INTERSECTIONS OF PARTICLE AND NUCLEAR PHYSICS

INTERSECTIONS OF PARTICLE AND NUCLEAR PHYSICS

8th Conference
CIPANP2003

New York, New York   19-24 May 2003

EDITOR
Zohreh Parsa
Brookhaven National Laboratory
Upton, New York

SPONSORING ORGANIZATIONS
Fermi National Laboratory
National Science Foundation
Brookhaven National Laboratory
Thomas Jefferson National Laboratory
Los Alamos National Laboratory

CD-ROM INCLUDED

Melville, New York, 2003
AIP CONFERENCE PROCEEDINGS ■ VOLUME 698

Editor:

Zohreh Parsa
Brookhaven National Laboratory
Physics Department
Building 510A
Upton, NY 11973-5000
USA
E-mail: parsa@bnl.gov

Authorization to photocopy items for internal or personal use, beyond the free copying permitted under the 1978 U.S. Copyright Law (see statement below), is granted by the American Institute of Physics for users registered with the Copyright Clearance Center (CCC) Transactional Reporting Service, provided that the base fee of $20.00 per copy is paid directly to CCC, 222 Rosewood Drive, Danvers, MA 01923. For those organizations that have been granted a photocopy license by CCC, a separate system of payment has been arranged. The fee code for users of the Transactional Reporting Service is: 0-7354-0169-1/03/$20.00.

© 2003 American Institute of Physics

Individual readers of this volume and nonprofit libraries, acting for them, are permitted to make fair use of the material in it, such as copying an article for use in teaching or research. Permission is granted to quote from this volume in scientific work with the customary acknowledgment of the source. To reprint a figure, table, or other excerpt requires the consent of one of the original authors and notification to AIP. Republication or systematic or multiple reproduction of any material in this volume is permitted only under license from AIP. Address inquiries to Office of Rights and Permissions, Suite 1NO1, 2 Huntington Quadrangle, Melville, N.Y. 11747-4502; phone: 516-576-2268; fax: 516-576-2450; e-mail: rights@aip.org.

L.C. Catalog Card No. 2003114758
ISBN 0-7354-0169-1
ISSN 0094-243X
Printed in the United States of America

Contents

Foreword xvii
Organizing Committee and Sponsors xix

Electroweak Physics 1
    P. Langacker
Lepton Dipole Moments 13
    B. L. Roberts
Electron Scattering and Hadron Structure 23
    E. J. Beise
Structure Functions — Status and Prospect 33
    J. C. Peng
QCD and RHIC 44
    D. Kharzeev
Neutron Spin Structure Results from JLab Hall A 54
    Z.-E. Meziani
Recent E158 Results 59
    P. A. Souder for the E158 Collaboration
**Do We Live in a Vanilla Universe? Theoretical Perspectives
on WMAP** 64
    R. Easther
Recent Results from KLOE at DAΦNE 74
    M. Moulson for the KLOE Collaboration
Recent Results from NA48, CERN's North Area Experiment 48 80
    P. D. Rubin for the NA48 Collaboration
Future Accelerators 85
    J. Womersley

LEPTON HADRON & HAD-HAD SCATTERING

**Cross Section Measurements and Charm Production in the
NuTeV Experiment** 95
    S. Boyd, T. Adams, A. Alton, S. Avvakumov, L. de Barbaro, P. de Barbaro,
    R. H. Bernstein, A. Bodek, T. Bolton, J. Brau, D. Buchholz, H. Budd,
    L. Bugel, J. Conrad, R. B. Drucker, B. T. Fleming, J. Formaggio, R. Frey,
    J. Goldman, M. Goncharov, D. A. Harris, R. A. Johnson, J. H. Kim,
    S. Koutsoliotas, M. J. Lamm, W. Marsh, D. Mason, J. McDonald,
    K. S. McFarland, C. McNulty, D. Naples, P. Nienaber, V. Radescu,
    A. Romosan, W. K. Sakumoto, H. Schellman, M. H. Shaevitz,
    P. Spentzouris, E. G. Stern, N. Suwonjandee, N. Tobien, A. Vaitaitis,
    M. Vakili, U. K. Yang, J. Yu, G. P. Zeller, and E. D. Zimmerman
**Measurement of the Absolute Drell-Yan Dimuon Cross Section in 800
GeV/c Proton-Proton and Proton-Deuterium Collisions** 100
    P. E. Reimer, J. C. Webb, T. C. Awes, M. E. Beddo, M. L. Brooks,
    C. N. Brown, J. D. Bush, T. A. Carey, T. H. Chang, W. E. Cooper,

C. A. Gagliardi, G. T. Garvey, D. F. Geesaman, E. A. Hawker, X. C. He,
L. D. Isenhower, D. M. Kaplan, S. B. Kaufman, P. N. Kirk, D. D. Koetke,
G. Kyle, D. M. Lee, W. M. Lee, M. J. Leitch, N. Makins,
P. L. McGaughey, J. M. Moss, B. A. Mueller, P. M. Nord, V. Papavassiliou,
B. K. Park, J. C. Peng, G. Petitt, M. E. Sadler, W. E. Sondheim,
P. W. Stankus, T. N. Thompson, R. S. Towell, R. E. Tribble, M. A. Vasiliev,
Y. C. Wang, Z. F. Wang, J. L. Willis, D. K. Wise, and G. R. Young

Hard QCD at the Tevatron 105
C. Mesropian

Aspects of Diffraction at the Tevatron: Review and Phenomenological Interpretation of CDF Results on Diffraction 110
K. Goulianos for the CDF Collaboration

Leading Baryon Production at HERA 115
I. Gialas

Nuclear Transparency in Exclusive ρ^0 Production at HERMES 119
W. Lorenzon

DVCS Results from HERMES 124
J. Volmer

DVCS with CLAS 129
E. S. Smith for the CLAS Collaboration

Pion-Pion Scattering Lengths from $\pi^+\pi^-$ Atom with the DIRAC Experiment 134
D. Goldin for the DIRAC Collaboration

Measurement of Invariant Mass Spectra of Vector Meson Decaying in Nuclear Matter at KEK-PS 138
R. Muto, J. Chiba, H. En'yo, Y. Fukao, H. Funahashi, H. Hamagaki,
M. Ieiri, M. Ishino, H. Kanda, M. Kitaguchi, M. Mihara, K. Miwa,
T. Miyashita, T. Murakami, T. Nakura, M. Naruki, M. Nomachi, K. Ozawa,
F. Sakuma, O. Sasaki, M. Sekimoto, T. Tabaru, K. H. Tanaka, M. Togawa,
S. Yamada, S. Yokkaichi, and Y. Yoshimura

***pp* Elastic Scattering at LHC and Nucleon Structure** 142
M. M. Islam, R. J. Luddy, and A. V. Prokudin

Modeling Neutrino Quasielastic Cross Sections on Nucleons and Nuclei 148
A. Bodek, H. Budd, and J. Arrington

Using Electron Scattering Data to Tune Neutrino Monte Carlos 153
H. Gallagher

FUNDAMENTAL SYMMETRY

Recent Results from the SAMPLE Experiment 161
T. M. Ito for the SAMPLE Collaboration

Neutrino Scattering in Perturbative QCD and Implications for the Weinberg Angle 165
S. Kretzer and M. Hall Reno

Probing Supersymmetry with Neutral Current
Scattering Experiments 168
    A. Kurylov, M. J. Ramsey-Musolf, and S. Su
Qweak: A Precision Measurement of the Proton's Weak Charge 172
    G. S. Mitchell for the Qweak Collaboration
Hadronic Parity Violation: Past, Present, and Future. 176
    B. R. Holstein
Progress at the WITCH Experiment 180
    M. Beck, S. Coeck, B. Delauré, V. V. Golovko, V. Y. Kozlov, I. S. Kraev,
    A. Lindroth, T. Phalet, N. Severijns, S. Versyck, D. Beck, W. Quint,
    F. Ames, P. Delahaye, C. Guenaut, and the NIPNET Collaboration
New Results in Superallowed Nuclear Beta Decay 184
    J. C. Hardy
Charged Current Universality Problem and NuTeV Anomaly:
Is SUSY to Blame? 188
    A. Kurylov, M. Ramsey-Musolf, and S. Su
Progress Towards Measuring the Electric Dipole Moment of the
Electron in Metastable PbO 192
    D. Kawall, F. Bay, S. Bickman, Y. Jiang, and D. DeMille
A New Experiment to Measure the Muon Electric Dipole Moment 196
    J. P. Miller, R. M. Carey, V. Logashenko, K. R. Lynch, B. L. Roberts,
    A. Silenko, G. Bennett, D. M. Lazarus, L. B. Leipuner, W. Marciano,
    W. Meng, W. M. Morse, R. Prigl, Y. K. Semertzidis, V. Balakin,
    A. Bazhan, A. Dunikov, B. Khazin, I. B. Khriplovich, G. Sylvestrov,
    Y. Orlov, K. Jungmann, P. T. Debevec, D. W. Hertzog, C. J. G. Onderwater,
    C. S. Ozben, E. Stephenson, M. Auzinsh, P. Cushman, R. McNabb,
    N. Shafer-Ray, K. Yoshimura, A. Aoki, Y. Kuno, A. Sato, M. Iwasaki, and
    F. J. M. Farley
A New Method for a Sensitive Deuteron EDM Experiment. 200
    Y. K. Semertzidis, M. Aoki, M. Auzinsh, V. Balakin, A. Bazhan,
    G. W. Bennett, R. M. Carey, P. Cushman, P. T. Debevec, A. Dudnikov,
    F. J. M. Farley, D. W. Hertzog, M. Iwasaki, K. Jungmann, D. Kawall,
    B. Khazin, I. B. Khriplovich, B. Kirk, Y. Kuno, D. M. Lazarus,
    L. B. Leipuner, V. Logashenko, K. R. Lynch, W. J. Marciano, R. McNabb,
    W. Meng, J. P. Miller, W. M. Morse, C. J. G. Onderwater, Y. F. Orlov,
    C. S. Ozben, R. Prigl, S. Rescia, B. L. Roberts, N. Shafer-Ray, A. Silenko,
    E. J. Stephenson, and K. Yoshimura
Production and Detection of Cold Anti-Hydrogen Atoms. A First Step
Towards High Precision CPT Test 205
    A. Variola, M. Amoretti, G. Bonomi, A. Boutcha, P. Bowe, C. Carraro,
    C. L. Cesar, M. Charlton, M. Doser, V. Filippini, A. Fontana, M. Fujiwara,
    R. Funakoshi, P. Genova, J. S. Hangst, R. S. Hayano, L. V. Jorgensen,
    V. Lagomarsino, R. Landua, D. Lindelof, E. Lodi Rizzini, M. Macrì,
    N. Madsen, G. Manuzio, P. Montagna, H. Pruys, C. Regenfus, A. Rotondi,
    P. Riedler, G. Testera, and D. P. Van der Werf
Phenomenology of the Littlest Higgs Model 209
    T. Han
Testing V-A in Top Decay at CDF at $\sqrt{s}=1.8$ TeV 218
    B. Kilminster

Symmetric Textures in SO(10) and LMA Solution for Solar Neutrinos 222
    M.-C. Chen and K. T. Mahanthappa
Analysis Details from Muon (g-2) 226
    C. C. Polly for the Muon (g-2) Collaboration
Muon Lifetime and Muon Capture 230
    B. Lauss for the MuCAP[1] and MuLAN[2] Collaborations
TWIST: Measuring the Space-Time Structure of Muon Decay 234
    C. A. Gagliardi and the TWIST Collaboration
Searches at the Run II Tevatron Collider 238
    L. S. Groer for the CDF and DØ Collaborations
Hadronic Matrix Elements of Proton Decay on the Lattice 243
    Y. Aoki for the RBC Collaboration
**Calculation of CP Violation in Non-Leptonic Kaon Decay
on the Lattice** 247
    J. Noaki for the RBC Collaboration

NEUTRINOS

SNO Past, Present and Future 253
    N. McCauley for the SNO Collaboration
First Results from KamLAND 258
    B. E. Berger for the KamLAND Collaboration
**The KARMEN Final Results on the Search for $\bar{\nu}_e$ from μ^+
Decay at Rest** 265
    K. Eitel
Neutrino Oscillation Search at MiniBooNE 270
    T. Hart for the MiniBooNE Collaboration
$\bar{\nu}_e$ and the Sudbury Neutrino Observatory 275
    J. L. Orrell for the SNO Collaboration
**Inelastic Cross-Sections for Neutrinos and Two Nucleons at Low
Energy in Effective Field Theory** 278
    M. N. Butler
Neutrino Interactions at Low Energy 283
    E. A. Hawker
**The Majorana Project: Determining $m\beta\beta$ to the 50 meV Scale
through $\beta\beta$ Measurements in ^{76}Ge** 288
    A. R. Young for the Majorana Collaboration
The MINOS Experiment 293
    H. Gallagher for the MINOS Experiment
Performance of the ICARUS T600 Detector 298
    S. Navas-Concha for the ICARUS Collaboration
**The Future of Reactor Neutrino Experiments: A Novel Approach to
Measuring θ_{13}** 303
    K. M. Heeger, S. J. Freedman, and K.-B. Luk
**Physics of an Intense Neutrino Beam from BNL to a Very Long
Baseline Detector** 307
    Z. Parsa

Neutrinos Parallel Session—A Summary 314
    B. E. Berger and B. T. Fleming

NUCLEAR AND PARTICLE ASTROPHYSICS

Probing Dark Energy in the Accelerating Universe with SNAP 323
    M. Schubnell for the SNAP Collaboration
Results from DAMA 328
    R. Bernabei, P. Belli, F. Cappella, F. Montecchia, F. Nozzoli, A. Incicchitti,
    D. Prosperi, R. Cerulli, C. J. Dai, H. H. Kuang, J. M. Ma, and Z. P. Ye
ADMX Dark-Matter Axion Search 332
    L. J. Rosenberg
An Experiment to Measure the Air Fluorescence Yield in
Electromagnetic Showers 341
    P. Hüntemeyer for the FLASH Collaboration
Big Bang Nucleosynthesis and the Missing Hydrogen Mass
in the Universe 345
    D. C. Choudhury and D. W. Kraft
Progress Towards a FLUKA Based Simulation Tool Aimed at the
Evaluation of Space Radiation Environments 349
    V. Andersen, F. Ballarini, G. Battistoni, M. Campanella, M. Carboni,
    F. Cerutti, A. Empl, A. Fassò, A. Ferrari, E. Gadioli, M. V. Garzelli,
    K. Lee, A. Ottolenghi, M. Pelliccioni, L. S. Pinsky, J. Ranft, S. Roesler,
    P. R. Sala, and T. L. Wilson
Cluster Structure of Atomic Nuclei and Nucleosynthesis 353
    R. Y. Kezerashvili
Ultra High Energy Cosmic Rays 357
    T. Stanev
AMS-02 on the International Space Station 362
    K. Scholberg for the AMS Collaboration
The Pierre Auger Observatory 366
    J. Swain
HiRes—Searching for the Origins of Ultra High Energy Cosmic Rays 370
    S. Westerhoff for the HiRes Collaboration
Nuclear and Particle Astrophysics at CIPANP 2003 374
    E.A. Baltz and J. Stone

LIGHT QUARKS AND LEPTONS

New, High Statistics Measurement of the $K^+ \to \pi^0 e^+ \nu (K^+_{e3})$
Branching Ratio 381
    A. Sher for the E865 Collaboration
Radiative Corrections and the Universality of the Weak Interactions 385
    A. Sirlin
Can an Amended Standard Model Account for Cold Dark Matter? 390
    M. Goldhaber

Nucleon Electromagnetic Form Factors and Densities 393
    J. J. Kelly
Nucleon Axial Charge from Quenched Lattice QCD with Domain
Wall Fermions... 398
    S. Ohta for the RBCK Collaboration
Hadronic Effects in Theory of g-2 403
    A. Vainshtein
Chiral Dynamics in the Meson Sector at Two Loops 407
    J. Bijnens
The Weak Production of Λ Particles in Muon and Tau Scattering
from Protons 411
    S. L. Mintz

HEAVY QUARKS AND LEPTONS

CP Asymmetries at *BABAR*.. 417
    F. Blanc for the BABAR Collaboration
$\sin 2\beta$ from Pure Penguins and Signs of New Physics 422
    M. Ciuchini, E. Franco, A. Masiero, and L. Silvestrini
A Mixing-Independent Construction of the Unitarity Triangle 427
    M. Neubert
CLEO Results on $|V_{cb}|$ and $|V_{ub}|$ 432
    K. M. Ecklund
Charged to Neutral B Meson Yield Ratio Across the
$\Upsilon(4S)$ Resonance.. 436
    M. B. Voloshin
Rare Hadronic B Decays: Probing Deeper into the Standard Model 440
    A. J. Schwartz
Rare Charm Decays at Colliders 452
    D. Cinabro
D^0–$\bar{D}^0$ Mixing, Rare Decays, and New Physics 456
    A. A. Petrov
The Future of Charm Physics.. 461
    T. E. Coan
Heavy-Quark Recombination in Z^0 Decay 465
    Y. Jia
Top Quark Mass Measurements at the Tevatron.......................... 469
    M. F. Canelli for the CDF and DØ Collaborations
Charm and Beauty at the Tevatron 474
    J. Cranshaw for the CDF and D0 Collaborations
Tau-Mu Flavor Violation and the Scale of New Physics..................... 478
    D. Black
Rare Decays of Tau Leptons: An Experimental Review 482
    J. Urheim

QCD SPECTROSCOPY, STRUCTURE AND DYNAMICS

Observation of a Narrow Resonance in the $D_s^+ \pi^0$ System at
2.32 GeV/c^2 with BABAR 489
    R. F. Cowan for the BABAR Collaboration

A DK Molecule or Other 4q Model for the $D_s \pi$ Resonance
at 2.32 GeV 493
    H. J. Lipkin

Observation of the D_{sJ} (2463) and Confirmation of the D^*_{sJ} (2317) 497
    J. Urheim and S. Stone for the CLEO Collaboration

Photoproduction of Charm Pairs 503
    E. E. Gottschalk for the Focus Collaboration

Charm Production Asymmetries from Heavy-Quark Recombination 508
    T. Mehen

Large-N_c Selection Rules for Decay of J^{PC} Exotic Hybrid Mesons 513
    P. R. Page

Meson Spectroscopy in Photo-Production at CLAS 517
    M. Nozar for the CLAS Collaboration

Effect of Light Scalar Mesons in $\eta \to 3\pi$ 522
    A. Abdel-Rehim, D. Black, A. H. Fariborz, and J. Schechter

Measurement of the Angular Distribution of $\Psi'(3686) \to e^+ e^-$
from $\bar{p}p$ Annihilation 526
    S. H. Seo for the E835 Collaboration

Baryon Spectroscopy on the Lattice: Recent Results 530
    C. Morningstar

An Analysis of $\gamma p \to p \pi^+ \pi^-$ Using the CLAS Detector 535
    M. Bellis and the CLAS Collaboration

Evidence for an Exotic S=+1 Baryon Resonance at a Mass
of 1540 MeV 539
    K. H. Hicks and the LEPS Collaboration

Evidence for an Exotic Baryon State, $\Theta^+(1540)$, in Photoproduction
Reactions from Protons and Deuterons with CLAS 543
    V. Kubarovsky and S. Stepanyan

Excited $L=1$ Baryons in Large N_c QCD 548
    D. Pirjol and C. Schat

An Experimental Overview of Gluonic Mesons 554
    C. A. Meyer

Observation of a 1750 MeV/c^2 State in the Peripheral
Photoproduction of $K^+ K^-$ 559
    R. E. Mitchell for the FOCUS Collaboration

Confinement Dynamics 562
    M. G. Olsson and T. J. Allen

Review of Two-Photon Interactions 566
    D. Urner

An Interferometric Study of the χ_{c0} (1^3P_0) in the Reactions
$\bar{p}p \to \pi^0 \pi^0, \eta\eta, \pi^0 \eta$ 571
    J. L. Rosen for the Fermilab E835 Collaboration

CLEO Results in Upsilon Spectroscopy.................................... 575
    R. S. Galik
The Resummed Photon Spectrum in Radiative Upsilon Decays
(And More) 579
    S. Fleming
The Enhancement by the Axial Anomaly of the
Decay $Y(1D) \rightarrow \eta Y(1S)$... 583
    M. B. Voloshin
QCD Spectroscopy, Structure, and Dynamics 587
    J. Napolitano and J. Russ

SPIN

Polarized Parton Distributions Measured at the
HERMES Experiment 595
    J. Wendland for the HERMES Collaboration
The Status of the COMPASS Experiment................................. 599
    E. M. Kabuß
Fragmentation Functions and Implications for Spin Physics.................. 603
    S. Kretzer
Renormalons in Exclusive Meson Electroproduction 607
    A. V. Belitsky
Single Spin Asymmetries at CLAS.. 612
    H. Avakian and L. Elouadrhiri for the CLAS Collaboration
Novel Transversity Properties in SIDIS................................... 617
    L. Gamberg, G. R. Goldstein, and K. A. Oganessyan
Single Spin Azimuthal Asymmetries and Transversity 621
    R. Seidl for the HERMES Collaboration
Single Transverse Spin Asymmetries 624
    D. S. Hwang
Future Measurements of Spin Dependent Fragmentation Functions
in e^+e^- Annihilation at Belle.. 628
    K. Hasuko, M. Grosse Perdekamp, A. Ogawa, J. Söeren Lange, and
    V. Siegle
Proton and Neutron Spin Structure Functions in and Near
the Resonance Region 632
    R. C. Minehart
Double-Transverse Spin Asymmetries at NLO............................ 636
    M. Mukherjee, M. Stratmann, and W. Vogelsang
Status of the Michigan Ultra-Cold Spin-Polarized Hydrogen Jet 639
    K. Yonehara, B. K. Harris, M. C. Kandes, B. H. Kienman, A. D. Krisch,
    M. A. Leonova, V. G. Luppov, V. S. Morozov, J. B. Olson, C. C. Peters,
    R. S. Raymond, D. L. Sisco, N. S. Borisov, V. V. Fimushkin, and
    A. F. Prudkoglyad

Spin Dependence in Polarized $pC \to pC$ Scattering at Low Momentum Transfer and Polarimetry at RHIC 643
    A. Bravar, I. Alekseev, L. Ahrens, M. Bai, G. Bunce, S. Dhawan,
    H. Huang, V. Hughes, G. Igo, O. Jinnouchi, K. Kurita, Z. Li,
    W. W. MacKay, S. Rescia, T. Roser, N. Saito, H. Spinka, D. Svirida,
    D. Underwood, C. Whitten, and J. Wood

A Study of Heavy Quark and Quarkonium Production in Polarized p-p Collisions at RHIC 647
    M. X. Liu for the PHENIX Collaboration

Single-Spin Asymmetries in Exclusive Electroproduction of Pseudoscalar and Vector Mesons at HERMES 651
    D. Hasch for the HERMES Collaboration

Flavor Decomposition of Nucleon Spin Structure: A Proposed Experiment at Jefferson Lab Hall C 655
    X. Jiang, D. B. Day, and M. K. Jones

Summary of Spin Physics Parallel Sessions 659
    J. Qiu and M. Grosse Perdekamp

RELATIVISTIC HEAVY IONS

Jets and High p_T Hadrons in Dense Matter: Recent Results from STAR 667
    P. Jacobs and J. Klay for the STAR Collaboration

PHENIX High Pt Results 673
    L. Aphecetche for the PHENIX Collaboration

First Results on d+Au Collisions from PHOBOS 677
    B. B. Back, M. D. Baker, M. Ballintijn, D. S. Barton, B. Becker,
    R. R. Betts, A. A. Bickley, R. Bindel, A. Budzanowski, W. Busza,
    A. Carroll, M. P. Decowski, E. García, T. Gburek, N. George,
    K. Gulbrandsen, S. Gushue, C. Halliwell, J. Hamblen, A. S. Harrington,
    C. Henderson, D. J. Hofman, R. S. Hollis, R. Hołyński, B. Holzman,
    A. Iordanova, E. Johnson, J. L. Kane, N. Khan, P. Kulinich, C. M. Kuo,
    J. W. Lee, W. T. Lin, S. Manly, A. C. Mignerey, A. Noell, R. Nouicer,
    A. Olszewski, R. Pak, I. C. Park, H. Pernegger, C. Reed, L. P. Remsberg,
    C. Roland, G. Roland, J. Sagerer, P. Sarin, P. Sawicki, I. Sedykh,
    W. Skulski, C. E. Smith, P. Steinberg, G. S. F. Stephans, A. Sukhanov,
    R. Teng, M. B. Tonjes, A. Trzupek, C. Vale, G. J. van Nieuwenhuizen,
    R. Verdier, G. I. Veres, B. Wadsworth, F. L. H. Wolfs, B. Wosiek,
    K. Woźniak, A. H. Wuosmaa, B. Wysłouch, and J. Zhang

Dynamics of Soft Particle Production in Heavy Ion Collisions 680
    P. Steinberg

Saturation Physics in Heavy Ion Collisions 685
    Y. V. Kovchegov

Particle Production at RHIC Energies 690
    R. Debbe for the Brahms Collaboration

What Did We Learn and What Will We Learn from Hydrodynamics at RHIC? 694
    P. F. Kolb

Particle Correlations at High Partonic Density 698
    K. Tuchin
Elliptic Flow from Au+Au Collisions at $\sqrt{s_{NN}}$=200 GeV 701
    A. Tang for the STAR Collaboration
Heavy Flavour Hadro-Production Cross-Sections 704
    H. K. Wöhri and C. Lourenço
J/Ψ and Open Charm Measurements at RHIC/PHENIX 709
    D. Silvermyr for the PHENIX Collaboration
The NA60 Experiment: Results and Perspectives 713
    J. M. Heuser, R. Arnaldi, K. Banicz, K. Borer, J. Buytaert, J. Castor,
    B. Chaurand, W. Chen, B. Cheynis, C. Cicalo, C. Colla, P. Cortese,
    A. David, A. de Falco, N. de Marco, A. Devaux, A. Drees, L. Ducroux,
    H. En'yo, A. Ferretti, M. Floris, P. Force, A. Grigorian, J.-Y. Grossiord,
    N. Guettet, A. Guichard, H. Gulkanian, M. Keil, L. Kluberg, Z. Li,
    C. Lourenço, J. Lozano, F. Manso, A. Masoni, A. Neves, H. Ohnishi,
    C. Oppedisano, P. Parracho, G. Puddu, E. Radermacher, P. Rosinsky,
    E. Scomparin, J. Seixas, S. Serci, R. Shahoyan, P. Sonderegger, R. Tieulent,
    G. Usai, H. Vardanyan, R. Veenhof, and H. Wöhri
Charm Flow versus Fragmentation in RHIC 718
    S. Batsouli
Multiparton Tomography of Hot and Cold Nuclear Matter 721
    I. Vitev
The dA Collisions at Forward Rapidities at RHIC 725
    J. Jalilian-Marian
Charged Particle Jet Studies in $\sqrt{s_{NN}}$=200 GeV p+p, d+Au, and Au+Au Collisions at RHIC 729
    M. L. Miller and D. H. Hardtke
Particle Composition at High p_T in Au+Au Collisions at $\sqrt{s_{NN}}$=200 GeV 732
    T. Chujo
Conference on the Intersections of Particle and Nuclear Physics 2003: Relativistic Heavy Ion Parallel Session Summary 735
    J. L. Nagle and T. Hallman

ACCELERATORS, FACILITIES AND DETECTORS

The CMS Experiment—Status and Physics 741
    N. Neumeister
Overview of the PHENIX Experiment 745
    E. J. O'Brien for the PHENIX Collaboration
Recent Developments for Experiments in the MIT-Bates South Hall Ring 751
    W. A. Franklin
Detection of High-Z Objects using Multiple Scattering of Cosmic Ray Muons 755
    G. E. Hogan, K. N. Borozdin, J. Gomez, C. Morris, W. C. Priedhorsky,
    A. Saunders, L. J. Schultz, and M. E. Teasdale

Development of Photon Collider and Solid-State Modulator Technology for the NLC 759
    J. Gronberg

Spin-Flipping Polarized Deuterons at COSY 763
    K. Yonehara, A. D. Krisch, V. S. Morozov, R. S. Raymond, V. K. Wong,
    U. Bechstedt, R. Gebel, A. Lehrach, B. Lorenz, R. Maier, D. Prasuhn,
    A. Schnase, H. Stockhorst, D. Eversheim, F. Hinterberger, H. Rohdjess,
    K. Ulbrich, and W. Scobel

LENS—the Low-Energy Neutrino Spectrometer 767
    R. L. Hahn for the International LENS R&D Collaboration

Future Neutrino Experiments at Nuclear Reactors 771
    J. M. Link

Particle Identification in the PHENIX Experiment at RHIC (Present and Future) 775
    E. Kistenev for the PHENIX Collaboration

Silicon Vertex Detector Upgrade for the PHENIX Experiment at RHIC 785
    Y. Akiba for the PHENIX Collaboration

A Large Tracking Detector in Vacuum Consisting of Self-Supporting Straw Tubes 789
    P. Wintz for the COSY-TOF Collaboration

Proposed Detector Upgrade for Measuring Low-Mass Lepton Pairs in PHENIX 793
    C. Aidala for the PHENIX Collaboration

Electron Beam Polarimetry for EIC/eRHIC 797
    W. Lorenzon

Progress on the Concept and Design of the Rare Isotope Accelerator 801
    D. F. Geesaman

The Physics of eRHIC 806
    R. G. Milner

ELIC: A High Luminosity and Efficient Spin Manipulation Electron-Light Ion Collider Based at CEBAF 811
    L. Merminga and Y. Derbenev

A New Detector for Physics at HERA 816
    I. Abt for the HERA-III Detector Group

CIPANP2003 Program 827
List of Registered Participants 831
Photos 839
Author Index 849

FOREWORD

The Eighth Conference on the Intersection of Particles and Nuclear Physics (CIPANP2003) was held from May 19 to May 24, 2003, in New York City, New York, USA. The purpose of this meeting, as with the seven previous conferences in this series, was to bring together particle and nuclear physicists in a pleasant setting where they could hear up-to-date scientific reports and discuss areas of research which overlap both their disciplines. The conference was overwhelmingly successful and was attended by more than 400 participants from the U.S. and abroad.

The success of the standard model has provided a common underpinning for both disciplines as well as similar fundamental goals. Indeed, Quantum Chromodynamics (QCD) has proven to be "the" theory of strong interactions. As such, it forms the basis for nuclear physics as well as high energy hadronic interactions. QCD is a perfect theory. It is parameter free. Similarly, all known electroweak phenomena can be described by the $SU(2)_L \times U(1)_Y$ sector of the standard model. It has been tested at the incredible ± 0.1 % in reactions ranging from nuclear beta decay through Z-pole studies. Nevertheless, important outstanding questions remain. Why are there three generations of fermions? What is the true origin of electroweak symmetry breaking, mass generation and CP violation? Why is parity violated? These and other important issues provide valuable input for ongoing discussions and in making decisions regarding the future direction of the field.

The conference on Intersections between Particle and Nuclear Physics started with opening remarks by Zohreh Parsa followed by a welcome by Bill Marciano (Chair) and ended with a closing presentation by F. Wilczek. We would like to thank them, the conference speakers, and all the authors for providing their contributions, given in the following chapters (1–11), arranged starting with overviews (plenary) followed by (parallel) contributions arranged by topics respectively. Plenary sessions were held in the mornings and ten groups were held in parallel sessions in the afternoons daily during the week of the conference. The general program is included in the back. Aside from the many interesting physics sessions, the exhibits, reception and conference dinner provided a stimulating environment for participants to interact and discuss physics. For more information see the conference web page www.cipanp2003.bnl.gov.

We would like to thank the conference chair, organizing committee, session coordinators, staff, sponsors, exhibitors, and others who contributed to the success of the conference. Our special thanks to Stanley Kowalski for his interest and assistance. Also thanks to those who provided photos used in these proceedings, and to Keith Lally, Tim Hallman, and Wayne Betts for computer support. Special thanks to the City of New York for their hospitality and for providing the photo used on the inside the cover of these proceedings, and also Alan D. Krisch and Malcolm H. MacFarlane, founding fathers of the CIPANP series.

Zohreh Parsa, Editor
Co-Organizer, Organizing Committee
Chair, Local Organizing Committee

Dr. Z. Parsa, Brookhaven National Laboratory, Physics Department, Upton, NY 11973

INTERSECTIONS ORGANIZING COMMITTEE

William J. Marciano, Chairman (BNL), David Hertzog, Vice Chair (Illinois),
Jeffrey A. Appel (FNAL), Edmond L. Berger (ANL), Karl Berkelman (Cornell),
Thomas J. Bowles (LANL), Wick Haxton (Washington),
Barry Holstein (Massachusetts), Kees de Jager,(JLAB), Yoshitaka Kuno (KEK),
Stanley Brodsky (SLAC), Frank Maas (Johanes-Gutenberg),
Costas Papanicolas (Athens), Zohreh Parsa (BNL), Heidi Schellman (Northwestern),
Alan Shotter, (TRIUMF).

LOCAL ORGANIZING COMMITTEE

Zohreh Parsa, Chair (BNL), Alan Mincer (NYU), Michael Tannenbaum (BNL).

EXHIBITORS

American Institute of Physics,
Apple Computer, Inc.,
Dell Computer Corporation,
Elsevier,
GovConnection, Inc.,
Institute of Physics Publishing.

SPONSORS

We thank the following Laboratories and Agencies for their support of the 8th Conference on Intersections of Particle and Nuclear Physics (CIPANP2003)

Fermi National Laboratory,
National Science Foundation,
Brookhaven National Laboratory,
Thomas Jefferson National Laboratory,
Los Alamos National Laboratory,
Institute of Physics Publishing/Journal of Physics G.

Electroweak Physics

Paul Langacker

Department of Physics and Astronomy, University of Pennsylvania
Philadelphia, PA 19104, USA

Abstract. The results of high precision weak neutral current (WNC), Z-pole, and high energy collider electroweak experiments have been the primary prediction and test of electroweak unification. The electroweak program is briefly reviewed from a historical perspective. Current changes, anomalies, and things to watch are summarized, and the implications for the standard model and beyond discussed.

THE Z, THE W, AND THE WEAK NEUTRAL CURRENT

The weak neutral current was a critical prediction of the electroweak standard model (SM) [1, 2]. Following its discovery in 1973 by the Gargamelle and HPW experiments, there were generations of ever more precise WNC experiments, typically at the few % level. These included pure weak νN and νe scattering processes, and weak-electromagnetic interference processes such as polarized $e^{\uparrow\downarrow}D$ or μN, $e^+e^- \to$ (hadron or charged lepton) cross sections and asymmetries below the Z pole, and parity-violating effects in heavy atoms (APV). There were also early direct observations of the W and Z by UA1 and UA2. The early 1990's witnessed the very precise Z-pole experiments at LEP and the SLC, in which the lineshape, decay modes, and various asymmetries were measured at the 0.1% level. The subsequent LEP 2 program at higher energies measured M_W, searched for the Higgs and other new particles, and constrained anomalous gauge self-interactions. Parallel efforts at the Tevatron by CDF and DØ led to the direct discovery of the t and measurements of m_t and M_W, while a fourth generation of weak neutral current experiments continued to search for new physics to which the (more precise) Z-pole experiments were blind. The program was supported by theoretical efforts in the calculation of QCD and electroweak radiative corrections; the expectations for observables in the standard model, large classes of extensions, and alternative models; and global analyses of the data.

The precision program has established that the standard model (SM) is correct and unique to first approximation, establishing the gauge principle as well as the SM gauge group and representations; shown that the SM is correct at loop level, confirming the basic principles of renormalizable gauge theory and allowing the successful prediction or constraint on m_t, α_s, and the Higgs mass M_H; severely constrained new physics at the TeV scale, with the ideas of unification strongly favored over TeV-scale compositeness; and yielded precise values for the gauge couplings, consistent with (supersymmetric) gauge unification.

RESULTS BEFORE THE LEP/SLD ERA

Even before the beginning of the Z-pole experiments at LEP and SLC in 1989, the precision program had established [2]-[5]:

- Global analyses of all data carried more information than the analysis of individual experiments, but care has to be taken with systematic and theoretical uncertainties.
- The SM is correct to first approximation. The four-fermion operators for vq, ve, and eq were uniquely determined, in agreement with the standard model, in model (i.e., gauge group) independent analyses. The W and Z masses agreed with the expectations of the $SU(2) \times U(1)$ gauge group and canonical Higgs mechanism, eliminating contrived alternative models with the same four-fermi interactions as the standard model.
- QCD evolved structure functions and electroweak radiative corrections were necessary for the agreement of theory and experiment.
- The weak mixing angle (in the on-shell renormalization scheme) was determined to be $\sin^2 \theta_W = 0.230 \pm 0.007$; consistency of the various observations, including radiative corrections, required $m_t < 200$ GeV.
- Theoretical uncertainties, especially in the c threshold in deep inelastic weak charge current (WCC) scattering, dominated.
- The combination of WNC and WCC data uniquely determined the $SU(2)$ representations of all of the known fermions, i.e., v_e and v_μ, as well as the L and R components of the e, μ, τ, d, s, b, u, and c [6]. In particular, the left-handed b and τ were the lower components of $SU(2)$ doublets, implying unambiguously that the t quark and v_τ had to exist. This was independent of theoretical arguments based on anomaly cancellation (which could have been evaded in alternative models involving a vector-like third family), and of constraints on m_t from electroweak loops.
- The electroweak gauge couplings were well-determined, allowing a detailed comparison with the gauge unification predictions of the simplest grand unified theories (GUT). Ordinary $SU(5)$ was excluded (consistent with the non-observation of proton decay), but the supersymmetric extension was allowed, "perhaps even the first harbinger of supersymmetry" [4].
- There were stringent limits on new physics at the TeV scale, including additional Z' bosons, exotic fermions (for which both WNC and WCC constraints were crucial), exotic Higgs representations, leptoquarks, and new four-fermion operators.

THE LEP/SLC ERA

The LEP/SLC era greatly improved the precision of the electroweak program. It allowed the differentiation between non-decoupling extensions to the SM (such as most forms of dynamical symmetry breaking and other types of TeV-scale compositeness), which typically predicted several % deviations, and decoupling extensions (such as most of the parameter space for supersymmetry), for which the deviations are typically 0.1%.

The first phase of the LEP/SLC program involved running at the Z pole, $e^+e^- \to Z \to \ell^+\ell^-$, $q\bar{q}$, and $\nu\bar{\nu}$. During the period 1989-1995 the four LEP experiments ALEPH, DELPHI, L3, and OPAL at CERN observed $\sim 2 \times 10^7 Z's$. The SLD experiment at the SLC at SLAC observed some 5×10^5 events. Despite the much lower statistics, the SLC had the considerable advantage of a highly polarized e^- beam, with $P_{e^-} \sim 75\%$. There were quite a few Z pole observables, including:

- The lineshape: M_Z, Γ_Z, and the peak cross section σ.
- The branching ratios for e^+e^-, $\mu^+\mu^-$, $\tau^+\tau^-$, $q\bar{q}$, $c\bar{c}$, $b\bar{b}$, and $s\bar{s}$. One could also determine the invisible width, $\Gamma(\text{inv})$, from which one can derive the number $N_\nu = 2.986 \pm 0.007$ of active (weak doublet) neutrinos with $m_\nu < M_Z/2$, i.e., there are only 3 conventional families with light neutrinos. $\Gamma(\text{inv})$ also constrains other invisible particles, such as light sneutrinos and the light majorons associated with some models of neutrino mass.
- A number of asymmetries, including forward-backward (FB) asymmetries; the τ polarization, P_τ; the polarization asymmetry A_{LR} associated with P_{e^-}; and mixed polarization-FB asymmetries.

The expressions for the observables are summarized in [1, 2], and the experimental values and SM predictions in Table 1. The precision of the Z mass determination was extraordinary for a high energy experiment. These combinations of observables could be used to isolate many Z-fermion couplings, verify lepton family universality, determine $\sin^2\theta_W$ in numerous ways, and determine or constrain m_t, α_s, and M_H. LEP and SLC simultaneously carried out other programs, most notably studies and tests of QCD, and heavy quark physics.

LEP 2 ran from 1995-2000, with energies gradually increasing from ~ 140 to ~ 209 GeV. The principal electroweak results were precise measurements of the W mass, as well as its width and branching ratios (these were measured independently at the Tevatron); a measurement of $e^+e^- \to W^+W^-$, ZZ, and single W, as a function of center of mass (CM) energy, which tests the cancellations between diagrams that is characteristic of a renormalizable gauge field theory, or, equivalently, probes the triple gauge vertices; limits on anomalous quartic gauge vertices; measurements of various cross sections and asymmetries for $e^+e^- \to f\bar{f}$ for $f = \mu^-, \tau^-, q, b$ and c, in reasonable agreement with SM predictions; a stringent lower limit of 114.4 GeV on the Higgs mass, and even hints of an observation at ~ 116 GeV; and searches for supersymmetric or other exotic particles.

In parallel with the LEP/SLC program, there were precise ($< 1\%$) measurements of atomic parity violation (APV) in cesium at Boulder, along with the atomic calculations and related measurements needed for the interpretation; precise new measurements of deep inelastic scattering by the NuTeV collaboration at Fermilab, with a sign-selected beam which allowed them to minimize the effects of the c threshold and reduce uncertainties to around 1%; and few % measurements of $\stackrel{(-)}{\nu}_\mu e$ by CHARM II at CERN. Although the precision of these WNC processes was lower than the Z pole measurements, they are still of considerable importance: the Z pole experiments are blind to types of new physics that do not directly affect the Z, such as a heavy Z' if there is no $Z-Z'$ mixing, while the WNC experiments are often very sensitive. During the same period there were important electroweak results from CDF and DØ at the Tevatron, most notably a

TABLE 1. Principal Z-pole observables, their experimental values, theoretical predictions using the SM parameters from the global best fit as of 1/03 (updated from [2]), and pull (difference from the prediction divided by the uncertainty). See [1] for definitions of the quantitites. Γ(had), Γ(inv), and $\Gamma(\ell^+\ell^-)$ are not independent.

Quantity	Group(s)	Value	Standard Model	pull
M_Z [GeV]	LEP	91.1876 ± 0.0021	91.1874 ± 0.0021	0.1
Γ_Z [GeV]	LEP	2.4952 ± 0.0023	2.4972 ± 0.0011	-0.9
Γ(had) [GeV]	LEP	1.7444 ± 0.0020	1.7436 ± 0.0011	—
Γ(inv) [MeV]	LEP	499.0 ± 1.5	501.74 ± 0.15	—
$\Gamma(\ell^+\ell^-)$ [MeV]	LEP	83.984 ± 0.086	84.015 ± 0.027	—
σ_{had} [nb]	LEP	41.541 ± 0.037	41.470 ± 0.010	1.9
R_e	LEP	20.804 ± 0.050	20.753 ± 0.012	1.0
R_μ	LEP	20.785 ± 0.033	20.753 ± 0.012	1.0
R_τ	LEP	20.764 ± 0.045	20.799 ± 0.012	-0.8
$A_{FB}(e)$	LEP	0.0145 ± 0.0025	0.01639 ± 0.00026	-0.8
$A_{FB}(\mu)$	LEP	0.0169 ± 0.0013		0.4
$A_{FB}(\tau)$	LEP	0.0188 ± 0.0017		1.4
R_b	LEP/SLD	0.21664 ± 0.00065	0.21572 ± 0.00015	1.1
R_c	LEP/SLD	0.1718 ± 0.0031	0.17231 ± 0.00006	-0.2
$R_{s,d}/R_{(d+u+s)}$	OPAL	0.371 ± 0.023	0.35918 ± 0.00004	0.5
$A_{FB}(b)$	LEP	0.0995 ± 0.0017	0.1036 ± 0.0008	-2.4
$A_{FB}(c)$	LEP	0.0713 ± 0.0036	0.0741 ± 0.0007	-0.8
$A_{FB}(s)$	DELPHI/OPAL	0.0976 ± 0.0114	0.1037 ± 0.0008	-0.5
A_b	SLD	0.922 ± 0.020	0.93476 ± 0.00012	-0.6
A_c	SLD	0.670 ± 0.026	0.6681 ± 0.0005	0.1
A_s	SLD	0.895 ± 0.091	0.93571 ± 0.00010	-0.4
A_{LR} (hadrons)	SLD	0.15138 ± 0.00216	0.1478 ± 0.0012	1.7
A_{LR} (leptons)	SLD	0.1544 ± 0.0060		1.1
A_μ	SLD	0.142 ± 0.015		-0.4
A_τ	SLD	0.136 ± 0.015		-0.8
$A_e(Q_{LR})$	SLD	0.162 ± 0.043		0.3
$A_\tau(\mathcal{P}_\tau)$	LEP	0.1439 ± 0.0043		-0.9
$A_e(\mathcal{P}_\tau)$	LEP	0.1498 ± 0.0048		0.4
Q_{FB}	LEP	0.0403 ± 0.0026	0.0424 ± 0.0003	-0.8

precise value for M_W, competitive with and complementary to the LEP 2 value; a direct measure of m_t, and direct searches for Z', W', exotic fermions, and supersymmetric particles. Many of these non-Z pole results are summarized in Table 2.

The effort required the calculation of the needed electromagnetic, electroweak, QCD, and mixed radiative corrections to the predictions of the SM. Careful consideration of the competing definitions of the renormalized $\sin^2\theta_W$ was needed. The principal theoretical uncertainty is the hadronic contribution $\Delta\alpha_{had}^{(5)}(M_Z)$ to the running of α from its precisely known value at low energies to the Z-pole, where it is needed to compare the Z mass with the asymmetries and other observables. The radiative corrections, renormalization schemes, and running of α are further discussed in [1, 2]. The LEP Electroweak Working Group (LEPEWWG) [7] combined the results of the four LEP experiments, and also those of SLD and some WNC and Tevatron results, taking proper account of common systematic and theoretical uncertainties. Much theoretical effort also

TABLE 2. Non-Z-pole observables, 1/03. The SM values are updated from [2].

Quantity	Group(s)	Value	Standard Model	pull
m_t [GeV]	Tevatron	174.3 ± 5.1	174.4 ± 4.4	0.0
M_W [GeV]	LEP	80.447 ± 0.042	80.391 ± 0.018	1.3
M_W [GeV]	Tevatron/UA2	80.454 ± 0.059		1.1
g_L^2	NuTeV	0.30005 ± 0.00137	0.30396 ± 0.00023	-2.9
g_R^2	NuTeV	0.03076 ± 0.00110	0.03005 ± 0.00004	0.6
R^ν	CCFR	$0.5820 \pm 0.0027 \pm 0.0031$	0.5833 ± 0.0004	-0.3
R^ν	CDHS	$0.3096 \pm 0.0033 \pm 0.0028$	0.3092 ± 0.0002	0.1
R^ν	CHARM	$0.3021 \pm 0.0031 \pm 0.0026$		-1.7
$R^{\bar\nu}$	CDHS	$0.384 \pm 0.016 \pm 0.007$	0.3862 ± 0.0002	-0.1
$R^{\bar\nu}$	CHARM	$0.403 \pm 0.014 \pm 0.007$		1.0
$R^{\bar\nu}$	CDHS 1979	$0.365 \pm 0.015 \pm 0.007$	0.3816 ± 0.0002	-1.0
$g_V^{\nu e}$	CHARM II	-0.035 ± 0.017	-0.0398 ± 0.0003	—
$g_V^{\nu e}$	all	-0.041 ± 0.015		-0.1
$g_A^{\nu e}$	CHARM II	-0.503 ± 0.017	-0.5065 ± 0.0001	—
$g_A^{\nu e}$	all	-0.507 ± 0.014		0.0
$Q_W(\text{Cs})$	Boulder	-72.69 ± 0.44	-73.10 ± 0.04	0.8
$Q_W(\text{Tl})$	Oxford/Seattle	-116.6 ± 3.7	-116.7 ± 0.1	0.0
$10^3 \frac{\Gamma(b \to s\gamma)}{\Gamma_{SL}}$	BaBar/Belle/CLEO	$3.48^{+0.65}_{-0.54}$	3.20 ± 0.09	0.5
τ_τ [fs]	direct/$\mathcal{B}_e$/$\mathcal{B}_\mu$	$290.96 \pm 0.59 \pm 5.66$	291.90 ± 1.81	-0.4
$10^4 \Delta\alpha_{had}^{(3)}$	e^+e^-/τ decays	$56.53 \pm 0.83 \pm 0.64$	57.52 ± 1.31	-0.9
$10^9 (a_\mu - \frac{\alpha}{2\pi})$	BNL/CERN	$4510.64 \pm 0.79 \pm 0.51$	4508.30 ± 0.33	2.5

went into the development, testing, and comparison of radiative corrections packages, and into the study of how various classes of new physics would modify the observables, and how they could most efficiently be parametrized.

NEW INPUTS, ANOMALIES, THINGS TO WATCH

The results in Tables 1 and 2 are from 1/03, while the fit results to be presented in the next Section are from June 2002, updated from [2]. Jens Erler and I are currently performing a new analysis for the next edition of the *Review Of Particle Physics*; it is useful to list here some of the things that have or will change or to watch for.

- As of 3/03, the LEP 2 value for the W mass, 80.412(42) GeV, is smaller than the previous value of 80.447(42) GeV (used in Table 2) due to a revised ALEPH analysis [7]. This is closer to the SM best fit prediction of 80.391(18) GeV and will lead to a small increase in the predicted M_H. The Tevatron (CDF, DØ) Run I/UA2 value of 80.454(59) GeV is also slightly high. A new Run II value is expected.
- The direct lower limit on the SM Higgs mass from LEP 2 is $M_H > 114.4$ GeV (95% cl). The hints for events around 116 GeV were weakened in the final analysis.
- A more precise m_t from the Tevatron Run II is awaited. The preliminary CDF and DØ values still have large uncertainties [8]. A new preliminary DØ analysis of their

Run I data yields 180.1 ± 5.4 GeV [8], about 1σ above the previous combined value of 174.3 ± 5.1 GeV. This will again lead to an increase in the M_H prediction.

- There is a new estimate of α_s from the τ lifetime [9], which is quite precise though theory-error dominated, yielding $\alpha_s(M_\tau) = 0.356^{+0.027}_{-0.021}$, corresponding to $\alpha_s(M_Z) = 0.1221^{+0.0026}_{-0.0023}$.

- $A_{FB}(b)$, the forward-backward asymmetry into b quarks, has the value $0.0995(17)$, 2.4σ below the standard model global fit value of $0.1036(8)$. However, the SLD value for the related quantity $A_b = 0.922(20)$ is only 0.6σ below the expected $0.9348(1)$, and the hadronic branching fraction $R_b = 0.2166(7)$, which at one time appeared anomalous, is now only 1.1σ above the expectation $0.2157(2)$. If not just a statistical fluctuation or systematic problem, $A_{FB}(b)$ could be a hint of new physics. However, any such effect should not contribute too much to R_b. The deviation is only around 5%, but if the new physics involved a radiative correction to the coefficient κ of $\sin^2\theta_W$, the change would have to be around 25%. Hence, the new physics would most likely be at the tree level, mainly increasing the magnitude of the right-handed coupling to the b. This could be due to a heavy Z' boson with non-universal couplings to the third family [10, 11]; or to the mixing of the b_R with exotic quarks [2, 12], such as with an $SU(2)$ doublet involving a heavy B_R quark and a charge $-4/3$ partner [12]. There is a strong correlation between $A_{FB}(b)$ and the predicted Higgs mass M_H in the global fits. It has been emphasized [13] that if one eliminated $A_{FB}(b)$ from the fit (e.g., because it is affected by new physics) then the M_H prediction would be lower, with the central value well below the lower limit from the direct searches at LEP 2. One resolution, assuming $A_{FB}(b)$ is due to an experimental problem or fluctuation, is to invoke a supersymmetric extension of the standard model with light sneutrinos, sleptons, and possibly gauginos [14], which modify the radiative corrections and allow an acceptable M_H.

- The NuTeV collaboration at Fermilab [15] have reported the results of their deep inelastic measurements of $\dfrac{\overset{(-)}{\nu}_\mu N \to \overset{(-)}{\nu}_\mu X}{\overset{(-)}{\nu}_\mu N \to \mu^{\mp} X}$. They greatly reduce the uncertainty in the charm quark threshold in the charged current denominator by taking appropriate combinations of ν_μ and $\bar{\nu}_\mu$. They find a value for the on-shell weak angle s_W^2 of $0.2277(16)$, which is 3.0σ above the global fit value of $0.2228(4)$. The corresponding values for the left and right handed neutral current couplings [2] are $g_L^2 = 0.3001(14)$ and $g_R^2 = 0.0308(11)$, which are respectively 2.9σ below and 0.7σ above the expected $0.3040(2)$ and $0.0300(0)$. Possible standard model explanations include an unexpectedly large violation of isospin in the quark sea [15]; an asymmetric strange sea [16], though NuTeV's data seems to favor the wrong sign for this effect; nuclear shadowing effects [17]; or next to leading order QCD effects [16].

More exotic interpretations could include a heavy Z' boson [10, 16], although the standard GUT-type Z's do not significantly improve the fits, suggesting the need for a Z' with "designer" couplings. Mixing of the ν_μ with a heavy neutrino could account for the effect [18, 19], and also for the slightly low value for the number of light neutrinos $N_\nu = 2.986(7)$ from the Z line shape when N_ν is allowed to deviate from 3 (this shows up as a slightly high hadronic peak cross section in the standard model fit with $N_\nu = 3$) [2, 10]. This mixing would also affect muon decay, leading to an apparent Fermi constant smaller than the true value. This would be problematic for the other Z-pole observables, but could be compensated by a large negative T parameter [19]. However, such mixings would also lead to a lower value for $|V_{ud}|$, significantly aggravating the universality problem discussed below.

- The Brookhaven $g_\mu - 2$ experiment has reported a precise new value [20] using positive muons, leading to a new world average $a_\mu = 11659203(8) \times 10^{-10}$. Improvements in the statistical error from negative muon runs are anticipated. Using the theoretical value quoted by the experimenters for the hadronic vacuum polarization contribution a_μ^{had}, there was a small discrepancy, with $a_\mu(exp) - a_\mu(SM) = (26 \pm 11) \times 10^{-10}$, a 2.6σ effect. The value and uncertainty in a_μ^{had} are still controversial[1]: subsequent analyses based on e^+e^- data [21, 22] found a 3σ discrepancy, while an analysis using τ decay data [21] found a smaller 1σ effect. Recently the CDM-2 collaboration found a mistake in their theoretical code for the $e^+e^- \to e^+e^-$ cross section, used to determine the luminosity in the hadronic cross section [23]. This should lower the discrepancy from e^+e^- data to around 2σ, closer to the τ value. New data from KLOE is anticipated.

Because of the confused situation with the vacuum polarization, it is hard to know how seriously to take the discrepancy. Nevertheless, a_μ is more sensitive than the electron moment to most types of new physics, so it is important. One obvious candidate for a new physics explanation would be supersymmetry [24], with relatively low masses for the relevant sparticles and high $\tan\beta$ (roughly, one requires an effective mass scale of $\tilde{m} \sim 55$ GeV $\sqrt{\tan\beta}$). There is a correlation between the theoretical uncertainty in the vacuum polarization and in the hadronic contribution to the running of α to the Z pole [25], leading to a slight reduction in the predicted Higgs mass when a_μ is included in the global fit assuming the standard model.

- $\Delta\alpha_{had}^{(5)}(M_Z)$, the hadronic contribution to the running of α up to the Z-pole, introduces the largest theoretical uncertainty into the precision program, in particular to the relation between M_Z and the $\overline{MS}$ weak angle $\hat{s}_Z^2$ (extracted mainly from the asymmetries). The uncertainty is closely related to that in a_μ^{had}. There has been much recent progress using improved QCD calculations for the high energy part and more precise e^+e^- data from BES and elsewhere for the low energy part.

- A few years ago there was an apparent 2.3σ discrepancy between the measured value of the effective (parity-violating) weak charge $Q_W(Cs)$ measured in cesium [26], and the expected value. Cesium has a single electron outside a tightly bound core, so the atomic matrix elements could be reliably calculated, leading (it was thought) to a combined theoretical and experimental uncertainty of around 0.6%. However, it turns out that there are surprisingly large (O(1%)) radiative cor-

---

[1] There are also uncertainties in the smaller hadronic light by light diagram. An unfortunate sign error increased the apparent discrepancy with experiment at an earlier stage, but this has now been corrected.

rections, including Breit (magnetic) interactions, vacuum polarization, vertex, and self-energy corrections [27, 28]. After a somewhat confusing period, the situation has apparently stabilized, with the current value [28], $Q_W(\text{Cs}) = -72.84(46)$, in excellent agreement with the SM expectation, $-73.10(4)$. (An earlier $-72.69(44)$ is listed in Table 1.)

- The unitarity of the CKM matrix can be partially tested by the universality prediction that $\Delta \equiv 1 - |V_{ud}|^2 - |V_{us}|^2 - |V_{ub}|^2$ should vanish. In particular $|V_{ud}|$ can be determined by the ratio of G_β^V/G_μ, where G_β^V and G_μ are respectively the vector coupling in β decay and the μ decay constant. The most precise determination of $|V_{ud}|$ is from superallowed $0^+ \to 0^+$ transitions, currently yielding $|V_{ud}| = 0.9740(5)$ [29]. Combining with the PDG values for $|V_{us}|$ from kaon and hyperon decays and $|V_{ub}|$ from b decays, this yields a 2.3σ discrepancy $\Delta = 0.0032(14)$, suggesting either the presence of unaccounted-for new physics, or, possibly, effects from higher order isospin violation such as nuclear overlap corrections. However, the latter have been carefully studied, so the effect may be real. This problem has been around for some time, but until recently less precise determinations from neutron decay were consistent with universality. Recently, a more precise measurement of the neutron decay asymmetry has been made by the PERKEO-II group at ILL [30]. When combined with the accurately known neutron lifetime, this allowed the new determination $|V_{ud}| = 0.9713(13)$, implying $\Delta = 0.0083(28)$, i.e., a 3σ violation of unitarity. Note, however, that this value is only marginally consistent with the value obtained from superallowed transitions.

  Mixing of the ν_μ with a heavy neutrino, suggested as a solution of the NuTeV anomaly, would mean that G_μ is larger than the apparent value and would aggravate this discrepancy. (ν_e mixing would affect G_β^V and G_μ in the same way and have no effect.) However, a very small mixing of the W boson with a heavy W' coupling to right handed currents, as in left-right symmetric models, could easily account for the discrepancy for the appropriate sign for the mixing [31], especially if the right-handed neutrinos are Majorana and too heavy to be produced in the decays.

  The situation has recently become more complicated, by the suggestion that the culprit may be in the long accepted value of $|V_{us}|$. The BNL E865 experiment has recently performed a high statistics measurement of the K_{e3}^+ branching ratio [32], obtaining a result 2.3σ higher than the old measurements. This would be sufficient to account for the entire discrepancy, but must be confirmed by new analyses and measurements from KLOE, CMD-2 and NA48.

- The LEP and SLC Z-pole experiments are the most precise tests of the standard electroweak theory, but they are insensitive to any new physics that doesn't affect the Z or its couplings. Non-Z-pole experiments are therefore extremely important, especially given the possible NuTeV anomaly. In the near future we can expect new results in polarized Møller scattering from SLAC [33], and in the QWEAK polarized electron experiment at Jefferson Lab [34].

- Although the Z-pole program has ended for the time being, there are prospects for future programs using the Giga-Z option at a linear collider, which might yield a factor 10^2 more events. This would enormously improve the sensitivity [35], but would also require a large theoretical effort to improve the radiative correction calculations.

FIT RESULTS (06/02)

As of June, 2002, the result of the global fit was

$$\begin{align}
M_H &= 86^{+49}_{-32} \text{ GeV}, \\
m_t &= 174.2 \pm 4.4 \text{ GeV}, \\
\alpha_s &= 0.1210 \pm 0.0018, \\
\hat{\alpha}(M_Z)^{-1} &= 127.922 \pm 0.020 \\
\hat{s}_Z^2 &= 0.23110 \pm 0.00015, \\
\chi^2/\text{d.o.f.} &= 49.0/40(15\%)
\end{align} \tag{1}$$

The precision data alone yield $m_t = 174.0^{+9.9}_{-7.4}$ GeV from loop corrections, in impressive agreement with the direct Tevatron value 174.3 ± 5.1. The result $\alpha_s = 0.1210 \pm 0.0018$ for the strong coupling is somewhat above the previous world average $\alpha_s = 0.1172(20)$, which includes other determinations, most of which are dominated by theoretical uncertainties [36]. This is due in part to the inclusion of the new τ lifetime result [9]. (Without it, one would obtain $\alpha_s=0.1200 \pm 0.0028$.) The Z-pole value is insensitive to oblique (propagator) new physics, but is very sensitive to non-universal new physics, such as those which affect the $Zb\bar{b}$ vertex.

The prediction for the Higgs mass from indirect data, $M_H = 86^{+49}_{-32}$ GeV, should be compared with the direct LEP 2 limit $M_H \gtrsim 114.4(95\%)$ GeV. The theoretical range in the standard model is $115 \text{ GeV} \lesssim M_H \lesssim 750 \text{ GeV}$, where the lower (upper) bound is from vacuum stability (triviality). In the MSSM, one has $M_H \lesssim 130$ GeV, while M_H can be as high as 150 GeV in generalizations. Including the direct LEP 2 exclusion results, one finds $M_H < 215$ GeV at 95%. M_H enters the expressions for the radiative corrections logarithmically. It is fairly robust to many types of new physics, with some exceptions. In particular, a much larger M_H would be allowed for negative values for the S parameter or positive values for T. The predicted value would decrease if new physics accounted for the value of $A_{FB}(b)$ [13].

BEYOND THE STANDARD MODEL

The ρ_0 or S, T, and U parameters describe the tree level effects of Higgs triplets, or the loop effects on the W and Z propagators due to such new physics as nondegenerate fermions or scalars, or chiral families (expected, for example, in extended technicolor). The current values are:

$$\begin{align}
S &= -0.14 \pm 0.10(-0.08) \\
T &= -0.15 \pm 0.12(+0.09) \\
U &= 0.32 \pm 0.12(+0.01) \quad (2.6\sigma)
\end{align} \tag{2}$$

for $M_H = 115.6$ (300) GeV, where these represent the effects of new physics only (the m_t and M_H effects are treated separately). Similarly, $\rho_0 \sim 1 + \alpha T = 0.9997^{+0.0011}_{-0.0008}$ for $M_H = 73^{+106}_{-34}$ GeV and $S = U = 0$. If one constrains $T = U = 0$, then $S = 0.10^{+0.12}_{-0.30}$. There is a strong negative $S - M_H$ correlation, so that the Higgs mass constraint is relaxed to $M_H < 570$ GeV at 95%. For M_H fixed at 115.6 GeV, one finds $S = -0.040(62)$, which implies that the number of ordinary plus degenerate heavy families is constrained to be $N_{\text{fam}} = 2.81 \pm 0.29$. This is complementary to the lineshape constraint, $N_\nu = 2.986 \pm 0.007$, which only applies to neutrinos less massive than $M_Z/2$. One can also restrict additional nondegenerate families by allowing both S and T to be nonzero, yielding $N_{\text{fam}} = 2.79 \pm 0.43$ for $T = -0.01 \pm 0.11$.

In the decoupling limit of supersymmetry, in which the sparticles are heavier than $\gtrsim 200 - 300$ GeV, there is little effect on the precision observables, other than that there is necessarily a light SM-like Higgs, consistent with the data. There is little improvement on the SM fit, and in fact one can somewhat constrain the supersymmetry breaking parameters [37].

Heavy Z' bosons are predicted by many grand unified and string theories [2]. Limits on the Z' mass are model dependent, but are typically around $M_{Z'} > 500 - 800$ GeV from indirect constraints from WNC and LEP 2 data, with comparable limits from direct searches at the Tevatron. Z-pole data severely constrains the $Z - Z'$ mixing, typically $|\theta_{Z-Z'}| < \text{few} \times 10^{-3}$. A heavy Z' would have many other theoretical and experimental implications [38].

Precision data constrains mixings between ordinary and exotic fermions, large extra dimensions, new four-fermion operators, and leptoquark bosons [2].

Gauge unification is predicted in GUTs and some string theories. The simplest non-supersymmetric unification is excluded by the precision data. For the MSSM, and assuming no new thresholds between 1 TeV and the unification scale, one can use the precisely known α and $\hat{s}_Z^2$ to predict $\alpha_s = 0.130 \pm 0.010$ and a unification scale $M_G \sim 3 \times 10^{16}$ GeV [39]. The α_s uncertainties are mainly theoretical, from the TeV and GUT thresholds, etc. α_s is high compared to the experimental value, but barely consistent given the uncertainties. M_G is reasonable for a GUT (and is consistent with simple seesaw models of neutrino mass), but is somewhat below the expectations $\sim 5 \times 10^{17}$ GeV of the simplest perturbative heterotic string models. However, this is only a 10% effect in the appropriate variable $\ln M_G$. The new exotic particles often present in such models (or higher Kač-Moody levels) can easily shift the $\ln M_G$ and α_s predictions significantly, so the problem is really why the gauge unification works so well. It is always possible that the apparent success is accidental (cf., the discovery of Pluto).

CONCLUSIONS

The precision Z-pole, LEP 2, WNC, and Tevatron experiments have successfully tested the SM at the 0.1% level, including electroweak loops, thus confirming the gauge principle, SM group, representations, and the basic structure of renormalizable field theory. The standard model parameters $\sin^2 \theta_W$, m_t, and α_s were precisely determined. In fact, m_t was successfully predicted from its indirect loop effects prior to the direct discovery at the Tevatron, while the indirect value of α_s, mainly from the Z-lineshape, agreed with more direct QCD determinations. Similarly, $\Delta\alpha_{\text{had}}^{(5)}(M_Z)$ and M_H were constrained. The indirect (loop) effects implied $M_H \lesssim 215$ GeV, while direct searches at LEP 2 yielded $M_H > 114.5$ GeV, with a hint of a signal at 116 GeV. This range is consistent with, but does not prove, the expectations of the supersymmetric extension of the SM (MSSM), which predicts a light SM-like Higgs for much of its parameter space. The agreement of the data with the SM imposes a severe constraint on possible new physics at the TeV scale, and points towards decoupling theories (such as most versions of supersymmetry and unification), which typically lead to 0.1% effects, rather than TeV-scale compositeness (e.g., dynamical symmetry breaking or composite fermions), which usually imply deviations of several % (and often large flavor changing neutral currents). Finally, the precisely measured gauge couplings were consistent with the simplest form of grand unification if the SM is extended to the MSSM.

ACKNOWLEDGMENTS

It is a pleasure to thank my collaborators, especially Jens Erler, for fruitful interactions. This work was supported in part by a Department of Energy grant DOE-EY-76-02-3071.

REFERENCES

1. The early era is described in more detail in P. Langacker, *J. Phys.* G **29**, 1, 35 (2003) and in *eConf* **C010630**, P107 (2001), hep-ph/0110129.
2. For complete references, see J. Erler and P. Langacker, *Phys. Rev.* D **52**, 441 (1995), and *Electroweak Model and Constraints on New Physics*, in K. Hagiwara et al. [Particle Data Group Collaboration], *Phys. Rev.* D **66**, 010001 (2002); *Precision Tests of the Standard Electroweak Model*, ed. P. Langacker (Singapore, World, 1995); P. Langacker, M. Luo and A. K. Mann, *Rev. Mod. Phys.* **64**, 87 (1992).
3. J. E. Kim, P. Langacker, M. Levine and H. H. Williams, *Rev. Mod. Phys.* **53**, 211 (1981).
4. U. Amaldi et al., *Phys. Rev.* D **36**, 1385 (1987).
5. G. Costa, J. R. Ellis, G. L. Fogli, D. V. Nanopoulos and F. Zwirner, *Nucl. Phys.* **B297**, 244 (1988).
6. P. Langacker, *Comments Nucl. Part. Phys.* **19**, 1 (1989).
7. The LEP Collaborations ALEPH, DELPHI, L3, OPAL, the LEP Electroweak Working Group and the SLD Heavy Flavour Group: D. Abbanco et al., LEPEWWG/2003-01.
8. See the talk by P. Azzi at the *XXI International Symposium on Lepton and Photon Interactions at High Energies*, Fermilab, 8/03.
9. J. Erler and M. x. Luo, *Phys. Lett.* B **558**, 125 (2003).
10. J. Erler and P. Langacker, *Phys. Rev. Lett.* **84**, 212 (2000) and references theirin.
11. P. Langacker and M. Plümacher, *Phys. Rev.* D **62**, 013006 (2000).
12. D. Choudhury, T. M. Tait and C. E. Wagner, *Phys. Rev.* D **65**, 053002 (2002).
13. M. S. Chanowitz, *Phys. Rev. Lett.* **87**, 231802 (2001); *Phys. Rev.* D **66**, 073002 (2002).
14. G. Altarelli, F. Caravaglios, G. F. Giudice, P. Gambino and G. Ridolfi, *JHEP* **0106**, 018 (2001).
15. G. P. Zeller et al. [NuTeV Collaboration], *Phys. Rev. Lett.* **88**, 091802 (2002); *Phys. Rev.* D **65**, 111103 (2002); hep-ex/0207037, 0207052, 0210010; R. H. Bernstein, *J. Phys.* G **29**, 1919 (2003).
16. S. Davidson, S. Forte, P. Gambino, N. Rius and A. Strumia, *JHEP* **0202**, 037 (2002); S. Davidson, *J. Phys.* G **29**, 2001 (2003).
17. G. A. Miller and A. W. Thomas, hep-ex/0204007; W. Melnitchouk and A. W. Thomas, *Phys. Rev.* C **67**, 038201 (2003); S. Kovalenko, I. Schmidt and J. J. Yang, *Phys. Lett.* B **546**, 68 (2002); S. Kumano, *Phys. Rev.* D **66**, 111301 (2002). Some of these papers are commented on in [15].

18. K. S. Babu and J. C. Pati, hep-ph/0203029.
19. W. Loinaz, N. Okamura, T. Takeuchi and L. C. Wijewardhana, *Phys. Rev.* D **67**, 073012 (2003).
20. G. W. Bennett *et al.*, *Phys. Rev. Lett.* **89**, 101804 (2002) [Erratum-ibid. **89**, 129903 (2002)].
21. M. Davier, S. Eidelman, A. Hocker and Z. Zhang, *Eur. Phys. J.* C **27**, 497 (2003).
22. K. Hagiwara, A. D. Martin, D. Nomura and T. Teubner, *Phys. Lett.* B **557**, 69 (2003).
23. R. R. Akhmetshin *et al.* [the CMD-2 Collaboration], hep-ex/0308008.
24. See, for example, A. Czarnecki and W. J. Marciano, *Phys. Rev.* D **64**, 013014 (2001).
25. J. Erler and M. x. Luo, *Phys. Rev. Lett.* **87**, 071804 (2001).
26. S.C. Bennett and C.E. Wieman, *Phys. Rev. Lett.* **82**, 2484 (1999).
27. A. Derevianko, *Phys. Rev. Lett.* **85**, 1618 (2000); W. R. Johnson, I. Bednyakov and G. Soff, *Phys. Rev. Lett.* **87**, 233001 (2001) [Erratum-ibid. **88**, 079903 (2002)]; V. A. Dzuba, V. V. Flambaum and J. S. Ginges, *Phys. Rev.* D **66**, 076013 (2002); M. Y. Kuchiev, *J. Phys.* B **35**, L503 (2002).
28. M. Y. Kuchiev and V. V. Flambaum, hep-ph/0305053.
29. I. S. Towner and J. C. Hardy, *Phys. Rev.* C **66**, 035501 (2002).
30. H. Abele *et al.*, *Phys. Rev. Lett.* **88**, 211801 (2002).
31. See the articles by J. Deutsch and P. Quin, p. 706, and by A. Sirlin, p. 766, in *Precision Tests of the Standard Electroweak Model*, ed. P. Langacker (World, Singapore, 1995).
32. BNL E865, A. Sher *et al.*, hep-ex/0305042.
33. SLAC E158: http://www.slac.stanford.edu/exp/e158/.
34. The QWEAK experiment, http://www.jlab.org/qweak/.
35. J. Erler, S. Heinemeyer, W. Hollik, G. Weiglein and P. M. Zerwas, *Phys. Lett.* B **486**, 125 (2000)
36. See the review *Quantum chromodynamics*, I. Hinchliffe, in K. Hagiwara *et al.* [Particle Data Group Collaboration], *Phys. Rev.* D **66**, 010001 (2002).
37. J. Erler and D. M. Pierce, *Nucl. Phys.* B**526**, 53 (1998).
38. For a recent discussion, see P. Langacker, hep-ph/0308033.
39. P. Langacker and N. Polonsky, *Phys. Rev.* D **52**, 3081 (1995) and references theirin.

Lepton Dipole Moments

B. Lee Roberts

roberts@bu.edu
Department of Physics
Boston University
Boston, MA 02215 USA

Abstract. From the famous experiments of Stern and Gerlach to the present, measurements of magnetic dipole moments, and searches for electric dipole moments of "elementary" particles have played a major role in our understanding of sub-atomic physics. In this talk I discuss the progress on measurements and theory of the magnetic dipole moments of the electron and muon. I also discuss a new proposal to search for a permanent electric dipole moment (EDM) of the muon and put it into the more general context of other EDM searches.

INTRODUCTION AND THEORY OF THE LEPTON ANOMALIES

Over the past 82 years, the study of dipole moments of elementary particles has provided a wealth of information on subatomic physics. From the pioneering work of Stern[1] through the discovery of the large anomalous magnetic moments of the proton[2] and neutron[3], the ground work was laid for the discovery of spin, of radiative corrections and the renormalizable theory of QED, of the quark structure of baryons and the development of QCD.

A charged particle with spin $\vec{s}$ has a magnetic moment

$$\vec{\mu}_s = g_s(\frac{e}{2m})\vec{s}; \quad \mu = (1+a)\frac{e\hbar}{2m}; \quad a \equiv \frac{(g_s-2)}{2}; \qquad (1)$$

where g_s is called the gyromagnetic ratio. The expression in the middle is the quantity one finds listed in the Particle Data Tables.[4] The quantity a is the anomalous magnetic dipole moment (or simply the anomaly) which is related to the g-factor in the right-hand equation.

The Dirac equation tells us that $g \equiv 2$ for spin angular momentum, and is unity for orbital angular momentum (the latter having been verified experimentally[6]). This can be seen from the non-relativistic reduction for an electron in a weak magnetic field:

$$i\hbar\frac{\partial \psi}{\partial t} = \left[\frac{p^2}{2m} - \frac{e}{2m}(\vec{L}+2\vec{S})\cdot\vec{B}\right]\psi, \qquad (2)$$

and the subscript on g is dropped in the following discussion. For point particles, the anomaly arises from radiative corrections, three examples of which are shown in Fig. 1. The "vertex" correction and vacuum polarization are important in the lepton anomaly, while the self-energy term is included in the dressed mass. The situation for baryons is quite different, since their internal quark structure gives them large anomalies.

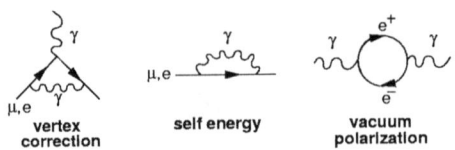

FIGURE 1. Three examples of radiative corrections. The middle term is absorbed into the dressed mass, but the vertex correction and vacuum polarization play an important role in the anomaly.

The vertex correction in lowest-order gives the famous Schwinger[5] result, $a = \alpha/2\pi$, which was verified experimentally by Foley and Kusch.[6] To lowest order, the S-matrix element[7] for a charged particle in a magnetic field is given by

$$-ie\bar{u}(p')\left[\frac{(p+p')_\lambda}{2m} + \left(1 + \frac{\alpha}{2\pi}\right)\frac{i\sigma_{\lambda\nu}q^\nu}{2m}\right]u(p)A^\lambda(q) \qquad (3)$$

In general a (or g) is an expansion in $\left(\frac{\alpha}{\pi}\right)$,

$$a = C_1\left(\frac{\alpha}{\pi}\right) + C_2\left(\frac{\alpha}{\pi}\right)^2 + C_3\left(\frac{\alpha}{\pi}\right)^3 + C_4\left(\frac{\alpha}{\pi}\right)^4 + \cdots \qquad (4)$$

with 1 diagram for the Schwinger (second-order) contribution, 5 for the fourth order, 40 for the sixth order, 891 for the eighth order.

The QED contributions to electron and muon $(g-2)$ have now been calculated through eighth order, $(\alpha/\pi)^4$ and the the tenth-order contribution has been estimated.[8] The first few orders are shown schematically below in Fig. 2.

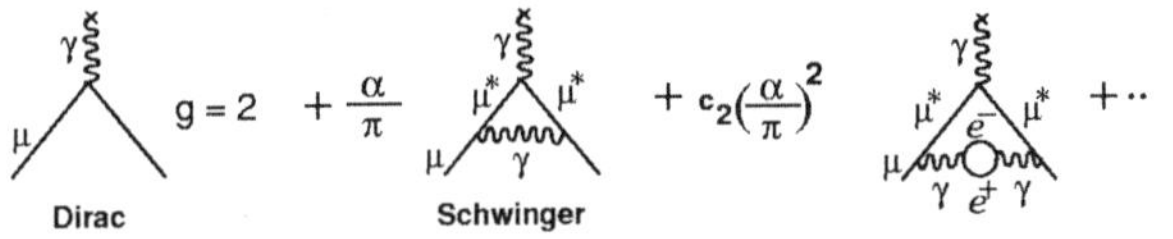

FIGURE 2. A schematic of the first few terms in the QED expansion for the muon. The vacuum polarization term shown is one of five of order $(\alpha/\pi)^2$.

While magnetic dipole moments (MDMs) are a natural property of charged particles with spin, electric dipole moments (EDMs) are forbidden both by parity and by time reversal symmetry. Interestingly enough, Purcell and Ramsey[9] suggested in 1950 that a measurement of the neutron EDM would be a good way to search for parity violation, well in advance of the paper by Lee and Yang. After the discovery of parity violation, Landau[10] and Ramsey[11] pointed out that an EDM would violate both P and T symmetries. This can be seen by examining the Hamiltonian for a spin one-half particle in the presence of both an electric and magnetic field, $\mathcal{H} = -\vec{\mu}\cdot\vec{B} - \vec{d}\cdot\vec{E}$. The transformation properties of $\vec{E}$, $\vec{B}$, $\vec{\mu}$ and $\vec{d}$ are given in the Table 1, and we see that while $\vec{\mu}\cdot\vec{B}$ is even under all three, $\vec{d}\cdot\vec{E}$ is odd under both P and T. Thus the existence of an EDM implies that both P and T are violated. In the context of CPT symmetry, an EDM implies CP violation. The standard model value for the electron and muon EDM is $\leq 10^{-35}$ e-cm, well beyond the reach of experiments (which are at the 10^{-26} e-cm level). Observation of a non-zero e or μ EDM would be a clear signal for new physics.

TABLE 1. Transformation properties of the magnetic and electric fields and dipole moments.

	$\vec{E}$	$\vec{B}$	$\vec{\mu}$ or $\vec{d}$
P	-	+	+
C	-	-	-
T	+	-	-

The connection between the magnetic and electric dipole moments can be seen by writing the interaction Lagrangian as

$$\mathscr{L}_{dm} = \frac{1}{2}\left[D\bar{\mu}\sigma^{\alpha\beta}\frac{1+\gamma_5}{2} + D^*\bar{\mu}\sigma^{\alpha\beta}\frac{1-\gamma_5}{2}\right]\mu F_{\alpha\beta} \qquad (5)$$

where the dipole operator D has $\mathrm{Re}\,D = a_\mu \frac{e}{2m_\mu}$ and $\mathrm{Im}\,D = d_\mu$.

The standard model value of a has three contributions from radiative processes: QED loops containing leptons (e, μ, τ) and photons; hadronic loops containing hadrons in vacuum polarization loops; and weak loops involving the weak gauge bosons W, Z, and Higgs. Thus $a_{e,\mu}(\mathrm{SM}) = a_{e,\mu}(\mathrm{QED}) + a_{e,\mu}(\mathrm{hadronic}) + a_{e,\mu}(\mathrm{weak})$. A difference between the experimental value and the standard model prediction would signify the presence of new physics beyond the standard model. Examples of such potential contributions are lepton substructure, extra gauge bosons, anomalous $W - \gamma$ couplings, or the existence of supersymmetric partners of the leptons and gauge bosons.

The electron anomaly is now measured to a relative precision of about four parts in a billion (ppb),[14] and the muon is measured to 0.7 parts per million (ppm).[15] The relative contributions of heavier particles to a scales as $(m_e/m_\mu)^2$, and the muon has a sensitivity factor of about 40,000 over the electron to higher mass scale radiative corrections. This gives the muon an overall advantage of 230 in measurable sensitivity to larger mass scales, including new physics. At a precision of 0.7 ppm, the muon anomaly is sensitive to ≥ 100 GeV scale physics.

In fact, the contribution of anything heavier than an electron to the electron anomaly is at the level of about 3 ppb. So while the the electron $(g-2)$ experiments are triumphs of experimental and theoretical physics, they are purely a test of QED. Since the independent measurements of α are less precise (7.4 ppb) than the present accuracy on the electron anomaly, the measurement of a_e has been used to determine the best measurement of the fine-structure constant.[16] The uncertainty in α is not an issue for a_μ.

The CERN experiment[12] observed the contribution of hadronic vacuum polarization shown in Fig. 3(a) at the 10 standard deviation level. Unfortunately, the hadronic contribution cannot be calculated directly from QCD, since the energy scale is very low $(m_\mu c^2)$, although Blum[13] has performed a proof of principle calculation on the lattice. Fortunately dispersion theory gives a relationship between the vacuum polarization loop and the cross section for $e^+e^- \to$ hadrons,

$$a_\mu(\mathrm{Had};1) = (\frac{\alpha m_\mu}{3\pi})^2 \int_{4m_\pi^2}^{\infty} \frac{ds}{s^2} K(s) R(s), \quad \text{where} \quad R \equiv \frac{\sigma_{\mathrm{tot}}(e^+e^- \to \mathrm{hadrons})}{\sigma_{\mathrm{tot}}(e^+e^- \to \mu^+\mu^-)} \qquad (6)$$

and experimental data are used as input. The factor s^{-2} in the dispersion relation, means that values of $R(s)$ at low energies (the ρ resonance) dominate the determination of $a_\mu(\text{Had};1)$. This information can also be obtained from hadronic τ^- decays such as $\tau^- \to \pi^-\pi^0\nu_\tau$, which can be related to e^+e^- annihilation through the CVC hypothesis and isospin conservation.[17, 18]

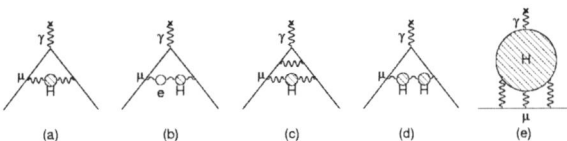

FIGURE 3. The hadronic contribution to the muon anomaly, where the dominant contribution comes from (a). The hadronic light-by-light contribution is shown in (e).

The quantity R is not directly measured. The cross-section for e^+e^- is determined using some other normalization, followed by careful subtractions for initial state radiation, vacuum polarization etc., and the denominator is calculated from QED. Most of these effects would cancel in the ratio if $R(s)$ were measured directly.

Knowledge of the hadronic contribution has steadily improved over the past 15 years. When the muon $(g-2)$ experiment E821 began at Brookhaven in the early 1980s, $a_\mu(\text{Had};1)$ was known to about 5 ppm. Now the uncertainty is about 0.5 ppm. This progress has not come without some pain. The hadronic light-by-light contribution has changed sign twice, with the positive sign now having been confirmed by a number of authors.[19] New high-precision data from Novosibirsk on $e^+e^- \to$ hadrons lowered both the value and the error on $a_\mu(\text{Had};1)$.[20, 18] The value obtained from τ-decay differed from that using e^+e^-.[18] Recently a normalization error was uncovered, which moves the value of $a_\mu(\text{Had};1)$ obtained from e^+e^- closer to that obtained from τ-decay,[21] and a revised value[22] of $a_\mu(\text{Had};1)$ shows a difference between the standard model value and the experimental result from E821 to be: $(22.1 \pm 7.2 \pm 3.5 \pm 8.0) \times 10^{-10}$ (1.9σ) and $(7.4 \pm 5.8 \pm 3.5 \pm 8.0) \times 10^{-10}$ (0.7σ) for the e^+e^- and τ-based estimates respectively. The second error is from the hadronic light-by-light contribution and the third one from E821. Further measurements of the hadronic cross section are underway at Frascati and BaBar, using initial state radiation to lower $\sqrt{s}$ below the beam energy (sometimes called radiative return).

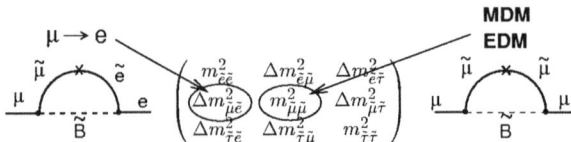

FIGURE 4. The supersymmetric contributions to the anomaly, and to $\mu \to e$ conversion, showing the relevant slepton mixing matrix elements. The MDM and EDM give the real and imaginary parts of the matrix element, respectively.

One of the very useful roles measurements of a_μ have played in the past is placing serious restrictions on physics beyond the standard model. With the development of supersymmetric theories as a favored scheme of physics beyond the standard model, interest in the experimental and theoretical value of a_μ has grown substantially. SUSY contributions to a_μ could be at a measurable level in a broad range of models. Furthermore, there is a complementarity between the SUSY contributions to the MDM, EDM and transition moment for $\mu \to e$. The MDM and EDM are related to the real and imaginary parts of the diagonal element of the slepton mixing matrix, and the transition moment is related to the off diagonal one, as shown in Fig. 4.

MEASUREMENT OF THE MUON ANOMALY

The method used in the third CERN experiment and the BNL experiment are very similar, save the use of direct muon injection[23] into the storage ring,[24, 25] which was developed by the E821 collaboration. These experiments are based on the fact that for $a_\mu > 0$ the spin precesses faster than the momentum vector when a muon travels transversely to a magnetic field. The spin precession frequency ω_S consists of the Larmor and Thomas spin-precession terms. The spin frequency ω_S, the momentum precession (cyclotron) frequency ω_C, and the difference frequency ω_a are given by

$$\omega_S = \frac{geB}{2mc} + (1-\gamma)\frac{eB}{\gamma mc}; \qquad \omega_C = \frac{eB}{mc\gamma}; \qquad \omega_a = \omega_S - \omega_C = \left(\frac{g-2}{2}\right)\frac{eB}{mc}. \qquad (7)$$

The difference frequency is the frequency with which the spin precesses relative to the momentum, which is proportional to the anomaly, rather than to g. A precision measurement of a_μ requires precision measurements of the precession frequency ω_a and the magnetic field, which is expressed as the free-proton precession frequency ω_p in the storage ring magnetic field.

The muon frequency can be measured as accurately as the counting statistics and detector apparatus permit. The design goal for the NMR magnetometer and calibration system was a field accuracy of 0.1 ppm. The B which enters in Eq. 7 is the average field seen by the ensemble of muons in the storage ring. The need for vertical focusing implies that a gradient field is needed, but the usual magnetic gradient used in storage rings is ruled out, since a sufficient magnetic gradient for vertical focusing would spoil the ability to use NMR to measure the magnetic field to the necessary accuracy. An electric quadrupole is used for vertical focusing, taking advantage of the "magic" $\gamma = 29.3$ at which an electric field does not contribute to the spin motion relative to the momentum. With both an electric and a magnetic field, the spin difference frequency is given by

$$\vec{\omega}_a = -\frac{e}{mc}\left[a_\mu \vec{B} - \left(a_\mu - \frac{1}{\gamma^2 - 1}\right)\vec{\beta} \times \vec{E}\right], \qquad (8)$$

which reduces to Eq. 7 in the absence of an electric field. For muons with $\gamma = 29.3$ in an electric field alone, the spin would follow the momentum vector.

The experimental signal is the $e^\pm$ from $\mu^\pm$ decay, which were detected by lead-scintillating fiber calorimeters.[26] The time and energy of each event was stored for analysis offline. Muon decay is a three-body decay, so the 3.1 GeV muons produce a continuum of positrons (electrons) from the end-point energy down. Since the highest energy $e^\pm$ are correlated with the muon spin, if one counts high-energy $e^\pm$ as a function of time, one gets an exponential from muon decay modulated by the $(g-2)$ precession. The expected form for the positron time spectrum is $f(t) = N_0 e^{-\lambda t}[1 + A\cos(\omega_a t + \phi)]$, however in analyzing the data it is necessary to take a number of small effects into account in order to obtain a satisfactory χ^2 for the fit.[15]

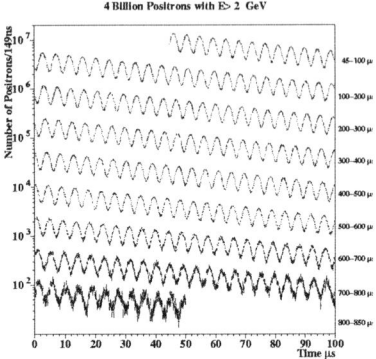

FIGURE 5. The time spectrum of positrons with energy greater than 2.0 GeV from the Y2000 run. The endpoint energy is 3.1 GeV. The time interval for each of the diagonal "wiggles" is given on the right.

The experimental results thus far are shown in Fig. 6, with the average

$$a_\mu(\exp) = 11\,659\,203(8) \times 10^{-10} \qquad (\pm 0.7 \text{ ppm}) \qquad (9)$$

being dominated by results from E821. The theory value does not reflect the re-analysis just made available,[22] but rather $a_\mu(\text{Had};1)$ determined from older data[17] is shown. One additional data set from E821 is being analyzed; the only data set obtained with negative muons. The result should be finalized in Fall '03 with an expected uncertainty between 0.7 and 0.8 ppm. We are exploring the possibility of an upgraded experiment at Brookhaven or at J-PARC[27] which could reach 0.1 ppm (BNL) or 0.06 ppm (J-PARC) precision.

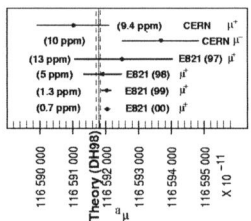

FIGURE 6. Measurements of a_μ. The theory comes from Davier and Höcker[17] with the sign of the hadronic light by light corrected.

MEASUREMENT OF THE ELECTRON ANOMALY

While measurements of the electron anomaly have a history stretching back to Kusch,[6] a major breakthrough in precision came with the trap experiments of Dehmelt.[14]

Single electrons were captured in a Penning trap, and the difference (beat) frequency ω_a was measured (see Eq. 7). Conventional resonance techniques were employed to measure ω_c, and this work was carried out using a hyperbolic trap where the cavity shifts of the measured magnetic moments placed serious limitations on the precision available.[28]

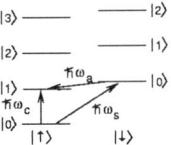

FIGURE 7. The energy levels of the quantum cyclotron. The energy difference between the spin-flip transition and the cyclotron excitation arises because $g \neq 2$ ($a \neq 0$).

Recently Gabrielse has developed a cooled cylindrical trap which he calls a quantum cyclotron.[29] The trap is cooled to mK temperatures, so that thermal excitation of the quantum cyclotron levels is very improbable, and the axial motion is reduced. The quantum levels of this system are shown in Fig. 7, where the three frequencies now become quantum transitions. The spin flip transition is slightly different in energy from the cyclotron energy since $a \neq 0$. The anomaly is determined by the ratio of two frequencies

$$\frac{\omega_a}{\omega_c} = \frac{g-2}{2} = a; \quad \text{where} \quad \omega_a \simeq 170 \text{ MHz}; \quad \omega_{s,c} \simeq 160 \text{ GHz} \qquad (10)$$

without the need for fundamental constants. The goal of is an order of magnitude improvement on the precision of a_e, or $\sim \pm 0.3$ ppb.

In the cylindrical trap, spontaneous emission is suppressed by two orders of magnitude. Thus one can stimulate these quantum transitions and measure the two frequencies. The benefits of cooling the trap can be seen in Fig. 8 where transitions to the first and second energy states are observed until the temperature is lowered to 0.08 K. The trap is built and working, and the first results can be expected in late 2003.

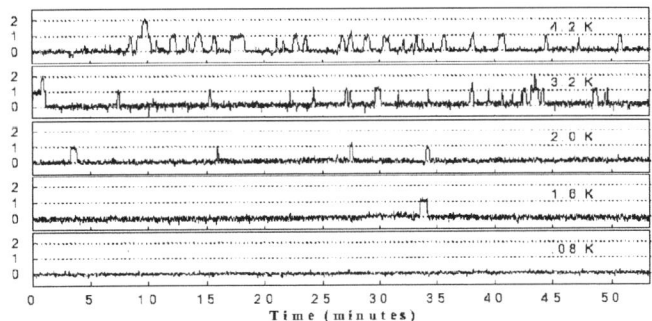

FIGURE 8. Thermal excitation of the levels of the quantum cyclotron. As the trap is cooled, the electron spends more time in the ground state. At 0.08 K the electron can spend on the order of two hours without a thermal excitation (courtesy of G. Gabrielse).

EDM SEARCHES, ESPECIALLY FOR THE MUON

While the MDM has a substantial standard model value, the predicted EDMs for the leptons are unmeasurably small and lie orders of magnitude below the present experimental limits given in Table 2. Thus the presence of an EDM at a measurable level would signify physics beyond the standard model. Since the presently known *CP* violation is inadequate to describe the baryon asymmetry in the universe, additional sources of *CP* violation should be present. SUSY models do predict an EDM.[33] For a new physics contribution to a_μ of 3×10^{-9} (of the order which might have been seen in E821 before the CDM2 normalization error was found).

$$d_\mu^{NP} \simeq 3 \times 10^{-22} \left(\frac{a_\mu^{NP}}{3 \times 10^{-9}} \right) \tan \phi_{CP} \ e\text{--cm} \qquad (11)$$

where ϕ_{CP} is a *CP* violating phase.

TABLE 2. Measured limits on electric dipole moments, and their standard model values

Particle	Present EDM Limit (e-cm)	Standard Model Value (e-cm)
n[30]	6.3×10^{-26}	10^{-31}
e^-[31]	$\sim 1.6 \times 10^{-27}$	10^{-38}
μ[12]	$< 10^{-18}$ (CERN) $\sim 10^{-19}$ (E821)* $\sim 10^{-24}$ †	10^{-35}

* Estimated limit, work in progress.
† Proposed new dedicated experiment.[32]

A new experiment to search for a permanent EDM of the muon with a design sensitivity of 10^{-24} *e*-cm is now being planned for J-PARC.[32] This sensitivity lies well within values predicted by SUSY models.[33] Feng, et al.,[34] have calculated the range of ϕ_{CP} available to such an experiment, which is shown in Fig. 9.

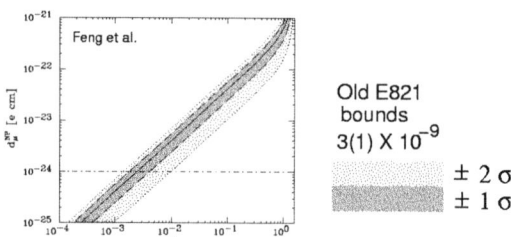

FIGURE 9. The range of ϕ_{CP} available to a dedicated muon EDM experiment.[34] The two bands show the one and two standard-deviation ranges if a_μ differs from the standard model value by $(3 \pm 1) \times 10^{-9}$.

With an EDM present, the spin precession relative to the momentum is given by

$$\vec{\omega} = -\frac{e}{m}\left[a_\mu \vec{B} - \left(a_\mu - \frac{1}{\gamma^2-1}\right)\frac{\vec{\beta}\times\vec{E}}{c}\right] + \frac{e}{m}\left[\frac{\eta}{2}\left(\frac{\vec{E}}{c} + \vec{\beta}\times\vec{B}\right)\right] \quad (12)$$

where $d_\mu = \frac{\eta}{2}(\frac{e\hbar}{2mc}) \simeq \eta \times 4.7 \times 10^{-14}$ $e-$cm and $a_\mu = (\frac{g-2}{2})$. For reasonable values of β, the motional electric field $\vec{\beta}\times\vec{B}$ is much larger than electric fields that can be obtained in the laboratory, and the two vector frequencies are orthogonal to each other. The EDM has two effects on the precession: the magnitude of the observed frequency is increased, and the precession plane is tipped relative to the magnetic field.

E821 was operated at the magic γ so that the focusing electric field did not cause a spin precession. The EDM signal in E821 is very difficult to observe, since the tipping of the precession plane is very small (≤ 5 mrad). The dedicated experiment will be operated at 500 MeV/c, off of the magic γ, and will use a radial electric field to stop the $(g-2)$ precession. Thus the EDM would cause a steady build-up of the spin out of the plane with time. Detectors would be placed above and below the storage region, and a time-dependent up-down asymmetry would be the signal of an EDM.

SUMMARY AND CONCLUSIONS

Measurements of the muon and electron anomalies played an important role in our understanding of sub-atomic physics in the 20th century. The electron anomaly was tied closely to the development of QED. The subsequent measurement of the muon anomaly showed that the muon was indeed a "heavy electron" which obeyed QED.[12] With the sub-ppm accuracy now available for the muon anomaly,[15] there may be indications that new physics is beginning to appear. Marciano[35] has pointed out that using a few very well measured standard model parameters, rather than a global fit to all electroweak measurements, predicts a Higgs mass which is much smaller than the present experimental limit from LEP. If one argues that any discrepancy between the standard model value of a_μ and the experimental one is an indication that the hadronic contribution has been underestimated, and uses this discrepancy to "determine the hadronic contribution", the Higgs mass limit gets even smaller. Marciano concludes that "hints of 'New Physics' may be starting to appear in quantum loop effects."

The non observation of an electron EDM is becoming an issue for supersymmetry, just as the non-observation of a neutron EDM implies such a mysteriously (some would say un-naturally) small θ-parameter for QCD. The search for EDMs will continue, and if one is observed, the motivation for further searches in other systems will be even stronger. The muon presents a unique opportunity to observe an EDM in a second-generation particle, where the CP phase might be different from the first generation, or the scaling with mass might be quadratic rather than linear.

It is clear that the study of lepton moments (and neutron EDM searches) will continue to be a topic of great importance in the first part of the 21st century. Both the theoretical and experimental situations are evolving. Stay tuned for further developments.

ACKNOWLEDGMENTS

I wish to thank my colleagues on the muon $(g-2)$ experiment, as well as M. Davier, E. de Rafael, G. Gabrielse, W. Marciano, and B. Odum for helpful discussions. Special thanks go to R. Carey and D. Hertzog for their helpful comments on this manuscript.

REFERENCES

1. O. Stern, Z. Phys. **7**, 249 (1921), W. Gerlach and O. Stern, Ann. Physik, **74**, 1924.
2. R. Frisch and O. Stern, Z. Phys.**85**, 4 (1933).
3. Luis W. Alvarez and F. Bloch, Phys. Rev. **57**, 111 (1940).
4. Particle Data Group, Phys. Rev. **D66**, Part 1 (2002).
5. J. Schwinger, Phys. Rev. **73**, 416L (1948).
6. H.M. Foley and P. Kusch, Phys. Rev. **73**, 4121 (1948).
7. J.D. Bjorken and S.D. Drell, *Relativistic Quantum Mechanics*, McGraw-Hill, (1964), p172.
8. T. Kinoshita and M. Nio, Phys. Rev. Lett. **90**, 021803-1 and references therein.
9. E.M. Purcell and N.F. Ramsey, Phys. Rev. **78**, 807 (1950)
10. L. Landau, Nucl. Phys. **3**, 127 (1957).
11. N.F. Ramsey Phys. Rev. **109**, 225 (1958).
12. J. Bailey, et. al, Nucl. Phys. **B150**, 1 (1979).
13. T. Blum, Phys. Rev. Lett. **91**, 052001-1 (2003), and at this conference.
14. R.S. Van Dyck et al., Phys. Rev. Lett., **59**, 26(1987) and in *Quantum Electrodynamics*, (Directions in High Energy Physics Vol. 7) T. Kinoshita ed., World Scientific, 1990, p.322.
15. G.W. Bennett, et al., (Muon $(g-2)$ Collaboration), Phys. Rev. Lett. **89**, 101804 (2002)
16. T. Kinoshita, Rep. Prog. Phys.**59**,1459 (1996) which is updated in Ref. [8].
17. M. Davier and A. Höcker, Phys. Lett.**B435**, 427 (1998).
18. M. Davier, S. Eidelman, A. Höcker, and Z. Zhang, Eur. Phys J. **C 27** 497 (2003). See K. Hagiwara, A.D. Martin, D. Nomura, and T. Teubner Phys. Lett. **B557**, 69 (2003) for an independent analysis of the e^+e^- data.
19. Marc Knecht, Andreas Nyffeler, Phys. Rev. **D65**, 073034 (2002), M. Knecht, A. Nyffeler, M. Perrottet, E. De Rafael, Phys. Rev. Lett. **88**, 071802 (2002), I. Blokland, A. Czarnecki and K. Melinkov Phys. Rev. Lett. **88**, 071803 (2002), and references therin.
20. R.R. Akhmetshin, et al, CMD2 collaboration, Phys. Lett. **B 527**, 161 (2002).
21. R.R. Akhmetshin, et al, CMD2 collaboration arXiv:hep-ex/0308008, (2003), submitted to Phys. Lett.
22. M. Davier, S. Eidelman, A. Höcker, and Z. Zhang, Xiv.org/abs/hep-ph/0308213, (2003)
23. E. Efstathiadis, et al., Nucl. Inst. and Meth. **A496**,8-25 (2002).
24. G.T. Danby, et al., Nucl. Instr. and Methods, **A 457**, 151-174 (2001).
25. A. Yamamoto, et al., Nucl. Inst. and Meth. **A496**,8-25 (2002).
26. S.A. Sedyk, et al., Nucl. Inst.and Meth., **A455** 346, (2000).
27. J-PARC Letter of Intent L17, *An Improved Muon $(g-2)$ Experiment at J-PARC*, B.L. Roberts contact person.
28. G. Gabrielse, J. Tan and L.S. Brown in *Quantum Electrodynamics*, (Directions in High Energy Physics Vol. 7) T. Kinoshita ed., World Scientific, 1990, p.389. See also Ref. [14].
29. See B. D'Urso, B. Odom and G. Gabrielse, Phys. Rev. Lett. **90**, 043001-1 (2003) for a description of the trap.
30. P.G. Harris, et al., Phys. Rev. Lett. **82**, 904 (1999).
31. B.C. Regan, et al., Phys. Rev. Lett. **88**, 071805-1 (2002).
32. J-PARC Letter of Intent L22, *Search for a permanent muon electric dipole moment at the 10^{-24} e cm level.*, Y. Kuno, J.Miller, Y. Semertzidis spokespersons.
33. K.S. Babu, B. Datta, and R.N. Mohapatra, Phys. Rev. Lett.**85**, 5064 (2000).
34. J.L. Feng, K.T. Matchev .Y, Shadmi, Nucl. Phys. **B 613**, 366 (2001), and Phys. Lett. **B555**, 89 (2003).
35. W.J. Marciano, J. Phys. **G29**, 23 (2003), and J. Phys. **G29**, 225 (2003).

Electron Scattering and Hadron Structure

E. J. Beise

University of Maryland, College Park, MD, USA

Abstract. Electron Scattering from nucleons and light mesons has provides precise information about the ground state structure of matter. In this talk I present a survey of recent experimental progress on the electromagnetic and weak structure of nucleons as well as the structure of pions and kaons.

INTRODUCTION

One of the fundamental challenges of nuclear physics today is to construct a quantitative description of the lightest hadrons in terms of their underlying constituents. Such a description should be able to account not only for their static properties, such as overall charge, magnetism, mass and spin which can largely be accounted for by constituent quarks, but also their dynamical properties such as contributions from gluons and the quark-antiquark sea and the relativistic motion of quarks to charge and magnetization distributions of hadrons in their ground state. By nature such a description is non-perturbative and is therefore tremendously challenging theoretically. Lattice calculations of these properties are beginning to provide information [1], but are still in their early stages, so quantitative interpretation of the data is still to come. From the experimental side there has been significant progress since the last Intersections meeting, largely the result of technical advances in polarized beams, targets and recoil polarimetry that have occurred over the last decade. In this paper I will focus on an overview of some of the recent experimental progress at the various electron scattering laboratories on nucleon electromagnetic form factors, as well as neutral weak nucleon structure and the charge structure of pions and kaons. For more detailed discussion on these topics, see the talks by others at this conference [2, 3, 4, 5].

PION AND KAON CHARGE FORM FACTORS

From a theoretical point of view, the simplest multi-quark system that can be studied in its ground state is the pion, the lightest $q\bar{q}$ object. Its electromagnetic structure is determined by a single charge form factor F_π, which has a well-defined behavior as $Q^2 \longrightarrow \infty$ to lowest order in perturbative QCD [6], depending only on the QCD coupling constant α_s and the pion decay constant f_π:

$$F_\pi \longrightarrow 8\pi \frac{\alpha_s f_\pi^2}{Q^2}. \tag{1}$$

Experimental determination of F_π is complicated by the lack of a target. The charge radius is well determined from π-e scattering [7], but anything other than the leading Q^2 behavior requires the use of pion electroproduction where a virtual photon is absorbed by the nucleon's pion cloud. The unpolarized cross section for pion electroproduction is

$$\frac{d^5\sigma}{d\Omega_{\pi^*}d\Omega_e dE_e} = 2\pi\Gamma(Q^2)\frac{d\sigma_v}{dt\,d\phi} \quad (2)$$

where $\Gamma(Q^2)$ is the virtual photon flux and

$$\frac{d\sigma_v}{dt\,d\phi} = \frac{d\sigma_T}{dt\,d\phi} + \varepsilon\frac{d\sigma_L}{dt\,d\phi} + \sqrt{2\varepsilon(1+\varepsilon)}\cos\phi\frac{d\sigma_{LT}}{dt\,d\phi} + \varepsilon\cos 2\phi\frac{d\sigma_{TT}}{dt\,d\phi}. \quad (3)$$

Integrating over the pion out-of-plane angle and varying the photon polarization $\varepsilon = \left[1+2(1+\tau)\tan^2\frac{\theta}{2}\right]^{-1}$ allows one to isolate the longitudinal component σ_L which is dominated by the photon coupling directly to the virtual pion cloud of the nucleon. There is also a strong t-dependence from the π-N vertex. Extrapolating to the mininum allowable t allows extraction of F_π. The t-dependence of σ_L is reasonably well described by the Regge model of Vanderhaeghen, Guidal and Laget [8], as shown in Figure 1(a), for the kinematics of a recent pion form factor measurement at Jefferson Laboratory [9]. The transverse component is not as well described: it is speculated that baryon resonance contributions to the cross section, which are not included in the model, may be the cause. In Figure 1(b) is shown the results from [9] along with earlier measurements, two future projections, and a variety of models. At moderate momentum transfers the JLab data give the first determination of F_π from a single experiment, providing a significant improvement in precision, even with the model dependence of the t-behavior which is included in the experimental uncertainties. One can see that the data to date are far from the pQCD limit. The calculation that best describes the measured F_π is that of Maris and Tandy [10], who use the Bethe Salpeter and Schwinger-Dyson equations to reproduce the pion mass, decay constant and chiral condensate and then predict the Q^2-dependence $F_\pi(Q^2)$. The data are also reasonably well described by the QCD sum-rules calculation of Nestarenko and Radyushkin [11]. A new measurement is presently underway that will provide a modest increase in dynamic range, perhaps ruling out some calculations and giving a hint about the higher Q^2 behavior. Beyond this a higher energy electron beam, such as that proposed for a 12-GeV upgrade to Jefferson Lab, would be required to provide the kinematic reach where one might begin to approach the expectation from pQCD.

While the charge radius of the K^+ is experimentally well determined [16], its charge form factor away from $Q^2 = 0$ is less well known than that of the pion. Until recently, very little electroproduction data have been available with which to establish either that the longitudinal portion of the cross section is large or to determine its t-dependence. Many more resonances are involved in K electroproduction, and models of the process are not yet well constrained by data. Recent measurements at JLab of kaon photo- and electro-production over a wide kinematic range [17, 18], as well as precise measurements of L/T separated cross sections at specific kinematics [19] are now finally beginning to constrain the models. New measurements of the t-dependence of the longitudinal

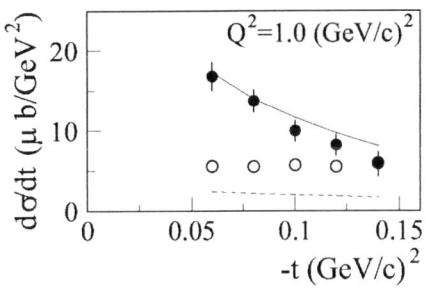 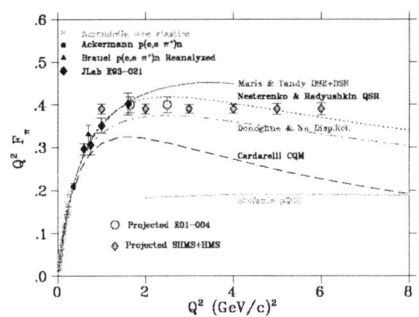

FIGURE 1. (a) Pion Electroproduction cross sections from [9], separated into longitudinal (solid circles) and transverse (open circles) components, and compared to a calculation based on Regge trajectories for π- and ρ-like particles [8]. (b) The pion's charge form factor. The published data are from (solid diamonds) [9], ($\times$) [7], (solid triangles) [12], and (solid circles) [13]. Also shown are several calculations and projected uncertainties for a new measurement in progress (open circles) [14], and for a possible measurement at Jefferson Lab with 12 GeV beam (open diamonds) [15].

cross section are presently being analyzed, with a goal of extracting the kaon form factor at two values of Q^2 near 2 (GeV/c)^2 [20]. Thus, the first determination of the kaon form factor at significant momentum transfer from electron scattering may be available by the time of the next Intersections Conference.

NUCLEON ELECTROMAGNETIC FORM FACTORS

Precise determination of nucleon electromagnetic form factors has been a topic of intense activity at all of the electron scattering laboratories over the last decade, and significant progress has been made in recent years. The majority of the information on the proton charge and magnetic form factors comes from unpolarized electron scattering experiments. In the one-photon exchange approximation, the electron-nucleon elastic scattering cross section in the laboratory frame is

$$\frac{d\sigma}{d\Omega} = \left(\frac{\alpha^2}{4E^2 \sin^4 \frac{\theta}{2}}\right) \cos^2 \frac{\theta}{2} \frac{E'}{E} \frac{1}{\varepsilon(1+\tau)} \left[\varepsilon G_E^2 + \tau G_M^2\right], \tag{4}$$

where E and E' are the incident and scattered electron energy and θ is the electron's scattering angle in the lab frame. Through measurements at various scattering angles one separates the electric and magnetic contributions. Very recently, this Rosenbluth technique has come under some degree of scrutiny because of new results from polarization experiments that are in conflict with the results from unpolarized electron scattering at momentum transfers above 1 (GeV/c)^2.

With a polarized beam and either a polarized target or detection of the polarization of the recoiling nucleon it is possible to determine directly either the ratio of G_E/G_M, or the product $G_E G_M$ in a single measurement. These techniques do not require absolute determination of the experimental cross section and are thus less susceptible to experimental systematic uncertainties related to knowledge of detector acceptance, absolute beam flux, or detector efficiencies. With a longitudinally polarized beam and detection of the recoil polarization, the ratio of polarization components perpendicular and parallel to the nucleon's momentum determines the ratio [21, 22, 23, 24]

$$\frac{G_E}{G_M} = -\frac{P_T}{P_L}\frac{E+E'}{2M_N}\tan\frac{\theta}{2}. \qquad (5)$$

In the last few years a series of data has been taken at Jefferson Laboratory using the recoil polarization technique for e-p scattering [25, 26, 27], in which a monotonic decrease of G_E/G_M with increasing Q^2 was found, in constrast to the world's set of Rosenbluth data that seemed to indicate a more constant ratio. In an attempt to understand this discrepancy, Arrington carried out a new fit to the world's cross section data [28], but did not uncover significant discrepancies or outlying data sets that would bring them into agreement with the polarization measurements. Recently, several theoretical groups have revisited the radiative corrections to the e-p cross section and found that 2γ contributions can account for about half of the discrepancy in the two data sets [29]. At the time of this writing, the remaining difference has yet to be reconciled. A new measurement at JLab [30] using a third experimental technique, presently under analysis, may provide some clues.

If one assumes that the polarization data are correct, the implication is that the proton's Pauli form factor $F_2 = \frac{G_M - G_E}{\kappa(1+\tau)}$, which comes from a quark helicity flip, is both nonzero and falls off more slowly than the $1/Q^4$ expected from leading order pQCD. Both facts imply a significant contribution to the proton's wave function from quark orbital angular momentum. Miller [31] has reproduced the experimental behavior of F_2 using a cloudy bag model of the proton with a relativistic wave function that includes nonspherical contributions generated by the wave function's small components. Belitsky, Ji and Yuan [32] computed the ratio F_2/F_1 to higher order in pQCD, and found an asymptotic scaling of $F_2(Q^2)/F_1(Q^2) \sim ln^2(Q^2/\Lambda^2)/Q^2$ which describes the data remarkably well for $\Lambda \sim 300$ MeV.

The state of the G_E/G_M data and the comparison of QF_2/F_1 extracted from the polarization data along with calculations, are shown in Figure 2. For additional experimental and theoretical detail on the proton's electromagnetic form factors, see [2] and [4], respectively, in these proceedings.

Achieving comparable information on the neutron's electromagnetic structure is more problematic. Nuclear targets are required, which introduces uncertainties having to do with the nuclear wave function. The neutron's charge form factor is expected to be small so uncertainties are typically magnified. On the other hand because of its small magnitude, the role of the pion cloud and/or $\bar{q}q$ components is enhanced relative to that in the proton. There has been substantial effort and significant progress on both the theoretical and experimental fronts in the last several years. State of the art two- and three-nucleon calculations have reduced the theoretical uncertainties associated with the target.

Two new sets of data on the neutron's magnetic form factor have recently become available, in one case from asymmetry data with a polarized ^{3}He [33] target, in the other

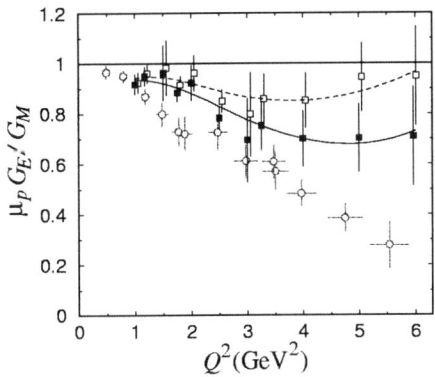 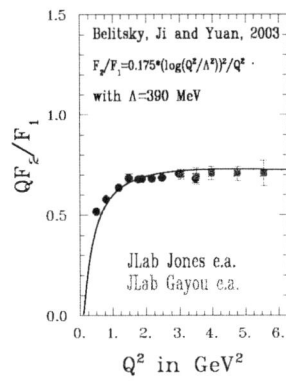

FIGURE 2. Left: Summary of world's data on the ratio $\mu_p G_E/G_M$ for the proton. The open circles are from the recoil polarization method (see text) and the open squares are the global fit to cross section data from [28]. The solid squares are the fit to the cross section data with a correction for 2γ-exchange as computed in [29] (from which the figure is taken). Right: The ratio $QF_2/F1$ from the polarization data (see text), compared with the pQCD calculation of [32] that includes leading logarithms.

from the ratio of $D(e,e'n)/D(e,e'p)$ unpolarized cross sections [34]. In the former, the asymmetry in the spin-dependent $^3\vec{\text{He}}(\vec{e},e')$ process was measured at kinematics where scattering from the nucleon dominates. The asymmetry when one flips the incident electron's helicity is

$$A = \frac{\sigma^+ - \sigma^-}{\sigma^+ + \sigma^-} = \frac{-(\cos\theta^* v_{T'} R_{T'} + 2\sin\theta^* \cos\phi^* v_{TL'} R_{TL'})}{v_L R_L + v_T R_T} \quad (6)$$

where the v_k are kinematic factors and (θ^*, ϕ^*) determine the direction of the target polarization relative to the momentum transfer vector $\vec{q}$. The response functions R_k contain the form factors of interest, convoluted with the spin-dependent components of the ^{3}He wave function. At momentum transfers below $Q^2 \sim 0.3$ (GeV/c)^2, where final state interaction effects modify the response functions from the free-nucleon case, a non-relativistic Fadeev calculation [35] was used to extract G_M^n. Above $Q^2 = 0.3$ (GeV/c)^2 the FSI effects become negligible but relativistic effects begin to enter, so a PWIA calculation that includes relativistic corrections is used.

In determining G_M^n from measurements of $D(e,e'n)/D(e,e'p)$, nuclear structure effects largely cancel in the ratio, and dominant systematic uncertainty comes from absolute knowledge of the neutron detection efficiency, and this has been a source of disagreement between previous data sets. In [34], the neutron detector efficiency was measured using tagged and monoenergetic neutron beams with a claimed accuracy of 1%, resulting in uncertainties in G_M^n of less than 2%. While these two methods have very different systematic uncertainties, the agreement between the two data sets is remarkable. A summary of all data, compared with several model calculations, is shown in Figure 3. A new experiment using the cross section ratio technique has recently been completed at JLab, which will push the measurements to higher momentum transfer [36]. While precise information at high momentum transer is still in the future, the low-momentum transfer data have now reached a level of precision comparable to our knowledge of G_M^p.

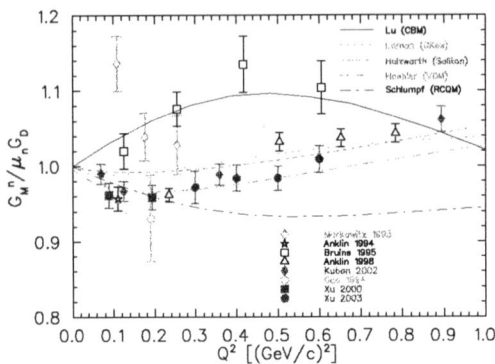

FIGURE 3. Summary of measurements the neutron's magnetic form factor, compared with several model calculations. See [33] for description of the calculations. The two most recent measurements from [34] (solid diamonds) and from [33] (solid circles) use very different experimental techniques but are in excellent agreement with each other.

Finally, the most significant challenge of the four electromagnetic distributions is the neutron's charge form factor. Until the mid-1990's the only detailed information came from electron-deuteron elastic scattering combined with a deuteron wave function to remove the (much larger) proton contributions. Initial polarization measurements demonstrated the feasibility of extracting G_E^n in a less model-dependent way, but they were still limited by statistical precision. Schiavilla and Sick [37] showed that one can reduce the model uncertainties in extracting G_E^n from elastic e-d scattering data by using only the deuteron's quadrupole form factor, which has been experimentally determined from the deuteron's tensor polarization [38] and which is much less sensitive to uncertain short-range two-body currents in the deuteron. Two recent JLab experiments, one using a polarized target and the other using recoil polarimetry, now provide the first double polarization measurements at momentum transfers above 1 (GeV/c)^2. These two experiments and their results are discussed in more detail in [2]. The two G_E^n measurements are in good agreement with each other, and, combined with recent results from Mainz at lower momentum transfer using $^3\vec{\text{He}}$ and $\vec{d}$ targets [39, 40], have greatly improved our knowledge of G_E^n. The BLAST collaboration at MIT-Bates [41] will take data using both $^3\vec{\text{He}}$ and $\vec{d}$ targets at low momentum transfer over the next year, providing additional systematic checks of the nuclear structure contributions.

As yet no single theoretical calculation can reproduce all four of the nucleon form factors over their measured ranges. However, the first lattice calculations are becoming available [1], and we can expect much new theoretical activity on the form factors in the next few years.

NEUTRAL WEAK FORM FACTORS AND STRANGE QUARKS

While, at least at the experimental level, a coherent picture of nucleon's electromagnetic form factors is becoming possible, very little is known about the comparable form factors resulting from a neutral weak probe. The neutral weak contributions can be isolated through measurements of parity-violating electron scattering, and select individual measurements now exist from MIT-Bates, Jefferson Laboratory and Mainz. These provide, when combined with the existing electromagnetic data, another degree of freedom that allows separation of the up, down and strange quark contributions to the nucleon's charge and magnetism in a model independent way. As a result, parity-violating electron scattering is a direct probe of the sea quark contributions to nucleon electromagnetic structure.

For elastic scattering from a spin-1/2 target, the parity-violating (PV) asymmetry is

$$A_{PV} = \frac{d\sigma_R - d\sigma_L}{d\sigma_R + d\sigma_L} = -\frac{G_F Q^2}{4\pi\alpha\sqrt{2}} \frac{A_E + A_M + A_A}{\left[\varepsilon\left(G_M^\gamma\right)^2 + \tau\left(G_M^\gamma\right)^2\right]} \quad (7)$$

where

$$A_E = \varepsilon G_E^Z G_E^\gamma, \quad A_M = \tau G_M^Z G_M^\gamma,$$
$$A_A = -\left(1 - 4\sin^2\theta_W\right)\sqrt{\tau(1+\tau)(1-\varepsilon^2)} G_A^e G_M^\gamma, \quad (8)$$

τ and ε are kinematic factors, and R,L correspond to the two helicity states of a longitudinally polarized beam. Assuming only that the difference between a proton and neutron is the interchange of u and d quarks, the weak vector form factors can be written in terms of the nucleon's EM form factors since they involve the same matrix element $\bar{q}\gamma_\mu q$:

$$G_{E,M}^Z = \left(1 - 4\sin^2\theta_W\right)\left(1 + R_V^p\right) G_{E,M}^{\gamma,p} - G_{E,M}^{\gamma,n}\left(1 + R_V^n\right) - G_{E,M}^s. \quad (9)$$

There is an additional axial form factor G_A^e arising from the matrix element $\bar{q}\gamma_\mu\gamma_5 q$, and PV e-p scattering is primarily sensitive to its isovector component. Unfortunately, G_A^e contains relatively large and somewhat uncertain higher order corrections that can come, for example, from γ-Z box diagrams or from an electromagnetic e-p interaction but with PV interactions between quarks. These nonperturbative effects are challenging to calculate since they involve nonperturbative physics, but have been estimated in [42].

The first experimental limits on the strange vector form factors were from the SAMPLE experiment at MIT-Bates [43, 44, 45]. The experiment measured the PV asymmetry from both hydrogen and deuterium targets to determine G_M^s and G_A^e at $Q^2 = 0.1$ (GeV/c)^2. While the data indicated that G_M^s is likely small, there was significant disagreement between experiment and theory regarding G_A^e. Subsequent improved analysis of the deuterium data that includes better simulation of background and radiative processes, along with improved calculations of *PV* effects in deuterium, have brought the experiment into agreement with the calculations, as shown on the left side of Figure 4.

The right side of Figure 4 shows a summary of the three parity-violation experiments that have been completed to date, shown as the fractional deviation from the asymmetry that would be expected if $G_E^s = G_M^s = 0$. While each of the measurements is consistent with no significant strange quark contributions, taken collectively one sees a possible trend in the same direction: in the SAMPLE case this corresponds to a slightly positive value for G_M^s. While little or no contribution may seem surprising in light of the

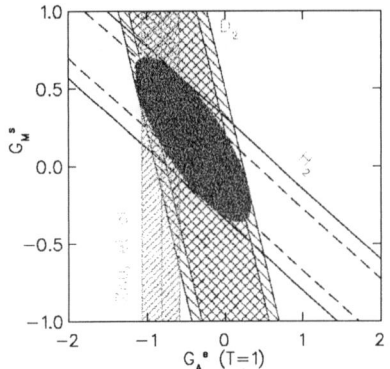
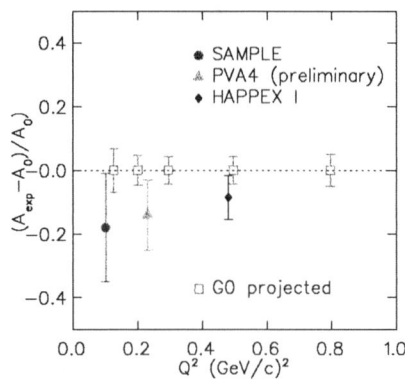

FIGURE 4. Left: Updated results from the SAMPLE result, showing agreement between the deuterium data and the calculation of Zhu et al. [42], for the proton's axial form factor $G_A^e(T=1)$. Right: summary of existing PV asymmetry measurements, showing the fraction deviation of the experimental measurements from the asymmetry that would be expected with no strange quarks from the SAMPLE experiment [46], HAPPEX [47], and the PVA4 experiment at Mainz [48].

somewhat larger role that strange quarks seem to play in the nucleon's mass and momentum, chirality arguments may cause the $\bar{s}\gamma_\mu s$ matrix elements to be small (see [49] for further discussion). On the other hand, each experiment is sensitive to different linear combinations of G_E^s and G_M^s, so if G_E^s and G_M^s were to have opposite sign, as predicted by some theoretical models, cancellations could cause a net null result. So the present state of the data is as yet inconclusive. Both the HAPPEX and PVA4 collaborations plan to carry out measurements at forward angle and low momentum transfer, which can be combined with the SAMPLE data to constrain the relative signs of G_E^s and G_M^s. The G0 experiment at Jefferson Laboratory [50] is dedicated to providing the first complete separation of all three form factors (G_E^s, G_M^s and G_A^e) from a single experimental apparatus. Also shown on Figure 4 is the level precision that would be achieved in the forward angle measurement of the G0 experiment, which is presently scheduled for early 2004.

The level of precision in parity violation experiments has improved to the point that, once knowledge of the hadron structure can be brought under control, precision tests of the Standard Model are possible, and experiments are presently being planned at Jefferson Lab. For a more detailed summary of the experimental parity violation programs, see [3], and for future possibilities for placing new constraints on the Standard Model, see [5].

CONCLUSIONS

Since the last Intersections Conference, there has been significant experimental progress in developing a quantititative understanding of the electromagnetic structure of the lightest nucleons and mesons, largely due to technological advances in polarization techniques in electron scattering that have taken place over the last decade. While we still await a thorough theoretical understanding of the nonperturbative aspects of nucleon structure, these data will provide important constraints both for models of hadron structure, and for lattice calculations for which these data will be an important benchmark. New puzzles have also emerged as a result of the improved level of precision of the data. At the next conference we can look forward to results in several areas: new pion form factor measurements and possibly the first determination of the kaon charge form factor at large momentum transfer, a deeper understanding of the proton's magnetism and the role of quark orbital angular momentum, perhaps a better theoretical understanding of neutron structure now that precise G_E^n and G_M^n data are available, and more comprehensive information on the role of strange quarks in the nucleon's charge and magnetism from parity violation experiments.

The author is supported by NSF contract PHY-0140010. Much of the work discussed in this paper has been supported both by the National Science Foundation and the U.S. Dept. of Energy.

REFERENCES

1. Göckeler, M., et al., *hep-lat/0303019* (2003), see also references therein.
2. Kelly, J. J., *these proceedings* (2003).
3. Beck, D. H., *these proceedings* (2003).
4. Ji, X., *these proceedings* (2003).
5. Ramsey-Musolf, M. J., *these proceedings* (2003).
6. Farrar, G., and Jackson, D., *Phys. Rev. D*, **43**, 246 (1979).
7. Amendolia, S. R., et al., *Nucl. Phys. B*, **227**, 168 (1986).
8. Vanderhaeghen, M., Guidal, M., and Laget, J.-M., *Phys. Rev. C*, **57**, 1454 (1998).
9. Volmer, J., et al., *Phys. Rev. Lett.*, **86**, 1713 (2001).
10. Maris, P., and Tandy, P., *Phys. Rev. C*, **62**, 055204 (2000).
11. Nestarenko, V., and Radyushkin, A., *Phys. Lett. B*, **115**, 410 (1982).
12. Brauel, P., et al., *Z. Phys. C*, **3**, 101 (1979).
13. Ackermann, H., et al., *Nucl. Phys. B*, **137**, 294 (1978).
14. Huber, G., Mack, D., and Blok, H., *JLab proposal E01-004* (2003).
15. Huber, G., and Mack, D., *private communication* (2003).
16. Amendolia, S. R., et al., *Phys. Lett. B*, **178**, 435 (1986).
17. McNabb, J., et al., *nucl-ex/0305028, submitted to Phys. Rev. Lett.* (2003).
18. Feuerbach, R., *Ph.D. thesis, Carnegie Mellon University* (2002).
19. Mohring, R. M., et al., *Phys. Rev. C*, **67**, 055205 (2003), this paper supercedes results from the same experiment previously published in G. Niculescu *et al.*, Phys. Rev. Lett. **81**,1805 (1998).
20. Markowitz, P., *JLab experiment E98-108* (2003), private communication.
21. Akhiezer, A., and Rekalo, M., *Sov. Phys.-Diklady*, **13**, 572 (1968).
22. Dombey, N., *Rev. Mod. Phys.*, **41**, 236 (1969).
23. Akhiezer, A., and Rekalo, M., *Sov. J. Part. Nucl.*, **3**, 277 (1974).
24. Arnold, R., Carlson, C., and Gross, F., *Phys. Rev. C*, **23**, 363 (1981).
25. Jones, M. K., et al., *Phys. Rev. Lett.*, **84**, 1398 (2000).
26. Gayou, O., et al., *Phys. Rev. Lett.*, **88**, 092301 (2002).
27. Punjabi, V., et al., *submitted to Phys. Rev. C* (2003).
28. Arrington, J., *nucl-ex/0305009* (2003), submitted to Phys. Rev. C.
29. Blunden, P., Melnitchouk, W., and Tjon, J., *nucl-th/0306076* (2003).
30. Arrington, J., and Segal, R., *JLab experiment E01-001* (2003).
31. Miller, G., *Phys. Rev. C*, **66**, 032201 (2002), see also G. Miller, nucl-th/0304076.
32. Belitsky, A., Ji, X., and Yuan, F., *hep-ph/0212351* (2002).
33. Xu, W., et al., *Phys. Rev. C*, **67**, 012201 (2003).
34. Kubon, G., et al., *Phys. Lett. B*, **524**, 26 (2002).

35. Golak, J., et al., *Phys. Rev. C*, **51**, 1638 (1995).
36. Brooks, W., et al., *JLab experiment E94-017* (2003).
37. Schiavilla, R., and Sick, I., *Phys. Rev. C*, **64**, 041002 (2001).
38. Abbott, D., et al., *Euro. Phys. Jour. A*, **7**, 421 (2000).
39. Ostrick, M., et al., *Phys. Rev. Lett.*, **83**, 276 (1999).
40. Bermuth, J., et al., *Phys. Lett. B*, **564**, 199 (2003).
41. Alarcon, R., et al., *BLAST collaboration* (2003), URL http://blast.lns.mit.edu.
42. Zhu, S.-L., et al., *Phys. Rev. D*, **62**, 033008 (2000).
43. Mueller, B., et al., *Phys. Rev. Lett.*, **78**, 3824 (1997).
44. Spayde, D., et al., *Phys. Rev. Lett.*, **84**, 1106 (2000).
45. Hasty, R., et al., *Science*, **290**, 2117 (2000).
46. Spayde, D. T., *Ph.D thesis, University of Maryland* (2001).
47. Aniol, K., et al., *Phys. Lett. B*, **509**, 211 (2001).
48. Maas, F., *private communication* (2003).
49. Jaffe, R. L., *these proceedings* (2003).
50. Beck, D., et al., *JLab experiment E00-006 ("G0")* (2003).

Structure Functions - Status and Prospect

J. C. Peng

University of Illinois, Urbana, Illinois, 61801, U.S.A.

Abstract. Current status and future prospects of the structure functions and parton distribution studies are presented.

INTRODUCTION

The study of nucleon's structure functions and parton distributions is an active area of research in nuclear and particle physics. The parton distributions address both the perturbative and nonperturbative aspects of QCD, and they also provide an essential input for describing hard processes in high-energy hadron collisions. As a result of several decades's intense effort, the unpolarized proton structure functions have been well mapped out over a broad range of Q^2 and Bjorken-x. While these data are invaluable for testing QCD and for extracting various parton distributions, several questions remain unanswered. For example, the unexpected finding of the flavor asymmetry of the light-quark sea ($\bar{u}, \bar{d}$) suggests that other aspects of the flavor structure, such as possible asymmetry between the s and $\bar{s}$ sea quark distributions and the bahavior of valence d/u ratio at large x, need to be examined. The issue of quark-hadron duality, as reflected in the intriguing similarity between the structure functions measured at the resonance region and at the DIS region, also requires further studies.

Remarkable progress in the study of spin-dependent structure functions has been made since the discovery of the "proton spin puzzle" in the late 1980's. Very active spin-physics programs have been pursued at many facilities including SLAC, CERN, HERA, JLab, and RHIC. The polarized DIS data now cover a sufficiently broad Q^2 range for scaling-violation to be observed. In recent years, new experimental tools such as semi-inclusive polarized DIS, polarized proton-proton collision, and deeply exclusive reactions have been employed to address the major unresolved question in spin physics: How is the proton's spin distributed among its various constituents?

On the theory front, the formulation of the generalized parton distributions as well as the identification of various k_T (intrinsic transverse momentum of partons)-dependent structure and fragmentation functions have opened exciting new directions of research. Furthermore, important progress in the Lattice calculations for the moments of various parton distributions and in the extrapolations to their chiral limits has been made.

In this review I will focus on recent progress in the following areas:

- Flavor structure of parton distributions
- Transition from high-Q^2 to low-Q^2
- Novel distribution and fragmentation functions; Generalized parton distributions

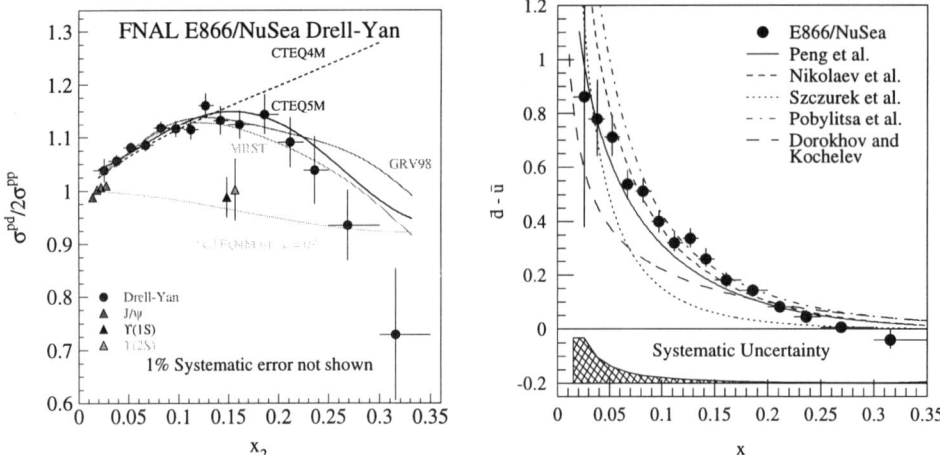

FIGURE 1. Left panel: Cross section ratios of $p+d$ over $2(p+p)$ for Drell-Yan, J/Ψ, and Υ production from FNAL E866. Right panel: Comparison of E866 $\bar{d}-\bar{u}$ data with calculations from various models [2].

FLAVOR STRUCTURES OF PARTON DISTRIBUTIONS

$\bar{d}/\bar{u}$ flavor asymmetry

The earliest parton models assumed that the proton sea was flavor symmetric, even though the valence quark distributions are clearly flavor asymmetric. The flavor symmetry assumption was not based on any known physics, and it remained to be tested. Under the assumption of a $\bar{u}$, $\bar{d}$ flavor-symmetric sea in the nucleon, the Gottfried Sum Rule [1], $I_G = \int_0^1 (F_2^p(x,Q^2) - F_2^n(x,Q^2))/x\, dx = 1/3$, is obtained. The NMC collaboration determined the Gottfried integral to be 0.235 ± 0.026, significantly below 1/3. This surprising result can be explained by a large flavor asymmetry between the $\bar{u}$ and the $\bar{d}$.

The x dependence of $\bar{d}/\bar{u}$ asymmetry has been determined by proton-induced Drell-Yan (DY) as well as semi-inclusive DIS measurements. Figure 1 shows that the Fermilab E866 [2] DY cross section per nucleon for $p+d$ clearly exceeds $p+p$, and it indicates an excess of $\bar{d}$ with respect to $\bar{u}$ over an appreciable range in x. In contrast, the $\sigma(p+d)/2\sigma(p+p)$ ratios for J/Ψ and Υ production, also shown in Fig. 1, are very close to unity. This reflects the dominance of gluon-gluon fusion process for quarkonium production and the expectation that the gluon distributions in the proton and in the neutron are identical.

Many theoretical models, including meson cloud model, chiral-quark model, Pauli-blocking model, instanton model, chiral-quark soliton model, and statistical model, have been proposed to explain the $\bar{d}/\bar{u}$ asymmetry. For recent reviews, see [3, 4]. These models can describe the $\bar{d}-\bar{u}$ data very well, as shown in Fig. 1. However, they all have difficulties explaining the $\bar{d}/\bar{u}$ data at large x ($x > 0.2$). The new 120 GeV Fermilab Main Injector and the proposed 50 GeV Japanese Hadron Facility present opportunities for extending the $\bar{d}/\bar{u}$ measurement to larger x ($0.25 < x < 0.7$).

Models in which virtual mesons are admitted as degrees of freedom have implications that extend beyond the $\bar{d}, \bar{u}$ flavor asymmetry addressed above. They create hidden strangeness in the nucleon via such virtual processes as $p \to \Lambda + K^+, \Sigma + K$, etc. Such processes are of considerable interest as they imply different s and $\bar{s}$ parton distributions in the nucleon, a feature not found in gluonic production of $s\bar{s}$ pairs.

A difference between the s and $\bar{s}$ distribution can be made manifest by direct measurements of the s and $\bar{s}$ parton distribution functions in neutrino DIS. A fit to the CDHS neutrino charged-current inclusive data together with charged lepton DIS data found evidence for $\int_0^1 s(x)dx > \int_0^1 \bar{s}(x)dx$ [5]. However, an analysis [6] of the recent CCFR and NuTeV $\nu(\bar{\nu})N \to \mu^+\mu^-x$ dimuon production data [7] favored $\int_0^1 s(x)dx < \int_0^1 \bar{s}(x)dx$ ($\int_0^1 (s(x) - \bar{s}(x))dx = -0.0027 \pm 0.0013$). To better determine the $s/\bar{s}$ asymmetry, an NLO analysis is currently underway [8]. Violation of the $s/\bar{s}$ symmetry would have impact on the recent extraction [9] of $\sin^2\theta_W$ from the CCFR/NuTeV νN scattering data.

Asymmetry in the $s, \bar{s}$ distributions can also be revealed in the measurements of the strange quark's contribution to the nucleon's electromagnetic and axial form factors. These "strange" form factors can be measured in neutrino elastic scattering [10] from the nucleon, or by selecting the parity-violating component of electron-nucleon elastic scattering. Two completed parity-violating experiments [11, 12] suggest small contributions of strange quarks to nucleon form factors. Several new experiments are underway at JLab and MAMI to measure parity-violating asymmetry at various kinematic regions.

Flavor structure of polarized nucleon sea

The flavor structure and the spin structure of the nucleon sea are closely connected. Many theoretical models originally proposed to explain the $\bar{d}/\bar{u}$ flavor asymmetry also have specific implications for the spin structure of the nucleon sea. In the meson-cloud model, for example, a quark would undergo a spin flip upon an emission of a pseudoscalar meson ($u^\uparrow \to \pi^0(u\bar{u}, d\bar{d}) + u^\downarrow$, $u^\uparrow \to \pi^+(u\bar{d}) + d^\downarrow$, $u^\uparrow \to K^+ + s^\downarrow$, etc.). The antiquarks ($\bar{u}, \bar{d}, \bar{s}$) are unpolarized ($\Delta\bar{u} = \Delta\bar{d} = \Delta\bar{s} = 0$) since they reside in spin-0 mesons. The strange quarks (s), on the other hand, would have a negative polarization.

In the chiral-quark soliton model [13, 14], the polarized isovector distributions $\Delta\bar{u}(x) - \Delta\bar{d}(x)$ appears in leading-order (N_c^2) in a $1/N_c$ expansion, while the unpolarized isovector distributions $\bar{u}(x) - \bar{d}(x)$ appear in next-to-leading order (N_c). Therefore, this model predicts a large flavor asymmetry for the polarized sea $[\Delta\bar{u}(x) - \Delta\bar{d}(x)] > [\bar{d}(x) - \bar{u}(x)]$.

The HERMES collaboration has recently reported the extraction of $\Delta\bar{u}(x)$, $\Delta\bar{d}(x)$, and $\Delta\bar{s}(x)(= \Delta s(x))$ using polarized semi-inclusive DIS (SIDIS) data [15]. Although the statistics are still limited, the HERMES results for $\Delta\bar{u}, \Delta\bar{d}, \Delta\bar{u} - \Delta\bar{d}$, as shown in Fig. 2, are all consistent with being zero. In particular, there is no evidence for a large positive $\Delta\bar{u}(x) - \Delta\bar{d}(x)$ asymmetry as was predicted [16] by the chiral quark soliton model. Figure 2 also shows that Δs tends to be positive, in contrast to the predictions of a negative polarization of the strange sea in the analysis of inclusive DIS and hyperon decay data assuming SU(3) symmetry. However, the HERMES result of $\Delta s = 0.03 \pm 0.03 \pm 0.01$ over $0.023 < x < 0.3$ is not in disagreement with the inclusive

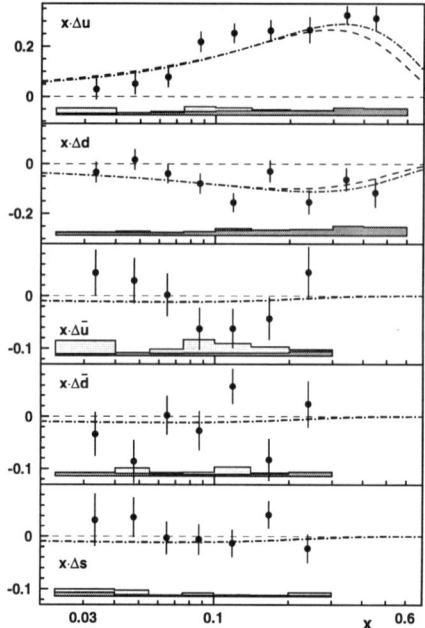

FIGURE 2. Quark and antiquark polarizations extracted from the HERMES SIDIS data [15].

DIS result of $(\Delta s + \Delta \bar{s})/2 \simeq -0.02$ [17].

Another promising technique for measuring sea-quark polarization is W-boson production [18] at RHIC. The longitudinal single-spin asymmetry for W production in polarized $p + p \to W^{\pm} + x$ gives a direct measure of sea-quark polarization. The RHIC W-production and the HERMES SIDIS measurements are clearly complementary tools for determining polarized sea quark distributions.

d/u ratio at large x

Another quantity related to the flavor symmetry of the proton is the d/u ratio at large x. Assuming $SU(2)_{spin} \times SU(2)_{flavor}$ symmetry, the proton wave function is given as

$$|p>\uparrow \; = \; \frac{1}{\sqrt{2}} u\uparrow (ud)_{S=0,S_z=0} + \frac{1}{\sqrt{18}} u\uparrow (ud)_{S=1,S_z=0} - \frac{1}{3} u\downarrow (ud)_{S=1,S_z=1}$$
$$-\frac{1}{3} d\uparrow (uu)_{S=1,S_z=0} + \frac{\sqrt{2}}{3} d\downarrow (uu)_{S=1,S_z=1} \qquad (1)$$

The neutron wave function is readily obtained from $u \leftrightarrow d$ interchange. In nature, the $SU(2)_{spin} \times SU(2)_{flavor}$ symmetry is clearly broken, as evidenced by the large $N - \Delta$ mass splitting. The dynamic origins of this symmetry breaking remains unclear. Close and Carlitz [19, 20] argued that the dominance of the $S = 0$ diquark configuration over

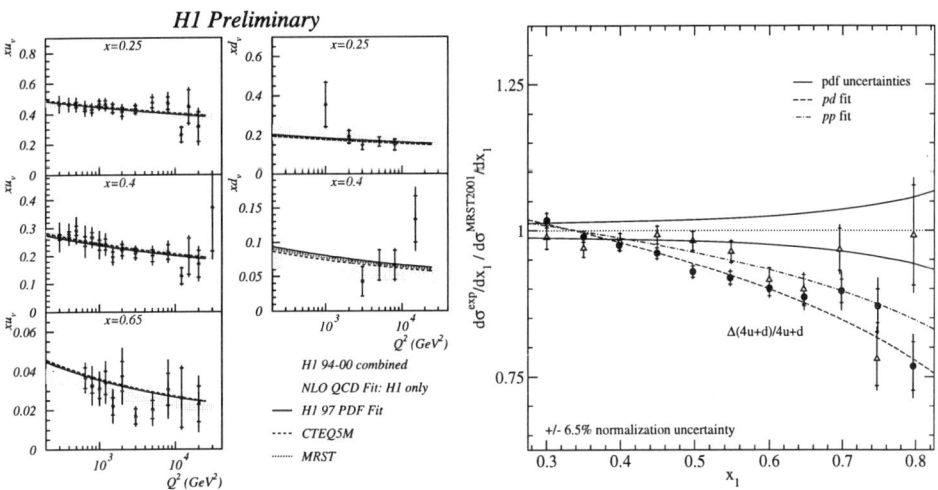

FIGURE 3. Left panel: u and d valence quark densities obtained from H1 charged-current measurements [25]. Right panel: Comparison of the E866 pp and pd DY cross sections with PDF calculations [26].

the $S = 1$ configuration would account for the $N - \Delta$ mass splitting as well as the SU(2) × SU(2) symmetry breaking. An alternative suggestion, based on perturbative QCD, was offered by Farrar and Jackson [21]. They pointed out that the spin-aligned diquark configuration with $S_z = 1$ is suppressed since only longitudinal gluons can be exchanged. A similar result was also obtained by Brodsky et al. [22] using counting rule argument. It is straightforward to show that in the $x \to 1$ limit, the different models predict the folllowing values for various ratios:

- SU(2)$_{spin}$× SU(2)$_{flavor}$ symmetry: $\frac{d}{u} = \frac{1}{2}$, $\frac{\Delta u}{u} = \frac{2}{3}$, $\frac{\Delta d}{d} = -\frac{1}{3}$, $\frac{F_2^n}{F_2^p} = \frac{2}{3}$.
- $S = 0$ diquark dominance: $\frac{d}{u} = 0$, $\frac{\Delta u}{u} = 1$, $\frac{\Delta d}{d} = -\frac{1}{3}$, $\frac{F_2^n}{F_2^p} = \frac{1}{4}$.
- $S_z = 0$ diquark dominance: $\frac{d}{u} = \frac{1}{5}$, $\frac{\Delta u}{u} = 1$, $\frac{\Delta d}{d} = 1$, $\frac{F_2^n}{F_2^p} = \frac{3}{7}$.

The distinct predictions for F_2^n/F_2^p from various models could be tested against DIS experiments. However, there exist considerable uncertainties in the extraction of F_2^n from the measurement of F_2^d. Depending on the treatment of the nuclear effects in the deuteron, very different values for F_2^n/F_2^p (and d/u) were obtained at large x [23]. It is clearly desirable to measure d/u without the need to model nuclear effects in the deuteron. One method is to measure the charge asymmetry of W production in $p - \bar{p}$ collision. Indeed, the CDF data [24] on the W charge asymmetry have already provided useful constraints on the d/u ratio.

The d/u ratio can also be probed by measuring the $e^- p \to v_e x$ and $e^+ p \to \bar{v}_e x$ charged-current DIS, where the underlying processes are $e^- u \to v_e d$ and $e^+ d \to \bar{v}_e u$, respectively. The recent H1 charged-current data [25], shown in Fig. 3, indicate that the u quark density at large $x(x = 0.65)$ is smaller than expected from the current

PDF parametrization. Very recently, the Fermilab E866/NuSea collaboration reported the absolute Drell-Yan cross sections of 800 GeV $p+p$ and $p+d$ [26]. As shown in Fig. 3, the data fall below the PDF predictions at large x (up to $x = 0.8$). The H1 and the E866 results suggest that u quark density at large x might be smaller than expected from current PDFs. This clearly would impact on the d/u ratio at large x, as shown in a recent global PDF analysis [27].

The uncertainties involved in the extraction of F_2^n from F_2^d data can be greatly reduced using the technique of neutron-tagging. A new experiment [28] has been proposed at the JLab Hall-B to detect $e^- d \to e^- px$, where a low-energy recoiled proton will be measured in coincidence with the (e,e') scattering. Using this method, the F_2^n/F_2^p ratio over the range $0.2 < x < 0.7$ could be determined with small systematic uncertainties.

TRANSITION FROM HIGH-Q^2 TO LOW-Q^2

Quark-hadron duality

The recent studies at JLab of the spin-averaged and spin-dependent structure functions at low Q^2 region have shed new light on the subject of quark-hadron duality. Thirty years ago, Bloom and Gilman [29] noticed that the structure functions obtained from deep-inelastic scattering experiments, where the substructures of the nucleon are probed, are very similar to the averaged structure functions measured at lower energy, where effects of nucleon resonances dominate. This surprising similarity between the resonance electroproduction and the deep inelastic scattering suggests a common origin for these two phenomena, called local duality.

Recently, high precision data [30] from JLab have verified the quark-hadron duality for spin-averaged scattering on proton and deuteron targets. For Q^2 as low as 0.5 GeV^2, the resonance data are within 10% of the DIS results. When the mean F_2 curve from the resonance data is plotted as a function of the Nachtmann variable, $\xi = 2x/(1+\sqrt{1+4M^2x^2/Q^2})$, it resembles the xF_3 structure function obtained in neutrino scattering experiments. Since xF_3 is a measure of the valence quark distributions, this suggests that the F_2 structure function at low Q^2 originates from valence quarks only.

The study of quark-hadron duality was recently extended to other structure functions. Results from HERMES [31] show that duality is also observed for the spin-dependent quantity A_1^p. Another recent result from JLab shows that the nuclear modifications to the unpolarized structure functions in the resonance region are in surprisingly good agreement with those measured in DIS [32].

$\Gamma_1(Q^2)$ at low Q^2 and the generalized GDH integral

The extensive data on $g_1(x,Q^2)$ allow accurate determinations of the integrals $\Gamma_1^{p,n}(Q^2) = \int_0^1 g_1^{p,n}(x,Q^2)dx$ for the proton and the neutron, as well as $\Gamma_1^p(Q^2) - \Gamma_1^n(Q^2)$. While the values of Γ_1^p and Γ_1^n are different from the predictions of Ellis and Jaffe who

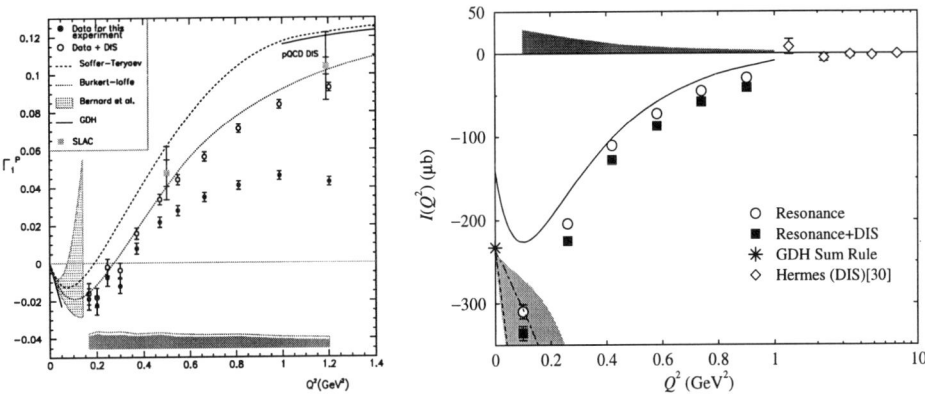

FIGURE 4. Left panel: $\Gamma_1^p(Q^2)$ from CLAS [33]. Right panel: Generalized GDH integral from JLab Hall-A experiment [34].

assumed SU(3) flavor symmetry and an unpolarized strange sea, the data are in good agreement with the prediction of the Bjorken sum rule.

How does $\Gamma_1(Q^2)$ evolve as $Q^2 \to 0$? This question is closely related to the Gerasimov-Drell-Hearn (GDH) sum rule:

$$\int_{v_0}^{\infty} [\sigma_{1/2}(v) - \sigma_{3/2}(v)] \frac{dv}{v} = -\frac{2\pi^2 \alpha}{M^2} \kappa^2. \quad (2)$$

The GDH sum rule, based on general physics principles (causality, unitarity, Lorentz and gauge invariances) and dispersion relation, relates the total absorption cross sections of circularly polarized photons on longitudinally polarized nucleons to the static properties of the nucleons. In Eq. 2, $\sigma_{1/2}$ and $\sigma_{3/2}$ are the photo-nucleon absorption cross sections of total helicity of 1/2 and 3/2, v is the photon energy and v_0 is the pion production threshold, M is the nucleon mass and κ is the nucleon anomalous magnetic moment. The GDH integral in Eq. 2 can be generalized from real photon absorption to virtual photon absorption with non-zero Q^2:

$$I_{GDH}(Q^2) \equiv \int_{v_0}^{\infty} [\sigma_{1/2}(v, Q^2) - \sigma_{3/2}(v, Q^2)] \frac{dv}{v} = \frac{16\pi^2 \alpha}{Q^2} \Gamma_1(Q^2). \quad (3)$$

Eq. 3 shows that the Q^2-dependence of the generalized GDH integral is directly related to the Q^2-dependence of Γ_1. The GDH sum rule (Eq. 2) predicts $\Gamma_1^p = 0$ at $Q^2 = 0$ with a negative slope for $d\Gamma_1^p(Q^2)/dQ^2$ and Γ_1^p is known to be positive at high Q^2, therefore, $\Gamma_1^p(Q^2)$ must become negative at low Q^2.

The GDH integrals at low Q^2 have recently been measured in several experiments at JLab [34, 33] and HERMES [35]. Results from a JLab Hall-B measurement [33] of $\Gamma_1^p(Q^2)$ are shown in Fig. 4. These data indeed show that Γ_1^p changes sign around $Q^2 = 0.3$ GeV^2. The origin of the sign-change can be attributed to the competition between $\Delta(1232)$ and higher nucleon resonances. At the lowest Q^2, the $\Delta(1232)$ has a dominant negative contribution to Γ_1^p. However, at larger Q^2, higher mass nucleon resonances take over to have a net positive Γ_1^p.

Results [34] from a JLab Hall-A measurement of the generalized GDH integral for neutron using a polarized ^{3}He target are shown in Fig. 4. In contrast to the proton case, the strong negative contribution to the GDH integral from the $\Delta(1232)$ resonance now dominates the entire measured Q^2 range. Future experiments at JLab will extend the measurements down to $Q^2 = 0.02$ GeV^2 in order to map out the low Q^2 behavior of the neutron and proton generalized GDH integrals.

NOVEL DISTRIBUTION AND FRAGMENTATION FUNCTIONS

In addition to the unpolarized and polarized quark distributions, $q(x, Q^2)$ and $\Delta q(x, Q^2)$, a third quark distribution, called transversity, is the remaining twist-2 distribution yet to be measured. This helicity-flip quark distribution, $\delta q(x, Q^2)$, can be described in quark-parton model as the net transverse polarization of quarks in a transversely polarized nucleon. Due to the chiral-odd nature of the transversity distribution, it can not be measured in inclusive DIS experiments. In order to measure $\delta q(x, Q^2)$, an additional chiral-odd object is required. For example, the double spin asymmetry, A_{TT}, for Drell-Yan cross section in transversely polarized pp collision, is sensitive to transversity since $A_{TT} \sim \Sigma_i e_i^2 \delta q_i(x_1) \delta \bar{q}_i(x_2)$. Such a measurement could be carried out at RHIC [18], although the anticipated effect is small, on the order of $1-2\%$.

Several other methods for measuring transversity have been proposed for semi-inclusive DIS. In particular, Collins suggested [36] that a chiral-odd fragmentation function in conjunction with the chiral-odd transversity distribution would lead to a single-spin azimuthal asymmetry in semi-inclusive pion production.

The HERMES collaboration recently reported [37] observation of single-spin azimuthal asymmetry for charged and neutral hadron electroproduction. Using unpolarized positron beam on a longitudinally polarized hydrogen and deuterium targets, the cross section was found to have a $\sin\phi$ dependence correlating with the target spin direction. ϕ is the azimuthal angle between the pion and the (e, e') scattering plane. This Single-Spin-Asymmetries (SSA) can be expressed as the analyzing power in the $\sin\phi$ moment, and the result is shown in Fig. 5. The $\sin\phi$ moment for an unpolarized (U) positron scattered off a longitudinally (L) polarized target contains two main contributions

$$\langle \sin\phi \rangle \quad \alpha \quad S_L \frac{2(2-y)}{Q\sqrt{1-y}} \sum_q e_q^2 x h_L^q(x) H_1^{\perp,q}(z) + S_T(1-y) \sum_q e_q^2 x h_1^q(x) H_1^{\perp,q}(z), \quad (4)$$

where S_L and S_T are the longitudinal and transverse components of the target spin orientation with respect to the virtual photon direction. For the HERMES experiment with a longitudinally polarized target, the transverse component is nonzero with a mean value of $S_T \approx 0.15$. The observed azimuthal asymmetry could be a combined effect of the h_1 transversity and the twist-3 h_L distribution. Recently, another mechanism involving a chiral-even T-odd Sivers distribution function [38] was shown to contribute

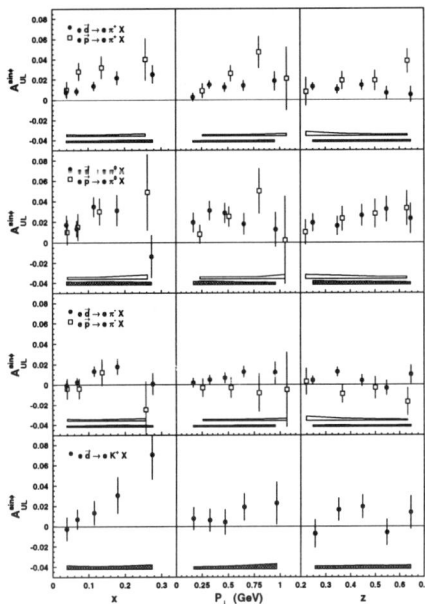

FIGURE 5. Analyzing power in the sinϕ moment from HERMES [35].

to azimuthal asymmetry [39, 40]. For a longitudinally polarized target the Collins and the Sivers mechanisms can not be distinguished.

If the azimuthal asymmetry observed by HERMES is indeed caused by the h_1 transversity, a much larger asymmetry is expected for a transversely polarized target. The HERMES and COMPASS collaborations have collected polarized SIDIS using transversely polarized hydrogen and ^{6}LiD targets, respectively. These data would shed much light on the origins of the SSA and could also disentangle the Sivers effect from the Collins effect. The Collins effect has a $\sin(\phi_h^l + \phi_s^l)$ dependence while the Sivers effect is proportional to $\sin(\phi_h^l - \phi_s^l)$, where $\phi_s^l = \phi_s - \phi^l$ is the angle between target spin and the lepton scattering plane. For longitudinally polarized target $\phi_s^l = 0$ and the two effects have identical ϕ dependence. For transversely polarized target, however, $\phi_s^l \neq 0$ and the two effects can be separated.

The Collins fragmentation function represents a correlation between the quark's transverse spin and the transverse momentum of the leading hadron formed in the fragmentation process. The Sivers distribution function reflects the correlation between the quark's transverse spin and its transverse momentum within the proton. It has been shown [41, 42] that both the Collins and the Sivers effects can contribute to the analysing power A_N observed in the Fermilab E704 $p \uparrow p \to \pi x$ reaction [43]. Very recently, A_N was measured [44] at RHIC at a much higher energy of $\sqrt{s} = 200$ GeV using transversely polarized proton beams. The RHIC data could provide new information on the Collins and Sivers functions.

GENERALIZED PARTON DISTRIBUTIONS

There has been intense theoretical and experimental activities in recent years on the subject of Generalized Parton Distribution (GPD). In the Bjorken scaling regime, exclusive leptoproduction reactions can be factorized into a hard-scattering part calculable in QCD, and a non-perturbative part parameterized by the GPDs. The GPD takes into account dynamical correlations between partons with different momenta. In addition to the dependence on Q^2 and x, the GPD also depends on two more parameters, the skewedness ξ and the momentum transfer to the baryon, t. Of particular interest is the connection between GPD and the nucleon's orbital angular momentum [45].

The deeply virtual Compton scattering (DVCS), in which an energetic photon is produced in the reaction $ep \to ep\gamma$, is most suitable for studying GPD. Unlike the exclusive meson productions, DVCS avoids the complication associated with mesons in the final state and can be cleanly interpreted in terms of GPDs. An important experimental challenge, however, is to separate the relatively rare DVCS events from the abundant electromagnetic Bethe-Heitler (BH) background. From the collision of 800 GeV protons with 27.5 GeV positrons, both the ZEUS [46] and the H1 [47] collaborations at DESY observed an excess of $e^+ + p \to e^+ + \gamma + p$ events in a kinematic region where the BH cross section is largely suppressed. The excess events were attributed to the DVCS process and the ZEUS collaboration further determined [46] the DVCS cross section over the kinematic range $5 < Q^2 < 100$ GeV^2, $40 < W < 140$ GeV. Both the W and Q^2 dependences of the ZEUS DVCS cross section data are well described by calculations based on GPD and on the color-dipole model.

At lower c.m. energies, the HERMES [48] and the CLAS [49] collaborations observed the interference between the DVCS and the BH processes, which manifests itself as a pronounced $\sin\phi$ azimuthal asymmetry correlated with the beam helicity. Another observable sensitive to the interference between the DVCS and the BH processes is the azimuthal asymmetry between unpolarized e^+ and e^- beams. In contrast to the Beam Spin Asymmetry (BSA) which is sensitive to the imaginary part of the DVCS amplitudes, the Beam Charge Asymmetry (BCA) is probing the real part of the DVCS amplitudes [50]. Analysis of the HERMES e^- data in 98-99 and the e^+ data in 99-00 has shown a positive effect for BSA [51].

QCD factorization was proved to be valid for exclusive meson production with longitudinal virtual photons [52]. Such factorization allowed new means to extract the unpolarized and polarized GPD. In particular, unpolarized GPDs can be measured with exclusive vector meson production, while polarized GPDs can be probed via exclusive pseudoscalar meson production. A broad program of DVCS and hard exclusive processes has been proposed [53] for the 12 GeV upgrade at JLab.

ACKNOWLEDGMENTS

I would like to thank V. Burkert, J. P. Chen, C. Keppel, N. Makins, and W. K. Tung for helpful discussion.

REFERENCES

1. Gottfried, K., *Phys. Rev. Lett.*, **18**, 1174 (1967).
2. Towell, R. S., et al., *Phys. Rev.*, **D64**, 052002 (2001).
3. Kumano, S., *Phys. Rep.*, **303**, 183 (1998).
4. Garvey, G. T., and Peng, J. C., *Prog. Part. Nucl. Phys.*, **47**, 203 (2001).
5. Barone, V., et al., *Eur. Phys. J.*, **C12**, 243 (2000).
6. Zeller, G. P., et al., *Phys. Rev.*, **D18**, 111103 (2002).
7. Goncharov, M., et al., *Phys. Rev.*, **D64**, 112006 (2001).
8. Olness, F., *Talk presented at DIS03* (2003).
9. Zeller, G. P., et al., *Phys. Rev. Lett.*, **88**, 091802 (2002).
10. Garvey, G., Louis, W., and White, H., *Phys. Rev.*, **C48**, 761 (1993).
11. Spayde, D. T., et al., *Phys. Rev. Lett.*, **84**, 1106 (2000).
12. Aniol, K. A., et al., *Phys. Lett.*, **B509**, 211 (2001).
13. Diakonov, D. I., et al., *Nucl. Phys.*, **B480**, 341 (1996).
14. Wakamatsu, M., and Kubota, T., *Phys. Rev.*, **D57**, 5755 (1998).
15. Airapetian, A., et al., *hep-ex/0307064* (2003).
16. Dressler, B., et al., *Eur. Phys. J.*, **C14**, 147 (2000).
17. Adeva, B., et al., *Phys. Rev.*, **D58**, 112002 (1998).
18. Bunce, G., et al., *Ann. Rev. Nucl. Part. Sci.*, **50**, 525 (2000).
19. Close, F. E., *Phys. Lett.*, **B43**, 422 (1973).
20. Carlitz, R., *Phys. Lett.*, **B58**, 345 (1975).
21. Farrar, G. R., and Jackson, D. R., *Phys. Rev. Lett.*, **35**, 1416 (1975).
22. Brodsky, S., Burkardt, M., and Schmidt, I., *Nucl. Phys.*, **B441**, 197 (1995).
23. W., M., and Thomas, A. W., *Phys. Lett.*, **B377**, 11 (1996).
24. Abe, F., et al., *Phys. Rev. Lett.*, **81**, 5754 (1998).
25. Zhang, Z., *hep-ph/0110231* (2001).
26. Webb, J. C., et al., *hep-ex/0302019* (2003).
27. Tung, W. K., *Talk presented at DIS03* (2003).
28. Kuhn, S., et al., *JLab proposal E-03-012* (2003).
29. Bloom, E. D., and Gilman, F. J., *Phys. Rev. Lett.*, **25**, 1140 (1970).
30. Niculescu, I., et al., *Phys. Rev. Lett.*, **85**, 1182,1186 (2000).
31. Airapetian, A., et al., *Phys. Rev. Lett.*, **90**, 092002 (2003).
32. Arrington, J., et al., *nucl-ex/0307012* (2003).
33. Fatemi, R., et al., *nucl-ex/0306019* (2003).
34. Amarian, M., et al., *Phys. Rev. Lett.*, **89**, 242301 (2002).
35. Airapetian, A., et al., *hep-ex/0210047* (2002).
36. Collins, J., *Nucl. Phys.*, **B396**, 161 (1993).
37. Airapetian, A., et al., *Phys. Lett.*, **B562**, 182 (2003).
38. Sivers, D. W., *Phys. Rev.*, **D41**, 83 (1990).
39. Brodsky, S., Hwang, D., and Schimidt, I., *Phys. Lett.*, **B530**, 99 (2002).
40. Collins, J., *Phys. Lett.*, **B536**, 43 (2002).
41. Anselmino, M., Boglione, M., and Murgia, F., *Phys. Lett.*, **B362**, 164 (1995).
42. Anselmino, M., and Murgia, F., *Phys. Lett.*, **B442**, 470 (1998).
43. Adams, D. L., et al., *Phys. Lett.*, **B261**, 201 (1991).
44. Bland, L. C., *hep-ex/0212013* (2002).
45. Ji, X., *Phys. Rev. Lett.*, **78**, 610 (1997).
46. Chekanov, S., et al., *hep-ex/0305028* (2003).
47. Adloff, C., et al., *Phys. Lett.*, **B517**, 47 (2001).
48. Airapetian, A., et al., *Phys. Rev. Lett.*, **87**, 182001 (2001).
49. Stepanyan, S., et al., *Phys. Rev. Lett.*, **87**, 182002 (2001).
50. Diehl, M., et al., *Phys. Lett.*, **B411**, 193 (1997).
51. Bianchi, N., *Int. J. Mod. Phys.*, **A18**, 1311 (2003).
52. Collins, J. C., et al., *Phys. Rev.*, **D56**, 2982 (1997).
53. Burkert, V., *hep-ph/0303006* (2003).

QCD and RHIC

D. Kharzeev

*Physics Department
Brookhaven National Laboratory
Upton, New York 11973, USA*

Abstract. In this talk I discuss recent advances in Quantum Chromo-Dynamics, in particular the progress in understanding the collective dynamics of the theory. I emphasise the significance of the RHIC program for establishing the properties of hot and dense QCD matter and for understanding the dynamics of the theory at the high parton density, strong color field frontier. Hopes and expectations for the future are discussed as well.

INTRODUCTION

It is hard to deny the importance of Quantum Chromo–Dynamics – strong interaction is, indeed, the strongest force of Nature, responsible for over 80% of the baryon masses, and thus for most of the mass of everything on Earth and in the Universe. Strong interactions bind nucleons in nuclei, which, being then bound into molecules by much weaker electro-magnetic forces, give rise to the variety of the physical World. Quantum Chromo–Dynamics is *the* theory of strong interactions, and its practical importance is thus undeniable. But QCD is more than a useful tool – it is a consistent and very rich field theory, which continues to serve as a stimulus for, and testing ground of, many exciting ideas and new methods in theoretical physics.

QCD: SUCCESSES AND UNSOLVED PROBLEMS

So what is QCD? From the early days of the accelerator experiments it has become clear that the number of hadronic resonances is very large, suggesting that all hadrons may be classified in terms of a smaller number of (more) fundamental constituents. A convenient classification was offered by the quark model, but QCD was not born until the hypothetical existence of quarks was not supplemented by the principle of local gauge invariance, previously established as the basis of electromagnetism. The resulting Lagrangian has the form

$$\mathscr{L} = -\frac{1}{4} G^a_{\mu\nu} G^a_{\mu\nu} + \sum_f \bar{q}^a_f (i\gamma_\mu D_\mu - m_f) q^a_f; \qquad (1)$$

the sum is over different colors a and quark flavors f; the covariant derivative is $D_\mu = \partial_\mu - igA^a_\mu t^a$, where t^a is the generator of the color group $SU(3)$, A^a_μ is the gauge (gluon) field and g is the coupling constant. The gluon field strength tensor is given by

$$G^a_{\mu\nu} = \partial_\mu A^a_\nu - \partial_\nu A^a_\mu + g f^{abc} A^b_\mu A^c_\nu, \qquad (2)$$

where f^{abc} is the structure constant of $SU(3)$: $[t^a, t^b] = i f^{abc} t^c$.

Asymptotic freedom

Due to the quantum effects of vacuum polarization, the charge in field theory can vary with the distance. In electrodynamics, summation of the electron–positron loops in the photon propagator leads to the following expression for the effective charge, valid at $r \gg r_0$:

$$\alpha_{em}(r) \simeq \frac{3\pi}{2\ln(r/r_0)}. \qquad (3)$$

This formula clearly exhibits the "zero charge" problem [1] of QED: in the local limit $r_0 \to 0$ the effective charge vanishes at any finite distance away from the bare charge due to the screening. Fortunately, because of the smallness of the physical coupling, this apparent inconsistency of the theory manifests itself only at very short distances $\sim exp\{-3\pi/[2\alpha_{em}]\}$, $\alpha_{em} \simeq 1/137$. Such short distances are (and probably will always remain) beyond the reach of experiments, and one can safely use QED as a truly effective theory.

As it has been established long time ago [2], QCD is drastically different from electrodynamics in possessing the remarkable property of "asymptotic freedom" – due to the fact that gluons carry color, the behavior of the effective charge $\alpha_s = g^2/4\pi$ changes from the familiar from QED screening to anti-screening:

$$\alpha_s(r) \simeq \frac{3\pi}{(11N_c/2 - N_f)\ln(r_0/r)}; \qquad (4)$$

as long as the number of flavors does not exceed 16 ($N_c = 3$), the anti-screening originating from gluon loops overcomes the screening due to quark–antiquark pairs, and the theory, unlike electrodynamics, is weakly coupled at short distances: $\alpha_s(r) \to 0$ when $r \to 0$.

Chiral symmetry and $U_A(1)$ problem

In the limit of massless quarks, QCD Lagrangian (1) possesses an additional symmetry $U_L(N_f) \times U_R(N_f)$ with respect to the independent transformation of left– and right–handed quark fields $q_{L,R} = \frac{1}{2}(1 \pm \gamma_5)q$:

$$q_L \to V_L q_L; \quad q_R \to V_R q_R; \quad V_L, V_R \in U(N_f); \qquad (5)$$

this means that left– and right–handed quarks are not correlated. Even a brief look into the Particle Data tables, or simply in the mirror, can convince anyone that there is no symmetry between left and right in the physical World. One thus has to assume that the symmetry (5) is spontaneously broken in the vacuum. The flavor composition of the existing eight Goldstone bosons (3 pions, 4 kaons, and the η) suggests that the $U_A(1)$ part of $U_L(3) \times U_R(3) = SU_L(3) \times SU_R(3) \times U_V(1) \times U_A(1)$ does not exist. This constitutes the famous "$U_A(1)$ problem".

The origin of mass

There is yet another problem with the chiral limit in QCD. Indeed, as the quark masses are put to zero, the Lagrangian (1) does not contain a single dimensionful scale – the only parameters are pure numbers N_c and N_f. The theory is thus apparently invariant with respect to scale transformations, and the corresponding scale current is conserved: $\partial_\mu s_\mu = 0$. However, the absence of a mass scale would imply that all physical states in the theory should be massless!

Quantum anomalies and classical solutions

Both apparent problems – the missing $U_A(1)$ symmetry and the origin of hadron masses – are related to quantum anomalies. Once the coupling to gluons is included, both flavor singlet axial current and the scale current cease to be conserved; their divergences become proportional to the $\alpha_s G^a_{\mu\nu} \tilde{G}^a_{\mu\nu}$ and $\alpha_s G^a_{\mu\nu} G^a_{\mu\nu}$ gluon operators, correspondingly. This fact by itself would not have dramatic consequences if the gluonic vacuum were "empty", with $G^a_{\mu\nu} = 0$. However, it appears that due to non-trivial topology of the $SU(3)$ gauge group, QCD equations of motion allow classical solutions even in the absence of external color source, i.e. in the vacuum. The well-known example of a classical solution is the instanton, corresponding to the mapping of a three-dimensional sphere S^3 into the $SU(2)$ subgroup of $SU(3)$; its existence was shown to solve the $U_A(1)$ problem.

Confinement

The list of the problems facing us in the study of QCD would not be complete without the most important problem of all – why are the colored quarks and gluons excluded from the physical spectrum of the theory? Since confinement does not appear in perturbative treatment of the theory, the solution of this problem, again, must lie in the properties of the QCD vacuum.

Understanding the Vacuum

As was repeatedly stated above, the most important problem facing us in the study of all aspects of QCD is understanding the structure of the vacuum, which does not at all behave as an empty space, but as a physical entity with a complicated structure. As such, the vacuum can be excited, altered and modified in physical processes [3] (for a review, see [4]), such as heavy ion collisions.

RECENT ADVANCES IN QCD

Recent years have seen very significant developments in QCD; most of them are discussed in other talks at this Conference. Some of these developments are listed below:

- Perturbative QCD
  - Higher order computations, necessary for the precision descriptions of hard processes, and for ensuring theoretical control over the reliability of perturbative calculations.
  - resummations (multiple parton effects) – even though at large momentum transfers Q^2 the coupling $\alpha_s(Q^2)$ becomes weak, this does not mean that the probability of parton emission is small – this is because the phase space factors of $\ln Q^2$ and/or $\ln(1/x)$ can become large in certain kinematical regions. This leads to the dominance of multi–parton effects in many processes, including high–energy heavy ion collisions.
- Lattice QCD – numerical Monte–Carlo calculations on the lattice have become one of the main sources of information about the dynamics of QCD in the non–perturbative domain [5], and in particular about the phase structure of the theory at finite temperature [6].
- Low energy QCD, as represented by the effective chiral theories. Chiral symmetry of the QCD lagrangean allows for a rigorous formulation of an effective theory describing interactions of pions and nucleons at the energies below the hadron "compositeness scale" of $\Lambda_\chi \simeq 1$ GeV.
- Extensions of QCD and related theories: it is often possible to glean a better understanding of QCD by looking at the properties of related and simpler theories; the examples include:
  - Large N_c, N_f
  - Chiral gauge theories
  - SUSY QCD, AdS correspondence, ...
- QCD of collective phenomena – this is the main topic of this talk; here I will just list some of the exciting recent developments:
  - Large baryon density, small T: color superconductivity (for reviews, see [7, 8, 9])
  - Phase diagram: tricritical point [10]
  - Small x / large A: Color Glass Condensate ([11, 12]; for reviews, see [13, 14, 15, 16, 17])
  - Jet propagation in dense QCD matter: energy loss (see [18, 19, 20] and references therein)

PHASE DIAGRAM OF STATISTICAL QCD

It has been clearly established by now that at high temperature QCD exhibits non-trivial critical behavior, corresponding to the transition from the hadron gas to the phase dominated by the color degrees of freedom ("the quark–gluon plasma"). The deconfinement transition to quark–gluon plasma is accompanied, in the presence of light quarks, by the restoration of chiral symmetry, and, perhaps at still higher temperatures, by the restoration of $U_A(1)$.

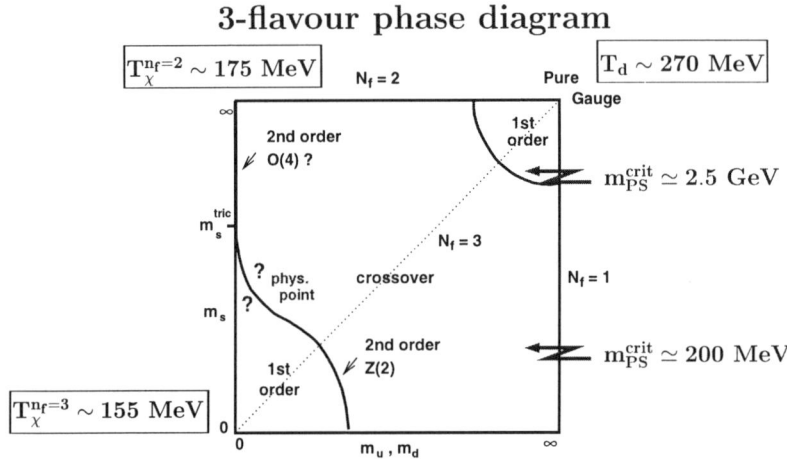

FIGURE 1. Phase diagram of finite–temperature QCD for three quark flavors. From [6].

The dynamical properties of this transition in some limiting cases (e.g., deconfinement phase transition in pure gauge theory, or the chiral phase transition in the limit of massless quarks [21]) are constrained by the symmetries of QCD but in general have to be deduced from numerical calculations on the lattice.

It has been established that the order of phase transition, or whether there is a phase transition at all or a rapid cross–over, depends on the values of the quark masses as shown in Fig.1 – see [6] for a recent review. Since, despite a very impressive progress, the physical values of light quark masses are still beyond the reach of current lattice calculations, the properties of the phase transition in the real physical world still have to be determined. Many unexpected phenomena may appear close to the phase transition; one example is the formation of $\mathscr{P}$ and $\mathscr{CP}$ odd metastable vacua [22].

At finite baryon number density, lattice calculations meet with significant difficulties. Despite recent progress in the extension of lattice results to (small) finite baryon density, to understand the dynamics of QCD in this domain one has to rely on analytical methods and models. This is why recent rigorous results (for reviews, see [7, 8, 9]) in understanding the properties of dense and cold QCD matter in the region of large baryon density, or corresponding chemical potential μ, and small μ/T are so important – and interesting. It appears that the new state of matter is formed in this domain – Color Super–Conductor. The properties of this matter depend on the masses of quarks; for three light quarks an

EXPLORING the PHASES of QCD

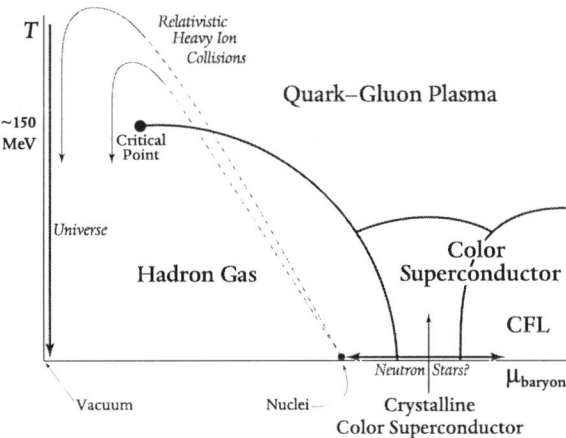

FIGURE 2. Phase diagram of QCD in the $(T-\mu)$ plane for two massless flavors; high temperature phase is the Quark–Gluon Plasma and the high baryon density phase is the Color Super–Conductor. From [23].

interesting symmetry breaking pattern of "Color–Flavor Locking" has been uncovered [7]. This progress in understanding the properties of cold dense QCD matter leads to important astrophysical applications, in particular in the physics of neutron stars.

QCD OF STRONG COLOR FIELDS

Most of the applications of QCD so far have been limited to the short distance regime of high momentum transfer, where the theory becomes weakly coupled and can be linearized. While this is the only domain where our theoretical tools based on perturbation theory are adequate, this is also the domain in which the beautiful non–linear structure of QCD does not yet reveal itself fully. On the other hand, as soon as we decrease the momentum transfer in a process, the dynamics rapidly becomes non–linear, but our understanding is hindered by the large coupling.

Being perplexed by this problem, one is tempted to dream about an environment in which the coupling is weak, allowing a systematic theoretical treatment, but the fields are strong, revealing the full non–linear nature of QCD. Fortunately such an environment can be created on Earth with the help of colliders – high collision energy ensures access to the domain of small Bjorken x, where the density of partons and the coherence length of hard processes $\sim 1/(mx)$ (m is the nucleon mass) are both large. If nuclear beams of atomic number A are used, one can further increase the parton density by a large factor of $\sim A^{1/3}$ – when the coherence length starts to exceed the size of the nucleus, all $\sim A^{1/3}$ partons contribute coherently to the hard scattering. Once the density of partons in the transverse plane ρ reaches the limit $\sim 1/\alpha_s$, the dynamics becomes classical

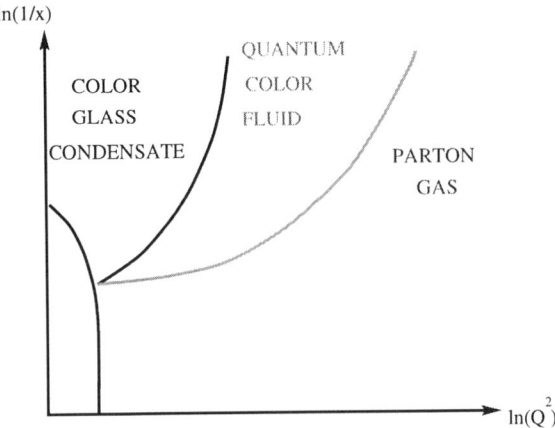

FIGURE 3. Phase diagram of high–energy QCD in the $(x - Q^2)$ plane; at small Bjorken x and moderate Q^2 partons form the Color Glass Condensate, which affects the properties of quantum evolution at larger Q^2 and smaller x within the Quantm Color Fluid domain.

– this is the domain of parton saturation [24, 25, 26] and "Color Glass Condensate" [13, 14, 15, 16, 17], see Fig.3). This correspondence allowed to formulate an effective quasi–classical theory [27], which is a subject of vigorous investigations at present.

The influence of classical dynamics on the properties of quantum evolution in fact extends significantly beyond the "saturation boundary" – formally, this corresponds to the domain in which the gluon density acquires an anomalous dimension close to $\gamma \simeq 1/2$ [28, 29]. This domain is marked as "Quantum Color Fluid" in Fig.3. Finally, the domain in which the parton density is not very large and their dynamics can thus be described by linear DGLAP evolution equations corresponds to the "Parton Gas" phase.

In nuclear collisions, the transverse density of partons becomes a function of centrality; a generic feature of the quasi–classical approach – the proportionality of the number of gluons to the inverse of the coupling constant – thus leads to definite predictions [30] on the centrality dependence of multiplicity, which are so far in accord with the data coming from RHIC [31, 32, 33].

QCD: NEW FRONTIERS AT RHIC

The arguments presented above show how research with relativistic heavy ions can, and does already, advance the understanding of the phase diagram of statistical QCD and the non–linear classical behavior of the theory. But is this enough to justify the heavy ion program? Indeed, QCD has already firmly occupied its place as part of the Standard Model. However, understanding the physical World does not mean only establishing its fundamental constituents; it means, mostly, understanding how these constituents interact and bring to the existence the entire variety of physical objects composing the Universe. Think of electrodynamics – the simplest of all gauge theories –

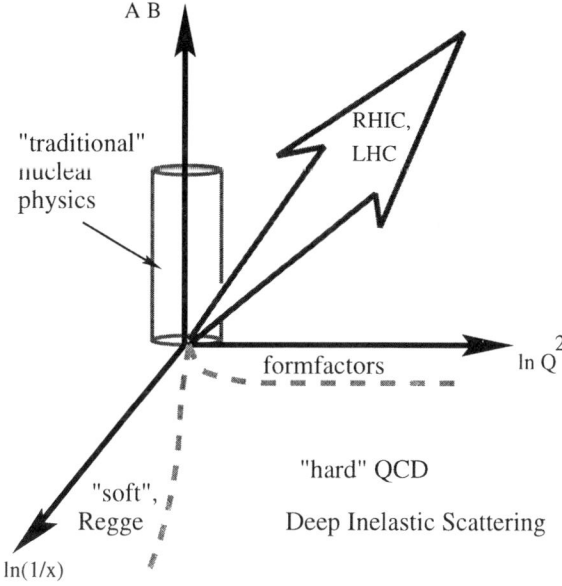

FIGURE 4. The place of relativistic heavy ion physics in the study of QCD; the vertical axis is the product of atomic numbers of projectile and target, and the horizontal axes are the momentum transfer Q^2 and rapidity $y = \ln(1/x)$ (x is the Bjorken scaling variable).

which is responsible for an enormous assortment of materials and substances of different structure. Now try to imagine the beauty and complexity of collective phenomena made possible in the theory where "electrons" carry three different "charges", "photons" carry eight, and they are all bound by the force two orders of magnitude stronger than electromagnetic forces! Just as the research in condensed matter physics is driven by the ability to perform experiments with different number of atoms, under different conditions of low and high temperature and pressure, further progress in QCD will be largely driven by the studies of hadronic matter under extreme conditions. By increasing the atomic number of the colliding systems and by raising the energy of the collision, we get access to the high parton density, high field strength QCD (see Fig.4).

A GLIMPSE OF RHIC RESULTS

RHIC began operation in 2000, culminating over ten years of development and construction and a much longer period of theoretical speculations about the properties of hot QCD matter produced in nuclear collisions in the collider regime. The results from the first years of RHIC operation have been summarized at this Conference [34, 35], so I will just briefly mention some highlights:

- Hadron multiplicity tells us what fraction of the collision energy is inelastically transferred to secondary particles. If nucleus–nucleus collisions were an incoherent superposition of nucleon–nucleon collisions, the multiplicity would scale with the number N_{coll} of individual *NN* collisions. This has not been observed; the multiplicity is reduced by factor of $\simeq 6(!)$ with respect to N_{coll} scaling. Final state interactions, by second law of thermodynamics, cannot decrease the number of particles, so the reduction of multiplicity has to originate in the initial state. The multiplicity measurements [31, 32, 33] thus provide a proof of coherence in parton interactions at high energy.

- Azimuthal asymmetry with respect to the reconstructed reaction plane is an important characteristic of collectivity in partonic interactions. Indeed, if all of the *NN* collisions were independent, there would be no reason to expect asymmetry in the distribution of the produced hadrons in the azimuthal angle. Experimentally, the asymmetry of the azimuthal distribution is quite sizable, and for peripheral collisions reaches about 35% [36].

- Suppression of high p_t particles – the yield of high p_t hadrons has been found [37, 38] to be drastically reduced with respect to what is expected for incoherent production in *NN* collisions. Does this important discovery signal jet energy loss in the quark–gluon plasma? We are close to answering this question since the first results of *dA* measurements have been reported at this Conference [39, 40, 41]. The comparison of these results to the $Au-Au$ data indicates the presence of final-state interaction in the intermediate range of p_t, up to $5 \div 6$ GeV, consistent with the expectations based on strong collective parton interactions driving the system towards the equilibrium. There is a hint of differing attenuation factors for neutral pions and heavier particles (baryons and kaons) for $p_t \leq 5 \div 6$ GeV [40]. If this is confirmed, the difference in attenuation factors will point towards a production mechanism beyond the independent fragmentation of perturbative jets at $p_t \leq 5 \div 6$ GeV. It remains to be checked if there is a statistically significant suppression of high p_t particles in $d-Au$ collisions at $p_t \geq 5 \div 6$ GeV [39, 40] . It will also be extremely important to check the dependence of high p_t particle production on rapidity, since moving into the deuterium fragmentation region allows access to the higher parton density in the initial wave function of the Au nucleus and at the same time to smaller density of final state hadrons. Measurements at forward rapidity thus allow to separate clearly the effects of the final state interaction in cold nuclear matter from the effects related to high parton densities in the initial state. In any case, it is clear that once the analysis of $d-Au$ data is complete, the origin of high p_t suppression can be clarified.

It is too early to fully assess the implications of RHIC results; however, we can already conclude that many of the observed phenomena clearly manifest collective behavior; nuclear collisions at RHIC are not an incoherent superposition of nucleon–nucleon collisions. The measured particle multiplicities and transverse momentum spectra allow to estimate initial energy density at the early moments of the collision; a typical value inferred in this way is about 20 GeV/fm^3. The dynamics of strongly interacting matter at such energy density (exceeding the energy density in a nucleus by over two orders of magnitude!) should be described in terms of quarks and gluons, and the collective phenomena observed at RHIC thus directly reflect the properties of high density QCD.

REFERENCES

1. L.D. Landau and I.Ya. Pomeranchuk, Dokl.Akad.Nauk 102 (1955) 489.
2. D.J. Gross and F. Wilczek, Phys. Rev. Lett. **30** (1973) 1343;
   H.D. Politzer, Phys. Rev. Lett. **30** (1973) 1346.
3. T.D. Lee and G.C. Wick, Phys. Rev. **D9** (1974) 2291.
4. H. Satz, Nucl. Phys. A **715**, 3 (2003) [arXiv:hep-ph/0209181].
5. R. Mawhinney, in these Proceedings.
6. F. Karsch and E. Laermann, arXiv:hep-lat/0305025.
7. K. Rajagopal and F. Wilczek, arXiv:hep-ph/0011333.
8. T. Schafer and E. V. Shuryak, Lect. Notes Phys. **578**, 203 (2001) [arXiv:nucl-th/0010049].
9. D. H. Rischke and R. D. Pisarski, arXiv:nucl-th/0004016.
10. M. A. Stephanov, K. Rajagopal and E. V. Shuryak, Phys. Rev. Lett. **81**, 4816 (1998) [arXiv:hep-ph/9806219].
11. L.D. McLerran, in these Proceedings.
12. Y. V. Kovchegov, in these Proceedings; arXiv:hep-ph/0308076.
13. L. D. McLerran, Lect. Notes Phys. **583**, 291 (2002) [arXiv:hep-ph/0104285].
14. A. H. Mueller, arXiv:hep-ph/0111244.
15. E. Iancu and R. Venugopalan, arXiv:hep-ph/0303204.
16. E. Levin, arXiv:hep-ph/0105205.
17. D. Kharzeev, arXiv:hep-ph/0204014.
18. M. Gyulassy, I. Vitev, X. N. Wang and B. W. Zhang, arXiv:nucl-th/0302077.
19. R. Baier, Y. L. Dokshitzer, A. H. Mueller and D. Schiff, JHEP **0109**, 033 (2001) [arXiv:hep-ph/0106347].
20. R. Baier, D. Schiff and B. G. Zakharov, Ann. Rev. Nucl. Part. Sci. **50**, 37 (2000) [arXiv:hep-ph/0002198].
21. R. D. Pisarski and F. Wilczek, Phys. Rev. D **29**, 338 (1984).
22. D. Kharzeev, R. D. Pisarski and M. H. Tytgat, Phys. Rev. Lett. **81**, 512 (1998) [arXiv:hep-ph/9804221].
23. K. Rajagopal, private communication.
24. L. V. Gribov, E. M. Levin and M. G. Ryskin, Phys. Rept. 100, 1 (1983).
25. A. H. Mueller and J. w. Qiu, Nucl. Phys. B 268, 427 (1986).
26. J. P. Blaizot and A. H. Mueller, Nucl. Phys. B 289, 847 (1987).
27. L. McLerran and R. Venugopalan, Phys. Rev. D 49 (1994) 2233, 3352; D 50 (1994) 2225.
28. E. Iancu, K. Itakura and L. McLerran, Nucl. Phys. A **708**, 327 (2002) [arXiv:hep-ph/0203137].
29. D. Kharzeev, E. Levin and L. McLerran, Phys. Lett. B **561**, 93 (2003) [arXiv:hep-ph/0210332].
30. D. Kharzeev and M. Nardi, Phys. Lett. B 507 (2001) 121 [arXiv:nucl-th/0012025]; D. Kharzeev and E. Levin, Phys. Lett. B 523 (2001) 79 [arXiv:nucl-th/0108006]; D. Kharzeev, E. Levin and M. Nardi, arXiv:hep-ph/0111315.
31. B. B. Back et al. [PHOBOS Collaboration], Phys. Rev. Lett. 88 (2002) 022302; B. B. Back et al. [PHOBOS Collaboration], Phys. Rev. C 65 (2002) 061901.
32. A. Bazilevsky [PHENIX Collaboration], arXiv:nucl-ex/0209025.
33. I. G. Bearden et al. [BRAHMS Collaboration], Phys. Rev. Lett. 88 (2002) 202301.
34. B. Jacak, in these Proceedings.
35. J. L. Nagle and T. Hallman, in these Proceedings; arXiv:nucl-ex/0308019.
36. C. Adler et al. [STAR Collaboration], Phys. Rev. C 66 (2002) 034904.
37. K. Adcox et al. [PHENIX Collaboration], Phys. Rev. Lett. 88, 022301 (2002) [arXiv:nucl-ex/0109003].
38. C. Adler et al. [STAR Collaboration], arXiv:nucl-ex/0206006.
39. P. Jacobs [STAR Collaboration], in these Proceedings.
40. L. Alphecetche [PHENIX Collaboration], in these Proceedings.
41. G. Roland [PHOBOS Collaboration], in these Proceedings.

Neutron spin structure results from JLab Hall A

Zein-Eddine Meziani

Department of Physics, Temple University, Philadelphia, PA 19122

Abstract. My presentation will focus on some of the latest results of the neutron spin physics program at Jefferson Laboratory in Hall A using a polarized ^{3}He target. This program includes several completed experiments in which the spin structure functions of ^{3}He were measured. The covered kinematic regions were these measurements were performed include the low Q^2 resonance and inelastic regions and the high Q^2 deep inelastic region. These experiments offer a ground for testing our understanding of the strong regime of quantum chromodynamics (QCD) through the determination of the neutron spin-dependent structure functions and their moments.

INTRODUCTION

After 25 years of spin structure measurements at the high energy physics laboratories (i.e. CERN, SLAC and DESY) leading to the determination of the quark spin content of the nucleon [1, 2] and culminating with the test of the Bjorken sum rule [3], the spin structure of the neutron in the deep inelastic large x region at Q^2 (above 1 GeV^2) is still poorly measured. This region does not contribute much to the first moment of spin structure functions, but it is crucial for evaluating higher moments of these functions. These moments offer a testing ground of QCD because they are connected with specific matrix elements [4, 5] that are directly calculable in lattice QCD [6]. The spin structure function in this valence quark region can also be tested through quark model calculations. At momentum transfers below Q^2=1 GeV^2 the large x region is dominated by the resonance contributions and information on the neutron spin structure is also very scarce. This region is needed to determine the higher twists corrections in deep inelastic scattering (DIS) using the operator product expansion (OPE) technique to express moments of structure functions.

Here we present results of a precision measurement (JLab E99-117) of the neutron asymmetry A_1^n in the valence region (large x and Q^2). We also show results of moments of the neutron spin structure g_2 (JLab E94-010) at Q^2's ranging from 1 GeV^2 to 0.1 GeV^2. The above experiments share the same experimental setup. They were carried out at Jefferson Lab in Hall A using a highly polarized electron beam (70-80%) with an average current up to 15μA and a high pressure polarized (on average between 30% and 40% in-beam) ^{3}He target with the highest polarized luminosity in the world. Details on these experiments can be found at[7].

SPIN AND FLAVOR DECOMPOSITION IN THE VALENCE QUARK REGION

The virtual photon-nucleon asymmetry A_1^n and spin structure function g_1^n are the most poorly known in the valence quark region ($x > 0.3$). This shortcoming is due to the small scattering cross sections at large x and Q^2 combined with a lack of high polarized luminosity facilities. This region, however, is clean and unambiguous since it is not polluted by sea quarks and gluons offering thus a unique opportunity to test predictions that are difficult if not impossible at low x. The set of predictions of A_1^n in the valence quark region fall into two categories, those of RCQM's which break SU(6) symmetry in the ground state wave function by hyperfine interaction [8, 9], and those of pQCD with a hadron helicity conservation (HHC) constraint [10, 11] as $x \to 1$ which break SU(6) symmetry dynamically.

The difference between these approaches is dramatic when the constituents flavor-spin decomposition is performed. For a proton and in the case of pQCD with HHC, we have $\Delta u(x)/u(x) \to 1$ and $\Delta d(x)/d(x) \to 1$, while for the case of RCQM's $\Delta u/u \to 1$, $\Delta d/d \to -2/3$. We notice that in leading order pQCD with HHC $\Delta d/d$ changes sign from negative at low x to positive at large x.

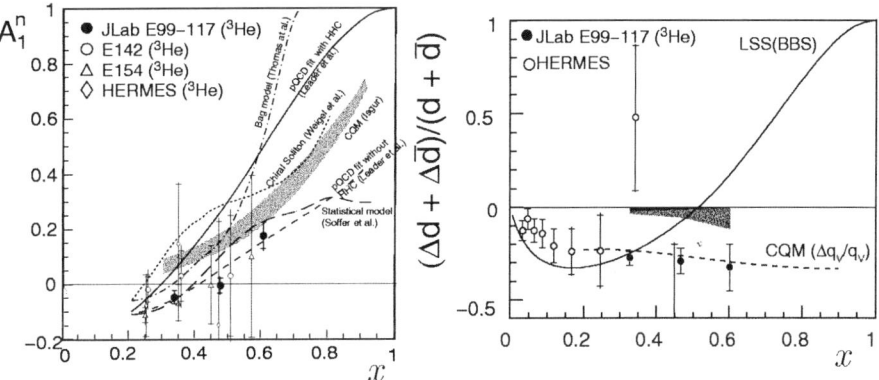

FIGURE 1. Left panel: Preliminary results of Jefferson Lab experiment E99-117 (solid circles) along with the world data (open symbols). The curves are predictions described in the text. Right panel: Spin-flavor dependent down-quark distribution extracted from this experiment. The light error band is an estimate of the difference between the valence and the total down-quark distribution including the sea.

In Figure 1 (left panel) we show preliminary results of A_1^n. The first data point at $x = 0.33$ is in good agreement with previous measurements. The data points show a clear change of sign of A_1^n as x increases and are compared with theoretical predictions. The total error in each point is dominated by the statistical error. The solid line is a prediction using HHC based on LSS(BBS) parameterization of g_1^n/F_1^n [12], the long-dashed line is a prediction of g_1^n/F_1^n from LSS 2001 parametrization at $Q^2 = 5$ GeV^2 without HHC constraints [13]. The shaded area is a range of predictions of A_1^n from the constituent quark model [9] while the dashed line is a calculation of the statistical model at $Q^2 = 4$ GeV^2 by Bourrely et. al. [14]. The short-dashed line is the chiral soliton model prediction at $Q^2 = 3$ GeV^2 by Weigel, Gamberg and Reinhardt [15]. Finally, the dot-dashed line is a bag model calculation but without meson cloud by Boros and Thomas [16] Data from Hermes and SLAC are original values without being re-analyzed for the Δ contribution of the nuclear corrections.

We used the quark parton model interpretation of g_1 and F_1 to perform a flavor decomposition of the spin dependent quark distributions assuming a negligible strange quark contribution above $x = 0.3$. The down-quark distribution obtained in E99-117 (filled squares) along with preliminary results of the HERMES semi-inclusive measurements (open circles) [17] are also shown in Fig. 1 (left panel). The solid line is a pQCD fit to the world data using the HHC constraint as $x \to 1$. The dashed line correspond to an RCQM prediction. It is clear that up to $x = 0.6$ the data favor the RCQM rather than the HHC pQCD based calculations. In the latter no orbital angular momentum (OAM) in considered while in the RCQM some OAM is included through the small components of the nucleon wavefunction. These results, perhaps point towards the importance of considering the orbital momentum or quarks in the nucleon wave function. In the meantime, one has to wait for a more complete QCD calculation.

BURKHARDT-COTTHIGHAM SUM RULE AND HIGHER MOMENTS OF STRUCTURE FUNCTIONS

The g_2 structure function obeys the Burkhardt-Cotthigham sum rule

$$\Gamma_2(Q^2) = \int_0^1 g_2(x, Q^2)\, dx = 0 \qquad (1)$$

which was derived from the dispersion relation and the asymptotic behavior of the corresponding Compton amplitude [18]. This sum rule is true t all Q^2 and does not follow from the OPE. It is rather a super-convergence relation based on Regge asymptotics as discussed in the review paper by Jaffe [19]. Many scenarios which invalidate this sum rule have been discussed in the literature. Surprisingly a first precision measurement of g_2 of the proton at SLAC at $Q^2 = 5$ GeV^2 but with a limited range of x has revealed a deviation of this sum rule for the proton at the level of three standard deviations [20]. In contrast, the neutron sum rule is poorly measured but consistent with zero at the one standard deviation.

IN Fig. 2 (left panel) we show Γ_2^n in the measured region (solid circles) and after adding the elastic contribution evaluated using Mergell et al. parametrization of G_M^n and G_E^n [22] (open circles). The solid line is the resonances contribution evaluated using MAID [21]. The positive light grey band corresponds to the total experimental systematic error while the dark negative band is an estimated deep inelastic contribution (DIS) assuming $g_2 = g_2^{WW}$ following the same method as in [23]. The data show that the BC sum rule is verified within uncertainties over the Q^2 range measured. Our result is at odds with the observed violation of this sum rule for the proton at high Q^2, where the elastic contribution is negligible. The neutron result of SLAC E155x [20] is consistent with the sum rule but with a rather large error bar.

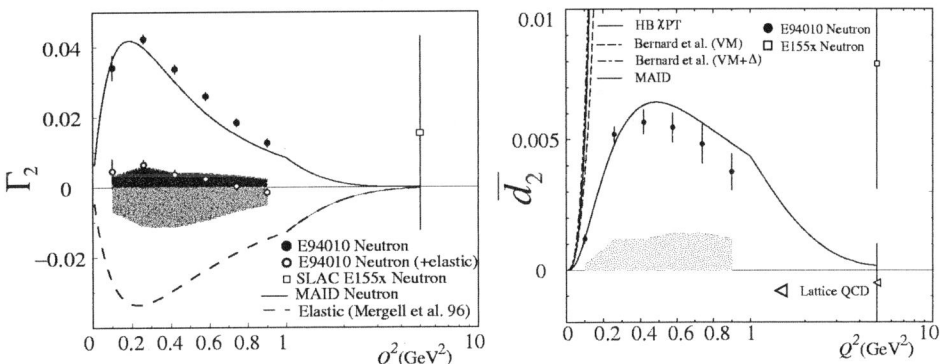

FIGURE 2. Left panel: Preliminary results of the Q^2 evolution of Γ_2^n from JLab experiment E94-010 along with the world data from DIS and theoretical calculations (see text). Right panel: Preliminary results of $\bar{d}_2^n$ compared to SLAC E155X results and χPT prediction at low Q^2 and Lattice QCD calculation at $Q^2 = 5$ GeV^2. (see text).

In the high Q^2 regime, the quantity

$$d_2(Q^2) = \int_0^1 x^2 [2g_1(x,Q^2) + 3g_2(x,Q^2)]dx, \quad (2)$$

coincides with a twist three-matrix element and is related to the electric and magnetic polarizabilities of the color field [2]. At low Q^2, a region covered by our data, its conventional interpretation in terms of higher twist is not obvious. However a interpretation in terms of the electromagnetic polarizabilities is possible [24]. These polarizabilities can be evaluated using chiral perturbation theory (χPT). Thus, the Q^2 evolution of d_2 is a quantity that offers a dual interpretation depending on the resolution of the probe. At low Q^2 it is sensitive to the a direct polarization of the electromagnetic field present in the nucleon while at high Q^2 it is sensitive to the polarization of the color field in the nucleon.

In Fig. 2 (right panel), $\bar{d}_2$ where the integration in equation (2) excludes the elastic peak is shown at several values of Q^2. The results of this experiment are the solid circles and the grey band represents their corresponding systematic uncertainty. The SLAC E155x neutron (open square) results is also shown. The solid line is the MAID calculation [21] while the solid line is a Heavy Baryon χPT calculation [24] and the covariant χPT [25]. The Lattice prediction [6] at $Q^2 = 5$ GeV^2 for the neutron d_2 matrix element is negative but close to zero. We note that all models (not shown here) predict d_2^n to be negative or zero at large Q^2. At moderate Q^2 the data of E94-010 show a positive d_2^n but decreasing perhaps to zero at high Q^2. The SLAC data also show a positive d_2^n value but with a rather large error bar. More measurements are needed to have a complete determination of the transition from low to very high Q^2 of this important quantity.

CONCLUSION

In summary, we took advantage of the highly polarized beam and high pressure polarized ^{3}He target at Jefferson Lab Hall A to investigate the internal spin structure of the neutron in the perturbative and the strong regimes of QCD. In E99-117 we have determined the world most precise down-quark helicity distribution in the valence region. The results agree with the constituent quark model rather than the HHC constrained pQCD prediction and perhaps point to the importancce of the quark OAM in this region. In experiment E94-010 we measured both g_1 and g_2 in the resonances region. We find the BC sum rule to be verified within uncertainties. The quantity $\bar{d}_2^n$ related to the electromagnetic polarizabilities at low Q^2 and the twist-three matrix element at large Q^2 was also measured and found to be small but finite.

ACKNOWLEDGMENTS

The work presented here was supported in part with funds provided to the Nuclear and Particle group at Temple University by the U.S. Department of Energy (DOE) under contract number DE-FG-02-94ER40844. The Southeastern Universities Research Association operates the Thomas Jefferson Accelerator Facility for the DOE under contractDE-AC05-84ER40150

REFERENCES

1. Hughes, E. W. and Voss, V., *Ann. Rev. Nucl. Part. Sci.* **49**, 303 (1999).
2. Filippone, B. W. and Ji, X., *Adv. in Nucl. Phys.* **26**, 1 (2001).
3. Bjorken, J. D. *Phys. Rev.* **148**, 1467 (1966); *Phys. Rev.* **D 1**, 1376 (1970).
4. Shuryak, E.V. and Vainshtein, A.I., *Nucl. Phys.* **201**, 141 (1982).
5. Jaffe, R. and Ji, X. *Phys. Rev. Lett.* **67**, 552 (1991).
6. Gockeler, M., et al., *Phys. Rev.* **D 63**, 074506 (2001).
7. Details of experiments at www.jlab.org/e99117/ and www.jlab.org/e94010/.
8. Close, F. and Thomas, A. W., *Phys. Lett.* **B 212**, 227 (1988).
9. Isgur, N., *Phys. Rev.* **D 59**, 034013 (1999).
10. Farrar, G. R. and Jackson, A. D., *Phys. Rev. Lett.* **35**, 1416 (1975).
11. Brodsky, S. J., Burkhardt, M., Schmidt, I., *Nucl. Phys.* **B 441**, 197 (1995).
12. Leader, E., Sidorov, A. V. and Stamenov, D. B., *Int. J. Mod. Phys.* **A13**, 5573 (1998).
13. Leader, E., Sidorov, A. V. and Stamenov, D. B., *Eur. Phys. J.* **C 23**, 479 (2002).
14. Bourrely, C. Soffer, J. and Bucella, F., *Eur. Phys. J.* **C23**, 479 (2002).
15. Weigel, H. and Gamberg, L., *Nuc. Phys.* **A 680**, 48 (2000) and references therein.
16. Boros, C. and Thomas, A. W., *Phys. Rev.* **D60**, 074017 (1999)
17. Wendland, J., http://hermes.desy.de/notes/pub/TRANS/Deltaq.5p.ps.gz.
18. Burkhardt, H. and Cottingham, W. N., *Ann. Phys.* **56**, 453 (1970).
19. Jaffe, R., *Comments Nucl. Part. Phys.* **19**, 239 (1990)
20. SLAC E155x, Anthony, P. L., et. al., *Phys. Lett.* **B553**, 18 (2003).
21. Drechsel, D., Kamalov, S. and Tiator L., *Phys. Rev.* **D 63**, 114010 (2001).
22. Mergell, P., Meissner, Ulf-G. and Drechsel, D., *Nucl. Phys.* **A 596**, 367 (1996).
23. Wandzura, S. and Wilczek, F., *Phys. Lett.* **B 72** 195, (1977).
24. Kao, C. W., Spitzenberg, T. and Vanderhaeghen, M., *Phys. Rev.* **D 67**, 016001 (2003).
25. Bernard, V. Hemmert, T. and Meissner, Ulf-G., *Phys. Lett.* **B 545**, 105 (2002).

Recent E158 Results[1]

P. A. Souder[2]

*Department of Physics,
201 Physics Building,
Syracuse, NY 13244, USA
E-mail: souder@physics.syr.edu*

Abstract. Experiment E158 has recently observed parity violation in the scattering of polarized electrons from the electrons in a liquid hydrogen target at SLAC. The results are consistent with the predictions of the Standard Model. We plan to obtain additional data by the end of the summer of 2003 to provide the most precise measurement of $\sin^2 \theta_W$ at low energy.

INTRODUCTION AND MOTIVATION

Experiment E158 at SLAC has measured the parity-violating electroweak interference in the elastic scattering of 48 GeV polarized electrons from the electrons in a liquid hydrogen target. The observed quantity

$$A_{LR} = \frac{\sigma_L - \sigma_R}{\sigma_L + \sigma_R}. \qquad (1)$$

is 3×10^{-7} at tree level in the Standard Model for our Q^2 value of 0.03 (GeV/c)^2. The measurement is especially sensitive to $\sin^2 \theta_W$ because

$$A_{LR} \propto (1 - 4\sin^2 \theta_W), \qquad (2)$$

and $\sin^2 \theta_W \approx 0.23$. [1] Indeed, the electroweak radiative corrections to the process [2] including the effects of hard photons [3] reduce the size of the asymmetry by almost a factor of 2. Observing this large effect is one of the goals of the experiment.

Møller scattering is also sensitive to possible extensions of the Standard Model. [4, 5] The advantage of low Q^2 data over the precise data from LEP and SLAC at the Z-pole is that the Z and new physics amplitudes are both real and thus add coherently. As pointed out by Eichten, Lane, and Peskin in 1983[6], the amplitudes at the Z-pole are imaginary and any small contribution of new physics would add in quadrature. Specific examples include extra Z-bosons motivated by Grand Unification Theories [7] or extra dimensions [8] as well as theories in which the electron is composite.

---

[1] For the E158 Collaboration
[2] Work supported by the United States Department of Energy under Grant No. DE-FG02-84ER40146

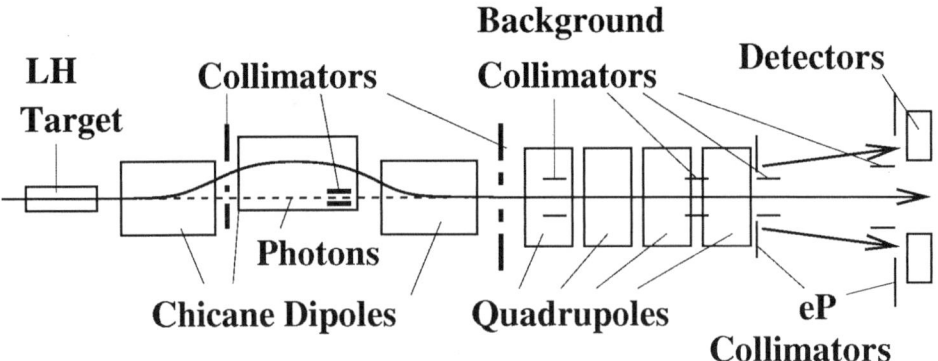

FIGURE 1. Schematic diagram of the E158 apparatus.

APPARATUS

The signature feature of E185 is the smallness of the asymmetry. Achieving this level of precision requires an apparatus that can collect immense statistics while maintaining small systematic errors. Special techniques for achieving these goals in parity experiments with polarized electrons have been developed during recent years. [9]

Controlling the Helicity of the Beam

The polarized electrons are produced by photoemission from polarized laser light striking a GaAsP photocathode. By using a gradient-doped structure, [10] we can achieve high peak current and a polarization of about 85%. The beam is delivered in bunches a few hundred ns long at a rate of 120 Hz.

The helicity of the beam is changed between pulses by changing the voltage on a Pockels' cell. The sign of the helicity is reversed from time to time by inserting a half-wave plate in the beam. The helicity of the beam reverses several times in the bend between the accelerator and the experimental hall due to the $g-2$ effect. By taking half our data 48 GeV and the rest at 45 GeV, we obtain an additional sign flip. These sign flips provide a powerful means to detect and cancel possible systematic errors.

Beam Monitors

It is essential that the data be corrected for any helicity-correlated differences in the parameters of the beam such as energy, position, or angle. To achieve this, we measure the position of the beam with a precision of a few microns every beam pulse with a set of beam monitors that are sensitive to all of the important parameters.

TABLE 1. Corrections and backgrounds to the measured asymmetries. The dilution factor is given by f and corrections to the asymmetry are given by A. Errors in these quantities are df and dA respectively

Issue	f	df	A	dA
Beam first order	-	-	-	18
Beam spot size	-	-	0	5
Transverse asymmetry	-	-	0	5
High energy photons	0.004	0.002	0	0
Synchrotron	0.002	0.002	0	5
Neutrons	0.003	0.001	-5	3
eP elastic	0.080	0.020	-11	4
eP inelastic	0.017	0.005	-31	10
Soft photons	0.001	0.001	0	9
Pions	0.002	0.002	0	5
Total	0.109	0.021	-42	24

Møller Spectrometer

Møller electrons scattered in our 2m-long liquid hydrogen target are detected in the Møller spectrometer shown in Fig. 1. The main elements are an achromatic chicane, focusing quadrupoles, and an integrating detector. Since there is in excess of 10^7 detected particles per pulse, the signal must be integrated. The spectrometer with its collimators then must deliver a pure sample of scattered particles. The collimator in front of the quadrupoles accepts electrons from Møller scattering near 90° in the CM frame. The quadrupoles bend the Møller electrons onto the detector while the higher energy electrons scattered from the protons in the target miss. The chicane absorbs low energy particles produced in the target while passing unscattered electrons and forward angle photons to the high power beam dump. The "background collimators" prevent energy from slit scattering from striking the detector.

The high flux in the detector creates problems with radiation damage. Thus the detector was made of Cu sheets with about 10% of the volume filled with quartz fibers to serve as the optical medium for Čerenkov light. The light is detected by photomultiplier tubes and digitized by custom 16-bit ADC's.

RESULTS

We report here the preliminary results from our run in the spring of 2002. The precision of a single measurement, comprised of one beam pulse of each helicity, is about 200 parts per million. To search for possible systematic effects, data were grouped onto about 24 "slugs" which contained data with a constant beam energy and half-wave plate setting. The state for the half-wave plate was changed between slugs and the energy was changed in the middle of the run. Since helicity-correlated beam differences tend to be approximately constant during a slug but differ significantly between slugs, studying the

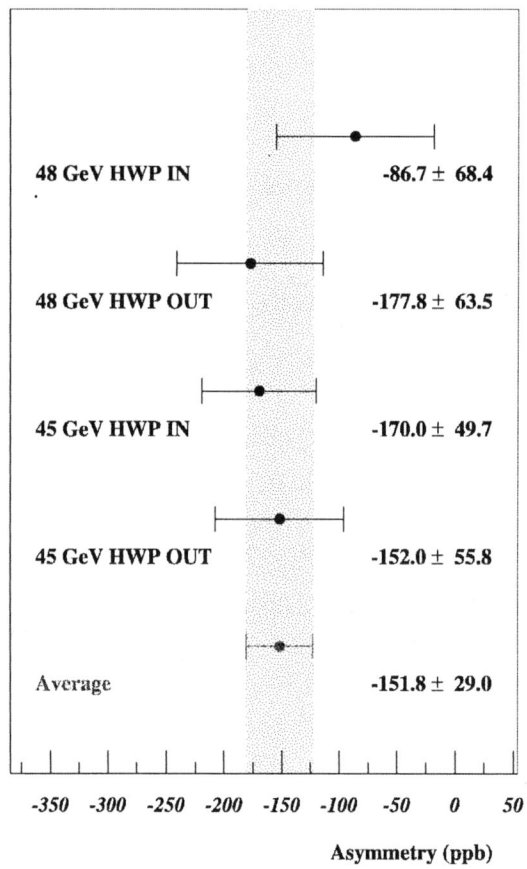

FIGURE 2. Møller asymmetry for different beam conditions.

TABLE 2. Normalization Factors and Errors

Issue	f	df
Polarimetry	.85	0.05
Dilutions	.89	0.02
Energy Scale	1.00	0.01
Geometry	1.00	0.01
Linearity	.99	0.02

data at the slug level provides the greatest sensitivity to unwanted beam systematics. The preliminary result is an asymmetry of $-152 \pm 29 \pm 33$ parts per billion (ppb), where the first error is statistical and the second systematic. The χ^2 is excellent. The significance for the observation of parity violation is 3.6σ. As a further check, the data for each half wave plate setting and for both energies are presented in Fig. 2. All results are consistent.

The corrections to the asymmetries and their associated systematic errors are given in Table I. We believe that they are conservative and that they will be reduced as we understand the experiment better. The normalization factors and errors are given in Table II. The line "Dilutions" in the second line is a summary of the dilution factors in Table I. The errors in Table II do not enter into the estimation of the statistical significance of the asymmetry.

In terms of the weak mixing angle, the preliminary result is

$$\sin^2 \theta_{eff}(Q^2 = 0.027(GeV/c)^2) = 0.2371 \pm 0.0025(\text{stat}) \pm 0.0027(\text{syst}) \quad (3)$$

For comparison with other experiments, we extrapolate the result to the Z-pole to get $\sin^2 \theta_W^{\overline{MS}}(M_Z) = 0.2296 \pm 0.0038$. The result is consistent with the Standard Model.

FUTURE RUNNING

We have completed a second run in the fall of 2002. The analysis of that data are nearing completion. We are obtaining our final data sample in the summer of 2003. We anticipate that our final result for A_{PV} will have an error on the order of 10 ppb.

REFERENCES

1. K. S. Kumar, E. W. Hughes, R. Holmes, and P. A. Souder, *Mod. Phys. Lett.* **A10**, 2979 (1995).
2. A. Czarnecki and W. J. Marciano, *Phys. Rev. D* **53**, 1066 (1996).
3. F. J. Petriello, hep-ph/0210259.
4. A. Czarnecki and W. J. Marciano, *Int. J. Mod. Phys.* **A15**, 2365 (2000).
5. J. Ramsey-Musolf, *Phys. Rev. C* **60**, 015501 (1999).
6. E. J. Eichten, K. D. Lane, and M. E. Peskin, *Phys. Rev. Lett.* **50**, 811 (1983).
7. P. Langacker, M. Luo, and A. K. Mann, Rev. Mod. Phys., **64**, 87 (1992).
8. T. G. Rizzo and J. D. Wells, *Phys. Rev. D* **61** 016007 (1999).
9. For a general review, see K. S. Kumar and P. A. Souder, *Prog. Part. Nucl. Phys.* **45** S333 (2000).
10. T. Maruyama, *et al.*, *Nucl.Instrum.Meth.* **A492**, 199 (2002).

Do We Live in a Vanilla Universe? Theoretical Perspectives on WMAP

Richard Easther

ISCAP, Columbia Astrophysics Lab, Mailcode 5247, 550 W 120th Street, New York, NY10027

Abstract. I discuss the theoretical implications of the WMAP results, stressing WMAP's detection of a correlation between the E-mode polarization and temperature anisotropies, which provides strong support for the overall inflationary paradigm. I point out that almost all inflationary models have a "vanilla limit," where their parameters cannot be distinguished from a genuinely de Sitter inflationary phase. Because its findings are consistent with vanilla inflation, WMAP cannot exclude entire classes of inflationary models. Finally, I summarize hints in the current dataset that the CMB contains relics of new physics, and the possibility that we can use observational data to reconstruct the inflaton potential.

INTRODUCTION

On February 12, 2003, the Wilkinson Microwave Anisotropy Probe [WMAP] reported results based on its first year of observations (e.g. [1, 2, 3, 4, 5]), and cosmology took a giant step towards its long promised "golden age." Ref. [2] lists the values of 22 cosmological parameters determined using the WMAP data and other recent observational information. Many of these values are quoted with several significant figures, whereas a decade ago they were either completely undetermined or had massive uncertainties.

For the theoretical cosmologist, the WMAP results are more tantalizing than revolutionary. On the one hand, WMAP confirms that there is a significant contribution from dark energy in the present epoch, puts tight constraints on the parameter space open to inflation, and provides strong support to the overall inflationary paradigm. However, this does not surprise most theoreticians, and nothing in the current WMAP dataset puts genuinely nontrivial constraints on the physics of inflation and the very early universe.

To explain further, consider *vanilla inflation* – an almost exactly de Sitter inflationary epoch which lasts long enough to deliver a primordial universe – whose measurable parameters all have their "default values". After vanilla inflation, scalar perturbations are Gaussian and scale-free, and there is no discernible contribution from tensor modes or curvature in the present epoch. Consequently, all parameters which constrain the primordial universe would be measured as upper bounds, rather than definite values.

Almost all inflationary models have parameters which can be tuned to provide a vanilla limit. In some cases these tunings may appear so unnatural, and one may want to exclude the model on aesthetic grounds. For instance, the perturbation spectrum associated with a ϕ^n potential becomes more strongly scale dependent as n increases, and one may argue that n should be an even, positive number. The vanilla limit of this model is $n \to 0$, so if $n \geq 2$ is experimentally excluded, the model becomes less attractive. However, this prejudice is far less compelling than the observation of an unambiguously blue spectrum (one with more power on short scales than on long scales): in this case *all* positive values of n are excluded. The one type of inflation which is hard to tune is de Sitter inflation, where the inflaton is trapped in a local minimum of the potential and does not evolve at all. However, in this case the spectrum is precisely scale invariant, which is the vanilla result.

Vanilla inflation has an ambiguous position in theoretical cosmology. It has the most "natural" set of parameter values but offers the smallest leverage for discriminating between different realisations of inflation, and thus the least insight into the early universe. Consequently, theoreticians tend to seize on any hints that the inflationary epoch contains non-vanilla flavorings, since these significantly constrain the inflationary parameter space and, in extreme cases, challenge the overall paradigm. The first year WMAP dataset has two tantalizing features: the apparent lack of power at long wavelengths, and the suggestion that the scalar spectral index itself is a function of the perturbation's wavelength.

THE CMB AND INFLATION

Almost any possible inflationary epoch can be described in terms of a scalar field ϕ moving in a potential, $V(\phi)$.[1] From the Einstein field equations and the energy momentum tensor for a minimally coupled scalar field once can deduce

$$H^2 = \left(\frac{\dot{a}}{a}\right)^2 = \frac{8\pi}{3m_{\rm Pl}^2}\left[\frac{1}{2}\dot{\phi}^2 + V(\phi)\right], \qquad (1)$$

where $a(t)$ is the spacetime scale factor, and a dot denotes differentiation with respect to time. We are implicitly assuming a spatially flat, homogeneous and isotropic universe where the inflaton field is the only contribution to the energy-momentum tensor. The motion of the field is given by

$$\ddot{\phi} + 3H\dot{\phi} + V'(\phi) = 0, \qquad (2)$$

where the dash notes differentiation with respect to ϕ.

Guth's original paper on inflation [6] addressed problems associated with the dynamics of the cosmological background. Inflation can be implemented in a multitude of different ways, all of which solve the cosmological problems addressed by Guth. Consequently, inflationary model builders do not focus directly on the expansion history of the universe. In addition to the zero-order dynamics needed to set the stage for a hot bigbang universe, inflation also predicts the first order perturbations about this background solution. These perturbations determine both the clustering properties of galaxies, and the anisotropies in the microwave background.

---

[1] I am restricting myself to single field models, but all of the statements below have an analogous (although often weaker) form for multi-field models.

Inflation produces primordial perturbations by magnifying quantum fluctuations until their wavelength is equal to or larger than the present size of the observable universe. The properties of these fluctuations differ markedly between different inflationary models and specific inflationary scenarios can thus be distinguished from one another and tested via their perturbation spectra. Consequently, putting tight experimental constraints on the perturbation spectrum is of prime importance, since the theoretical cosmologist can use this data to eliminate specific inflationary models. However, almost all inflationary models have a vanilla limit, so as long as vanilla inflation remains consistent with the observational data we cannot exclude entire classes of models.

Given the functional form of the potential and fairly mild assumptions about the dynamics of inflation, we express the perturbation spectra as a function of the potential and its derivatives. A general perturbation to the background $g_{\mu\nu}$ is a symmetric tensor, $h_{\mu\nu}$, where the perturbed spacetime is $g_{\mu\nu} + h_{\mu\nu}$ [7]. We decompose h into scalar, vector and anti-symmetric tensor components, where the decomposition reflects the transformation properties of the different pieces under (small) transformations. Cosmologically, we need consider only the scalar and tensor modes, as the vector modes decay with time. The scalar modes are associated with a gravitational potential and are the source of density fluctuations in the universe. The tensor modes are effectively gravity waves (and are often referred to as such) and do not contribute to the formation of structure in the universe, but do contribute to the microwave background anisotropies, especially at large angular scales.

The perturbations are described in terms of their power spectra,

$$P_S^{1/2} \sim k^{n_S - 1}, \quad P_T^{1/2} \sim k^{n_T} \qquad (3)$$

where k is the comoving wavenumber of the perturbation. If $n_S = 1$ or $n_T = 0$ the amount of power in each mode is independent of k and the resulting spectra are *scale invariant*.[2] We know that the underlying spectra must be roughly scale invariant, but the question is whether the difference between n_S and n_T from their "natural" values of 1 and 0 is detectable observationally.

Using the slow roll approximation, we can write n_S and n_T in terms of derivatives of the potential [8],

$$n_S = 1 + 2\eta - 4\varepsilon, \qquad n_T = -2\varepsilon \qquad (4)$$

where

$$\varepsilon = \frac{m_p^2}{16\pi}\left(\frac{V'}{V}\right)^2, \quad \eta = \frac{m_p^2}{8\pi}\left[\frac{V'}{V} - \frac{1}{2}\left(\frac{V'}{V}\right)^2\right]. \qquad (5)$$

Finally, the amplitudes of the two spectra must obey a consistency condition,

$$\frac{P_T^{1/2}}{P_S^{1/2}} = 16\varepsilon \qquad (6)$$

---

[2] The differing definitions of P_S and P_T are an historical anomaly.

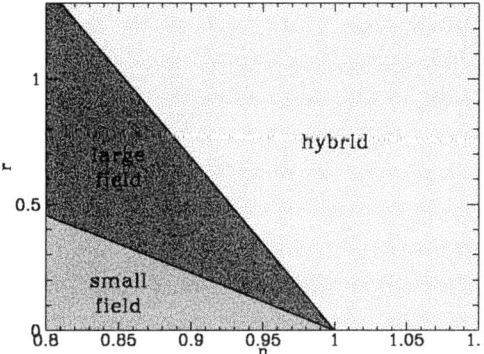

FIGURE 1. The three classes of single field inflationary models are displayed. The hybrid models have $\eta > \varepsilon > 0$, and the small field models have $\eta < 0$. Vanilla inflation corresponds to $r = 0$ and $n = 1$, the point at which the three wedges intersect. (Plot from [9].)

where the numerical coefficient is, to some extent, a matter of definition. We can divide inflationary models into three general classes, summarized in Figure 1 [9]. In the hybrid case, the $V''/V \gg (V'/V)^2$, and the field is evolving towards a local minimum of the potential. Conversely, if we find $n_S < 1$ and no detectable tensor contribution, $V'' < 0$ and $V' \sim 0$, and we have *small field inflation*. Finally, if we observe a significant tensor component, we must have $V'/V \neq 0$. From the equation of motion for ϕ, we see that $\dot{\phi} \neq 0$, and the total evolution of ϕ during inflation can be substantial – leading to the moniker *large field inflation*.

While a non-zero scalar spectrum is needed to provide the primordial density fluctuations that seed the formation of structure, a primordial tensor spectrum is optional. We can show that the $P_T^{1/2}$ is proproportional to the value of H (and thus the square root of the energy density) during inflation and, unless H is GUT scale or above, the tensor signal will most likely be forever undetectable. Since a detectable tensor signal is produced by a limited range of inflationary models the vanilla prediction is that the CMB contains no detectable contribution from tensors. However, if we *do* observe a primordial tensor spectrum, then we can immediately deduce the energy scale at which inflation occurred. Moreover, if we can measure both its amplitude and index, we are in the pleasant position of having four observable quantities (the amplitudes and indices of the tensor and scalar spectrum) which are specified in terms of three parameters. This leads to a "consistency condition" which must be satisfied (to first order in slow-roll) by all single field inflationary models.

THE WMAP RESULTS: SUMMARY

Conceptually, the WMAP mission is very simple: over a period of several years, it makes repeated observations of the microwave sky, and is sensitive to both temperature and po-

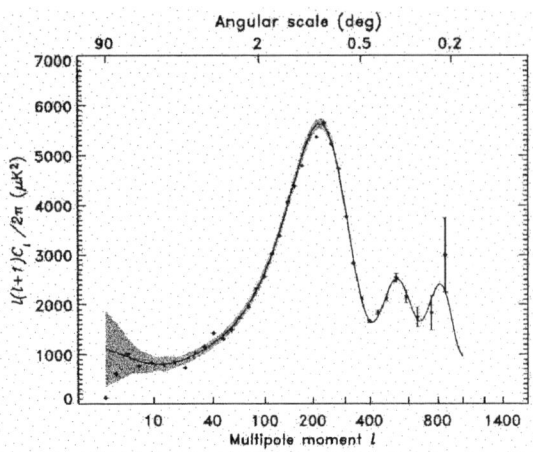

FIGURE 2. The WMAP power spectrum, where the shaded region represents the expected cosmic variance about the best fit spectrum [3].

larization. It observes in five frequency bands, since the main foreground contaminants scale differently with frequency from the underlying black body of the CMB.

Information can be extracted from the maps directly (for instance, the topology of isotemperature contours is a function of the underlying cosmological model [10]), but the maps are frequently distilled into a power spectrum by expanding them in spherical harmonics, and the WMAP power spectrum is shown in Figure 2. Different theoretical models of the early universe predict different scalar and tensor spectra, but the CMB also depends on parameters such as the cosmological constant and the Hubble constant via their influence on the evolution of the perturbations as they propagate in an expanding universe. This is crucial, since it turns the CMB into a tool for estimating a wide range of cosmological parameters which, taken together, put tight constraints on the composition and history of our universe.

Given a set of parameters, the theoretical spectrum is estimated using a tool such as CMBFAST. Armed with high quality CMB data (and data from other sources) one can find the "best fit" model by varying the parameters until the underlying spectrum is matched as accurately as possible. Spergel *et al.* [2] describes this process. The WMAP team concludes that the age of the universe is 13.7 ± 0.2Gyr, the total mass-energy of the universe is $\Omega_{tot} = 1.02 \pm 0.02$ (where a value of unity corresponds to a universe with no spatial curvature), the parameterized Hubble constant $h = 0.71^{+0.04}_{-0.03}$, and baryon density (as a fraction of the total energy density) $\Omega_b h^2 = 0.0224 \pm 0.0009$. Little more than a decade ago, the value of all of these numbers was the subject of significant controversy, and yet here they are all quoted to two or three significant figures.

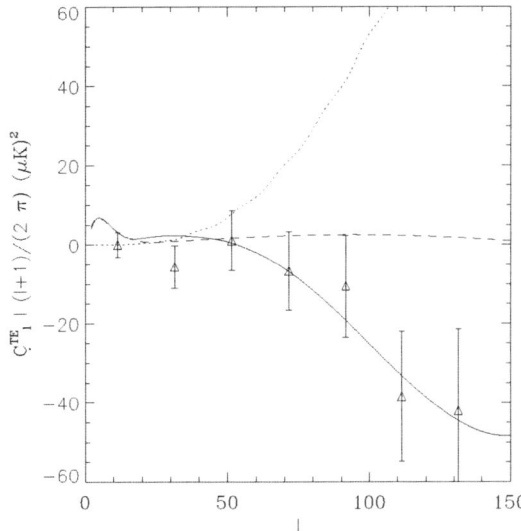

FIGURE 3. The cross-correlation between the E-mode polarization and temperature anisotropies in the CMB. The solid line represents the inflationary prediction, while the defect prediction is shown by the dotted line [3].

POLARIZATION

Since the first detection of the CMB anisotropies, attention has focussed on the temperature variations. The polarization also varies from point to point, but this anisotropy is intrinsically smaller and more difficult to measure. WMAP is the first full-sky mission to return a non-zero polarization measurement [5].[3]

In a universe where inflation occurs, the primordial perturbations may be correlated on scales far larger than the present size of the observable universe, and were definitely correlated on super-horizon scales when the microwave background photons decoupled from the rest of the universe some 300,000 years after the big bang. While other mechanisms for generating the primordial density perturbations have been explored (particularly "defect models" where objects such as cosmic strings generate perturbations as they move through the universe), these do not produce super-horizon correlations.[4] A frequent critique of inflation is that the overall paradigm makes no generic predictions, and that there is no one property that all inflationary models share. However, a universe

---

[3] The polarization itself was first detected by the DASI mission [11].

[4] The ekpyrotic [12] and pre big bang [13] scenarios both produce super-horizon correlations, but in a universe which is initially contracting. While the proponents of these models typically define them in contrast to inflation, they resemble inflation more than they resemble defect models of structure formation. In particular, like inflation, they possess an era in which $\ddot{a} > 0$ and modes with fixed comoving size are continually leaving the horizon.

with super-horizon perturbations at decoupling has a characteristic correlation between the temperature anisotropies and the E-mode polarization signal. WMAP was observed this correlation, and found that it closely matches the inflationary prediction.

It is worth pausing to reflect what a significant achievement this is: from my perspective the observed correlation between temperature and E-mode polarization is most important single result in the WMAP dataset. Prior to inflation, no-one had predicted super-horizon correlations, so this observation is a stunning verification of one of the key features of inflation. Moreover, if the data had not confirmed the inflationary prediction, almost all models of inflation would have been ruled out in a single stroke.[5] Consequently, this is a key test of the inflationary paradigm. The only downside (and perhaps the reason why more is not being said about it) is that because all inflationary models make this prediction, it only eliminates models such as defect scenarios which have already fallen from favor in the theoretical community. However, I suspect that this observation will be regarded as one of the lasting achievements of WMAP, and that observations of the $\langle TE \rangle$ cross-correlation will be followed carefully as the data improves.

SURPRISES: NON-VANILLA FEATURES IN WMAP?

The two obvious peculiarities in the WMAP dataset are that the spectrum has anomalously low power on small scales, and that the fit to the data improves if the scalar index is allowed to be scale-dependent (that is, n_s is a function of k). Either of these results *could* provide a dramatic non-vanilla flavor to the early universe. However, in both cases their physical significance is hard to quantify and await both further data and theoretical analysis.

Low Quadrupole

The low quadrupole is visible in the power spectrum of Figure 2, as the first data points lie well below the best-fit spectrum. Due to cosmic variance[6], we don't expect an exact match between theory and experiment at small values of l, but the discrepancy is, on the face of it, unexpectedly large.

While it is clear that the observed CMB sky has less power at large wavelengths (low k) than that suggested by the "best fit" ΛCDM model, it is not clear is whether this is something we *need* to explain, since the result could be produced by cosmic variance alone. Spergel *et al.* [2] generated multiple realizations of the microwave background with the parameters estimated by WMAP, and found that the probability that the low power at small l is due to cosmic variance is 1.5×10^{-3}. However, other authors ([14],

---

[5] Inflation may still have occurred, but it could not have produced the observed primordial perturbations.
[6] The C_l values plotted in Figure 2 represent averages over the $2l+1$ m values of the corresponding Y_{lm}. Roughly speaking, this average will have a sampling uncertainty of $1/\sqrt{2l+1}$ – but since we can observe only one sky we cannot reduce this uncertainty by gathering more data. This intrinsic uncertainty is the cosmic variance.

for example) argue that this calculation underestimates cosmic variance, and that the discrepancy is not large enough to be a signal of new physics.

If this result is significant, the possible modifications to the standard paradigm that would suppress power at large scales take a variety of forms. For example, a comparatively conservate approach is provided by Contaldi et al. [15], who look at inflationary models which are tuned to suppress P_S for values of k which dominate the low l terms in the CMB spectrum. Carefully tuned inflationary models violate the spirit of the inflationary paradigm, but they are less radical departures from the standard cosmology than (for instance) advocating a toroidal universe with a "cell size" that is smaller than the current size of the visible universe [16], or that the universe possesses detectable (and positive) spatial curvature [17], both of which would tend to cut off the power spectrum at long wavelengths.

Simply measuring the microwave sky more accurately will not reduce the cosmic variance. However, it is possible that some of lack of power at low l could be explained by an over-aggressive foreground subtraction, and this is amenable to testing and improvement. Conversely, if the suppression of power at low l is a real effect, evidence for it will appear in other places. For example, Kesden et al. [18] show that the shear produced by gravitational inhomogeneities, which distorts the correlation between temperature and polarization anisotropies, would be measurably different in a universe where the spectrum lacked power at small values of k, and this can be tested by future experiments.

Running Index

The low quadrupole seen by WMAP is suggestive of unsuspected physics that plays a role at large angular scales. However, WMAP also hints that the standard assumption of an underlying spectrum described by a constant index n_S may be too simplistic [4]. In this case, the principal evidence is found in CMB data at small angular scales. This is currently dominated by observational uncertainty, rather than cosmic variance. In fact, the first-year WMAP dataset alone does cover a large enough range of l-values to provide any evidence for a running (k dependent) n_S. The evidence for running (between the 1 and 2 σ level) appears when the WMAP power spectrum is combined with that derived from galaxy surveys and Lyman-α forest data, both of which provide data on the primordial spectrum at smaller scales than is possible with the CMB alone [4].

If the running index is confirmed, it will put tight constraints on inflation. The slow roll expansion for the spectral index given by equation (3) can be extended to beyond the leading order result given here, and $dn_S/d\ln k$ is dependent on the third derivative of the potential. However, having $V'''(\phi)$ large enough to produce a $dn_S/d\ln k$ observable by WMAP rules out almost all standard models of inflation. This is not, in itself, a drawback. Moreover, it would comprehensively rule out vanilla inflation, which would be very a welcome development indeed. Moreover, if the potential has a number of "features" then it may not need to be carefully tuned in order to ensure that one of these features is found within the range of ϕ covered by the inflaton field as the cosmologically relevant perturbations are generated [19]. Indeed, a potential of this sort was considered by the WMAP team, and there is weak support in the WMAP dataset (in combination with other survey information) for this type of feature.

The major caveat about this possible non-vanilla flavoring of the early universe is simply that the data is inconclusive. The result hinges on the merger of several datasets, which increases the complexity of any statistical analysis, and the level of significance is small enough for it to simply be the result of a statistical fluctuation. This uncertainty will soon be resolved – the next year of WMAP data will significantly improve the sensitivity of the measurements of the C_l for larger values of l, and the SDSS [Sloan Digital Sky Survey] will supplant the galaxy surveys used by the WMAP team in their previous papers.

RECONSTRUCTING THE POTENTIAL

One of the principal dreams of the theoretical cosmologist is to reconstruct the underlying physical mechanism of inflation. In general, this amounts to recovering the functional form of the potential. Even in a "golden age" this inverse problem remains enormously difficult. Several efforts have been made to develop a methodology for reconstructing the potential from its Taylor series, but these appear to be best by observational difficulties [8]. More recently, Easther and Kinney developed *Monte Carlo reconstruction*, a stochastic approach to the problem based on generating a large number of "trial" inflationary models and isolating those for which the observable parameters coincide with the window in parameter space permitted by the available data[20]. This approach builds on a thorough understanding of the "flow equations" [21], a consistent expansion of the inflationary dynamics.[7] If the permitted window of parameter space is sufficiently narrow the class of allowed potentials will be sufficiently well-defined that one can then proceed to estimate the functional form of the potential.

The WMAP team used a variant of Monte Carlo reconstruction [4]. This problem has been tackled in more detail by Kinney *et al.* [23], who find three classes of reconstructed potential, corresponding to the three subdivisions of the "zoo plot" shown in Figure 1. This is of course expected, given that vanilla inflation remains consistent with the observational data. However, it is only with the release of WMAP data that the observational constraints on inflationary theories are tight enough for this sort of calculation to return any non-trivial limits on the possible range of potentials which could have driven inflation.

CONCLUSION

This paper has given a quick overview of theoretical cosmologists' response to the WMAP data. WMAP confirms what we already believed we knew – that the perturbations are correlated on super-horizon scales as a result of an inflation(like) mechanism, thanks to the observed anti-correlation between the temperature anisotropies and

---

[7] See also Liddle's recent paper [22].

the E-mode polarization signal. However, at present the observational evidence is not tight enough to rule out whole classes of inflationary models. Crucially, vanilla inflation – which is a limit of almost all inflationary models – remains viable. Since vanilla inflation is allowed, it follows that no classes of model can be excluded, even if certain parameter values can be ruled out within each class. I have reviewed the two hints in the WMAP data for a non-vanilla universe – the low quadrupole and the possible running scalar index – and sketched how these effects could change our understanding of the early universe, and how future data is likely to constrain them more closely. In conclusion, though, it is clear that WMAP marks a profound change in the theoretical debate, and that the "golden age" of cosmology is upon us.

ACKNOWLEDGEMENTS

I thank my Columbia colleagues Ted Baltz, Brian Greene and Will Kinney for many useful conversations which helped me form the viewpoints expressed here, and I thank Hiranya Peiris for several useful discussions about the WMAP results and their interpretation.

REFERENCES

1. C. L. Bennett et al., arXiv:astro-ph/0302207.
2. D. N. Spergel et al., arXiv:astro-ph/0302209.
3. G. Hinshaw et al., arXiv:astro-ph/0302217.
4. H. V. Peiris et al., arXiv:astro-ph/0302225.
5. A. Kogut et al., arXiv:astro-ph/0302213.
6. A. H. Guth, Phys. Rev. D **23**, 347 (1981).
7. V. F. Mukhanov, H. A. Feldman and R. H. Brandenberger, Phys. Rept. **215**, 203 (1992).
8. J. E. Lidsey, A. R. Liddle, E. W. Kolb, E. J. Copeland, T. Barreiro and M. Abney, Rev. Mod. Phys. **69**, 373 (1997) [arXiv:astro-ph/9508078].
9. W. H. Kinney, Phys. Rev. D **58**, 123506 (1998) [arXiv:astro-ph/9806259].
10. W. N. Colley and J. R. Gott, arXiv:astro-ph/0303020.
11. J. Kovac, E. M. Leitch, P. C., J. E. Carlstrom, H. N. W. and W. L. Holzapfel, Nature **420**, 772 (2002) [arXiv:astro-ph/0209478].
12. J. Khoury, B. A. Ovrut, P. J. Steinhardt and N. Turok, Phys. Rev. D **64**, 123522 (2001) [arXiv:hep-th/0103239].
13. M. Gasperini and G. Veneziano, Astropart. Phys. **1**, 317 (1993) [arXiv:hep-th/9211021].
14. G. Efstathiou, arXiv:astro-ph/0306431.
15. C. R. Contaldi, M. Peloso, L. Kofman and A. Linde, arXiv:astro-ph/0303636.
16. A. de Oliveira-Costa, M. Tegmark, M. Zaldarriaga and A. Hamilton, arXiv:astro-ph/0307282.
17. G. Efstathiou, arXiv:astro-ph/0303127.
18. M. H. Kesden, M. Kamionkowski and A. Cooray, arXiv:astro-ph/0306597.
19. J. Adams, B. Cresswell and R. Easther, Phys. Rev. D **64**, 123514 (2001) [arXiv:astro-ph/0102236].
20. R. Easther and W. H. Kinney, Phys. Rev. D **67**, 043511 (2003) [arXiv:astro-ph/0210345].
21. W. H. Kinney, Phys. Rev. D **66**, 083508 (2002) [arXiv:astro-ph/0206032].
22. A. R. Liddle, arXiv:astro-ph/0307286.
23. W. H. Kinney, E. W. Kolb, A. Melchiorri and A. Riotto, arXiv:hep-ph/0305130.

Recent results from KLOE at DAΦNE

The KLOE collaboration[1]
Presented by Matthew Moulson

Laboratori Nazionali di Frascati, 00044 Frascati RM, Italy

Abstract. The KLOE experiment at DAΦNE collected about 450 pb^{-1} of data in 2001–2002. Much of this data set has been analyzed and has yielded definitive results on K_S and radiative ϕ decays, as well as studies concerning a wide range of topics in kaon and hadronic physics.

KLOE is a large, general-purpose detector with optimizations for the study of discrete symmetries in the neutral kaon system. The experiment is permanently installed at DAΦNE, the Frascati ϕ factory, an e^+e^- machine with $W \approx m_\phi \approx 1.02$ GeV. The DAΦNE design luminosity is $5.3 \cdot 10^{32}$ cm^{-2}s^{-1}. ϕ's are produced with a cross section of ~ 3.2 μb, and decay into K^+K^- and $K_S K_L$ pairs with branching ratios (BR's) of $\sim 49\%$ and $\sim 34\%$. These pairs are produced in a pure $J^{PC} = 1^{--}$ quantum state, so observation of a K_S in an event signals the presence of a K_L and vice versa. With an appropriate tagging technique, highly pure and nearly monochromatic K_S, K_L, K^+, or K^- beams can be obtained.

The KLOE detector consists essentially of a large drift chamber surrounded by an electromagnetic calorimeter. The drift chamber [1] is 4 m in diameter and 3.3 m in length, which results in a fiducial volume for K_L decays that extends to about half of a decay length. The momentum resolution for tracks with $p \gtrsim 100$ MeV and $\theta > 45°$ is $\sigma_p/p \leq 0.4\%$. The lead/scintillating-fiber calorimeter [2] consists of a barrel and two endcaps and covers 98% of the solid angle. The energy resolution is $\sigma_E/E = 5.7\%/\sqrt{E(\text{GeV})}$. The intrinsic timing resolution is $\sigma_t = 54$ ps$/\sqrt{E(\text{GeV})} \oplus 50$ ps, which allows photon vertices from π^0 decays to be reconstructed with a resolution of

---

[1] The KLOE collaboration: A. Aloisio, F. Ambrosino, A. Antonelli, M. Antonelli, C. Bacci, G. Bencivenni, S. Bertolucci, C. Bini, C. Bloise, V. Bocci, F. Bossi, P. Branchini, S. A. Bulychjov, R. Caloi, P. Campana, G. Capon, T. Capussela, G. Carboni, G. Cataldi, F. Ceradini, F. Cervelli, F. Cevenini, G. Chiefari, P. Ciambrone, S. Conetti, E. De Lucia, P. De Simone, G. De Zorzi, S. Dell'Agnello, A. Denig, A. Di Domenico, C. Di Donato, S. Di Falco, B. Di Micco, A. Doria, M. Dreucci, O. Erriquez, A. Farilla, G. Felici, A. Ferrari, M. L. Ferrer, G. Finocchiaro, C. Forti, A. Franceschi, P. Franzini, C. Gatti, P. Gauzzi, S. Giovannella, E. Gorini, E. Graziani, M. Incagli, W. Kluge, V. Kulikov, F. Lacava, G. Lanfranchi, J. Lee-Franzini, D. Leone, F. Lu, M. Martemianov, M. Matsyuk, W. Mei, L. Merola, R. Messi, S. Miscetti, M. Moulson, S. Müller, F. Murtas, M. Napolitano, A. Nedosekin, F. Nguyen, M. Palutan, E. Pasqualucci, L. Passalacqua, A. Passeri, V. Patera, F. Perfetto, E. Petrolo, L. Pontecorvo, M. Primavera, F. Ruggieri, P. Santangelo, E. Santovetti, G. Saracino, R. D. Schamberger, B. Sciascia, A. Sciubba, F. Scuri, I. Sfiligoi, A. Sibidanov, T. Spadaro, E. Spiriti, M. Testa, L. Tortora, P. Valente, B. Valeriani, G. Venanzoni, S. Veneziano, A. Ventura, S. Ventura, R. Versaci, I. Villella, G. Xu

∼ 1.5 cm. A superconducting coil surrounding the calorimeter barrel provides a 0.52 T magnetic field.

During 2002 data taking, the maximum luminosity sustained by DAΦNE was $7.5 \cdot 10^{31}$ cm^{-2}s^{-1}. Although this is lower than the design value, the performance of the machine during 2002 was much improved with respect to previous years, and the KLOE experiment was able to collect as much as 4.5 pb^{-1} per day. The combined KLOE 2001–2002 data set amounts to about 450 pb^{-1}, or 1.4 billion ϕ decays.

A series of recent upgrades to the machine, including an overhaul of the interaction region inside the KLOE detector, is expected to bring the design luminosity to within reach. Data taking with KLOE is scheduled to restart during the fall of 2003.

Kaon physics with KLOE. The tagging of K_L and K_S decays is fundamental to all KLOE studies of the $K_S K_L$ system. The $K_S \to \pi^+ \pi^-$ decay provides an efficient tag for K_L decays. K_S's can be tagged by identifying a K_L interaction in the calorimeter. Since the neutral kaons from ϕ decays have $\beta = 0.22$ at KLOE, the signature of such an interaction, or "K_L crash," is a late, high-energy cluster that is not associated to any track in the drift chamber. In either case, reconstruction of one kaon establishes the trajectory of the other with an angular resolution of $\sim 1°$ and a momentum resolution of ~ 2 MeV.

Using the K_L crash to tag K_S decays, KLOE has measured the ratio of the partial widths for the dominant K_S decay modes: $\Gamma(K_S \to \pi^+\pi^-(\gamma))/\Gamma(K_S \to \pi^0\pi^0) = 2.236 \pm 0.003 \pm 0.015$ [3]. This value was obtained using just 17 pb^{-1} of data from the 2000 run. In addition to providing the first part of the double ratio for Re$(\varepsilon'/\varepsilon)$, the measurement of this value allows determination of $\chi_0 - \chi_2$, the difference in $\pi\pi$ phase shifts in $K \to \pi\pi$ transitions with $I = 0$ and 2. As argued in [4], the extraction of the $K \to \pi\pi$ amplitudes from the measured widths must take into account the effective cutoff for the detection of final state photons from $K_S \to \pi^+\pi^-\gamma$ decays. Due to the tagging technique used at KLOE, the detection efficiency for such decays is good out to high values of the photon energy, which allows a fully inclusive measurement to be made. By the evaluation of [4], the previously existing data give $\chi_0 - \chi_2 = (56 \pm 8)°$, which is in somewhat poor agreement with the estimates of the difference in strong $\pi\pi$ phase shifts, $\delta_0 - \delta_2$, from chiral perturbation theory, $(45 \pm 6)°$ [5], and from the phenomenological analysis of $\pi\pi$ scattering data, $(45 \pm 6)°$ [6].[2] The KLOE measurement gives $\chi_0 - \chi_2 = (48 \pm 3)°$, which considerably improves the agreement with the predictions from phenomenology.

KLOE has also measured the BR's for the K_{e3} decays of the K_S. The π and e assignments are made using time-of-flight measurements, so the BR's to final states of each lepton charge are measured independently. Based on 170 pb^{-1} of 2001 data, KLOE obtains the preliminary values for the BR's to $\pi^- e^+ \nu$, $(3.46 \pm 0.09 \pm 0.06) \cdot 10^{-4}$, and to $\pi^+ e^- \bar{\nu}$, $(3.33 \pm 0.08 \pm 0.05) \cdot 10^{-4}$. A nonzero value for $A_S - A_L$, the difference in the semileptonic charge asymmetries for the K_S and K_L, would signal *CPT* violation, either in the K_S-K_L mixing or in direct transitions that also violate the $\Delta S = \Delta Q$ rule. While A_L

---

[2] The difference between $\chi_0 - \chi_2$ and $\delta_0 - \delta_2$ is that the former quantity includes an additional phase shift difference from isospin-breaking electromagnetic effects, estimated to be about 3° [4].

has recently been measured with precision [7], these preliminary KLOE results give the first-ever measurement of A_S: $(19 \pm 17 \pm 6) \cdot 10^{-3}$. When the semileptonic final states are not distinguished by charge, KLOE obtains $(6.81 \pm 0.12 \pm 0.10) \cdot 10^{-4}$ for the BR. With respect to the previous KLOE measurement [8], this represents a tenfold increase in statistics accompanied by a reduction of the systematic error by a third. The difference between the partial widths for the K_{e3} decays of the K_S and K_L can be related to Re x_+, the parameter which quantifies violations of the $\Delta S = \Delta Q$ rule in CPT-conserving transitions. The preliminary KLOE measurement of the charge-undifferentiated K_{e3} branching ratio of the K_S gives Re $x_+ = (3.3 \pm 5.2 \pm 3.5) \cdot 10^{-3}$, which is comparable in significance to the CPLEAR result [9]. KLOE has an additional 280 pb^{-1} of data under analysis, and forthcoming KLOE measurements of the K_L lifetime and K_{e3} BR will allow further improvements on the value of Re x_+.

Using the $K_S \to \pi^+\pi^-$ decay to tag K_L decays, KLOE has recently completed a measurement of $\Gamma(K_L \to \gamma\gamma)/\Gamma(K_L \to 3\pi^0)$. The BR for the decay $K_L \to \gamma\gamma$ provides interesting tests of chiral perturbation theory; additionally, this decay dominates the long-distance contribution to $K_L \to \mu^+\mu^-$ [10]. For the ratio of BR's, KLOE obtains $(2.793 \pm 0.022 \pm 0.024) \cdot 10^{-3}$ from 362 pb^{-1} of 2001–2002 data [11]. This value is comparable in significance to that from NA48 [12]. The normalization sample of $K_L \to 3\pi^0$ decays (which is downscaled by a factor of ten) provides a value for the K_L lifetime that is comparable in statistical significance to the world-average value.

KLOE is currently studying the BR's of the various K_L decays to charged particles. As a proof of principle, an analysis of the tagged K_L vertices in 78 pb^{-1} of 2002 data has been performed, and gives values for the K_L BR's to $\pi^+\pi^-\pi^0$, $\pi\mu\nu$, and $\pi e\nu$ that are consistent with, and which have statistical significance comparable to, the world-average values. When a similar, dedicated analysis of BR$(K_L \to \pi^+\pi^-)$ is performed (on 429 pb^{-1} of 2001–2002 data), the value $(2.04 \pm 0.04) \cdot 10^{-3}$ is obtained. The systematic errors on the above values have not been fully evaluated yet, but are thought to be at the 1–2% level.

In the longer term, KLOE intends to measure Re $(\varepsilon'/\varepsilon)$ via the double ratio:

$$1 - 6\mathrm{Re}\,(\varepsilon'/\varepsilon) = \frac{K_L \to \pi^0\pi^0}{K_L \to \pi^+\pi^-} \cdot \frac{K_S \to \pi^+\pi^-}{K_S \to \pi^0\pi^0}.$$

The manner in which this expression is written calls attention to the fact that at KLOE, cancellations of experimental systematics are sought principally in the ratios of the BR's for the charged and neutral decay modes. KLOE prospects for the measurement of Re $(\varepsilon'/\varepsilon)$ can be summarized as follows. The current KLOE measurement of the ratio of $K_S \to \pi\pi$ BR's has a negligible statistical error and a systematic error of 0.7%. This error is expected to be reduced to the 0.1% level when the 2001–2002 data are analyzed, both because of changes to the tagging algorithm already implemented, and because the errors on the various corrections are determined by the statistics of the control samples used to obtain them. The statistical errors on the $K_L \to \pi\pi$ BR measurements obtained using the entire 2001–2002 data set are currently at the 1.5% level; systematic errors are at about the same level and work is in progress to significantly reduce them. A measurement of the ratio of the $K_L \to \pi\pi$ BR's with an overall error at the level of a few per mil will

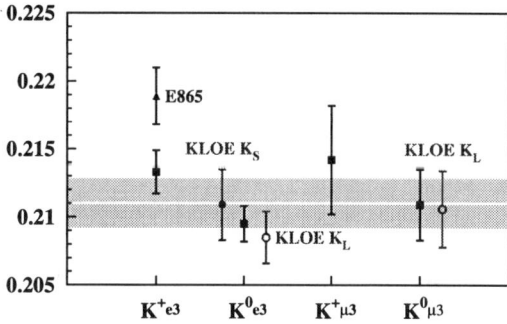

FIGURE 1. Current status of $|V_{us}|f_+^{K^0\pi^-}(0)$. Evaluations of $|V_{us}|f_+^{K^0\pi^-}(0)$ from the published world data on $K_{\ell3}$ decays are shown as the squares [15]. The average over the K_{e3}^+ and K_{e3}^0 modes (with its error) is shown as the horizontal band. The value obtained [15] from the recent measurement of the K_{e3}^+ BR by BNL-E865 [16] is shown as the triangle. The values from the preliminary KLOE measurements of the $K_{\ell3}$ decays of the K_S and K_L are shown as the solid and open circles.

require at least a factor of ten more data.

KLOE is also undertaking a comprehensive program for the study of the decays of the charged kaons. The most advanced analysis concerns the BR and Dalitz plot for the decay $K^\pm \to \pi^\pm \pi^0 \pi^0$. Charge asymmetries in the rates and Dalitz plot slopes for this decay would signal direct *CP* violation (see *e.g.* [13]). Based on 188 pb^{-1} of 2001–2002 data, the KLOE value for the BR is $(1.781 \pm 0.013 \pm 0.016)\%$ [14].

The measurement of the CKM matrix element $|V_{us}|$ represents a point of convergence for KLOE studies of charged and neutral kaon decays. Fig. 1 summarizes values for the quantity $|V_{us}|f_+^{K^0\pi^-}(0)$ obtained from measurements of the $K_{\ell3}$ BR's and kaon lifetimes. The errors on these values are completely dominated by the experimental inputs. The four essentially independent evaluations of $|V_{us}|f_+^{K^0\pi^-}(0)$ from published world data are in agreement. The recent measurement of the K_{e3}^+ BR by BNL-E865, on the other hand, gives a discrepant value. This is intriguing, because the value for $|V_{us}|$ obtained from the E865 measurement agrees with current determinations of $|V_{ud}|$ given the first-row unitarity constraint on the CKM matrix elements. Preliminary KLOE results, on the other hand, decisively weigh in on the side of the existing value for $|V_{us}|$. A KLOE measurement of the K_{e3}^+ BR would offer direct comparison with the E865 result, and is forthcoming. In the longer term, KLOE should be able to measure all four $K_{\ell3}$ BR's to much better than 1%, and to significantly improve the determinations of the lifetimes of the K_L and $K^\pm$, as well as the form factor slopes λ_+ and λ_-.

Hadronic physics with KLOE. KLOE has recently completed an analysis of the decay $\phi \to \pi^+\pi^-\pi^0$, which proceeds mainly through $\rho\pi$ intermediate states [17]. A fit to the Dalitz plot containing 2 million events from 17 pb^{-1} of 2000 data gives values for the masses and widths of the ρ^+, ρ^-, and ρ^0. When the fit is performed such that all three ρ charge states are described by the same mass and width, KLOE obtains $m_\rho = 775.8 \pm 0.5 \pm 0.3$ MeV (in agreement with other recent e^+e^- measurements; see [18]) and $\Gamma_\rho = 143.9 \pm 1.3 \pm 1.1$ MeV (confirming the result of [19]). When this constraint is relaxed, no significant differences are observed between the masses of the charged and neutral ρ's, or between the masses of the ρ^+ and ρ^-.

The ratio of BR's for the decays $\phi \to \eta'\gamma$ and $\phi \to \eta\gamma$ provides information on the η-η' mixing angle and on the gluonium content of the η' [20, 21]. KLOE has measured this ratio using 17 pb^{-1} of 2000 data and obtained a value for the pseudoscalar mixing angle in the flavor basis, $\varphi_P = (41.8^{+1.9}_{-1.6})°$, as well as a limit on the gg content of the η' [22]. An extension of the analysis based on the 2001-2002 data set is in progress.

KLOE has also analyzed the ϕ decays to $\pi^0\pi^0\gamma$ and $\eta\pi^0\gamma$, where the dominant contributions involve production and decay of the scalar mesons f_0 and a_0, respectively. Fits based on a kaon-loop model to the $\pi\pi$ and $\eta\pi$ invariant-mass spectra obtained from 17 pb^{-1} of 2000 data have been used to obtain values of the coupling constants g_{f_0KK} and g_{a_0KK}. The value obtained for g_{f_0KK} is compatible with predictions that assume a $q\bar{q}q\bar{q}$ model for the f_0 structure, while that obtained for g_{a_0KK} is in poor agreement with these predictions [23, 24]. The factor-of-ten increase in statistics from the 2001–2002 data has allowed KLOE to undertake a model-independent analysis of these decays featuring a complete study of the contributions to the Dalitz plots.

Finally, KLOE is concluding a determination of $\sigma(e^+e^- \to \pi^+\pi^-)$ as a function of s_π, the squared CM energy of the $\pi\pi$ system, for $0.3 < s_\pi < 1$ GeV^2. DAΦNE operates exclusively at $W \approx m_\phi$, so the actual measurement is the cross section for the process $e^+e^- \to \pi^+\pi^-\gamma$, where the photon is radiated from the initial state (ISR). The PHOKHARA generator [25] is then used to relate $\sigma(e^+e^- \to \pi^+\pi^-\gamma)$ to $\sigma(e^+e^- \to \pi^+\pi^-)$. Complications from processes with final-state radiation (FSR) are avoided by restricting the current study to events with low-angle photons ($\theta < 15°$); in this kinematic region ISR events completely dominate the sample. Detection of the photon is unnecessary; s_π and the photon angle are reconstructed in the drift chamber. The preliminary KLOE data presented in Fig. 2 provide valuable comparison to the energy-scan data from CMD-2 [19] with respect to the determination of the hadronic contribution to the anomalous magnetic moment of the muon, a_μ^{had}. Evaluations of a_μ^{had} based on e^+e^--annihilation and τ-decay data differ by -3.0σ and -0.9σ from consistency with the a_μ measurement from BNL-E821 [26]; at present, the CMD-2 results dominate the value of a_μ^{had} from e^+e^- data [27]. The KLOE data in Fig. 2 give $a_\mu^{\text{had}}(0.37 < s_\pi < 0.93 \text{ GeV}^2) = 374.1 \cdot 10^{-10}$, with a negligible statistical error and a 1.6% systematic error including all experimental and theoretical sources except for the lack of ISR+FSR events in the simulation (these are already included in a new version of PHOKHARA). In this same interval for s_π, the CMD-2 data give $a_\mu^{\text{had}} = (368.1 \pm 2.6 \pm 2.2) \cdot 10^{-10}$. The discrepancy between the KLOE and CMD-2 results is concentrated below the ρ peak; the values of a_μ^{had} from the two experiments agree for the interval $0.6 < s_\pi < 0.97$ GeV^2, where the $\pi^+\pi^-$ spectral functions for e^+e^- and τ data disagree by 10–15% [28].

Conclusions. KLOE is currently analyzing a unique data set consisting of 500 pb^{-1} of ϕ decays. Extensions of previous KLOE results on K_S and radiative ϕ decays are nearly ready, and important new measurements such as those of $|V_{us}|$ and $\sigma(e^+e^- \to$ hadrons) are forthcoming. In 2003–2004 running, KLOE expects to collect a few fb^{-1} of data, which will allow the broad physics reach of the experiment to be further extended.

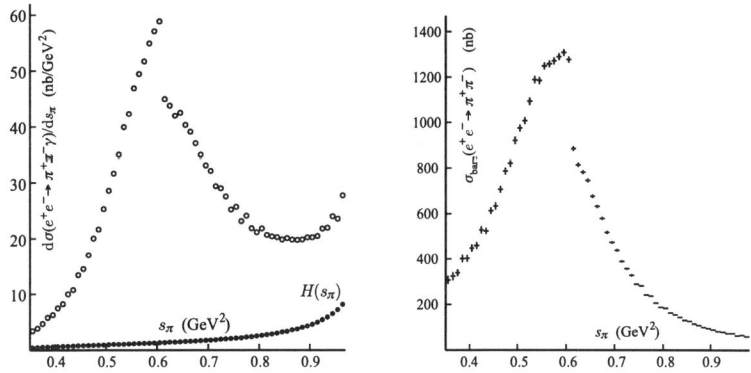

FIGURE 2. Left: Open markers show preliminary KLOE measurements of $d\sigma(e^+e^- \to \pi^+\pi^-\gamma)/ds_\pi$ from 1.5 million events (140 pb^{-1}) of 2001 data; solid markers show the radiation function used to extract $\sigma(e^+e^- \to \pi^+\pi^-)$. Right: KLOE determination of the bare cross section for the process $e^+e^- \to \pi^+\pi^-$.

REFERENCES

1. Adinolfi, M., et al. (KLOE Collaboration), *Nucl. Instr. Meth.*, **A488**, 51 (2002).
2. Adinolfi, M., et al. (KLOE Collaboration), *Nucl. Instr. Meth.*, **A482**, 364 (2002).
3. Aloisio, A., et al. (KLOE Collaboration), *Phys. Lett.*, **B538**, 21 (2002).
4. Cirigliano, V., Donoghue, J., and Golowich, E., *Eur. Phys. J.*, **C18**, 83 (2000).
5. Gasser, J., and Meissner, U.-G., "On the phase of ε'," in *Proceedings of the Joint International Lepton-Photon Symposium and Europhysics Conference on High Energy Physics, Geneva, 25 July–1 August 1991*, edited by S. Hegarty et al., World Scientific, 1992, p. 202.
6. Colangelo, G., Gasser, J., and Leutwyler, H., *Nucl. Phys.*, **B603**, 125 (2001).
7. Alavi-Havarti, A., et al. (KTeV Collaboration), *Phys. Rev. Lett.*, **88**, 181601 (2002).
8. Aloisio, A., et al. (KLOE Collaboration), *Phys. Lett.*, **B535**, 37 (2002).
9. Angelopoulos, A., et al. (CPLEAR Collaboration), *Phys. Lett.*, **B444**, 38 (1998).
10. D'Ambrosino, G., et al., "Radiative Non-Leptonic Kaon Decays," in [29], p. 265.
11. Aloisio, A., et al. (KLOE Collaboration), *Phys. Lett.*, **B566**, 61 (2003).
12. Lai, A., et al. (NA48 Collaboration), *Phys. Lett.*, **B551**, 7 (2003).
13. Maiani, L., and Paver, N., "*CP* Violation in $K \to 3\pi$ decays," in [29], p. 51.
14. Aloisio, A., et al. (KLOE Collaboration), hep-ex/0307054 (2003).
15. Isidori, G., et al., "Determination of the Cabibbo Angle," in *The CKM Matrix and the Unitary Triangle*, edited by M. Battaglia et al., CERN, 2003, p. 43, hep-ph/0304132.
16. Sher, A., et al., hep-ex/0305042 (2003).
17. Aloisio, A., et al. (KLOE Collaboration), *Phys. Lett.*, **B561**, 55 (2003).
18. Hagiwara, K. et al. (Particle Data Group), *Phys. Rev.*, **D66**, 010001 (2002).
19. Akhmetshin, R.R., et al. (CMD-2 Collaboration), *Phys. Lett.*, **B527**, 161 (2002).
20. Feldmann, T., *Int. J. Mod. Phys.*, **A15**, 159 (2000).
21. Rosner, J., *Phys. Rev.*, **D27**, 1101 (1983).
22. Aloisio, A., et al. (KLOE Collaboration), *Phys. Lett.*, **B541**, 45 (2002).
23. Aloisio, A., et al. (KLOE Collaboration), *Phys. Lett.*, **B537**, 21 (2002).
24. Aloisio, A., et al. (KLOE Collaboration), *Phys. Lett.*, **B536**, 209 (2002).
25. Czyż, H., et al., *Eur. Phys. J.*, **C27**, 563 (2003).
26. Bennett, G.W., et al. (Muon g-2 Collaboration), *Phys. Rev. Lett.*, **89**, 101804 (2002).
27. Davier, M., et al., *Eur. Phys. J.*, **C27**, 497 (2003).
28. Aloisio, A., et al. (KLOE Collaboration), hep-ex/0307051 (2003).
29. Maiani, L., Pancheri, G., and Paver, N., editors, *The Second DAΦNE Physics Handbook*, Frascati, 1995.

Recent Results from NA48, CERN's North Area Experiment 48

P. D. Rubin[1]

for the NA48 Collaboration

CERN, CH-1211, Genève 23, Switzerland

Abstract. Recent results, of particular interest in the context of chiral perturbation theory (χ_{PT}), from the NA48 collaboration are presented.

Introduction

The NA48 experiment was designed and constructed to search for evidence of direct CP-violation in neutral kaon decays. In particular, it measured the the relative fraction of direct to indirect CP-violation through the double ratio, $R = \frac{\Gamma(K_L \to \pi^0 \pi^0)}{\Gamma(K_S \to \pi^0 \pi^0)} / \frac{\Gamma(K_L \to \pi^+ \pi^-)}{\Gamma(K_S \to \pi^+ \pi^-)} \approx 1 - 6 \times Re(\varepsilon'/\varepsilon)$. In order to minimize net systematic effects on this measurement, the beamline was constructed to provide simultaneous K_L (far-target) and K_S (near-target) beams. Data-taking with simultaneous beams was the standard configuration, but it was possible to run with each beam separately, as was the case for dilution measurements and other studies.

Beam running conditions for the results described in this note are summarized in Table 1.

TABLE 1. Beam Characteristics

	1998-1999	2000*	2001
Proton Momentum [GeV/c]	450	400	400
SPS Cycle Time [s]	14.4	14.4	16.8 s
Spill Length (Effective) [s]	2.4 (1.7)	3.2 (2.2)	5.2 (3.6)
Duty Cycle	0.17	0.22	0.31
K_L **Beam Intensity [($\times 10^{12}$) ppp]**	1.5	1.0	2.4
K_S **Beam Intensity [($\times 10^7$) ppp]**	3.0	1.0×10^2	5.0

* No drift chambers

---

[1] Email: philip.rubin@cern.ch; on leave from University of Richmond, Richmond, VA 23173 USA; supported in part by US NSF grant #0140230.

χ_{PT} is a low energy, effective field theory, which ought to describe kaon decays. We discuss six relevant decay processes: $K_{L,S} \to \pi^+\pi^- e^+ e^-$, $K_{L,S} \to \pi^0 \gamma\gamma$, and $K_{L,S} \to \gamma\gamma$.

$K_{L,S} \to \pi^+\pi^- e^+ e^-$

Four diagrams contribute to each of these processes: inner bremsstrahlung, $M1$ and $E1$ direct emission, and direct CP-violation. For the K_L, interference between the first two diagrams leads to CP-violating circularly polarized γ^*s, which in turn implies a decay asymmetry in the angle ϕ between the $\pi^+\pi^-$ and e^+e^- planes. The predicted value of this asymmetry,

$$\mathscr{A}_\phi = \frac{N_{\pi\pi ee}(\sin\phi\cos\phi > 0) - N_{\pi\pi ee}(\sin\phi\cos\phi < 0)}{N_{\pi\pi ee}(\sin\phi\cos\phi > 0) + N_{\pi\pi ee}(\sin\phi\cos\phi < 0)},$$

from the interference term of the differential rate, $\frac{d\Gamma}{d\phi} = \Gamma_1 \cos^2\phi + \Gamma_2 \sin^2\phi + \Gamma_3 \sin\phi\cos\phi$, is 14%.[1] No such effect is expected for the K_S.

Data collected during the 1998 and 1999 ε'/ε runs and a 40-hour high-intensity K_S^0 run in 1999 were analyzed and normalized to reconstructed $K_L^0 \to \pi^+\pi^-\pi^0_{Dalitz}$ decays.

In the K_L channel, 1161 candidates were found in the signal region, with a signal:background ratio of 31. This gives a branching fraction of $(3.08 \pm 0.20) \times 10^{-7}$, where statistical and systematic errors have been combined in quadrature. In the K_S channel, 621 events were found in the signal region, with no background, for a branching fraction of $(4.69 \pm 0.30) \times 10^{-5}$. The acceptance-corrected angular asymmetries for the K_L is $\mathscr{A}_\phi = (14.2 \pm 3.0_{stat})\%$ and for the K_S is $\mathscr{A}_\phi = (0.5 \pm 4.0_{stat})\%$, both in good agreement with theory. The finite value of the K_L angular asymmetry indicates CP-violation.[2]

$K_L^0 \to \pi^0 \gamma\gamma$

χ_{PT} predicts a branchng fraction of 0.6×10^{-6}, for this mode, to better than 5%, where $\mathscr{O}(p^2)$ contributions are zero and $\mathscr{O}(p^4)$ contributions are unambiguous.[3, 5, 4, 2] The measured fraction, however, is $(1.68 \pm 0.10) \times 10^{-6}$, which χ_{PT} manages to approximate (1.5×10^{-6}) by including $\mathscr{O}(p^6)$ terms (including vector meson exchange) with a coupling (so-called counter) term, $a_v = -0.7$. This latter compensates for higher-order loop divergences and, in this case, can be extracted from the $m_{\gamma\gamma}$ distribution. This counter term appears in χ_{PT} calculations of the CP-conserving amplitude in $K_L \to \pi^0 e^+ e^-$. A determination of its value, then, leads to an estimation of $BR(K_L \to \pi^0 e^+ e^-)_{CP-C}$.

After cuts to reject $2\pi^0$ and $3\pi^0$ events, and subtraction of pile-up, some 2500 candidate events were found. The branching fraction is determined to be $[1.36 \pm 0.03(\text{stat}) \pm$

---

[2] Accepted for publication in EPJ C.

$0.03(\text{syst}) \pm 0.03(\text{norm})] \times 10^{-6}$, and a likelihood fit gives $a_v = -0.46 \pm 0.03(\text{stat}) \pm 0.03(\text{syst}) \pm 0.02(\text{theo})$. This implies $BR(K_L \to \pi^0 e^+ e^-)_{CP-C} = (4.7 \pm 2.2) \times 10^{-13}$.[3]

$K_S^0 \to \pi^0 \gamma\gamma$

This mode is dominated by the π-pole of $K_S^0 \to \pi^0 \pi^0$, and the analysis therefore is limited to a squared mass-ratio $m_{\gamma\gamma}^2/m_K^2 \geq 0.2$. The shape of this ratio's distribution tests the chiral structure of weak vertex. The χ_{PT} prediction[6] for the branching fraction is 3.8×10^{-8} ($m_{\gamma\gamma}^2/m_K^2 \geq 0.2$). The best published limit was given in a recent paper due to NA48[4]: $< 3.3 \times 10^{-7}$ (90% CL, $m_{\gamma\gamma}^2/m_K^2 \geq 0.2$), on the basis of data collected in 1998 and 1999.

Preliminarily, NA48 has found 31.0 ± 5.6 events in the signal region from the 2000 high-intensity near-target run. Of these, 13.6 ± 2.8 events have been determined to be either beam-related backgrounds, $K_S^0 \to \pi^0 \pi_{Dalitz}^0$, $K_L^0 \to \pi^0 \gamma\gamma$, or acceptance-influenced. The spectrum in the signal region has a 9×10^{-4} probability of being due to backgrounds such as these. The preliminary result reached is $BR(K_S \to \pi^0 \gamma\gamma, z_q \geq 0.2) = [4.9 \pm 1.6(\text{stat}) \pm 0.8(\text{syst})] \times 10^{-8}$ (preliminary). Unfortunately, the sample is inadequate to test the chiral structure.

$K_{L,S} \to \gamma\gamma$

The channel $K_L^0 \to \gamma\gamma$ presents an irreducible background to $K_S^0 \to \gamma\gamma$. Knowledge of the level of this contamination is limited, however, by the 3% uncertainty in the K_L^0 branching fraction. By taking far-target and near-target data separately, it was possible to determine the K_L branch to about 1% relative to $K_L^0 \to \pi^0 \pi^0 \pi^0$ and subsequently obtain a K_S result that could challenge χ_{PT} predictions. The data for this two-stage analysis were collected in 2000. Five samples were gleaned from the data: $\gamma\gamma$ and $3\pi^0$ from the far target and $\gamma\gamma$, $2\pi^0$, and $3\pi^0$ from the near target.

The initial result, $\frac{\Gamma(K_L^0 \to \gamma\gamma)}{\Gamma(K_L^0 \to 3\pi^0)} = [2.81 \pm 0.01(\text{stat}) \pm 0.02(\text{syst})] \times 10^{-3}$, came after reducing hadronic and other backgrounds to $(0.6 \pm 0.3)\%$ in the far-target data. Then, by counting the number of $K_L^0 \to 3\pi^0$ events in the near-target data and reducing hadronic, accidental, and $K_L^0 \to 2\pi_{Dalitz}^0$ backgrounds to about 2.5% of the total 20,000 $K_{L,S} \to \gamma\gamma$ events found, the $K_S^0 \to \gamma\gamma$ branch was determined to be $[2.78 \pm 0.06(\text{stat}) \pm 0.03(\text{syst}) \pm 0.02(\text{ext})] \times 10^{-6}$.[5] While compatible with previous results, the error is small enough to show a 30% discrepency with the $\mathcal{O}(p^4)$ prediction

---

[3] Published in Phys.Lett. B 536 (2002) 229; note that these results differ signficantly from those of KTeV.
[4] Phys. Lett. B 556 (2003) 105.
[5] Published in Phys. Lett. B 551 (2003) 7.

of 2.1×10^{-6}.[7, 8]

Two other results:

K_{e3} Charge Asymmetry

A rate difference of sign in kaon semi-leptonic decays indicates *CP*-violation:

$$\delta_\ell(e) = \frac{BR(K_L^0 \to \pi^- e^+ \nu_e) - BR(K_L^0 \to \pi^+ e^- \overline{\nu}_e)}{BR(K_L^0 \to \pi^- e^+ \nu_e) + BR(K_L^0 \to \pi^+ e^- \overline{\nu}_e)}$$
$$= 2\Re e(\varepsilon)$$

In 2001, NA48 collected about 2×10^8 K_{e3} events, with a background fraction of less than 10^{-5}. After correcting for trigger efficiency, particle identification, and punch-through, the asymmetry may be measured in bins of pion momentum. The weighted-average preliminary result is:

$$\delta_\ell(e) = [3.317 \pm 0.070(\text{stat}) \pm 0.072(\text{syst})] \times 10^{-3} \text{ (preliminary)}.$$

η_{000}

In 2000, NA48 had no drift chambers, but exploited the situation to take far-target data for dilution measurements and near-target data at high intensity to test for the K_S program of 2002. The near-target running period yielded 5.9×10^6 $\pi^0\pi^0\pi^0$ events, which were used, after acceptance corrections determined from the far-target data, to measure

$$\eta_{000} = \frac{A(K_S^0 \to \pi^0\pi^0\pi^0)}{A(K_L^0 \to \pi^0\pi^0\pi^0)}.$$

A function

$$f(E,t) = \frac{I^{\text{near}}_{\pi^0\pi^0\pi^0}}{I^{\text{far}}_{\pi^0\pi^0\pi^0}} =$$

$$A(E) \left[1 + |\eta_{000}|^2 e^{t(\frac{1}{\tau_L} - \frac{1}{\tau_S})} + 2D(E) e^{\frac{t}{2}(\frac{1}{\tau_L} - \frac{1}{\tau_S})} (\Re e\eta_{000} \cos \Delta mt - \Im m\eta_{000} \sin \Delta mt) \right]$$

was fit in energy bins between 70 and 170 GeV to determine $A(E)$, $\Re e\eta_{000}$, and $\Im m\eta_{000}$. The preliminary results are

$$\Re e\eta_{000} = [-2.6 \pm 1.0(\text{stat}) \pm 0.5(\text{syst})] \times 10^{-2}$$
$$\Im m\eta_{000} = [-3.4 \pm 1.0(\text{stat}) \pm 1.1(\text{syst})] \times 10^{-2}.^6$$

If we assume CPT good, then $\Re e\eta_{000} = \Re e\varepsilon$, and $\Im m\eta_{000}$ is sensitive to direct CP-violation. Under this assumption, the preliminary result is $\Im m\eta_{000} = (-1.2 \pm 1.3) \times 10^{-2}$ (CPT). From this, it is possible to calculate that $BR(K_S^0 \to \pi^0\pi^0\pi^0) < 3.0 \times 10^{-7}$ (preliminary).

REFERENCES

1. Heiliger, P., and Sehgal, L., *Phys. Rev. D*, **47**, 4920 (1993).
2. D'Ambrosio, G., and Portoles, J., *Nucl. Phys. B*, **492**, 417 (1997).
3. Ecker, G., Pich, A., and Rafael, E. D., *Phys. Lett. B*, **237**, 481 (1990).
4. Cohen, A., Ecker, G., and Pich, A., *Phys. Lett. B*, **304**, 347 (1993).
5. Cappiello, L., D'Ambrosio, G., and Miragliuolo, M., *Phys. Lett. B*, **298**, 423 (1993).
6. Ecker, G., Pich, A., and Rafael, E. D., *Phys. Lett. B*, **189**, 363 (1987).
7. D'Ambrosio, G., and Espriu, D., *Phys. Lett. B*, **175**, 237 (1986).
8. Goity, J. L., *Z. Phys. C*, **34**, 341 (1987).

---

[6] For comparison, CPLEAR found $\Re e\eta_{000} = [18 \pm 14(\text{stat}) \pm 6(\text{syst})] \times 10^{-2}$, $\Im m\eta_{000} = [15 \pm 20(\text{stat}) \pm 2(\text{syst})] \times 10^{-2}$.

Future Accelerators

John Womersley

Fermi National Accelerator Laboratory, Batavia, IL 60510

Abstract. I describe the future accelerator facilities that are currently foreseen for electroweak scale physics, neutrino physics, and nuclear structure. I will explore the physics justification for these machines, and suggest how the case for future accelerators can be made.

In asking me to give one of the closing presentations at this meeting, I imagine the conference organizers may expect me to impart some inspiration as well as information. The inspirational part will explore why I have added a question mark to the title. The informational part will describe future accelerators aimed at understanding electroweak symmetry breaking (TeV scale physics), neutrino physics, and nuclear physics.

WHY THE ?

The "?" indicates that the existence of future accelerators is far from assured. In fact, the climate is arguably rather hostile. In recent years we seem to have done a poor job of making the case for future machines, at least where particle physics is concerned. Here are two examples of statements from representatives of the administration that show how far the case is from being made:

- Michael Holland of the White House Office of Management and Budget, at Snowmass 2001: *"How much importance do scientists outside your immediate community attach to yoru fervent quest for the Higgs boson? How else would you expect us to evaluate your priorities? What would you do if the government refused to fund any big accelerator?"* [1]
- Dr. John Marburger, Director of the Office of Science and Technology Policy, at SLAC, October 2002: *"At some point we will simply have to stop building accelerators. [...] we must start thinking about what fundamental physics will be like when it happens. [...] experimental physics at the frontier will no longer be able to produce direct excitations of increasingly massive parts of nature's spectrum [...] There are two alternatives. The first is to use the existing accelerators to measure parameters of the standard model with ever-increasing accuracy so as to capture the indirect effects of higher energy features of the theory [...] The second is to turn to the laboratory of the cosmos, as physics did in the cosmic ray era before accelerators became available more than fifty years ago."* [2]

With all due respect, I have to assert that Dr. Marburger is wrong on both counts. At some point, yes, any given accelerator technology becomes too expensive to pursue.

That does not mean that we have to stop building accelerators; it means that we need to develop new accelerator technologies. Secondly, the very richness of the "laboratory of the cosmos" is exactly the reason why we need to keep building accelerators. There is a universe full of weird stuff out there — the more we look, the more weird stuff we find. Do we really think we can understand it all without making these new quanta in the laboratory and studying their properties under controlled conditions?

How might we then start to better make the case? I have a couple of suggestions.

1. Emphasize the Unknown

As Shakespeare had Hamlet point out, "there are more things in Heaven and Earth than are dreamt of in our philosophy." In justifying and describing the potential of new facilities, I believe that we have tended much too far in the direction of 'one last piece of the puzzle' or 'we know what we're doing and we know what we'll find.' This reinforces the mistaken idea that we are close to 'the end of science' and is rather hard to justify given that 95% of the universe is not made of quarks and leptons. In fact, exploring the unknown has a lot more resonance with the public. We have to search for new phenomena in ways that are not constrained by our preconceptions of what may be 'out there.' The Tevatron collider experiments have done just that. The DØ collaboration has published [3] a model-independent search for deviations from the standard model in the 1992–95 data. Only two channels had any hint of disagreement and overall the confidence level for the standard model—in this small dataset—was 89%. CDF has also pursued signature-based searches. Such approaches are good science but also good tools for publicity and outreach.

2. It's all about the Cosmos

The composition of the universe is a powerful unifying theme for particle and nuclear physics. Mass shapes the universe through gravity, the only force that is important over astronomical distances. The masses of stars and planets arise largely through QCD (binding energies of protons and neutrons), but it has long been known that there is substantial invisible (dark) matter and that (from primordial D/He abundances) that this matter is not baryons. Recent measurements of the multipole moments of the cosmic microwave background such as that from WMAP[4] have allowed the dark matter density to be extracted quite precisely. There seems to be about six to seven times more mass ($27 \pm 4\%$) than baryons ($4.4 \pm 0.4\%$). The most likely explanation is that the dark matter is a new kind of particle: weakly interacting, massive relics from the early universe. There are two complementary experimental approaches that should be pursued: to search for dark matter particles impinging on Earth, and to try to create such particles in our accelerators.

Supersymmetry (SUSY) is an attractive idea theoretically; it can unify couplings, cancel divergences in the Higgs mass, and provides a path to the incorporation of gravity and string theory. It also predicts a particle, the lightest neutralino, which is a good explanation for cosmic dark matter and which could be discovered at the Tevatron or LHC, and studied in detail at a linear collider. In fact the search for dark matter is underway now, in Run II at the Tevatron collider. Neutralinos would be produced in cascade decays of squarks or gluinos and could be detected through their escape from the detector, as missing transverse energy.

The same cosmic microwave background data, together with supernova measurements of the velocity of distant galaxies, suggest that two-thirds of the energy density of the universe is in the form of dark energy—some kind of field that expands along with the universe. Again, there are two complementary approaches to learn more. We should refine our cosmologically-based understanding of the properties of dark energy in bulk (its 'equation of state') through new projects such as SNAP. We should also understand what we can do under controlled conditions in the laboratory. Ultimately I am sure we will want to make dark energy quanta in accelerators. For now, we should explore the only other example of a 'mysterious field that fills the universe,' namely the Higgs field. The Standard Model Higgs field would produce something like 54 orders of magnitude too much dark energy compared with the cosmological observations, but surely it cannot be totally unrelated.

We know that photons and W and Z bosons couple to particles with the same strength—this is electroweak unification. Yet while the whole universe is filled with photons, the W's and Z's only mediate a weak force that occurs inside nuclei in radioactive beta decay. This is because the W and Z are massive particles, and the unification is thus broken. This mass (the electroweak symmetry breaking) appears to arise because the universe is filled with an energy field, called the Higgs field, with which the W and Z interact (and in fact mix). We want to excite the quanta of this field and measure their properties. The field need not result from a single, elementary scalar boson: there can be more than one particle (as is the case in supersymmetry), or composite particles can play the role of the Higgs (e.g. in technicolor or topcolor models). We do know that electroweak symmetry breaking occurs, so there is something out there coupling to the W and Z. Precision electroweak measurements imply that this thing looks very much like a standard model Higgs (though its couplings to fermions are less constrained). We also know that WW cross sections would violate unitarity at ~ 1 TeV without it, and this is a real process that will be seen at the LHC. For all of these reasons, electroweak symmetry breaking remains a focus of the experimental high energy physics program.

This naturally leads me to the second part of my presentation, where I shall review future accelerator intitiatives, starting with those aimed at the electroweak scale.

FUTURE ACCELERATORS FOR ELECTROWEAK SCALE PHYSICS

The flagship future facility for TeV-scale physics will be the **Large Hadron Collider** (LHC) at CERN. The LHC is a 14 TeV proton-proton collider. It will serve two large general purpose detectors, ATLAS and CMS, together with a heavy-ion and B-physics program. Underground construction is well advanced and the detectors are making good progress. Accelerator dipole magnet production is the overall pacing item; if all goes well, first beam will be circulated in 2007.

The LHC will be able to discover a standard model Higgs over the entire range of allowed masses (115 GeV – 1 TeV). Beyond discovery, we will need to verify that the observed state actually provides both vector bosons and fermions with their masses. The LHC will be able to start this job by measuring various ratios of Higgs couplings and branching fractions (at the 25% level) by comparing rates in different Higgs production and decay channels.

The more complex Higgs sector in supersymmetric models can also be quite thoroughly explored. Tau decay modes are very important over a large region of parameter space at moderate to large $\tan\beta$. At least one Higgs state is visible no matter what; the most problematic region of parameter space is where one light state h is discoverable, but looks very much like the standard model H.

To elucidate this case, and of course in general too, one would use the LHC to search for supersymmetry through sparticle production. The mass range covered for squarks and gluinos is huge (up to ~ 2.5 TeV) and a signal to background ratio as high as ten can be achieved even with simple cuts. Exclusive mass reconstruction of SUSY cascade decays has been demonstrated for several benchmark points. New Higgs signals also appear in such decays.

The combination of high energy (14 TeV) and luminosity (100 fb^{-1}) means that the LHC will have the potential to observe almost any other new physics associated with the TeV scale. Extra dimensions of space-time and/or TeV-scale gravity could have subtle, indirect effects—or direct, spectacular signatures like the production of black holes. The LHC would also be sensitive to compositeness, excited quarks, leptoquarks, technicolor, strong WW interactions, new gauge bosons, and heavy neutrinos.

In summary, by the year $201x$, if all goes well, we should have observed at least one and maybe several Higgs bosons, and will have tested their properties at the 25% level. We will not always have been able to distinguish a Standard Model from a SUSY Higgs, but we almost always expect to have discovered SUSY in other ways. If we don't see a Higgs, we will have observed some other signal of electroweak symmetry breaking (technicolor, or strong WW scattering, for example). In addition, we will have learned a great deal more about the physics landscape at the TeV scale: is there supersymmetry? Are there extra dimensions?

There is an international consensus[5] that the highest priority facility to follow the LHC should be an **Electron-positron Linear Collider** (LC). This would collide e^+e^- beams at a center-of-mass energy between 500 GeV and 1 TeV and deliver a few hundred inverse femtobarns per year. The cost is perhaps \$5–7B and will require an international effort to build; it could be in operation by 2015–20.

The physics of the Linear Collider is no longer about discovery, it is about precision. (In this sense, it plays a similar role to the one that LEP did, after the W and Z had been discovered at the SPS collider). The LC program aims to exploit aggressive detector technology such as displaced vertex charm-tagging and energy-flow calorimetry, and also make use of highly polarized beams to reduce backgrounds.

Higgs production at a LC occurs through both $e^+e^- \to HZ$ and $e^+e^- \to \nu\bar{\nu}H$ processes. The HZ process can be used to reconstruct the Higgs (actually whatever the Z recoils against) even if it decays invisibly, and permits the g_{HZZ} coupling to be determined to a few percent. This in turn provides a simple test of whether the observed particle is actually the only Higgs: namely, does it account for all the mass of the Z? For example, in minimal SUSY the h couples $g_{hZZ} \sim g_Z M_Z \sin(\beta - \alpha)$ and the H couples $g_{HZZ} \sim g_Z M_Z \cos(\beta - \alpha)$ and together they create the full M_Z that we observe. The $\nu\bar{\nu}H$ process, with $H \to b\bar{b}$, allows the g_{HWW} coupling to be extracted with a precision of a few percent.

The couplings of the Higgs to fermions determine whether the Higgs field is indeed responsible for fermion masses as well as for electroweak symmetry breaking. With 500 fb^{-}1 at $\sqrt{s} = 500$ GeV, the Yukawa couplings of a 120 GeV Higgs could be determined at the level of $\Delta g_{Hbb} = 4\%$, $\Delta g_{Hcc} = 7\%$, $\Delta g_{H\tau\tau} = 7\%$, and $\Delta g_{H\mu\mu} = 30\%$. At $\sqrt{s} = 800$ GeV, it would also be possible to measure g_{Htt}, through $t\bar{t}H$ production, at the 10% level. We could thus determine whether the top quark's unexpectedly large mass arises from the Higgs or from some other mechanism.

The quantum numbers of the Higgs itself can be excplored. The angular dependence of $e^+e^- \to ZH$ and of the $Z \to f\bar{f}$ decay products can cleanly separate CP-even and odd Higgs states (H and A in minimal supersymmetry). One would be sensitive to a 3% admixture of CP-odd A in the "H" signal. This could be a window to CP violation in the Higgs sector. With sufficient luminosity, the Higgs self-coupling can be probed through ZHH production (six jets in the final state). The cross section is tiny, about 0.2 fb, so of order 1 ab (1000 fb) is required for a 20-30% measurement of g_{HHH}. Such a measurement would constrain the Higgs potential and, compared with the expectation from the Higgs mass, would give a self-consistency test for the Higgs.

There are very clean signals for light superpartner production at a LC. For example, chargino pair production occurs through s-channel annihilation or through t-channel sneutrino exchange. One can select the mixture of processes by polarizing the electron beam: since a right-handed electron has no coupling to the sneutrino, one suppresses the t-channel process. In this way the "Wino" and "Higgsino" parts of the chargino can be separated. The Wino coupling to $e\tilde{\nu}$ can then be compared to the W coupling to $e\nu$ — if it is truly supersymmetry, they must be equal. The chargino decays to neutralinos, and at the LC all the masses can be measured. This would enable the expected dark matter abundance and properties to be calculated.

In summary, we are planning a relay race at the electroweak scale. The Tevatron will discover new TeV-scale physics if we are lucky. The LHC is "guaranteed" discovery and will start to measure and constrain. The Linear Collider will measure, measure, measure — and build the physics case for the next accelerator to follow.

FUTURE ACCELERATORS FOR NEUTRINO PHYSICS

We now have three distinct signals for neutrino oscillation:

- **Solar neutrinos:** missing ν_e, as observed by Homestake, GALLEX, SAGE, Kamiokande, SuperK, SNO and KamLAND.
- **Atmospheric neutrinos:** missing ν_μ, as observed by Kamiokande, SuperK and K2K.
- **LSND signal:** a $\nu_\mu \leftrightarrow \nu_e$ oscillation, as seen by the LSND experiment at Los Alamos.

Parenthetically, we may note (and point out to Dr. Marburger) that while the "laboratory of the solar system" gave us the first two signals, it required terrestrial beams (at KamLAND and K2K) to really understand and have confidence in what we were seeing.

The solar and atmospheric signals form a consistent picture in which three neutrino mass eigenstates each contain admixtures of the flavor states. The ν_1 and ν_2 states are separated by $\Delta m^2 \sim 5 \times 10^{-5}$ eV (the solar oscillation signal) while ν_3 is split from these two states by $\Delta m^2 \sim 3 \times 10^{-3}$ eV (the atmospheric oscillation signal). The overall mass scale and ordering in mass is not known. Unlike quarks, there is a lot of mixing; the mass eigenstates do not correspond "mostly" to any single flavor. If the LSND result is confirmed, it would require drastic extensions to this picture: either additional neutrino states, or new physics (CPT violation, for example).

There are a significant number of neutrino experiments now running. At Fermilab, miniBooNE is seeking to confirm LSND's signal for $\nu_\mu \to \nu_e$ (and also $\bar{\nu}_\mu \to \bar{\nu}_e$). In Japan, K2K is pursuing the "atmospheric" oscillation using an accelerator neutrino beam, and KamLAND is exploring the "solar" signal using reactor neutrinos. SNO continues to detect solar neutrinos with flavor selection. It will soon be joined by Borexino, a solar neutrino detector with a very low energy threshold. Two new long-baseline projects are also under construction: MINOS, with a beam from Fermilab to Soudan to measure the atmospheric oscillation and search for $\nu_\mu \to \nu_e$; and the CERN Neutrinos to Gran Sasso project which will focus on $\nu_\mu \to \nu_\tau$ using the OPERA (emulsion) and ICANOE (liquid argon) detectors.

These experiments will tell us whether the LSND result is correct (if yes, confirming that there is new physics). They will better pin down the mass-squared splittings and mixing angles in the solar and atmospheric oscillations. Most importantly, they will give some information on the critical parameter θ_{13}, which describes how much electron-neutrino there is in the ν_3 eigenstate. It is θ_{13} which governs the size of possible CP violation in the neutrino sector, which is of great interest in understanding the baryon asymmetry of the universe. Currently, θ_{13} is known to be less than about 0.10. If it is large enough (where large enough means greater than 0.05 or so), a rich program of next generation experiments opens up. The goal would be to search for electron neutrinos in the "atmospheric" distance/energy regime, to observe matter effects (to resolve the mass ordering) and ultimately CP violation. This would require any or all of the following:

- Bigger detectors, 20–100 kt compared with MINOS's 3 kt fiducial mass;
- Better instrumentation (for example, calorimetry);
- Higher intensity neutrino beams ("superbeams").

There are a number of concepts that exploit new beams to existing detectors, or new detectors in existing beams, or entirely new projects: Fermilab to Minnesota or Canada, Brookhaven to Homestake or WIPP, and JPARC to Kamioka. One could also access the physics through ν_e disappearance using a very high precision reactor experiment.

If θ_{13} is small, things become much more challenging. Baselines of thousands of kilometers become optimal, and low rates require new technology for neutrino beams. In this scenario, a muon storage ring neutrino factory may be essential.

No matter what we learn in the next few years, it is clear that we will need major new accelerator and detector facilities for neutrino physics. There is no complete consensus—

yet—on just what those facilities should be, but there are lots of good ideas, and lots more data are coming.

FUTURE ACCELERATORS FOR NUCLEAR PHYSICS

The nuclear physics community has developed a long range plan for the next decade[6], and recently the Facilities Subcommittee of the Nuclear Science Advisory Committee reported on the importance of the science and readiness for construction of new facilities[7]. The following three projects were the highest ranked in the two categories:

- The Rare Isotope Accelerator (RIA)
- A new gamma-ray detector array GRETA (instrumentation for RIA)
- CEBAF energy upgrade (from 6 to 12 GeV).

(RHIC upgrades and an underground detector were also highly ranked but not judged to be immediately ready for construction.)

RIA is a facility to produce rare isotopes. It is driven by a linac (400 MeV/u U, 900 MeV p) which feeds production targets followed by online isotope separation, possible re-acceleration, trapping or isotope recovery. So why do we need such a major ($\sim$\$900M) new facility for nuclear physics now? The science case is based on:

- Nuclear struture;
- Astrophysics — the origin of elements heavier than iron. Creation of such elements in supernovae is believed to occur through a complex series of reactions involving unstable, neutron-rich nuclei that could be explored at RIA;
- Low energy tests of standard model symmetries.

As well as the science, RIA would offer "collateral benefits" through the production of medical isotopes and the understanding of processes relevant to nuclear stockpile stewardship.

In preparing this talk I discussed the RIA science case with several of my high energy physics colleagues. Their initial skepticism generally turned to interest once they heard the astrophysics aspects (and, implicitly, how much had been glossed over in the undergraduate astronomy classes they had taken). This observed resonance is a good lesson for all of us in how to explain the relevance and interest of future facilities to those outside our immediate field.

CONCLUSIONS

Accelerators are the key to understanding this weird and wonderful universe that we inhabit. Only accelerators can provide the controlled conditions, known particle species, high rates and high energies that we need to make sense of cosmological observations. Recent progress in astroparticle physics and cosmology does not weaken the case for new accelerators, it strengthens it; and there is no shame in exploiting public interest in these discoveries. The major problems are political. As Joe Lykken stated at the Lepton–Photon

Symposium at Stanford in 1999, "It is much more likely that we will fail to build new accelerators than that these accelerators will fail to find interesting physics." It will take a concerted effort to overcome the political obstacles, but if we work together we can do it.

ACKNOWLEDGMENTS

I would like to thank Peter Meyers for allowing me to use material from his excellent presentation at the 2003 meeting of the American Physical Society in Philadelphia.

REFERENCES

1. as quoted in Physics Today, September 2001
2. the full text is available at www.ostp.gov
3. DØ Collaboration (V. Abazov et al.), Phys. Rev. D **64**, 012004 (2001).
4. C.L. Bennett et al., astro-ph/0302207; D.N. Spergel et al., astro-ph/0302209.
5. Science Ahead: the Way to Discovery, report of the HEPAP subpanel on Long Range Planning for US High Energy Physics, January 2002.
6. 2002 NSAC Long-Range Plan: Opportunities in Nuclear Science, a long-Range Plan for the next decade, April 2002.
7. The Nuclear Physics Scientific Horizon: Projects for the Next Twenty Years. Report of the ad-hoc Facilities Subcomittee of the Nuclear Science Advisory Committee, March 2003.

LEPTON HADRON & HAD-HAD SCATTERING

Cross Section Measurements and Charm Production in the NuTeV Experiment

S. Boyd*, T. Adams†, A. Alton†, S. Avvakumov**, L. de Barbaro‡,
P. de Barbaro**, R. H. Bernstein§, A. Bodek**, T. Bolton†, J. Brau¶,
D. Buchholz‡, H. Budd**, L. Bugel§, J. Conrad‖, R. B. Drucker¶,
B. T. Fleming‖, J. Formaggio‖, R. Frey¶, J. Goldman†, M. Goncharov†,
D. A. Harris**, R. A. Johnson††, J. H. Kim‖, S. Koutsoliotas‖,
M. J. Lamm§, W. Marsh§, D. Mason¶, J. McDonald*, K. S. McFarland**,
C. McNulty‖, D. Naples*, P. Nienaber§, V. Radescu*, A. Romosan‖,
W. K. Sakumoto**, H. Schellman‡, M. H. Shaevitz‖, P. Spentzouris‖,
E. G. Stern‖, N. Suwonjandee††, N. Tobien§, A. Vaitaitis‖, M. Vakili††,
U. K. Yang**, J. Yu§, G. P. Zeller‡ and E. D. Zimmerman‖

*University of Pittsburgh, Pittsburgh, PA
†Kansas State University, Manhattan, KS
**University of Rochester, Rochester, NY
‡Northwestern University, Evanston, IL
§Fermi National Accelerator Laboratory, Batavia, IL
¶University of Oregon, Eugene, OR
‖Columbia University, New York,NY
††University of Cincinnati, Cincinnati, OH

Abstract. The NuTeV experiment at Fermilab obtained pure high statistics samples of neutrino and antineutrino interactions using its sign-selected beam. Preliminary inclusive charged current differential cross sections of neutrino and antineutrino interactions on iron are presented, along with preliminary measurements of $F_2(x,Q^2)$ and $xF_3(x,Q^2)$. Preliminary results from the next-to-leading order QCD analysis of deep inelastic charged current dimuon data are also shown.

1. INTRODUCTION

Neutrino nucleon deep inelastic scattering provides a precise probe of the nucleon structure and QCD. Neutrinos are uniquely sensitive to the parity violating structure function, $xF_3(x,Q^2)$. The NuTeV experiment is a high-energy fixed target neutrino-nucleon scattering experiment. An iron-scintillator sampling calorimeter, the experiment used the Fermilab Sign Selected Quadrupole Train (SSQT) beam to produce high purity neutrino and antineutrino beams A continuous calibration beam was also installed with which NuTeV was able to determine muon and hadron energy scales to a precision of 0.7% and 0.43% respectively. A detailed description of the experiment, the neutrino beam and the calibration procedure can be found in [1].

2. ν-FE CC DIFFERENTIAL CROSS SECTION AND ν-FE STRUCTURE FUNCTIONS

The neutrino-nucleon differential scattering cross section in x, the Bjorken scaling variable, and y, the inelasticity, is determined from the relation

$$\frac{d^2\sigma^{\nu(\bar{\nu})}}{dxdy} = \frac{1}{\Phi(E)^{\nu(\bar{\nu})}} \frac{d^2N^{\nu(\bar{\nu})}(E)}{dxdy} \quad (1)$$

where $\Phi(E)^{\nu(\bar{\nu})}$ is the $\nu(\bar{\nu})$ flux in energy bins. The events in the data sample used for the cross section analysis are required to pass fiducial volume cuts and have a momentum analyzed muon in the toroid spectrometer which passes reconstruction quality cuts. In addition, the muon is required to have an energy greater than 15 GeV, the hadronic energy is required to be greater than 10 GeV and the reconstructed energy of the neutrino is required to be greater than 30 GeV. The square of the 4-momentum transfer of each event, Q^2, must be greater than 1 (GeV/c)^2 to minimize the non-perturbative contribution to the cross section. The NuTeV kinematic range extends from 10^{-3} to 0.95 in x, 0.05 to 0.95 in y, and from 30 GeV to 360 GeV in E_ν. The data sample contains a total of 8.6×10^5 ν and 2.3×10^5 $\bar{\nu}$ events after selection.

The neutrino flux is determined from a sample of events with $E_{HAD} < 20$ GeV using the "fixed ν_0" relative flux extraction method [2]. The integrated number of events in this sample as $y = \frac{E_{HAD}}{E_\nu} \to 0$ is proportional to the flux. Corrections up to order y^2, determined from the data sample, are applied to determine the relative flux to about the 1% level. This procedure can only determine the relative flux. The absolute normalisation of the flux is obtained using the world average ν-Fe cross section [3] $\frac{\sigma^\nu}{E_\nu} = 0.677 \times 10^{-38} cm^2/GeV$.

The Monte Carlo simulation, used to estimate acceptance and resolution effects, employs an input cross section model which is iteratively tuned to the data. The initial cross section and flux inputs are extracted from CCFR[4] data. The cross section is extracted and then fit to empirically determine a set of parton distribution functions[5]. The new parton distribution functions are used as inputs to the Monte Carlo, and the procedure is iterated until convergence.

Figure 1 shows the preliminary extracted cross section in x bins at $E_\nu = 85$ GeV compared to CCFR results.

2.1. Measurement of $F_2(x, Q^2)$ and $xF_3(x, Q^2)$

The structure functions $F_2(x, Q^2)$ and $xF_3(x, Q^2)$ may be obtained from the sum and differences of the neutrino and antineutrino differential scattering cross sections :

$$\left(\frac{d^2\sigma^\nu}{dxdy} + \frac{d^2\sigma^{\bar{\nu}}}{dxdy} \right) = \frac{y^2 G_F^2 ME}{\pi(1-\varepsilon)} \left[(1+\varepsilon R_L)2xF_1 + \frac{y(1-y/2)}{1+(1-y)^2} \Delta xF_3 \right] \quad (2)$$

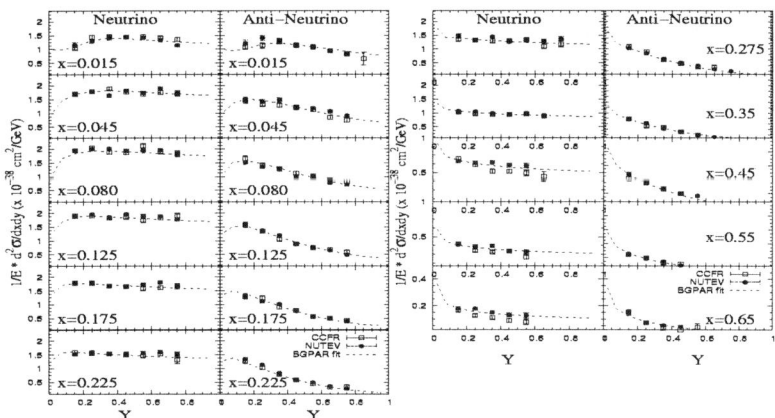

FIGURE 1. NuTeV preliminary measurement of $\frac{d^2\sigma^\nu}{dxdy}$ and $\frac{d^2\sigma^{\bar\nu}}{dxdy}$ for (left) $x < 0.225$ and (right) $x > 0.225$ compared with CCFR at a neutrino energy of $E = 85$ GeV.

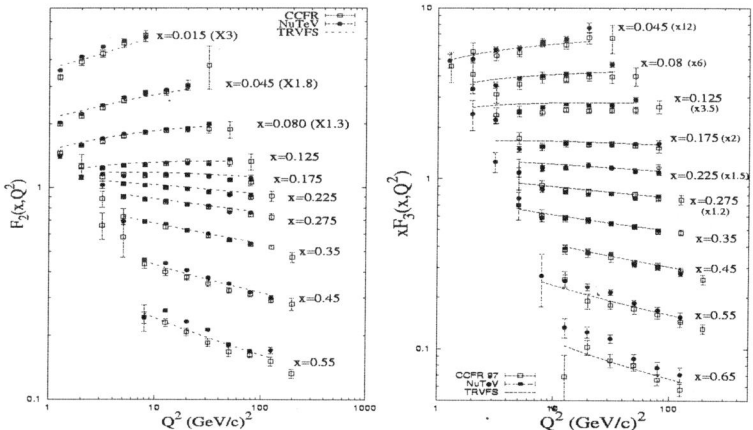

FIGURE 2. Preliminary NuTeV measurement of (left) $F_2(x,Q^2)$ and (right) $xF_3(x,Q^2)$ compared with data from CCFR[4]. Curves show predictions from the Thorne-Roberts model[6].

$$\left(\frac{d^2\sigma^\nu}{dxdy} - \frac{d^2\sigma^{\bar\nu}}{dxdy}\right) = \frac{2G_F^2 ME}{\pi} y\left(1 - \frac{y}{2}\right) xF_3^{AVG}(x,Q^2) \quad (3)$$

In Equation 2, $\varepsilon = \frac{2(1-y)-Mxy/E}{1+(1-y)^2+Mxy/E}$ is the polarisation of the virtual W-boson and R_L is the ratio of the longitudinally polarised W-boson absorption cross section to the transversely polarised W-boson absorption cross section. The term $\Delta xF_3 = xF_3^\nu - xF_3^{\bar\nu} \sim 4x(s-c)$ is sensitive to heavy flavours. A fit to the sum of differential cross sections is used to extract $F_2(x,Q^2)$ using, as input, a model prediction for ΔxF_3[6] and $R_{WORLD}(x,Q^2)$[7] for $R_L(x,Q^2)$.

Figure 2 compares the NuTeV measurements of $F_2(x,Q^2)$ and $xF_3(x,Q^2)$ with those from CCFR[3] and the predictions from the a Next-to-Leading Order QCD model[6].

3. CHARM PRODUCTION IN NEXT-TO-LEADING ORDER QCD

A charged current interaction of a $\nu(\bar{\nu})$ with a $d(\bar{d})$ or a $s(\bar{s})$ quark may produce a $c(\bar{c})$ quark, resulting in an event with a charmed hadron (usually a D meson). Ten percent of the time, the charmed hadron decays semi-muonically, resulting in a distinctive event with an exclusive final state containing two muons of opposite charge and a hadronic shower. In ν interactions, approximately half of these dimuon events arise from scattering off d quarks, and half from scattering off s quarks. In $\bar{\nu}$ interactions, since the scattering is solely off the sea quarks and the $\bar{\nu}-\bar{d}$ channel is Cabibbo suppressed, more than 90 of the events come from $\bar{s}$ quarks. Such events are a powerful probe of the strange component of the nucleon sea. In many experiments determining which muon arose from the leptonic vertex and which from the charmed meson decay is made difficult due to the presence of $\bar{\nu}$ contamination in a ν beam (and vice versa). NuTeV is unique in its use of the sign-selected beam, eliminating such ambiguity.

Dimuon production has been studied within the context of Leading Order (LO) QCD in many experiments[8]. However the presence of a large gluon component in the nucleon, and the fact that a large fraction of dimuon production occurs solely off the quark sea, really requires a Next-to-Leading-Order (NLO) analysis. A measurement of the differential cross section for forward dimuon production, in which the muon from the decay of the charmed meson is required to have an energy greater than 5 GeV, has been published[9]. The NLO analysis proceeds by fitting an NLO dimuon production cross section model to the measured forward dimuon cross section using,

$$\frac{d^2\sigma^{2\mu}(E,x,y)}{dxdy} = \frac{d^2\sigma^{charm}(E,x,y;m_c,\kappa,\bar{\kappa},\alpha,\bar{\alpha},\varepsilon)}{dxdy} \times EMC \times B_c \times A(E,x,y;\varepsilon,m_c) \quad (4)$$

The charm production cross section model is a function of the charm quark mass, m_c, strange sea shape parameters (see Eq. 5), $\kappa,\bar{\kappa},\alpha$ and $\bar{\alpha}$, and the Collins-Spiller[10] fragmentation parameter, ε. In this preliminary analysis the model is computed using the DISCO[11] Monte Carlo program. Nuclear shadowing and the EMC effect[3] are taken into account by the EMC term. The semi-muonic branching ratio, B_c, is set to 0.093[12] and $A(E,x,y;\varepsilon,m_c)$ is an acceptance correction accounting for the minimum muon energy cut of 5 GeV. It is a function of the fragmentation parameter, ε, and the mass of the charm quark. The usual dependencies on the Bjorken-scaling variable, x, the inelasticity, y, and the neutrino energy, E have also been shown.

A preliminary fit of the NLO model to the forward dimuon cross section for ε, m_c, and the strange sea shape parameters has been performed. The strange sea is parameterised in terms of the light quark sea distributions by

$$s(\bar{s}) = \kappa(\bar{\kappa})\frac{\bar{u}(x)+\bar{d}(x)}{2}(1-x)^{\alpha(\bar{\alpha})} \quad (5)$$

TABLE 1. Preliminary 6-parameter fit to the NuTeV forward dimuon cross section.

Parameter	Fit value	Stat. Error	Sys. Error
m_c	1.46	0.22	0.08
κ	0.579	0.036	0.038
$\bar{\kappa}$	0.620	0.040	0.048
α	1.37	0.57	0.72
$\bar{\alpha}$	1.22	0.44	0.08
ε	0.176	0.111	0.09
χ^2/dof	36.2/38		

The CTEQ6[13] parton distribution set was used in the fit. The best fit values for the fit parameters are shown in Table 1. The NLO fit describes the data well.

CONCLUSIONS AND PROSPECTS

NuTeV has extracted preliminary differential cross sections for ν-Fe and $\bar{\nu}$-Fe inelastic scattering. The NuTeV result has improved the systematic precision on these measurements due to a precise understanding of the muonic and hadronic energy scales and to the unique SSQT beam. Preliminary determinations of $F_2(x, Q^2)$ and $xF_3(x, Q^2)$ have been carried out and furthur work on the extraction of Λ_{QCD} is underway. NuTeV has also provided new results on Next-to-Leading Order charm quark production. The charm quark mass, and $s(\bar{s})$ sea shape parameters have been extracted from the NLO fit.

REFERENCES

1. D. A. Harris et al., Nucl. Inst. Methods. **A 447**, p. 337 (2000).
2. J. M. Conrad, M. H. Shaevitz, and T. Bolton, Rev. Mod. Phys. **70**, p. 4 (1998).
3. W. Seligman, Ph. D. Thesis (Columbia University), Nevis Reports 292, (1997).
4. U. K. Yang et al., Phys. Rev. Lett. **86**, p. 2742 (2001).
5. A. J. Buras and K. L. F. Gaemers, Nucl. Phys. **B 132**, p. 2109 (1978).
6. R. S. Thorne and R. G. Roberts, Phys. Lett. **B 421**, p. 303 (1998). A. D. Martin, R. G. Roberts, W. J. Stirling, and R. S. Thorne, Eur. Phys. J. **C 18**, p. 117 (2000).
7. L. W. Whitlow et al., Phys. Lett. **B 250**, p. 193 (1990).
8. S. A. Rabinowitz et al., Phys. Rev. Lett. **70**, p. 134 (1993), H. Abramowicz et. al., Z. Phys. **C 15**, p. 19 (1982), P. Vilain et. al., Eur. Phys. J. **C 11**, p. 19 (1999), P. Asteir et. al., Phys. Lett. **B 486**, p. 35 (2000).
9. M. Goncharov et al., Phys. Rev. **D 62**, 112006 (2001).
10. P. Collins and T. Spiller et al., J. Phys. **G 11**, p. 1289 (1985).
11. S. Kretzer, D. Mason and F. Olness, Phys. Rev. **D 65**, 074010 (2002).
12. T. Bolton, KSU-HEP-97-04, e-print hep-ex/9708014 (1997).
13. J. Pumplin, D.R. Stump, J. Huston, H.L. Lai, P. Nadolsky and W.K. Tung, JHEP **207**, 012 (2002).

Measurement of the Absolute Drell-Yan Dimuon Cross Section in 800 GeV/c Proton-Proton and Proton-Deuterium Collisions

P.E. Reimer*[†], J.C. Webb**, T.C. Awes[‡], M.E. Beddo**, M.L. Brooks*,
C.N. Brown[§], J.D. Bush[¶], T.A. Carey*, T.H. Chang**, W.E. Cooper[§],
C.A. Gagliardi[∥], G.T. Garvey*, D.F. Geesaman[†], E.A. Hawker[∥*],
X.C. He[††], L.D. Isenhower[¶], D.M. Kaplan[‡‡], S.B. Kaufman[†], P.N. Kirk[§§],
D.D. Koetke[¶¶], G. Kyle**, D.M. Lee*, W.M. Lee[††], M.J. Leitch*,
N. Makins[†], P.L. McGaughey*, J.M. Moss*, B.A. Mueller[†], P.M. Nord[¶¶],
V. Papavassiliou**, B.K. Park*, J.C. Peng*, G. Petitt[††], M.E. Sadler[¶],
W.E. Sondheim*, P.W. Stankus[‡], T.N. Thompson*, R.S. Towell[¶*],
R.E. Tribble[∥], M.A. Vasiliev[∥], Y.C. Wang[§§], Z.F. Wang[§§], J.L. Willis[¶],
D.K. Wise[¶] and G.R. Young[‡]

Los Alamos National Laboratory, Los Alamos, NM 87545
[†]*Argonne National Laboratory, Argonne, IL 60439*
**New Mexico State University, Las Cruces, NM 88003*
[‡]*Oak Ridge National Laboratory, Oak Ridge, TN 37831*
[§]*Fermi National Accelerator Laboratory, Batavia, IL 60510*
[¶]*Abilene Christian University, Abilene, TX 79699*
[∥]*Texas A&M University, College Station, TX 77843*
[††]*Georgia State University, Atlanta, GA 30303*
[‡‡]*Illinois Institute of Technology, Chicago, IL 60616*
[§§]*Louisiana State University, Baton Rouge, LA 70803*
[¶¶]*Valparaiso University, Valparaiso, IN 46383*

Abstract. The Fermilab E866/NuSea Collaboration has measured the Drell-Yan dimuon cross sections in 800 GeV/c proton-proton and proton-deuterium collisions. This is the first measurement of the absolute Drell-Yan cross section in proton-proton collisions over a broad kinematic region and the most extensive study to date of the Drell-Yan cross section in proton-deuterium collisions. The Drell-Yan mechanism is sensitive to both the beam and target parton distributions. In particular, with the kinematics of the E866/NuSea data, the Drell-Yan mechanism is sensitive to the target antiquark distributions at low and intermediate Bjorken-x and to the beam quark distributions at high-x. Approximately 55K proton-proton and 121K proton-deuterium Drell-Yan events over the longitudinal momentum fraction (Feynman-x) range $-0.05 < x_F < 0.8$ and the mass ranges $4.2 < M_{\mu^+\mu^-} < 8.7$ GeV and $10.85 < M < 16.85$ GeV are included. The data analysis will be described, and the doubly-differential $M^3 d^2\sigma/dM dx_F$, and triply-differential cross sections $E d^3\sigma/dp^3$ will be presented. These results will be compared with previous measurements by E605 and E772 and to predictions based upon next-to-leading order calculations utilizing the MRST2001 and CTEQ6 global parton distribution function fits. The results indicate that recent global parton distribution fits provide a good description of the light antiquark sea in the nucleon over the Bjorken-x range $0.03 < x < 0.15$. In contrast, the valence quark distributions appear to be overestimated by the current parton distribution fits as $x \to 1$; a region in which, prior to this data, there was very little proton data to constrain the global fits.

The quark distributions of the proton have been extensively studied experimentally, and the data from these experiments have been incorporated into global fits of the parton distribution functions (PDFs) [1, 2, 3]. Much of these data are from charged lepton deep inelastic scattering (DIS) experiments, which are not able to distinguish between the quark and antiquark distributions. Complementary to DIS experiments are Drell-Yan experiments that, through kinematics, offer the ability to distinguish between the quark and antiquark distributions.

In leading order, the Drell-Yan mechanism represents the annihilation of a beam (or target) quark with an antiquark from the target (or beam), and hence is sensitive to the PDFs of both the beam and target. The Drell-Yan cross section for $pA \to \mu^+\mu^- X$ written in terms of the PDFs of the colliding hadrons in leading order (LO) is

$$M^3 \frac{d^2\sigma}{dM\,dx_F} = \frac{8\pi\alpha^2}{9} \frac{x_1 x_2}{x_1 + x_2} \sum_q e_q^2 \left[q_1(x_1)\bar{q}_2(x_2) + \bar{q}_1(x_1)q_2(x_2) \right] \qquad (1)$$

where the subscript 1(2) denotes the beam (target) hadron. Here, $q(x)$ denotes the PDF of quark flavor q and x represents the fraction of the hadron's momentum carried by the interacting quark (Bjorken-x); M is the mass of the decaying photon and $x_F = x_1 - x_2$ is Feynman-x. The general features of Eq. 1 are preserved in next-to-leading-order (NLO) calculations.

Fermilab Experiment E866/NuSea has measured the Drell-Yan cross sections in 800 GeV/c proton-proton (pp) and proton-deuterium (pd) collisions. In the fixed target environment of E866/NuSea, the acceptance selects lepton pairs with large longitudinal momenta, and one generally has $x_1 \gg x_2$. Thus the first term in the sum of Eq. 1 dominates the cross section. Under these conditions, the cross section is sensitive to the valence distributions of the beam proton at high-x and the antiquark distributions of the target's sea quarks at low- and intermediate-x. The cross section data described here cover the range $4.2 < M < 8.7$ GeV or $10.85 < M < 16.85$ and $-0.05 < x_F < 0.8$.

E866 used a 3-dipole magnet spectrometer [4, 5] employed previously in experiments E605, E772 and E789, modified by the addition of new detectors at the first tracking station. An 800 GeV/c proton beam bombarded 50.8-cm long target flasks containing liquid hydrogen, liquid deuterium and vacuum that were alternated every few minutes. After passing through the target, the remaining beam was intercepted by a copper beam dump, which was followed by a thick absorber to remove hadrons produced in the target and the dump, ensuring that only muons traversed the spectrometer's active elements. The detector consisted of four tracking stations and a momentum analyzing magnet.

Previous publications [6, 5] have described the E866 determination of the $\bar{d}/\bar{u}$ ratio in the proton, based on the observed ratio of the Drell-Yan cross sections $\sigma^{pd}/2\sigma^{pp}$. The results presented here are based on the same data. The present analysis used 55,000 pp and 121,000 pd Drell-Yan events, approximately half the statistics of the E866 $\bar{d}/\bar{u}$ study [5]. The $\bar{d}/\bar{u}$ analysis was optimized to achieve minimum *relative* uncertainties in $\sigma^{pd}/2\sigma^{pp}$. The present analysis adopted more stringent fiducial cuts which eliminated events where the absolute acceptance of the spectrometer could not be reliably determined–trading statistical precision for smaller systematic uncertainty.

Detailed Monte Carlo (MC) studies were performed to determine the spectrometer's acceptance. Realistic kinematic distributions were used in the MC. To provide a precise

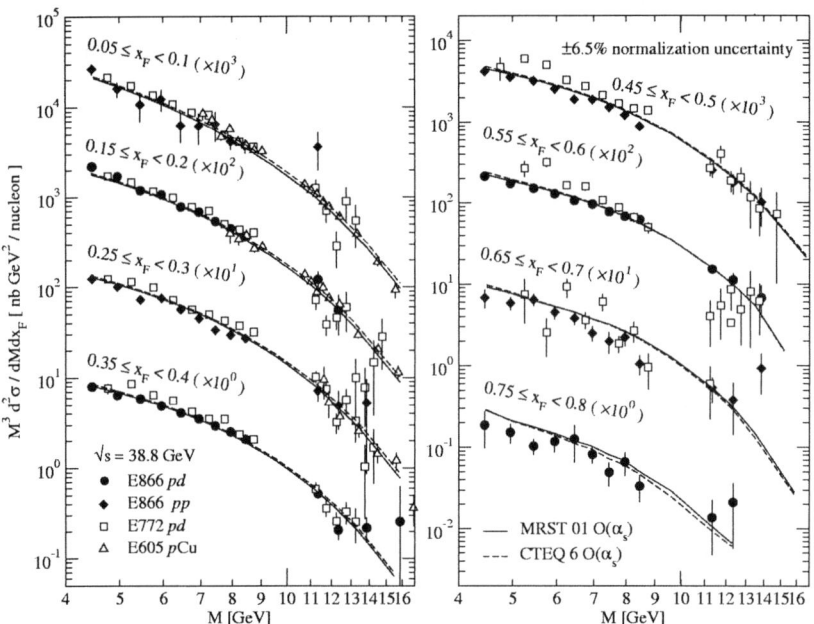

FIGURE 1. FNAL E866 Drell-Yan cross sections per nucleon. The E866 pd (solid circles) and pp (solid diamonds) are shown in alternate decades, compared with previous pCu results from E605 [4] (open triangles) and pd results from E772 [8, 9] (open squares). NLO cross section calculations based on the CTEQ6 [1] (dashed curves) and MRST2001 [2] (solid curves) PDFs are also shown. The E605 and E772 uncertainties are statistical only. The E866 data points show both the statistical and point-to-point systematic uncertainties added in quadrature. An additional $\pm 6.5\%$ normalization uncertainty is common to all E866 data points.

match to the acceptance, the MC events were then reweighted to match the observed distributions. The effect of the reweighting on the acceptance calculation was approximately 1%. The overall and point-to-point systematic uncertainties were also studied carefully. The dominant contributions to the systematic uncertainties were due to limited MC statistics and uncertainties in the absolute field strength of the spectrometer magnets. An additional possible source of uncertainty is due to radiative corrections, which are currently under study. In addition to the point-to-point systematic uncertainties, there is a $\pm 6.5\%$ overall normalization uncertainty, associated with the integration of the proton beam, which is common to all data points [7].

The Drell-Yan pp and pd cross sections per nucleon are shown in Fig. 1 for selected x_F bins. The results agree with previous measurements in pCu collisions by E605 [4] and, for $x_F < 0.3$, in pd collisions by E772 [8, 9]. At larger x_F and small M, the E772 cross sections are systematically larger by more than expected based on the combined systematic uncertainties, with the largest differences in the region $0.4 < x_F < 0.6$ and $4.2 < M < 7$ GeV. Here inconsistencies between the E772 results and expectations from global PDF fits were noted previously [10]. This region was studied during E866 by both the low- and high-mass spectrometer settings. The two measurements, which involve

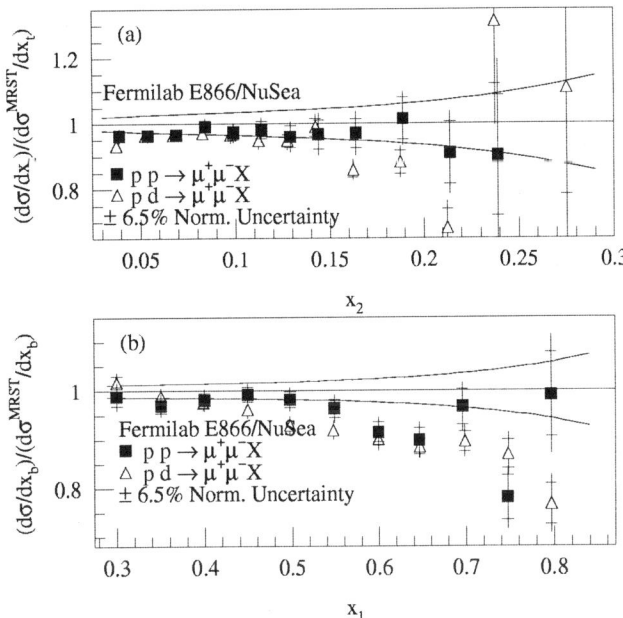

FIGURE 2. Ratios of the measured Drell-Yan pp (solid squares) and pd (open triangles) cross sections to NLO calculations based on the MRST2001 [2] PDF fit plotted vs a) x_2 and b) x_1, while averaging over the other variable. The inner error bars represent the statistical uncertainty in the ratio, the outer error bar the sum in quadrature of the statistical and point-to-point systematic uncertainties. Solid lines represent the uncertainties on a) $\bar{u}+\bar{d}$ and b) $4u+d$ in the MRST2001 PDF fit.

very different acceptances, are consistent.

Figure 1 also shows the results of next-to-leading order calculations of the Drell-Yan cross sections based on the CTEQ6 [1] and MRST2001 [2] global PDF fits. The agreement with the global fits is very good over the entire kinematic region. This agreement may be quantified by computing a K'-factor, which we define to be the ratio of the experimental cross section to a NLO prediction. Table 1 shows K'-factors for several recent global PDF fits. With the exception of GRV98, all of the recent PDF fits predict the absolute magnitude of the Drell-Yan cross sections to within the $\pm 6.5\%$ normalization uncertainty.

TABLE 1. $K' = \sigma^{\text{exp}}/\sigma^{\text{NLO}}$ factors obtained with various PDFs. The χ^2 includes statistical and point-to-point systematic uncertainties, but not the $\pm 6.5\%$ global normalization uncertainty.

PDF	K'_{pp}	χ^2/dof	K'_{pd}	χ^2/dof
CTEQ6 [1]	1.016	1.39	1.001	2.56
MRST2001 [2]	0.980	1.45	0.966	2.44
GRV98 [3]	0.811	2.04	0.808	4.15

While the overall normalization is well reproduced, there are systematic deviations between the measurements and the predictions that are reflected in the large χ^2 values. To elucidate these deviations, it is useful to examine the experimental cross sections separately as functions of x_1 and x_2. As noted, most of the events have $x_1 \gg x_2$, which implies that the x_1 dependence is primarily sensitive to the valence quarks in the proton beam while the x_2 dependence measures the antiquarks in the target. Figure 2 shows the ratios of the measured cross sections to NLO calculations using the MRST2001 PDFs, separately as functions of x_2 and x_1. The uncertainties in the NLO calculations from the PDF fit are also shown. The MRST2001 partons provide a good description of the x_2 dependence of both the pp and pd cross sections. The CTEQ6 fits describe the x_2 dependence equally well.

As a function of x_1, MRST2001 and CTEQ6 have similar behavior: both overestimate the valence quarks by 15-20% at large x_1. The MRST2001 and CTEQ6 partons show essentially the same x_1 dependence, and the discrepancy between the data and current PDFs appears to be larger for the pd cross sections than the pp cross sections. Some of this discrepancy may be accounted for by radiative corrections, which are presently being studied. The pp and pd cross sections at large x_1 constrain two slightly different linear combinations of u_V and d_V. The results imply that the u quark distributions in CTEQ6 and MRST2001 are overestimated as $x \to 1$. The Drell-Yan cross sections may also point to problems with the d/u ratio as $x \to 1$ [11]. However, determination of the d/u ratio at large x from these data will require a global fit to the full two-dimensional invariant cross sections together with the rest of the current world's data.

It is important to recognize that this discrepancy is nearly within the quoted uncertainties on the valence quark distributions in the current PDFs [1, 12] and the $\pm 6.5\%$ normalization uncertainty on the cross sections. Therefore, future global PDF fits that include these new results should provide a much better description of the E866 Drell-Yan cross sections, together with improved determinations of the antiquark distributions for $0.03 \lesssim x < 0.15$ and of the valence quark distributions for $x \to 1$, without significant degradation to the quality of the fit for the rest of the world's current data.

This work was supported in part by the U.S. Department of Energy.

REFERENCES

1. Pumplin, J., et al., *JHEP*, **07**, 012 (2002).
2. Martin, A. D., Roberts, R. G., Stirling, W. J., and Thorne, R. S., *Eur. Phys. J.*, **C23**, 73–87 (2002).
3. Gluck, M., Reya, E., and Vogt, A., *Eur. Phys. J.*, **C5**, 461–470 (1998).
4. Moreno, G., et al., *Phys. Rev.*, **D43**, 2815–2836 (1991).
5. Towell, R. S., et al., *Phys. Rev.*, **D64**, 052002 (2001).
6. Hawker, E. A., et al., *Phys. Rev. Lett.*, **80**, 3715–3718 (1998).
7. Webb, J. C., *Measurement of continuum dimuon production in 800-GeV/c proton nucleon collisions*, Ph.D. thesis, New Mexico State University (2003), hep-ex/0301031.
8. McGaughey, P. L., et al., *Phys. Rev.*, **D50**, 3038–3045 (1994).
9. McGaughey, P. L., et al., *Phys. Rev.*, **D60**, 119903 (1999).
10. Martin, A. D., Roberts, R. G., Stirling, W. J., and Thorne, R. S., *Eur. Phys. J.*, **C4**, 463–496 (1998).
11. Kuhlmann, S., et al., *Phys. Lett.*, **B476**, 291–296 (2000).
12. Martin, A. D., Roberts, R. G., Stirling, W. J., and Thorne, R. S., *Eur. Phys. J.*, **C28**, 455–473 (2003).

Hard QCD at the Tevatron

Christina Mesropian

The Rockefeller University
for CDF and DØ Collaborations

Abstract. Results from QCD studies at the Tevatron from new Run 2 data are presented. The inclusive jet cross section and dijet mass spectrum are measured at $\sqrt{s} = 1960$ GeV by the CDF and DØ collaborations. CDF also reports results of searches for new particles decaying into dijets, and a study of jet shapes.

INTRODUCTION

The measurement of the inclusive jet cross section represents one of the basic tests of QCD at hadron-hadron colliders. The Tevatron Run I results, in particular the discrepancy between the CDF inclusive jet cross section and NLO QCD calculations, generated substantial interest in the particle physics community and resulted in the revision of existing parton distribution functions (PDFs) [1]. New PDF sets now exploit the flexibility of gluon distributions at high x values, which could account for the excess observed in the data at high E_T. Other important outcomes of jet measurements at the Tevatron, were the measurement of the strong coupling constant from the inclusive jet cross section [2] and inclusion of both CDF and DØ jet results in the global fit by PDF collaborations [3].

The increase in center-of-mass energy and increased luminosity in Run 2 resulted in a dramatically larger kinematic range for measuring jet production. With a data sample similar to that obtained in Run I, both CDF and DØ collaborations are able to measure jet production at much higher E_T than those possible in the previous run. All preliminary results from CDF and DØ are based on data samples collected during the limited time period from February 2002 to January 2003 at Fermilab Tevatron Collider at $\sqrt{s} = 1.96$ TeV.

INCLUSIVE JET CROSS SECTION

The inclusive jet cross section measurement from CDF is based on a data sample of integrated luminosity 85 pb^{-1}. To obtain results in a prompt fashion, the CDF collaboration utilized the same techniques as in Run I inclusive jet analyses [4], when possible. Briefly, jets are reconstructed using the iterative fixed cone algorithm with cone radius $R = 0.7$. The inclusive jet cross section includes all jets in an event in the pseudorapidity range $0.1 < |\eta| < 0.7$. The following data quality cuts are used: events with large missing E_T are excluded to avoid background from cosmic rays, and the event vertex is required to be within 60 cm of the center of the detector, to ensure a good

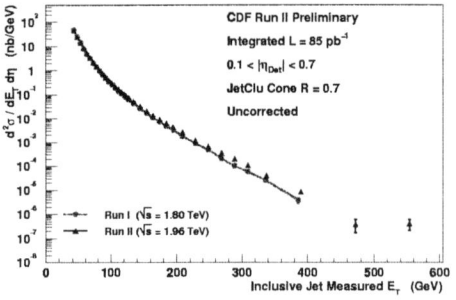 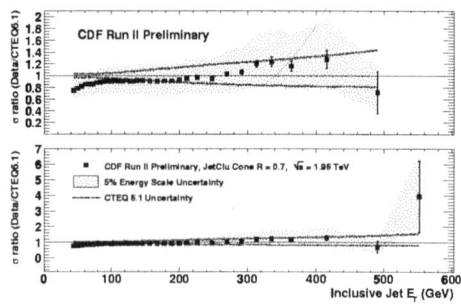

FIGURE 1. (left) The measured inclusive jet E_T distribution for the Run I and Run 2 data sets; (right) the ratio of the measured inclusive jet cross section from Run 2 data to the NLO QCD calculation with CTEQ6.1 PDF.

jet energy measurement. The measured spectrum is corrected for calorimeter response, resolution and underlying event energy using an iterative unsmearing procedure. The absolute energy scale of jets in the central region is calibrated to the known energy of jets in Run I by requiring the p_T balance of central photons to central jets to be the same in Run 2 and Run I. Fig. 1(left) shows CDF Run 2 and Run I jet spectra. The new results are in good agreement with the previous measurement. As one can see, the E_T reach due to the increased $\sqrt{s}$ is rather dramatic, spanning from 40 to 568 GeV. Fig. 1(right) shows the corrected Run 2 cross section compared to a QCD prediction with CTEQ6.1 PDF as an input to the calculations. The shaded area represents 5% energy scale uncertainty, which

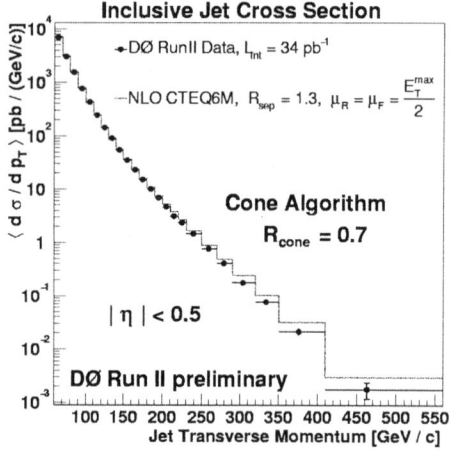 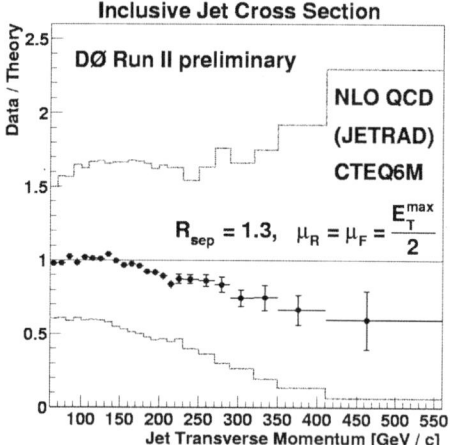

FIGURE 2. (left) DØ inclusive jet cross section as a function of p_T; (right) the inclusive jet cross section shown as data/theory as a function of p_T for CTEQ6M PDF.

is the dominant source of experimental systematic error. The data are in good agreement with the NLO QCD predictions, within the theoretical and experimental uncertainties.

The DØ collaboration measured the inclusive jet cross section with 34 pb^{-1} of data collected during Run 2. Jets are reconstructed with the Run 2 iterative cone algorithm [5] with a cone radius $R = 0.7$. The analysis is restricted to the central pseudorapidity region of $|\eta| < 0.5$. The data quality cuts for jet events are similar to those of CDF. The calorimeter energy is corrected to particle level using information from γ+jet events, low bias triggers and Monte Carlo simulations. Figure 2(left) shows the jet spectrum falling almost by seven orders of magnitude and covering the E_T range from 60 to 560 GeV. In order to see the level of agreement with the theoretical calculations, Fig. 2(right) presents the data to theory ratio. There is agreement within rather large uncertainties. The overall uncertainty is dominated by the jet energy scale.

DIJET MASS CROSS SECTION

Another important probe of QCD is the measurement of the dijet cross section. It provides a handle on parton structure functions at large values of x and also can be used for searches by identifying resonances at high mass.

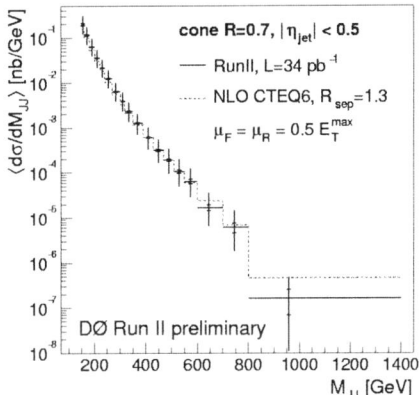

FIGURE 3. DØ dijet cross section as a function of dijet mass. Overlayed on the data are the predictions of a NLO pQCD calculation.

The DØ collaboration measured the dijet mass cross section as a function of the dijet invariant mass. The size of the data sample, requirements for the events and the algorithm used for jet reconstruction are the same as for the inclusive jet cross section measurement. The measured cross section is presented in Fig. 3. The inner bars represent experimental statistical errors and the outer bars are systematic uncertainties; the 10% luminosity uncertainty is not shown. There is an agreement within the errors with the theoretical predictions using CTEQ6M PDF.

The CDF collaboration used 75 pb^{-1} of jet data to search for new particles decaying to dijets by using a general search for narrow resonances and a direct search for several kinds of new particles: axigluons ($A \rightarrow q\bar{q}$), excited states of composite quarks ($q^* \to qg$), and E_6 diquarks ($D(D^c) \to \bar{q}\bar{q}(qq)$). Fitting the mass spectrum with a simple background parametrized function and a mass resonance allows to obtain a 95% confidence level upper limit on the cross section for new particles as a function of mass. A dijet event is defined as an event with the two largest E_T jets, restricted to the pseudorapidity region of $|\eta|<2.0$. In addition, dijets are required to satisfy the condition $|\cos\theta^*|<2/3$[1] to supress QCD background. Figures 4(top and bottom) show Run 2 dijet mass spectra compared to Run I data and the ratio of both CDF results compared with lowest order parton level calculations, respectably. CDF excludes at 95% confidence limit axigluons for $200<M_A<1130$ GeV/c^2, excited quarks for $200<M^*<760$ GeV/c^2, and E_6 diquarks for $280<M_{E6}<420$ GeV/c^2.

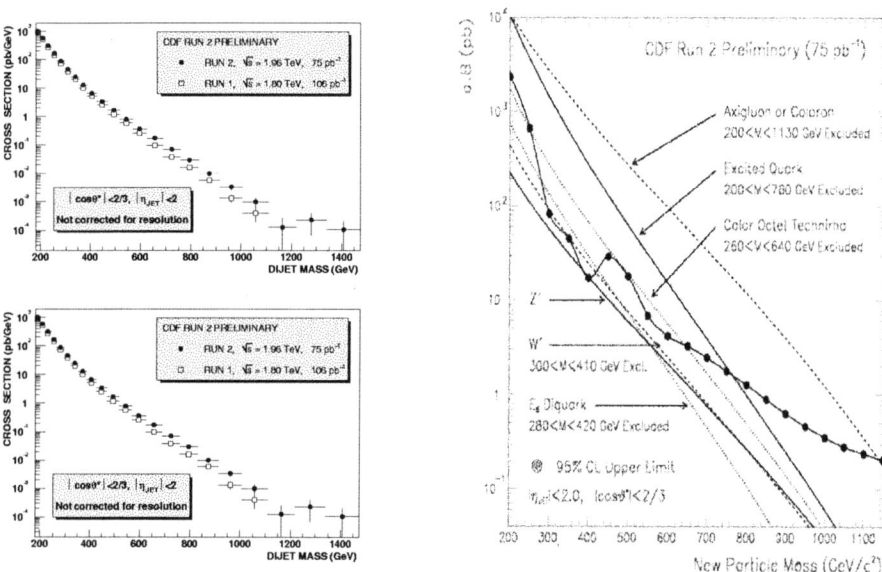

FIGURE 4. (top) CDF dijet mass distribution in Run 2 and Run I presented as a differential cross section in the same mass bins; (bottom) the dijet mass distribution in Run 2 divided by that in Run I is compared to a lowest order parton level calculation(curve); (right) the 95% confodence level upeer linit on the cross section times branching ratio for new particles decaying to dijets.

JET SHAPE STUDIES

The CDF collaboration reports an analysis of jet shapes in inclusive dijet events measured using calorimeter towers. To study the internal structure of jets, integrated jet shape, $\Psi(r,R)$, is defined as the average fraction of jet E_T that lies inside a subcone

---

[1] $\cos\theta^* = \tanh(\eta^*) = \tanh((\eta_1 - \eta_2)/2)$.

with radius $r < R$. In Fig. 5(left), the jet shape Ψ is shown and compared with the results of a PYTHIA Monte Carlo simulation. The error bars represent the statistical and experimental systematic uncertainties added in quadrature. The MC simulation provides good description of measured jet shapes, but produces jets slightly narrower than the data for low E_T and forward η regions. The low E_T discrepancy can be partially attributed to the fact that the MC simulation underestimates the underlying event component. The measured data distributions also show that jets become narrower as the E_T of the jet increases. This can be observed in Fig. 5(right), where Ψ is shown for fixed $r = 0.4$ in different regions of E_T^{jet} and η^{jet}.

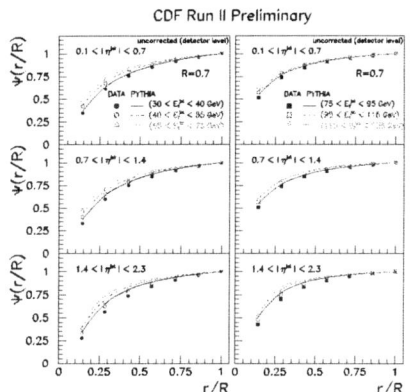
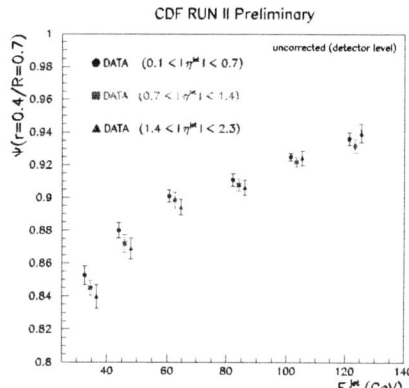

FIGURE 5. (left) CDF measurement of the integrated jet shape in different regions of E_T^{jet} and η^{jet}; (right) measured uncorrected integrated jet shapes, $\Psi(r = 0.4)$. The outer error bars indicate statistical and systematic uncertainties added in quadratures.

CONCLUSIONS AND ACKNOWLEDGMENTS

We presented first Run 2 jet results from the Tevatron. Both CDF and DØ collaborations have accumulated larger data samples and are currently working on reduction of the jet energy scale uncertainties and application of different jet reconstruction algorithms.

We would like to acknowledge the work of all CDF and DØ collaborators that made these results possible.

REFERENCES

1. J. Huston et al., *Phys. Rev. Lett.* **77**, 444 (1996).
2. T. Affolder et al., *Phys. Rev. Lett.* **88**, 042001 (2002).
3. J. Pumplin et al., *JHEP* **0207**, 012 (2002).
4. T. Affolder et al., *Phys. Rev.* D **64**, 032001 (2001).
5. G. Blazey, et al., Proc. of Physics at Run 2: QCD and Weak Boson Physics Workshop, Batavia, IL, FERMILAB-CONF-00-092-E.

Aspects of Diffraction at the Tevatron

Review and Phenomenological Interpretation of CDF Results on Diffraction

Konstantin Goulianos

The Rockefeller University
1230 York Avenue, New York, NY 10021
(Experimental results are presented on behalf of the CDF Collaboration)

Abstract. Results on soft and hard diffraction obtained by the CDF Collaboration at the Fermilab Tevatron $\bar{p}p$ Collider are reviewed with emphasis on aspects of the data that point to the underlying QCD mechanism for diffraction. The results are interpreted in terms of a phenomenological approach in which diffraction is due to an exchange of low-x partons subject to color constraints.

INTRODUCTION

Diffractive interactions between hadrons are characterized by the presence of one or more large rapidity gaps in an event. Processes which (*do not*) incorporate a hard partonic scattering in addition to the rapidity gap signature of diffraction are referred to as (*soft*) hard diffractive. A rapidity gap is a region of pseudorapidity[1] devoid of particles. Rapidity gaps may be formed in non-diffractive (ND) interactions by multiplicity fluctuations. However, from Poisson statistics, the probability for a ND gap of width $\Delta\eta$ is expected to be $P(\Delta\eta) = \exp[-\rho\Delta\eta]$, where ρ is the average particle density per unit η. Thus, ND gaps are exponentially suppressed with increasing $\Delta\eta$. In contrast, diffractive gaps do not exhibit such a suppression. This aspect of diffraction could be explained if the exchange across the gap were a color singlet quark/gluon object with vacuum quantum numbers. For historical reasons, this object is referred to as Pomeron [1]. In this paper, we briefly review the results on soft and hard diffraction reported by the Collider Detector at Fermilab (CDF), present new results from Run II, and "interrogate" the data to learn about the partonic structure and factorization properties of Pomeron exchange. The information obtained is compared with expectations form the "renormalized gap probability" phenomenological model (RENORM), in which the Pomeron is formed from the underlying partonic structure of the interacting hadrons subject to the color-matching requirements appropriate for "vacuum exchange" [2, 3].

The paper is organized in two sections: soft diffraction and hard diffraction. For pedagogical reasons, experimental results and RENORM model expectations are presented concurrently and conclusions are interspersed within the main body of the presentation.

---

[1] We use *rapidity* and *pseudorapidity* interchangeably, since in the kinematic region of interest in this paper the pseudorapidity of a particle, defined as $\eta = -\ln\tan\frac{\theta}{2}$, where θ is the polar angle, is numerically very close to its rapidity, $y = \frac{1}{2}\frac{E+p_L}{E-p_L}$, where p_L is the longitudinal momentum of the particle.

SOFT DIFFRACTION

The following soft $\bar{p}p$ processes have been studied by CDF:
- **ND**   Non-Diffractive $\bar{p}p \to X$
- **SD**   Single Diffraction [4] $\bar{p}p \to \bar{p} + \text{gap} + X$
- **DD**   Double Diffraction [5] $\bar{p}p \to X + \text{gap} + Y$
- **DPE**   Double Pomeron Exchange [6] $\bar{p}p \to \bar{p} + \text{gap} + X + \text{gap} + p$
- **SDD**   Single $\oplus$ Double Diffraction [7] $\bar{p}p \to \bar{p} + \text{gap} + X + \text{gap} + Y$

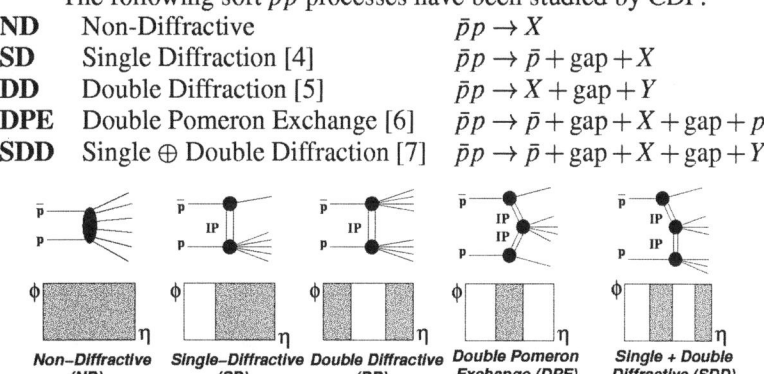

FIGURE 1. Diagrams and η-ϕ topologies of soft processes studied by CDF; the shaded areas are regions where particle production occurs and are referred to in this paper as diffractive clusters.

Diffraction has been traditionally treated phenomenologically in the framework of Regge theory. The connection of the theory to QCD is best seen by expressing cross sections in terms of rapidity gap and "diffractive cluster" variables, with the latter defined as regions of pseudorapidity where particle production occurs. The SDD process, for example, has two rapidity gaps and two diffractive clusters, which we designate, from left to right in Fig. 1, as $\Delta\eta_1$, $\Delta\eta_1'$, $\Delta\eta_2$ and $\Delta\eta_2'$. The gap $\Delta\eta_1$ can be thought of as being formed by the elastic scattering between the $\bar{p}$ and the cluster $\Delta\eta_1'$, and the gap $\Delta\eta_2$ by the elastic scattering between the two diffractive clusters. Each gap is associated with a four-momentum transfer squared, t. There are 5 independent variables in SDD: the two rapidity gaps with their associated t-values and the center of the "floating" gap (non-adjacent to the $\bar{p}$), η_c. The Regge theory SDD differential cross section is given by

$$\frac{d^5\sigma}{dt_1 dt_2 d\Delta\eta_1 d\Delta\eta_2 d\eta_c} = P_{gap}(t_1, t_2, \Delta\eta_1, \Delta\eta_2, \eta_c) \times \kappa^2 \times \sigma_{tot}(s') \quad (1)$$

$$P_{gap}(t_1, t_2, \Delta\eta_1, \Delta\eta_2, \eta_c) = C \times F_{\bar{p}}^2(t_1) \times \left[e^{(\varepsilon + \alpha' t_1)\Delta\eta_1}\right]^2 \times \left[e^{(\varepsilon + \alpha' t_2)\Delta\eta_2}\right]^2 \quad (2)$$

$$\sigma_{tot}(s') = \beta(0)^2 (s')^\varepsilon = \beta(0)^2 e^{\varepsilon \ln s'} = \beta(0)^2 e^{\varepsilon(\Delta\eta_1' + \Delta\eta_2')} \quad (3)$$

where $\beta(0)$ is the $I\!Pp$ coupling at $t = 0$, ε and α' the parameters of the Pomeron trajectory, $\alpha(t) = 1 + \varepsilon + \alpha' t$, $\kappa = g^{I\!PI\!PI\!P}/\beta^{I\!Pp}$ the ratio of the triple-Pomeron to the Pomeron-proton couplings, s' the diffractive cluster sub-energy defined by $\ln s' = \Delta\eta' = \Delta\eta_1' + \Delta\eta_2'$, and C a constant [3]. The parameter κ has been measured to be $\kappa = 0.17 \pm 0.02$ [8].

The QCD connection. There are three factors in Eq. (1): P_{gap}, κ^2 and σ_{tot}. Recalling that the total $\bar{p}p$ cross section is $\beta(0)^2 e^{\varepsilon \ln s}$, the last factor is identified as the $\bar{p}p$ cross section at the diffractive sub-energy squared, s'. From the optical theorem, the term $e^{\varepsilon \Delta\eta'}$ is proportional to the forward elastic scattering amplitude at s'. The fact that the two diffractive clusters are not contiguous does not present a conceptual problem in the parton model, in which the amplitude is $\sim e^{\varepsilon \Delta \eta_i'}$ for each cluster [9] and thus the regions $\Delta \eta_1'$ and $\Delta \eta_2'$ add in the exponent. The full t-dependent parton model amplitude is:

$$f_{\bar{p}p}(t, \Delta \eta) \propto e^{(\varepsilon + \alpha' t) \Delta \eta} \quad \text{Parton Model Amplitude} \quad (4)$$

Thus, in Eq. (2), the terms in the square brackets are identified as the amplitudes for elastic scattering between the diffractive clusters on either side of each gap, while $F_{\bar{p}}(t_1)$ is the $\bar{p}$ form factor. Finally, the parameter κ is identified as the color factor required to produce a color singlet exchange; two such factors are needed in SDD, one for each gap.

Similar equations can be written for SD, DD and DPE [3]. In all cases the cross section factorizes into $P_{gap}(\Delta \eta)$ and $\sigma_{tot}(\Delta \eta')$ terms. The predicted shapes of the differential cross sections for all four processes agree with the CDF data [5, 6, 7, 8]. However, as seen in Fig. 2 (a,b), the s-dependence of the SD and DD cross sections is approximately flat at high energies, contrary to the Regge theory expectation of $\sim s^{2\varepsilon}$. The culprit for this problem was identified [2] as the normalization of the $P_{gap}(\Delta \eta)$ term, which is obtained from the elastic and total cross sections using factorization *independently* from the normalization of the $\sigma_{tot}(\Delta \eta')$ term. Interpreting $P_{gap}(\Delta \eta)$ as a gap probability distribution and renormalizing it to unity by dividing it by its integral over all phase space [2, 3] yields excellent agreement with the all data (see Fig. 2).

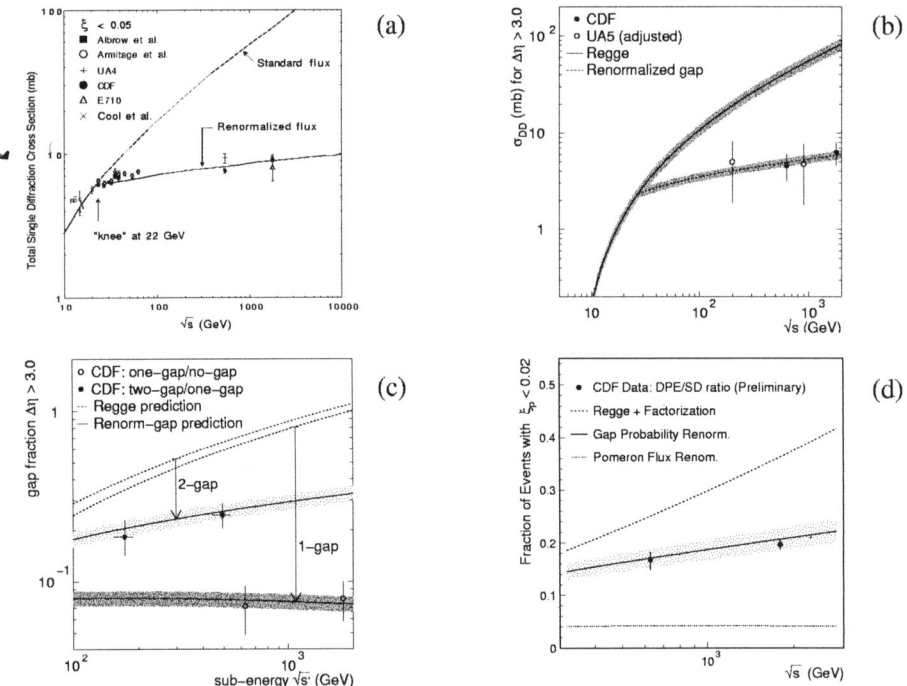

FIGURE 2. Soft diffraction cross sections compared with Regge theory and RENORM model predictions: (a) SD, (b) DD, (c) ratios of SDD to SD and DD to TOTAL, (d) ratio of DPE to SD.

HARD DIFFRACTION

Hard diffraction processes studied by CDF include SD (dijet, W, b-quark and J/ψ), DD (dijet) and DPE (dijet) production, corresponding to the topologies shown in Fig. 3.

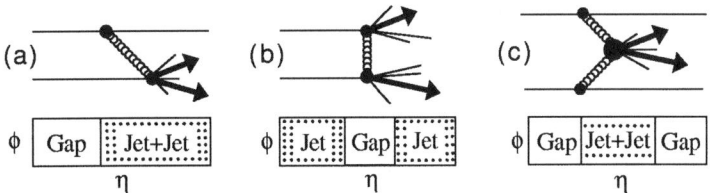

FIGURE 3. Topologies in η-ϕ space for hard (a) SD, (b) DD and (d) DPE processes.

Two types of results have been obtained: diffractive to non-diffractive cross section ratios (using the rapidity gap signature to select diffractive events), and diffractive to non-diffractive structure function ratios (using a Roman Pot Spectrometer to trigger on leading antiprotons). For a recent review of CDF Run I hard diffraction result see Ref. [10]. Here we summarize the aspects of the Run I results that point to the QCD structure of the Pomeron and present new results from Run II.

Run I rapidity gap results. (a) At $\sqrt{s} = 1800$ GeV, the SD/ND ratios for dijet, W, b-quark and J/ψ production, as well the ratio of DD/ND dijet production, are all $\approx 1\%$. This "gap fraction" is suppressed relative to QCD inspired theoretical expectations (*e.g.* 2-gluon exchange) by a factor of ~ 10, which is comparable to the suppression factor observed in soft diffraction relative to Regge theory expectations based on factorization. (b) The gluon fraction of the diffractive exchange was determined from dijet, W and b-quark production to be $0.54 \pm 0.15\%$, which is similar to the ND fraction.
The above results indicate that (i) the diffractive structure function is similar to the ND one, apart from an overall suppression in normalization, and (ii) at fixed $\bar{p}p$ collision energy QCD factorization approximately holds within the diffractive sector.

Run I Roman Pot Results. (a) The diffractive structure function determined from SD dijet production is suppressed by a factor of ~ 10 relative to expectations based on extrapolations from parton densities determined from diffractive DIS at HERA. This suppression is approximately the same as that observed in soft diffraction. (b) The ratio of SD to ND structure functions behaves approximately as $x_{Bj}^{-0.5}$. For a prediction of such behavior by the RENORM model see Ref. [11]. (c) The double-ratio of (DPE/SD)/(SD/ND) structure functions was found to be 5.3 ± 2.0, which is equal within errors to the ratio of (two-gap/one-gap)/(one-gap/no-gap) in soft diffraction (Fig. 3c).

Conclusions from Run I results. In both soft and hard diffraction processes cross sections factorize into two terms, one containing the cross section at the sub-energy of the diffractive cluster and another representing the gap probability distribution, which must be normalized to unity. A color factor is required for each gap. Diffraction appears as the interaction between low-x partons subject to color-matching constraints imposed by the rapidity gap requirement, as prescribed by the RENORM model.

Run II results. In Run II, diffractive data have been collected by CDF at $\sqrt{s} = 1.96$ TeV and results obtained on the $Q^2 \equiv (E_T^{jet})^2$ dependence of the diffractive structure function and on exclusive dijet production in hard DPE. Results are shown in Fig. 4.

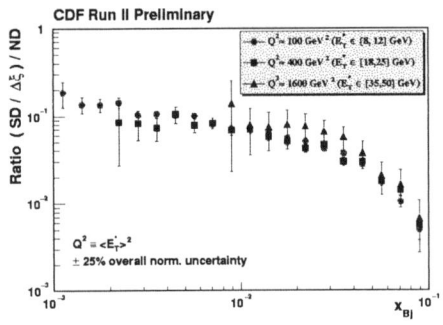
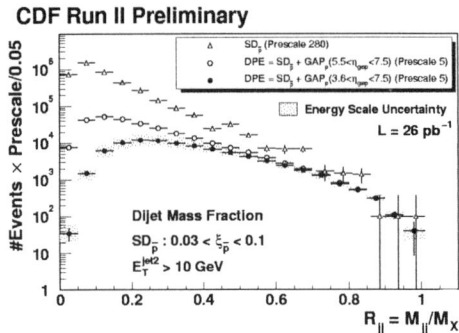

FIGURE 4. (*left*) Ratio of SD/$\Delta \xi_{\bar{p}}$ over ND rates obtained from dijet data at various Q^2 ranges; (*right*) ratio of dijet mass to total mass "visible" in the calorimeters for dijet production in events with a leading antiproton within $0.3 < \xi_{\bar{p}} < 0.1$ and various gap requirements on the proton side: (*triangles*) no gap requirement, (*open circles*) gap in $5.5 < \eta < 7.5$, and (*filled circles*) gap in $3.5 < \eta < 7.5$.

The ratio of SD/ND rates, which in LO QCD is equal to the ratio of the corresponding structure functions at a given x_{Bj}, shows no appreciable Q^2 dependence. This result supports the RENORM model, in which the diffractive structure function is basically *extracted* from the non-diffractive one.

Exclusive dijet production in DPE, which has been proposed as a process on which to *calibrate* models of diffractive Higgs production [12], would appear in Fig. 4 as a peak in the vicinity of $R_{jj} = 1$. No such peak is observed in the data. For dijets of minimum E_T^{jet} of 10 GeV [25 GeV], the cross section for $R_{jj} > 0.8$ is measured to be $970 \pm 65(\text{stat}) \pm 272(\text{syst})$ [$34 \pm 5 \pm 10$] pb. Although similar values are obtained in Ref. [12] for exclusive dijets, we emphasize that no exclusive signal is seen in the data.

REFERENCES

1. V. Barone and E. Predazzi, High-Energy Particle Diffraction, Springer Press (2001).
2. K. Goulianos, Phys. Lett. B **358**, 379 (1995); Erratum-*ib*. **363**, 268 (1995).
3. K. Goulianos, "Diffraction in QCD," Presented at Corfu Summer Institute on Elementary Particle Physics (Corfu 2001), Corfu, Greece, 31 Aug - 20 Sep 2001; e-print Archive: hep-ph/0203141.
4. F. Abe et al. (CDF Collaboration), Phys. Rev D **50**, 5535 (1994).
5. T. Affolder et al. (CDF Collaboration), Phys. Rev Lett. **87**, 141802 (2001).
6. D. Acosta *et al.* (CDF Collaboration), "Inclusive Double-Pomeron Exchange at the Fermilab Tevatron $\bar{p}p$ Collider," to be submitted to Phys. Rev. Letters.
7. D. Acosta et al. (CDF Collaboration), Phys. Rev Lett. **91**, 011802 (2003).
8. K. Goulianos and J. Montanha, Phys. Rev. D **59**, 114017 (1999).
9. E. Levin, "An Introduction to Pomerons," Preprint DESY 98-120.
10. K. Goulianos, "Diffraction at the Tevatron in Perspective," Presented at Workshop on Diffraction 2002, Alushta, Ukraine, 31 Aug - 6 Sep 2002; e-Print Archive: hep-ph/0306085.
11. K. Goulianos, J.Phys.G**26**, 716 (2000).
12. V.A. Khoze, A.D. Martin and M.G. Ryskin, Eur. Phys. J. **C23**, 211 (2001); *ib*. **C26**, 229 (2002).

Leading Baryon Production at HERA

Ioannis Gialas

University of Aegean, Chios, Greece

Abstract.
We have studied the production of fast neutrons and protons in proton positron collisions at the HERA collider. Results from experiments ZEUS and H1 are presented. Special detectors were used by the two experiments to detect particles emerging very close to the beam. The results support the assumption of vertex factorization. Measurements of the F_2 structure function was made as well as the pion structure function F_2^π. Both of these structure functions behave as the electron proton F_2 structure function.

INTRODUCTION

In the Regge picture, a proton in the final state is produced via the exchange of neutral states. In case of exchange of a Pomeron, the final state proton has a momentum very close to that of the incoming proton ($x_L \approx 1$), where x_L is the ratio of the baryon momentum to the beam proton momentum. Other neutral particle exchanges produce protons with lower momenta. The exchange of charged particles, notably pions, gives rise to final state neutrons. Within the QCD picture leading baryons are produced in the target (proton) fragmentation process. One question to be answered is if the assumption of Projectile-Target factorization is valid. In such case the positron proton cross section $\sigma_{ep \to e'pX}$ can be written as

$$\sigma_{ep \to e'pX} = f(x_L, t) \sigma_{e_i \to e'X}(\beta, Q^2)$$

where i is any exchanged meson, t is the square of the momentum transfer from the initial state proton to the final state proton, and $\beta \equiv \frac{x_{BJ}}{1-x_L}$.

Under limiting fragmentation, leading baryon production in the proton fragmentation region will be independent of the photon four momentum squared Q^2 and the photon proton system mass W. If factorization of vertices is valid, the dependence of the cross section on the lepton variables (x_{BJ} and Q^2) should be independent of the baryon variables (x_L and p_T^2), where x_{BJ} is the Bjorken x.

Special detectors were used. The ZEUS experiment used the Leading Proton Spectrometer, a system of retractable silicon detectors, to detect fast protons and the Forward Neutron Calorimeter to detect neutrons.

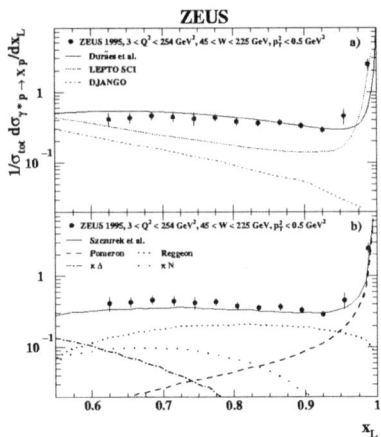

FIGURE 1. Leading Proton x_L spectrum. The lines represent predictions by various models

RESULTS

The spectrum of leading protons is plotted in figure 1. The plot shows there is not dependence of the cross section on x_L, for $x_L \leq 0.97$. Above $x_L \simeq 0.97$ one sees the diffractive peak. An analogous plot (without the diffractive peak) one can derive for leading neutrons. For $x_L \leq 0.97$, the fraction of events with a leading proton is consistent for pp and $\gamma*p$ data sets, in accord with vertex factorization. Standard fragmentation models do not describe the data. Models based on virtual particle exchange can describe the production rate and energy spectrum.

The leading baryon transverse momentum distributions can be described well by a single exponential of the form $\exp(-bp_T^2)$. The slopes parameter b is plotted in figure 2 versus x_L. The BPC and DIS data together indicate that b is independent of Q^2 and x_L. The present results are compatible with the $pp \to pX$ data also shown in figure 2. This, together with the fact that b is approximately Q^2 independent, provides additional support for vertex factorization.

Several results have been produced in terms of the ratio of cross sections for production of leading baryons to the cross section for inclusive e^+p scattering, $r^{LB(2)}(Q^2)$. In these ratios, the systematic uncertainties coming from the positron selection procedure cancel. In figure 3 the $r^{LB(2)}(Q^2)$ data are presented for $0.6 < x_L < 0.97$ and $p_T^2 < 0.5 GeV^2$, averaged over x for different Q^2 ranges. The leading baryon yield increases by approximately 20%, from about $r \simeq 0.12$ to $r \simeq 0.15$, when Q^2 varies from 0.25 GeV^2 to 100 GeV^2, indicating a modest but definite breakdown of vertex factorization. The effect is of similar size for leading protons and neutrons. The neutron data are measured for scattering angles less than 0.8 mrad, corresponding to $p_T^2 < 0.43 x_L^2 GeV^2$. The ZEUS and H1 data agree very well.

When the ratio $r^{LP(3)}(x, Q^2, x_L)$ is plotted in several x and Q^2 bins only a weak dependence on these variables is apparent, indicating that F_2^{LP} has approximately the

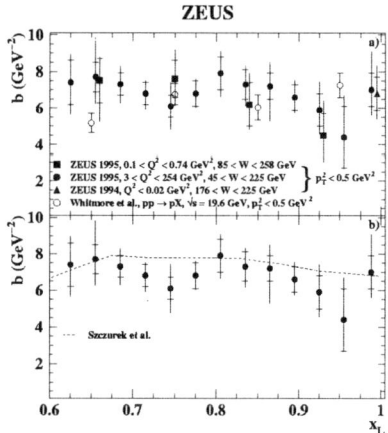

FIGURE 2. The transverse momentum slope parameter b as a function of x_L for events with a leading proton.

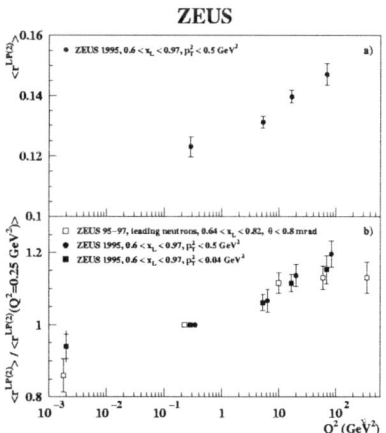

FIGURE 3. The leading baryon production ratio as a function of Q^2.

same x and Q^2 dependence as $F_2(x,Q^2)$. It is straightforward to calculate the $F_2^{LP}(x,Q^2)$ from the measured ratios using the known inclusive electron proton F_2. The leading neutron structure function $F_2^{LN}(x,Q^2)$ has been plotted in figure 4 as a function of x in different Q^2 bins.

Having established that the leading neutron data are dominated by pion exchange in the range $0.64 < x_L < 0.82$, the OPE model can be used to determine the structure function of the pion, F_2^π. In the DIS region, F_2^π is approximately proportional to $F_2^{LN}(x,Q^2,x_L)$ for a fixed x_L. The evaluation of $F_2^\pi(x_\pi,Q^2)$ has been made for that x_L range where OPE is expected to dominate and where the fraction of events with a leading neutron is observed to be approximately independent of x and Q^2. Although the

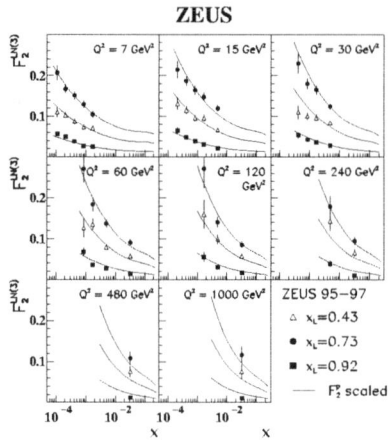

FIGURE 4. The F_2 structure function of events with a leading neutron as a function of x for different Q^2 bins

normalization of F_2^π is uncertain to a factor of two, the x and Q^2 dependence is well measured. It is striking that F_2^π has approximately the same x and Q^2 dependence as the F_2 of the proton.

Summary

Measurements of leading neutron and proton production were performed in a wide Q^2 range (0.1-250 GeV^2) in a variety of processes. The data can be described by a combination of virtual particle exchanges. One pion exchange describes the neutron data well. For the leading proton production more exchanges are needed. The ratio F_2^{LB}/F_2^{ep} which minimizes the systematic uncertainties is largely independent of x_L. F_2^{LB}/F_2^{ep} has little dependence on Q^2 and x_{BJ} indicating that F_2^{LP} and F_2^{LN} exhibit the same behavior as F_2^{ep}. However, a modest violation of vertex factorization is observed. Finally, a model dependent F_2^π was extracted and it was found to be proportional to the F_2^{ep}.

ACKNOWLEDGMENTS

The author would like to express his thanks to the University of Hamburg Institute of Experimental Physics.

Nuclear Transparency in Exclusive ρ^0 Production at HERMES

W. Lorenzon[†]

Randall Laboratory of Physics, University of Michigan, Ann Arbor, Michigan 48109-1120, USA
(on behalf of the HERMES collaboration)

Abstract. Exclusive coherent and incoherent electroproduction of the ρ^0 meson from ^{1}H and ^{14}N targets has been studied at the HERMES experiment as a function of coherence length (l_c), corresponding to the lifetime of hadronic fluctuations of the virtual photon, and squared four-momentum of the virtual photon ($-Q^2$). The ratio of ^{14}N to ^{1}H cross sections per nucleon, *called* nuclear transparency, was found to increase (decrease) with increasing coherence length for coherent (incoherent) ρ^0 electroproduction. For fixed coherence length, a rise of nuclear transparency with Q^2 is observed for both coherent and incoherent ρ^0 production, which is in agreement with theoretical calculations of color transparency.

INTRODUCTION

Exclusive electroproduction of ρ^0 mesons from nuclei is considered to be an excellent tool to investigate the properties of elementary particles interacting with the nuclear medium, such as the phenomena of a "shrinking photon" [1, 2, 3] and Color Transparency (CT) [4, 5, 6]. The latter phenomenon is a prediction of perturbative QCD. It suggests that, due to their reduced transverse size, particles produced with high virtuality in exclusive reactions should exhibit a reduced interaction with other hadrons. In particular, the "size" of the hadronic components of the virtual photon at high negative four-momentum transfer squared, Q^2, is conjectured to be smaller than the size of a normal hadron. This would account for the pointlike behavior and the diminished absorption of virtual photons in nuclear interactions, as compared to real photons. In QCD, the reaction amplitudes for exclusive interactions at large momentum transfer are expected to be dominated by components of the photon wave function with small transverse size, which give rise to diminished final state interactions in the nuclear medium. Theoretical models typically describe the exclusive production of light vector mesons as occurring via the fluctuation of the virtual photon into a quark-antiquark pair (or off-shell vector meson), which is scattered onto the mass shell by a diffractive interaction with the target. The corresponding tree level diagram is shown in Fig. 1.

Several experiments in search of CT have been carried out in the past. Although none of these experiments is in conflict with CT, no unambiguous signature for the onset of CT has been found yet. The pioneering searches of CT found an oscillation in the nuclear transparency in quasielastic proton scattering [7], and a nuclear transparency that is compatible with the Glauber model in quasielastic electron scattering [8, 9]. The interpretation of these results in terms of CT is still debatable. First evidence for CT

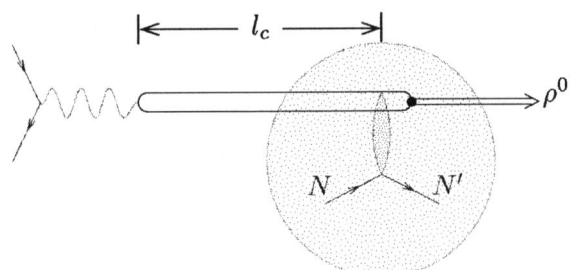

FIGURE 1. Cartoon describing exclusive ρ^0 meson production and illustrating the importance of coherence length, which is the propagation distance of the short-lived quark-antiquark state.

came from quasifree charge exchange scattering data of 40 GeV/c negative pions on carbon [10] as suggested in Ref. [11]. Further evidence for CT comes from Fermilab experiment E791 on the A-dependence of coherent diffractive dissociation of 500 GeV/c pions into di-jets [12]. This result shows a platinum to carbon cross section about ten times larger than expected if soft processes would dominate, which is qualitatively consistent with theoretical calculations of CT effects [13, 14]. Also experiment E665 on exclusive incoherent ρ^0 muoproduction from nuclei [15] gives an indication of CT. However, that signal is of indecisive statistical significance.

EXPERIMENT

The nuclear transparency data were obtained during the 1996-1997 running periods of the HERMES experiment using ^{1}H and ^{14}N gas targets in the 27.5 GeV HERA positron storage ring at DESY. The HERMES detector is described in detail in Ref. [16]. The scattered positron and the h^+h^- hadron pair arising from the decay $\rho^0 \to \pi^+\pi^-$ were detected and identified in the HERMES forward spectrometer. The ρ^0 production sample was extracted from events with exactly these three detected tracks. A more detailed and comprehensive description of the extraction procedure for exclusive diffractive ρ^0 can be found in Refs. [17, 18]. Here, we only describe the method for extracting the ratio of coherent to incoherent cross sections based on nuclear data. The coherent component of the cross section, where the scattering occurs from the nucleus as a whole, dominates at low $|t'|$ and is absent for the hydrogen target, while the incoherent part, where the ρ^0 meson scatters from a single quasifree nucleon within the target, dominates starting at $|t'| \approx 0.09$ GeV^2. Here, $-t' = -(t - t_{min})$ with t being the four-momentum transfer between the vector meson and target nucleon and $|t_{min}|$ the minimum $|t|$ allowed by the kinematics.

The cross section is therefore approximated with the sum of coherent and incoherent contributions as

$$\frac{d\sigma}{dt} = b_N e^{b_N t'} + R_A b_A e^{b_A t'}, \tag{1}$$

where b_N and b_A are the slope parameters for the nucleon and nucleus, respectively. This yields the first observable, the coherent to incoherent full cross section ratio $R_A = \frac{\sigma_c}{\sigma_{inc}}$. The Monte Carlo generator DIPSI [19] was used to calculate the detector acceptance. There, different diffractive slope parameters and relativistic (non relativistic) Breit-Wigner mass distributions were used as an input parameters. Finally, corrections ($\approx 15\%$) were applied to the ratios due to the "Pauli blocking" effect [20, 21] for incoherent scattering.

The coherent (incoherent) nuclear transparency is defined as

$$T_{c(inc)} = \frac{\sigma^A_{c(inc)}(Q^2)}{A\sigma^p(Q^2)}, \qquad (2)$$

where σ^p refers to scattering from the proton, and A is the atomic number of the nuclear target. For the nuclear transparency measurement, the DIS positron cross section was used as a luminosity measure in addition to the standard luminosity measurement based on Bhabha scattering from atomic electrons. The ratio of the integrated luminosities represents the largest source of kinematics-independent uncertainties. The total estimated systematic uncertainty from all normalization factors is 11%. For incoherent ρ^0 production, nuclear transparency is associated with the probability that the produced ρ^0 meson escapes the nucleus without interaction. For coherent production, measured for the first time at HERMES, such a probabilistic interpretation is not applicable, though T_c is still sensitive to coherence length and color transparency effects.

RESULTS

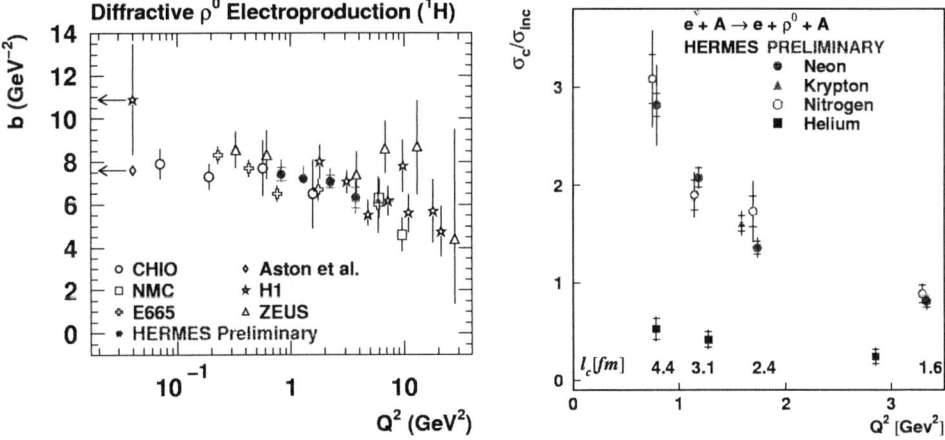

FIGURE 2. *Left panel:* The measured Q^2 dependence of the diffractive slope parameter in exclusive ρ^0 production from a hydrogen target. The compilation of all results including HERMES data is taken from Ref. [22]. *Right panel:* Q^2 dependence of coherent to incoherent cross section ratio (as described by Eq. (1).

In the left panel of Fig. 2 data on the diffractive slope parameter b versus Q^2 for diffractive ρ^0 production from a hydrogen target are presented. This parameter characterizes the rate of exponential decay of the cross section with t. Physically, b is a measure of the transverse size of the interaction region. Fig. 2 (left panel) demonstrates that the virtual photon "shrinks" with increasing virtuality Q^2. When ρ^0 mesons are produced from a nuclear target rather than hydrogen, this shrinkage is one possible source of reduced final state interactions. Another arises from the coherence length $l_c = \frac{2\nu}{Q^2 + M_{q\bar{q}}^2}$, which describes the propagation distance of the short-lived quark-antiquark state (see Fig. 1). In the right panel of Fig. 2, the coherent-to-incoherent cross section ratio versus Q^2 is presented. A strong Q^2 dependence is observed, which is likely due to the variation of the coherence length l_c and the nuclear form factor's Q^2 dependence.

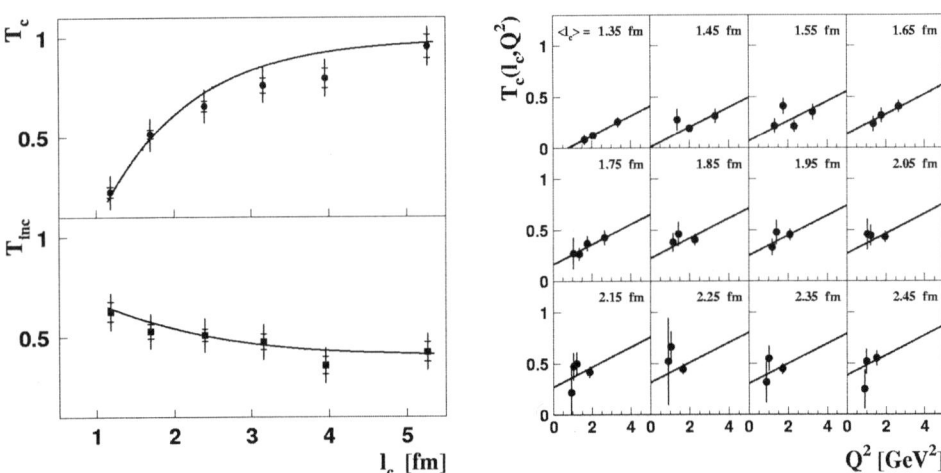

FIGURE 3. *Left panel:* Coherence length dependence of nuclear transparency, for coherent (T_c, top) and incoherent (T_{inc}, bottom) ρ^0 production on nitrogen. The curves are from Ref. [23] and include CT effects. The inner error bars include only statistical uncertainties, while the outer error bars present the statistical and systematic uncertainties added in quadrature. *Right panel:* Q^2 dependence at fixed l_c on nitrogen. The straight line is the result of the common fit.

In Fig. 3 (left panel) results on coherent and incoherent transparency versus l_c [17] are shown. As one can see, both the coherent and incoherent transparencies show a distinct l_c dependence. The incoherent data decrease with increasing l_c, as expected from the effects of initial state interactions. The nuclear transparency for coherent ρ^0 production increases with coherence length as expected from the effects of the nuclear form factor. Good agreement is found between the measured nuclear transparencies, integrated over the available Q^2 region, and calculations including both the coherence length and CT effects [23].

To separate the color transparency effect, which is purely Q^2 dependent, from coherence length effects a two-dimensional analysis was performed: the slope of the data with Q^2 was extracted at fixed l_c. This two-dimensional analysis represents a new approach in the search of CT. Since the combination of statistical significance and Q^2 coverage is largest near $l_c = 2.0$ fm, the region $1.3 < l_c < 2.5$ fm has been chosen for this two-dimensional analysis. To deconvolute the CT and coherence length effects, coherence length bins of 0.1 fm were used. In order to extract the Q^2 dependence, each l_c bin was split into three or four Q^2 bins. The nuclear transparency was extracted in each (l_c, Q^2) sub-bin, and is shown in the right panel of Fig. 3 for coherent ρ^0 production. The data have been fitted with a common Q^2 dependent slope of transparency ratio, presented as lines in Fig. 3 (right panel), resulting in reduced chi-square values close to unity. The Q^2 slopes were found to be 0.070 ± 0.021(stat.) ± 0.017(syst.) for coherent and 0.089 ± 0.046(stat.) ± 0.020(syst.) for incoherent production [18]. According to Ref. [23], a positive slope of the transparency with Q^2, for fixed coherence length, is evidence for CT. Indeed, the results presented here support the CT prediction. If the results are combined, the measured Q^2 slope $(0.074 \pm 0.023(\text{tot.}))$ GeV^{-2} is found to differ from zero by more than three standard deviations.

ACKNOWLEDGMENTS

I wish to thank my colleagues in the HERMES collaboration. I acknowledge Avetik Airapetian and Harold Jackson for critical reading of the manuscript. The author's research is supported in part by the U.S. National Science Foundation, Intermediate Energy Nuclear Science Division under grant No. PHY-0072297 and PHY-0244842.

REFERENCES

1. Bauer, T.H. et al., *Rev. Mod. Phys.*, **50**, 261 (1978).
2. Cheng, H., and Wu, T. T., *Phys. Rev.*, **183**, 1324 (1969).
3. Bjorken, J. D., and Kogut, J. B., *Phys. Rev. D*, **5**, 1152 (1972).
4. Bertsch, G. et al., *Phys. Rev. Lett.*, **47**, 297 (1981).
5. Brodsky, S. J., and Mueller, A. H., *Phys. Lett. B*, **206**, 685 (1988).
6. Kopeliovich, B. Z., and Hüfner, J., *Phys. Lett. B*, **309**, 179 (1993).
7. Caroll, A.S. et al., *Phys. Rev. Lett.*, **61**, 1698 (1988).
8. Makins, N.C.R. et al., *Phys. Rev. Lett.*, **72**, 1986 (1994).
9. O'Neill, T.G. et al., *Phys. Lett. B*, **351**, 87 (1995).
10. Apokin, V.D. et al., *Sov. J. Nucl. Phys.*, **46**, 877 (1987).
11. Kopeliovich, B. Z., and Zakharov, B. G., *Phys. Lett. B*, **264**, 434 (1991).
12. E791 Collaboration, Aitala, E.M. et al., *Phys. Rev. Lett.*, **86**, 4773 (2001).
13. Frankfurt, L.L. et al., *Phys. Lett. B*, **304**, 1 (1993).
14. Frankfurt, L.L. et al., *Found. Phys.*, **30**, 533 (2000).
15. E665 Collaboration, Adams, M.R. et al., *Phys. Rev. Lett.*, **74**, 1525 (1995).
16. HERMES Collaboration, Ackerstaff, K. et al., *Nucl. Instr. Meth. A*, **417**, 230 (1998).
17. HERMES Collaboration, Ackerstaff, K. et al., *Phys. Rev. Lett.*, **82**, 3025 (1999).
18. HERMES Collaboration, Airapetian, A. et al., *Phys. Rev. Lett.*, **90**, 052501 (2003).
19. Arneodo, M. et al., Tech. Rep. DESY96-149 (1996).
20. Renk, T., Piller, G., and Weise, W., *Nucl. Phys. A*, **689**, 869 (2001).
21. Trefil, J. S., *Nucl. Phys. B*, **11**, 330 (1969).
22. Tytgat, M., Ph.D. thesis, Gent University (2001), DESY-THESIS-2001-018.
23. Kopeliovich, B.Z. et al., *Phys. Rev. C*, **65**, 035201 (2002).

DVCS results from HERMES

Jochen Volmer[†]

for the HERMES collaboration
DESY Zeuthen, Platanenallee 6, 15738 Zeuthen

Abstract. The HERMES experiment studies the spin structure of the nucleon using the 27.6 GeV longitudinally polarized positron beam of HERA and internal targets of pure gases. Recently, HERMES has published measurements of azimuthal single-spin asymmetries in the hard electroproduction of photons on an unpolarized hydrogen target arising from the interference of the amplitudes of the Bethe-Heitler (BH) and the Deeply Virtual Compton Scattering (DVCS) processes. The measurements constitute an effort towards accessing generalized parton distributions, and open a new approach to study the spin structure of the nucleon.

INTRODUCTION

Lepton–nucleon scattering experiments have long been an important tool in the study of nucleon structure. Elastic lepton–nucleon scattering has been used to extract nucleon form factors, whereas from inclusive and semi–inclusive deeply inelastic scattering (DIS) parton distribution functions (PDFs) were extracted. The PDFs describe the momentum distribution of partons in the nucleon along the direction of the momentum exchange.

PDFs and form factors alone cannot render a complete picture of the nucleon. For example, polarized DIS has determined that a large fraction of the nucleon spin does not lie in the sum of the simple helicities of the quarks; however, there is no PDF defined to describe the mixture of transverse and longitudinal motion inherent in orbital angular momentum (see e.g. Ref. [1] and references therein).

A more comprehensive, non–perturbative description of the nucleon has emerged in the framework of generalized parton distributions (GPDs), which may provide access to the orbital momentum contribution to the nucleon spin via the Ji relation [2]. PDFs and nucleon form factors appear as limiting cases of the GPDs. Moreover, a recent interpretation has linked the GPDs to the three dimensional picture of the nucleon (see e.g. Ref. [3]).

GENERALIZED PARTON DISTRIBUTIONS

The DVCS process, the electroproduction of a real photon on a proton in the Bjorken limit, is thought to be described by the leading handbag diagram shown in the left panel of Fig. 1. It describes the factorization of the process into a hard photon–quark scattering part calculable in perturbative QCD, and a soft part describing the nucleon structure, which can be expressed in terms of GPDs. PDFs depend on the longitudinal momentum

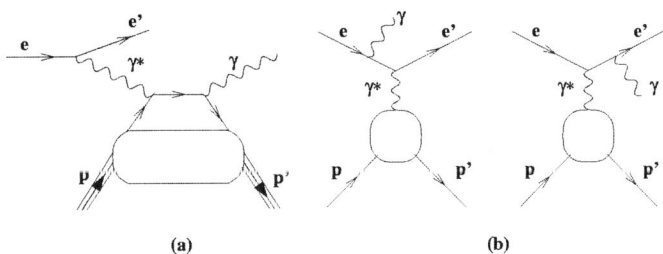

FIGURE 1. DVCS (a) and BH (b) diagrams.

fraction x carried by the struck quark in the proton. GPDs additionally depend on the Mandelstam variable $t = (p-p')^2$, and a skewedness parameter ξ.

There are four leading twist quark–chirality conserving GPDs for each quark flavour q, namely the unpolarized GPDs H^q and E^q, and the polarized ones $\tilde{H}^q$ and $\tilde{E}^q$. Of these, H^q and $\tilde{H}^q$ occur in amplitudes where nucleon helicity is conserved, while E^q and $\tilde{E}^q$ occur in amplitudes where the nucleon helicity is flipped. In the forward limit ($t \to 0$, implying also $\xi \to 0$), the GPD $H^q(x,0,0)$ becomes the quark number density distribution $q(x)$, whereas $\tilde{H}^q(x,0,0)$ becomes $\Delta q(x)$, the quark helicity distribution. The nucleon helicity-odd GPDs E and $\tilde{E}$ vanish in the forward limit. Furthermore, the GPDs are related to the elastic form factors of the nucleon via their first moments, whereas the second moment of the sum of the unpolarized GPDs H^q and E^q in the limit $t \to 0$ is related to the total quark angular momentum J^q by the Ji relation [2]:

$$J^q = \lim_{t \to 0} \int_{-1}^{1} dx\, x [H^q(x,\xi,t) + E^q(x,\xi,t)].$$

An analogous relation exists for the gluon case. In principle, this set of quantities is sufficient to describe all contributions to the nucleon spin.

DVCS AND BETHE-HEITLER AMPLITUDES

The final state of the DVCS process, as shown in the left panel of Fig. 1 is indistinguishable from the Bethe–Heitler process (right panel), in which the real photon is radiated off the lepton instead of the quark. The square of the total process amplitude τ is given as

$$|\tau|^2 = |\tau_{BH} + \tau_{DVCS}|^2 = |\tau_{BH}|^2 + |\tau_{DVCS}|^2 + \underbrace{\tau_{DVCS}\tau^*_{BH} + \tau^*_{DVCS}\tau_{BH}}_{I} \quad (1)$$

At HERMES kinematics with average values of $\langle Q^2 \rangle = 2.5$ GeV^2, $\langle x_B \rangle = 0.1$ and $\langle -t \rangle = 0.25$ GeV^2, the BH process largely dominates over the DVCS process, making it difficult to extract the DVCS cross section. However, the interference terms I in Eq. (1) allow the DVCS process amplitude to be addressed indirectly by measuring various cross section asymmetries in the azimuthal angle ϕ between the lepton scattering plane and the photon production plane [4]. As has been shown in Ref. [3], each of the three terms of Eq. (1) can be written in a series of Fourier moments in ϕ, the amplitude of which depends on linear combinations of the GPDs.

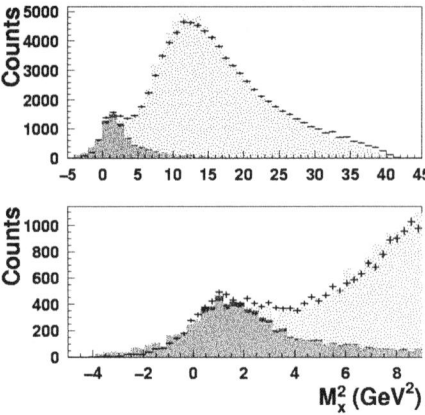

FIGURE 2. Crosses: M_x distribution of experimental data; histograms: Monte Carlo: exclusive events (grey), resonance contribution (dark grey), semi-inclusive background (light grey).

The data collected at HERMES have been taken with longitudinally polarized electron and positron beams at 27.6 GeV and with longitudinally polarized deuterium and unpolarized proton, neon and krypton targets. The data have been used to measure the single beam spin asymmetry A_{LU}, the single target–spin asymmetry A_{UL}, and the beam charge asymmetry A_C:

$$A_{LU}(\phi) = \frac{d\sigma^\uparrow(\phi) - d\sigma^\downarrow(\phi)}{d\sigma^\uparrow(\phi) + d\sigma^\downarrow(\phi)} \propto \Im(H)\sin(\phi), \quad (2)$$

$$A_{UL}(\phi) = \frac{d\sigma^\Uparrow(\phi) - d\sigma^\Downarrow(\phi)}{d\sigma^\Uparrow(\phi) + d\sigma^\Downarrow(\phi)} \propto \Im(\tilde{H})\sin(\phi), \quad (3)$$

$$A_C(\phi) = \frac{d\sigma^+(\phi) - d\sigma^-(\phi)}{d\sigma^+(\phi) + d\sigma^-(\phi)} \propto \Re(H)\cos(\phi). \quad (4)$$

The subscripts denote the polarization states of the beam and the target, respectively, with possible values L and U for longitudinally polarized and unpolarized. The superscripts stand for the polarization direction of the beam (↑, ↓), the target (⇑, ⇓), and for the sign of the lepton charge (+,-), respectively. These azimuthal asymmetries provide access to different combinations of the real and imaginary parts of the interfering amplitude in Eq. (1), and of the GPDs embedded therein. In all three asymmetries, higher moments are present, but suppressed by powers of Q^2.

RESULTS

The results of the HERMES DVCS analysis presented here are based on data taken between 1998 and 2000 with 27.6 GeV polarized electron and positron beams incident on gaseous nucleon and nuclear targets. In the analysis of single-photon events in the HERMES spectrometer [5], events were accepted if exactly one positron and one real photon were identified. Such events were accepted if the virtual photon momentum squared $Q^2 > 1$ GeV^2 and the invariant mass of the virtual photon-nucleon system $W^2 >$

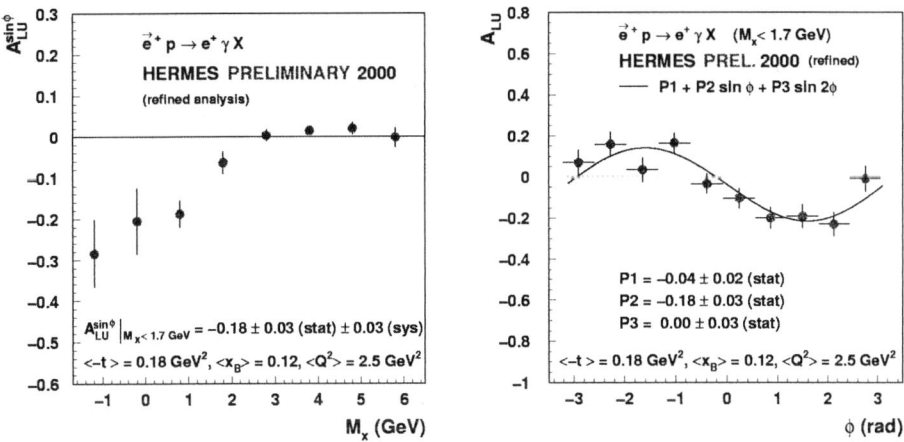

FIGURE 3. Left: $\sin\phi$ moment $A_{LU}^{\sin\phi}$ of the beam spin asymmetry on the proton as function of the missing mass M_X. Right: beam spin asymmetry on the proton in the exclusive region as function of ϕ.

4 GeV². Since the recoiling proton was not detected, the determination of the exclusivity of the event (*i.e.*, no other particles were produced) rested on the missing mass M_X of the process (see Fig. 2). The M_X resolution was limited by the energy resolution of the electromagnetic calorimeter. In order to reduce the amount of background in the exclusive sample, a cut was placed at $M_X < 1.7$ GeV, defining the exclusive region. The remaining contribution of the resonant and semi-inclusive background was then studied using a Monte Carlo simulation of the experiment and the effect was included in the systematic uncertainty. Shown in Fig. 3 is the beam spin asymmetry A_{LU} on a proton target. The left panel shows the $\sin\phi$-weighted moment as a function of M_X, demonstrating that there is a signal only in the exclusive region; the right panel shows the ϕ distribution of the asymmetry (see Eq. 2) in the exclusive region. The beam charge asymmetry on the proton is shown in a similar fashion in Fig. 4. As expected, a $\cos\phi$ signal is observed in the exclusive region. Recently, theoretical interest has been focused on coherent DVCS on nuclei [6],[7],[8]. This process competes against the incoherent one, in which the virtual photon interacts with one nucleon only. Due to the steep t-dependence of the nuclear elastic form factor, the coherent part is enhanced at forward angles. Shown in Fig. 5 are the $\sin\phi$ moments $A_{LU}^{\sin\phi}$ of the beam spin asymmetries on deuterium, neon and krypton targets, as functions of M_X. In all three cases, signals are observed in the exclusive region. The fact that for heavier nuclei the coherent part plays a stronger role is reflected in the trend of a decreasing average $\langle -t \rangle$ value with increasing mass and size of the nucleus.

CONCLUSION AND OUTLOOK

The status of the DVCS analysis on a proton target at HERMES was discussed along with first preliminary results on the beam spin asymmetry from coherent DVCS on nuclei. The measurements will be enhanced by a better way to ensure the exclusivity

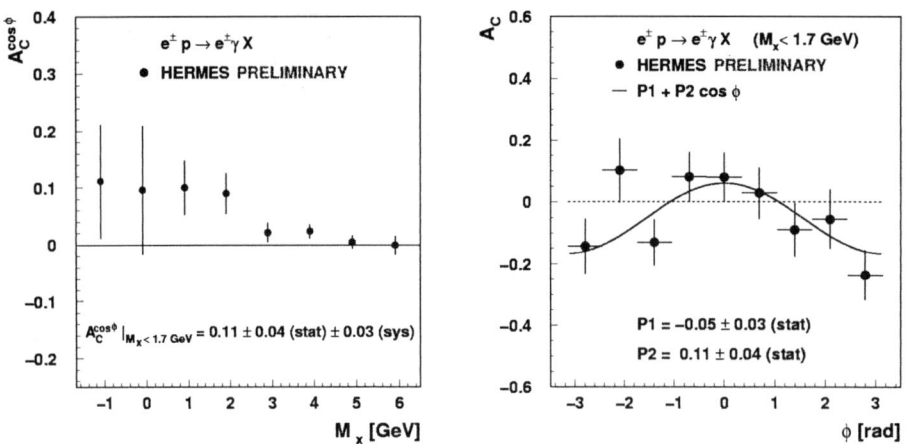

FIGURE 4. Left: $\cos\phi$ moment $A_C^{\cos\phi}$ of the beam charge asymmetry on the proton as function of M_X. Right: beam charge asymmetry A_C on the proton in the exclusive region as function of ϕ.

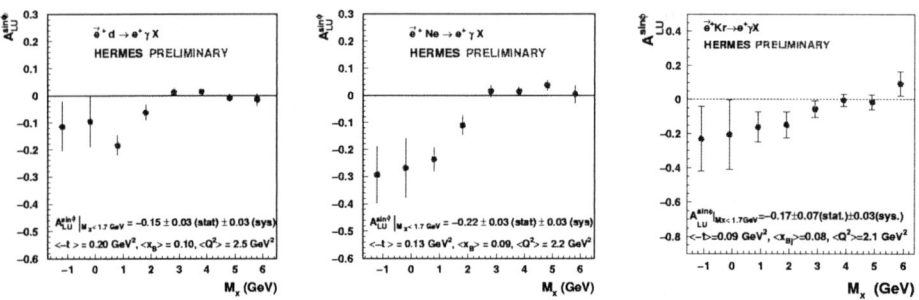

FIGURE 5. $\sin\phi$ moment $A_{LU}^{\sin\phi}$ of the beam spin asymmetry on the proton as function of the missing mass M_X, for deuterium, neon and krypton targets, respectively.

of the process. To this end, a recoil detector will be installed around the HERMES gas target, removing background contributions to the exclusive sample and their effect on the measured asymmetries.

REFERENCES

1. Rith, K., *Prog. Part. Nucl. Phys.*, **49**, 245–324 (2002).
2. Ji, X., *Phys. Rev. Lett.*, **78**, 610–613 (1997).
3. Belitsky, A., and Müller, D., *Nucl. Phys. A*, **711**, 118–126 (2002).
4. Diehl, M., Gousset, T., Pire, B., and Ralston, J. P., *Phys. Lett. B*, **411**, 193–202 (1997).
5. Ackerstaff, K., *Nucl. Instr. Meth. A*, **417**, 230–265 (1998).
6. Polyakov, M., *Phys. Lett. B*, **555**, 57–62 (2003).
7. Guzey, V., and Strikman, M., *hep-ph/0301216* (2003).
8. Kirchner, A., and Müller, D., *hep-ph/0302007* (2003).

DVCS with CLAS

Elton S. Smith
for the CLAS Collaboration

*Thomas Jefferson National Accelerator Facility,
12000 Jefferson Avenue, Newport News, VA 23606, USA*

Abstract. Generalized parton distributions provide a unifying framework for the interpretation of exclusive reactions at high Q^2. The most promising reaction for the investigation of these distributions is the hard production of photons using Deeply Virtual Compton Scattering (DVCS). This reaction can be accessed experimentally by determining the production asymmetry using polarized electrons on a proton target. Pioneering experiments with CLAS and HERMES have produced the first measurements of this asymmetry. We will review the current experimental program to study DVCS at Jefferson Lab. Recent high statistics data taken with CLAS at 5.75 GeV allows us to determine this asymmetry at low -t in the valence region (x_B=0.1-0.5) up to a Q^2 of 4 GeV^2/c^2.

INTRODUCTION

Historically electron scattering experiments have focussed either on the measurements of form factors using exclusive processes or on measurements of inclusive processes to extract deep inelastic structure functions. Elastic processes measure the momentum transfer dependence of the form factors, while the latter ones probe the quark's longitudinal momentum and helicity distributions in the infinite momentum frame. Form factors and deep inelastic structure functions measure two different one-dimensional slices of the proton structure. While it is clear that the two pictures must be connected, a common framework for the interpretation of these data have only recently been developed using Generalized Parton Distribution (GPD) functions [1, 2, 3]. The GPD's are two-parton correlation functions that encode both the transverse spatial dependence and the longitudinal momentum dependence. At the twist-2 level, for each quark species there are two spin-dependent GPD's, $\tilde{E}(x,\xi,t)$ and $\tilde{H}(x,\xi,t)$, and two spin averaged GPD's, $E(x,\xi,t)$ and $H(x,\xi,t)$. The four GPD's are each functions of the longitudinal momentum fraction x, the longitudinal momentum transfer ξ, and the four-momentum transfer t. The first x moment of the GPD's links them to the proton's form factors, while at t=0, the GPD's H and $\tilde{H}$ reduce to the quark longitudinal momentum $q(x)$ and the quark helicity distributions $\Delta q(x)$, respectively. The new physics which can be accessed using exclusive reactions is contained in the dependence on ξ and t. The t dependence is directly related to the distribution of the parton densities as a function of impact parameter [4]. Mapping out the GPD's will allow, for the first time, to obtain a 3-dimensional picture of the nucleon [5].

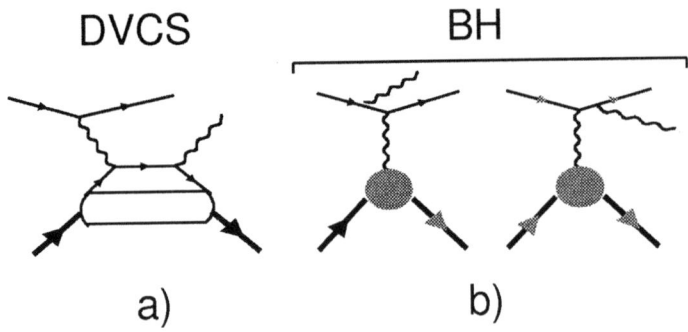

FIGURE 1. a) Deeply virtual Compton scattering (DVCS) and b) Bethe-Heitler processes contributing to $ep \to ep\gamma$ scattering.

DEEPLY VIRTUAL COMPTON SCATTERING

Deeply Virtual Compton Scattering (DVCS) is one of the key reactions to determine the GPD's experimentally, as it is the simplest process that can be described in terms of the GPD's. This reaction is also expected to enter the Bjorken regime at relatively low photon virtuality compared to exclusive meson production. Fig. 1 shows two contributions to the production of $ep \to ep\gamma$ events. At present CEBAF energies, the reaction is dominated by the Bethe-Heitler (BH) amplitudes. However, the interference term between DVCS and BH can be probed using polarized electron beams. The interference term is given by

$$\mathscr{I} = \mathscr{T}^*_{DVCS}\mathscr{T}_{BH} + \mathscr{T}_{DVCS}\mathscr{T}^*_{BH}, \qquad (1)$$

Only the imaginary part of the DVCS amplitude survives in the single beam asymmetry A_{LU} which is accessible experimentally. In this case, the small DVCS amplitude which depends on the GPD's, is amplified by the larger but well-known BH amplitude. The interference term is dominated by the $\sin\phi$ moment, where ϕ is the angle between the $\gamma^*\gamma$ plane and the electron scattering plane. At this level of approximation, the interference term is a linear combination of Compton Form Factors ($\mathscr{H}, \widetilde{\mathscr{H}}, \mathscr{E}$)

$$A_{LU}(\phi) \propto \Im m \left(F_1 \mathscr{H} + \frac{x_B}{2-x_B}(F_1+F_2)\widetilde{\mathscr{H}} - \frac{\Delta^2}{4M^2} F_2 \mathscr{E} \right) \sin\phi \qquad (2)$$

where $F_1(t)$ and $F_2(t)$ represents the elastic Dirac and Pauli form factors, respectively. The Compton Form Factors are related to the corresponding GPD's via convolution integrals [6]. Thus the magnitude of the azimuthal asymmetry probes a linear combination of GPD's.

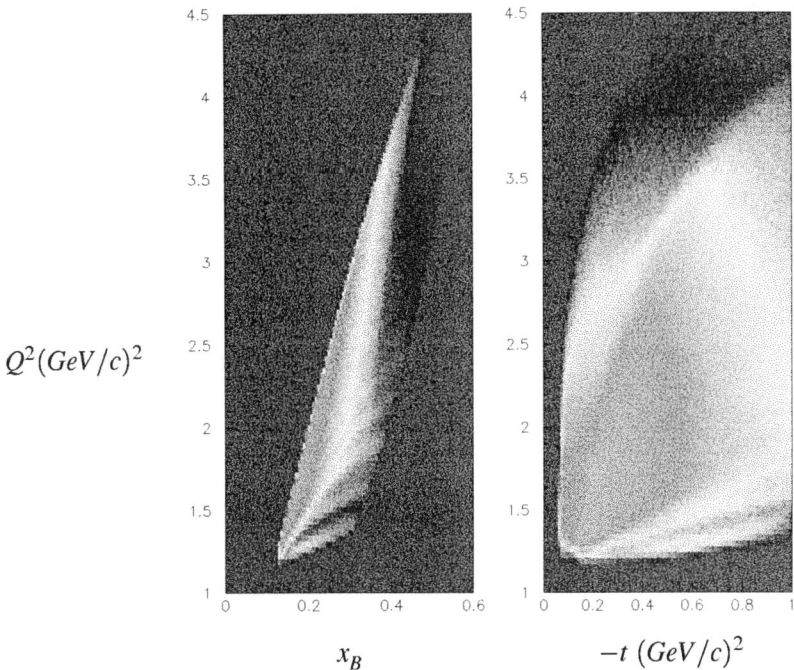

FIGURE 2. Kinematic coverage of data taken at 5.75 GeV showing a) Q^2 vs. x_B and b) Q^2 vs. $-t$.

DATA SAMPLE AND EVENT SELECTION

Preliminary results are now available from an extensive data set taken with the CLAS detector [7], which was taken at a beam energy of 5.75 GeV. The data set corresponds to an integral luminosity of $2.6 \times 10^{40} cm^{-2}$ accumulated on a 5-cm-long unpolarized hydrogen target. The data were taken at a typical luminosity of $0.9 \times 10^{34} cm^{-2} s^{-1}$ and the average electron beam polarization was 70%. The kinematic coverage for W > 2 GeV is shown in Fig. 2 with x_B ranging from 0.1 to 0.5 and Q^2 up to 4 GeV$^2/c^2$ at large x_B. The data also covered a large range in momentum transfer $-t$, but analysis of DVCS was restricted to $-t < 0.5$ GeV$^2/c^2$.

Events are selected with positive identification of the electron in the Cerenkov counter and electromagnetic calorimeter (EC), and the proton by time-of-flight and momentum reconstruction. The ideal experiment would also detect the scattered photon, but the EC has limited coverage down to only 8° and therefore generally misses the scattered photon from the DVCS process. The missing mass resolution of CLAS cannot cleanly separate the $ep\pi^0$ from $ep\gamma$ reactions, and in previous analyses we have extracted the yield by fitting the missing mass distribution to the sum of those two distributions [8]. At higher energies, this procedure becomes more difficult and we use the following selection criteria to obtain a clean sample of $ep\gamma$ events: veto photons from π^0 decay in the EC, select data at low $-t$ using the lab angle of the scattered photon relative to the virtual photon $\theta_{\gamma^*\gamma} \leq 0.12$ rad, and require that the missing mass squared $M_X^2 \leq 0.025$

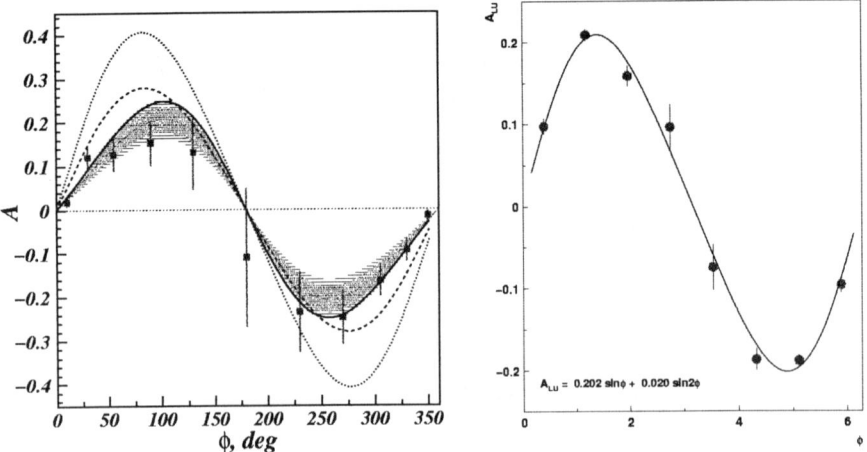

FIGURE 3. The beam spin asymmetry as a function of azimuthal angle for a) the published data at beam energy of 4.25 GeV and b) the new data sample at 5.75 GeV averaged over the entire kinematic range of the data. The increased statistics available at the higher energy will allow mapping out the asymmetry as a function of Q^2, x_B and t.

GeV^2 [9].

The beam spin asymmetry at 5.75 GeV averaged over the entire data sample is shown in Fig. 3 compared to our published data at 4.25 GeV [8]. The beam spin asymmetry is well described by a $\sin\phi$ distribution, which is expected if the handbag diagram dominates. The small $\sin 2\phi$ component is of order 10%, and can be used to place limits on the size of twist-3 contributions. The increased statistical power of the new event sample is reflected in the size of error bars. This will allow us to map out the dependence of the asymmetry as a function of x and $-t$ up to a Q^2 of about 4 (GeV/c)^2.

FUTURE PROSPECTS

The data presented so far is only the first step toward probing GPD's at Jefferson Lab. There are two dedicated DVCS experiments approved to run using 6 GeV polarized electron beams [10]. The first experiment E00-110 is scheduled to be run in Hall A [12] and is designed to determine both beam spin asymmetries and cross section differences in three Q^2 intervals, for fixed values of x_B. The second experiment, to take data with CLAS, will use dedicated apparatus to over-determine the reaction kinematics [11]. This experiment is building a a forward calorimeter to directly detect the scattered photon and a solenoidal magnet to improve the shielding of Møller backgrounds from the target to achieve higher luminosity. The study of GPD's to higher precision and higher Q^2 will continue as one of the main research programs driving the CEBAF upgrade to 12 GeV [13].

ACKNOWLEDGMENTS

I would like to thank L. Elouadrhiri, H. Avagyan, M. Garçon for useful discussions and suggestions. This work was supported in part by the U.S. Department of Energy, including DOE Contract No. DE-AC05-84ER40150.

REFERENCES

1. D. Müller *et al.*, Fortschr. Phys. **42** (1994) 101.
2. X. Ji, Phys. Rev. D **55** (1997) 7114.
3. A.V. Radyushkin, Phys. Lett. B **380** (1996) 417; Phys. Rev. D **56** (1997) 5524.
4. M. Burkardt, Nucl. Phys. A **711** (2002) 127.
5. A.V. Belitsky, talk at this conference.
6. A.V. Belitsky, D. Müller, A. Kirchner, Nucl. Phys. **B**629 (2002) 323.
7. B.A. Mecking *et al.*, Nucl. Instr. and Meth. A **503** (2003) 513.
8. S. Stepanyan *et al.*, Phys. Rev. Lett. **87** (2001) 182002.
9. H. Avagyan, talk at this conference.
10. L. Elouadrhiri, Nucl. Phys. A **711** (2002) 154.
11. V. Burkert, L. Elouadrhiri, M. Garçon, S. Stepanyan *et al.*, Jefferson Lab Experiment E01-113.
12. P. Bertin, C. Hyde-Wright, F. Sabatié *et al.*, Jefferson Lab Experiment E00-110.
13. "The Hall B 12 GeV Upgrade Preconceptual Design Report," December, 2002., and references therein.

Pion-Pion Scattering Lengths from $\pi^+\pi^-$ Atom with the DIRAC experiment

Daniel Goldin

University of Basel
Klingelbergstrasse 82, CH-4056 Basel, Switzerland
FOR THE DIRAC COLLABORATION

Abstract. The DIRAC experiment at CERN is presently measuring the lifetime of the pionium atom, a bound state of a π^+ and π^- meson. The isospin 0 and 2 scattering length difference will be determined unambiguously once the value of the lifetime is known. In what follows we describe the present status of the experiment, the experimental goals and the first results.

MOTIVATION

The goal of the DIRAC experiment is to find the difference of isospin 0 and 2 pion-pion S-wave scattering lengths a_0 and a_2 by measuring the lifetime of the $\pi^+\pi^-$ bound state ($A_{2\pi}$). The principal decay of such a state is into two neutral pions ($BR = 99.6\%$). The lifetime and difference of scattering lengths are related through [1][2]:

$$\Gamma_{\pi^0\pi^0} = 1/\tau_{A_{2\pi}} = \frac{2}{9}\alpha^3 p^* \cdot (a_0 - a_2)^2 (1+\delta_\Gamma), \qquad (1)$$

where α is the fine structure constant, and $p^* = (M_{\pi^+}^2 - M_{\pi^0}^2 - M_{\pi^+}^2\alpha^2/4)^{1/2}$ and $\delta_\Gamma = (5.8 \pm 1.2) \cdot 10^{-2}$. The Chiral Perturbation theory (ChPT) currently predicts $|a_0 - a_2| = 0.265 \pm 0.004\,[m_\pi^{-1}]$ yielding the $A_{2\pi}$ lifetime of $\tau_{A_{2\pi}} = 2.9 \pm 0.1\,fs$ [2].

DIRAC's goal is to determine the pionium lifetime with a precision of $0.3\,fs$. Once the lifetime is known, the determination of the scattering lengths will follow from (1).

Our scattering length measurement is model-independent in a sense that we are detecting pion pairs at close to 0 relative momenta, and, hence, do not need to rely on the precise knowledge of the scattering matrix at higher momenta.

DESCRIPTION OF THE SETUP

DIRAC's experimental setup is located on the 24 GeV PS beamline at CERN. Secondary particles resulting from collisions between proton beam and a thin target (of the order of $100\,\mu m$ in thickness) are registered by 3 coordinate detectors. Charged secondaries are subsequently separated by the 1.65 Tesla magnet into the positive and negative arm, where they traverse another set of coordinate detectors, time-of-flight and identification counters.(See Ref. [3] for more details.)

The spectrometer has an excellent time and momentum resolution allowing us to measure low relative momenta (chosen to be recorded on tape by the multilevel trigger system described in detail Ref. [4]) at high laboratory momenta.

SIGNAL AND BACKGROUND

Coherent $\pi^+\pi^-$ pairs originating from proton-target collisions may sometimes form a $\pi^+\pi^-$ bound state. After evolving in the target material pionium atoms either annihilate or dissociate. Dissociated atoms constitute the signal for the DIRAC experiment.

The background events are formed by two types of pion pairs: Coulomb-correlated and non-correlated ones. Coulomb correlated pairs are similar in nature to the pionium atoms, with the exception that the interactions between the former are characterized by a continuous, rather than a discrete, energy spectrum, as is the case for the latter. Non-correlated pairs are formed by the *incoherent* pion pairs.

The breakup probability is defined as a ratio of the number of pionium atoms n_A that have dissociated in the target over the number of the initially produced pionium atoms $N_{A2\pi}$: $P_{br} = n_A/N_{A2\pi}$. When combined with the relationship between the number of atoms and the Coulomb pairs $N_{A2\pi} = 0.615 \cdot N_c(Q < 2MeV/c)$ [5][6], the expression for the breakup probability becomes

$$P_{br} = n_A/(0.615 \cdot N_c(Q < 2MeV/c)). \qquad (2)$$

Hence, if the number of Coulomb pairs with relative momenta below $2MeV/c$ and number dissociated atomic pairs is known, the breakup probability is determined unambiguously. The $A_{2\pi}$ lifetime is then found from the breakup probability vs. lifetime relationship calculated using Glauber and Born approximations [7][8][9].

SINGLE LAYER TARGET SIGNAL EXTRACTION

We analyzed the results of the 2001 run with the single layer $98\,\mu m$ Ni target. Before proceeding with the signal extraction we applied the following set of cuts: (a) prompt events selected by imposing a narrow time window of -0.5 to $0.5\,ns$, (b) fast protons, electrons and muons coming from the upstream region were rejected, (c) only the pion pairs with relative momenta $Q_{trans} = (Q_x^2 + Q_y^2)^{1/2} < 4MeV/c$ and $|Q_l| < 22MeV/c$ were taken.

The background was constructed using Monte Carlo methods. Both Q_l and Q_{tot} background spectra were fitted to the *signal-free regions of the measured data*, i.e. those with $Q_l > 2MeV$ and $Q_{tot} > 4MeV$. Multiplication of the Coulomb-correlated and non-correlated data by the fit parameters in the *entire relative momentum range* gives us the overall background in the -0.5 to $0.5\,ns$ time interval. Subtracting the background from the experimental data yields the signal (shown in Fig. 1).

6800 ± 400 atomic pairs were found in 2001. We note that the solid-shaded histograms on the bottom two plots corresponding to the simulated and subsequently reconstructed atomic pairs replicate well the extracted signal. Additionally, the flatness and the 0 mean value of the signal-free parts of the distributions indicate the accuracy of the simulated background.

Due to the sizable effects of multiple scattering [10] and, to a lesser extent, the effects of the detector response, the proportionality constant 0.615 in Eq. 2 has a value different from the model-based 0.615. A dedicated multiple scattering measurement to quantify this effect is under way.

Once the breakup probability is determined with sufficient accuracy, the pionium lifetime may be obtained from the functional dependence as described at the end of the previous section.

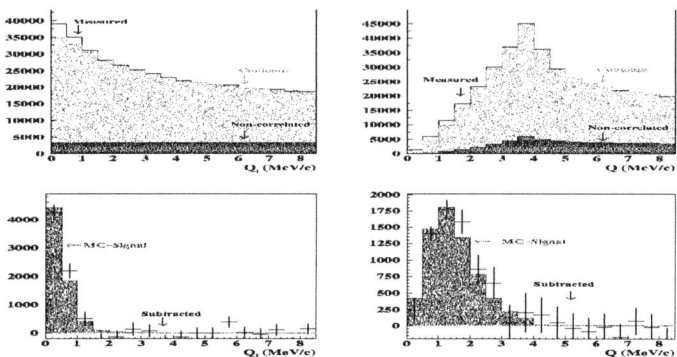

FIGURE 1. Top row: relative momenta distributions of the experimental data and background. Bottom row: signal resulting from subtracting the background from the experimental data.

COMBINED METHOD OF SINGLE AND MULTILAYER TARGET MEASUREMENTS

We have an additional way to measure the pionium lifetime at our disposal. In the 2002 run we have introduced a segmented Ni target consisting of 12 planes with 1 mm gap between each [10]. The combined thickness of all planes is approximately equal to that of the single layer $98\,\mu m$ Ni target.

Single and multilayer target event distributions are identical in all but one respects: the multilayer target yields a *lower number* of dissociated pairs due to the annihilations in the interlayer gaps. Hence, without resorting to a Monte Carlo simulation we can at once obtain the single/multilayer signal difference (Fig. 2). We observe that the subtraction yields a flat spread of relative momenta in the signal-free regions ($Q_l > 2\,MeV$ and $Q_{tot} > 4\,MeV$) centered around 0 indicating that the backgrounds for both targets are indeed identical. **The signal difference between the two targets is found to be 825 ± 239 events** and its shape again accurately replicates that of the modeled atomic pairs.

The signal extraction from both targets individually is accomplished in the same way as described in the previous section. The ratio of the atomic pair signals from each target is equal to the ratio of breakup probabilities [11]. $A_{2\pi}$ lifetime is then found from the latter and the functional dependence similar to the one described in section 3.

The single/multilayer target method offers the following advantages: (a) As evidenced by Fig. 2 the signal shape can be determined relying only on the experimental data. (b) It allows us to bypass the procedure of finding the number of pionium atoms through the Coulomb-correlated background (Eq. 2). (c) We will be able to obtain a "pure" background distribution and use it as a cross-check of the Monte Carlo-generated background: $N_{bckgrnd} = \varepsilon \cdot N^{exp}_{multi} - N^{exp}_{single}$, where ε is the ratio of atomic pair signals from the single layer over the multilayer target, respectively [10].

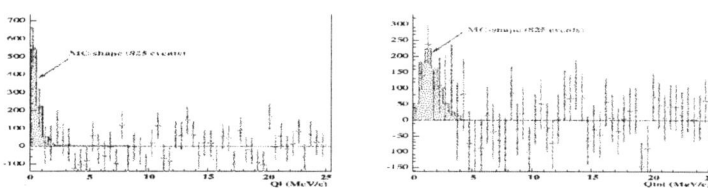

FIGURE 2. Signal resulting from subtracting normalized single from multilayer relative momenta distributions.

SUMMARY AND OUTLOOK

The DIRAC experiment has observed and quantified the atomic pair signal, which will be used in determining the $A_{2\pi}$ lifetime and pion-pion scattering length difference. Our Monte Carlo-simulated background was found to be consistent with the normalization in Q_{tot} and Q_l and the extracted signal to be in good agreement with the simulated atomic pairs.

In 2003 we are planning to get a better handle on multiple scattering by performing a dedicated measurement to determine it with a 1% accuracy in order to improve on the current 5% value. More multilayer and single layer data taking is planned to improve on the statistical accuracy in obtaining the $A_{2\pi}$ lifetime.

REFERENCES

1. S. Weinberg, Phys. Rev. Lett. **17** (1966) 616.
2. J. Gasser, H.Leutwyler, Ann. Phys. (NY), **158**, 142 (1984); J. Gasser, V. E. Lyubovitskij, A. Rusetsky and A. Gall, Phys. Rev. D **64**, 016008 (2001).
3. B. Adeva et al., to be published in Nucl. Instrum. Meth. A. [arXiv:hep-ex/0305022 v1]
4. L. Afanasyev et al., Nucl. Instrum. Meth. A **491**, 376 (2002).
5. L. L. Nemenov, Sov. J. Nucl. Phys. **41**, 629 (1985) [Yad. Fiz. **41**, 980 (1985)].
6. L. Afanasyev, Observation of $\pi^+\pi^-$ Atoms, Ph.D. thesis, JINR Dubna (1997).
7. L. Afanasyev, A. Tarasov, Phys. At. Nucl. **59** 2130 (1996); L. Afanasyev et al., J. Phys. G **25** (1999) B7.
8. Z. Halabuka et al., Nucl. Phys. B **554** (1999) 86; T. Heim et al., J. Phys. B **33** (2000) 3583; T. Heim et al., J. Phys. B **34** (2001) 3763; M. Schumann et al., J. Phys. B **35** 2683 (2002).
9. C. Santamarina et al., Article submitted to J. Phys. B: At. Mol Phys. [arXiv:physics/0306161].
10. A. Benelli et al., DIRAC Note 03-01 (http://dirac.web.cern.ch/DIRAC/i_notes.html).
11. C. Santamarina, A. Kuptsov, DIRAC Note 02-11.

Measurement of Invariant Mass Spectra of Vector Meson Decaying in Nuclear Matter at KEK-PS

R. Muto*, J. Chiba†, H. En'yo*, Y. Fukao**, H. Funahashi**, H. Hamagaki‡,
M. Ieiri†, M. Ishino§, H. Kanda¶, M. Kitaguchi**, M. Mihara§, K. Miwa**,
T. Miyashita**, T. Murakami**, T. Nakura**, M. Naruki**, M. Nomachi[1]†,
K. Ozawa‡, F. Sakuma**, O. Sasaki†, M. Sekimoto†, T. Tabaru*,
K.H. Tanaka†, M. Togawa**, S. Yamada**, S. Yokkaichi* and
Y. Yoshimura**

RIKEN, 2-1 Hirosawa, Wako, Saitama 351-0198, Japan
†*KEK, 1-1 Oho, Tsukuba, Ibaraki 305-0801, Japan*
**Department of Physics, Kyoto University, Kitashirakawa Sakyo-Ku, Kyoto 606-8502, Japan*
‡*Center for Nuclear Study, Graduate School of Science, University of Tokyo, 7-3-1 Hongo, Tokyo 113-0033, Japan*
§*ICEPP, University of Tokyo, 7-3-1 Hongo, Tokyo 113-0033, Japan*
¶*Physics Department, Graduate School of Science, Tohoku University, Sendai 980-8578, Japan*

Abstract. We measured invariant mass spectra of e^+e^- pairs produced in 12-GeV p+A interaction to investigate in-medium mass modification of vector mesons. We observed the excess over the known hadronic sources below the ω meson mass peak. The obtained ρ/ω ratio and velocity dependence of excess/ω ratio imply that this excess mainly results from ρ meson whose mass was modified in nuclear matter.

INTRODUCTION

In quantum chromodynamics, light quarks originally have only a few MeV/c^2 bare mass, but as a consequence of spontaneous chiral symmetry breaking, they obtain an effective mass of about a few hundred MeV/c^2. At a very high density and/or temperature, this broken symmetry will be restored and quark mass will return to its original value. Various theories predict hadron spectral modifications even at a normal nuclear density, as a precursor of chiral phase transition. For example, one model based on the QCD sum rule has been proposed by Hatsuda and Lee[1], which predicted that the mass decrease at a normal nuclear density is about 120~180 MeV/c^2 for the ρ/ω mesons and about 20~40 MeV/c^2 for the ϕ meson. We have carried out an experiment to detect such vector-meson modification in nuclear matter and to investigate the origin of the quark mass in hadron.

---

[1] Present Address: Department of Physics, Osaka University, 1-1 Machikaneyama, Toyonaka, Osaka 560-0043 Japan

EXPERIMENT

The experiment, KEK-PS E325, was designed to detect vector mesons, ρ, ω and ϕ, produced in 12-GeV p+A interactions, by measuring invariant mass spectra of e^+e^- and K^+K^- pairs which are decay products of vector mesons. We constructed a large acceptance spectrometer in a dedicated primary beam line EP1-B at KEK 12-GeV Proton-Synchrotron. The spectrometer acceptance was optimized to detect the slowly moving mesons which have a larger probability to decay inside a target nucleus. For such kinematical region, several tens of percent of the ρ mesons and several percent of the ω/ϕ mesons are expected to decay inside a copper-size nucleus without no width broadening effect. Thus, for example, we can expect the original ϕ meson peak decaying outside a nucleus with some tail or a second peak which consists of ϕ decaying inside a nucleus.

The layout of experimental setup is shown in FIGURE 1. The spectrometer had two electron arms and two kaon arms, which share the dipole magnet and the tracking devices. The electron arms covered from $\pm 12°$ to $\pm 90°$ horizontally and $\pm 22°$ vertically. The kaon arms covered from $\pm 12°$ to $\pm 54°$ horizontally and $\pm 6°$ vertically. Primary proton beam (typical intensity is 5×10^8 Hz) was delivered to the targets located at the center of the magnet. During the data taking in 2002, five targets: four copper targets (0.05% interaction length / 0.6% radiation length each) and one carbon target (0.2% interaction length / 0.4% radiation length) were aligned in-line along the beam axis, separated each other by 23 mm. The magnetic field at the center of the magnet was 0.71 T and the field integral was 0.81 T·m from the center to the radius of 1600 mm, where the last tracking devices, the barrel drift chambers, were located.

The mass resolution and the absolute mass scale were evaluated through a comparison of the observed spectra of known resonances with the Monte Carlo simulation by taking into account the chamber resolutions, multiple scattering and energy losses. The observed peak widths of $\Lambda \rightarrow p\pi^-$ and $K_s \rightarrow \pi^+\pi^-$ were well reproduced with the simulation. The mass resolutions for $\omega \rightarrow e^+e^-$ and $\phi \rightarrow e^+e^-$ were estimated to be 9.6 and 12.0 MeV/c^2, respectively [2].

The data acquisition started in 1997 and ended in March 2002. In the last physics run in 2002, we accumulated statistics of about 100 times as large as our earlier publication [2] with upgraded drift chambers and electron identification counters. In the next section, preliminary results of the e^+e^- invariant mass spectra from the 2002 data are discussed.

RESULT AND DISCUSSION

The invariant mass spectra of e^+e^- pairs are shown in FIGURE 2 for carbon and copper targets. The gray lines are the best-fit results of the cocktails of the known hadronic sources with the combinatorial background. As the hadronic sources the decay modes, $\rho \rightarrow e^+e^-$, $\omega \rightarrow e^+e^-$, $\phi \rightarrow e^+e^-$, $\eta \rightarrow e^+e^-\gamma$ and $\omega \rightarrow \pi^0 e^+e^-$ were taken into account, and the combinatorial background was evaluated by the event mixing method. The relative abundances of these components were determined by the fitting.

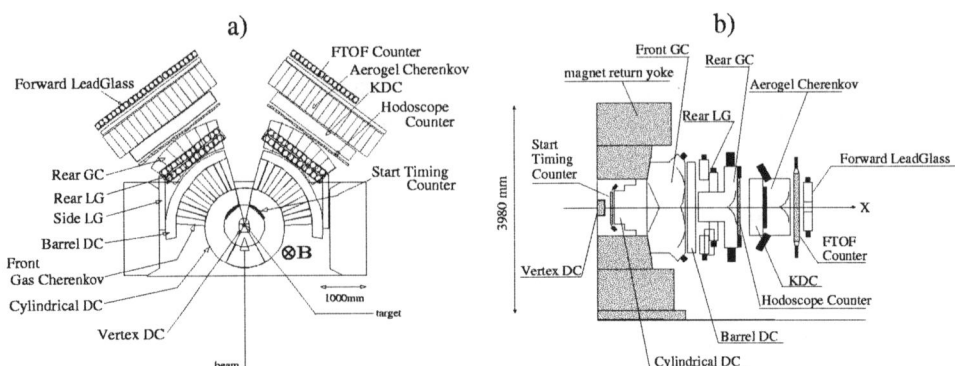

FIGURE 1. Schematic view of the E325 spectrometer. a) is a top view and b) is a vertical cross section along the 33 degree line from the beam line, which is the center of the kaon arm (see text).

The significant excess can be seen at the low mass side of the clear peak from decays of ω mesons in free space, which is located at 0.78 GeV/c^2.

Besides the excess, the ρ/ω ratio is determined mainly from the tail of high-mass side of the ω peak. The obtained values are statistically consistent with zero for both targets, which is much smaller than the known ρ/ω ratio, unity, in pp interactions [3]. This fact implies that most of the ρ mesons are modified to the lower mass region in nuclear matter.

Next, we examined the velocity and nuclear size dependences of this excess. We divided the data into halves at $\beta\gamma = 2.2$ of the mesons. We fitted each divided data in the same scheme and obtained the abundances of ω and the excess. Here, the excess was integrated in the mass region from 0.5 to 0.8 GeV/c^2. FIGURE 3 shows the velocity dependence of the excess to ω ratio for each target. Following tendency can be observed; First, the excess to ω ratio in larger (copper) target is larger than that in smaller (carbon) target. Second, the ratio is larger in the lower velocity region than in the larger velocity region. We also found that the amount of ρ meson decaying in vacuum is consistent with zero in the both velocity regions. A possible scenario is that most of the ρ mesons decay inside a nucleus due to the in-media broadening, and show up as the excess together with unknown contribution from the ω mesons.

ACKNOWLEDGMENTS

We would like to thank all the staff members of KEK-PS, especially the beam channel group for their helpful support. This work was partly supported by the Japan Society for the Promotion of Science, RIKEN Special Postdoctoral Researchers Program and a Grant-in-Aid for Scientific Research of the Japan Ministry of Education, Culture, Sports, Science and Technology (MEXT). Finally, we would like to thank the staff members of CC-J system at RIKEN.

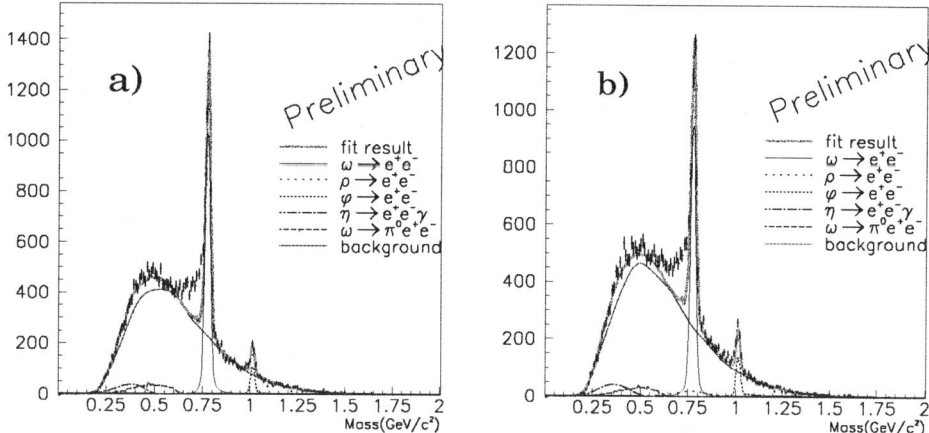

FIGURE 2. Invariant mass spectra of the 2002 e^+e^- data. a) is for the carbon target and b) is for the copper target. The gray lines are the best-fit results by the mixture of the known hadronic sources and the combinatorial background.

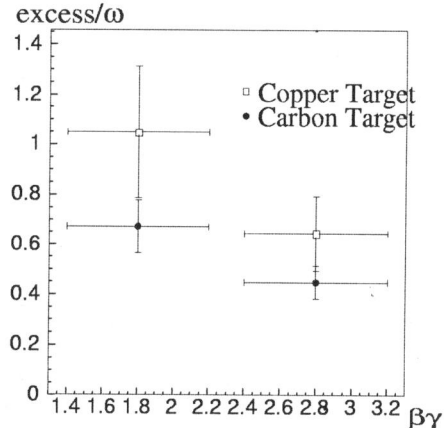

FIGURE 3. The velocity and target nucleus dependences of the excess to ω ratio. The vertical axis is the ratio, and horizontal is average $\beta\gamma$ value.

REFERENCES

1. T. Hatsuda and S.H.Lee, Phys. Rev. **C46** (1992) R34-R38.
2. K. Ozawa *et. al.*, Phys. Rev. Lett. **86** 22(2001) 5019-5022.
3. V. Blobel *et. al.*, Phys. Lett. **B48**:73 (1974)

pp Elastic Scattering at LHC and Nucleon Structure[1]

M. M. Islam*, R. J. Luddy* and A. V. Prokudin[†]

Department of Physics, University of Connecticut, USA
†*Department of Theoretical Physics, University of Torino and INFN, Torino, Italy*

Abstract. High energy elastic pp differential cross section at LHC at the c.m. energy 14 TeV is predicted using the asymptotic behavior of $\sigma_{tot}(s)$ and $\rho(s)$, and the measured $\bar{p}p$ differential cross section at $\sqrt{s}$ =546 GeV. The phenomenological investigation has progressively led to an effective field theory model that describes the nucleon as a chiral bag embedded in a quark-antiquark condensed ground state. The measurement of pp elastic scattering at LHC up to large |t| $\gtrsim$ 10 GeV^2 by the TOTEM group will be crucial to test this structure of the nucleon.

High energy pp and $\bar{p}p$ elastic scattering have been measured at the CERN ISR[1] and SPS Collider[2-3] over a wide range of energy and momentum transfer: $\sqrt{s}$ = 23-630 GeV and |t| = 0-10 GeV^2. These measurements have been followed by Fermilab Tevatron measurement[4-5] of $\bar{p}p$ at $\sqrt{s}$ = 1.8 TeV and |t| = 0-0.5 GeV^2. Such a large experimental effort naturally presents us with the following questions:
1. What do we learn from these experiments about NN interactions at high energies?
2. What insight do we get from them about the physical structure of the nucleon?

These questions have now assumed greater significance because of the Large Hadron Collider (LHC) currently being built at CERN. One of the first experiments planned at LHC called TOTEM (Total and Elastic Measurement) will measure pp elastic $\frac{d\sigma}{dt}$ in the near forward direction at an unprecedented c.m. energy $\sqrt{s}$ = 14 TeV.

My collaborators and I have been studying high energy pp, $\bar{p}p$ elastic scattering for some time[6-8]. Our initial phenomenological investigation led us to the following description. The nucleon has an outer cloud and an inner core (Fig. 1). High energy elastic scattering is primarily due to two processes (Fig. 2): 1) a glancing collision where the outer cloud of one nucleon interacts with that of the other giving rise to diffraction scattering; 2) a hard (or large |t|) collision where one nucleon core scatters off the other core via vector meson ω exchange, while their outer clouds overlap and interact independently. In the small |t| region diffraction dominates, but the hard scattering takes over as |t| increases.

Let me present an example from our recent calculations. The solid curve in Fig. 3 is our calculated $\frac{d\sigma}{dt}$ for $\bar{p}p$ scattering at $\sqrt{s}$ =546 GeV. The dotted curve is the differential cross section due to diffraction alone, while the dot-dashed curve is that due to the hard scattering alone. As we can see, diffraction dominates in the small |t| region, but falls off rapidly as |t| increases, and the hard scattering takes over. The interference between

---

[1] Presented by M. M. Islam. Details have appeared in: Mod. Phys. Lett. 18(2003)743; hep-ph/0210437.

the diffraction and the hard scattering produces the dip. The experimental data are from SPS Collider[2]. The thick dashed curve in Fig. 3 is our calculated pp elastic $\frac{d\sigma}{dt}$ at $\sqrt{s}$ = 500 GeV, which is currently being measured at RHIC in the small |t| region[9].

We describe diffraction scattering using the impact parameter representation:

$$T_D(s,t) - ipW \int_0^\infty b\, db\, J_0(bq)\Gamma_D(s,b), \tag{1}$$

q is the momentum transfer ($q = \sqrt{|t|}$) and $\Gamma_D(s,b)$ is the profile function, which is related to the eikonal function $\chi_D(s,b)$: $\Gamma_D(s,b) = 1 - exp(i\chi_D(s,b))$. We take $\Gamma_D(s,b)$ to be an even Fermi profile function:

$$\Gamma_D(s,b) = g(s)\left[\frac{1}{1+exp((b-R)/a)} + \frac{1}{1+exp(-(b+R)/a)} - 1\right]. \tag{2}$$

The parameters R and a are energy dependent: $R = R_0 + R_1(lns - \frac{i\pi}{2})$, $a = a_0 + a_1(lns - \frac{i\pi}{2})$; $g(s)$ is a complex crossing even energy-dependent coupling strength.

Our hard scattering amplitude is of the form

$$T_H(s,t) \sim exp[i\,\chi_D(s,0)]\, s\, \frac{F^2(t)}{m_\omega^2 - t}. \tag{3}$$

The t-dependence is the product of two form factors and the ω propagator. It shows that ω probes two density distributions corresponding to the two form factors. The density distributions represent the nucleon cores. The factor of s originates from spin 1 of ω. The factor $e^{i\chi_D(s,0)}$ represents absorptive correction due to diffraction scattering. The diffraction amplitude obtained by us satisfies a number of general properties associated with the phenomenon of diffraction:

1. $\sigma_{tot}(s) \sim (a_0 + a_1 lns)^2$ (Froissart-Martin bound)
2. $\rho(s) \simeq \frac{\pi a_1}{a_0 + a_1 lns}$ (derivative dispersion relation)
3. $T_D(s,t) \sim i\, s\, ln^2 s\, f(|t|ln^2 s)$ (AKM scaling)
4. $T_D^{\bar{p}p}(s,t) = T_D^{pp}(s,t)$ (crossing even)

Our present approach is different from our earlier one[8], where we fitted known $\frac{d\sigma}{dt}$ at different energies using complex energy-dependent parameters. Our goal now is to obtain the asymptotic behavior and the approach to the asymptotic behavior of the elastic scattering amplitude, so that we can predict the pp differential cross section at $\sqrt{s}$ = 14 TeV. To this end, we require the energy-dependent parameters to describe quantitatively the asymptotic behavior and the approach to the asymptotic behavior of total cross section $\sigma_{tot}(s)$ and $\rho(s) = \frac{ReT(s,0)}{ImT(s,0)}$ as known from dispersion relation calculations. Furthermore, we require them to describe well the measured $\bar{p}p$ elastic differential cross section at 546 GeV[2]. Here are the results of our calculations of $\sigma_{tot}(s)$ (Fig. 4), $\rho(s)$ (Fig. 5), and $\frac{d\sigma}{dt}$ at $\sqrt{s}$ = 546 GeV (Fig. 3) shown earlier. We find a satisfactory description. Once the parameters are determined, we can test our model by predicting $\frac{d\sigma}{dt}$ at higher energies where experimental data are available. Fig. 6 shows our prediction at $\sqrt{s}$ = 1.8 TeV for $\bar{p}p$ compared with the Tevatron data[4-5]. Fig. 7 shows our prediction for $\bar{p}p$ elastic scattering at $\sqrt{s}$ = 630 GeV, where large |t| data are available from SPS Collider[3]. These tests indicate that the model provides a reasonably quantitative description of high energy elastic scattering.

We now proceed to predict pp elastic $\frac{d\sigma}{dt}$ at LHC at the c.m. energy 14 TeV (Fig. 8). The solid curve is our predicted differential cross section. The dashed curve represents the prediction by the impact-picture model of Bourrely et al. and the dot-dashed curve represents that by the Regge pole-cut model of Desgrolard et al.[10-13]. The latter models predict typical diffraction oscillations in the large |t| region, while our model predicts smooth fall-off of $\frac{d\sigma}{dt}$ for |t| > 1.5 GeV^2. The dotted line in Fig. 8 represents schematically the expected change in our model in the behavior of $\frac{d\sigma}{dt}$ from Orear fall-off: $\frac{d\sigma}{dt} \sim e^{-a\sqrt{|t|}}$ to a power fall-off: $\frac{d\sigma}{dt} \sim t^{-10}$ due to quark-quark scattering.

Our phenomenological investigation progressively led us to an effective field theory model that describes the nucleon structure. This development began with a criticism of our model which was the following: The hard scattering amplitude in our model (Eq.(3)) has a factor of s from spin 1 of ω, and the s and t dependence of this amplitude shows that ω behaves as an elementary vector meson. On the other hand, at such high energies one would expect ω to Reggeize and s be replaced by $s^{\alpha_\omega(t)}$, where $\alpha_\omega(t)$ is the ω trajectory. $\alpha_\omega(t)$ is considerably less than 1 at large |t| and therefore this amplitude should give negligible contribution contrary to our calculations. However, we noticed that in the non-linear σ-model of the nucleon, ω couples to the baryonic current like a gauge boson: $g\omega_\mu J_B^\mu$, and the baryonic current is topological:

$$J_B^\mu = \frac{\varepsilon^{\mu\nu\rho\sigma}}{24\pi^2} tr[U^\dagger \partial_\nu U\, U^\dagger \partial_\rho U\, U^\dagger \partial_\sigma U] \qquad (4)$$

What this model says is that it is an effective field theory model. But, as long as it holds, baryonic current continues to behave as a topological current and ω coupled to it as a gauge boson continues to behave as a gauge boson, i.e. as an elementary vector meson. And we seem to be seeing this behavior.

Fortunately, there was a way of testing this conclusion. From our ωNN form factor $F(t)$, we can obtain by Fourier transform the baryonic charge distribution and then derive the pion field that gives rise to this baryonic charge distribution. We can compare this pion field with the pion field obtained in the n.l. σ-model, which describes the nucleon as a topological soliton or Skyrmion. Here is the result of our analysis: Fig. 9. The solid curve is our calculated pion field configuration, or pion profile function $\theta(r)$, while the dotted and the dashed curves are the pion profile functions from the n.l. σ-model. The curves are consistent with each other, if we bear in mind that our curve is coming from c.m. energy region $\gtrsim$ 23 GeV, while the other curves are coming from an energy region of order 1 GeV. Furthermore, the r.m.s. radius for the baryonic charge distribution obtained by us is 0.44 F, while that from the n.l. σ-model is about 0.5 F.

We faced another problem at this point. Even though the n.l. σ-model when gauged describes the low energy properties of the nucleon quite well, it typically predicts a soliton mass $m_{sol} \sim$ 1500 MeV compared to the nucleon mass m_N = 939 MeV (see, for example,[14]). We obviously had to confront this problem of large soliton mass as we were claiming evidence in favor of the soliton model. To this end, we examined a model more general than the n.l. σ-model. The model turns out to be the linear σ-model of Gell-Mann and Levy, which is described by the Lagrangian:

$$\mathscr{L} = \bar{\psi}i\gamma^\mu \partial_\mu \psi + \tfrac{1}{2}(\partial_\mu \sigma \partial^\mu \sigma + \partial_\mu \vec{\pi} \partial^\mu \vec{\pi}) - g\bar{\psi}[\sigma + i\vec{\tau}\cdot\vec{\pi}\gamma^5]\psi - \lambda(\sigma^2 + \vec{\pi}^2 - f_\pi^2)^2. \quad (5)$$

The model has $SU(2)_L \times SU(2)_R \times U(1)_V$ global symmetry and spontaneous breakdown of chiral symmetry. ψ is the quark field, σ is an isospin-zero scalar field, and $\vec{\pi}$ is an isovector pseudoscalar field. The model can be expressed in terms of right

and left quark fields $\psi_{R,L} = \frac{1}{2}(1 \pm \gamma^5)\psi$ by introducing a scalar field ζ and a unitary field U in the following way: $\sigma + i\vec{\tau} \cdot \vec{\pi} = \zeta U, \zeta = \sqrt{\sigma^2 + \vec{\pi}^2}, U = e^{\frac{i\vec{\tau} \cdot \vec{\phi}}{f_\pi}}$.
The field $\vec{\phi}$ is the massless Goldstone pion field; U is the Skyrmion field that gives rise to the topological baryonic current. In terms of these fields, Eq.(5) takes the form

$$\mathscr{L} = \bar{\psi}_R i\gamma^\mu \partial_\mu \psi_R + \bar{\psi}_L i\gamma^\mu \partial_\mu \psi_L + \frac{1}{2}\partial_\mu \zeta \partial^\mu \zeta + \frac{1}{4}\zeta^2 tr[\partial_\mu U \partial^\mu U^\dagger]$$
$$- g \zeta \left(\bar{\psi}_L U \psi_R + \bar{\psi}_R U^\dagger \psi_L \right) - \lambda \left(\zeta^2 - f_\pi^2 \right)^2. \tag{6}$$

In the conventional n.l. σ-model, one replaces from the very beginning the scalar field ζ by its vacuum value f_π. Furthermore, one introduces a Weiss-Zumino-Witten anomalous action term[15], which arises from the underlying quark structure of the model. It is this action that contains the term $g\omega_\mu J_B^\mu$, which couples ω to the topological baryonic current. The n.l. σ-model also assumes that all the important low-energy interactions are in the meson sector. The only important interaction coming from the quark sector is that given by the WZW action and no further interaction in the quark sector needs to be included. This, of course, leads to a Skyrmion lying in a non-interacting Dirac sea (Fig. 10). On the other hand, we notice from the linear σ-model that even though replacing ζ by its vacuum value f_π may be reasonable in the meson sector, completely neglecting it in the quark sector is questionable, because the ζ field provides an interaction between left and right quarks (Eq.(6)). The latter makes the quarks massive and leads to the spontaneous breakdown of chiral symmetry. This, of course, means that we have a soliton lying in an interacting Dirac sea (Fig. 10). What one finds is that if the scalar field has a critical behavior, and by this I mean a scalar field that is zero for small distances, but rises sharply at some distance R to its vacuum value f_π (Fig. 11), then the energy of the interacting Dirac sea together with that of the scalar field is considerably less than that of the non-interacting Dirac sea[16]. The system therefore makes a transition to this lower ground state and significantly reduces its total energy or mass. This condensation phenomenon solves the soliton mass problem and is analogous to superconductivity. Instead of spin up and down electrons, we have left and right quarks forming a $q\bar{q}$ condensate.

The behavior of the ζ field shown in Fig. 11 has significant implications. First, for $r < R$, $\zeta = 0$ — quarks are massless, and we are in a perturbative regime. Therefore, we end up with the nonperturbative structure of the nucleon shown in Fig. 12, which shows that the nucleon is a chiral bag[15,17] embedded in a $q\bar{q}$ condensed ground state. Second, for momentum transfer $Q = \sqrt{|t|}$ sufficiently large, one nucleon probes the other nucleon at an impact parameter $b \approx \frac{1}{Q} < R$, and therefore in the perturbative regime where pp elastic scattering originates from valence quark-quark scattering. The latter has been investigated by Sotiropoulos and Sterman[18] who concluded that at very large $|t|$, $\frac{d\sigma}{dt} \sim t^{-10}$ (same as power counting rules). From our point of view, this means that as momentum transfer Q increases, there will be a critical value $Q_0 \approx \frac{1}{R}$ beyond which $\frac{d\sigma}{dt}$ will tend to a power fall-off. Schematically, the dotted line in Fig. 8 represents this transition from the nonperturbative regime to the perturbative regime and, in fact, will be a signature of the chiral phase transition.

Concluding remarks

1. Our phenomenological investigation has led us to physical aspects of the nucleon which have been proposed and studied by other authors in different contexts.

2. We find that the nucleon is a chiral bag embedded in a quark-antiquark ground state, and this ground state is analogous to a superconducting ground state. We also find that this structure is described by an effective field theory model — a gauged Gell-Mann-Levy linear σ-model.

3. The experimental study of pp elastic scattering at LHC at $\sqrt{s}$ = 14 TeV by the TOTEM group up to large $|t|$ will be crucial to test this structure of the nucleon.

REFERENCES

1. Nagy, E., et al., *Nucl. Phys.*, **B150**, 221 (1979).
2. Bozzo, M., et al., *Phys. Lett.*, **B147**, 385 (1984); **B155**, 197 (1985).
3. Bernard, D., et al., *Phys. Lett.*, **B171**, 142 (1986).
4. Amos, N., et al., *Phys. Lett.*, **B247**, 127 (1990).
5. Abe, F., et al., *Phys. Rev.*, **D50**, 5518 (1994).
6. Heines, G. W., and Islam, M. M., *Nuovo Cimento*, **61A**, 149 (1981).
7. Islam, M. M., Fearnley, T., and Guillaud, J. P., *Nuovo Cimento*, **81A**, 737 (1984).
8. Islam, M. M., Innocente, V., Fearnley, T., and Sanguinetti, G., *Europhys. Lett.*, **4**, 189 (1987).
9. Guryn, W., *Nucl. Phys. (Proc. Suppl.)*, **B99**, 299 (2001).
10. Bourrely, C., Soffer, J., and Wu, T. T., *Phys. Rev. Lett.*, **54**, 757 (1985).
11. Bourrely, C., et al., "Quarter Century of Rising Total Cross Sections," in *Frontiers in Strong Interactions*, edited by P. Chiappetta, M. Haguenauer, and J. T. T. Van, Editions Frontieres, 1996, p.15.
12. Desgrolard, P., Giffon, M., and Predazzi, E., *Z. Phys.*, **C63**, 241 (1994).
13. Buenerd, M., "The TOTEM project at LHC," in *Frontiers in Strong Interactions*, edited by P. Chiappetta, M. Haguenauer, and J. T. T. Van, Editions Frontieres, 1996, p.437.
14. Zhang, L., and Mukhopadhyay, N. C., *Phys. Rev.*, **D50**, 4668 (1994).
15. Bhaduri, R. K., *Models of the Nucleon: from Quarks to Soliton*, Addison-Wesley, Reading, Massachusetts, 1988.
16. Islam, M. M., *Z. Phys.*, **C53**, 253 (1992).
17. Hosaka, A., and Toki, H., *Quarks, Baryons, and Chiral Symmetry*, World Scientific, 2001.
18. Sotiropoulos, M. G., and Sterman, G., *Nucl. Phys.*, **B425**, 489 (1994).

Figures 1 - 12: Explanations of the figures are given in the main body of the paper.

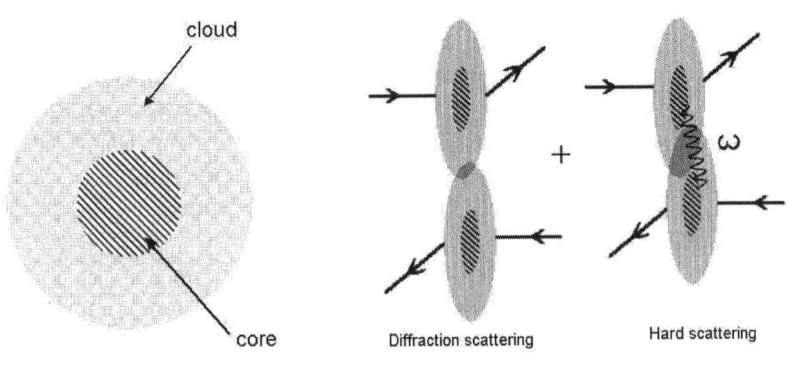

FIGURE 1.

FIGURE 2.

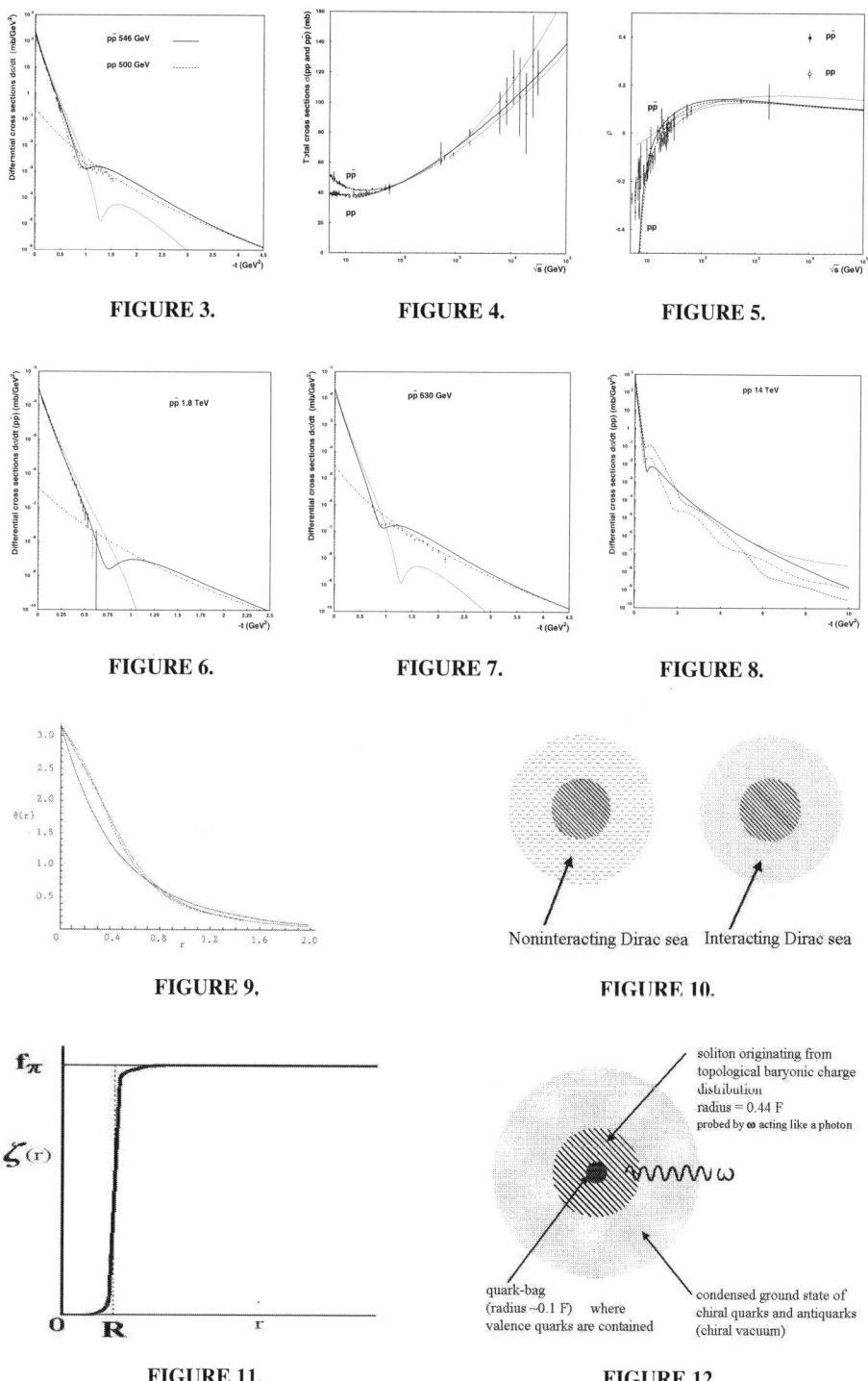

Modeling Neutrino Quasielastic Cross Sections on Nucleons and Nuclei

A. Bodek*, H. Budd* and J. Arrington[†]

*University of Rochester, Rochester, NY
[†]Argonne National Laboratory, Argonne, Illinois 60439, USA

Abstract. We calculate the total and differential quasielastic cross sections for neutrino and antineutrino scattering on nucleons using up to date fits to the nucleon elastic electromagnetic form factors G_E^p, G_E^n, G_M^p, G_M^n, and G_A and pseudoscalar form factors. We compare predictions of the cross sections for nucleons and nuclei to experimental data. (Presented by Arie Bodek at CIPANP2003, New York City, NY 2003)

1. INTRODUCTION

Experimental evidence for oscillations among the three neutrino generations has been recently reported [1]. Since quasielastic (QE) scattering forms an important component of neutrino scattering at low energies, we have undertaken to investigate QE neutrino scattering using the latest information on nucleon form factors.

Recent experiments at SLAC and Jefferson Lab (JLab) have given precise measurements of the vector electromagnetic form factors for the proton and neutron. These form factors can be related to the form factors for QE neutrino scattering by conserved vector current hypothesis, CVC. These more recent form factors can be used to give better predictions for QE neutrino scattering.

The hadronic current for QE neutrino scattering is given by [2]

$$< p(p_2)|J_\lambda^+|n(p_1) > = $$
$$\bar{u}(p_2)\left[\gamma_\lambda F_V^1(q^2) + \frac{i\sigma_{\lambda\nu}q^\nu \xi F_V^2(q^2)}{2M} + \gamma_\lambda \gamma_5 F_A(q^2) + \frac{q_\lambda \gamma_5 F_P(q^2)}{M}\right] u(p_1),$$

where $q = k_\nu - k_\mu$, $\xi = (\mu_p - 1) - \mu_n$, and $M = (m_p + m_n)/2$. Here, μ_p and μ_n are the proton and neutron magnetic moments. We assume that there are no second class currents, so the scalar form factor F_V^3 and the tensor form factor F_A^3 need not be included. Using the above current, the cross section is

$$\frac{d\sigma^{\nu,\bar{\nu}}}{dq^2} = \frac{M^2 G_F^2 \cos^2\theta_c}{8\pi E_\nu^2} \times \left[A(q^2) \mp \frac{(s-u)B(q^2)}{M^2} + \frac{C(q^2)(s-u)^2}{M^4}\right],$$

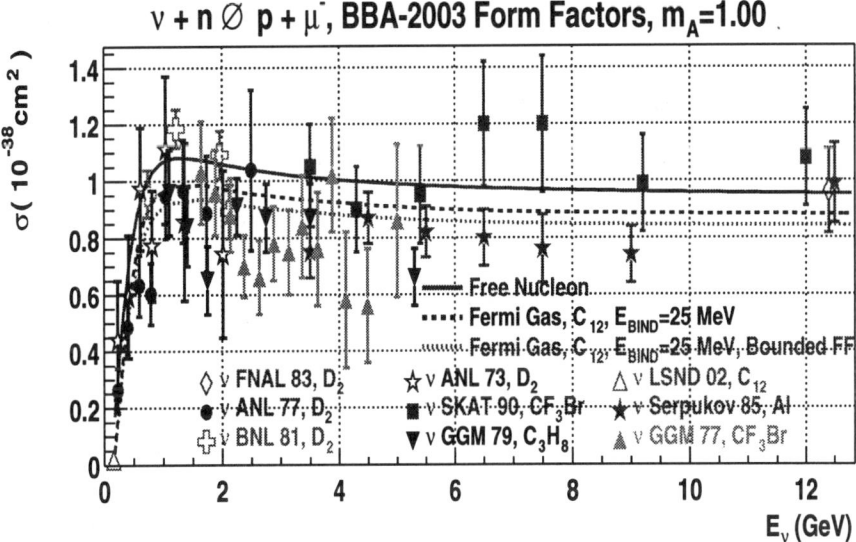

FIGURE 1. The QE neutrino cross section along with data from various experiments. The calculation uses M_A=1.00 GeV, g_A=−1.267, M_V^2=0.71 GeV^2 and BBA-2003 Form Factors. The solid curve uses no nuclear correction, while the dashed curve [5] uses a Fermi gas model for carbon with a 25 MeV binding energy and 220 Fermi momentum. The dotted curve is the prediction for Carbon including both Fermi gas Pauli blocking as wel effect of nuclear binding on the nucleon form factors. [8] The lower plot is identical to the upper plot with the E_ν axis limit changed to 2 GeV. The data shown [3] are from FNAL 1983, ANL 1977, BNL 1981, ANL 1973, SKAT 1990, GGM 1979, LSND 2002, Serpukov 1985, and GGM 1977.

where

$$A(q^2) = \frac{m^2 - q^2}{4M^2}\left[\left(4 - \frac{q^2}{M^2}\right)|F_A|^2 \right.$$
$$\left. - \left(4 + \frac{q^2}{M^2}\right)|F_V^1|^2 - \frac{q^2}{M^2}|\xi F_V^2|^2\left(1 + \frac{q^2}{4M^2}\right) - \frac{4q^2 Re F_V^{1*} \xi F_V^2}{M^2}\right],$$

$$B(q^2) = -\frac{q^2}{M^2} Re F_A^*(F_V^1 + \xi F_V^2), \quad C(q^2) = \frac{1}{4}\left(|F_A|^2 + |F_V^1|^2 - \frac{q^2}{M^2}\left|\frac{\xi F_V^2}{2}\right|^2\right).$$

Although we have have not shown terms of order $(m_l/M)^2$, and terms including $F_P(q^2)$ (which is multiplied by $(m_l/M)^2$), these terms are included in our calculations [2].) The form factors $F_V^1(q^2)$ and $\xi F_V^2(q^2)$ are given by:

$$F_V^1(q^2) = \frac{G_E^V(q^2) - \frac{q^2}{4M^2}G_M^V(q^2)}{1 - \frac{q^2}{4M^2}}, \quad \xi F_V^2(q^2) = \frac{G_M^V(q^2) - G_E^V(q^2)}{1 - \frac{q^2}{4M^2}}.$$

We use the CVC to determine $G_E^V(q^2)$ and $G_M^V(q^2)$ from the electron scattering form factors $G_E^p(q^2)$, $G_E^n(q^2)$, $G_M^p(q^2)$, and $G_M^n(q^2)$:

$$G_E^V(q^2) = G_E^p(q^2) - G_E^n(q^2), \quad G_M^V(q^2) = G_M^p(q^2) - G_M^n(q^2).$$

The axial form factor F_A and the pseudoscalar form factor F_P (related to F_A by PCAX) are given by

$$F_A(q^2) = \frac{g_A}{\left(1 - \frac{q^2}{M_A^2}\right)^2} \cdot F_P(q^2) = \frac{2M^2 F_A(q^2)}{M_\pi^2 - q^2}.$$

In the expression for the cross section, $F_P(q^2)$ is multiplied by $(m_l/M)^2$. Therefore, in muon neutrino interactions, this effect is very small except at very low energy, below 0.2 GeV. $F_A(q^2)$ needs to be extracted from QE neutrino scattering. A low Q^2, $F_A(q^2)$r can also be extracted from pion electroproduction data.

Previously, people have assumed that the vector form factors are described by the dipole approximation.

$$G_D(q^2) = \frac{1}{\left(1 - \frac{q^2}{M_V^2}\right)^2}, \quad M_V^2 = 0.71 \; GeV^2$$

$$G_E^p = G_D(q^2), \quad G_E^n = 0, \quad G_M^p = \mu_p G_D(q^2), \quad G_M^n = \mu_n G_D(q^2).$$

We refer to the above combination of form factors as 'Dipole Form Factors'. It is an approximation that has been improved by us in a previous publication [3]. We use our updated form factors to which we refer as 'BBA-2003 Form Factors' (Budd, Bodek, Arrington). We also use our updated value [3] of M_A 1.00 ± 0.020 GeV which is in good agreement with the theoretically corrected value from pion electroproduction [4] of 1.014 ± 0.016 GeV.

2. COMPARISON TO EXPERIMETAL DATA

Figures 1, 2 and 3 show the QE cross section for ν and $\bar{\nu}$ with BBA-2003 Form Factors and M_A=1.00 GeV. The normalization uncertainty in the data is approximately 10%. The solid curve uses no nuclear correction, while the dashed curve [5] uses a NUANCE [6] calculation of a Smith and Moniz [7] based Fermi gas model for carbon. This nuclear model includes Pauli blocking (see Figure 4(a)) and Fermi motion, but not final state interactions. The Fermi gas model was run with a 25 MeV binding energy and 220 MeV Fermi momentum. The dotted curve is the prediction for Carbon including both Fermi gas Pauli blocking and effect of nuclear binding on the nucleon form factors as modeled by Tsushima et al [8] (see Figure 4(b)). Note that this model is only valid for Q^2 less than 1 GeV^2, and that the binding effects on the form factors are expected to be very small at higher Q^2. Both the Pauli blocking and the the nuclear modifications to bound nucleon form factors reduce the cross section relative to the cross section with free nucleons.

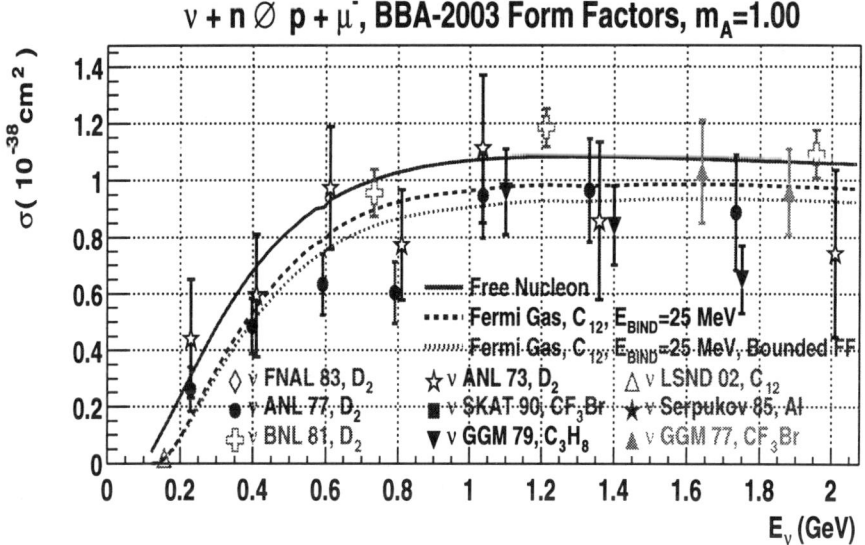

FIGURE 2. Same as Figure 1 with the E_ν axis limit changed to 2 GeV.

The updated form factors improve the agreement with neutrino QE cross section data and give a reasonable description of the cross sections from deuterium.

We plan to continue to study the nuclear corrections, adopting models which have been used in precision electron scattering measurements from nuclei at SLAC and JLab. For example, we plan to study the Pauli blocking correction using an improved Fermi Gas model with a high momentum tail [9], as well as more sophisticated nuclear spectral functions. In addition, we will continue to update the extraction of M_A from previous neutrino experiments, using the updated versions of the input parameters and electromagnetic form factors.

This work is supported in part by the U. S. Department of Energy, Nuclear Physics Division, under contract W-31-109-ENG-38 (Argonne) and High Energy Physics Division under grant DE-FG02-91ER40685 (Rochester).

REFERENCES

1. Y. Fukada *et al.*, Phys. Rev. Lett. 81 (1998) 1562.
2. C.H. Llewellyn Smith, Phys. Rep. 3C (1972).
3. H. Budd, A. Bodek and J. Arrington, hep-ex[0308005].
4. V. Bernard, L. Elouadrhiri, U.G. Meissner, J.Phys.G28 (2002), hep-ph[0107088].
5. G. Zeller, private communication.
6. D. Casper, Nucl. Phys. Proc. Suppl. 112 (2002) 161.
7. R.A. Smith and E.J. Moniz, Nucl. Phys. B43 (1972) 605.
8. K. Tsushima, Hungchong Kim, K. Saito, nucl-th[0307013].
9. A. Bodek and J. L. Ritchie, Phys. Rev. D23 (1981) 1070; ibid Phys. Rev. D24 (1981) 1400.

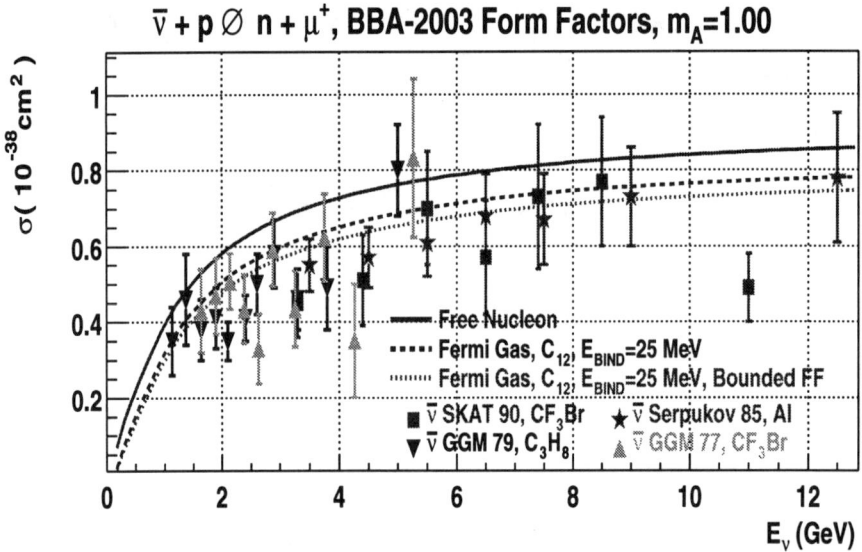

FIGURE 3. The QE antineutrino cross section along with data from various experiments. The calculation uses M_A=1.00 GeV, g_A=−1.267, M_V^2=0.71 GeV^2 and BBA-2003 Form Factors. The solid curve uses no nuclear correction, while the dashed curve [5] uses a Fermi gas model for carbon with a 25 MeV binding energy and 220 MeV Fermi momentum. The dotted curve is the prediction for Carbon including both Fermi gas Pauli blocking and the effect of nuclear binding on the nucleon form factors. [8] The data shown are from SKAT 1990, GGM 1979, Serpukov 1985, and GGM 1977.

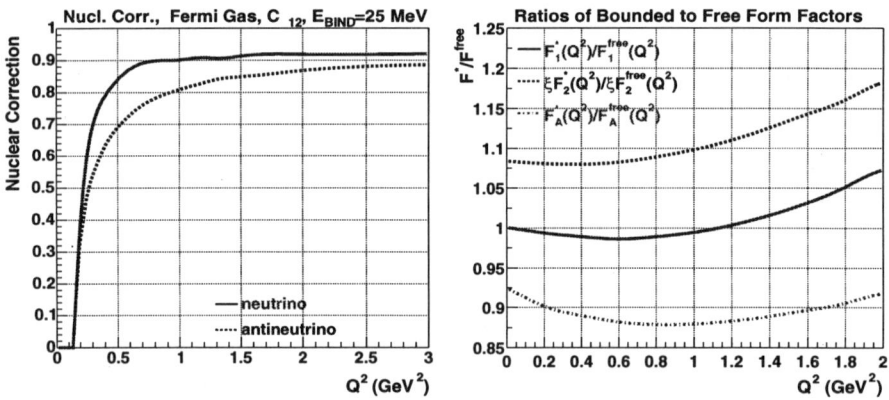

FIGURE 4. (a) The Pauli blocking suppression for a Fermi gas model for carbon with a 25 MeV binding energy and 220 MeV Fermi momentum. (b) The ratio of bound to free nucleon form factors for F_1, F_2 and F_A from ref. [8]

Using Electron Scattering Data to Tune Neutrino Monte Carlos

Hugh Gallagher

Tufts University

Abstract. Monte Carlo programs have played a major role in the analysis of data from neutrino experiments. These programs have been largely calibrated on bubble chamber data which does not fully cover the range of energies and nuclear targets of interest to current and future neutrino oscillation experiments. The possibilities of using electron scattering data to address areas of uncertainty in the simulation are presented.

MOTIVATION

Monte Carlo simulations play a variety of important roles in neutrino experiments. They are used to: evaluate the feasibility of proposed projects and estimate their physics impact, make decisions about detector design and optimization, and ultimately analyze the collected data samples. These simulations are important whether neutrinos are themselves the object of study, as in atmospheric and long-baseline experiments, or are a source of background, as in proton decay searches in underground detectors. With the advent of high-intensity neutrino beams from proton colliders, we are entering a new era of high-statistics, precision neutrino experiments which will require a new level of accuracy in our knowledge (and simulation) of neutrino interaction physics.

Neutrino interactions in the energy range of interest to current and near-future experiments (1 to 10 GeV), pose particular problems. In this energy range, bridging the perturbative and non-perturbative pictures of the nucleon, a variety of scattering mechanisms are important. Nuclear effects, both in the initial state (nuclear structure, fermi motion, short-range correlations), and the final state (Pauli blocking, nuclear re-interactions, nuclear transparency), play a role. The data in this region comes primarily from bubble chamber experiments which ran in the 1960s - 1980s. Data samples were, at most, in the tens of thousands of events. Although nuclear physics plays a role in the analysis of data from heavy targets in bubble chambers (typically freon or neon), nuclear effects were not themselves the subject of study. Figure 1 shows one such comparison, between the predictions of the NEUGEN event generator [1] and data from the BNL bubble chamber [2].

The models incorporated into neutrino simulations at these energies have been tuned primarily to this bubble chamber data. This data is not sufficient to completely constrain the models, particularly with regards to the simulation of nuclear effects. A logical place to turn for guidance are electron scattering experiments.

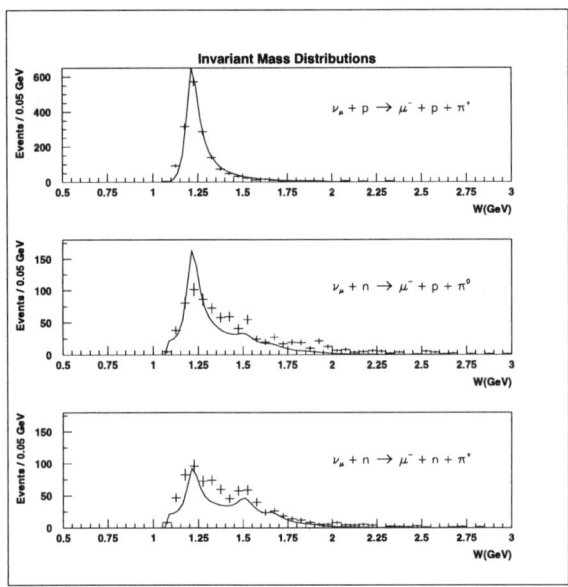

FIGURE 1. Invariant mass distributions of single pion reactions measured by the BNL bubble chamber [2] compared with the predictions of the NEUGEN event generator. Errors shown are statistical only.

SOURCES OF UNCERTAINTY

Several neutrino Monte Carlo codes, incorporating quite different models and approaches, have been independently developed and are described elsewhere [3, 1, 4, 5]. The assembly of a neutrino Monte Carlo can be broken down into a number of distinct steps.

1. Collecting a variety of calculations for neutrino scattering off of free nucleons / nuclei for the interactions considered to be important.
2. Devise a scheme for combining the different contributions to determine the total cross section at an arbitrary point in phase space.
3. Decide on a fragmentation / hadronization scheme.
4. Determine how nuclear effects modify each of the above.

Substantial uncertainties exist at each of these phases. For high energy experiments fragmentation is often simulated using one of the standard packages like JETSET [6], however for low invariant masses alternative algorithms, often based on KNO scaling [7], need to be used. Numerous studies have explored the universality of hadronic systems produced in neutrino, e^+e^-, and hadron scattering experiments [8, 9], the results of which can be used to build a hadronization model over the full (W, Q^2) range. In each of the other areas, measurements from electron scattering experiments are the strongest, and often only, way to provide useful constraints.

Free Nucleons: The validity of the resonance models and the treatment of the overlap region between resonance and DIS scattering processes

While elastic and deep-inelastic scattering processes are well understood theoretically, in the resonance region a variety of different models have been proposed or used in different programs. The Rein-Seghal model, used by several neutrino codes, has the advantage of being relatively straightforward to implement and gives predictions over a broad range of kinematics [10]. This model is based on the Feynman-Kislinger-Ravndal (FKR) model of the baryon resonances [11].

The DIS / resonance overlap issue has also been handled very differently in the different Monte Carlos. New proposals motivated by duality considerations and incorporating new scaling variables have been put forward which produce good agreement with the cross section averaged over the resonance region and are applicable down to the lowest values of Q^2 [12]. Many measurements have been made on electron scattering with hydrogen and deuterium targets in the resonance region. Measurements of F2 and differential distributions in (e,e') scattering can be used to provide stringent constraints on the validity of the Rein-Seghal model and the scheme employed to combine resonance and deep inelastic contributions at each point in phase space.

Nuclear Effects

While one might expect that in some kinematic regimes neutrinos interact differently with the nuclear medium due to the V-A nature of the interaction, many other details of the nuclear environment should affect neutrino and electron scattering in the same way. In particular the effects of Fermi motion and the binding of the nucleon in the nuclear material is very cleanly explored in electron scattering measuerements at quasi-free kinematics and missing energy measurements from (e,e'p) measurements.

Models of electron scattering in the GeV range are typically derived in the Plane Wave Impulse Approximation (PWIA) using spectral functions derived from nuclear many-body theory and scattering measurements to give the probability of finding a nucleon within the nucleus with a particular value of momentum and binding energy [13]. The full theoretical framework for describing the scattering kinematics in terms of these spectral functions is well developed and has been shown to produce good agreement with data. Extending neutrino Monte Carlos to use spectral functions for quasi-elastic and inelastic reactions is a topic of current activity [14], and the validity of the various assumptions of the nuclear model can be evaluated through direct electron scattering experiments. Previous calculations have often relied on Fermi Gas model calculations, where the Fermi momentum and binding energy of the nucleus are measured directly in electron scattering [15]

A variety of calculations and studies have shown that intra-nuclear rescattering and absorption of the final state hadrons produces large effects on the observed distributions in neutrino detectors [16]. Data from high-statistics 4π detectors like CLAS [17] could provide the answers to many of the questions that plague neutrino simulations here.

Studies could be carried out of charged particle multiplicities for different nuclear targets, and changes in energy and angular distributions. Of particular interest would be information to confirm baryon re-scattering models in quasi-elastic interactions. Nuclear transparency results from coincidence experiments provide one set of useful data, but the final destination of scattered protons, for instance, could only be determined from a detailed study in CLAS of correlated low-energy particle production in events with quasi-free kinematics.

MECHANICS OF NEUTRINO MONTE CARLO / ELECTRON DATA COMPARISONS

There are a number of different ways in which one can compare the predictions of neutrino simulations to electron scattering data. One approach involves generating events weighted by the neutrino interaction cross section. The corresponding weight for an electron scattering event with the same kinematics can then be calculated and distributions generated with neutrinos can be re-weighted to approximate electron distributions. This approach has been used to perform a number of comparisons to the NUANCE Monte Carlo [18].

An alternative approach is to fully replace the fundamental interaction cross sections in the neutrino simulations by their electromagnetic counterparts. If unweighted events are being produced it is generally necessary to limit the phase space selection to small regions which correspond to the acceptance of the spectrometer for the particular electron experiment one wishes to compare to.

A further complication involves the inclusion of radiative corrections, which although calculated for neutrino scattering are generally not included as part of event generation. For the level of accuracy required in current studies, radiative corrections can be handled in a number of ways. The simplest thing is of course to compare only to data corrected for radiative effects. In many cases the uncertainties related to the neutrino MC predictions are quite large, in which case one can gain useful information as long as the magnitude of the radiative corrections are significantly smaller than the uncertainties under investigation. Such an approach requires assistance from radiative corrections experts in determining distributions which are less affected by radiative corrections.

In addition to comparing to published cross sections and other fully corrected measurements, the possibility also exists to compare to data in a somewhat more raw form. Such comparisons could be carried out if one were to wish to study a topic of particular interest or concern for neutrino scattering experiments but for which data from electron scattering is typically not fully analyzed. One example is how, in detail, the hadronic final state is degraded (in terms of particle multiplicities, energy and angular distributions) due to nuclear re-interactions for few-Gev scattering off of different nuclear targets. Such a study could be very useful for future long-baseline experiments searching for sub-dominant $v_\mu \to v_e$ appearance. These experiments often base their search on cuts on shower energy and angle, and have as a main background contributor single π^o NC events.

CONCLUSION

Data from electron scattering experiments, particularly data from the CLAS experiment on nuclear targets, provides an excellent testing ground for many of the models that are currently being used in neutrino interaction Monte Carlos. Obtaining agreement with this data is crucial for demonstrating that the vector part of the hadronic current is being modeled correctly and that nuclear effects are being appropriately handled. Correctly modeling neutrino interactions then becomes a question of incorporating the axial current. Quantitative comparisons to electron data are thus a crucial first step in establishing the validity of these simulations for the role they will play in analyzing data from future high precision neutrino experiments.

ACKNOWLEDGMENTS

The author would like to acknowledge Steve Wood, Omar Benhar, Jorge Morfin, Thia Keppel, Dave Casper, Rolf Ent, Will Brooks, Andrei Afanasev, Arie Bodek and John Arrington for their helpful comments and suggestions.

REFERENCES

1. Gallagher, H., *Nucl. Phys. Proc. Suppl.*, **112**, 188–194 (2002).
2. Kitagaki, T., et al., *Phys. Rev.*, **D34**, 2554–2565 (1986).
3. Casper, D., *Nucl. Phys. Proc. Suppl.*, **112**, 161–170 (2002).
4. Hayato, Y., *Nucl. Phys. Proc. Suppl.*, **112**, 171–176 (2002).
5. Cavanna, F., and Palamara, O., *Nucl. Phys. Proc. Suppl.*, **112**, 183–187 (2002).
6. Sjostrand, T., *Comput. Phys. Commun.*, **82**, 74–90 (1994).
7. Koba, Z., Nielsen, H. B., and Olesen, P., *Nucl. Phys.*, **B40**, 317–334 (1972).
8. Bonvicini, G., et al., *Nuovo Cim. Lett.*, **36**, 555 (1983).
9. Schmitz, N. (1981), rapporteur's talk given at 1981 Int. Symp. on Lepton and Photon Interactions at High Energy, Bonn, West Germany, Aug 24-29, 1981.
10. Rein, D., and Sehgal, L. M., *Ann. Phys.*, **133**, 79 (1981).
11. Feynman, R. P., Kislinger, M., and Ravndal, F., *Phys. Rev.*, **D3**, 2706–2732 (1971).
12. Bodek, A., and Yang, U. K., *J. Phys.*, **G29**, 1899–1906 (2003).
13. Benhar, O., Fabrocini, A., Fantoni, S., and Sick, I., *Nucl. Phys.*, **A579**, 493–517 (1994).
14. Nakamura, H., and Seki, R., *Nucl. Phys. Proc. Suppl.*, **112**, 197–202 (2002).
15. Moniz, E. J., et al., *Phys. Rev. Lett.*, **26**, 445–448 (1971).
16. Battistoni, G., Lipari, P., Ranft, J., and Scapparone, E., hep-ph/9801426 (1998).
17. Mecking, B. A., et al., *Nucl. Instrum. Meth.*, **A503**, 513–553 (2003).
18. Wood, S., Talk at the Second International Workshop on Neutrino-Nucleus Interactions in the Few-GeV Region, U. California Irvine, Dec. 2002.

FUNDAMENTAL SYMMETRY

Recent Results from the SAMPLE Experiment

Takeyasu M. Ito
for the SAMPLE Collaboration

W.K.Kellogg Laboratory, California Institute of Technology, Pasadena, CA 91125

Abstract. The previous two SAMPLE experiments yielded a measurement of the axial *e-N* form factor G_A^e substantially different from the theoretical estimate. In order to confirm this observation, a third SAMPLE experiment was carried out at a lower beam energy of 125 MeV ($Q^2 = 0.038$ (GeV/c)^2) on a deuterium target. The data analysis is now at the final stage and the results are consistent with the theoretical prediction of the axial form factor G_A^e. Also, reevaluation of the background dilution factor and the electromagnetic radiative correction for the 200 MeV deuterium data lead to updated results, which are also consistent with the theoretical prediction.

INTRODUCTION

It is well established that parity-violating (PV) electron scattering can provide a measurement of the neutral weak form factors of the nucleon [1, 2]. When combined with the information on the known electromagnetic form factors, the information on the neutral weak form factors would allow a separation of the nucleon's form factors into the three contributing flavors of quarks: up, down, and strange [3]. Measuring the contribution from strange quark-antiquark pairs is of special interest because it relates directly to the proton's virtual sea that apparently determines the bulk of its mass. This provided the basis for the SAMPLE project (and other experiments that followed). The primary goal of SAMPLE is to determine the proton's strange magnetic form factor G_M^s through PV electron scattering at backward angles.

PV electron scattering from the proton at backward angles is not only sensitive to G_M^s, but is also sensitive to the proton's neutral weak axial form factor. The neutral weak axial form factor as measured in electron scattering G_A^e can potentially receive large electroweak corrections, including the anapole moment, that are absent in neutrino scattering. Determining G_A^e is not only important for a reliable extraction of G_M^s, but also is interesting on its own. Parity violating quasielastic electron-deuteron scattering is mainly sensitive to G_A^e.

The SAMPLE collaboration performed an experiment on a deuterium target as well as on a hydrogen target at 200 MeV ($Q^2 = 0.1$ (GeV/c)^2). Combining the results from these two experiments allows separate determination of G_M^s and G_A^e. The results from these measurements were published in Ref. [4]. Our data indicated that, while the overall contribution from strange quarks to the proton's magnetic form factor is small, the size of the electroweak radiative corrections (potentially including the anapole moment) to the axial form factor is significantly larger than anticipated from theory [5].

These results stimulated considerable interest among theorists. Many different processes and effects were studied for their potential contributions to the axial form factor, including the anapole moment [5–8], nuclear effects including two body currents [9], and the parity violating hadronic interaction [10, 11]. Despite intensive efforts, the discrepancy between the theoretical prediction and the experimental value was not reconciled.

SAMPLE III EXPERIMENT

In order to experimentally confirm these results, we performed a third SAMPLE experiment, with a deuterium target at a lower beam energy of 125 MeV ($Q^2 = 0.038$ (GeV/c)^2). Just like the 200 MeV experiment with a deuterium target, this experiment would mainly be sensitive to G_A^e. Since the PV asymmetry in the cross section is (to first order) proportional to Q^2, the expected asymmetry would be roughly 3 times smaller than that for 200 MeV. The cross section, however, is larger roughly by a factor of 2, and since it was expected (and was later experimentally confirmed) that the background level for 125 MeV would be the same as that for 200 MeV, we would expect roughly the same figure-of-merit for the 125 MeV experiment as for the 200 MeV experiment. Therefore, this experiment at 125 MeV is an experiment that is sensitive to the same physics but with very different systematics.

The experiment was performed at the MIT Bates Linear Accelerator Center. The apparatus used for the SAMPLE III was identical to the one used for the previous SAMPLE experiments. A 125 MeV longitudinally polarized electron beam was incident on a 40 cm long liquid deuterium target, and electrons that were scattered at backward angles were detected by a large solid angle air Čerenkov detector covering angles between 130° and 170°. The beam was pulsed at 600 Hz, and the average beam current was 40 µA. The polarized electron beam was generated at the Bates Polarized Electron Source by shining a circularly polarized laser beam onto a GaAs crystal. The helicity of the beam was randomly chosen for each pulse. The helicity of the beam (with respect to the electronics signal in the Polarized Source) was manually reversed every 2-3 days by inserting and removing a halfwave plate in the laser beam path to check and reduce possible systematic effects.

After a successful test run in April-May 2001 in which the expected background conditions were achieved, the production data were taken in November 2001-February 2002. About 96 C of beam charge worth data was written on tape. In order to measure background components, 11% of the data were taken with the shutters (made of think aluminum sheet) in front of the PMT's closed to block the light coming into the PMT's.

SAMPLE II ANALYSIS UPDATE

In the mean time, the data analysis on the previous SAMPLE experiments (200 MeV run on hydrogen and deuterium targets) was continued. The background dilution factor was carefully reevaluated with the contribution from the coherent photoproduction of

π^0 on the deuteron taken into account using new experimental results. Moreover, electromagnetic radiative effects were reevaluated with an improved detector acceptance model, and the theoretical value for the asymmetry with which the experimental value is to be compared was reevaluated with full nuclear calculation [12] and an improved detector acceptance model. The overall effect is such that the final physics asymmetry is increased (closer to the theoretical prediction) by 10% and the theoretical prediction for the asymmetry is decreased (closer to the experimental value) by 2%. The implication of these updated results will be discussed at the end of the next section.

SAMPLE III ANALYSIS PRESENT STATUS

The data analysis was performed in a similar manner to the previous SAMPLE experiments. One of the differences from the analysis of the previous SAMPLE data is that the transmission (defined as the ratio of the beam intensity at the target to that at the end of the accelerator) was used as one of the parameters for the corrections procedure to correct for the intensity asymmetry due to the differential scraping at the energy defining slit.

In Fig. 1, the detector asymmetry is plotted as a function of time. The sign of the IN data is reversed. Correcting this raw asymmetry for the background dilution, the beam polarization, and electromagnetic radiative effects, we obtain the physics asymmetry. We then compare this physics asymmetry to the theoretical prediction, which was generated by averaging over the detector acceptance the results of the full nuclear calculation [12]. The comparison between the experimental asymmetry with the theoretical prediction is plotted in Fig. 2 for both SAMPLE II (updated results) and SAMPLE III. The data for both energies are consistent with the theoretical prediction with the value of G_A^e taken from Ref. [5].

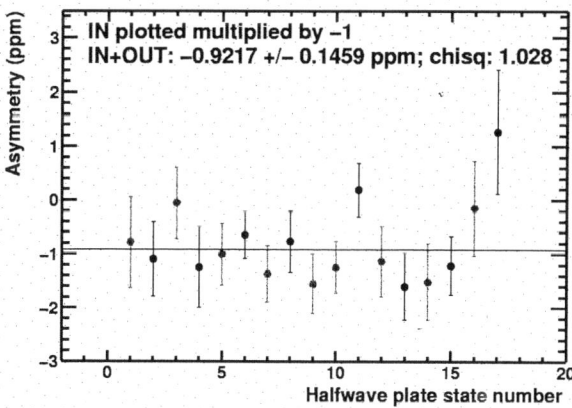

FIGURE 1. Results for the detector signal asymmetry for SAMPLE III. Each data point represents the average of the detector asymmetry between two halfwave plate state changes. The OUT data are plotted in red and the IN data are plotted in blue. The sign of the IN data is reversed.

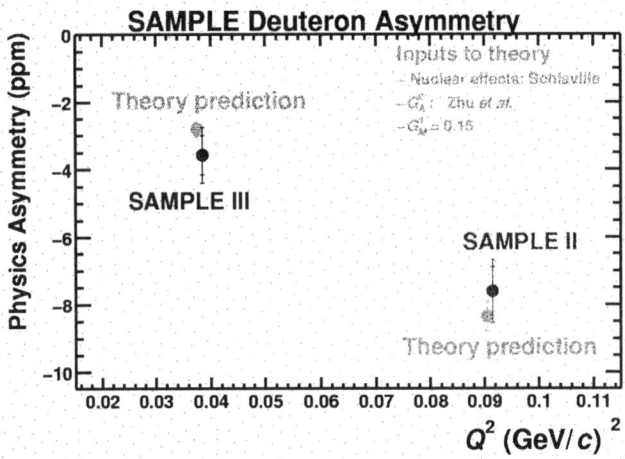

FIGURE 2. The physics asymmetry measured in SAMPLE II (updated results) and SAMPLE III are plotted as a function of Q^2. Also plotted are the theoretical prediction with the value of G_A^e taken from [5], and $G_M^s = 0.15$. The dependence of the theoretical values on the value of G_M^s is small. The experimental values are consistent with the theoretical prediction.

SUMMARY

The updated results from the 200 MeV SAMPLE deuterium run (SAMPLE II) and the results from the 125 MeV SAMPLE deuterium run (SAMPLE III) both agree with the theoretical prediction on the electroweak radiative correction on the neutral weak axial form factor of the nucleon. In addition to these two experimental results, various theoretical efforts also support the theoretical predition on the electroweak radiative corrections on the axial form factor by Zhu et al. [5]. With these confirmation on the theoretical value of G_A^e, the theoretical value of G_A^e can be used to extract G_M^s from the data from the SAMPLE hydrogen run (SAMPLE I).

REFERENCES

1. R. D. McKeown, Phys. Lett. B **219**, 140 (1989).
2. D. H. Beck, Phys. Rev. D **39**, 3248 (1989).
3. D. B. Kaplan and A. Manohar, Nucl. Phys. **B310**, 527 (1988).
4. SAMPLE Collaboration, R. Hasty et al., Science, **290**, 2021 (2000).
5. Shi-lin Zhu, S. J. Puglia, B. R. Holstein, and M. J. Ramsey-Musolf, Phys. Rev. D **62**, 033008 (2000).
6. C. M. Maekawa, U. van Kolck, Phys. Lett. B **478**, 73 (2000).
7. C. M. Maekawa, J. S. Veiga, U. van Kolck, Phys. Lett. B **488**, 167 (2000).
8. D. O. Riska, Nucl. Phys. **A678**, 79 (2000).
9. L. Diaconescu, R. Schiavilla, U. van Kolck, Phys. Rev. C, **63**, 044007 (2001).
10. R. Schiavilla, J. Carlson and M. Pais, Phys. Rev. C **67**, 032501(R) (2003).
11. C.-P. Liu, G. Prézeau, and M. J. Ramsey-Musolf, Phys. Rev. C **67**, 035501 (2003).
12. R. Schiavilla, *Private communication* (2003).

Neutrino Scattering in Perturbative QCD and Implications for the Weinberg Angle

Stefan Kretzer* and Mary Hall Reno[†]

*Physics Department and RIKEN-BNL Research Center, Brookhaven National Laboratory,
Upton, New York 11973, U.S.A.
[†]Department of Physics and Astronomy, University of Iowa
Iowa City, Iowa 52242 USA

Abstract. We summarize a recent calculation of perturbative neutrino cross sections that includes NLO and mass corrections. We provide numerical results for quantities that are related to the extraction of the weak mixing angle from neutrino deep inelastic scattering.

NEUTRINO CROSS SECTIONS IN PERTURBATIVE QCD

At neutrino energies above a few GeV, neutrino nucleon cross sections are dominated by deep inelastic interactions that are perturbatively assessable in QCD [1]. In Ref. [2] we calculated

- $\mathcal{O}(\alpha_s)$ perturbative next-to-leading order (NLO) corrections
- $\mathcal{O}(M^n/Q^n)$ target mass effects
- $\mathcal{O}(m^n/Q^n)$ and $\mathcal{O}(\ln m^2/Q^2)$ heavy quark mass effects
- $\mathcal{O}(m_l^{2n}/(M^n E_l^n))$ heavy lepton mass effects (mostly for τ production)

by combining the OPE technique of Georgi & Politzer [3] with the 1-loop corrections in Refs. [4]. One could summarize the above list as ξ-scaling for weak structure functions merged with NLO QCD for light and heavy quarks: While the heavy quark and perturbative corrections come in through the expansion of the Wilson coefficients in the operator product expansion, the target mass corrections enter through the Lorentz structure of the corresponding (non-reduced) operators. In this contribution to the proceedings, we will apply the calculation in Ref. [2] to quantities that are related to the weak mixing angle analysis in Ref. [5] which found a value of $\sin^2\Theta_W$ deviating by $\sim 3\sigma$ from the standard model expectation.

THE PASCHOS–WOLFENSTEIN RELATION AND $R^{\nu,\bar{\nu}}$

In the following, integrated neutral current (NC) and charged current (CC) cross sections

$$\sigma_{NC,CC}^{\nu,\bar{\nu}} = \frac{\int dE_{\nu,\bar{\nu}}\, d\sigma_{NC,CC}^{\nu,\bar{\nu}}\, \Phi(E_{\nu\bar{\nu}})\,|_{20\text{ GeV}<yE_{\nu,\bar{\nu}}<180\text{ GeV}}}{\int dE_{\nu,\bar{\nu}}\, \Phi(E_{\nu\bar{\nu}})} \qquad (1)$$

will refer to flux-averaged integrals with a cut on hadronic energy as in the experimental analysis [5]. We will consider the *counting experiment* observables

$$R^{\nu,\bar{\nu}} \equiv \frac{\sigma_{NC}^{\nu,\bar{\nu}}}{\sigma_{CC}^{\nu,\bar{\nu}}} \qquad (2)$$

as well as the Paschos-Wolfenstein [6] relation

$$R^{-} \equiv \frac{\sigma_{NC}^{\nu} - \sigma_{NC}^{\bar{\nu}}}{\sigma_{CC}^{\nu} - \sigma_{CC}^{\bar{\nu}}} \simeq \frac{1}{2} - \sin^{2}\Theta_{W} \qquad (3)$$

For an ideally iso-scalar target and under the neglect of charm production components one has that the approximation Eq. (3) is exact at arbitrary order in QCD as long as isospin symmetry is exact and $(s - \bar{s})(x) = 0$ holds for the nucleon's (anti-)strange quark parton distribution functions [7]. For definiteness, we will consider the scattering of neutrinos on an $Z = 26, A = 56$ iron target and charm production with $m_c = 1.3$ GeV. We assume $(s - \bar{s})(x) = 0$ to focus on the NLO and mass corrections.

Our results for $R^{\nu,\bar{\nu}}$ are summarized in Table 1 for two sets of parton distribution functions [8, 9], for a LO or NLO evaluation and for the standard model and the anomalous value of the Weinberg angle. In the first line, the numbers in parentheses refer to a perturbative expansion of the ratios $R^{\nu,\bar{\nu}}$ directly (instead of the ratios of perturbatively expanded cross sections in Eq. (2)); i.e. schematically $R^{\nu} = R_{(0)}^{\nu} + \alpha_s R_{(1)}^{\nu}$ inside the parentheses. These results can be summarized as follows

- $R^{\bar{\nu}}$ is insensitive to the Weinberg angle and sensitive to NLO corrections
- R^{ν} is insensitive to NLO corrections within its sensitivity to the Weinberg angle
- The impact of PDF uncertainties from pre-determined PDF fits is inconclusive, e.g. the error estimate of CTEQ6 for R^{ν} does not overlap with the evaluation based on GRV.

From these results, one cannot derive a conclusive estimate of the impact of NLO corrections on the analysis [5] and further work will be required. For now, we will restrict ourselves to playing the game to treat Eq. (3) as a would-be identity and to solve it for the Weinberg angle:

$$\left.\frac{1}{2} - R^{-}\right|_{\{\text{GRV LO, }\sin^2\Theta_W=0.2227\}} = 0.2192(3) \qquad (4)$$

$$\left.\frac{1}{2} - R^{-}\right|_{\{\text{GRV NLO, }\sin^2\Theta_W=0.2227\}} = 0.2192(2) \qquad (5)$$

$$\left.\frac{1}{2} - R^{-}\right|_{\{\text{CTEQ NLO, }\sin^2\Theta_W=0.2227\}} = 0.2196(9) \pm 0.0005(1) \qquad (6)$$

The difference between the numerical values in Eqs. (4) and Eqs. (5) reflects the impact of a LO or NLO evaluation of the cross sections entering R^{-}. The error quoted with the NLO evaluation using CTEQ6M refers to the master formula (3) in Ref. [9]

TABLE 1. The ratios $R^{\nu,\bar{\nu}}$ as defined in Eq. (2). Details in the text.

PDF ($\sin^2 \Theta_W$)	R^ν	$R^{\bar{\nu}}$
GRV NLO (0.2227)	0.3120 (0.3115)	0.3844 (0.3838)
GRV LO (0.2227)	0.3125	0.3860
GRV NLO (0.2277)	0.3088	0.3839
CTEQ6 NLO (0.2227)	0.3105 ± 0.0006	0.3841 ± 0.0038

and has to be understood as explained in detail in this reference. For the observable R^- we find a very robust stability under NLO corrections. We also find a similar stability with regards to PDF variations as long as they do not exploit any new physical degree of freedom such as isospin violations or $(s - \bar{s})(x) \neq 0$. Note, however, that $(s - \bar{s})(x) \neq 0$ has to be expected in general because there is no symmetry that would protect $(s - \bar{s})(x) = 0$. The consequences of this strange sea asymmetry will be discussed in [10].

ACKNOWLEDGMENTS

S.K. is grateful to RIKEN, Brookhaven National Laboratory and the U.S. Department of Energy (contract No. DE-AC02-98CH10886) for providing the facilities essential for the completion of this work. The work is also supported by the U.S. Department of Energy under Contract No. FG02-91ER40664.

REFERENCES

1. E. A. Paschos and J. Y. Yu, Phys. Rev. D **65**, 033002 (2002); S. Kretzer and M. H. Reno, Phys. Rev. D **66**, 113007 (2002); H. M. Gallagher and M. C. Goodman, NuMI note NuMI-112 (1995), http://www.hep.anl.gov/ndk/hypertext/numi_notes.html; and references in these articels.
2. S. Kretzer and M. H. Reno, hep-ph/0307023.
3. H. Georgi and H. D. Politzer, Phys. Rev. D **14**, 1829 (1976); A. De Rújula, H. Georgi, and H. D. Politzer, Ann. Phys. **103**, 315 (1977).
4. G. Altarelli, R. K. Ellis and G. Martinelli, Nucl. Phys. B **157**, 461 (1979); T. Gottschalk, Phys. Rev. D **23** (1981) 56; W. Furmanski and R. Petronzio, Z. Phys. C **11**, 293 (1982); M. Glück, R.M. Godbole and E. Reya, Z. Phys. C**38** (1988) 441; **39** (1988) 590 (E). M. Glück, S. Kretzer and E. Reya, Phys. Lett. B **380**, 171 (1996); B **405**, 391 (1996) (E).
5. NuTeV Collaboration, G.P. Zeller *et al.*, Phys. Rev.Lett. **88**, 091802, 2002; and K. McFarland's contribution to these proceedings.
6. E.A. Paschos and L. Wolfenstein, Phys. Rev. D **7**, 91 (1973).
7. S. Davidson, S. Forte, P. Gambino, N. Rius, A. Strumia, JHEP **0202**, 037, 2002.
8. M. Glück, E. Reya and A. Vogt, Eur. Phys. J. C **5**, 461 (1998) [arXiv:hep-ph/9806404].
9. J. Pumplin, D. R. Stump, J. Huston, H. L. Lai, P. Nadolsky and W. K. Tung, JHEP **0207**, 012 (2002).
10. S. Kretzer, F. Olness, J. Pumplin, M.H. Reno, D. Stump and W.-K. Tung, article under preparation.

Probing Supersymmetry with Neutral Current Scattering Experiments

A. Kurylov*, M.J. Ramsey-Musolf*[†] and S.Su*

California Institute of Technology, Pasadena, CA 91125 USA
[†]*Department of Physics, University of Connecticut, Storrs, CT 06269 USA*

Abstract. We compute the supersymmetric contributions to the weak charges of the electron (Q_W^e) and proton (Q_W^p) in the framework of Minimal Supersymmetric Standard Model. We also consider the ratio of neutral current to charged current cross sections, R_ν and $R_{\bar\nu}$ at ν ($\bar\nu$)-nucleus deep inelastic scattering, and compare the supersymmetric corrections with the deviations of these quantities from the Standard Model predictions implied by the recent NuTeV measurement.

INTRODUCTION

In the Standard Model (SM) of particle physics, the predicted running of $\sin^2\theta_W$ from Z-pole to low energy: $\sin^2\theta_W(0) - \sin^2\theta_W(M_Z) = 0.007$, has never been established experimentally to a high precision. $\sin^2\theta_W(M_Z)$ can be obtained through the Z-pole precision measurements with very small error. However, no determination of $\sin^2\theta_W$ at low energy with similar precision is available. More recently, the results of cesium atomic parity-violation (APV) [1] and ν- ($\bar\nu$-) nucleus deep inelastic scattering (DIS)[2] have been interpreted as determinations of the scale-dependence of $\sin^2\theta_W$. The cesium APV result appears to be consistent with the SM prediction for $q^2 \approx 0$, whereas the neutrino DIS measurement implies a $+3\sigma$ deviation at $|q^2| \approx 100\,\text{GeV}^2$. If conventional hadron structure effects are ultimately unable to account for the NuTeV "anomaly", the results of this precision measurement would point to new physics.

In light of this situation, two new measurements involving polarized electron scattering have taken on added interest: parity-violating (PV) Möller (ee) scattering at SLAC[3] and elastic, PV ep scattering at the Jefferson Lab (JLab)[4]. In the absence of new physics, both measurements could be used to determine $\sin^2\theta_W$ at the same scale: $|q^2| \approx 0.03\,\text{GeV}^2$, with comparable precision in each case: $\delta\sin^2\theta_W = 0.0007$. Furthermore, the precision needed to probe new physics effects, e.g. supersymmetry (SUSY), is roughly an order of magnitude less stringent, owing to a fortuitous suppression of the SM electron and proton weak charge: $Q_W^p = -Q_W^e = 1 - 4\sin^2\theta_W \approx 0.1$ at tree-level. Consequently, experimental precision of order a few percent, rather than a few tenths of a percent, is needed to probe new physics corrections.

The goal of our study is to develop consistency check for theories of new physics using the low energy neutral current scattering measurements. In particular, we will consider the Minimal Supersymmetric Extension of SM (MSSM)[5], which is the most promising candidate for new physics beyond SM. For R-parity conserved MSSM, low-energy precision observables experience SUSY only via loop effects involving virtual super-symmetric particles. Tree level corrections appear once R-parity is broken explicitly. We studies both the PV electron scattering (PVES) and ν ($\bar{\nu}$)-nucleus DIS processes. Details of the calculations presented here can be found in Ref. [6] and [7].

RADIATIVE CONTRIBUTION TO WEAK CHARGE

The weak charge of a particle f is defined as the strength of the effective $A(e) \times V(f)$ interaction: $\mathscr{L}_{EFF}^{ef} = -\frac{G_\mu}{2\sqrt{2}} Q_W^f \bar{e}\gamma_\mu \gamma_5 e \bar{f} \gamma_\mu f$. With higher-order corrections included, the weak charge can be written as $Q_W^f = \rho_{PV}\left[2T_3^f - 4Q_f \kappa_{PV} \sin^2\theta_W \right] + \lambda_f$. The quantities ρ_{PV} and κ_{PV} are universal, while the correction λ_f, on the other hand, does depend on the fermion species. At tree-level, one has $\rho_{PV} = 1 = \kappa_{PV}$ and $\lambda_f = 0$, while at one-loop order $\rho_{PV} = 1 + \delta\rho_{PV}^{SM} + \delta\rho_{PV}^{SUSY}$, and similar formulae apply to κ_{PV} and λ_f.

The counterterm $\delta\hat{G}_\mu$ determined by muon life time and the Z^0 boson self-energy are combined into ρ_{PV} (expressed in terms of the oblique parameters S, T [8]):

$$\rho_{PV} = 1 + \frac{\delta\hat{G}_\mu}{G_\mu} + \frac{\hat{\Pi}_{ZZ}(0)}{M_Z^2} = 1 - \frac{\hat{\Pi}_{WW}(0)}{M_W^2} + \frac{\hat{\Pi}_{ZZ}(0)}{M_Z^2} - \hat{\delta}_{VB}^\mu = 1 + \hat{\alpha}T - \hat{\delta}_{VB}^\mu. \quad (1)$$

The quantity $\hat{\delta}_{VB}^\mu$ denotes the the electroweak vertex, external leg, and box graph corrections to the muon decay amplitude.

$Z - \gamma$ mixing and parity-violating electron-photon coupling $F_A^e(0)$ contribute to κ_{PV}:

$$\kappa_{PV} = 1 + \frac{\hat{c}}{\hat{s}}\frac{\hat{\Pi}_{\gamma Z}(q^2)}{q^2} + 4\hat{c}^2 F_A^e(0) + \frac{\delta\hat{s}_{new}^2}{\hat{s}^2} = 1 + \left(\frac{\hat{c}^2}{\hat{c}^2 - \hat{s}^2}\right)\left(\frac{\hat{\alpha}}{4\hat{s}^2\hat{c}^2}S - \hat{\alpha}T + \hat{\delta}_{VB}^\mu\right)$$

$$+ \frac{\hat{c}}{\hat{s}}\left[\frac{\hat{\Pi}_{Z\gamma}(q^2)}{q^2} - \frac{\hat{\Pi}_{Z\gamma}(M_Z^2)}{M_Z^2}\right] + \left(\frac{\hat{c}^2}{\hat{c}^2 - \hat{s}^2}\right)\left[-\frac{\hat{\Pi}_{\gamma\gamma}(M_Z^2)}{M_Z^2} + \frac{\Delta\hat{\alpha}}{\alpha}\right] + 4\hat{c}^2 F_A^e(q^2) \quad (2)$$

The shift $\delta\hat{s}_{new}^2$ in $\hat{s}^2$ follows from the definition of $\hat{s}^2$ in terms of α, G_μ, and M_Z[6].

The non-universal contribution λ_f to the weak charge is determined by the sum of the the renormalized vertex corrections and the box graphs.

SUSY CORRECTION TO WEAK CHARGES

In order to evaluate the potential size of SUSY loop corrections, a set of about 3000 different combinations of SUSY-breaking parameters was generated. Fig. 1(a) shows the shift in the weak charge of the proton, $\delta Q_W^p = 2\delta Q_W^u + \delta Q_W^d$, versus the corresponding shift in the electron's weak charge, δQ_W^e, normalized to the respective SM values. The corrections in the MSSM (with R-parity conserved) can be as large as $\sim 4\%$ (Q_W^p) and $\sim 8\%$ (Q_W^e) – roughly the size of the proposed experimental errors for the two PVES

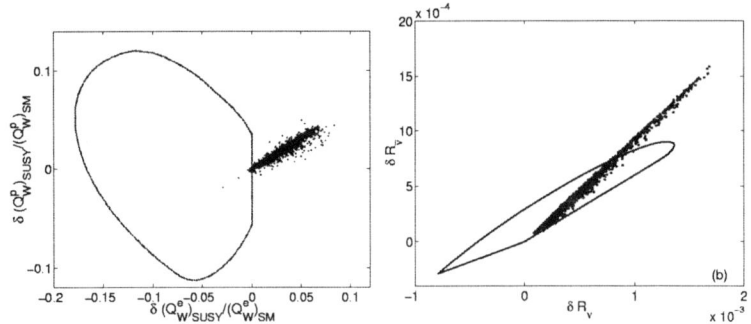

FIGURE 1. Plot(a) shows the relative shifts in electron and proton weak charges due to SUSY effects. Plot(b) shows the MSSM contribution to R_V and $R_{\bar{\nu}}$. Dots indicate MSSM loop corrections for ~ 3000 randomly-generated SUSY-breaking parameters. Interior of truncated elliptical region gives possible shifts due to RPV SUSY interactions (95% confidence).

measurements. The shifts $\delta Q_W^{e,p}$ are dominated by $\delta \kappa_{PV}^{SUSY}$, which is nearly always negative, corresponding to a reduction in the value of $\sin^2 \theta_W^{eff}(q^2) = \kappa_{PV}(q^2) \sin^2 \theta_W$ for the PVES experiments. Since this effect is identical for both Q_W^e and Q_W^p, the dominant effect of $\delta \kappa_{PV}$ produces a linear correlation between the two weak charges.

As evident from Fig. 1 (a), the relative sign of the loop corrections to both Q_W^p and Q_W^e is nearly always the same and positive. This correlation is significant, since the effects of other new physics scenarios can display different signatures. For example, for the general class of theories based on E_6 gauge group, with neutral gauge bosons Z' having mass < 1000 GeV, the effects on Q_W^p and Q_W^e also correlate, but $\delta Q_W^{e,p}/Q_W^{e,p}$ can have either sign in this case[9, 10]. In contrast, leptoquark interactions would not lead to discernible effects in Q_W^e but could induce sizable shifts in Q_W^p [9, 10].

As a corollary, we also find that SUSY loop corrections to the weak charge of cesium is suppressed: $\delta Q_W^{Cs}/Q_W^{Cs} < 0.2\%$ and is equally likely to have either sign, which is smaller than the presently quoted uncertainty for the cesium nuclear weak charge of about 0.6% [11]. Therefore, the present agreement of Q_W^{Cs} with the SM prediction does not preclude significant shifts in $Q_W^{e,p}$ arising from SUSY. The situation is rather different, for example, in the E_6 Z' scenario, where sizable shifts in $Q_W^{e,p}$ would also imply observable deviations of Q_W^{Cs} from the SM prediction.

New tree-level SUSY contributions to the weak charges can be generated when the R parity in MSSM is not conserved. The effects of R-parity violating (RPV) contribution can be parametrized by positive, semi-definite, dimensionless quantities $\Delta_{ijk}(\tilde{f})$ and $\Delta'_{ijk}(\tilde{f})$[12], which are constrained from the existing precision data [12]. The 95% CL region allowed in the $\delta Q_W^p/Q_W^p$ vs. $\delta Q_W^e/Q_W^e$ plane is shown by the closed curve in Fig. 1 (a). We observe that the prospective effects of RPV are quite distinct from SUSY loops. The value of $\delta Q_W^e/Q_W^e$ is never positive in contrast to the situation for SUSY loop effects, whereas $\delta Q_W^p/Q_W^p$ can have either sign.

Thus, a comparison of the two PVES measurements could help determine which extension of the MSSM is to be favored over other new physics scenarios [10].

NUTEV MEASUREMENT

Recently, the NuTeV collaboration has performed a precise determination of the ratio R_ν ($R_{\bar{\nu}}$) of neutral and charged current deep-inelastic ν_μ ($\bar{\nu}_\mu$)-nucleus cross sections[2], which can be expressed in terms of the effective $\nu - q$ hadronic couplings $(g_{L,R}^{\text{eff}})^2$:

$$R_{\nu(\bar{\nu})} = \frac{\sigma(\nu(\bar{\nu})N \to \nu(\bar{\nu})X)}{\sigma(\nu(\bar{\nu})N \to l^{-(+)}X)} = (g_L^{\text{eff}})^2 + r^{(-1)}(g_R^{\text{eff}})^2 \quad , \tag{3}$$

where $r = \sigma_{\bar{\nu}N}^{CC}/\sigma_{\nu N}^{CC}$. Comparing the SM predictions[13] for $(g_{L,R}^{\text{eff}})^2$ with the values obtained by the NuTeV Collaboration yields deviations $\delta R_{\nu(\bar{\nu})} = R_{\nu(\bar{\nu})}^{\text{exp}} - R_{\nu(\bar{\nu})}^{\text{SM}}$, $\delta R_\nu = -0.0029 \pm 0.0015$, $\delta R_{\bar{\nu}} = -0.0015 \pm 0.0026$.

The numerical results for SUSY contributions to R_ν and $R_{\bar{\nu}}$ via the correction to the effective hadronic couplings $(g_{L,R}^{\text{eff}})^2$ are shown in Fig. 1 (b). For detailed analysis, see Ref. [7]. SUSY loop contributions to R_ν and $R_{\bar{\nu}}$ are smaller than the observed deviations. More significantly, the sign of the SUSY loop corrections is nearly always positive, in contrast to the sign of the NuTeV anomaly. Tree-level RPV contributions to R_ν and $R_{\bar{\nu}}$ are by and large positive. While small negative corrections are also possible, they are numerically too small to be interesting.

CONCLUSION

In summary, we have studied the SUSY corrections to the weak charge of the electron and proton, which could be measured at PV ee and ep scattering experiments. The correlation between these two quantities could be used to distinguish various new physics. We also examined the SUSY contributions to the NuTeV measurements and found that it is hard to explain the NuTeV anomaly in the framework of MSSM.

REFERENCES

1. S.C. Bennett and C.E. Wieman, Phys. Rev. Lett. **82**, 2484 (1999); C.S. Wood et al., Science **275**, 1759 (1997).
2. G.P. Zeller et al, NuTeV Collaboration, Phys. Rev. Lett. **88**:091802 (2002).
3. SLAC Experiment E-158, E. W. Hughes, K. Kumar, P.A. Souder, spokespersons.
4. JLab experiment E-02-020, R. Carlini, J.M. Finn, S. Kowalski, and S. Page, spokespersons.
5. H.E. Haber, G.L Kane, Phys. Rep. **117**, 74 (1985).
6. A. Kurylov, M. J. Ramsey-Musolf, and S. Su, hep-ph/0303026, to appear in Phys. Rev. **D**.
7. A. Kurylov, M. J. Ramsey-Musolf, and S. Su, hep-ph/0301208, to appear in Nucl. Phys. **B**.
8. G. Degrassi, B. A. Kniehl, and A. Sirlin, Phys. Rev. D **48**, 3963 (1993).
9. M.J. Ramsey-Musolf, Phys. Rev. C **60**:015501 (1999).
10. J. Erler, M.J. Ramsey-Musolf, and A. Kurylov, [hep-ph/0302149] (2003).
11. A. I .Milstein, O.P. Sushkov, and I.S. Terekhov, hep-ph/0212072.
12. M.J. Ramsey-Musolf, Phys. Rev. D **62**:056009 (2000).
13. K. Hagiwara et. al., Review of Particle Physics, Phys. Rev. D **66**, 10001 (2002); J. Erler, private communication.

Qweak: A Precision Measurement of the Proton's Weak Charge

Gregory S. Mitchell, for the Qweak Collaboration[1]

Physics Division, Los Alamos National Laboratory, Los Alamos, NM 87545, USA

Abstract.
The Qweak experiment at Jefferson Lab aims to make a 4% measurement of the parity-violating asymmetry in elastic scattering at very low Q^2 of a longitudinally polarized electron beam on a proton target. The experiment will measure the weak charge of the proton, and thus the weak mixing angle at low energy scale, providing a precision test of the Standard Model. Since the value of the weak mixing angle is approximately 1/4, the weak charge of the proton $Q_w^p = 1 - 4\sin^2\theta_w$ is suppressed in the Standard Model, making it especially sensitive to the value of the mixing angle and also to possible new physics. The experiment is approved to run at JLab, and the construction plan calls for the hardware to be ready to install in Hall C in 2007. The experiment will be a 2200 hour measurement, employing: an 80% polarized, 180 μA, 1.2 GeV electron beam; a 35 cm liquid hydrogen target; and a toroidal magnet to focus electrons scattered at 9°, a small forward angle corresponding to $Q^2 = 0.03$ (GeV/c)^2. With these kinematics the systematic uncertainties from hadronic processes are strongly suppressed. To obtain the necessary statistics the experiment must run at an event rate of over 6 GHz. This requires current mode detection of the scattered electrons, which will be achieved with synthetic quartz Čerenkov detectors. A tracking system will be used in a low-rate counting mode to determine average Q^2 and the dilution factor of background events. The theoretical context of the experiment and the status of its design are discussed.

INTRODUCTION

There are strong theoretical reasons to expect that the Standard Model is a low-energy effective theory of some more fundamental description of nature. To identify new physics, one method is to observe directly new particles and interactions at large energy scale. Alternatively, an indirect search can be made at low energy, where a precision measurement may observe small effects due to the new physics.

The Qweak experiment [1] at Jefferson Lab (JLab) will make such a precision measurement of the asymmetry between cross-sections for positive and negative helicity electrons in polarized elastic electron-proton scattering. The asymmetry violates parity, and arises from the interference of electromagnetic and weak amplitudes (photon and Z boson exchange). At this low energy scale, the asymmetry is a measure of the weak

---

[1] The Qweak Collaboration: California Institute of Technology, University of Connecticut, Los Alamos National Laboratory, Louisiana Tech University, University of Manitoba, Massachusetts Institute of Technology, Mississippi State University, Universidad Nacional Autonoma de Mexico, University of New Hampshire, University of Northern British Columbia, Ohio University, Thomas Jefferson National Accelerator Facility, TRIUMF, Virginia Polytechnic Institute, College of William & Mary, Yerevan Physics Institute.

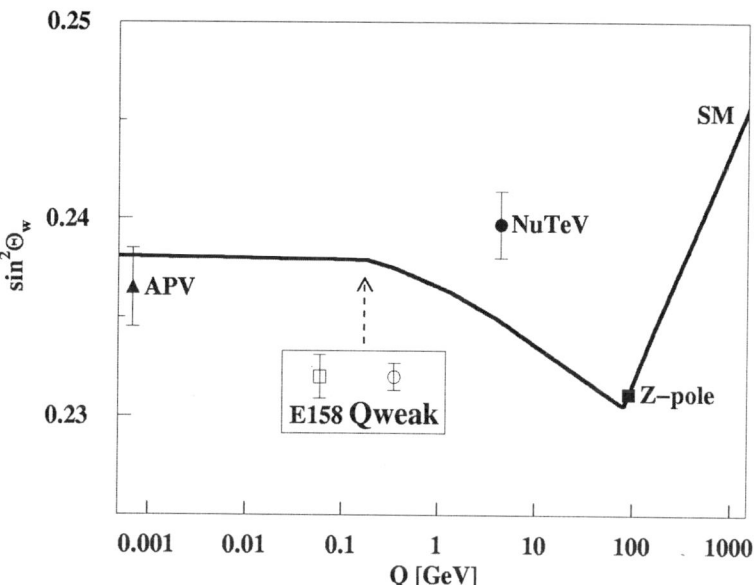

FIGURE 1. Running of the weak mixing angle in the Standard Model, calculated in the $\overline{MS}$ scheme. Shown are results from atomic parity violation [6, 7, 8], NuTeV [11], and the Z pole [5]. E158 [12] is a currently running experiment at SLAC, and the error bar shown is the E158 proposal goal. Figure courtesy of J. Erler, A. Kurylov, and M.J. Ramsey-Musolf [3].

charge of the proton, Q_w^p, which is the strength of the weak vector coupling of the Z boson to the proton. In the limit of small scattering angle and small momentum transfer ($Q^2 \to 0$), the asymmetry is given by [2]:

$$\frac{\sigma_+ - \sigma_-}{\sigma_+ + \sigma_-} = \left[\frac{-G_F}{4\pi\alpha\sqrt{2}}\right][Q^2 Q_w^p + Q^4 B(Q^2)] \approx -0.3 \text{ ppm at } Q^2 = 0.03 \text{ GeV}^2$$

where $B(Q^2)$ is a contribution from electromagnetic and weak form factors. To lowest order, the weak charge is $Q_w^p = 1 - 4\sin^2\theta_w$, where $\sin^2\theta_w \approx 0.23$ is the weak mixing angle. The goal of the Qweak experiment is a 4% measurement of Q_w^p, which corresponds to a 0.3% measurement of $\sin^2\theta_w$.

The weak mixing angle is the single most important parameter of the Standard Model. The Qweak experiment will be a low-energy measurement of the weak mixing angle, which is a test of the Standard Model running of the angle [3, 4]. As shown in Fig. 1 the value of the weak mixing angle is predicted to change (in the $\overline{MS}$ renormalization scheme) by $\sim 4\%$ from the energy scale of the Z pole (where a decade of precision measurements has been made at high energy colliders at SLAC and CERN [5]) to the low energy scale of Qweak. With a 0.3% measurement of the weak mixing angle, the Qweak experiment will make a $\sim 10\sigma$ verification of this effect.

Weak charge and mixing angle results can be obtained in other types of experiments. In atomic parity violation experiments [6, 7, 8], weak charge values are complicated to extract due to atomic structure and many-body nuclear effects (see, for example [9,

10]). The NuTeV experiment [11] at Fermilab has seen evidence of a non-Standard Model value for the weak mixing angle at a lower energy scale than the Z pole, but its result is sensitive to choice of parton distributions and radiative corrections. A currently running experiment at SLAC, E158 [12], is measuring parity violation in electron-electron scattering, and is complementary to Qweak in its sensitivity to different types of new physics (for example: SUSY, leptoquarks, Z' boson, R-parity violation, ...) [3, 4]. The scientific impact of the Qweak experiment will be: a direct measurement of the proton's weak charge in a simple system; a theoretically clean method of measuring precisely the weak mixing angle at low energy scale, a relatively unexplored regime of the Standard Model; and a possible unambiguous indication of new physics.

THE QWEAK EXPERIMENT

An illustration of the conceptual design of the Qweak experiment is shown in Fig. 2. The experiment consists of scattering of a longitudinally polarized 1.2 GeV electron beam by a 35 cm liquid hydrogen target. Elastically scattered electrons at 9 ± 2 degrees are selected by a collimation system, and the electrons are focused by a large toroidal magnet onto a set of eight synthetic quartz Čerenkov detectors. At the average experimental momentum transfer of $Q^2 = 0.03$ GeV^2 the expected Qweak asymmetry is small, -0.3 parts per million (ppm). The expected event rate for scattered electrons of ~ 6 GHz (760 MHz per detector octant) precludes counting the individual events. Instead, the experiment will use current mode detection and low noise front end electronics.

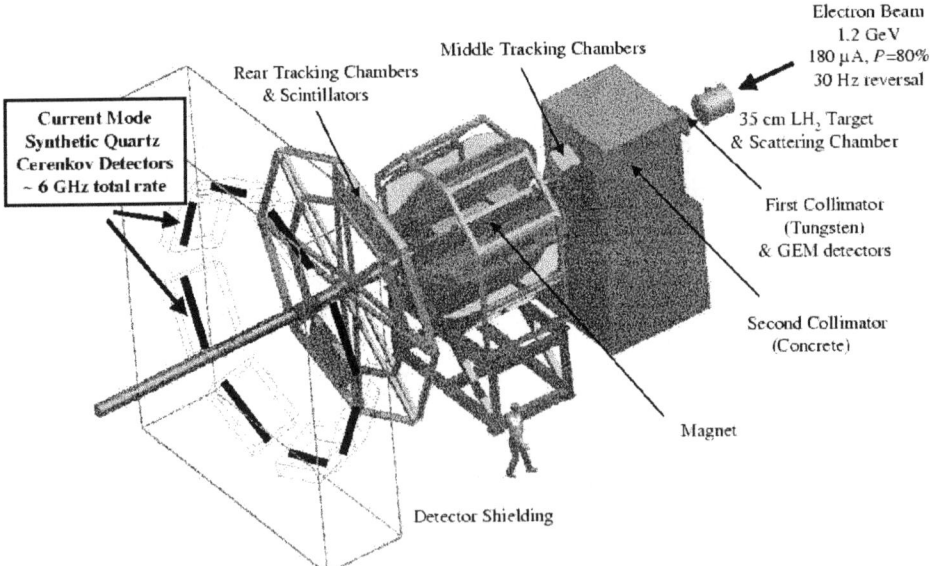

FIGURE 2. Conceptual design for the Qweak experimental setup, in Hall C at Jefferson Lab. The eight quartz detectors are each 2 m × 12 cm × 2.5 cm. The spectrometer provides clean separation of elastic and inelastic electrons at its focal plane.

A tracking system will be used with the beam current reduced by four orders of magnitude, allowing individual events to be observed. This will enable both a measurement of the dilution of the Čerenkov detector signal by background and a precise determination of the average Q^2. The tracking system components will rotate to cover all octants, and will include the following three sets of detectors: a gas electron multiplier as a front vertex detector; wire chambers near the magnet entrance as a measure of scattering angle; and, at the focal plane, vertical drift chambers to allow mapping of the analog response of the Čerenkov system, and large scintillators as a charged particle trigger.

To extract the physics of interest for Qweak requires an extrapolation to low Q^2 of hadronic form factors, $B(Q^2)$ above, which will be the dominant systematic error for the experiment at 2%. These form factors will contribute approximately one third of the experimental asymmetry. They are known with reasonable precision at higher Q^2, and are suppressed by a factor of Q^2 with respect to the weak charge contribution for Qweak. Other experiments completed or underway at JLab (G^0, HAPPEX, HAPPEX II), Bates at MIT (SAMPLE), and MAMI at Mainz (A4), will be used to constrain these form factors. If necessary, in the future the Qweak collaboration could pursue an independent measure of $B(Q^2)$ by running at a Q^2 different from 0.03 GeV^2.

Precision beam polarimetry is required in order to have a polarization contribution to systematic error of less than 1.5%, and a Compton polarimeter will be constructed in JLab Hall C. Radiative corrections for the polarized $e - p$ scattering process at these kinematics have been calculated and are known with a precision of better than 0.7% [3].

The Qweak physics proposal was approved by the January 2002 JLab PAC with an 'A' scientific rating, and Qweak has become an important new thrust of the JLab scientific program. The experiment presented a successful technical design review in January 2003. The Qweak experiment will proceed in two stages: a statistics limited run with a low power target to achieve an 8% or better result on the asymmetry in 2007, followed by a run of 2200 hours at 180 μA to achieve a 4% result. The absolute limits of the technique are under study. This research was supported in part by the U.S. Department of Energy, the National Science Foundation, and the Natural Sciences and Engineering Research Council of Canada.

REFERENCES

1. Qweak Collaboration: Carlini, R. et al., Jefferson Lab Proposal E-02-020.
   Proposal and collaboration member list available at http://www.jlab.org/qweak/.
2. Musolf, M.J. et al., Phys. Rep. **239**, 1 (1994).
3. Erler, J., Kurylov, A., and Ramsey-Musolf, M.J., Phys. Rev. **D68**, 016006 (2003).
4. Kurylov, A., Ramsey-Musolf, M.J., and Su, S., hep-ph/0303026, to appear in Phys. Rev. D.
5. Particle Data Group: Hagiwara, K. et al., Phys. Rev. **D66**, 010001 (2002).
   In particular, the section *Electroweak Model and Constraints on New Physics*.
6. Edwards, N.H. et al., Phys. Rev. Lett. **74**, 2654 (1995).
7. Vetter, P.A. et al., Phys. Rev. Lett. **74**, 2658 (1995).
8. Wood, C.S. et al., Science **275**, 1759 (1997).
9. Kuchiev, M.Y. and Flambaum, V.V., Phys. Rev. Lett. **89**, 283002 (2002).
10. Milstein, A.I., Sushkov, O.P., and Terekhov, I.S., Phys. Rev. Lett. **89**, 283003 (2002).
11. NuTeV Collaboration: Zeller, G.P. et al., Phys. Rev. Lett. **88**, 091802 (2002).
12. SLAC-E-158 Collaboration: Hughes, E.W., Kumar, K.S., Souder, P.A., et al., SLAC-Proposal-E-158.

Hadronic Parity Violation: Past, Present, and Future

Barry R. Holstein

Department of Physics-LGRT
University of Massachusetts
Amherst, MA 01003

Abstract. The history of hadronic parity violation is reviewed, from its beginning in 1957 to the present time. The current state of the subject is shown to be confused and a program to resolve this dilemma, including theoretical calculations using effective theoretic methods together with new low energy experiments in low A systems, is proposed.

1. HADRONIC PARITY VIOLATION: PAST AND PRESENT

The subject of hadronic parity violation began soon after the 1957 publication of the experiment by Wu et al.[1] wherein maximal violation of parity invariance was discovered in nuclear beta decay. Indeed the initial search for parity violation in the NN interaction was carried out by Tanner in in the same year[2]. However, it was not until a decade later that Lobashov et al., using integration techniques, was able to identify a $(-6\pm1) \times 10^{-6}$ signal among the much larger parity conserving background in radiative neutron capture from ^{181}Ta. On the theoretical side the initial systematic study was carried out in 1964 by F.Curtis Michel[3], who who pictured the parity-violating NN interaction as occurring via meson exchange, with the parity-violating weak NNM couplings being estimated via the so-called factorization hypothesis. This paper was followed by other works on the subject, but the seminal contribution in this area was the 1980 paper bv Desplanques, Donoghue, and Holstein which has come to be called simply DDH[4]. In this work, the parity violating NNM couplings were estimated using quark model and symmetry ideas, enabling bounds to be placed as well as estimated "best values" for each coupling, as shown in Table 1. However, the bounds are rather generous and the "best values" are really "best guesses," and the goal of the program is to come up with reliable *experimental* values for these quantities.

Unfortunately, carrying out this program has proved to be exceptionally challenging. Since the natural size of such effects is

$$\sim G_F m_\pi^2 = 10^{-7}$$

early experiments tried to use nuclei wherein the presence of closely spaced parity doublets acted to enhance the size of the effect and produced parity violating signals as large as 1%! However, interpreting such signals, even in P-shell and S,D-shell nuclei wherein one has believable nuclear wavefunctions, has not produced a consistent picture.

For a detailed summary of such efforts, see, e.g., review articles by Adelberger and Haxton[5] and by Haeberli and Holstein[8]. Without going into detail, the present state

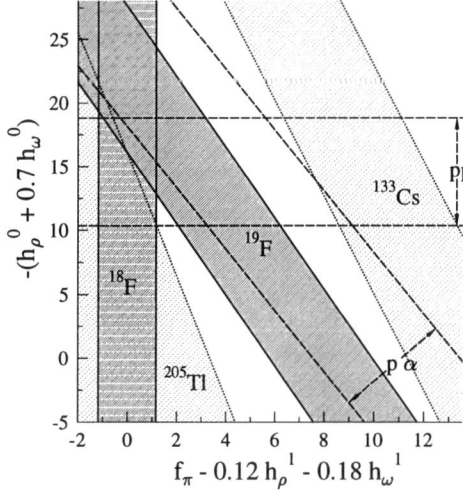

FIGURE 1. Weak parity violating pion and isoscalar vector meson couplings in units of 3.8×10^{-7}.

of affairs is summarized in Figure 1 and shows that such an approach does *not* at the present time yield a consistent set of couplings. (Note: although there exist seven weak couplings in general, the most important by far are the isovector pion coupling, because of its long range, and a linear combination of isoscalar rho and omega couplings. This is the reason that one can represent the experimental situation in terms of a simple two-dimensional plot.) The various allowed ranges shown here come from:

i) experiments in ^{18}F which place limits on the isovector pion coupling[9];

ii) experiments in ^{19}F which place limits on a linear combination of the pion and vector meson couplings[10];

iii) $\vec{p}p$ scattering experiments, which measure the size of the vector meson couplings[11];

iv) $\vec{p}\alpha$ scattering experiments, which measure a linear combination of the pion and vector meson couplings[12];

v) anapole moment measurements in ^{133}Cs and ^{205}Th, which constrain a combination of pion and isoscalar vector meson couplings[13].

There are a number of reasons for this situation, as explored in the above review articles, but the most likely problem is the inadequacy of existing nuclear wavefunctions when dealing with parity violating observables. A recent paper by Miller is intriguing in this regard[15].

TABLE 1. Weak NNM couplings as calculated in Refs. [4, 6, 7]. All numbers are quoted in units of the "sum rule" value $g_\pi = 3.8 \cdot 10^{-8}$.

Coupling	DDH[4] Reasonable Range	DDH[4] "Best" Value	DZ[6]	FCDH[7]
f_π	$0 \to 30$	$+12$	$+3$	$+7$
h_ρ^0	$30 \to -81$	-30	-22	-10
h_ρ^1	$-1 \to 0$	-0.5	$+1$	-1
h_ρ^2	$-20 \to -29$	-25	-18	-18
h_ω^0	$15 \to -27$	-5	-10	-13
h_ω^1	$-5 \to -2$	-3	-6	-6

2. HADRONIC PARITY VIOLATION: FUTURE

In order to bring order out of this chaos, a new approach is needed and this has been suggested in recent theoretical work, which proposes approaching the problem from the perspective of effective theory as well as removing some of the nuclear uncertainty from the problem[16]. In the effective theory picture the parity-violating NN interaction is described in terms of its long range pion exchange component plus a group of short range contact interactions, without reference to vector meson exchange, although the basic form of the operators is similar to that used by DDH. Nevertheless, there is more freedom in choosing the contact couplings here.

In order to remove some of the nuclear uncertainty it is proposed to utilize a series of experiments on systems which are no heavier than the alpha particle, in order that nuclear effects be completely under control. At low energy one need consider only S-P-wave mixing and it is easy to see that there can exist only five independent such couplings:

i) $^3S_1 - {}^1P_1$ mixing with $\Delta I = 0$
ii) $^3S_1 - {}^3P_1$ mixing with $\Delta I = 1$
iii) $^1S_0 - {}^3P_0$ mixing with $\Delta I = 0, 1, 2$

In order to determine these five weak couplings, five independent experiments are required. Two of these are already in hand—

i) low energy $\vec{p} - p$ scattering, which has been performed at LANL, PSI, and Bonn[11]
ii) low energy $\vec{p} - \alpha$ scattering, which has been performed at PSI[12]

but additional results are desperately needed. Two are already on the horizon:

iii) neutron spin rotation in He is an approved experiment at NIST and is proposed at the SNS.
iv) near threshold radiative capture of polarized neutrons by protons is a LANSCE experiment which is now being mounted by Dave Bowman and collaborators[14]. The capture asymmetry is sensitive to the parity-violating pion coupling h_π.

However, other experiments are needed in order to bring this program to completion. Some examples include

- v) measurement of the asymmetry in low energy photodisintegration of deuterium by polarized photons, which could be accomplished at HIGS.
- vi) neutron spin rotation in hydrogen, which could take place at SNS.

The combination of five such precise experiments should allow the extraction of the five independent weak couplings. At this point a reanalysis of the experiments on heavier systems should take place, using the best nuclear wavefunctions, since any discrepancy should now be due solely to nuclear effects. A second front once this program is completed should be to attempt to confront the measured couplings with a fundamental theory, such as QCD, which could be accomplished via lattice or other methods. In any case, after nearly a half century of effort these developments should allow at last a consistent picture of the parity-violating NN force to be developed.

Acknowledgement

This work was supported in part by the National Science Foundation under award PHY-98-01875.

REFERENCES

1. C.S. Wu et al., Phys. Rev. **105**, 1413 (1957).
2. N. Tanner, Phys. Rev. **107**, 1203 (1957).
3. F.C. Michel, Phys. Rev. **133**, B329 (1964).
4. B. Desplanques, J.F. Donoghue, and B.R. Holstein, Ann. Phys. (NY), **124**, 449 (1980).
5. E.G. Adelberger and W.C. Haxton, Ann. Rev. Nuc. Part. Sci. **35**, 501 (1985).
6. V.M. Dubovik and S.V. Zenkin, Ann. Phys. (NY) **172**, 100 (1986).
7. G.B. Feldman, G.A. Crawford, J. Dubach, and B.R. Holstein, Phys. Rev. **C43**, 863 (1991).
8. W. Haeberli and B.R. Holstein, in "Symmetries and Fundamental Interactions in Nuclei," ed. W.C. Haxton and E.M. Henley, World Scientific, Singapore (1995), p.17.
9. C.A. Barnes et al., Phys. Rev. Lett. **40**, 840 (1978); M. Bini et al., Phys. Rev. Lett. **55**, 795 (1985); G. Ahrens et al., Nucl. Phys. **A390**, 486 (1982); H.C. Evans et al., Phys. Rev. Lett. **55**, 791 (1965); P.G. Bizetti et al., Lett. Nuovo Cim. **29**, 167 (1980).
10. E.G. Adelberger in *Polarization Phenomena in Nuclear Physics*, AIP Conf. Proc NO. 69, AIP, New York (1981), p. 1367; E.G. Adelberger et al., Phys. Rev. **C27**, 2833 (1983); K. Elsener et al, Phys. Lett. **B117**, 167 (1982); Phys Rev. Lett. **52**, 1476 (1984).
11. S. Kistryn, J. Lang, J. Liechti, Th. Maier, R. Müller, F. Nessi-Tedaldi, M. Simonius, J. Smyrski, S. Jaccard, W. Haeberli, and J. Sromicki, *Phys. Rev. Lett.* **58** (1987) 1616; P.D. Eversheim, W. Schmitt, S. Kuhn, F. Hinterberger, P. von Rossen, J. Chlebek, R. Gebel, U. Lahr, B. von Przewoski, M. Wiemer, and V. Zell, Phys. Lett. **B256**, 11 (1991).
12. J. Lang et al., Phys. Rev. **C34**, 1545 (1986) and Phys. Rev. Letters **54**, 170 (1985).
13. C.S. Wood et al., Science **275**, 1759 (1997); P. Vetter et al., Phys. Rev. Letters **74**, 2658 (1995); N.H. Edwards et al., Phys. Rev. Letters **74**, 2654 (1995).
14. W.M. Snow et al., Nucl. Inst. Meth. **A440**, 729 (2000).
15. G.A. Miller, [arXiv: nucl-th/0301012].
16. S.-L. Zhu et al., in preparation.

Progress at the WITCH Experiment

M. Beck*, S. Coeck*, B. Delauré*, V. V. Golovko*, V. Yu. Kozlov*, I. S. Kraev*, A. Lindroth*, T. Phalet*, N. Severijns*, S. Versyck*, D. Beck†, W. Quint†, F. Ames**, P. Delahaye**, C. Guenaut‡ and The NIPNET Collaboration

**Instituut voor Kern- en Stralingsfysica, Katholieke Universiteit Leuven, Celestijnenlaan 200D, B-3001 Leuven, Belgium*
†*GSI-Darmstadt, Planckstr. 1, D-64291 Darmstadt, Germany*
***Ludwig-Maximilians-Universität, Sektion Pysik, Schellingstraße 4/IV, D-80799 München, Germany*
‡*Centre de Spectrométrie Nucléaire et de Spectrométrie de Masse, F-91405 Orsay, France*

Abstract. The WITCH-experiment will measure the energy spectrum of the recoiling daughter ions in beta decay to search for non-standard scalar and tensor type interaction. To facilitate this a Penning trap is used to store the radioactive ions. Thus the recoil ions can leave the source without any energy loss and their energy can be probed by the subsequent retardation spectrometer. The experiment is being set up at present at ISOLDE/CERN. The principle and the status of the WITCH-experiment will be presented.

INTRODUCTION

The WITCH experiment will search for scalar (S) and tensor (T) interactions in nuclear beta decay. These interaction types are not present in the standard model of the electroweak interaction, that incorporates just vector (V) and axial-vector (A) interactions. However, there is no fundamental reason why S- and T-interaction should not be present. Moreover, their present experimental limits are at best 6% (68% CL) of the V-interaction strength (see e.g. [1, 2] and references therein).

In β-decay any S(T)-interaction will modify the beta-neutrino angular correlation in pure Fermi (Gamov-Teller) decays. This β-ν angular correlation can be determined by measuring the recoil energy spectrum of the daughter ions after β-decay (Fig. 1).

The recoil energy spectrum after β-decay is difficult to measure due to the low energies involved (typically <500eV). At these energies the recoil ions usually get stopped already in the source. At the WITCH experiment a Penning trap [3, 4] is used to store the radioactive ions in vacuum (Fig. 2). Due to the cylindrical structure of the Penning trap and the magnetic field of 9T necessary to operate it the recoil ions leave the trap through the open ends without any energy loss in surrounding matter.

Any isotope that may be produced can be stored in a Penning trap. For this reason the WITCH experiment is being set up at the isotope separator ISOLDE at CERN [5], where a multitude of different isotopes can be produced. Thus the most suitable isotope can be selected for a given purpose and systematic checks using different isotopes become possible. Eventually it is intended to measure the β-ν angular correlation with the

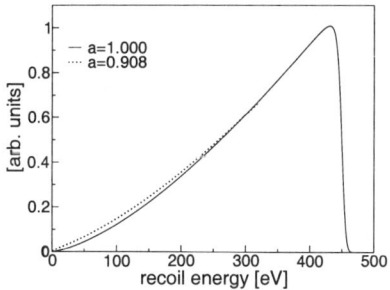

FIGURE 1. Typical recoil energy spectra (calculated). Two different β-ν angular correlation coefficients are compared here (a=1 → pure Fermi decay, a=0.908 → mixed decay of ^{35}Ar).

WITCH experiment with a presicion exceeding 0.5% (68%CL), corresponding to limits well beyond the best present limits on S- and T-interaction. At these levels of precision additional effects will have to be taken into account which are presently negligible in this type of experiment [6, 7, 8, 9].

THE EXPERIMENT

The beam of radioactive ions from ISOLDE gets captured and pre-cooled by REX-TRAP [10]. From REXTRAP the ions are sent in bunches to the cooler trap of WITCH. Here the ions are cooled and centered using buffer gas. Then they get transmitted to the decay trap where they are stored as an ion cloud. The recoil ions from β-decays in

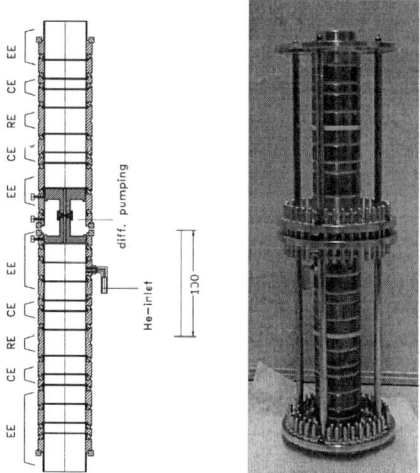

FIGURE 2. The Penning traps (left: schematic, right: assembled). The lower electrodes belong to the cooler trap, the upper ones to the decay trap. The traps are separated by a differential pumping barrier.

this ion cloud spiral adiabatically from the magnetic field of 9T in the decay trap to a field of 0.1T. Due to the adiabatic invariance of the magnetic flux contained in the ion motion [11] a fraction of $1 - 0.1T/9T \approx 98.9\%$ of the radial ion energy gets converted into axial energy. This axial energy is probed by a retardation potential that is applied between the trap region and the analysis region at 0.1T. Those ions that pass the analysis region get re-accelerated to $\approx 10 keV$ and focused onto a micro-channel plate detector where they are counted. By varying the retardation potential the integral recoil energy spectrum is determined. A schematic of this retardation spectrometer is shown on the left side of Fig. 3. The superconducting magnets are shown on the right side. They got mounted in a common cryostat which is shown in Fig. 4 while it was inserted into its support structure. A more detailed description of the experiment can be found in [12].

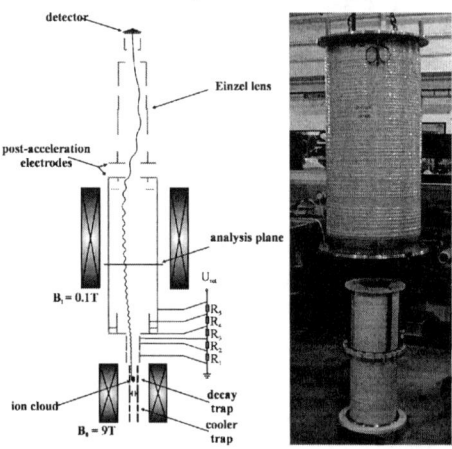

FIGURE 3. The retardation spectrometer (left: schematic, right: the superconducting magnets).

The WITCH experiment is at present being set up at ISOLDE at CERN. The beamlines from REXTRAP to WITCH are operational. The traps are mounted. The retardation section and the top chamber with the detector are under construction. A first measurement of a recoil energy spectrum is expected for spring 2004.

CONCLUSION

The WITCH experiment is well on its way towards a first measurement of a recoil energy spectrum in nuclear beta decay. After a thorough investigation of the systematic uncertainties the sensitivity to scalar and tensor contributions to the standard model vector and axial vector interactions will be greatly enhanced compared to existing experiments.

FIGURE 4. The superconducting magnet system in place. The Penning traps and the superconducting magnets are located inside the large cylinder, the cryostat.

ACKNOWLEDGMENTS

This work is supported by the European Union grant HPRI-CT-2001-50034 (the NIP-NET RTD network), by the Flemish Fund for Scientific Research FWO and by the project GOA 99-02 of the K.U.Leuven. It was partly funded with a fellowship of the Flemish Institute for the stimulation of Scientific-Technological Research in the Industry (IWT). The authors want to thank J. Deutsch for discussions and suggestions.

REFERENCES

1. Severijns, N., Deutsch, J., Beck, D., Beck, M., Delauré, B., Prieels, R., Schuurmans, P., Vereecke, B., and Versyck, S., *Hyp. Int.*, **129**, 223 (2000).
2. Herczeg, P., *Prog. Part. Nucl. Phys.*, **46**, 413 (2001).
3. Penning, F. M., *Physica*, **3**, 873 (1936).
4. Raimbault-Hartmann, H., Beck, D., Bollen, G., König, M., Kluge, H.-J., Schark, E., Stein, J., Schwarz, S., and Szerypo, J., *Nucl. Instr. and Meth.*, **B 126**, 378 (1997).
5. Kugler, E., *Hyp. Int.*, **129**, 23 (2000).
6. Gell-Mann, M., *Phys. Rev.*, **111**, 363 (1958).
7. Sirlin, A., *Phys. Rev.*, **164**, 1767 (1967).
8. Holstein, B. R., *Rev. Mod. Phys.*, **46**, 789 (1974).
9. Grenacs, L., *Ann. Rev. Nucl. Part. Sci.*, **35**, 455 (1985).
10. Ames, F., Beck, D., Bollen, G., Forstner, O., Habs, D., Huber, G., Reisinger, K., and Schmidt, P., "Bunching and cooling of radioactive ions with REXTRAP," in *AIP Conference Proceedings*, AIP, 2002, vol. 606, p. 609.
11. Jackson, J. D., *Classical Electrodynamics*, Wiley, 1962.
12. Beck, M., Ames, F., Beck, D., Bollen, G., Delauré, B., Golovko, V. V., Kozlov, V. Y., Kraev, I. S., Lindroth, A., Phalet, T., Quint, W., Schuurmans, P., Severijns, N., Vereecke, B., and Versyck, S., *Nucl. Instr. and Meth. A*, **503**, 567 (2003).

New results in Superallowed Nuclear Beta Decay

J.C. Hardy

Cyclotron Institute, Texas A & M University,
College Station, TX 77843

Abstract. The value of the V_{ud} matrix element of the Cabibbo-Kobayashi-Maskawa matrix can be derived from nuclear superallowed beta decays, neutron decay and pion beta decay. Today, the most precise value of V_{ud} (0.05% precision) comes from the nuclear decays; and its precision is limited not by experimental error but by the estimated uncertainty in theoretical corrections, which themselves are of order 1%. When combined with the accepted values for V_{us} and V_{ub}, the result differs at the 98% confidence limit from the unitarity condition for the CKM matrix. The focus of recent work on superallowed beta decay has been to test and refine the small nuclear-structure-dependent corrections so that the precision on the value of V_{ud} can be improved and the unitarity test sharpened. A consistent set of calculated corrections has been produced not only for the nine superallowed transitions currently measured to high precision, but for a number of additional superallowed transitions in nuclei not previously accessible to exacting experiments. Now experiments are beginning to achieve the precision required on these previously inaccessible transitions: already in print or in press are results for the superallowed decays of ^{22}Mg, ^{34}Ar, ^{62}Ga and ^{74}Rb. Only the ^{22}Mg and ^{34}Ar transitions have so far provided a test of the structure-dependent corrections, but before long one can expect to have a meaningful comparison between experiment and theory over a range of new cases. Some improvement in the precision of V_{ud} can be expected to follow.

OVERVIEW

Superallowed $0^+ \rightarrow 0^+$ nuclear β-decay depends uniquely on the vector part of the weak interaction. Measurement of the ft-value for such a transition yields a direct determination of the vector coupling constant, G_V, provided that small radiative corrections are properly accounted for. To date, the ft-values for nine $0^+ \rightarrow 0^+$ transitions – the decays of ^{10}C, ^{14}O, ^{26m}Al, ^{34}Cl, ^{38m}K, ^{42}Sc, ^{46}V, ^{50}Mn and ^{54}Co – have been measured with $\sim 0.1\%$ precision or better. Though these nuclei span a wide range of nuclear masses, their ft-values yield fully consistent values for G_V as anticipated by the Conserved Vector Current hypothesis. With G_V thus determined, it is possible to establish a very precise value for the up-down element of the Cabibbo-Kobayashi-Maskawa (CKM) quark-mixing matrix:

$$V_{ud} = 0.9740 \pm 0.0005 \tag{1}$$

Not only is this the most precise determination of V_{ud}, it is the most precise result for any element in the CKM matrix. It also leads to the most demanding test available of CKM unitarity, a fundamental tenet of the minimal standard model.

Strikingly, this test on the top row of the matrix fails by more than two standard deviations [1, 2]: viz.

$$V_{ud}^2 + V_{us}^2 + V_{ub}^2 = 0.9968 \pm 0.0014. \tag{2}$$

In obtaining this result, we have used the Particle Data Group's [3] recommended values for the much smaller matrix elements, V_{us} and V_{ub}.

Because this deviation is not completely definitive statistically, and because any conclusive violation of unitarity would have profound implications for the standard model, there is now considerable interest in sharpening this unitarity test. Recent experimental results are already suggesting that the value of V_{us} may need to be revised [4]. The status of V_{ud} is quite different. The body of experimental data that contributes to its determination includes well over 100 independent – mostly concordant – measurements on nine different transitions, and overall it is certainly very robust; however, the small calculated correction terms that must be applied to the experimental ft-values in order to extract G_V are less well determined. Specifically, G_V is obtained from each ft-value via the relationship [2]

$$\mathcal{F}t \equiv ft(1 + \delta'_R + \delta_{NS})(1 - \delta_C) = \frac{K}{2G_V^2(1 + \Delta_R)}, \tag{3}$$

where K is a known constant, f is the statistical rate function and t is the partial half-life for the transition. The correction terms – all of order 1% or less – comprise δ_C, the isospin-symmetry-breaking correction, δ'_R and δ_{NS}, the transition-dependent parts of the radiative correction and Δ_R, the transition-independent part. Here we have also defined $\mathcal{F}t$ as the "corrected" ft-value. Note that, of the four calculated correction terms, two – δ_C and δ_{NS} – depend on nuclear structure and their influence in Eq.(3) is effectively in the form $(\delta_C - \delta_{NS})$.

The overall ± 0.0005 uncertainty obtained for V_{ud} in Eq.(1) comprises contributions of 0.0001 from experiment, 0.0001 from δ'_R, 0.0003 from $(\delta_C - \delta_{NS})$, and 0.0004 from Δ_R. Thus, to sharpen the unitarity test, the most pressing objective is to reduce the uncertainties on Δ_R and $(\delta_C - \delta_{NS})$. The former is especially important because it also appears in the extraction of G_V from neutron and pion decay, and thus it will also ultimately limit the precision achievable from these decays to approximately the same level as the current nuclear result, regardless of the experimental precision achieved. Improvements in Δ_R are a purely theoretical challenge, however, the solution of which will not depend on further experiments. Experiments can play a role in improving the next most important contributor to the uncertainty on V_{ud}, namely $(\delta_C - \delta_{NS})$. Recently, a new set of consistent calculations for $(\delta_C - \delta_{NS})$ has appeared [1] not only for the nine well known superallowed transitions but for eleven others that are potentially accessible to precise measurements. It is to these new unexplored cases that experiments have turned, seeking validation for the structure-dependent corrections. Success would lead to a significant reduction in the uncertainty applied to V_{ud} from this source.

NEW RESULTS

Although there continue to be fresh measurements that refine and improve the ft-values for the nine well-known cases of superallowed decay, the recent experimental thrust is towards two new series of 0^+ nuclei: the even-Z, $T_z = -1$ nuclei with $18 \leq A \leq 42$, and the odd-Z, $T_z = 0$ nuclei with $A \geq 62$. The main attraction of these new regions is that the calculated values of $(\delta_C - \delta_{NS})$ for the superallowed transitions [1] are larger, or show larger variations from nuclide to nuclide, than the values applied to the familiar cases. This means that their uncorrected ft values are expected to exhibit these larger variations since, according to CVC, the corrected $\mathcal{F}t$ values should all have the same value. This is just what is needed to test the accuracy of the calculations. If the calculations reproduce large experimental variations where they occur, then that must surely verify their reliability for the original nine transitions whose $\delta_C - \delta_{NS}$ values are considerably smaller.

Both new regions present substantial experimental challenges. The decays of the heavier $T_z = 0$ nuclei are of higher energy than any previously studied and each therefore involves numerous weak Gamow-Teller transitions in addition to the superallowed one [5]. Likely there are hundreds of these weak branches, which in total can carry significant strength ($\sim 1\%$) but individually may be too weak to unobserve. Branching-ratio measurements are thus very demanding, particularly with the limited intensities often available for these rather exotic nuclei. Nevertheless, a combination of favorable production rates, meticulous experiments [6, 7, 8] and shell-model calculations [5] have made it possible to obtain rather precise branching ratios for the decays of ^{62}Ga and ^{74}Rb. Precise half-lives have also been measured recently for these two isotopes [8, 9, 10]. Unfortunately, since their half-lives (~ 0.1 s) are considerably shorter than those of the lighter superallowed emitters, high-precision mass measurements (± 2 keV) may be some time in coming. Without them, precise ft values cannot be extracted nor correction terms validated.

More accessible in the short term are the $T_z = -1$ superallowed emitters with $18 \leq A \leq 42$. Their experimental challenge arises from the fact that they have competing *strong* Gamow-Teller transitions. To achieve the $\sim 0.1\%$ precision required for a meaningful superallowed branching-ratio result, all these transitions must be measured with compatible precision, a goal that requires efficiency calibration of a γ-ray detector to unprecedented precision [11]. So far, at Texas A&M we have obtained precise branching ratios for two of these decays: ^{22}Mg [12] and ^{34}Ar [13]. Half-lives and Q-values are also known for both decays, although the half-life for ^{34}Ar and the Q-value for ^{22}Mg have not yet reached the precision that one might ultimately hope for. Even so, ft-values can be derived and used in a meaningful test of the calculated structure-dependent corrections.

The ft-value results for the two new $T_z = -1$ superallowed emitters are shown in Fig. 1, along with those from the nine previously well-known cases. The shaded bands show calculated results that assume CVC, by taking a unique value for $\mathcal{F}t$ (shown at the top of the figure), and obtain ft from Eq.(3) by applying the carrection terms from Ref. [1]. The agreement is excellent, and clearly demonstrates – within the uncertainties of the new measurements – that the calculations closely follow the nucleus-to-nucleus variations in ft-values. As these uncertainties

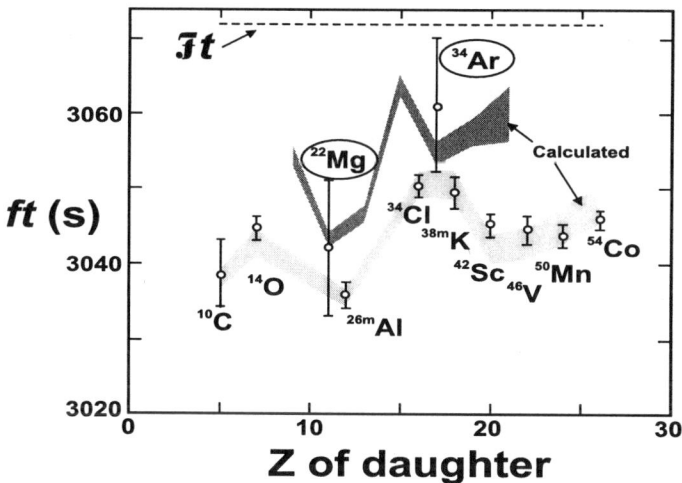

FIGURE 1. Uncorrected ft-values measured for the 11 most precisely measured superallowed transitions. The shaded bands show the results of calculations for each series of nuclei.

are improved, and other new superallowed transitions are added, the calculated structure-dependent corrections will come under even greater scrutiny. If the current level of agreement is maintained, then the precision on the extracted value of V_{ud} will be significantly improved.

ACKNOWLEDGMENTS

This work involved many valuable collaborators, among them I.S. Towner, V.E. Iacob, M. Sanchez-Vega, N. Nica and R.G. Neilson. It was supported by the U.S. Department of Energy under Grant No. DE-FG03-93ER40773 and by the Robert A. Welch Foundation.

REFERENCES

1. I.S. Towner and J.C. Hardy, Phys. Rev. C **66**, 035501 (2002).
2. I.S. Towner and J.C. Hardy, J. Phys. G: Nucl. Part. Phys. **29**, 197 (2003).
3. K. Hagiwara et al., Phys. Rev. D **66**, 010001 (2002).
4. A. Sher et al., arXiv:hep-ex/0305042.
5. J.C. Hardy and I.S. Towner, Phys. Rev. Lett. **88** 252501 (2002).
6. B. Blank, Eur. Phys. J A **15**, 121 (2002).
7. A. Piechaczek et al., Phys. Rev. C **67**, 051305(R) (2003).
8. B.C. Hyman et al., Phys. Rev. C **68**, 015501 (2003).
9. B. Blank et al., to be published.
10. G.C. Ball et al., Phys. Rev. Lett. **86**, 1454 (2001).
11. R.G. Helmer et al., Nucl. Instrum. Meth. Phys. Res. A, to be published.
12. J.C. Hardy et al., Phys. Rev. Lett., in press.
13. V.E. Iacob et al., to be published.

Charged current universality problem and NuTeV anomaly: is SUSY to blame?

Andriy Kurylov*, Michael Ramsey-Musolf*† and Shufang Su*

*California Institute of Technology, Pasadena, CA 91125, USA
†Department of Physics, University of Connecticut, Storrs, CT 06269, USA

Abstract.
We compute the complete one-loop contributions to low-energy charged current weak interaction observables in the Minimal Supersymmetric Standard Model (MSSM). We obtain the constraints on the MSSM parameter space which arise when precision low-energy charged current (CC) data are analyzed in tandem with measurements of the muon anomaly. The data imply a pattern of mass splittings among first and second generation sleptons and squarks which contradicts predictions of widely used models for supersymmetry breaking mediation. We also discuss the implications of these constraints on the SUSY one-loop contributions to the (anti)neutrino-nucleus deep inelastic scattering. We consider the ratios of neutral current to charged current cross sections, and compare with the deviations of these quantities from the Standard Model predictions implied by the recent NuTeV measurement. We discuss one scenario in which a right-sign effect arises, and show that it is ruled out by the CC data. We also study R parity-violating contributions. Although such effects can account for the violation of the first row CKM unitarity, they can not reproduce the NuTeV anomaly. If NuTeV anomaly is ultimately explained within the SM, R parity-violating resolution of the CKM unitarity problem can be tested in parity-violating electron scattering experiments at SLAC and TJNAF.

INTRODUCTION

The universality of the charged current weak interaction (CCWI) is an important feature of the Standard Model (SM). The presence of a common coupling strength and $(V-A) \times (V-A)$ current-current interaction structure for all CCWI processes has been tested with high precision in a number of leptonic and semileptonic experiments. The results place significant limits on scenarios for physics beyond the SM which may generate breakdowns of CCWI universality. In the first part of this talk, we report our findings regarding implications of universality tests for minimal supersymmetric extension of the SM (MSSM) with R-parity conservation–one of the leading candidates for "new physics". Supersymmetric theories which break R-parity conservation (equivalently, baryon minus lepton number, $B-L$, conservation) have been considered elsewhere [1, 2, 3].

Of particular interest for our analysis are the results of superallowed nuclear β-decays, from which one extracts the Cabibbo-Kobayashi-Maskawa (CKM) quark mixing matrix element $|V_{ud}|$. When $|V_{ud}|$ is considered along with the values of $|V_{us}|$ and $|V_{ub}|$ determined from K_{e3} and B-meson decays, respectively, one obtains for the sum of the squares a result falling below the unitarity requirement by 2.2σ[4]. In what follows, we discuss the implications of this deviation for the MSSM spectrum.

In the second part of the talk, we discuss the recent result by NuTeV collaboration, which has performed a precise determination of the ratio R_ν ($R_{\bar\nu}$) of neutral current (NC) and charged current (CC) deep-inelastic ν_μ ($\bar\nu_\mu$)-nucleus cross sections [5], which can be expressed as $R_{\nu(\bar\nu)} = (g_L^{\text{eff}})^2 + r^{(-1)}(g_R^{\text{eff}})^2$, where $r = \sigma_{\nu N}^{CC}/\sigma_{\nu N}^{CC}$ and $(g_{L,R}^{\text{eff}})^2$ are effective hadronic couplings (defined below). Comparing the SM predictions [6, 7] for $(g_{L,R}^{\text{eff}})^2$ with the values obtained by the NuTeV Collaboration yields deviations[1] $\delta R_{\nu(\bar\nu)} = R_{\nu(\bar\nu)}^{\text{exp}} - R_{\nu(\bar\nu)}^{\text{SM}}$: $\delta R_\nu = -0.0033 \pm 0.0007$, $\delta R_{\bar\nu} = -0.0019 \pm 0.0016$.

Within the SM, these results may be interpreted as a test of the scale-dependence of the $\sin^2\theta_W$ since the $(g_{L,R}^{\text{eff}})^2$ depend on the weak mixing angle. While the SM prediction for $\sin^2\theta_W$ at $\mu = M_z$ has been confirmed with high precision at LEP and SLC, the predicted running of this parameter to lower scales has yet to be studied systematically. The results from the NuTeV measurement imply a $+3\sigma$ deviation at $\mu \sim 10$ GeV. This interpretation of the NuTeV results has been the subject of some debate. Unaccounted for QCD effects, such as charge symmetry-breaking in parton distributions or nuclear shadowing[8], have been proposed as possible remedies for the anomaly. Alternatively, one may consider physics beyond the SM, as reviewed in Ref. [9]. In what follows, we focus on one new physics scenario, namely, supersymmetry (SUSY). Here, we carry out a model-independent treatment, avoiding the choice of a specific mechanism for SUSY-breaking mediation. We find that it is difficult – if not impossible – to choose MSSM parameters so as to improve agreement with the NuTeV result.

MSSM IN A NUTSHELL

Low energy SUSY is an attractive scenario from a number of standpoints. By introducing a superpartner for every SM particle, it provides a solution to the hierarchy problem associated with Higgs mass renormalization; it produces coupling unification at the GUT scale; and it is a prediction of superstring theory. It remains to be seen, however, which version of SUSY correctly describes electroweak phenomena. In particular, details of the superpartner spectrum (*e.g.*, masses and mixing angles) are largely unknown. Limits on branching ratios obtained from collider data provide, in general, only weak lower bounds.

If R-parity (and, thus, $B - L$) is conserved, the SUSY corrections to low energy observables arise only via tiny loop effects. In order to become sensitive to such contributions, one generally requires a precision of $\sim (\alpha/\pi) \times (M/\tilde{M})^2$, where M is the relevant mass of a SM particle and $\tilde{M}$ is a superpartner mass. For CCWI observables and NuTeV, one has $M \sim M_W$, so that only a few $\times 10^{-3}$ precision is needed to achieve sensitivity. If R-parity is not conserved, new tree-level SUSY effects appear for virtually all observables. A comprehensive review of supersymmetry can be found in Ref. [10].

---

[1] We use the quoted experimental errors on R_ν and $R_{\bar\nu}$, rather than adding the errors on $(g_{L,R}^{\text{eff}})^2$ in quadrature, since the latter are correlated and derived from the experimental cross section ratios.

CHARGED CURRENT UNIVBERSALITY

Any CCWI amplitude is properly normalized to $G_\mu = \sqrt{2}g^2/(8M_W^2)[1+\Delta r_\mu]$, the Fermi constant measured in μ-decay, which is one of the three most precise inputs for the gauge sector of the MSSM. Here, g is the universal weak coupling, M_W is the W-boson mass, and Δr_μ includes the effects of weak, radiative corrections in the MSSM as well as other possible new physics contributions to μ-decay. The Fermi constant relevant for light quark β-decay is $G_F^\beta = G_\mu V_{ud}(1 - \Delta r_\mu + \Delta r_\beta)$, where Δr_β is the analogue of Δr_μ. The difference $\Delta r_\beta - \Delta r_\mu$ is sensitive only to non-universal effects, such as vertex corrections, box diagrams, and external leg corrections. Because it compares these corrections as they appear in leptonic and semileptonic decay amplitudes, G_F^β is essentially a measure of slepton-squark universality in the MSSM. Moreover, since both Δr_β and Δr_μ pertain to processes with $e^+\nu_e$ ($e^-\bar{\nu}_e$) in the final state, the difference $\Delta r_\beta - \Delta r_\mu$ is considerably more sensitive to effects produced by second generation sleptons than to those produced by the first generation.

Requiring that the non-universal SUSY corrections produce no additional deviation from CKM unitarity (on top of existing 2.2σ) together with constraints from muon anomalous magnetic moment [11] implies that $\Delta r_\beta^{SUSY} - \Delta r_\mu^{SUSY} < 0$ at 95% confidence level. We found that this constraint leads to a non-trival relationship between the masses of the superpartners of the muon $\tilde{\mu}_L$ and the first generation squarks $\tilde{q}_L$. In particular, one always has $M_{\tilde{\mu}_L} > M_{\tilde{q}_L}$. This phenomenological solution is particularly interesting from the standpoint of both gauge-mediated and mSUGRA models of SUSY-breaking mediation, which generally predict $M_{\tilde{q}} > M_{\tilde{l}}$. In mSUGRA, this hierarchy results from gluino contributions to the renormalization group running of the masses down from the GUT scale. Inverting this hierarchy would presumably require modifying the universality assumptions made for the parameters of $\mathcal{L}_{\text{soft}}$ at the GUT scale.

NUTEV ANOMALY

We represent the charged and neutral current neutrino-quark interactions by an effective four fermion Lagrangian:

$$\mathcal{L}_{\nu q}^{NC} = -\frac{G_\mu \rho_{\nu N}^{NC}}{\sqrt{2}} \bar{\nu}_\mu \gamma^\lambda (1-\gamma_5)\nu_\mu \sum_q \bar{q}\gamma_\lambda [2\epsilon_L^q(1-\gamma_5)/2 + 2\epsilon_R^q(1+\gamma_5)/2]q \quad (1)$$

$$\mathcal{L}_{\nu q}^{CC} = -\frac{G_\mu \rho_{\nu N}^{CC}}{\sqrt{2}} \bar{\mu}\gamma^\lambda(1-\gamma_5)\nu_\mu \bar{u}\gamma_\lambda(1-\gamma_5)d + \text{h.c.} \; , \quad (2)$$

where

$$\epsilon_L^q = I_L^3 - Q_q \kappa_\nu \sin^2\theta_W + \lambda_L^q, \; \epsilon_R^q = -Q_q \kappa_\nu \sin^2\theta_W + \lambda_R^q. \quad (3)$$

The parameters $\rho_{\nu N}^{NC} = \rho_{\nu N}^{CC} = \kappa_\nu = 1$ and $\lambda_{L,R}^q = 0$ at tree-level in the SM. These quantities differ from their tree-level values when $\mathcal{O}(\alpha)$ corrections in the SM or MSSM are included or when other new physics contributions arise. The NC to CC cross section ratios R_ν and $R_{\bar\nu}$ can be expressed in terms of the above parameters via the effective couplings $(g_{L,R}^{\text{eff}})^2$ in a straightforward way:

$$(g_{L,R}^{\text{eff}})^2 = \left(\frac{\hat{M}_Z^2}{\hat{M}_W^2}\right)^2 \left(\frac{\hat{M}_W^2 - q^2}{\hat{M}_Z^2 - q^2}\right)^2 \left(\frac{\rho_{\nu N}^{NC}}{\rho_{\nu N}^{CC}}\right)^2 \sum_q (\epsilon_{L,R}^q)^2 \qquad (4)$$

We calculate the MSSM contributions to $\rho_{\nu N}^{NC,CC}$, κ_ν, $\lambda_{L,R}^q$, R_ν, and $R_{\bar\nu}$ both with and without R-parity conservation. In the former case we randomly choose values for these parameters discarding any points that yield SUSY particle masses below present collider lower bounds or violate constraints from the Z-pole electroweak precision measurements. In the latter case we utilize constraints on R-parity violating parameters from superallowed nuclear β-decay, atomic PV measurements of the cesium weak charge, the e/μ ratio $R_{e/\mu}$ in π_{l2} decays, and a comparison of the Fermi constant G_μ with the appropriate combination of α, M_Z, and $\sin^2\theta_W$. In both cases, SUSY corrections to R_ν and $R_{\bar\nu}$ generally have too small a magnitude and the wrong sign to account for the effect.

CONCLUSIONS

SUSY loop corrections to CC universality are generically small. They can alleviate the unitarity problem by no more that 0.5σ but required parameters are difficult to accommodate in all known models of SUSY breaking. SUSY corrections also tend to worsen the NuTeV anomaly. If the CKM unitarity problem and NuTeV anomaly are not resolved within the SM (by, *e.g.* new value of V_{us} for the former and nuclear effects for the latter), they will pose serious difficulties for considering the MSSM as the true extension of the SM.

REFERENCES

1. V. Barger and K. Cheung, Phys. Lett. B480, 149 (2000).
2. M.J. Ramsey-Musolf, Phys. Rev. D62:056009 (2000).
3. B.C. Allanach, A. Dedes, and H. Dreiner, Phys. Rev. D60:075014 (1999).
4. I.S. Towner and J.C. Hardy, nucl-th/9809087.
5. G.P. Zeller, *et al.*, the NuTeV Collaboration, Phys. Rev. Lett. **88**, 091802 (2002).
6. Review of Particle Properties, Phys. Rev. **D 66**, 010001 (2002).
7. J. Erler, private communication.
8. G.A. Miller and A.W. Thomas, hep-ex/0204007; see also K.S. McFarland, hep-cx/021001.
9. S. Davidson, *et al.*, JHEP **02**, 037 (2002).
10. For a review, see Perspectives on Supersymmetry, G.L. Kane, Ed., World Scientific, Singapore, 1998.
11. Muon g-2 Collaboration, H.N. Brown *et al.*, Phys. Rev. Lett. **86**, 2227 (2001).

Progress towards measuring the electric dipole moment of the electron in metastable PbO

D. Kawall*, F. Bay*, S. Bickman*, Y. Jiang* and D. DeMille*

Department of Physics, Yale University, New Haven, CT 06520-8120

Abstract. A permanent electric dipole moment (EDM) of an elementary particle would violate parity and time-reversal symmetries. This has motivated an experiment to search for the EDM of the electron in the metastable excited state $a(1)[^3\Sigma^+]$ of the polar diatomic molecule PbO. A sensitivity to an electron EDM d_e of 10^{-29} e·cm appears possible in the short term, representing two orders of magnitude improvement over current limits, with the possibility of 10^{-31} e·cm sensitivity in the future. Since $|d_e| > 10^{-31}$ e·cm is predicted by many theories of physics beyond the standard model, the experiment has the potential to find or exclude new physics. The properties of PbO and techniques required to achieve this sensitivity to new physics are discussed.

INTRODUCTION

A permanent electric dipole moment (EDM) of an elementary particle, atom or molecule would violate both parity and time-reversal symmetries, and has not been observed in any system[1, 2]. However, through the CPT theorem, T violation is equivalent to CP violation, which has been observed in the K and B systems. This violation is accommodated in the standard model (SM) by a single complex, CP-violating phase in the CKM matrix. Thus radiative corrections involving CP violation and a P-violating weak interaction could generate an electron EDM entirely from SM physics, but this requires more than three loops[3], and the SM upper bound $d_e < 10^{-38}$ e·cm is fantastically small and far below current or projected experimental sensitivities.

There are reasons to expect there might be other non-SM sources of CP violation, so d_e could be larger than the SM prediction. The CP violation observed in K and B mesons is orders of magnitude too small to explain the baryon asymmetry in the universe. Also, most extensions of the SM, in particular supersymmetric models, predict many new particles and more complex phases, generating EDMs in fewer loops and consequently with larger magnitudes (for a review see [2]), many within a factor of 10^{-1} to 10^{-4} of the current bound $|d_e| < 1.6 \times 10^{-27}$ e·cm[4]. Thus improving the limits on d_e by a few orders of magnitude has the potential to find or exclude these new theories of physics beyond the standard model, and motivates the experiment in PbO, suggested more than a decade ago by V. Flambaum[5], and initiated by D. DeMille[6].

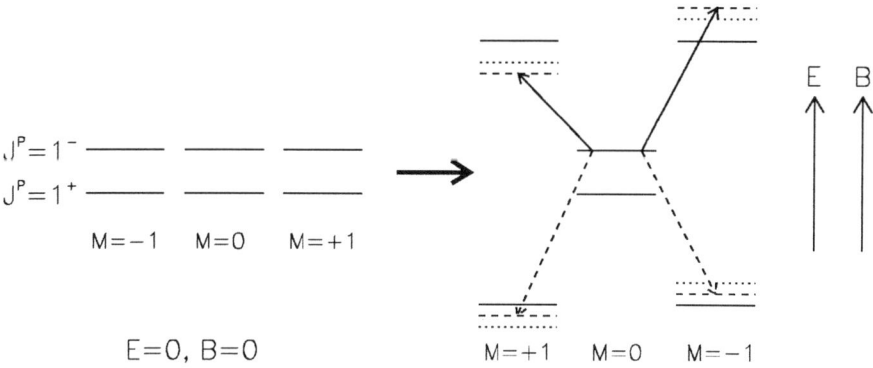

FIGURE 1. The level structure of the $J = 1$ Ω-doublet of the $a(1)$ state is shown in the absence of external fields on the left. The Ω-doublet splitting between the $J^P = 1^+$ and 1^- states is 11.2 MHz. The right diagram indicates the levels in a static vertical electric field, $E\hat{z} \sim 50$ V/cm, with solid lines. The Zeeman shift from a parallel $\vec{B}$ field is indicated with dotted lines, and the shifts from an electron EDM are indicated with dashed lines, where both shifts are highly exaggerated. As indicated, an electron EDM would change the energy splitting between the $M = +1$ and $M = -1$ sublevels of the upper doublet in a manner opposite to its change on the lower $M = \pm 1$ doublet sublevels. This is because the internal electric field is oriented oppositely in the two cases. The EDM search can be made by coherently driving the RF transition (solid arrows) from the upper $M = 0$ level (populated with the laser) to the upper $M = \pm 1$ doublet sublevels, and then measuring the difference in the $M = \pm 1$ energies using quantum beat spectroscopy. An EDM can be detected by changing the RF frequency by 11.2 MHz and driving the transition to the lower doublet sublevels (dashed arrows), and looking for the change in the $M = \pm 1$ energy difference induced by the electron EDM and visible as a change in the quantum beat frequency.

EXPERIMENTAL TECHNIQUE

An electron EDM is expected to lie along the spin vector and is manifested as a linear Stark shift, so EDM searches require an unpaired electron spin in an external electric field. Polar diatomic paramagnetic molecules containing a heavy atom can have large internal electric fields ($\gg 10$ GV/cm) which can be fully polarized along much smaller laboratory fields[7, 8]. However, these radicals (such as PbF, HgF, YbF) are difficult to produce and chemically reactive (so vapor cells cannot be used), making large counting rates difficult to achieve. Still, great technical progress and an electron EDM limit have been made by an experiment using YbF[9]. To obviate these problems we use PbO which is easily vaporized, thermodynamically and chemically stable in its ground state and a spin singlet, but which has a metastable paramagnetic excited state $a(1)[^3\Sigma^+]$ with a lifetime of 80 μs accessible via laser excitation.

This excited state, described by Hund's case (c), can have projections of electronic angular momentum along or against the internuclear axis. Even and odd parity states which are nominally degenerate are formed from even and odd combinations of these two projections[10]. The degeneracy is broken by the Coriolis coupling between electronic and rotational angular momentum, leading to a small splitting (Ω-doubling) of

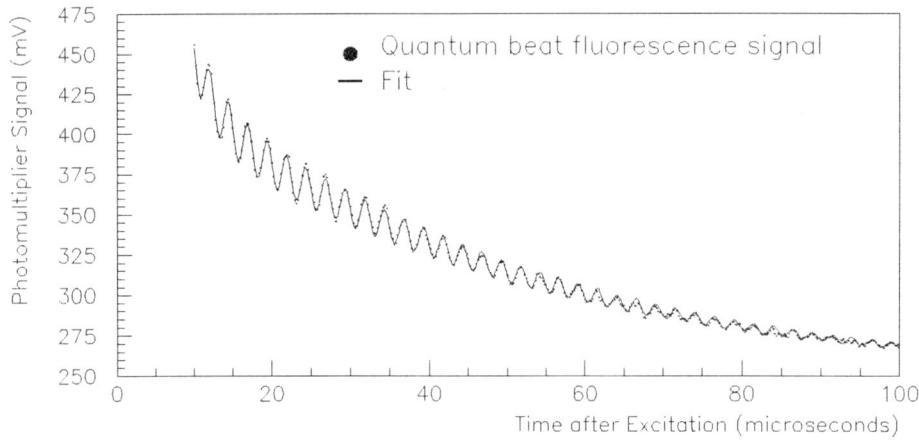

FIGURE 2. The frequency of the $\Delta M = 2$ quantum beat in the fluorescence decay of the excited $a(1)$ state is determined by the $M = \pm 1$ energy difference and is sensitive to an EDM.

11.2 MHz between the $J^P = 1^{\pm}$ states of opposite parity [11]. This small splitting allows the $J = 1$ $a(1)$ state to be fully polarized in electric fields as small as 50 V/cm. Semi-empirical[12] and *ab initio* calculations[13] suggest an electron EDM would perturb these $a(1)$ energy levels by $3 - 12 \times 10^{24} \left[\frac{\text{Hz}}{\text{e·cm}}\right] \times d_e[\text{e·cm}]$. We try to detect this small EDM-induced shift ($d_e = 10^{-29}$ e·cm corresponds to a shift of 30-120 μHz) by exploiting the changes in energy level structure and driving the transitions shown in Fig.1. The small shifts in energy are detected using quantum beat spectroscopy[14]. Briefly, the $M = \pm 1$ states have different energies in the magnetic field so their phases evolve at different rates. Interference between these two state reveals itself as a beat note in the exponential decay of the excited $a(1)$ state, where the beat frequency is determined by the $M = \pm 1$ energy difference (see Fig.2). As outlined in the Fig.1 caption, by sensitively comparing the beat frequency in the upper and lower doublet levels we can detect an EDM.

Spurious EDM-like signals will be identified by looking for consistency in the results upon reversals of the electric and magnetic fields. The upper and lower Ω-doublets should exhibit opposite level shifts due to an EDM, and changing the RF frequency to toggle between the doublet levels is a unique and powerful technique to reject a false EDM without requiring any field reversals. Systematic effects from motional magnetic fields should be highly suppressed since the magnetic moment is coupled to the internuclear axis, which is parallel to the external electric field and perpendicular to any motional magnetic fields.

RESULTS

By heating PbO to 700°C in a sapphire and alumina vapor cell and populating the $a(1)$ state via laser excitation, high counting rates $> 1 \times 10^7$/s, and long coherence times have been demonstrated. We have achieved shot-noise limited sensitivity in the extraction of Zeeman quantum beat frequencies in the short term of 35 Hz/$\sqrt{\text{Hz}}$. Precision measurements have been made of the Ω-doublet splitting, g-factors and coherence quenching cross-sections, and we have successfully applied electric fields in the cell large enough to completely polarize the molecule with no adverse effects.

Small, easily implemented changes in our apparatus should improve our sensitivity to 100 mHz/$\sqrt{\text{Hz}}$, and allow a limit of 10^{-29} e·cm to be achieved in less than a month of integration.

A review of EDM experiments and techniques can be found in [1], and in these proceedings from M. Romalis, S. Lamoreaux, J. Miller and Y. Semertzidis. More details regarding the PbO experiment may be found in [6].

ACKNOWLEDGMENTS

We gratefully acknowledge support from the David and Lucile Packard Foundation, National Science Foundation, a NIST Precision Measurements grant, Yale University and Research Corporation.

REFERENCES

1. I.B. Khriplovich and S.K. Lamoreaux, *CP Violation Without Strangeness: Electric Dipole Moments of Particles, Atoms, and Molecules*, New York: Springer, 1997.
2. E.D. Commins, Adv. At. Mol. Opt. Phys. **40**, 1 (1999).
3. M.E. Pospelov and I.B. Khriplovich, Sov. J. Nucl. Phys. **53**, 638 (1991).
4. B.C. Regan, E.D. Commins, C.J. Schmidt and D. DeMille, Phys. Rev. Lett. **88**, 071805 (2002).
5. V.V. Flambaum, Ph.D. thesis, Institute of Nuclear Physics, Novosibirsk, 1987; see also L. Barkov, M. Zolotorev, and D. Melik-Pashaev, Kvant. Elektron. **15**, 1106 (1988) [Sov. J. Quantum Electron. **18**, 710 (1988)].
6. D. DeMille *et al.*, Phys. Rev. A **61**, 052507 (2000); D. DeMille *et al.*, in *Art and Symmetry in Experimental Physics*, edited by D. Budker *et al.* (AIP, New York 2001).
7. P.G.H. Sandars, Phys. Rev. Lett. **19**, 1396 (1967).
8. M.G. Kozlov and L.N. Labzowsky, J. Phys. B. **28**, 1933 (1995).
9. J.J. Hudson, B.E. Sauer, M.R. Tarbutt, and E.A. Hinds, Phys. Rev. Lett. **89**, 023003 (2002).
10. G. Herzberg, *Molecular Spectra and Molecular Structure Volume I. Spectra of Diatomic Molecules*, Florida: Krieger Publishing, 1950.
11. L.R. Hunter *et al.*, Phys. Rev. A **65**, 030501(R) (2002).
12. M.G. Kozlov and D. DeMille, Phys. Rev. Lett. **89**, 133001 (2002).
13. T.A. Isaev *et al.*, physics/0306071, submitted to Phys. Rev. Lett.
14. A. Corney, *Atomic and Laser Spectroscopy*, Oxford: Clarendon Press, 1977.

A New Experiment to Measure the Muon Electric Dipole Moment

The EDM Collaboration: J.P. Miller, R.M. Carey, V. Logashenko, K.R. Lynch, B.L. Roberts, *Boston U.*, A. Silenko, *Belarus*, G. Bennett, D.M. Lazarus, L.B. Leipuner, W. Marciano, W. Meng, W.M. Morse, R. Prigl, Y.K. Semertzidis, *BNL*, V. Balakin, A. Bazhan, A. Dunikov, B. Khazin, I.B. Khriplovich, G. Sylvestrov, *BINP Novosibirsk*, Y. Orlov, *Cornell U.*, K. Jungmann, *KVI*, P.T. Debevec, D.W. Hertzog, C.J.G. Onderwater, C.S. Özben, *U. of Illinois*, E. Stephenson, *Indiana U.*, M. Auzinsh, *U. of Latvia*, P. Cushman, R. McNabb, *U. of Minnesota*, N. Shafer-Ray, *U. of Oklahoma*, K. Yoshimura, *KEK*, A. Aoki, Y. Kuno, A. Sato, *Osaka*, M. Iwasaki, *RIKEN*, F.J.M. Farley, *Yale U.*

Abstract. A description is given of a new experiment to measure the muon electric dipole moment (EDM) to between $\sigma = 10^{-24}$ $e-cm$ and 10^{-25} $e-cm$, which would be 5 to 6 orders of magnitude improvement over the current world average. Muons are stored in a magnetic ring. Precession due to Thomas precession and the magnetic moment are canceled with the proper combination of applied E and B fields. Only precession due to a non-vanishing EDM remains, resulting in a large amplification of the EDM signal. The method has general applicability to charged particles.

A new technique is being developed to measure the permanent electric dipole moments (EDM) of charged elementary particles and nuclei. The precession of the spin is observed in a magnetic storage ring, in which the background precession due to the magnetic moment is canceled by means of an externally applied radial electric field. Only spin precession due to a non-vanishing electric dipole moment remains, resulting in much improved signal size relative to background. This technique represents the only practical means to measure the muon EDM. While there is no large EDM enhancement as in electron EDM measurements with atoms or molecules, there is also no uncertainty due to cancellations.

The immediate plan is to develop experiments to measure the muon and the deuteron EDMs. Such measurements will be highly competitive with and complimentary to other planned EDM experiments. This note will concentrate discussion of the muon case, where the goal is to measure the EDM with $\sigma = 10^{-24} - 10^{-25}$ $e-cm$. The deuteron case is discussed in these proceedings by Y. Semertzidis.

There has been increased interest in the measurements of the EDMs of the electron, neutron, atoms, and complex nuclei. The reasons are persuasive:

1) The permanent EDMs of elementary particles, atoms and nuclei are predicted in the Standard Model to be extremely small and far below current and foreseeable experimental sensitivities. Therefore the experimental detection of a non-vanishing EDM

would be an *unambiguous* indication of physics beyond the Standard Model.

2) A non-vanishing permanent EDM violates both P and T symmetries. Under the assumption of CPT invariance, it also violates CP symmetry. While the Standard model is able to accommodate the observed violation of CP in the K and B systems, it offers no explanation as to its origin. Observation of CP violation in other systems is essential to understanding its source, and to explain, for example, baryogenesis of the universe.

3) A number of candidate models beyond the Standard Model predict or allow for very large CP violations, a notable example being supersymmetry.

4) The current experimental limits on the EDMs of the electron, neutron and the ^{199}Hg nucleus have placed some of the strongest limits or eliminated altogether many theories beyond the Standard Model. In particular, it is easy to find combinations of supersymmetry parameters which predict large EDMs, and many of the most stringent constraints on supersymmetric models come from EDM data.

The Standard model, as well as degenerate minimal supersymmetric models in which only the masses change from one lepton to another, predict that the lepton EDMs scale with the mass. Using the current experimental limit on the electron EDM leads to an estimate for the muon, $d_\mu < 1.5 \pm 1.4 \times 10^{-25}$ $e-cm$. This is less than or equal to the proposed experimental limit, although only a relatively small deviation from linear mass scaling would bring the EDM into the measurable range. If one or more of the large number of CP violating phases in supersymmetry are non-vanishing, or if some of the parameters of the model follow a non-scaling behavior, then it is easy to get large deviations from mass scaling. One model (Babu, *et al.*, **PRL** 85, 5064(2000)), using left-right symmetry with a see-saw mechanism which accounts for the large observed neutrino mixing, predicts $d_\mu \approx 5 \times 10^{-23}$ $e-cm$, at least 50 times larger than our proposed experimental precision, while at the same time predicting an electron EDM which is ten times smaller than the current experimental limit. There are many other models, involving lepto-quarks, multiple Higgs, etc., which also predict d_μ greater than 10^{-24} $e-cm$. Interestingly, the muon's EDM is likely the only experimentally accessible EDM outside the first generation of particles for the foreseeable future, likely making its measurement especially important to understanding its origins.

The present value of the muon EDM, $d_\mu = 3.7 \pm 3.4 \times 10^{-19}$ $e-cm$, which is consistent with zero, was obtained as a secondary measurement by the CERN muon (g-2) experiment (Bailey, *et al.*, **NP B**150, 1(1979)). The current BNL (g-2) experiment expects to report a result with 3-5 times improved precision, using the same method as CERN. It is now recognized that further improvements in the CERN/BNL method are limited by severe systematic problems. The new technique described in this document promises to greatly reduce these systematic errors. The proposed precision represents 5 to 6 orders of magnitude improvement over the current value.

Following is a brief description of the method used by the (g-2) experiments to determine the EDM. A description of the new method is then given.

In a storage ring possessing both E- and B- fields, the angular precession rate of the spin relative to the momentum, in the approximation $\vec{\beta} \cdot \vec{B} = \vec{\beta} \cdot \vec{E} = 0$, is given by

$$\vec{\omega} = -\frac{e}{m}[a_\mu \vec{B} + (-a_\mu + \frac{1}{\gamma^2 - 1})\vec{\beta} \times \frac{\vec{E}}{c} + \frac{\eta}{2}(\vec{\beta} \times \vec{B} + \frac{\vec{E}}{c})]$$

where the magnetic dipole moment is given by $\frac{g}{2}\frac{e\hbar}{2m}=(1+a_\mu)(\frac{e\hbar}{2m})$, and the electric dipole moment is given by $d_\mu = \frac{\eta}{2}(\frac{e\hbar}{2mc})$.

The (g-2) experiments chose the 'magic' $\gamma = 29.2$ so that on average the coefficient of $\vec{\beta} \times \vec{E}$, $-a_\mu + \frac{1}{\gamma^2-1} = 0$, thereby minimizing the contribution of the E-field to the spin precession. At the magic γ, a non-vanishing EDM would cause the precession vector to 1) tip in the radial direction by an angle $\delta = \tan^{-1}(\frac{\eta}{2a_\mu})$ and to 2) increase in magnitude according to $\omega = \sqrt{\omega_a^2 + \omega_{EDM}^2}$. Effect 1) causes the plane of spin precession to tilt relative to the orbit plane. Recalling that the decay positrons are emitted preferentially along the muon spin direction, the average position of the decay positrons at the detectors oscillates vertically with frequency $\frac{\omega}{2\pi}$. CERN obtained its measurement by observing the oscillations in the ratio $R_N(t) = \frac{N_{up}-N_{dn}}{N_{up}+N_{dn}}$, whose amplitude is proportional to the EDM. Here, N_{up} and N_{dn} refer to the number of positrons detected above and below the orbit plane, respectively. However, any misalignment between the average vertical position of the stored muons and the detector mid-plane led directly to a false EDM signal. It was difficult to monitor this alignment, and as a result, the CERN measurement had a large systematic error which was comparable in size to the statistical error. The BNL (g-2) experiment expects to improve upon the CERN EDM limit with increased statistics and by the use of detectors with multiple vertical segments to reduce uncertainties in the beam-detector alignment.

It was realized that significant further improvements in the EDM limit using this method are not possible, therefore a collaboration was formed to develop a dedicated EDM experiment using a new technique. Particles are stored in a magnetic ring as before. An alternating gradient B-field would be employed to store the beam in the ring which increases the number of stored muons compared to weak focusing. γ, $\vec{E}$ and $\vec{B}$ are chosen so that the sum of the first two terms in the expression for $\vec{\omega}$ is zero. Only precession due to the EDM from the last term remains. Instead of the tilt angle observed by CERN, $\delta < 5 \times 10^{-3}$, one would observe a constant rotation rate of the spin about the radial axis, resulting in steady accumulation of a vertical component of the spin. It is necessary to move away from the magic γ in order to use the E-field to cancel (g-2) precession.

To minimize the statistical error, one needs to maximize the polarization, B-field, and momentum subject to the constraint that $\vec{B}$, $\vec{E}$, γ be chosen to cancel (g-2), $E_r = \frac{a_\mu B_z}{(\frac{1}{\gamma^2-1}-a_\mu)\beta_\theta}$. The conservative maximum is $E_r < 2$ MV/m. Simulations show that with $E_r = 2$ MV/m, $p \approx 500$ MeV/c ($\gamma \approx 5$), and $B \approx 0.25$ T (radius of curvature ≈ 7 m) produces an optimal EDM signal relative to backgrounds. To achieve $\sigma = 10^{-24}$ $e-cm$, assuming conservative values $A = 0.3$ and $P = 0.4$, requires $N \approx 4 \times 10^{16}$. With the installation of the proposed PRISM-II beam line, the needed muons would be available at the Japanese facility J-PARC in one year of running, using conservative running parameters. Currently being investigated are a) increasing the radial E-field by a factor of two or three, b) increasing the internal current available at JPARC, and c) increasing the polarization of the muons. A combination of improvements could lead to an order of magnitude improvement in the precision, to 10^{-25} $e-cm$.

The most serious systematic error to be tackled arises if the applied radial E-field has a small non-vanishing average component perpendicular to the orbit plane. The spin

precession due to the EDM is proportional to the Lorentz force on the particle:
$$\vec{\omega}_{EDM} = -\frac{e\eta}{2m}(\vec{\beta} \times \vec{B} + \frac{\vec{E}}{c}) = -\frac{e\eta}{2m}(\beta B_z + \frac{E_r}{c})\hat{e}_r \propto \vec{F}$$
The radially directed centripetal force produces the desired EDM signal (with the $\vec{\beta} \times \vec{B}$ term being, in the case considered here, far larger than the $\vec{E}$ term). In the vertical direction, the average force must be zero for a stable stored beam, $\langle F_z \rangle - 0 \rightarrow \beta_\theta B_r = \frac{E_z}{c}$. A non-zero $\langle E_z \rangle$ leads to a non-zero $\langle B_r \rangle$, each of which cause a false EDM signal (e.g. spin precession about a radial axis) through the first two terms in $\vec{\omega}$,
$$(a_\mu \vec{B} + (-a_\mu + \frac{1}{\gamma^2-1})\vec{\beta} \times \vec{E}) = (a_\mu B_r + (-a_\mu + \frac{1}{\gamma^2-1})\beta_\theta E_z)\hat{e}_r \approx \frac{E_z}{\beta_\theta \gamma^2}\hat{e}_r$$

In order to measure the muon EDM to 10^{-24} $e-cm$, the direction of the electric field must be controlled to 10 nR. This is a very stringent requirement. Fortunately, there are factors which greatly ease the difficulty of the problem:

1) The electric field does not need to be in one plane or perpendicular to B with high accuracy. It is only the average component of $\vec{E}$ perpendicular to the average plane of motion which needs to be controlled well. (See Farley, et al., hep-ex/0307006).

2) The false EDM signal from an out-of-plane E-field can be canceled by combining data with particles stored in the ring in both the clockwise and counter-clockwise directions. The EDM precession will reverse sign while the false EDM precession will have the same sign.

3) Y. Orlov is developing a promising technique to determine the initial alignment of the electric field by exploiting the effect of an out of plane electric field component on the particles' betatron oscillations.

4) A team at Brookhaven National Laboratory has developed an inclinometer capable of monitoring the stability of the electric field plates to better than 1 nR, or much better than needed for the muon measurement.

Because of the very high rates of decays in the proposed experiment, the EDM will be deduced from the ratio of energies deposited in the up and down counters, $R_E(t) = \frac{E_{up} - E_{dn}}{E_{up} + E_{dn}}$, rather than the ratio of counts. One design places lead-glass calorimeters above and below the storage region. Calorimeters would be added to the left and right of the stored beam in order to monitor how well the spin precession in the orbit plane has been canceled by the applied electric field.

The proposed muon beam line at JPARC would consist of a pion capture solenoid in the target region followed by a long pion decay and muon collection solenoid. The muons would then be injected into a Fixed Phase Alternating Gradient (FFAG) phase rotator. The polarized muons would then be injected into the storage ring. This beam line is now being actively considered by JPARC. A decision on implementation should be forthcoming in a year or two, and the beam-line would be ready in about FY 2008.

In the preliminary design, the storage ring would be divided into sixteen identical sectors, arranged in a ring of radius 11 m. Each sector would consist of a 2.7 m region of homogeneous B- field plus a (g-2) canceling radial electric field. This would be followed by a 1.7 m region, containing a pair of quadrupoles and the detectors.

This work is supported in part by a grant from the US National Science Foundation.

A New Method For A Sensitive Deuteron EDM Experiment

Y.K. Semertzidis[4], M. Aoki[13], M. Auzinsh[9], V. Balakin[2], A. Bazhan[2], G.W. Bennett[4], R.M. Carey[3], P. Cushman[10], P.T. Debevec[7], A. Dudnikov[2], F.J.M. Farley[15], D.W. Hertzog[7], M. Iwasaki[14], K. Jungmann[5], D. Kawall[15], B. Khazin[2], I.B. Khriplovich[2], B. Kirk[4], Y. Kuno[13], D.M. Lazarus[4], L.B. Leipuner[4], V. Logashenko[2,3], K.R. Lynch[3], W.J. Marciano[4], R. McNabb[10], W. Meng[4], J.P. Miller[3], W.M. Morse[4], C.J.G. Onderwater[7], Y.F. Orlov[6], C.S. Ozben[7], R. Prigl[4], S. Rescia[4], B.L. Roberts[3], N. Shafer-Ray[11], A. Silenko[1], E.J. Stephenson[8], K. Yoshimura[12]

EDM Collaboration

1. Belarusian State University, Belarus; 2. BINP, Novosibirsk; 3. Boston University; 4. Brookhaven National Laboratory; 5. Kernfysisch Versneller Instituut, Groningen; 6. Newman Laboratory, Cornell University, Ithaca; 7. University of Illinois, at Urbana-Champaign; 8. Indiana University; 9. University of Latvia; 10. University of Minnesota; 11. University of Oklahoma; 12. KEK, Japan; 13. Osaka University, Japan; 14. RIKEN, Japan; 15. Department of Physics, Yale University, New Haven.

Abstract. In this paper a new method is presented for particles in storage rings which could reach a statistical sensitivity of 10^{-27} e·cm for the deuteron EDM. This implies an improvement of two orders of magnitude over the present best limits on the T-odd nuclear forces ξ parameter.

INTRODUCTION

Sensitive searches for electric dipole moments (EDM) of fundamental particles are excellent probes of physics beyond the standard model (SM) [1]. The reason for this is because physics beyond the SM allows for values that are well within reach of present technology while the backgrounds from the SM predictions are many orders of magnitude below them.

In this paper a new method is presented for particles in storage rings which could reach a statistical sensitivity of 10^{-27} e·cm for the deuteron EDM. The method uses a radial electric field to cancel the g-2 precession of particles in order to maximize the EDM signal strength by several orders of magnitude [2,3]. The same technique is used to improve upon the muon EDM sensitivity level; see contribution by J. Miller elsewhere in these proceedings.

There are three candidate laboratories that the collaboration is considering at this point for the deuteron EDM experiment: Brookhaven National Laboratory (BNL),

Indiana University Cyclotron Facility (IUCF) and Kernfysisch Vernsneller Instituut (KVI) at Groningen of The Netherlands. In the numerical examples of this paper we will assume parameters which are relevant at the BNL site. The systematic errors are still under consideration.

SPIN PRECESSION

The spin s of a particle with a magnetic moment μ and an electric dipole moment d, located in magnetic and electric fields B and E, precesses according to

$$\frac{d\vec{s}}{dt} = \vec{\mu} \times \vec{B} + \vec{d} \times \vec{E} \qquad (1)$$

In a storage ring with only B-field present and $d=0$, the spin precession rate relative to the momentum vector is given by

$$\vec{\omega}_a = \frac{e}{m} a \vec{B} \qquad (2)$$

with $a=(g-2)/2$, where g is the gyromagnetic ratio. The spin precession plane is in the same plane as the momentum precession. If $d \neq 0$ the above Equation 2 becomes

$$\vec{\omega}_T = \frac{e}{m} a \vec{B} + \frac{1}{\hbar} \vec{d} \times (\vec{u} \times \vec{B}) \qquad (3)$$

for spin 1 particles, where $(\vec{u} \times \vec{B})$ is the motional electric field, i.e. the E-field the particle feels in its own rest frame, $\vec{u} = \vec{\beta}/c$ the particle's velocity. The effect of the EDM component is to tilt the spin precession plane relative to the momentum plane at *every* azimuthal location. For a small value of d and a relatively large value of a the tilt is very small, e.g. for the deuteron, a=-0.143, and assuming a value for $d = 10^{-27}$ e·cm and $\beta=0.7$, then the tilt angle is 5×10^{-13} rad which is very small. If the g-2 precession term was zero, the only precession would be due to the EDM and it would be entirely out-of-plane. Moreover, the spin phase angle would accumulate with time, proportional to the EDM. One way to eliminate the g-2 term is to apply an electric field equal to

$$E = \frac{aBc\beta}{1-(1+a)\beta} \approx aBc\beta\gamma^2 (1 + a\beta^2\gamma^2) \qquad (4)$$

where c is the speed of light and γ the relativistic Lorentz factor [2,3]. If the g-2 cancellation is achieved at the 10^{-7} level then the tilt angle becomes 5×10^{-6} rad, i.e. seven orders of magnitude higher than without the cancellation.

EXPERIMENTAL PARAMETER VALUES

EDM Signal Strength

In the absence of the g-2 precession, Equation (3) can be re-written as

$$\omega_d = \frac{1}{\hbar} duB \approx \frac{d}{\hbar a \gamma^2 (1 + a\beta^2 \gamma^2)} E \quad (5)$$

i.e. the EDM signal is proportional to the electric field strength E, inversely proportional to the anomalous magnetic moment a, and approximately inversely proportional to γ^2. Assuming a 5 cm plate separation, then an electric field of 4 MV/m is reasonable. For $d = 10^{-27}$ e·cm, and $\beta = 0.7$ the EDM signal becomes $\omega_d = 2.5 \times 10^{-7}$ rad/s.

Storage Ring Parameters

At BNL one possible location for the deuteron EDM experiment is the AGS tunnel. Assuming 80% ring coverage with magnetic and electric fields, an electric field of 4MV/m and $\beta = 0.7$, $\gamma = 1.4$ then the required magnetic field is $B \approx 0.8$ KG, a rather low field. The radius of the tunnel is 127 m and there is space for another ring in the tunnel. The deuteron momentum is 2 GeV/c and its kinetic energy 0.9 GeV.

Signal Detection

The out of plane spin precession can be detected by scattering the deuterons off a proton or carbon target and looking for a left-right asymmetry versus time. We will assume a carbon target here. The deuteron cross section on carbon targets has been measured and analyzed [5,6,7,8] at 1.69 GeV/c or 650 MeV kinetic energy of deuteron, close to the parameters we are considering here.

The analyzing power for deuterons on carbon has been measured with the POMME polarimeter in the semi-inclusive [9] and inclusive [10] mode in the energy range of 0.175 to 1.6 GeV using the polarized deuteron beam of the Laboratoire National Saturn in France. In the interesting range 3^0-13^0 the elastic cross section is estimated from ref. [5] to be (40±10) mb or about 30% of the total elastic.

Expected Rates in the AGS Tunnel and Estimated Statistical Error

In our case we want to use a thin carbon target and detect in-time coincidence between the deuteron and recoiling carbon as a function of storage time. The asymmetry is expected to be at least as much as measured by the semi-inclusive method [9].

Preliminary M.C. studies showed that the useful elastic scattering rate ratio is about $f \approx 3 \times 10^{-4}$. If we have 100 carbon target stations distributed around the ring, the counting rate on the DAQ will be reduced by the same factor. The estimated polarized deuteron beam intensity is 12×10^{11} per cycle [11], and making the reasonable assumption that the distributed DAQ can take the rate, the useful rate is then $\approx 3 \times 10^8$ per cycle with an average asymmetry of at least 0.35. The beam polarization is expected to be 85% of the maximum 2/3 for a "pure" vector polarized beam, i.e. 0.56 [11].

The deuteron beam will de-polarize as a function of time due to E, B-field multipoles, betatron oscillations, momentum dispersion, etc. At this point we believe we can achieve a polarization lifetime of 10 s. One can then optimize the beam lifetime (the beam is lost as a result of multiple scattering in the target), and the beam measuring time per cycle for best statistical sensitivity. The statistical error on the deuteron EDM when the above parameters are optimized is given by

$$\sigma_d = 4.6 \frac{\hbar a \gamma^2 \left(1 + a \beta^2 \gamma^2\right)}{E\left[1 + a\gamma^2\left(1 + a\beta^2\gamma^2\right)\right]AP\sqrt{NfT_{tot}\tau}} \qquad (6)$$

where τ is the beam lifetime, E is the radial electric field value, A is the polarimeter asymmetry, P is the deuteron beam polarization, $\hbar = 6.58 \times 10^{-22}$ MeV s, $a = -0.143$ the deuteron anomalous magnetic moment, $\gamma^2 \approx 2$, N is the average number of stored particles per cycle, f the fraction of useful events, and T_{tot} the total running time of the experiment. Assuming $\tau = 5$ s, $E_R = 4$ MV/m with 80% coverage of the ring with the E-field, and a total 10^7 s running time, then $\sigma_d \approx 1.1 \times 10^{-27}$ e·cm statistical error. Equation (6) is valid for the deuteron EDM experiment when the polarization lifetime is 10 s, the beam lifetime is 5 s and the measuring time is 10 s.

Systematic Errors

The main systematic error is due to an out of plane electric field component [2,3] and the background spin precession rate is given by

$$\omega_B = \frac{geE}{2m\beta c \gamma^2}\theta_E \qquad (7)$$

with g the gyromagnetic ratio and θ_E the misalignment angle. We are planning to inject the particles into the ring [3] clockwise (CW) and counter clockwise (CCW) to cancel the effect. We are also studying the possibility to store polarized protons in the same ring to study and optimize the electric field directional stability since their sensitivity to the misalignment angle is more than 10 times greater than that of the deuterons. Using the same electric field with opposite polarity the protons would have a momentum of about 0.5 GeV/c and the required magnetic field would be about 50 Gauss. The intensity of the polarized protons is expected to be similar to the intensity of the polarized deuterons [11] and the results would be easier to interpret since it is a spin ½ particle and has no tensor polarization components.

THEORETICAL IMPLICATIONS

Improving the deuteron EDM sensitivity to the level of low 10^{-27} e·cm, would be a very interesting experiment. The parameter ξ of the T-odd nuclear forces is related [12] to the electric dipole moment of the deuteron according to the equation $d = 2 \times 10^{-22} \xi$ e·cm. The best limit on ξ of 0.5×10^{-3} currently comes form the ^{199}Hg EDM experiment [13]. A deuteron EDM experiment at the intended sensitivity level would be an improvement of two orders of magnitude over that limit. Also, since the deuteron consists of a proton and a neutron, at that level it would be improving upon the neutron EDM by a factor of 60 to 100 and more than four orders of magnitude upon the proton EDM.

REFERENCES

1. Y. Nir, Plenary talk at the Amsterdam ICHEP02, July 2002.
2. Y.K. Semertzidis et al., hep-ph/0012087 and references therein.
3. F.J.M. Farley et al., hep-ex/0307006.
4. L.M.C. Dutton, et al., Phys. Lett. **16** (1965) 331.
5. L.M.C. Dutton, et al., Phys. Lett. **25B** (1967) 245.
6. K.S. Chadha and V.S. Varma, Phys. Rev. **C13** (1976) 715.
7. L.M.C. Dutton, et al., Nucl. Phys. **A343** (1980) 356.
8. B. Bonin, et al., NIM **A288** (1990) 389.
9. V.P. Ladygin, et al., NIM **A404** (1998) 129.
10. J. Arvieux, et al., NIM **A273** (1988) 48.
11. T. Roser, private communication.
12. V.V. Flambaum, I.B. Khriplovich, and O.P. Sushkov, Phys. Lett. **B162** (1985) 213.
13. I.B. Khriplovich and R.A. Korkin, Nucl. Phys. **A665** (2000) 365.
14. M.V. Romalis, et al., Phys. Rev. Lett. **86** (2001) 2505.

Production and Detection of Cold Anti-Hydrogen Atoms. A First Step Towards High Precision CPT Test

A. Variola*, M.Amoretti*, G. Bonomi[†], A. Boutcha[†], P. Bowe**, C. Carraro*[‡], C. L. Cesar[§¶], M. Charlton[∥], M. Doser[†], V. Filippini[††‡‡], A. Fontana[††‡‡], M. Fujiwara[§§], R. Funakoshi[§§], P. Genova[††‡‡], J. S. Hangst**, R. S. Hayano[§§], L. V. Jorgensen[∥], V. Lagomarsino*[‡], R. Landua[†], D.Lindelof[¶¶], E. Lodi Rizzini[††***], M. Macrì*, N. Madsen[¶¶], G. Manuzio*[‡], P. Montagna[††‡‡], H. Pruys[¶¶], C. Regenfus[¶¶], A. Rotondi[††‡‡], P. Riedler[†], G. Testera* and D.P. Van der Werf[∥]

*Istituto Nazionale di Fisica Nucleare, Sezione di Genova, 16146 Genova Italy
[†]EP Division, CERN, CH-1211 Geneva 23, Switzerland
**Department of Physics and Astronomy, University of Aarhus, DK-8000 Aarhus C, Denmark
[‡]Dipartimento di Fisica, Universita di Genova, 16146 Genova, Italy
[§]Instituto de Fisica, Univesidade Federal do Rio de Janeiro, Rio de Janeiro 21945-970
[¶]Centro Federal de Educacao Tecnologica do Ceara, Fortaleza 60040-531, Brazil
[∥]Department of Physics, University of Wales Swansea, Swansea SA2 8PP, UK
[††]Istituto Nazionale di Fisica Nucleare, Universita di Pavia, 27100 Pavia, Italy
[‡‡]Dipartimento di Fisica Nucleare e Teorica, Universita di Pavia, 27100, Pavia, Italy
[§§]Department of Physics, University of Tokyo, Tokyo 113-0033, Japan
[¶¶]Physik-Institut, Zurich University, CH-1211 Zurich, Switzerland
***Dipartimento di Chimica e Fisica per l'Ingegneria e per i Materiali, Università di Brescia, 25123 Brescia, Italy

Abstract.
Observations of anti-hydrogen in small quantities have been reported at CERN and at FermiLab, but these experiments were not suited to spectroscopy experiments. In 2002 the ATHENA collaboration reported the production and detection of very low energy anti-hydrogen atoms produced in cryogenic environment. This is the first major step in the study of antiatom's internal structure and it can lead to a high precision test of the CPT fundamental symmetry. The method of production and detection of cold anti-hydrogen will be introduced. The absolute rate of anti-hydrogen production and the signal to background ratio in the ATHENA experiment will be discussed.

Testing fundamental symmetries is of great interest in modern physics. A major role is played by the tests of the CPT theorem which ensures the physical invariance of systems simultaneously subjected to charge conjugation, parity inversion and time reversal. This happens for point particles in a flat space-time under the assumptions of Lorentz invariance and unitarity. Different theories that are not based on these assumptions suggest CPT violations. The one put forward by Colladay and Kostelecky[1] is worth mentioning. Another CPT violation mechanism has been suggested by Ellis et al. [2] who invoke quantum gravity.

Why antihydrogen?

A number of measurements have already been carried out as CPT tests comparing the characteristics of particles with those of antiparticles. The precision attained in the neutron kaon mass measurement (10^{-18}) is impressive, even though it was obtained in a theoretically dependent manner. In future an high degree of precision could be obtained in a direct measurement by means of the spectroscopy of antihydrogen atoms in the 1S-2S transition [3]. The first antihydrogen atoms formed in flight have been detected at CERN [4]. Their low number and high velocity made them unsuitable for spectroscopy which requires a high number of antiatoms at low temperatures. Cryogenic temperatures are needed for three main reasons: the inverse dependence of temperature on radiative and three-body recombination; the inverse dependence of velocity for neutral atom trapping; the Doppler effect determining the ultimate resolution of the spectroscopy measurement. In the summer of 2002 the ATHENA experiment made a first breakthrough producing and detecting a large quantity of antihydrogen atoms at cryogenic temperatures [5]. The experimental apparatus, the technique used for antihydrogen production and detection, as well as the first experimental results are illustrated in this paper.

The ATHENA experiment

The Athena apparatus is made up of three main devices, each performing a specific function: the main cryostat, the positron accumulator and the antihydrogen detector. In the main cryostat antiprotons are caught, cooled, transferred and mixed by means of Penning Malberg traps. Bunches of some 10^7 antiprotons are supplied by the CERN AD facility [6] in cycles of $\sim$ two minutes at $\sim$ 5 MeV energy. Once extracted, they are injected into the main cryostat where a 3T solenoidal magnetic field is imposed for the radial confinement of charged particles. The low energy tail is subsequently trapped in the so-called catching trap by means of a high voltage switch. Once caught, the antiprotons are cooled by means of collisions with a preloaded electron cloud. The energy released by antiprotons to the electron cloud is rapidly radiated ($\tau \sim$400 msec) by cyclotron radiation into the 3T magnetic field. At the end both the $\bar{p}$s and the electron populations reach the thermal equilibrium with the cryogenic environment. The cooled antiprotons are then transferred into the so-called mixing trap where recombination takes place. The efficiency of all these processes is relatively low and in the end roughly 10000 antiprotons are injected in the mixing cycle for recombination. The mixing trap electrodes' arrangement allows a nested potential configuration (see fig. 1). This configuration allows for the si-

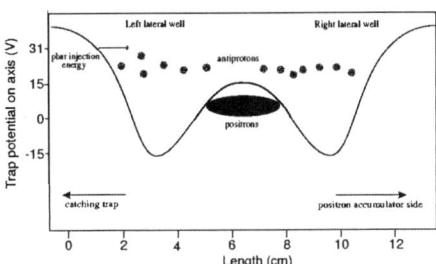

FIGURE 1. Nested potential configuration for the simultaneous axial confinement of $\bar{p}$s and positrons. $\bar{p}$s are injected at high energy and passing through the positron cloud they cool down. Once thermalized, they diffuse into the positron plasma increasing the probability of recombination.

multaneous axial confinement of oppositely charged particles The antiprotons injected in the nested trap's outer well cross the positrons and cool, thus starting the interaction process which leads to recombination. The positron plasma utilized for recombination is originally produced in the positron accumulator. A radioactive Na^{22} source emits positrons, which are then cooled by collisions with a buffer gas in differential pressure sections. The positron accumulation process continues for about two minutes. At the end roughly $1.5 \cdot 10^8$ e^+ are ready to be transferred to the mixing region with an efficiency of about 50 %. The final result is a plasma of ~ 75 million positron ready for recombination. Plasma characteristics such as radius, density and temperature are monitored by an innovative diagnostic system [7]. The antihydrogen atom, which is electrically neutral, is no longer confined once the recombination has taken place. Therefore, it drifts toward the trap walls causing the simultaneous and localized annihilations of both the antiproton and the positron. The antiproton-proton annihilation produces mainly pions, whereas the positron-electron annihilation provides two characteristic back-to-back 511 keV photons. They are all detected by an innovative antihydrogen detector operating in a cryogenic environment at ~ 140 K. Two layers of doubled sided silicon micro strip detectors are used to localize the vertex of the charged pions tracks (fig. 2). A microsec-

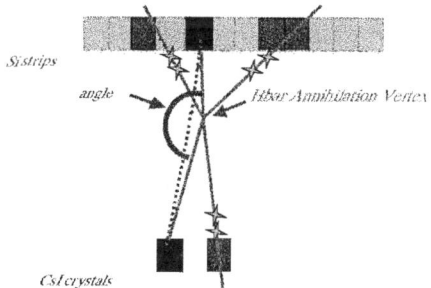

FIGURE 2. Identification of $\bar{H}$. The drawing shows the meaning of the opening angle. If $\cos(\theta) = -1$ the vertex and the two 511 keV line are collinear.

ond window is provided for the coincidence of the two back-to-back 511 keV photons from the e^+-e^- annihilation with the associated vertex.. The gammas are identified and the energy measured by means of an external layer of CsI crystals read out by APD's.

Antihydrogen identification: experimental results

Events associating a vertex with a back-to-back 511 keV signal are plotted in a histogram illustrating the cosine of the opening angle θ. This is the angle that is subtended from the vertex to the geometrical center of the two crystals identifying the two 511 keV gamma lines (see fig. 2). The antihydrogen "golden events" are obtained for $\cos(\theta)=-1$ i.e when there is a perfect overlap between the vertex and the back to back 511 keV line. The events characterized by $\cos\theta \neq -1$ can still identify an $\bar{H}$ event where the vertex misalignment is possibly due to background or noise events. A full Monte Carlo simulation that takes into account the showers produced in the magnet coils has modeled this. The measurement results are shown in fig. 3 a,b. In fig. 3a the continuous line represents the spectrum when the mixing takes place at cryogenic temperature (cold mixing). The $\cos \theta=-1$ peak is the signature of antihydrogen. The rest of the spectrum is partially composed by the $\bar{H}$ signal and the background. To rule out any possible misleading mechanism in antihydrogen detection in-depth background studies and measurements have been carried out (see fig. 3 a, b). In fig. 3a the triangles show the

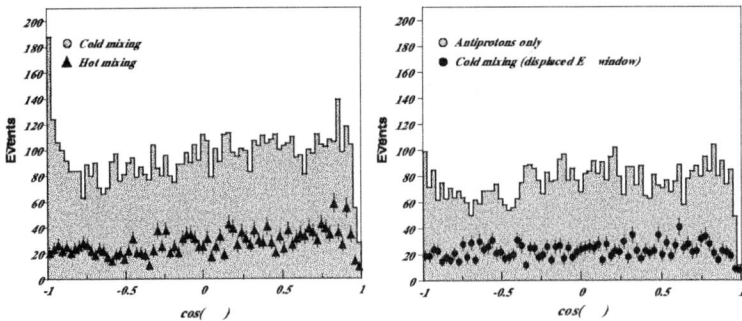

FIGURE 3. The experimental results proving antihydrogen production are displayed

spectrum acquired when the positron plasma was heated by applying RF (hot mixing). In fig. 3b the continuous line illustrates the spectrum resulting from the annihilation of $\bar{p}$s in the absence of positrons, whereas the circles show the cold mixing results when the crystal energy window is shifted away from the 511 keV range. In all three cases the absence of the $\cos\theta = -1$ peak is noticeable. This is the final and decisive evidence of antihydrogen atom production. Considering only the so-called golden events, at least 50000 antihydrogen atoms have been produced, but the analysis of the signal to background ratio shows that the production is much larger, which will be the subject of a forthcoming publication. Our heating techniques and the use of the detector have enabled us to assess the production dependence on temperature and the spatial distribution of the $\bar{H}$s produced. These results are also under study and will be submitted for publication.

Conclusions

In this article the main recent results of the ATHENA collaboration have been briefly summarized. The importance of antihydrogen physics in relation to the CPT tests has been highlighted. The experimental apparatus and the techniques used have been rapidly shown. The criteria for antihydrogen identification and the experimental results that have enabled us to provide strong evidence of the first production of antihydrogen atoms at cryogenic temperature have been illustrated. The next goal of the ATHENA collaboration is the detailed study and characterization of the antihydrogen produced. Therefore the production rate needs to be optimized to allow for the first interaction between anthydrogen atoms and laser.

REFERENCES

1. D. Colladay, V.A. Kostelecky "CPT Violation and the Standard Model", Phys. Rev. D 55 (1997), 6760-6774.
2. J. Ellis, J. Lopez, N. E. Mavromatos, D. V. Nanopulos, Phys. Rev. D 53 (1996), 3846.
3. M.H. Holzscheiter and M. Charlton, Rep. Prog. Phys. 62 (1999) 1-60.
4. B. Baur et al, Phys.Lett. B 368 (1996), 251.
5. M. Amoretti *et al.*, Nature 419 (2002) 456.
6. S. Maury, Hyperfine Interact. 109 (1997), 43.
7. M. Amoretti *et al.*, Phys. Rev. Lett. **91** (2003) 055001.

Phenomenology of the Littlest Higgs Model

Tao Han

Department of Physics, University of Wisconsin,
1150 University Avenue, Madison, Wisconsin 53706
E-mail: than@pheno.physics.wisc.edu

Abstract. The little Higgs idea is a new way to solve the "little hierarchy" problem by protecting the Higgs mass from quadratically divergent one-loop corrections. I describe the phenomenology of one particular realization of the little Higgs idea: the "Littlest Higgs" model.

INTRODUCTION

One of the major motivations for physics beyond the Standard Model (SM) is to resolve the hierarchy and fine-tuning problems between the electroweak scale (v) and the Planck scale (M_{pl}), a big hierarchy. The Higgs boson mass in the SM is quadratically sensitive to the cutoff scale Λ of the SM effective theory via radiative corrections. The quantum-corrected Higgs mass is given at one-loop by

$$m_h^2 = (m_h^2)_{\text{bare}} + \frac{3g^2\Lambda^2}{32\pi^2 m_W^2}\left(m_h^2 + 2m_W^2 + m_Z^2 - \frac{4}{3}N_c \sum_f m_f^2\right). \quad (1)$$

For a high cutoff scale Λ, this cancellation must be fine-tuned; for example, for $\Lambda = 10$ TeV, $(m_h^2)_{\text{bare}}$ must be tuned at the 1% level to cancel the radiative corrections. In fact, requiring that the one-loop contributions to the Higgs mass-squared parameter are no more than 10 times the size of the renormalized Higgs mass-squared term (i.e., no more than 10% fine-tuning), leads to the requirement that $\Lambda_t \lesssim 2$ TeV, $\Lambda_{W,Z} \lesssim 5$ TeV, $\Lambda_H \lesssim 10$ TeV. If we accept this "naturalness" argument and believe the existence of the new physics scale $\Lambda \lesssim 10$ TeV, what could the natural cancellation mechanism be? This is referred as the "little hierarchy", between v and Λ.

The classic solution is supersymmetry. The Higgs mass is protected by supersymmetry to be one loop factor below the soft supersymmetry breaking scale. Thus weak scale supersymmetry is natural if $M_{SUSY} \sim \mathcal{O}(1 \text{ TeV})$.

The little Higgs idea [1] is an alternative way to keep the Higgs boson naturally light. The basic idea is as follows:

(i) The Higgs field is a pseudo-Nambu-Goldstone boson [2] of a global symmetry that is spontaneously broken at a scale $\Lambda \sim 4\pi f \sim 10 - 30$ TeV;

(ii) The quadratic divergences in the Higgs mass are canceled at the one-loop level by new particles with masses $M \sim gf \sim 1 - 3$ TeV;

(iii) The Higgs acquires a mass radiatively at the electroweak scale $v \sim g^2 f/4\pi \sim 100 - 300$ GeV.

Little Higgs models [1, 3, 4, 5, 6, 7, 8, 9] are constructed so that at least two operators are needed to explicitly break all of the global symmetry that protects the Higgs mass. This forbids quadratic divergences at one-loop; the Higgs mass is then smaller than Λ by not one but two loop factors, leading to the little hierarchy $\Lambda \gg f \gg v$.

In this talk, I start by summarizing the Littlest Higgs model [4], and give an overview of its phenomenology, then present an outlook and conclusions. This talk is based on [10, 11].

THE LITTLEST HIGGS MODEL

We consider here a specific realization of the little Higgs idea called the "Littlest Higgs" model [4], introduced last year by Arkani-Hamed, Cohen, Katz and Nelson.

The Littlest Higgs model is a nonlinear sigma model with a global SU(5) symmetry group broken down to SO(5) by the vacuum expectation value (vev) Σ_0. An $[SU(2) \times U(1)]^2$ subgroup of SU(5) is gauged; Σ_0 breaks this gauge symmetry down to the diagonal $SU(2) \times U(1)$ subgroup, which is identified with the SM gauge group. The symmetry breaking leads to $24 - 10 = 14$ Goldstone bosons, four of which are eaten by the broken gauge generators. The remaining ten Goldstone bosons transform under the SM gauge symmetry as a complex doublet h (which will become the SM Higgs doublet) and a complex triplet ϕ. The Goldstone bosons can be written as $\Sigma = \exp(2i\Pi/f)\Sigma_0$, where $f \sim \Lambda/4\pi$ is the "pion decay constant" that will set the scale of the new particle masses. The uneaten Goldstone bosons are given by

$$\Pi = \begin{pmatrix} & h^\dagger/\sqrt{2} & \phi^\dagger \\ h/\sqrt{2} & & h^\star/\sqrt{2} \\ \phi & h^T/\sqrt{2} & \end{pmatrix}, \quad h = (h^+, h^0), \quad \phi = \begin{pmatrix} \phi^{++} & \phi^+/\sqrt{2} \\ \phi^+/\sqrt{2} & \phi^0 \end{pmatrix}. \quad (2)$$

Gauge sector

The gauge couplings $g_1, g_2, g'_1,$ and g'_2 break the global symmetry explicitly. However, the model is constructed such that no single interaction breaks all the global symmetry protecting the Higgs mass. This implements the little Higgs mechanism: at least two interactions are required to break all the global symmetry and give mass to the Higgs, thus forbidding quadratically divergent radiative corrections at the one loop level. The Σ_0 vev gives mass to one linear combination of the two SU(2) gauge bosons W_1 and W_2 and to one linear combination of the two U(1) gauge bosons B_1 and B_2 as listed in Table 1, where we have defined the mixing angles $c, s \equiv \cos\theta, \sin\theta$ and $c', s' \equiv \cos\theta', \sin\theta'$ in terms of the gauge couplings by $g = g_1 s = g_2 c$, $g' = g'_1 s' = g'_2 c'$. After electroweak symmetry breaking which will be discussed later, the heavy gauge bosons mix with the SM gauge bosons at order v^2/f^2 and form the mass eigenstates W_H, Z_H, A_H.

TABLE 1. The new particles of the Littlest Higgs model and their masses to leading order in v/f. The masses given all receive corrections of order v^2/f. For M_ϕ, we obtain the lower bound by assuming $m_H \geq 115$ GeV.

Particle	A_H	Z_H, W_H	$\Phi^0, \Phi^P, \Phi^+, \Phi^{++}$	T
Mass	$f\frac{m_Z s_W}{v\sqrt{5}s'c'}$	$f\frac{m_W}{vsc}$	$f\frac{\sqrt{2}m_H}{v\sqrt{1-(4v'f/v^2)^2}}$	$f\sqrt{\lambda_1^2+\lambda_2^2}$
Mass lower bound	$0.16f$	$0.65f$	$0.66f$	$1.42f$

Top sector

To cancel the quadratic divergence due to the top quark loop, the Higgs coupling to the top quark must also be generated through collective breaking. This can be done by introducing a vector-like pair of colored Weyl fermions $\tilde{t}$ and $\tilde{t}'^c$, and writing the following couplings: $\mathscr{L}_{\text{Yuk}} = (\lambda_1/2)f\varepsilon_{ijk}\varepsilon_{xy}Q_i\Sigma_{jx}\Sigma_{ky}u_3'^c + \lambda_2 f\tilde{t}\tilde{t}'^c + \text{h.c.}$, where the third-generation quark doublet is expanded to $Q = (b,t,\tilde{t})$ [4]. Here $i,j,k = 1,2,3$ and $x,y = 4,5$. The first term generates the Higgs couplings to fermions when Σ is expanded in powers of the Goldstone bosons. It is symmetric under the global $SU(3)_2$ symmetry, thus ensuring that the quadratic divergences cancel between the top loop and a loop of the new heavy fermion. The second term in the Lagrangian is a mass term for the vector-like quark. Inserting the Σ_0 vev, $\tilde{t}$ marries a linear combination of $\tilde{t}'^c$ and $u_3'^c$ and gets a mass of order f, as listed in Table 1. The remaining linear combination becomes the right-handed top quark. The top quark mass is given by $m_t = \frac{\lambda_1\lambda_2}{\sqrt{\lambda_1^2+\lambda_2^2}}v$. Note that m_t vanishes if either λ_1 or λ_2 is zero: this is a manifestation of the collective breaking.

The rest of the fermions

There is no need to cancel the quadratic divergences in the Higgs mass due to light fermion loops because they do not become important until much higher cutoff scales. Thus we can generate masses for the rest of the fermions by writing terms of the same form as $\mathscr{L}_{\text{Yuk}}$ but without the extra vector-like quarks. If we make this choice for the light fermion masses, then gauge invariance of $\mathscr{L}_{\text{Yuk}}$ fixes the charges of the fermions under the two U(1) gauge symmetries up to only two free continuous parameters y_u and y_e per generation. Imposing anomaly cancellation then fixes the U(1) charges uniquely $y_u = -2/5$, $y_e = 3/5$. Different U(1) charges are possible for the light fermions if their masses are instead generated by higher-dimensional operators.

Higgs potential and electroweak symmetry breaking

The Higgs potential is generated radiatively by integrating out the heavy gauge bosons and heavy top-partner. It can be written in the general form of Coleman-Weinberg, $V = \lambda_{\phi^2}f^2\text{Tr}(\phi^\dagger\phi) + i\lambda_{h\phi h}f\left(h\phi^\dagger h^T - h^*\phi h^\dagger\right) - \mu^2 hh^\dagger + \lambda_{h^4}(hh^\dagger)^2$. The parameter μ^2 gets log-divergent contributions at one-loop and quadratically divergent contributions at two-loop, which are both generically one loop factor smaller than f^2, leading to a Higgs mass of order $g^2 f/4\pi$. Since there will be additional cutoff-scale freedom in the two-loop contributions to μ^2, we regard it as a free parameter. For $\mu^2 > 0$, electroweak symmetry is broken, and both h and ϕ get vevs:

$$2\langle h^0\rangle^2 \equiv v^2 = \frac{\mu^2}{\lambda_{h^4} - \lambda_{h\phi h}^2/\lambda_{\phi^2}}, \qquad \langle i\phi^0\rangle \equiv v' = \frac{\lambda_{h\phi h}}{2\lambda_{\phi^2}}\frac{v^2}{f}. \qquad (3)$$

Note that $v' \sim v^2/f$ is much smaller than v. The scalar masses, to leading order in v/f, are $M_\phi^2 = \lambda_{\phi^2} f^2$, $m_H^2 = 2\mu^2$.

Summary of new parameters and particles

The Littlest Higgs model contains six new free parameters, which we can choose as follows:
(1) $\tan\theta = s/c = g_1/g_2$: new SU(2) gauge coupling.
(2) $\tan\theta' = s'/c' = g_1'/g_2'$: new U(1) gauge coupling.
(3) f: symmetry breaking scale, $\mathcal{O}$(TeV).
(4) v': triplet vev; $v' < v^2/4f$.
(5) m_H: SM Higgs mass.
(6) M_T: top-partner mass (together with m_t, this fixes λ_1 and λ_2).

The new particles and their masses are summarized in Table 1. For more detailed discussions, see *e.g.* [10].

PHENOMENOLOGY

There are by now quite a number of little Higgs models in the literature [1, 3, 4, 5, 6, 7, 8, 9]. It thus behooves us to look for generic features of the phenomenology. All little Higgs models must contain the following features at the TeV scale: new heavy gauge bosons to cancel the W and Z loops; a new heavy fermion to cancel the top quark loop; and new heavy scalars to cancel the Higgs loop.

Electroweak precision constraints

The constraints on little Higgs models from electroweak precision data have been examined in detail in [12, 13, 14, 15]. The constraints come from Z pole data, low-energy neutrino-nucleon scattering, and the W mass measurement. Together, these measurements probe little Higgs model contributions from the exchange of the heavy gauge bosons between fermion pairs, mixing between the heavy and light gauge bosons that modifies the Z-boson couplings to fermions, and a shift in the mass ratio of the W and

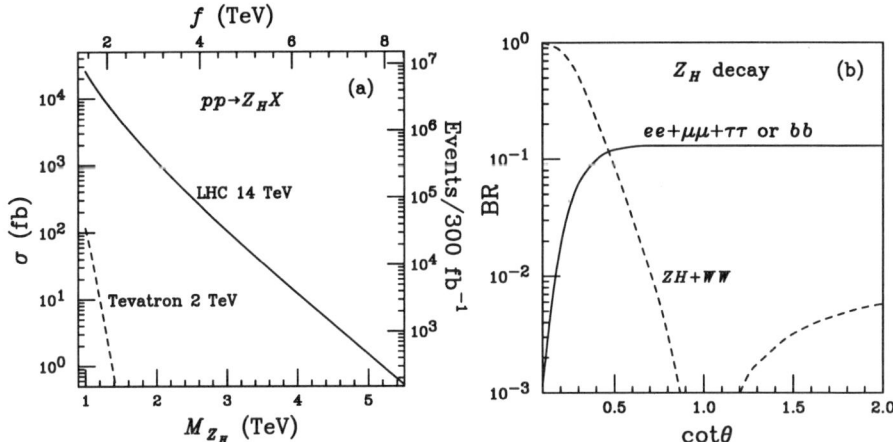

FIGURE 1. (a). Cross section for Z_H production in Drell-Yan at the LHC and Tevatron, for $\cot\theta = 1$; and (b). Branching ratios of Z_H into SM particles as a function of $\cot\theta$, neglecting final-state mass effects.

Z. Examining these contributions yields a strategy for reducing the impact of the Littlest Higgs model on electroweak observables:

(1) Reduce the triplet vev: $v' \ll v$;
(2) Reduce the heavy SU(2) gauge boson contributions: $c \ll 1$;
(3) Reduce the heavy U(1) gauge boson contributions to M_Z: $c' \approx s'$;
(4) Reduce the heavy U(1) gauge boson contributions to neutral current couplings: $c'^2 Y_1 \approx s'^2 Y_2$.

The lower bounds on the masses of the new heavy gauge bosons are generally in the 1.5 − 2 TeV range [14, 15]. The electroweak precision measurements do not directly constrain the mass of the top-partner. However, the mass of the top-partner is related to the heavy gauge boson masses by the structure of the model. For naturalness, the top-partner should be as light as possible. The lower bounds on the top-partner mass are generally in the 1 − 2 TeV range.

Collider signatures

Z_H and W_H: The heavy SU(2) gauge bosons Z_H and W_H can be produced via the Drell-Yan process at hadron colliders. The cross section is proportional to $\cot^2\theta$ because of the Z_H and W_H couplings to fermion pairs. In Fig. 1(a) we show the cross section for Z_H production at the Tevatron and LHC for $\cot\theta = 1$. Even with this suppression factor $\cot^2\theta \simeq 0.04$, a cross section of 40 fb is expected at the LHC for $M_{Z_H} \simeq 2$ TeV, leading to 4,000 events in 100 fb^{-1} of data.

The decay branching fractions of Z_H are shown in Fig. 1(b). The partial widths to fermion pairs are proportional to $\cot^2\theta$, while the partial widths to ZH and W^+W^- are

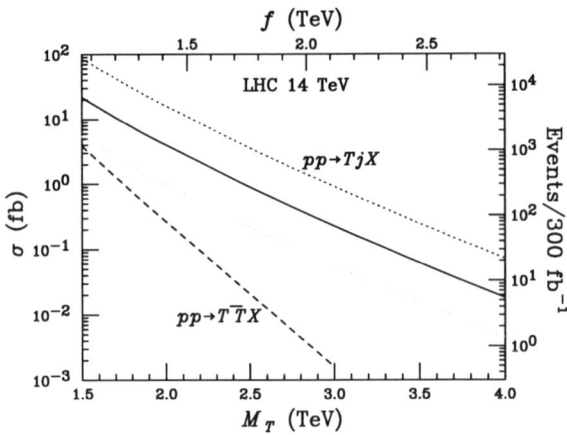

FIGURE 2. Cross sections for T production at the LHC. The single T cross section is shown for $\lambda_1/\lambda_2 = 1$ (solid) and $\lambda_1/\lambda_2 = 2$ (upper dotted) and $1/2$ (lower dotted). The QCD pair production cross section is shown for comparison (dashed).

proportional to $\cot^2 2\theta$. This offers a method to distinguish the Littlest Higgs model from a "big Higgs" model with the same gauge group in which the Higgs doublet transforms under only one of the SU(2) groups, in which case the ZH and W^+W^- partial widths would also be proportional to $\cot^2 \theta$. Neglecting final-state particle masses, the branching fraction into three flavors of charged leptons is equal to that into one flavor of quark ($\simeq 1/8$ for $\cot\theta \gtrsim 0.5$), due to the universal coupling of Z_H to all SU(2) fermion doublets. The branching ratio into ZH is equal to that into W^+W^-. The total width of Z_H depends on $\cot\theta$; for $\cot\theta \sim 0.2$ the Z_H width is about 1% of the Z_H mass.

The $W_H^\pm$ couplings to fermion doublets are larger by a factor of $\sqrt{2}$ than the Z_H couplings; this together with the parton distribution of the proton leads to a $W_H^\pm$ cross section at the LHC about 1.5 times that of Z_H [16]. As for the W_H decays, the branching fraction into three lepton flavors is equal to that into one generation of quarks ($\simeq 1/4$ for $\cot\theta \gtrsim 0.5$). At low $\cot\theta$, $W_H^\pm$ decays predominantly into $W^\pm H$ and $W^\pm Z$ with partial widths proportional to $\cot^2 2\theta$.

The heavy U(1) gauge boson A_H is the lightest new particle in the Littlest Higgs model, but its couplings to fermions are more model dependent than those of the heavy SU(2) gauge bosons, since they depend on the U(1) charges of the fermions. We will not discuss its collider phenomenology here.

T: The heavy top-partner T can be pair produced by QCD interactions with model-independent couplings. However, this production mode is suppressed by phase space due to the high mass of T. The single T production mode, $W^+b \to T$, is dominant for M_T above about a TeV. The cross section for single T production depends on the ratio of couplings λ_1/λ_2, which relates M_T to the scale f. The cross sections are shown in Fig. 2. The top-partner T decays into tH, tZ and bW with partial widths in the ratio $1 : 1 : 2$. The top sector is quite similar in many of the other little Higgs models in the literature, so that these general features of T production and decay should apply.

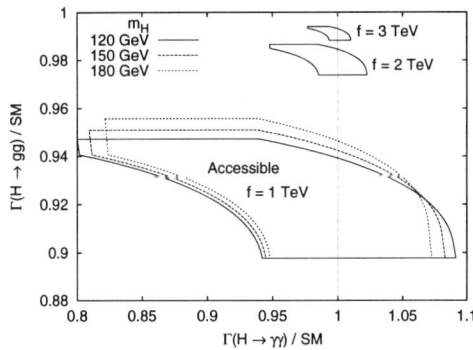

FIGURE 3. Range of values of $\Gamma(H \to gg)$ versus $\Gamma(H \to \gamma\gamma)$ accessible in the Littlest Higgs model normalized to the SM value, for $m_H = 120, 150, 180$ GeV and $f = 1, 2, 3$ TeV. From [12].

Φ^{++}: The doubly charged Higgs triplet state Φ^{++} can be singly produced at resonance $W^+W^+ \to \Phi^{++} \to W^+W^+$. The cross section for this process is proportional to v'^2/v^2, which may make it difficult to see due to lack of rate. The doubly charged Higgs can in principle decay to a pair of like-sign charged leptons via the dimension-four operator $L\Phi L$, offering a more distinctive signature; however, the coupling is highly model dependent and care must be taken to avoid generating too large a neutrino mass from v'.

Loop-induced Higgs decays: A natural question regarding the Little Higgs phenomenology would be if the "little Higgs" itself can be distinguishable from the SM one. The Higgs decays into gluon pairs or photon pairs will be modified in the Littlest Higgs model by the new particles running in the loop and by the shifts in the Higgs couplings to the SM W boson and top quark due to the structure of the model. The range of partial widths as a function of f accessible by varying the other model parameters are shown in Fig. 3.

For $f \geq 1$ TeV, the correction to $\Gamma(H \to gg)$ is always less than 10%. This is already smaller than the remaining theoretical and experimental uncertainties on the gluon fusion cross section at the LHC. For the partial width to photons, the situation is more promising because the QCD corrections are well under control. At the LHC, the $H \to \gamma\gamma$ decay rate can be measured to 15-20%; this probes $f < 1.0$ TeV at 1σ. A linear e^+e^- collider has only comparable precision since the $H \to \gamma\gamma$ branching ratio measurement is limited by statistics. The most promising measurement would be done at a photon collider, in which the $\gamma\gamma \to H \to b\bar{b}$ rate could be measured to about 2%. Combining this with a measurement of the branching ratio of $H \to bb$ to about 1.5-2% at the e^+e^- collider allows the extraction of $\Gamma(H \to \gamma\gamma)$ with a precision of about 3%. Such a measurement would be sensitive to $f < 2.7$ TeV at the 1σ level, or $f < 1.8$ TeV at the 2σ level. A 5σ deviation is possible for $f < 1.2$ TeV.

CONCLUSIONS

The little Higgs idea provides a new way to address the little hierarchy problem of the Standard Model by making the Higgs a pseudo-Nambu-Goldstone boson of a spontaneously broken global symmetry. The global symmetry is explicitly broken by gauge and Yukawa interactions; however, no single interaction breaks all the symmetry protecting the Higgs mass. This prevents quadratically divergent radiative corrections to the Higgs mass from appearing at the one-loop level, and thus allows the cutoff scale to be pushed higher by one loop factor, to ~ 10 TeV.

The details of the phenomenology depend on the specific model. Very generically, there must be new gauge bosons, fermions and scalars to cancel the quadratic divergences in the Higgs mass. Less generically, models with product gauge groups of the form $[SU(2) \times U(1)]^2$ contain an $SU(2)$ triplet of new heavy gauge bosons, $Z_H, W_H^\pm$.

There is some tension between the precision electroweak constraints pushing up the new particle masses and the requirement that the new particles be light to avoid fine tuning. However, by tuning the parameters of the models appropriately one can satisfy both constraints. This tuning of the parameters should be explained in the ultraviolet completion of the nonlinear sigma model. Our developing understanding of the effects of little Higgs models on the electroweak precision observables is now driving model building to incorporate features that loosen the constraints. Taking these constraints into account, the new particles should live in the $1 - 2$ TeV mass range and should be accessible at the LHC.

ACKNOWLEDGMENTS

I would like to thank Heather Logan, Bob McElrath and Lian-Tao Wang for collaborations leading to the papers [10, 11] on which this talk was based. I also thank the organizers of the CIPANP for their invitation to the conference. The research is supported in part by the US Department of Energy under grant DE-FG02-95ER40896 and in part by the Wisconsin Alumni Research Foundation.

REFERENCES

1. N. Arkani-Hamed, A. G. Cohen and H. Georgi, Phys. Lett. B **513**, 232 (2001) [arXiv:hep-ph/0105239].
2. S. Dimopoulos and J. Preskill, Nucl. Phys. B **199**, 206 (1982); D. B. Kaplan and H. Georgi, Phys. Lett. B **136**, 183 (1984); D. B. Kaplan, H. Georgi and S. Dimopoulos, Phys. Lett. B **136**, 187 (1984); H. Georgi and D. B. Kaplan, Phys. Lett. B **145**, 216 (1984); H. Georgi, D. B. Kaplan and P. Galison, Phys. Lett. B **143**, 152 (1984); M. J. Dugan, H. Georgi and D. B. Kaplan, Nucl. Phys. B **254**, 299 (1985); T. Banks, Nucl. Phys. B **243**, 125 (1984).
3. N. Arkani-Hamed, A. G. Cohen, E. Katz, A. E. Nelson, T. Gregoire and J. G. Wacker, JHEP **0208**, 021 (2002) [arXiv:hep-ph/0206020].
4. N. Arkani-Hamed, A. G. Cohen, E. Katz and A. E. Nelson, JHEP **0207**, 034 (2002) [arXiv:hep-ph/0206021].
5. I. Low, W. Skiba and D. Smith, Phys. Rev. D **66**, 072001 (2002) [arXiv:hep-ph/0207243].
6. D. E. Kaplan and M. Schmaltz, arXiv:hep-ph/0302049.

7. S. Chang and J. G. Wacker, arXiv:hep-ph/0303001.
8. W. Skiba and J. Terning, arXiv:hep-ph/0305302.
9. S. Chang, arXiv:hep-ph/0306034.
10. T. Han, H. E. Logan, B. McElrath and L. T. Wang, Phys. Rev. D **67**, 095004 (2003) [arXiv:hep-ph/0301040].
11. T. Han, H. E. Logan, B. McElrath and L. T. Wang, Phys. Lett. B **563**, 191 (2003) [arXiv:hep-ph/0302188].
12. C. Csaki, J. Hubisz, G. D. Kribs, P. Meade and J. Terning, Phys. Rev. D **67**, 115002 (2003) [arXiv:hep-ph/0211124].
13. J. L. Hewett, F. J. Petriello and T. G. Rizzo, arXiv:hep-ph/0211218.
14. C. Csaki, J. Hubisz, G. D. Kribs, P. Meade and J. Terning, arXiv:hep-ph/0303236.
15. T. Gregoire, D. R. Smith and J. G. Wacker, arXiv:hep-ph/0305275.
16. G. Burdman, M. Perelstein and A. Pierce, Phys. Rev. Lett. **90**, 241802 (2003) [arXiv:hep-ph/0212228].

Testing V-A in Top Decay at CDF at $\sqrt{s}$ = 1.8 TeV

Ben Kilminster*

*University of Rochester
for the CDF Collaboration*

Abstract. The structure of the tbW vertex can be probed by measuring the polarization of the W in $t \to W + b \to l + \nu + b$. The invariant mass of the the lepton and b quark measures the W decay angle which in turn allows a comparison with polarizations expected from a V-A and V+A tbW vertex. We measure the fraction by rate of Ws produced with a V+A coupling in lieu of the Standard Model V-A to be $f_{V+A} = -0.21^{+0.42}_{-0.24}$ (stat) $\pm$ 0.21 (sys). We assign a limit of $f_{V+A} <$ 0.80 @ 95% CL. By combining this result with a complementary observable in the same data, we assign a limit of $f_{V+A} <$ 0.61 @ 95% CL. From this CDF Run I preliminary result, we find no evidence for a non-standard Model tbW vertex.

INTRODUCTION

The heavy top quark mass has led to speculation that the top quark may have a unique relationship with the electroweak force which might modify the V-A structure of the decay of the top to a W boson and b quark. This could in turn lead to non-standard model W polarization evident from the study of angular distributions of leptons from W decay [1]. This analysis exploits the relationship between these angular distributions and the invariant mass of $l - b$ combinations from the top decay $t \to W + b \to l + \nu + b$ to fit for a non-standard model V+A contribution. To analyze the polarization of the W, it is necessary to know the weak isospin of the W decay products, therefore we use the leptonic decay of the W rather than the hadronic decay since quark jets cannot be distinguished from anti-quark jets. Scenarios introducing a V+A contribution include mirror fermions having a right-handed weak interaction either mixing with the top, or faking the top if having similar mass [2].

The spin-one W has three possible helicities for the W^+: -1 (left-handed), 0 (longitudinal), and +1 (right-handed). V-A theory predicts the probability of each W helicity distribution in top decay. Because $M_t > M_W$, a significant number of Ws will be longitudinally polarized with rate, $F_0 = 0.70$ for $M_t = 174.3$ GeV and $M_W = 80.4$ GeV [1]. Leptons from the decay of longitudinally polarized Ws have a symmetric angular distribution of the form $1 - (cos\psi_l^\star)^2$, where $cos\psi_l^\star$ is defined as the angle between the lepton in the W rest frame and the boost vector from the top to the W rest frame. Maximal parity violation in the Standard Model V-A weak theory predicts that the remaining W helicity rate is left-handed, creating an asymmetric angular distribution of the form $(1 - cos\psi_l^\star)^2$ [1]. Since the angle $\psi_l^\star$ can be related to the $l - b$ invariant mass

*Current E-Mail: bjk@fnal.gov

combination by Equation 1,

$$M_{lb}^2 = \frac{1}{2}(M_T^2 - M_W^2)(1 + cos\psi_l^*) \quad (1)$$

the distribution of M_{lb}^2 in $t\bar{t}$ data can be studied to determine the polarization of the non-longitudinal Ws. M_{lb}^2 is a good choice because no information about the top or W rest frames are required, and so the neutrino momentum need not be reconstructed. In a V+A theory, M_{lb}^2 would be larger on average than for a V-A theory.

DATA SAMPLES

The $t\bar{t}$ data samples are obtained from $p\bar{p}$ collisions at $\sqrt{s} = 1.8$ TeV in the CDF detector [3]. Three sub-samples of $t\bar{t}$ data are chosen for their low background and high efficiency for b jet identification. The samples and backgrounds are defined in detail elsewhere [4]. Each sample is classified by the number of leptons and identified b jets in the final state.

The dilepton sample is composed of $t\bar{t}$ in which both Ws decay to an electron or muon, and are distinguished by a signal of $\not{E}_T > 25 GeV$ from the combination of $W \to l\nu$ decays, two opposite sign and flavor leptons with $P_T > 20 GeV$ in the central region $|\eta| < 1.0$, and two jets with $E_T > 10 GeV$ and $|\eta| < 2.0$. This sample is consistent with previous CDF analyses, but here only the $e + \mu + jets$ channel is used to eliminate Drell Yan production which is the dominant source of background in the dilepton sample. Initial and final state radiation can result in extra jets, so the b jets are chosen to be the two highest E_T jets. There are four M_{lb} pairings in each dilepton event.

Two lepton+jets samples are defined where only one W decays into an electron or muon, the other decaying into two jets. The lepton is required to have $P_T > 20 GeV$, in the central region $|\eta| < 1.0$. Four jets are required, three with $E_T > 15 GeV$, $|\eta| < 2.0$, and the fourth having $E_T > 8 GeV$ and $|\eta| < 2.4$. One of these samples is required to have a jet "b-tagged" with a displaced vertex using the silicon vertex detector [4], reducing background greatly, and decreasing the number of M_{lb} pairings from at least four to only one per event. This sample is referred to as the "single-tagged" sample. The other lepton+jets sample is known as "double-tagged" and is required to have two b tags, further reducing background and providing two M_{lb} pairings. The b-tagging algorithm does not assign a charge to the b-jet, therefore half of all pairings incorrectly match leptons with b jets from the other top decay.

In the actual data, 7 events were found in the dilepton $e\mu$ sample with an expected background of 0.76 ± 0.21 events, for the single-tagged sample 15 events were found with a background 2.0 ± 0.7, and in the double-tagged sample there were 5 events with a 0.2 ± 0.2 background. Interesting to note is that since right-handed leptons have higher P_T, an increase in events passing the lepton P_T trigger requirement could also indicate a V+A theory. The actual number of events found is not used in this analysis, only the shape of the M_{lb}^2 distributions. When two pairings of the lepton are possible, the two variables are used simultaneously in the fit. While correct pairings are limited kinematically by $1/2(M_t^2 - M_W^2)$, incorrect pairings may have significantly higher mass.

METHOD

The M^2 data shape is fit to be a linear combination of V-A $t\bar{t}$, V+A $t\bar{t}$ [5], and background using a log likelihood method. Each data sample: dilepton, single-tagged, and double-tagged, is fit individually, and a combined log likelihood is used to determine f_{V+A} for the total data sample. f_{V+A} is not constrained to be in the physical region in the fit. Background is allowed to fluctuate within its uncertainties. The background and V-A $t\bar{t}$ samples are weighted by the relative efficiency for the events to pass the lepton P_T requirements as compared to the V+A sample which has a harder lepton P_T distribution. The likelihood fit technique is evaluated using Monte Carlo experiments and determined to be consistent with Gaussian central value and errors by including a 4% scale factor to the error of the fit.

Since this analysis fits to V-A and V+A rates, it does not take into account the interference effects resulting from left and right-handed W polarizations being involved in the decay. For $f_{V+A} = 0.5$, the interference is maximal and the matrix element fails to account for $m_b/E_b \sim 10\%$ of the decays. The actual uncertainty introduced in the fit is expected to be less than 10% in this case[6]. There is no interference in the case of all V-A or all V+A. This error is not significant compared to statistical and systematic uncertainties.

The largest systematic uncertainties are the top mass and jet energy scale. The top mass enters into the definition of M_{lb}^2, resulting in a shift of f_{V+A} of +0.21 for a top mass shift of 5 GeV. The jet energy scale uncertainty [4] results in a shift of +0.14 for a shift of the jet energy scale upwards within its uncertainty. Since the top mass has a large uncertainty due to the CDF jet energy scale [4], these systematics are highly correlated. Accounting for their correlations results in a top mass systematic shift of 0.19, independent of jet energy scale, and a jet energy scale systematic shift of 0.04, independent of top mass. The total systematic uncertainties amount to a shift in f_{V+A} of 0.21, and are listed in Table 1.

RESULTS

The results of the likelihood fit are $f_{V+A} = -0.21^{+0.42}_{-0.24}$ (stat) ± 0.21 (sys) for the combined data sample found in 109 pb^{-1}. The individual fits are shown in figure 1. Constructing a Neyman confidence band [7], we find an upper limit of $f_{V+A} < 0.80$ @ 95% CL. The result is combined with a previous CDF analysis using the lepton P_T [8] to discriminate between left-handed and right-handed Ws for a fixed longitudinal component, and determined to be $f_{V+A} < 0.61$ @ 95%. The combined result is inconsistent with a pure V+A theory at the 2.67 σ level. All results are preliminary. For a data sample of 2 fb^{-1} expected in the first part of Run II, this analysis technique is expected to result in total systematic and statistical uncertainties of 0.14 in f_{V+A}.

TABLE 1. Summary of systematic uncertainties in terms of shift in measurement of V+A fraction.

Systematic Uncertainties	
Top mass Uncertainty	0.19 *(0.21 w/out jet energy correlation)*
Jet energy Scale	0.04 *(0.14 w/out top mass correlation)*
Background shape uncertainty	0.05
Background normalization	0.05
ISR Gluon radiation	0.04
FSR Gluon radiation	0.03
B tagging efficiency	0.03
Parton distribution Functions	0.02
Monte Carlo Statistics	0.01
Relative acceptance uncertainty	0.005
Total systematic	0.21

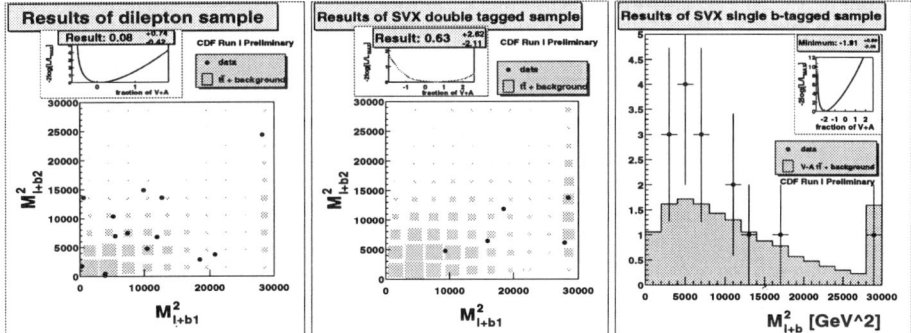

FIGURE 1. Data and Standard Model Monte Carlo distributions for each sample with $-2\log\mathcal{L}$ as a function of f_{V+A}.

REFERENCES

1. R. Peccei and X. Zhang, Nuc. Phys. B **337**, 269, 1990; G. Kane, C.P. Yuan, and D. Ladinsky, Phys. Rev. D **45**, 1531 (1992); M. Jezabek and J.H. Kuhn, Phys. Lett. B **329**, 317 (1994); C. A. Nelson, B. T. Kress, M. Lopes, and T.McCauley, Phys. Rev. D **56**, 5928 (1997).
2. D. Choudhury, T. M. Tait and C. E. Wagner, Phys. Rev. D **65**, 053002 (2002).
3. F. Abe *et al.*, Nucl. Instr. Meth. Phys. Res. A **271**, 387 (1988); D. Amidei*et al., ibid.* **350**, 73 (1994); P. Azzi *et al., ibid.* **360**, 137 (1995).
4. F. Abe *et al.* [CDF Collaboration], Phys. Rev. Lett. **80**, 2779 (1998); *ibid.* **80**, 2767 (1998); *ibid.* **74**, 2626 (1995).
5. G. Corcella *et al.*, JHEP **0101**, 010 (2001). Both $t\bar{t}$ samples were generated with HERWIG, the latter using a custom version with adjustable W helicity amplitudes.
6. Private communication with T. Tait.
7. Particle Data Group, K. Hagiwara et al., Phys. Rev. D **66**, 010001 (2002).
8. T. Affolder *et al.* [CDF Collaboration], Phys. Rev. Lett. **84**, 216 (2000).

Symmetric Textures in SO(10) and LMA Solution for Solar Neutrinos

Mu-Chun Chen* and K.T. Mahanthappa[†]

*HET Group, Physics Department, Brookhaven National Laboratory, Upton, NY 11973
[†]Department of Physics, University of Colorado, Boulder, CO 80309

Abstract. A model based on SUSY SO(10) combined with SU(2) family symmetry is constructed. In contrast with the commonly used effective operator approach, 126-dimensional Higgs fields are utilized to construct the Yukawa sector. R-parity symmetry is thus preserved at low energies. The symmetric mass textures arising from the left-right symmetry breaking chain of SO(10) give rise to very good predictions for quark and lepton masses and mixings. The prediction for $\sin 2\beta$ agrees with the average of current bounds from BaBar and Belle. In the neutrino sector, our predictions are in good agreement with results from atmospheric neutrino experiments. Our model accommodates the LMA solution to the solar neutrino anomaly. The prediction of our model for the $|U_{e v_3}|$ element in the MNS matrix is close to the sensitivity of current experiments; thus the validity of our model can be tested in the near future. We also investigate the correlation between the $|U_{e v_3}|$ element and $\tan^2 \theta_\odot$ in a general two-zero neutrino mass texture.

SO(10) has long been thought to be an attractive candidate for a grand unified theory (GUT) for a number of reasons: First of all, it unifies all the 15 known fermions with the right-handed neutrino for each family into one 16-dimensional spinor representation. The seesaw mechanism then arises very naturally, and the non-zero neutrino masses can thus be explained. Since a complete quark-lepton symmetry is achieved, it has the promise for explaining the pattern of fermion masses and mixing. Because B-L contained in SO(10) is broken in symmetry breaking chain to the SM, it also has the promise for baryogenesis. Recent atmospheric neutrino oscillation data from Super-Kamiokande indicates non-zero neutrino masses. This in turn gives very strong support to the viability of SO(10) as a GUT group. Models based on SO(10) combined with discrete or continuous family symmetry have been constructed to understand the flavor problem. Most of the models utilize "lopsided" mass textures which usually require more parameters and therefore are less constrained. Furthermore, the right-handed neutrino Majorana mass operators in most of these models are made out of $16_H \times 16_H$ which breaks the R-parity at a very high scale. The aim of this talk, based on Ref.[1, 2, 3], is to present a realistic model based on supersymmetric SO(10) combined with SU(2) family symmetry which successfully predicts the low energy fermion masses and mixings. Since we utilize *symmetric* mass textures and $\overline{126}$-dimensional Higgs representations for the right-handed neutrino Majorana mass operator, our model is more constrained in addition to having R-parity conserved. We first discuss the viable phenomenology of mass textures followed by the model which accounts for it, and then the implications of the model for neutrino mixing, CP violation, and neutrinoless double beta decay are presented.

The set of up- and down-quark mass matrix combination is given by, at the GUT scale,

$$M_u = \begin{pmatrix} 0 & 0 & ae^{i\gamma_a} \\ 0 & be^{i\gamma_b} & ce^{i\gamma_c} \\ ae^{i\gamma_a} & ce^{i\gamma_c} & e^{i\gamma_d} \end{pmatrix} dv_u, \quad M_d = \begin{pmatrix} 0 & ee^{i\gamma_e} & 0 \\ ee^{i\gamma_e} & fe^{i\gamma_f} & 0 \\ 0 & 0 & e^{i\gamma_h} \end{pmatrix} hv_d \quad (1)$$

with $a \sim b \ll c \ll 1$, and $e \ll f \ll 1$. Symmetric mass textures arise naturally if SO(10) breaks down to the SM through the left-right symmetric breaking chain $SU(4) \times SU(2)_L \times SU(2)_R$. SO(10) relates the up-quark mass matrix to the Dirac neutrino mass matrix, and the down-quark mass matrix to the charged lepton mass matrix. To achieve the Georgi-Jarlskog relations, a factor of -3 is needed in the (2,2) entry of the charged lepton mass matrix,

$$M_e = \begin{pmatrix} 0 & ee^{i\gamma_e} & 0 \\ ee^{i\gamma_e} & -3fe^{i\gamma_f} & 0 \\ 0 & 0 & e^{i\gamma_h} \end{pmatrix} hv_d \quad (2)$$

This factor of -3 can be accounted for by the SO(10) CG coefficients associated with $\overline{126}$-dimensional Higgs representations. The smallness of the neutrino masses is explained by the type I seesaw mechanism. The Dirac neutrino mass matrix is identical to the mass matrix of the up-quarks in the framework of SO(10)

$$M_{v_{LR}} = \begin{pmatrix} 0 & 0 & ae^{i\gamma_a} \\ 0 & be^{i\gamma_b} & ce^{i\gamma_c} \\ ae^{i\gamma_a} & ce^{i\gamma_c} & e^{i\gamma_d} \end{pmatrix} dv_u \quad (3)$$

The right-handed neutrino sector is an unknown sector. It is only constrained by the requirement that it gives rise to a bi-maximal mixing pattern and a hierarchical mass spectrum at low energies. To achieve this, we consider an effective neutrino mass matrix of the form

$$M_{v_{LL}} = M_{v_{LR}}^T M_{v_{RR}}^{-1} M_{v_{LR}} = \begin{pmatrix} 0 & 0 & t \\ 0 & 1 & 1+t^n \\ t & 1+t^n & 1 \end{pmatrix} \frac{d^2 v_u^2}{M_R} \quad (4)$$

The effective neutrino mass matrix of this form is obtained if the right-handed neutrino mass matrix has the same texture as that of the Dirac neutrino mass matrix,

$$M_{v_{RR}} = \begin{pmatrix} 0 & 0 & \delta_1 \\ 0 & \delta_2 & \delta_3 \\ \delta_1 & \delta_3 & 1 \end{pmatrix} M_R \quad (5)$$

and if the elements δ_i are of the right orders of magnitudes, determined by $\delta_i = f_i(a,b,c,t,\theta)$, where $\theta \equiv (\gamma_b - 2\gamma_c - \gamma_d)$. Note that $M_{v_{LL}}$ has the same texture as that of $M_{v_{LR}}$ and $M_{v_{RR}}$, thus the seesaw mechanism is form invariant. A generic feature of mass matrices of the type given in Eq.(4) is that they give rise to bi-maximal mixing pattern. After diagonalizing this mass matrix, one can see immediately that the squared mass difference between $m_{v_1}^2$ and $m_{v_2}^2$ is of the order of $O(t^3)$, while the squared mass difference between $m_{v_2}^2$ and $m_{v_3}^2$

is of the order of $O(1)$, in units of Λ. For $t \ll 1$, the phenomenologically favored relation $\Delta m^2_{atm} \gg \Delta m^2_\odot$ is thus obtained.

The SU(2) family symmetry is implemented á la the Froggatt-Nielsen mechanism. The heaviness of the top quark and to suppress the SUSY FCNC together suggest that the third family of matter fields transform as a singlet and the lighter two families of matter fields transform as a doublet under SU(2). In the family symmetric limit, only the third family has non-vanishing Yukawa couplings. SU(2) breaks down in two steps: $SU(2) \xrightarrow{\varepsilon M} U(1) \xrightarrow{\varepsilon' M} nothing$, where $\varepsilon' \ll \varepsilon \ll 1$ and M is the family symmetry scale. These small parameters ε and ε' are the ratios of the vacuum expectation values of the flavon fields to the family symmetry scale. A discrete symmetry $(Z_2)^3$ is needed to avoid unwanted couplings. The field content of our model is then given by

– matter fields

$$\psi_a \sim (16,2)^{-++} \quad (a=1,2), \quad \psi_3 \sim (16,1)^{+++}$$

– Higgs fields:

$$(10,1): \quad T_1^{+++}, \quad T_2^{-+-}, \quad T_3^{--+}, \quad T_4^{---}, \quad T_5^{+--}$$
$$(\overline{126},1): \quad \overline{C}^{---}, \quad \overline{C}_1^{+++}, \quad \overline{C}_2^{++-}$$

– Flavon fields:

$$(1,2): \quad \phi_{(1)}^{++-}, \quad \phi_{(2)}^{+-+}, \quad \Phi^{-+-}$$
$$(1,3): \quad S_{(1)}^{+--}, \quad S_{(2)}^{---}, \quad \Sigma^{++-}$$

and the superpotential of our model which generates fermion masses is given by

$$W = W_{D(irac)} + W_{M(ajorana)}$$
$$W_D = \psi_3 \psi_3 T_1 + \frac{1}{M} \psi_3 \psi_a (T_2 \phi_{(1)} + T_3 \phi_{(2)}) + \frac{1}{M} \psi_a \psi_b (T_4 + \overline{C}) S_{(2)} + \frac{1}{M} \psi_a \psi_b T_5 S_{(1)} \quad (6)$$
$$W_M = \psi_3 \psi_3 \overline{C}_1 + \frac{1}{M} \psi_3 \psi_a \Phi \overline{C}_2 + \frac{1}{M} \psi_a \psi_b \Sigma \overline{C}_2$$

where T_i's and $\overline{C}_i$'s are the 10 and $\overline{126}$ dimensional Higgs representations of SO(10) respectively, and Φ and Σ are the doublet and triplet of SU(2), respectively. Detailed quantum number assignment and the VEVs acquired by various scalar fields are given in Ref.[1]. This superpotential gives rise to the mass textures given in Eq.(1)-(5):

$$M_{u,\nu_{LR}} = \begin{pmatrix} 0 & 0 & \langle 10_2^+ \rangle \varepsilon' \\ 0 & \langle 10_4^+ \rangle \varepsilon & \langle 10_3^+ \rangle \varepsilon \\ \langle 10_2^+ \rangle \varepsilon' & \langle 10_3^+ \rangle \varepsilon & \langle 10_1^+ \rangle \end{pmatrix} = \begin{pmatrix} 0 & 0 & r_2 \varepsilon' \\ 0 & r_4 \varepsilon & \varepsilon \\ r_2 \varepsilon' & \varepsilon & 1 \end{pmatrix} M_U \quad (7)$$

$$M_{d,e} = \begin{pmatrix} 0 & \langle 10_5^- \rangle \varepsilon' & 0 \\ \langle 10_5^- \rangle \varepsilon' & (1,-3) \langle \overline{126}^- \rangle \varepsilon & 0 \\ 0 & 0 & \langle 10_1^- \rangle \end{pmatrix} = \begin{pmatrix} 0 & \varepsilon' & 0 \\ \varepsilon' & (1,-3) p \varepsilon & 0 \\ 0 & 0 & 1 \end{pmatrix} M_D \quad (8)$$

where $M_U \equiv \langle 10_1^+ \rangle$, $M_D \equiv \langle 10_1^- \rangle$, $r_2 \equiv \langle 10_2^+ \rangle / \langle 10_1^+ \rangle$, $r_4 \equiv \langle 10_4^+ \rangle / \langle 10_1^+ \rangle$ and $p \equiv \langle \overline{126}^- \rangle / \langle 10_1^- \rangle$. The right-handed neutrino mass matrix is

$$M_{\nu_{RR}} = \begin{pmatrix} 0 & 0 & \langle \overline{126}_2'^0 \rangle \delta_1 \\ 0 & \langle \overline{126}_2'^0 \rangle \delta_2 & \langle \overline{126}_2'^0 \rangle \delta_3 \\ \langle \overline{126}_2'^0 \rangle \delta_1 & \langle \overline{126}_2'^0 \rangle \delta_3 & \langle \overline{126}_1'^0 \rangle \end{pmatrix} = \begin{pmatrix} 0 & 0 & \delta_1 \\ 0 & \delta_2 & \delta_3 \\ \delta_1 & \delta_3 & 1 \end{pmatrix} M_R \quad (9)$$

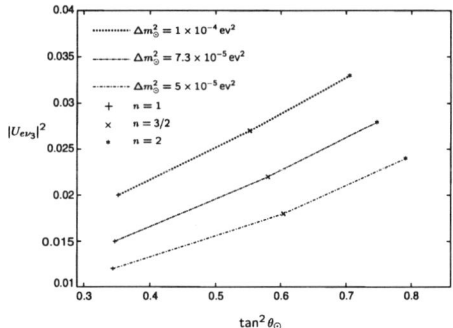

FIGURE 1. Correlation between $|U_{ev_3}|^2$ and $\tan^2 \theta_\odot$. Δm_{atm}^2 is $2.8 \times 10^{-3} eV^3$. The dotted line corresponds to the upper bound $\Delta m_\odot^2 = 10^{-4} eV^2$; the dotted-long-dashed line corresponds to the best fit value $\Delta m_\odot^2 = 7.3 \times 10^{-5} eV^2$; the dotted-short-dashed line corresponds to the lower bound $\Delta m_\odot^2 = 5 \times 10^{-5} eV^2$.

with $M_R \equiv \langle \overline{126}_1^{'0} \rangle$. Note that, since we use $\overline{126}$-dimensional representations of Higgses to generate the heavy Majorana neutrino mass terms, R-parity is preserved at all energies.

With values of $m_f, (f=u,c,t,e,\mu,\tau)$ and those of $|V_{us,ub,cb}|$ at the weak scale, the input parameters at the GUT scale are determined. The predictions for the charged fermion masses and CKM mixing of our model at M_Z which are summarized below including 2-loop RGE effects are in good agreements with the experimental values:

$$\frac{m_s}{m_d}=25, \quad m_s=85.66 MeV, \quad m_b=3.147 GeV, \quad |V_{ud}|=0.9751, \quad |V_{cd}|=0.2218$$
$$|V_{cs}|=0.9744, \quad |V_{td}|=0.005358, \quad |V_{ts}|=0.03611, \quad |V_{tb}|=0.9993, \quad J_{CP}^q=1.748 \times 10^{-5} \quad (10)$$
$$\sin 2\alpha = -0.8913, \quad \sin 2\beta = 0.7416, \quad \gamma = 34.55^0.$$

Using the mass square differences $\Delta m_{atm}^2 = 2.78 \times 10^{-3}$ eV^2 and $\Delta m_\odot^2 = 7.25 \times 10^{-5}$ eV^2 for the LOW solution as input parameters, we determine $(t, M_R) = (0.35, 5.94 \times 10^{12} GeV)$, and correspondingly $(\delta_1, \delta_2, \delta_3) = (0.00119, 0.000841 e^{i(0.220)}, 0.0211 e^{-i(0.029)})$. The three mass eigenvalues are predicted to be $(m_{v_1}, m_{v_2}, m_{v_3}) = (0.00363, 0.00926, 0.0535)$ eV, and the mixing angles are predicted to be $\sin^2 2\theta_{atm} = 1$, $\tan^2 \theta_\odot = 0.58$, $\sin^2 \theta_{13} = 0.022$. These predictions agree with current bounds from experiments within 1 σ. The strengths of CP violation in the lepton sector are $(J_{CP}^l, \alpha_{31}, \alpha_{21}) = (-0.00690, 0.490, -2.29)$, and the matrix element for the neutrinoless double β decay is given by $|<m>| = 2.22 \times 10^{-3}$ eV. The correlation between $|U_{ev_3}|^2$ and $\tan^2 \theta_\odot$ is plotted in Fig. 1. A comparison of the predictions for $\sin^2 2\theta_{13}$ and (α, β, γ) from different SO(10) models is given in Ref.[4]. This work was supported by US DOE under Grant No. DE-AC02-98CH10886 and DE-FG03-95ER40894.

REFERENCES

1. M. -C. Chen and K. T. Mahanthappa, *Phys. Rev.* **D62**, 113007 (2000).
2. M. -C. Chen and K. T. Mahanthappa, *Phys. Rev.* **D65**, 053010 (2002).
3. M. -C. Chen and K. T. Mahanthappa, *Phys. Rev.* **D68**, 017301 (2003).
4. M. -C. Chen and K. T. Mahanthappa, arXiv:hep-ph/0305088.

Analysis Details from Muon $(g-2)$

C.C. Polly for the Muon $(g-2)$ Collaboration [1]

University of Illinois at Urbana-Champaign

Abstract. The emphasis of this document is to illustrate a few of the analysis issues that arise in detemining the spin precession frequency of muons in the $g-2$ storage ring. Due to the precise nature of the experiment, effects from pileup of decay electrons in the calorimeters, muons prematurely lost from the storage ring, and coherent betatron motion must be considered.

THE SIGNAL

When placed in a magnetic field, a muon will experience a torque, which rotates its spin in a plane perpendicular to the magnetic field. The rate at which the spin turns is determined by the field strength, intrinsic properties of the muon, and quantum field effects. Parity violation in the muon decay dictates that the decay electrons are preferentially emitted in the direction of the muon spin. Therefore in the muon rest frame, the precession frequency of an ensemble of initially polarized muons could be measured by placing a detector to one side of the ensemble and observing the time interval between peaks in the flux of decay electrons. When boosted back into the lab frame, one observes a modulation in the energy spectrum where the maximum energy decay electrons tend to come from decays in which the spin direction was parallel to the momentum vector. The number of decays measured above a certain energy threshold is then given by,

$$N(t) = \frac{N_0}{\tau} e^{-t/\tau}[1 - A\cos(\omega_a t + \phi)]. \qquad (1)$$

Additional detail on the experimental technique and and latest results may be found in Refs. [1], [2], [3], [4], and [5].

---

[1] "R.M. Carey, E. Efstathiadis, M.F. Hare, X. Huang, F. Krienen, A. Lam, I. Logashenko, J.P. Miller, J. Paley, Q. Peng, O. Rind, B.L. Roberts, L.R. Sulak, and A. Trofimov (Boston University); G.W. Bennett, H.N. Brown, G. Bunce, G.T. Danby, R. Larsen, Y.Y. Lee, W. Meng, J. Mi, W.M. Morse, D. Nikas, C. Özben, R. Prigl, Y.K. Semertzidis, and D. Warburton (Brookhaven National Laboratory); Y. Orlov (Cornell University); A. Grossmann, G. zu Putlitz, and P. von Walter (Universität Heidelberg); P.T. Debevec, W. Deninger, F.E. Gray, D.W. Hertzog, C.J.G. Onderwater, C. Polly, M. Sossong, and D. Urner (University of Illinois at Urbana-Champaign); A. Yamamoto (KEK); K. Jungmann (Kernfysisch Versneller Instituut); B. Bousquet, P. Cushman, L. Duong, S. Giron, J. Kindem, I. Kronkvist, R. McNabb, T. Qian, and P. Shagin (University of Minnesota); V.P. Druzhinin, G.V. Fedotovich, D. Grigoriev, B.I. Khazin, N.M. Ryskulov, Yu.M. Shatunov, and E. Solodov (Budker Institute of Nuclear Physics); M. Iwasaki (Tokyo Institute of Technology); M. Deile, H. Deng, S.K. Dhawan, F.J.M. Farley, V.W. Hughes (deceased), D. Kawall, M. Grosse-Perdekamp, J. Pretz, S.I. Redin, E. Sichtermann, and A. Steinmetz (Yale University).

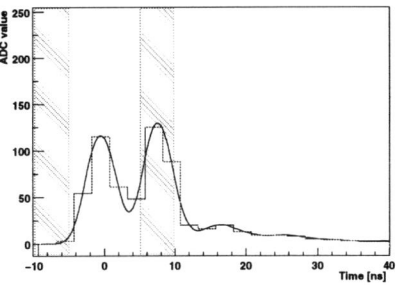 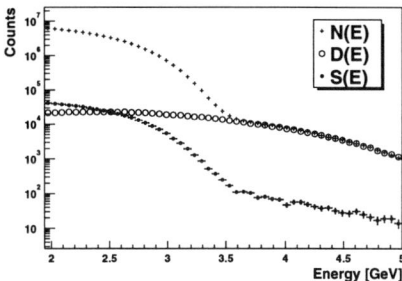

FIGURE 1. The plot on the left shows a sample of the WFD data with two pulses separated by 8 ns. The hashed regions indicate where random windows are drawn for constructing pileup. The plot on the right contains energy spectra from the raw data and the constructed pileup.

DECAY ELECTRON PILEUP

One of the primary deviations from the ideal fitting function, Eq. 1, is caused by pileup. Each raw event is recorded by a 400 MHz waveform digitizer (WFD) and then fit with a minimization technique to determine an energy and time. Pileup occurs as a result of two decay electrons entering a calorimeter within the deadtime of the pulse fitter. The left panel in Fig. 1 shows two pulses that enter a calorimeter with a time separation of about 8 ns. Clearly these two pulses are discernable, but once the pulse separation falls below 5 ns, the efficiency of the pulse fitter to find two pulses degrades rapidly. Instead, one pulse of a higher energy is constructed, which causes a perturbation to the ideal function. One could try to fit with a pileup term, but complications arise due to the hardware thresholds and correlations between the pileup phase and the $g - 2$ frequency. An alternate solution is to use the periods of time directly before and after a pulse has triggered the WFD to construct a statistically equivalent pileup spectrum. Every time an event is found in the pileup window it is combined with the trigger pulse and an entry is recorded in a doubles histogram, $D(E,t)$. These events are subtracted from the raw data, and the singles $S(E,t)$ from which they are constructed are added back in. The efficiency of the method is illustrated in the right panel of Fig. 1, which shows the energy spectrum of the events before pileup subtraction, $N(E)$, along with the energy spectra from the doubles and singles. The events in the right shoulder of $N(E)$ arise from pileup in the raw data. The constructed doubles, $D(E)$, agree to within a few percent.

COHERENT BETATRON MOTION

A Fourier transform of the residuals after fitting the pileup-subtracted data with Eq. 1 reveals another deviation from the standard form. As shown in the left panel of Fig. 2, there is a strong peak at 420 kHz. This peak is a result of the coherent betatron motion of the stored muons. As the muon population oscillates radially, the number of counts observed in the calorimeters also fluctuates. The dominant peak stems from the difference

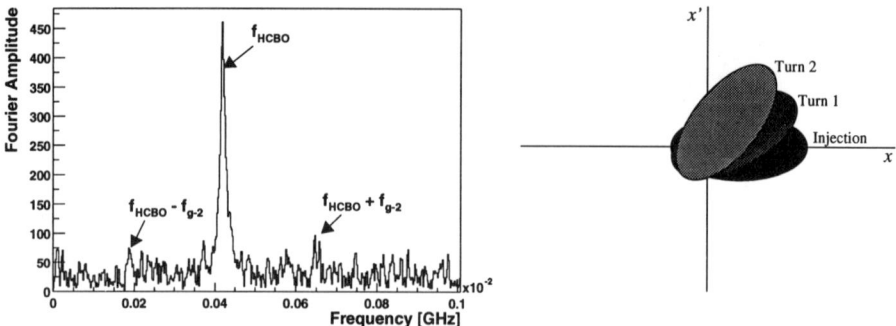

FIGURE 2. The plot on the left shows the Fourier transform of the residuals after a fit with Eq. 1. The right panel is an illustration of the phase space rotation of the stored muons.

or 'beating' between the horizontal betatron oscillations and the cyclotron frequency. The coherent betatron osciallations (CBO) arise from the emittance of the inflector not being perfectly matched to the acceptance of the storage ring, as illustrated in the right panel of Fig. 2. As the ellipse rotates in phase space, vertical and horizontal oscillations, as well as breathing modes are generated. At very early times after injection other modes are visible, but after 30 μs only the 420 kHz peak is relevant. To account for CBO, an overall multiplicitive term, $C(t)$, is included in the fitting function.

$$N(t) = \frac{N_0}{\tau} C(t)[1 - A\cos(\omega_a t + \phi)], \text{ where} \qquad (2)$$

$$C(t) = 1 - A_{cbo} e^{-t/\tau_{cbo}} [1 - \cos(\omega_{cbo} t + \phi_{cbo})]. \qquad (3)$$

MUON LOSSES

Muons with eccentric orbits are lost from the storage ring through interactions with collimators. Since the losses are not a constant fraction of the stored population, the result is another perturbation to Eq. 1. The number of muons being lost per unit time, $L(t)$, can be constructed from data taken with hodoscopes placed on the face of the $g-2$ calorimeters. Decay electrons are absorbed, but muons can pass through several detectors in a row. The loss distribution is constructed by looking for coincidences in the hodoscopes on the front side of three consecutive calorimeters with an absence of energy in the calorimeters. Examples of the muon loss construction are shown in the left panel of Fig 3. The high losses at early times are purposely incurred by manipulating the quad voltages so that muons on the periphery of phase space are 'scraped' by the collimators before the fitting period begins. Once $L(t)$ has been constructed, the fitting function is modified by a term $\Lambda(t)$,

$$N(t) = \frac{N_0}{\tau}\Lambda(t)C(t)[1 - A\cos(\omega_a t + \phi)], \text{ where} \quad (4)$$

$$\Lambda(t) = 1 - A_{loss}\exp\frac{-t_0}{\tau}\int_{t_0}^{t} L(t')\exp\frac{t'}{\tau}dt'. \quad (5)$$

CONCLUSIONS

Although the effects discussed in this article are by no means a comprehensive picture of the analysis, they are the effects that have the largest impact on the quality of the fit. The right panel of Fig. 3 shows the reduced χ^2 as a function of when the fit is started. At early start times, there is a dramatic improvement in the quality of the fit when the aforementioned terms are included. By producing an acceptable fit at early times, the statistical power of the $\mu-$ dataset is increased to a precision of 0.7 ppm, while still maintaining systematic errors below 0.2 ppm.

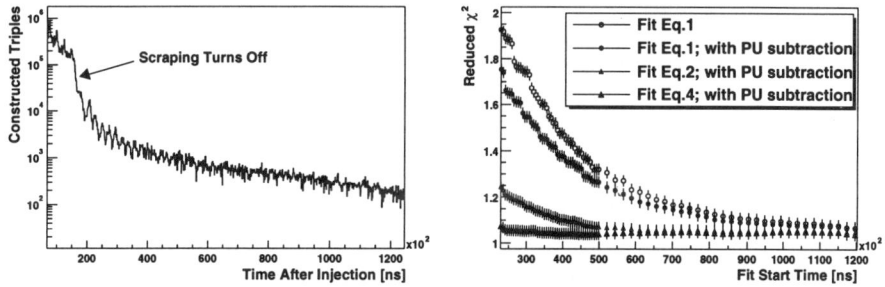

FIGURE 3. The plot on the left shows the muon loss construction, $L(t)$. The plot on the right shows the improvement in the reduced χ^2 as pileup, CBO, and lost muons are incorporated.

ACKNOWLEDGMENTS

We acknowledge support from the U.S. Department of Energy, the National Science Foundation, the German Bundesminister fur Bildung und Forschung, the Russian Ministry of Science, and the U.S.-Japan Agreement in High Energy Physics.

REFERENCES

1. Bennett, G. W., et al., *Phys. Rev. Lett.*, **89**, 101804 (2002).
2. Sedykh, S. A., et al., *Nucl. Instrum. Meth.*, **A455**, 346–360 (2000).
3. Semertzidis, Y. K., et al., *Nucl. Instrum. Meth.*, **A503**, 458–484 (2003).
4. Prigl, R., et al., *Nucl. Instrum. Meth.*, **A374**, 118–126 (1996).
5. Efstathiadis, E., et al., *Nucl. Instrum. Meth.*, **A496**, 8–25 (2003).

Muon Lifetime and Muon Capture

Bernhard Lauss

on behalf of the MuCAP[1] and MuLAN[2] Collaborations

University of California at Berkeley, Physics Department, 366 LeConte Hall, and Lawrence Berkeley National Laboratory, Berkeley, CA, 94720, USA

Abstract. We present an introduction to the MuLAN and MuCAP experiments at PSI, which aim at high precision determinations of two fundamental Weak Interactions parameters: the Fermi constant G_F and the induced pseudoscalar form factor g_p, respectively.

MULAN - MUON LIFETIME

The Fermi coupling constant G_F is one of the fundamental constants of the Standard Model. G_F is obtained from the muon lifetime via a calculation in the Fermi Model, in which weak interactions are represented by a contact interaction.

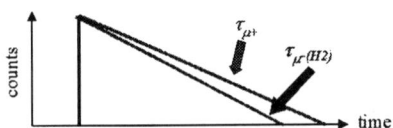

FIGURE 1. The experimental principle of the lifetime method is to measure the time difference between a muon stopping in a target and its decay electron or positron. In the case of μ^+ this ideally results in a single exponential with $\lambda_0 = 1/\tau_{\mu^+}$ (MuLAN); for μ^- in hydrogen the rate is increased due to the capture process to be $\lambda = \lambda_0 + \lambda_{capture}$. The capture rate consequently follows by comparing the lifetimes of both muon charge states in hydrogen (MuCAP).

The goal of the MuLAN (**Mu**on **L**ifetime **An**alysis) experiment [3] is the determination of the positive muon lifetime, τ_{μ^+}, with a precision of 1 ppm; the method is sketched in Fig.1. In order to achieve this higher precision than all combined existing experimental results, the statistics and systematics of the measurement have to be dramatically improved.

The ingredients to achieve this challenging goal are: 1) Construction of a multisegment (170 tiles) detector, read out via fast (500 MHz, 8 bit) waveform digitizers; both are crucial in order to separate pile-up events of two simultaneous decay electron hits in one detector module. The layout of the detector and its elements is shown in Fig.2. A depolarizing and dephasing target material - e.g. sulfur - will be used in a 70 Gauss magnetic field to control the muon spin rotation from residual muon polarization.

2) Construction of a new kicked muon beam line in the high intensity $\pi E3$ muon channel at the Paul Scherrer Institute (PSI). This will allow us to collect 10^{12} events in

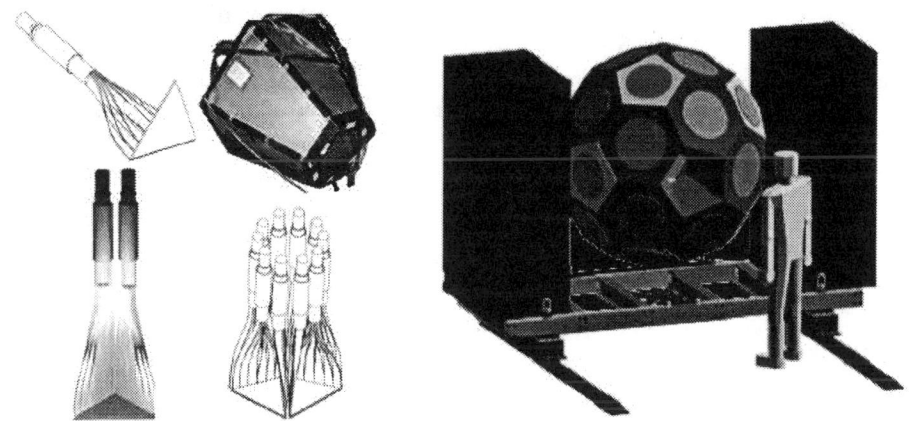

FIGURE 2. Individual scintillation counters, double layered scintillator tiles; their arrangement to a detector module; schematic view of the full **MuLAN** soccer-ball detector and electronic racks.

a few weeks. Muons will be stored in the target in a $\sim 5\mu s$ period followed by a $22\mu s$ ($10 \times \tau_\mu$) detection period with the electrostatic kicker off.

In a recent run we successfully installed and tested the kicker in the $\pi E3$ beam line, which demonstrated the feasibility of the measurements timing. A offline run with the detector is under way to test the full setup with LEDs simulating hits. Fall'03 will see a first beam test.

MUCAP - MUON CAPTURE

FIGURE 3. View of the **MuCAP** experiment located in PSI's $\mu E4$ area. From left to right: the final beamline quadrupoles; the cylindrical scintillator hodoscope (eSc) with photomultiplier tubes; the TPC with the surrounding magnet rolled back from its center position inside the hodoscope; the hydrogen purification and filling apparatus.

The goal of MuCAP[3] is the high precision measurement of the singlet **Mu**on **Cap**ture rate on the proton in low-density, ultra-clean H_2 gas. The muon capture rate on the proton λ_{cap} can be directly related to the induced pseudoscalar form factor g_p of the nucleus, which was calculated very accurately with recent heavy baryon chiral perturbation theory approaches. Existing experimental data are outdated, lack accuracy, and show a unresolved discrepancy between results from ordinary muon capture and radiative muon capture. MuCAP determines the capture rate via the lifetime method as sketched in Fig.1. Goal of MuCAP is a 1% precision on λ_{cap}, which in turn yields a 7% error on g_p. This is very challenging, due to the large difference in involved rates, ($\lambda_{cap} \sim 700 s^{-1}$, $\lambda_0 = 455000 s^{-1}$, $\lambda_{transfer\ to\ Z>1} \sim 10^9 s^{-1}$), and due to the complex chemistry of negative muons in hydrogen.

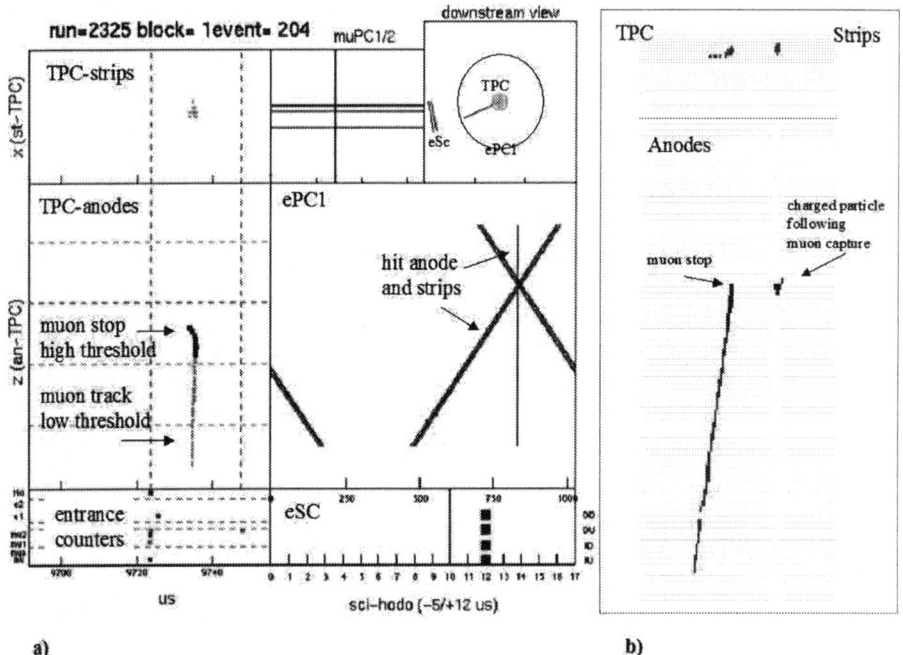

FIGURE 4. a) Typical MuCAP event: Hits in the entrance scintillator (mu) and wire chambers (muPC1/2) are followed by a muon stop in the TPC. The stopping muon triggers a higher threshold as it deposits more energy at the end of the Bragg range. The dashed lines demonstrate the allowed range of 24 μs drift time in the TPC after the initial entrance scintillator hit. The muon decay electron is observed in one wire chamber (ePC1) and in the four-fold hodoscope coincidence (eSC). b) Detected impurity capture event in the TPC, most likely on a nitrogen nucleus. The muon stops and after a short time a very large signal from a charged particle occurs. The time difference is defined by the muon transfer rate times the impurity concentration.

As part of the effort to control the molecular processes, ultra-clean target conditions are selected which enhance only muonic atomic singlet states and suppress muonic molecular formation. A unique high pressure (10 bar) pure hydrogen time projection chamber (TPC) serves as an active target detector. To meet stringent purity conditions it is made out of quartz-glass and bakeable up to 130 degree C. It is surrounded by a μSR-

controlling saddle-coil magnet which provides a 70 Gauss field (relevant only for the μ^+ measurement, as negative muons are effectively depolarized in the atomic cascade); two large cylindrical wire chambers (ePC1/2); and a 16-tile, two-layer scintillator hodoscope (eSc), which sees approximately 2/3 of all decay electrons in coincidence. A view of the setup is presented in Fig.3. The timing start with a hit in the entrance scintillator (mu) and stops with a hit in the scintillator hodoscope counters, both with excellent timing resolution. The eSc is read out with fast waveform digitizers.

The TPC is essential for control of the systematics, for several reasons: 1) It allows the unambiguous identification in 3D of the muon stopping positions in hydrogen (and consequently excludes wall stops). 2) It can detect impurity captures (our high Z contamination in the hydrogen after passing the palladium filter is smaller than 0.1 ppm and can actively be monitored via muon capture events on the contaminant nuclei - Fig.4b). 3) Muon transfer to deuterium can be observed: A mismatch between muon stopping position and back-tracked decay electron in the two surrounding wire chambers indicate μd diffusion events.

Fig.4a shows a typical event where a muon, after being seen in all entrance counters, stops in the TPC. Most of the time muons only trigger a low threshold, but in the final part of the track, where there is a large energy deposition, the high threshold is also fired. The decay electron is observed in surrounding counters. Fig.4b shows a detected impurity event: A stopping muon followed shortly after by a very high threshold trigger.

As of summer'03 a MuCAP commissioning run is in progress and first physics data are expected soon.

REFERENCES

1. **The MuCAP Collaboration:** V.A.Andreev, A.A.Fetisov, V.A.Ganzha, V.I.Jatsoura, A.G.Krivshich, E.M.Maev, O.E.Maev, G.E.Petrov, S.Sadetsky, G.N.Schapkin, G.G.Semenchuk, M.Soroka, A.A.Vorobyov, Petersburg Nuclear Physics Institute, PNPI, Gatchina, Russia; P.U.Dick, A.Dijksman, J.Egger, D.Fahrni, M.Hildebrandt, A.Hofer, L.Meier, C.Petitjean, R.Schmidt, Paul Scherrer Institute, PSI, Villigen, Switzerland; T.I.Banks, T.A.Case, K.M.Crowe, S.J.Freedman, F.E.Gray, B.Lauss, University of California, UCB and LBNL, Berkeley, USA; D.B.Chitwood, S.Clayton, P.Debevec, D.W.Hertzog, P.Kammel, B.Kiburg, C.J.G.Onderwater, C.Ozben, C.C.Polly, A.Sharp, University of Illinois at Urbana-Champaign, Urbana, USA; L.Bonnet, J.Deutsch, J.Govaerts, D.Michotte, R.Prieels, Universite Catholique de Louvain la Neuve, Belgium; R.M.Carey, J.Paley, Boston University, USA; T.Gorringe, M.Ojha, P.Zolnierzcuk, University of Kentucky, Lexington, USA; F.J.Hartmann, TU München, Germany.
2. **The MuLAN Collaboration:** R.M.Carey, A.Gafarov, I.Logachenko, K.Lynch, J.Miller, L.Roberts, Boston University, USA; D.B.Chitwood, S.Clayton, P.Debevec, D.W.Hertzog, P.Kammel, B.Kiburg, C.J.G.Onderwater, C.Ozben, C.C.Polly, A.Sharp, S.Williamson, University of Illinois at Urbana-Champaign, Urbana, USA; M.Deka, T.Gorringe, M.Ojha, University of Kentucky, Lexington, USA; K.Giovanetti, James Madison University, Harrisonburg, USA; K.M.Crowe, F.E.Gray, B.Lauss, University of California, UCB and LBNL, Berkeley, USA.
3. http://www.npl.uiuc.edu/exp/mulan/
4. http://weak0.physics.berkeley.edu/weakint/research/muons/mucap_home.html
   http://www.npl.uiuc.edu/exp/mucapture/

TWIST: Measuring the Space-Time Structure of Muon Decay

C. A. Gagliardi* and the TWIST Collaboration[†]

*Cyclotron Institute, Texas A&M University, College Station, TX 77843, U.S.A.
[†]http://twist.triumf.ca

Abstract. TWIST, the TRIUMF Weak Interaction Symmetry Test, is a precision measurement of the energy and angular distributions of the positrons emitted in polarized muon decay. The goal is to search for new physics that leads to deviations of the Michel parameters ρ, δ, and $P_\mu \xi$ from their Standard Model values. TWIST will determine these parameters to a few parts in 10^4, an improvement of at least an order of magnitude in each case. At this level, TWIST will confront several proposed extensions to the Standard Model.

PHYSICS OF TWIST

The energy and angular distributions of the positrons emitted in the decay of polarized muons may be written in a number of equivalent forms. One convenient form [1, 2] describes the distributions in terms of four parameters ρ, η, δ, and ξ, commonly referred to as the Michel parameters. Neglecting the electron and neutrino masses and radiative corrections, the differential decay rate for positive muon decay is given in terms of the decay parameters ρ, δ, and ξ by

$$\frac{d^2\Gamma}{x^2 dx d(\cos\theta)} \propto 3 - 3x + \frac{2}{3}\rho(4x-3) + P_\mu \xi \cos\theta \left[(1-x) + \frac{2}{3}\delta(4x-3) \right], \quad (1)$$

where P_μ is the polarization of the muon, x is the outgoing positron energy as a fraction of the maximum possible value, and θ is the angle between the muon polarization axis and the positron decay direction. The fourth decay parameter η contributes to the angle-independent part of the distribution if one includes the finite electron mass.

The decay parameters characterize the space-time structure of muon decay. Thus, within the Standard Model they take on precise values: $\rho = \frac{3}{4}$, $\eta = 0$, $\delta = \frac{3}{4}$, $\xi = 1$. Any deviations from these values would signify new physics. For example, in left-right symmetric models [3], deviations of ρ from $\frac{3}{4}$ imply mixing between the left- and right-handed W bosons, and deviations of ξ from 1 primarily measure the ratio of the squares of the two W boson masses. In more general extensions that include scalar and tensor interactions in addition to vector and axial vector currents, the linear combination

$$Q_R^\mu = \frac{1}{2}\left[1 + \frac{1}{3}\xi - \frac{16}{9}\xi\delta \right] \quad (2)$$

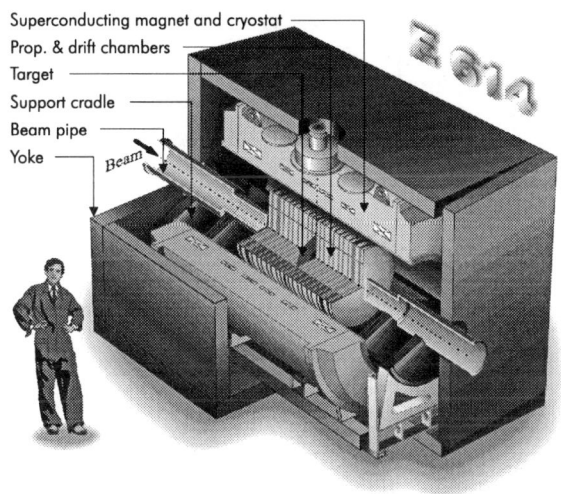

FIGURE 1. The TWIST spectrometer.

provides a model-independent measure of the total right-handed contributions to muon decay [4]. In Eq. (1), ξ appears only in the product $P_\mu \xi$. Conventionally, P_μ represents the polarization of the muons produced in pion decay. Thus, $P_\mu = 1$ in the Standard Model and any deviation would also provide a signature for new physics.

The best current measurements of the Michel parameters are [5]: $\rho = 0.7518 \pm 0.0026$, $\eta = -0.007 \pm 0.013$, $\delta = 0.7486 \pm 0.0026 \pm 0.0028$, $P_\mu \xi = 1.0027 \pm 0.0079 \pm 0.0030$, $P_\mu \xi \delta/\rho > 0.99682$. The ultimate goal of TWIST is to determine ρ, δ, and $P_\mu \xi$ to a few parts in 10^4, over an order of magnitude improvement in each case. At this level, TWIST will be sensitive to the existence of right-handed W bosons with masses up to 800 GeV/c^2, without needing to make assumptions about the form of the right-handed CKM matrix, and will be sensitive to mixing angles between left- and right-handed W bosons as small as 0.01.

EXPERIMENT

Figure 1 shows the TWIST spectrometer. It consists of an array of very thin, high precision planar wire chambers [6] located within a 2 T uniform magnetic field. The spectrometer includes 44 drift chambers and 12 MWPCs. The ~5000 sense wires are positioned with 3 μm accuracy, and longitudinal and transverse dimensions are known to better than 5 parts in 10^5. A highly-polarized beam of surface muons from the M13 beam line at TRIUMF enters the spectrometer. The muons are tracked as they pass through the detector planes until they stop in a thin target in the center of the spectrometer. The decay positrons then follow helical trajectories through the detectors, permitting their energies and angles to be measured precisely.

Equation (1) shows that the differential decay rate is linear in the decay parameters.

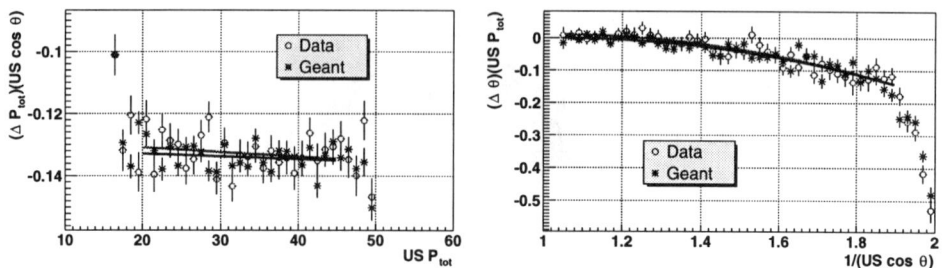

FIGURE 2. The energy loss in MeV as a function of energy (left) and the change in the polar angle in radians as a function of relative path length (right) when positrons pass through the stopping target and half of the detector planes. The energy loss is corrected event-by-event for the observed path length, yielding the typical energy loss for positrons moving parallel to the magnetic field. Similarly, the scattering is rescaled event-by-event according to the momentum of the positron, yielding the average scattering angle that would be experienced by a 1 MeV/c positron. Monte Carlo predictions are shown for comparison.

This provides the basis for the blind analysis scheme. The measured energy-angle spectrum will be compared to the sum of a Monte Carlo 'standard' spectrum produced with unknown Michel parameters, together with additional Monte Carlo distributions that describe the dependence on $\Delta\rho$, $\Delta\eta$, $\Delta\delta$, and $\Delta P_\mu \xi$. These Monte Carlo spectra are generated including the effects of the electron mass, plus first- and many second-order radiative corrections, in contrast to Eq. (1).

CURRENT STATUS

TWIST had its first physics run during Fall, 2002. The goal was to determine ρ and δ to 10^{-3}. TWIST is a systematics-dominated experiment. Thus, most of the running time was devoted to exploring the possible systematic effects by amplifying them as much as practical, then measuring their impact. 3×10^8 muon decay events are sufficient to determine ρ and δ with a statistical precision of $\sim 6 \times 10^{-4}$, but a total of 6×10^9 events were recorded to tape. Independent data sets were taken to explore the sensitivity to the beam properties (polarization, stopping distribution, steering, focus, intensity), the detector performance (efficiencies, gas pressure), the magnetic field, the upstream-downstream symmetry of the system, and the overall system stability. In addition, special runs were taken to provide additional data to validate the quality of our GEANT-based Monte Carlo simulation.

These data are now being analyzed, and the initial results are very encouraging. The Monte Carlo simulation has been shown to provide an excellent description of the muon stopping distribution. Special runs during which the muon beam was stopped near the upstream end of the detector have been used to study the overall detector response. For these runs, positrons which are emitted in the downstream direction pass through the upstream half of the detector, the stopping target, then the downstream half of the detector. Each half may be treated as an independent detector, permitting the two measurements of the helix parameters at different points along the trajectory to be

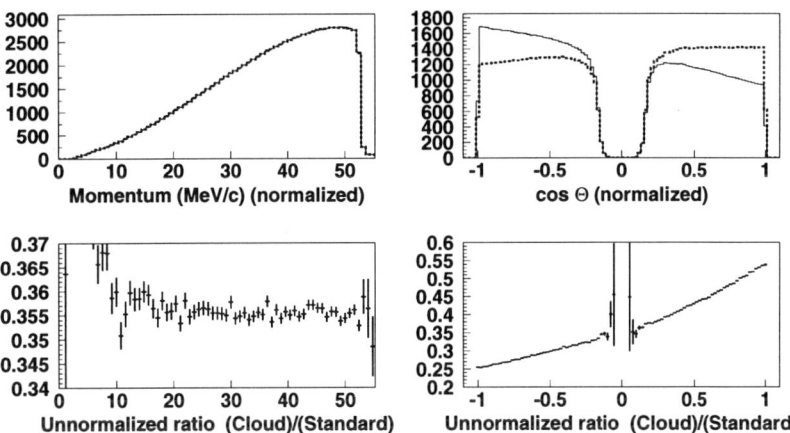

FIGURE 3. Comparison of momentum and angular distributions between surface muon (solid curves, $P_\mu \approx -1.0$) and cloud muon (dashed curves, $P_\mu \approx +0.3$) data sets. In each case, fiducial cuts have been applied to the other variable. The upper panels show the spectra normalized to the same total number of events. The lower panels show the ratio of the yields for cloud muons relative to surface muons.

compared. Figure 2 shows that the Monte Carlo simulates the positron energy loss and scattering very well.

The various independent data sets are also being analyzed. Figure 3 shows a comparison between spectra obtained with a surface muon beam under nominal conditions and with a cloud muon beam. The angular distributions are quite different, but the overall momentum distributions are very similar within the fiducial range, demonstrating the uniformity and upstream-downstream symmetry of the spectrometer. At present, work continues to optimize the pattern recognition and track fitting algorithms and to validate the Monte Carlo. The goal is to have the first physics results available near the end of 2003.

ACKNOWLEDGMENTS

This work was supported in part by the Natural Sciences and Engineering Research Council and the National Research Council of Canada and by the U.S. Department of Energy.

REFERENCES

1. L. Michel, Proc. Phys. Soc. **A63**, 514 (1950); C. Bouchiat and L. Michel, Phys. Rev. **106**, 170 (1957).
2. T. Kinoshita and A. Sirlin, Phys. Rev. **108**, 844 (1957).
3. P. Herczeg, Phys. Rev. D **34**, 3449 (1986).
4. W. Fetscher, H.-J. Gerber, and K.F. Johnson, Phys. Lett. **B173**, 102 (1986).
5. K. Hagiwara et al., Phys. Rev. D **66**, 010001 (2002).
6. Yu. Davydov et al., Nucl. Instrum. Methods **A461**, 68 (2001).

Searches at the Run II Tevatron Collider

Leslie S. Groer
for the CDF and DØ Collaborations.

Department of Physics, Columbia University, New York, New York 10027, USA.
E-mail: groer@fnal.gov

Abstract.
Some initial results based on 40-90 pb^{-1} of integrated luminosity from Run II are presented for searches for new phenomena beyond the Standard Model, conducted at the upgraded Tevatron CDF and DØ experiments.

Run II commenced at the upgraded Tevatron proton-antiproton collider complex in 2001 with a center-of-mass collision energy of 1.96 TeV. Both the CDF and DØ detectors underwent major upgrades to take full advantage of the increased instantaneous luminosity, with the installation of new inner tracking systems within solenoidal fields, including silicon vertex detectors, upgrades and extensions to the muon detection systems and new electronics, trigger and data acquisition systems for the higher rate with the reduced bunch crossing time of 396 ns. The Tevatron is delivering approximately 7 pb^{-1} a week with a peak instantaneous luminosity achieved of 4.5×10^{31} cm^{-2}s^{-1} and has delivered a total of 240 pb^{-1}.

The general purpose detectors with very good azimuthal and forward coverage allow the identification of many final states that could provide hints of new physics. Some of these searches, from approximately 40–90 pb^{-1} of integrated luminosity that has been analyzed, are presented below.

Searches for Higgs Bosons

One of the main interests in Run II is the search for Higgs bosons, either from the Standard Model (SM) or one of the many varieties predicted by extensions to the SM. The predicted inclusive production cross section for a SM Higgs boson at 1.96 TeV is of order 1 pb, with gluon-gluon fusion being the dominant process. For a SM Higgs mass less than 135 GeV/c^2 the dominant decay is to $b\bar{b}$ quark pairs. Searches in this channel however are swamped by the large production cross section of b-quark pairs from QCD jet production; a more fruitful search is for the production of a Higgs boson in association with a heavy vector boson ($p\bar{p} \to W^{\pm *}/Z^{0*} \to W^{\pm}H^0/Z^0H^0$), where the vector boson decays leptonically. For SM Higgs masses greater than 135 GeV/c^2, the Higgs boson is predicted to decay predominantly to W^+W^- pairs. Again the leptonic decay channels yield the cleanest signatures.

Studies in the Tevatron Supersymmetry and Higgs Working Groups[1] indicated that a lower mass SM Higgs may be within the reach of the Tevatron Run II program provided sufficient luminosity can be delivered ($\sim 5-10$ fb^{-1}) and by combining all the decay channels from both experiments.

Both CDF and DØ have preliminary investigations in the associated production channels looking for either electron and muon pairs in the case of the Z or a high-p_T electron or muon and a large imbalance in the transverse momentum (missing E_T) in the case of a W. These channels are not yet competitive for setting limits but do prove the capabilities of the detectors and the analysis techniques.

DØ has also searched for higgs decays to WW pairs in the leptonic decay modes and sees zero events in the ee channel in 44 pb^{-1}, and one event each in the $e\mu$ (34 pb^{-1}) and $\mu\mu$ (48 pb^{-1}) channels with expected backgrounds of 0.7 ± 1.4, 0.9 ± 1.5 and 0.3 ± 0.1 events respectively. These searches exploit the fact that the Higgs boson is a fundamental spin-zero scalar and hence angular spin correlations can be used to reduce the backgrounds from fakes and diboson and $t\bar{t}$ production. Combining these results sets an upper limit on the production of a Higgs boson times branching ratio to WW pairs of 3 pb at the 95% confidence level (CL), about two orders of magnitude above the SM predictions, but only a factor of ten above the predictions of more exotic models such as a fermiophobic or topcolor Higgs. Searches for Higgs decaying to pairs of high-p_T photons also set limits on the production times branching ratio of about 10-1 pb for Higgs masses running from 60 to 120 GeV/c^2. For Higgs masses of about 120 GeV/c^2, the cross section limits are comparable to the theoretical expectations if $B(h \to \gamma\gamma) = 1$.

CDF has searched for signatures of decays of doubly-charged higgs (H^{++}) which are predicted in various classes of left-right symmetry breaking models [2]. These decays provide strong experimental signatures as there are very low backgrounds for same-sign high-p_T leptons in hadron collisions. Zero same-sign central electron events are observed in 91$\pm$5.3 pb^{-1}, based on an inclusive electron trigger, with a signal acceptance of about 20-35% and an expected background of 0.6$\pm$0.5 coming mainly from Z, QCD and $W+$ jets. The production cross section limit established at 95% CL using a sliding mass window of M$_{H^{++}} \pm 10\%$ is 125-90 pb for a H^{++} mass going from 105 to 145 GeV/c^2.

Searches for Heavy Neutral Gauge Bosons and Extra Dimensions

Both CDF and DØ have searched for evidence of new heavy neutral gauge bosons assuming SM couplings to both dielectron and dimuon lepton pairs. Neither experiment observes any excess with backgrounds dominated by Drell-Yan lepton pair production and smaller contributions from misidentified electrons and diboson and $t\bar{t}$-pair production. The Run I limits were 690 and 670 GeV/c^2 from CDF and DØ respectively. The current limits are 650 (ee) and 455 ($\mu\mu$) GeV/c^2 from CDF in 72 pb^{-1} giving a combined limit of 665 GeV/c^2. The limit from DØ in the dielectron channel in 50 pb^{-1} of integrated luminosity is 620 GeV/c^2. The Run II sensitivity, which is enhanced by the increased cross sections at the higher center of mass energies (1.8 TeV in Run I), is approaching that of Run I.

Similar experimental signatures are indicated in the recently developed extra dimension models. DØ has searched for Large Extra Dimensions (LED) in the ADD [3] model assuming SM particles are confined to a 3-brane, but gravity propagates in the extra dimensions which, if large, lead to an effective Planck scale M_S in the TeV range. The expected signatures would be excesses of high-mass dilepton or diphoton pairs coupling to the the excited states of the graviton, the so-called Kaluza-Klein modes. DØ has searched in 50 pb^{-1} of well-balanced high-p_T di"em" events where dielectron and diphoton events are treated identically (i.e. with no track requirements) and in 30 pb^{-1} of dimuon events. The data distributions are fitted to a signal plus background hypothesis to evaluate upper limits on the coupling η_G which in turn can be translated into 95% CL lower limits on M_S in various formalisms. The limits on M_S are of order 1 TeV in the diem channel, very competitive with the Run I results. The new collider limit from the dimuon channel is of order 0.7 TeV. The Tevatron is expected to probe up to 1.6 (2) TeV with 0.3 (2) fb^{-1} in Run II.

CDF has also searched for anomalous dilepton production at large mass within the context of Randall-Sundrum (RS) graviton models which predict observable discrete spin-2 resonances [4]. The Kaluza Klein excitations of the graviton can be separately produced as resonances, enhancing the Drell-Yan cross section at large mass. CDF has compared the Drell Yan spectrum in the ee and $\mu\mu$ channel to the SM expectations in 72 pb^{-1} and no excess has been observed. A 95% CL upper limit on the cross section as a function of the RS graviton mass can be derived, which is 535 GeV/c^2 in the dielectron and 370 GeV/c^2 in the dimuon channel, giving a combined limit of 555 GeV/c^2.

Searches in Diphoton Events

In gauge mediated supersymmetry the lightest supersymmetric particle (LSP) is a very light gravitino ($\widetilde{G}$) so the phenomenology is driven by the nature of the next to lightest supersymmetric particle (NLSP), the neutralino, which decays as $\overline{\chi}_1^0 \to \gamma\widetilde{G} \to \gamma\gamma\widetilde{G}$. DØ has searched in this context for diphotons with $p_T > 20$ GeV/c^2 and large missing E_T. No events are observed in 50 pb^{-1} with expected backgrounds of 1.6 ± 0.4 coming from QCD fakes. A 95% CL limit of 66 GeV/c^2 for the mass of the NLSP is derived in the Snowmass model where M = 2Λ, $N_5 = 1$, $\tan\beta = 15$ and $\mu = 0$ [5]. CDF has also not seen any anomalous high-p_T diphoton plus missing E_T events in 84 pb^{-1}.

Searches in Dilepton and Trilepton Events

Both CDF and DØ conduct model-independent searches for anomalous production of high-p_T dilepton and trilepton events with large missing E_T, especially of different lepton flavors. These channels have very low backgrounds, dominated by $Z \to \tau\tau$ at low missing E_T and heavy vector-boson production with fake leptons and with smaller contributions from diboson and $t\bar{t}$ production at high missing E_T. DØ sees 13 events in the $e\mu$ channel in 33 pb^{-1} with expected backgrounds of 9.6 ± 2.7 and sets 95% CL limits of acceptance times branching ratio for new physics of 0.4 − 0.1 pb with missing E_T going from 0 to 35 GeV. DØ also sees no events in the $ee\ell\nu$ channel in 42 pb^{-1} with 0 ± 2 events expected from mostly heavy vector boson backgrounds and sets a 95% CL

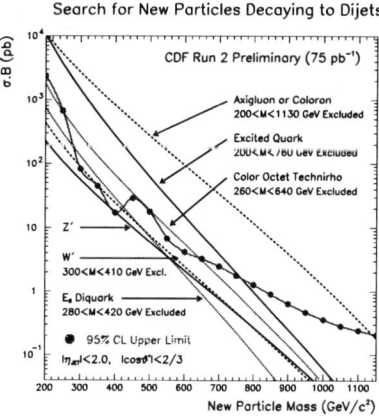

FIGURE 1. Preliminary CDF 95% CL upper limits on $\sigma \cdot B$ for narrow dijet resonances for a variety of non-Standard Model models.

limit on the production cross section times branching ratio of 3.5 pb.

Searches in Dijet Events

The dijet mass spectrum has been investigated at both experiments as a test of QCD and for new mass resonances. CDF fits the mass spectrum in 75 pb^{-1} with a simple background parameterization and searches for bumps comparable to the mass resolution. No significant evidence of any new resonance is found. Several 95% CL upper limits can be derived (shown in Figure 1) for the cross section times branching ratio for a host of non-SM models which predict narrow dijet resonances: axigluons, flavor universal colorons, excited quarks, color octet technirhos, E6 diquarks and W' and Z' bosons. All these limits surpass the Run I results.

Searches for Leptoquarks

Extended gauge sector and composite models imply a direct coupling between the lepton and quark sectors via scalar leptoquarks (LQ) that can be pair produced from quark or gluon fusion. Models are characterized by the coupling $\beta = B(\text{LQ} \to \ell q) = 1 - B(\text{LQ} \to \nu q)$. Strategies involve searching for dilepton and jets events and reconstructing the LQ mass or searching for events with missing E_T and dijets. CDF and DØ have seen no evidence for first (e) or second (μ) generation leptoquarks with 95% CL limits from CDF of 230 GeV/c^2 for LQ$_1$ from 72 pb^{-1}, an improvement over the 220 GeV/c^2 limit from Run I. DØ's result for LQ$_2$ is 157 GeV/c^2 from 40 pb^{-1}, not yet competitive with the Run I result of 200 GeV/c^2. CDF also conducts a flavor-blind search for LQ$\to \nu q$ in 76 pb^{-1} with limits of $M_{LQ} < 60$ or $M_{LQ} > 107$ GeV/c^2.

Search for Excited Electrons

CDF has searched its $ee\gamma$ data for evidence of the decay of an excited or exotic electron ($p\bar{p} \to e^* + e \to e\gamma + e$) which would indicate compositeness. No events are observed in 72 pb^{-1} with expected background contributions coming mainly from $Z\gamma$ events and smaller contributions from Z+jet, multijet and W+jet, so a new mass limit can be set of $M_{e^*} > 785$ GeV/c^2 at 95% CL, assuming a contact interaction model [6] and the compositeness scale $\Lambda = M_{e^*}$. This limit exceeds the previous published limit from H1 of 223 GeV/c^2 [7].

Charged Massive Stable Particles

CDF has performed a search for charged massive stable particles that escape the detector without decaying and hence present themselves as slow-moving high-p_T muon-like objects. The time-of-flight detector (TOF) encasing the tracking volume at 140 cm from the beampipe with time resolution of 100 ps allows detection of these long-lived particles. In 53 pb^{-1} of data, seven events are observed with $\Delta_{TOF} > 2.5$ ns and with $2.9 \pm 0.7_{\text{stat}} \pm 3.1_{\text{sys}}$ events expected.

The results can be interpreted in the context of several SUSY-based models (e.g. gauge-mediated supersymmetry breaking) where the LSP is a nearly massless Goldstino and the stop or stau is the NLSP, which can be long-lived, depending on the supersymmetry breaking scale. Interpreting the results assuming the stop as the stable NLSP gives a lower limit on the stop mass of 107 GeV/c^2 for an isolated decay or 96 GeV/c^2 for a stop decay associated with a jet. The previous limit on these models is from ALEPH at 95 GeV/c^2 [8].

CONCLUSION

The Tevatron Run II is progressing steadily with very well performing detectors. Many of the results are already competitive or surpass those from Run I due to improved detectors and the higher center of mass collision energy. Sensitivity will increase at the 0.1 pb and 1 TeV scale in many channels as the integrated luminosity accumulates with the exciting potential for discoveries.

REFERENCES

1. Carena, M., et al., *hep-ph/0010338* (2000).
2. Gunion, J. F., Loomis, C., and Pitts, K. T., *hep-ph/9610237* (1996).
3. Arkani-Hamed, N., Dimopoulos, S., and Dvali, G., *Phys. Lett. B*, **429**, 263 (1998).
4. Randall, L., and Sundrum, R., *Phys. Rev. Lett.*, **83**, 3370 (1999).
5. Allanach, B. C., et al., *Eur. Phys. J.*, **C25**, 113–123 (2002).
6. Baur, U., *Phys. Rev. D*, **42**, 3 (1990).
7. Adloff, C., et al., *Eur. Phys. J.*, **C17**, 567–581 (2000).
8. Heister, A., et al., *Phys. Lett.*, **B537**, 5–20 (2002).

Hadronic matrix elements of proton decay on the lattice

Yasumichi Aoki[1] for RBC collaboration[†]

RIKEN BNL Research Center, Brookhaven National Laboratory, Upton, NY 11973, USA
Physics Department, Columbia University, New York, NY 10027, USA

Abstract. We report on our on-going project to calculate proton decay matrix elements using domain-wall fermions on the lattice. By summarizing the history of the proton decay calculation on the lattice, we reveal the systematic errors of those calculations. Then we discuss our approach to tackle those uncertainties and show our preliminary results on the matrix elements.

Nucleon decay is one of the most important aspect that any (SUSY) GUT model has. At low energy dimension-six operators are dominant contribution to the proton decay while higher dimensional operators are suppressed by the inverse power of the heavy mass (M_X). The dimension-six operators consist of three quark and one lepton fields. While the lepton part is treated trivially, the matrix element of the three-quark part

$$\mathcal{O}^{\beta}_{R/L;L} \equiv \varepsilon^{ijk}(u^{iT} C P_{R/L} d^j) P_L u^k \tag{1}$$

between the initial proton and final pion (K or η meson) states receives a highly non-perturbative contribution from QCD, which we want to tackle in this study. The matrix element has a tensor structure [1],

$$\langle \pi; \vec{p} | \mathcal{O}^{\beta}_{R/L;L} | p; \vec{k} \rangle = P_L[W_0 - i\slashed{q} W_q] u_p, \tag{2}$$

where $q = k - p$ is the momentum transfer, u_p is the proton spinor. The relevant form factor W_0 is what we need since the $\slashed{q}$ is practically zero by the on-shell condition of the outgoing lepton.

The lattice gauge theory gives the first principle computational ground for the hadronic quantities like this matrix element of proton decay. In the first two calculations on the lattice [2, 3], the tree level chiral perturbation theory [4] was used to evaluate W_0 from the low energy constant α and β,

$$\alpha P_L u_p \equiv \langle 0 | \mathcal{O}^{\beta}_{R;L} | p \rangle, \quad \beta P_L u_p \equiv \langle 0 | \mathcal{O}^{\beta}_{L;L} | p \rangle, \tag{3}$$

which are calculated on the lattice. This method is sometimes called the indirect method. Few years ago, JLQCD published on their large simulation of the nucleon decay matrix

---

[1] Present address: Physics Department, University of Wuppertal, 42097 Wuppertal, Germany

element [1], where they employed both direct and indirect methods. The direct method was first used by the authors of ref. [5]. However, the treatment of the form factors was improper, which led a large discrepancy in the results from direct and indirect methods. Once the direct method is treated properly, JLQCD [1] found the discrepancy not so large, yet, 30 − 40% in most of the cases. Their results of the matrix elements are 3–5 times larger than those from a model calculation commonly used, which pushes down the theoretical estimate of the life time of the proton, and makes much severe constraint on the GUT models.

The existing calculations are all done with the Wilson fermion at a single lattice spacing. The Wilson fermion has an $O(a)$ discretization error, where a is the lattice spacing. There are two sources of error propagating to the matrix elements. One is the measurement of the matrix element in lattice unit. The other is the estimate of the lattice scale a^{-1}. Even for the state of the art calculation by JLQCD, the systematic error of the a^{-1} is as much as 30% [2]. Of course there should be a scaling violation for W_0, α and β, too, which could diminish the overall violation by compensating that from the scale. But it is unknown until it is studied. Also the Wilson fermion breaks chiral symmetry explicitly. Thus, the applicability of the chiral perturbation theory at a finite lattice spacing is not guaranteed. One has to take the continuum limit of quantities of interest.

The second problem is that up to now the operator renormalization has been done by one-loop (tadpole-improved) perturbation theory. This should be improved by employing a non-perturbative technique [3].

Finally the calculations are all done in the quenched approximation, where all quark loop effects are neglected. This approximation is commonly used in the lattice calculation as the unquenched simulation is much more expensive. One has to check how large is the effect of quenching by doing the unquenched simulation.

We use the domain-wall fermions [10, 11, 12] in our simulation. This fermion discretization has almost exact chiral symmetry and exact flavor symmetry. Hence there practically is no mixing of the operators with different chiral structure, making the data cleaner. Moreover, there is no $O(a)$ discretization error. This second point has been demonstrated in the simulation results for the hadron spectrum [13] and the kaon B parameter [14, 15]. The chiral symmetry can be further improved dramatically by improving the gauge action [16, 13]. We use DBW2 gauge action which reduce the residual chiral symmetry breaking by factor 100 from that for the Wilson gauge action at a typical lattice spacing [13].

---

[2] The dimensionless quantity, the product of the Sommer scale [6, 7] and the ρ mass $r_0 m_\rho$ is about 30% off from its continuum limit [8] at the simulation point of JLQCD. Note that the decay width is proportional to the square of W_0 (dimension two), or α and β (dimension three).

[3] For the recent summary of the non-perturbative renormalization on the lattice, see [9].

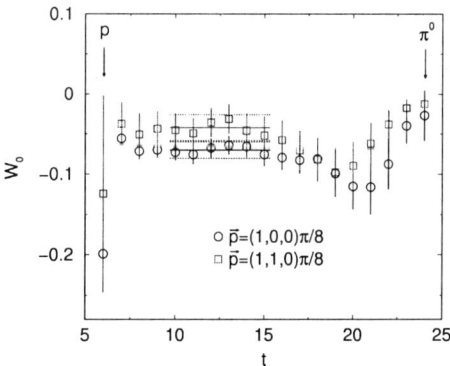

FIGURE 1. Ratio $R(t)$ for the relevant form factor W_0 for $\langle \pi^0 | \mathcal{O}^\beta_{R;L} | p \rangle$ at $m_1 = m_2 = 0.04$.

We use the $16^3 \times 32$ lattice with $a^{-1} \simeq 1.3$ GeV [4]. The direct method uses the ratio of the three- and two-point functions

$$R(t) \equiv \frac{\langle J_\pi(t_1) \mathcal{O}^\beta_{R/L;L}(t) \bar{J}_p(t_0) \rangle}{\langle J_\pi(t_1) J_\pi^\dagger(t) \rangle \langle J_p(t) \bar{J}_p(t_0) \rangle} \sqrt{Z_\pi Z_p}, \qquad (4)$$

where the proton and pion interpolating fields are located at $t_0 = 6$ and $t_1 = 24$ respectively. Momentum $\pm \vec{p}$ with $\vec{p}a = (1,0,0)\pi/8$ or $(1,1,0)\pi/8$ is injected to the pion and the operator in the three point function, as well as in the pion two point function in the denominator. $\sqrt{Z_\pi}$ and $\sqrt{Z_p}$ are overlap of J_π and J_p to the corresponding pion and proton states, which is estimated from the fit of two point functions.

Figure 1 shows ratio at a parameter point with the particular projection and subtraction to get W_0, which is taken from the fit to the plateau. In addition to the data shown in the ref. [17], we have further performed the calculation for the non-degenerate quark mass m_1, m_2 in the final pseudoscalar state, where the initial proton state is made up of quarks with m_1. Then we get W_0 as a function of m_1, m_2, and q^2. The chiral perturbation [1] helps to fit W_0 to get to the physical point. The results for various decay amplitudes are shown in Fig. 2. We are assuming the SU(2) symmetry for the u and d quarks. There are other possible matrix elements, but they can be calculated with the matrix elements in the figure when the SU(2) symmetry is intact. We are yet to have the renormalization factor for the operators by a non-perturbative renormalization. Preliminary value using the perturbative estimate of the renormalization factor [18] is listed in ref. [17].

The results of the indirect method are also shown in Fig. 2. The direct and indirect calculations give consistent results within the error, in contrast to the result of JLQCD. However, this could be caused by larger statistical error in our calculation. We need to have more statistics to judge it. Nevertheless, the relative size of the matrix element in our calculation for each decay mode is similar to that obtained by JLQCD.

---

[4] The more precise description of our simulation is given in ref. [17].

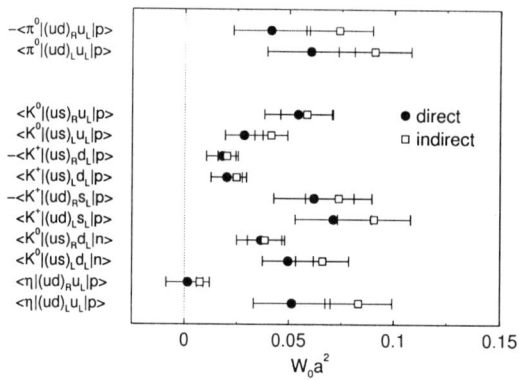

FIGURE 2. Summary of the relevant form factor of the nucleon decay with both direct and indirect methods in lattice unit.

We have investigated the proton decay matrix elements at a lattice cut off of $a^{-1} \simeq 1.3$ GeV with the domain-wall fermion in the quenched approximation. The direct and indirect methods give consistent results within our precision. The non-perturbative renormalization program [19, 20] is underway to get the continuum matrix elements. Also we are performing the two flavor dynamical domain-wall fermion simulation, which will give us an idea of the size of the quenching error.

We thank RIKEN, Brookhaven National Laboratory and the U.S. Department of Energy for providing the facilities essential for the completion of this work.

REFERENCES

1. Aoki, S., et al., *Phys. Rev.*, **D62**, 014506 (2000).
2. Hara, Y., Itoh, S., Iwasaki, Y., and Yoshie, T., *Phys. Rev.*, **D34**, 3399 (1986).
3. Bowler, K. C., Daniel, D., Kieu, T. D., Richards, D. G., and Scott, C. J., *Nucl. Phys.*, **B296**, 431 (1988).
4. Claudson, M., Wise, M. B., and Hall, L. J., *Nucl. Phys.*, **B195**, 297 (1982).
5. Gavela, M. B., et al., *Nucl. Phys.*, **B312**, 269 (1989).
6. Sommer, R., *Nucl. Phys.*, **B411**, 839–854 (1994).
7. Guagnelli, M., Sommer, R., and Wittig, H., *Nucl. Phys.*, **B535**, 389–402 (1998).
8. Aoki, S., et al., *Phys. Rev.*, **D67**, 034503 (2003).
9. Sommer, R., *Nucl. Phys. Proc. Suppl.*, **119**, 185–197 (2003).
10. Kaplan, D. B., *Phys. Lett.*, **B288**, 342–347 (1992).
11. Shamir, Y., *Nucl. Phys.*, **B406**, 90–106 (1993).
12. Furman, V., and Shamir, Y., *Nucl. Phys.*, **B439**, 54–78 (1995).
13. Aoki, Y., et al., "Domain wall fermions with improved gauge actions", hep-lat/0211023.
14. Blum, T., and Soni, A., *Phys. Rev. Lett.*, **79**, 3595–3598 (1997).
15. Ali Khan, A., et al., *Phys. Rev.*, **D64**, 114506 (2001).
16. Ali Khan, A., et al., *Phys. Rev.*, **D63**, 114504 (2001).
17. Aoki, Y., *Nucl. Phys. Proc. Suppl.*, **119**, 380–382 (2003).
18. Aoki, S., Izubuchi, T., Kuramashi, Y., and Taniguchi, Y., *Phys. Rev.*, **D67**, 094502 (2003).
19. Martinelli, G., Pittori, C., Sachrajda, C. T., Testa, M., and Vladikas, A., *Nucl. Phys.*, **B445**, 81–108 (1995).
20. Blum, T., et al., *Phys. Rev.*, **D66**, 014504 (2002).

Calculation of CP Violation in Non-leptonic Kaon Decay on the Lattice

J. Noaki for RBC Collaboration

RIKEN BNL Research Center, Bldg. 510A, Brookhaven National Laboratory, Upton NY, 11973

Abstract. We give a progress report of our lattice calculation of direct and indirect CP violation in kaon decays, parametrized as ε'/ε and B_K, which require non-perturbative calculation of the matrix elements of the Standard Model effective Hamiltonian.

INTRODUCTION

In the investigations of the Standard Model, a very important issue is the theoretical treatment of $K \to \pi\pi$ decay to the accuracy such that comparison with the experimental results is possible. In particular, the ratio of direct and indirect CP violation, ε'/ε, has been determined experimentally [1] in recent years and theoretical calculation is desired to test Kobayashi-Maskawa theory. In the theoretical calculation, numerical simulation of lattice QCD is the most systematic method to estimate the non-perturbative effect of QCD which is the main source of the error. Using the operator product expansion, the interaction in this decay is written as $H_W = \frac{G_F V_{us} V_{ud}^*}{\sqrt{2}} \sum_i W_i(\mu) Q_i$, where the coefficients W_i contain the effects of the energy scales higher than the matching point μ and can be obtained perturbatively [2]. Non-perturbative QCD effects will appear in $K \to \pi\pi$ matrix elements of the local operators $\langle \pi\pi | Q_i | K \rangle$, which should be calculated on the lattice. A couple of years ago, CP-PACS and RBC Collaboration [3, 4] calculated all of the matrix elements using the domain-wall fermion formalism [5, 6] to realize the chiral symmetry required in this calculation and reported small and negative values of ε'/ε in conflict with the experimental result. Another work using staggered fermion has obtained a larger negative value [7]. In these works, however, there are several uncontrolled systematic errors such as 1) the effect of the small, but non-zero, chiral symmetry breaking, 2) the effect of finite lattice spacing, 3) the effect of the perturbative treatment of the charmed quark in the matrix elements, 4) quenching effect, and 5) $K \to \pi\pi$ matrix elements are obtained from $K \to \pi$ and $K \to 0$ (vacuum) by using lowest order chiral perurbation theory [8].

In order to examine all of these systematic errors except the fifth one, we are performing two types of numerical simulation with domain-wall fermion and the DBW2 gluonic action [9] to improve the chiral symmetry on the lattice. "Numerical Simulation I" is the quenched calculation including directly the effect of the charm quark on the lattice. The degree of chiral symmetry breaking is decreased by a factor 1/10 compared with the previous work of RBC Collaboration. In addtion, we are generating gauge configurations with $N_f = 2$ dynamical quarks [10] in "Numerical Simulation II." In the rest of this article, we present the contents of these numerical simulations and report preliminary results of the matrix elements which numerically dominate ε'/ε and kaon B-parameter B_K.

NUMERICAL SIMULATION I

We are generating gauge configurations on a relatively fine $24^3 \times 48$ lattice with the scale $a^{-1} = 2.86(9)$. The residual quark mass m_{res} which measures the chiral symmetry breaking is as small as $\lesssim 0.3$ MeV. Since quark mass $m_f a$ is introduced as a parameter of the boundary condition in the fifth dimension in domain-wall QCD, the localization of chiral modes on domain-walls in the fifth dimension tends to fail for a heavy quark mass m_f. However, our small lattice spacing made the value of $m_c a$ acceptable as a domain-wall fermion: $m_c a \approx 0.45$. We found that, around this value, the behavior of wave function in the fifth dimension is qualitatively same as the case of much smaller quark mass.

At the lowest order of chiral perturbation theory, $K \to \pi\pi$ matrix elements are in proportion to $K \to \pi$ matrix elements calculated on the lattice. For $i = 1-6, 9, 10$, these matrix elements are related as,

$$\langle \pi^+\pi^- | Q_i^{(I)} | K^0 \rangle = \frac{m_K^2 - m_\pi^2}{\sqrt{2}f} \left[\frac{1}{m_{\text{PS}}^2} \langle \pi^+ | Q_i^{(I)} | K^+ \rangle \bigg|_{\text{(subt)}} + \mathcal{O}(p^2) \right], \quad (1)$$

in particular, for $\Delta I = 1/2$, or $I = 0$, subtraction of a lower dimension operator is needed:

$$\langle \pi^+ | Q_i^{(0)} | K^+ \rangle \bigg|_{\text{subt}} = \langle \pi^+ | Q_i^{(0)} - \alpha_i Q_{\text{sub}} | K^+ \rangle, \quad (2)$$

$$Q_{\text{sub}} = (m_s + m_d)\bar{s}d - (m_s - m_d)\bar{s}\gamma_5 d, \quad \alpha_i = \langle 0 | Q_i^{(0)} | K^0 \rangle / \langle 0 | Q_{\text{sub}} | K^0 \rangle \quad (3)$$

For $i = 7, 8$, we have the simpler relation

$$\langle \pi^+\pi^- | Q_i^{(I)} | K^0 \rangle = -\frac{1}{\sqrt{2}f} \langle \pi^+ | Q_i^{(I)} | K^+ \rangle + \mathcal{O}(p^2). \quad (4)$$

In particular, $Q_6^{(0)}$ and $Q_8^{(2)}$ have the largest contribution to ε'/ε, numerically. In FIG. 1, results of $K \to \pi$ matrix elements of these operators are plotted. In particular, the left panel, which is the example with $m_c a = 0.40$, shows that there is a severe cancellation in the subtraction in (2) for $Q_6^{(0)}$. Since the slope of the subtracted matrix elements have the error of $\sim 200\%$, we cannot quote a result of $K \to \pi\pi$ matrix element with the current statistics. And its depencence on $m_c a$ is not visible, so far.

Lattice value of kaon B parameter which is defined by

$$B_K = \frac{\langle \bar{K} | Q_{\Delta S=2} | K \rangle}{8/3 \langle \bar{K} | A_\mu | 0 \rangle \langle 0 | A_\mu | K \rangle}, \quad (5)$$

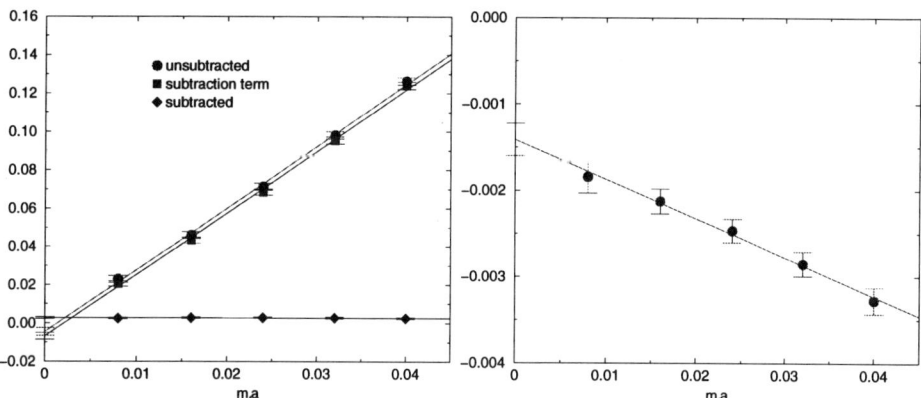

FIGURE 1. $K \to \pi$ matrix element of $Q_6^{(0)}$ (left) and $Q_8^{(2)}$ (right) as a function of $m_f a$ from Numerical Simulation I. In the left panel, data for matrix elements before (circle) and after (diamond) the subtraction and the subtraction term $-\alpha_6 \langle \pi | Q_{\text{sub}} | K \rangle$ (square) are plotted from 50 statistics. Linear extrapolation was used for all of plots.

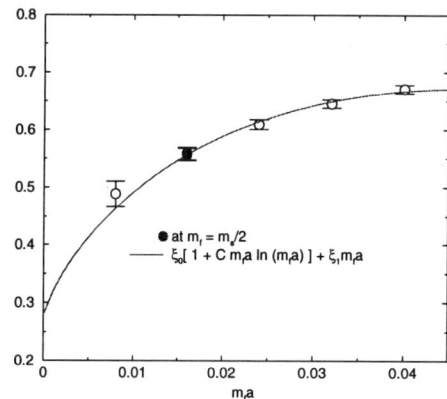

FIGURE 2. Lattice value of B_K as a function of $m_f a$ from 77 configurations.

is plotted in FIG. 2 as a function of $m_f a$. In this figure, the fit function used is $B_K = \xi_0[1 + Cm_f a \ln(m_f a)] + \xi_1 m_f a$ with C taken from analytic result [11]. The physical result for B_K can be obtained at $m_f = m_s/2$ (the filled symbol). To obtain the physical value, we are now calculating the Z factor for B_K by non-perturbative renormalization proposed in [12] and the preliminary result is roughly consistent with the previous works [13, 4].

NUMERICAL SIMULATION II

Since the dynamical simulation demands much more resources than a quenched one, dynamical domain-wall QCD has been explored by only our collaboration [10], so far.

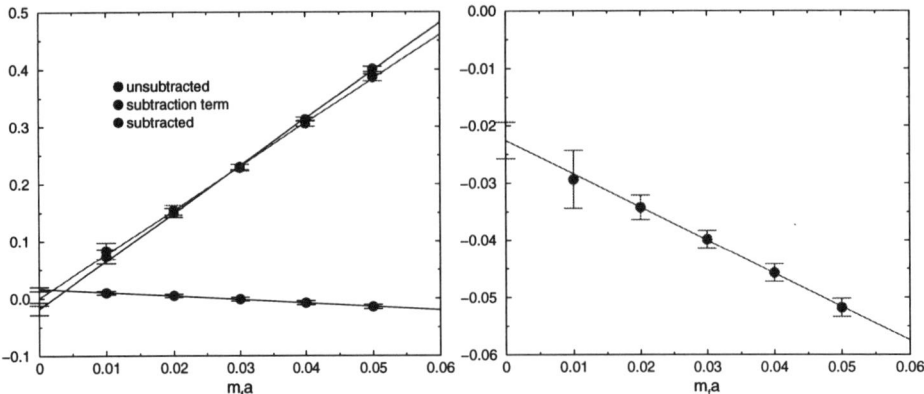

FIGURE 3. Same as FIG. 1 but from Numerical Simulation II with $m_{sea}a = 0.03$. 72 configurations were used.

In this calculation, we generated three kinds of gauge configuration on a $16^3 \times 32$ lattice with the mass of the sea quark (u or d quark) being $m_{sea}a = 0.02, 0.03$ and 0.04. For each series of configurations, $K \to \pi$ and $K \to 0$ matrix elements are calculated in the same way as Simulation I with the five valence quark masses $m_{val} = 0.01$–0.05, and basic parameters $a^{-1} \approx 1.8$ GeV and $m_{res} \approx 3$ MeV are obtained. FIG. 3 shows the same matrix elements as in FIG. 1 as an example of the case of $m_{sea} = 0.03$. Although the signal seems to be reasonable, we need much more statistics to take correct chiral limit $m_{sea}a = m_{val} \to 0$ using three data points with $m_{sea}a = m_{val} = 0.02, 0.03$ and 0.04. B_K at the physical point such that $m_{sea} = m_{u/d}$ and $m_s \sim 120$ MeV will be also obtained after a careful treatment of our data [14] and non-perturbative renormalization which is now under calculation.

We thank RIKEN, BNL and the U.S. DOE for providing the facilities essential for the completion of this work.

REFERENCES

1. Alavi-Harati, A. *et al.*, Phys. Rev. Lett. **83**, 22, 1999; Fanti, A. *et al.*, Phys. Lett. **B465**, 335, 1999.
2. For a review, see Buchalla, G., Buras, A. J., Leutenbacher, M E., Rev. Mod. Phys **68**, 1125, 1995.
3. CP-PACS Collaboration, Noaki, J.*et al.*, Phys. Rev. **D68**, 014501, 2003.
4. RBC Collaboration, Blum, T.*et al.*, hep-lat/0110075.
5. Kaplan, D., Phys. Lett. **B288**, 342, 1992.
6. Shamir, Y., Nucl. Phys. **B406**, 90, 1993; Furman, V. and Shamir, Y., Nucl. Phys. **B439**, 54, 1995.
7. Pekurovsky, D. and Kilcup, G., Phys. Rev. **D64**, 074502, 2001.
8. Bernard, C., Draper, T., Soni, A., Politzer, H. D. and Wise, M. B., Phys. Rev. **D32**, 2343, 1985.
9. RBC Collaboration, Aoki, Y.*et al.*, hep-lat/0211023.
10. RBC Collaboration, Izubuchi, T.*et al.*, hep-lat/0210011.
11. Sharpe, S., Phys. Rev. **D46**, 3146, 1992.
12. Martinelli, G. *et al.*, Nucl. Phys. **B445**, 81, 1995.
13. CP-PACS Collaboration, Ali Khan, A.*et al.*, Phys. Rev. **D64**, 114506, 2001.
14. Golterman, M. and Leung, K-C., Phys. Rev. **D57**, 5703, 1998.

NEUTRINOS

SNO Past, Present and Future

N. McCauley. On behalf of the SNO collaboration.

Department of Physics and Astronomy, University of Pennsylvania, Philadelphia, PA 19104-6396

Abstract. The Sudbury Neutrino Observatory (SNO) is a one kiloton heavy water Cherenkov detector sensitive to the flavor content of the ^{8}B neutrinos from the sun. Results from phase 1, the pure D_2O phase, of the experiment show evidence for neutrino flavor change as the solution of the solar neutrino problem at the 5.3 sigma level. The second phase of the experiment, where 0.2% NaCl solution is added to the D_2O, is underway. The benefits and drawbacks of adding salt are discussed. Finally plans for phase three of the experiment, where ^{3}He proportional counters will be deployed inside the D_2O, are reviewed.

INTRODUCTION.

Solar neutrino experiments over the past 30 years[1],[2],[3],[4],[5],[6] have observed fewer neutrinos than predicted by solar models[7]. One possible explanation for this effect is that neutrinos transform from one flavor to another in transit from creation at the center of the Sun to detection on Earth. The Sudbury Neutrino Observatory (SNO)[8] was designed to measure the flavor content of the ^{8}B solar neutrino flux and to test the model of neutrino flavor change.

The SNO[8] detector is a water Cherenkov detector, located at 6010 m of water equivalent in the INCO Ltd. Creighton mine near Sudbury, Ontario. The detector uses 1 kiloton of ultra pure heavy water contained in a transparent acrylic shell of 12m diameter as the solar neutrino target. Cherenkov photons produced in the neutrino interactions are detected by 9456 photo-multiplier tubes (PMTs) mounted on a 17.8m diameter stainless steel geodesic sphere (PSUP). The detector is immersed in ultra-pure light water to provide shielding from radioactivity in both the rock wall of the cavity and the PMTs and support structure.

Sensitivity to the flavor content of the solar ^{8}B neutrino flux is obtained by comparison of three different reaction rates with different sensitives to electron neutrinos (v_e) and other active neutrinos ($v_{\mu,\tau}$). These reactions are:

$$v_e + d \rightarrow p + p + e^- \quad \text{(CC)}$$
$$v_{e,\mu,\tau} + d \rightarrow p + n + v_{e,\mu,\tau} \quad \text{(NC)}$$
$$v_{e,\mu,\tau} + e^- \rightarrow v_{e,\mu,\tau} + e^- \quad \text{(ES)}$$

The Charged Current (CC) reaction is sensitive only to the electron neutrino flux, while the Neutral Current (NC) reaction has equal sensitivity to each active neutrino flavor. The Elastic Scattering reaction (ES) has enhanced sensitivity to electron neutrinos, due to the additional possibility for this reaction to occur via a charged current process.

Both the elastic scattering reaction and the charged current interaction are detected by SNO using the Cherenkov light emitted by the relativistic electrons produced in the reaction. The neutral current is detected via the neutron produced in the deuteron breakup. SNO uses a variety of techniques to detect these neutrons, each technique being a different phase of the experiment.

THE PURE D$_2$O PHASE.

The pure D$_2$O phase of SNO ran from November 2 1999 until May 28 2001, providing 306.4 live days of data. Using pure D$_2$O neutrons are detected via the capture on deuterons.

$$d + n \to {}^3H + \gamma, \; E_\gamma = 6.25 MeV \quad (1)$$

The gamma ray produced is detected via Cherenkov light from Compton scattered electrons. The low energy of the neutron signal means that it lies close the background from radioactive decays in the detector. These provide a steeply climbing background wall at low energy that is removed by imposing a kinetic energy (T) threshold at 5MeV. This means that the peak of the neutron signal occurs very close to the energy threshold. After imposing the energy threshold and fiducial volume cut (at r=550cm) the neutron detection efficiency is 14.4±0.5%.

The separate rates of the three reactions are determined by a signal extraction procedure. Each reaction type has distinctive distributions in at least one of three variables: energy, radius and direction with respect to the sun. The elastic scattering reaction for example is strongly peaked away from the sun. The NC signal has a strong radial fall off as neutrons produced close to the acrylic vessel are more likely to capture on H in the acrylic or light water and produce a signal below the energy threshold. The signal extraction procedure uses these distributions in a maximum likelihood fit to the data to provide the number of CC, NC and ES events along with their uncertainties and correlations. As the sensitivity to both the ν_e and $\nu_{\mu,\tau}$ fluxes is known for each reaction we can instead use signal extraction to directly extract these fluxes. We find that:

$$\phi_e = 1.76^{+0.05}_{-0.05}(stat.)^{+0.09}_{-0.09}(sys.) \times 10^6 cm^{-2}s^{-1}$$
$$\phi_{\mu,\tau} = 3.41^{+0.45}_{-0.45}(stat.)^{+0.48}_{-0.45}(sys.) \times 10^6 cm^{-2}s^{-1}$$

The flux $\phi_{\mu,\tau}$ is greater than 0 at the 5.3σ level, strongly favoring models of neutrino flavor change. This is shown graphically in figure 1. More details on the pure D$_2$O analysis can be found in [9],[10] and [11].

THE SALT PHASE

On May 28 2001 two tons of salt (NaCl) were added to the heavy water. Production data taking began on July 27 2001. Adding salt provides a new channel for neutron capture in SNO. Neutrons now predominantly capture on ^{35}Cl:

$$^{35}Cl + n \to {}^{36}Cl + multiple \gamma s, \; \Sigma E_\gamma = 8.6 MeV \quad (2)$$

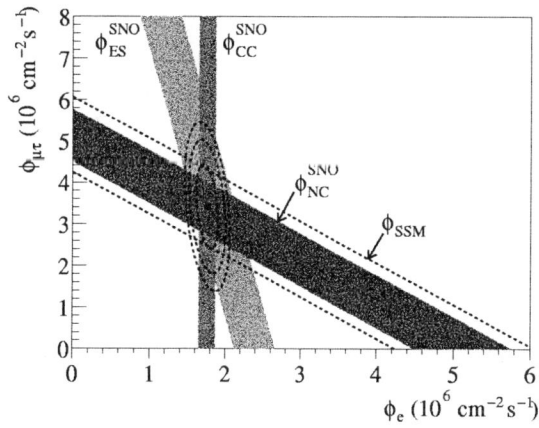

FIGURE 1. The three neutrino fluxes measured by SNO and their flavor composition. The hypothesis that $\phi_{\mu,\tau} = 0$ is strongly disfavored.

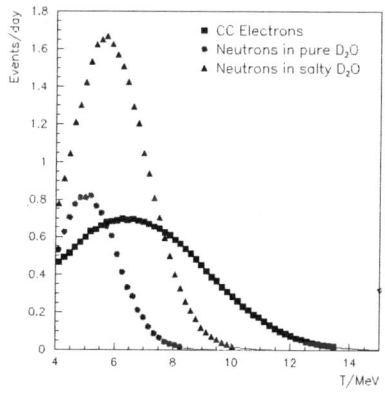

FIGURE 2. The kinetic energy distribution of neutrons in salty and pure D_2O and the charged current signal. The signals are normalized to their expected signal rates.

The addition of salt gives three distinct improvements upon the situation in pure D_2O.

- The cross section for capture on chlorine is greater than that for deuterons. This means that more neutrons capture on an isotope that provides a Cherenkov signal that is in the analysis region.
- The overall energy of the gamma rays in the neutron capture is greater than for captures on deuterons, moving the neutron signal away from the low energy background wall. The shift in the neutron energy in comparison to pure D_2O and the charged current signal is shown in figure 2. With the increase in cross section and energy the neutron detection efficiency is expected to increase three-or four-fold

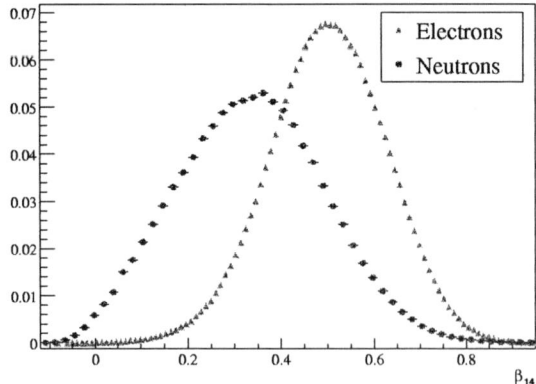

FIGURE 3. The isotropy as measured by β_{14} of electrons and neutrons in salt.

over that seen in pure D_2O.
- The neutron signal is provided by multiple particles and is therefore more isotropic than the single electrons produced in the CC and ES reactions. We parameterize the isotropy of an event using a parameter known as β_{14}, which a combination of spherical harmonic parameters of the hit distribution. The β_{14} distribution for neutrons and electrons is shown in figure 3. This gives us an additional handle to use in the signal extraction, in particular allowing signal extraction to be carried out without a constraint on the energy distribution of the signals. This provides a more model independent measure of the solar ^8B neutrino flux and flavor content.

The addition of salt also provides some obstacles that must be overcome. The chemistry of the water is changed which results in difference in radiochemical assays and optical attenuation. In addition ^{24}Na, which has a half life of 15 hours, can be created by neutron capture during calibration or outside the detector, in the pipes of the water systems. This provides an additional low energy radio-isotope that must be accounted for. The salt phase is expected to end some time in 2003.

THE NCD PHASE

The third phase of SNO, expected to commence in late 2003, is the Neutron Capture Detector (NCD) phase. Unlike the previous two phases a direct neutron detection technique will be used. Proportional counters containing ^3He will be deployed in the D_2O volume. The counters utilize the reaction:

$$n + ^3He \rightarrow p + ^3H \qquad (3)$$

Neutrons that capture on ^3He are then directly detected by the charged tracks in the counter from the proton and triton. The energy and time of the charge deposition then allow the neutron signal to be separated from the background, principally α decays.

The use of the NCDs breaks the covariance between the NC and CC signals and provides a neutral current measurement with very different systematics from the previous phases of the experiment.

CONCLUSIONS

SNO has measured the flavor content of the solar ^{8}B neutrino flux and sees evidence at the 5.3σ level for neutrino flavor change. The salt phase of SNO is currently underway, where the addition of chlorine ions to the heavy water provides additional sensitivity to neutral current interactions, by enhancing neutron detection. The final phase of SNO, expected to commence late in 2003, has NCDs deployed in the heavy water to provide a neutron detection channel that does not use Cherenkov light and therefore has very different systematics to the previous two phases. SNO has demonstrated that neutrino flavor change is the solution to the solar neutrino problem. The future phases of SNO will help us test the physics behind the flavor change.

ACKNOWLEDGMENTS

This work is presented on behalf of the SNO collaboration. This research was supported by: Canada: NSERC, Industry Canada, NRC, Northern Ontario Heritage Fund Corporation, INCO, AECL, Ontario Power Generation; US: Dept. of Energy; UK: PPARC. The SNO collaboration wishes to thank the SNO technical staff for their strong contribution.

REFERENCES

1. Cleveland, B., et al., *The Astrophysical Journal*, **496**, 505 (1998).
2. Fukuda, Y., et al., *Physical Review Letters*, **77**, 1683 (1996).
3. Abdurashitov, J., et al., *Physical Review C*, **60**, 055801 (1999).
4. Hampel, W., et al., *Physics Letters B*, **447**, 127 (1999).
5. Fukuda, S., et al., *Physical Review Letters*, **86**, 5651 (2001).
6. Altmann, W., et al., *Physics Letters B*, **490**, 16 (2000).
7. Bahcall, J. N., Pinsonneault, M. H., and Basu, S., *The Astrophysical Journal*, **555**, 990–1012 (2001).
8. Dragowsky, M. R., et al., *Nuclear Instruments and Methods in Physics Research*, **A449**, 172 (2000).
9. Ahmed, Q. R., et al., *Physical Review Letters*, **87**, 071301 (2001).
10. Ahmed, Q. R., et al., *Physical Review Letters*, **89**, 011301 (2002).
11. Ahmed, Q. R., et al., *Physical Review Letters*, **89**, 011302 (2002).

First Results from KamLAND

B. E. Berger for the KamLAND Collaboration

Lawrence Berkeley National Laboratory, Berkeley, CA 94720

Abstract. The KamLAND collaboration recently published the first evidence for reactor antineutrino disappearance [1]. The measured ratio of observed inverse β-decay events to the number expected from standard assumptions about $\bar{\nu}_e$ propagation is 0.611±0.085(stat)±0.041(syst). Fewer $\bar{\nu}_e$'s were seen than expected at the 99.95% confidence level. In the context of two-flavor neutrino oscillations and CPT invariance, this measurement rules out all but the LMA region of solar neutrino oscillation parameter space. I summarize these first results and discuss the prospects for future oscillation measurements at KamLAND.

INTRODUCTION

The primary goal of KamLAND, the Kamioka Liquid-Scintillator Anti-Neutrino Detector, is to search for the oscillation of $\bar{\nu}_e$'s produced by nuclear reactors at long baseline distances. KamLAND is located in central Japan at the site of the earlier Kamiokande [2] experiment, where there is a large flux of reactor $\bar{\nu}_e$'s from the full ensemble of Japanese nuclear power plants, with a flux-weighted average distance of ~180 km. This distance is over two orders of magnitude greater than the previous generation of reactor neutrino experiments, CHOOZ [3] and Palo Verde [4]. These earlier experiments saw no evidence for oscillations at baseline distances of ~1 km. The longer baseline allows KamLAND to reach much lower values of neutrino mass difference Δm^2 than previous reactor experiments. This extended reach allows KamLAND to probe the large mixing angle (LMA) region of solar neutrino oscillation (Δm_{12}^2, θ_{12}) parameter space. This region is the one favored by global analysis of all solar neutrino measurements [5].

The KamLAND collaboration published its first results [1] earlier this year, based on 145.1 days of data corresponding to a 162 ton-year exposure. In this paper, I summarize the first published results and their implications, and I discuss the prospects for future neutrino measurements at KamLAND.

THE KAMLAND EXPERIMENT

KamLAND detects $\bar{\nu}_e$'s through the inverse β-decay interaction $\bar{\nu}_e + p \rightarrow e^+ + n$. This reaction produces two energy depositions in delayed coincidence. The positron deposits energy both through ionization and by annihilation with a detector electron. All this energy is detected by KamLAND as a single "prompt" event with total energy $E_{\text{prompt}} \simeq E_{\bar{\nu}_e} - 0.8$ MeV. The 0.8 MeV includes the proton-neutron mass difference, the electron rest mass, and the small average recoil energy of the neutron. The neutron first

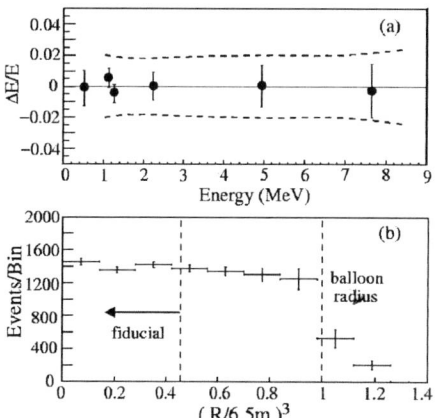

FIGURE 1. Energy reconstruction and vertex fitter performance. (a) The fractional differences between reconstructed energy and known γ energy for various γ calibration sources. The dashed curve represents the assigned systematic error. (b) The R^3 vertex distribution of 2.2 MeV γ's from spallation neutron capture. The 6.5 m balloon radius and 5 m fiducial volume are marked.

thermalizes through multiple elastic scattering and is eventually captured on a proton. The capture produces a 2.2 MeV γ, which KamLAND detects as a second "delayed" event. The delay time between these two events is a characteristic of the scintillator; the measured capture time in KamLAND is $188 \pm 23 \, \mu$sec.

The KamLAND detector is described in detail elsewhere [1, 6]. Briefly, the $\bar{\nu}_e$ target is a kiloton of liquid scintillator (LS); this target is surrounded by a 2.5 m mineral oil buffer. An array of 1325 17-inch photomultiplier tubes (PMT's) detects scintillation light produced when energy is deposited in the detector. (An additional 554 20-inch PMT's are not used in this first analysis.) The detector is read out whenever 200 or more PMT's fire in coincidence, corresponding to an energy threshold of $\sim$0.7 MeV. The trigger threshold is reduced to 120 PMT hits for 1 msec after each event, so the energy threshold for delayed events is $\sim$0.4 MeV. The liquid-scintillator inner detector (ID) is surrounded by a 3.2 kton water-Čerenkov outer detector (OD) that tags cosmic-ray muons.

DATA ANALYSIS

The raw KamLAND data are the arrival times and sizes of PMT pulses. We fit the timing information to determine the location of the energy deposit for each event. Our vertex fitting algorithm has a typical position resolution for single events of $\sim$25 cm; it reconstructs the positions of calibration sources to $\sim$5 cm.

We reconstruct event energies based on the total charge collected by all PMT's, after corrections for gain variations, solid angle, shadowing by detector components, and the light transport properties of both the liquid scintillator and buffer oil. Further corrections are applied to account for quenching and for Čerenkov light contributions.

TABLE 1. Background summary.

Background	Number of Events
Accidental	0.0086 ± 0.0005
^{9}Li/^{8}He	0.94 ± 0.85
Fast neutron	< 0.5
Total B.G. events	1 ± 1

We calibrate the energy response with both calibration sources, including ^{68}Ge, ^{65}Zn, ^{60}Co, and Am-Be γ sources, and radioactive backgrounds in the scintillator, in particular the 2.2 MeV γ from spallation neutron capture. The error in reconstructed energy versus calibration source energy is shown in Figure 1(a). The detector energy resolution is $\sim 7.5\%/\sqrt{E(\text{MeV})}$.

We select candidate $\bar{\nu}_e$ events with the following cuts. Both the prompt and delayed events must be reconstructed within a fiducial volume defined to be within 5 m of the detector center. The time difference Δt between the prompt and delayed events must fall in the delayed coincidence window $0.5\,\mu\text{sec} < \Delta t < 660\,\mu\text{sec}$. We apply a vertex correlation cut; the distance ΔR between the prompt and delayed events must satisfy $\Delta R < 1.6\,\text{m}$. The reconstructed energy of the delayed event must be within 0.4 MeV of the 2.2 MeV capture γ energy. Finally, we veto events in which the delayed vertex is within 1.2 m of the detector's central vertical axis to reduce backgrounds due to thermometers deployed down the axis. The total efficiency of these cuts is $78.3 \pm 1.6\%$.

Table 1 gives the backgrounds we estimate for prompt event energies above 2.6 MeV. The delayed-coincidence signal is a powerful tool for suppressing the accidental background; we find this background to be only 0.0086 ± 0.0005 events. Our primary backgrounds are due to muon spallation. Inelastic interactions of high-energy cosmic-ray muons liberate neutrons from the nuclei of detector atoms. We veto all events within 2 msec of any muon event to eliminate this neutron background. Spallation also produces ^{8}He ($T_{1/2} = 119$ msec) and ^{9}Li ($T_{1/2} = 178$ msec), both of which produce delayed-coincidence backgrounds. We veto most of these events with a 2 sec veto on events within 3 m of all muon tracks and a 2 sec veto on the entire detector for the highest-energy muons, which can produce showers well away from the muon track. We estimate that 0.94 ± 0.85 background events still pass these cuts. The three muon spallation cuts together reduce the detector livetime by 11.4%. A final background is due to fast neutrons produced outside the detector that shower in the scintillator before being captured. We estimate this background to be less than 0.5 events based on the OD reconstruction

TABLE 2. Estimated systematic uncertainties (%).

Total scintillator mass	2.1	Reactor power	2.0
Fiducial mass ratio	4.1	Fuel composition	1.0
Energy threshold	2.1	Time lag	0.28
Efficiency of cuts	2.1	$\bar{\nu}$ spectra	2.5
Live time	0.07	Cross section	0.2
Total systematic error			6.4%

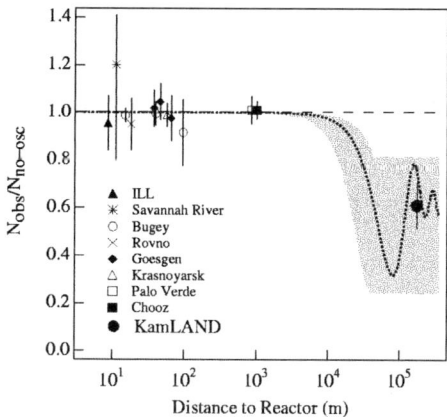

FIGURE 2. The ratio of measured $\bar{\nu}_e$ flux to that expected with no oscillations for reactor neutrino experiments [7]. The KamLAND result is plotted at the flux-weighted mean distance of ∼180 km. The dashed line is the expectation for no oscillations. The dotted curve is the best-fit LMA prediction from a global analysis of solar neutrino data [5] ($\sin^2 2\theta = 0.883$, $\Delta m^2 = 5.5 \times 10^{-5}$ eV^2), while the shaded region is the 95% C.L. LMA region from that fit.

efficiency. The total background above 2.6 MeV comes to 1±1 events.

The systematic uncertainties listed in Table 2 are described in detail in the first results paper [1]. The largest uncertainty is on the number of target protons in the fiducial volume. This uncertainty depends directly on how well we understand the performance of the vertex fitter, a challenging task in the absence of off-axis calibration sources. We have tested the performance of this fitter with the reconstructed distribution of spallation neutrons, shown in Figure 1(b). Based on the uniformity of this distribution, we calculate that the fraction of LS events reconstructed within the fiducial volume is consistent with the geometric fraction to within 4.1%. The other major sources of systematic error are the energy resolution at the 2.6 MeV threshold, the efficiencies of the analysis cuts, and our understanding of the neutrino flux from the nuclear reactors.

After applying all analysis cuts, we find a total of 54 $\bar{\nu}_e$ candidate coincidences with prompt energy greater than 2.6 MeV. The expected number in the absence of neutrino disappearance is 86.8±5.6, including the systematic uncertainties in Table 2. With the 1±1 event background, the ratio of observed to expected events is:

$$\frac{N_{\text{obs}} - N_{\text{BG}}}{N_{\text{no-osc}}} = 0.611 \pm 0.085(\text{stat}) \pm 0.041(\text{syst}). \quad (1)$$

Accounting for the background and systematic errors, the probability of a fluctuation from 86.8 down to 54 events is <0.05% by Poisson statistics, so this result rules out standard neutrino propagation at the 95% confidence level. Figure 2 plots the ratio of measured to expected $\bar{\nu}_e$ flux for KamLAND and previous reactor neutrino experiments as a function of baseline distance.

We have also performed a two-flavor oscillation fit to the data that incorporates both the observed $\bar{\nu}_e$ rate and the energy distribution of the observed events. This

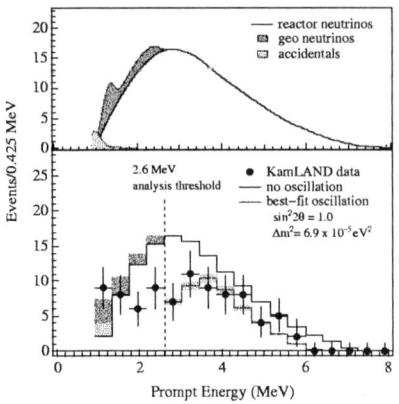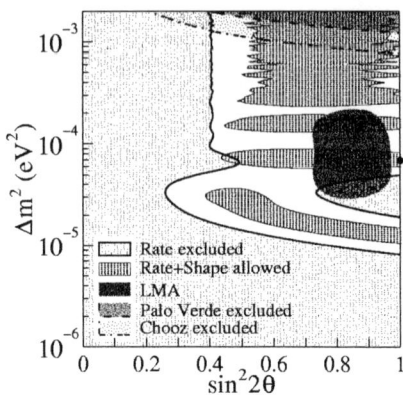

FIGURE 3. Upper left: The expected reactor $\bar{\nu}_e$ spectrum, along with geoneutrino (model Ia of [8]) and accidental background spectra. Lower left: The energy spectrum of observed prompt events (points with statistical errorbars), compared with the no-oscillation expectation (upper histogram) and the best fit including oscillations (lower histogram). The shaded bands indicate the systematic errors. The geoneutrino and background histograms and the 2.6 MeV analysis threshold are also shown. Right: The 95% C.L. excluded region from the KamLAND rate analysis and the 95% C.L. allowed region from the rate+shape analysis. The 95% C.L. region from a global analysis of solar neutrino data [5] is also shown, as are the 95% excluded regions from the CHOOZ [3] and Palo Verde [4] experiments. The point indicates the best-fit values from the rate+shape analysis.

"rate+shape" analysis is described in detail in the first results paper [1]. The fit itself is shown in the lower left of Figure 3. The best-fit oscillation parameters are $\sin^2 2\theta = 1.0$, $\Delta m^2 = 6.9 \times 10^{-5}$ eV^2, while the reduced $\chi^2 = 0.31$ for 8 degrees of freedom.

The 95% C.L. excluded region from the rate analysis and the 95% C.L. allowed region from the rate+shape analysis are both shown in the right half of Figure 3. In the context of two-flavor oscillations and CPT invariance, the KamLAND results can be compared directly to the LMA region preferred by a global analysis of solar neutrino data [5]. The rate analysis is consistent with the global fit, eliminating a region at the smaller values of Δm^2, while the rate+shape analysis places tighter constraints on the mixing parameters. The KamLAND results rule out all other regions of the solar neutrino oscillation parameter space. The lower and upper overlap regions between the solar global fit and the KamLAND rate+shape analysis have recently come to be called LMA1 and LMA2, respectively.

It should be noted that KamLAND has not directly observed an oscillation, a signal that rises and falls with either energy or distance. The measured $\bar{\nu}_e$ flux as a function of energy is consistent with a flat suppression of flux vs. energy. In fact, the rate+shape analysis rules out regions of the Δm^2_{12}, θ_{12} parameter space precisely because oscillations are *not* seen: mixing parameters in those regions correspond to large oscillations inconsistent with the KamLAND data.

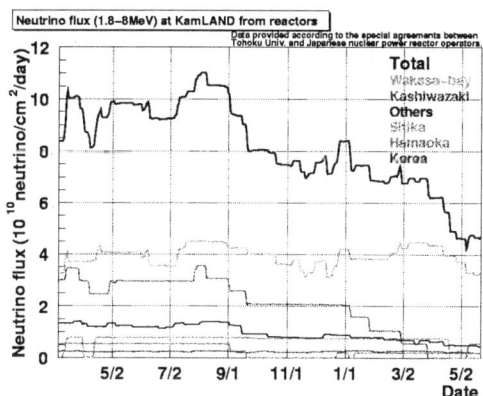

FIGURE 4. The reactor antineutrino flux at KamLAND from early 2002 to mid-2003.

OUTLOOK

KamLAND continues to accumulate reactor antineutrino data. We have already more than doubled the 145.1 day dataset used in the first publication. However, we cannot simply extrapolate our final sensitivity from our first results. Our rate+shape analysis has an effective reduced $\chi^2 = 0.31$. This low value, less than unity, means that our fit is "too good"—that statistical fluctuations in the data have worked to make our allowed region tighter than we would expect. We cannot expect future fluctuations in the data be so favorable.

The reactor antineutrino flux at KamLAND has also fallen significantly, as illustrated by Figure 4. Some Japanese reactors have been shut down for inspection, and it is not clear when they will return to operation. KamLAND may have to run longer to collect the same antineutrino sample. In addition, the shutdowns alter the distribution of baseline distances, which affects the experiment's sensitivity to oscillation parameters. However, these shutdowns give us an unexpected opportunity to measure the observed electron antineutrino event rate at different neutrino fluxes.

The expected KamLAND sensitivity also depends strongly on the actual values of the the oscillation parameters. Based on Monte Carlo studies with an assumed 500 event sample, we can hope to measure Δm_{12}^2 to the 4-5% level if the true value is in the LMA1 region, but only to the 5-7% level in LMA2. The LMA2 expectations are particularly sensitive to how long reactors remain off in Japan. The KamLAND reactor neutrino experiment is less sensitive to the mixing angle θ_{12} than it is to Δm_{12}^2. For $\tan^2 \theta_{12} \simeq 0.4$, close to the current best-fit values, we expect uncertainties at the 0.2-0.3 level.

The KamLAND collaboration is also considering upgrades to allow the experiment to detect ^{7}Be solar neutrinos through elastic scattering from electrons. This is an extremely challenging measurement, as the signal events have visible energies well below 1 MeV, where radioactive backgrounds are high. The KamLAND scintillator already meets the radiopurity levels of ^{238}U and ^{232}Th that would be necessary for ^{7}Be neutrino detection. However, the levels of other contaminants, such as ^{85}Kr and ^{210}Pb, are currently far too high. KamLAND will have to make substantial upgrades to the scintillator purification system for this measurement to be possible.

CONCLUSIONS

KamLAND has demonstrated reactor $\bar{\nu}_e$ disappearance for the first time. In the context of two-flavor neutrino oscillation and CPT invariance, our results rule out all the solar neutrino oscillation parameter space except the LMA region. The rate+shape analysis restricts the allowed region. Future KamLAND measurements will provide a more precise determination of neutrino oscillation parameters.

REFERENCES

1. K. Eguchi et al. (KamLAND Collaboration), Phys. Rev. Lett. **90**, 021802 (2003)
2. K. S. Hitata et al., Phys. Rev. D **38**, 448 (1988).
3. M. Apollonio et al. (CHOOZ Collaboration), Eur. Phys. J. C **27**, 331 (2003).
4. F. Boehm et al. (Palo Verde Collaboration), Phys. Rev. D **62**, 072002 (2002).
5. G. L. Fogli et al., Phys. Rev. D **66**, 053010 (2002).
6. A. Suzuki et al. (KamLAND Collaboration), Nucl. Phys. B (Proc. Suppl.) **77**, 171 (1999); http://www.awa.tohoku.ac.jp/KamLAND/; http://kamland.lbl.gov; and references therein.
7. Particle Data Group, Phys. Rev. D **66**, 010001 (2002).
8. R. S. Raghavan et al., Phys. Rev. Lett. **80**, 635 (1998).

The KARMEN final results on the search for $\bar{\nu}_e$ from μ^+ decay at rest

Klaus Eitel

Institut für Kernphysik, Forschungszentrum Karlsruhe, 76021 Karlsruhe, Germany

Abstract. We review the search for $\bar{\nu}_e$ from μ^+ decays at rest with the KARMEN experiment. A potential excess of $\bar{\nu}_e$-induced events as seen in the LSND experiment can be interpreted as evidence for $\bar{\nu}_\mu \to \bar{\nu}_e$ or as indication of lepton flavor (LF) violation in μ^+ decays. The KARMEN results show no hint for extra $\bar{\nu}_e$, thereby excluding large parameter areas for an oscillation signal as well as the LF violating μ^+ decay at more 90% C.L. as source of the LSND $\bar{\nu}_e$ excess.

In the last years, tremendous progress has been achieved to firmly establish the nature of neutrino oscillations using neutrinos from the sun as well as neutrinos produced in the earth's atmosphere. However, with oscillation parameters accessible to accelerator based experiments, the situation remains unsettled. There is one evidence for oscillations in the appearance mode $\bar{\nu}_\mu \to \bar{\nu}_e$ from the LSND experiment [1] which is in contrast to results from the KARMEN experiment, but also partly excluded by accelerator-based experiments such as NOMAD [2] and NuTeV [3]. In this paper, we summarize the results of the KARMEN search for appearance of $\bar{\nu}_e$ within a very pure $\bar{\nu}_\mu$ flux from μ^+ decays at rest (DAR) indicating either the flavor oscillation $\bar{\nu}_\mu \to \bar{\nu}_e$ or other non-SM processes, e.g. lepton family number violating μ^+ decays $\mu^+ \to e^+ + \bar{\nu}_e + {}^{(\bar{\nu})}$.

EXPERIMENTAL CONFIGURATION AND DATA REDUCTION

The experiment was performed at the neutrino source of the ISIS synchrotron accelerating protons to an energy of 800 MeV before striking a massive beam stop target. Neutrinos emerge isotropically from the consecutive decays at rest (DAR) $\pi^+ \to \mu^+ + \nu_\mu$ and $\mu^+ \to e^+ + \nu_e + \bar{\nu}_\mu$ assuming the ν–flavors of the SM decay channels. Neutrinos from μ^+ DAR have a continuous energy spectrum up to 52.83 MeV. Due to the narrow time structure of 525 ns of the proton pulses muons are produced in a short time window compared to their lifetime of 2.2 μs.

The neutrinos are detected in a 56 t scintillation calorimeter [4] at a mean distance of 17.6 m from the ISIS target (fig. 1). The calorimeter is a mineral oil based scintillator segmented into 512 independent modules. Gadolinium within the module walls allows effective neutron detection via Gd(n,γ) with on average 3 γ's of energy $\Sigma E_\gamma = 8$ MeV in addition to the capture on the hydrogen of the scintillator via p(n,γ). The scintillation detector provides an almost pure target of ^{12}C and ^{1}H for ν-interactions. Three veto layers ensure a search for LF violating μ^+ decays almost free of cosmic background.

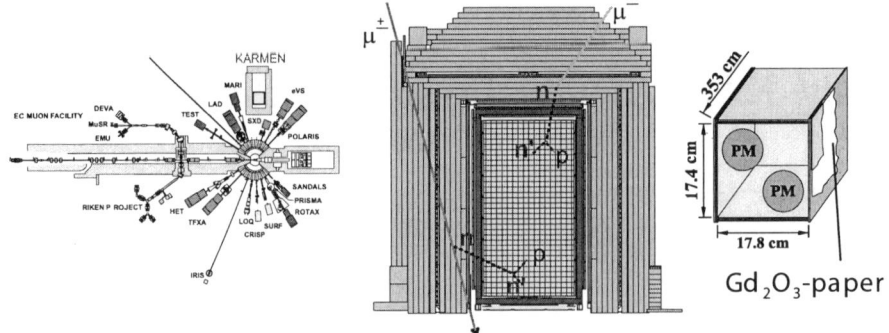

FIGURE 1. Schematic view of the KARMEN detector located at the ISIS target; front view of detector tank with its layers of active vetos and the passive iron shielding; cross section of an individual scintillator module with the photomultiplier pair at each module end.

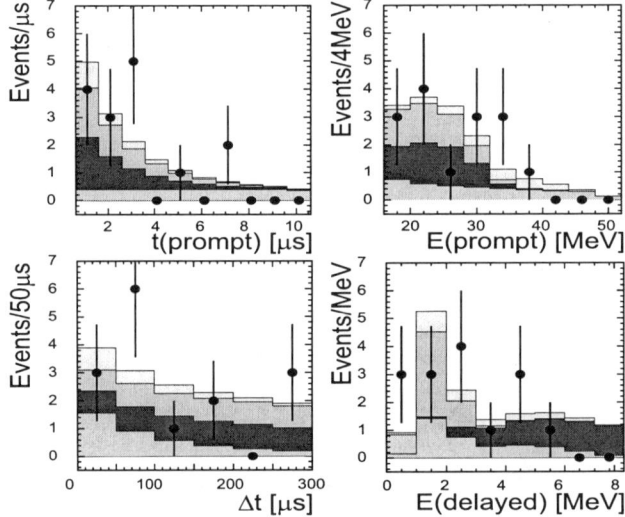

FIGURE 2. KARMEN 15 coincidence candidates after final cuts.

After a substantial upgrade consisting of a third active muon veto system of 300 m^2 plastic scintillators to reduce cosmic induced neutron background [5], the KARMEN 2 experiment took data from February 1997 to March 2001. During this time, protons equivalent to a total charge of 9425 Coulombs have been accumulated on the ISIS target.

A $\bar{\nu}_e$ signal consists of a spatially correlated delayed (e$^+$,n) sequence. The requirements for event sequences in KARMEN are described in detail in [6] and references therein. Applying all cuts to the data, 15 (e$^+$,n) candidate sequences were finally reduced. Figure 2 shows the remaining sequences in the appropriate energy and time windows. The background components are also given with their distributions. The expected

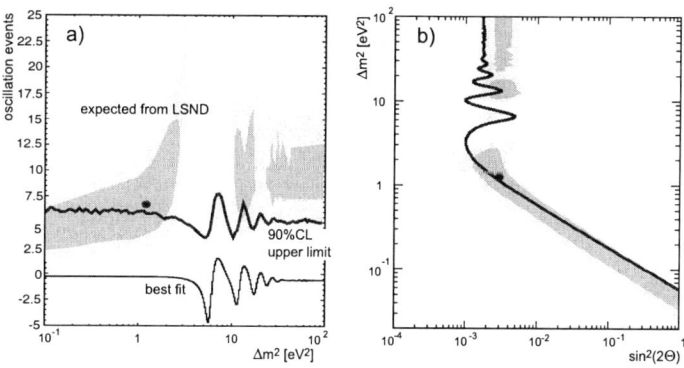

FIGURE 3. *(a) KARMEN limit and expected event rate following the LSND signal strength. (b) oscillation plot with LSND favored regions and KARMEN exclusion curve. The dot denotes the LSND best fit values.*

background amounts to 15.8 ± 0.5 events. This number comprises 3.9 ± 0.2 events from cosmic induced sequences as well as ν induced reactions such as intrinsic source contamination of $\bar{\nu}_e$ (2.0 ± 0.2), ν_e induced random coincidences (4.8 ± 0.3) and (e^-, e^+) sequences from $^{12}C(\nu_e, e^-)^{12}N_{g.s.}$ with subsequent ^{12}N decay (5.1 ± 0.2). Except for the intrinsic $\bar{\nu}_e$ contamination, deduced from detailed MC simulations, all the background components have been measured in different time and energy regimes with the KARMEN detector and extrapolated into the evaluation cuts applied for the $\bar{\nu}_e$ search. The extracted number of sequences is in excellent agreement with the background expectation, consistent with no additional $\bar{\nu}_e$ signal.

LIMITS ON NEUTRINO OSCILLATIONS $\bar{\nu}_\mu \to \bar{\nu}_e$

To analyze the extracted sequences with respect to a potential contribution of $\bar{\nu}_e$ from oscillations $\bar{\nu}_\mu \to \bar{\nu}_e$, an event-based maximum likelihood method is applied. This method includes the detailed spectral information of each individual event as well as the expected event parameters for a specific oscillation hypothesis with the free parameters Δm^2 and the oscillation amplitude, usually expressed in the simplified 2-dim scheme as $\sin^2(2\Theta)$. Fig. 3(a) shows the results of this analysis in terms of oscillation events as function of Δm^2. The best fit line is compatible with zero. Applying a unified frequentist approach [7] leads to the given 90% C.L. upper limit which can be compared to what one would expect as signal strength in KARMEN taken the evidence from LSND. These KARMEN results are in clear contrast to the LSND evidence. However, KARMEN does not rule out completely the parameters favored by LSND. Therefore, two problems have to be addressed: What is the level of compatibility of both experiments? Assuming statistical compatibility, what are the oscillation parameters accepted by both experiments? A detailed combined statistical analysis [8] based on an earlier study with intermediate data sets [9] has been performed to answer these questions. To summarize, LSND and KARMEN are incompatible at a level of 36% confidence. Assuming statistical compatibility, all parameter combinations with $\Delta m^2 > 1\,\text{eV}^2$ are excluded apart from a little 'island' at $\approx 7\,\text{eV}^2$.

LIMITS ON LF VIOLATING μ^+ DECAYS

The LF number violating decay mode $\mu^+ \to e^+ + \bar{\nu}_e + {}^{(-)}_{\nu}$ is allowed in many extensions of the SM, e.g. the decay mode $\mu^+ \to e^+ + \bar{\nu}_e + \nu_\mu$ in left–right (LR) symmetric models [11], supersymmetric models with R parity violation [12], or the decay mode $\mu^+ \to e^+ + \bar{\nu}_e + \bar{\nu}$ in extensions involving additional scalar multiplets [13] (fig. 4). With the rest masses much smaller than the energy of all neutrinos emitted in $\mu^+ \to e^+ + \bar{\nu}_e + {}^{(-)}_{\nu}$, an analytical description of the neutrino spectra similar to the SM one can be applied, with the spectral parameter $\tilde{\rho}$ to be specified, replacing the SM Michel parameter ρ.

Although the energy scale of LR symmetry of weak interactions or the appearance of supersymmetric particles is expected to be in the range of 0.1–1 TeV, these extensions could manifest in small branching ratios at energy scales of muon decays at rest and lead to an excess of $\bar{\nu}_e$-induced events compared to SM predictions, i.e. explain the $\bar{\nu}_e$ signal seen by LSND. For details of the KARMEN analysis towards such μ^+ decays, we refer to [10].

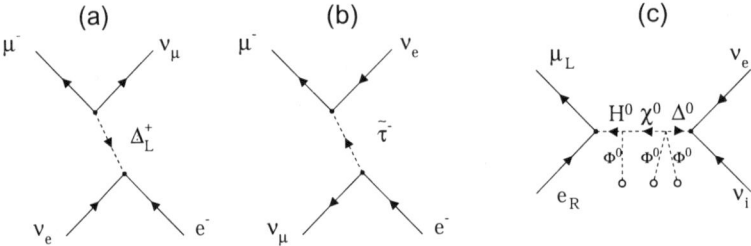

FIGURE 4. LF violating μ^+ decays in different extensions of the SM, with the exchange of (a) charged Higgs triplet fields [11] in left-right symmetric models, (b) supersymmetric partners of the τ lepton [12] in Supersymmetry or (c) scalar multiplets [13]

Figure 5 shows the prompt energy of the 15 reduced sequences together with the results of the maximum likelihood analysis for different kinematical parameters $\tilde{\rho}$ (as expected for models [11] or [13]) for the emitted $\bar{\nu}_e$. Table 1 shows the results of the likelihood method. They are consistent with no $\bar{\nu}_e$ emission from μ^+ decay with upper limits extracted within a unified frequentist analysis near the physical boundary $N(\bar{\nu}_e)=0$ following [7]. The above limits on the branching ratio BR on μ^+ decays emitting $\bar{\nu}_e$ improve by more than an order of magnitude the most sensitive limit so far of $BR(\mu^+ \to e^+ + \bar{\nu}_e + \nu_\mu) < 0.012$ obtained by the E645 experiment at LAMPF [14]. The most conservative upper limit of $BR < 1.7 \cdot 10^{-3}$ for any parameter $\tilde{\rho}$ is also in direct experimental disagreement with the possibility that the beam excess of $\bar{\nu}_e$ seen in the LSND experiment with a branching ratio or probability of $P = (2.64 \pm 0.67 \pm 0.45) \cdot 10^{-3}$ [1] is due to μ^+ decays with $\bar{\nu}_e$ emission.

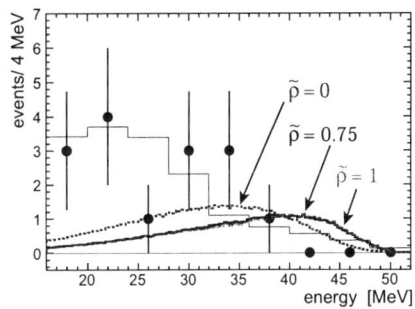

FIGURE 5. Visible energy distribution of candidate events with background expectation (shaded area). The lines show the 90% C.L. limit for an additional $\bar{\nu}_e$ signal with different spectral parameters $\tilde{\rho}$.

TABLE 1. Flux averaged cross section $\langle\sigma\rangle$ for $p(\bar{\nu}_e,e^+)n$ and $^{12}C(\bar{\nu}_e,e^+n)^{11}B$, expected (e^+,n) sequences for μ^+ decaying entirely via $\mu^+ \to e^+ + \bar{\nu}_e + {}^{(-)}{\nu}$, experimental results for potential $\bar{\nu}_e$-induced events and deduced upper limits for the branching ratio.

$\tilde{\rho}$	$\langle\sigma\rangle[10^{-42}\text{cm}^2]$		$N(\bar{\nu}_e)_{BR=1}$	$N(\bar{\nu}_e)_{\text{bestfit}}$	$N(\bar{\nu}_e)_{90\%CL}$	BR (90% C.L.)
	$\bar{\nu}_e + p$	$\bar{\nu}_e + {}^{12}C$				
0.0	72.0	4.5	4304 ± 403	+0.3	< 7.1	< 1.7 · 10^{-3}
0.25	78.8	5.8	4773 ± 445	−0.1	< 6.2	< 1.3 · 10^{-3}
0.5	86.0	7.2	5273 ± 489	−0.4	< 6.0	< 1.1 · 10^{-3}
0.75	93.5	8.5	5828 ± 538	−0.8	< 5.3	< 0.9 · 10^{-3}

The data presented here are the results of years of experimental work and data analysis. I am deeply indebted to all the colleagues of the KARMEN collaboration for their expertise and enthusiasm in running a long term experiment successfully. It is a great pleasure to thank the organizers of the CIPANP conference, especially the coordinators of an exciting session on neutrinos, Bruce Berger and Bonnie Fleming, for a very stimulating and interesting meeting.

REFERENCES

1. A. Aguilar et al., Phys. Rev. D **64**, 112007 (2001).
2. P. Astier et al., hep-ex/0306037.
3. S. Avvakumov et al., Phys. Rev. Lett. **89**, 011804 (2002).
4. G. Drexlin et al., Nucl. Instrum. Methods A **289**, 490 (1990).
5. G. Drexlin, Prog. Part. Nucl. Phys. **40**, 193 (1998).
6. B. Armbruster et al., Phys. Rev. D **65**, 112001 (2002).
7. G.J. Feldman and R.D. Cousins, Phys. Rev. D **57**, 3873 (1998).
8. E. Church et al., Phys. Rev. D **66**, 013001 (2002).
9. K. Eitel, New Jour. Phys. **2**, 1.1 (2000).
10. B. Armbruster et al., Phys. Rev. Lett. **90**, 181804 (2003).
11. P. Herczeg and R.N. Mohapatra, Phys. Rev. Lett **69**, 2475 (1992).
12. A. Halprin and A. Masiero, Phys. Rev. D **48**, R2987 (1993).
13. K.S. Babu and S. Pakvasa, arXiv:hep-ph/0204236.
14. K. Hagiwara et al., Phys. Rev. D **66**, 010001 (2002).

Neutrino Oscillation Search at MiniBooNE

T. Hart* for the MiniBooNE Collaboration

University of Colorado, Department of Physics, Campus Box 390, Boulder, CO 80309

Abstract. The MiniBooNE experiment at Fermilab will search for $\nu_\mu \to \nu_e$ oscillations, and is designed specifically to confirm or refute the LSND observation of neutrino oscillations. We present an overview of the MiniBooNE beamline and detector and conclude with future analysis plans.

MOTIVATION AND OVERVIEW

The latest neutrino experiments have provided strong evidence for the phenomena of neutrino oscillations in which a neutrino will change type while in flight. In the approximation of oscillation from one neutrino type to another, the probability of oscillation is given by

$$Prob(\nu_\alpha \to \nu_\beta) = (sin^2 2\theta)(sin^2 \frac{1.27 \Delta m^2 L}{E}) \qquad (1)$$

where θ is the mixing angle between the weak and mass eigenstates, Δm^2 is the difference in the square of the masses of the two neutrino types, and L and E are the flight path length and energy of the oscillating neutrino.

The most well established neutrino oscillation results are from solar neutrino experiments (SNO [1, 2], KamLAND [3]) and atmospheric neutrino experiments (Super-Kamiokande [4], Soudan [5]) which measure Δm^2 values of approximately $10^{-5} eV^2$ and $10^{-3} eV^2$ respectively. These experiments establish two independent Δm^2 values consistent with three neutrino types. However, the LSND experimental result [6], which to date has not yet been independently verified, yields a third Δm^2 value of approximately $1 eV^2$. Possible explanations of an extra Δm^2 include that one or more experiments is observing something that's not neutrino oscillation, the existence of extra neutrinos that do not feel the weak force ('sterile' neutrinos), or perhaps even CPT violation.

The MiniBooNE experiment is designed to confirm or refute the LSND result. MiniBooNE will search for the oscillation of muon neutrinos to electron neutrinos (LSND reported an anti-muon neutrino to anti-electron neutrino oscillation measurement). This cross-check will offer an important clarification to the understanding of neutrino behavior. Figure 1 shows the LSND result and our expected sensitivities for both a neutrino beam (our current running condition) and for a possible future anti-neutrino beam.

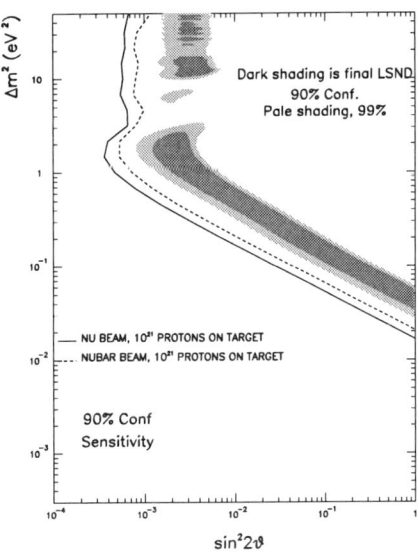

FIGURE 1. Expected Δm^2-$\sin^2 2\theta$ sensitivity for MiniBooNE and the LSND oscillation result.

THE BEAMLINE

MiniBooNE detects neutrinos from decays of pions and kaons from proton-beryllium collisions and neutrinos from decays of muons from pions and kaons. The 8 GeV kinetic energy proton beam (the primary beam) used by MiniBooNE is extracted from the Booster, one of the accelerators in the chain of Fermilab accelerators. The protons arrive in 1.6 μs wide pulses at a maximum rate of 15 Hz. Each pulse is subdivided into 2 ns wide buckets spaced 19 ns apart. The beryllium target is located in a magnetic horn which focuses the pions and kaons (the secondary beam) from the proton-beryllium collisions. As the neutrinos come from the secondary beam, the focusing of the pions and kaons increases the neutrino flux at the MiniBooNE detector (541 m from the target) approximately seven-fold. Between the target and detector is a permanent absorber of the pions, kaons, and muons (so that only neutrinos pass through) about 50 m from the target and a removable absorber located 25 m from the target. Figure 2 shows the main components of the MiniBooNE beamline.

Our neutrino flux is almost entirely of muon neutrinos with an approximately 0.3% electron neutrino component (from muon and kaon decays). However, the expected number of electron neutrinos from oscillations is comparable to the number of electron neutrinos expected from kaon and muon decays. Accounting for these background electron neutrinos to the oscillation signal is of vital importance in obtaining a reliable oscillation measurement.

The measurement of the intrinsic electron neutrino content will be approached in several ways. Measurement of the muon neutrino rate at the detector will provide kinematic constraints on the intrinsic electron neutrino rate. MiniBooNE has the option

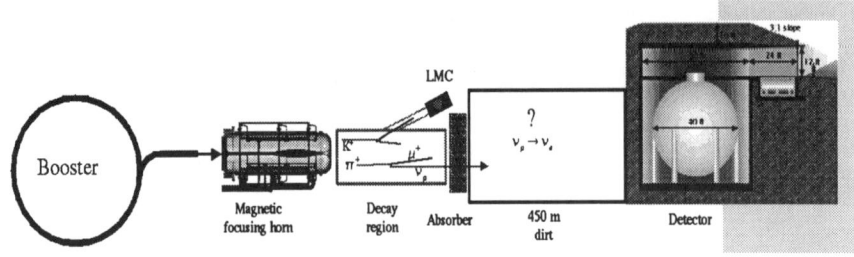

FIGURE 2. The MiniBooNE beamline.

of stopping the secondary beam with a 25 m (instead of the nominal 50 m) absorber which should reduce the muon neutrino and electron neutrino rates at the detector in a predictable way. We will also take advantage of improved particle production modeling using GEANT4 [7] and cross section measurements from HARP [8] and BNL910 (latest publication is [9]).

MiniBooNE is also commissioning a permanent magnet fiber spectrometer located 7° off-axis from the secondary beam line to constrain the kaon content of the secondary beam which will, in turn, provide valuable information about the number of electron neutrinos from kaons. The spectrometer, named the Little Muon Counter or LMC, will measure the momenta of muons decaying from pions and kaons. As kaons have a much higher mass than pions, muons from kaons have a much higher maximum transverse momentum than pionic muons. Therefore, the parentage of muons which reach the LMC after traveling at a large angle off the beam axis will be ascertained by their momentum; muons from kaons have a momentum of about 2 GeV, and muons from pions have a momentum of about 0.2 GeV. Currently, a temporary detector consisting of photomultiplier tubes behind various amounts of steel is providing good tests of our data readout system as well as a preliminary measurement of the overall muon rate at the LMC site.

THE DETECTOR

The MiniBooNE detector consists of a 40' diameter sphere filled with 250,000 gallons of mineral oil and lined with 1280 optically isolated 8" photomultiplier tubes pointing inward and a veto region consisting of 240 photomultiplier tubes pointing outward. Figure 3 shows the MiniBooNE detector. Neutrinos are detected indirectly by the Čerenkov light that comes from neutrino interactions with carbon nuclei of the oil. Electron neutrino, muon neutrino, and neutral pion events are distinguished by characteristic Čerenkov light signatures.

The calibration of MiniBooNE is ongoing along with development and testing of track reconstruction and particle identification algorithms. Laser pulses provide light of known duration and intensity to calibrate the time and charge measurements of the photomultiplier tubes. Our timing resolution is about 1.8 ns for the photomultiplier tubes that were previously used in the LSND experiment and about 1.2 ns for newer tubes. Our

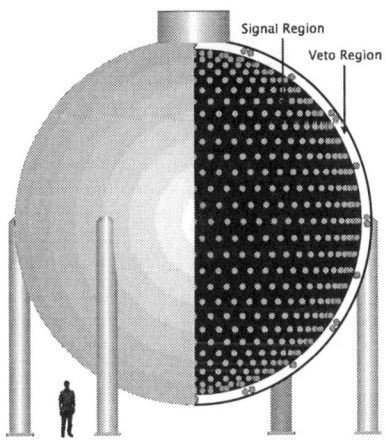

FIGURE 3. The MiniBooNE Detector.

charge resolution is about 15% at one photoelectron. Electrons from muon decays help to develop and test low energy track measurements and electron particle identification routines. From energy measurements of electrons from muon decays, we know that the energy resolution is about 15%. Further work in our energy reconstruction routines will improve this resolution. MiniBooNE has demonstrated good agreement between Monte Carlo simulations and data measurements of vertex locations and energy measurements, and we have made preliminary measurements of the neutral pion mass and the muon lifetime in oil. Figure 4 shows our preliminary mass reconstruction.

FUTURE PLANS

To date, MiniBooNE has accumulated about 1.3×10^{20} protons on target, roughly 13% of the 10^{21} protons on target goal. We expect to reach full proton intensity this Fall. Our analysis goals include cross sections and exotic process measurements, a preliminary muon neutrino disappearance measurement by this Fall, and a ν_e oscillation result sometime in 2005.

REFERENCES

1. Q. R. Ahmad *et al.* [SNO Collaboration], Phys. Rev. Lett. **89**, 011301 (2002) [arXiv:nucl-ex/0204008].
2. Q. R. Ahmad *et al.* [SNO Collaboration], Phys. Rev. Lett. **89**, 011302 (2002) [arXiv:nucl-ex/0204009].

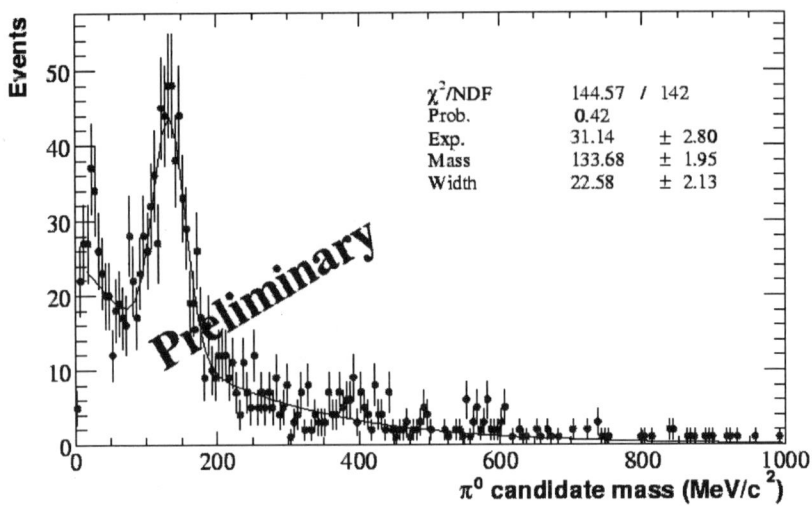

FIGURE 4. Preliminary π^0 mass from reconstruction of MiniBooNE detector events.

3. K. Eguchi et al. [KamLAND Collaboration], Phys. Rev. Lett. **90**, 021802 (2003) [arXiv:hep-ex/0212021].
4. S. Fukuda et al. [Super-Kamiokande Collaboration], Phys. Lett. B **539**, 179 (2002) [arXiv:hep-ex/0205075].
5. W. A. Mann [Soudan-2 Collaboration], Nucl. Phys. Proc. Suppl. **91**, 134 (2000) [arXiv:hep-ex/0007031].
6. A. Aguilar et al. [LSND Collaboration], Phys. Rev. D **64**, 112007 (2001) [arXiv:hep-ex/0104049].
7. S. Agostinelli et al. [GEANT4 Collaboration], Nucl. Instrum. Meth. A **506**, 250 (2003).
8. E. Radicioni [HARP Collaboration], arXiv:hep-ex/0206028.
9. S. Mioduszewski [E910 Collaboration], Nucl. Phys. A **698**, 595 (2002) [arXiv:nucl-ex/0105005].

$\bar{\nu}_e$ and the Sudbury Neutrino Observatory

J. L. Orrell on behalf of the SNO Collaboration

jorrell@u.washington.edu   http://www.sno.phy.queensu.ca/

Abstract. Neutrino oscillation results from KamLAND [1], the Sudbury Neutrino Observatory (SNO) [2], and Super-Kamiokande [3] provide evidence for neutrino mass. Determination of the Dirac or Majorana nature of neutrinos is an important next step in neutrino physics. An electron antineutrino, $\bar{\nu}_e$, component of the solar neutrino flux would provide a telltale sign neutrinos are Majorana particles. The SNO Collaboration is currently searching for an $\bar{\nu}_e$ signal, intending to measure or limit the flux of $\bar{\nu}_e$ in the solar neutrino energy range. A method for increasing the fiducial volume and lowering the analysis energy threshold using the time coincidence signature of the product particles of the charged current weak interaction of a $\bar{\nu}_e$ with a deuterium nucleus, $\bar{\nu}_e + d \rightarrow e^+ + n + n$, is presented.

HISTORY AND LIMITS ON SOLAR $\nu_e \rightarrow \bar{\nu}_e$ CONVERSION

Massive neutrinos may have non-zero magnetic moments, $\mu_\nu < 1.5 \times 10^{-10} \mu_B$ [4]. If neutrinos are Dirac particles, a magnetic moment interaction with a magnetic field transverse to the neutrino's momentum induces spin precession from a left-handed state, ν_{xL}, to a sterile right-handed state, ν_{xR}, where $x = e$, μ, τ. If neutrinos are Majorana particles, only a transition magnetic moment is possible for which a magnetic field interaction both induces spin precession and changes the neutrino's flavor, $\nu_{xL} \rightarrow \nu_{yR}$. The Majorana ν_{xR} is active as it is identified as the antineutrino, $\nu_{xR} \equiv \bar{\nu}_{xR}$. These ideas have guided researchers for over 20 years in contemplating solar neutrino interactions with solar magnetic fields.

The hypothesis of solar magnetic field induced conversion of solar ν_{eL} to sterile ν_{eR} was first proposed as a solution to the neutrino deficit measured by the Homestake ^{37}Cl experiment [5]. Later, it was proposed to account for an apparent time (*anti*)correlation between the ^{37}Cl experiment's neutrino signal and solar magnetic field activity as quantified by sunspot number [6]. It was realized the combination of matter enhanced neutrino flavor oscillation (MSW) and spin precession (SP) (Dirac case) or spin-flavor precession (SFP) (Majorana case) provides a single description including both significant ν_e depletion and a time varying signal [7]. The combination of these effects are known as *resonant* conversion schemes (RSP & RSFP) [8]. More realistic descriptions include a varying magnetic field direction along the neutrino's path [9]. Current conversion schemes have five characteristics:

- Focus on Majorana neutrino signature rather than Dirac sterile admixture.
- Matter enhanced flavor oscillations (MSW).
- Hypothetical solar magnetic field configurations with varying field direction.
- Consideration of two conversion scenarios: $\nu_e \xrightarrow{\text{SFP}} \bar{\nu}_x \xrightarrow{\text{Osc}} \bar{\nu}_e$ OR $\nu_e \xrightarrow{\text{Osc}} \nu_x \xrightarrow{\text{SFP}} \bar{\nu}_e$

- Two experimentally unconfirmed assumptions:
  - A neutrino magnetic moment on the order of $\mu_V \approx 10^{-10} \mu_B$.
  - Solar magnetic field strengths of 10-100 kGauss. The actual interior field strength is not known. Predictions range from 10-10^6 Gauss [10].

Several authors have used available experimental data to search for a solar $\bar{\nu}_e$ flux, $\Phi_{\bar{\nu}_e}$. Table 1 shows limits on $\Phi_{\bar{\nu}_e}$ assuming the $\nu_e \to \bar{\nu}_e$ conversion does not alter the initial ^{8}B neutrino energy spectrum. Note the 2003 limit is 0.38% of the total

TABLE 1. Limits on the solar $\bar{\nu}_e$ flux, $\Phi_{\bar{\nu}_e}$, assuming a ^{8}B flux spectrum. References in brackets.

Year	$\Phi_{\bar{\nu}_e} \left(\frac{10^4}{cm^2 s}\right)$	CL	Note	Year	$\Phi_{\bar{\nu}_e} \left(\frac{10^4}{cm^2 s}\right)$	CL	Note
2003	≤ 1.92 [11]	95%	KamLAND data	1997	≤ 19.8 [14]	95%	Super-K data
2003	≤ 4.04 [12]	90%	Super-K Coll.	1996	≤ 10 [15]	90%	LSD Coll.
2000	≤ 18 [13]	95%	Super-K data	1991	≤ 30.3 [16]	99%	Kamiokande data

produced ^{8}B electron neutrino flux. Limits on $\Phi_{\bar{\nu}_e}$ and neutrino oscillation results from KamLAND [1] and SNO [2] show $\nu_e \to \bar{\nu}_e$ conversion is not a dominant process in the Sun. However, it remains true that detection of solar $\bar{\nu}_e$'s is a sign neutrinos are Majorana particles. In light of the discovery of massive neutrinos, determination of the Dirac or Majorana nature of the neutrino is a next major step in understanding the properties of the neutrino, thus searching for solar $\bar{\nu}_e$'s remains a valuable endeavor.

THE SUDBURY NEUTRINO OBSERVATORY'S $\bar{\nu}_e$ SENSITIVITY

SNO detects $\bar{\nu}_e$ through charged current weak interactions on deuterons, $\bar{\nu}_e + d \to e^+ + n + n$. This $\overline{CC}$ reaction is distinguishable via the time coincidence between the

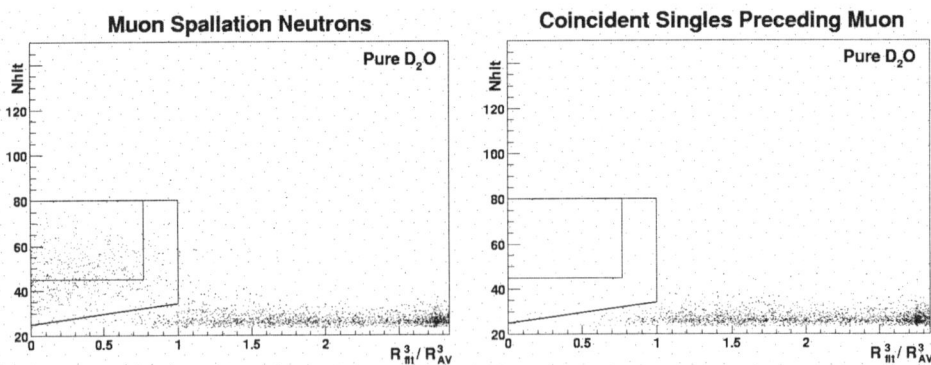

FIGURE 1. Example neutron analysis regions for a $\bar{\nu}_e$ search. N_{hit} measures an event's energy. $R^3_{\text{fit}}/R^3_{\text{AV}}$ is an event's reconstructed radial position divided by the radius of the acrylic vessel (AV) containing the D_2O target volume. Here muon spallation neutrons are used as a "calibration" source. The small rectangle is a neutron selection window derived from the analysis presented in Ref. [2] (*standard* analysis). The large trapezoid is proposed to increase SNO's $\bar{\nu}_e$ sensitivity (*enlarged* analysis). The coincidence singles show the background when no neutrons are present. Both plots show events within 0.5 seconds of the muon's passage. Solar $\overline{CC}$ product positrons will fill the region to the left of $R^3_{\text{fit}}/R^3_{\text{AV}} = 1$.

prompt Cherenkov radiating positron, e^+, and the millisecond delayed captures of the product neutrons, n. Figure 1 describes *standard* and *enlarged* analysis regions in an energy and volume phase-space. During the D_2O phase of SNO, the *standard* analysis's neutron and positron detection efficiencies are $\varepsilon_n \simeq 0.14$ and $\varepsilon_{e^+} \simeq 0.40$ (ε_{e^+} is based on a ^8B neutrino spectrum) leading to a $\overline{CC}$ detection efficiency of $\varepsilon_{\overline{CC}} \simeq 0.12$. The proposed *enlarged* analysis would increase $\varepsilon_{e^+} \simeq 0.67$. During the Salt phase of SNO, ε_n increases significantly and for the *enlarged* analysis may reach $\varepsilon_n \simeq 0.76$. Thus, during the Salt phase, the *enlarged* analysis's $\overline{CC}$ detection efficiency is potentially $\varepsilon_{\overline{CC}} \simeq 0.82$. A no-signal, no-background calculation of the limit on $\Phi_{\bar{\nu}_e}$ set by one year of data, estimates SNO's sensitivity to solar $\bar{\nu}_e$. This limit for a D_2O phase, *standard* analysis is $\Phi_{\bar{\nu}_e} \leq 2.68$ ($10^4/cm^2 s$). This limit for a Salt phase, *enlarged* analysis is $\Phi_{\bar{\nu}_e} \leq 0.39$ ($10^4/cm^2 s$). Accounting for $\overline{CC}$ backgrounds (not discussed here) raises these upper limits, however it is clear SNO is competitive with current limits on $\Phi_{\bar{\nu}_e}$.

ACKNOWLEDGMENTS

This work is presented on behalf of the SNO Collaboration under the auspices of NSERC, Industry Canada, NRC, Northern Ontario Heritage Fund Corporation, Inco, AECL, Ontario Power Generation, US Dept. of Energy, and PPARC.

REFERENCES

1. K. Eguchi et al., Phys. Rev. Lett. **90**, 021802 (2003).
2. Q. R. Ahmad et al., Phys. Rev. Lett. **89**, 011301 (2002); Q. R. Ahmad et al., Phys. Rev. Lett. **89**, 011302 (2002).
3. Y. Fukuda et al., Phys. Rev. Lett. **81**, 1562 (1998); Y. Fukuda et al. Phys. Rev. Lett. **82**, 2644 (1999); S. Fukuda et al., Phys. Rev. Lett. **85**, 3999 (2000).
4. K. Hagiwara et al., Phys. Rev. D **66**, 010001 (2002).
5. A. Cisneros, Astrophys. Space Sci. **10**, 87 (1971).
6. M. B. Voloshin and M. I. Vysotskii, Sov. J. Nucl. Phys. **44**, 544 (1986); L. B. Okun, Sov. J. Nucl. Phys. **44**, 546 (1986).
7. M. B. Voloshin, M. I. Vysotskii and L. B. Okun, JETP **64**, 446 (1986).
8. E. Kh. Akhmedov and O. V. Bychuk, JETP **68**, 250 (1989); C.-S. Lim and W. J. Marciano, Phys. Rev. D **37**, 1368 (1988); A. B. Balantekin, P. J. Hatchell and F. Loreti, Phys. Rev. D **41**, 3583 (1990).
9. J. Vidal and J. Wudka, Phys. Lett. B **249**, 473 (1990); A. Yu. Smirnov, Phys. Lett. B **260**, 161 (1991); S. Toshev, Phys. Lett. B **271**, 179 (1991); C. Aneziris and J. Schechter, Phys. Rev. D **45**, 1053 (1992); E. Kh. Akhmedov, S. T. Petcov and A. Yu. Smirnov, Phys. Rev. D **48**, 2167 (1993); A. B. Balantekin and F. Loreti, Phys. Rev. D **48**, 5496 (1993); T. Kubota, T. Kurimoto and T. Eiichi, Phys. Rev. D **49**, 2462 (1994); E. Torrente-Lujan, Phys. Rev. D **59**, 093006 (1999).
10. C. P. Burgess, N. S. Dzhalilov, T. I. Rashba, V. B. Semikoz and J. W. F. Valle, arXiv:astro-ph/0304462 (2003).
11. B. C. Chauhan, J. Pulido and E. Torrente-Lujan, arXiv:hep-ph/0304297 (2003).
12. Y. Gando et al., Phys. Rev. Lett. **90**, 171302 (2003).
13. E. Torrente-Lujan, Phys. Lett. B **494**, 255 (2000).
14. G. Fiorentini, M. Moretti and F. L. Villante, arXiv:astro-ph/9707097 (1997).
15. M. Aglietta et al., JETP **63**, 791 (1996).
16. R. Barbieri, G. Fiorentini, G. Mezzorano and M. Moretti, Phys. Lett. B **259**, 119 (1991).

Inelastic Cross-Sections for Neutrinos and Two Nucleons at Low Energy in Effective Field Theory

Malcolm N. Butler

Department of Astronomy and Physics, Saint Mary's University, Halifax NS Canada B3H 3C3

Abstract. Proton-proton fusion and neutrino-deuteron breakup are key reactions in the standard solar model and for the resolution of the solar neutrino problem. However, cross-sections for both processes rely on our understanding of the theory of two-nucleon dynamics. Here I discuss how well we understand these cross-sections, and the implications of any uncertainty therein.

INTRODUCTION

The proton-proton (pp) fusion reaction $pp \to De^+\nu_e$ is the most fundamental process in the standard solar model. Neutrino-deuteron breakup $\nu_x D \to lNN$ has been used by the Sudbury Neutrino Observatory (SNO) to confirm the hypothesis of neutrino oscillations as the solution to the solar neutrino problem [1]. pp-fusion and neutrino-deuteron breakup are related processes, and share a common problem; neither has been measured to any reasonable precision in the laboratory. As a result, both the standard solar model and SNO rely on theory to provide them with values for these cross-sections. The most difficult question for theory to answer is how precise the calculations are. This work reviews how effective field theory can be used to gain some insight into this question.

The standard approach to studying both the fusion and breakup processes is to use the potential model [2, 3, 4, 5, 6, 7, 9]. Here, potentials extracted from high-precision fits to nucleon-nucleon scattering data are used to construct initial and final-state wavefunctions for the two-nucleon system. The scattering matrix elements are calculated using these wavefunctions and weak-interaction current operators. These operators have two components. First, the 'one-body' contribution comes from the weak interaction coupling to one nucleon while the other nucleon is a passive spectator (Fig. 1a)). The second 'two-body' contribution comes from processes where both nucleons are involved in the coupling to the external weak current. Examples of such contributions are shown in Fig. 1, where the weak current couples to the $NN\pi$ vertex directly (1b), or the weak current excites a virtual Δ, which decays through the exchange of a pion with the spectator nucleon (1c). Both diagrams involve degrees of freedom not represented in the NN wavefunctions and are thus part of the irreducible two-body operator.

The one-body contributions at low energy are well-constrained, and little uncertainty lies there. The issue is with the two-body contributions. These are constructed within models and contain unknown parameters that must be constrained from other processes.

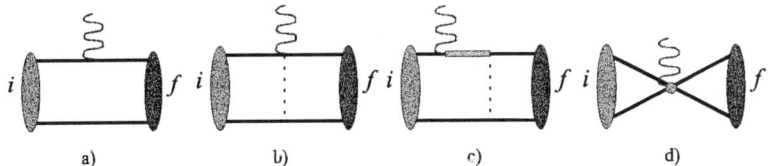

FIGURE 1. One and two-body contributions to weak interactions involving deuterium. i and f represent the initial and final two-body states, respectively, calculated either in a potential model or a field-theoretical framework. a) The one-body current. b) Two-body currents arising from gauging the derivative-coupling to pions. c) Two-body currents arising from excitation of the Δ resonance. d) Examples of a contact operator arising in effective field theory.

So, uncertainties have two sources. The first, inherent (and difficult to quantify) uncertainties in the model; and the second, uncertainties (experimental and systematic) associated with constraining parameters in the two-body operator.

The unknown parameters in the two-body operator are usually constrained by Tritium β-decay [8, 9]. Here, two-body matrix elements are extracted from the three-body system using the well-determined half-life of Tritium. In principle, this is the most precise data available that has any sensitivity to the two-body operators. However, the precision of constraint of the two-body operator is usually taken to the precision of the half-life measurement. This ignores two concerns: firstly, that the two-body matrix elements are off-shell in the three-body system, and there might be a three-body operator present that complicates the extraction; and secondly, that we do not know the systematic uncertainties introduced due to our less detailed understanding of the three-body system (the role of three-body forces, etc.).

EFFECTIVE FIELD THEORY AND CONSTRAINTS

Here, we use a particular formulation of two-nucleon effective field theory (EFT) to study weak interactions involving the deuteron [10]. The advantage is that the EFT is perturbative, so that there is a systematic power-counting scheme for the theory, ensuring convergence at low energies. Uncertainties in a calculation can be determined from the convergence of the expansion and the intrinsic uncertainties in the parameters used. Theoretical uncertainties are now a well-defined quantity.

Two-nucleon scattering is described by momentum-dependent contact interactions that lead to a systematic perturbative expansion at low energy that matches directly onto the effective range expansion. External currents are introduced easily at the one-nucleon level, and two-body current operators appear naturally as counterterms in the perturbative expansion [11] (see Fig. 1d)). For the processes studied here, there is one counterterm of import corresponding to an axial isovector two-body current, whose strength we parameterize by $L_{1,A}$. It first appears at NLO. To NNLO, the cross-sections for process i can be parameterized as

$$\sigma_i(E) = A_i(E) + B_i(E)L_{1,A} + C_i(E)L_{1,A}^2 + \mathcal{O}(3\%), \qquad (1)$$

where $A_i(E)$, $B_i(E)$, and $C_i(E)$ are all defined in terms of known parameters [12, 13, 14]. E is the energy of the incident neutrino or antineutrino for breakup reactions. $E = 0$ for pp-fusion. Dimensional analysis of $L_{1,A}$ suggests that it is of order 6 fm^3. At this scale, two-body physics represents a 5-8% contribution to fusion and breakup. However, we have no true measure of $L_{1,A}$, or its uncertainty, as yet.

Reactor antineutrino scattering

A number of experiments probe antineutrino breakup of deuterium using reactor antineutrinos [15, 16, 17]. Here, the flux is well-calibrated through the power output of the reactor. Both charged and neutral current cross-sections can be measured. The experiments themselves never provide uncertainties better than 10%, however, they can be used to represent a global dataset and we can extract a 'world' value for $L_{1,A}$ of 3.6±5.5 fm^3, including both experimental uncertainties and an intrinsic theoretical uncertainty from truncating our calculation at NNLO [18].

Helioseismology

Another interesting constraint on $L_{1,A}$ comes from the solar model itself. Helioseismology serves as a precise constraint on the dynamics of the solar interior, and could thus be sensitive to variations in the pp-fusion cross-section. We have varied that cross-section by varying $L_{1,A}$, while keeping the static properties of the sun fixed. Fig. 2 shows the deviation of these models relative to the data (from the BiSON collaboration [20], and the variation shown represents that which maintains acceptable agreement [19]. The corresponding constraint on $L_{1,A}$ is 4.8±6.7 fm^3.

SNO self-calibration

One of the most interesting constraints on $L_{1,A}$ comes from using the SNO data to calibrate the breakup cross-sections [21]. Here, the active neutrino flux detected in the neutral current channel is assumed to have an undistorted ^{8}B spectral shape, meaning that there are three unknowns: the total charged current flux, the total neutral current flux, and $L_{1,A}$. There are three distinct sets of data corresponding to this (charged current events, neutral current events, and elastic scattering events), meaning that $L_{1,A}$ can indeed be constrained from the data. This fit produces a value of $L_{1,A} = 4.0 \pm 6.3$ fm^3, while maintaining the significance of SNO's detection of neutrino oscillations. A more recent work again considers this possibility, and also the idea of using SNO solely as a measure of the $L_{1,A}$ [22].

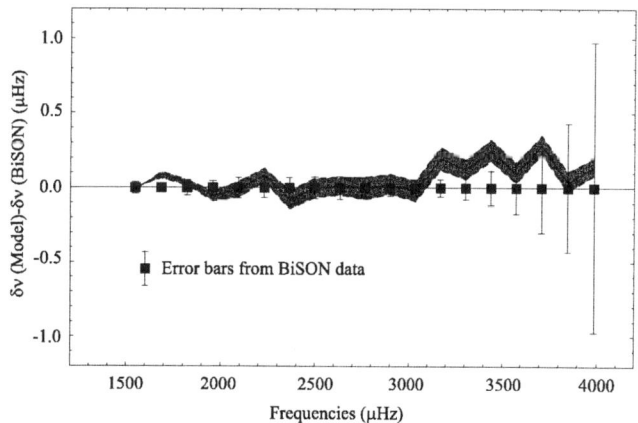

FIGURE 2. Comparison of standard solar model results to experiment for $l = 0, 1$ p-modes. The error bars represent uncertainties in the measured small frequency spacings $\delta v(n,l) = v(n,l) - v(n-1, l+2)$ from the BiSON collaboration [20]. The shaded range represents acceptable models for various values of $L_{1,A}$.

Potential Model Comparisons

Finally, we calibrate the potential model calculations to our EFT expressions. If we treat a recent potential model calculation as data [7], we find that $L_{1,A} = 4.0$ fm^3 fits all four breakup channels to within 1%. Further, we can fit a recent pp-fusion calculation [9] using $L_{1,A} = 4.2$ fm^3. The results are similar if we compare to potential models where the two-body physics is parameterized using heavy baryon chiral perturbation theory [23, 24].

At these low energies, the potential model calculations are truly indistinguishable from our work in terms of physics input - the difference is that the potential model calculations have no means of calibrating their uncertainties.

SUMMARY

In summary, we find a set of independent constraints on the two-body counterterm that are all consistent, as shown in Table 1. The standard solar model is clearly robust against variations in the pp-fusion cross-section within these constraints because of the helioseismology analysis. Further, the SNO self-calibration shows that the uncertainty in the breakup cross-sections does not impact on their verification of solar neutrino oscillations in any significant way. However, as SNO's statistical precision improves, and more subtle analysis of the data is desired, then these uncertainties could become problematic. Thus, there is still a need to improve upon the theoretical uncertainties. A calculation of Tritium β-decay in EFT might resolve this, as the systematic questions can be addressed in a clean and unambiguous fashion.

TABLE 1. Summary of constraints on $L_{1,A}$ from different data and techniques.

Process	$L_{1,A}$ (fm^3)
Dimensional analysis	± 6
Reactor antineutrino scattering	3.6 ± 5.5
Helioseismology	4.8 ± 6.7
SNO self-calibration	4.0 ± 6.3
Potential model neutrino-deuteron breakup	4.0
Potential model pp-fusion	4.2

ACKNOWLEDGMENTS

The work presented here would not have been accomplished without the collaboration with Jiunn-Wei Chen. The author's work is supported by the Natural Sciences and Engineering Research Council (NSERC) of Canada.

REFERENCES

1. Q.R. Ahmad et al., *Phys. Rev. Lett.* **89**, 011301 (2002).
2. J.N. Bahcall, K. Kubodera, and S. Nozawa, *Phys. Rev.* D **38**, 1030 (1987).
3. S. Ying, W.C. Haxton, and E. M. Henley, *Phys. Rev.* C **45**, 1982 (1992); *Phys. Rev.* D **40**, 3211 (1989).
4. N. Tatara, Y. Kohyama and K. Kubodera, *Phys. Rev.* C **42**, 1694 (1990).
5. M. Doi and K. Kubodera, *Phys. Rev.* C **45**, 1988 (1992).
6. K. Kubodera and S. Nozawa, *Int. J. Mod. Phys.* **E3**, 101 (1994); Y. Kohyama and K. Kubodera, USC(NT)-Report-92-1, 1992, unpublished.
7. S. Nakamura, T. Sato, V. Gudkov, and K. Kubodera, *Phys. Rev.* C **63**, 034617 (2001).
8. S. Nakamura et al., *Nucl. Phys.* **A707**, 561 (2002).
9. R. Schiavilla et al., *Phys. Rev.* C **58**, 1263 (1998).
10. D.B. Kaplan, M.J. Savage and M.B. Wise, *Phys. Lett.* **B424**, 390 (1998); *Nucl. Phys.* **B534**, 329 (1998); *Phys.Rev.* C **59**, 617 (1999).
11. J.W. Chen, G. Rupak and M.J. Savage, *Nucl. Phys.* **A653**, 386 (1999).
12. M.N. Butler and J.W. Chen, *Nucl. Phys.* **A675**, 575 (2000).
13. M.N. Butler, J.W. Chen and X. Kong, *Phys. Rev.* C **63**, 035501 (2001).
14. M.N. Butler and J.-W. Chen, *Phys. Lett.* **B520**, 87 (2001).
15. A.G. Vershinsky et al., *JETF Lett.* **53**, 513 (1991).
16. Yu.V. Kozlov et al., *Phys. Atom. Nucl.* **63**, 1016 (2000).
17. S.P. Riley, Z.D. Greenwood, W.R. Kropp, L.R. Price, F. Reines, H.W. Sobel, Y. Declais, A. Etenko, and M. Skorokhatov, *Phys. Rev.* C **59**, 1780 (1998); Erratum, October 22, 2001 (unpublished).
18. M.N. Butler, J.-W. Chen, P. Vogel, *Phys. Lett.* **B549**, 26 (2002).
19. K.I.T. Brown, M.N. Butler, and D.B. Guenther, nucl-th/0207008.
20. W.J. Chaplin et al., *MNRAS* **308**, 424 (1999).
21. J.W. Chen, K.M. Heeger, and R.G.H. Robertson, *Phys. Rev.* C **67**, 025801 (2003).
22. A.B. Balantekin and H. Yuksel, hep-ph/0307227
23. T. Park et al., *Phys. Rev.* C **67**, 055206 (2003).
24. S. Ando et al., *Phys. Lett.* **B555**, 49 (2003).

Neutrino Interactions at Low Energy

E. A. Hawker

University of Cincinnati, P.O. Box 210011, Cincinnati, Ohio 45221, USA

Abstract.
  Understanding neutrino interactions at energies of a few GeV and lower is vitally important to understand many current and future neutrino experiments. Presented here are brief overviews of what is currently known about neutrino cross sections, what present day experiments such as K2K and MiniBooNE can teach us, and some future experiments that are being planned.

MODELS OF NEUTRINO INTERACTIONS

Neutrinos have become an important tool used to study a large variety of different topics in physics. Several of these topics, such as neutrino masses and mixings, CP violation in neutrino oscillations, and proton decay, could very well shed light on physics beyond the Standard Model. Many current experiments that are investigating these topics, such as SNO, Super Kamiokande, K2K, MiniBooNE, and MINOS, are sensitive to processes that occur at low neutrino energies (from a half to a few GeV). In order to more fully understand their data these experiments use theoretical models and Monte Carlo simulations of neutrino interactions in their analyses. Most of these Monte Carlos are proprietary, but two notable ones, NUANCE [1] and NEUGEN [2], have been made available to the general public. Most Monte Carlo simulations used by neutrino experiments contain common theoretical models for some processes, such as

- the Llewellyn Smith [3] model for free nucleon charged current quasi-elastic scattering,
- the Rein and Sehgal [4] model for resonant single pion production, and
- a standard PDF formula for DIS at high W and Q^2.

However for some processes, such as final state interactions, there are no generally accepted models that can be used in Monte Carlos.

Important processes

At low neutrino energies there are four types of dominant processes: charged current quasi-elastic (CCQE) scattering, resonant single pion production, deep inelastic scattering (DIS), and nuclear effects. While nuclear effects are not a type of neutrino interaction, they can be very important processes to understand in order to analyze neutrino data. The three neutrino processes can be seen in Figure 1.

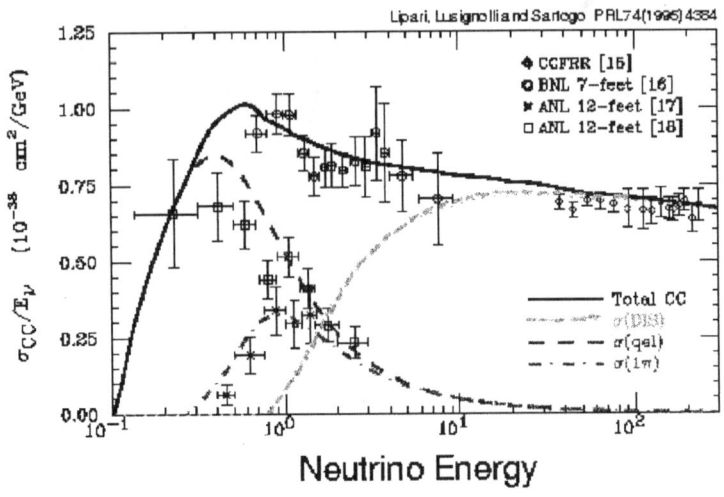

FIGURE 1. Neutrino charged current cross sections [5].

- Charged current quasi-elastic scattering is probably the best understood of these processes. From the spread in the experimental data there is about a 20% uncertainty in the cross section, however the uncertainty in the theoretical model is much smaller (several percent). Form factors and other constants in the model, such as the axial and vector masses M_A and M_V, in the model are determined from electron and neutrino scattering experiments, and it is the uncertainty in these constants that determines the uncertainty in the CCQE model. Recently a group of physicists, including A. Bodek, H. Budd, and J. Arrington, have started to extract more precise values of the CCQE form factors from updated electron and neutrino data [6].
- Resonant single pion production as modeled by Rein and Sehgal also agrees with experimental data to within 20% or so, but this comparison can only be made in the charged current channels. There is very little data available in the neutral current resonant single pion production channels so the uncertainties are larger [7].
- Resonant single pion production models at low Q^2 and DIS models at higher Q^2 both agree with data in their kinematic ranges, however the intersection of these two regions has been more difficult to model. One idea put forth by A. Bodek and U.K. Yang [8] is a model that extends parton distribution functions (PDFs) used for modeling DIS to much lower Q^2. Currently work is underway to try to incorporate this new method into both the NUANCE and NEUGEN Monte Carlos.
- Nuclear effects play an important role in understanding neutrino interactions because most current experiments use water, mineral oil, iron, or argon as a neutrino target in their detectors. Fermi motion and Pauli blocking have a larger effect at lower energy, near the threshold of processes, however final state interactions can be

important anywhere in this energy range, and they are more difficult to understand. These re-interactions can change the kinematics and types of particles observed in the final state, and therefore seriously affect the reconstruction and analysis of events.

CURRENT NEUTRINO EXPERIMENTS

MiniBooNE is a neutrino oscillation experiment at the Fermi National Accelerator Laboratory looking for the appearance of an excess of electron neutrinos in a beam of primarily muon neutrinos. MiniBooNE uses the 8 GeV proton beam from the Fermilab booster, a beryllium target, a pulsed toroidal field magnet (a horn) and a 50 meter decay region to form a neutrino beam with an average neutrino energy of about 1.0 GeV. The MiniBooNE detector is located 540 meters downstream of the target. The detector is a 12.2 meter diameter steel sphere filled with pure mineral oil as a neutrino target with a fiducial mass of 445 tons, and is instrumented with 1520 photomultiplier tubes.

A large number of different types of neutrino interactions should be seen at MiniBooNE, so there will be opportunities for measurements to be made on several processes. One process that MiniBooNE is especially interested in is neutral current single π^0 production, which is a background to the oscillation search. A preliminary result from this analysis is shown in Figure 2.

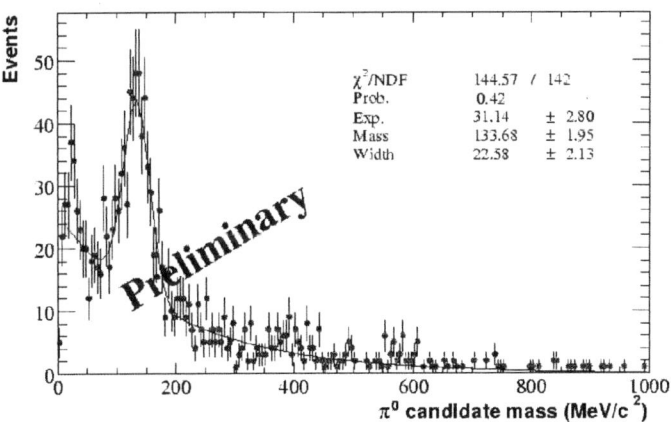

FIGURE 2. Preliminary results from MiniBooNE: the reconstructed invariant mass for candidate neutral current single π^0 events.

The K2K experiment is a neutrino oscillation experiment in Japan looking for the "disappearance" of muon neutrinos between the near detector at KEK and the far detector - Super Kamiokande. K2K uses the 12 GeV proton beam from the KEK PS, an aluminum target, a pair of pulsed horns, and a 200 meter decay region to produce a neutrino beam with an average neutrino energy of about 1.3 GeV. The K2K near detector array is located 300 meters downstream of the target. The near detector consists of a 1 kiloton (25 ton fiducial) water Cherenkov detector, a scintillating fiber detector, and a muon range stack.

In addition to their oscillation search, data collected with the K2K near detector could possibly allow several interesting measurements to be made on neutrino interactions. One example of this is an analysis the K2K collaboration has done on π^0 production. The ratio of the neutral current π^0 production rate to CCQE event rate was measured at the K2K near detector, and then compared to a Monte Carlo prediction [9]. The result that was found,

$$R_{\pi^0} = \frac{(\pi^0/\mu)_{data}}{(\pi^0/\mu)_{MC}} = 1.03 \pm 0.02 \,(stat) \pm 0.09 \,(syst), \qquad (1)$$

has allowed K2K to reduce the uncertainty on neutral current π^0 production in Super Kamiokande analyses allowing a better determination if atmospheric ν_μs are oscillating to ν_τ or to ν_s. The result of this analysis is shown in Figure 3.

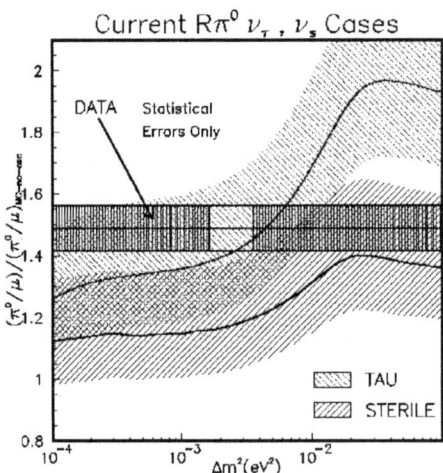

FIGURE 3. An example on how the uncertainty in neutral current single π^0 production affects an oscillation analysis for Super Kamiokande. The open box in the center is the allowed region [9], © 2002, reprinted with permission from Elsevier.

FUTURE NEUTRINO EXPERIMENTS

Two possible future experiments have submitted Expressions of Interest to the Fermilab PAC.

FINeSE (Fermilab Intense Neutrino Scattering Experiment) proposes to put another detector in the MiniBooNE neutrino beam closer to the neutrino source. The main goal of this experiment would be to use neutrino-nucleon elastic scattering to measure the strange quark contribution to the spin of the nucleon. However, the highly segmented target/detector and the high numbers of events that are being proposed would be ideal to make precision measurements of other processes [10].

MINeRVA (Main INjector ExpeRiment ν-A) proposes to put another detector in the NuMI neutrino beam, just in front of the MINOS near detector. The main goal of this experiment would be to measure total and differential cross sections and also to study nuclear effects in neutrino interactions. This experiment would also have a highly segmented target/detector, like FINeSE, but would also have the ability to insert into their detector a variety of different nuclear targets such as iron, carbon, aluminum, or others.

CONCLUSIONS

Understanding neutrino interactions is important for studying several important physics topics. At low neutrino energies modeling these interactions can be difficult and complex. However, as shown above there is a great amount work currently being done to improve the understanding of neutrino interactions both by acquiring more experimental data, and by improving the Monte Carlo models.

ACKNOWLEDGMENTS

This work was supported by the National Science Foundation under grant PHY-0070413

REFERENCES

1. D. Casper, proceedings of the First Workshop on Neutrino–Nucleon Interactions in the Few-GeV Region (NuInt01), Nucl. Phys. Proc. Suppl. 112 (2002) 161.
2. H. M. Gallagher and M. C. Goodman, NuMI-112 (1995), also proceedings of the First Workshop on Neutrino–Nucleon Interactions in the Few-GeV Region (NuInt01), Nucl. Phys. Proc. Suppl. 112 (2002) 188.
3. C. H. Llewellyn Smith, Phys. Rept. 3C, 261 (1972).
4. D. Rein and L. M. Sehgal, Annals of Physics, 133, 79 (1981).
5. P. Lipari, M. Lusignoli, and F. Sartogo, Phys. Rev. Lett. 74 (1995) 4384.
6. H. Budd, A. Bodek, and J. Arrington, hep-ex/0308005, Second International Workshop on Neutrino-Nucleus Interactions in the Few GeV Region (NuInt02), Dec. 2002. To be published in Nucl. Phys. B, Proceedings Suppl.
7. E. Hawker, Second International Workshop on Neutrino-Nucleus Interactions in the Few GeV Region (NuInt02), Dec. 2002. To be published in Nucl. Phys. B, Proceedings Suppl.
8. A. Bodek, and U.K. Yang, hep-ex/0308007, Second International Workshop on Neutrino-Nucleus Interactions in the Few GeV Region (NuInt02), Dec. 2002. To be published in Nucl. Phys. B, Proceedings Suppl.
9. C. Mauger, proceedings of the First Workshop on Neutrino–Nucleon Interactions in the Few-GeV Region (NuInt01), Nucl. Phys. Proc. Suppl. 112 (2002) 146.
10. M.O. Wascko, Neutrino Physics at FINeSE, these proceedings.

The Majorana project: determining $m_{\beta\beta}$ to the 50 meV scale through $\beta\beta$ measurements in ^{76}Ge

A. R. Young for the Majorana Collaboration

Physics Dept., Box 8202, NCState University, Raleigh, NC 27606

Abstract. The goal of the Majorana project is to provide limits on the effective neutrino mass at the level of 50 meV through measurements of neutrinoless double beta-decay in ^{76}Ge. The planned experiment relies entirely on proven technology and uses materials and techniques demonstrated to produce the lowest background per kilogram of fiducial germanium. It will utilize 500 kg of germanium detector material enriched to 86% ^{76}Ge. Each detector subunit will be segmented and instrumented for digital pulse acquisition, with both the segmentation and pulse digitization providing us with powerful signatures for background signals coming from processes other than neutrinoless double beta-decay. We present an overview of the Majorana project and include some details characterizing a prototype for our segmented detector elements: SEGA (the Segmented, Enriched Germanium Assembly).

INTRODUCTION

The Majorana experiment's proposed goal is to measure the effective mass of the electron neutrino to roughly the 0.05 eV level by measuring the rate for neutrinoless double beta-decay in ^{76}Ge to the ground state of ^{76}Se using well-established, Ge ionization detector technology. Our proposed mass limit corresponds to a half-life of about 4×10^{27} y for neutrinoless double beta-decay and can be reached by operating 500 kg of germanium enriched to 86% in ^{76}Ge in a well-shielded environment, deep underground. To achieve this goal, our proposed increase in mass over previous experiments of about a factor of 50 must be accompanied by a signficant decrease in radiological backgrounds. These backgrounds will be controlled and reduced by ultra-low-background screening and chemical processing of materials, by a combination of active and passive shielding, and by identification and "electronic" rejection of multi-site background events in the detector. Because all of the materials processing and shielding techniques have already been demonstrated by members of our collaboration, and the detector technology we plan to exploit is commercially available, we are confident that the physics goals of this project can be met in a timely fashion.

In ^{76}Ge, the neutrinoless double beta-decay process, ^{76}Ge$\rightarrow ^{76}$Se+2β proceeds via the exchange of a virtual Majorana neutrino. The decay rate's sensitivity to the neutrino mass can be written[1]:

$$\Gamma_{0\nu} = G_{0\nu}|M_{0\nu}|^2 \langle m_{\beta\beta}\rangle^2, \tag{1}$$

with $\Gamma_{0\nu}$ the neutrinoless double beta-decay rate, $|M_{0\nu}|^2$ a function of nuclear physics matrix elements and $m_{\beta\beta}$ the effective Majorana electron neutrino mass. From Eq. 1, it

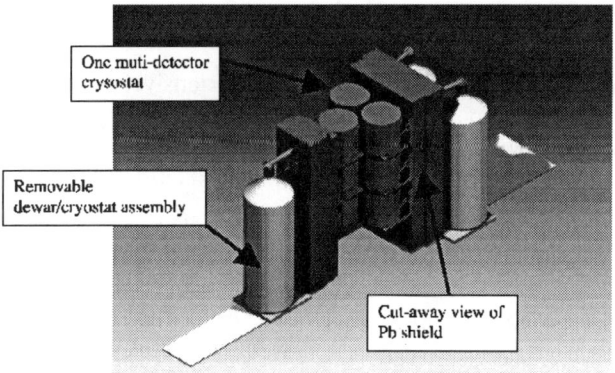

FIGURE 1. The Majorana baseline design, consisting of a set of Cu cryostats, each containing 57, approximately 1 kg segmented Ge ionization detectors, enriched to 86% ^{76}Ge, with a total Ge mass of about 500 kg.

is evident that a measurement of a finite half-life for the neutrinoless double beta-decay mode provides, in a straightforward way, a value for the effective neutrino mass. Because the nuclear matrix elements that appear in Eq. 1 are difficult to directly determine, the measured value of the half-life corresponds to a range of values for $m_{\beta\beta}$ consistent with the range of theoretical values available for $|M_{0\nu}|^2$.

The Majorana experiment will consist of a large number of high-resolution ionization detectors enriched to 86% ^{76}Ge. The signature for the 0ν mode will be a sharp peak around the double beta-decay endpoint energy of 2038.6 keV. The most stringent limits, at present, for the mass of the Majorana neutrino come from measurements of ^{76}Ge in the form of a collection of several large, single-element ionization detectors of roughly 11 kg total Ge mass. Our strategy to improve this limit is to increase the mass of Ge to 500 kg and to dramatically reduce the backgrounds in the 5 keV region of interest around the endpoint energy.

The availability of segmented Ge ionization detectors commercially in 1 to 2 kg masses provides the Majorana experiment with a great deal of flexibility. Our baseline design utilizes ultra-low background Cu cryostats, each containing 57 roughly 1 kg detectors, see Fig. 1. We have used this baseline design to evaluate the impact of various sources of background, and from these studies we have derived a conservative assessment of our baseline performance. We expect, however, to further optimize this design and the modular nature of the detectors will make it straightforward to realize whatever repackaging of the individual detectors provides us with the best projected performance.

RADIOLOGICAL BACKGROUNDS

Our approach to the backgrounds is based on the extensive, concrete experience within our collaboration with Ge ionization detectors operating underground[2]. One of the critical lessons learned from this experience is that with great care one can ensure that the dominant source of backgrounds is cosmogenic production of ^{68}Ge ($T_{1/2} = 271$d) and ^{60}Co ($T_{1/2} = 5.2$y).[3] Given the relatively long half-life of ^{60}Co, waiting for these backgrounds to decay is not a practical solution. The contributions of these two backgrounds, however, can be mitigated in two reasonable ways: reduced cosmogenic exposure before introduction underground and application of electronic techniques to identify and reject their signals. Reduced exposure can be achieved by carefully orchestrating the logistics of detector production, and in particular by performing some of the detector manufacturing underground.

The remaining contaminants in the crystals can be further suppressed by exploiting the fact neutrinoless double beta-decay occurs as a "single-site" process. Two beta particles are emitted that create short, very localized ionization paths in the Ge crystal and producing a single site from which charge is detected. The situation is quite different for many background processes, such as the decay of both ^{68}Ge and ^{60}Co, in which one anticipates multiple, well-seperated sites at which ionization occurs in the Ge crystal. This is true because these decays are associated with the emission of multiple gamma-rays, and essentially all of the emitted higher energy gamma-rays must undergo several Compton scatters to create an event near the range of interest. The presence of multiple sites from which charge is collected is therefore a very effective signature for background events.

The Majorana experiment uses two state-of-the-art techniques in Ge ionization detector technology to reject multiple-site events. The first and most straightforward technique is contact segmentation. It is now possible to purchase Ge detectors subdivided into a collection of segments, permitting the experimenter to optimize the volume of the detector subelements in a practical fashion. Segmentation provides a powerful tool for discriminating between multi-site events, simply by identifying events which simultaneously deposit energy in more than one segment. More sophisticated options for how to use the information from multiple-segment events are also available, in which the presence of "image" charges induced in detector elements adjacent to the segment in which the actual ionization occurs can also be used to characterize the events. To assess the impact of segmentation to the sensitivity of the Majorana experiment, we only use the multiple-segment "hit" criterion to determine multiple versus single-site events, but we still see a reduction in the efficiency for relevant gamma backgrounds of a factor of 7, with a loss of only 10% in the single-site efficiency.

The second technique we can use to discriminate multi-site from single-site events is pulse-shape analysis (PSA). Our experiment will utilize the capability of digitizing a time-record of the charge collected from each segment during an event. The pulse-shape analysis recently demonstrated by members of the collaboration[4] allows three pulse parameters to be calculated for each pulse, resulting in a three-dimensional space populated by pulses of different origin. Sources such as the 1592.47 keV double escape of the ^{208}Tl transition at 2614.47 keV produce a characteristic single site spatial population in the three dimensional parameter space. Events encountered outside this distribution can be discarded as backgrounds, with an appropriate efficiency correction. Applying the single-site parameterization to the ^{212}Bi gamma-ray line at 1620.6 keV reduces the efficiency for photopeaks by 74%, and the reduces the efficiency for single-site events by 20%. Since the multiplicity of internal ^{68}Ge and ^{60}Co events is expected to be even larger than full-energy external single gamma ray events such as those in the 1620.6 keV ^{212}Bi peak, the efficacy of the cut for rejecting ordinary gamma ray peak events is conservative.

Although previous experiments indicate the cosmogenic activity in the Ge dominates the background, there is still some possibility of a primoridal or other contribution at the level of the previous experimental uncertainties. Of particular concern is the copper used for our cryostats, which is chemically purified and electroformed into components to control radiological backgrounds. A specialized measurement of copper produced in 1995 with this procedure[5] estimated that the electroformed copper contained <25 mBq/kg ^{226}Ra and $\approx$9 mBq/kg ^{228}Th. Since then these results have been reanalyzed, suggesting the possibility that the activity was neither in the copper or the lead shielding. In addition, the copper production technique has been refined since 1995. For instance, cleaner reagents are available, multiple bath re-crystallizations and sequential electro-forming can all reduce the radioactive contaminants of the copper. If the copper is truly starting at 9 mBq/kg of either primordial decay chain, our estimates of suppression lead us to believe that we need no lower than about 1 mBq/kg, so the copper is already not far from acceptability. Note that the cosmogenic production of ^{60}Co in copper can also be greatly reduced by sequential electroforming underground. Ultimately, we will use very sensitive methods of screening to determine that the copper, electronic components, cabling, wires and bonding materials all have acceptable contaminant levels.

SENSITIVITY ESTIMATE

The two electronic background rejection methods, pulse shape analysis and segmentation, reduce background by a factor of 4 and 7, respectively, while decreasing the signal by 20% and 10%, respectively. If we compute a figure of merit (FOM) based on the remaining signal fraction divided by the square root of the remaining background, we can compute a factor that shows the effective multiplication of the half-life obtainable if the previously mentioned cosmogenic backgrounds are the limiting contribution. The FOM resulting from PSA is 1.6 and the FOM from segmentation is 2.4, resulting in a total FOM of 4 for the electronic rejection techniques for cosmogenic backgrounds. Coupled with a decay factor of about 20, weighted by isotope and based on previous experience with cosmogenics in germanium, the total reduction in cosmogenic background would equal about 480. This background level produces about 10 counts in the analysis region in 5000 kg-y of operation, corresponding to a half-life limit of 4×10^{27} y and an effective Majorana neutrino mass of $\langle m_{\beta\beta} \rangle = [0.02 \text{ eV to } 0.07 \text{ eV}]$.

CURRENT RESEARCH AND DEVELOPMENT: THE MEGA ARRAY AND THE SEGA DETECTOR

The MEGA array, currently under construction, will consist of 18 70% germanium detectors surrounding a central sample volume. This system should be able to screen items or materials to below the potential to contribute even one background count in the region of interest during the Majorana Experiment, and should provide a useful tool for exploring double beta-decay to excited states, placing new limits on solar axions, and evaluating the efficacy of active and passive shielding techniques proposed for the full Majorana experiment.

SEGA is a 1.4 kg, N-type Ge ionization detector, with 6-fold azimuthal and 2-fold axial contact segmentation. This detector is unique, in that it is also enriched to 85% ^{76}Ge; making it to our knowledge the first enriched, segmented Ge detector to be produced. SEGA was installed in a test cryostat and its performance evaluated by the collaboration. The detector's measured leakage current and energy resolution meet our expectations. Each contact of the detector is currently coupled through a room temperature preamplifier to a data acquisition system which digitizes each pulse, permitting us to test the efficacy of PSA for this detector. In the near future, SEGA will be transfered to a low-background cryostat and moved to an underground laboratory for futher tests.

ACKNOWLEDGMENTS

Pre-approval work on the Majorana Project is supported at Collaboration institutions by: the U.S. Department of Energy Office of Science, the National Nuclear Security Administration Office of Research and Engineering, several US National Laboratory Laboratory-Directed Research and Development offices, the National Science Foundation, several university research offices, and international research support institutions. The presenter also received generous support for this work from the Triangle Universities Nuclear Laboratory.

REFERENCES

1. Boehm, F., and Vogel, P., *Physics of Massive Neutrinos*, Cambridge University Press, Cambridge, 1987, pp. 154–155.
2. Avignone, F. T. *et al.*, *Nucl. Phys. B (Proc. Suppl.)*, **28A**, 280–285 (1992).
3. Brodzinski, R. L., *Nucl. Instr. and Meth.*, **A292**, 337–342 (1990).
4. Aalseth, C. E., Ph.D. thesis, University of South Carolina (2000).
5. Brodzinski, R. L., *Journal of Radioanalytical and Nuclear Chemistry*, **193**, 61–70 (1995).

The MINOS Experiment

Hugh Gallagher
for the MINOS Experiment

Tufts University

Abstract. The MINOS (Main Injector Neutrino Oscillation Search) Experiment is a long baseline neutrino oscillation experiment using the Fermilab Main Injector neutrino beam and two detectors, one located at Fermilab and a second located 730 km away in Soudan, MN. The experiment is designed to probe neutrino oscillations at the value of Δm^2 suggested by atmospheric neutrino results, and will measure the oscillation parameters with high precision, search for sub-dominant $\nu_\mu \to \nu_e$ mixing with unprecedented accuracy, and study possible CPT-violating oscillation scenarios with atmospheric neutrinos.

PHYSICS GOALS

One of the most significant discoveries in particle physics in the past decade has been the observation of neutrino oscillations in the atmospheric neutrino flux [1, 2, 3, 4]. The current generation of experiments will probe the oscillation phenomenology uncovered by these non-accelerator experiments using intense, well-understood neutrino beams. The experiment has a number of physics objectives:

1. Precision measurements of the oscillation parameters Δm^2_{23} and $\sin^2 2\theta$ from the analysis of fully reconstructed ν_μ CC events. The right hand plots of Figure 1 shows the expected sensitivity (statistical + systematic) for the given numbers of protons on target. In particular MINOS will be able to address the question of whether the mixing strength differs from unity by a small amount. A related goal of the experiment is to observe the 'dip' in the energy distribution corresponding to the first minimum in the neutrino survival probability. The left hand plots of Figure 1 show the reconstructed neutrino energy distribution for ν_μ CC events which have oscillated with $\Delta m^2 = 0.0025$ eV^2.

2. Precision measurements of sub-dominant $\nu_\mu \to \nu_e$ mixing through observation of electron events in the far detector. MINOS will be able to extend the search for $\nu_\mu \to \nu_e$ mixing below the existing Chooz limit [5]. Backgrounds arise primarily from ν_e in the beam and NC interactions in which most of the hadronic energy goes into a single π^0, both of which will be measured with high precision in the MINOS near detector.

3. Atmospheric neutrino studies. MINOS is the first atmospheric neutrino detector capable of separating ν_μ and $\bar\nu_\mu$ interactions, due to the presence of the magnetic field. This leads to the possibility of performing sensitive tests of CPT violation in the neutrino sector, which has been proposed as a means of reconciling the solar,

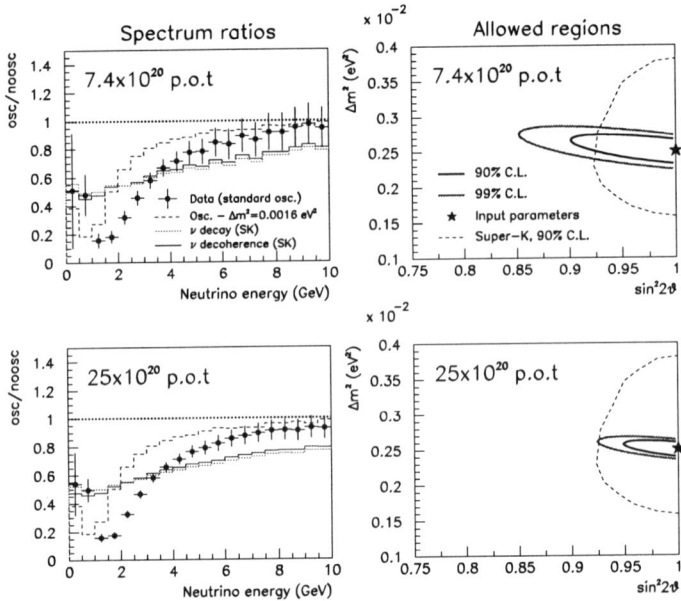

FIGURE 1. MINOS measurements for 2 different numbers of protons on target assuming true values for the oscillation parameters of $\Delta m^2 = 0.0025$ eV^2 and $\sin^2 2\theta = 1.0$. Plots on the left show the expected CC energy distribution divided by the no-oscillation prediction. Plots on the right show the corresponding allowed regions. Also shown at the left are the predicted energy distributions for $\Delta m^2 = 0.0016$ eV^2 and several alternative explanations - neutrino decay and decoherence.

atmospheric, and LSND results [6]. The nanosecond level timing in the scintillator makes it possible to identify several categories of neutrino interactions: contained events, upgoing stopping muons, and upward throughgoing muons. Early analyses indicated that cosmic ray particles entering the detector through the large gaps between scintillator planes would pose an insurmountable problem for contained event atmospheric neutrino studies, for this reason a veto shield covering the top and sides of the detector was built using existing scintillator modules. By the end of 2007 over 1000 neutrino interactions will be recorded, with several hundred coming from anti-neutrinos.

Crucial to the success of the MINOS program is the proton intensity of the Fermilab Main Injector. While previous studies have been based on assumptions of 7.4×10^{20} p.o.t. in the first two year's running, both the current performance of the main injector and the difficulties in making precision measurements at low values of Δm^2 have necessitated consideration of run plans longer than two years. One proposal put forward by the collaboration calls for a 5 year run with total intensity of 25×10^{20} p.o.t. [7]. Such a goal, which would require a significant investment in the beam facilities over the next several years, would greatly increase the physics reach of the experiment [7].

THE NUMI BEAM

The MINOS experiment will utilize the high-intensity neutrino beam created at the NuMI facility at Fermilab. 120 GeV/c protons are extracted from the Main Injector using a single-turn fast extraction technique, which produces a 10 μs beam spill with an internal structure of 2 ns buckets every 19 ns. These protons are then directed onto a graphite target, after which produced pions and kaons are focussed using a two-horn system. Following the horns is a 2m diameter by 700 m long decay pipe in which the produced mesons decay.

One of the early recommendations to the experiment from the Fermilab PAC was to build into the experimental program enough flexibility to be able to respond to new developments in this rapidly evolving field. One of the main ways this was done was through the design of the NuMI beamline, which allows the possibility of moving the horn and target elements, thereby changing the 'tune' of the beam and the mean energy of the focused pions. For the current best fit values to atmospheric neutrino data, the maximum of the oscillation probability for the 730 Fermilab-Soudan baseline occurs at around 2 GeV. Of the beam options available the 'low-energy' configuration maximizes the proportion of the flux in the energy range below 4 GeV and the experiment will begin data-taking with this beam. It is expected that over the course of a possible five-year run, other beam configurations, including changing to anti-neutrinos, will also be used.

THE MINOS DETECTORS

MINOS is a two-detector experiment with a 0.98 kton tracking calorimeter (the 'near detector') located at Fermilab and a 5.4 kton tracking calorimeter (the 'far detector') located 730 km away in the Soudan Underground State Park in Soudan, MN. The purpose of the near detector is to provide a snapshot of the beam before oscillations have occurred. This approach substantially reduces systematic errors which would otherwise arise from uncertainties in the beam predictions and neutrino interaction cross sections.

The completed MINOS detector is made of two 2.7 kton 'super-modules' with alternating 8m wide octagonal planes of steel and plastic scintillator. The steel planes are one inch thick and are magnetized to 1.3 T. The detector was completed in August 2003, nearly nine months ahead of schedule (Figure 2 shows a picture of the completed detector). The scintillator planes are composed of 192 4.1 cm wide x 1 cm thick strips of plastic scintillator of varying lengths. The scintillator strips are co-extruded with a TiO_2 outer layer to provide internal reflectivity. A 1.2 mm scintillating fiber is glued with optical epoxy into a groove which runs the length of the strip. Strips are routed through clear fiber cables to boxes which hold 16-pixel Hammamatsu PMTs. Strips are optically multiplexed eight to a pixel. Since light is read out from both ends of the strip, using a different multiplexing scheme for the two ends allows for unambiguous de-multiplexing of through-going tracks.

Scintillator strips are packed in 20 or 28 strip light-tight aluminum modules which provide structural support and ease of installation. Measurements of the light output of the MINOS modules (both in the fabrication process and with cosmic ray muons at the

FIGURE 2. The completed MINOS far detector. The veto shield, consisting of standard MINOS scintillator modules, is visible above the detector.

far detector) show light levels a factor of 2 higher than the performance requirement of 4.7 photoelectrons / MIP which was set at the design stage.

The MINOS near detector, while in principle identical to the far detector, operates in a significantly different environment; the beam at the near detector is smaller in transverse size and more intense than that seen at the far detector. Because of this, the near detector is a 'squashed octagon' design, which provides a region with identical acceptance and magnetic field characteristics while reducing the overall transverse size. Because of the high rate, events are not multiplexed, which allows readout on 64-channel PMTs. The electronics at the near detector are also significantly faster to cope with the high rates produced in a single spill. The planes for the near detector have been assembled and are stored at Fermilab awaiting installation, scheduled to begin with the completion of the NuMI MINOS Hall in December, 2003. Installation of the near detector is scheduled to be completed in October 2004.

Detector Calibration

The physics requirements of the experiment require the measurement of hadronic energy to be consistent to 2% between the near and far detectors and the absolute energy scale to be known to 5%. To obtain this level of calibration accuracy substantial activity over the past several years has gone into several test beam runs at the CERN PS. The MINOS calibration detector consists of 60 1 m x 1m planes, alternating steel and scintillator identical to the MINOS detectors. The detector has been exposed to pions, electrons, muons and protons of various energies and angles chosen to represent particles characteristically produced in neutrino interactions. This individual particle data provides a number of uses: studying particle topologies in the detector allows for tuning of the Monte Carlo, measurement of the energy resolution – around $22\%/\sqrt{E}$ for electromagnetic showers and $55\%/\sqrt{E}$ for hadrons, and cross-calibration of the near and far detector electronics. Running periods with both the near and far detector electronics have been taken. The final calibration run is scheduled to be completed in the fall of 2003.

The strip-to-strip calibration of the detectors is obtained with cosmic muons, there are about 530 hits/strip/month in the far detector and 25k hits/strip/month in the near detector, and the relative calibrations near-far can be determined using stopping muons in both detectors. In addition, a light-injection system using pulsed blue LEDs monitored by PIN photodiodes [8] operates in situ to determine the non-linearity of the PMT gain curves, monitor short term gain drifts, and provide an independent means of determining PMT gains by pulsing at low light levels to obtain single photoelectron spectra.

CONCLUSION

All aspects of the MINOS and NuMI experimental program are proceeding at a rapid pace. The MINOS far detector has been completed nine months ahead of schedule. Outfitting of the NuMI beamline facilities continues with installation of the near detector scheduled to begin in late 2003 with first beam in late 2004. In the meantime, analysis of atmospheric neutrino and cosmic ray data continues with over 1000 atmospheric neutrino-induced events expected by 2007. This data should provide a strong test of CPT-violating models of neutrino oscillations. Studies of cosmic ray muons show that the light levels in scintillator are at least a factor of two higher than the design requirements, and that the timing resolution of several nanoseconds, which is dominated by the decay time of the fluor in the WLS fiber, is sufficient to determine the direction of throughgoing muon tracks.

REFERENCES

1. Fukuda, Y., et al., *Phys. Rev. Lett.*, **81**, 1562–1567 (1998).
2. Fukuda, Y., et al., *Phys. Lett.*, **B436**, 33–41 (1998).
3. Ambrosio, M., et al., *Phys. Lett.*, **B517**, 59–66 (2001).
4. Allison, W. W. M., et al., *Phys. Lett.*, **B449**, 137–144 (1999).
5. Apollonio, M., et al., *Eur. Phys. J.*, **C27**, 331–374 (2003).
6. Barenboim, G., and Lykken, J., *Phys. Lett.*, **B554**, 73–80 (2003).
7. The MINOS Collaboration, Proposal For a Five Year Run Plan For MINOS, Tech. Rep. Fermilab report NuMI-GEN-0930, Fermilab (2003).
8. Adamson, P., et al., *Nucl. Instrum. Meth.*, **A492**, 325–343 (2002).

Performance of the ICARUS T600 detector

S. Navas-Concha

(On behalf of the ICARUS Collaboration)
Dpto. de Física Teórica y del Cósmos & C.A.F.P.E., University of Granada, c/ Severo Ochoa s/n, Granada, Spain

Abstract. The ICARUS detector is a liquid argon time projection chamber. It provides three dimensional imaging and calorimetry of ionizing particles over a large volume, with high granularity. This multipurpose detector opens up unique opportunities to look for phenomena beyond the Standard Model through the study of atmospheric, solar and supernova neutrinos, nucleon decay searches and neutrinos from the CERN to Gran Sasso beam. The ICARUS technology has reached maturity with the construction and test (during summer 2001) of a 600 ton detector, demonstrating the feasibility of building large mass devices relevant for non-accelerator physics. The collected data allow to assess the detector performance, i.e. the spatial reconstruction, calorimetry and particle identification, as well as the test of all technical aspects of the system (cryogenics, electronics, purification, etc). A summary of the detector performances and an overview of the general physics program of the ICARUS experiment is reported.

THE ICARUS T600 DETECTOR

The ICARUS T600 liquid Argon (LAr) detector [1] (see figure 1) consists of a large cryostat split in two identical, adjacent half-modules, each of $3.6 \times 3.9 \times 19.9$ m^3 internal dimensions. Each half-module is an independent unit housing an internal detector composed by two Time Projection Chambers (TPC), a field shaping system, monitors and probes, and by two arrays of photo-multipliers. Externally the cryostat is surrounded by a set of thermal insulation layers. The TPC wire read-out electronics is located on the top side of the cryostat. The detector layout is completed by a cryogenic plant made of a liquid Nitrogen cooling circuit to maintain uniform the LAr temperature, and of a system of LAr purifiers.

A liquid Argon TPC allows to detect the ionization charge released at the passage of charged particles in the volume of LAr, for three dimensional image reconstruction and calorimetric measurement of ionizing events. The detector, equipped with an electronic read-out system, works as an "electronic bubble chamber" employing LAr as ionization medium. Unlike traditional bubble chambers, limited by a short window of sensitivity after expansion, the LAr TPC detector remains fully and continuously sensitive, self-triggerable and without read-out dead time.

A uniform electric field applied to the medium makes the ionization electrons drift onto the anode, following the electric field lines; thanks to the low transverse diffusion of the ionization charge, the electron images of ionizing tracks are preserved. Successive anode wire planes, biased at a different potential and oriented at different angles, make possible the three dimensional reconstruction of the track image. While approaching a plane, the electrons induce a current on the wires near which they are drifting; when moving away after crossing the plane, a current of opposite sign is induced. By appropriate biasing, the first planes can be made non-destructive (*Induction* planes), whereas the charge is finally collected in the last plane (*Collection* plane). Each wire plane provides an independent view of the event.

In each T600 half-module, the two identical TPC's are separated by a common cathode and are referred to as *left* and *right* chambers. Each TPC consists of three parallel wire planes: the first, facing the drift region, with horizontal wires (Induction plane); the other two with the wires at $\pm 60^0$ from the horizontal direction (Induction and Collection planes, respectively). The wire pitch is 3 mm. The maximum drift path (distance between the cathode and the wire planes) is 1.5 m and the nominal drift field 500 V/cm (a 3 meter drift is foreseen for future ICARUS modules).

Each wire of the chamber is independently digitized every 400 ns. The electronics was designed to allow continuous read-out, digitization and independent waveform recording of signals from each wire of the TPC. Measurement of the time when the ionizing event occurred (so called "T_0 time" of the event), together with the electron drift velocity information, provides the absolute position of the tracks along the drift coordinate. The T_0 can be determined by detection of the prompt scintillation light produced by ionizing particles in LAr [2].

A very detailed description of the T600 physics program at Gran Sasso can be found in Ref.[3]. In this phase the available mass is limited, however the high efficiency and the detailed information which can be collected for each event will allow to initiate the study of some of the fundamental issues of underground physics: the study of neutrino physics, with solar, atmospheric and supernova neutrinos, and the study of nucleon decay.

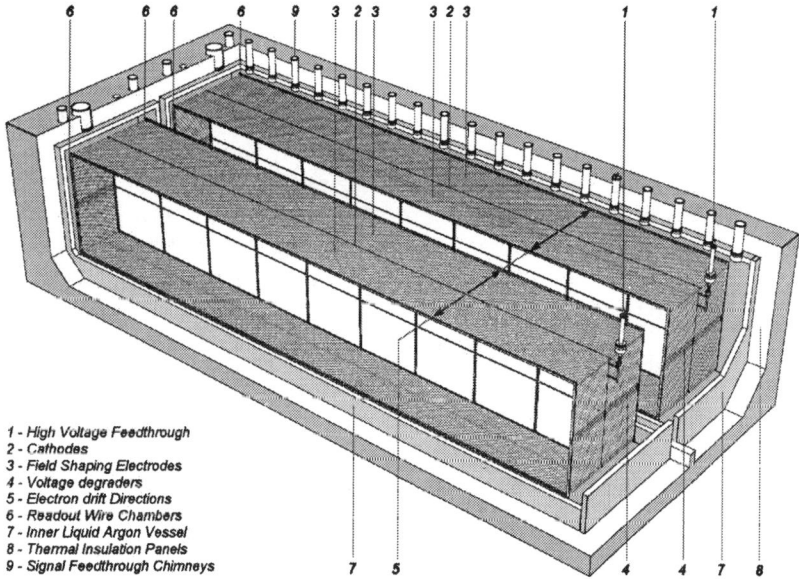

1 - High Voltage Feedthrough
2 - Cathodes
3 - Field Shaping Electrodes
4 - Voltage degraders
5 - Electron drift Directions
6 - Readout Wire Chambers
7 - Inner Liquid Argon Vessel
8 - Thermal Insulation Panels
9 - Signal Feedthrough Chimneys

FIGURE 1. Cut view of the T600 module.

THE T600 TEST RUN

A full above-ground test of the T600 experimental set-up was carried out in Pavia (Italy) during summer 2001. One T600 half-module was fully instrumented to allow a complete test in real experimental conditions. All technical aspects of the system, cryogenics, mechanics, LAr purification, read-out chambers, scintillation light detection, electronics, and DAQ were tested, and found to be satisfactorily in agreement with expectations.

During the test run, a very large amount of cosmic ray events was recorded [4] with different configurations of a dedicated trigger system: long, penetrating muons and spectacular, high multiplicity muon bundles, electromagnetic and hadronic showers and low energy events among other categories of events. An example of a large electromagnetic shower event is shown in figure 2.

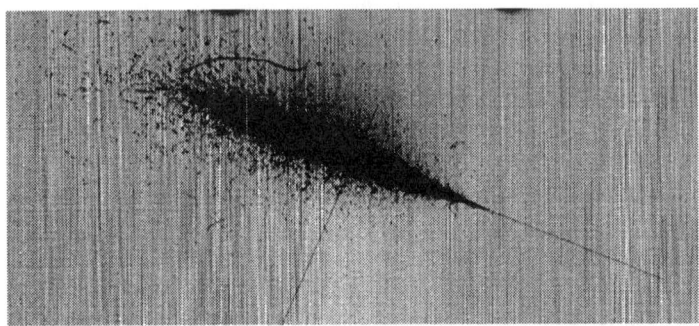

FIGURE 2. Electromagnetic shower detected on the T600 first half module (event size 3.7×1.7 m^2).

Like a bubble chamber, the ICARUS detector provides a measurement of the total ionization loss of a track with very high sampling. By extracting the physical information contained in the wires output signal, i.e. the energy deposited by the different particles and the point where such a deposition has occurred, it is possible to build a complete three dimensional spatial and calorimetric picture of the event. The measurement of the dE/dx and positions for a large number of points along a given track provides a way of estimating the particle momentum from range (for stopping particles) or multiple scattering, providing, in addition, a method for particle identification. As an example, figure 3 shows a kaon decay candidate event acquired during the T600 test.

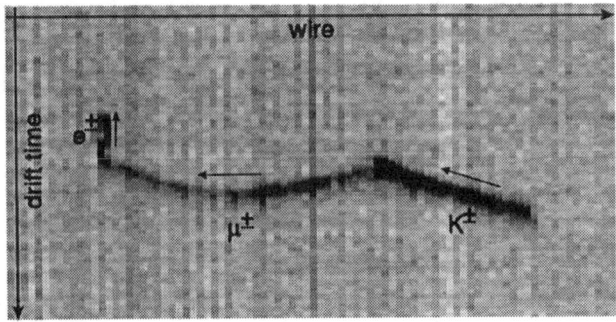

FIGURE 3. Charged Kaon decay candidate from the T600 test run.

A fundamental requirement for the ICARUS technology is that electrons produced by ionizing particles might travel unperturbed in LAr from the point of production to the collecting wire planes. To this extend, the concentration of any kind of electro-negative impurity diluted in the liquid must be reduced at extraordinarily low levels. At the end of the run, the measurements provided by the *Purity Monitors* (dedicated devices immersed in the LAr) and from crossing muon tracks shown a maximum value of the drift electron lifetime of about 1.8 ms, equivalent to an electron mean free path greater than 280 cm, and was still increasing [5].

THE ICARUS PHYSICS PROGRAM

A liquid Argon time projection chamber (TPC), working as a electronic bubble chamber, continuously sensitive, self-triggering, with the ability to provide 3D imaging of any ionizing event, together with excellent calorimetric response, offers the possibility to perform complementary and simultaneous measurements of neutrinos, as those of the CERN to Gran Sasso beam (CNGS), those from cosmic ray events, and even those from the sun and from supernovae. The same class of detector can also be envisaged for high precision measurements at a neutrino factory and can be used to perform background-free searches for nucleon decays. Hence an extremely rich and broad physics program, encompassing both accelerator and non-accelerator physics, will be addressed. These will answer fundamental questions about neutrino properties and about the possible physics of the nucleon decay.

The performance of a neutrino detector is proportional to its total mass but also to its geometrical granularity, which determines the quality with which signal events can be reconstructed and hence separated from backgrounds.

Though it has a physics program on its own [3], the T600 construction was mainly motivated by technical issues. Since the concept of modularity was inserted starting from the very initial phases of the detector design, growing the detector mass to the several kton scale can be accomplished by "cloning" the actual T600 to the required number of units. An instrumented target of 5 ktons would provide the following:

- *Atmospheric neutrinos*: about one thousand atmospheric electron and muon charged current (CC) events per year and 500 neutral current (NC) atmospheric events. In addition, about 5 tau CC events per year coming from neutrino oscillations are expected, assuming $v_\mu \to v_\tau$ oscillations ($\sin^2 2\theta = 1, \Delta m^2 \approx 10^{-3}$ eV^2).
- *Solar and supernovae neutrinos*: about $16200 \times f_B$ solar neutrino electron scattering and absorption events per year with an electron detection threshold of 5 MeV, where f_B is the Boron flux suppression factor. A supernova at a distance of 10 kpc would produce about 360 events, in a similar energy range.
- *CNGS neutrinos*: a total of 13600 v_μ CC events per year (4.5×10^{19} pots) and accompanying smaller samples of $\bar{v}_\mu$, v_e and $\bar{v}_e$ flavors. In addition, 455 v_τ CC events for $v_\mu \to v_\tau$ oscillations are expected ($\Delta m^2 = 3 \times 10^{-3}$eV^2, $\sin^2 2\theta = 1$). Concerning $v_\mu \to v_e$ studies, after 5 years of data taking (SPS dedicated mode) a limit of $\sin^2 2\theta_{13} < 2 \times 10^{-2}$ for $\Delta m^2_{23} = 3 \times 10^{-3}$eV^2 could be set in case of

negative search.
- *Nucleon decays*: 3×10^{33} nucleons, which translates into background-free searches up to 2×10^{32} years for $p \to e^+\pi^0$ and 6×10^{32} years for $p \to K^+\bar{\nu}$ after 1 year of running. ICARUS takes in advantage of the excellent tracking and particle identification capabilities to access inclusive analyses with more than two visible particles on the final state.
- *Neutrino factory*: 6150 ν_μ CC events for 10^{21} decays of 5 GeV μ^- at a distance of $L = 730$ km and 1130 ν_μ CC events per 10^{20} decays of 30 GeV μ^- at a distance of $L = 7400$ km (FNAL-LNGS).

CONCLUSIONS

In this paper we have presented a detailed description of the ICARUS T600 detector. All technical aspects of the system, namely cryogenics, LAr purification, readout chambers, detection of LAr scintillation light, electronics and DAQ were tested and performed as expected. The ICARUS T600 successful test provides the proof that its technology, developed over many years, is now mature. The statistically significant sample of cosmic ray events recorded has been used to assess the performance of the reconstruction tools.

A broad physics program, including accelerator and non-accelerator physics, will be addressed by ICARUS in the next years. The first phase will include the detection of atmospheric and solar neutrino events, supernovae and the search for proton decay. On the basis of the experience of the T600 module, the design and assembly of "clones" of the present prototype in a series of units will allow to reach a mass of the order of 3000 tons for the time of arrival of the CNGS beam, starting the multi-kton physics program.

ACKNOWLEDGMENTS

The author wishes to thank the CIPANP'03 organizers for preparing such an excellent conference. The credit of the work presented on this report is due to the ICARUS Collaboration members that contributed to the detector construction, operation and physics analysis. Special thanks to A.Bueno, I.Gil and J.Rico for helping on the elaboration of this document. The author gratefully acknowledges support from the Spanish Ministry of Science and Technology.

REFERENCES

1. ICARUS Collab., LNGS-94/99 (1994); LNGS-95/10 (1995); LNGS-EXP 13/89 add.2/01 (2001) and references therein.
2. ICARUS Collab., *Nucl. Instr. and Meth.* **A 432** (1999) 240.
3. ICARUS Collab., "Initial Physics Program", LNGS-P28/2001 (2001).
4. ICARUS Collab., "Observation of long ionizing tracks with the ICARUS T600 first half-module", accepted for publication in *Nucl. Instr. and Meth. A*
5. ICARUS Collab., "Analysis of the liquid Argon purity in the ICARUS T600 TPC", submitted to NIM.

The Future of Reactor Neutrino Experiments: A Novel Approach to Measuring θ_{13}

Karsten M. Heeger[*], Stuart J. Freedman[†*] and K.-B. Luk[†*]

[*]*Physics Division, Lawrence Berkeley National Laboratory, Berkeley, CA 94720, USA*
[†]*University of California at Berkeley, Berkeley, CA 94720, USA*

Abstract. Results from non-accelerator neutrino oscillation experiments have provided evidence for the oscillation of massive neutrinos. The subdominant oscillation, the coupling of the electron neutrino flavor to the third mass eigenstate, has not been measured yet. The size of this coupling U_{e3} and its corresponding mixing angle θ_{13} are critical for CP violation searches in the lepton sector and will define the future of accelerator neutrino physics. The current best limit on U_{e3} comes from the CHOOZ reactor neutrino disappearance experiment. In this talk we review proposals for future measurements of θ_{13} with reactor antineutrinos.

Recent results from atmospheric, solar, and reactor neutrino experiments [1] have provided evidence for the mixing of massive neutrinos. The phenomenon of neutrino mixing is characterized by the coupling between the neutrino flavors ($v_{e,\mu,\tau}$) and mass eigenstates ($v_{1,2,3}$), and the associated mixing angles. Past and present neutrino oscillation experiments have determined two of the three mixing angles in the neutrino mixing matrix, U_{MNSP}, of three active species. The coupling of the electron neutrino flavor to the third mass eigenstate, U_{e3}, is yet to be determined. Equation 2 shows a parametrization of U_{MNSP} and the experimental input to this matrix.

$$U_{MNSP} = \begin{pmatrix} U_{e1} & U_{e2} & U_{e3} \\ U_{\mu 1} & U_{\mu 2} & U_{\mu 3} \\ U_{\tau 1} & U_{\tau 2} & U_{\tau 3} \end{pmatrix} \quad (1)$$

$$= \begin{pmatrix} 1 & 0 & 0 \\ 0 & c_{23} & s_{23} \\ 0 & -s_{23} & c_{23} \end{pmatrix} \begin{pmatrix} c_{13} & 0 & s_{13}e^{-i\delta_D} \\ 0 & 1 & 0 \\ -s_{13}e^{i\delta_D} & 0 & c_{13} \end{pmatrix} \begin{pmatrix} c_{12} & s_{12} & 0 \\ -s_{12} & c_{12} & 0 \\ 0 & 0 & 1 \end{pmatrix} \begin{pmatrix} 1 & 0 & 0 \\ 0 & e^{i\alpha/2} & 0 \\ 0 & 0 & e^{i\alpha/2+i\beta/2} \end{pmatrix} \quad (2)$$

PRESENT: atmospheric v — reactor v — solar v — $0\nu\beta\beta$
FUTURE: accelerator v — reactor and accelerator v — solar v — $0\nu\beta\beta$

The current best upper limit on U_{e3} comes from the CHOOZ reactor neutrino disappearance experiment [4]. In contrast to the surprisingly large mixing of the other neutrino states, the U_{e3} coupling was found to be small [4]. The discovery of subdominant effects in $\bar{v}_e \to \bar{v}_{\mu,\tau}$ oscillations and a non-zero U_{e3} coupling would have a profound impact on neutrino physics. It determines whether CP violation can play a significant role in lepton mixing; if U_{e3} is zero, CP is conserved.

CP violation is a well-established phenomenon in the quark sector but leptonic CP violation is as yet unknown. CP violation in the lepton sector could have cosmological implications far beyond the phenomenon on neutrino oscillations. It may be the only way to explain the observed matter-antimatter asymmetry in the Universe. In this context a successful θ_{13} experiment has the potential to define the direction of neutrino research and the neutrino program at accelerators for the next decade and beyond. The small size of θ_{13} compared to the other neutrino mixing angles may also point us to an underlying symmetry in theoretical neutrino mass models.

Nuclear reactors are an abundant source of $\overline{\nu}_e$ and have been the site of several experiments. From the discovery of the free antineutrino by Reines and Cowan in 1956 [3], to the first discovery of reactor $\overline{\nu}_e$ disappearance at KamLAND in Japan in 2002 [2], reactor neutrino experiments have played a central role in the history of neutrino physics. Almost five decades after the discovery of the neutrino, reactor neutrino experiments have – together with solar neutrino studies – provided evidence for the mixing of massive neutrinos. Reactor neutrino experiments study antineutrinos with an average energy of 4 MeV produced in the fission reactions in the core of a nuclear reactor. Reactor antineutrinos are usually detected through the inverse β-decay reaction on protons $\overline{\nu}_e + p \to e^+ + n$. The coincidence signal from the prompt positron and the delayed neutron capture allows the unique identification of electron antineutrinos.

The observation of neutrino flavor transformation in the atmospheric and solar neutrino experiments have allowed us to measure the oscillation parameters, including the square of the mass differences between the neutrino mass eigenstates. We expect to observe the signatures of neutrino oscillations associated with these mass states in accelerator and reactor neutrino experiments. In 2002, the KamLAND experiment made the first measurement of the disappearance of reactor $\overline{\nu}_e$ at an average distance of 180 km from the source and confirmed the dominant oscillation effect observed in solar neutrinos. Past reactor experiments such as CHOOZ and Palo Verde have studied the flux of reactor $\overline{\nu}_e$ at distances of ~ 1 km from the reactor cores. These experiments did not observe a flux suppression and limited the subdominant mixing associated with θ_{13} to $sin^2 2\theta_{13} < 0.9$ [4]. Combined analyses of the result from reactor and solar neutrino experiments now place a limit of $sin^2 2\theta_{13} < 0.6$ on the subdominant $\overline{\nu}_e \to \overline{\nu}_{\mu,\tau}$ oscillation. The goal of the next generation reactor neutrino oscillation experiment is the discovery of this subdominant neutrino oscillation and the first measurement of θ_{13}.

Over a distance of a few kilometers the survival probability of reactor $\overline{\nu}_e$ is well described by the approximate expression

$$P_{ee} \simeq 1 - \sin^2 2\theta_{13} \sin^2 \frac{\Delta m_{31}^2 L}{4E_\nu} + \left(\frac{\Delta m_{21}^2 L}{4E_\nu} \right) \cos^4 \theta_{13} \sin^2 2\theta_{12} \qquad (3)$$

which is dominated by the frequency associated with the atmospheric mass splitting $\Delta m_{13} = \Delta m_{12} + \Delta m_{23} \simeq \Delta m_{23}$. A next-generation reactor experiment with 2 or more antineutrino detectors can make a precise measurement of the relative change in rate and shape of the energy spectrum and detect the signature of θ_{13} in the relative spectral distortion. The relative measurement between multiple detectors will make this experiment largely independent of the absolute reactor $\overline{\nu}_e$ flux, absolute detector systematics, and uncertainties in the cross-section.

Principle features of a suitable reactor site are a powerful reactor (or multiple cores) and overburden in excess of 300 mwe to shield the antineutrino detectors from cosmic

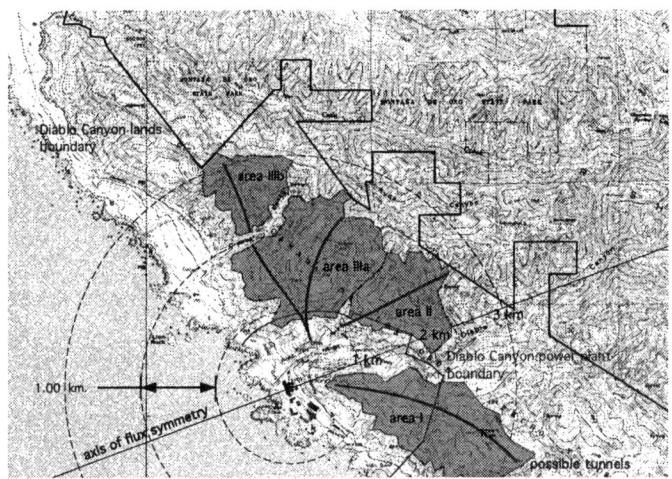

FIGURE 1. Topographic map of the site of the Diablo Canyon nuclear power plant in San Luis Obispo County, CA, USA. The land boundary as well as the power plant site boundary are indicated. The shaded areas identify regions with sufficient overburden for the placement of antineutrino detectors. The green lines illustrate the location of possible horizontal tunnels ranging in distance between 600 m to 3.2 km from the reactor cores of the power plant. At these locations we find an overburden of 300-600 mwe.

rays. Multiple reactor cores may lead to interference effects and reduce the ultimate sensitivity of an experiment. The construction of a horizontal tunnels or a vertical shafts are usually required to obtain this overburden. The variable baseline provided by horizontal tunnels is useful for optimizing the baseline of the two detectors and to demonstrate the neutrino oscillation effect with measurements at various locations along the tunnel. A tunnel may also be used to facilitate the relative calibration of the detectors at one location.

A number of reactor sites worldwide are currently under evaluation by different experimental groups [6]. For an overview of the current activities see [7] and references therein. One of the promising sites in the US is the Diablo Canyon nuclear power plant in central California. Two cores separated by ~ 100 m provide a total thermal energy of 6.2 GW_{th}. Nearby coastal mountains provide good overburden and make the plant an almost ideal site for a reactor neutrino experiment. Horizontal tunnels in the coastal mountains may provide overburden of 300-400 mwe at distances of 0.6-1 km and 600-800 mwe at < 3.2 km. The general layout and topography of the site allows the construction of a kilometer-long tunnel for two movable detectors. An overview map of the site is shown in Figure 1.

The physics potential of a next-generation reactor neutrino measurement has been discussed in [5] and references therein. The ultimate sensitivity of such an experiment will strongly depend on the layout of the experiment including the distances from the reactor cores, the overburden, the size of the detectors and their mobility. With a sensitivity of up to $\sin^2 2\theta_{13} = 0.01$ [5] a reactor neutrino measurement of θ_{13} is comparable to next-generation accelerator experiments. The difference between a low-energy, short-baseline reactor neutrino experiment and a longer-baseline accelerator

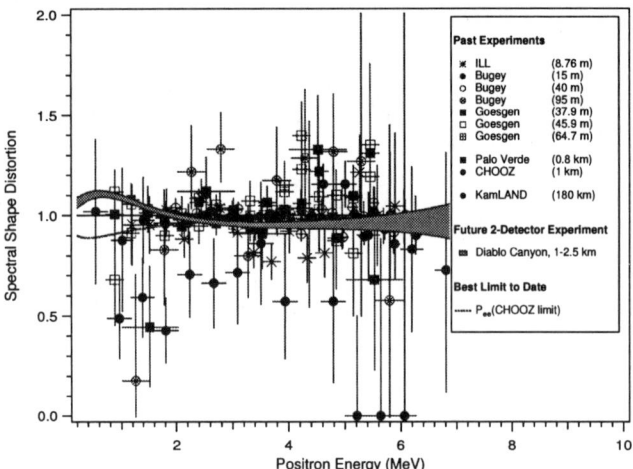

FIGURE 2. Expected statistical precision of a next-generation 2-detector reactor neutrino oscillation experiment compared to the results from past and present single-detector reactor neutrino experiments. This figure shows the ratio of the measured positron spectrum to the unoscillation spectrum for single-detector experiments and the ratio between the near and far detector for the proposed 2-detector experiment at Diablo Canyon. The errors are statistical.

experiment in which matter effects play a role make these approaches complementary. The combined interpretation of the reactor and accelerator experiments will provide the best sensitivity to θ_{13} and δ_{CP}. The expected statistical accuracy of a next-generation reactor neutrino experiment is illustrated in Figure 2.

In summary, next-generation reactor neutrino experiments have the potential to discover the neutrino mixing angle θ_{13} and define the roadmap of accelerator neutrino studies for the next decade and beyond. A discovery or a new limit on θ_{13} would determine the prospects for measuring leptonic CP violation. Through this effect we might find that neutrinos may play a significant role in the evolution of the early Universe and can in fact explain the long-standing mystery of the baryon asymmetry in the Universe.

REFERENCES

1. Fukuda, Y. et al. (Super-Kamiokande Collaboration), *Phys.Rev.Lett.82* (1999); Ahmad, Q.R. et al. (SNO Collaboration), *Phys.Rev.Lett.89:011301* (2002); Eguchi, K. et al. (KamLAND Collaboration), *Phys.Rev.Lett.90:021802* (2003)
2. Eguchi, K. et al. (KamLAND Collaboration), *Phys.Rev.Lett.90:021802* (2003)
3. Cowan, C.L. et al., *Science 124, 103* (1956)
4. Appollonio, M. et al., *Eur.Phys.J.C27:331-374* (2003)
5. Huber, P. et al., *arXive: hep-ph/030323* (2003)
6. Martemianov, V. et al., *arXive: hep-ph/0305295* (2003); Yasuda, O., *arXive: hep-ph/0305295* (2003); Suekane, F. et al., *arXive: hep-ex/0306029* (2003); Shaevitz, M.H. and Link, J.M., *arXive: hep-ex/0306031* (2003); Goodman, M., *arXive: hep-ex/0307017* (2003)
7. http://theta13.lbl.gov/ (2003)

Physics of an Intense Neutrino Beam from BNL to a Very Long Baseline Detector

Zohreh Parsa[1]

Brookhaven National Laboratory, Physics Dept., 510 A, Upton, NY 11973, USA.

Abstract. An intense neutrino facility allows probing of the neutrino mixing angles, mass hierarchy, and leptonic CP violation. Physics potential, for making precision measurements of all neutrino oscillation parameters (θ_{ij}, Δm_{ij}^2, δ) using a wide band ν_μ beam, to a (very long baseline) detector is presented. Potential of a Neutrino beam from Brookhaven National Laboratory to a 2540 km baseline (with 0.5 megaton) detector at Homestake Mine in South Dakota, is (under study by our neutrino working group) discussed. Schemaics of the beam facility for the AGS upgrade to 1 MW with a cycle time of 2.5 and 10^{14} protons on target at 28 GeV; and a map with possible detector sites are also included.

INTRODUCTION

Success of the atmospheric and solar neutrino experiments that has provided evidence for non - zero neutrino masses and mixing has increased our interest in neutrino oscillation searches using accelerator created neutrinos. Protons from an accelerator (e.g. AGS) would hit a target (e.g. Mercury Jet, or graphite), and produce bursts of particles e.g., pions, that decay to muons, which then decay to neutrinos. To focus the beam a magnetic horn (and/ or solenoid) can be used to keep the particles from spreading and to direct the beam in the detector(s) direction. After leaving the horn pions decay into neutrinos.

Upgraded conventional Neutrino horn beams (Superbeams) are being considered (at BNL) for probing of the neutrino masses, mixing angles, leptonic CP violation, matter effects, new interactions, etc. We discuss in the following sections: Physics & Extra long baseline experiment; AGS Upgrade; Neutrino Superbeam; Detector; and Outlook.

---

[1] Supported by US Department of Energy contract Number DE-AC02-98CH10886. E-mail: parsa@bnl.gov, URL: http://www.neutrinos.bnl.gov, member of BNL Neutrino Working Group

PHYSICS

The Atmospheric Neutrino "Anomaly" suggests that GeV ν_μ's (from $p + N \to \pi \to \mu\nu_\mu$) disappear while traversing the Earth's diameter, indicating $\Delta m_{32}^2 = m_3^2 - m_2^2 = \pm 2.0^{+1.0}_{-0.7} \times 10^{-3}(eV)^2$ for $\sin^2 2\theta_{23} \simeq 0.85 - 1.0$. The value of Δm_{32}^2 has decreased over the years, with recent reductions from [3] $3.0 \to 2.5 \to 2.0 \times 10^{-3} eV^2$. Fortunately this change is good, for experiments with very long baselines ($L \simeq$ 2000–4000) such as our BNL to HomStake, (WIPP or Henderson) proposal.

Solar neutrino ($\nu_e \to \nu_e$ and $\nu_e \to \nu_k$) oscillation experiments and the Kamland reactor study of $\bar{\nu}_e$ disappearance prefer [4] $\Delta m_{21}^2 = m_2^2 - m_1^2 = 7.3 \pm 1 \times 10^{-5} eV^2$, and $\sin^2 2\theta_{12} \simeq 0.84 \pm 0.10$.

Increased interest in the Neutrino oscillation physics span from the solar neutrino deficit and some evidence for $\nu_\mu \to \nu_e$, oscillations (from the LSND experiment), as well as the exciting atmospheric neutrino results including measurements of the atmospheric Muon - Neutrino deficit from the SuperK (Superkamiokande) experiment that has provided convincing evidence for lepton flavor violation. The experimental results interpeted is based on oscillation of one neutrino flavor ν_e, ν_μ and ν_τ, (state $|\nu_\ell >, \ell = e, \mu, \tau$) into others and are related to the neutrino mass eigenstates $|\nu_i >, i = 1, 2, 3$ (with masses m_i) by U a 3×3 unitary matrix, with $c_{ij} = \cos\theta_{ij}$, and $s_{ij} = \sin\theta_{ij}$:

$$U = \begin{pmatrix} c_{12}c_{13} & s_{12}c_{13} & s_{13}e^{-i\delta} \\ -s_{12}c_{23} - c_{12}s_{23}s_{13}e^{i\delta} & c_{12}c_{23} - s_{12}s_{23}s_{13}e^{i\delta} & s_{23}c_{13} \\ s_{12}s_{23} - c_{12}c_{23}s_{13}e^{i\delta} & -c_{12}s_{23} - s_{12}c_{23}s_{13}e^{i\delta} & c_{23}c_{13} \end{pmatrix}$$

Extra Long-Baseline Physics

Extra-long neutrino flight paths provide the possibility of observing multiple nodes of the neutrino oscillation (probability) in appearance and disappearance experiments. Observation of such a pattern will directly demonstrate the oscillatory nature of the flavor changing phenomenon. For fixed distance L, the oscillation maxima will occur roughly at energies of

$$E_\nu(n) = \frac{\Delta m_{32}^2 L}{2(2n-1)\pi},$$
$$n = 1, 2, 3, \ldots \quad (1)$$

For a given E_ν and L, the oscillation of $\nu_\mu \to \nu_e$ appearance can be described by:

$$P(\nu_\mu \to \nu_e) = 4(s_2^2 s_3^2 c_3^2 + J_{CP} \sin\Delta_{21})\sin^2\frac{\Delta_{21}}{2} \quad (2)$$

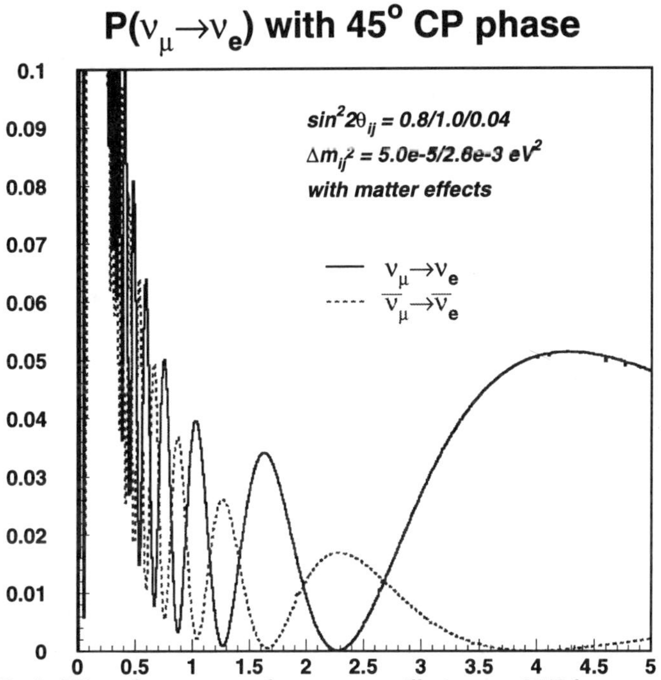

P($\nu_\mu \to \nu_e$) with 45° CP phase

$\sin^2 2\theta_{ij} = 0.8/1.0/0.04$
$\Delta m_{ij}^2 = 5.0e$-$5/2.6e$-3 eV^2
with matter effects

—— $\nu_\mu \to \nu_e$
----- $\bar\nu_\mu \to \bar\nu_e$

FIGURE 1. Probability of $\nu_\mu \to \nu_e$ and $\bar\nu_\mu \to \bar\nu_e$ oscillations at 2540 km assuming a 45° CP violation phase, including matter effect.

$$+2(s_1 s_2 s_3 c_1 c_2 c_3^2 \cos\delta - s_1^2 s_2^2 s_3^2 c_3^2) \sin\Delta_{31} \sin\Delta_{21}$$
$$+4(s_1^2 c_1^2 c_2^2 c_3^2 + s_1^4 s_2^2 s_3^2 c_3^2 - 2s_1^3 s_2 s_3 c_1 c_2 c_3^2 \cos\delta$$
$$-J_{CP} \sin\Delta_{31}) \sin^2 \frac{\Delta_{21}}{2} + 8(s_1 s_2 s_3 c_1 c_2 c_3^2 \cos\delta$$
$$-s_1^2 s_2^2 s_3^2 c_3^2) \sin^2 \frac{\Delta_{31}}{2} \sin^2 \frac{\Delta_{21}}{2} + \text{matter effects}$$

Where, $c_i \equiv \cos\theta_i$, $s_i \equiv \sin\theta_i$, $J_{CP} \equiv s_1 s_2 s_3 c_1 c_2 c_3^2 \sin\delta$, $\Delta_{31} \equiv \Delta m_{31}^2 L/2E_\nu$, and $\Delta_{21} \equiv \Delta m_{21}^2 L/2E_\nu$.

J_{CP} is an invariant that quantifies CP violation in the neutrino sector; Δ_{31} is the atmospheric term and Δ_{21} is the solar term [7]. For $P(\bar\nu_\mu \to \bar\nu_e)$ the above formula holds except J_{CP} terms will have opposite sign and matter effect will change.

The oscillation is primarily due to the first term linear in $\sin^2 \frac{\Delta_{31}}{2}$, and oscillation probability rises for lower energies due to the terms linear in $\sin^2 \frac{\Delta_{21}}{2}$.

The interference terms involve CP violation and they create an asymmetry between neutrinos and anti-neutrinos. The CP asymmetry grows linearly with distance:

$$A_{\text{CP}} = \frac{P(\nu_\mu \to \nu_e) - P(\bar\nu_\mu \to \bar\nu_e)}{P(\nu_\mu \to \nu_e) + P(\bar\nu_\mu \to \bar\nu_e)} \quad (3)$$

$$\simeq \frac{2s_1c_1c_2\sin\delta}{s_2s_3} \frac{\Delta m_{21}^2}{\Delta m_{31}^2} \frac{\Delta m_{31}^2 L}{4E_\nu} + O(\Delta_{21}^2) \quad (4)$$
+matter effects

or is given by (to order of Δm_{21}^2 assuming $\sin^2 2\theta_{13}$ is not too small)

$$A_{CP} \simeq \frac{\cos\theta_{23} \sin 2\theta_{12} \sin\delta}{\sin\theta_{23} \sin\theta_{13}} \left(\frac{\Delta m_{21}^2 L}{4E_\nu} \right)$$
+matter effects (5)

In this expression, the asymmetry grows linearly with distance and increases as θ_{13} gets smaller, noted earlier.

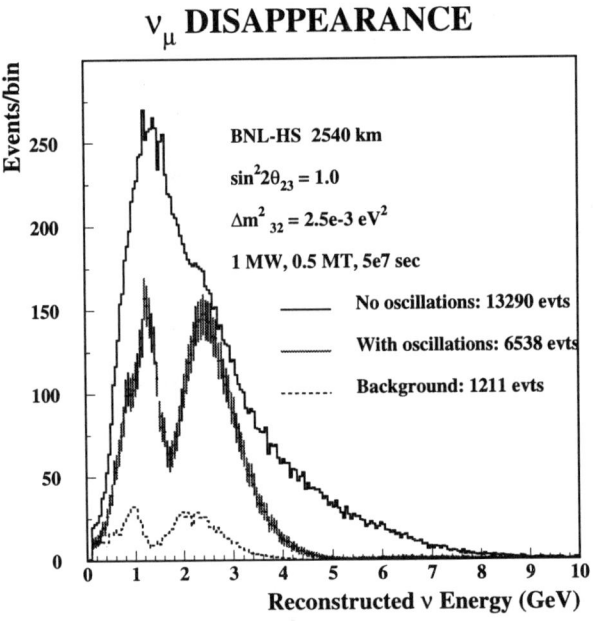

FIGURE 2. Example of expectected ν_μ disappearance spectra, without oscillations; (the middle) with oscillation and (the bottom histogram shows) the background contribution to the oscillating spectrum. This spectrum has improved with decrease in Δm_{32}^2, with recent reductions from [3] $3.0 \to 2.5 \to 2.0 \times 10^{-3} \text{eV}^2$.

Fig. 1 includes the matter effect since matter will enhance (suppress) neutrino (anti-neutrino) conversion at high energies and will lower (increase) the energy at which the oscillation maximum occurs, detection of matter enhancement effect can be made by measuring the asymmetry between neutrino and anti-neutrino oscillations (or by measuring the spectrum of electron neutrinos which also provide the sign of Δm_{32}^2).

If both the CP violation and the signal to $\nu_\mu \to \nu_e$ is large then effects of CP violation can be measured with only the (ν_μ) neutrino beam. It grows linearly with decrease in energy or the increase in baseline. For extra-long baseline experiments, comparison of the signal strength in the $\pi/2$ node versus the $3\pi/2$ (or higher) nodes will provide measurements of CP violation.

AGS UPGRADE AND νBEAM

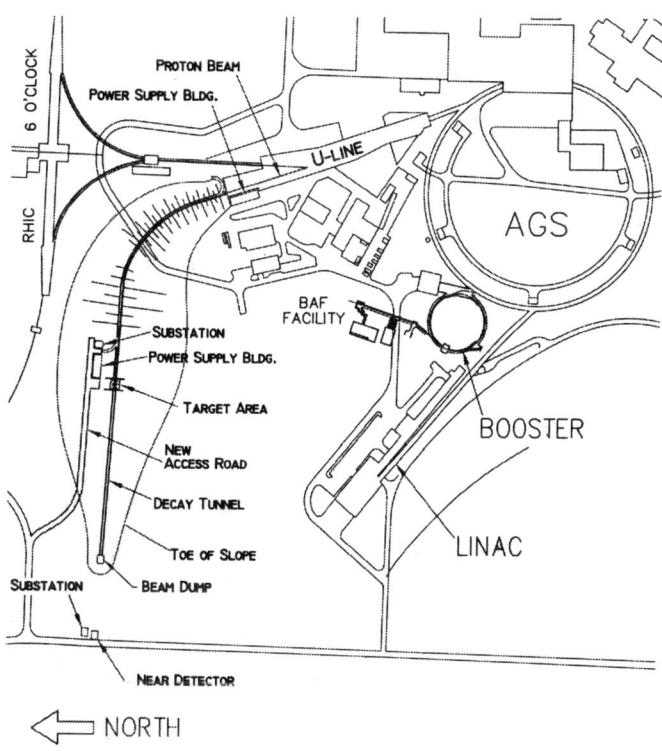

FIGURE 3. Schematic of the BNL-AGS RHIC facility showing location of the new beam-line for sending a neutrino beam to Homestake mine in South Dakota, and any detector in the Western direction.

The preliminary design of the BNL-AGS upgrades and the new neutrino beam has been produced by the AGS department [8] to reach an AGS power of e.g. 0.53 MW ($1.2 \times 10^{21} ppp$) in its first phase and 1.3 MW ($1.2 \times 10^{21} ppp$) in the second phase. In the first phase the LINAC will be improved to inject protons to the booster at 400 MeV (at present it is 200 MeV), and the booster energy increases to 2.5 GeV from 1.8 GeV. The addition of a fixed field accumulator storage ring between the booster and the AGS main ring will increase the AGS input beam from the present 4 booster pulses per AGS acceleration to 6 booster pulses per AGS acceleration and, at the same time, increase the AGS frequency from 0.6 Hz to 1.0 Hz. The AGS power increase would be from 0.14 to 0.53 MW. The new accumulator will be in the same tunnel as the AGS. In the second phase of the upgrades the AGS repetition rate will be increased to 2.5 Hz to reach a total beam power of 1.3 MW.

The proton beam is to be elevated to a target station on top of the hill. And the new proposed fast extracted proton beam line in the U-line tunnel will come off the line feeding RHIC. And will turn west, a few hundred meters before the horn-target building. In addition to its 90 degree bend, the extracted proton beam will be bent upward through 13.76 degrees to strike the proton target. The downward 11.30 degree angle of the 667.8 ft meson decay region will then be aimed at the 2500 meter level of the Homestake Laboratory. This will require the construction of a 39 meter hill to support the target-horn building, so as to avoid any penetration of the water table. At its midpoint (about Lake Michigan) the center of the neutrino beam will be roughly 120 km below the Earth's surface. (For a shorter baseline e.g., to Lansing NY in approximately the same direction as Homestake the hill won't be needed. Various combinations of the proton transport and the target station for the extra-long, (short/intermediate) baselines are being considered.)

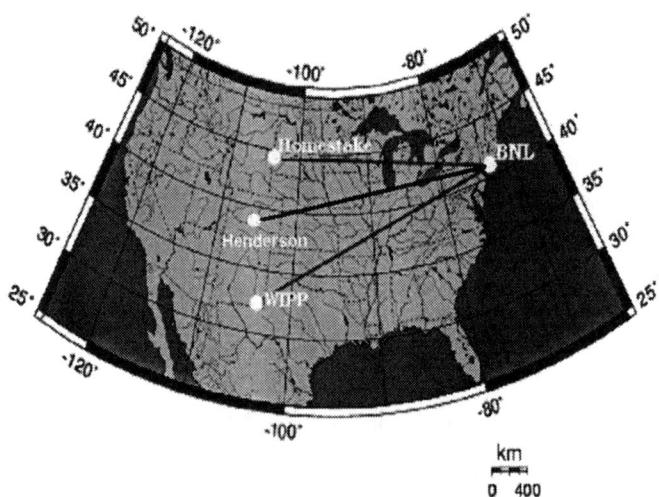

FIGURE 4. Possible extra long neutrino baselines from BNL to Lead (Homestake) SD ($\sim 2540 Km$, 11.5 degrees dip angle), to Carlsbad (WIPP) NM ($\sim 2900 km$, 13.0 degrees), and to the Henderson Mine in Colorado.

DETECTORS FOR THE VERY LONG BASELINE EXPERIMENT

There is an interest to convert the Homestake Gold Mine in Lead, South Dakota into a National Underground Science Laboratory (NUSL). This will provide unique opportunity for an extra-long baseline neutrino oscillation experiments from BNL. The extra-long baseline is 2540 km from the (Brookhaven National Laboratory) BNL to Lead, South Dakota. The proposed NUSL facility is to accommodate an array of detectors with about 1 Megaton total mass. Most of these will be water Cerenkov detectors that can observe neutrino interactions in the desired energy range with sufficient energy and time resolution.

Other detector types (e.g. Liquid Argon), and sites are also being considered, e.g.,Henderson Mine in Colorado, the Waste Isolation Pilot Plant (WIPP) located in an ancient salt bed at a depth of $\sim 700m$ near Carlsbad, New Mexico, etc. The distance from BNL to WIPP is about 2880 km,. The cosmic ray background will be higher at WIPP because the facility is not as deep as Homestake (with levels as deep as $\sim 2500m$).

OUTLOOK

Four goals of neutrino physics: precise determination of Δm_{32}^2, observation of $\nu_\mu \to \nu_e$ appearance, measurement of matter effects, and detection of CP violation are all possible with an intense neutrino broad band beam, very long distance baseline, and large detector. Both very long O(2500 km) and intermediate O(400 km) baseline experiments can be staged (from Brookhaven) as the AGS is upgraded to .5 MW, as much as 2.5 MW or higher (4 MW needed for a Neutrino Factory). AGS improvements will also allow rare muon and kaon decay studies, muon EDM measurements, etc. Thus providing additional windows for discovery.

REFERENCES

1. Z. Parsa, ν Superbeam for BNL-sec Procds. PAC20001, Chicago, Ill. ; ibid, Procds. EPAC2002, Paris France (2002). Presentation here are, from [1]-[8] and Refs therein.
2. BNL Neutrino Working Group: M. Diwan et al., hep-ex/0211001.ibid, Phys. Rev. **D68**, 012002; LOI submitted to BNL (April 2002) and Refs therein.
3. Y. Hayato, HEP2003 talk, Aachen Germany July 2003. For a review see V. Barger, D. Marfatia and K. Whisnant, Int. J. Mod. Phys. **E12**, 569 (2003).
4. SNO Collaboration, Q. Ahmad et al., nucl-ex/0309004; A. Balantekin and H. Yuksel, hep-ph/0309079.
5. CHOOZ: M. Apollonio et al., Phys. Lett. **B466**, 415 (1999), Palo Verde: F. Boehm et al., Nucl. Phys. Proc. Suppl. **91**, 91 (2001)
6. C. Jarlskog, Z. Phys. **C29**, 491 (1985); Phys. Rev. **D35**, 1685 (1987).
7. W. Marciano, "Extra Long Baseline Neutrino Oscillations and CP Violation", hep-ph/0108181, (22 Aug, 2001).
8. D Beavis et al., BNL Report 52459, AGS proposal 889.

Neutrinos Parallel Session—A Summary

B. E. Berger* and B. T. Fleming[†]

Lawrence Berkeley National Laboratory, Berkeley, CA 94720
[†]*Fermi National Accelerator Laboratory, Batavia, IL 60510*

Abstract. We summarize the presentations on neutrino research made during the *Neutrinos* parallel sessions at CIPANP2003.

INTRODUCTION

The last ten years have seen an enormous growth of knowledge about neutrinos and their properties. We now have strong evidence for two different neutrino flavor oscillations, called "atmospheric" and "solar" oscillations after the sources of neutrinos with which these oscillations were first observed. Where neutrinos were once held to be massless, we now know they have small but finite mass. A consistent picture has emerged that explains these two oscillations in terms of an "MNSP" (Maki-Nakagawa-Suzuki-Pontecorvo) matrix, analogous to the CKM matrix in the quark sector. There are are two independent mass differences, Δm_{12}^2 and Δm_{23}^2; three mixing angles θ_{ij}; and one CP phase δ_{CP}.

Even with all the progress in neutrino research in the last decade, our field is best defined by the questions that remain to be answered, so we organized our sessions accordingly. The atmospheric and solar neutrino oscillation parameters still have large uncertainties; these parameters will be measured much better in the next few years. The next round of neutrinoless double-beta decay measurements hope to answer several fundamental questions: What is the neutrino mass hierarchy? What is the absolute neutrino mass scale? Are neutrinos Dirac or Majorana particles? Future neutrino experiments will also attempt to measure the other MNSP matrix elements, θ_{13} and the CP phase δ_{CP}. Another open question is how to interpret the LSND [1] measurement, as it does not fit into the MNSP framework. The MiniBooNE [2] experiment will validate or contradict the LSND result. Several beyond-the-standard-model approaches have been proposed to incorporate LSND with the other two observed neutrino oscillations. Finally, high flux neutrino sources and improved detection techniques have allowed for high precision neutrino scattering measurements.

"SOLAR" NEUTRINO OSCILLATIONS: Δm_{12}^2 AND θ_{12}

The biggest advances in our understanding of neutrino properties in recent years have come from measurements of the solar neutrino oscillations. Our sessions included pre-

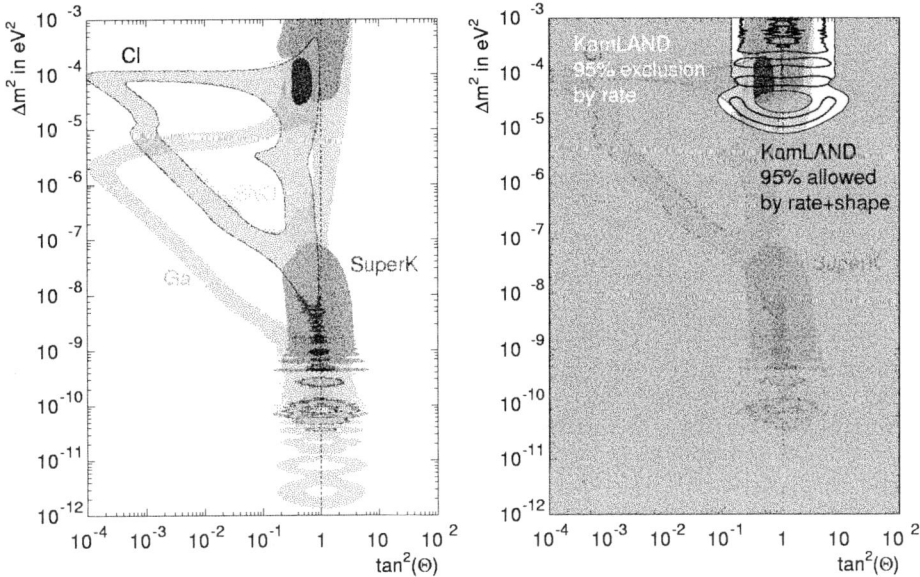

FIGURE 1. The constraints from "solar" neutrino oscillation experiments in Δm_{12}^2, $\tan^2 \theta_{12}$ parameter space. The left plot shows the allowed regions from the gallium and chlorine radiochemical experiments, from Super-K, and from SNO, plus the allowed region from a combined fit to these experiments. The right plot overlays the additional constraints from the KamLAND rate and rate+shape analyses.

sentations of new measurements from the SNO [3] and KamLAND [4] collaborations that were published within the last year. Neil McCauley and John Orrell presented talks on SNO, while Bruce Berger spoke about KamLAND.

In the MNSP framework, the solar neutrino oscillation is that between ν_1 and ν_2. Radiochemical experiments with chlorine and gallium plus water-Čerenkov detectors like Super-Kamiokande [5] have all measured the deficit of solar ν_e's relative to the number expected from the SSM (solar standard model). The deficits measured in these experiments are consistent with a wide range of parameters in very different regions of Δm_{12}^2, θ_{12} parameter space.

SNO is also a water-Čerenkov detector, but it has a heavy water (D_2O) target. It is sensitive to not only to ν_e's through charged-current (CC) interactions but also to the other neutrino flavors through elastic-scattering (ES) and neutral-current (NC) interactions. By measuring all three processes, SNO showed that the total neutrino flux, summed over flavors, is consistent with the SSM prediction. Furthermore, the results show an inferred 5.3σ appearance of $\nu_{\mu,\tau}$ in a ν_e beam, clear evidence for neutrino flavor change. The combination of the SNO results with the previous solar neutrino results rules out much of the previously-available oscillation parameter space, as shown in the left half of Figure 1.

KamLAND, by contrast, studies "solar" neutrino oscillations with a very different source: antineutrinos produced in nuclear power plants. KamLAND is located in a unique position in central Japan that allows the experiment to detect neutrinos produced by an ensemble of power plants. At a mean baseline of 180 km, the experiment detected $0.611 \pm 0.085(\text{stat}) \pm 0.041(\text{syst})$ of the number of neutrinos expected if there were no oscillations. KamLAND is sensitive only to one of the previously-possible solar oscillation parameter regions, the so-called LMA (large mixing angle) region, so the KamLAND result rules out all the other solar oscillation solutions. A "shape" analysis of the energy dependence of the observed signal further constrains the LMA region. Some LMA parameters would have produced large shape distortions in the KamLAND data, but such distortions were not observed. The additional constraints due to the KamLAND results are shown in the right half of Figure 1.

Both SNO and KamLAND will provide additional measurements of the solar neutrino oscillation parameters in the next few years. SNO will measure the day/night flux asymmetry, which is sensitive to Δm_{12}^2, and the CC/NC ratio, sensitive to θ_{12}. KamLAND continues to add statistics to the reactor antineutrino measurement. With more data KamLAND will be able to make a much better Δm_{12}^2 measurement, and it has some sensitivity to θ_{12}. Further down the road, a potential KamLAND solar phase could measure θ_{12} better with solar ^{7}Be neutrinos.

"ATMOSPHERIC" NEUTRINO OSCILLATIONS: Δm_{23}^2 AND θ_{23}

In the MNSP framework, the atmospheric oscillation is that between ν_2 and ν_3. Cosmic-ray showers in the earth's atmosphere produce both ν_μ's and ν_e's; the ν_μ's oscillate as the neutrinos propagate through the earth, principally into ν_τ's. The best current measurement of atmospheric neutrino oscillations comes from Super-Kamiokande [5] measurements of both the ν_μ and ν_e flux as a function of zenith angle.

The "atmospheric" oscillation will be measured to higher precision with accelerator neutrinos by the MINOS [6] experiment, as described in a talk by Hugh Gallagher. MINOS will look for the appearance of ν_τ in a ν_μ beam. The neutrino beam will be produced at Fermilab with the NuMI beamline, while the MINOS experiment itself is located 735 kilometers away in the Soudan mine in northern Minnesota.

NEUTRINOLESS DOUBLE-BETA DECAY

We devoted a full session to the topic of neutrinoless double-beta decay. Rabi Mohapatra began with a theoretical introduction. Neutrinoless double-beta decay is possible if the neutrino is a Majorana rather than a Dirac fermion, in which case it is its own antiparticle. The decay rate depends on the absolute neutrino mass scale, or more precisely an effective mass $\langle m_{\beta\beta} \rangle$ that depends on all the neutrino masses and mixing angles—plus new Majorana phases. Thanks to this dependence, a measurement of neutrinoless double-beta decay could probe the neutrino mass hierarchy. However, under the "normal" mass hierarchy $m_1 < m_2 < m_3$, the phases can conspire to suppress $\langle m_{\beta\beta} \rangle$ such that a null result in the search for neutrinoless double-beta decay cannot rule out Majorana neutrinos.

Neutrinoless double-beta decay experiments face difficult challenges. The signal itself is tiny. The two-neutrino double-beta decay is a background, so good energy resolution is required to distinguish the neutrinoless double-beta decay peak at the endpoint of the two-neutrino spectrum. Backgrounds from natural radioactivity demand the use of very clean materials, while cosmogenic backgrounds require the experimental site to be underground. In addition, the necessary nuclear matrix elements are difficult to calculate.

Our sessions included talks on three neutrinoless double-beta decay projects that have chosen very different experimental techniques. Rick Norman described the CUORE [7] experiment, which will use a bolometric technique to search for decays of ^{130}Te in tellurium crystals having the natural 33.9% ^{130}Te abundance. A pilot program, Cuoricino, is already running at Gran Sasso. Albert Young described the Majorana [8] project, which will take advantage of the excellent energy resolution of germanium ionization detectors, a well-established technology. Majorana will search for neutrinoless double-beta decay of ^{76}Ge at 85% isotopic enrichment. Finally, Peter Rowson described the EXO [9] experiment, based on 80% enriched ^{136}Xe. EXO will use a liquid xenon TPC, and this ambitious experiment will attempt to extract and identify the barium daughter of the xenon decay on an event-by-event basis, a unique method to reject background.

FUTURE MEASUREMENTS OF MNSP ELEMENTS

Solar and atmospheric oscillations probe a subset of the MNSP matrix elements. They cannot determine the final two parameters, namely θ_{13} and the CP phase δ_{CP}. Our sessions included three talks on future projects to measure these parameters.

Karsten Heeger spoke about the possibility of measuring θ_{13} with nuclear reactors as a source. The CHOOZ [10] and Palo Verde [11] experiments saw no flux deficit to the 3% level at a baseline of 1 km, while KamLAND has seen a $39 \pm 12\%$ deficit due to the solar neutrino (θ_{12}) oscillation at a mean distance of 180 km. The θ_{13} oscillation should give a small subdominant oscillation in the $\bar{\nu}_e$ flux on top of the large θ_{12} oscillation as a function of distance. The absolute magnitude of this subdominant term depends on the unknown θ_{13}, but the locations of the maxima and minima depend on Δm^2_{13}, which can be inferred from the known solar and atmospheric Δm^2's. A two-detector experiment with systematic errors at the 1% level could either measure this subdominant oscillation or set much tighter limits on θ_{13}. This is a very interesting idea, and an experiment could be running within a few years. Groups in the US, Japan, Russia, and Europe are all actively pursuing this idea.

Adam Para presented a talk on the possibility of using a long-baseline off-axis neutrino beam to measure MNSP matrix elements. An off-axis NuMI beam could be used to provide a narrow-band 2 GeV beam for a $\nu_\mu \to \nu_e$ counting experiment. The probability for this oscillation depends on multiple parameters: both θ_{13} and δ_{CP}, but also on matter effects. Runs with both neutrino and antineutrino beams would give complementary measurements to partially resolve the parameter degeneracy. In addition, other experiments at different baseline distances, for example JPARC to Super-K, would also give complementary information.

Zohreh Parsa presented a very ambitious idea [12] for the ultimate neutrino oscillation experiment, one that could measure all MNSP parameters in a single ultra-long-baseline experiment. One proposal is a 2540 km baseline from Brookhaven to the Homestake mine in South Dakota, with a 500 kiloton water Čerenkov detector such as UNO [13]. An upgrade of the AGS to 1 MW is proposed to provide a wideband (.5–5 GeV) on-axis beam, aimed down into the ground. The ultra-long baseline provides oscillations versus energy across the beam energy spread, which allows the mixing angles to be measured well. In addition, the shape of the observed oscillation depends on both δ_{CP} and matter effects, so this experiment would also be sensitive to the CP phase and the neutrino mass hierarchy.

SHORT BASELINE OSCILLATION RESULTS AND THEIR IMPLICATIONS

In addition to the strong evidence for solar and atmospheric neutrino oscillations, there is also evidence for neutrino oscillations at shorter baselines from the LSND [1] experiment. This result implies oscillations at high Δm^2 and small mixing. Because of the large Δm^2, this signal, along with the solar and atmospheric oscillation interpretations, cannot be explained with the three standard model neutrinos.

The LSND experiment observed $\bar{\nu}_e$ appearance in a $\bar{\nu}_\mu$ beam created from muon decay at rest. This signal is typically interpreted via a neutrino oscillation model, but it could also be described by a rare lepton-number-violating μ^+ decay: $\mu^+ \to e^+ \bar{\nu}\nu_e$. Klaus Eitel from Karmen, another short baseline neutrino oscillation experiment, presented new results on a search for this μ^+ decay [14]. Karmen does not observe this signal and therefore rules it out as a possible explanation for the LSND result. Karmen is sensitive to some of the LSND neutrino oscillation signal, as is Bugey, at higher mixing, but there is still a substantial amount of the LSND allowed region which must be addressed.

The MiniBooNE experiment, presented by Terry Hart, is designed to confirm or rule out the entire LSND signal to 5σ [2]. It runs at higher energy and at a longer baseline than the LSND experiment to preserve the oscillation parameter L/E. MiniBooNE tests the signal in an independent way, with different detection techniques, different systematic errors, and different backgrounds. MiniBooNE began data taking in August of 2002, expects first results on ν_μ disappearance and cross sections by Fall 2003, and expects ν_e appearance results by 2005.

If the LSND signal is due to oscillations, there are several beyond the standard model theories which can accommodate LSND along with solar and atmospheric neutrino oscillations. Gabriela Barenboim presented one such theory in which CPT is violated in the neutrino sector [15]. In this case, the different neutrino and antineutrino mass spectra can reconcile all three signals. By tagging the neutrino sign in atmospheric oscillations, the MINOS experiment will be able to address this theory in the near future. The combined results of Borexino and SNO compared to KamLAND can also address this model. Likewise, the MiniBooNE experiment will also test this theory.

A number of beyond the standard model theories, such as GUT's, SUSY, and those involving large extra dimensions, predict the existence of sterile neutrinos which cannot interact via the weak interaction but can oscillate with the usual three neutrinos. Theories involving one sterile neutrino, invoked to accommodate all three oscillation signals, in either a 3+1 or a 2+2 mass hierarchy, are increasingly disfavored by the latest neutrino oscillation data. However, Michel Sorel presented recent work showing that a fifth neutrino in a 3+2 mass hierarchy opens possibilities for such sterile neutrino theories [16]. Upcoming v_μ disappearance results from MiniBooNE and FINeSE [17] can directly address the mass hierarchy and what mixing parameters for these 3+2 models.

NEUTRINO SCATTERING PHYSICS

Intensive work in neutrino and accelerator physics over the last 30 years has led to high flux neutrino sources and much improved detection techniques. These advances have paved the way for a new generation of neutrino scattering physics experiments that probe other physics and contribute to understanding neutrinos.

M. Komatsu presented new results from CHORUS on charm hadron production measurements from vN deep inelastic scattering (DIS) interactions [18]. In the 1-20 GeV range there is rekindled interest in neutrino scattering physics in order to understand the DIS-to-resonance crossover region and to study nucleon structure at low Q^2. Eric Hawker gave an overview of low energy neutrino cross section data at these energies. Thia Keppel and Arie Bodek presented work on understanding the DIS-to-resonance region from charged lepton and neutrino scattering data [19]. These studies are crucial for the next generation of neutrino experiments that need good neutrino cross section models. Hugh Gallagher presented work on one such model called NEUGEN [20].

In addition to cross section measurements important as input for oscillation experiments, neutrino scattering can probe other physics. For example, neutrinos can pick out the strange spin of the nucleon, Δs, through measurement of the neutral weak current extrapolated to $Q^2 = 0$. Charged lepton experiments measuring Δs suffer from model dependence [21, 22], and the results from different experiments disagree even on the sign of Δs. Neutrino scattering cleanly picks out Δs and can provide an independent measurement. Morgan Wasko presented a soon-to-be-proposed experiment called FINeSE [17], designed to measure Δs at a near detector on the Fermilab Booster Neutrino Beamline. With a fine-grained detector followed by a muon range-out, FINeSE will also be able to measure neutrino cross sections to high precision and study v_μ disappearance in conjunction with the MiniBooNE experiment.

Neutrino cross sections at even lower energies, in the tens of MeV, are of interest to solar oscillation experiments and astrophysics. Malcolm Butler discussed modeling of inelastic vN cross sections in these regions in order to understand v-deuteron breakup reactions at SNO energies as well as pp fusion at threshold in the sun [23]. Measurements of these cross sections can be made at a high-precision low-energy neutrino scattering experiment at the Spallation Neutrino Source (SNS), presented by Bill Bugg [24]. SNS will be a copious source of neutrinos from muon decay at rest, allowing for cross section measurements crucial to astrophysics and nuclear theory.

CONCLUSIONS

This is an exciting period of neutrino physics, one in which new discoveries are being made rapidly. The field has changed dramatically in the three years since CIPANP2000, especially due to new measurements of solar neutrino oscillations. We expect many new results in the next three years, in time for CIPANP2006.

ACKNOWLEDGMENTS

We thank all the speakers who participated in our sessions. Their efforts and enthusiasm made our sessions enjoyable and interesting for us. We also thank Bill Marciano and Zohreh Parsa for the opportunity to organize the *Neutrinos* parallel sessions at CIPANP2003 and for their hard work in organizing such a successful conference.

REFERENCES

1. A. Aguilar *et al.* (LSND Collaboration), Phys. Rev. D **64**, 112007 (2001); http://www.neutrino.lanl.gov/LSND/.
2. http://www-boone.fnal.gov/.
3. Q. R. Ahmad *et al.* (SNO Collaboration), Phys. Rev. Lett. **87**, 071301 (2001); Phys. Rev. Lett. **89**, 011301 (2002); Phys. Rev. Lett. **89**, 011302 (2002); http://www.sno.phy.queensu.ca/.
4. K. Eguchi, *et al.* (KamLAND Collaboration), Phys. Rev. Lett. **90**, 021802 (2003); http://www.awa.tohoku.ac.jp/KamLAND/; http://kamland.lbl.gov/.
5. S. Fukuda *et al.* (Super-Kamiokande Collaboration), Phys. Lett. B **539**, 179 (2002); Y. Fukuda *et al.* (Super-Kamiokande Collaboration), Phys. Rev. Lett. **81**, 1562 (1998); Phys. Rev. Lett. **82**, 2644 (1999); http://www-sk.icrr.u-tokyo.ac.jp/doc/sk/index.html; http://www.phys.washington.edu/~superk/.
6. http://www-numi.fnal.gov/.
7. S. Pirro *et al.*, Nucl. Instrum. Methods **A44**, 71 (2000); http://crio.mib.infn.it/wig/halla/index_CUORE.html.
8. C. E. Aalseth *et al.*, hep-ex/0201021; http://majorana.pnl.gov/.
9. M. Danilov *et al.*, Phys. Lett. B **480**, 12 (2000); http://grattalab3.stanford.edu/exo/.
10. M. Apollonio *et al.* (CHOOZ Collaboration), Eur. Phys. J. C **27**, 331 (2003); http://www.pi.infn.it/chooz/.
11. F. Boehm *et al.* (Palo Verde Collaboration), Phys. Rev. D **62**, 072002 (2002).
12. M. V. Diwan *et al.*, Phys. Rev. D **68**, 012002 (2003); http://www.neutrino.bnl.gov/.
13. http://ale.physics.sunysb.edu/uno/.
14. B. Armbruster *et al.*, Phys. Rev. Lett. **90**, 181804 (2003).
15. G. Barenboim, *et al.*, hep-ph/0212116.
16. M. Sorel, *et al.*, hep-ph/0305255.
17. http://home.fnal.gov/~bfleming/finese.html
18. A. Kayis-Topaksu *et al.* (CHORUS Collaboration), Phys. Lett. B **555**, 156 (2003); Phys. Lett. B **549**, 48 (2002); Phys. Lett. B **527**, 173 (2002).
19. A. Bodek and U. K. Yang, hep-ex/0308007; H. Budd, *et al.*, hep-ex/0308005.
20. H. Gallagher, Nucl. Phys. Proc. Suppl. **112**, 188 (2002).
21. D. Adams *et al.*, Phys. Rev. D **56**, 5330 (1997) and references therein.
22. H. E. Jackson, Int. J. Mod. Phys. A17, 3551 (2002).
23. K. I. Brown, M. N. Butler and D. B. Guenther, nucl-th/0207008.
24. http://www.phy.ornl.gov/sns2/

NUCLEAR AND PARTICLE ASTROPHYSICS

Probing Dark Energy in the Accelerating Universe with SNAP

Michael Schubnell (for the SNAP Collaboration) [1]

Physics Department, University of Michigan, Ann Arbor, MI 48109

Abstract. It has now been firmly established that the Universe is expanding at an accelerated rate, driven by a presently unknown form of dark energy that appears to dominate our Universe today. A dedicated satellite mission has been designed to precisely map out the cosmological expansion history of the Universe and thereby determine the properties of the dark energy. The SuperNova / Acceleration Probe (SNAP) will study thousands of distant supernovae, each with unprecedented precision, using a 2-meter aperture telescope with a wide field, large-area optical-to-near-IR imager and high-throughput spectrograph. SNAP can not only determine the amount of dark energy with high precision, but test the nature of the dark energy by examining how its equation of state evolves. The images produced by SNAP will have an unprecedented combination of depth, solid-angle, angular resolution, and temporal sampling and will provide a rich program of auxiliary science.

INTRODUCTION

Recent measurements of luminosity distance versus redshift of nearby Type Ia supernovae by the Supernova Cosmology Project and the High-z Supernova Team have determined that the expansion of the Universe is accelerating [1, 2]. Furthermore, the results constrain the mass density, Ω_M, and the density of an unknown form of negative pressure energy, Ω_Λ, characterized by an equation of state $w \equiv p/\rho < -1/3$ causing the acceleration. This additional energy component, coined dark energy, appears to dominate energy density and dynamics of the Universe at the present epoch.

The evidence for dark energy is in remarkable concordance with other observations. Measurements of small scale fluctuations in the cosmic microwave background (CMB) radiation support the supernova results and have determined that the Universe is nearly flat [3]. Observations of galaxy clustering [4] have shown that the fraction of the critical density consisting of matter is $\Omega_M \approx 0.3$, also consistent with the results obtained from the supernova measurements (Figure 1). Combined, these results strongly suggest that - at the present epoch - at least 70% of the Universe's density is in the form of dark energy and only approximately 30% in some form of matter (which is mostly dark).

In its simplest form, dark energy might well be Einstein's cosmological constant in the form of a vacuum energy but numerous other theories have been proposed including the possibility of slowly evolving scalar fields (so-called quintessence models [5, 6].) When combined with CMB and galaxy cluster measurements, a tight bound on the dark energy equation of state w can be extracted (assuming it is constant over the expansion

---

[1] for a list of SNAP collaboration members see http://snap.lbl.gov

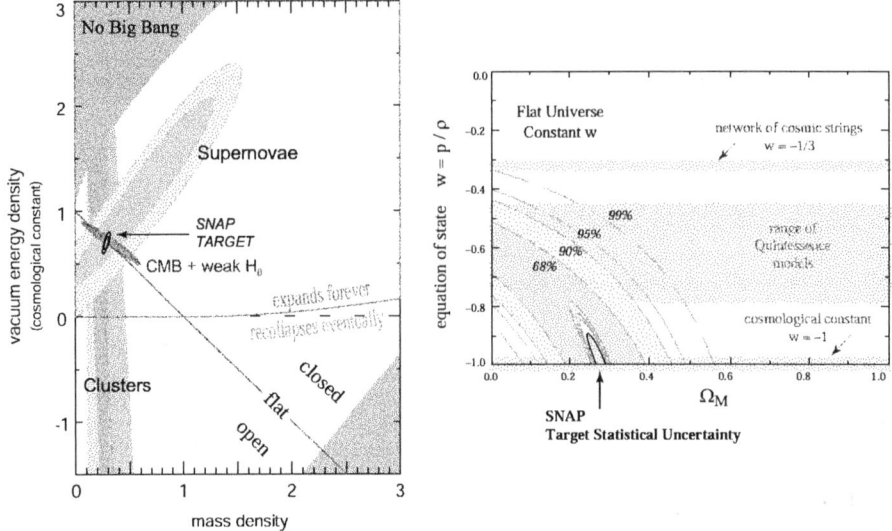

FIGURE 1. Left figure – Confidence regions in the $\Omega_\Lambda - \Omega_M$ plane for supernovae [1], galaxy cluster [4], and CMB [3] data. The consistent overlap is compelling evidence for a geometrically flat, dark energy dominated Universe. Also shown (in both figures) is the expected confidence region from the SNAP satellite for a flat $\Omega_M = 0.28$ Universe. Right figure – Constraints on the equation-of-state parameter w from the Supernova Cosmology Project [1]. Shown are confidence regions in the $\Omega_\Lambda - w$ plane for an energy density component Ω_Λ characterized by $w = p/\rho$. If the dark energy is Einstein's cosmological constant, then $w = -1$. Also shown are w predictions for other dark energy models.

time, i.e. $dw/dz = 0$). Current constraints on w are consistent with dark energy being a cosmological constant but also allow for a wide range of alternate models, including those with a time dependent value of w (Figure 1).

A precise measurement of dark energy properties requires a much larger data set of supernovae than currently available and a significant improvement of the systematic uncertainties in the measurements over current experiments. A definitive program to study dark energy with supernovae must provide a high degree of statistical and systematic rigor [7]. Furthermore, the greatest sensitivity to cosmological parameters is obtained with measurements extending from the present epoch of acceleration into the matter dominated deceleration phase [8]. Because measurements of the highly redshifted light from very distant supernovae require sensitivity into the near infrared (NIR), such a program can only be achieved in space, unhindered by absorption in the earth's atmosphere.

SATELLITE AND MISSION

The primary goal of the SNAP mission is to measure cosmological parameters with a precision that will allow to distinguish between different dark energy models. For this supernova observations provide a proven and well understood cosmological tool. The

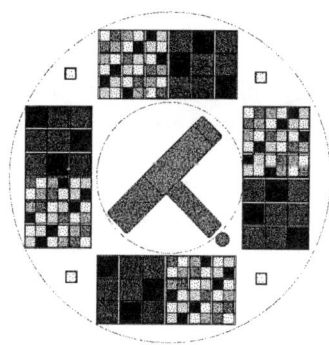

FIGURE 2. Left figure – Cross-sectional view of the SNAP satellite. Right figure – The SNAP focal plane. For detailed description of satellite and anstrument see [9, 10]

essential measurement for this purpose is a comparison of luminosity distance to redshift providing information on the scale size as a function of expansion time. With a precisely calibrated data set of several thousand Type Ia SNe with $z = 0.1 - 1.7$ the expansion history of the Universe can be reconstructed back to more than 70% of its age.

It has been shown that Type Ia supernovae have uniform peak B-band brightness when their light curves are corrected for a stretch factor which describes the relation between absolute brightness and explosion [1]. However, to fully standardize the SN peak brightness, a variety of additional observations must be made. Color measurements throughout the light curve for instance provide constraints on host-galaxy environment and galactic extinction and spectra obtained near maximum brightness allow identification of the explosion as a Type Ia through characteristic features (e.g. SiII at 6150 Å).

The SNAP satellite and mission design has been optimized for efficient supernova detection and high quality follow-up measurements. The combination of a three mirror 2-meter telescope and a $\approx$ 600 million pixel optical to near infrared imager with a large 0.7 square degree field of view will allow discovery and follow up of many supernovae at once. The imaging system comprises 36 large format (3.5k × 3.5k) CCDs and the same number of 2k × 2k HgCdTe infrared sensors. Both the CCDs and the NIR detectors are placed in four symmetric 3 × 3 arrangements as shown in Figure 2. Both the imager and a low resolution (R$\approx$100) high-throughput spectrograph cover the waveband from 350 to 1700 nm, allowing detailed characterization of supernovae out to $z = 1.7$. This deep reach in redshift is essential to the mission as it will allow to resolve degeneracies in cosmological parameters and to discriminate between models of dark energy.

Nine special filters fixed above the imaging sensors will provide overlapping redshifted B-band coverage from 350 – 1700 nm. As SNAP repeatedly steps across its target fields in the north and south ecliptic poles, every supernova will be seen in every filter in both the visible and NIR. Because of their larger linear size, each NIR filter will be visited with twice the exposure time of the visible filters. This, combined with the time-dilated light curve, will ensure that Type Ia supernovae out to redshift 1.7 will be detected with a S/N > 6 at least 2 magnitudes below peak brightness [9, 10].

SNAP SCIENCE

SNAP will conduct two primary surveys, a ~ 15 square degree ultra-deep ($m_{AB} \sim 30$ for point sources) supernova survey, and a ~ 300 square degree deep ($m_{AB} \sim 27.8$ for point sources) weak lensing survey. With this wealth of detailed data, SNAP will construct a Hubble diagram with unprecedented control over systematic uncertainties, addressing all known and proposed sources of error. The first goal is to provide precision measurements of the cosmological parameters: the matter density, Ω_M, will be measured to ± 0.02, while Ω_Λ, and the curvature parameter, Ω_k, will both be determined to an accuracy of ± 0.04. The SNAP measurements will be largely orthogonal to the CMB measurements in the $\Omega_M - \Omega_\Lambda$ plane, and the curvature measurement at $z \approx 1$ will test cosmological models by comparison with the CMB determination at $z \approx 1000$. SNAP's science reach will then extend to an exploration of the nature of the dark energy, measuring the present equation of state, w, to 5%. Of even more interest is a determination of w as a function of redshift. SNAP's tight control of systematics and high statistics in each redshift bin allows determination of the dynamical variation of w.

To complement its supernova cosmology observations, SNAP will conduct a wide-area weak lensing survey. These weak lensing observations provide important independent measurements and complementary determinations of the dark matter and dark energy content of the Universe. They will substantially enhance SNAP's ability to constrain the nature of dark energy[11]. SNAP weak lensing observations benefit enormously from the high spatial resolution, the accurate photometric redshifts, and the very high surface density of resolved galaxies available in these deep observations.

Although the SNAP mission is tailored for supernova and weak lensing observations, with the large survey field, depth, spatial resolution, temporal sampling and wavelength coverage into the infrared, the resulting data sets will provide a rich program of auxiliary science. Here we highlight a few selected areas where the large area deep-field observations are expected to significantly impact our understanding of the Universe:

- Galaxies – Within the ultra-deep 15 square degree survey area, SNAP will obtain accurate photometric redshift measurements for at least 5×10^7 galaxies to $z=3.5$. This will provide a unique opportunity to study the evolution of galaxies through more than 90% of the age of the Universe.
- Galaxy clusters – Galaxy clusters, the most massive bound objects and probably largest structures in the Universe, provide important probes of our understanding of structure formation. The SNAP surveys will provide detailed information on roughly 15,000 galaxy clusters with masses above 5×10^{13} M$_\odot$.
- Quasars – NIR photometry extends the redshift range for quasar discovery using colors and dropout surveys. SNAP will be able to detect quasars beyond redshift 10, and to probe the quasar luminosity function to 100 times fainter than the brightest quasars.
- GRBs – Gamma-ray bursts continue to pose a great mystery. Recent observations point to GRBs as the product of core-collapse of super-massive short lived stars. If so, then GRBs may trace the star formation rate and thus GRBs coincident with the epoch of first stars formation are expected. The most distant GRB currently known occurred at redshift of 3.4. SNAP will be able to identify GRB afterglows to $z = 10$.

- Re-ionisation – Most likely the re-ionisation of the Universe did not occur as a single instant in time, but rather as a complex process happening at slightly different epochs in different parts of the Universe. By identifying many quasars and galaxies to $z = 10$, SNAP will map the epoch of re-ionization in unprecedented detail.
- Gravitational lensing – The high spatial resolution of SNAP NIR observations will enable the discovery of a large number of new strong lenses. In weak lensing measurements, SNAP spatial resolution and NIR sensitivity will allow the use of a huge number of faint, high-redshift background galaxies. With these galaxies, it will be possible to extend weak lensing studies to lower mass objects, and to study lens objects beyond $z = 1$ – measurements which are impossible from the ground.

CONCLUSION

SNAP presents a unique opportunity to probe the dark energy and advance our understanding of the Universe. It will discover and precisely measure thousands of supernovae of Type Ia and will provide a combination of depth, solid angle and angular resolution heretofore unachieved. From the data collected, it will be possible to precisely measure the equation of state of the Universe, measure the history of its accelerations and decelerations and to study the nature of dark energy, which is causing the current acceleration of the expansion of the Universe. SNAP will be able to measure the equation of state, w, of the Universe as well as its variation over time, w'. This detailed knowledge will allow to distinguish between different models for the nature of the dark energy and lead to deeper understanding of the Universe.

ACKNOWLEDGMENTS

This work was supported by the U.S. Department of Energy.

REFERENCES

1. Perlmutter, S., et al., *Astrophysical Journal*, **517**, 565–586 (1999).
2. Riess, A., et al., *Astronomical Journal*, **116**, 1009–1038 (1998).
3. Spergel, D. N., et al., *astro-ph/0302209* (2003).
4. Bahcall, N. A., et al., *Science*, **284**, 1481 (1999).
5. Ratra, B., and Peebles, P. J. E., *Phys. Rev D*, **37**, 3406–3427 (1988).
6. Caldwell, R. R., Dave, R., and Steinhardt, P., *Phys. Rev. Letters*, **80**, 1582–1585 (1999).
7. Kim, A., et al., *astro-ph/0305286* (2003).
8. Linder, E. V., and Huterer, D., *Phys. Rev. D*, **67**, 081303 (2003).
9. Lampton, M., et al., *astro-ph/0209549* (2002).
10. Lampton, M., et al., *astro-ph/0210003* (2003).
11. Rhodes, J., et al., *astro-ph/0304417* (2003).

Results from DAMA

R. Bernabei*, P. Belli*, F. Cappella*, F. Montecchia*, F. Nozzoli*, A. Incicchitti†, D. Prosperi†, R. Cerulli**, C.J. Dai‡, H.H. Kuang‡, J.M. Ma‡ and Z.P. Ye‡

*Dip. di Fisica, Università di Roma "Tor Vergata" and INFN, sez. Roma2 I-00133 Rome, Italy
†Dip. di Fisica, Università di Roma "La Sapienza" and INFN, sez. Roma I-00185 Rome, Italy
**INFN, LNGS I-67010 Assergi (AQ), Italy
‡IHEP, Chinese Academy - P.O. Box 918/3, Beijing 100039, China

Abstract. DAMA is an observatory for rare processes based on the development and use of various kinds of radiopure scintillators. Several low background set-ups have been realized with time passing and many rare processes have been investigated. Results achieved in the first four annual cycles investigating the WIMP annual modulation signature will be mainly addressed. Next perspectives are also mentioned.

The DAMA experiment – in the Gran Sasso underground laboratory of I.N.F.N. – is an observatory for rare processes (such as WIMPs direct detection, $\beta\beta$ processes, charge-non-conserving processes, Pauli exclusion principle violating processes, nucleon instability, detection of solar axions and search for exotics [1, 2, 3, 4, 5, 6, 7, 8, 9, 10, 11]) by developing and using low radioactive scintillators; the main activity field is the investigation on WIMPs in the galactic halo.

The main experimental set-ups are: i) the $\simeq$ 100 kg NaI(Tl) set-up [4], which has completed its data taking in July 2002; ii) the new 250 kg NaI(Tl) DAMA/LIBRA (Large sodium Iodide Bulk for RAre processes) set-up, operative since March 2003; iii) the $\simeq$ 6.5 kg liquid Xenon (LXe) pure scintillator [2]; iv) the R&D installation for tests on prototypes and small scale experiments. Moreover, in the framework of devoted R&D for higher radiopure detectors and PMTs, sample measurements are regularly carried out by means of the low background DAMA/Ge detector, installed deep underground since about a decade and, in some cases, at Ispra.

In the following, the results obtained by the highly radiopure $\simeq$ 100 kg NaI(Tl) (DAMA/NaI) set-up are summarized. In particular, its main goal has been the investigation on WIMPs in the galactic halo by the annual modulation signature. The set-up – which has completed its data taking in July 2002 after collecting data during seven annual cycles – and its performances have been described in details in [4]. As we have already pointed out, the annual modulation signature is very distinctive[3, 4, 5, 7, 8, 9] because a WIMP induced seasonal effect must simultaneously satisfy several requirements. At the time of this Conference, results obtained by analysing the data of the first four annual cycles (57986 kg·day statistics) were already released [3, 4, 5, 6, 7, 8, 9, 10, 11] and are summarized in the following.

A model independent analysis has given evidence for the presence of an annual modulation of the rate of the single hit events in the lowest energy interval (see Fig. 1).

The χ^2 test on the data disfavors the hypothesis of unmodulated behaviour (probability: $4 \cdot 10^{-4}$), while a behaviour compatible with the one expected for a WIMP signal is found by fitting the residuals with a cosine function [7, 8]. No known systematic effect or side

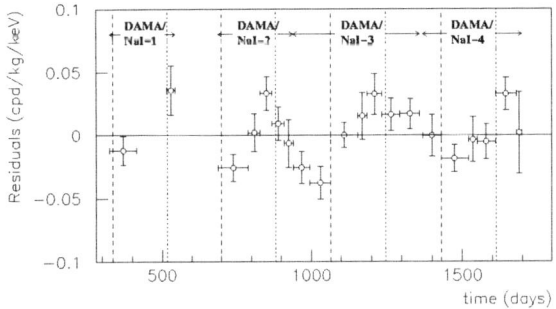

FIGURE 1. *A model independent analysis:* model independent residual rate for single hit events, in the 2–6 keV cumulative energy interval. The expected behaviour of a WIMP signal is a cosine function with minimum roughly at the dashed vertical lines and with maximum roughly at the dotted ones[7, 8].

reactions able to mimic such a signature has been identified (see ref. [8] for details). Thus, a WIMP contribution to the measured rate is candidate by the data independently on the nature and coupling with ordinary matter of the possible WIMP particle.

To investigate the nature and coupling with ordinary matter of a possible candidate, an energy and time correlation analysis of the data should be carried out in given model framework. This is identified not only by the general astrophysical, nuclear and particle physics assumptions, but also by the chosen values for all the parameters needed in the model itself and in related quantities, such as for example WIMP local velocity, form factor parameters, etc.

Firstly, a model dependent analysis has been carried out on the cumulative data in a purely spin-independent model framework[7, 11]; the region allowed at 3σ C.L. given by the superposition of all the allowed regions obtained in this given model framework considering several possible non-rotating halo models [11] is shown in Fig. 2. The shaded region (which corresponds to the particular case of the approximate and non-consistent isothermal sphere halo model when assuming also $v_0 = 220$ km/s and $\rho_0 = 0.3$ GeV/cm^3 for the astrophysical, nuclear and particle physics assumptions and fixed parameters of ref. [11]) is shown only to point out the effect due to the poor knowledge of the right halo model.

Secondly, since the ^{23}Na and ^{127}I nuclei are indeed fully sensitive to both spin-independent (SI) and spin-dependent (SD) couplings we have also extended the analysis to the more general (SI/SD) framework [9]. In this case an allowed volume in the ($\xi \sigma_{SI}$, $\xi \sigma_{SD}$, m_W, θ) four-dimensional space is obtained by the correlation analysis, where $\sigma_{SI(SD)}$ is the point-like SI (SD) WIMP cross section on nucleon and $tg\theta$ is the ratio between the effective SD coupling constants on neutron, a_n, and on proton, a_p; therefore, θ can assume values between 0 and π depending on the SD coupling. For simplicity, Fig. 3 on the left shows slices for some m_W of this volume allowed at 3σ C.L. for some fixed θ value in model frameworks of ref. [9].

Finally, the model independent annual modulation effect observed by DAMA during four annual cycles has been analysed, by energy and time correlation analysis, also in

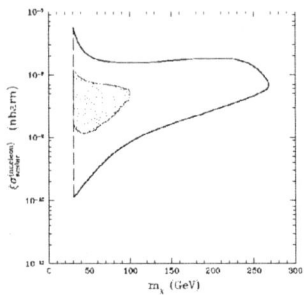

FIGURE 2. *A model dependent analysis:* region allowed at 3σ C.L. given by the superposition of all the allowed regions obtained, in given model frameworks, considering several possible non-rotating halo models [11]. In these calculations only uncertainties on the halo model have been considered; the inclusion of the other existing uncertainties would further enlarge it. This cumulative region, as those given e.g. in ref. [6, 7], accounts for a large set of best fit values for WIMP mass and cross section. The shaded region (which corresponds to the particular case of the approximate and non-consistent isothermal sphere halo model when assuming also $v_0 = 220$ km/s and $\rho_0 = 0.3$ GeV/cm^3 for the astrophysical, nuclear and particle physics assumptions and fixed parameters of ref. [11]) is shown only to point out the effect due to the poor knowledge of the right halo model. We note that no result directly comparable with the present one is available.

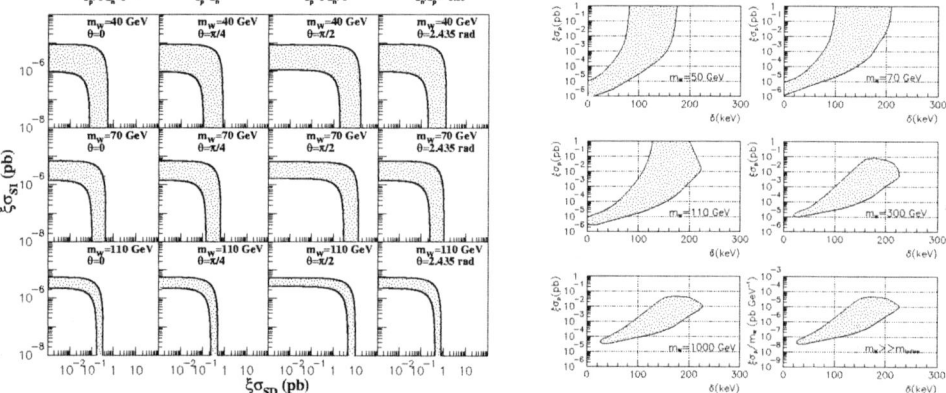

FIGURE 3. *Some model dependent analyses.* **On the left: a mixed SI/SD case.** Example of slices of the region allowed at 3 σ C.L. in the ($\xi\sigma_{SI}$, $\xi\sigma_{SD}$, m_W) space for some m_W and θ values in the model frameworks considered in ref. [9]. Only four particular couplings are reported here for simplicity: i) $\theta = 0$; ii) $\theta = \pi/4$ iii) $\theta = \pi/2$; iv) $\theta = 2.435$ rad. Note that e.g. Ge experiments are sensitive mainly only to SI coupling and, therefore, cannot explore most of the DAMA allowed regions in this scenario; the same is in most cases also for ^{nat}Xe since the odd-spin isotopes have the neutron as unpaired nucleon. **On the right: an inelastic case.** Slices at fixed WIMP masses of the volume allowed at 3 σ C.L. in the space ($\xi\sigma_p$, δ, m_W) obtained for the model frameworks considered in ref. [10]; some of the uncertainties on used parameters have been included. Note that e.g. Ge experiments cannot explore most of the DAMA allowed regions in this scenario. These allowed regions would be further enlarged by taking into account the uncertainties existing on the halo models and their parameters and on some other experimental and theoretical parameters.

terms of a particle Dark Matter candidate with a preferred inelastic scattering by making a transition to a slightly heavier state [12]. The found allowed volume [10] in the space

($\xi\sigma_p$, m_W, δ), where δ is the mass splitting, largely lies in δ regions where experiments with light nuclei (such as e.g. Ge) are disfavoured; see Fig. 3 on the right.

We remark that in all the presented model frameworks the inclusion of present uncertainties on other astrophysical, nuclear and particle physics parameters would enlarge these regions (increasing the sets of best fit values).

Finally, comparison with other model dependent results can be found in ref. [13]. Here we just mention that no model independent comparison with the DAMA/NaI effect is available. Only few model dependent approaches have been used in direct searches to claim for a particular model dependent comparison, which appears in addition neither based on solid procedures nor fully correct nor complete. On the other hand, the indirect search approaches, which can also offer only model dependent comparisons, are either not in contradiction or in substantial agreement with the DAMA/NaI observed effect [13].

As mentioned, the data taking of the $\simeq$ 100 kg NaI(Tl) set-up has been completed on July 2002. For the sake of completeness we mention that during the writing of these Proceedings the cumulative data analysis of the seven annual cycles has been released (total statistics of 107731 kg · d). The data over the seven annual cycles confirm the model independent effect observed in the first four annual cycles summarized at this Conference and favour the presence of WIMPs in the galactic halo at 6.3 σ C.L. [14].

After the completion of the data taking of the DAMA/NaI set-up, the procedures to install the new DAMA/LIBRA set-up (consisting of $\simeq$ 250 kg of radiopure NaI(Tl)) have been carried out. Many improvements have been realized and DAMA/LIBRA is operative since March 2003.

REFERENCES

1. R. Bernabei et al., Astrop. Phys.5 (1996) 217; *Phys. Lett.* B **389** (1996) 757; Il Nuovo Cimento C**19** (1996) 537; Astrop. Phys. **7** (1997) 73; Il Nuovo Cimento A **110** (1997) 189; *Phys. Lett.* B **408** (1997) 439 ; Astrop. Phys. **10** (1999) 115; Nucl. Phys. B**563** (1999) 97; Phys. Lett. B**465** (1999) 315; *Il Nuovo Cimento* A **112** (1999) 1541; *Phys. Rev. Lett.* **83** (1999) 4918 ; Phys. Rev. D**61** (2000) 117301; Phys. Lett. B**493** (2000) 12; Eur. Phys. J. direct C **11** (2001) 1; Phys. Lett. B**527** (2002) 182; Nucl. Phys. A**705** (2002) 29; Phys. Lett. B**546** (2002) 23; *Phys. Lett.* B **387** (1996) 222; *Phys. Lett.* B **436** (1998) 379; *Phys. Rev.* C **60** (1999) 065501; *Phys. Lett.* B **460** (1999) 236; *New Journal of Physics* **2** (2000) 15.1; *Phys. Lett.* B **515** (2001) 6 ; *Eur. Phys. J.-direct* C **14** (2002) 1; *Nucl. Instr. & Meth.* A **498** (2003) 352
2. R. Bernabei et al., *Nucl. Instr. & Meth.* A **482** (2002) 728
3. R. Bernabei et al., *Phys. Lett.* B **424** (1998) 195
4. R. Bernabei et al., *Il Nuovo Cim.* A **112** (1999) 545
5. R. Bernabei et al., *Phys. Lett.* B **450** (1999) 448
6. P. Belli et al., *Phys. Rev.* D **61** (1999) 023512
7. R. Bernabei et al., *Phys. Lett.* B **480** (2000) 23
8. R. Bernabei et al., *Eur. Phys. J.* C **18** (2000) 283
9. R. Bernabei et al., *Phys. Lett.* B **509** (2001) 197
10. R. Bernabei el al., *Eur. Phys. J.* C **23** (2002) 61
11. P. Belli et al., *Phys. Rev.* D **66** (2002) 043503
12. D. Smith and N. Weiner, *Phys. Rev.* D **64** (2001) 043502
13. R. Bernabei et al., ROM2F/2003/9 to appear on the Proceed. of "Neutrino Telescopes, Venice 2003; *astro-ph/0305542*.
14. R. Bernabei et al., *Riv. N. Cim.* **26** N.1 (2003) 1-73; *astro-ph/0307403*

ADMX Dark-Matter Axion Search

Leslie J Rosenberg
Lawrence Livermore National Laboratory
PO Box 808, Livermore CA 84551

August 22, 2003

Abstract

The axion, a hypothetical elementary particle, emerged from a compelling solution to the Strong-CP Problem in QCD. Subsequently, the axion was recognized to be a good Cold Dark Matter candidate. Although dark-matter axions have only feeble couplings to matter and radiation, these axions may be detected through resonant conversion of axions into microwave photons in a high-Q cavity threaded by a strong static magnetic field. This technique is at present the only means whereby dark-matter axions with plausible couplings may be detected at the required sensitivity. This talk describes recent results from the Axion Dark Matter Experiment (ADMX), now the world's most sensitive search for axions. There will also be a short overview of the ADMX upgrade, which promises sensitivity to even the more feebly coupled dark matter axions even should they make up only a minority fraction of the local dark matter halo.

1 Introduction

Peccei-Quinn (PQ) symmetry, from which the axion arises, still stands after more than two decades as the most minimal and elegant extension of the Standard Model to enforce Strong-CP in particle physics.[Peccei and Quinn(1977), Weinberg(1978), Wilczek(1978)] Within the presently allowed mass range $10^{-(6-3)}$ eV, the axion is also a good particle cold dark matter candidate. While the couplings of such light axions to matter and radiation would be exceedingly weak, it is realistic to expect that the status of the dark-matter axion may be definitively resolved within the next decade.

The original axion with the PQ symmetry-breaking scale $f_{PQ} \sim f_{EW} \sim$ 250 GeV implied an axion mass of a few hunderd keV, and couplings with matter and radiation that would have made it readily observable in reactor- and accelerator-based experiments.[Rosenberg and van Bibber(2000)]. It was quickly ruled-out, and axion models were constructed with $f_{PQ} \gg f_{EW}$, rendering the mass and all couplings extremely small and giving rise to the name 'invisible axion'.

In 1983, Pierre Sikivie of the University of Florida, among others, showed how halo dark matter axions could be detected by their Primakov conversion into microwave photons by a high-Q electromagnetic resonator permeated by a strong static magnetic field.[Sikivie(1983)] First generation experiments at BNL[DePanfilis(1987)] and the University of Florida[Hagmann(1990)] demonstrated the experimental technique and laid the groundwork for second-generation efforts in the US[Hagmann(1998)] and Japan[Ogawa(1997)]. More recently, the US effort "Axion Dark-Matter Experiment" (ADMX) has reached the sensitivity where realistic dark-matter halo axions could finally be detected. Upgrades to ADMX are underway that will increase the sensitivity to where even the more weakly coupled axions would be detected even if they comprised a minority fraction of the local halo density.

This talk will focus on ADMX present results and their plans for the upgrade.

2 The Cavity Microwave Experiment

Halo axions could be detected through their resonant conversion into photons in a high-Q microwave cavity threaded by a static magnetic field. In practice, a tunable helium-cooled high-Q cavity is placed within the bore of a superconducting solenoid, and the resonant frequency of the lowest cavity TM mode is slowly changed while the power spectrum within the cavity is monitored for excess power from axion-to-photon conversions.[Sikivie(1983)] The excess power for realistic axions in this type of experiment is expected to be less than 10^{-21} watts.

The axion mass is *a priori* unknown, as is the corresponding cavity resonant frequency. However, for axions comprising a significant fraction of the primordial dark matter energy density, the most likely axion mass is in the 1–10 μeV range; this is what makes the first decade of the axion mass search window so attractive. As the cavity radius for the TM_{010} mode of interest is R=0.115 m/ν[GeV], and 1 GHz = 4.136 μeV, this mass range also corresponds to an accessible range of cavity diameters, 5–50 cm.

3 The ADMX Detector

The Axion Dark-Matter Experiment (ADMX) has been designed, constructed and operated by a collaboration of Lawrence Livermore Laboratory (LLNL), the University of Florida, Lawrence Berkeley Laboratory, and the National Radio Astronomy Observatory (NRAO). The experiment is sited at LLNL. The key goals of this effort are (i) to achieve a power sensitivity to realistic axions that would constitute our halo; (ii) have sensitivity over the mass range over the decade 1.3–13. μeV. The sensitivity was achieved by employing a large-field (8 tesla) high volume (1 m long by 1/2 m diameter bore) magnet, as well as employing state-of-the-art low-noise amplifiers built by NRAO.

Figure 1 shows a sketch of the ADMX experiment. The magnet bore, 1 m

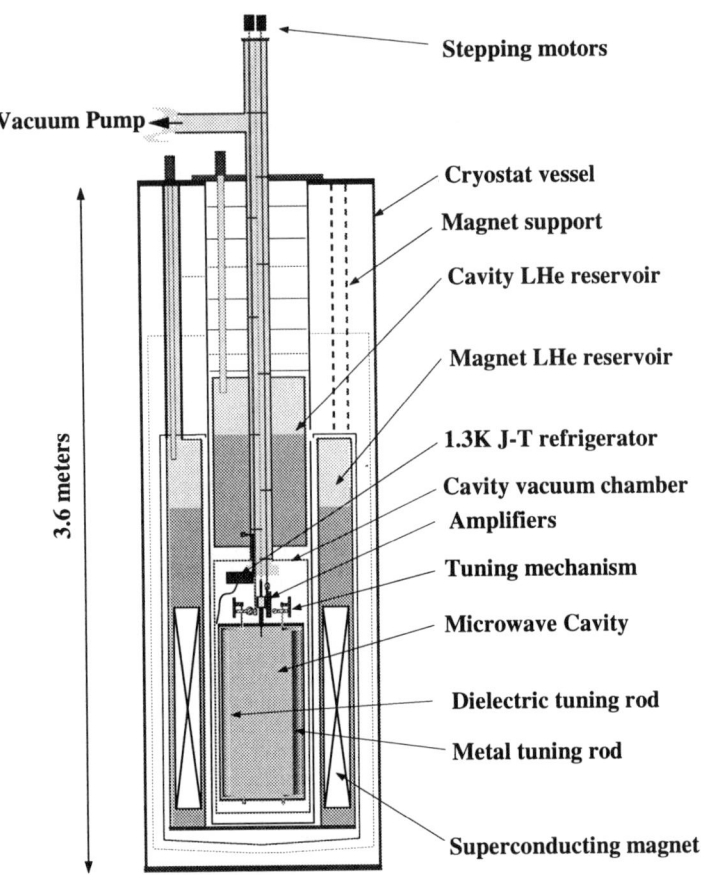

Figure 1: Sketch of the ADMX experiment.

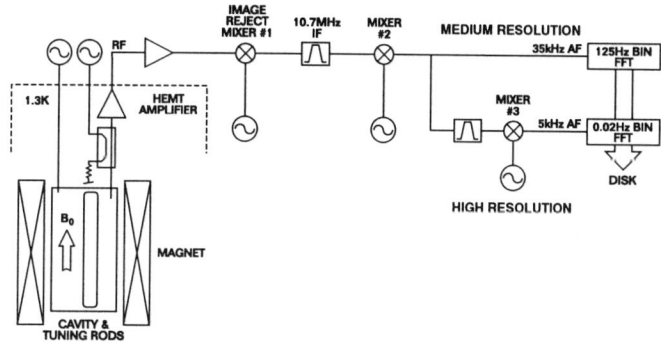

Figure 2: Block diagram of the axion receiver.

long by 1/2 m diameter, contains the resonant cavity. The cavity is tuned by moving tuning rods from near the cavity central axis to the cavity wall. The tuning range is approximately 300–900 MHz. The cavity and cryogenic electronics are operated at ∼1.3 K.

Some cavity power is extracted by a adjustable antenna inserted through the top-plate of the cavity. A second weakly-coupled port allows applying diagnostic signals. The extracted cavity power is amplified by two HFET cryogenic amplifiers in series. The combined cryogenic gain is around 34 dB. The noise temperature of these amplifiers is now below 2 K. Figure 2 shows the block diagram of the axion receiver. The power leaving the cryostat on a coaxial cable is mixed-down to a 10.7 MHz intermediate frequency, then applied to a 35 kHz wide filter. A second mixing stage places the cavity center frequency at 35 kHz. This near-audio (AF) power then splits into two hardware processing paths. In the first path, the AF signal is sent to a commercial fast Fourier transform instrument which computes 10,000 averages of a single-sided 8 ms power spectrum of 400 channels, each channel 125 Hz wide. This is the "medium resolution" path for which a virialized axion would appear as a peak approximately six channels wide. A second path records a 5 kHz wide single-sided power spectrum, of channel width 0.02 Hz, by an on-board ADC/FFT from a single 50 second integration. This is the "high resolution" path that is optimized for narrow peaks resulting from recent infall axions that have not yet virialized. We believe the ADMX microwave receiver is the world's lowest noise receiver. This is supported by Figure 3, which shows the single channel (125 Hz wide) relative power fluctuation about the mean versus the number of averages. The log-log slope falls as the expected inverse square-root of the number of averages up to about a month of continuous integration. This corresponds to a power sensitivity in this single channel of 10^{-26} watts.

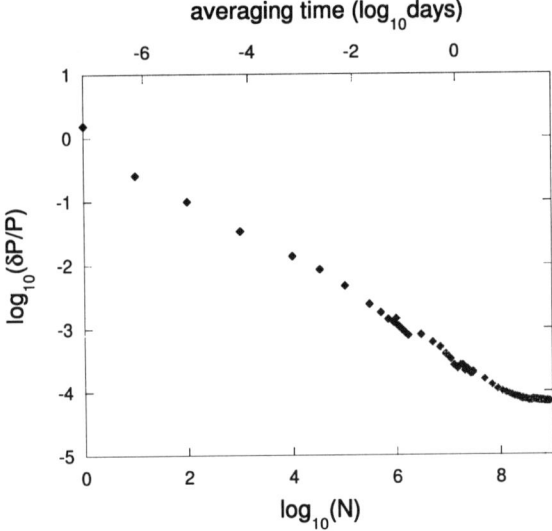

Figure 3: Single channel (125 Hz wide) relative power fluctuation about the mean versus the number of averages. The log-log slope falls as the expected inverse square-root of the number of averages up to about a month of continuous integration.

4 Recent Results

This experiment has analyzed data over the frequency range 460–812 MHz, corresponding to a range of axion masses 1.9–3.4 μeV. For this whole range we have reported limits from the medium resolution path. The high resolution path is only recently commissioned and for this path we have data only in the range 460–550 MHz; in this talk I am not reporting results from the high resolution path.

Figure 4 shows the limits we have placed on axions over the full mass range from the medium resolution path. The lower horizontal axis is frequency, the upper horizontal axis is the corresponding axion mass. The left vertical axis shows the upper limit to the power, in units of one benchmark KSVZ axion model, produced in the cavity at a particular frequency. The right vertical axis is the corresponding limit to the effective axion-to-two-photon coupling constant. The solid line is the limit from the very conservative assumption that axion power can be produced with any shape within a 6-channel window. The dotted line corresponds to the upper limit when we require any developed axion power have the expected virial line shape. These KSVZ axions are therefore excluded in the "prime hunting ground" of this region of the first mass decade. The more weakly-coupled GUT-inspired axions develop about a factor of 7 less power than this KSVZ axion and are not excluded. These results represent the first experiment sensitive to plausible dark-matter axions.

Figure 5 shows the limits we have placed on the contribution of axionic dark matter to the local dark-matter halo density. The horizontal axes are the same as for Figure 4, as are the meanings of the solid and dotted lines. The vertical axis is the upper limit to the contribution of axions to the local dark matter halo density, which is approximately 1/2 GeV/cc. This shows that the KSVZ benchmark axion is excluded as saturating the local halo dark matter, while GUT-inspired DFSZ axions remain candidates.

5 The ADMX Upgrade

There is a high premium on noise temperature in the ADMX experiment. For a fixed axion model, the search rate of the experiment improves as the inverse square of the system noise temperature. Our group has embarked on a program of incorporating DC SQUIDs as microwave amplifiers. These devices are typically not used at microwave frequencies as parasitic capacitance dramatically reduces the device gain. A recent breakthrough in coupling the input signal to the SQUID has significantly increased the useful upper frequency reach of these devices.[Mück(1998)] The recent incarnation of these devices are achieving noise temperatures of around 100 mK, within a factor of two of the quantum limit and a factor of 20 lower in noise than our present HFET amplifiers.

We are at present upgrading ADMX to incorporate these DC SQUID amplifiers. With the dilution refrigerator in place, this upgraded experiment will be sensitive to even the more pessimistically coupled axions even if they formed

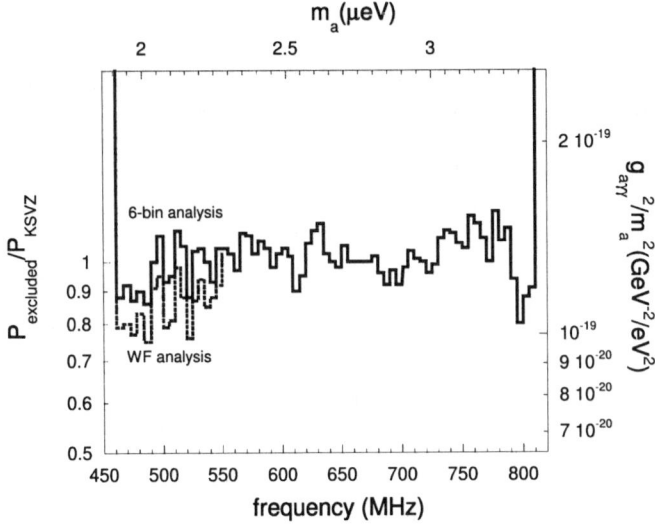

Figure 4: Limits on axion power and coupling. The lower horizontal axis is frequency, the upper horizontal axis is the corresponding axion mass. The left vertical axis shows the upper limit to the power, in units of one benchmark KSVZ axion model, produced within the cavity at a particular frequency. The right vertical axis is the corresponding limit to the effective axion-to-two-photon coupling constant. The solid line is the limit with the very conservative assumption that axion power can be produced with any shape within a 6-channel window. The dotted line corresponds to the upper limit when we require any developed axion power to have the expected virial line shape.

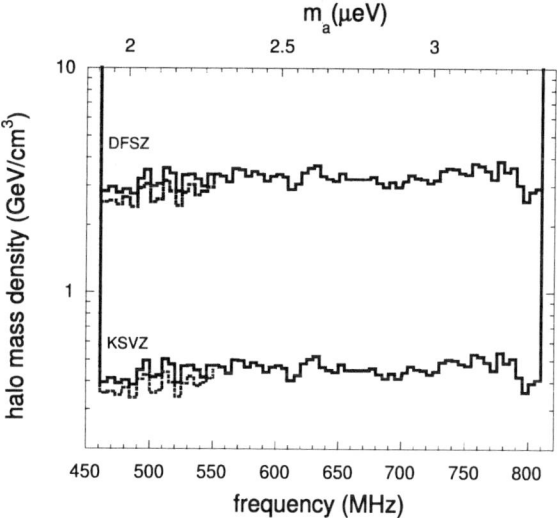

Figure 5: Limits on the contribution of axionic dark matter to the local dark-matter halo density. The horizontal axes are the same as for Figure 4, as are the meanings of the solid and dotted lines. The vertical axis is the upper limit to the contribution of axions to the local dark matter halo density.

a minority fraction of the local halo density.

6 Conclusions

The axion, a hypothetical elementary particle and good Cold Dark Matter candidate, once thought to be 'invisible' can now be detected by exquisitely sensitive RF cavity detectors. The most sensitive of the present axion detectors, ADMX, is sensitive enough to detect the more optimistically coupled darkmatter axions should they form a dominant fraction of the local halo density. ADMX is currently undergoing an upgrade to DC SQUID microwave amplifiers that promises to ultimately deliver sensitivity to even the more pessimistically coupled axions even if they formed a minority fraction of the local dark matter halo. Much more information on axion searches may be found in recent reviews.[Rosenberg and van Bibber(2000), Bradley(2003)]

7 Acknowledgments

Work performed under the auspices of the U.S. Department of Energy by the University of California, Lawrence Livermore National Laboratory under Contact W-7405-Eng-48.

References

[Peccei and Quinn(1977)] Peccei, R., and Quinn, H., *Phys. Rev. Lett.*, **38** (1977).

[Weinberg(1978)] Weinberg, S., *Phys. Rev. Lett.*, **40** (1978).

[Wilczek(1978)] Wilczek, F., *Phys. Rev. Lett.*, **40** (1978).

[Rosenberg and van Bibber(2000)] Rosenberg, L. J., and van Bibber, K. A., *Physics Reports*, **325**, 1–39 (2000).

[Sikivie(1983)] Sikivie, P., *Phys. Rev. Lett.*, **51** (1983).

[DePanfilis(1987)] DePanfilis, S. e. a., *Phys. Rev. Lett.*, **59** (1987).

[Hagmann(1990)] Hagmann, C. e. a., *Phys. Rev.*, **D90** (1990).

[Hagmann(1998)] Hagmann, C. e. a., *Phys. Rev. Lett.*, **80** (1998).

[Ogawa(1997)] Ogawa, I. e. a., *Proc. 2nd RESCEU Symp. on Dark Matter* (1997).

[Mück(1998)] Mück, M. e. a., *Appl. Phys. Lett.*, **72** (1998).

[Bradley(2003)] Bradley, R. e. a., *Rev. Mod. Physics*, **75**, 777–817 (2003).

An Experiment to Measure the Air Fluorescence Yield in Electromagnetic Showers

Petra Hüntemeyer for the FLASH Collaboration

Department of Physics, University of Utah, Salt Lake City, Utah 84112-0830, United States

Abstract. FLuorescence in Air from SHowers (FLASH) or E-165 is an experiment to be carried out at the Stanford Linear Accelerator Center (SLAC). It aims to measure the total and the spectrally resolved air fluorescence yield of electromagnetic showers with an accuracy of better than 10%. The experiment explores the energy dependence of the yield down to the lowest energies effective in air showers, $\sim$ 100 keV. For this experiment, the SLAC linac will deliver a 28.5 GeV electron beam at intensities of 10^7 to 10^9 particles per pulse.

A thin target run will allow us to measure the fluorescence yield per beam track depending on pressure and atmospheric impurities. Later, the interaction of the beam in a thick target will mimic the distribution of electron energies found deep in cosmic ray induced air showers. In June 2002, a test experiment at SLAC measuring the total fluorescence of air and nitrogen between 300 and 400 nm in a thin target mode has proven the feasibility of such an experiment. Results of this test run will be presented.

INTRODUCTION

To measure the energy spectrum of ultra high energy cosmic rays (UHECR) is the goal of several past, present, and future experiments using the air fluorescence technique. Perhaps the most interesting question in this regard today is if there is a GZK cut-off[1, 2] as predicted by well-established physical laws or if there is no end point and if the spectrum extends beyond 10^{20} eV. The High Resolution Fly's Eye (HiRes) detector utilizes air fluorescence to detect cosmic rays and to measure their energy. HiRes began operation in 1997 and has since then observed one event above 10^{20} eV[3]. AGASA, the world largest ground array located in Japan, utilizes the particle sampling technique. AGASA has accumulated data for a decade and has found seven super-GZK events with an aperture similar to that of HiRes[4]. The HiRes flux measurement is systematically lower than that of AGASA[5]. Several future experiments using the fluorescence technique are planned or are beginning construction, e.g. the hybrid detector of the Pierre Auger Observatory[6] in Argentina or the space-based fluorescence detectors EUSO[7] and OWL[8, 9]. They are designed to increase the detection aperture above the GZK cut-off and to search for point sources. The importance of the physics involved, the disagreement between HiRes and AGASA, and the expectation that other systematic uncertainties in existing and proposed future fluorescence-based cosmic ray experiments will be reduced significantly makes careful studies of the fluorescence yield necessary.

AN EXPERIMENT TO MEASURE THE AIR FLUORESCENCE YIELD

In his thesis from 1967 Bunner[10] summarized all existing data and quoted errors of around 30% on the listed fluorescence efficiencies. In a more recent experiment Kakimoto et al.[11] measured the total fluorescence yield between 300 and 400 nm with an uncertainty of >10%. The newest measurement using a ^{90}Sr β source by Nagano et al.[12], determined the total fluorescence yield between 300 nm and 406 nm with a systematic uncertainty of 13.2%. Both experiments also measured a number of individual spectral lines. The proposed FLASH experiment using the final focus test beam at SLAC aims for a systematic uncertainty of less than 10% in the net fluorescence yield and also in the yield of the individual spectral lines. SLAC is an ideal site for this study because the SLAC beams interacting in a thick target produce secondary electron energy distributions similar to those generated in extensive air showers (EAS). In addition, all the relevant N_2 fluorescence transitions are only accessible by electron excitation. Finally, the final focus test beam (FFTB) pulse has an energy equivalent to that of an EAS in the range of 10^{15} to 10^{20} eV.

The experimental program of FLASH is designed in view of the basic fluorescence issues:

- The gas composition will be varied, starting with a measurement of the yield in dry air. Then the yield in pure nitrogen and in a range of intermediate N_2/O_2 concentrations will be measured. In addition it is planned to introduce small quantities of argon, water vapor and possibly carbon dioxide to check for possible enhanced de-excitation effects from trace contaminants that are known to be present in the atmosphere.
- The pressure dependence of the yield will be studied with great precision in pressure sweeps between 10 Torr and 760 Torr.
- The fluorescence yield Y is expected to vary linearly with the energy deposited in the gas by the passing electrons. This will be studied by showering a beam in blocks of material of different thicknesses, and measuring the air fluorescence produced by the emerging particles.
- A detailed measurement of the fluorescence spectrum resolved to 10 nm is important in limiting the systematic uncertainty of fluorescence-based cosmic ray experiments, since the λ^{-4} dependence of Rayleigh scattering in the atmosphere leads to a significant distortion in the differential spectrum of light arriving at a detector from that emitted at the air shower.
- The decay time of fluorescence is a function of the transition involved and of the gas pressure. A measurement of the pressure dependence of the decay time for each spectral line is an important cross-check confirming that indeed air or N_2 fluorescence is measured and that possible backgrounds are understood.

The implementation of FLASH will be done in two stages. The first stage of the experiment will be run in a thin target mode. The electron beam will pass largely undisturbed through a small volume of gas at a rate of 10-30 Hz. In a test run in June 2002 (T-461), two PMTs located at 40 cm from the beam and viewing a 1 cm

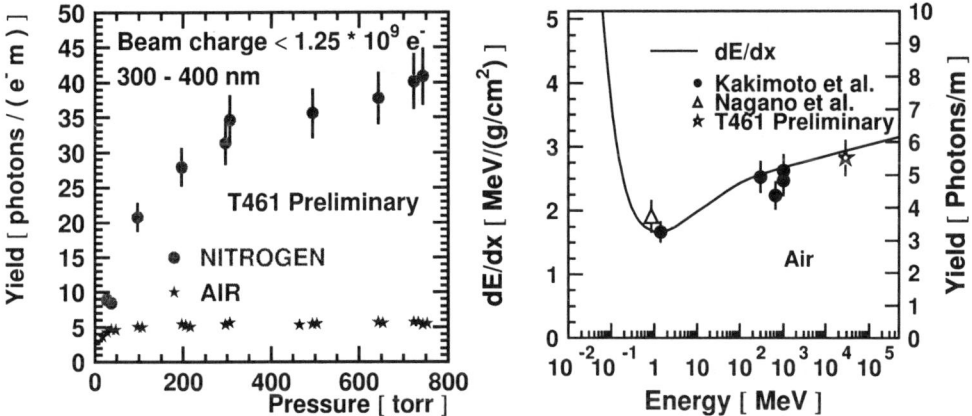

FIGURE 1. Left plot: Fluorescence yield in dry air and in nitrogen versus pressure. Right plot: Energy dependence of N_2 fluorescence in dry air at atmospheric pressure.

length of the beam line measured the fluorescence light directly produced by the beam electrons. T-461 has shown that Y is linear with respect to beam intensity below $\sim 10^9$ e^- / pulse. Like in T-461, HiRes wide-band filters will be used for FLASH to measure the total fluorescence yield between 300 nm and 400 nm. The PMT response stability will be monitored with UV LEDs fired between beam pulses and the background will be measured by filling the system with a non-fluorescing gas and by a installing a shutter in front of the detectors. Narrow band filters will be used to measure emission spectrum between 300 nm and 450 nm.

In the second stage of the implementation, the experiment will be run in a thick target mode to study the relation of Y to dE/dx. An electromagnetic shower is generated by the introduction of varying thicknesses of radiator into the beam. Al_2O_3 is the preferred material and will be assembled in remotely removable layers of square cross section, approximately 2 to 4 radiation length thick. A fluorescence chamber will be installed behind the shower material, with a thickness of only 4 cm along the beam to reduce the effect of the fast lateral spread of the showers in air.

RESULTS OF THE TEST RUN

In the left plot of figure 1, the fluorescence yield in dry air and in nitrogen versus pressure as measured by T-461 is shown in units of number of photons per electron per travelled distance in meters. The results are preliminary. The shape of the distribution roughly agrees with the measurements of Kakimoto and Nagano *et al.* but more detailed studies are necessary, and are currently being performed. T-461 obtained a ratio of the fluorescence yield in pure nitrogen and in dry air of around seven. In the right plot of figure 1, the photon yields versus electron energy as measured by Kakimoto and Nagano *et al.* are shown together with the preliminary result of T-461 at atmospheric pressure. The dE/dx curve is normalized to the Kakimoto measurement at an electron energy of 1.4 MeV. Assuming that Y scales as dE/dx, a good agreement of the preliminary result of T-461 with the preceding measurements by Kakimoto *et al.* and Nagano *et al.* is observed. The assigned systematic uncertainties of T-461 of around 10% (with the largest contribution, of around 7%, coming from the PMT calibration) are preliminary, and subject to ongoing studies. A refinement of the analysis is in progress.

CONCLUSIONS

In November 2002 the experiment FLASH or E-165 was approved by the Experimental Program Advisory Committee (EPAC) at SLAC. Since then the FLASH collaboration has been preparing the implementation of both the thin and thick target modes of the experiment in the test beam area of SLAC where we will measure the total and spectrally resolved fluorescence yield in air. The preliminary results of a test run conducted at SLAC in June 2002 are consistent with measurements of the total fluorescence yield by earlier experiments and serve as a valuable input to the experimental design and program of FLASH.

ACKNOWLEDGMENTS

This work is supported by US NSF grants PHY-9322298, PHY-9321949 PHY-9974537, PHY-0098826, PHY-0071069, by the DOE grant FG03-92ER40732, and by the Australian Research Council. The cooperation of Colonel E. Fischer, the US Army, and the Dugway Proving Ground staff is greatly appreciated.

REFERENCES

1. Greisen, K., *Phys. Rev. Lett.*, **16**, 748–750 (1966).
2. Zatsepin, G. T., and Kuzmin, V. A., *JETP Lett.*, **4**, 78–80 (1966).
3. Abu-Zayyad, T., et al. (2002), astro-ph/0208243.
4. Hayashida, N., et al., *Astrophys. J.*, **522**, 225 (1999).
5. Takeda, M., et al., *Phys. Rev. Lett.*, **81**, 1163–1166 (1998).
6. Mantsch, P., *SLAC Beam Line*, **28N3**, 4–11 (1998).
7. Scarsi, L., Catalano, ., Maccarone, M., and Sacco, B., *Proceedings of the 27th International Cosmic Ray Conference, Hamburg, Germany, 07-15 Aug 2001, Edited by K.-H. Kampert, G. Heinzelmann, C. Spiering, (Copernicus Gesellschaft e.V., Hamburg 2001)*, **HE**, 175 (2001).
8. Scarsi, L., *Proceedings of the 26th International Cosmic Ray Conference, Salt Lake City, Utah, 17-25 Aug 1999, Edited by B.L. Dingus, D.B. Kieda, M.H. Salamon*, **2**, 384 (1999).
9. Linsley, J., *Proceedings of the 26th International Cosmic Ray Conference, Salt Lake City, Utah, 17-25 Aug 1999, Edited by B.L. Dingus, D.B. Kieda, M.H. Salamon*, **2**, 423 (1999).
10. Bunner, A. N. (1967), Ph.D. Thesis, Cornell Univ., Ithaca, N. Y.
11. Kakimoto, F., et al., *Nucl. Instrum. Meth.*, **A372**, 527–533 (1996).
12. Nagano, M., Kobayakawa, K., Sakaki, N., and Ando, K. (2003), astro-ph/0303193.

Big Bang Nucleosynthesis and the Missing Hydrogen Mass in the Universe

D. C. Choudhury* and David W. Kraft[†]

*Department of Physics, Polytechnic University, Brooklyn, New York 11201, USA
E-mail: dchoudhu@duke-poly.edu
[†]Dana Hall, University of Bridgeport, Bridgeport, CT 06601, USA

Abstract. It is proposed that when the era of the big-bang nucleosynthesis ended, almost all of the 75 percent of the observed total baryonic matter remained in the form of hydrogen and continued to exist in the form of protons and electrons. They are present today as baryonic dark matter in the form of intergalactic hydrogen plasma. To test our hypothesis we have investigated the effects of Thomson scattering by free electrons on the reported dimming of Type Ia supernovae. The quantitative results of our calculation suggest that the dimming of these supernovae, which are dimmer than expected and hence more distant than predicted by Hubble expansion, is a result of Thomson scattering without cosmic acceleration.

Recent observations [1-2] of Type Ia supernovae (SNe Ia) appear to suggest that the universe is accelerating. The basic idea proposed to account for the acceleration is dark energy or quintessence [3-6]. However, as yet there is no direct confirmation of their existence or exact nature. The present work examines whether the question of acceleration can be resolved within the limits of established laws of physics. Consequently, we have investigated the effects of Thomson scattering of photons by free electrons present in the form of H-plasma while propagating from the supernova source to the point of observation. This particular scattering process has been chosen for the following reasons: the scattering cross-section is pure elastic in nonrelativistic region, the total cross-section is a universal constant and it is independent of the incident frequency [7]. Hence characteristics of the atomic spectra, which are relevant in the present work, remain unchanged. The calculation is performed within the framework of Friedmann-Robertson-Walker (FRW) cosmology [8] for the special case of a flat universe, consistent with the recent cosmic microwave background anisotropy measurements [9,10] indicating a spatially flat, critical density universe with $\Omega = 1$ and with the inflationary model [11-13] of cosmology. Hence in the present investigation we consider a universe, for large-scale structure, consisting only of matter and negligible radiation without cosmological constant in accord with all popular cosmological models prior to the late 1990s. Therefore in terms of standard notation,

$$\Omega = \frac{\rho(z)}{\rho_c(z)} = 1 \qquad (1)$$

and

$$\rho_c(z) = \frac{3H^2(z)}{8\pi G} = \frac{3H_0^2}{8\pi G}(1+z)^3 \qquad (2)$$

where ρ is the total mass density including radiation, ρ_c is the critical density, G is the gravitational constant, H is the Hubble constant and in which we have incorporated the flat universe relation between $H(z)$ and the current Hubble constant H_0.

In light of the above considerations, the total matter density of universe consists of ordinary baryonic matter, baryonic dark matter and nonbaryonic dark matter; thus we have $\Omega = \Omega_{bm} + \Omega_{bdm} + \Omega_{nbdm}$ where each term is in units of ρ_c, and where the subscripts refer, respectively, to the three aforementioned components. The observed mass density $\Omega_m \approx 0.30 \pm 0.10$ as estimated from its gravitational pull on visible matter and is assumed to consist of ordinary baryonic matter and baryonic dark matter instead of exotic dark matter [14]; thus $\Omega_m = \Omega_{bm} + \Omega_{bdm}$. Big bang nucleosynthesis and synthesis of heavy elements in stars have been extensively investigated for more than half a century. The quantitative results of atomic abundances of various groups of elements [15-16] by mass-fraction of the total are hydrogen $\cong 0.75$ and rest of the elements $\cong 0.25$. From the above value of Ω_m we obtain the magnitude of Ω_{bm} which includes all the elements formed during big bang nucleosynthesis, and of Ω_{bdm} which remained as free hydrogen (in the form of mostly protons and equal number of electrons, neutrons having decayed into protons and electrons) within half an hour after the era of big bang nucleosynthesis ended. These magnitudes are $0.05 \leq \Omega_{bm} \leq 0.10$ and $0.15 \leq \Omega_{bdm} \leq 0.30$. From this analysis we conclude that when the era of the big bang nucleosynthesis ended, the universe continued expanding for several thousand years or so until the temperature dropped low enough to form neutral atoms. We believe that it is in this time interval that most of the free protons and electrons (contained in Ω_{bdm}) escaped into cosmic space and are most likely present in the form of an intergalactic hydrogen plasma [17]. They are dark because they cannot emit light.

The intensity of the radiation lost to Thomson scattering depends critically on the density of free electrons in the path from the source to the observer. We take the free electron number density at redshift z as

$$n(z) = \frac{\Omega_{bdm} \rho_c(z)}{m_h} \qquad (3)$$

where $\Omega_{bdm} \rho_c$ is the hydrogen mass density and m_h is the hydrogen mass. We consider radiation emitted by a Type Ia supernova at redshift z_S and received by an infinitesimal volume element of length cdt at redshift z. The fractional reduction of intensity owing to Thomson scattering by free electrons within this volume is

$$\frac{-dI}{I(z)} = \sigma_T n(z) cdt \qquad (4)$$

where $I(z)$ is the incident intensity and σ_T is the total Thomson scattering cross-section by a free electron. The total attenuation by scattering is the integral of eq. (7) over the entire path taken by the light from the supernova. We perform the integration by identifying dt with $dT_H(z)$ where $T_H(z)$ is the Hubble time $H^{-1}(z)$. Thus

$$dt = dT_H(z) = -\frac{3}{2} H_0^{-1} (1+z)^{-5/2} dz. \qquad (5)$$

TABLE 1. Distance moduli are those reported in ref. [1]. Data at $z = 0.43$ and 0.48 are the mean values of two observations at each of these redshifts. Distances in columns 3-6 are computed from eq. (8). Column 3 lists distances computed from observed values of $m-M$ and the values in columns 4-6 result from distance moduli corrected for Thomson scattering; these three columns correspond to the extreme values of the hydrogen mass fraction Ω_{bdm} cited above and to their midpoint, $\Omega_{bdm} \simeq 0.23$. Values of the theoretical luminosity distance in the last column are computed from eq. (9). The best agreement with the theoretical values in column 7 are those listed for $\Omega_{bdm} \simeq 0.23$ in column 5.

Redshift z	Distance modulus $m-M$	Distance R_{obs} (Mpc)	With Thomson scattering correction			Theoretical distance D_L (Mpc)
			$\Omega_{bdm}=0.15$ (Mpc)	$\Omega_{bdm}=0.23$ (Mpc)	$\Omega_{bdm}=0.30$ (Mpc)	
0.0043	31.72	22	22	22	22	20
0.0077	32.81	36	36	36	36	36
0.025	35.35	117	117	117	117	115
0.052	36.72	221	219	218	217	241
0.053	37.12	265	263	262	261	248
0.068	37.58	328	324	322	320	318
0.090	38.51	504	495	491	487	424
0.17	39.95	977	945	929	915	825
0.30	41.38	1888	1775	1718	1670	1474
0.38	41.63	2118	1956	1874	1807	1893
0.43	42.15	2692	2454	2336	2238	2160
0.44	41.95	2455	2234	2296	2035	2214
0.48	42.39	3006	2783	2560	2438	2430
0.50	42.40	3020	2707	2553	2428	2539
0.57	42.76	3565	3138	2931	2766	2924
0.62	42.98	3945	3428	3179	2982	3203
0.83	43.67	5420	4447	4000	3656	4400
0.97	44.39	7551	5937	5221	4681	5225

Combining eqs. (2)-(5) then yields

$$\frac{dI}{I} = \sigma_T \Omega_{bdm} \frac{9cH_0}{16\pi Gm_h}(1+z)^{1/2}dz. \qquad (6)$$

Integrating over a path from the source to the observer at $z=0$ yields

$$I(0) = I_S \exp\left\{-\sigma_T \Omega_{bdm} \frac{9cH_0}{16\pi Gm_h}[(1+z)^{3/2}-1]\right\} \qquad (7)$$

where I_S is the light intensity at the source and in which the exponential factor represents the loss of intensity by Thomson scattering.

Observations of redshift and distance modulus $m-M$ for Type Ia supernovae reported by Riess et al. [1] are listed in Table 1 in columns 1 and 2. Column 3 contains the distance computed for each distance modulus according to [18]

$$R_{obs} = 10^{(m-M-25)/5} \qquad (8)$$

where R_{obs} is the distance in Mpc. Corrections for Thomson scattering are effected by multiplying the observed values of $m-M$ by the ratio $I(0)/I_S$ from eq. (7) and columns

4-6 contain the corrected distances as computed from eq. (8); these three columns correspond to the extreme values of the hydrogen mass fraction Ω_{bdm} listed above and to their midpoint, $\Omega_{bdm} \cong 0.23$. The calculations employ $H_0 = 65$ km-s^{-1}-Mpc^{-1}. The last column lists values of the theoretical luminosity distance D_L which, for a flat universe with deceleration parameter $q_0 = 1/2$, is given by [8]

$$D_L = \frac{2c}{H_0}(1+z-\sqrt{1+z}). \qquad (9)$$

Conclusion. The effects of Thomson scattering on enhanced dimming of SNe Ia presented in Table 1 suggests that (i) the recently observed supernovae data can be understood without dark energy; (ii) the amounts of baryonic ordinary matter, baryonic dark matter and nonbaryonic dark matter are in the ranges of 5-10%, 15-30% and 60-80%, respectively; and (iii) the total matter density Ω consists of 5-10% baryonic ordinary matter and 90-95% total dark matter, consistent with the understanding of most cosmologists prior to late 1990s. Further details of the present investigation will be published elsewhere.

REFERENCES

1. Riess, A.G. et al, *Astron. J.* **116**, 1009-1038 (1998).
2. Perlmutter, et al, *Astrophys. J.* **517**, 565-586 (1999).
3. Wang, L. and Steinhardt, P. J., *Astrophys. J.* **508**, 483-490 (1998).
4. Wang, L., Caldwell, R. R., Ostriker, J. P. and Steinhardt, P. J., *Astrophys. J.* **530**, 17-35 (2000).
5. Perlmutter, S., Turner, M. S. and White, M., *Phys. Rev. Lett.* **83**, 670-673 (1999).
6. Peebles, P. J. E. and Ratra, B., *Rev. Mod. Phys.* **75**, 559-606 (2003).
7. Heitler, W., *The Quantum Theory of Radiation*, 3rd ed.,(Dover, New York, 1984), pp. 34-35.
8. Kolb, E. W. and Turner, M. S., *The Early Universe*, paperback ed., (Addison-Wesley, Massachusetts, 1994), pp. 47-86.
9. de Bernardis, P. et al., *Nature* **404**, 955-959 (2000).
10. Netterfield, C. B. et al., *Astrophys. J.* **571**, 604-614 (2002).
11. Guth, A. H., *Phys. Rev.* **D23**, 347-356 (1981).
12. Linde, A. D., *Phys. Lett.* **B108**, 389-393 (1982).
13. Albrecht, A. and Steinhardt, P. J., *Phys. Rev. Lett.* **48**, 1220-1223 (1982).
14. Bahcall, N. A., Ostriker, J. P., Perlmutter, S. and Steinhardt, P. J., *Science* **284**, 1-16 (1999).
15. Burbidge, E. M., Burbidge, G. B., Fowler, W. A. and Hoyle, F., *Rev. Mod. Phys.* **29**, 547-650 (1957).
16. Boesgaard, A. M. and Steigman, G., *Ann. Rev. Astron. Astrophys.* **23**, 319-378 (1985).
17. Weinberg, S., *Gravitation and Cosmology* (Wiley, New York, 1972), p. 500.
18. Peebles, P. J. E., *Principles of Physical Cosmology*, (Princeton Univ. Press, Princeton, 1993), p.21.

Progress towards a FLUKA based simulation tool aimed at the evaluation of space radiation environments

V. Andersen*, F. Ballarini†, G. Battistoni**, M. Campanella‡, M. Carboni‡, F. Cerutti**, A. Empl*, A. Fassò§, A. Ferrari**¶, E. Gadioli**, M.V. Garzelli**, K. Lee*, A. Ottolenghi†, M. Pelliccioni‡, L.S. Pinsky*, J. Ranft‖, S. Roesler¶, P.R. Sala††** and T.L. Wilson‡‡

*Houston University, Texas, USA
†University of Pavia and INFN, Italy
**University of Milan and INFN, Italy
‡Laboratori Nazionali di Frascati, INFN, Italy
§SLAC, California, USA
¶CERN, Geneva, CH-1211 Switzerland
‖Siegen University, Germany
††ETH Zurich, Switzerland
‡‡NASA/JSC, Houston, USA

Abstract. Goal of the NASA funded FLEUR project is to develop a simulation tool to predict the impact of radiation environments, in particular to evaluate the effect of shielding in space applications. The heart of this tool is the FLUKA Monte Carlo transport code which is traditionally used in related areas of research such as radio-protection and dosimetry, cosmic ray physics and modeling of biological effects of radiation on DNA (in connection with further external micro codes). An important aspect in this context are heavy ion nuclear interactions which at this point have been implemented in FLUKA for high and medium energies while work is proceeding to cover the low energy range. Further information is available at http://www.fluka.org and http://fleur.cern.ch

INTRODUCTION

The Monte Carlo program FLUKA [1, 2] has traditionally been employed for radiation protection and dosimetry simulation, owing to its sound and internally consistent framework which relies primarily on microscopic physics models. In order to accurately predict effects of cosmic radiation in space, particularly in connection with shielding questions, the functionality of FLUKA is being extended to include heavy ion nuclear interaction by implementing suitable, available models.

For high energies, E from 5-10 GeV/n up to 10^9-10^{11} GeV/n in the laboratory frame, the DPMJET model [4, 5] was implemented[1]. DPMJET is a successful high energy nucleus-nucleus event generator based on the dual parton model which has been well

---

[1] The version of DPMJET currently implemented is DPMJET-II.53.

bench-marked against available experimental data. To cover the energy range from 5-10 GeV/n down to about 100 MeV/n the implementation of a relativistic Quantum-Molecular-Dynamics (QMD) model RQMD-2.4 [6, 7] was modified and adapted as an interim solution and is operational. To describe the final state of the interaction, specifically for possible heavy residual objects, both models profit from the FLUKA infrastructure by utilizing its pre-equilibrium and evaporation modules.

Further technical improvements to the FLUKA code relevant to this project include the extension of the employed combinatorial geometry model to include voxel geometries and the introduction of a new geometry input mode to facilitate a modular approach to complex configurations.

HEAVY ION NUCLEAR INTERACTIONS

FLUKA requires nucleus-nucleus reaction cross sections internally in order to select nucleus-nucleus interactions appropriately. Nucleus-nucleus reaction cross sections were prepared for a full matrix of projectile-target combinations up to a mass number of A=246. For projectile kinetic energies above 1 GeV/n these were simulated using DPMJET (based on the robust Glauber approach) while a modified version of a nucleus-nucleus cross section parameterization [8] was adopted for energies down to the Coulomb barrier. Figure 1 demonstrates the available reaction cross section as a function of projectile kinetic energy for one projectile-target combination.

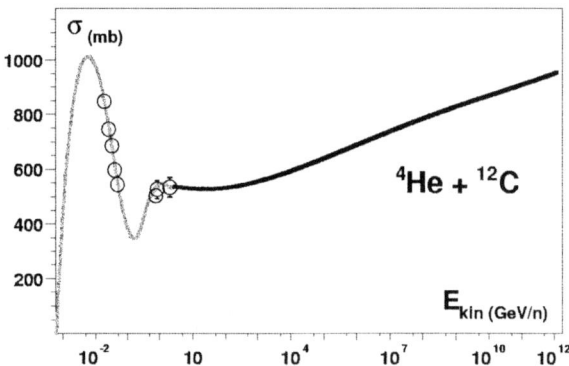

FIGURE 1. Combined reaction cross section prediction for ^{4}He + ^{12}C as a function of kinetic energy given by the adopted parameterization (grey curve) and the DPMJET model (black curve). Open circles indicate available experimental data. For a reference to the experimental data see [8] and therein.

The interface to the nucleus-nucleus event generators was first developed for the DPMJET model and had to be modified substantially in case of RQMD which does not identify nuclei in the final state and exhibits serious energy non-conservation issues. The adopted solution was to modify the code, reworking the nuclear final state out of the available information on spectators, correlating the excitation energy to the actual hole depth of hit nucleons. Remaining energy-momentum conservation issues were resolved taking into account experimental binding energies. After these improvements a meaningful excitation energy could be computed and the FLUKA evaporation model was used to produce the low energy (in their respective rest frames) particles emitted by the excited projectile and target residuals. Figure 2 gives an example of the performance of FLUKA making use of the modified RQMD-2.4 model simulating double differential neutron yields by 400 MeV/n Ar ions on thick Al targets.

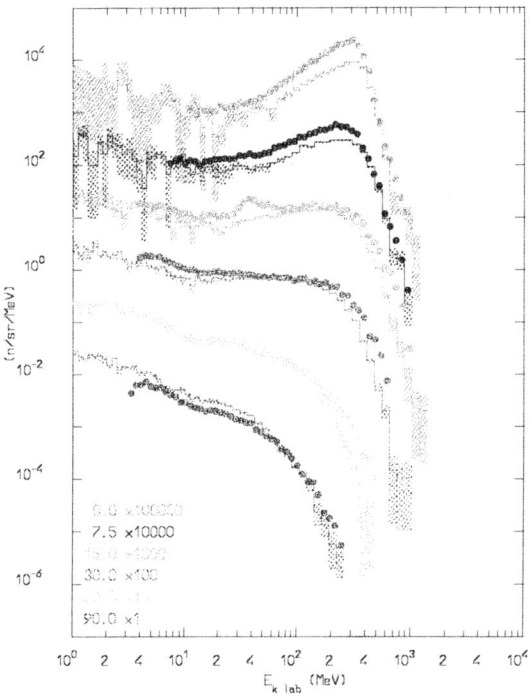

FIGURE 2. Double differential neutron yield by 400 MeV/n Ar ions on thick Al targets. Data are shown for six different laboratory emission angles, with the most forward on top: histograms are FLUKA results, dots experimental data [11].

While the implemented nucleus-nucleus models are available now, further benchmarking and testing is carried out in particular in case of the RQMD model implementation. In parallel, first steps have been accomplished to develop a (r)QMD model within the FLUKA framework [3] as a replacement for RQMD-2.4. In contrast to RQMD-2.4, the new model is geared towards medium/low energies, with particular attention for energy conservation and the nuclear structure (i.e. binding and excitation energies) from the beginning.

Because of the importance of light nuclear projectiles, the functionality of the hadron-nucleus interaction code within FLUKA (PEANUT) is being upgraded to include the treatment of systems with light nuclear projectiles (up to carbon) for energies below about 5 GeV/n.

Finally, work was started to interface a code that implements the Boltzmann master equation theory (BME) suitable for the Monte Carlo approach [9, 10]. The BME theory has been successfully used to describe nucleus-nucleus interactions from threshold up to an incident energy of 50 MeV/n and the intent is to extend its range up to about 100 MeV/n.

CONCLUSION

Heavy ion event generators have been interfaced with the FLUKA Monte Carlo code to implement nucleus-nucleus interaction for energies above 100 MeV/n. Work is underway to improve this new functionality and extend the applicable energy range down to the Coulomb barrier. The initial results are very encouraging and show that the code can already be applied to practical problems. Important progress to include all physics processes relevant to the detailed evaluation of space radiation environments in FLUKA has been made.

ACKNOWLEDGMENTS

This work was partially supported under NASA Grant NAG8-1658, ASI contract 1/R/320/02, and EC contract FIGH-CT1999-00005. In addition, research support from NASA Marshall Space Flight Center under grant no. NAG8-1901 is gratefully acknowledged.

REFERENCES

1. Fassò, A., A. Ferrari, P.R. Sala, Electron-photon transport in FLUKA: status, Proc. MonteCarlo 2000 Conference, Lisbon, October 23-26 2000, Springer-Verlag Berlin, 159-164, 2001.
2. Fassò, A., A. Ferrari, J. Ranft, P.R. Sala, FLUKA: Status and Prospective for Hadronic Applications, Proc. MonteCarlo 2000 Conference, Lisbon, October 23-26 2000, Springer-Verlag Berlin, 955-960, 2001.
3. Anderson, V. et al., The FLUKA Code for Space Applications: Recent Developments, in *Proc. World Space Congress 2002, Adv.Space Res.*, to be published.
4. Ranft, J., Dual Parton Model at cosmic ray energies, *Phys. Rev.* **D 51**, 64-84, 1995.
5. Roesler, S., R. Engel and J. Ranft, The Monte Carlo event generator DPMJET-III, Proc. Monte-Carlo 2000 Conference, Lisbon, October 23–26 2000, Springer-Verlag Berlin, 1033-1038, 2001.
6. Sorge, H., Flavor production in Pb(160 AGeV) on Pb collision: Effect of color ropes and hadronic rescattering, *Phys. Rev.* **C 52**, 3291-3314, 1995.
7. Sorge, H., H. Stocker and W. Greiner, Poincaré Invariant Hamiltonian Dynamics: Modelling Multi-hadronic Interactions in a Phase Space approach, *Ann. of. Phys.* **192**, 266-306, 1989.
8. Tripathi, R. K., F. A. Cucinotta and J. W. Wilson, Accurate universal parameterization of absorption cross sections III - light systems, *Nucl. Instr. Meth.* **B 155**, 349-356, 1999.
9. Cavinato, M., E. Fabrici, E. Gadioli, E. Gadioli Erba, and E. Galbiati, Monte Carlo calculations using the Boltzmann master equation theory of nuclear reactions, *Phys. Lett.*, **B382**, 1-5, 1996.
10. Cavinato, M., E. Fabrici, E. Gadioli, E. Gadioli Erba, and G. Riva, Monte Carlo calculations of heavy ion cross-sections based on the Boltzmann master equation theory, *Nucl. Phys.*, **A679**, 753-764, 2001.
11. Kurosawa, T., N. Nakao, T. Nakamura, et al., Neutron yields from thick C, Al, Cu and Pb targets bombarded by 400 MeV/nucleon Ar, Fe, Xe and 800 MeV/nucleon Si ions, *Phys. Rev.* **C 62**, 044615-1, 11, 2000.

Cluster Structure of Atomic Nuclei and Nucleosynthesis

Roman Ya. Kezerashvili

New York City College of Technology, City University of New York
300 Jay Street, Brooklyn, NY 11201.
Email: rkezerashvili@citytech.cuny.edu

Abstract. It is shown that the static and dynamic α–cluster models of nuclei, which describe an elastic electron scattering, photodisintegration reactions and pion double charge exchange reactions on α-cluster nuclei are in favor of the α-capture and α process of the formation of these nuclei.

One of the fundamental problems of astrophysics is the problem of the origin of chemical elements. There are different theories of the origin of the elements. Here we concentrated our attention on the formation of α–cluster or α-particle nuclei. The helium nucleus is the fundamental building block, which build up all these nuclei. There are basically two approaches to explain the origin of helium in its measured abundance through hydrogen burning. In the 1950s Gamow, Alpher and Herman (cf. Ref.1) proposed that ^{4}He produced by hydrogen burning in the early stage of a big bang some 10^{10} years ago. In Ref. 2 proposed that the stars are the seats on origin of the elements, and only hydrogen is primeval. In contrast with the theories, which demand matter in a particular primordial state, this approach is intimately related to the known fact that nuclear transformations are currently taking place inside stars, and many of detail of the observed abundances of the elements were explained in terms of stellar processes. Yet one major problem remained, the origin of helium. In the late 60th interest to a hot big bang cosmological model of synthesizes has been stimulated by the discovery of Penzias and Wilson [3] of the cosmic microwave background. In Refs. 4 and 5 were demonstrated that the existence of ^{4}He and the other light elements, together with the cosmic microwave background radiation, as a primary evidence in favor of the a big bang cosmological model. New measurements of the wavelength of the cosmic microwave background radiation have shown that it corresponds to the temperature of 2.728 K. Just a few years later, based on this fact Burbidge and Hoyle [6] found that the energy released in the synthesis of cosmic ^{4}He from hydrogen is almost exactly equal to the energy contained in the cosmic microwave background radiation, and this result strongly suggests that the ^{4}He was produced by hydrogen burning in stars and not in the early stage of a big bang.

Let's now begin by reviewing the nuclear reaction leading to α-particle element production at various stages of stellar evolution. When hydrogen burning in a star's main sequence stage leads eventually to hydrogen exhaustion, a helium core remains at the star's center. As helium builds up in the core of a star, the burning ceases, and the core contracts, and part of gravitational energy converted into thermal energy. When the temperature exceeds 10^8 K, and densities of about 10^5 g/cm^3 helium nuclei can overcome their mutual electrical repulsion, leading to the fusion processes: $^4He + {}^4He \rightarrow {}^8Be$ and then $^8Be + {}^4He \rightarrow {}^{12}C + \gamma$. The net result of these reactions is that three ^{4}He nuclei are combined into one carbon-^{12}C nucleus. At temperature above $2 \cdot 10^8$ K the ^{12}C produced in the helium fusion can capture an α particle to form ^{16}O and thus, continue the element synthesis: $^4He + {}^{12}C \rightarrow {}^{16}O + \gamma$. Further α-capture reactions depend critically on the existing excited states and parity. For example the rate of the $^{16}O(\alpha,\gamma)^{20}$Ne process depends on the excited states in ^{20}Ne at 4.95 and 5.62 MeV have a proper spin and parity to formed by ^{16}O and the α particles in their ground states. A resonance can also be expected to occur through helium capture by ^{20}Ne. But in the reaction ^{20}Ne$(\alpha,\gamma)^{24}$Mg the ^{24}Mg production will be small because of the large Coulomb barrier factor for α–particle. It is understandable that as the star evolves, heavier elements tend to form through helium capture rather than fusion of like nuclei, like the fusion of two ^{12}C nuclei to form ^{24}Mg or fusion of two ^{16}O to form ^{32}S. Because the repulsive force between two carbon nuclei is three times greater than the repulsion between carbon and helium, carbon-helium fusion occurs at a lower temperature than that at which carbon-carbon fusion occurs. Similarly the ^{16}O, which produced through helium capture by ^{12}C, may fuse with other ^{16}O to form ^{32}S, but it is much more probable that the capture process $^{16}O(\alpha,\gamma)^{20}$Ne to form ^{20}Ne. As a result, elements ^{4}He, ^{12}C, ^{16}O, ^{20}Ne, and ^{24}Mg stand out as prominent peaks in the chart of cosmic abundances. α-capture processes occur at temperatures between 10^8 and $2 \cdot 10^8$ K and as results in the exhaustion of the helium produced in hydrogen burning. An inner core of ^{12}C, ^{16}O, ^{20}Ne and perhaps a little ^{24}Mg develops in the star and eventually undergoes gravitation contraction and as a result conversion of the gravitation energy into heat just as occurred previously in the case of helium core. Gravitation is "a built-in" mechanism in stars, which leads to the development of high temperature in the ashes of exhausted nuclear fuel. Gravitation takes over whenever nuclear reaction stops; it raises the temperature to the point where the ashes of the previous process begin to burn. Implicit in this argument is the assumption that mixing of core and surrounding zones does not occur.

Around the time ^{28}Si appears in the core of a star, a competitive struggle begins between the continued capture of helium to produce even heavier nuclei and the tendency of more complex nuclei to break down into simpler ones. The cause of this breakdown is heat. By now the star's core temperature has reached the unimaginably large value of $3 \cdot 10^9$ K, and the gamma rays associated with that temperature have enough energy to promote photodisintegration reactions. α-particle binding energy in ^{12}C, ^{16}O, and ^{20}Ne nuclei are 7.15 MeV, 7.37 MeV and 4.75 MeV, respectively. This means that the sequence of photodisintegration reactions will be the following: ^{20}Ne$(\gamma,\alpha)^{16}$O, ^{16}O$(\gamma,\alpha)^{12}$C and $\gamma + {}^{12}C \rightarrow 3\alpha$. The photodisintegration reaction (γ,α) precedes (γ,p) and (γ,n) processes on ^{12}C, ^{16}O, and ^{20}Ne nuclei, because the proton and neutron binding energies in these nuclei are larger, therefore photo-dissociation threshold falls higher. α-particle released in the process ^{20}Ne$(\gamma,\alpha)^{16}$O, can now penetrate the Coulomb barrier of the other nearby ^{20}Ne nuclei that have not yet photodisintegrated may capture some or all of these ^{4}He nuclei, leading to the formation of still heavier elements: ^{20}Ne$(\alpha,\gamma)^{24}$Mg. Thus, once some ^{24}Mg is produced we also expect ^{24}Mg$(\alpha,\gamma)^{28}$Si to take place, since it is possible to penetrate the Coulomb barrier of ^{24}Mg at that temperature. Once an appreciable concentration of ^{28}Si is built up, the reaction ^{28}Si$(\alpha,\gamma)^{32}$S take place, and so on for the production of ^{36}A and ^{40}Ca. This two-step process—photodisintegration followed by the direct capture of some or all of the resulting ^{4}He nuclei—is called the α process [2]. The α process is responsible for building, in decreasing proportion, the α-particle nuclei from ^{24}Mg to ^{40}Ca. All these elements stand out as prominent peak in the relative abundances of cosmic matter. Of course, a proportion of the releasing α–particles is consumed in scouring out the previous α-particle nuclei. Thus, α–particle nuclei are synthesis through two sources: helium burning (α-capture) process and α process. The α process is very similar to helium burning. However, these processes are different from a point of view that α particle sources are quite different in the two cases and they occur in the different range of temperatures.

In atomic nuclei, as is well known, there are two type of correlations: short range correlation due to the strong repulsive part of nucleon-nucleon interaction and long range correlation, which lead to the formation of the nucleon associations or clusters within atomic nuclei. Let's consider the static cluster model of atomic nuclei. Following Ref. 6, we represent the spatial configurations of alpha-cluster nuclei of ^{12}C, ^{16}O, ^{20}Ne ^{24}Mg, ^{28}Si, and ^{32}S respectively in the form of an equilateral triangle, tetrahedron, and regular triangular, quadrangular, pentagonal, and hexagonal bipyramids. The nuclei ^{12}C, ^{16}O are characterized by one parameter R_1, namely the distance from the center of nucleus to the center of the α-particle formation, which is located in one the vertices of the triangle or tetrahedron. The nuclei of 2s-1d shells are characterized by two parameters: the distance R_1 from the center of nucleus to the center of the α-particle formation located at one of the vertices of the bipyramid, and the distance R_2 from the center of nucleus to the center of the α-particle formation located at the base of the bipyramid. Maximum-symmetry consideration [7] indicate that the centers of the α-particles formation, the distances between which are fixed, are singled out in the α-cluster model. Since the four nucleons are in the 1s states relative these centers, the density of a nucleus with 4N nucleons can be expressed in the form

$$\rho(r, R_k) = \frac{1}{N}\sum_{k=1}^{N}\rho(r - R_k), \quad (1)$$

where N is the number of α–clusters, $\rho(r - R_k)$ is the density distribution in the α-particle formation, and will be calculated using shell model wave function with Jastrow factor. Under these assumptions, the charge form factor for the elastic scattering of electron by an α-cluster nucleus takes a form

$$f(q) = \frac{1}{N}f_\alpha(q)[Aj_0(qR_1) + Bj_0(qR_2)], \quad (2)$$

where $f_\alpha(q)$ is the elastic form factor of ^{4}He, q is a transferred momentum and the coefficients A and B are related to the number of α–clusters. In our calculations the Jastrow short-range correlation are included into the ^{4}He charge form factors and charge form factors for the elastic scattering of electrons on the ^{12}C, ^{16}O, ^{20}Ne ^{24}Mg, ^{28}Si, and ^{32}S nuclei are calculated. Figure presents the results of our calculation.

Consideration of the short-range correlation in description of the charge form factor $f_\alpha(q)$ of ^{4}He, results an essential improvements of the agreements between the theoretical and experimental form factors. The best-fit values for R_1 and R_2 also predict the reasonable values for the size of the α–cluster nuclei. The considered model enables one to conclude on the extent of the cluster separation on the mass number. We also observe that the size of α–cluster increases with the increment of atomic number of a nucleus. The static cluster model of atomic nuclei also describes the quasielastic electron scattering on α–cluster nuclei under the assumption that the corresponding elementary process proceeds by the α–particle formation within the nucleus.

In Ref. 8 a dynamic model of α–cluster nuclei have been suggested and developed using the method of hyperspherical functions. A theory of complete α–particle photodisintegration of light nuclei was developed, assuming that the elementary process occurs through quasi-α–particle formation inside the nucleus. One of the examples of reactions with several α-particles in the final state is $\gamma + ^{12}C \to N\alpha$. The general reversibility of nuclear reactions makes it possible to obtain the information about reverse reaction $3\alpha \to ^{12}C + \gamma$ and understand its role in the evolution of the stars and production of the ^{12}C. We consider the internal structure of the α particle and studied the process of 3α photodisintegration by the method of hyperspherical functions in the coordinate representation. Expanding the wave function in the basis of four-particle hyperspherical functions and using NN potential from Ref. 9, calculate the internal structure of the α–particle. In the region of photon energies from the threshold up to 30 MeV the photodisintegration of ^{12}C into three α particles can proceed by two mechanisms: *i*. direct disintegration of ^{12}C into three interacted α particles as a result of the interaction of photon with a quasi-α-particle cluster inside the nucleus; *ii*. Two step disintegration of ^{12}C, with formation of ^{8}Be subsystem in an excited state as a first step, and its subsequent decay into two α particles as a second step. As our calculations show, the maximum in the cross section can be explained by the first mechanism, and there is no need that the process goes through the intermediate 2^+ state of ^{8}Be. The success of the α-cluster model of atomic nuclei was also demonstrated in Ref. 10, where the quasi-α-particles mechanism of pion double charge exchange reaction on light nuclei has been suggested and developed.

Thus, assumption of α-cluster structure of atomic nuclei is in very good agreement with varieties of observing experimental phenomena, and is just a reflection of the history of matter, on which we can make observations today.

REFERENCES

1. Alpher, R. A., & Herman R.C., *Rev. Mod. Phys.* **22**, 153 (1950).
2. Burbidge, E. M., Burbidge, G. R., Fowler, W. A. & Hoyle, F., *Rev. Mod. Phys.* **29**, 547 (1957).
3. Penzias, A., & Wilson, R., *ApJ.* **142**, 419 (1965).
4. Peebles, P. J. E., *ApJ.* **146**, 419 (1966).
5. Wagoner, R., Fowler, W. A., & Hoyle, F., *ApJ.* **148**, 3 (1967).
6. Dzhibuti, R. I., Kezerashvili, R. Ya., *Sov. J. Nucl. Phys.* **20**, 181 (1975).
7. Mattheis, Z., Neudachin, V. G., & Smirnov, Yu. F., *Sov. Phys., JETP* **18**, 281 (1964).
8. Dzhibuti, R. I., Kezerashvili, R. Ya., & Shubitidze, N. I., *J. Nucl. & Part. Phys.* **55**, 1801 (1992).
9. Gogny, D., Pires, P., & De Tourreil, R., *Phys. Lett.* **32B**, 591 (1970).
10. Jibuti, R.I., & Kezerashvili, R. Ya., *Nucl. Phys.* **A430**, 573 (1984).

Ultra High Energy Cosmic Rays

Todor Stanev

Bartol Research Institute, University of Delaware, Newark, DE 19716

Abstract. We discuss briefly the phenomenon of the Ultra High Energy Cosmic Rays (UHECR), particles of energy approaching and exceeding 10^{11} GeV. The world experimental statistics contains a small number of events, but their existence is a puzzle. Its solution may lead to exciting discoveries in high energy particle astrophysics, as well as in particle physics and astronomy.

INTRODUCTION

Soon after the discovery of the microwave background two almost simultaneous papers [1, 2] predicted that the radiation should cut-off the cosmic ray spectrum at energy about 5×10^{10} GeV - the GZK cut-off. Such high energy protons can interact with the 3°K microwave photons and there is enough room in CMS to contain a a secondary pion. Protons lose on the average about 1/5 of their energy in a single interaction and this limits their propagation in the Universe to 50 Mpc, a cosmologically negligible distance.

This prediction is correct and is universally accepted. The only problem is that a cosmic ray particle of energy above 10^{20} eV was detected three years earlier [3]. During the 40 years since such events were reported every couple of years. These rare events kept some interest in the problem until several years ago two powerful detectors were built and increased the world statistics relatively fast.

In the same time new bold projects were proposed and one of them, The Auger Southern Observatory [4] is now being constructed in Argentina. Even more powerful are the proposed satellite experiments for observation of the interactions of UHECR in the atmosphere, EUSO [5] and OWL [6].

These interactions are observed from the Earth's surface in two ways:
a) The interaction of the primary particles generates a huge atmospheric cascade, consisting of electrons, photons, muons and hadrons. The rule of thumb is that in the maximum shower development every electron carries about 1.5 GeV, i.e. the number of electrons in a 10^{20} eV shower approaches 10^{11}. At such energy the shower maximum is close to the surface of the Earth. The shower particles arrive almost simultaneously and trigger in coincidence particle detectors spread over many kilometers. Such detector arrangements are called air shower arrays.
b) The charged particles in the cascade induce nitrogen fluorescence that can be observed at distances of tens of kilometers by optical detectors. The rough number for the production of fluorescence light is 4 photons per meter pathlength of a charged particle. This is the atmospheric fluorescence technique, pioneered by the Fly's Eye detector.

The flux of particles of energy above 10^{20} eV is roughly 1 *per* 100 square kilometers *per* year *per* steradian. The world statistics above the GZK cut-off is dominated by the

Akeno [7] giant air shower array in Japan and by the new generation HiRes Fly's Eye detector [8].

EXPERIMENTAL DATA

Fig. 1 presents the recent results of the two detectors with the highest statistics above 10^{19} eV - AGASA [7] and HiRes [8]. The differential cosmic ray flux is multiplied by E^3. The energy assignment by the experiments affects strongly the normalization and overemphasizes the differences between the two groups. Still, the energy assignment is different by about 40%, which is the sum of the unceratinties defined by the experimental groups. There is no obvious energy dependence of these differences.

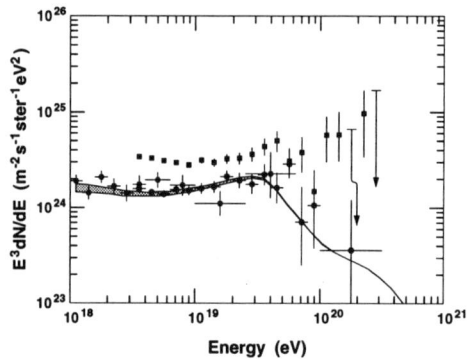

FIGURE 1. The energy spectra of UHECR derived by the AGASA(squares) and the HiRes experiments. The lines show the expected energy spectra if the UHECR sources were distributed isotropically in the Universe.

The other difference, which is more interesting from astrophysical point of view is the end of the cosmic rays spectrum. The HiRes data are consistent with a GZK cut-off [9], shown in the figure for $E^{-2.5}$ acceleration spectrum and $(1+z)^n$ source evolutions for $n = 3$ and $n = 4$ (upper edge). The curves are normalized to HiRes data at 10^{19} eV. The AGASA data set, on the other hand, indicates spectrum which may be even flatter in the highest energy range. The statistics is such, though, that the inconsistency with the GZK cut-off is less than 3σ [10] after the spectra are renormalized. All other experiments, with lower statistics, have intermediate normalization between these two extremes.

The nature of UHECR is not well known. The highest energy shower detected by the Fly's Eye [11] does not look like a gamma ray shower [12]. There are also two analyzes that limit the fraction of gamma rays in UHECR to about 1/2 of the flux.

Fly's Eye group reached a conclusion of cosmic ray composition that changes from heavy nuclei to protons [13] at about 3×10^{18} eV. A change of the cosmic ray nuclear composition is expected if there is a change from Galactic to extragalactic cosmic ray sources. AGASA does not confirm this conclusion

POSSIBLE COSMIC RAY SOURCES

There are two general scenarios for the acceleration of cosmic rays to and above 10^{20} eV. The conventional model is the extension of the astrophysical shock acceleration that is fully consistent with the Galactic cosmic rays of lower energy. We believe that in the Galaxy cosmic rays are accelerated at supernova blast shocks.

Extending the shock acceleration to very high energy is not trivial. The minimum requirement for particle acceleration is to contain the particle at the acceleration site. If we use this requirement to estimate the maximum energy, it becomes

$$E_{max} \leq \gamma e Z B R,$$

where B is the magnetic field strength and R is the length of the acceleration region. The factor γ accounts for the bulk Lorentz factor of the shock. To accelerate a proton to 10^{20} eV one needs the product $B/\text{G} \times R/\text{pc} > 0.2$ [14] without any account for efficiency. For typical Galactic field strength of 5 μG the extension of the acceleration region should exceed 40 Kpc. For extragalactic fields of order 1 nG one needs extensions of 200 Mpc. UHECR sources that have been studied to certain extend include:
- *Hot Spots* in giant radio galaxies
- Active Galactic Nuclei
- Colliding Galaxies
- Quiet black holes
- Gamma Ray bursts

The problem with most of these scenarios is that the detected UHECR do not point of any of the cosmologically nearby energetic astrophysical systems that could in principle power shocks of the required strength. Protons of such energy are not expected to deflect from the source direction by more than 2-3°. A source should be identified by the direction of one or more UHECR and surrounded by a halo of lower energy cosmic rays. The observed arrival direction distribution is however almost isotropic. There might be only a slight preference for directions associated with large scale concentration of matter. The AGASA date however show a small scale anisotropy and clustering on the 2° scale [15]. The chance probability of this clustering is on the 1% level.

The UHECR luminosity of their sources is estimated [16] to be 4.5×10^{44} erg/Mpc/yr for energies above 10^{19} eV and E^{-2} acceleration spectrum in the assumption that the sources are isotropically and homo. If one accounts for particles of lower energy the luminosity estimate exceeds 10^{45} erg/Mpc/yr. The luminosity estimate depends very strongly on the assumed acceleration spectrum. For steeper spectra, which seem to fit better the detected events it could increase by two orders of magnitude [17].

The other general class of models (called TopDown scenarios to be distinct from the BottomUp acceleration ones) was inspired by the difficulties to identify suitable astrophysical sources for UHECR. Its basic idea is that UHECR are decay products of very massive X particles of mass M_X as high as 10^{16} GeV/c^2. The X particles may, or may not, be a significant part of the dark matter. The different TD scenarios are described and reviewed in [18]. A common feature is that the decay (QCD) spectrum is flat: $E^{-3/2}$. Another distinctive feature is that most of the stable decay products are γ–rays and neutrinos, while nucleons are only about 3% of the final decay products with the possible exception of the highest end of the spectrum. The detected UHECR are expected to be UHE γ–rays.

TD scenarios are very exciting from the point of view of particle physics and cosmology. If the future observations prove them correct they will lead to a better understanding of the cosmological evolution of the Universe or/and to further development of the *Standard Model* of particle physics.

There are also scenarios that do not fit in these two classes. Especially interesting is the Z burst proposal [19]. It exploits the sharp peak in $\sigma_{\nu\nu}$ at the Z_0 mass and the indications for non zero neutrino mass from the observations of neutrino oscillations. If the primordial neutrinos have mass of order 1 eV they would be gravitationally attracted to dense regions including the local Supercluster and their local density will exceed their average density in the Universe. UHE astrophysical neutrinos can than interact of this dense neutrino cloud and produce Z-bosons. For m_ν of 1 eV the resonance energy is 4×10^{21} eV. Among the Z_0 decay products there are 2 nucleons and 20 photons of very high energy, that could be the detected UHECR. Another possibility [20] is that the Lorentz invariance is violated and as a result the photoproduction cross section is changed and the GZK cut-off is moved to significantly higher energy.

MAGNETIC FIELDS

The general astrophysical parameter that affects the most the UHECR (in case they are charged nuclei) is the possible existence of extragalactic magnetic fields. Microgauss fields have been observed with Faraday rotation measurements in many clusters of galaxies [21] together with some magnetic bridges between them. The technique is not sensitive enough to identify lower fields,

It has been shown that even 1 to 10 nG fields [22], if they are organized on the scale of 20 Mpc (the extension of the local Supercluster), would affect not only the direction, but also the energy spectrum of the observed UHECR. As an example we show in Fig. 2 the arrival directions of protons of energy above 4×10^{19} eV isotropically emitted by M87 if there were a 10 nG magnetic field along the direction of the Supergalactic plane.

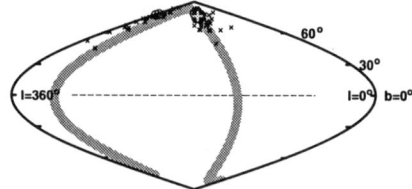

FIGURE 2. The arrival direction in galactic coordinates of protons emitted by M87 (cross) if there were an organized 10 nG field along the Supergalactic plane (shown with wide gray curve). Particles above 10^{20} eV are shown with circles. The Galactic plane is indicated with a dashed line. The map is centered at the Galactic anticenter.

The statistics of events in Fig. 2 is the same as of the AGASA experiment. Nine out of 11 events above 10^{20} eV are clustered around the source. The lower energy particles are however spread around the Supergalactic plane, which is the center of mass plane of the galaxies within redshift of 0.04. If a galaxy in the local Supercluster were the source of a fraction of the detected UHECR and organized magnetic fields of 10 nG indeed existed, this direction distribution would be consistent with the experimental.

Some authors relate the *clustering* observed by AGASA to effects induced by the Galactic magnetic field. Protons above 10^{20} eV are not affected much by the strength and extension of the Galactic field, but particles of lower energy arriving at our Galaxy from different direction could be systematically bent to arrive in clusters. The efficiency of this process is however low.

It is now too early to explain the UHECR phenomena. We need the much higher experimental statistics which is expected after the completion of Auger and the flights of EUSO and OWL. The solution of these puzzles will have a strong impact in our understanding of the dynamics of the powerful astrophysical systems, the evolution of the Universe as well as on the extension of the particle physics models to ultra high energy.

Acknowledgments The research on the topic is performed in collaboration with R. Engel, T.K. Gaisser, R.J. Protheroe, D. Seckel and other colleagues. TS is supported in part by the DOE grant DE-FG02 91ER 4062 and by NASA grant NAG5-10919.

REFERENCES

1. K. Greisen, Phys. Rev. Lett. **16**, 748 (1966)
2. G.T. Zatsepin & V.A. Kuzmin, JETP Lett. **4** 78 (1966).
3. J. Linsley, Phys. Rev. Lett., **10**, 146 (1963)
4. http://www.auger.org
5. http://www.euso-mission.org
6. http://owl.gsfc.nasa.gov
7. Takeda, M. et al., Phys. Rev. Lett., **81**, 1163 (1998); for updates see http://www-akeno.icrr.u-tokio.ac.jp
8. Abu-Zayyad, T. et al., *astro-ph/0208301*
9. J.N. Bahcall & E. Waxman, *astro-ph/0206217*
10. D. DeMarco, P. Blasi & A.V. Olinto, *astro-ph/0301497*
11. D.J. Bird et al., Ap.J., **441**, 144 (1995)
12. F. Halzen et al., Astropart. Phys., **3**, 151 (1995)
13. D.J. Bird et al., Phys. Rev. Lett., **71**, 3401 (1993)
14. A.M. Hillas, Ann. Rev. Astron. Astrophys., **22**, 425 (1984)
15. Y. Uchihori et al., Astropart. Phys., **13**, 151 (2000)
16. E. Waxman, Ap. J., Phys. Rev. Lett., **386** (1995)
17. V.S. Berezinsky, A.Z. Gazizov & S.I. Grigorieva *astro-ph/0204357*; *astro-ph/0210095*
18. P. Bhattacharjee & G. Sigl, Phys. Reports, **327**, 109 (2000)
19. T. Weiler, Astropart. Phys. **11**, 303 (1999); D. Fargion, B. Mele & A. Salis, Ap. J., **517**, 517 (1999)
20. S. Coleman & S. Glashow, Phys. Rev. D**59**:116008 (1999); L. Gonzales-Mestres, physics/9704017
21. P.P. Kronberg, Rep. Prog. Phys., **57**, 325 (1994)
22. T.Stanev, D. Seckel & R. Engel, *astro-ph/0108338, v.2*

AMS-02 on the International Space Station

K. Scholberg
for the AMS Collaboration

Massachusetts Institute of Technology, Cambridge, MA, 02139.

Abstract. AMS-02 is the main phase of the Alpha Magnetic Spectrometer experiment and is to be installed on the International Space Station for a three-year exposure. I will review motivations for the experiment and capabilities of the instrument.

INTRODUCTION

The Alpha Magnetic Spectrometer (AMS) is a charged particle in space, with a main goal of studying cosmic rays with energies up to TeV. The basic idea is simple: a high dipole magnetic field provided by a superconducting magnet allows momentum and charge sign measurements in a precision silicon tracker. Combining tracking information with dE/dx and velocity measurement, one can identify masses and charges of particles traversing the detector. Several other sub-detectors, including a transition radiator detector, a ring-imaging Cherenkov detector and an electromagnetic calorimeter, provide additional and redundant information to improve characterization of the fluxes of the different charged particles species.

PHYSICS MOTIVATIONS

There are several pmotivations for studying the composition and spectra of cosmic rays above the atmosphere:

- **Search for primordial antimatter:** The apparent asymmetry between matter and antimatter is a long-standing mystery of cosmology. A possible solution to the puzzle is a universe which does in fact contain distant domains of antimatter. Even for an asymmetric universe, small nearby pockets of antimatter created by an early phase transition are not ruled out. A smoking-gun signature of primordial antimatter would be antinuclei such as $\overline{\mathrm{He}}$ (or heavier elements) observed above the atmosphere.

- **Indirect search for SUSY dark matter:** Another long-standing puzzle is the nature of the dark matter of the universe. Some sort of non-baryonic dark matter is now thought to make up to $\sim 25\%$ of the critical density of the universe. Neutralinos (χ), heavy, stable, neutral particles predicted by supersymmetric theories, are prime candidates for the dark matter. If neutralinos comprise the our galaxy's dark halo, they may annihilate, with antimatter ($\bar{p}$, e^+ or $\bar{d}$) among the direct or indirect annihilation products[1]. Such "primary" χ annihilation antimatter could be distinguished from "secondary" antimatter produced in cosmic ray collisions by an anomalous energy spectrum. For instance, a bump in the observed e^+ spectrum at around 10 GeV/c could be the signature of $\chi\bar{\chi}$ annihilation. Gamma-rays may also be among the annihilation products.

- **Cosmic Ray Propagation:** A precision, high-statistics measurement of the ^{10}Be to ^{9}Be ratio in the cosmic ray flux, and its energy dependence, would be a powerful method of distinguishing between Galactic cosmic ray propagation models. These isotopes are "clocks" which measure the confinement time of charged cosmic rays in the galaxy.
- **Exotic Particles:** Exotic matter such as "strangelets" (a possibly stable state of matter consisting of u, d, and s quarks) or fractionally charged particles may manifest itself as particles with anomalous charge-to-mass ratio in a spectrometer[2].
- **The Unexpected?** Finally, since high-statistics cosmic ray measurements of this type have never been made before above the atmosphere, one can never rule out the possibility of surprising new observations.

THE AMS-01 PRECURSOR MISSION

The AMS-01 precursor experiment flew on Space Shuttle Discovery in June of 1998 for a period of 10 days, recording 100 hours of data and 10^8 particles. The orbit was 51.7°, and the altitude 320-390 km. The precursor mission employed a permanent Nd-Fe-B magnet with a 0.15 T field, in addition to six planes of silicon tracker, time-of-flight scintillator counters, and a threshold Cherenkov counter. This successful flight produced a number of new results. In particular the limit on the $\overline{\text{He}}$/He ratio was pushed down to nearly 10^{-6} for rigidities up to 20 GV[3]. No $|Z| > 2$ nuclei were found. Unprecedented high-statistics measurements of protons[4, 5], leptons[6, 7], and helium isotopes[8] were made. A full reporting of physics results from AMS-01 can be found in Reference [9].

THE AMS-02 EXPERIMENT

The AMS-02 experiment is to be installed on ISS in 2006 for a three-year exposure. A significant upgrade with respect to AMS-01 is the superconducting magnet with a field of 0.87 T, allowing spectral measurement up to TeV energies[10]. The acceptance of the tracker will be about 0.5 m^{2}sr. A few notes on the major AMS-02 sub-detectors are given below.

- **Silicon tracker**[11]: there will be eight layers of Si strip tracking planes, with a total of 196,000 channels covering 6.45 m^2. Maximum detectable rigidity is approximately 1 TV. The silicon tracker provides dE/dx information as well as a rigidity measurement.
- **Time of flight scintillator counters**[12]: four layers of scintillator counters provide time of flight ($\sim$140 picoseconds) and dE/dx information. The TOF also provides

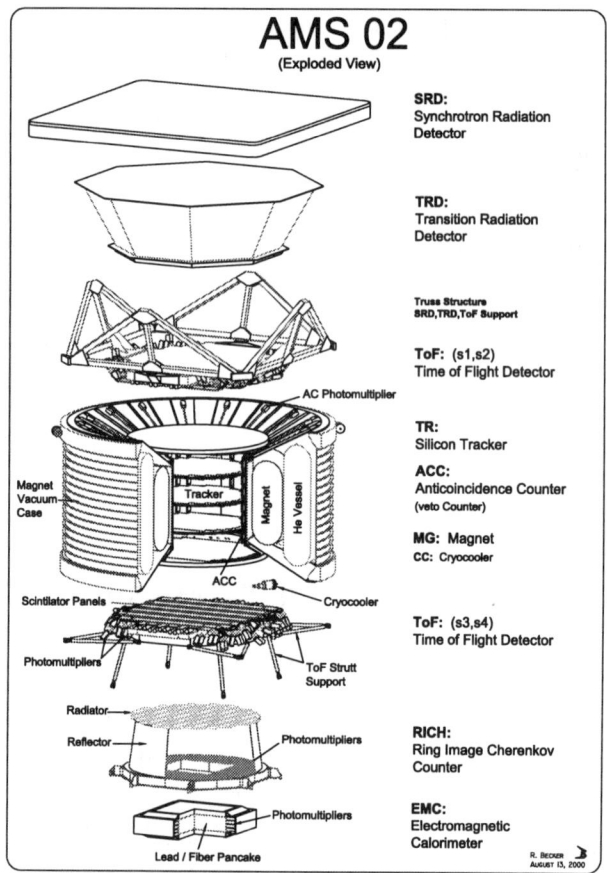

FIGURE 1. View of AMS-02 showing the subdetector components.

the fast trigger for the detector.
- **Transition radiation detector**[13]: there will be 20 layers of polypropylene radiator interspersed with Xe/CO$_2$ drift tubes to detect the transition radiation resulting from charged particles traversing the interfaces between materials of differing refractive index. Since transition radiation depends on the relativistic γ of the charged particle, the TRD improves p/e^+ separation with a proton rejection factor of $\sim 10^2 - 10^3$, up to about 300 GeV/c.
- **Ring-imaging Cherenkov detector**[14]: the RICH consists of a layer of aerogel (and possible some NaF) radiator, a conical reflector, and a layer of photomultiplier tubes, is sensitive to charge and velocity, via the intensity and angle, respectively, of the Cherenkov ring produced. The RICH will provide nuclear isotope identification up to ~ 10 GeV/n for isotopes up to approximately carbon.
- **Electromagnetic calorimeter**[15]: the 3D sampling ECAL, comprising 9 superlayers of lead and scintillating fibers, can measure energies and improve p/e^+ sep-

aration up to TeV energies, with a proton rejection factor of $\sim 10^3$. The combined rejections of the ECAL and TRD are essential for the measurement of positron spectral properties, because in the regime of interest for dark matter at around 10-100 GeV, protons outnumber positrons by a factor of $\sim 10^4$ to 1.

The requirements for building a detector for space are extremely challenging. The detector has a strict weight limit of 14809 lb, and must have a power consumption of less than 2 kW. It must withstand temperature variations between $-180°$ and $50°C$, and of course it must work in vacuum. It must survive accelerations up to 9 G during shuttle launch. The data rate is limited to 2 Mbits per second, which constrains trigger configurations. Finally, the detector must function without intervention for three years.

AMS-02 will be able to study e^+, e^-, p, $\bar{p}$, d, t, ^{3}He and ^{4}He with statistics three or four orders of magnitude greater than previous measurements. It can search for anti-ions, such as $\overline{\text{He}}$ and $\overline{\text{C}}$; in particular the $\overline{\text{He}}$/He sensitivity will be 1 in 10^9, giving limits some orders of magnitude beyond current ones.

In addition, AMS-02 may have some γ-ray sensitivity[16, 17] in the $\sim$10-100 GeV range, via pair conversions in the upper layers of the detector, and shower production in the ECAL. This capability will permit studies of GRBs, blazars and other sources, as well as SUSY dark matter annihilation γ's.

SUMMARY

In summary, the AMS-02 experiment will measure cosmic rays with momenta between 300 MeV/c and 3 TeV/c with unprecented statistics and precision over a three-year period starting in 2006. This will allow an antihelium search with $\overline{\text{He}}$/He sensitivity of 10^{-9}, a SUSY dark matter annihilation product search, tests of cosmic ray propagation models, exotic matter searches and more.

REFERENCES

1. K. Griest and M. Kamionkowski, *Phys. Rept.*, **333**, 167 (2000).
2. J. Madsen, *J. Phys. G*, **28**, 1737 (2000).
3. The AMS collaboration, J. Alcaraz et al., *Phys. Lett. B*, **461**, 387 (1999).
4. The AMS collaboration, J. Alcaraz et al., *Phys. Lett. B*, **472**, 215 (2000).
5. The AMS collaboration, J. Alcaraz et al., *Phys. Lett. B*, **490**, 23 (2000).
6. The AMS collaboration, J. Alcaraz et al., *Phys. Lett. B*, **484**, 10 (2000).
7. The AMS collaboration, J. Alcaraz et al., *Phys. Lett. B*, **495**, 440 (2000).
8. The AMS collaboration, J. Alcaraz et al., *Phys. Lett. B*, **494**, 193 (2000).
9. The AMS collaboration, M. Aguilar et al., *Physics Reports*, **366/6**, 331 (2002).
10. B. Blau et al., *Nucl. Phys. Proc. Suppl.*, **113**, 125 (2002).
11. W. J. Burger et al., *Nucl. Phys. Proc. Suppl.*, **113**, 139 (2002).
12. V. Bindi et al., *ICRC Proceedings, 2003* (2003).
13. T. Siedenburg et al., *Nucl. Phys. Proc. Suppl.*, **113**, 154 (2002).
14. J. Casaus et al., *Nucl. Phys. Proc. Suppl.*, **113**, 147 (2002).
15. F. Cadoux et al., *Nucl. Phys. Proc. Suppl.*, **113**, 159 (2002).
16. R. Battiston, *Astropart. Phys.*, **13**, 51 (2000).
17. M. Pohl, *Int. J. Mod. Phys.*, **A17**, 1809 (2002).

The Pierre Auger Observatory

John Swain, for the Pierre Auger Observatory

Department of Physics, Northeastern University, Boston, MA02115, USA

Abstract. One of the most fascinating puzzles in particle astrophysics today is that of the origin and nature of the highest energy cosmic rays. The Pierre Auger Observatory (PAO), currently under construction in Province of Mendoza, Argentina, and with another site planned in the Northern hemisphere, is a major international effort to make precise, high statistics studies of the highest energy cosmic rays. It is the first experiment designed to work in a hybrid mode incorporating both a ground-based array of 1600 particle detectors spread over 3000 km^2 with fluorescence telescopes placed on the boundaries of the surface array. The current status of the observatory is presented and prospects for the future discussed.

INTRODUCTION

The observation of ultra high energy cosmic rays (UHECR) with energies above 10^{20} eV [1], is puzzling in at least two ways. First of all, it is quite difficult to conceive of acceleration mechanisms which are adequate to impart such enormous energies to cosmic ray particles[2, 3] – energies comparable to those carried by everyday objects like tennis balls or golf balls! Secondly, even if such a mechanism is found, it is difficult to see how such high energy particles would make it through the background radiation: the celebrated GZK cutoff[4] predicts that protons over about 5×10^{19} eV should rapidly lose energy in inelastic collisions with the cosmic microwave background photons with similar energy degradation mechanisms being present for most other particles, including heavy nuclei.

A very important point to make early in this talk is that while there is a puzzle concerning so-called "super-GZK events", the Pierre Auger Obseratory will provide interesting information regardless of how well the GZK cutoff holds out. If there is no GZK cutoff, this will be an unambiguous sign of new physics. If the GZK cutoff is found, then we can rest assured of that piece of the physics and confidently use the data to try to understand the nature of the sources. This point is often missed by people who see the whole physics motivation as "is there a GZK cutoff or not?".

Above 10^{15} eV, cosmic ray primaries are not detected directly, but rather through the effects that such particles produce when they strike the upper atmosphere. There they initiate a cascade of reactions, some nuclear, but most forming an electromagnetic shower made of repeated bremsstrahlung and e^+e^- pair creation events. This shower can be detected experimentally through the fluorescence it produces in the atmosphere (due to excited nitrogen) or via the particles that reach the ground.

Our understanding of the highest energy part of the spectrum above the so-called "ankle" (5×10^{18} eV) is poor due to a combination of low statistics, uncertain energy resolution, uncertainties in energy conversion arising from models, and a lack of knowledge of the mass composition and the fluorescence yield efficiency. Clearly there is something interesting going on, but to fully understand the situation we need more statistics, better control over systematic uncertainties, and full sky coverage: enter the Pierre Auger Obervatory. A comprehensive review of the state of the art prior to the Pierre Auger Observatory can be found in [5].

THE PIERRE AUGER OBSERVATORY

The Pierre Auger Observatory (PAO) is actually comprised of two sub-observatories, one currently under construction in Mendoza, Argentina since 2000. This site is especially interesting, since in addition to being in the wine-growing district, it offers a view of the centre of the galaxy. Another is planned for the Northern hemisphere, and while for the remainder of this talk I will concentrate on the Southern site it is important to understand that the full observatory is comprised of two sites. This will eventually allow full-sky coverage which is very important to allow good studies on anisotropies[6].

The PAO is designed to measure the energy, arrival direction and primary species with excellent precision and very high statistics. A unique feature of the design is the combination of both fluorescence detection and ground-based particle detectors which can be operated independently as well as together in "hybrid" mode.

The scale of the observatory was determined by the requirement that we can collect high statistics in and beyond the region of the expected GZK cutoff, with 1600 particle detectors separated from each other by 1.5 km and covering an area of 3000 km^2, overlooked by four fluorescence detectors which can only operate when ambient conditions offer a clear, dark sky, which leads to a roughly 10% duty cycle. Figure 1 shows, together with one of the surface detectors, a photograph of one the fluorescence detectors where both the mirror and the box of 440 photomultipliers which register the fluorescence light can be seen. The fluorescence measurements are complemented by a a very comprehensive atmospheric monitoring system. The surface array stations are water Čerenkov detectors and can operate continuously.

The fact that about 10% of the showers detected by PAO will be observed by *both* surface and fluorescence detectors, offers the possibility of doing calibrations and understanding systematic errors in a manner that has never been possible before. Access to a large-dimensional parameter space of observables should allow not only determination of the direction and energy of incoming primaries, but also the disentanglement of information about composition from the notoriously difficult systematic errors associated with the choice of hadronic interaction models.

Each ground-based detector is a cylindrical, opaque tank of 10 m^2 and a water depth of 1.2 m, where particles produce light by Čerenkov radiation. The filtered water is contained in a bag which diffusely reflects the light collected by three photomultipliers (PMT's) installed on the top. The large diameter PMT's ($\approx$ 20 cm) are mounted facing down and look at the water through sealed polyethylene windows that are integral part of the internal liner. The signals are processed locally and a second level trigger is identified before transmitting the data to the central acquisition system [7]. The fact that the tanks are quite deep enables showers to be detected efficiently over a wide angular range. Due to the size of the array, the stations have to be able to function independently and yet in communication with the central data acquisition system. The stations operate on battery-backed solar power and communicate with a central station by using wireless LAN radio links [8]. Absolute timing information is obtained from the Global Positioning Satellite (GPS) system [9] and is used to reconstruct the direction from which the primary came. Figure 1 shows a water Čerenkov detector installed in the Southern Observatory as well as one of the fluorescence detectors. Mounted on top of the tank are the solar panel, electronic enclosure, mast, radio antenna and GPS antenna. The battery is housed in a box attached to the tank.

FIGURE 1. Photographs of a typical surface detector and one of the fluoresence detectors showing the mirrors and the array of phototubes onto which they direct the collected light.

The expected angular resolution for the ground array of the Southern Auger Observatory is less than 1° for all energies, and better for large events above 10^{20}eV. The expected energy resolution is estimated to be 12%, averaged over all energies (assuming a proton-iron primary mixture), falling to 10% at 10^{20}eV. The limiting aperture for the full Southern Observatory array and for zenith angle less than 60° is 7350 km^{2}sr. The detection efficiency at the trigger level should reach 100% for energies above 10^{19}eV [10]. Additionally, if events above 60 degrees can be analyzed effectively, the aperture will increase by about 50%.

In hybrid mode, the Pierre Auger Observatory is expected to have 6% energy resolution and an angular precision of 0.5° at 10^{20}eV where only statistical errors are taken into account in these estimates. The detector is optimized for energies above 10^{19}eV, with good reconstruction expected at energies down to 1 EeV. The hybrid data set will provide the best evaluation of primary species, allowing a simultaneous fit to all parameters sensitive to mass composition.

The first cosmic ray event detected by one of the two prototype telescopes installed at Los Leones is displayed in Figure 2. A twenty pixel track, produced by light from a shower, with a length of 8 μs can be seen. The angular velocity of the shower image across the sky allows the distance of the shower core to be established as 5 km. The time duration of the signal in the field of view of the telescope corresponds to a track length of 2.4 km. The mirror inverts the image, and particles from the sky appear to be going upwards as seen by the camera.

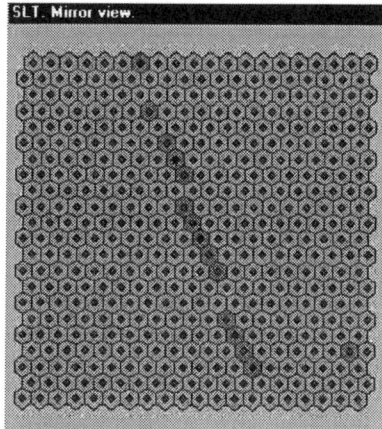

FIGURE 2. First high energy cosmic ray observed by one of the prototype telescopes at Los Leones.

Construction continues, and we look forward to the large amount of very high quality which will appear in the years to come, and the light it will shed on one of nature's great mysteries.

ACKNOWLEDGMENTS

I would like to thank the organizers of this conference for a most interesting meeting, as well as all my collaborators in the Pierre Auger Observatory. Special thanks are due to Maria Teresa Dova for assistance in the preparation of this talk, and, as always, I would like to thank the National Science Foundation for its continued support.

REFERENCES

1. J. Linsley, Phys. Rev. Lett. **10** (1963) 146; M. A. Lawrence, R. J. O. Reid, A. A. Watson, J. Phys **G17** (1991) 733; N. Hayashida et al., Phys. Rev. Lett. **74** (1994) 3491; D. J. Bird et al., Astrophys. J. **441** (1995) 144; N. Sakaki et al. [AGASA Coll.] Proc. of 27th ICRC (Hamburg) **1** (2001) 333.
2. P. Bhattacharjee and G. Sigl, Phys. Rep. **327** 2000.
3. A. V. Olinto, Phys. Rep. **333-334** (2000) 329.
4. K. Greisen, Phys Rev Letters **16** (1966) 748, G. T. Zatsepin and V. A. Kuzmin, JETP Lett. **4** (1966) 78.
5. L. Anchordoqui, T. Paul, S. Reucroft, and J. Swain, Int. J. Mod. Phys. **A18** (2003) 2229.
6. L. A. Anchordoqui, C. Hojvat, T. P. McCauley, T. C. Paul, S. Reucroft, J. D. Swain and A. Widom, arXiv:astro-ph/0305158; P. Sommers, Astropart. Phys. **14** (2001) 271.
7. T. Suomijärvi [Pierre Auger Coll.], Proc. of 27th ICRC (Hamburg) (2001).
8. P. D. J.Clark and D. Nitz, [Pierre Auger Coll.], Proc. of 27th ICRC (Hamburg), (2001).
9. C. Pryke et al Nucl. Inst. Methods **A354** (1995) 354.
10. M. Ave, J. Lloyd-Evans, A. A. Watson, [Pierre Auger Coll.], Proc. of 27th ICRC (Hamburg), (2001).

HiRes – Searching For the Origins of Ultra High Energy Cosmic Rays

Stefan Westerhoff (for the HiRes Collaboration) [1]

Department of Physics, Columbia University, New York, NY, USA

Abstract. The High Resolution Fly's Eye Experiment (HiRes) in Utah is an air fluorescence telescope mapping the northern sky in cosmic rays at energies above 10^{18} eV. Since November 1999, HiRes has been operated in stereo mode, *i.e.* with two sites separated by 13 km to provide cosmic ray data of unprecedented quality of the northern sky. This paper focuses on first results from the stereoscopic data. We present a measurement of the primary chemical composition above 10^{18} eV, and results on the search for small-scale anisotropies in the cosmic ray arrival distribution, with emphasis on the highest energies, where previous experiments have observed clustering.

INTRODUCTION

Cosmic ray particles with energies in excess of 10^{20} eV, more than 11 orders of magnitude greater than the equivalent rest mass of the proton, have been detected on Earth. At this point, we do not know where these particles come from, how they are accelerated, and how they travel astronomical distances without substantial loss of energy. The High Resolution Fly's Eye (HiRes) Experiment [1] in Dugway, Utah, measures the energy spectrum, chemical composition and arrival directions of cosmic rays above 10^{18} eV, often referred to as *ultra-high energy* cosmic rays. By mapping the sky in ultra-high energy cosmic rays, HiRes tries to shed some light on the origin of these most energetic particles of the known Universe.

The flux of particles above 10^{18} eV is extremely low, so detectors need to probe a large effective area to detect a statistically convincing number of events during their lifetime. This requires earth-bound detectors rather than balloon- or satellite-bound experiments. On Earth, the cosmic ray primary is not detected directly, but rather by means of the air shower it induces in the Earth's atmosphere. Two techniques are commonly used to detect this air shower cascade and extract the direction and the energy of the primary particle, ground-based air shower arrays and air fluorescence detectors. Ground arrays like the Akeno Giant Air Shower Array (AGASA) [2] in Japan are sparsely instrumented arrays of counters which measure the secondary particles reaching the ground. Air fluorescence detectors like HiRes make use of the fact that the air shower dissipates much of its energy exciting and ionizing air molecules, which subsequently fluoresce in the ultraviolet. This fluorescence light is emitted isotropically, and air fluorescence detectors consist of arrays of telescopes that image fluorescence light from distant air

---

[1] see http://www.cosmic-ray.org for a complete list of authors

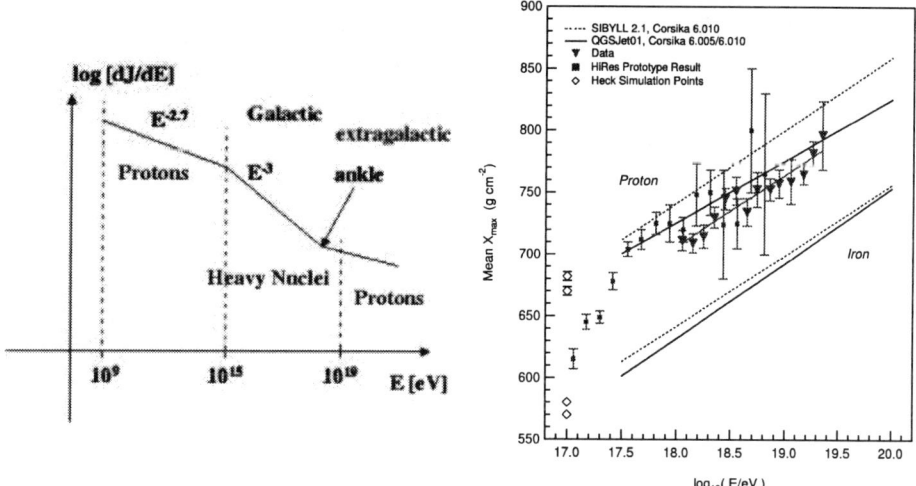

FIGURE 1. (Left) Scheme of the cosmic ray flux, composition, and presumed origin as a function of energy. (Right) Mean shower maximum as a function of energy for HiRes stereo data (preliminary, taken from [4]) and HiRes prototype data (taken from [5]), together with Monte Carlo predictions for proton- and iron-initiated showers.

showers onto arrays of photomultiplier tubes. As this fluorescence light is extremely weak, the operation of these detectors is limited to dark, moonless nights, or about 10 % of the "on time" of ground arrays.

With its two detector sites, about 13 km apart, HiRes is able to make *stereoscopic* observations of cosmic ray induced air showers with unprecedented energy resolution and angular resolution. HiRes has been taking data in stereo mode since November 1999. Here, we report on recent stereo results on the *chemical composition* and the *arrival directions* of cosmic rays.

CHEMICAL COMPOSITION ABOVE 10^{18} EV

The study of the chemical composition of the incoming flux of cosmic rays is an important tool to reveal their origin. In our current understanding, the origin of cosmic rays changes between $10^{17.5}$ and $10^{18.5}$ eV. Galactic sources are responsible for the cosmic ray flux below 10^{18} eV, but quickly run out of steam at higher energies, where extragalactic sources take over. One indication supporting this picture is the change in the spectral index of the cosmic ray energy spectrum around 10^{18} eV, often referred to as the "ankle" (see Fig. 1 (a)). Another indication is the chemical composition. Since the Galactic flux is expected to be dominated by heavier nuclei like iron, and the extragalactic flux should be mainly protons, we expect the chemical composition to change at energies around the "ankle." This has been observed with an earlier prototype of the HiRes detector [3], HiRes-MIA. New data from the HiRes stereo detector extends

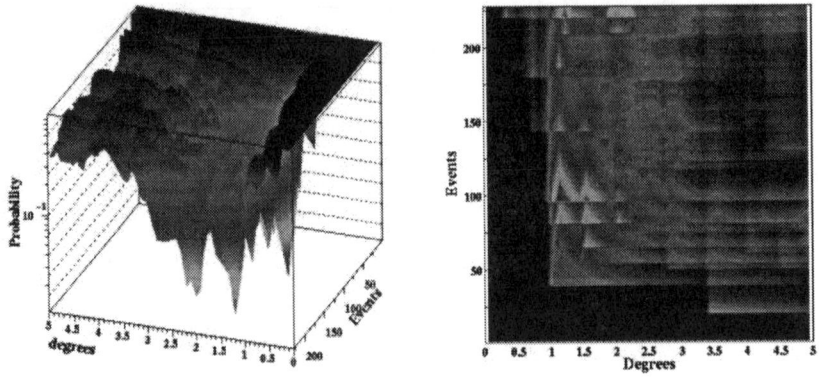

FIGURE 2. Chance probability for a clustering signal in HiRes stereo events above 10^{19} eV in a scan over minimum energy E (or events above E) and angular distance (degrees). Probabilities do not account for the scanning. Results are preliminary.

this analysis to energies above 10^{18} eV.

Measuring the chemical composition of the cosmic ray flux is one of the strengths of air fluorescence detectors. With this detector type, we directly measure the atmospheric depth of the shower maximum, X_{max}, which is an indicator of the nature of the primary particle. Heavier nuclei induce an earlier shower development, so their shower maximum is higher up in the atmosphere than it is for protons. However, due to the large intrinsic fluctuations in the depth of the shower maximum, it is not possible to identify the chemical nature of the incoming cosmic ray on a shower-by-shower basis with any reasonable certainty. Only quantities averaged over a large number of showers can be used as indicators for the cosmic ray composition at a given energy. Simulations predict (Fig. 1(b)) that the mean atmospheric depth of the shower maximum increases logarithmically with primary energy and differs by about $100\,\mathrm{g\,cm^{-2}}$ for proton- and iron-induced showers at all energies.

As shown in Fig. 1(b), HiRes stereo results are consistent with a constant light composition [4] above 10^{18} eV, consistent with the rough scheme shown in Fig. 1(a).

ARRIVAL DIRECTIONS ABOVE 10^{19} EV

A direct way to search for sources of cosmic ray particles is to analyze the distribution of their arrival directions. Sky maps of cosmic ray arrival directions at all energies are amazingly isotropic, with no obvious source or region standing out. However, clustering of cosmic ray arrival directions at small angles and high energies has been claimed [6, 7]. If confirmed, any clustering signal would favor compact sources. Unfortunately, the significance of the clustering is small and tainted by an unknown number of statistical trials.

Arrival directions of cosmic ray particles do not necessarily point back to their sources. As charged particles, they are subject to deflections in Galactic and intergalactic magnetic fields of unknown strength and orientation. However, the Larmor radius is proportional to the energy of the particle. The highest energy cosmic rays, at energies above 10^{19} eV, have suffered the smallest deflections, Therefore, we anticipate that any clustering may only be apparent at the highest energies. However, the statistical power of the available data quickly weakens as the sample is limited to higher energy events. Given these two competing forces, the optimal value for the minimum energy is hard to determine *a priori*. This is also true for the maximum angular separation which two events can have in order to be considered a cluster. We therefore adopt a scanning technique which searches for the separation angle θ and the minimum energy E which *maximizes* the clustering signal. This method requires a careful evaluation of the final significance of any clustering signal, since the significance attached to this signal has to account for the scanning process itself. In order to correctly evaluate the true chance probability, we perform identical scans over simulated data.

Using data taken between November 1999 and June 2003, we search for anisotropy in the arrival direction of cosmic ray particles at small scales ($< 5°$) and the highest energies ($E > 10^{19}$). This data sample contains 228 events with an energy above 10^{19} eV. All events have RMS energy uncertainties less than 20 %, and the typical angular uncertainty is between 0.5° and 1.0°, with 67 % of all events having an uncertainty less than 0.75°. The error in arrival direction caused by the angular resolution of the detector is therefore smaller than the smearing expected from deflections in magnetic fields.

The maximum signal in the HiRes stereo data set occurs at energies above $1.7 \cdot 10^{19}$ eV and an angular separation of 1.2° (see Fig. 2). The sample of 97 events above this energy contains 4 doublets. The corresponding probability is 1.1 %, but after accounting for the scanning process, the final chance probability is 39 %. We conclude that at this point, HiRes stereo data does not support the hypothesis that the arrival directions of ultra-high energy cosmic rays cluster [8].

Acknowledgments: The HiRes project is supported by the National Science Foundation under contract numbers NSF-PHY-9321949, NSF-PHY-9322298, NSF-PHY-9974537, NSF-PHY-0098826, NSF-PHY-0071069, by the Department of Energy Grant FG03-92ER40732, and by the Australian Research Council. The cooperation of Colonels E. Fisher and G. Harter, the US Army and Dugway Proving Ground staff is appreciated.

REFERENCES

1. Matthews, J. N., *et al.*, Proc. 27th Int. Cosmic Ray Conf. (ICRC), Hamburg (2001) 350.
2. Chiba, N., *et al.*, Nucl. Instr. Meth. A 311 (1992) 338.
3. Abu-Zayyad, T., *et al.*, Phys. Rev. Lett. 84 (2000) 4276.
4. Archbold, G., PhD thesis, University of Utah (2003).
5. Song, C., PhD thesis, Columbia University (2001).
6. Takeda, M., *et al.*, ApJ 522 (1999) 225.
7. Tinyakov, P. G., and Tkachev, I. I., JETP Lett. 74 (2001) 1 (astro-ph/0102101).
8. Finley, C. B., *et al.*, Proc. 28th Int. Cosmic Ray Conf. (ICRC), Tsukuba (2003) 425.

Nuclear and Particle Astrophysics at CIPANP 2003

Edward A. Baltz*† and James Stone**

*ISCAP, Columbia Astrophysics Laboratory, MC 5247, 550 W 120th St., New York, NY 10027
†KIPAC, SLAC, M/S 78, 2575 Sand Hill Rd., Menlo Park, CA 94025 (from Sept. 2003)
**Physics Department, Boston University, Room 255, 590 Commonwealth Ave., Boston, MA 02215

Abstract. In the nuclear and particle astrophysics session of CIPANP 2003 we heard talks on a number of topics, focused for the most part into four broad areas. Here we outline the discussions of the standard cosmological model, dark matter searches, cosmic rays, and neutrino astrophysics. The robustness of theoretical and experimental programs in all of these areas is very encouraging, and we expect to have many questions answered, and new ones asked, in time for CIPANP 2006.

1. THE STANDARD COSMOLOGY

In recent months the WMAP satellite has provided the clearest picture of temperature fluctuations in cosmic microwave background (CMB) to date [1]. The power spectrum of these fluctuations is a sensitive function of the cosmological parameters, such as the total energy density, matter density, baryon density, Hubble constant and others. The emerging model was convincingly confirmed: the universe is spatially flat, consisting mostly of "dark energy" that behaves like a cosmological constant, about 23% dark matter, and about 4% baryons. The sum of the neutrino masses is limited to be less than 0.7 eV, much more strict than the bound from terrestrial laboratories. The Compton scattering optical depth implies an early reionization at redshift ~ 17. These results will improve, as the WMAP team plans to take at least 4 years of data.

Several profound questions arise from these results. To start, the nature of the dark matter and the dark energy are completely unknown. We understand the 4% in baryons, but the other 96% is a mystery. Furthermore, there is an indication of a deficit of power on the largest scales (in e.g. the quadrupole and octupole moments), but the meaning of this unclear. A finite universe is one among many unlikely possibilities.

The SNAP satellite[2] has been proposed to study the nature of the dark energy by measuring the Hubble diagram (redshift – distance relation) using type Ia supernovae (SNIa) as standard candles, as has been done from the ground. Ground based measurements of this kind have already shown that there is a large density of dark energy. The SNAP team hopes to study the equation of state $w : (p = w\rho)$ of this material. A satellite is required to significantly extend the ground–based results, as a large number of SNIa with redshifts $z > 1$ is required: this tests the universe in its decelerating phase. The wavelengths of interest are redshifted into the infrared, thus the necessity of a space–based instrument. Furthermore, the exact nature of the progenitors of SNIa is unknown. A large sample can be split into many subsamples to study systematic effects. SNAP is fundamentally a large survey telescope, and should have a broad science reach.

Big Bang Nucleosynthesis (BBN)[3] tries to explain the primordial abundances of light elements, in particular the stable isotopes of hydrogen, helium and lithium. As a function of only the baryon to photon ratio, these abundances can be calculated by tracking the network of nuclear reactions in the hot big bang. The primordial abundance of deuterium depends sensitively on the one free parameter, thus deuterium measurements can provide an accurate assessment of the cosmological baryon density. Lyman-α clouds obscuring distant quasars presumably consist of predominantly unprocessed gas, and thus reflect primordial abundances. With the right column depth of hydrogen (not so small that deuterium is unobservable, and not so large so that the damping wings of hydrogen cover the deuterium line), the primordial deuterium abundance is measured. At present, the implied baryon density is fully consistent with the WMAP value, with comparable errors. To go further, many more such systems would be needed. Suitable systems are currently found at the rate of one per year, so for now progress on the baryon density will come from CMB measurements.

Cosmological observations have the potential to probe Planck-scale physics[4]. Inflationary theories usually predict that observable wavelengths (e.g. galaxy and cluster scales) originated during inflation as sub-Planck fluctuations. Thus, inflation can in principle probe quantum theories of gravity such as superstring theory. With not overly optimistic assumptions, a 1% modulation of the CMB fluctuations might be produced by Planck-scale physics during inflation, detectable in the next decade or two.

2. DARK MATTER SEARCHES

The Standard Cosmological Model requires that 23% of the energy density in the universe is some form of non–baryonic nearly collisionless clustering matter. A new stable particle would fit the bill, as has been known for several decades. In this regard the dark matter problem is more tractable than the dark energy problem — dark matter "looks like" something we understand, while dark energy is completely mysterious.

Two possible candidates for dark matter have survived numerous tests and remain viable. The first is the lightest superpartner in supersymmetric extensions to the Standard Model, which is naturally stable, weakly interacting, and electrically neutral. The second is the axion, arising in a compelling solution to the strong CP problem.

Weakly Interacting Massive Particles (WIMPs), such as those in supersymmetric models can be detected in sensitive low–background experiments by their rare scattering from atomic nuclei. The nuclear recoil deposits energy, which in principle is detectable. Two such detectors currently running are CDMS and DAMA.

CDMS[5] uses germanium and silicon detectors. These are sensitive to both phonons and ionization. The ratio of these two signals powerfully discriminates against background, as nuclear recoils exhibit much lower ionization that most backgrounds (electrons and gamma rays). CDMS-I ran in a shallow site at Stanford, and the final WIMP exclusion results are now available. CDMS-II is currently running in the Soudan mine, with new results anticipated by the end of 2003.

The DAMA[6] detector uses NaI scintillators. They do not have the background rejection capabilities of CDMS, but instead rely on the annual modulation of the WIMP signal: as the Earth orbits the Sun, its relative velocity with respect to the WIMP "wind" is modulated by several tens of km s^{-1}, leaving a rate modulation of a few percent. In the 4-year data such a modulation is seen, though the implied mass and cross section are nearly ruled out by other experiments (CDMS, EDELWEISS, ZEPLIN). Three more years of data have been released since the conference, and the modulation signal is strengthened. Furthermore, the successor experiment LIBRA is being installed now.

The future of WIMP searches requires that ton-scale detectors be constructed. The XENON proposal[7] to use a two-phase detector for both scintillation light and ionization is a promising possibility for scalability to a one ton target mass. Background can again be rejected by the lower levels of ionization from nuclear recoils relative to electronic processes. Small prototypes are currently being tested. The construction of a 10 kg prototype is well underway. The goal is to build a 100 kg module; with this a ton-scale detector could feasibly be built in the next decade.

Axions in the range micro- to milli-eV remain a viable dark matter candidate. They require a vastly different experimental approach: conversion to microwave photons in a magnetic field. These photons are then detected with with what is essentially a very sensitive radio receiver. The axion dark matter experiment ADMX is ongoing, already scanning deep in the allowed model range[8]. Upgraded receivers using SQUID amplifiers are on the way. As light pseudoscalars, axions have a bounded parameter space, as their interactions are essentially the same as neutral pions. The lowest allowed axion-photon couplings are within reach.

3. COSMIC RAYS

Energetic cosmic ray nuclei are an important probe of high energy processes in astrophysics[9]. Supernova blast waves are capable of accelerating protons to energies of 10^{15} eV, the "knee" in the spectrum where the power law shifts to a steeper value. Very puzzling are the ultra high energy cosmic rays with energies in excess of 10^{20} eV. Various acceleration mechanisms have been proposed, involving FR II galaxies, interacting galaxies, jets in radio sources (with Lorentz factor 10), gamma ray bursts (with Lorentz factor 300), and others. Top-down models have also been considered: UHECRs may arise in the decay chains of supermassive particles. Whichever mechanism produces UHECRs, they must originate cosmologically nearby, within roughly 100 Mpc. The GZK cutoff operates for nuclei above 10^{20} eV / A, where the threshold for pion photoproduction on the CMB is exceeded. UHE photons have a similar cutoff at lower energy, at the pair production threshold. The nature of the observed events above 10^{20} eV is unknown, and difficult to determine experimentally. Nuclei such as iron, protons, photons, and neutrinos are all possibilities.

The experimental situation in UHECRs is quite promising. The HiRes experiment[10] consists of two air fluorescence detectors situated 12.6 km apart, allowing stereoscopic viewing of the air showers induced in the atmosphere. The dataset exhibits the GZK cutoff, though at low significance. The data are in agreement with AGASA, the Japanese particle detector, and the AGASA data does not exhibit the cutoff. The relative calibration between air fluorescence and particle detectors is uncertain; the FLASH collaboration[11] at SLAC expects to measure this to better than 10% accuracy. HiRes does not see the clustering visible in the AGASA data, though again the experiments are consistent. This of course dilutes the AGASA evidence for clustering. The collaboration expects three more years of data to be taken, which may clear up the situation.

The plan for the Pierre Auger project[12] is to use both particle detectors (with 100% duty cycle) and air fluorescence detectors (with 10% duty cycle). The project promises greatly enhanced statistics and cross-correlation between the two methods. Two sites are planned: one in Argentina to be fully operational in 2004, and one in the northern hemisphere. 3000 events per year per site above 10^{19} eV are expected, with 30 per year per site above 10^{20} eV. The comparison of the two detector types will allow more accurate energy measurements, and furthermore the identification of the primary will be more certain (e.g. proton vs. Fe). At the southern site, a mountain range will act as a neutrino converter for low altitude primaries, so UHE neutrinos can be studied as well.

At energies below 1 TeV, the Alpha Magnetic Spectrometer (AMS)[13] will measure the spectra of cosmic ray species with higher accuracy that has been possible. Of particular interest are several exotic possibilities, including antinuclei, dark matter, and quark matter "strangelets". AMS-01, which flew on the space shuttle, placed limits on the antihelium to helium ratio of 10^{-6}. AMS-02 will be installed on the international space station as early as 2005. With three years of data, the sensitivity to antihelium will be improved to 10^{-9}. AMS is sensitive to dark matter since annihilations in the galactic halo may produce anomalous levels of antiprotons and positrons at low to moderate energies, though very low energy particles are difficult to study because of the geomagnetic cutoff. AMS will have some sensitivity to photons in the sub-TeV range as well. A final possibility is the search for stable strangelets. They would be easily identified by their anomalous charge to mass ratios. Overall, AMS is a high statistics cosmic ray experiment, and will advance our knowledge considerably.

4. NEUTRINO ASTROPHYSICS

The advent of large neutrino telescopes is exciting for high energy astrophysics. AGNs, GRBs, microquasars, and dark matter annihilations are all interesting possible sources of neutrinos at GeV energies and higher.

The AMANDA array[14] at the south pole uses strings of photomultipliers (PMTs) deep in the ice to detect the tracks from (primarily muon) neutrinos. At 200m in diameter and 400m tall, it detects about 4 neutrinos per day, primarily atmospheric. From the 2000 dataset, no point sources were detected, but the photon flux limit was reached (based on the expected relationship between photon and neutrino fluxes from hadronic processes). With timing information, GRBs are within a single order of magnitude of detectability. Over the next decade, AMANDA will be upgraded to IceCube, with 80 strings totaling 4800 PMTs, with an effective volume of a cubic km. Neutrino energies from 50 GeV to more than a PeV will be studied. At high enough energies, the three neutrino flavors can be disentangled: electron from the shower characteristics, and tau from the "double bang" signature of the the recoil track and subsequent decay. IceCube will be installed starting in 2004, with completion expected in 2010.

ANTARES[15] is a complementary neutrino telescope to be built at the floor of the Mediterranean. 12 cables totaling 1000 PMTs will cover 0.1 km^2, spaced at intervals of 60m. Seawater has a longer scattering length but a shorter absorption length than ice. Thus, the PMTs must be closer together, but the angular resolution is superior: above 10 TeV, a resolution of 0.2° is expected. Between AMANDA / IceCube and ANTARES, most of the sky will be covered. One string is currently running, and completion is planned for 2005. The design for a km^3 upgrade is also underway.

The Super Kamiokande[16] neutrino detector has also begun to do extra-solar neutrino astronomy. A search for the diffuse neutrino background from supernovae has been performed, and interesting limits have already been set.

5. FUTURE OUTLOOK

The future of nuclear and particle astrophysics looks bright. We have a Standard Cosmological Model, but we understand very little about its matter and energy contents. We have observed very high energy processes, beyond the reach of terrestrial accelerators, but the results are puzzling. Several programs of extra-solar neutrino astronomy are underway. The level of experimental activity is very encouraging. We fully expect many new and interesting results to be reported at CIPANP 2006.

ACKNOWLEDGMENTS

We thank all of the speakers in the nuclear and particle astrophysics session of CIPANP 2003 for such interesting and informative talks, and we thank the organizers for putting on such a good conference.

REFERENCES

1. Spergel, D. N. et al., Astrophys. J., in press (2003).
2. http://snap.lbl.gov
3. Burles, S., Nollett, K. M., & Turner, M. S., Astrophys. J. Lett. **552**, L1 (2001).
4. Easther, R., Greene, B. R., Kinney W. H., and Shiu, G., Phys. Rev. D **64**, 103502 (2001).
5. Akerib, D. S., et al., hep-ex/0306001
6. Bernabei R., et al. Phys. Lett. B **480**, 23 (2000); Bernabei R., et al. Riv. N. Cim. **26**, 1 (2003).
7. Aprile, E., et al., astro-ph/0207670
8. Asztalos, S. J., et al., Astrophys. J. **571**, L27 (2002).
9. Gaisser, T. K., and Stanev, T., "Cosmic Rays," Phys. Rev. D **66**, 010001 (2002).
10. Abu-Zayyad, T., et al., astro-ph/0208301
11. http://www.slac.stanford.edu/grp/rd/epac/Meeting/200211/sokolsky.pdf
12. http://www.auger.org
13. http://ams.cern.ch/AMS/ams_homepage.html
14. http://amanda.uci.edu
15. http://antares.in2p3.fr
16. Malek, M., et al., Phys. Rev. Lett. **90**, 061101 (2003).

LIGHT QUARKS AND LEPTONS

New, high statistics measurement of the $K^+ \to \pi^0 e^+ \nu$ (K_{e3}^+) branching ratio

A. Sher for the E865 collaboration[†]

University of Pittsburgh, Pittsburgh, PA 15260, USA
Present address: SCIPP UC Santa Cruz, Santa Cruz, CA 95064

Abstract. E865 at the Brookhaven National Laboratory AGS collected about 70,000 K_{e3}^+ events to measure the K_{e3}^+ branching ratio relative to the observed $K^+ \to \pi^+ \pi^0$, $K^+ \to \pi^0 \mu^+ \nu$, and $K^+ \to \pi^+ \pi^0 \pi^0$ decays. The π^0 in all the decays was detected using the $e^+ e^-$ pair from $\pi^0 \to e^+ e^- \gamma$ decay and no photons were required. Using the Particle Data Group branching ratios [1] for the normalization decays we obtain $BR(K_{e3(\gamma)}^+) = (5.13 \pm 0.02_{stat} \pm 0.09_{sys} \pm 0.04_{norm})\%$, where $K_{e3(\gamma)}^+$ includes the effect of virtual and real photons. This result is $\approx 2.3\sigma$ higher than the current Particle Data Group value. Implications for the V_{us} element of the CKM matrix, and the matrix's unitarity are discussed.

The experimentally determined Cabibbo-Kobayashi-Maskawa (CKM) matrix describes quark mixing in the Standard Model framework. Any deviation from the matrix's unitarity would undermine the validity of the Standard Model. One unitarity condition involves the first row elements:

$$|V_{ud}|^2 + |V_{us}|^2 + |V_{ub}|^2 = 1 - \delta \qquad (1)$$

where a non-zero δ indicates a deviation from unitarity. V_{ub}, obtained from the semileptonic decays of B mesons [1], is too small to affect Eqn. 1. The V_{us} with the smallest theoretical uncertainty is obtained from K_{e3} decays[1, 2, 3]. The most precise value of V_{ud} obtained from the nuclear superallowed Fermi beta decays leads to $\delta = (3.2 \pm 1.4) \cdot 10^{-3}$ [4], a 2.3σ deviation from unitarity. Theoretical contributions to V_{us} were reevaluated recently[5, 6, 7, 8], but there has been little new experimental input on the K_{e3}^+ branching ratio. Since the V_{ud}^2 and V_{us}^2 uncertainties are comparable, a high statistics measurement of the K_{e3}^+ B.R. with good control of systematic errors is useful.

The bare (without QED contributions) K_{e3}^+ decay rate [2, 5, 6, 9] is:

$$d\Gamma(K_{e3}^+) = C(t)|V_{us}|^2 |f_+(0)|^2 [1 + \lambda_+ \frac{t}{M_\pi^2}]^2 dt \qquad (2)$$

where $t = (P_K - P_\pi)^2$, C(t) is a known kinematic function, and $f_+(0)$ is the vector form factor value at $t = 0$, determined theoretically [2, 5]. Two recent experiments[10, 11] both produced λ_+ (the form factor slope) consistent with previous measurements. An omitted negligible term in Eq. 2 containing the form factor f_- is proportional to M_e^2/M_π^2.

E865 [12] searched for the lepton flavor violating decay $K^+ \to \pi^+ \mu^+ e^-$. The detector resided in a 6 GeV/c positive beam. For the Ke3 running, the intensity was reduced by a factor of 10, to 10^7 kaons, 2×10^8 protons, and $2\times 10^8 \pi$ per 2.8 second pulse. The spectrometer consisted of four multiwire proportional chambers (MWPCs) and a dipole magnet. The particle identification used the threshold multichannel Čerenkov detectors (C1 and C2, each separated into left and right volumes, for four independent counters) filled with gaseous methane with $\gamma_t \approx 30$ and electron detection efficiency $\varepsilon_e \approx 0.98$ [14], followed by an electromagnetic calorimeter, and a muon detector (not used for K_{e3}^+).

The π^0 from the kaon decays was detected through the e^+e^- from the $\pi^0 \to e^+e^-\gamma$ decay, with the γ detected in some cases. To eliminate the uncertainty (2.7%) of the $\pi^0 \to e^+e^-\gamma$ B.R., and to reduce systematic uncertainty we used the other three major decay modes with a π^0 in the final state ($K^+ \to \pi^+\pi^0(K_{\pi 2})$, $K_{\mu 3}^+$, $K^+ \to \pi^+\pi^0\pi^0(K_{\pi 3})$) for the normalization sample ("Kdal").

The K_{e3}^+ data was collected in a one-week dedicated run in 1998, with special online trigger logic [13, 14]. The Kdal and K_{e3}^+ data were collected by the same trigger, designed to detect e^+e^- pairs. Each detector element's efficiency was measured as a function of the relevant phase space using either the built in detection and reconstruction redundancy or the data collected with the additional prescaled triggers designed specifically for this purpose[13, 14].

All relevant kaon decay chains [14] were simulated with GEANT [15] including decays of secondary pions and muons. The radiative corrections to the K_{e3}^+ decay phase-space density [5] were used. The $K_{e3\gamma}^+$ (inner bremsstrahlung) decays outside the K_{e3}^+ Dalitz plot boundary were explicitly simulated [9]. For $\pi^0 \to e^+e^-\gamma$ decay, radiative corrections were taken into account according to Ref. [16]. Measured efficiencies were applied[14], and accidental detector hits obtained from the data[14] were added.

Selection criteria[14], common to K_{e3}^+ and Kdal, required good quality three track events with the low ($M_{ee} < 0.05$ GeV) mass e^+e^- pair identified in the Čerenkov counters. All tracks were required to have less than 3.4 GeV/c momentum corresponding to the muon Čerenkov threshold. A geometric Čerenkov ambiguity cut rejected events where the Čerenkov counter response could not be unambiguously assigned to separate tracks[14]. The K_{e3}^+ sample was then selected by requiring the second positive track to be identified as e^+ in 2 of the 3 electron detectors: C1, C2, or the calorimeter, each with efficiency $\varepsilon_e \approx 98\%$. Events entering the Kdal sample had no response in at least one of the two Čerenkov counters. The $K_{\pi 2}$ acceptance is $\approx 1.2\%$. The K_{e3}^+ acceptance $\approx 0.7\%$ [14], somewhat lower because of the lower average e^+ momentum in the K_{e3}^+ decay. The final K_{e3} and Kdal samples were 71,204 and 558,186, respectively.

Contamination of the K_{e3}^+ sample by other K^+ decays occurred when π^+ or μ^+ from Kdal decays were misidentified as e^+, as a result of $\pi^0 \to e^+e^-e^+e^-$, or due to secondary particle decays. Total contamination of the Ke3 sample was estimated to be $(2.49 \pm 0.05_{stat} \pm 0.32_{sys})\%$[14]. Contamination due to overlapping events was $(0.25\pm 0.07)\%$ and $(0.12\pm 0.05)\%$ of the Kdal and K_{e3}, respectively. Figure 1 shows the energy distribution in the calorimeter from the e^+ in the K_{e3}^+ sample. The contamination is manifest in the minimum ionization spike at 250 MeV. The small excess of data in the spike agrees with our contamination uncertainty estimate.

The final K_{e3}^+ sample included $\approx 30\%$ events with fully reconstructed π^0s. Not requiring π^0s in our main analysis minimized the uncertainty arising from photon detection

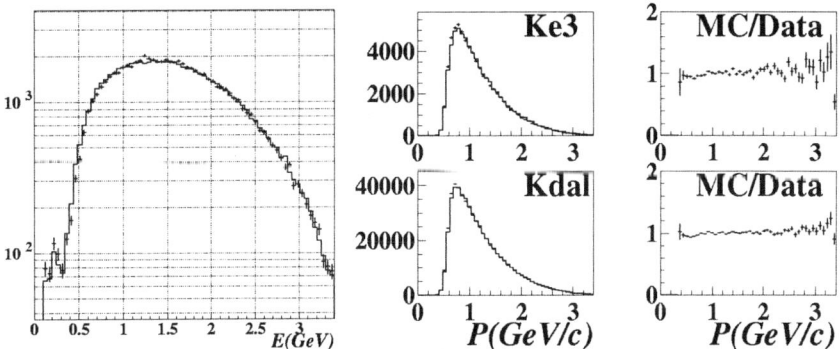

FIGURE 1. Plot on the left: Energy deposited in the calorimeter by the second positive track from the selected K_{e3}^+ sample (e^+ which is not from the low mass e^+e^- pair). No calorimeter information was used for the e^+ identification. Four plots on the right: Reconstructed momentum of the e^+ from the low mass e^+e^- pair from the selected K_{e3}^+ and Kdal samples. Histograms represent Monte Carlo; points with errors represent data. Two plots on the right show the bin by bin Monte Carlo to data ratio.

and reconstruction in the calorimeter, but increased vulnerability to contamination from upstream decays and photon conversion. The decay volume was evacuated to about 10^{-8} nuclear interaction length, which, in combination with the selection criteria applied, largely inhibited this contamination source. It was verified by the absence of the significant variation in the result after requirement for the fully reconstructed π^0[14].

The K_{e3}^+ statistical precision is 0.4%. The systematic error estimate was determined from the B.R. stability under variation of reconstruction procedure, selection criteria, assumed detector efficiencies, and subdivision of both K_{e3} and Kdal samples[14]. No significant correlations between any of the different systematic uncertainties were observed. The total systematic error was estimated to be 1.8%. Individual contributions to it are discussed in detail in Ref. [14]. The two largest contributions to the systematic error come from the discrepancies [14] between data and Monte Carlo in the momentum (Figure 1) and spatial distributions, 1.3% and 0.8%, respectively. These errors were determined by dividing K_{e3}^+ and Kdal events in roughly equal samples, using the relevant parameter and observing the result variation[14]; the errors were uncorrelated. The sensitivity of the vertical spatial discrepancy to the MWPC alignment and of the momentum discrepancy to the spectrometer parameters [14] indicate possible origins.

As an additional consistency check, we estimated the $K^+ \to \pi^+\pi^+\pi^-$/Kdal B.R.. The result was $(1.01 \pm 0.02) \times R_{PDG}$[1], where the theoretical prediction [17] was used for the $\pi^0 \to e^+e^-\gamma$ decay rate. The 2% error was dominated by the uncertainty in the prescale factor of the $K^+ \to \pi^+\pi^+\pi^-$ trigger. We also compared the K_{e3}^+ B.R. from 1998 and 1997 data. The 1997 K_{e3}^+ data used a trigger that required calorimeter hits, and A and D-counters, which neither allowed measurement of these detector efficiencies, nor of the trigger efficiency. While we did not use the 1997 data for our final result, the 1997 K_{e3}^+ branching ratio was statistically consistent with that from 1998. We estimated the form factor slope λ_+ from both 1998 and 1997 K_{e3}^+ data[14], and obtained: $\lambda_+ = 0.0324 \pm 0.0044_{stat}$ for 1998, and $\lambda_+ = 0.0290 \pm 0.0044_{stat}$ for the 1997 data, both consistent with the current PDG fit.

After contamination subtraction[14], our result is $BR(K^+_{e3(\gamma)})/(BR(K_{\pi 2})+BR(K_{\mu 3})+BR(K_{\pi 3}))=0.1962\pm 0.0008_{stat}\pm 0.0035_{sys}$, where $K^+_{e3(\gamma)}$ includes all QED contributions (loops and inner bremsstrahlung).

Using current[1] Kdal B.R.'s we infer $BR(K^+_{e3(\gamma)})=(5.13\pm 0.02_{stat}\pm 0.09_{sys}\pm 0.04_{norm})\%$. Where the normalization error is determined by the PDG estimate of the Kdal B.R. uncertainties. The PDG fit to the previous K^+ decay experiments yields $BR(K^+\to\pi^0 e^+\nu)=(4.87\pm 0.06)\%$ [1], $\approx 2.3\sigma$ lower than our result.

Radiative corrections for decays inside the K^+_{e3} Dalitz plot boundary were estimated to be -1.3% using the procedure of Ref. [5]; $K^+_{e3\gamma}$ decays outside the Dalitz plot boundary gave $+0.5\%$. Thus the total radiative correction was -0.8% resulting in the bare $BR(K^+_{e3})=(5.17\pm 0.02_{stat}\pm 0.09_{sys}\pm 0.04_{norm})\%$.

Using the PDG value for G_F, the short-distance enhancement factor $S_{EW}(M_\rho,M_Z)=1.0232$[5, 18], and our result for the bare K^+_{e3} rate we obtain $|V_{us}f_+(0)|=0.2239\pm 0.0022_{rate}\pm 0.0007_{\lambda_+}$, which gives $|V_{us}|=0.2272\pm 0.0023_{rate}\pm 0.0007_{\lambda_+}\pm 0.0018_{f_+(0)}$ if $f_+(0)=0.9874\pm 0.0084$[2, 5]. With this value of V_{us} and V_{ud} from superallowed nuclear Fermi beta decays[4], $\delta=0.0001\pm 0.0016$.

This result is consistent with CKM unitarity, but increases the discrepancy with the V_{us} from K^0_{e3} decay if extracted under conventional theoretical assumptions about symmetry breaking. K_{e3} measurements in progress (CMD2, NA48, KLOE)[3] should help to clarify the experimental situation.

We thank V. Cirigliano for the K^+_{e3} radiative corrections code. We gratefully acknowledge the contributions by the staffs of the AGS, and participating institutions. This work was supported in part by the U.S. Department of Energy, the National Science Foundations of the USA, Russia and Switzerland, and the Research Corporation.

REFERENCES

1. K. Hagiwara et al., Phys. Rev. D **66**, 010001 (2002).
2. H. Leutwyler, M. Roos, Z. Phys. C **25**, 91 (1984).
3. Workshop on the CKM Unitarity Triangle Durham, UK, 5th-9th April 2003
4. J.C. Hardy and I.S. Towner, J. Phys. G **29**, 197 (2003).
5. V. Cirigliano et al., Eur. Phys. J. C **23**, 121 (2002).
6. A. Bytev et al., Eur. Phys. J. C **27**, 57 (2003).
7. G. Calderon and G. Lopez Castro, Phys. Rev. D **65**, 073032 (2002)
8. J. Bijnens and P. Talavera, e-Print Archive: hep-ph/0303103
9. J. Bijnens et al., Nucl. Phys. **B396**, 81 (1993).
10. S. Shimizu et al., Phys. Lett. **B495**, 33 (2000).
11. I.V. Ajinenko et al., Phys. Atom. Nucl. **66**, 105 (2003).
12. R. Appel et al., Nucl. Instr. and Meth. A **479**, 349 (2002).
13. A. Sher et al., hep-ex/0305042 (2003).
14. A. Sher, Ph.D. thesis, University of Pittsburgh (2002) http://scipp.ucsc.edu/~sasha/thesis/th_ke3.ps.
15. R. Brun et al., "GEANT, Detector Description and Simulation Tool", CERN, Geneva (1994).
16. K.O. Mikaelian, J. Smith, Phys. Rev. **D5**, 1763 (1972)
17. B.E. Lautrup, J. Smith, Phys. Rev. **D3**, 1122 (1971).
18. W.J. Marciano, A. Sirlin, Phys. Rev. Lett. **71**, 3629 (1993).

Radiative Corrections and the Universality of the Weak Interactions

Alberto Sirlin

*Department of Physics, New York University,
4 Washington Place, New York, NY 10003, USA*

Abstract. We review the radiative corrections to superallowed Fermi transitions and neutron β decay in the Standard Model, and their relevance for the universality of the Weak Interactions.

The study of radiative corrections (R.C.) to μ and β decays has played an important role in the analysis of weak interactions before and after the emergence of the Standard Model (SM).

In the framework of the local V-A theory that preceded the SM, the R.C. to μ decay are convergent, to first order in G_F and all orders in α, after charge and mass renormalization [1]. The corrections to the electron spectrum in μ decay are large and play an important role in verifying that the parameters ρ and δ equal 3/4, a major prediction of the V-A theory [2].

The expression for the μ lifetime is

$$\frac{1}{\tau_\mu} = \frac{G_F^2 m_\mu^5}{192\pi^3} f\left(\frac{m_e^2}{m_\mu^2}\right)[1+\delta_\mu], \tag{1}$$

where $f(x) = 1 - 8x - 12x^2 \ln x + 8x^3 - x^4$ and δ_μ is the R.C.. One finds $\delta_\mu = -4.1995 \times 10^{-3} + 1.5 \times 10^{-6} + \ldots$, where the first and second terms are the $\mathcal{O}(\alpha)$ and $\mathcal{O}(\alpha^2)$ contributions [2, 3]. This leads to $G_F = 1.16637(1) \times 10^{-5}/\text{GeV}^2$.

Instead, the R.C. to β decay in the V-A theory are logarithmically divergent. For some time it was thought that form factor effects from the strong interactions (S.I.) can give rise to an effective cutoff. However, using current algebra (C.A.) techniques, Bjorken, and Abers, Dicus, Norton, and Quinn [4] studied the short distance behavior of the R.C. to β decay and reached the conclusion that the S.I. cannot tame their logarithmic divergence!

In the SM with three generations, the interactions of $W^\pm$ with fermions are given by

$$-(g_0/\sqrt{2})W_\mu \left[\overline{U}\gamma^\mu a_- D' + \overline{N}'\gamma^\mu a_- L\right] + \text{h.c.},$$

where $a_- = (1-\gamma_5)/2$, $D' = VD$, V is the unitary 3×3 CKM matrix, and U, D, N, and L are column vectors representing the up and down quarks, the neutrinos, and the charged leptons. The principle of non-abelian gauge invariance tells us that g_0 is a universal parameter, independent of the nature of the fermions involved. That fundamental property of g_0 and the unitarity of V may be regarded as the present statement of universality.

Since the SM is renormalizable, it should provide a convergent answer for the R.C. to β decay! In fact, using a simplified version of the SM, and neglecting the S.I., it was found in 1974 that, to very good approximation, the corrections are the same as in the V-A theory, with $\Lambda \to M_Z$! [5] During 1974-1978 the Current Algebra Formulation was extended to the real SM, including the effect of the S.I.[6]. This leads to the following expression for pure Fermi β decay transitions:

$$\mathscr{P}d^3p = \mathscr{P}^0 d^3p \left\{ 1 + \frac{\alpha}{2\pi} \left[3\ln\left(\frac{M_Z}{m_p}\right) + g(E, E_m) + 6\overline{Q}\ln\left(\frac{M_Z}{M}\right) + 2C + \mathscr{A}_{\bar{g}} \right] \right\}, \quad (2)$$

$$\mathscr{P}^0 d^3p = \frac{G_F^2 (V_{ud})^2}{8\pi^4} |M_F|^2 F(Z, E) (E_m - E)^2 d^3p. \quad (3)$$

The first two terms between square brackets in Eq. (2) arise from the vector current and are independent of the S.I.. In fact, the proton mass m_p cancels in the sum. The function $g(E, E_m)$, where E is the energy of the electron or positron and E_m its end-point energy, describes the R.C. to the spectrum in β decay in the presence of S.I.. It was first derived using the so-called "$1/k$" method [7].

The third term between the square brackets in Eq. (2) is a short distance contribution to the Fermi amplitude arising from the axial vector current. $\overline{Q}$ is the average charge of the fundamental doublet involved in the transition. In the SM this is the u-d doublet and we have $\overline{Q} = (2/3 - 1/3)/2 = 1/6$. The $2C$ term is a corresponding non-asymptotic part while $\mathscr{A}_{\bar{g}} \sim -0.34$ is a very small asymptotic contribution from QCD.

The R.C. to β decay are dominated by a large logarithmic term: $(3\alpha/2\pi)\ln(M_Z/2E_m)$. As an example, for the superallowed ^{14}O Fermi transition $E_m = 2.3$MeV, and this correction amounts to 3.4%. It turns out that such a large correction is phenomenologically crucial to verify the unitarity of the CKM matrix. Early smoking gun for the SM at the level of the quantum corrections?

Contributions of $\mathscr{O}(Z\alpha^2)$ and $\mathscr{O}(Z^2\alpha^3)$ are denoted by δ_2 and δ_3. One finds that δ_2 varies from 0.22% for ^{14}O to 0.50% for ^{54}Co, while δ_3 is much smaller [8].

There is also a correction δ_c that reflects the lack of perfect overlap between the wavefunctions of the parent and daughter nuclei due to Coulomb forces and configuration mixing effects in the shell-model wavefunctions. It has been extensively discussed in the literature [9, 10].

Leading logarithms of $\mathscr{O}[(\alpha/\pi)\ln(M_Z/m_p)]^n$ ($n \geq 2$) have been incorporated by means of a renormalization group analysis [11].

Putting these various contributions together, and integrating over the positron momentum one obtains

$$\Gamma = \Gamma^0 \left\{ 1 + \frac{\alpha(m_p)}{2\pi} \left[\overline{g}(E_m) + \mathscr{A}_{\bar{g}} \right] + \frac{\alpha}{2\pi} \left[\ln\left(\frac{m_p}{M}\right) + 2C \right] + \delta_2 + \delta_3 \right\}$$
$$\times S(m_p, M_Z)(1 - \delta_c), \quad (4)$$

where $S(m_p, M_Z) = 1.0225$ is the short distance contribution and $\overline{g}(E_m)$ is the average of $g(E, E_m)$ over the positron spectrum. The term $(\alpha/2\pi)[\ln(m_p/M) + 2C]$ from the axial vector current is model dependent. In recent discussions the mass M, that represents the onset of the asymptotic behavior, is allowed to vary in the range $m_{A1}/2 \leq M \leq 2m_{A1}$, with a central value $M_c = m_{A1} = 1.26\text{GeV}$, the mass of the A_1 resonance, which has the correct quantum numbers to mediate that contribution [9, 12]. Jaus and Rasche proposed to split $C = C_{Born} + C_{NS}$, where the first term is identified with the Born approximation calculation of the diagram where the insertions of the axial vector and electromagnetic currents involves the same nucleon, while C_{NS} corresponds to the contributions in which the insertions occur in different ones [13]. One obtains $C_{Born} = 0.881 \pm 0.030$.

In order to verify CVC, it is advantageous to factor out the nuclear-dependent part of the R.C.. A simple way of doing this is to factor out the expression in Eq. (4) in the form $(1+\delta_R)(1+\Delta_R)(1-\delta_c)$, where

$$1 + \delta_R = 1 + \frac{\alpha(m_p)}{2\pi} \bar{g}(E_m) + \frac{\alpha}{\pi} C_{NS} + \delta_2 + \delta_3, \qquad (5)$$

$$1 + \Delta_R = \left\{ 1 + \frac{\alpha}{2\pi} \left[\ln\left(\frac{m_p}{M}\right) + 2C_{Born} + \mathscr{A}_{\bar{g}} \right] \right\} S(m_p, M_Z). \qquad (6)$$

One can then introduce a radiatively corrected $\mathscr{F}t$ value

$$\mathscr{F}t = ft(1+\delta_R)(1-\delta_c) = K/2G_V'^2, \qquad (7)$$

$$K = 2\pi^3 \ln 2 \, \hbar^7 / m_e^5 c^4 = 8.12027 \times 10^{-7} (\hbar c)^6 \text{GeV}^{-4} s; \qquad (8)$$

$$G_V'^2 = G_V^2(1+\Delta_R) \, ; \, G_V = G_F V_{ud}. \qquad (9)$$

The test of CVC consists in checking the constancy of the $\mathscr{F}t$ values. Using then the average $\mathscr{F}t$, one obtains $G_V'^2$. Inserting the calculated Δ_R one obtains G_V^2 and therefore V_{ud}. A recent determination by Towner and Hardy is $V_{ud} = 0.9740 \pm 0.0005$ [9] (nuclear β decay). It is important to note that the error is mainly theoretical ($\pm 4 \times 10^{-4}$ from Δ_R, $\pm 3 \times 10^{-4}$ from δ_c).

In the case of neutron β decay, we avoid nuclear physics complexities, but this is not a pure Fermi transition! However, we can apply C.A. in combination with the $1/k$ method [7]. The latter allows the calculation of some important observables in the presence of the S.I. in terms of effective coupling constants G'_V and G'_A, neglecting small contributions of $\mathscr{O}((\alpha/\pi)(E/M)\ln(M/E), (\alpha/\pi)(q/M))$, where M is a hadronic mass. The observables include the correction to the electron spectrum (given by $(\alpha/2\pi)g(E, E_m)$), the longitudinal polarization of electrons, and the electron asymmetry from polarized neutrons.

We use the $1/k$ method to express the lifetime and the electron asymmetry in terms of G'_V and G'_A. The inverse lifetime is proportional to $G_V'^2 + 3G_A'^2$. The asymmetry gives us G'_A/G'_V. Combining the two observables, we can find G'_V. Using $G_V'^2 = G_F^2 V_{ud}^2 (1+\Delta_R)$, we extract V_{ud}.

Employing $G'_A/G'_V = -1.2690 \pm 0.0022$ and $\tau_n = 885.6 \pm 0.8$ s, a recent analysis by Towner and Hardy [9] gives $|V_{ud}| = 0.9745 \pm 0.0016$ (neutron β decay), which is consistent with the nuclear result but has considerably larger error. Combining with

$|V_{us}| = 0.2196 \pm 0.0026$ and $|V_{ub}| = 0.0036 \pm 0.0007$, recommended by PDG02 [14], one obtains

$$\sum_i |V_{ui}|^2 = 0.9969 \pm 0.0015 \quad \text{(nuclear } \beta \text{ decay [10])}, \tag{10}$$

$$\sum_i |V_{ui}|^2 = 0.9979 \pm 0.0033 \quad \text{(neutron } \beta \text{ decay [11])}. \tag{11}$$

The first test is short by 2.1σ, while the second one is in agreement, but has a larger error. On the other hand, PDG02 averages only over recent asymmetry experiments with polarization $> 90\%$, leading to $G'_A/G'_V = -1.2720 \pm 0.0018$ and $|V_{ud}| = 0.9725 \pm 0.0013$ (neutron) [14], and a 2.2σ shortfall.

Based on a recent high statistics experiment [15], a preliminary value Br $(K^+ \to \pi^0 e^+ \nu) = (5.13 \pm 0.2 \pm 0.08 \pm 0.04)\%$ has been reported, which is higher than the PDG02 entry $(4.82 \pm 0.06)\%$. If the result is confirmed and the lifetime is not modified, it may lead to a solution of the unitarity deviation. In fact, the central value in the unitarity test would become $(0.9740)^2 + (0.2196)^2 \, 5.13/4.82 = 1.000002!$ Of course, it would be important to check the experimental status of $K^0 \to \pi^- e^+ \nu_e$, as well as the $K_{\mu 3}$ modes.

It is also interesting to remember that the deviation in Eq. (10) can be removed in "manifest" left-right symmetric models [16] by choosing $2\zeta = 0.0031 \pm 0.0015$, where ζ is the mixing angle [9, 12].

The determination of V_{us} is derived mainly from K_{l3} decays applying R.C. and chiral perturbation theory (ChPT). One considers $K^+ \to \pi^0 e^+ \nu$, $K^0 \to \pi^- e^+ \nu$, and $K_{\mu 3}$ modes. After applying R.C. the experiments determine $f_+(0) V_{us}$. To get V_{us}, we need $f_+(0)$. For $K^0 \to \pi^- l^+ \nu$, the non-renormalization theorem tells us that $f_+(0)$ differs from 1 by terms of second order in the mass splittings [17]. Expanding $f_+(0) = 1 + f_1 + f_2 + \cdots$, where $f_1 = \mathcal{O}(m_q \ln m_q)$, $f_2 = \mathcal{O}(m_q^2 \ln m_q)$, and m_q are generic quark masses, f_1 was obtained in a model independent manner [18] and lowers $f_+^{K^0 \pi^-}(0)$ to 0.977, while an estimate for f_2 gives $f_2 = -0.016 \pm 0.008$ [19]. Combining the two results, one has $f_+^{K^0 \pi^-}(0) = 0.961 \pm 0.008$ [19]. For $K^+ \to \pi^0 l^+ \nu$, there is a complication. One finds $|\pi^0 > = \cos\varepsilon |3> + \sin\varepsilon |8>$ where $|\pi^0>$ is the physical state and $\varepsilon = (\sqrt{3}/4)(m_d - m_u)/(m_s - \hat{m}) \approx 0.01$. As a consequence, to zeroth order in m_q, there is a breaking of isospin invariance and $f_{K^+\pi^0}(0)/f_{K^0\pi^-}(0) = 1.0172$. Including terms of $\mathcal{O}(\varepsilon m_q)$, the ratio becomes 1.022 [18, 19]. Thus, there is an interesting isospin breaking effect that enhances the $K^+ \to \pi^0 e^+ \nu$ rate by 4.45% relative to $K^0 \to \pi^- e^+ \nu$. Using the above results, the experimental data, some of which had been corrected by long distance R.C., and including the short distance R.C., Leutwyler and Roos obtained $|V_{us}| = 0.2196 \pm 0.0023$ [19], while PDG02 recommends $|V_{us}| = 0.2196 \pm 0.0026$ [14].

Very recently, the R.C. to K_{l3} decays have been studied in the ChPT framework [20], leading to $|V_{us}| = 0.2201 \pm 0.0024$, very close to the other determinations.

Also very recently Bijnens and Talavera have discussed the evaluation of K_{l3} decays to two-loop order in ChPT in the isospin limit [21]. Their expression for $f_+(0)$ depends on two unknown constants that can in principle be determined by accurate measurements of the scalar form factor $f_0(t) = f_+(t) + f_-(t) t/(M_K^2 - M_\pi^2)$, specifically its slope and curvature.

Bill Marciano tells me that precise lattice calculations of $f_+(0)$ are possible. Lattice practitioners should be encouraged to carry out this important calculation!

This work was supported in part by NSF Grant PHY-0245068.

REFERENCES

1. S. M. Berman and A. Sirlin, Ann. Phys. **20**, 20 (1962).
2. T. Kinoshita and A. Sirlin, Phys. Rev. **113**, 1652 (1959); S. M. Berman, Phys. Rev. **112**, 267 (1958).
3. T. van Ritbergen and R. G. Stuart, Phys. Rev. Lett. **82**, 488 (1999); M. Steinhauser and T. Seidensticker, Phys. Lett. **B467**, 271 (1999); A. Ferroglia, G. Ossola and A. Sirlin, Nucl. Phys. B **560**, 23 (1999).
4. J. D. Bjorken, Phys. Rev. **148**, 1467 (1966); E. S. Abers, D. A. Dicus, R. E. Norton and H. R. Quinn, Phys. Rev. **167**, 1461 (1968).
5. A. Sirlin, Nucl. Phys. **B71**, 29 (1974).
6. A. Sirlin, Rev. Mod. Phys. **50**, 573 (1978).
7. A. Sirlin, Phys. Rev. **164**, 1767 (1967); A. Sirlin, in *Particles, Currents, Symmetries*, ed. P. Urban, Acta Physics Austriaca, Suppl. V (Springer-Verlag, New York 1968) p. 353.
8. A. Sirlin and R. Zucchini, Phys. Rev. Lett. **57**, 1994 (1986); A. Sirlin, Phys. Rev. D **35**, 3423 (1987).
9. I. S. Towner and J. C. Hardy, J. Phys. G **29**, 197 (2003) and references cited therein.
10. D. H. Wilkinson, J. Phys. G **29**, 189 (2003) and references cited therein.
11. W. J. Marciano and A. Sirlin, Phys. Rev. Lett. **56**, 22 (1986).
12. A. Sirlin, in *Precision Tests of the Standard Electroweak Model*, ed. P. Langacker (World Scientific Publishing Co., Singapore 1995), p. 766.
13. W. Jaus and G. Rasche, Phys. Rev. D **41**, 166 (1990).
14. K. Hagiwara *et al.* [Particle Data Group Collaboration], Phys. Rev. D **66**, 010001 (2002).
15. E865 Experiment at BNL.
16. M. A. Bég, R. V. Budny, R. N. Mohapatra and A. Sirlin, Phys. Rev. Lett. **38**, 1252 (1977) [Erratum-ibid. **39**, 54 (1977)].
17. R. E. Behrends and A. Sirlin, Phys. Rev. Lett. **4**, 186 (1960); M. V. Terent'ev, Zhur. Eksptl. i Teort. Fiz. **44**, 1320 (1963) [Phys. JHTP **17**, 890 (1963)]; M. Ademollo and R. Gatto, Phys. Rev. Lett. **13**, 264 (1964).
18. J. Gasser and H. Leutwyler, Nucl. Phys. B **250**, 517 (1985).
19. H. Leutwyler and M. Roos, Z. Phys. C **25**, 91 (1984).
20. V. Cirigliano, arXiv:hep-ph/0305154.
21. J. Bijnens and P. Talavera, arXiv:hep-ph/0303103.

Can An Amended Standard Model Account For Cold Dark Matter?

Maurice Goldhaber

Physics Department, Brookhaven National Laboratory, Upton, NY 11973

Abstract

It is generally believed that one has to invoke theories beyond the Standard Model to account for cold dark matter particles. However, there may be undiscovered universal interactions that, if added to the Standard Model, would lead to new members of the three generations of elementary fermions that might be candidates for cold dark matter particles.

The Standard Model (SM) of the elementary fermions and their interactions can accommodate hot dark matter, believed to consist of zero mass or lightweight particles: photons (the micro-wave background), gravitons and neutrinos. However, it cannot account for cold dark matter, particles that may have considerable mass and are long-lived or stable. This has often been regarded as evidence for the need to invoke a theory beyond the SM, usually Supersymmetry and its favorite candidate for a cold dark matter particle, the lightest supersymmetrical particle that is believed to be stable, the neutralino. Several experimental searches are designed to detect cold dark matter particles that have weak interactions with "our" matter. Such particles have been dubbed WIMPS (weakly interacting massive particles) which would include the neutralino.

In a recent review, Ostriker and Steinhardt [1] consider many possible properties of candidates for cold dark matter particles, including the conclusion by Sperling and Steinhardt [2] that these particles strongly interact with each other, which they deduced from the work by Tyson et al. [3] who studied the matter distribution in a large gravitational lens, consisting of many galaxies as well as cold dark matter.

It seems worthwhile to ponder whether an extension of the SM might be able to accommodate cold dark matter particles. For this purpose I make use of rules -- some long known -- based on the known properties of elementary fermions, discussed in a recent paper, entitled *"A closer look at the elementary fermions"* [4]. Rules have often been considered as proto-theories from which conclusions of predictive value can be drawn by interpolation or extrapolation. I have recently refined the rules concerning the elementary fermions (to be published). For the present purpose I make use of two rules:

Rule A. In each generation $(i = 1\text{-}3)$ a correlation exists between the masses of the elementary fermions and the strength of their *dominant* interactions.

Rule B. The masses of elementary fermions with equal dominant interactions increase as i increases, known as the *hierarchical mass problem*.

In the following I shall refer to cold dark matter particles as δ_i's. To include fairly massive δ's that strongly interact with each other, we would have to assume, according to Rule A, the existence of a new, fairly strong, interaction, and according to Rule B, the three δ_i would have hierarchical masses. Elementary fermions possessing this new interaction would interact with "our" elementary fermions through the only interaction they have in common, the gravitational one, which is too weak to be detected in experiments designed to detect WIMPS. This also implies that cross sections for *producing* δ's in collisions between "our" elementary fermions would be too small to be observable at attainable accelerator energies. The decay of the heavier δ_i ($i > 1$) into δ_1 by gravitational interaction in one or two steps would be too slow to have happened to a noticeable degree in the time that has elapsed since their creation. A faster decay of the heavier δ's might exist, if they could e.g. decay to the lighter δ's by emission of pairs of lighter $\delta_i + \overline{\delta}_i$ or, if the new interaction has a gauge boson of suitable mass, by emissions of the gauge boson. Thus we have to leave the possibility open that cold dark matter may consist not only of δ_1 and $\overline{\delta}_1$ and gauge bosons but also of the heavier δ_i and $\overline{\delta}_i$ and, depending on the forces between them, they may be able to bind to each other, leading perhaps to combinations of $\delta_i \delta_j$, $\overline{\delta}_i \overline{\delta}_j$ and $\delta_i \overline{\delta}_j$ [i (1-3), j (1-3)] that might grow into "clumps" containing many δ_i and $\overline{\delta}_i$. Therefore, we cannot exclude the possibility that the mass range for cold dark matter particles may be an average of several masses.

If we want to speculate further, we might consider the possibility that larger compounds containing many combinations of δ_i and $\overline{\delta}_i$ might have also been formed, and that "δ-clumps" might have grown into congregations of astrophysical interest, perhaps even into early black holes around which "our" galaxies may have later formed.

Wandelt et al [5] consider the further possibility that δ's may also have strong interactions with "our" matter. This would imply that they share QCD interactions with either u or d quarks or both, and thus would carry electric charges like quarks do, which would not be compatible with the properties of δ's. The existence of unusually heavy isotopes, if cold dark matter particles could bind to "our" nuclei, that has often been contemplated, thus should not be expected.

Since the ratio of visible matter to dark matter is ~ *1:6*, the number of δ_i's + $\overline{\delta}_i$'s appear to be of the same order as the number of nucleons (or even less if they are considerably heavier than nucleons). This leads to an interesting paradox: Why should the δ_i's + $\overline{\delta}_i$'s that survived, have only been produced in numbers comparable to the number of surviving nucleons while the origin of the nucleon—anti-nucleon asymmetry is ascribed to CP non-conservation in the origin of nucleons and anti-nucleons, with only ~10^{-9} nucleons estimated to exceed the number of anti-nucleons and surviving mutual annihilation.

Even in the unlikely case that a similar CP non-conservation had played a role for δ's, mutual $\delta_t + \bar{\delta}_t$ annihilation at the end of inflation would be too slow and the amount of cold dark matter would still be overwhelming! How can we understand the paradox of apparently too few δ's and $\bar{\delta}$'s by many orders of magnitude?

While some cold dark matter may be "hidden," partly inside "our" stars, partly in δ-stars and partly in black holes of δ origin, and some may exist in voids, the total could not possibly explain the big deficiency, though parenthetically, it might nevertheless be of interest, whether a void can be found that acts as a gravitational lens.

In the to-and-fro of reactions in the early inflationary period, any δ_i's $+\bar{\delta}_i$'s produced at the highest energy might, when still energetic enough, be able to annihilate each other into gravitons, or by graviton stimulated emission, if no other channels were available. But as energies decrease further δ_i's $+\bar{\delta}_i$'s pairs would be poorly recreated and thus end up finally in smaller numbers than the expected equilibrium number known for other particles.

ACKNOWLEDGMENTS

I wish to thank A.S. Goldhaber and W.J. Marciano for valuable discussions.

REFERENCES

1. J.P. Ostriker and P. Steinhardt, Science, 300, 1909 (2003).
2. D.N. Sperling and P.J. Steinhardt, Phys. Rev. Lett., 84, 3760 (2000).
3. J.A. Tyson, G.P. Kochanski, and Dell'antonio, ApJ, 498, L107 (1998).
4. M. Goldhaber, Proc. Nat. Acad. Sci. 99, 33 (2002) and with minor corrections hep-ph/0201208.
5. B. D. Wandelt, R. Davé, G.R. Farrar, P.C. Mcguire, D.N. Spergel and P.J. Steinhardt, Proc. of Dark Matter (2000), astro-ph/0006344.

Nucleon Electromagnetic Form Factors and Densities

James J. Kelly

Department of Physics, University of Maryland, College Park, MD 20742

Abstract. We review data for nucleon electromagnetic form factors, emphasizing recent measurements of G_E/G_M that use recoil or target polarization to minimize systematic errors and model dependence. The data are parametrized in terms of densities that are consistent with the Lorentz contraction of the Breit frame and with pQCD.

The electromagnetic structure of nucleons provides fundamental tests of the QCD confinement mechanism, as calculated on the lattice or interpreted with the aid of models. From elastic electron scattering one obtains the Sachs electric and magnetic form factors, which are closely related to the charge and magnetization densities. Dramatic improvements in the quality of these measurements have recently been achieved by using beams that combine high polarization with high intensity and energy together with either polarized targets or measurements of recoil polarization. In this paper we review the current status of nucleon elastic form factors, emphasizing recent polarization measurements, and parametrize these data using a model that permits visualization of the underlying charge and magnetization densities. Recent reviews of the data and experimental techniques may be found in [1, 2].

It is difficult to extract the proton electric form factor, G_{Ep}, at large Q^2 using Rosenbluth separation of the differential cross section because the magnetic contribution, G_{Mp}, is much stronger. Nevertheless, a series of Rosenbluth experiments performed at SLAC with modest precision suggested that $\mu_p G_{Ep}/G_{Mp} \approx 1$ remains a reasonable approximation for $Q^2 < 7$ (GeV/c)^2. Fortunately, it is now possible to measure the electromagnetic ratio, $g_p = \mu_p G_{Ep}/G_{Mp}$, much more accurately using the polarization transferred to the recoil proton instead. For the proton, both the longitudinal and transverse components can be measured simultaneously using a polarimeter in the focal plane of a magnetic spectrometer, thereby minimizing systematic uncertainties due to beam polarization, analyzing power, and kinematic parameters. The systematic uncertainty due to precession of the proton spin in the magnetic spectrometer is usually much smaller than the uncertainties in comparing the cross sections obtained with different kinematical conditions and acceptances needed for the Rosenbluth method. Similarly, it is difficult to extract G_{En} from unpolarized elastic electron-deuteron scattering because the model dependence of corrections for G_{Mn}, G_{Ep}, G_{Mp}, final-state interactions, and two-body currents is prohibitive. Schiavilla and Sick [3] demonstrated that use of the quadrupole form factor extracted from t_{20} and T_{20} reduces this model dependence, but the uncertainties remain appreciable. Much more accurate results can be obtained using either neutron recoil polarization in the quasifree $d(\vec{e},e'\vec{n})p$ reaction or the target asymmetry for the $\vec{d}(\vec{e},e'n)p$ or $^3\vec{He}(\vec{e},e'n)$ reactions.

Recoil-polarization measurements for proton elastic scattering were performed at Jefferson Laboratory. Although we expected to improve the precision of g_p for $Q^2 > 1$ $(GeV/c)^2$, I doubt that anyone expected to observe a significant departure from unity — the experiment was undertaken primarily to commission and calibrate the new focal-plane polarimeter. Nevertheless, the high-precision measurements [4, 5, 6] shown in Figure 1 revealed a surprisingly strong linear decrease in g_p for $Q^2 > 1$ $(GeV/c)^2$, with very small systematic uncertainties. This marked difference from the SLAC observations prompted Arrington [7] to re-analyze the Rosenbluth data with careful analysis of relative normalization between various subsets. His results, also shown in Fig. 1, tend to confirm the original finding that $g_p \approx 1$ — the uncertainties are large enough to permit a small slope but are clearly inconsistent with the recoil-polarization data. Furthermore, Arrington finds that restoring consistency between the two types of experiments would require modification of the ε dependence of the Rosenbluth formula by about 5%.

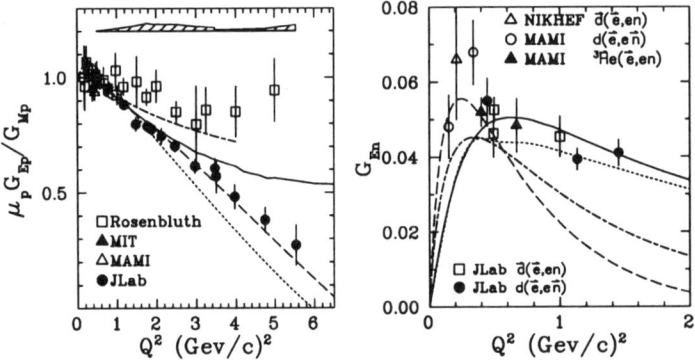

FIGURE 1. Recent data for g_p and G_{En} are compared with selected models. The hatched region indicates the systematic uncertainty in JLab measurements for the proton. Rosenbluth data were taken from the global analysis of Arrington [7]. The proton polarization-transfer data are from [8, 9, 6, 5]. The neutron data are from [10, 11, 12, 13, 14, 15, 16]. See text for explanation of curves.

Reconciliation between these techniques is now crucial. If we accept that both analyses are free of fundamental error and that both propagate uncertainties properly, then we must re-examine the validity of the one-photon exchange approximation. Given that the cross section falls rapidly with Q^2, it is clear that two-step scattering processes should begin to compete favorably with the direct mechanism for sufficiently large Q^2. Several recent papers [17, 18, 19, 20] have discussed mechanisms in which two virtual photons share Q^2 approximately equally, but none have provided rigorous calculations that integrate over intermediate states. Guichon and Vanderhaeghen [18] employed a semi-empirical argument to demonstrate that the two-photon contribution has negligible effect upon the recoil-polarization ratio but that its effect upon the Rosenbluth separation increases with Q^2 without destroying the linearity with respect to ε expected for the Rosenbluth plot at fixed Q^2. However, Rekalo and Tomasi-Gustafsson [20] criticized the neglect of C-invariance and emphasized the importance of crossing symmetry. The calculation of Blunden et al. [19] using only the nucleon intermediate state accounts for part of the discrepancy between the two techniques but does not achieve consistency; clearly there is room for the Δ and higher contributions to the box diagrams. Similarly, a measurement of the parity-conserving asymmetry for elastic scattering of transversely polarized electrons gave a result at $Q^2 = 0.1$ (GeV/c)^2 that is of order 10^{-5} and somewhat larger than the calculation based upon the nucleon box diagrams [21].

Experimentally, one can investigate corrections to one-photon exchange by a) more accurate ε dependence of eN elastic scattering, b) measurement of the C-odd difference between e^-N and e^+N elastic scattering, or c) T-odd parity-conserving single-spin asymmetry (SSA). The data from a recently completed *super Rosenbluth* experiment [22] that was designed to minimize systematic errors are presently under analysis. It has been argued that SLAC data [23] for b) are consistent with $g_p \approx 1$ but, although the statistical precision is modest, the fact that the data for σ^+/σ^- are actually systematically above unity for large Q^2 appears to leave room for a two-photon contribution to positron scattering opposite to that for electron scattering. Proposals to measure P_N for recoil-polarization or the asymmetry A_T for a transversely polarized target are under development but are technically challenging – the former must compensate accurately for instrumental asymmetries in the polarimeter while a new target must be developed for the latter. Although SSA would measure the imaginary part of the doubly virtual Compton scattering amplitude while it is the real part that is implicated in the present controversy, we should demand consistency before declaring victory.

For the neutron one must use quasifree scattering from a neutron in a nuclear target and correct for the effects of Fermi motion, meson-exchange currents, and final-state interactions also. Both recoil and target polarization for quasifree knockout give similar PWIA formulas for the form factor ratio, but the systematic errors and the nuclear physics corrections are appreciably different. Therefore, confident extraction of G_{En}/G_{Mn} benefits from comparison of data for both recoil and target polarization. Data from recent experiments of these types are also shown in Fig. 1, where we now find good consistency between several methods.

Results for selected models are also shown in Fig. 1. The chiral soliton model of Holzwarth (dashed) predicted the linear behavior of G_{Ep}/G_{Mp} but fails to reproduce neutron form factors [24, 25]. The relativistic cloudy bag model (dotted) of Miller et al. [26] reproduces G_{En} well and also predicts a linear form factor ratio for the proton, but the slope is too steep. The point-form spectator approximation (PFSA) using pointlike constituent quarks and a Goldstone boson exchange interaction fitted to spectroscopic data (dash-dot) lies well above the g_p and well below the G_{En} data for large Q^2 [27]. Finally, a light-front calculation using one-gluon exchange and constituent-quark form factors fitted to $Q^2 < 1$ (GeV/c)^2 provides a good fit (solid) up to about 4 (GeV/c)^2 [28]. However, none of the available theoretical calculations provides a truly quantitative description for all four form factors over a wide range of Q^2. The differences between these models are largest for G_{En}, which is especially sensitive to small mixed-symmetry and deformed components of the nucleon wave function. Clearly it will be very important to extend the G_{En} data to larger Q^2.

Several parametrizations of these data have been proposed. Lomon fit the data with a hybrid model that interpolates between vector-meson dominance at low Q^2 and perturbative QCD at high Q^2 [29]. The model has the advantage of fitting all four elastic form factors in a consistent framework with a clear physical interpretation and can be

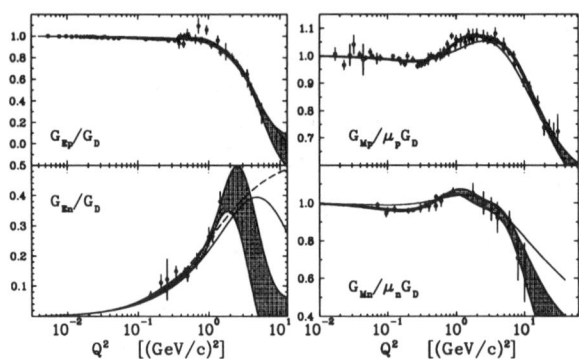

FIGURE 2. LGE fits to selected data for nucleon form factors using $\lambda_E = \lambda_M = 2$ are shown as error bands. The solid lines represent the VMD+pQCD fit of Lomon. For G_{En} we also show a two-parameter Galster fit as a dashed line.

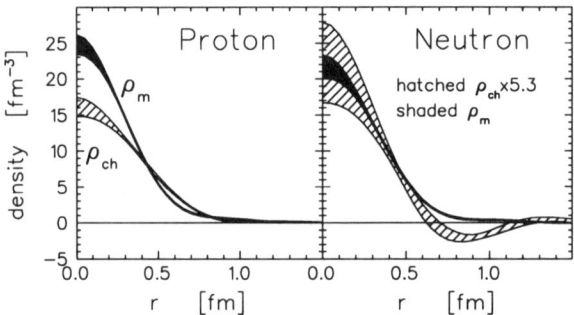

FIGURE 3. Nucleon electromagnetic densities using $\lambda_E = \lambda_M = 2$. The neutron charge density has been scaled for comparison with its magnetization density.

extended naturally to the timelike regime. The fits shown in Fig. 2 are fairly accurate for the proton but for G_{Mn}/G_D have difficulty describing the dip around $Q^2 \sim 0.2$ and the slope for large Q^2. The deficiency in G_{En} for $Q^2 \sim 1.5$ may simply reflect the lack of data at the time of that analysis.

Alternatively, I proposed a parametrization that was motivated by the interpretation of form factors in terms of densities and accounts for Lorentz contraction while enforcing pQCD constraints at high Q^2 [1]. Furthermore, by expanding the density or, equivalently, the form factor in a complete set of basis functions, one can construct error bands for fitted quantities and evaluate the uncertainties involved in extrapolation to higher Q^2. The density interpretation is admittedly model dependent, but the Q^2 parametrization is robust. Fits to the form factors are shown as bands in Fig. 2 and the resulting densities are shown in Fig. 3, where the error bands include the uncertainty due to the limited experimental range of Q^2. Details of the method may be found in Ref. [1], but we have included some newer data, rejected data for G_{Mn} using the associated-particle method for neutron detector efficiency, and have enforced constraints at $Q^2 = 0$ explicitly. The observation that g_p decreases rapidly for $Q^2 > 1$ (GeV/c)^2 is interpreted as a broader distribution of charge than of magnetization. The magnetization density is slightly broader for the neutron than for the proton. For the purposes of comparing shapes, the neutron charge density is shown scaled to the interior magnetization. Despite the limited range of Q^2 and larger uncertainties in the G_{En} data, the neutron charge density is determined with useful precision. The neutron charge density results from incomplete cancellation between u and d quark densities with slightly different shapes that leaves a positive core surrounded by negative surface charge; alternatively, this feature can be interpreted in terms of the pion cloud [30, 31].

In summary, recent advances in polarized electron beams, nucleon polarimetry, and polarized targets have rekindled interest in the Sachs form factors. The surprising discrepancy between Rosenbluth and recoil-polarization measurements of the g_p may show that long-neglected dispersive effects become important $Q^2 > 1$ (GeV/c)^2. Precise data for G_{En} are finally available for $Q^2 < 1.5$ (GeV/c)^2 and data for higher Q^2 are expected in a few years.

The support of the U.S. National Science Foundation under grant PHY-0140010 is gratefully acknowledged.

REFERENCES

1. Kelly, J. J., *Phys. Rev.* **C**, **66**, 065203 (2002).
2. Gao, H., *Int. Journ. Mod. Phys.* **E**, **12**, 1–40 (2003).
3. Schiavilla, R., and Sick, I., *Phys. Rev.* **C**, **64**, 041002(R) (2001).
4. Jones, M. K., et al., *Phys. Rev. Lett.*, **84**, 1398–1402 (2000).
5. Gayou, O., et al., *Phys. Rev. Lett.*, **88**, 092301 (2002).
6. Punjabi, V., et al., submitted to Phys. Rev. C (2003).
7. Arrington, J., arXiv:nucl-ex/0305009 (2003).
8. Milbrath, B. D., et al., *Phys. Rev. Lett.*, **82**, 2221 (1999).
9. Pospischil, T., et al., *Eur. Phys. J. A*, **12**, 125–127 (2001).
10. Herberg, C., et al., *Eur. Phys. J. A*, **5**, 131–135 (1999).
11. Golak, J., et al., *Phys. Rev.* **C**, **63**, 034006 (2001).
12. Passchier, I., et al., *Phys. Rev. Lett.*, **82**, 4988–4991 (1999).
13. Bermuth, J., et al., *Phys. Lett.*, **B564**, 199–204 (2003).
14. Zhu, H., et al., *Phys. Rev. Lett.*, **87**, 081801 (2001).
15. Warren, G., et al., submitted to Phys. Rev. Lett (2003).
16. Madey, R., et al., accepted by Phys. Rev. Lett (2003).
17. Afanasev, A., Akushevich, I., and Merenkov, N. P., arXiv:hep-ph/0208260 (2002).
18. Guichon, P. A. M., and Vanderhaeghen, M., arXiv:hep-ph/0306007 (2003).
19. Blunden, P. G., Melnitchouk, W., and Tjon, J. A., arXiv:nucl-th/0306076 (2003).
20. Rekalo, M. P., and Tomasi-Gustafsson, E., arXiv:nucl-th/0307066 (2003).
21. Wells, S. P., et al., *Phys. Rev.* **C**, **63**, 064001 (2001).
22. Arrington, J., et al., Jefferson Laboratory proposal e01-001 (2001).
23. Mar, J., et al., *Phys. Rev. Lett.*, **21**, 482–484 (1968).
24. Holzwarth, G., *Zeit. Phys. A*, **356**, 339 (1996).
25. Holzwarth, G., arXiv:hep-ph/0201138 (2002).
26. Miller, G. A., *Phys. Rev.* **C**, **66**, 032201(R) (2002).
27. Wagenbrunn, R., et al., *Phys. Lett. B*, **511**, 33–39 (2001).
28. Simula, S., arXiv:nucl-th/0105024 (2001).
29. Lomon, E. L., arXiv:nucl-th/0203081 (2002).
30. Kaskulov, M. M., and Grabmayr, P., arXiv:nucl-th/0308015 (2003).
31. Friedrich, J., and Walcher, T., arXiv:hep-ph/0303054 (2003).

Nucleon Axial Charge from Quenched Lattice QCD with Domain Wall Fermions

Shigemi Ohta for RBCK Collaboration[†]

Institute of Particle and Nuclear Studies, KEK, Oho 1-1, Tsukuba, Ibaraki 305-0801, Japan
RIKEN BNL Research Center, Brookhaven National Laboratory, Upton, NY 11973, USA

Abstract. The chiral symmetry of domain wall fermions makes the calculation of the nucleon axial charge particularly easy since the Ward-Takahashi identity requires the vector and axial-vector currents to have the same renormalization, up to lattice spacing errors of order $O(a^2)$. The DBW2 gauge action provides enhancement of the good chiral symmetry properties of domain wall fermions at larger lattice spacing than the conventional Wilson gauge action. Taking advantage of these methods and performing a high statistics simulation, we find a significant finite volume effect between the nucleon axial charges calculated on lattices with $(1.2\,\text{fm})^3$ and $(2.4\,\text{fm})^3$ volumes ($a \approx 0.15$ fm). On the large volume we find $g_A = 1.212 \pm 0.027(\text{stat}) \pm 0.024(\text{norm})$. The quoted systematic error is the dominant one, corresponding to current renormalization. This theoretical first principles calculation, which does not yet include isospin breaking effects, yields a value of g_A only a little bit below the experimental one, 1.2670 ± 0.0030.

Four form factors appear in neutron β decay: the vector and induced tensor form factors from the vector current,

$$\langle p|V_\mu^+(x)|n\rangle = \bar{u}_p[\gamma_\mu g_V(q^2) - q_\lambda \sigma_{\lambda\mu} g_T(q^2)]u_n e^{-iq\cdot x}, \qquad (1)$$

and the axial-vector and induced pseudo-scalar form factors from the axial-vector current,

$$\langle p|A_\mu^+(x)|n\rangle = \bar{u}_p[\gamma_\mu \gamma_5 g_A(q^2) - iq_\mu \gamma_5 g_P(q^2)]u_n e^{-iq\cdot x}. \qquad (2)$$

Here $q = p_n - p_p$ is the momentum transfer between the proton (p) and neutron (n). In the limit $|\vec{q}| \to 0$, the momentum transfer should be small because the mass difference of the neutron and proton is only about 1.3 MeV. This makes the limit $q^2 \to 0$, where the vector and axial-vector form factors dominate, a good approximation. Their values in this limit are called the vector and axial charges of the nucleon: $g_V = g_V(q^2 = 0)$ and $g_A = g_A(q^2 = 0)$. Experimentally [1], $g_V = \cos\theta_C$ (with the Cabibbo mixing angle θ_C), and $g_A = 1.2670(30) \times g_V$ [1]. Among the nucleon form factors or moments of structure functions, it is technically the simplest from the point of view of a lattice QCD numerical calculation.

Since they are defined at zero momentum transfer, a naive expectation is that g_V and g_A are easier to calculate on the lattice than form factors which require non-zero momentum

---

[1] Note that the Particle Data Group defines g_A to be negative because no assumption about the structure of the weak interaction is made. In this article, assuming the $V - A$ structure of the weak interaction, the axial form factor in Eq. 2 is defined to make g_A positive.

transfer. Despite this, quenched QCD lattice calculations with Wilson fermions at finite lattice cutoff ($a^{-1} \sim 2$ GeV) have underestimated g_A by about 20% [2, 3, 4]. This suggests systematic errors, which may arise from (1) the quenched approximation, (2) operator renormalization, (3) non-zero-lattice-spacing a and loss of chiral symmetry for Wilson and Kogut-Susskind fermions, and (4) finite volume, remain in the lattice calculation.

The first three errors have been addressed in previous calculations. The SESAM and LHPC collaborations found that unquenching does not solve the problem as the estimated value g_A decreases by 5-10% [5, 6]. On the other hand, reducing the lattice spacing error seems to increase the value, but only by a small amount, $\sim 5\%$ [7, 8]. Perhaps more important is the calculation of the renormalization factor Z_A for the axial current. The one-loop perturbative renormalization factor, used in the case of Wilson fermions [2, 3, 4, 6, 5], was probably overestimated. The QCDSF-UKQCD collaboration reported that the non-perturbatively calculated renormalization factor ($Z_A^{\text{nonpert}} \sim 0.8$) is roughly 10% smaller than the one-loop one ($Z_A^{\text{pert}} \sim 0.9$) in the case of the non-perturbatively $O(a)$ improved Wilson fermions [8] at $a^{-1} \sim 2-3$ GeV. Thus, the systematic error in the determination of the renormalization factor appears to be more important than the first two effects mentioned. The first two systematic errors listed above likely cannot resolve the issue that previous lattice calculations of g_A underestimate the experimental value.

The loss of chiral symmetry on the lattice is potentially significant. As is well known, $g_A/g_V = 1$ in the absence of chiral symmetry breaking in QCD. Further, in the realistic case of spontaneously broken chiral symmetry, the ratio is still constrained by the axial Ward-Takahashi identity; $\partial_\mu A^a_\mu(x) = 2mP^a(x)$. The Goldberger-Treiman relation derives from the nucleon matrix elements of the currents on both sides of this identity in the soft pion limit [9]. We can easily understand the deviation of the ratio from unity in the context of the Gell-Mann-Oakes-Renner relation [10] which is also related to the axial Ward-Takahashi identity. Thus, the explicit breaking of chiral symmetry at non-zero lattice spacing a for Wilson fermions may induce significant errors which are only removed in the continuum limit.

In this work we use domain wall fermions (DWF), a fermion discretization scheme with almost perfectly preserved chiral symmetry [11, 12, 13]. This scheme introduces a fictitious fifth dimension in addition to the four dimensions of space-time. In the limit where the fifth-dimensional extent L_s is taken to ∞, DWF preserve the axial Ward Takahashi identity [14] at non-zero lattice spacing. With finite L_s the suppression of explicit chiral symmetry breaking is effectively exponential in quenched simulations if the gauge field is sufficiently smooth [15, 16, 17, 18, 19, 20]. This is always true if the lattice spacing is sufficiently small. In low energy cases like the one investigated here, the small breaking of the symmetry at finite L_s is parametrized by a single universal "residual mass" parameter, m_{res}, acting as an additive quark mass and which is defined from the axial Ward-Takahashi identity [21, 18]. Furthermore, the DWF scheme greatly simplifies the non-perturbative determination of the renormalization of quark bilinear currents [22]. For example the renormalization factor of local vector and axial-vector current operators should be equal, $Z_A = Z_V$ [22]. This means the ratio of the nucleon axial and vector charges calculated on the lattice directly yields the continuum value, *i.e.* it is not renormalized [23, 24]. By employing the DWF scheme, the ambiguity in the renormalization of quark currents which may be present and problematic in other fermion discretization schemes is eliminated. We emphasize that the DWF calculation of the nucleon axial charge should not suffer from the systematic errors due to the operator renormalization and loss of chiral symmetry [23, 24].

However, in our first DWF calculation with the single-plaquette Wilson gauge action at $\beta = 6.0$ and lattice volume $16^3 \times 32 \times 16$ (which correspond to $a^{-1} \approx 2$ GeV and spacial volume $\sim (1.6\,\text{fm})^3$), we found that g_A exhibits a fairly strong dependence on the quark mass [24]. A simple linear extrapolation of g_A to the chiral limit yielded a value that was almost a factor of two smaller than the experiment [24]. This implied the presence of a large finite volume effect. To our surprise, we found no systematic study of such an effect in the literature. Note also that there is no volume dependence in the naive quark model [25] nor in the MIT bag model [26]. In the former the ratio is determined by a simple spin-isospin algebra, and in the latter it arises from a simple overlap integral of the upper and lower component of the bag Dirac wave function.

To address the finite volume issue we need to have at the same time a sufficiently high lattice cutoff to preserve chiral symmetry reasonably well and at least two lattice volumes, preferably ones that are large compared to the charge radius of the proton. The Wilson gauge action will not work for this purpose since the chiral symmetry of DWF in the quenched case degrades rapidly as lattice spacing a increases, while the computational cost necessitated by a very large lattice volume would be prohibitive. Fortunately various "renormalization-group-inspired" improved gauge actions preserve the chiral symmetry of DWF well while not demanding a large cutoff [19, 20]. Thus both requirements, chiral symmetry and large physical volume, can be met at reasonable computational cost. Of the relatively well-established candidates in this class of improved gauge actions, we choose the "doubly-blocked Wilson 2 (DBW2)" action [27, 20]. We refer our full paper [29] for the details of these calculations.

We plot the value of $(g_A/g_V)^{\text{lattice}}$ as a function of $(m_\pi/m_\rho)^2$ in Figure 1. The smaller volume results using the DBW2 gauge action ($\beta = 0.87$) are the same (within statistical errors) as our previous results using the Wilson gauge action ($\beta = 6.0$) on a slightly larger volume. The large volume DBW2 results exhibit mild quark mass dependence while both smaller volume results show a marked decrease toward the chiral limit. We conclude that our previous DWF-Wilson-gauge-action results were significantly adversely affected by finite volume.

Finally, we extrapolate g_A^{ren} to the chiral limit. For this purpose, we have two methods. One is to extrapolate the charge ratios $(g_A/g_V)^{\text{lattice}}$ to the chiral limit where the relation $Z_V = Z_A$ is valid. The second method was the conventional one utilized in all other calculations [2, 3, 4, 5, 6, 7, 8]. The chiral extrapolation is performed on $g_A^{\text{lattice}} \times Z_A$. Recall that the latter requires the value of Z_A, whether nonperturbatively or perturbatively calculated, while the former does not. In the present case, we use the nonperturbative value of Z_A from [20]. We plot $(g_A/g_V)^{\text{lattice}}$ and $Z_A \times g_A^{\text{lattice}}$ together in Figure 1 (right pane) and perform a simple linear extrapolation in each case. The two methods provide consistent results in the chiral limit: the ratio method gives $g_A^{\text{ren}} = 1.212(27)$ while the conventional method gives $g_A^{\text{ren}} = 1.188(25)$. The systematic difference if there is one, is related to our choice of renormalization. A two percent error stemming from $Z_V \neq Z_A$

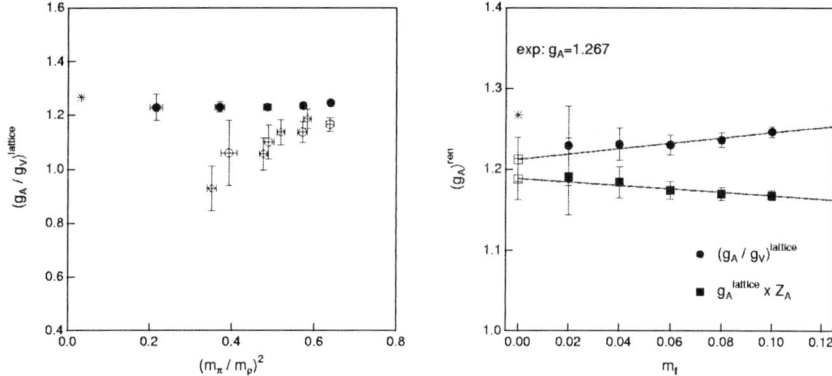

FIGURE 1. Left: The physical ratio of nucleon charges. DBW2 gauge action results on two different physical volumes, $(2.4\,\text{fm})^3$ (solid circles) and $(1.2\,\text{fm})^3$ (open circles), reveal the existence of a significant finite volume effect. Wilson gauge action results (diamonds), $V \approx (1.6\,\text{fm})^3$, also appear to be affected by finite volume. **Right**: Two methods to obtain the physical nucleon axial charge, the ratio of axial-vector to vector charge (circles), and the lattice axial-vector charge times the axial-vector current renormalization factor Z_A in the chiral limit from Ref. [20]. They show slightly different quark mass dependence, but exptrapolate to consistent values. Each underestimates the experimental value (burst) by roughly five percent.

yields 0.024. This is also the difference in the central values just obtained. Thus, we quote

$$g_A^{\text{ren}} = 1.212 \pm 0.027(\text{stat}) \pm 0.024(\text{norm}) \qquad (3)$$

which underestimates the experimental value of 1.267 by less than five percent. We have not attempted to estimate residual non-zero lattice spacing, finite volume, explicit chiral symmetry breaking, and quenching effects. The first three are probably small [18, 20, 19]. The only remaining error not under good control is the quenching one which does not appear to be large [5, 6], in light of the relatively good agreement with experiment shown above. This view does not change unless significant non-analytic behavior, which we did not detect here, arises near the chiral limit.

As mentioned above, this calculation of g_A is performed for relatively heavy quark masses; the quenching error at this unphysically large mass scale is probably small. However, one may worry that such a calculation does not capture relevant physics in the region where the quark mass is much lighter, and the so-called "pion cloud" surrounding the nucleon becomes important. Nevertheless the values of g_A^{ren} at these heavier quark masses already lie just a few percent below the experimental value and show little dependence on the quark mass. This presents an important question concerning the role of the pion cloud: is it a few percent effect, as seems plausible from our first principles calculation, or is it larger, as estimated from phenomenological models [28].

The dependence of the product $m_N g_A$ on the lattice volume is of interest. While the smaller volume results always lie below the larger volume ones, within one standard deviation they almost always agree. There is only one exception at $am_f = 0.08$ in the bare lattice result. No volume dependence is detected. This is in clear contrast to the situation of the axial charge alone. Since the product is the one that appears in the Goldberger-Treiman relation, it would be interesting to see how its counterpart, the induced pseudo-scalar form factor, behaves at small momentum transfer.

After the CIPANP 2003 meeting, Shoichi Sasaki, Kostas Orginos, Shigemi Ohta and Tom Blum, for the RIKEN-BNL-Columbia-KEK QCD Project, posted a fuller account of this work [29]. In the meeting results on the moments of nucleon structure functions were also presented. These will appear elsewhere. We thank RIKEN, Brookhaven National Laboratory and the U.S. Department of Energy for providing the facilities essential for the completion of this work.

REFERENCES

1. K. Hagiwara et al. [Particle Data Group Collaboration], Phys. Rev. D **66**, 010001 (2002).
2. M. Fukugita, Y. Kuramashi, M. Okawa and A. Ukawa, Phys. Rev. Lett. **75**, 2092 (1995) [arXiv:hep-lat/9501010].
3. K. F. Liu, S. J. Dong, T. Draper, J. M. Wu and W. Wilcox, Phys. Rev. D **49**, 4755 (1994) [arXiv:hep-lat/9305025].
4. M. Göckeler, R. Horsley, E. M. Ilgenfritz, H. Perlt, P. Rakow, G. Schierholz and A. Schiller, Phys. Rev. D **53**, 2317 (1996) [arXiv:hep-lat/9508004].
5. S. Güsken et al. [TXL Collaboration], Phys. Rev. D **59**, 114502 (1999).
6. D. Dolgov et al. [LHPC collaboration], Phys. Rev. D **66**, 034506 (2002) [arXiv:hep-lat/0201021].
7. S. Capitani et al., Nucl. Phys. Proc. Suppl. **79**, 548 (1999) [arXiv:hep-ph/9905573].
8. R. Horsley [UKQCD Collaboration], Nucl. Phys. Proc. Suppl. **94**, 307 (2001) [arXiv:hep-lat/0010059].
9. M. L. Goldberger and S. B. Treiman, Phys. Rev. **110**, 1178 (1958).
10. M. Gell-Mann, R. J. Oakes and B. Renner, Phys. Rev. **175**, 2195 (1968).
11. D. B. Kaplan, Phys. Lett. B **288**, 342 (1992) [arXiv:hep-lat/9206013].
12. Y. Shamir, Nucl. Phys. B **406**, 90 (1993) [arXiv:hep-lat/9303005].
13. R. Narayanan and H. Neuberger, Phys. Lett. B **302**, 62 (1993) [arXiv:hep-lat/9212019].
14. V. Furman and Y. Shamir, Nucl. Phys. B **439**, 54 (1995) [arXiv:hep-lat/9405004].
15. Y. Kikukawa, Nucl. Phys. B **584**, 511 (2000) [arXiv:hep-lat/9912056].
16. P. Hernandez, K. Jansen and M. Luscher, Nucl. Phys. B **552**, 363 (1999) [arXiv:hep-lat/9808010].
17. P. Hernandez, K. Jansen and M. Luscher, arXiv:hep-lat/0007015.
18. T. Blum et al., [RBC Collaboration], to appear in PRD, [arXiv:hep-lat/0007038].
19. A. Ali Khan et al. [CP-PACS Collaboration], Phys. Rev. D **63**, 114504 (2001) [arXiv:hep-lat/0007014].
20. Y. Aoki et al., [RBC Collaboration], [arXiv:hep-lat/0211023].
21. T. Blum, Nucl. Phys. Proc. Suppl. **73**, 167 (1999) [arXiv:hep-lat/9810017].
22. T. Blum et al., Phys. Rev. D **66**, 014504 (2002) [arXiv:hep-lat/0102005].
23. T. Blum and S. Sasaki, arXiv:hep-lat/0002019.
24. T. Blum, S. Ohta and S. Sasaki, Nucl. Phys. Proc. Suppl. **94**, 295 (2001) [arXiv:hep-lat/0011011].
25. N. Isgur and G. Karl, Phys. Rev. D **20**, 1191 (1979).
26. A. Chodos, R. L. Jaffe, K. Johnson and C. B. Thorn, Phys. Rev. D **10**, 2599 (1974).
27. T. Takaishi, Phys. Rev. D **54**, 1050 (1996).
28. W. Detmold, W. Melnitchouk and A.W. Thomas, Phys. Rev. D **66**, 054501 (2002) [arXiv:hep-lat/0206001].
29. S. Sasaki, K. Orginos, S. Ohta and T. Blum [RBCK Collaboration], arXiv:hep-lat/0306007, to appear in Phys. Rev. D.

Hadronic effects in theory of $g-2$

Arkady Vainshtein

*William I. Fine Theoretical Physics Institute, University of Minnesota,
116 Church St SE, Minneapolis, MN 55455*

Abstract. Hadronic effects and corresponding uncertainties in theory of the muon anomalous magnetic moment are discussed.

MUON ANOMALOUS MAGNETIC MOMENT

The anomalous magnetic moment of muon is measured with a very high precision [1]

$$a_\mu^{\exp} = \frac{g_\mu - 2}{2} = 116\,592\,030(80) \times 10^{-11}. \quad (1)$$

The Standard Model prediction for a_μ can be represented as a sum

$$a_\mu^{\rm SM} = a_\mu^{\rm QED} + a_\mu^{\rm had} + a_\mu^{\rm EW}. \quad (2)$$

The QED part involving only leptons and photons is the main one [2],
$a_\mu^{\rm QED} = 116\,584\,706(3) \times 10^{-11}$.

HADRONIC CONTRIBUTION

Next is the hadronic contribution which can be broken in three pieces,

$$a_\mu^{\rm had} = a_\mu^{\rm had, LO} + a_\mu^{\rm had, HO} + a_\mu^{\rm LBL}. \quad (3)$$

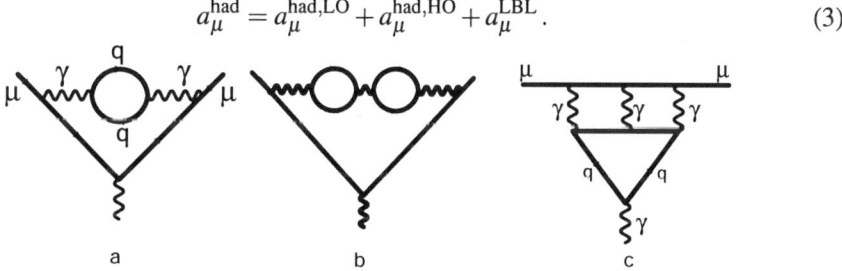

FIGURE 1. Hadronic contributions represented by quark loops

The leading part, $a_\mu^{\rm had, LO}$, presented by the diagram a in Fig. 1 is determined by experimental data from e^+e^- annihilation to hadrons and from semileptonic decays of τ inclusive over hadrons. The e^+e^- and τ based analyses in Ref. [3] led to different results:

$$a_\mu^{\rm had, LO} = \begin{cases} 6847(60)(36) \times 10^{-11} & e^+e^- \text{ based} \\ 7090(51)(12)(28) \times 10^{-11} & \tau \text{ based} \end{cases} \quad (4)$$

Recently the Novosibirsk collaboration discovered an error in the calibration [4]. The resulting corrections increase the e^+e^- result and leave less than a half of the quoted above difference of 243×10^{-11}. One can hope that the e^+e^- and τ data will converge eventually.

Higher order hadronic corrections of the type presented by the diagram b in Fig. 1 are known with a good accuracy, $a_\mu^{h,HO} = -100(6) \times 10^{-11}$, see Refs.[5]. More theoretically challenging is a computation of the light-by-light hadronic contribution to a_μ, see the diagram c in Fig. 1. The theoretical error in the usually quoted number $a_\mu^{LBL} = 86(35) \times 10^{-11}$ could be even larger.

The theoretical expression for $a_\mu^{had,LO}$ is

$$a_\mu^{had,LO} = \left(\frac{\alpha m_\mu}{3\pi}\right)^2 \int_{4m_\pi^2}^{\infty} \frac{ds}{s^2} K(s) R(s) \qquad (5)$$

where $K(s)$ is the known function and $R(s)$ is the cross section of e^+e^- annihilation into hadrons in units of $\sigma(e^+e^- \to \mu^+\mu^-)$. The threshold region where $R(s) \approx (1/4)(1 - 4m_\pi^2/s)^{3/2}$ gives a parametrical enhanced in the chiral limit contribution, $a_\mu^{had,LO} \propto 1/m_\pi^2$. Numerically, however, this is not a leading contribution, it constitutes

$$a_\mu^{had,LO}(4m_\pi^2 \leq s \leq m_\rho^2/2) \approx 320 \times 10^{-11}. \qquad (6)$$

Compare with an estimate for the ρ peak which contains no chiral enhancement,

$$a_\mu^{had,LO}(\rho) = \frac{m_\mu^2 \Gamma(\rho \to e^+e^-)}{\pi m_\rho^3} \approx 5170 \times 10^{-11}. \qquad (7)$$

The ρ peak constitutes 74% of the total. Together with remaining ω, ϕ and large s parts we get the main bulk of $a^{had,LO}$.

What is a lesson from this exercise? The ρ contribution is enhanced by another theoretical parameter, namely, by number of colors N_c. We see that this enhancement prevails over chiral one. We will use this observation in consideration of the light-by-light effects.

LIGHT-BY-LIGHT

The $\gamma^*\gamma^* \to \gamma^*\gamma$ amplitude is not accessible experimentally, a challenge for theorists. Parametrically the LbL contribution to a_μ has the form

$$a_\mu^{LbL} \sim C \left(\frac{\alpha}{\pi}\right)^3 \frac{m_\mu^2}{M_{had}^2} \qquad (8)$$

where M_{had} is some hadronic mass parameter. This parameter is minimal, $\sim 2m_\pi$, in case of the charged pion loop, see the diagram c in Fig. 1 with the quark substituted by pion in the loop. Similarly to the polarization operator case above this chirally enhanced contribution again does not result in large number, it is actually quite small [6]

$$a_\mu^{LbL}(\text{pion box}) \approx -4 \times 10^{-11}. \qquad (9)$$

The π^0 pole part of LbL, Fig. 2, contains the chiral enhancement in the logarithmic form only [7]

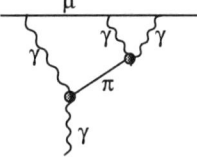

$$a_\mu^{LbL}(\pi^0) = \left(\frac{\alpha}{\pi}\right)^3 N_c \frac{m_\mu^2 N_c}{48\pi^2 F_\pi^2} \ln^2 \frac{m_\rho}{m_\pi} + \ldots \quad (10)$$

but it contains the N_c enhancement absent for the charged pion box. Numerically

FIGURE 2. The π^0 part in the LbL

$$a_\mu^{LbL}(\pi^0) = 56 \times 10^{-11} \quad (11)$$

The number increases to 63 by applying the old sum rules analysis of the $\pi^0 \gamma^* \gamma^*$ form factor [8].

Rather uncertain is an estimate of terms without any chiral enhancement — they are seemingly positive. This can probably be improved along within the OPE based approach discussed in the next section in application to the hadronic part of the electroweak correction.

ELECTROWEAK CONTRIBUTIONS

The presentation in this section follows our work with A. Czarnecki and B. Marciano [9] where we resolved, in particular, problems raised earlier in Ref. [10]. In the Standard Model the one-loop electroweak contributions were calculated about 30 years ago

$$a_\mu^{EW}(\text{1-loop}) = \frac{5G_\mu m_\mu^2}{24\sqrt{2}\pi^2}\left[1 + \frac{1}{5}(1-4\sin^2\theta_W)^2 + \mathscr{O}\left(\frac{m_\mu^2}{m_{W,H}^2}\right)\right] = 194.8 \times 10^{-11} \quad (12)$$

Two-loop corrections are more involved ($F = \tau, u, d, s, c, b$)

$$a_\mu^{EW}(\text{2-loop})_{LL} = \frac{5G_\mu m_\mu^2}{24\sqrt{2}\pi^2} \cdot \frac{\alpha}{\pi}\left\{-\frac{43}{3}\ln\frac{m_Z}{m_\mu} + \frac{36}{5}\sum_{f\in F} N_f Q_f^2 I_f^3 \ln\frac{m_Z}{m_f}\right\} \approx -37 \times 10^{-11} \quad (13)$$

These corrections include the fermion triangles, see Fig. 3. The numbers above are obtained for $m_{u,d} = 0.3\,\text{GeV}$, $m_s = 0.5\,\text{GeV}$, $m_c = 1.5\,\text{GeV}$, $m_b = 4.5\,\text{GeV}$ with strong corrections neglected. This approach is justifiable for the heavy quarks where the result is defined by short distances $\sim 1/m_Q$ but is questionable in case of light u,d,s quarks.

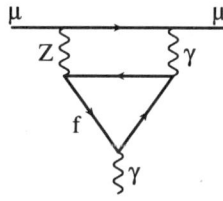

The $Z^* \gamma \gamma^*$ vertex containing one soft photon

FIGURE 3. Fermion triangles in a_μ

$$T_{\mu\nu} = i\int d^4x e^{iqx}\langle 0|T\{j_\mu(x)j_\nu^5(0)\}|\gamma(k)\rangle, \quad j_\mu = \sum_f Q_f \bar{f}\gamma_\mu f, \quad j_\nu^5 = \sum_f 2I_f^3 \bar{f}\gamma_\nu\gamma_5 f \quad (14)$$

is given by two invariant functions $w_{T,L}(q^2)$,

$$T_{\mu\nu} = -\frac{ie}{4\pi^2}\left[w_T(q^2)\left(-q^2\tilde{f}_{\mu\nu} + q_\mu q^\sigma \tilde{f}_{\sigma\nu} - q_\nu q^\sigma \tilde{f}_{\sigma\mu}\right) + w_L(q^2) q_\nu q^\sigma \tilde{f}_{\sigma\mu}\right] \quad (15)$$

where $\tilde{f}_{\mu\nu} = \frac{1}{2}\varepsilon_{\mu\nu\gamma\delta}f^{\gamma\delta}$, $f_{\mu\nu} = k_\mu e_\nu - k_\nu e_\mu$. The corresponding piece in a_μ is

$$\Delta a_\mu^{EW} = \frac{\alpha}{\pi}\frac{G_\mu m_\mu^2}{8\pi^2\sqrt{2}}\int_0^\infty dQ^2 \left(w_L(Q^2) + \frac{m_Z^2}{m_Z^2 + Q^2}w_T(Q^2)\right) \quad (16)$$

The well-known one-loop result can be written as

$$w_L^{1-\text{loop}} = 2w_T^{1-\text{loop}} = \sum_f 4I_f^3 N_f Q_f^2 \int_0^1 \frac{d\alpha\,\alpha(1-\alpha)}{\alpha(1-\alpha)Q^2 + m_f^2} \xrightarrow{Q^2 \to \infty} \sum_f \frac{4I_f^3 N_f Q_f^2}{Q^2} \quad (17)$$

The leading $1/Q^2$ term leads to the UV divergent integral $\int dQ^2/Q^2$ in a_μ. These divergences are canceled within a given generation (anomaly cancellation), $\sum_f I_f^3 N_f Q_f^2 = 0$. First, about QCD perturbative corrections to quark loops at $Q \gg m_q$. The longitudinal function w_L is protected by Adler-Bardeen nonrenormalization theorem. Moreover, perturbative corrections to the transversal function w_T are also absent at $Q \gg m_q$ due to the new nonrenormalization theorem [11] based on the relation $w_T[m_q = 0] = \frac{1}{2} w_L[m_q = 0]$. No α_s corrections in chiral limit!

In the chiral limit $w_L[u,d]$ is protected against nonperturbative corrections as well, the $1/Q^2$ pole corresponds to the exact duality between the pion and quarks: 't Hooft matching condition. Deviations from the chiral limit are readily accounted for by a shift $1/Q^2 \to 1/(Q^2 + m_\pi^2)$.

For the transversal function $w_T[u,d]$ it is only spin one hadrons, like $\rho(770)$, $\omega(770)$ and $a_1(1230)$, which contribute. Thus, nonperturbative effects shift $1/Q^2$ at $Q \sim m_{\rho,a_1}$. The model of $w_T[u,d]$ consistent with the OPE is

$$w_T[u,d] = \frac{1}{m_{a_1}^2 - m_\rho^2} \left[\frac{m_{a_1}^2 - m_\pi^2}{Q^2 + m_\rho^2} - \frac{m_\rho^2 - m_\pi^2}{Q^2 + m_{a_1}^2} \right]. \quad (18)$$

Overall, the approach results in

$$a_\mu^{EW} = 154(1)(2) \times 10^{-11} \quad (19)$$

which is a slightly larger (but consistent) with the previous value. It is important that the hadronic uncertainties is shown to be small (first error). The second error is due to an allowed Higgs mass range of 114 GeV $< m_H <$ 250 GeV, the current top mass uncertainty and unknown three-loop effects.

REFERENCES

1. G. W. Bennett et al. [Muon g-2 Collaboration], Phys. Rev. Lett. **89**, 101804 (2002) [hep-ex/0208001].
2. V. W. Hughes and T. Kinoshita, Rev. Mod. Phys. **71**, S133 (1999).
3. M. Davier, S. Eidelman, A. Hocker and Z. Zhang, Eur. Phys. J. C **27**, 497 (2003) [hep-ph/0208177].
4. R. R. Akhmetshin et al. [the CMD-2 Collaboration], hep-ex/0308008.
5. B. Krause, Phys. Lett. B **390**, 392 (1997) [hep-ph/9607259].
6. M. Hayakawa, T. Kinoshita and A. I. Sanda, Phys. Rev. D **54**, 3137 (1996) [hep-ph/9601310]; K. Melnikov, Int. J. Mod. Phys. A **16**, 4591 (2001) [hep-ph/0105267].
7. M. Knecht and A. Nyffeler, Phys. Rev. D **65**, 073034 (2002) [hep-ph/0111058]; M. Hayakawa and T. Kinoshita, Phys. Rev. D **57**, 465 (1998) [Erratum-ibid. D **66**, 019902 (2002)] [hep-ph/9708227]. J. Bijnens, E. Pallante and J. Prades, Nucl. Phys. B **626**, 410 (2002) [hep-ph/0112255]; I. Blokland, A. Czarnecki and K. Melnikov, Phys. Rev. Lett. **88**, 071803 (2002) [hep-ph/0112117].
8. V. A. Novikov et al Nucl. Phys. B **237**, 525 (1984).
9. A. Czarnecki, W. J. Marciano and A. Vainshtein, Phys. Rev. D **67**, 073006 (2003) [hep-ph/0212229].
10. M. Knecht, S. Peris, M. Perrottet and E. de Rafael, JHEP **0211**, 003 (2002) [hep-ph/0205102].
11. A. Vainshtein, hep-ph/0212231.

Chiral Dynamics in the Meson Sector at two Loops

Johan Bijnens

*Department of Theoretical Physics 2, Lund University,
Sölvegatan 14A, S 22362 Lund, Sweden*

Abstract. I give a very short introduction to Chiral Perturbation Theory and an overview of the next-to-next-to-leading order three-flavour calculations done. I discuss those relevant for an improvement in the accuracy of the measurement of V_{us} in more detail. One major conclusion is that all needed p^6 low energy constants can be obtained from experiment via the scalar form-factor in $K_{\ell 3}$ decays.

INTRODUCTION

The problem of dynamics of mesons at low energies is important. It plays a major role in the precise determination of the elements of the Cabibbo-Kobayashi-Maskawa matrix (CKM) which is a main part of the study of the standard model flavour sector [1]. In this talk I will concentrate on the theory behind the measurement of V_{us} from $K_{\ell 3}$ ($K \to \pi \ell \nu$) decays and in particular on the recent work of P. Talavera and myself on the two-loop calculation and the determination of the relevant low-energy constants [2, 3].

I review shortly Chiral Perturbation Theory (ChPT) and the relevant two-loop calculations done up to now, followed by a discussion of $K_{\ell 3}$ and the present theory situation. I include a short discussion of the validity of the linear approximation of the form factors normally used in the data analysis. The main relevant results from ChPT [2] can be summarized as follows. The curvatures are important in the analysis but can be predicted using ChPT from the pion electromagnetic form-factor [2] and all order p^6 parameters needed to determine V_{us} can be experimentally obtained via the scalar form factor, $f_0(t)$, in $K_{\ell 3}$ [2]. The curvature of $f_0(t)$ can be predicted as well from knowledge about scalar form factors of the pion [4], albeit only at fairly low precision at present.

CHIRAL PERTURBATION THEORY

ChPT is an effective field theory valid as an approximation to Quantum Chromodynamics (QCD) at low energies. Its modern form has was introduced by Weinberg, Gasser and Leutwyler [5, 6]. The global chiral symmetry, $SU(3)_L \times SU(3)_R$, of QCD in the limit of massless quarks is spontaneously broken down to the diagonal subgroup $SU(3)_V$ by a nonzero quark condensate, $\langle \bar{q}q \rangle = \langle \bar{q}_L q_R + \bar{q}_R q_L \rangle \neq 0$. The eight broken generators lead to eight Goldstone bosons. These are massless *and* their interactions vanish at zero momentum. The latter allows the construction of a well defined perturbative expansion in terms of momenta, generically referred to as an expansion in p^2. Quark masses are usually counted as order p^2 since $p_\pi^2 = m_\pi^2 \sim m_q \langle \bar{q}q \rangle / F_\pi^2$. Inserting an external photon or $W^\pm$-bosons counts as order p since these are included via covariant derivatives. Recent lectures, much more detailed than what is included here are [7]. ChPT being an effective field theory implies that the number of parameters increases order by order. In the purely mesonic strong and semi-leptonic sector there are two parameters at lowest order (p^2), ten at NLO (p^4) [6], and 90 at NNLO (p^6) [8]. The renormalization procedure and the divergences are worked out in general to NNLO [9] and provide a good check on all calculations. One problem shared with other high order loop calculations in comparing different calculations is the use of different renormalization schemes. The calculations that were used to determine all the needed parameters are those of the masses and decay constants [10], $K_{\ell 4}$ [11] and the electromagnetic form factors [12].

$K_{\ell 3}$: DEFINITIONS, V_{us} AND FORM FACTOR LINEARITY

The neutral and charged $K_{\ell 3}$ decays amplitudes, $K^{+,0}(p) \to \pi^{0,-}(p')\ell^+(p_\ell)\nu_\ell(p_\nu)$, are

$$T^{(+,0)} = \frac{G_F}{\sqrt{2}} V_{us}^* \ell^\mu F_\mu^{(+,0)}(p',p), \qquad \ell^\mu = \bar{u}(p_\nu)\gamma^\mu(1-\gamma_5)v(p_\ell),$$

$$F_\mu^{+,0}(p',p) = \left(1/\sqrt{2}, 1\right)\left[(p'+p)_\mu f_+^{K^+\pi^0, K^0\pi^-}(t) + (p-p')_\mu f_-^{K^+\pi^0, K^0\pi^-}(t)\right]. \quad (1)$$

Isospin leads to the relations

$$f_+^{K^0\pi^-}(t) = f_+^{K^+\pi^0}(t) = f_+(t) \quad \text{and} \quad f_-^{K^0\pi^-}(t) = f_-^{K^+\pi^0}(t) = f_-(t), \quad (2)$$

The scalar form factor and the usual linear parameterizations are defined as

$$f_0(t) = f_+(t) + t/(m_K^2 - m_\pi^2)f_-(t), \qquad f_{+,0}(t) = f_+(0)\left(1 + \lambda_{+,0} t/m_\pi^2\right). \quad (3)$$

To determine $|V_{us}|$ we need $f_+(0)$ theoretically and experimentally. There are three main theoretical effects. There is a well-known short-distance correction from G_μ to G_F calculated by Marciano and Sirlin. The corrections of order $(m_s - \hat{m})^2$ allowed by the Behrends-Sirlin-Ademollo-Gatto theorem are discussed at p^6 here and in [2, 3]. The sizable isospin breaking found by Leutwyler and Roos [13] is in the process of being evaluated at order p^6 too. On the experimental side, the old radiative correction calculations used in [13] are updated in [14] where a clean procedure with generalized form-factors is proposed. The experimental data are mostly analyzed using a linear form factor $f_+(t)$. The recent precise CPLEAR data [15] allow to test this assumption [2]. Using a linear fit to their data and *neglecting* systematic errors we get a normalized $f_+(0) = 1$ and $\lambda_+ = 0.0245 \pm 0.0006$. Allowing for curvature we obtain a sizable curvature, $f_+(0) = 1.008 \pm 0.009$ and $\lambda_+ = 0.0181 \pm 0.0068$. The fitted curvature is compatible with zero, the central value is precisely at the ChPT prediction given below. In order to obtain $|V_{us}|$ with an error of 1% it is therefore important to include the effect of curvature in the analysis. The central value of λ_+ is outside the errors quoted for the

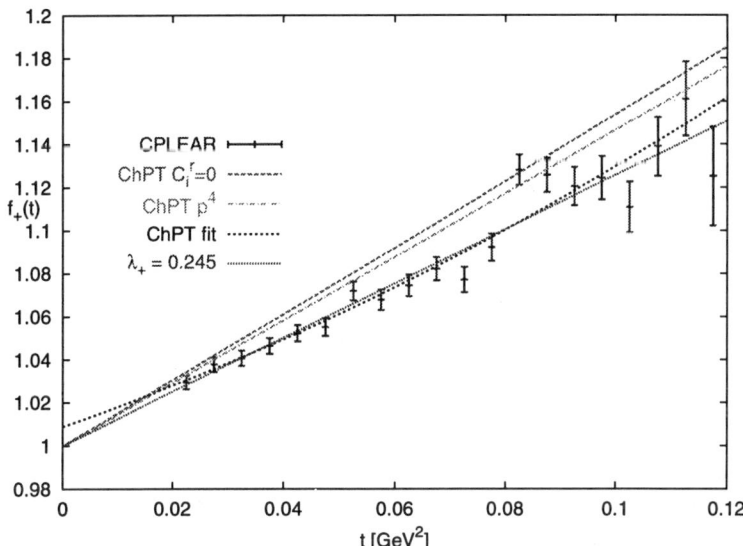

FIGURE 1. ChPT fits to the CPLEAR data showing the effect of the curvature compared to linear fits.

linear fit [15]. After the meeting we obtained similar conclusions for the KEK-PS E246 data [16], see v2 of [2].

$f_+(t)$ AND $f_0(t)$: THEORY

The $f_+(t)$ ChPT calculation for is rather cumbersome. I will use our work, [2] but an independent calculation exists [17] and agrees reasonably well, the difference is discussed in [2]. We write the amplitude as

$$f_+(t) = 1 + f_+^{(4)}(t)/F_\pi^2 + f_+^{(6)}(t)/F_\pi^4 \quad \text{with} \quad f_+^{(4)}(t) = tL_9^r/2 + \text{loops},$$
$$f_+^{(6)}(t) = -8(C_{12}^r + C_{34}^r)(m_K^2 - m_\pi^2)^2 + tR_{+1}^{K\pi} + t^2(-4C_{88}^r + 4C_{90}^r) + \text{loops}(L_i^r). \quad (4)$$

The pion electromagnetic form factor data yield [12] $L_9^r = 0.00593 \pm 0.00043$ and $-4C_{88}^r + 4C_{90}^r = 0.00022 \pm 0.00002$. With this input we fit the CPLEAR data and obtain

$$R_{+1}^{K\pi} = -(4.7 \pm 0.5)\,10^{-5}\,\text{GeV}^2 \quad \text{and} \quad \lambda_+ = 0.0170 \pm 0.0015. \quad (5)$$

The first agrees with the VMD estimate [2] $R_{+1}^{K\pi}|_{VMD} \approx -4\,10^{-5}\,\text{GeV}^2$. The latter comes from ChPT as

$$\lambda_+ = 0.0283\,(p^4) + 0.0011\,(\text{loops } p^6) - 0.0124(C_i^r). \quad (6)$$

Our main conclusion follows from rewriting the full p^6 result for $f_0(t)$ as

$$f_0(t) = 1 - (8/F_\pi^4)(C_{12}^r + C_{34}^r)(m_K^2 - m_\pi^2)^2 + (8t/F_\pi^4)(2C_{12}^r + C_{34}^r)(m_K^2 + m_\pi^2)$$

$$+t/\left(m_K^2-m_\pi^2\right)(F_K/F_\pi-1)-(8t^2/F_\pi^4)C_{12}^r+\overline{\Delta}(t)+\Delta(0)\,. \tag{7}$$

$\overline{\Delta}(t)$ and $\Delta(0)$ contain NO C_i^r and only depend on the L_i^r at order p^6 thus ALL needed parameters can be determined experimentally.

$\Delta(0) = -0.0080 \pm 0.0057 [\text{loops}] \pm 0.0028[L_i^r]$, is known and an expression for $\Delta(t)$ can be found in Ref. [2] The errors are an estimate of higher orders and using fits of the L_i^r using different assumptions. The p^6 estimate of [13] corresponds to $-(8/F_\pi^4)\left(C_{12}^r+C_{34}^r\right)(m_K^2-m_\pi^2)^2 \approx -0.016 \pm 0.008$.

CONCLUSIONS

I discussed the $K_{\ell 3}$ form factors in ChPT to order p^6. The main conclusions are that the curvatures for $f_+(t)$ and $f_0(t)$ can be predicted from the data on pion electromagnetic [12] and scalar [4] form-factors, the curvature in $f_+(t)$ and $f_0(t)$ should be taken into account in new precision experiments but from the slope and the curvature we can determine experimentally the needed parameters to calculate $f_+(0)$. A precision of better than one percent seems feasible for $|V_{us}|$.

ACKNOWLEDGMENTS

This work has been funded in part by the Swedish Research Council and the European Union RTN network, Contract No. HPRN-CT-2002-00311 (EURIDICE)

REFERENCES

1. M. Battaglia et al., hep-ph/0304132.
2. J. Bijnens and P. Talavera, hep-ph/0303103, to be published in Nucl. Phys. B.
3. J. Bijnens, Talk given at 38th Rencontres de Moriond on QCD and High-Energy Hadronic Interactions, Les Arcs, Savoie, France, 22-29 Mar 2003, hep-ph/0304284.
4. J. Bijnens and P. Dhonte, hep-ph/0307044
5. S. Weinberg, Physica A **96** (1979) 327; J. Gasser and H. Leutwyler, Annals Phys. **158** (1984) 142,
6. J. Gasser and H. Leutwyler, Nucl. Phys. B **250** (1985) 465.
7. A. Pich, A., hep-ph/9806303; G. Ecker, hep-ph/0011026; S. Scherer, hep-ph/0210398.
8. J. Bijnens, G. Colangelo and G. Ecker, JHEP **9902** (1999) 020 [hep-ph/9902437];
9. J. Bijnens, G. Colangelo and G. Ecker, Annals Phys. **280** (2000) 100 [hep-ph/9907333].
10. G. Amorós, J. Bijnens and P. Talavera, Nucl. Phys. B **568** (2000) 319 [hep-ph/9907264], Nucl. Phys. B **602** (2001) 87 [hep-ph/0101127].
11. G. Amorós, J. Bijnens and P. Talavera, Phys. Lett. B **480** (2000) 71 [hep-ph/9912398]; Nucl. Phys. B **585** (2000) 293 [Erratum-ibid. B **598** (2001) 665] [hep-ph/0003258].
12. J. Bijnens and P. Talavera, JHEP **0203** (2002) 046 [hep-ph/0203049].
13. H. Leutwyler and M. Roos, Z. Phys. C **25** (1984) 91.
14. V. Cirigliano et al., Eur. Phys. J. C **23** (2002) 121 [hep-ph/0110153].
15. A. Apostolakis et al. [CPLEAR Collaboration], Phys. Lett. B **473** (2000) 186.
16. A. S. Levchenko et al. [KEK-PS E246 Collaboration], Phys. Atom. Nucl. **65** (2002) 2232 [Yad. Fiz. **65** (2002) 2294] [hep-ex/0111048].
17. P. Post and K. Schilcher, Eur. Phys. J. C **25** (2002) 427 [hep-ph/0112352].

The Weak Production of Λ Particles in Muon and Tau Scattering From Protons

Stephan L. Mintz

Physics Department, Florida International University, Miami, Florida, 33199

Abstract. We calculate and discuss the differential cross section for the weak, strangeness changing, processes, $\mu^- + p \longrightarrow \nu_\mu + \Lambda$ and $\tau^- + p \longrightarrow \nu_\tau + \Lambda$, for incoming muon energies from threshold to 5.0 GeV and for an incoming tau of 20.0 GeV. The relatively high energy of the tau lepton is made necessary by its short lifetime. We obtain contributions from the form factors and particularly the interference terms with a view to observing those parts of the cross section which are suppressed in the corresponding electron induced process due to the small size of the electron mass. We make use of SU(3) relations and experimental data from electron scattering and Λ beta decay so that the calculation presented is phenomenological in nature.

INTRODUCTION

There is currently interest in the reaction $e^- + p \longrightarrow \Lambda + \nu_e$ by two groups centered at CEBAF. We have studied this reaction in some detail[1,2] and while very interesting, there are some disadvantages associated with using the electron as a probe. Because the contributions of the F_P and F_S form factors to the differential cross section are proportional to the charged lepton mass squared, this part of the interaction is suppressed in the electron induced process. Also because the weak leptonic charge changing currents contain the projector $(1 - \gamma_5)$ it is not possible to undertake an asymmetry measurement with electrons, as $(1-\gamma_5)\psi_R \propto (1 - p/(E+m_L))$ where L stands for the charged lepton and R stands for right handed. Because even at threshold, $E_{th} = 194.1753$ MeV for an electron, the right hand contribution is effectively suppressed. However at threshold for the muon induced process $1 - p/(E+m_L) = .47$ making asymmetry measurements a possibility. Since at threshold ($E_{th} = 188.226$ MeV) for the muon induced process, the muon would achieve a mean distance of 1200 meters from its point of production, such an experiment might be feasible.

MATRIX ELEMENTS

The process we are considering can be well described as a first order weak interaction. and may be written as:

$$< \nu\Lambda|H_w|L^- p > = \frac{G}{\sqrt{2}} \sin\theta_C \bar{u}_\nu \gamma^\lambda (1 - \gamma_5) u_L < \Lambda|J_\lambda^\dagger(0)|p > . \tag{1}$$

Using a notation very similar to that used in $p \leftrightarrow n$ transitions we may write the weak current matrix elements as follows:

$$< \Lambda |V_\mu^\dagger(0)|p> = \bar{u}_f[\gamma_\mu F_V(q^2) + i\frac{F_M(q^2)\sigma_{\mu\nu}q^\nu}{2m_p} - F_S(q^2)\frac{q_\mu}{2m_p}]u_i \qquad (2)$$

and

$$< \Lambda |A_\mu^\dagger(0)|p> = \bar{u}_f[\gamma_\mu\gamma_5 F_A(q^2) + \frac{q_\mu\gamma_5 F_P(q^2)}{m_\pi} + \frac{iF_E(q^2)\sigma_{\mu\nu}q^\nu\gamma_5}{2m_p}]u_i \qquad (3)$$

where i is the initial particle and f is the final particle, p and Λ respectively. The structure of the particles is contained, of course, in the six form factors, $F_V(q^2), F_M(q^2), F_S(q^2), F_A(q^2), F_P(q^2)$, and $F_E(q^2)$. We note that the form factors F_S and F_E would be due to contributions from second class currents and forbidden by G parity if this were a $p \leftrightarrow n$ transition but they are not forbidden in a $p \leftrightarrow \Lambda$ transition.

For the processes treated here the terms of the transition matrix element squared containing either F_P or F_S are proportional to the lepton mass squared. Thus we may unlike the electron case be able to observe these contributions. Therefore all form factors need to be determined. Because these are strangeness changing processes, SU(3) relations must be used to obtain the unknown form factors. These results are well known and lead to form factors:

$$F_r(q^2) = F_r(0)/(1 - q^2/M_r^2)^2 \qquad (4)$$

where $r = V, M, A, S, E$ but only $r = V$ and $r = M$ can be determined via SU(3) and electromagnetic data. The results are $F_V(0) = 1.2247$ with $M_V = .98 GeV/c^2$ and $F_M(0) = 1.793/2m_p$ with $M_M = .71 GeV/c^2$.

The axial current matrix element is more difficult to obtain. However there is very useful experimental data from Λ beta decay, $\Lambda \longrightarrow p + e^- + \bar{\nu}_e$, which gives, in the notation of Eq.(4), $M_A = 1.25 GeV/c^2$ and $F_A(0) = .8793$. For F_P we use a standard form first given by Nambu, namely, $F_P(q^2) = -F_A(q^2)m_k(m_i + m_f)/(q^2 - m_k^2)$. This completes the standard axial current form factors. Finally we estimate values for F_E and F_S. From a theoretical reference[1] we obtain an estimate of $F_E(0) = .705/2m_p$ and $F_S(0) = .344 F_E(0)$ in the notation used here. Making use of our experience that F_E, F_S and F_M have similar q^2 dependence, we write equations of the form of Eq.(4) with the M_M obtained earlier.

RESULTS AND DISCUSSION

We now are able to calculate the differential cross sections for incoming muons with energies of 0.265 GeV, (near threshold) and 5.0 GeV (highly relativisitc). We plot the former in figure 1. We also show the contributions of three interference terms of interest, $F_E F_A, F_P F_A$, and $F_E F_V$. We plot these because they contain form factors of interest, namely F_E and F_P. Because the contributions of F_E, F_S, and F_P are small, none of the squared terms from these form factors will be observable so that F_S will not

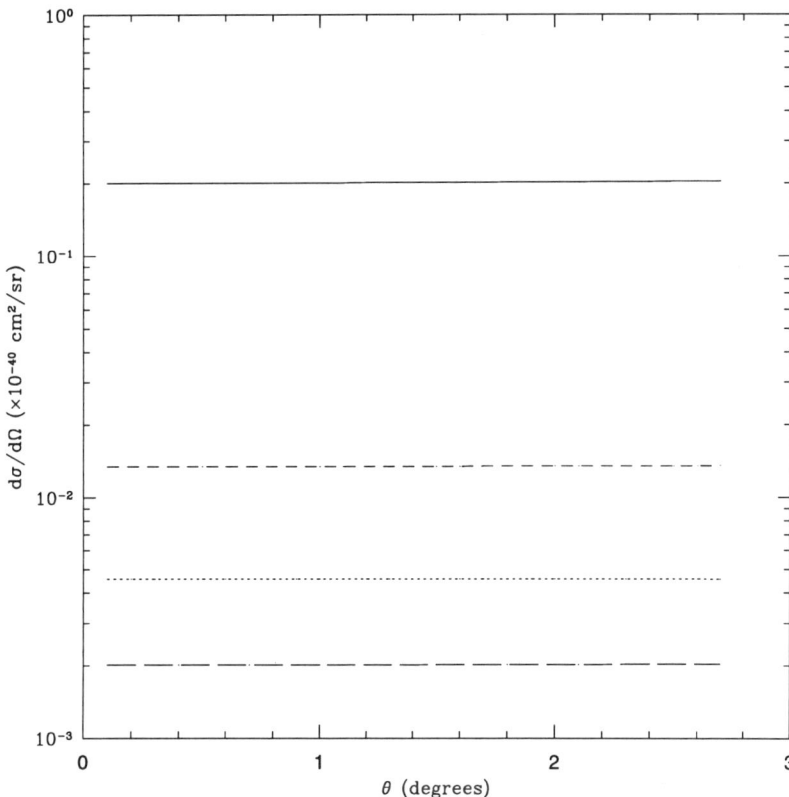

FIGURE 1. Plot of the differential cross section for the reaction, $\mu^- + p \longrightarrow \Lambda + \nu_\mu$ as a function of the laboratory angle of the outgoing Λ for an incident muon energy of 0.265 GeV. The solid curve indicates the whole differential cross section. The medium dashed, small dashed and large dashed curves represent the absolute values the absolute value of the $F_E F_A$, $F_P F_A$ interference terms respectively

be obtained in the muon or the tau induced induced reaction. From figure 1 we see that the $F_E F_A$ interference term contributes at near the 10 percent level which might be observable. The $F_P F_A$ interference term contributes at the 3 percent level which is probably too small to observe. We do not show the 5.0 GeV data due to space limitations. However the situation is considerably worse and it is unlikely that any of the interference terms could be observed. We plot these same quantities for the tau induced reaction in figure 2. Here the large mass of the tau lepton accentuates the contribution of the $F_P F_A$ interference term but it is still probably too small to observe.

There is also the possiblity of an asymmetry measurement particularly with the muon induced process near threshold. which might allow easier separation of the F_V and F_A form factors than could be done in an electron induced process. It might also allow a measurement of F_E if the $F_E F_V$ term is sufficiently large. Finally it might be

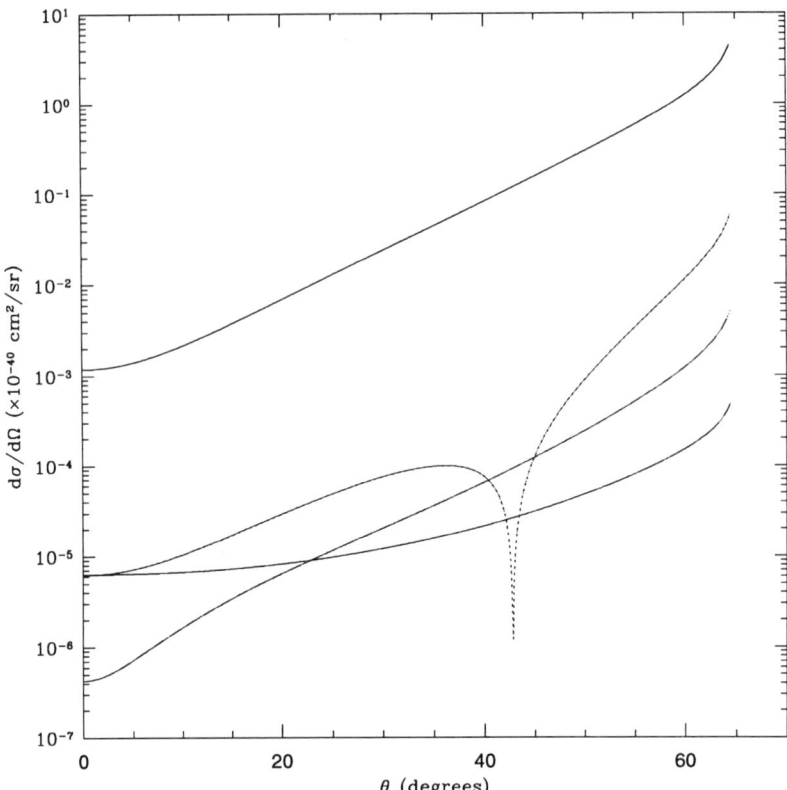

FIGURE 2. Plot of the differential cross section for the reaction, $\mu^- + p \longrightarrow \Lambda + \nu_\mu$ as a function of the laboratory angle of the outgoing Λ for an incident muon energy of 5.0 GeV. The solid curve indicates the whole differential cross section. The medium dashed, small dashed and large dashed curves represent the absolute values the absolute value of the $F_E F_A$, $F_P F_A$ interference terms respectively

possible eventually to perform the corresponding neutrino induced interactions, namely $\bar{\nu}_\mu + p \longrightarrow \Lambda + mu^+$ and $\bar{\nu}_\tau + p \longrightarrow \Lambda + \tau^+$. These could be run near threshold for producing the charged lepton. From figure 2 this should accentuate the contribution of F_P, the pseudoscalar form factor for the case of tau production. The pseudoscalar form factor is not well studied for strangeness changing weak processes. The opportunity to observe it would be very welcome indeed.

REFERENCES

1. Mintz,S.L., Nucl. Phys. **A 657**,303(1999).
2. Mintz,S.L., and Barnett,M.A.,Phys. Rev. **D 66**,117501(2002).

HEAVY QUARKS AND LEPTONS

CP Asymmetries at *BABAR*

Frederic Blanc

(on behalf of the *BABAR* Collaboration)

University of Colorado

Abstract. We present recent measurements of CP asymmetries in the B meson system, based on 92fb^{-1} collected with the *BABAR* detector at the PEP-II asymmetric-energy B factory at SLAC.

INTRODUCTION

CP violation was first observed in the K^0 system, almost 40 years ago [1]. The theoretical understanding of this phenomenon within the framework of the standard model (SM) led to the prediction of sizable CP violation in the B-meson system. In 2001, the *BABAR* and BELLE Collaborations announced the observation of CP violation in $B^0 \to J/\psi K_S$ decays [2, 3]. These results are in good agreement with the predictions of the standard model. Further constraints to the theory may come from the measurement of CP asymmetries in other B-decay channels. We present here several recent results of the measurement of CP asymmetries obtained by the *BABAR* Collaboration.

CP VIOLATION IN *B*-MESON DECAYS

In the standard model, CP violation arises from a single complex phase in the Cabibbo-Kobayashi-Maskawa (CKM) quark-mixing matrix [4]. In the Wolfenstein parametrization of the CKM matrix [5], only V_{ub} and V_{td} are complex, and we define $\gamma \equiv \arg(V_{ub})$, $\beta \equiv \arg(V_{td})$ and $\alpha = \pi - \beta - \gamma$, the angles of the CKM unitarity triangle.

At asymmetric B factories, the B mesons are produced in pairs from the decay of the $\Upsilon(4S)$ resonance, and evolve coherently until the decay of one of the B mesons, B_{tag}. At that time, the flavor of the other B meson, B_{rec}, is opposite to that of B_{tag}. We write the decay rate distribution F_+ (F_-) for B_{rec} decays to the final state f when $B_{tag} = B^0(\bar{B}^0)$

$$F_{\pm}(\Delta t) = \frac{e^{-|\Delta t|/\tau}}{4\tau}\left[1 \pm S_f \sin(\Delta m_d \Delta t) \mp C_f \cos(\Delta m_d \Delta t)\right], \quad (1)$$

where τ is the mean B^0 lifetime and Δm_d is the mixing frequency due to the B eigenstates mass difference. The CP parameters S_f and C_f are defined as $S_f = 2\mathrm{Im}\lambda_f/(1+|\lambda_f|^2)$ and $C_f = (1-|\lambda_f|^2)/(1+|\lambda_f|^2)$, where the convention-independent parameter $\lambda_f \equiv \eta_f \cdot q/p \cdot \bar{A}_{\bar{f}}/A_f$ [6]. In the definition of λ_f, η_f is the CP eigenvalue of f, q/p describes the $B^0 - \bar{B}^0$ mixing, and A_f ($\bar{A}_{\bar{f}}$) is the $B^0 \to f$ ($\bar{B}^0 \to \bar{f}$) amplitude. The standard model predicts negligible CP violation in the mixing and in the decay, but a sizable CP violation in the interference between the mixing and the decay is expected (i.e. $\text{Im}\lambda_{CP} \neq 0$).

In the processes of type $b \to c\bar{c}s$, $b \to c\bar{c}d$ and $b \to s\bar{s}s$, the dominant phase comes from mixing, and therefore the related decays are sensitive to the phase β. In addition to the phase β from mixing, the decays of type $b \to u\bar{u}d$ also depend on the phase γ from the $b \to u$ transition, and are therefore sensitive to the phase α.

Although direct CP violation is small in the standard model, it is expected to be larger in the B-meson system than it is in the kaon system (parameter ε'_K). We define the time-integrated CP asymmetry $\mathscr{A}_{CP} \equiv [\Gamma(\bar{B} \to \bar{f}) - (B \to f)] / [\Gamma(\bar{B} \to \bar{f}) + (B \to f)]$, which is sensitive to α and γ. Charmless B decays are of interest, because the penguin contribution can be large, allowing possible contributions from physics beyond the standard model. Unfortunately, the theoretical uncertainties are large due to the contribution of the strong phases in $\mathscr{A}_{CP}$.

DATA SAMPLE AND ANALYSIS TECHNIQUE

The results presented here are based on approximately $92\,\text{fb}^{-1}$, corresponding to about 89 million $B\bar{B}$ pairs. The data were taken with the BABAR detector, located at the e^+e^- PEP-II asymmetric B factory at SLAC. The PEP-II collider runs at the $\Upsilon(4S)$ resonance (center of mass energy $\sqrt{s} = 10.58\,\text{GeV}$). The $\Upsilon(4S)$ boost of $\beta\gamma = 0.56$ with respect to the laboratory frame allows time-dependent measurements of the B-meson decay rates. The BABAR detector is describe in details in Ref. [7].

Kinematic and events shape variables are used to separate signals from backgrounds, and the decay time difference Δt between the two B mesons produced in the $\Upsilon(4S)$ decay is used for the time-dependent measurements.

The kinematic variables are the energy difference $\Delta E = E_B^\star - \sqrt{s}/2$, the energy substituted mass $m_{ES} = \sqrt{(s/2 + \vec{p}_0 \cdot \vec{p}_B)^2/E_0^2 - |\vec{p}_B|}$, and the mass and decay angle of the resonances involved in the B-meson decay under study. In the above definitions, $(E_0, \vec{p}_0)$ and $(E_B, \vec{p}_B)$ are the 4-vectors of the initial $\Upsilon(4S)$ and B candidate, respectively, and $E_B^\star$ is the B candidate energy measured in the $\Upsilon(4S)$ frame.

The dominant background arises from the abundant $e^+e^- \to q\bar{q}$ ($q = u,d,s,c$) events. The topology of these events is jetty relative to the spherical $\Upsilon(4S) \to B\bar{B}$ events. This difference is used in constructing a set of event-shape variables.

For the time-dependent measurements, both B_{rec} and B_{tag} are reconstructed and Δt is calculated from the distance Δz between their decay vertices. The value of Δz is typically $260\,\mu\text{m}$, measured with a resolution of $180\,\mu\text{m}$. From the partial reconstruction of B_{tag}, we also determine its flavor at the time of the decay.

For each decay mode, the discriminating variables are used in a maximum likelihood (ML) fit, from which the signal and background yields, the charge asymmetries, $\mathscr{A}_{CP}$, and the time-dependent asymmetries are extracted. The ML fit is extensively tested and validated using Monte-Carlo simulated and real data control samples. A detailed description of the analysis technique can be found in Ref. [8].

TEST OF CP, T AND CPT CONSERVATION IN $B^0-\bar{B}^0$ MIXING

We tested the predicted (SM) very small lifetime difference $\Delta\Gamma \equiv \Gamma_H - \Gamma_L$ between the B mass eigenstates (B_H and B_L), the negligible T violation and the absence of CPT violation in $B^0-\bar{B}^0$ mixing. From a simultaneous time-dependent fit to the CP and flavor eigenstate samples, we measure $\Delta\Gamma/\Gamma$, $|q/p|$, the CPT-sensitive z parameter, and λ_{CP} [9]:

$$\begin{aligned}
\text{sign}(\text{Re}\lambda_{CP}) &= -0.008 \pm 0.037 \pm 0.018 & [-0.084, 0.068] \\
|q/p| &= 1.029 \pm 0.013 \pm 0.011 & [\;1.001, 1.057] \\
(\text{Re}\lambda_{CP}/|\lambda_{CP}|)\text{Re}z &= 0.014 \pm 0.035 \pm 0.034 & [-0.072, 0.101] \\
\text{Im}z &= 0.038 \pm 0.029 \pm 0.025 & [-0.028, 0.104]
\end{aligned}$$

where the first (second) uncertainty is statistical (systematic). The square brackets give the 90% confidence-level intervals. The results are compatible with the SM predictions.

MEASUREMENTS OF THE PHASE β

As a reminder, we give here the results for the measurement of $\sin 2\beta$ obtained from the theoretically well understood and experimentally clean B-meson decays involving $b \to c\bar{c}s$ transitions (golden channels). From a sample of 2641 events, we obtain [10]

$$\sin 2\beta = 0.741 \pm 0.067 \pm 0.034 \quad \text{and} \quad |\lambda| = 0.948 \pm 0.051 \pm 0.030,$$

compatible with the SM expectation of $|\lambda| \approx 1$. The above value of $\sin 2\beta$ is taken as a reference in the following discussion.

The cleanest example of a $b \to s\bar{s}s$ penguin dominated transitions comes from the $B \to \phi K$ decays. The value of $S_{\phi K_S}$ is expected to be a clean measurement of $\sin 2\beta$. Any significant deviation from $\sin 2\beta$ would be a sign of physics beyond the standard model. With $50 \pm 9\ B^0 \to \phi K_S$ and about 170 $B^+ \to \phi K^+$ signal events, we measure [11]

$$\begin{aligned}
\mathscr{A}_{CP}(B^+ \to \phi K^+) &= (3.9 \pm 8.6 \pm 1.1)\% \\
S_{\phi K_S} &= -0.18 \pm 0.51 \pm 0.07 \\
C_{\phi K_S} &= -0.80 \pm 0.38 \pm 0.12
\end{aligned}$$

$S_{\phi K_S}$ shows a two standard deviation discrepancy with the value of $\sin 2\beta$ obtained in the $b \to c\bar{c}s$ channels. Although not statistically significant, this may be a hint for non-standard model physics.

Similarly to $B \to \phi K$, the $B \to \eta' K$ decays are dominated by $b \to s\bar{s}s$ penguin transitions. However, due to the $u\bar{u}$ and $d\bar{d}$ content of the η' meson, a contamination from tree processes may exist, making the time-dependent asymmetry more difficult to interpret. On the other hand, the large branching fraction of this decay allows a more precise measurement of the CP parameters.

From $\approx 800\ B^+ \to \eta' K^+$ and $\approx 200\ B^0 \to \eta' K_S$ signal events reconstructed in two different decays of the η' ($\eta' \to \eta \pi^+ \pi^-$ and $\eta' \to \rho^0 \gamma$), we obtain [12]

$$\begin{aligned}
\mathscr{A}_{CP}(B^+ \to \eta' K^+) &= (3.7 \pm 4.5 \pm 1.1)\% \\
S_{\eta' K_S} &= 0.02 \pm 0.34 \pm 0.03 \\
C_{\eta' K_S} &= 0.10 \pm 0.22 \pm 0.04
\end{aligned}$$

$S_{\eta'K_S}$ shows a two standard deviation discrepancy with the expected value of $\sin 2\beta$, but the theoretical uncertainties are large, and therefore no firm conclusion can be drawn.

We also obtained preliminary CP results for the decay modes $B^0 \to J/\psi \pi^0$ and $B^0 \to D^{\star\pm}D^{\mp}$. These decays proceeds through a $b \to c\bar{c}d$ transition, with contributions from both tree and penguin amplitudes. However, we expect $S_{J/\psi\pi^0} = S_{D^{\star+}D^-} = S_{D^{\star-}D^+} = -\sin 2\beta$ if the penguin amplitude is negligible. From samples of 40 ± 7 $B^0 \to J/\psi\pi^0$ and 113 ± 13 $B^0 \to D^{\star\pm}D^{\mp}$ signal events, we obtain [13, 14]

$$S_{J/\psi\pi^0} = 0.05 \pm 0.49 \pm 0.16 \qquad C_{J/\psi\pi^0} = 0.38 \pm 0.41 \pm 0.09$$
$$S_{D^{\star+}D^-} = -0.82 \pm 0.75 \pm 0.14 \qquad C_{D^{\star+}D^-} = -0.47 \pm 0.40 \pm 0.12$$
$$S_{D^{\star-}D^+} = -0.24 \pm 0.69 \pm 0.12 \qquad C_{D^{\star-}D^+} = -0.22 \pm 0.37 \pm 0.10$$

Increased statistics will be necessary before these decay modes can be used for constraining the standard model.

MEASUREMENTS OF THE PHASE α

The $B^0 \to \pi^+\pi^-$ decay is dominated by the tree amplitude, but it is known from the large $B^0 \to K^+\pi^-$ branching fraction that the penguin amplitude contribution is large. As a consequence, the measured value of $\sin 2\alpha_{\text{eff}} \equiv S_{\pi^+\pi^-}$ is not expected to be equal to $\sin 2\alpha$. But it has been suggested to use an isospin analysis of the $B \to \pi\pi$ decay modes to extract the value of α from α_{eff} [15]. From our measurements of the $B \to \pi\pi$ branching fractions [16], we set the limit $|\alpha - \alpha_{\text{eff}}| < 51°$. From a simultaneous fit of both $B^0 \to \pi^+\pi^-$ and $B^0 \to K^+\pi^-$ decays, we obtain [17]

$$\mathcal{A}_{CP}(B^0 \to K^+\pi^-) = (-10.2 \pm 5.0 \pm 1.6)\%$$
$$S_{\pi^+\pi^-} = 0.02 \pm 0.34 \pm 0.05$$
$$C_{\pi^+\pi^-} = -0.30 \pm 0.25 \pm 0.04$$

The result for $S_{\pi^+\pi^-}$ is compatible with the naively expected value of $\alpha_{\text{eff}} \approx 90°$.

The phase α may also be extracted from a dalitz plot analysis of the decay $B \to \pi^+\pi^-\pi^0$. This channel has the advantage of a larger branching fraction ($\approx 4\times$) and a smaller contribution from penguin amplitude than the $B^0 \to \pi^+\pi^-$ channel. We performed this challenging analysis as a simpler pseudo two-body analysis, selecting a $\rho^{\pm} \to \pi^{\pm}\pi^0$ candidate recoiling against a $\pi^{\mp}$. From ≈ 430 $B^0 \to \rho^+\pi^-$ and ≈ 120 $B^0 \to \rho^+K^-$ signal events, we measure [18]

$$\mathcal{A}_{CP}(B^0 \to \rho^+K^-) = (28 \pm 17 \pm 8)\%$$
$$\mathcal{A}_{CP}(B^0 \to \rho^+\pi^-) = (-18 \pm 8 \pm 3)\%$$
$$S_{\rho^{\pm}\pi^{\mp}} = 0.19 \pm 0.24 \pm 0.03$$
$$C_{\rho^{\pm}\pi^{\mp}} = 0.36 \pm 0.18 \pm 0.04$$

The over two standard deviation discrepancy of $\mathcal{A}_{CP}(B^0 \to \rho^+\pi^-)$ from the no-direct CP violation hypothesis is not significant to allow any useful conclusion.

SEARCHES FOR DIRECT CP VIOLATION

Finally, we have searched for direct CP violation in 19 different charmless decay modes of the B mesons, including some of the results presented above. Only three of these modes show a deviation larger than two standard deviations from no asymmetry: $\mathcal{A}_{CP}(B^0 \to \rho^+\pi^-) = (-18 \pm 8 \pm 3)\%$, $\mathcal{A}_{CP}(B^+ \to \eta\pi^+) = (-51^{+20}_{-18} \pm 1)\%$ [19] and $C_{\phi K_S} = -0.80 \pm 0.38 \pm 0.12$. These deviations are compatible with the expectation of statistical fluctuations. However, it should be noted that theoretical calculations predict large asymmetries for both $\mathcal{A}_{CP}(B^0 \to \rho^+\pi^-)$ and $\mathcal{A}_{CP}(B^+ \to \eta\pi^+)$ [see references in [18] and [19]]. Increased statistics will help clarify the situation.

CONCLUSION

Besides the well measured $\sin 2\beta$ in the golden channels, the *BABAR* Collaboration has obtained preliminary results in several other decay modes sensitive to the CKM phases α and β. Some results show interesting deviations from the expected values, but more statistics will be required before any conclusion can be drawn. Pursuing the goal of better accuracy, the *BABAR* experiment is expecting to accumulate over $500\,\text{fb}^{-1}$ by 2006.

ACKNOWLEDGMENTS

I want to thank my *BABAR* colleagues for their help while preparing this talk, as well as the organizers for this beautiful conference.

REFERENCES

1. J. H. Christenson *et al.*, Phys. Rev. Lett. **13**, 138 (1964)
2. B. Aubert *et al.*, BABAR Collaboration, Phys. Rev. Lett. **87**, 091801 (2001)
3. K. Abe *et al.*, BELLE Collaboration, Phys. Rev. Lett. **87**, 091802 (2001)
4. N. Cabibbo, Phys. Rev. Lett. **10**, 531 (1963); M. Kobayashi and T. Maskawa, Prog. Theor. Phys. **49**, 652 (1973)
5. L. Wolfenstein, Phys. Rev. Lett. **51**, 1945 (1983)
6. For a detailed discussion, see e.g. "The *BABAR* Physics Book", SLAC-R-504 (1998)
7. B. Aubert *et al.*, BABAR Collaboration, Nucl. Instr. an Methods, A **479**, 1 (2002)
8. B. Aubert *et al.*, BABAR Collaboration, Phys. Rev. D **66**, 032003 (2002)
9. B. Aubert *et al.*, BABAR Collaboration, SLAC-PUB-9696 (2003), hep-ex/0303043
10. B. Aubert *et al.*, BABAR Collaboration, Phys. Rev. Lett. **89**, 201802 (2002)
11. B. Aubert *et al.*, BABAR Collaboration, SLAC-PUB-9297 (2002), hep-ex/0207070
12. B. Aubert *et al.*, BABAR Collaboration, SLAC-PUB-9698 (2003), hep-ex/0303046
13. B. Aubert *et al.*, BABAR Collaboration, SLAC-PUB-9668 (2003), hep-ex/0303018
14. B. Aubert *et al.*, BABAR Collaboration, Phys. Rev. Lett. **90**, 221801 (2003)
15. M. Gronau and D. London, Phys. Rev. Lett. **65**, 3381 (1990); see other related references in [17]
16. B. Aubert *et al.*, BABAR Collaboration, SLAC-PUB-9683 (2003), hep-ex/0303028
17. B. Aubert *et al.*, BABAR Collaboration, Phys. Rev. Lett. **89**, 281802 (2002)
18. B. Aubert *et al.*, BABAR Collaboration, SLAC-PUB-9923 (2003), hep-ex/0306030
19. B. Aubert *et al.*, BABAR Collaboration, SLAC-PUB-9692 (2003), hep-ex/0303039

sin 2β from pure penguins and signs of new physics [1]

M. Ciuchini*, E. Franco†, A. Masiero** and L. Silvestrini†

*INFN Sezione di Roma III and Dip. di Fisica, Univ. di Roma Tre, Via della Vasca Navale 84, I-00146 Rome, Italy.
†INFN Sezione di Roma and Dip. di Fisica, Univ. di Roma "La Sapienza", P.le A. Moro 2, I-00185 Rome, Italy.
**Dip. di Fisica "G. Galilei", Univ. di Padova and INFN, Sezione di Padova, Via Marzolo 8, I-35121 Padua, Italy.

Abstract. We discuss SUSY contributions to $b \to s$ processes in a generic MSSM, showing that the discrepancy between the SM prediction and the experimental value of the CP asymmetry in $B \to \phi K_s$ decays can be accounted for by SUSY. We then analyze in detail different classes of SUSY contributions, and identify other interesting channels for SUSY searches in $B_{(s)}$ decays.

A close look at the Unitarity Triangle (UT) fit reveals that New Physics (NP) contributions to $s \to d$ and $b \to d$ transitions are strongly constrained, while new contributions to $b \to s$ transitions affect the fit only if they interfere destructively with the SM amplitude for $B_s - \bar{B}_s$ mixing, bringing the mass difference below the present lower bound. Other processes not involved in the UT fit, for example the celebrated $B \to X_s \gamma$, can provide constraints on any NP in $b \to s$ transitions. However, $B \to X_s \gamma$ mostly constrains the helicity flipping contributions to the $b \to s$ transition. As we shall see in the following, plenty of room is left for NP contributions to interesting observables in this sector, in particular to the CP asymmetry in $B \to \phi K_S$ decays. Let us focus on the most realistic extension of the SM, the Minimal Supersymmetric Standard Model (MSSM), which in its general form can cause Flavour and CP violating processes to arise at a rate much higher than what is experimentally observed [1] and is therefore expected to give large contributions to $b \to s$ transitions.

We keep our analysis in the MSSM as general as possible [2]. Minimality refers here only to the minimal amount of superfields needed to supersymmetrize the SM and to the presence of R parity. Otherwise the soft breaking terms are left completely free and constrained only by phenomenology. Technically the best way we have to account for the SUSY FCNC contributions in such a general framework is via the mass insertion method using the leading gluino exchange contributions [3]. In the Super-CKM basis, SUSY FCNC and CP violation arise from off-diagonal terms in squark mass matrices only. These are conveniently expressed as $(\delta_{ij})_{AB} \equiv (\Delta_{ij})_{AB}/m_{\tilde{q}}^2$, where $(\Delta_{ij})_{AB}$ is the mass term connecting squarks of flavour i and j and "helicities" A and B, and $m_{\tilde{q}}$ is

---

[1] Talk given by L.S.

the average squark mass. Detailed analyses carried out in SUSY have shown that one must have $(\delta_{12}^d)_{AB}$ and $(\delta_{13}^d)_{AB}$ much smaller than what naively expected [4, 5]. It is therefore reasonable to assume that $(\delta_{12}^d)_{AB} \sim (\delta_{13}^d)_{AB} \sim 0$. Under this assumption, we present constraints on $(\delta_{23}^d)_{AB}$ from available data and possible effects in present and future measurements.

Our analysis aims at determining the allowed regions in the SUSY parameter space governing $b \to s$ transitions, studying the correlations among different observables and pointing out possible signals of SUSY. The constraints on the parameter space come from:

1. The BR$(B \to X_s \gamma) = (3.29 \pm 0.34) \times 10^{-4}$ (experimental results as reported in [6], rescaled according to ref. [7]).
2. The CP asymmetry $A_{CP}(B \to X_s \gamma) = -0.02 \pm 0.04$ [6].
3. The BR$(B \to X_s \ell^+ \ell^-) = (6.1 \pm 1.4 \pm 1.3) \times 10^{-6}$ [6].
4. The lower bound on the $B_s - \bar{B}_s$ mass difference $\Delta M_{B_s} > 14.4 \text{ ps}^{-1}$ [6].

We have also considered BR's and CP asymmetries for $B \to K\pi$ and found that, given the large theoretical uncertainties, they give no significant constraints on the δ's.

For $B \to \phi K_s$, we have studied the BR and the coefficients $C_{\phi K}$ and $S_{\phi K}$ of cosine and sine terms in the time-dependent CP asymmetry.

All the details concerning the treatment of the different amplitudes entering the analysis can be found in ref. [2]. In summary, we use:

i) $\Delta B = 2$ *amplitudes.* Full NLO SM and LO gluino-mediated matching condition, NLO QCD evolution and hadronic matrix elements from lattice calculations.

ii) $\Delta B = 1$ *amplitudes.* Full NLO SM and LO gluino-mediated matching condition and NLO QCD evolution. The matrix elements of semileptonic and radiative decays include α_s terms, Sudakov resummation, and the first corrections suppressed by powers of the heavy quark masses. For non-leptonic decays, such as $B \to K\pi$ and $B \to \phi K_s$, we adopt BBNS factorization [8], with an enlarged range for the annihilation parameter ρ_A, in the spirit of the criticism of ref. [9]. This choice maximizes the sensitivity of the factorized amplitudes to SUSY contributions, which is expected to be much lower if the power corrections are dominated by the "charming penguin" contributions [10].

Another source of potentially large SUSY effects in $B \to \phi K_s$ is the contribution of the chromomagnetic operator which can be substantially enhanced by SUSY without spoiling the experimental constraints from $B \to X_s \gamma$ [11]. Indeed, the time-dependent asymmetry in $B \to \phi K_s$ is more sensitive to the SUSY parameters in the case of chirality-flipping insertions which enter the amplitude in the coefficient of the chromomagnetic operator. One should keep in mind, however, that the corresponding matrix element, being of order α_s, has large uncertainties in QCD factorization.

We performed a MonteCarlo analysis, generating weighted random configurations of input parameters (see ref. [12] for details of this procedure) and computing for each configuration the processes listed above. We study the clustering induced by the constraints on various observables and parameters, assuming that each unconstrained δ_{23}^d fills uniformly a square $(-1\ldots 1, -1\ldots 1)$ in the complex plane. The ranges of CKM parameters have been taken from the UT fit ($\bar{\rho} = 0.178 \pm 0.046$, $\bar{\eta} = 0.341 \pm 0.028$), and hadronic parameter ranges are those used in ref. [2]. Concerning SUSY parameters, we

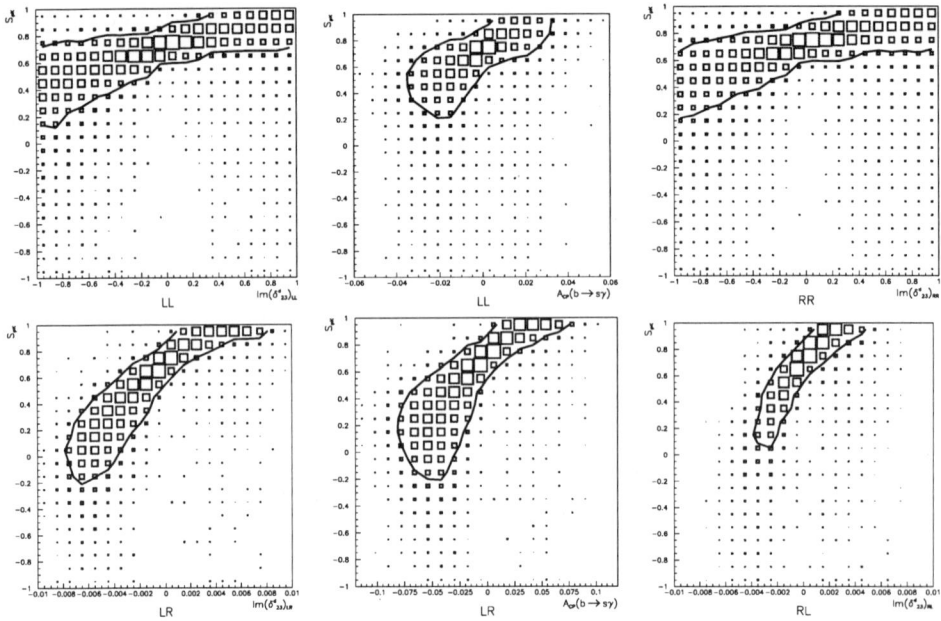

FIGURE 1. Correlations between $S_{\phi K}$ and $\text{Im}(\delta^d_{23})_{AB}$ or $A_{CP}(b \to s\gamma)$, for $AB = (LL, RR, LR, RL)$. The black line contains 68% of the weighted events.

fix $m_{\tilde{q}} = m_{\tilde{g}} = 350$ GeV and consider different possibilities for the mass insertions.

In fig. 1, we study the correlations of $S_{\phi K}$ with $\text{Im}(\delta^d_{23})_{AB}$ and $A_{CP}(B \to X_s\gamma)$ for various SUSY insertions. The reader should keep in mind that, in all the results reported in fig. 1, the hadronic uncertainties affecting the estimate of $S_{\phi K}$ are not completely under control. Low values of $S_{\phi K}$ can be more easily obtained with helicity flipping insertions. A deviation from the SM value for $S_{\phi K}$ requires a nonvanishing value of $\text{Im}(\delta^d_{23})_{AB}$ (see fig. 1, left and right), generating, for those channels in which the SUSY amplitude can interfere with the SM one, a $A_{CP}(B \to X_s\gamma)$ at the level of a few percents in the LL case, and up to the experimental upper bound in the LR case (see fig. 1, center).

A crucial question naturally arises at this point: what are the more promising processes to reveal some signal of low energy SUSY among the FCNCs involving $b \to s$ transitions? For this purpose, it is useful to classify different "classes of MSSM" according to the "helicities" LL, RR, etc, of the different δ^d_{23}'s.

The BaBar and BELLE Collaborations have recently reported the time-dependent CP asymmetry in $B_d(\bar{B}_d) \to \phi K_s$. While $\sin 2\beta$ as measured in the $B \to J/\psi K_s$ channel is 0.734 ± 0.054 (in agreement with the SM prediction [6]), the combined result from both collaborations for the corresponding $S_{\phi K}$ of $B_d \to \phi K_s$ is -0.39 ± 0.41 [13] with a 2.7σ discrepancy between the two results. In the SM, they should be the same up to doubly Cabibbo suppressed terms. Obviously, one should be very cautious before accepting such result as a genuine indication of NP. Nonetheless, the negative value of $S_{\phi K}$ could be due to large SUSY CP violating contributions. Then, one can wonder which δ's are relevant to produce such enhancement and, even more important, which other significant deviations from the SM could be detected.

We start discussing the RR case. As shown in Fig. 1 (upper right), although values of $S_{\phi K}$ in the range predicted by the SM are largely favoured, still pure δ_{RR} insertions are able to give rise to a negative $S_{\phi K}$ in agreement with the results of BaBar and BELLE quoted above. As for the $B_s - \bar{B}_s$ mixing, the distribution of ΔM_s is peaked at the SM value, but it has a long tail at larger values, up to 120 ps^{-1} for our choice of the range of δ_{RR}. In addition, we find that the expected correlation requiring large ΔM_s for negative $S_{\phi K}$ is totally wiped out by the large uncertainties. Hence, in the RR case it is possible to have a strong discrepancy between $\sin 2\beta$ and $S_{\phi K}$ whilst $B_s - \bar{B}_s$ oscillations proceed as expected in the SM (thus, being observable in the Run II of Tevatron). We expect the CP asymmetry in $B \to X_s \gamma$ to be as small as in the SM, while, differently from the SM, the time-dependent CP asymmetry in the decay channel $B_s \to J/\psi\phi$ is expected to be large.

We now move on to discuss the LL insertion. A major difference with the previous case concerns the SUSY contributions to $B \to X_s \gamma$. The LL insertion contributes to the same operator which is responsible for $B \to X_s \gamma$ in the SM and hence the SM and SUSY amplitudes interfere. As a consequence, the rate tends to be larger than the RR case and, moreover, a CP asymmetry can be generated up to 5% (see fig. 1, center). However, given the uncertainties, the correlation of $A_{CP}(B \to X_s \gamma)$ with $S_{\phi K}$ is not very stringent. As can be seen from the figure, negative values of $S_{\phi K}$ do not necessarily correspond to non-vanishing $A_{CP}(B \to X_s \gamma)$, although typical values are around 2%. The distribution of ΔM_s is quite similar to the RR case. Finally, one expects also in this case to observe CP violation in $B_s \to J/\psi\phi$ at hadron colliders.

Let us now consider helicity-flipping mass insertions. In these cases, negative values of $S_{\phi K}$ can be easily obtained (although a positive $S_{\phi K}$ is favoured, cfr. fig. 1, bottom row). The severe bound imposed by $BR(B \to X_s \gamma)$ (and $A_{CP}(B \to X_s \gamma)$ in the LR case) prevents any enhancement of the $B_s - \bar{B}_s$ mixing as well as any sizeable contribution to $A_{CP}(B_s \to J/\psi\phi)$. On the other hand, $A_{CP}(B \to X_s \gamma)$ as large as 5–10 % is attainable in the LR case (fig. 1, lower center), offering a potentially interesting hint for NP. Notice that the LR mass insertion contributes to $b_R \to s_L \gamma$, just as the SM. On the contrary, the RL mass insertion contributes to $b_L \to s_R \gamma$ and thus it does not interfere. Consequently, the CP asymmetry is as small as in the SM.

Our results confirm that FCNC and CP violation in physics involving $b \to s$ transitions still offer opportunities to disentangle effects genuinely due to NP. In particular the discrepancy between the amounts of CP violation in the two B_d decay channels $J/\psi K_s$ and ϕK_s can be accounted for in the MSSM while respecting all the existing constraints in B physics, first of all the $BR(B \to X_s \gamma)$. The relevant question is then which processes offer the best chances to provide other hints of the presence of low-energy SUSY.

First, it is mandatory to further assess the time-dependent CP asymmetry in the decay channel $B \to \phi K_s$. If the measurement will be confirmed, then this process would become decisive in discriminating among different MSSM realizations. Although, as we have seen, it is possible to reproduce the negative $S_{\phi K}$ in a variety of different options for the SUSY soft breaking down squark masses, the allowed regions in the SUSY parameter space are more or less tightly constrained according to the kind of δ^d_{23} mass insertion which dominates.

In order of importance, it then comes the measurement of the $B_s - \bar{B}_s$ mixing. Finding ΔM_s larger than 20 ps^{-1} would hint at NP. RR or LL could account for a ΔM_s up to 120 ps^{-1}. An interesting alternative would arise if ΔM_s is found as expected in the SM while, at the same time, $S_{\phi K}$ is confirmed to be negative. This scenario would favour the LR possibility, even though all other cases do not necessarily lead to large ΔM_s.

Keeping to B_d physics, we point out that the CP asymmetry in $B \to X_s \gamma$ remains of utmost interest. This asymmetry is so small in the SM that it should not be possible to detect it. We have seen that in particular with LR insertions such asymmetry can be enhanced up to 10 % making it possibly detectable in a not too distant future.

Finally, once we will have at disposal large amounts of B_s, it will be of great interest to study processes which are mostly CP conserving in the SM, while possibly receiving large contributions from SUSY. In the SM the amplitude for $B_s - \bar{B}_s$ mixing does not have an imaginary part up to doubly Cabibbo suppressed terms and decays like $B_s \to J/\psi\phi$ also have a negligible amount of CP violation. Quite on the contrary, if the measured negative $S_{\phi K}$ is due to a large, complex δ_{23}^d mass insertion, we expect some of the above processes to exhibit a significant amount of CP violation. In particular, in the case of RR insertions, both the $b \to s$ amplitudes and the B_s mixing would receive non negligible contributions from Imδ_{23}^d, while, if the LR insertions is dominant, we do not expect any sizable contribution to B_s mixing. Still, the SUSY contribution to CP violation in the $B_s \to J/\psi\phi$ decay amplitude could be fairly large.

REFERENCES

1. F. Gabbiani et al., Nucl. Phys. B **477** (1996) 321 [arXiv:hep-ph/9604387].
2. M. Ciuchini et al., Phys. Rev. D **67** (2003) 075016 [arXiv:hep-ph/0212397] (n.b.: versions lower than 3 are incorrect).
3. L. J. Hall, V. A. Kostelecky and S. Raby, Nucl. Phys. B **267** (1986) 415.
4. D. Becirevic et al., Nucl. Phys. B **634** (2002) 105 [arXiv:hep-ph/0112303].
5. M. Ciuchini et al., JHEP **9810** (1998) 008 [arXiv:hep-ph/9808328].
6. A. Stocchi, arXiv:hep-ph/0211245.
7. P. Gambino and M. Misiak, Nucl. Phys. B **611** (2001) 338 [arXiv:hep-ph/0104034].
8. M. Beneke, G. Buchalla, M. Neubert and C. T. Sachrajda, Phys. Rev. Lett. **83** (1999) 1914 [arXiv:hep-ph/9905312]; Nucl. Phys. B **606** (2001) 245 [arXiv:hep-ph/0104110].
9. M. Ciuchini et al., Phys. Lett. B **515** (2001) 33 [arXiv:hep-ph/0104126]; arXiv:hep-ph/0208048.
10. M. Ciuchini et al., Nucl. Phys. B **501** (1997) 271 [arXiv:hep-ph/9703353].
11. A. L. Kagan, Phys. Rev. D **51** (1995) 6196 [arXiv:hep-ph/9409215];
    M. Ciuchini, E. Gabrielli and G. F. Giudice, Phys. Lett. B **388** (1996) 353 [Erratum-ibid. B **393** (1997) 489] [arXiv:hep-ph/9604438].
12. M. Ciuchini et al., JHEP **0107** (2001) 013 [arXiv:hep-ph/0012308].
13. B. Aubert et al. [BABAR Collaboration], arXiv:hep-ex/0207070.

A mixing-independent construction of the unitarity triangle

Matthias Neubert

*F.R. Newman Laboratory for Elementary-Particle Physics
Cornell University, Ithaca, NY 14853, USA*

Abstract. The study of charmless hadronic two-body decays of B mesons is one of the most fascinating topics in B physics. A construction of the unitarity triangle based on such decays is presented, which is independent of B–$\bar B$ and K–$\bar K$ mixing. It provides stringent tests of the Standard Model with small theoretical uncertainties.

INTRODUCTION

Measurements of $|V_{ub}|$ in semileptonic decays, $|V_{td}|$ in B–$\bar B$ mixing, and $\text{Im}(V_{td}^2)$ from CP violation in K–$\bar K$ and B–$\bar B$ mixing have firmly established the existence of a CP-violating phase in the CKM matrix. The present situation, often referred to as the "standard analysis" of the unitarity triangle, is summarized in Figure 1.

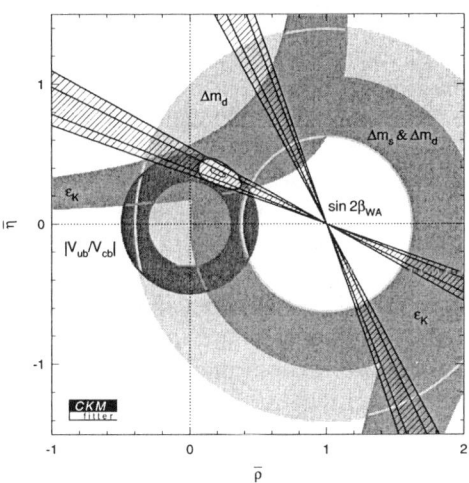

FIGURE 1. Standard constraints on the apex $(\bar\rho, \bar\eta)$ of the unitarity triangle [1].

Three comments are in order concerning this analysis:

1. The measurements of CP asymmetries in kaon physics (ε_K and ε'/ε) and B–$\bar B$ mixing ($\sin 2\beta$) probe the imaginary part of V_{td} and so establish CP violation in

the top sector of the CKM matrix. The Standard Model predicts that the imaginary part of V_{td} is related, by three-generation unitarity, to the imaginary part of V_{ub}, and that those two elements are (to an excellent approximation) the only sources of CP violation in flavor-changing processes. In order to test this prediction one must explore the phase $\gamma = \arg(V_{ub}^*)$ in the bottom sector of the CKM matrix.

2. With the exception of the $\sin 2\beta$ measurement the standard analysis is limited by large theoretical uncertainties, which dominate the widths of the various bands in the figure. These uncertainties enter via the calculation of hadronic matrix elements. Below I will discuss some novel methods to constrain the unitarity triangle using charmless hadronic B decays, which are afflicted by smaller hadronic uncertainties and hence provide powerful new tests of the Standard Model, which can complement the standard analysis.

3. With the exception of the measurement of $|V_{ub}|$ the standard constraints are sensitive to meson–antimeson mixing. Mixing amplitudes are of second order in weak interactions and hence might be most susceptible to effects from physics beyond the Standard Model. The new constraints on $(\bar{\rho}, \bar{\eta})$ discussed below allow a construction of the unitarity triangle that is over-constrained and independent of B–$\bar{B}$ and K–$\bar{K}$ mixing. It is in this sense complementary to the standard analysis.

The phase γ can be probed via tree–penguin interference in decays such as $B \to \pi K, \pi\pi$. Experiment shows that amplitude interference is sizable in these decays. Information about γ can be obtained from measurements of direct CP asymmetries ($\sim \sin\gamma$), but also from the study of CP-averaged branching fractions ($\sim \cos\gamma$). The challenge is, of course, to gain theoretical control over the hadronic physics entering the tree-to-penguin ratios in the various decays. Recently, much progress has been made toward achieving that goal.

Hadronic weak decays simplify greatly in the heavy-quark limit $m_b \gg \Lambda_{\rm QCD}$. The underlying physics is that a fast-moving light meson produced by a point-like source (the effective weak Hamiltonian) decouples from soft QCD interactions [2, 3]. A systematic implementation of this color transparency argument is provided by the QCD factorization approach [4, 5], which makes rigorous predictions for hadronic B-decay amplitudes in the heavy-quark limit. One can hardly overemphasize the importance of controlling nonleptonic decay amplitudes in the heavy-quark limit. While a few years ago reliable calculations of such amplitudes appeared to be out of reach, we are now in a situation where hadronic uncertainties enter only at the level of power corrections suppressed by the heavy b-quark mass.

In recent work, QCD factorization has been applied to the entire set of the 96 decays of B and B_s mesons into PP or PV final states ($P=$ pseudoscalar meson, $V =$ vector meson) [6]. It has been demonstrated that the approach correctly reproduces the main features seen in the data, such as the magnitudes of the various tree and penguin amplitudes, and the fact that they have small relative strong-interaction phases. In the future, when more data become available, this will allow us to extract much useful information about the flavor sector of the Standard Model either from global fits or from analysis of certain classes of decay modes such as $B \to \pi K$, $B \to \pi\pi$, and $B \to \pi\rho$. Detailed comparison with the data may also reveal limitations of the heavy-quark expansion in certain modes, perhaps hinting at the significance of some power corrections in $\Lambda_{\rm QCD}/m_b$.

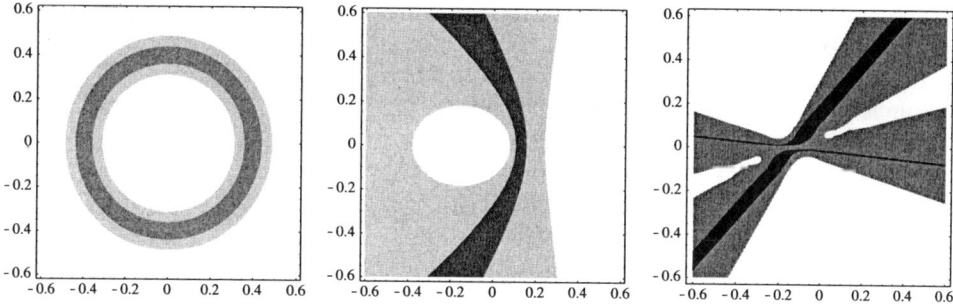

FIGURE 2. The three constraints in the $(\bar{\rho}, \bar{\eta})$ plane used in the construction of the CP-b triangle (see text for explanation). Experimental errors are shown at 95% CL. In each plot, the dark band shows the theoretical uncertainty, which is much smaller than the experimental error. This demonstrates the great potential of these methods once the data will become more precise.

THE CP-B TRIANGLE

Despite of the success of QCD factorization in describing the data, there is an interest in analyzing CKM parameters using methods that rely as little as possible on an underlying theoretical framework. In this talk I discuss a method for constructing the unitarity triangle from B physics using measurements whose theoretical interpretation is "clean" in the sense that it only relies on assumptions that can be tested using experimental data. I call this construction the CP-b triangle, because it probes the existence of a CP-violating phase in the b sector of the CKM matrix. The CP-b triangle is over-determined and can be constructed using already existing data. Most importantly, this construction is insensitive to potential New Physics effects in B–$\bar{B}$ or K–$\bar{K}$ mixing. The present analysis is an update of [7] using the most recent data as of summer 2003.

The first ingredient is the ratio $|V_{ub}/V_{cb}|$ extracted from semileptonic B decays, whose current value is $|V_{ub}/V_{cb}| = 0.09 \pm 0.02$. Several strategies have been proposed to determine $|V_{ub}|$ with an accuracy of about 10% [8, 9, 10, 11, 12], which would be a significant improvement. The first plot in Figure 2 shows the corresponding constraint in the $(\bar{\rho}, \bar{\eta})$ plane. Here and below the narrow, dark-colored band shows the theoretical uncertainty, while the lighter band gives the current experimental value.

The second ingredient is a constraint derived from the ratio of the CP-averaged branching fractions for the decays $B^\pm \to \pi^\pm K_S$ and $B^\pm \to \pi^0 K^\pm$, using a generalization of the method suggested in [13]. The experimental inputs to this analysis are a certain tree-to-penguin ratio $\varepsilon_{\exp} = 0.197 \pm 0.016$ and the ratio

$$R_* = \frac{\text{Br}(B^+ \to \pi^+ K^0) + \text{Br}(B^- \to \pi^- \bar{K}^0)}{2[\text{Br}(B^+ \to \pi^0 K^+) + \text{Br}(B^- \to \pi^0 K^-)]} = 0.804 \pm 0.085$$

of two CP-averaged $B \to \pi K$ branching fractions [14]. Without any recourse to QCD factorization this method provides a bound on $\cos \gamma$, which can be turned into a determination of $\cos \gamma$ (for fixed value of $|V_{ub}|/V_{cb}|$) when information on the relevant strong-interaction phase ϕ is available. The phase ϕ is bound by experimental data (and very general theoretical arguments) to be small, of order 10° [7]. (In the future, this phase can be determined directly from the direct CP asymmetry in $B^\pm \to \pi^0 K^\pm$ decays.) It is thus conservative to assume that $\cos\phi > 0.8$, corresponding to $|\phi| < 37°$. With this assumption the corresponding allowed region in the $(\bar{\rho}, \bar{\eta})$ plane was analyzed in [5]. The resulting constraint is shown in the second plot in Figure 2.

The third constraint comes from a measurement of the time-dependent CP asymmetry $S_{\pi\pi} = -\sin 2\alpha_{\text{eff}}$ in $B \to \pi^+\pi^-$ decays. The present experimental situation is still unclear, since the measurements by BaBar ($S_{\pi\pi} = -0.40 \pm 0.22 \pm 0.03$) and Belle ($S_{\pi\pi} = -1.23 \pm 0.41^{+0.08}_{-0.07}$) are not in good agreement with each other [15]. The naive average of these results gives $S_{\pi\pi} = -0.58 \pm 0.20$. (Inflating the error according to the PDG prescription would yield $S_{\pi\pi} = -0.58 \pm 0.34$, but for some reason the experimenters usually use the naive error without rescaling, and I will follow their example.) The theoretical expression for the asymmetry is

$$S_{\pi\pi} = -\frac{2\,\text{Im}\,\lambda_{\pi\pi}}{1+|\lambda_{\pi\pi}|^2}, \quad \text{where} \quad \lambda_{\pi\pi} = e^{-i\phi_d}\frac{e^{-i\gamma}+(P/T)_{\pi\pi}}{e^{+i\gamma}+(P/T)_{\pi\pi}}.$$

Here ϕ_d is the CP-violating phase of the B_d–$\bar{B}_d$ mixing amplitude, which in the Standard Model equals 2β. Usually it is argued that for small $(P/T)_{\pi\pi}$ ratio the quantity $\lambda_{\pi\pi}$ is approximately given by $e^{-2i(\beta+\gamma)} = e^{2i\alpha}$, and so apart from a "penguin pollution" the asymmetry $S_{\pi\pi} \approx -\sin 2\alpha$. In order to become insensitive to possible New Physics contributions to the mixing amplitude I adopt a different strategy [5]. I use the measurement $\sin\phi_d = 0.736 \pm 0.049$ [15] and write $e^{-i\phi_d} = \pm(1-\sin^2\phi_d)^{1/2} - i\sin\phi_d$, with a sign ambiguity in the real part. (The plus sign is suggested by the standard fit of the unitarity triangle.) A measurement of $S_{\pi\pi}$ can then be translated into a constraint on γ (or $\bar{\rho}$ and $\bar{\eta}$), which remains valid even if the $\sin\phi_d$ measurement is affected by New Physics. The result obtained with the current experimental values and assuming $\cos\phi_d > 0$ is shown in the third plot in Figure 2. The resulting bands for $\cos\phi_d < 0$ are obtained by a reflection about the $\bar{\rho}$ axis. This follows because the expression for $S_{\pi\pi}$ is invariant under the simultaneous replacements $e^{-i\phi_d} \to -e^{i\phi_d}$ and $\gamma \to -\gamma$.

Each of the three constraints in Figure 2 are, at present, limited by rather large experimental errors, while comparison with Figure 1 shows that the theoretical limitations are smaller than for the standard analysis. Yet, even at the present level of accuracy it is interesting to combine the three constraints and construct the resulting allowed regions for the apex of the unitarity triangle. The result is shown in the left-hand plot in Figure 3. Note that the lines corresponding to the new constraints intersect the circles representing the $|V_{ub}|$ constraint at large angles, indicating that the three measurements used in the construction of the CP-b triangle provide highly complementary information on $\bar{\rho}$ and $\bar{\eta}$. There are six (partially overlapping) allowed regions, three corresponding to $\cos\phi_d > 0$ (dark shading) and three to $\cos\phi_d < 0$ (light shading). If we use the information that the measured value of ε_K requires a positive value of $\bar{\eta}$, then only the solutions in the upper half-plane remain. Comparison with Figure 1 shows that one of these regions (corresponding to $\cos\phi_d > 0$) is in perfect agreement with the standard fit. This is highly nontrivial, since with the exception of $|V_{ub}|$ none of the standard constraints are used in this construction. Interestingly, there is a second allowed region (corresponding to $\cos\phi_d < 0$) which would be consistent with the constraint from ε_K but inconsistent

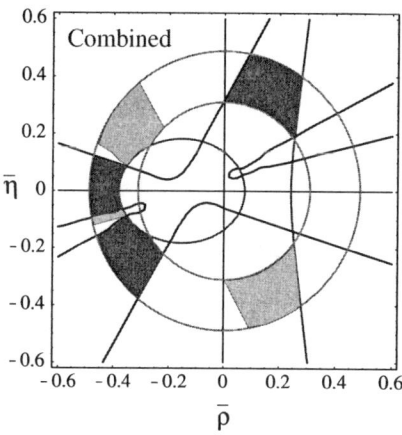
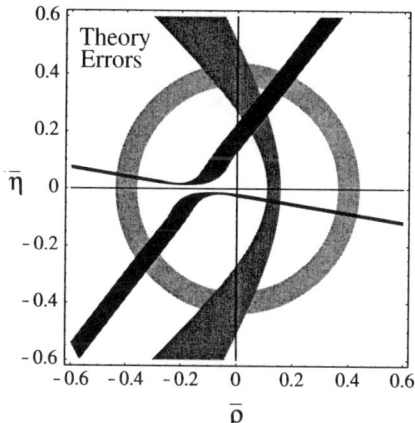

FIGURE 3. Left: Allowed regions in the $(\bar{\rho}, \bar{\eta})$ plane obtained from the construction of the CP-b triangle (at 95% CL). The light-shaded areas refer to $\cos\phi_d < 0$. Right: Theoretical error bands for the three constraints combined in the construction of the CP-b triangle.

with the constraints derived from $\sin 2\beta$ and $\Delta m_s/\Delta m_d$. Such a solution would require a significant New Physics contribution to B–$\bar{B}$ mixing.

Acknowledgment: This research was supported by the National Science Foundation under Grant PHY-0098631.

REFERENCES

1. A. Höcker, H. Lacker, S. Laplace and F. Le Diberder, Eur. Phys. J. C **21**, 225 (2001); for an updated analysis, see: http://www.slac.stanford.edu/laplace/ckmfitter.html.
2. J. D. Bjorken, Nucl. Phys. Proc. Suppl. **11**, 325 (1989).
3. H. D. Politzer and M. B. Wise, Phys. Lett. B **257**, 399 (1991).
4. M. Beneke, G. Buchalla, M. Neubert and C. T. Sachrajda, Phys. Rev. Lett. **83**, 1914 (1999); Nucl. Phys. B **591**, 313 (2000).
5. M. Beneke, G. Buchalla, M. Neubert and C. T. Sachrajda, Nucl. Phys. B **606**, 245 (2001).
6. M. Beneke and M. Neubert, Nucl. Phys. B **651**, 225 (2003); preprint hep-ph/0308039.
7. M. Neubert, preprint hep-ph/0207327.
8. M. Neubert, Phys. Rev. D **49**, 4623 (1994); Phys. Lett. B **543**, 269 (2002).
9. R. D. Dikeman and N. G. Uraltsev, Nucl. Phys. B **509**, 378 (1998); I. I. Bigi, R. D. Dikeman and N. Uraltsev, Eur. Phys. J. C **4**, 453 (1998).
10. A. F. Falk, Z. Ligeti and M. B. Wise, Phys. Lett. B **406**, 225 (1997).
11. C. W. Bauer, Z. Ligeti and M. E. Luke, Phys. Lett. B **479**, 395 (2000); Phys. Rev. D **64**, 113004 (2001).
12. M. Neubert, JHEP **0007**, 022 (2000); M. Neubert and T. Becher, Phys. Lett. B **535**, 127 (2002).
13. M. Neubert and J. L. Rosner, Phys. Lett. B **441**, 403 (1998); Phys. Rev. Lett. **81**, 5076 (1998); M. Neubert, JHEP **9902**, 014 (1999).
14. I use the average of BaBar, Belle, and CLEO data as compiled in the second paper in [6].
15. T. Browder and H. Jawahery, talks presented at the 21th International Symposium on Lepton and Photon Interactions at High Energies, Fermilab, Batavia, Illinois, 11–16 August 2003.

CLEO results on $|V_{cb}|$ and $|V_{ub}|$

Karl M. Ecklund

*Floyd R. Newman Laboratory for Elementary-Particle Physics
Cornell University, Ithaca, New York, 14853*

Abstract. I report results from the CLEO collaboration on semileptonic B decays, highlighting measurements of the Cabibbo-Kobayashi-Maskawa matrix elements $|V_{cb}|$ and $|V_{ub}|$. I describe the techniques used to obtain the recent improvements in precision for these measurements, including the use of the $b \to s\gamma$ photon spectrum to reduce hadronic uncertainties in semileptonic B decays.

The study of semileptonic B meson decays allows measurement of the Cabibbo-Kobayashi-Maskawa (CKM) matrix elements $|V_{cb}|$ and $|V_{ub}|$, providing important inputs to a test of the unitarity of the CKM matrix, which governs the weak charged current and gives rise to CP violation in the standard model. The rate for a b hadron to decay weakly to hadrons containing a c or u quark is proportional to $|V_{cb}|^2$ or $|V_{ub}|^2$ respectively. The absence of final-state interactions in semileptonic decay make the interpretation less dependent on hadronic matrix elements than fully hadronic B decays, although hadronic uncertainties still limit the precision of $|V_{ub}|$ and $|V_{cb}|$ measurements.

The current round of measurements from CLEO continues to test the hadronic calculations needed to disentangle weak matrix elements from strong interaction effects. For decays of B mesons to exclusive final states, the hadronic effects are expressed in terms of a form factor that depends only on the momentum transfer q^2 to the lepton neutrino pair. By measuring decay rates as a function of q^2 we have begun to test the form factors, particularly for $b \to u\ell\bar{\nu}$ transitions. In decays to inclusive final states, under the assumption of parton-hadron duality, quark-level calculations may be compared to inclusive measurements to extract CKM matrix elements. Measurement of spectral distributions in inclusive decays gives additional observables to overconstrain theory parameters and test how well the theory and parton-hadron duality works.

$|V_{cb}|$ Measurements

CLEO has measured $|V_{cb}|$ using the decay $\bar{B} \to D^* \ell \bar{\nu}$ [1], where the decay rate as a function of q^2 is extrapolated to maximum q^2 where the D^* is at rest in the frame of the initial B meson. At this kinematic point the form factor $\mathscr{F}$ is known to 4% of itself, owing to heavy quark symmetry considerations [2]. The differential decay rate is given by $\frac{d\Gamma}{dq^2} = \frac{G_F^2}{48\pi^3}|V_{cb}|^2[\mathscr{F}(q^2)]^2\mathscr{K}(q^2)$, where $\mathscr{K}$ is a known kinematic function. Using $\mathscr{F}(q^2_{\max}) = 0.91 \pm 0.04$ [3], we find $|V_{cb}| = (47.4 \pm 1.4_{\text{stat}} \pm 2.0_{\text{syst}} \pm 2.1_{\mathscr{F}}) \times 10^{-3}$, somewhat higher than other results from $\bar{B} \to D^* \ell \bar{\nu}$. The present world average from

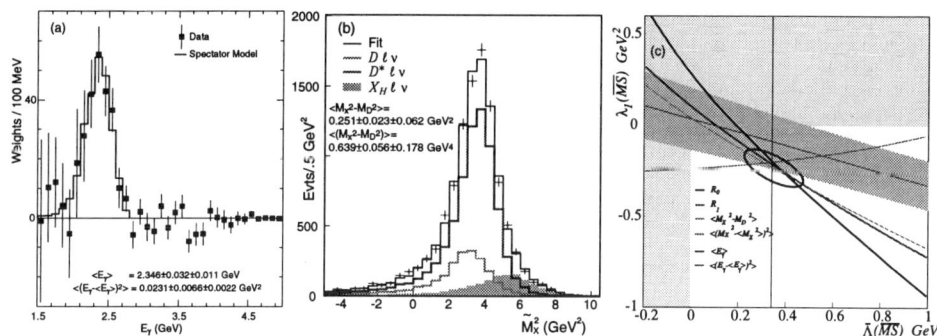

FIGURE 1. $B \to X_s \gamma$ photon spectrum (a), $\bar{B} \to X_c \ell \bar{\nu}\ M_X^2$ spectrum (b), and constraints on HQET parameters (c) from CLEO moment measurements. The shaded band includes $\mathcal{O}(1/M^3)$ theory uncertainties.

$\bar{B} \to D^* \ell \bar{\nu}$ is $|V_{cb}| = (42.4 \pm 1.2_{\text{expt}} \pm 1.9_{\text{theo}}) \times 10^{-3}$ [4].

A measurement of $|V_{cb}|$ using the inclusive semileptonic decay rate is also possible. Here the experimental inputs are the branching fraction for $\bar{B} \to X_c \ell \bar{\nu}$ and the B lifetime. The inclusive decay rate $\Gamma_c^{SL} = \gamma_c |V_{cb}|^2$, where γ_c comes from theory. Within the framework of heavy quark effective theory (HQET) [5], the inclusive semileptonic decay rate is expanded in a double series in α_s^n and $1/M^n$, where M is the heavy quark mass. Hadronic effects enter both in the perturbative expansion and as expansion parameters, matrix elements of non-perturbative QCD operators. At $\mathcal{O}(1/M^2)$ there are two parameters: λ_1, which is proportional to the kinetic energy of the b quark in the B meson, and λ_2, which comes from the chromomagnetic operator. An additional parameter $\bar{\Lambda}$ relates the B meson mass to the b quark mass. From the B-B^* mass difference $\lambda_2 = 0.128 \pm 0.010$ GeV^2. The other parameters can be estimated (e.g. in quark models) but they can also be measured using spectral moments in inclusive B decay. Moments, e.g. of the lepton energy spectrum, are also computed in HQET, allowing extraction of λ_1 and $\bar{\Lambda}$.

CLEO has a preliminary measurement of the inclusive semileptonic branching fraction using a high-momentum ($p > 1.5$ GeV/c) lepton tag. The analysis is an update of Ref. [6]. The tag identifies a sample of B decays with high purity (98%). Additional electrons may come from the decay chain of the same B or from the decay of the other B meson in the event ($e^+ e^- \to \Upsilon(4S) \to B\bar{B}$). Secondary leptons ($b \to c \to \ell$) and primary leptons are separated using kinematic and charge correlations, with a known correction from B^0-$\bar{B}^0$ mixing. The new semileptonic branching fraction is $10.88 \pm 0.08 \pm 0.33\%$. The spectrum of electrons above 600 MeV is also obtained, from which spectral moments will be measured.

CLEO has recently measured spectral moments in inclusive semileptonic decay and in $B \to X_s \gamma$. These are used to extract HQET parameters and reduce the theoretical uncertainty in inclusive $|V_{cb}|$ measurements. CLEO measured the $B \to X_s \gamma$ photon spectrum and moments (Fig. 1a) in [7]. In [8], CLEO measured the moments of the hadronic mass distribution in $\bar{B} \to X_c \ell \bar{\nu}$ decays (Fig. 1b). Combining the constraints on λ_1 and $\bar{\Lambda}$ from the first moments of the photon energy and hadronic mass spectra, we obtain a solution for λ_1 and $\bar{\Lambda}$ and extract $|V_{cb}| = (41.1 \pm 0.5_{\lambda_1, \bar{\Lambda}} \pm 0.7_\Gamma \pm 0.8_{HQET}) \times 10^{-3}$ using the new

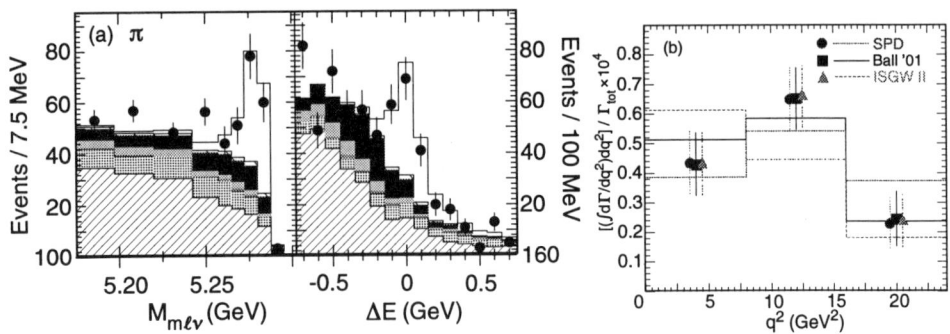

FIGURE 2. Exclusive $\bar{B} \to \pi \ell \bar{\nu}$: (a) projections of ML fit to $M_{m\ell\nu}$ and ΔE and (b) fit to $d\Gamma/dq^2$.

CLEO branching fraction and PDG2003 lifetime average as inputs. The uncertainties from unknown $\mathcal{O}(1/M^3)$ HQET parameters are dominant.

The lepton energy moments in $\bar{B} \to X_c \ell \bar{\nu}$ are also sensitive to the HQET parameters, and CLEO has measured the lepton spectrum [9] and moments [10] above 1.5 GeV. From all of the moment measurements, one can assemble the constraints on the HQET parameters λ_1 and $\bar{\Lambda}$. Figure 1c shows the remarkable consistency of these measurements, lending credibility to the inclusive $|V_{cb}|$ measurement.

At present the inclusive $|V_{cb}|$ measurement is more precise (3%) than that from $\bar{B} \to D^* \ell \bar{\nu}$, but with reliance on HQET for hadronic corrections. The first tests of HQET using spectral moments in inclusive B decays give us some confidence in the method, but additional tests with more inclusive moments are needed.

The agreement between inclusive and exclusive measurements is another test of our control of hadronic corrections. There is good agreement between inclusive and the world average exclusive $|V_{cb}|$ measurements, but CLEO's exclusive $|V_{cb}|$ is larger than the inclusive measurement and other measurements using $\bar{B} \to D^* \ell \bar{\nu}$.

$|V_{ub}|$ Measurements

Measurements of $b \to u \ell \bar{\nu}$ have to contend with a 50–100 times larger background from $b \to c \ell \bar{\nu}$. Requiring a lepton energy above the endpoint for $b \to c \ell \bar{\nu}$ (≈ 2.3 GeV) is the easiest strategy to reduce background, but this cut near the edge of the spectrum introduces sensitivity to the motion of b quark in the B meson. The sensitivity is reduced by using the $b \to s\gamma$ photon spectrum [7], which is sensitive to the same hadronic effects at leading order [11, 12, 13, 14].

CLEO measured the lepton spectrum from B decays in the endpoint region $E > 2.2$ GeV and extracted a partial branching fraction of $(2.30 \pm 0.15 \pm 0.35) \times 10^{-4}$ [15]. From the $b \to s\gamma$ photon spectrum, the fraction of $b \to u \ell \bar{\nu}$ events passing the lepton energy cut is $f_u = 0.130 \pm 0.024 \pm 0.015$. This gives $|V_{ub}| = (4.08 \pm 0.34_{\text{exp}} \pm 0.44_{f_u} \pm 0.16_\Gamma \pm 0.24_{NLO}) \times 10^{-3}$, where the theoretical uncertainties are Γ, from [16, 17], and *NLO*, from sub-leading terms relating hadronic effects in $b \to u \ell \bar{\nu}$ and $b \to s\gamma$.

CLEO has also measured $|V_{ub}|$ in the exclusive modes $\bar{B} \to [\pi/\rho/\omega/\eta]\ell\bar{\nu}$ [18], where kinematics from full reconstruction of the final state gives the needed suppression of $b \to c\ell\bar{\nu}$. The neutrino is reconstructed from the missing energy and momentum of the event, taking advantage of CLEO's large solid angle (95%). Combined with a lepton and light meson candidate, energy and momentum conservation leads to signal peaks in $\Delta E = E - E_{\text{beam}}$ and $M_{m\ell\nu}$, with $S/B \approx 1$. We perform a simultaneous maximum likelihood fit in ΔE and $M_{m\ell\nu}$ to seven sub-modes. Signals for π (Fig. 2a) and ρ are extracted separately in three q^2 bins. Given form factors from theory, we extract $|V_{ub}|$ from a fit to $d\Gamma/dq^2$ (Fig. 2b). Combining $\bar{B} \to \pi\ell\bar{\nu}$ and $\bar{B} \to \rho\ell\bar{\nu}$ results we find $|V_{ub}| = (3.17 \pm 0.17|_{\text{stat}} {}^{+0.16}_{-0.17}|_{\text{syst}} {}^{+0.53}_{-0.39}|_{\text{theo}} \pm 0.03|_{\text{FF}}) \times 10^{-3}$. This result uses form factors from Lattice QCD ($q^2 > 16$ GeV^2) and light cone sum rules ($q^2 > 16$ GeV^2) where each are most reliable. In a test of $\bar{B} \to \pi\ell\bar{\nu}$ form factors, ISGW2 [19] is disfavored (Fig. 2b).

We find good agreement between measurements of $|V_{ub}|$ using inclusive and exclusive techniques. The theoretical uncertainty on the form factor normalization currently limits the precision of the exclusive $|V_{ub}|$ measurement. In the future, unquenched Lattice QCD calculations can improve the $\bar{B} \to \pi\ell\bar{\nu}$ form factor in a limited region of q^2. The inclusive $b \to u\ell\bar{\nu}$ measurement can be further improved with increased $b \to s\gamma$ statistics and better phenomenological understanding of non-perturbative shape functions for the B meson [20, 21, 22]. Comparison between inclusive measurements that use different kinematic cuts (more inclusive and away from the endpoint region) will increase our confidence in inclusive $|V_{ub}|$ measurements. Since the principal background comes from $b \to c\ell\bar{\nu}$, better knowledge of the dominant semileptonic B decays will improve systematic errors for both inclusive and exclusive measurements.

REFERENCES

1. Adam, N. E., et al., *Phys. Rev.*, **D67**, 032001 (2003).
2. Falk, A. F., and Neubert, M., *Phys. Rev.*, **D47**, 2965–2981 (1993).
3. Hashimoto, S., et al., *Phys. Rev.*, **D66**, 014503 (2002).
4. Heavy Flavor Averaging Group, Updates of semileptonic results (2003), for summer conferences.
5. Manohar, A. V., and Wise, M. B., *Heavy quark physics*, Cambridge University Press, 2000, chap. 4.
6. Barish, B., et al., *Phys. Rev. Lett.*, **76**, 1570–1574 (1996).
7. Chen, S., et al., *Phys. Rev. Lett.*, **87**, 251807 (2001).
8. Cronin-Hennessy, D., et al., *Phys. Rev. Lett.*, **87**, 251808 (2001).
9. Mahmood, A. H., et al., *Phys. Rev.*, **D67**, 072001 (2003).
10. Gremm, M., Kapustin, A., Ligeti, Z., and Wise, M. B., *Phys. Rev. Lett.*, **77**, 20–23 (1996).
11. Neubert, M., *Phys. Rev.*, **D49**, 3392–3398 (1994).
12. Neubert, M., *Phys. Rev.*, **D49**, 4623–4633 (1994).
13. Bigi, I. I. Y., et al., *Int. J. Mod. Phys.*, **A9**, 2467–2504 (1994).
14. Leibovich, A. K., Low, I., and Rothstein, I. Z., *Phys. Rev.*, **D61**, 053006 (2000).
15. Bornheim, A., et al., *Phys. Rev. Lett.*, **88**, 231803 (2002).
16. Hoang, A. H., Ligeti, Z., and Manohar, A. V., *Phys. Rev.*, **D59**, 074017 (1999).
17. Uraltsev, N., *Int. J. Mod. Phys.*, **A14**, 4641–4652 (1999).
18. Athar, S. B., et al., *Phys. Rev.*, **D** (2003), hep-ex/0304019, to appear.
19. Scora, D., and Isgur, N., *Phys. Rev.*, **D52**, 2783–2812 (1995).
20. Leibovich, A. K., Ligeti, Z., and Wise, M. B., *Phys. Lett.*, **B539**, 242–248 (2002).
21. Bauer, C. W., Luke, M., and Mannel, T., *Phys. Lett.*, **B543**, 261–268 (2002).
22. Neubert, M., *Phys. Lett.*, **B543**, 269–275 (2002).

Charged to neutral B meson yield ratio across the $\Upsilon(4S)$ resonance

M.B. Voloshin

William I. Fine Theoretical Physics Institute, University of Minnesota, Minneapolis, MN 55455
and
Institute of Theoretical and Experimental Physics, Moscow, 117259

Abstract. It is shown that the relative yield of the pairs B^+B^- and $B^0\overline{B}^0$ should substantially and rapidly change with energy within the width of the $\Upsilon(4S)$ resonance, crossing the value of one near the center of the resonance. This behavior is due to an interference of the rapidly changing with energy Breit-Wigner phase with the phase introduced in the wave function of charged mesons by their Coulomb interaction.

The ratio of the production rates of charged and neutral B mesons in e^+e^- annihilation at the $\Upsilon(4S)$ resonance,

$$R^{c/n} = 1 + \delta R^{c/n} = \frac{\sigma(e^+e^- \to B^+B^-)}{\sigma(e^+e^- \to B^0\overline{B}^0)}, \quad (1)$$

is an important parameter in detailed studies of the properties of B mesons. Recent dedicated measurements[1, 2, 3] of $R^{c/n}$ at the maximum of the resonance report values ranging from $1.04 \pm 0.07 \pm 0.04$ [1] to $1.10 \pm 0.06 \pm 0.05$ [2], which leave enough room for further studies of the quantity of interest $\delta R^{c/n}$.

Theoretically the difference $\delta R^{c/n}$ of the discussed ratio from one arises as dominantly an effect of the Coulomb interaction, clearly different for charged and neutral B mesons, since the mass difference $m_{B^0} - m_{B^+} = 0.33 \pm 0.28\,MeV$ [4] is quite small, and its effect can be accounted separately. In the most simple approach[5], where the B mesons are treated as point particles, and the existence of the resonant interaction is ignored, the estimate of $\delta R^{c/n}$ can be expressed in terms of the c.m. velocity v of produced B mesons, using the textbook Coulomb wave functions:

$$\delta R^{c/n} = \frac{\pi\alpha}{2v} + O\left(\frac{\alpha^2}{v^2}\right). \quad (2)$$

At the excitation energy of the $\Upsilon(4S)$ resonance, $E_0 = M_{\Upsilon(4S)} - 2m_B \approx 20\,MeV$, one has $v \approx 0.06$, and the simple estimate (2) would yield $\delta R^{c/n} \approx 0.19$. It was subsequently argued[6] that the estimate of $\delta R^{c/n}$ can in fact be substantially reduced from eq.(2) if one accounts for a finite size of the B mesons (through their electromagnetic form factor) and also for the finite size of the $\Upsilon(4S)$. The latter effect was further discussed in a specific model of heavy quarkonium[7]. Recently the problem of calculation of the ratio $R^{c/n}$ was revisited[8] in the context of a chiral-type model for strong interaction of B mesons at short distances, including the $B^*B\pi$ vertex and the coupled channels with pairs of pseudoscalar and/or vector mesons, although still considering all the mesons as point-like with respect to the Coulomb interaction.

In all previous theoretical studies of the ratio $R^{c/n}$ the $\Upsilon(4S)$ was treated as a 'weak' resonance in the sense that it produced only a small perturbation in the P-wave wave function of the B meson pairs otherwise dominated by a non-resonant part. For this reason the results predicted a smooth behavior of $\delta R^{c/n}$ with energy in the region of the $\Upsilon(4S)$ resonance. There however are sound reasons to believe that such picture is not adequate for the actual $\Upsilon(4S)$ and that the behavior of the wavefunctions of the B mesons in the P wave is in fact *dominated by the resonance* at energies comparable to the excitation energy E_0. Indeed, the experimental width $\Gamma \approx 14\,MeV$ of the $\Upsilon(4S)$ corresponds to the value of its coupling g to the B meson channel (defined in the standard way) as large as $g^2/4\pi \approx 30$. For this reason the present analysis is done using the approach, where the wave function of the B meson pairs is described by only the Breit-Wigner resonant part, which is appropriate in the limit $g^2 \gg 1$.

The standard physical picture for considering the scattering in the resonance region (c.f. Ref.[9]) is that the strong interaction, responsible for the existence of the resonance has a short range a. At distances larger than a the motion is described by a known potential $V(r)$: either $V(r) = 0$, or a Coulomb potential (with a possible modification due to form factor at short distances), where the wave function of the scattering state can be found explicitly from the Schrödinger equation. The boundary (matching) conditions at $r \approx a$ for the 'outer' wave function are related to the measurable parameters of the resonance. Thus the 'outer' wave function at $r > a$ is described by the spherical wave with $L = 1$, whose radial part can be written as $R(r) = \chi(r)/r$, with a separate function $\chi(r)$ for each of the channels: $\chi_n(r)$ (for $B^0\bar{B}^0$) and $\chi_c(r)$ (for B^+B^-), each satisfying at energy $E = p^2/m$ the corresponding one-dimensional Schrödinger equation

$$\chi_n'' + \left(p^2 - \frac{2}{r^2}\right)\chi_n = 0, \quad \chi_c'' + \left(p^2 + m\frac{\alpha}{r} - \frac{2}{r^2}\right)\chi_c = 0, \qquad (3)$$

where the prime denotes derivative over r, and $m = m_B \approx 5280\,MeV$.

The coupling between the "neutral" and the "charged" channels takes place in the region of strong interaction at short distances. At those distances the isospin symmetry applies, so that the functions χ_n and χ_c evolve from one and the same function at a certain short distance $r = a$, i.e. that

$$\chi_c(a) = \chi_n(a) \quad \text{and} \quad \chi_c'(a) = \chi_n'(a), \qquad (4)$$

which boundary conditions can be viewed as our formal definition of the short distance parameter a.

According to the standard Breit-Wigner description of a resonance scattering[9] at energy E near the position E_0 of the resonance, the relevant 'outer' solution of the Schrödinger equations (3) for stationary wave functions has the form

$$\begin{aligned}\chi_n(r) &= (\Delta - i\gamma)\,b_n f_n(r) + (\Delta + i\gamma)\,b_n^* f_n^*(r), \\ \chi_c(r) &= (\Delta - i\gamma)\,b_c f_c(r) + (\Delta + i\gamma)\,b_c^* f_c^*(r),\end{aligned} \qquad (5)$$

where $\Delta = E - E_0$, $\gamma = \Gamma/2$ and the complex coefficients $b_{n(c)}$ are generally functions of the energy, which however have no zeros at $\Delta = i\gamma$. Finally, each of the functions f_n and f_c is the solution of the corresponding equation in (3), which contains only the outgoing wave, i.e. at $r \to \infty$ they contain only the factor $\exp(ipr)$ (while their complex conjugates $f^*_{n(c)}$ contain only the incoming wave factor $\exp(-ipr)$).

The function $f_n(r)$ specified by this condition is well known for the free motion with $L = 1$,

$$f_n(r) = \left(1 + \frac{i}{pr}\right) e^{ipr}, \qquad (6)$$

and with this condition for its phase, the phase of the coefficient b_n coincides with the non-resonant scattering phase δ_1 at $L = 1$: $\exp(2i\delta_1) = \frac{b_n}{b_n^*}$. The corresponding function $f_c(r)$ for the motion in the Coulomb potential is also well known (see e.g. in Ref.[9]), however for our present purpose it would be more convenient to make use of the perturbation theory in the Coulomb interaction, rather than to do an expansion of the explicit expression. The function $f_c(r)$ in eq.(5) can be chosen to exactly coincide (both in phase and in normalization) with $f_n(r)$ at asymptotically large distances. Under these conventions the ratio $R^{c/n}$ is determined by the coefficients b as $R^{c/n} = |b_n|^2/|b_c|^2$.

The latter ratio of the coefficents can be readily found from the solution of the boundary conditions (4), and the final result readrs as[10]

$$\delta R^{c/n} = -\frac{\alpha}{v} \operatorname{Im}\left[e^{2i\delta_{BW}+2i\delta_1} \int_a^\infty e^{2ipr}\left(1+\frac{i}{pr}\right)^2 \frac{dr}{r}\right] \qquad (7)$$

$$= \frac{\alpha}{v}\left[\frac{\Delta^2-\gamma^2}{\Delta^2+\gamma^2}(A\cos 2\delta_1 + B\sin 2\delta_1) - \frac{2\gamma\Delta}{\Delta^2+\gamma^2}(B\cos 2\delta_1 - A\sin 2\delta_1)\right],$$

where in the latter expression the coefficients A and B are given (with the oposite sign) by respectively the imaginary and the real part of the integral with complex exponent:

$$A = -\int_{pa}^\infty \left[\left(1-\frac{1}{u^2}\right)\sin 2u + \frac{2\cos 2u}{u}\right]\frac{du}{u},$$

$$B = \int_{pa}^\infty \left[\frac{2\sin 2u}{u} - \left(1-\frac{1}{u^2}\right)\cos 2u\right]\frac{du}{u}. \qquad (8)$$

At small values of the product pa the coefficients A and B have the expansion:

$$A = \frac{\pi}{2} - \frac{2pa}{3} + O(p^3 a^3), \quad B = \frac{1}{2p^2 a^2} - \ln 2pa - \gamma_E + 1 + O(p^4 a^4), \qquad (9)$$

where $\gamma_E = 0.577\ldots$ is the Euler constant.

A qualitative illustration of the expected variation of $R^{c/n}$ in the resonance region is provided by the plots in Fig.1. The curves in the plots are calculated with various values of a and δ_1 under the following assumptions: only the leading terms in the expansion of A and B at small pa, explicitly shown in eq.(9), are retained, the width parameter is parameterized as $\gamma = (\Gamma/2)(p/p_0)^3$, and the non-resonant phase as $\delta_1 = \delta_1(E_0)(p/p_0)^3$.

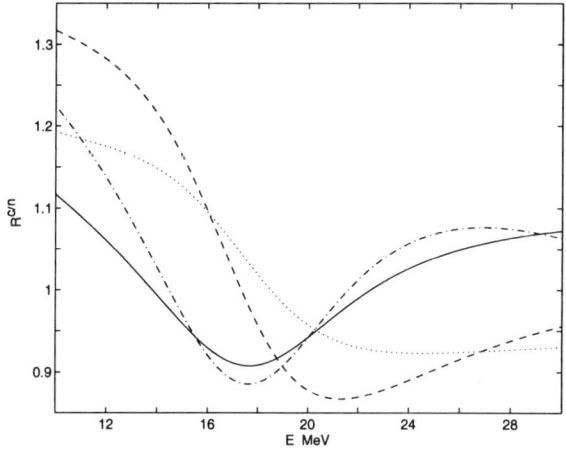

FIGURE 1. The dependence of the ratio $R^{c/n}$ on the excitation energy $E = \sqrt{s} - 2m_B$ in the region of the $\Upsilon(4S)$ resonance (the center position is assumed to be at $E_0 = 20\,MeV$) for some values of a and $\delta_1(E_0)$: $a^{-1} = 200\,MeV$, $\delta_1(E_0) = 0$ (solid), $a^{-1} = 400\,MeV$, $\delta_1(E_0) = 0$ (dashed), $a^{-1} = 300\,MeV$, $\delta_1(E_0) = 30^0$ (dashdot), and $a^{-1} = 300\,MeV$, $\delta_1(E_0) = -30^0$ (dotted).

(It should be mentioned that the range of $\delta_1(E_0)$ from -30^0 to $+30^0$ most likely is unrealistically broad, and is used here for an illustration of the effect of the phase under extreme assumptions.) As expected on general grounds from eq.(7), the very fact of a substantial and rapid variation of $\delta R^{c/n}$ within the resonance width stays robust under assumptions about the presently unknown parameters.

ACKNOWLEDGMENTS

A part of this text was prepared at the Aspen Center for Physics. This work is supported in part by the DOE grant DE-FG02-94ER40823.

REFERENCES

1. J.P. Alexander et. al. [CLEO Collaboration], *Phys. Rev. Lett.* **86**, 2737 (2001).
2. B. Aubert et. al. [BABAR Collaboration], *Phys. Rev.* **D65**, 032001 (2002).
3. S.B. Athar et. al. [CLEO Collaboration], *Phys. Rev.* **D66**, 052003 (2002).
4. K. Hagivara et.al (Particle Data Group), *Phys. Rev.* **D66**, 010001 (2002).
5. D. Atwood and W.J. Marciano, *Phys. Rev.* **D41**, 1736 (1990).
6. G.P. Lepage, *Phys. Rev.* **D42**, 3251 (1990).
7. N. Byers and E. Eichten, *Phys. Rev.* **D42**, 3885 (1990).
8. R. Kaiser, A.V. Manohar, and T. Mehen, Report hep-ph/0208194, Aug. 2002 (unpublished)
9. L.D. Landau and E.M. Lifshits, *Quantum Mechanics (Non-relativistic Theory)*, Third Edition, Pergamon, Oxford, 1977.
10. M.B. Voloshin *Mod. Phys. Lett. A* **18**, 1783 (2003)

Rare hadronic B decays: probing deeper into the Standard Model

A. J. Schwartz

Physics Department, University of Cincinnati, Cincinnati, Ohio 45221

Abstract. We present recent results from the Belle experiment on rare hadronic B meson decays. The results are based on a 78 fb^{-1} data sample and consist of branching fractions, CP asymmetries, and polarization amplitudes. The decays studied include two-body pseudoscalar-pseudoscalar final states ($B \to \pi\pi$, $K\pi$, KK, and $D^0 K^{\pm}$); pseudoscalar-vector final states ($B \to \omega\pi$, ωK, and ϕK); and vector-vector final states ($B \to \phi K^*$ and $\rho^+\rho^0$).

INTRODUCTION

Rare hadronic B decays are useful for probing physics beyond the Standard Model. Their amplitudes usually contain internal loops, which are sensitive to mass scales that cannot be accessed directly. Here we present recent measurements of such decays by the Belle experiment at KEK. This experiment runs at the KEKB asymmetric e^+e^- collider, which has a center-of-mass (CM) energy near the $\Upsilon(4S)$ resonance. The results are from 78 fb^{-1} of data, which corresponds to 85×10^6 $B\bar{B}$ pairs produced.

The Belle detector consists of a three-layer silicon vertex detector, a 50-layer central drift chamber (CDC) for charged-particle tracking, an array of silica aerogel threshold Čherenkov counters (ACC), time-of-flight scintillation counters (TOF), a CsI(Tl) electromagnetic calorimeter (ECL), and a superconducting solenoid providing a 1.5 T magnetic field. An iron flux-return located outside the coil is instrumented with resistive plate chambers to identify muons and K^0_L's. For details of the detector, see Ref. [1].

Charged tracks are identified as kaons or pions by the number of photoelectrons detected in the ACC, the specific ionization energy loss (dE/dx) in the CDC, and, if slow enough, their time-of-flight. This information is used to calculate kaon and pion relative likelihoods $\mathcal{L}_K$ and $\mathcal{L}_\pi$. Tracks are identified as pions or kaons based on the likelihood ratios $R_{K,\pi} \equiv \mathcal{L}_{K,\pi}/(\mathcal{L}_\pi + \mathcal{L}_K)$. The efficiency for kaons is typically 84% with a pion misidentification rate of 5%; the efficiency for pions is typically 91% with a kaon misidentification rate of 10%.

For most final states there is substantial background from $e^+e^- \to q\bar{q}$ continuum events ($q = u, d, s, c$). We distinguish this background from B decays by first combining five modified Fox-Wolfram moments into a Fisher discriminant. This is then combined with a likelihood function for θ_B, the polar angle of the B meson flight direction, and the resulting likelihood function is used to form a likelihood ratio $R_{q\bar{q}} \equiv \mathcal{L}_{sig}/(\mathcal{L}_{sig} + \mathcal{L}_{q\bar{q}})$. A mode-dependent cut on $R_{q\bar{q}}$ is made to significantly reduce background events. Typical cut values reject $\gtrsim 90\%$ of $q\bar{q}$ background with a signal efficiency of 40–70%.

The analyses presented here proceed in three steps: *(a)* selecting the final state of interest using $R_{K,\pi}$ to identify tracks as pions or kaons; *(b)* using $R_{q\bar{q}}$ to reject continuum background; and *(c)* selecting B decays by cutting on the variables $m_{bc} \equiv \sqrt{E^{*2}_{beam} - p^{*2}_B}$ and $\Delta E \equiv E^*_B - E^*_{beam}$, where E^*_{beam} denotes the beam energy and p^*_B and E^*_B denote the reconstructed momentum and energy of the candidate B meson, all evaluated in the e^+e^- CM frame. For correctly-identified B decays, $m_{bc} = M_B$ and $\Delta E = 0$. Throughout this paper, charge-conjugate modes are included unless stated otherwise. When two errors are listed for a measurement, the first one is statistical and the second one systematic.

$B \to \pi\pi/K\pi/KK$ DECAYS

These decays proceed via $b \to u$ tree and $b \to d,s$ loop diagrams, and the final states include both charged and neutral kaons and pions. Neutral kaons are identified via $K^0_S \to \pi^+\pi^-$, and neutral pions are identified via $\pi^0 \to \gamma\gamma$ (where the photons produce clusters in the ECL). The ΔE distributions after all selection cuts and a cut $5.27 < m_{bc} < 5.29$ GeV/c^2 are shown in Fig. 1. The event yields are obtained by fitting these distributions for signal, $q\bar{q}$ background, and other charmless B decay background. Possible reflections due to $K^\pm/\pi^\pm$ misidentification are included where applicable. All fit parameters other than the normalizations are fixed: most are determined from Monte Carlo (MC) simulation, while others are determined directly from the data, usually from events in a lower m_{bc} sideband. The resulting event yields are listed in Table 1. The statistical significance of signals is calculated as $\mathscr{S} = \sqrt{2\ln(L_{max}/L_0)}$, where L_0 is the likelihood obtained assuming no signal events and L_{max} is the (maximum) likelihood obtained with N_s signal events. For cases where no significant signal is observed, we quote a 90% C.L. upper limit using a Feldman-Cousins frequentist approach [2]. The corresponding branching fractions are also listed in Table 1 and show the theoretically-expected hierarchy: $B(B \to K\pi) > B(B \to \pi\pi) > B(B \to K\bar{K})$.

The branching fractions can be used to constrain the magnitudes of the CKM phases ϕ_2 and ϕ_3 [3]. For such constraints, ratios of partial widths are most useful because of their reduced hadronic uncertainties. We thus use the results in Table 1 and the lifetime ratio $\tau_{B^+}/\tau_{B^0} = 1.083 \pm 0.017$ [4] to calculate the partial width ratios listed in Table 2. The fact that $\Gamma(\pi^+\pi^-)/2\Gamma(\pi^+\pi^0) \neq 1$ implies that the penguin contribution to $B^0 \to \pi^+\pi^-$ is significant.

For the flavor-specific decays $B \to K^\pm\pi^\mp, K^\pm\pi^0, K^0\pi^\pm$, and $\pi^\pm\pi^0$, the ΔE distributions are fitted separately for B and $\bar{B}$ candidates to measure the CP asymmetry $A_{CP} \equiv [N(\bar{B} \to \bar{f}) - N(B \to f)] / [N(\bar{B} \to \bar{f}) + N(B \to f)]$, where $B(\bar{B})$ represents B^0 or B^+ ($\bar{B}^0$ or B^-). The results are listed in Table 3; no significant CP asymmetries are observed.

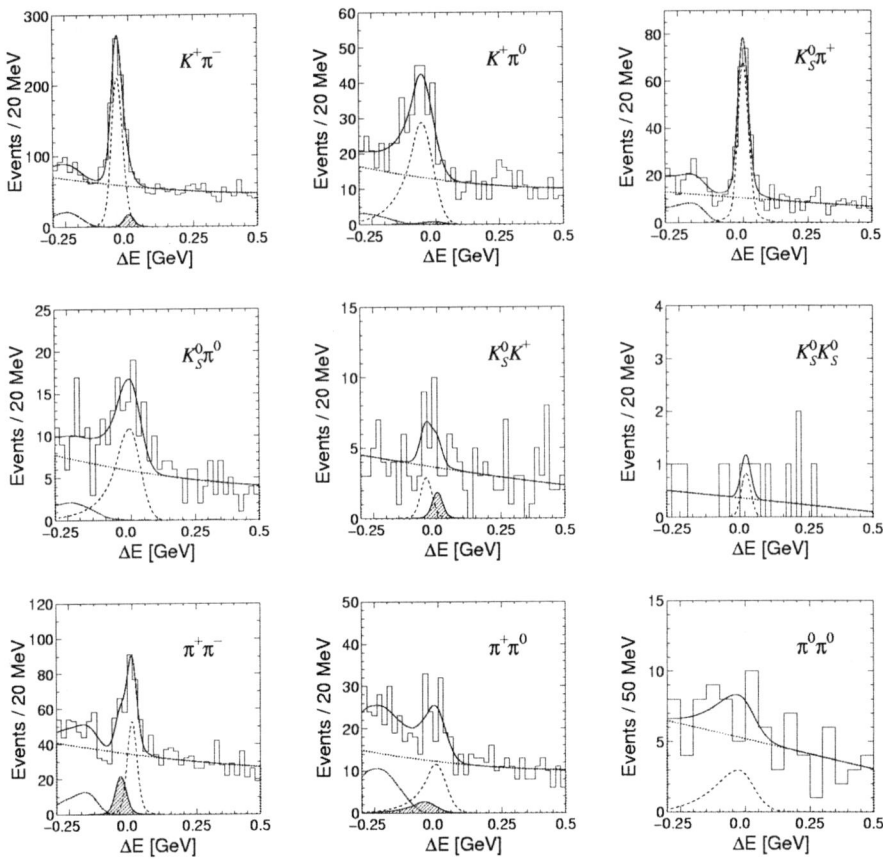

FIGURE 1. $B \to hh$ ΔE distributions for $5.27 < m_{bc} < 5.29$ GeV/c^2. The fit results are shown as the solid, dashed, dotted, and dash-dotted curves for the total, signal, $q\bar{q}$ background, and $B\bar{B}$ background, respectively. The hatched area indicates reflections resulting from $\pi^{\pm} \to K^{\pm}$ misidentification. All tracks are assigned the pion mass; this produces the shift of signal modes containing $K^{\pm}$'s towards $-\Delta E$ values.

$B^{\pm} \to D_{CP}K^{\pm}$ DECAYS

The decay $B^{\pm} \to D_{CP}K^{\pm}$, where D_{CP} represents a D^0 decaying to a CP eigenstate, proceeds via $b \to c$ and $b \to u$ transitions as shown in Fig. 2. Interference between the amplitudes gives rise to direct CP violation, and measuring A_{CP} allows one to constrain the CKM phase ϕ_3. The observables are [6]:

$$A_{1,2} \equiv \frac{B(B^- \to D_{1,2}K^-) - B(B^+ \to D_{1,2}K^+)}{B(B^- \to D_{1,2}K^-) + B(B^+ \to D_{1,2}K^+)} = \frac{2r\sin\delta' \sin\phi_3}{1 + r^2 + 2r\cos\delta' \cos\phi_3}$$

$$\mathscr{R}_{1,2} \equiv \frac{R^{D^{1,2}}}{R^{D^0}} = 1 + r^2 + 2r\cos\delta' \cos\phi_3, \qquad (1)$$

TABLE 1. Event yields, signal significance, efficiencies, and branching fractions (90% C.L. upper limits) for $B \to hh$ decays.

Mode	N_s	$\mathcal{S}$	ε (%)	$B \times 10^6$ (90% C.L. limit)
$\pi^+\pi^-$	133^{+19}_{-18}	8.5	35.2	$4.4 \pm 0.6 \pm 0.3$
$\pi^+\pi^0$	72.4 ± 17.4	4.5	16.1	$5.3 \pm 1.3 \pm 0.5$
$\pi^0\pi^0$	$12.0^{+9.1}_{-8.6}$	1.9	7.8	$1.8^{+1.4}_{-1.3}{}^{+0.5}_{-0.7}$ (< 4.4)
$K^+\pi^-$	596 ± 33	24.1	37.9	$18.5 \pm 1.0 \pm 0.7$
$K^+\pi^0$	199 ± 22	10.8	18.3	$12.8 \pm 1.4^{+1.4}_{-1.0}$
$K^0\pi^+$	187 ± 16	16.4	10.0	$22.0 \pm 1.9 \pm 1.1$
$K^0\pi^0$	72.6 ± 14.0	5.8	6.8	$12.6 \pm 2.4 \pm 1.4$
K^+K^-	$-1.0^{+6.6}_{-5.9}$	—	20.1	< 0.7
$K^+\overline{K^0}$	8.6 ± 5.9	1.6	5.9	$1.7 \pm 1.2 \pm 0.1$ (< 3.4)
$K^0\overline{K^0}$	2.0 ± 1.9	1.3	2.9	$0.8 \pm 0.8 \pm 0.1$ (< 3.2)

TABLE 2. Partial width ratios for $B \to hh$.

Ratio	Measured Value
$\Gamma(\pi^+\pi^-)/2\Gamma(\pi^+\pi^0)$	$0.45 \pm 0.13 \pm 0.05$
$\Gamma(\pi^+\pi^-)/\Gamma(K^+\pi^-)$	$0.24 \pm 0.04 \pm 0.02$
$\Gamma(\pi^0\pi^0)/\Gamma(\pi^+\pi^0)$	< 0.92 @ 90% C.L.
$2\Gamma(K^+\pi^0)/\Gamma(K^0\pi^+)$	$1.16 \pm 0.16^{+0.14}_{-0.11}$
$\Gamma(K^+\pi^-)/\Gamma(K^0\pi^+)$	$0.91 \pm 0.09 \pm 0.06$
$\Gamma(K^+\pi^-)/2\Gamma(K^0\pi^0)$	$0.74 \pm 0.15 \pm 0.09$

where $\delta' = \delta \, (\delta + \pi)$ for D_1 (D_2) and the ratios $R^{D^{1,2}}$ and R^{D^0} are:

$$R^{D^{1,2}} = \frac{B(B^- \to D_{1,2}K^-) + B(B^+ \to D_{1,2}K^+)}{B(B^- \to D_{1,2}\pi^-) + B(B^+ \to D_{1,2}\pi^+)}$$

$$R^{D^0} = \frac{B(B^- \to D^0 K^-) + B(B^+ \to \overline{D^0} K^+)}{B(B^- \to D^0 \pi^-) + B(B^+ \to \overline{D^0} \pi^+)}.$$

TABLE 3. CP asymmetries for $B \to hh$. For $B^0 \to \pi^+\pi^-$, see [5] (t-dependent analysis).

Mode	$N_s(\overline{B})$	$N_s(B)$	A_{CP}	90% C.L. Interval
$\pi^+\pi^0$	31.2 ± 11.9	41.3 ± 12.7	$-0.14 \pm 0.24^{+0.05}_{-0.04}$	$(-0.57, 0.30)$
$K^+\pi^-$	$235.4^{+19.8}_{-19.1}$	$270.2^{+19.7}_{-18.9}$	$-0.07 \pm 0.06 \pm 0.01$	$(-0.18, 0.04)$
$K^+\pi^0$	122.0 ± 15.8	76.5 ± 14.5	$0.23 \pm 0.11^{+0.01}_{-0.04}$	$(-0.01, 0.42)$
$K^0\pi^+$	$119.1^{+13.8}_{-13.1}$	$104.4^{+13.2}_{-12.5}$	$0.07^{+0.09}_{-0.08}{}^{+0.01}_{-0.03}$	$(-0.10, 0.22)$

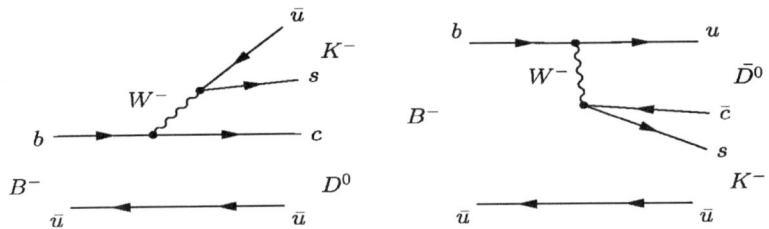

FIGURE 2. Feynman diagrams for $B^{\pm} \to D_{CP}K^{\pm}$: $b \to c$ tree (left) and $b \to u$ tree (right).

In these expressions, D_1 and D_2 are CP-even and CP-odd eigenstates, respectively, of the neutral D^0 meson; r is the ratio of the $b \to u$ and $b \to c$ amplitudes shown in Fig. 2; and δ is their strong phase difference. The ratio r is expected to be only ~ 0.1 due to a CKM suppression factor and a color suppression factor. The ratio R^{D^0} has been previously measured by Belle ($0.079 \pm 0.009 \pm 0.006$ [7]) and CLEO ($0.099^{+0.014}_{-0.012}{}^{+0.007}_{-0.006}$ [8]). The results are in agreement with naive factorization: $\tan^2\theta_C (f_K/f_\pi)^2 \approx 0.074$. Here we present new results for R^{D^0}, $R^{D^{1,2}}$, and $A_{1,2}$; these values can be inserted into Eq. (1) to determine the three unknowns r, δ, and ϕ_3.

For this analysis D^0 mesons are reconstructed as $K^-\pi^+$; D_1 mesons ($CP = +1$) as K^+K^- and $\pi^+\pi^-$; and D_2 mesons ($CP = -1$) as $K^0_S\pi^0$, $K^0_S\phi$, $K^0_S\omega$, $K^0_S\eta$, and $K^0_S\eta'$. The short-lived mesons are reconstructed as follows: $\phi \to K^+K^-$ with $1.008 < m_{KK} < 1.032$ GeV/c^2; $\omega \to \pi^+\pi^-\pi^0$ with $0.732 < m_{\pi\pi\pi} < 0.820$ GeV/c^2; $\eta \to \gamma\gamma$ with $0.495 < m_{\gamma\gamma} < 0.578$ GeV/c^2; and $\eta' \to \eta\pi^+\pi^-$ with $0.903 < m_{\eta\pi\pi} < 1.002$ GeV/c^2. The resulting D^0 candidates are required to have masses within 2.5σ of m_{D^0}, where σ is the measured mass resolution (4.5–18 MeV/c^2). The D^0 and π^+/K^+ candidates are combined to form B^+ candidates by selecting combinations with $5.27 < m_{bc} < 5.29$ GeV/c^2 and $|\Delta E| < 0.20$ GeV. The event yields are obtained from fits to the ΔE distributions. The results for R^{D^0}, $R^{D^{1,2}}$, and $A_{1,2}$ are listed in Table 4; all CP asymmetries are consistent with zero. The factor r can be calculated via $\mathcal{R}_1 + \mathcal{R}_2 = 2(1+r^2)$; the result is $r^2 = 0.31 \pm 0.21$, which is only 1.5σ from zero. Since, in Eq. (1), $\cos\phi_3$ and $\sin\phi_3$ are always multiplied by a factor of r, the value of r obtained precludes setting a stringent constraint upon ϕ_3 with the current statistics. The situation should improve with more data.

$B \to \omega K/\omega \pi$ DECAYS

The decays $B \to \omega K$ and $B \to \omega \pi$ also proceed via $b \to u$ tree and $b \to s$ loop diagrams. Theoretical calculations based on QCD factorization [9, 10, 11, 12] predict $B(B \to \omega\pi) \approx 2 \times B(B \to \omega K)$. A previous Belle measurement [13] based on 29 fb^{-1} of data did not agree with this prediction, and we update that result here.

Candidate events are selected by first selecting $\omega \to \pi^+\pi^-\pi^0$ decays. The π^0 is reconstructed from $\gamma\gamma$ pairs having $|m_{\gamma\gamma} - m_{\pi^0}| < 3\sigma$ ($\sigma = 5.4$ MeV/c^2); each γ must

TABLE 4. Results for A_{CP} (top) and R^{D^0}, $R^{D^{1,2}}$ (bottom).

Mode	A_{CP}	90% C.L. Interval
$D^0 K^\pm$	$0.04 \pm 0.06 \pm 0.03$	$(-0.07, 0.15)$
$D_1 K^\pm$	$0.06 \pm 0.19 \pm 0.04$	$(-0.26, 0.38)$
$D_2 K^\pm$	$-0.19 \pm 0.17 \pm 0.05$	$(-0.47, 0.11)$
R^{D^0}	$0.077 \pm 0.005 \pm 0.006$	
R^{D^1}	$0.093 \pm 0.018 \pm 0.008$	
R^{D^2}	$0.108 \pm 0.019 \pm 0.007$	

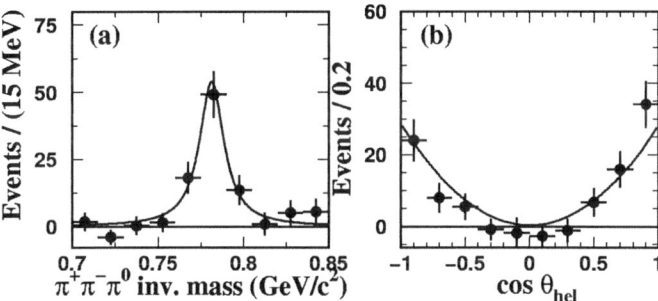

FIGURE 3. The fitted event yields in bins of $m_{\pi^+\pi^-\pi^0}$ (left) and $\cos\theta_h$ (right) for $B \to \omega K$ and $B \to \omega \pi$.

also satisfy $E_\gamma > 50$ MeV. We then require $|m_{\pi\pi\pi} - m_\omega| < 30$ MeV/c^2 (2σ), and the ω is paired with a $\pi^\pm$, π^0, $K^\pm$, or K_S^0 to form B candidates. Those candidates satisfying $5.20 < m_{bc} < 5.30$ GeV/c^2 and $|\Delta E| < 0.25$ GeV are subjected to an unbinned maximum likelihood (ML) fit using m_{bc} and ΔE as the independent variables. The event yields resulting from the fit and the corresponding branching fractions are listed in Table 5. We note that the central value for $B(B \to \omega K)$ is still greater than that for $B(B \to \omega \pi)$, in contrast with the theoretical prediction.

The main background is due to $q\bar{q}$ continuum events. To reduce this we cut on both $R_{q\bar{q}}$ and the helicity angle θ_h, which is defined as the angle between the B flight direction and the vector perpendicular to the ω decay plane, in the ω rest frame. The cut chosen is $|\cos\theta_h| > 0.5$. To confirm that signal candidates contain real ω decays, we relax the $m_{\pi^+\pi^-\pi^0}$ cut and repeat the fits for different $m_{\pi^+\pi^-\pi^0}$ bins. The resulting event yields are plotted in Fig. 3 and display a sharp peak at m_ω with negligible nonresonant background underneath.

Since the $B^\pm \to \omega h^\pm$ final states are self-tagging, we divide these samples into B^+ and B^- decays and search for a CP asymmetry. The quantity measured is $A_{CP} = [N(B^-) - N(B^+)]/[N(B^-) + N(B^+)]$. The event yields are determined from a two-dimensional binned fit in the m_{bc}-ΔE plane. The results are listed in Table 6. While

TABLE 5. $B \to \omega K$ and $B \to \omega \pi$ event yields, statistical significance, and branching fractions (90% C.L. limits).

Mode	N_s	$\mathscr{S}$	$B \times 10^6$ (90% C.L. limit)
ωK^-	$46.1^{+9.1}_{-8.4}$	7.8	$6.7^{+1.3}_{-1.2} \pm 0.6$
ωK^0	$11.1^{+5.2}_{-4.4}$	3.2	$4.0^{+1.9}_{-1.6} \pm 0.5$ (< 7.6)
$\omega \pi^-$	$42.1^{+10.1}_{-9.3}$	6.0	$5.7^{+1.4}_{-1.3} \pm 0.6$
$\omega \pi^0$	$0.0^{+2.1}_{-0.0}$	–	(< 1.9)

TABLE 6. $B^\pm \to \omega K^\pm$ and $B^\pm \to \omega \pi^\pm$ event yields separated by charge, and the resulting CP asymmetry.

Mode	$N(B^-)$	$N(B^+)$	A_{CP}	90% C.L. Interval
$\omega K^\pm$	$24.3^{+6.7}_{-5.9}$	$21.8^{+6.4}_{-5.7}$	$0.06^{+0.20}_{-0.18} \pm 0.01$	(−0.25, 0.40)
$\omega \pi^\pm$	$32.5^{+8.2}_{-7.5}$	$11.5^{+6.1}_{-5.3}$	$0.48^{+0.23}_{-0.20} \pm 0.02$	(0.14, 0.86)

$A_{CP}(\omega K)$ is consistent with zero, $A_{CP}(\omega \pi)$ is 2.4σ above zero, and a symmetric 90% C.L. interval excludes $A_{CP} = 0$.

$B \to \phi K / \phi K^*$ DECAYS AND POLARIZATION

The decays $B \to \phi K$ and $B \to \phi K^*$ proceed only via loop diagrams ($b \to s s \bar{s}$) and thus are especially sensitive to new physics. Because both the ϕ and K^* are spin-1, $B \to \phi K^*$ is a mixture of CP-even and CP-odd states; the individual components can be determined by measuring the ϕ polarization.

As a first step, $\phi \to K^+K^-$ decays are identified by requiring pairs of oppositely-charged tracks having $R_K > 0.1$ and $|m_{KK} - m_\phi| < 10$ MeV/c^2. K^* decays are reconstructed via $K^{*+} \to K^+\pi^0$, $K^{*+} \to K^0_S\pi^+$, and $K^{*0} \to K^+\pi^-$; the resulting two-body mass is required to be within 70 MeV/c^2 of m_{K^*}. $B \to \phi K$ ($B \to \phi K^*$) decays are selected by pairing a ϕ candidate with a K (K^*) candidate and requiring that they be within the signal region $5.271 < m_{bc} < 5.289$ (5.270–5.290) GeV/c^2 and $|\Delta E| < 0.64$ (0.60) GeV. The ΔE window is slightly larger for $K^{*+} \to K^+\pi^0$ decays due to shower leakage.

The dominant background is due to $q\bar{q}$ continuum events. There is also 5–9% contamination of $B \to \phi K^{(*)}$ decays from nonresonant $B \to K^+K^-K^{(*)}$, and 2–12% contamination from $B \to f(980)K^{(*)}$, $f(980) \to K^+K^-$. The uncertainty in the instrinsic width of the $f_0(980)$ is included in the systematic error.

The signal yields are obtained via an unbinned ML fit with m_{bc} and ΔE as the independent variables. The results and corresponding branching fractions are listed in Table 7. The projections of the fits are shown in Fig. 4. For the $B \to \phi K^*$ modes, there is an additional systematic error due to uncertainty in the K^* polarization and the

TABLE 7. Branching fractions and CP asymmetries for $B \to \phi K$ and $B \to \phi K^*$ decays.

Mode	N_s	$B \times 10^6$	A_{CP}	90% C.L. Interval
ϕK^+	136^{+16}_{-15}	$9.4 \pm 1.1 \pm 0.7$	$0.01 \pm 0.12 \pm 0.05$	$(-0.20, 0.22)$
ϕK^0	$35.6^{+8.4}_{-7.4}$	$9.0^{+2.2}_{-1.8} \pm 0.7$	–	–
ϕK^{*0}	$58.5^{+9.1}_{-8.1}$	$10.0^{+1.6}_{-1.5}{}^{+0.7}_{-1.8}$	$0.07 \pm 0.15^{+0.05}_{-0.03}$	$(-0.18, 0.33)$
ϕK^{*+}	$8.0^{+4.3}_{-3.5}$ $(K^+\pi^0)$ $11.3^{+4.5}_{-3.8}$ $(K^0_S\pi^+)$	$6.7^{+2.1}_{-1.9}{}^{+0.7}_{-1.0}$	$-0.13 \pm 0.29^{+0.08}_{-0.11}$	$(-0.64, 0.36)$

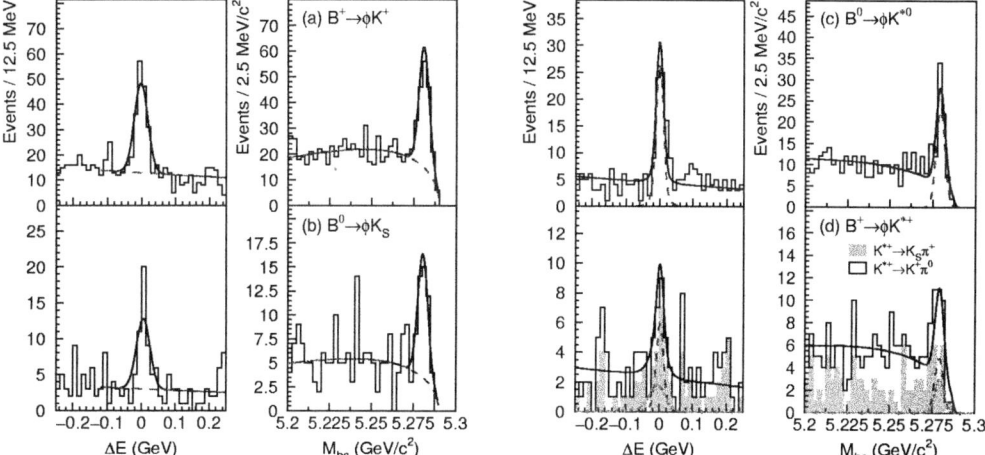

FIGURE 4. Projections of the unbinned ML fits for $B \to \phi K$ decays (left) and $B \to \phi K^*$ decays (right). The histograms show the data. Events in the m_{bc} plots are required to have $|\Delta E|$ within the signal region, and events in the ΔE plots are required to have m_{bc} within the signal region (see text).

corresponding uncertainty in the daughter π detection efficiency.

For the self-tagging modes $B^\pm \to \phi K^{(*)\pm}$ we measure $A_{CP} = [N(\overline{B}) - N(B)]/[N(\overline{B}) + N(B)]$, where B $(\overline{B})$ is B^0 or B^+ ($\overline{B^0}$ or B^-). The results are also listed in Table 7 and in all cases are consistent with zero.

The polarization of the ϕ in $B \to \phi K^*$ decays is measured using the transversity basis [14]. In this basis the ϕ is at rest. The x-y plane is defined by the K^{*0} daughters, with the $-x$ axis along the direction of the K^* (see Fig. 5). The angle θ_{K^*} is that between the K^{*0} direction and the K^+ daughter. The angles θ_{tr} and ϕ_{tr} are the polar and azimuthal angles, respectively, of the K^+ daughter of the ϕ. The decay distribution is given by [15]:

$$\frac{d^3\Gamma(\phi_{tr},\cos\theta_{tr},\cos\theta_{K^*})}{d\phi_{tr}\,d\cos\theta_{tr}\,d\cos\theta_{K^*}} = \frac{9}{32\pi}\left[|A_\perp|^2 2\cos^2\theta_{tr}\sin^2\theta_{K^*} \right.$$
$$\left. + |A_\parallel|^2 2\sin^2\theta_{tr}\sin^2\phi_{tr}\sin^2\theta_{K^*} \right.$$

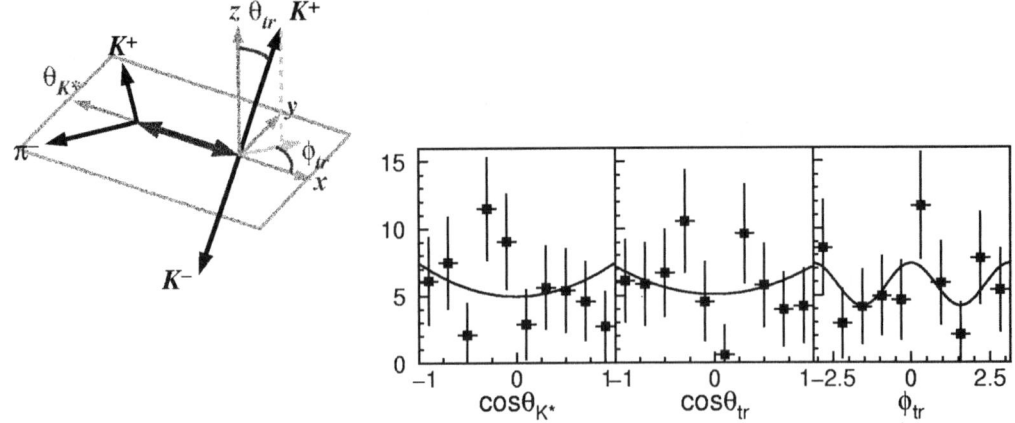

FIGURE 5. Definition of the angles θ_{K^*}, θ_{tr}, and ϕ_{tr} in the transversity basis (left), and projections of the unbinned fit for these angles (right).

$$\begin{aligned} &+\ |A_0|^2 4\sin^2\theta_{tr}\cos^2\phi_{tr}\cos^2\theta_{K^*} \\ &+\ \sqrt{2}\,\mathrm{Re}(A_\|^* A_0)\sin^2\theta_{tr}\sin 2\phi_{tr}\sin 2\theta_{K^*} \\ &-\ \eta\sqrt{2}\,\mathrm{Im}(A_0^* A_\perp)\sin 2\theta_{tr}\cos\phi_{tr}\sin 2\theta_{K^*} \\ &-\ 2\eta\,\mathrm{Im}(A_\|^* A_\perp)\sin 2\theta_{tr}\sin\phi_{tr}\sin^2\theta_{K^*} \Big], \end{aligned} \quad (2)$$

where A_0, $A_\|$, and $A_\perp$ are the complex amplitudes of the three helicity states, and $\eta = +1(-1)$ for B^0 ($\overline{B^0}$) decays. The amplitude A_0 denotes the longitudinal polarization of the final state, and $A_\perp$ ($A_\|$) denotes the transverse polarization along the $z(y)$ axis. Note that $|A_0|^2 + |A_\||^2 + |A_\perp|^2 = 1$. The value of $|A_\perp|^2$ $(1 - |A_\perp|^2 = |A_0|^2 + |A_\||^2)$ is the *CP*-odd (*CP*-even) fraction of the decay.

The complex amplitudes A_0, $A_\perp$, and $A_\|$ are determined via an unbinned ML fit to the candidates within the m_{bc}-ΔE signal region; the probability density function for signal is given by Eq. (2). By convention, the value of $\mathrm{Arg}(A_0)$ is set to zero and $|A_\||^2$ is calculated from the normalization constraint. The results of the fit are: $|A_0|^2 = 0.43 \pm 0.09 \pm 0.04$, $|A_\perp|^2 = 0.41 \pm 0.10 \pm 0.04$, $\mathrm{Arg}(A_\|) = -2.57 \pm 0.39 \pm 0.09$, and $\mathrm{Arg}(A_\perp) = 0.48 \pm 0.32 \pm 0.06$. The projections of the fit are shown in Fig. 5. The systematic errors include the (slow) pion detection efficiency (3–6%) and background from higher K^* states (6–9%). The value of $|A_\perp|^2$ obtained indicates that both *CP*-odd and *CP*-even components of $B \to \phi K^*$ are sizable.

$B^{\pm} \to \rho^{\pm}\rho^0$ DECAYS

The decay $B^+ \to \rho^+\rho^0$ proceeds via $b \to d$ loop and $b \to u$ tree diagrams and contains two vector mesons in the final state. Angular correlations among the decay products ($\pi^+\pi^0\pi^+\pi^-$) can be used to search for CP- and T-violating effects. In the final state, both ρ's are either longitudinally or transversely polarized; the corresponding amplitudes are denoted H_{00} and H_{11}, respectively.

In this analysis, $\rho^+\rho^0$ states are reconstructed by combining three charged pions with one neutral pion. The charged pions are required to have $p_T > 0.10$ GeV/c. Candidate π^0's are reconstructed from $\gamma\gamma$ pairs having $118 < m_{\gamma\gamma} < 150$ MeV/c^2; each γ must also satisfy $E_\gamma > 50(100)$ MeV in the barrel (endcap) region. Candidate ρ mesons are identified via $\pi^+\pi^-$ or $\pi^+\pi^0$ pairs having $0.65 < m_{\pi\pi} < 0.89$ GeV/c^2. $B^+ \to \rho^+\rho^0$ candidates are identified by requiring $5.272 < m_{bc} < 5.290$ GeV/c^2 and $-0.10 < \Delta E < 0.06$ GeV. The H_{00} amplitude gives rise to asymmetric $\rho \to \pi\pi$ decays, i.e., one pion has high momentum and the other has low momentum. The H_{11} amplitude gives rise to symmetric $\rho \to \pi\pi$ decays. Thus, the H_{00} state has a lower reconstruction efficiency and a ΔE resolution $\sim 15\%$ broader than that for H_{11}.

There are large backgrounds due to $q\bar{q}$ continuum events. To reduce these we cut on both $R_{q\bar{q}}$ and the thrust angle θ_{thr}, which is the angle between the thrust axis of tracks originating from the B candidate and that of the remaining tracks in the event. The cut chosen is $|\cos\theta_{thr}| < 0.80$. The overall rejection of continuum events is $> 99.5\%$, with a signal efficiency of 28%. There is also a small level of background from $b \to c$ processes and rarer B decays such as $B^+ \to \eta'\rho^+$, $K^{*+}\rho^0$, ρ^+K^{*0} and $\rho\pi$; these tend to be displaced in ΔE.

The resulting ΔE and m_{bc} distributions are shown in Fig. 6. The event yields are determined by fitting in ΔE. The fit yields 58.7 ± 13.2 events with a statistical significance ($\sqrt{2\ln[L_{max}/L_0]}$) of 5.3. Fitting the ΔE distributions for different $m_{\pi\pi}$ bins gives the event yields plotted in Fig. 7. These distributions agree well with MC expectations and show little nonresonant background beneath the ρ peaks.

The relative strengths of H_{00} and H_{11} are determined by studying distributions of the helicity angle θ_{hel}, which is the angle between the ρ flight direction in the B rest frame and the π^+ flight direction in the ρ rest frame. The signal yields determined from ΔE fits for different $\cos\theta_{hel}$ bins are plotted in Fig. 8 for both the ρ^0 and ρ^+. We perform simultaneous binned fits to these distributions using MC expectations for the H_{00} and H_{11} helicity states. The fit yields the fraction of $\rho^0\rho^+$ final states that are longitudinally polarized:

$$\frac{\Gamma_L}{\Gamma_{tot}} = (94.8 \pm 10.6 \pm 2.1)\%.$$

This result shows that the H_{00} state dominates, which is consistent with theoretical expectations [16]. The systematic error includes uncertainties in signal yield extraction and the polarization dependence of the detection efficiency. Based on this polarization ratio and the MC-determined reconstruction efficiencies of the two helicity states, we calculate $B(B^+ \to \rho^+\rho^0) = \left(3.17 \pm 0.71 ^{+0.38}_{-0.67}\right) \times 10^{-5}$.

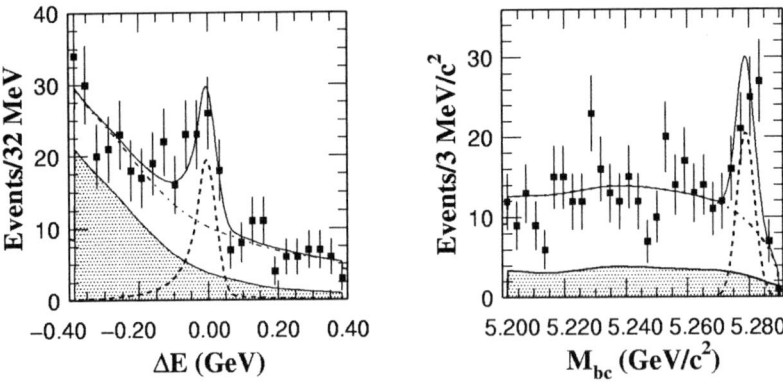

FIGURE 6. ΔE (left) and m_{bc} (right) fits for candidate $B^+ \to \rho^+\rho^0$ decays. The shaded curve represents $B\bar{B}$ background, the dash-dotted curve represents the sum of $B\bar{B}$ and continuum backgrounds, the dashed curve represents the $B^+ \to \rho^+\rho^0$ signal, and the solid curve represents the overall sum.

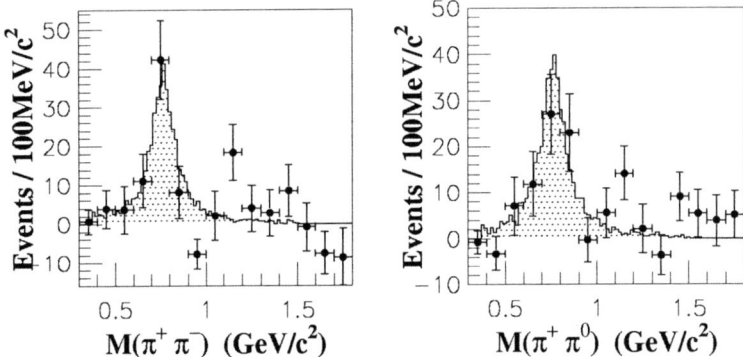

FIGURE 7. $m_{\pi^+\pi^-}$ (left) and $m_{\pi^+\pi^0}$ (right) distributions for candidate $B^+ \to \rho^+\rho^0$ decays. The shaded histogram shows MC signal.

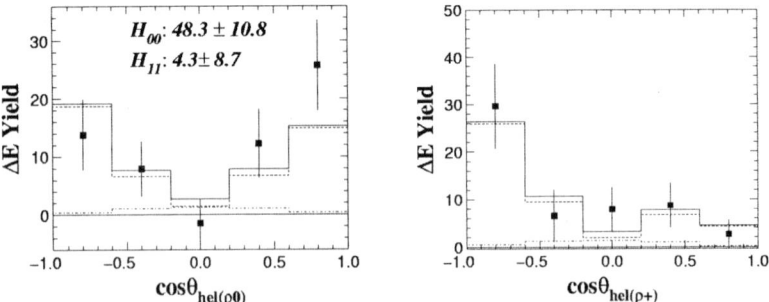

FIGURE 8. Helicity distributions for the ρ^0 (left) and ρ^+ (right) in $B^+ \to \rho^+\rho^0$ decays. The dashed (dash-dotted) histogram shows the H_{00} (H_{11}) component of the fit; the solid histogram is their sum. The bin size is 0.4. The low event yield near $\cos\theta_{hel(\rho+)} = 1$ is due to a requirement $p_{\pi^0}(\text{CM}) > 0.5$ GeV/c.

We separate the candidate events into $B^- \to \rho^-\rho^0$ and $B^+ \to \rho^+\rho^0$ subsamples and fit these subsamples individually. The resulting event yields are 29.3 ± 9.5 and 29.3 ± 9.1, respectively. The CP asymmetry is $A_{CP} = [N(\rho^-\rho^0) - N(\rho^+\rho^0)]/[N(\rho^-\rho^0) + N(\rho^+\rho^0)] = 0.00 \pm 0.22 \pm 0.03$, which is consistent with zero.

SUMMARY

With 78 fb^{-1} of data the Belle experiment has:

- updated the branching fractions and CP asymmetries for $B^0 \to \pi\pi$, $B^0 \to K\pi$, and $B^0 \to KK$ decays;
- measured the CP asymmetries in $B^{\pm} \to D_{CP}K^{\pm}$ decays, where D_{CP} represents a D^0 decaying to a $CP = +1$ or $CP = -1$ eigenstate, and investigated the possibility of using the measured asymmetries to constrain the CKM phase ϕ_3;
- updated the branching fractions and CP asymmetries for $B^{\pm} \to \omega\pi^{\pm}$ and $B^{\pm} \to \omega K^{\pm}$ decays;
- measured the branching fractions for $B \to \phi K$ and $B \to \phi K^*$ decays and, for the latter, measured the polarization amplitudes $A_\perp$, $A_\parallel$, and A_0 in the transversity basis;
- measured the branching fraction for $B^+ \to \rho^+\rho^0$ and the helicity amplitudes H_{00} and H_{11}. This is the first reported observation of this decay.

Most results are consistent with theoretical expectations, although some channels show interesting and possibly important discrepancies at the 2σ level. We look forward to investigating these further (and refining all of these measurements) with more data.

REFERENCES

1. Abashian et al. (Belle Collaboration), A., *Nucl. Instr. Meth. A*, **479**, 117 (2002).
2. Feldman, G. J., and Cousins, R. D., *Phys. Rev. D*, **57**, 3873 (1998).
3. Abbaneo et al., D., "The CKM Matrix and The Unitarity Triangle," in *Proceedings of the First Workshop on the CKM Unitarity Triangle, 13-16 Feb. 2002*, edited by M. Battaglia, A. J. Buras, P. Gambino, and A. Stocchi, CERN, hep-ph/0304132 (2003), pp. 291–308.
4. Hagiwara et al. (PDG), K., *Phys. Rev. D*, **66**, 010001 (2002).
5. Abe et al. (Belle Collaboration), K., *Phys. Rev. D*, **68**, 012001 (2003).
6. Quinn, H., and Sanda, A. I., *Eur. Phys. J.*, **C15**, 626 (2000).
7. Abe et al. (Belle Collaboration), K., *Phys. Rev. Lett.*, **87**, 111801 (2001).
8. Bornheim et al. (CLEO Collaboration), A., Tech. Rep. CLNS 03/1816, hep-ex/0302026 (2003).
9. Ali, A., Kramer, G., and Lu, C. D., *Phys. Rev. D*, **58**, 094009 (1998).
10. Chen, Y. H., Cheng, H. Y., Tseng, B., and Yang, K. C., *Phys. Rev. D*, **60**, 094014 (1999).
11. Du, D. S., Gong, H. J., Sun, J. F., and Zhu, G. H., *Phys. Rev. D*, **65**, 094025 (2002).
12. Lu, C. D., and Yang, M. Z., *Eur. Phys. J.*, **C23**, 275 (2002).
13. Lu et al. (Belle Collaboration), R. S., *Phys. Rev. Lett.*, **89**, 191801 (2002).
14. Dunietz et al., I., *Phys. Rev. D*, **43**, 2193 (1991).
15. Abe, K., Satpathy, M., and Yamamoto, H., Tech. Rep. UH-511-982-01, hep-ex/0103002 (2001).
16. Aleksan et al., R., *Phys. Lett. B*, **356**, 95 (1995).

Rare Charm Decays at Colliders

D. Cinabro

Wayne State University, Detroit, MI 481202

Abstract. I review recent results in rare charm decays from experiments at collider facilities.

INTRODUCTION

Rare charm decays are an excellent hunting ground for new physics. [1] Charm mixing is highly suppressed in the Standard Model since there are no heavy quarks, specifically the top is absent, in the charm to anti-charm box diagram, the b-quark contribution is suppressed by V_{ub} and is thus tiny, and the contributions from d and s cancel in the $SU(3)$ limit. Thus difficult to calculate long range effects are important and charm mixing can be enhanced by new physics.

CP violation in charm decays is also small in the Standard Model. Charm decays are dominated by the first two generations where the weak phase has only a small effect. The amount of CP-violation depends on a strong phase difference which is very difficult to calculate for charm. Any large CP effect are expected to show up in Cabibbo suppressed $c \to d$ decays. Large strong phases have been observed in charm decays and we expect CP asymmetries at the 10^{-3} level. CP violation in charm decays is another sensitive probe of new physics.

More generally it is possible that any new physics would couple differently to up-type and down-type quarks. For up-type quarks the Standard Model backgrounds to new physics effects are Cabibbo suppressed. Practically both precision measurements and probes sensitive to one in 10^6 effects are possible as experiments have clean samples of $\mathcal{O}(10^7)$ charm decays.

I summarize recent results in charm mixing, CP-violation, and searches for charm flavor changing neutral current decays from experiments at colliders.

MIXING AND DOUBLY CABIBBO SUPPRESSED DECAYS

Charm mixing is characterized by $y = \Delta\Gamma/2\Gamma$ which is from long range effects, and $x = \Delta M/\Gamma$ which is from short range effects. A large value of x would indicate new physics. In hadronic decays a D^0 has two paths to a "wrong sign" final state. One is via a doubly Cabibbo suppressed (DCS) decay, $c \to d$ and $W^+ \to \bar{s}$ with a rate $\propto \sin^4\theta_C$, with θ_C being the Cabibbo angle. The second is via mixing with $c \to \bar{c}$ followed by Cabibbo favored decay of the $\bar{c}$. The wrong sign signal can be observed by tagging the initial charm flavor with the charge of the pion in $D^* \to D^0\pi$ decay.

The DCS and mixing contributions can be separated using the lifetime distribution of the D^0 to wrong sign decay. The DCS contribution will be a pure exponential while mixing will cause terms modulated by the lifetime, due to interference between DCS and mixing decays, and the square of the lifetime to appear.

BABAR has a preliminary search for D^0 mixing in the $K\pi$ decay channel [2] They have 430 wrong sign events in a sample that is 73% pure. Fitting the lifetime distribution they observe no evidence of mixing and at the 95% C.L. limit $x' < 0.0022$, $-0.056 < y' < 0.039$, and $0.0023 < R_D < 0.0052$ where the primes represent the mixing parameters rotated by an unknown strong phase difference and R_D is the rate for DCS decays relative to the Cabibbo allowed decay. Under the assumption that there is no mixing or CP-violation in the decay they measure $R_D = 0.00357 \pm 0.00022 \pm 0.00027$.

Another route to mixing is to consider CP eigenstates such as KK or $\pi\pi$ which can have contributions from both mixed and unmixed neutral D decays. If there is no CP-violation, and x and y are both small, then $y = 1 - \tau_0/\tau_+$, where where τ_0 is the lifetime of a CP neutral state such as $K\pi$ and τ_+ is lifetime of a CP even state such as KK or $\pi\pi$. These decays have large statistics since they are only singly Cabibbo suppressed. BABAR has a preliminary analysis in these channels [3] which features large statistics ($K\pi$: 265k, KK: 26k, $\pi\pi$: 13k, and KK Untagged: 146k) and high purity ($K\pi$: 99.4%, KK: 97%, $\pi\pi$: 87.9%, and KK Untagged: 68.1%). They measure, combining all their samples, $y = 0.008 \pm 0.004^{+0.005}_{-0.004}$ which is consistent with no mixing. They see no hints of mixing in the sub-samples.

Since the 2002 PDG there have been two high quality measures of R_D, the relative rate of DCS decays to Cabibbo allowed in $D^0 \to K\pi$. One by BABAR is discussed above, and the second by Belle [4] of $R_D = 0.00372 \pm 0.00027$. Averaging these with the 2002 PDG value of 0.00390 ± 0.00060, yields $R_D = 0.00369 \pm 0.00021$. When this rate is normalized the expected ratio of CKM matrix elements squared that contribute to the two decays, it is expected to be one in the $SU(3)$ limit. Using the value above and the present values for the CKM elements I get 1.44 ± 0.23 which is an indication of violation of $SU(3)$ symmetry at the two standard deviation level.

CP VIOLATION

In charm CP asymmetries $A_{CP} \equiv (\text{Rate}(D \to f) - \text{Rate}(\overline{D} \to \overline{f}))/(\text{Rate}(D \to f) + \text{Rate}(\overline{D} \to \overline{f}))$ are expected at the 10^{-3} level. CLEO has been considering three body decays with the Dalitz technique facing the complication that it is not exactly clear what "rate" means when there are large interference effects. They have defined $\mathcal{A}_{CP}$ similar to A_{CP} by integrating all contributions over the entire Dalitz plot and alternatively measuring explicitly CP violating amplitudes and phases for observed two body modes. In a preliminary analysis of $D^0 \to K_S\pi\pi$ they see no evidence of CP-violation and measure $\mathcal{A}_{CP} = -0.039 \pm 0.034$ and for the two body modes, taking as an example $\overline{K^0}\rho^0$, CP-violating amplitudes over conserving of $0.00 \pm 0.02^{+0.02}_{-0.07}$ and CP-violating phases of $-3 \pm 16^{+6}_{18}$ degrees. Similarly in a preliminary analysis of $D^0 \to \pi\pi\pi^0$ [5] they see no hint of CP-violation and measure $\mathcal{A}_{CP} = 0.01^{+0.09}_{-0.07} \pm 0.09$ and in the two body modes, saturated by the ρ's, see consistency between D^0 and $\bar{D}^0$ amplitudes to 10% and phases to 7°. For D decays within the Standard Model this decay mode is expected to exhibit one of the largest violations of CP.

The mixing analyses from BABAR described above allow for CP violation. They measure A_{CP} in DCS $D \to K\pi$ at $0.095 \pm 0.061 \pm 0.083$ and Δy, the asymmetry in $D^0 \to$ CP eigenstates lifetimes in their combined sample, at $-0.008 \pm 0.006 \pm 0.002$. They see no hint of CP violations in any of the sub-samples.

CDF has a new detached vertex trigger designed for two body B decays that also enables them to do a great deal of charm physics. [6] They present a preliminary analysis of CP asymmetries in $D \to KK$ and $\pi\pi$. Even at this early stage of Run II they have large samples, 8300 in KK and 3400 in $\pi\pi$, in tagged D^0 decays. They measure $A(KK)_{CP} = 0.020 \pm 0.017 \pm 0.006$ and $A(\pi\pi)_{CP} = 0.030 \pm 0.019 \pm 0.006$ which are already competitive with the best measurements done previously.

There are no hints of C-violation in charm decays at our present level of sensitivity of a few percent.

FLAVOR CHANGING NEUTRAL CURRENTS

Flavor changing neutral currents for charm are expected at a very low level, $\mathcal{O}(10^{-8}$, in the Standard Model. New physics can enhance this by up to a factor of 100 making it an attractive hunting ground in rare charm decays.

CLEO has searched for $D^0 \to \gamma\gamma$ [7] using the $\pi^0\pi^0$ mode as normalization. They see no hint of a signal and limit $\mathcal{B}(D^0 \to \gamma\gamma)/\mathcal{B}(D^0 \to \pi^0\pi^0) < 0.033$ and $\mathcal{B}(D^0 \to \gamma\gamma) < 2.9 \times 10^{-5}$ at the 90% C.L. This is the first limit in this mode.

CDF has a preliminary search for $D^0 \to \mu\mu$. They normalize with $\pi\pi$, 1400 events, and limit $\mathcal{B}(D^0 \to \mu\mu) < 2.4 \times 10^{-6}$ at 90% C.L. This is the best limit in this mode and again shows the termendous potential for charm physics at CDF.

FUTURE PROSPECTS

The prospects for rare charm decays are very good in the near future. The B-Factories now have samples ten times CLEO, which currently leads the search for rare charm decays, and BABAR and Belle expect ten times more data by the end of their runs. CDF can clearly become a major contributer with twenty times more data expected from Run IIa alone. CLEO-c is turning on later this year. This experiment expect two orders of magnitude more data than previous charm threshold machines. Among other things this will allow them to have 1.5×10^6 tagged D^0 decays which allow for searches and measurements in an almost background free environment. Planned experiments such as BTeV, LHC-b, and BES-III, will have substantial charm capabilities. These experiments are discussed in the "Future of Flavor Physics" session.

CONCLUSION

Rare charm decay is a very rich field, with surprising diversity ranging from high statistics precision measurements to searches with one in a million sensitivity. As yet there is no hint of charm mixing, and systematics are starting to limit present techniques. There is no hint of CP violation in charm, but experiments running now are sensitive at the Standard Model predicted level. There is no hint of enhancements in flavor changing neutral current charm decays at the 10^6 level, but there is still two orders of magnitude to go before we test the prediction of the standard model. With many present experiments expecting more data and near future experiments likely to have excellent charm capabilities the future for this excellent hunting ground for physics beyond the standard model is very bright.

ACKNOWLEDGMENTS

Many thanks to Marino Artuso and Alexey Petrov. My work is supported by the National Science Foundation.

REFERENCES

1. A. Petrov, these proceedings for more on the theory of rare charm decays.
2. B. Aubert, *et al.* (The BABAR Collaboration), hep-ex/0304007.
3. B. Aubert, *et al.* (The BABAR Collaboration), hep-ex/0306003.
4. K. Abe, *et al.* (Belle Collaboration), hep-ex/0208051.
5. V. Frolov, *et al,* (CLEO Collaboration), hep-ex/0306048.
6. A. Korn, hep-ex/0305054.
7. T.E. Coan, *et al.* (CLEO Collaboration), Phys. Rev. Lett. **90** (2003) 101801.

$D^0 - \overline{D}^0$ mixing, rare decays, and New Physics

Alexey A. Petrov

Department of Physics and Astronomy
Wayne State University, Detroit, MI 48201, USA

Abstract. We review the current status of searches for new physics in mixing and CP violation in charmed mesons both at the currently operating and proposed facilities.

Charm physics plays a unique dual role in the modern investigation of flavor physics. Charm decay and production experiments provide valuable checks and supporting measurements for studies of CP-violation in measurements of CKM parameters in b-physics, as well as outstanding opportunities for searches for new physics. Historically, many methods of heavy quark physics have been first tested in charmed hadrons. The fact that a b-quark mainly decays into a charm quark makes charm physics an integral part of any b-physics program. In many cases, direct measurements of charm decay parameters directly affect the studies of fundamental physics in B decays.

One of the important areas of modern phenomenology where charm decays play an important role is the indirect search for physics beyond the Standard Model. Indeed, large statistics usually available in charm physics experiment makes it possible to probe small effects that might be generated by the presence of new physics particles and interactions. A program of searches for new physics in charm is complimentary to the corresponding programs in bottom or strange systems. This is in part due to the fact loop-dominated processes such as $D^0 - \overline{D}^0$ mixing or flavor-changing neutral current (FCNC) decays are sensitive to the dynamics of ultra-heavy *down-type particles*. Also, in many dynamical models, including the Standard Model, the effects in s, c, and b systems are correlated.

The low energy effect of new physics particles can be naturally written in terms of a series of local operators of increasing dimension generating $\Delta C = 1$ (decays) or $\Delta C = 2$ (mixing) transitions. For $D^0 - \overline{D}^0$ mixing these operators, as well as the one loop Standard Model effects, generate contributions to the effective operators that change D^0 state into $\overline{D}^0$ state leading to the mass eigenstates

$$|D_{\frac{1}{2}}\rangle = p|D^0\rangle \pm q|\bar{D}^0\rangle, \qquad (1)$$

where the complex parameters p and q are obtained from diagonalizing the $D^0 - \overline{D}^0$ mass matrix. The mass and width splittings between these eigenstates are given by

$$x \equiv \frac{m_2 - m_1}{\Gamma}, \quad y \equiv \frac{\Gamma_2 - \Gamma_1}{2\Gamma}. \qquad (2)$$

It is known experimentally that $D^0 - \overline{D}^0$ mixing proceeds extremely slowly, which in the Standard Model is usually attributed to the absence of superheavy quarks destroying GIM cancellations [1].

It is instructive to see how new physics can affect charm mixing. Since the lifetime difference y is constructed from the decays of D into physical states, it should be dominated by the Standard Model contributions, unless new physics significantly modifies $\Delta C = 1$ interactions. On the contrary, the mass difference x can receive contributions from all energy scales. Thus, it is usually conjectured that new physics can significantly modify x leading to the inequality $x \gg y$ [1]. The same considerations apply to FCNC decays as well, where new physics could possibly contribute to the decay rates of $D \to X_u \gamma$, $D \to X_u l^+ l^-$ (with X_u being exclusive or inclusive final state) as well as other observables [4]. One technical problem here is that in the standard model these decays are overwhelmingly dominated by long-distance effects, which makes them extremely difficult to predict model-independently. This problem can be turned into a virtue [5].

Another possible manifestation of new physics interactions in the charm system is associated with the observation of (large) CP-violation. This is due to the fact that all quarks that build up the hadronic states in weak decays of charm mesons belong to the first two generations. Since 2×2 Cabbibo quark mixing matrix is real, no CP-violation is possible in the dominant tree-level diagrams that describe the decay amplitudes. In the Standard Model CP-violating amplitudes can be introduced by including penguin or box operators induced by virtual b-quarks. However, their contributions are strongly suppressed by the small combination of CKM matrix elements $V_{cb} V_{ub}^*$. It is thus widely believed that the observation of (large) CP violation in charm decays or mixing would be an unambiguous sign for new physics. This fact makes charm decays a valuable tool in searching for new physics, since the statistics available in charm physics experiment is usually quite large.

As in B-physics, CP-violating contributions in charm can be generally classified by three different categories: (I) CP violation in the decay amplitudes. This type of CP violation occurs when the absolute value of the decay amplitude for D to decay to a final state f (A_f) is different from the one of corresponding CP-conjugated amplitude ("direct CP-violation"); (II) CP violation in $D^0 - \overline{D}^0$ mixing matrix. This type of CP violation is manifest when $R_m^2 = |p/q|^2 = (2M_{12} - i\Gamma_{12})/(2M_{12}^* - i\Gamma_{12}^*) \neq 1$; and (III) CP violation in the interference of decays with and without mixing. This type of CP violation is possible for a subset of final states to which both D^0 and $\overline{D}^0$ can decay.

For a given final state f, CP violating contributions can be summarized in the parameter

$$\lambda_f = \frac{q}{p} \frac{\overline{A}_f}{A_f} = R_m e^{i(\phi+\delta)} \left| \frac{\overline{A}_f}{A_f} \right|, \qquad (3)$$

where A_f and $\overline{A}_f$ are the amplitudes for $D^0 \to f$ and $\overline{D}^0 \to f$ transitions respectively and δ is the strong phase difference between A_f and $\overline{A}_f$. Here ϕ represents the convention-independent weak phase difference between the ratio of decay amplitudes and the

---

[1] This signal for new physics is lost if a relatively large y, of the order of a percent, is observed [2, 3].

mixing matrix.

Presently, experimental information about the $D^0 - \overline{D}^0$ mixing parameters x and y comes from the time-dependent analyses that can roughly be divided into two categories. First, more traditional studies look at the time dependence of $D \to f$ decays, where f is the final state that can be used to tag the flavor of the decayed meson. The most popular is the non-leptonic doubly Cabibbo suppressed decay $D^0 \to K^+\pi^-$. Time-dependent studies allow one to separate the DCSD from the mixing contribution $D^0 \to \overline{D}^0 \to K^+\pi^-$,

$$\Gamma[D^0 \to K^+\pi^-] = e^{-\Gamma t}|A_{K^-\pi^+}|^2 \left[R + \sqrt{R}R_m(y'\cos\phi - x'\sin\phi)\Gamma t + \frac{R_m^2}{4}(y^2+x^2)(\Gamma t)^2\right], \quad (4)$$

where R is the ratio of DCS and Cabibbo favored (CF) decay rates. Since x and y are small, the best constraint comes from the linear terms in t that are also *linear* in x and y. A direct extraction of x and y from Eq. (4) is not possible due to unknown relative strong phase δ_D of DCS and CF amplitudes [6], as $x' = x\cos\delta_D + y\sin\delta_D$, $y' = y\cos\delta_D - x\sin\delta_D$. This phase can be measured independently. The corresponding formula can also be written [2] for $\overline{D}^0$ decay with $x' \to -x'$ and $R_m \to R_m^{-1}$.

Second, D^0 mixing can be measured by comparing the lifetimes extracted from the analysis of D decays into the CP-even and CP-odd final states. This study is also sensitive to a *linear* function of y via

$$\frac{\tau(D \to K^-\pi^+)}{\tau(D \to K^+K^-)} - 1 = y\cos\phi - x\sin\phi\left[\frac{R_m^2-1}{2}\right]. \quad (5)$$

Time-integrated studies of the semileptonic transitions are sensitive to the *quadratic* form $x^2 + y^2$ and at the moment are not competitive with the analyses discussed above.

The construction of new tau-charm factories CLEO-c and BES-III will introduce new *time-independent* methods that are sensitive to a linear function of y. One can again use the fact that heavy meson pairs produced in the decays of heavy quarkonium resonances have the useful property that the two mesons are in the CP-correlated states [7].

By tagging one of the mesons as a CP eigenstate, a lifetime difference may be determined by measuring the leptonic branching ratio of the other meson. Its semileptonic *width* should be independent of the CP quantum number since it is flavor specific, yet its *branching ratio* will be inversely proportional to the total width of that meson. Since we know whether this $D(k_2)$ state is tagged as a (CP-eigenstate) $D_\pm$ from the decay of $D(k_1)$ to a final state S_σ of definite CP-parity $\sigma = \pm$, we can easily determine y in terms of the semileptonic branching ratios of $D_\pm$. This can be expressed simply by introducing the ratio

$$R_\sigma^L = \frac{\Gamma[\psi_L \to (H \to S_\sigma)(H \to Xl^\pm\nu)]}{\Gamma[\psi_L \to (H \to S_\sigma)(H \to X)] \, Br(H^0 \to Xl\nu)}, \quad (6)$$

where X in $H \to X$ stands for an inclusive set of all final states. A deviation from $R_\sigma^L = 1$ implies a lifetime difference. Keeping only the leading (linear) contributions due to mixing, y can be extracted from this experimentally obtained quantity,

$$y\cos\phi = (-1)^L \sigma \frac{R_\sigma^L - 1}{R_\sigma^L}. \quad (7)$$

The current experimental upper bounds on x and y are on the order of a few times 10^{-2}, and are expected to improve significantly in the coming years. To regard a future discovery of nonzero x or y as a signal for new physics, we would need high confidence that the Standard Model predictions lie well below the present limits. As was recently shown [3], in the Standard Model, x and y are generated only at second order in $SU(3)_F$ breaking,

$$x, y \sim \sin^2\theta_C \times [SU(3) \text{ breaking}]^2, \qquad (8)$$

where θ_C is the Cabibbo angle. Therefore, predicting the Standard Model values of x and y depends crucially on estimating the size of $SU(3)_F$ breaking. Although y is expected to be determined by the Standard Model processes, its value nevertheless affects significantly the sensitivity to new physics of experimental analyses of D mixing [2].

Theoretical predictions of x and y within and beyond the Standard Model span several orders of magnitude [8]. Roughly, there are two approaches, neither of which give very reliable results because m_c is in some sense intermediate between heavy and light. The "inclusive" approach is based on the operator product expansion (OPE). In the $m_c \gg \Lambda$ limit, where Λ is a scale characteristic of the strong interactions, ΔM and $\Delta \Gamma$ can be expanded in terms of matrix elements of local operators [9]. Such calculations yield $x, y < 10^{-3}$. The use of the OPE relies on local quark-hadron duality, and on Λ/m_c being small enough to allow a truncation of the series after the first few terms. The charm mass may not be large enough for these to be good approximations, especially for nonleptonic D decays. An observation of y of order 10^{-2} could be ascribed to a breakdown of the OPE or of duality, but such a large value of y is certainly not a generic prediction of OPE analyses. The "exclusive" approach sums over intermediate hadronic states, which may be modeled or fit to experimental data[10]. Since there are cancellations between states within a given $SU(3)$ multiplet, one needs to know the contribution of each state with high precision. However, the D is not light enough that its decays are dominated by a few final states. In the absence of sufficiently precise data on many decay rates and on strong phases, one is forced to use some assumptions. While most studies find $x, y < 10^{-3}$, Refs. [10] obtain x and y at the 10^{-2} level by arguing that $SU(3)_F$ violation is of order unity, but the source of the large $SU(3)_F$ breaking is not made explicit. It was also shown that phase space effects alone provide enough $SU(3)_F$ violation to induce $y \sim 10^{-2}$ [3]. Large effects in y appear for decays close to D threshold, where an analytic expansion in $SU(3)_F$ violation is no longer possible. Thus, theoretical calculations of x and y are quite uncertain, and the values near the current experimental bounds cannot be ruled out. Therefore, it will be difficult to find a clear indication of physics beyond the Standard Model in $D^0 - \overline{D}^0$ mixing measurements alone. The only robust potential signal of new physics in charm system at this stage is CP violation.

Acknowledgment: This research was supported by the National Science Foundation under Grant PHY-0244853 and by the US Department of Energy under grant DE-FG02-96ER41005.

REFERENCES

1. A. Datta, D. Kumbhakar, Z. Phys. C**27**, 515 (1985); A. A. Petrov, Phys. Rev. D**56**, 1685 (1997); E. Golowich and A. A. Petrov, Phys. Lett. B **427**, 172 (1998).
2. S. Bergmann, Y. Grossman, Z. Ligeti, Y. Nir, A. Petrov, Phys. Lett. B **486**, 418 (2000).
3. A. F. Falk, Y. Grossman, Z. Ligeti and A. A. Petrov, Phys. Rev. D **65**, 054034 (2002).
4. S. Fajfer, arXiv:hep-ph/0306263.
5. S. Fajfer, S. Prelovsek, P. Singer and D. Wyler, Phys. Lett. B **487**, 81 (2000).
6. A. F. Falk, Y. Nir and A. A. Petrov, JHEP **9912**, 019 (1999).
7. D. Atwood and A. A. Petrov, arXiv:hep-ph/0207165.
8. H. N. Nelson, in *Proc. of the 19th Intl. Symp. on Photon and Lepton Interactions at High Energy LP99* ed. J.A. Jaros and M.E. Peskin, arXiv:hep-ex/9908021.
9. H. Georgi, Phys. Lett. B297, 353 (1992); T. Ohl, G. Ricciardi and E. Simmons, Nucl. Phys. B403, 605 (1993); I. Bigi and N. Uraltsev, Nucl. Phys. B **592**, 92 (2001).
10. J. Donoghue, E. Golowich, B. Holstein and J. Trampetic, Phys. Rev. D33, 179 (1986); L. Wolfenstein, Phys. Lett. B164, 170 (1985); P. Colangelo, G. Nardulli and N. Paver, Phys. Lett. B242, 71 (1990); T.A. Kaeding, Phys. Lett. B357, 151 (1995). A. A. Anselm and Y. I. Azimov, Phys. Lett. B **85**, 72 (1979);
11. I. I. Bigi and A. I. Sanda, *CP violation* (Cambridge University Press, 2000).
12. D. Pedrini, J. Phys. G **27**, 1259 (2001).
13. I. I. Bigi and A. I. Sanda, Phys. Lett. B **171**, 320 (1986).
14. A. A. Petrov, Proc. of the *5th Workshop on Continuous Advances in QCD*, pp. 102-114; arXiv:hep-ph/0209049.

The Future of Charm Physics

Thomas E. Coan

Physics Department, Southern Methodist University, Dallas, TX, 75275 USA

Abstract. The CLEO-c and BESIII detectors at the CESR and BEPC accelerators, respectively, will collect in the near term large data sets of $e^+e^- \to c\bar{c}$ events in the energy range $\sqrt{s} = 3-5\,\text{GeV}$. These data sets will correspond to a huge fractional increase over the size of current ones. The physics goals and unique advantages of running at charm threshold production are discussed.

Charm Threshold Running

Over the next 3-5 years, the CLEO-c (CESR) and BESIII (BEPC) experiments will accumulate large statistics ($\int Ldt \geq 1\,\text{fb}^{-1}$) of $e^+e^- \to c\bar{c}$ events produced in charm threshold reactions in the energy range $\sqrt{s} = 3-5\,\text{GeV}$. Running at charm threshold production has distinct advantages over continuum $e^+e^- \to c\bar{c}$ production at colliders running at $\sqrt{s} = \Upsilon(4S)$. For example, the charged and neutral multiplicities in $\Psi(3770)$ events are only 5.0 and 2.4, respectively, reducing combinatorics and leading to high detection efficiencies and low systematic errors. Additionally, charm events at threshold are pure $D\bar{D}$, including the $\Psi(4140)$ decaying into $D\bar{D}^*$, $D_s\bar{D}_s$ and $D_s\bar{D}_s^*$. No additional particles from fragmentation are produced.

Low multiplicity events and pure $D\bar{D}$ states, coupled with the relatively high branching fractions typical of D decays, permit the efficient implementation of "double tag" studies where one D is fully reconstructed and the other is studied in a bias free fashion. This permits the determination of absolute branching fractions with very low backgrounds. The quantum coherence of the $D\bar{D}$ states produced in $\Psi(3770) \to D\bar{D}$ and $\Psi(4140) \to \gamma D\bar{D}$ decays permit relatively simple techniques[1] for measuring $D\bar{D}$ mixing parameters and direct CP violation.

This report summarizes estimates of the physics reach of the existing CLEO-c detector based on extensive monte carlo simulation. Similar physics topics can also be addressed by the soon-to-be-upgraded BESIII detector.

Absolute Branching Fractions

The combination of pure $D\bar{D}(D_s\bar{D}_s)$ at $\sqrt{s} = \Psi(3770)$ ($\sqrt{s} = 4140\,\text{MeV}$), typical charm branching fractions of $(1-15)\%$, and a high reconstruction efficiency for D-mesons, lead to a high net D-meson tagging efficiency of $\sim 15\%$ for CLEO-c. Key selection criteria for D candidates include constraints on the energy difference between the D candidate and the beam energy, the beam constrained mass of the D candidate, and particle identification cuts.

TABLE 1. Branching fraction precision of key D decay modes projected for $3\,fb^{-1}$ of CLEO-c data compared to PDG 2003 values.

Mode	$\sqrt{s}$ (GeV)	$(\frac{\delta Br}{Br})_{PDG}$	$(\frac{\delta Br}{Br})_{CLEO-c}$
$D^0 \to K^-\pi^+$	3770	2.4%	0.6%
$D^+ \to K^-\pi^+\pi^+$	3770	6.8%	0.7%
$D_s \to \phi\pi$	4140	25%	1.9%

TABLE 2. Charm decay constant precision expected with $3\,fb^{-1}$ of CLEO-c data compared to PDG 2002 values.

Decay Constant	Mode (GeV)	$(\frac{\delta f}{f})_{PDG}$	$(\frac{\delta f}{f})_{CLEO-c}$
f_{D_s}	$D_s^+ \to \mu\nu$	16%	1.9%
f_{D_s}	$D_s^+ \to \tau\nu$	17%	1.7%
f_D	$D^+ \to \mu\nu$	Upper limit	2.3%

The technique for tagging a single D candidate can clearly be extended to the second D-meson candidate in $e^+e^- \to c\bar{c}$ threshold production events by essentially applying the single-tag technique twice. Using a modified version of a technique developed at MARK III[2], CLEO-c can then precisely measure absolute hadronic charm meson branching fractions using double-tag events. Table 1 compares the branching fraction precision for some key D decay modes anticipated with $3\,fb^{-1}$ of CLEO-c data and the corresponding PDG 2003 values.

Leptonic and Semileptonic Decays

Precision measurements of leptonic and semileptonic decays in the charm sector are vital for determining CKM matrix elements that describe the mixing of flavors and generations induced by the weak interaction. The lowest order expression for the leptonic branching fraction of a D-meson is given by[3]

$$\mathcal{B}(D_q \to l\nu) = \frac{G_F^2}{8\pi} m_{D_q} m_l^2 (1 - \frac{m_l^2}{m_{D_q}^2}) f_{D_q}^2 |V_{cq}|^2 \tau_{D_q}, \qquad (1)$$

where f_{D_q} is the parameter that encapsulates the strong physics of the process and $|V_{cq}|$ is the CKM matrix parameter that encapsulates the weak physics and quantifies the amplitude for quark mixing. Measurements of leptonic branching fractions can then be used to extract f_{D_q} and, with additional semileptonic measurements, $|V_{cq}|$. Table 2 compares with PDG 2002 values the expected precision for D-meson decay constants with $3\,fb^{-1}$ of data and assuming 3 generation unitarity.

The differential semileptonic decay rate for a D-meson to a pseudoscalar P is given by[4]

TABLE 3. Expected precision in the branching fraction $\mathscr{B}$ for important semileptonic decays with CLEO-c and the comparison with PDG 2003 values.

Mode	$(\frac{\delta\mathscr{B}}{\mathscr{B}})_{PDG}$	$(\frac{\delta\mathscr{B}}{\mathscr{B}})_{CLEO-c}$
$D^0 \to K^-e^+\nu$	5%	0.4%
$D^0 \to \pi^-e^+\nu$	16%	1.0%
$D^+ \to \pi^0 e^+\nu$	48%	2.0%
$D_s \to \phi e^+\nu$	25%	3.1%

TABLE 4. Expected precision in V_{cq} matrix elements with CLEO-c and the comparison to PDG 2002 values.

$\frac{\delta V}{V}$	CLEO-c	PDG 2002
$\frac{\delta V_{cd}}{V_{cd}}$	1.6%	7%
$\frac{\delta V_{cs}}{V_{cs}}$	1.7%	11%

$$\frac{d\Gamma(D \to Pl\nu)}{dq^2} = \frac{G_F^2}{24\pi^3}|V_{cq}|^2 p_P^3 |f(q^2)|^2, \quad (2)$$

where the form factor $f(q^2)$ encapsulates the strong physics. Form factor measurements are a key means to test theory's description of heavy quark decays. Precision measurements in inclusive semileptonic decays can strenuously test heavy quark effective theory (HQET)[5] while exclusive decays are a rigorous testbed for Lattice QCD (LQCD) calculations[6]. Table 3 shows the expected precision in branching fraction $\mathscr{B}$ for some important semileptonic decays with CLEO-c for an integrated luminosity of $3\,\text{fb}^{-1}$ and the comparison with PDG 2003 values.

The absolute branching fraction for a semilptonic D decay to a pseudoscalar can be combined with a measurement of the D lifetime τ_D to yield the total decay width:

$$\Gamma(D \to Pl\nu) = \frac{\mathscr{B}(D^0 \to Pl\nu)}{\tau_D} = \beta_{cq} V_{cq}, \quad (3)$$

with β_{cq} given by theory. Using eqs. 1, 2 and 3 and combining measurements from leptonic and semileptonic decays make it possible to measure charm decays constants directly, without the assumption of 3-generation unitarity, and to then determine the CKM matrix elements $|V_{cd}|$ and $|V_{cs}|$, also without the unitarity assumption. Table 4 shows the expected precision in V_{cd} and V_{cs} for CLEO-c with $3\,\text{fb}^{-1}$ of integrated luminosity.

QCD Probes

BESIII and CLEO-c will probe the low-energy nonperturbative structure of QCD with new precision. QCD predicts the existence of bound hadronic states in the mass range $\sim (1.5-2.5)\,\text{GeV}/c^2$ in which gluons are both constituents and the source of the binding force. Both fully gluonic "glueballs" and quark-gluon "hybrids" are novel forms of matter whose existence has yet to be unambiguously demonstrated. Their detection and study will be a major focus of J/Ψ running. With an expected CESR luminosity of $\mathscr{L} = 2 \times 10^{32}\,\text{cm}^2/\text{sec}^{-1}$ at $\sqrt{s} = J/\Psi$, CLEO-c expects to collect 1×10^9 J/Ψ events. BESIII should collect even more.

Radiative J/Ψ decays are a fruitful environment to search for glue rich hadronic matter[7] and CLEO-c, for example, will collect roughly 60 million $J/\Psi \to \gamma X$ decays with its projected $1\,\text{fb}^{-1}$ of integrated luminosity from J/Ψ running. With this projected data set and if the branching fraction measurements from BES[8] are indeed correct for the putative glueball candidate $f_J(2220)$, then CLEO-c will see many thousands of events in a variety of exclusive $f_J(2220)$ decay modes, $J/\Psi \to \gamma f_J(2220), f_J(2220) \to \pi\pi, K K, p\bar{p}$. Firmly establishing or debunking the existence of the $f_J(2220)$ is a CLEO-c and BESIII priority.

The inclusive photon spectrum from radiative J/Ψ decays is also a powerful means to search for new glue-rich hadronic states. Due to its nearly hermetic structure (93% of 4π), the CLEO-c detector is highly efficient at rejecting events of the type $J/\Psi \to \pi^0 X$ where one of the photons from the π^0 gets lost. With its projected data set of 10^9 J/Ψ events, CLEO-c should be able to detect any narrow resonances in radiative J/Ψ decays with a branching fraction of $\mathscr{O}(10^{-4})$ or larger.

ACKNOWLEDGMENTS

The author would like to thank the conference organizers for the invitation and a pleasant environment, and the U.S. Department of Energy for its support under contract DE-FG03-95ER40908.

REFERENCES

1. M. Gronau *et al*, hep-ph/0103110.
2. MARK III Collaboration, R.M. Baltrusaitis *et al*, Phys. Rev. Lett. **56**, 89 (1988).
3. J.L. Rosner, Phys. Rev. D **42**, 3732 (1990).
4. B.Grinstein *et al*, Phys. Rev. Lett. **56**, 298 (1986); F.J. Gilman and R.L. Singleton, Phys Rev. D **41**, 142 (1990); K. Hagiwara *et al*, Nucl. Phys B **327** 569 (1989).
5. I.I. Bigi *et al*, Ann. Rev. Nucl. Part. Sci. **47**, 591 (1997) and references therein.
6. Cornell Workshop on High-Precision Lattice QCD, January 2001.
7. T. Appelquist *et al*, Phys. Rev. Lett. **34**, 365 (1975); M.S. Chanowitz, Phys. Rev. D **12**, 918 (1975).
8. BES Collaboration, J.Z. Bai *et al*, Phys. Rev. Lett. **76**, 3502 (1996).

Heavy-quark recombination in Z^0 decay

Yu Jia

*Department of Physics and Astronomy, Michigan State University
East Lansing, MI 48824*

Abstract. We briefly review the recent advances of heavy-quark recombination mechanism. This mechanism predicts a class of power-suppressed 3-jet events in Z^0 decay, such as $b\bar{b}q$ and $b\bar{b}\bar{q}$. Furthermore, heavy quark fragmentation function also receives a contribution from this mechanism. Some light can be shed on the scaling of the maximum of the fragmentation function for S-wave heavy hadrons. We finally comment on a new variant of this mechanism which has important impact on the precision electroweak physics.

Heavy flavor production serves as an excellent testing ground for perturbative QCD [1]. So far, the heavy quark cross sections in all different processes have been computed to at least the next-to-leading order. However, in order to compare with experimental data, a sound understanding of how a heavy quark turns into a heavy hadron is crucial. The standard strategy is to implement the heavy quark fragmentation as the sole hadronization mechanism.

Inspired by the Non-relativistic QCD factorization for the formation of heavy quarkonium [2], a new hadronization mechanism, dubbed *heavy-quark recombination* (HQR) was recently developed [3]. It was initially motivated as a "higher twist" mechanism, to supplement the usual fragmentation. The central picture of this mechanism is quite simple: after a hard scattering, a heavy quark may capture a nearby light *parton* which emerges from the hard scattering and happens to carry soft momentum in the heavy quark rest frame. Subsequently they can materialize into a heavy hadron, plus additional soft hadrons. A typical HQR process in hadron collision at $O(\alpha_s^3)$ is $\bar{q}g \to \bar{B} + \bar{b} + X$, where $\bar{B}$ is produced from the $b\bar{q}$ recombination. Similarly, cq recombination has later been introduced to account for Λ_c production [4].

HQR is drastically distinct from the other "higher-twist" mechanisms – conventional recombination model [5], intrinsic charm model [6], and so on. In all these cases, the beam remnants participate in the dynamics of forming a heavy hadron. In contrast, the beam remnants play no role in HQR processes, which leads to great simplifications. In fact, HQR respects a simple factorization formula. Namely, inclusive production of heavy hadron in HQR can be expressed as a product of hard-scattering parton cross section, which is calculable in perturbative QCD, and a nonperturbative parameter (*recombination factor*), which characterizes the probability for the heavy quark and the light parton to evolve into a state containing the heavy hadron [3].

One important achievement of HQR is that it can explain the charm meson and baryon production asymmetries observed in a number of fixed-target experiments, in a simpler, more coherent and controlled fashion than those aforementioned models [7, 4]. The charm asymmetry is simply attributed to the asymmetry between the densities of light

FIGURE 1. Diagrams for the $b\bar{q}$ recombination process $Z^0 \to b\bar{q}(n) + \bar{b} + q$. The shaded blob represents the hadronization of $b\bar{q}(n)$ into $\bar{B}$ meson plus anything else.

quark and anti-quark in the beam and target hadron.

Although this success constitutes a strong evidence for HQR, the complicated hadronic environment in fixed-target experiments prevents us from excluding other hadronization models. Most probably, the asymmetries arise from the interplay of several different mechanisms, one of which is HQR. A curious question thereby is, is there a cleaner playground where HQR can be unambiguously singled out?

The answer is yes, because the physical idea of HQR is quite general, so its applications are not only confined in the hadroproduction of heavy hadron. In fact, heavy flavor production in e^+e^- annihilation is an ideal place to test HQR [8, 9]. In particular, we will be interested in B production on the Z^0-pole, thanks to the huge statistics of Z^0 samples. Clearly, those "higher-twist" mechanisms which rely on the beam remnants in hadron collision, are simply absent here.

Let us consider B production at $O(\alpha_s^2)$ through $Z^0 \to b\bar{b}q\bar{q}$. If each quark independently fragments, then it represents a regular 4-jet event. Nonetheless, in a small corner of phase space where $\bar{q}$ is soft in the b rest frame, they can form a composite $b\bar{q}$ state with definite color and angular momentum. Subsequently this state hadronizes into a $\bar{B}$ meson plus soft hadrons. We thereby end up with a jet containing $\bar{B}$ from the recombination, the recoiling $\bar{b}$ jet and a light quark jet [8]. The corresponding Feynman diagrams are shown in Fig. 1. The inclusive $\bar{B}$ production rate from HQR can be written

$$d\Gamma[\bar{B}] = \sum_n d\hat{\Gamma}[Z^0 \to b\bar{q}(n) + \bar{b} + q] \rho[b\bar{q}(n) \to \bar{B}]. \quad (1)$$

where $d\hat{\Gamma}_n$ are the perturbatively calculable parton cross sections, and ρ_n are the recombination factors, and n denotes the color and angular momentum quantum numbers of $b\bar{q}$. These ρ_n parameters have recently been defined in terms of nonperturbative QCD matrix elements [10]. An important property of these parameters is that they scale as Λ_{QCD}/m_b. While $\bar{B}$ can be produced in four different recombination channels, the color-singlet, spin-matching channel is expected to dominate. Adopting the fitted value of ρ_1^c from Ref. [7, 4], and using its scaling property, we can obtain $\rho_1^b = 0.1$.

A striking signal of these novel 3-jets is that the third jet is initiated by a *light quark*, instead of by a *gluon*. However, distinguishing quark and gluon jets experimentally requires a large statistics. OPAL collaboration has selected 3,000 symmetric 3-jet events at the Z resonance, in which the most energetic jet is tagged to contain b, and the angle between this b-jet and each of the two low energy jets is roughly $150°$ [11]. These samples were assumed to all be the $b\bar{b}g$ events. However now we know there must be a small fraction of them are actually made of HQR 3-jets. Simple dimensional argument suggests these 3-jets are suppressed by a factor of $\alpha_s(M_Z)\Lambda_{QCD}m_b/M_Z^2 \sim 10^{-5}$ relative to $b\bar{b}g$. However, a more quantitative study indicates that the ratio of the yield for HQR 3-jet events to that for the $b\bar{b}g$ in such a topology is roughly 0.012. So there are about 36 new events out of 3,000 OPAL samples, seemingly not statistically important. We hope that prospective Giga-Z experiments with a much larger number of Z^0 samples will confirm the existences of these 3-jet events definitely. If true, it should be viewed as a decisive triumph of the HQR mechanism.

Though the HQR cross section is highly suppressed for 3-jet, its magnitude becomes much larger when $\bar{B}$ and q lie in the fragmentation region, i.e., with a small invariant mass, because the virtuality of the internal gluon that splits into $q\bar{q}$ (see Fig. 1) becomes much smaller in this region. This motivates us to examine if this $b\bar{q}$ recombination process also contributes to the b fragmentation function.

Fragmentation functions are nonperturbative objects and usually defy a tackle from perturbation theory. This is true for q, g to fragment into π, K. However, the fact that b is heavy ($m_b \gg \Lambda_{QCD}$) may allow us to proceed further. Armed with knowing how a heavy quark hadronizes in the recombination picture encoded in Eq. (1), we can readily derive the HQR contribution to b fragmentation function by integrating the inclusive $\bar{B}$ differential cross sections over some appropriate kinematic variables. For example, the HQR contribution to b fragmentation into $\bar{B}^*$ turns out to be

$$D^{HQR}_{b\to\bar{B}^*}(z) = \frac{32\rho_1^b \alpha_s^2(m_b)}{81} \frac{z(2-2z+3z^2)}{(1-z)^2}, \qquad (2)$$

where z is the energy fraction carried by $\bar{B}^*$ relative to b. This HQR fragmentation function is not away from zero until z becomes large, and finally diverges quadratically as $z \to 1$. This divergence is a symptom that perturbative calculation in the endpoint region becomes invalid. Yet, one can show that Eq. (2) is still valid as long as $1-z \gg \Lambda_{QCD}$. The z distribution in Eq. (2) is much harder than the widely-used Peterson fragmentation function [13]. This may suggest that b hadronizing via picking up a $\bar{q}$ from vacuum is still non-negligible, even at relatively small z. However, some model-independent extraction of the nonperturbative part of b fragmentation function shows also a harder spectrum than Peterson parametrization [12].

Insight may be gained if we assume that $z \sim 1 - \Lambda_{QCD}$ is where the peak of fragmentation function is located. While a perturbative QCD treatment from HQR is ceasing to work when close to the endpoint region, $D^{HQR}_{b\to\bar{B}^*}(1-\Lambda_{QCD}/m_b)$ may still betray the correct order of magnitude of the maximum of the "true" fragmentation function. If this is true, then the fragmentation function of b to B^{*-} is expected to peak around $z = 0.93$, with a height roughly $\frac{32}{27}\rho_1^b \alpha_s^2(m_b)(m_b/\Lambda_{QCD})^2 \approx 1.5$. If we approximate the "true" B^{*-} fragmentation function by Peterson function $D(z;\varepsilon_b)$ with $\varepsilon_b = 0.006$, and take the fragmentation probability $f_{b\to B^{*-}} \approx 0.3$, the "true" peak is also around $z \approx 0.93$ with a height about 1.7, in good agreement with our naive estimate. Since $\rho_1^b \propto \Lambda_{QCD}/m_b$, we thereby propose the maxima of the fragmentation functions for S-wave heavy hadrons scale as $\alpha_s^2(m)m/\Lambda_{QCD}$. For charmed hadrons, this scaling law doesn't hold so well, but still conveys the correct order of magnitude.

FIGURE 2. The diagrams for the bg recombination process $Z^0 \to bg(n) + \bar{b}$.

A comprehensive understanding of Z^0 decay to heavy flavor is important to precision electroweak physics [14]. If we were able to extract the finite power correction from the linearly divergent total HQR cross section, it would represent an $O(\alpha_s^2 \Lambda_{QCD} m_b / M_Z^2) \sim 10^{-6}$ correction to the partial width of Z^0 to $b\bar{b}$. The Z^0 width has been measured to per mille accuracy, thus the contribution associated with Fig. 1 can be neglected.

However, there is a new HQR process, as depicted in Fig. 2, occurring at order α_s only, with a genuine "higher twist" contribution of order Λ_{QCD}/m_b [9]. To accomplish this, bg recombination needs to be invoked. The net contribution of this new mechanism to the partial width of Z^0 to $b\bar{b}$ turns out to be $\Delta\Gamma[b\bar{b}] = 32\pi\alpha_s(M_Z)\xi_3^b/9\Gamma_0[b\bar{b}]$, where ξ_3^b is an unknown color-triplet recombination factor. Both ξ_3^b and ξ_3^c may be fitted from the global electroweak analysis, and consequently the Standard Model predictions of various electroweak observables will be updated.

All the three HQR mechanisms, $b\bar{q}$, bq and bg recombination have now been fulfilled.

ACKNOWLEDGMENTS

I thank E. Braaten and T. Mehen for the collaboration on laying down the foundation of heavy-quark recombination mechanism. This work is supported in part by the National Science Foundation under Grant No. PHY-0100677.

REFERENCES

1. For a review, see S. Frixione *et al.*, Adv. Ser. Direct. High Energy Phys. **15**, 609 (1998).
2. G. T. Bodwin, E. Braaten and G. P. Lepage, Phys. Rev. D **51**, 1125 (1995) [*Erratum ibid.* D **55**, 5853 (1997)].
3. E. Braaten, Y. Jia and T. Mehen, Phys. Rev. D **66**, 034003 (2002).
4. E. Braaten *et al.*, hep-ph/0304280.
5. K. P. Das and R. C. Hwa, Phys. Lett. B **68**, 459 (1977) [*Erratum ibid.* **73B**, 504 (1978)].
6. R. Vogt and S. J. Brodsky, Nucl. Phys. B **478**, 311 (1996).
7. E. Braaten, Y. Jia and T. Mehen, Phys. Rev. D **66**, 014003 (2002); Phys. Rev. Lett. **89**, 122002 (2002).
8. Y. Jia, hep-ph/0305172.
9. Y. Jia, hep-ph/0307072.
10. C. H. Chang, J. P. Ma and Z. G. Si, Phys. Rev. D **68**, 014018 (2003).
11. G. Alexander *et al.* [OPAL Collaboration], Z. Phys. C **69**, 543 (1996).
12. E. Ben-Haim *et al.*, hep-ph/0302157.
13. C. Peterson *et al.*, Phys. Rev. D **27**, 105 (1983).
14. K. Hagiwara *et al.* [Particle Data Group Collaboration], Phys. Rev. D **66**, 010001 (2002).

Top Quark Mass Measurements at the Tevatron

Maria Florencia Canelli

University of Rochester
On behalf of the CDF and DØ Collaborations

Abstract. We present two new measurements of the top-quark mass. Using the same methodology applied in Run I, the CDF experiment uses 72 event/pb of Run II data to measure $M_{top} = 171.2 \pm 13.4_{stat} \pm 9.9_{syst}$ GeV/c². On the other hand, the DØ experiment, using 125 event/pb from Run I, and appling a new method that extracts information from data through a direct calculation of a probability for each event, obtains $M_{top} = 180.1 \pm 3.6_{stat} \pm 4.0_{syst}$ GeV/c².

INTRODUCTION

In proton-antiproton collisions at Tevatron energies, top quarks are produced primarily in pairs, either via $q\bar{q}$ or gg fusion. At the Tevatron, the main contribution to the $t\bar{t}$ yield is from $q\bar{q}$ annihilation. This is purely the result of the fact that the parton distribution functions (PDFs) favor this channel at Run I $\sqrt{s}$=1.8 TeV and Run II $\sqrt{s}$=1.96 TeV. In fact, about 90% of the top quarks are produced through the quark interaction.

The top quark is detected indirectly via its decay products. It decays via the weak interaction, and according to the Standard Model is almost always expected to decay to a b quark and a W boson. This is followed by the W decaying into two quarks or a lepton and a neutrino. The final state of the $t\bar{t}$ system has different topological classifications that depend on the decay of the W. The results presented here use the lepton+jets channel, and corresponds to one W decaying leptonically (into a electron or a muon), while the other W decays hadronically. This channel has a branching fraction of about 30%.

Although its value is not predicted, M_{top} is a fundamental parameter in the Standard Model. The best value of the top quark mass found from combining all channels at the Tevatron is [1],

$$M_{top} = 174.4 \pm 5.1 \text{ GeV}/c^2 \qquad (1)$$

The top quark mass, along with the mass of the W boson, provides through radiative corrections the best indication for the value of the mass of the Higgs boson [2]. The measurement of M_W will improve significantly in the future, with an uncertainty of 27 GeV/c² being a realistic goal for Run II of the Tevatron. To be able to make maximum use of this precision measurement to constrain the mass of the Higgs, the top mass should be measured with an uncertainty of less than 3 GeV/c². This will yield a prediction for the Higgs mass with an uncertainty of 40%. It is therefore important to develop techniques for extracting the mass of the top quark that will optimize the use of the Run II data.

CDF RUN II TOP-QUARK MASS MEASUREMENT

This is the first measurement of the top-quark mass using data from Run II of the Tevatron. The luminosity in this analysis corresponds to 72 events/pb. The selection criteria applied and the method used to extract the top quark mass are the same as in the previous analysis of Run I data [3]. The selection criteria consist of requiring one isolated high-p_T electron or muon, $\not{E}_T > 20$ GeV, $E_T^{jets} > 15$ GeV and $|\eta^{jets}| < 2$, which reduces the sample to 33 candidates.

A constrained fitting technique is employed to reconstruct the mass of the top quark, requiring $M_{l\nu} = M_W$, $M_{jj} = M_W$; $M_{top} = M_{\bar{t}op}$. The inability to identify uniquely the four jets in the lepton+jets channel results in 12 posibilities to reconstruct the event. Since the longitudinal momentum of the neutrino is not known, every combination can have two possible solutions for the neutrino p_z. After the constrained fit, the combination with lowest χ^2 is chosen as a measure of the top mass. These reconstructed top masses are compared to parameterized templates of top and background Monte Carlos. The top mass is extracted using a maximum likelihood method comparing the data and template distributions. The top mass obtained (see Figure 1) is :

$$M_{top} = 171.2 \pm 13.4_{stat} \pm 9.9_{syst} \text{ GeV}/c^2 \tag{2}$$

The systematic error is dominated by the uncertainty on the jet energy. CDF aims to reduce this error to ≈ 2 GeV/c^2.

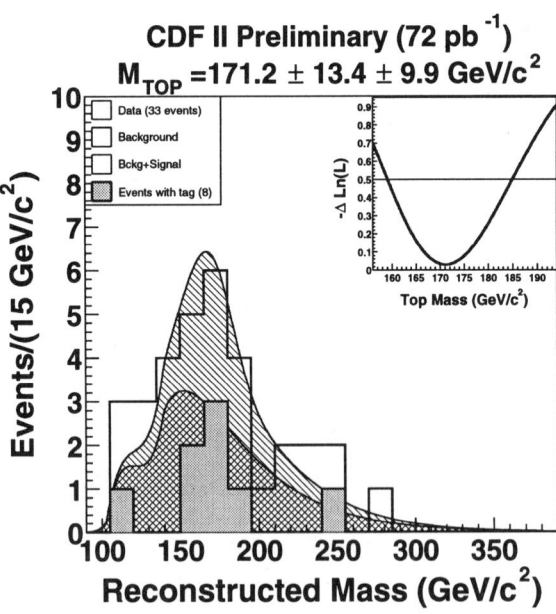

FIGURE 1. Mass of the top quark using lepton+jets events from Run II data. The white histogram consists of 33 events. The filled histogram is a subsample where at least one of the jets has been b-tagged. (This information is not used in reconstructing the top mass).

NEW DØ RUN I TOP QUARK MASS MEASUREMENT

This is a preliminary measurement of the mass of the top quark using a method that compares each individual event with the differential cross section for $t\bar{t}$ production and decay. This method is similar to that suggested for $t\bar{t}$ dilepton decay channels, and used in previous mass analyses of dilepton events [4]. The luminosity used in this analysis corresponds to 125 events/pb, and the data was accumulated by the DØ experiment during Run I of the Tevatron. This analysis is based on the same sample that was used to extract the mass of the top quark in the previous publication [5]. A set of selections was introduced to improve acceptance for lepton+jets from $t\bar{t}$ relative to background. The standard requirements were: $E_T^{lepton} > 20$ GeV, $|\eta_e| < 2$, $|\eta_\mu| < 1.7$, $E_T^{jets} > 15$ GeV, $|\eta_{jets}| < 2$, $\not{E}_T > 20$ GeV, $|E_T^{lepton}| + |\not{E}_T| > 60$ GeV ; $|\eta_{lepton+\not{E}_T}| < 2$. A total of 91 events remained after these selections.

The $t\bar{t}$ production probability is calculated as:

$$P_{t\bar{t}} = \frac{1}{12\sigma_{t\bar{t}}} \int d\rho_1 dm_1^2 dM_1^2 dm_2^2 dM_2^2 \sum_{\text{perm.},\nu} |M_{t\bar{t}}|^2 \frac{f(q_1)f(q_2)}{|q_1||q_2|} \Phi_6 W_{jet}(E_y, E_x) \qquad (3)$$

where $|M_{t\bar{t}}|^2$ is the leading-order matrix element, $f(q_1)$ and $f(q_2)$ are the CTEQ4M parton distribution functions for the incident quarks, Φ_6 is the phase-space factor for the 6-object final state, and the sum is over all 12 permutations of the jets (the permutation of the jets from W decay was performed by symmetrizing the matrix element), and all possible longitudinal momenta of the neutrino solutions. The integration variables used in the calculation are the top masses ($m_{1,2}$), the W masses ($M_{1,2}$), and the energy of one of the jets (ρ_1). Observed electron momenta are assumed to correspond to those of produced electrons. The angles of the jets are also assumed to reflect the angles of the partons on the final state, and we ignore any transverse momentum for the incident partons. $W_{jet}(E_y, E_x)$ corresponds to a function that parameterizes the mapping between parton-level energies E_y and energies measured in the detector E_x. A large Monte Carlo sample of $t\bar{t}$ events (generated with masses between 140–200 GeV in HERWIG [6], and processed through the DØ detector-simulation package) is used to determine $W_{jet}(E_y, E_x)$. For a final state with a muon, W_{jet} is expanded to include the known muon momentum resolution, and an integration over muon momentum is added to Eq. 3. Effects such as geometrical acceptance, trigger efficiencies, event selection, etc, are taken into account through a multiplicative function $A(x)$ that is independent of M_{top}. This function relates the production probability $P(x; M_{top})$ to the measured probability $P_m(x; M_{top})$: $P_m(x; M_{top}) = A(x)P(x; M_{top})$. All processes that can contribute to the observed final state must be included in the probability. Therefore the final probability is written as $c_1 P(x; M_{top}) + c_2 P_{background}(x)$. The VECBOS [7] W+jets matrix element is used to calculate the background probability, which is integrated over the four jet energies and the W-boson mass, and later summed over the 24 jet permutations and neutrino solutions.

Since the method involves a comparison of the data with a leading-order matrix element for the production and decay process, the sample is restricted to only four jets events, thereby reducing the sample to 71 events. In order to increase the purity of

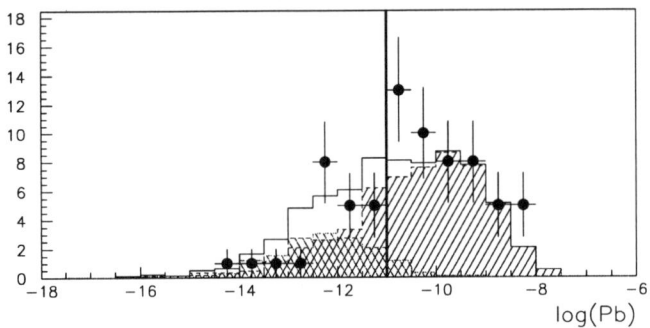

FIGURE 2. Distribution os probability of events being background. Only events which have $P_{background} < 10^{-11}$ are used in this analysis.

signal, a selection in the background probability, $P_{background} < 1.10^{-11}$, is applied. This selection is required to minimize a bias introduced by the presence of background, and its imposition leaves a sample of only 22 events. Figure 2 shows a comparison between the probability for a background interpretation of events calculated for a large sample of Monte Carlo events (solid histogram) and for the 71 $t\bar{t}$ candidates (data points). The left-hatched (right-hatched) histogram shows the contribution from $t\bar{t}$ (W+4 jets) MC events. The probabilities are inserted into a likelihood function for N observed events, which compares the Standard-Model prediction with the data. The best estimate of M_{top} is obtained by maximizing this likelihood function. Figure 3a) shows the value of $-lnL$ as a function of M_{top} for the 22 events that pass all the selection criteria, 12 of which are signal and 10 background. ($-lnL$ was minimized with respect to the parameters c_1 and c_2 at each mass point.) Figure 3b) shows the likelihood normalized to its maximum value. The Gaussian fit in the figure yields M_{top}=179.6 GeV/c², and an uncertainty δM_{top}=3.6 GeV/c². Monte Carlo studies show that there is a shift to 0.5 GeV/c² in the extracted mass. Applying this shift the new result yields:

$$M_{top} = 180.1 \pm 3.6(\text{stat}) \pm 4.0(\text{sys}) \text{ GeV}/c^2 \quad (4)$$

The main systematic uncertainties are due to the jet-energy scale (3.6 GeV/c²), model for $t\bar{t}$ (1.5 GeV/c²), model for background (1.0 GeV/c²), noise and multiple interactions (1.3 GeV/c²), parton distribution functions (0.2 GeV/c²), and acceptance corrections (0.5 GeV/c²).

CONCLUSIONS

Measuring the top quark mass with an uncertainty of less than 3 GeV/c² is important to restrict the mass of the Higgs boson. With the new data and improved techniques, this appears to be a realistic goal. CDF has made the first attempt to measure the top mass using Run II data. Meanwhile, DØ developed a method to maximize the use of information in each event to reduce the uncertainty on the top mass, which when applied

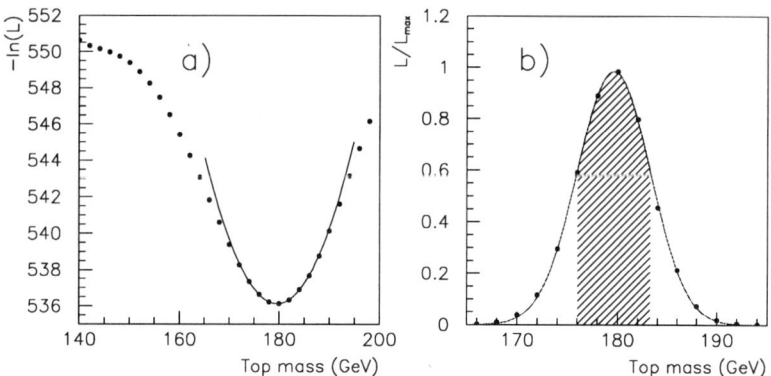

FIGURE 3. a) Negative of the log of the likelihood as a function of the mass of the top quark for the 22 $t\bar{t}$ candidates in our final sample. b) Likelihood normalized to the maximum value. The curves are Gaussian fits to the likelihood plot b). The hatched area corresponds to the 68.27% probability interval.

to Run I data, yields a result with an error comparable to all previous DØ and CDF measurements of the mass of the top quark, combined.

REFERENCES

1. Particle Data Group. *http://pdg.lbl.gov* (2002)
2. K. Hagiwara et al., Phys. Rev. D **66**, 01001 (2002).
   LEP Working Group for Higgs, LEPEWWG/2002-1 (2002).
3. T.Affolder et al., Phys. Rev. D **63**, 032003 (2001).
4. R. H. Dalitz and G. R. Goldstein, Proc. R. Soc. Lond. A **445**, 2803 (1999).
   K. Kondo et al., J. Phys. Soc. Jap. **62**, 1177 (1993).
   B. Abbott, et al., Phys. Rev D **60**, 052001 (1999).
5. DØ Collaboration, B. Abbott, et al., Phys. Rev. D **58**, 052001 (1998).
6. G. Marchesini et al., Comput. Phys. Commun. **67**, 467 (1992).
7. F. A. Berends, H. Kuijf, B. Tausk and W. T. Giele, Nucl. Phys. **B357**, 32 (1991).

Charm and Beauty at the Tevatron

Jack Cranshaw

Texas Tech University
for the CDF and D0 Collaborations

Abstract. The large heavy quark production cross section in $p\bar{p}$ collisions makes the Tevatron an excellent place to study charm and bottom physics. This allows for a rich program of spectroscopy, CP parameter measurements, and searches for new physics.

INTRODUCTION

The Tevatron is a $p\bar{p}$ collider at the Fermi National Accelerator Laboratory near Chicago Illinois. It operates at a center of mass energy of $\sqrt{s} = 1.96$ TeV and is currently the highest energy hadron collider in the world. Two detectors, the Collider Detector at Fermilab (CDF) and D0, are operated at the Tevatron.

The Tevatron is an excellent place to study bottom and charm quark physics for two main reasons reasons: the large cross section ($\sim 100 \mu b$) relative to e^+e^- colliders (few nb), and the production of a full spectrum of mesons and baryons with b and c quarks. Studies of the D^0, B_d^0 and B_s^0 systems give access to various CP parameters. Studies of rare decays give access to various CP effects as well as new physics[1]. In 2001, Fermilab began collider operations (Run 2) with a set of physics goals with a large emphasis on heavy quark (c,b,t) physics. The program in charm and and bottom physics has started with an emphasis on spectroscopy, and is starting to move on to searches for new physics and CP violation.

CDF and D0 Detectors

In order to pursue the previously described physics agenda after Run 1 the detectors at the Tevatron, CDF and D0, underwent extensive upgrades during the period 1996-2001. For the B physics programs at the two experiments, the upgrades addressed the following needs.

- Excellent tracking and vertexing.
- Ability to trigger on displaced vertices.
- Particle ID for flavor tagging.
- Efficient use of trigger bandwidth.

The upgraded CDF and D0 detectors have been described in more detail elsewhere[2].

Triggers and Datasets

Both D0 and CDF use three types of triggers for charm and beauty physics which roughly correspond to three different types of final states: di-lepton decays, semileptonic decays, and multi-hadron decays. The di-lepton triggers were used in Run 1 as well, but the single lepton trigger p_T thresholds have been lowered and the rapidity range extended. Semileptonic decays were measurable in Run 1 only by using single lepton triggers which had a low signal purity. The signal purity has been improved in Run 2 by adding a displaced vertex requirement which became possible due to the addition to the trigger of a programmable module which uses information from the silicon tracker to detect displaced vertices. This trigger has been in regular use at CDF, and is being commissioned at D0. The ability to trigger on displaced vertices also allows us to implement a trigger for charm and bottom decays to fully hadronic final states.

The Tevatron luminosity has been gradually improving since the beginning of 2002, and is approaching its initial design goals for instantaneous luminosity (5×10^{31} cm^{-2} s^{-1}). Since the end of commissioning in Spring 2002 this has allowed both experiments to accumulate over 150 pb^{-1} of integrated luminosity on tape. Due to data quality and ongoing detector studies the results presented here are based on up to 115 pb^{-1} for both CDF and D0.

RUN 2 RESULTS

Charm Spectroscopy

The displaced vertex trigger has opened up a large area of charm physics at the Tevatron which can be measured with relatively low levels of integrated luminosity. Figure 1 shows the ratio of the measured cross sections $d\sigma/p_T$ for $D^0 \rightarrow K^+\pi^-$, $D^+ \rightarrow K^\pm \pi^\mp \pi^\pm$, and $D^{*+} \rightarrow D^0 \pi^+$ to theoretical calculations[3] as a function of p_T. As the figure shows, all of the ratios are systematically above 1.0.

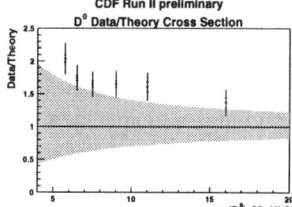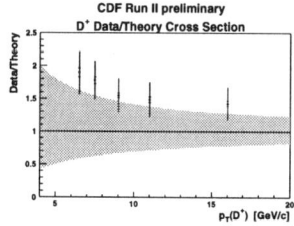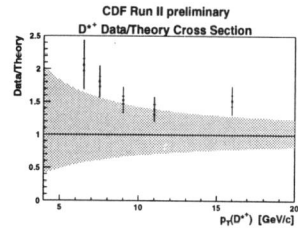

FIGURE 1. Ratio of Data to Theory for the exclusive charmed meson decays: $D^0 \rightarrow K^\pm \pi^\mp$, $D^\pm \rightarrow K^\pm \pi^\mp \pi^\pm$, and $D^{*\pm} \rightarrow D^0 \pi^\pm$.

Using the lower p_T thresholds for muons available in early running at CDF, they measured the J/ψ inclusive cross section in $p\bar{p}$ collisions down to $p_T^{J/\psi} = 0$. This is the first time this has been done at a hadron collider.

Bottom Spectroscopy

A variety of bottom mesons and baryons have been reconstructed at the Tevatron using all three datasets: dilepton, semileptonic, and hadronic. Other than extended kinematic coverage the dilepton dataset is similar to what was available in Run 1, so analyses using this dataset are the most advanced. So far all results use only the dimuon channel. The results for B meson and b baryon masses and lifetimes for decays to a final state with a J/ψ are summarized in table 1. For the B_s^0 and Λ_b^0 mass measurements, these are already the best available measurements. Currently the Tevatron is the only place that b baryons can be studied, and we have already collected the world's largest number of observed Λ_b^0 decays using all three datasets as shown in figure 2.

TABLE 1. Masses and Lifetimes of bottom mesons and baryons. Unless otherwise indicated, where multiple uncertainties are quoted, the first uncertainty is statistical and the second is systematic.

Decay		CDF	D0
J/ψ inclusive	life	1.526 ±0.034 ± 0.035	1.552 ±0.013 ± 0.028
$B_u^+ \to J/\psi K^+$	mass	5279.32 ± 0.68 ± 0.94	5271.2 ± 1.7(stat)
	life	1.63 ± 0.05 ± 0.04	1.65 ± 0.083 $^{+0.096}_{-.1233}$
$B_d^0 \to J/\psi K^{*0}$	mass	5280.30 ± 0.92 ± 0.96	5264.8 ± 2.6(stat)
	life	1.51 ± 0.06 ± 0.02	1.51 $^{+.19}_{-.17}$ ± 0.20
$B_s^0 \to J/\psi \phi$	mass	5365.50 ± 1.29 ± 0.94	5359.4 ± 3.8(stat)
	life	1.33 ± 0.14 ± 0.02	1.19 $^{+.19}_{-.16}$ ± 0.14
$\Lambda_b^0 \to J/\psi \Lambda$	mass	5620.4 ± 1.6 ± 1.2	5600 ± 25(stat)
	life	1.25 ± 0.26 ± 0.10	1.05 $^{+.21}_{-.18}$ ± 0.12

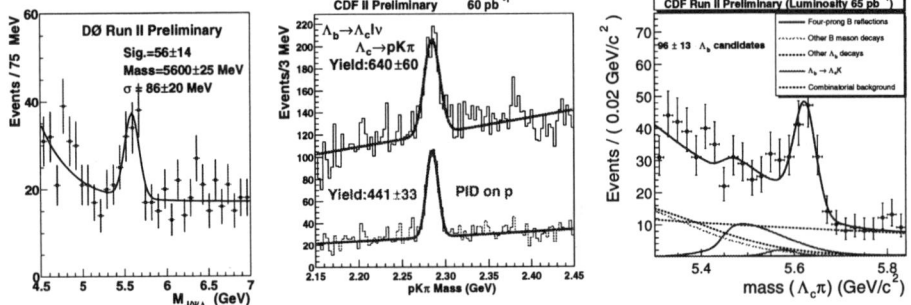

FIGURE 2. Reconstructed Λ_b^0 decays using the dimuon dataset, the semileptonic dataset, and the hadronic dataset.

Of particular interest for bottom physics at the Tevatron is the study of the properties of the B_s^0 meson. In a general sense it is less well measured than the B_d^0[4], and it also undergoes mixing in a similar way. In fact, there are a variety of CP measurements which benefit from a combined analysis of B_d^0 and B_s^0 decays which are possible at the Tevatron[5]. Figure 3 shows plots for the highest yield fully reconstructed B_s decay and an analogous B_d decay used for normalization and systematic studies.

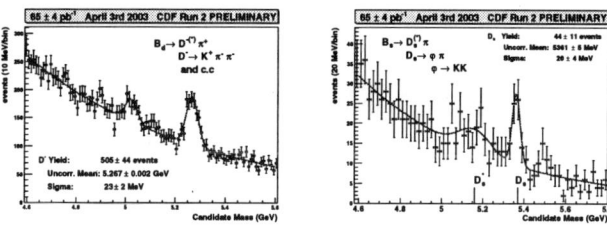

FIGURE 3. Reconstructed $B_d \to D^+\pi^-$ and $B_s \to D_s^+\pi^-$ decays.

Rare Decays

The large charm/bottom production rate has also produced results for the first rare decay searches from Run 2 at the Tevatron. The decays $D^0 \to \mu^+\mu^-$ and $B_s^0 \to \mu^+\mu^-$ are highly suppressed in the Standard Model $\sim 10^{-9}$[6], but can be enhanced by various types of new physics[7]. Figure 4 shows the result for searches for these two decays which gives values of $BR(D^0 \to \mu^+\mu^-) < 2.4 \times 10^{-6}$ at the 90% confidence level and $BR(B_s^0 \to \mu^+\mu^-) < 9.5 \times 10^{-7}$ at the 90% confidence level. Both of these measurements are significant improvements on the current limits[4].

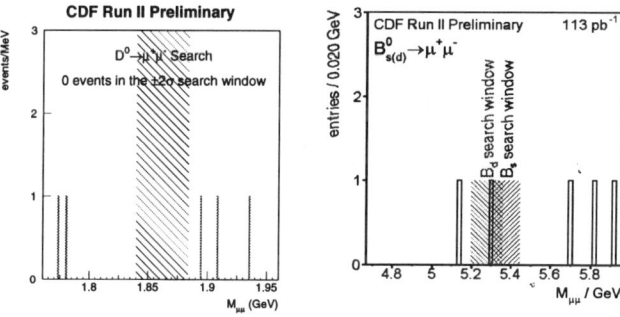

FIGURE 4. Search results for the rare decay $D^0 \to \mu^+\mu^-$ for a luminosity of 69 pb^{-1} and for the rare decay $B_s^0 \to \mu^+\mu^-$ for a luminosity of 113 pb^{-1}.

REFERENCES

1. K. Anikeev, *et. al.*, FERMILAB-Pub-01/197, 2001.
2. J. Cranshaw, *NIM* B115:302, 2003.
3. P. Nason, M. Cacciari, hep-ph/0306212..
4. M. Artuso, *et. al.*, *Phys. Rev.* D66, 2002.
5. Fleischer, *Phys Lett* B459:306, 1999.
6. M. Misiak and J. Urban, *Phys Lett* B451:161, 1999.
7. R. Arnowitt, *et. al.*, *Phys Lett* B538:121, 2002.

Tau-Mu Flavor Violation and the Scale of New Physics

Deirdre Black

Theory Group, Jefferson Lab, 12000 Jefferson Ave, Newport News, VA 23606

Abstract. Motivated by the strong experimental evidence of large $v_\mu - v_\tau$ neutrino oscillations, we study existing constraints for related $\mu - \tau$ flavor violation. Using a general bottom-up approach, we construct dimension-6 effective fermionic operators whose coefficients encode the scale of new physics associated with $\mu - \tau$ flavor violation, which is a piece in the puzzle of the origin of neutrino oscillations. We survey existing experimental bounds on this scale, which arise mostly from τ and B decays. In many cases the new physics scale is constrained to be above a few TeV. We also discuss the operators which are either weakly constrained or, at present, subject to no experimental bounds.

INTRODUCTION

In the past few years there has been a wealth of exciting results from solar, atmospheric, reactor and accelerator neutrino experiments [1] which provide strong evidence for neutrino oscillations. The atmospheric neutrino experiments point in particular to maximal $v_\mu - v_\tau$ mixing. In Ref. [2] we investigated bounds on related $\mu - \tau$ flavor violation in the charged lepton sector. This, together with other studies of lepton flavor violation, for example $\mu - e$ mixing, complements what we learn from neutrino oscillations and will help in eventually achieving an understanding of lepton flavor dynamics.

In order to carry out a systematic study, we use a model-independent effective operator approach. We consider an effective theory containing the Standard Model (SM) at dimension-4 and a class of dimension-6 four-fermion lepton flavor-violating operators of the form

$$O_{\mu\tau} = \frac{4\pi}{\Lambda^2}(\bar{\mu}\Gamma\tau)(\overline{q^\alpha}\Gamma q^\beta). \tag{1}$$

Here α and β are flavor indices and Γ denotes any of the four Dirac structures $\Gamma = 1, \gamma_5, \gamma^\mu, \gamma_5\gamma^\mu$. This effective interaction paramaterizes new physics effects associated with $\mu - \tau$ flavor-violation below an ultraviolet cutoff scale, Λ. As illustrated schematically in Fig. 1, such dimension-6 operators could be generated, for example, from exchange of new gauge bosons or scalar (pseudoscalar) particles in the underlying theory at the scale Λ. A careful discussion of the construction of these operators is given in Ref. [2].

In Ref. [2] we also considered purely leptonic effective interactions of the form $\frac{4\pi}{\Lambda^2}(\bar{\mu}\Gamma\tau)(\overline{l^\alpha}\Gamma l^\beta)$. Updated bounds using the latest upper limits from Belle [3] on the branching fractions for decays such as $\tau^- \to \mu^-\mu^+\mu^-$ are given in Ref. [4].

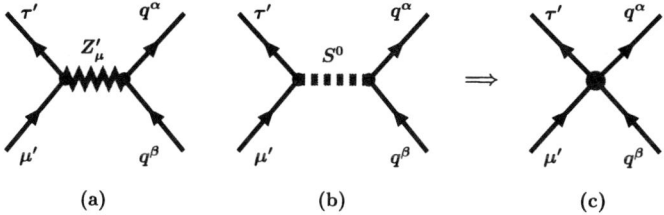

FIGURE 1. Illustration of generation of the effective four-Fermi interactions in Eq. (1), shown in diagram (c), from exchange of (a) heavy gauge boson Z'_μ or (b) new scalar or pseudoscalar particle S^0.

PHENOMENOLOGICAL CONSTRAINTS

Light quarks: τ decays

Experimental searches for lepton flavor violating (LFV) τ decay modes give strong bounds on many operators involving light quarks. The most recent experimental results[1] for branching ratios for τ decay to μ and pseudoscalar or vector mesons are given in Refs. [3, 5, 6]. In order to obtain bounds[2] on Λ, we calculate the appropriate matrix element of the effective operator leading to each decay using vacuum insertion and standard results for the hadronic matrix elements of the light quark bilinear current densities.

One b quark: B decays

The operators in Eq. (1) with $\Gamma = \gamma^5, \gamma^\mu \gamma^5$ and with quark-antiquark combinations $\bar{b}d$ and $\bar{b}s$ will give decays of B and B_s to $\mu\tau$. We calculate this width using vacuum insertion and a Heavy Quark Effective Theory estimate of $|\langle 0|\bar{q}\gamma^5 b|B\rangle|$. The experimental upper limits are $\text{Br}(B^0 \to \mu\tau) < 8.3 \times 10^{-4}$ [5] and $\text{Br}(B_s \to \mu\tau) < 10\%$ (a conservative estimate based on the observed B_s lifetime). Similarly, the operators with $\Gamma = 1, \gamma^\mu$ give the decays $B \to \pi\mu\tau, K\mu\tau$ which we calculate using a quark model estimate for the form factors in the heavy-light meson matrix elements. There is no experimental data at present on these branching ratios so we take conservative estimates $\text{Br}(B \to \pi\mu\tau, K\mu\tau) < 5\%$, which give $\Lambda > 2.5$ TeV.

---

[1] The speaker thanks Jon Urheim for drawing her attention to these new results [6, 3] in his talk "Rare tau decays: an experimental review" at this conference.

[2] We considered only one operator in Eq. (1) at a time, so treated the operators $\bar{\mu}\Gamma\tau\bar{d}\Gamma s + H.c.$ and $\bar{\mu}\Gamma\tau\bar{s}\Gamma d + H.c.$ independently and, using the new experimental limits [3] on the branching ratio for $\tau \to \mu K_s$, quote strong bounds on each one. If these operators occur with the same coefficients then their contributions to $\tau \to \mu K_s$ would actually cancel (up to presumably small CP violating effects).

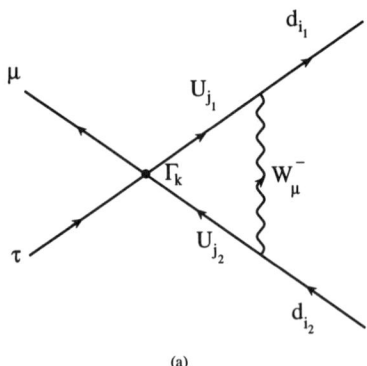

FIGURE 2. Diagram relating couplings involving heavy u-type quark bilinears to light d-type quark bilinears.

Loop-induced processes

We found that it is hard to constrain all operators $O_{\mu\tau}$ with heavy quarks directly from tree-level processes. Hence we consider one-loop processes which induce effective light-quark vertices via insertions of the relevant heavy quark operators, for example diagrams with $W^\pm$ exchange as shown in Fig. 2. As an example we see that through this mechanism, if the quarks in the loop are c and $\bar{c}$ then the final state quarks can be s and $\bar{s}$. This means that the operator $\bar{\mu}\Gamma\tau\bar{c}\Gamma c$ can lead to an amplitude at one loop for τ decay to μ and an $\bar{s}s$ state such as the ϕ vector meson. Other operators for which the strictest bounds come from similar one-loop processes are given in Table 2.

SUMMARY AND DISCUSSION

We have presented a survey based on Ref. [2] of existing constraints on flavor violation in the charged lepton sector due to effective interactions of the type $\frac{4\pi}{\Lambda^2}(\bar{\mu}\Gamma\tau)(\overline{q^\alpha}\Gamma q^\beta)$. Most of our lower bounds on the scale of new physics responsible for these lepton flavor-violating operators come from τ and B decays at tree or one-loop level and are of order 1-10 TeV. A complete summary of our results is given in Table 2.

It will be interesting to further apply the bounds presented in this study to constrain specific new physics scenarios which can generate the $\mu - \tau$ flavor violating interactions of the kind we have analyzed. We note that as ongoing experimental searches for rare τ decay modes reach higher levels of precision almost half of the bounds listed in Table 2 will become more stringent. Searches for $\mu\tau$ in charmonium and bottomonium decays may help to constrain some of the operators for which we could at present find no limits. Also we expect searches for $B_s \to \mu\tau$ and for B decays to $(\mu\tau + \text{meson})$ to improve the values listed in Table 2 for operators involving one b quark. It would also be interesting to look for the decay $t \to \mu\tau + \text{jet}$ at top-quark factories.

TABLE 1. Bounds at 90% C.L. on four-Fermi flavor-violation operators discussed in text. Asterisk indicates that no bounds have been found, otherwise we list the process which gives the strongest bound.

Bound	1	γ_5	γ_σ	$\gamma_\sigma \gamma_5$
$\bar{u}u$	2.6 TeV $(\tau \to \mu \pi^+ \pi^-)$	12 TeV $(\tau \to \mu \pi^0)$	12 TeV $(\tau \to \mu \rho)$	11 TeV $(\tau \to \mu \pi^0)$
$\bar{d}d$	2.6 TeV $(\tau \to \mu \pi^+ \pi^-)$	12 TeV $(\tau \to \mu \pi^0)$	12 TeV $(\tau \to \mu \rho)$	11 TeV $(\tau \to \mu \pi^0)$
$\bar{s}s$	1.5 TeV $(\tau \to \mu K^+ K^-)$	9.9 TeV $(\tau \to \mu \eta)$	14 TeV $(\tau \to \mu \phi)$	9.5 TeV $(\tau \to \mu \eta)$
$\bar{s}d$	2.3 TeV $(\tau \to \mu K^+ \pi^-)$	24.3 TeV $(\tau \to \mu K_s)$	13 TeV $(\tau \to \mu K^*)$	23.6 TeV $(\tau \to \mu K_s)$
$\bar{b}d$	2.2 TeV $(B \to \pi \mu \tau)$	9.3 TeV $(B \to \mu \tau)$	2.2 TeV $(B \to \pi \mu \tau)$	8.2 TeV $(B \to \mu \tau)$
$\bar{b}s$	2.6 TeV $(B \to K \mu \tau)$	2.8 TeV $(B_s \to \mu \tau)$	2.6 TeV $(B \to K \mu \tau)$	2.5 TeV $(B_s \to \mu \tau)$
$\bar{t}c$	190 GeV $(t \to c \mu \tau)$	190 GeV $(t \to c \mu \tau)$	310 GeV $(B \to \mu \tau)$	310 GeV $(B \to \mu \tau)$
$\bar{t}u$	190 GeV $(t \to u \mu \tau)$	190 GeV $(t \to u \mu \tau)$	650 GeV $(B \to \mu \tau)$	650 GeV $(B \to \mu \tau)$
$\bar{c}u$	★	★	550 GeV $(\tau \to \mu \phi)$	550 GeV $(\tau \to \mu \phi)$
$\bar{c}c$	★	★	1.1 TeV $(\tau \to \mu \phi)$	1.1 TeV $(\tau \to \mu \phi)$
$\bar{b}b$	★	★	180 GeV $(\Upsilon \to \mu \tau)$	★
$\bar{t}t$	★	★	75 GeV $(B \to \mu \tau)$	120 GeV $(B \to \mu \tau)$

ACKNOWLEDGMENTS

D.B. thanks M. Artuso and A. Soni for the invitation to speak in this interesting session and wishes to acknowledge support from the Thomas Jefferson National Accelerator Facility operated by the Southeastern Universities Research Association (SURA) under DOE Contract No. DE-AC05-84ER40150.

REFERENCES

1. For a recent review see M.C. Gonzalez-Garcia and Y. Nir, Rev. Mod Phys. 75 (2003) 345.
2. D. Black, T. Han, H.-J. He and M. Sher, Phys. Rev. D **66**, 053002 (2002); hep-ph/0206056.
3. Y. Yusa, H. Hayashii, T. Nagamine, A. Yamaguchi for the Belle Collaboration, hep-ex/0211017.
4. D. Black, T. Han, H.-J. He and M. Sher; Proceedings of MRST 2003; hep-ph/0307182
5. Particle Data Group, K. Hagiwara *et al*, Phys. Rev. D **66**, 1-I 010001 (2002) and http://pdg.lbl.gov.
6. CLEO Collaboration, S. Chen *et al*, Phys. Rev. D **66** 071101(R) (2002); hep-ex/0208019.

Rare Decays of Tau Leptons: an Experimental Review

Jon Urheim

School of Physics and Astronomy, University of Minnesota, Minneapolis, MN 55413

Abstract. I review the current experimental situation with regard to searches for rare decays of τ leptons. The recent progress in this area has been made by the BELLE and BaBar collaborations thanks to their impressive datasets and detection capabilities, attention to backgrounds, and careful statistical treatments of the data. After summarizing the present status, the review concludes with look at the future prospects for these searches.

1. INTRODUCTION

The τ lepton offers unique possibilities for searches for physics beyond the Standard Model (SM). By virtue of being a heavy lepton, neutrino-less decays to lighter charged leptons can be probed experimentally. These transitions are typically lepton flavor violating (LFV). That is, the additive quantum numbers (L_e, L_μ, L_τ) associated with the three lepton flavors are separately not all conserved in these decays, in contrast with all observations of known interactions. This apparent conservation law is not associated with a known symmetry, and thus remains an unexplained aspect of the SM. LFV processes are natural features of many models for new physics.

By virtue of being the heaviest lepton, the τ has more LFV final states accessible than the other leptons. On the other hand, searches for lepton flavor violation in rare decays of muons, pions and kaons benefit by virtue that these particles can be produced in copious quantities. However, it is possible that for example LFV interactions involving third generation particles are more prevalent, enhanced perhaps by mass-dependent couplings. At this conference, we heard from D. Black [1] on constraints on the scale of new physics from processes involving τ-μ flavor violation.

In this talk, I report on recent experimental progress on rare τ lepton decays, focussing on neutrino-less channels that are forbidden in the SM. Time limitations prevent me from discussing the status of searches for highly-suppressed channels (*i.e.*, those such as $\tau^- \to \pi^- \eta \nu_\tau$ in which G-parity is not conserved) where new physics might also appear.

2. THE CLASSIC DECAY $\tau^- \to \mu^- \gamma$

The decay $\tau \to \mu\gamma$ (see Fig. 1) is analogous to the $\mu \to e\gamma$ decay that has been the subject of a number of dedicated experiments. The current 90% CL upper limit on the branching ratio, $\mathcal{B}(\mu \to e\gamma) < 1.2 \times 10^{-11}$ from the MEGA experiment [2], is already

FIGURE 1. Feynman diagram for $\tau \to \mu\gamma$. Particles in the loop could include heavy neutrinos, supersymetric particles, heavy neutral bosons, etc.

quite impressive and future experiments are being planned [3]. Considering models with mass-dependent couplings, comparable sensitivity to the new physics can be attained in $\tau \to \mu\gamma$ at only the $10^{-6} - 10^{-8}$ level in branching fraction. Even at this meeting, we heard from R. Kitano [4] about interesting theoretical prospects for $\tau \to \mu\gamma$.

With τ-pair production in e^+e^- collisions, experimenters have two main kinematic handles on $\tau \to \mu\gamma$. First, the reconstructed $\mu\gamma$ invariant mass $M(\mu\gamma)$ must equal the τ mass. Second, the deviation $\Delta E = E_\mu + E_\gamma - E_{beam}$ where all quantities are evaluated in the e^+e^- CM frame, must be consistent with zero. The latter condition holds only approximately with the presence of initial state radiation (ISR). Additional requirements on the system recoiling against the $\mu\gamma$ system can be placed to ensure that the event is compatible with being due to τ-pair production.

Until recently, the most stringent limits on $\tau \to \mu\gamma$ came from the CLEO-II experiment operating at the Cornell Electron Storage Ring (CESR) at energies in the Υ energy region. An analysis in 1992 [5] of 1.6 fb^{-1} of data ($\sim 1.4 \times 10^6$ produced τ-pairs) yielded no candidate events and a 90% CL upper limit of 4.2×10^{-6} on $\mathscr{B}(\tau \to \mu\gamma)$.

This analysis was most recently updated in 1999 [6] with a sample of $N_{\tau\tau} \sim 12.6$ million τ-pairs, which yielded a limit of $\mathscr{B}(\tau \to \mu\gamma) < 1.1 \times 10^{-6}$, while six events were observed in the signal region in the $\Delta E - M(\mu\gamma)$ plane, consistent with 5.5 ± 0.5 expected background events. The improvement in the limit at a rate faster than $1/\sqrt{N_{\tau\tau}}$ despite the presence of background was facilitated by the use of an extended unbinned maximum likelihood fit (a "cut-and-count" analysis yielded a limit of 1.8×10^{-6}).

In the past year, BaBar and Belle have investigated this decay, both with considerably larger data samples. From the CLEO experience, the immediate question is whether $\tau \to \mu\gamma$ isn't already background-limited. The two main background sources are (1) radiative μ-pair events ($e^+e^- \to \mu^+\mu^-\gamma$), and (2) τ-pair events in which one τ decays to the common $\mu\nu\bar{\nu}$ final state. In the latter case, a radiative photon can be emitted in the decay, and thus, if the neutrinos are soft enough, one is left with the same kinematics as in $\tau \to \mu\gamma$ decay. Alternately, the muon from $\tau \to \mu\nu\bar{\nu}$ can be combined with an ISR photon. The τ backgrounds tend to give ΔE values less than zero, while the $\mu\mu\gamma$ background can yield ΔE values greater than zero.

The preliminary BaBar analysis [7], presented at ICHEP and at the Tau Lepton Workshop in summer 2002, is based on a sample of 56 million τ-pairs. A fully "blind" analysis was carried out, with signal and background control regions excluded from the determination of selection criteria. With careful attention to suppression of the $\mu\mu\gamma$ background, the total residual background was estimated based on extrapolation from sideband regions in the data to be 7.8 ± 1.4 events, with an acceptance for $\tau \to \mu\gamma$ decays of $(5.2 \pm 0.5)\%$ (the acceptance for the CLEO cut-and-count analysis of $\sim 12.7\%$). The background estimation method was validated by Monte Carlo simulations of background processes, as well as through comparisons of estimated yields in sideband regions with observed yields.

The final event sample from the BaBar analysis is shown in the left plot in Fig. 2. The signal region contains 13 events, a yield higher than, but consistent with, the background estimation. The 90% CL upper limit on the $\tau \to \mu\gamma$ yield of 11.5 events leads to a limit $\mathscr{B}(\tau \to \mu\gamma) < 2.0 \times 10^{-6}$, actually less stringent than the CLEO limit. This is due to the lower detection efficiency and the apparent upward fluctuation in the background.

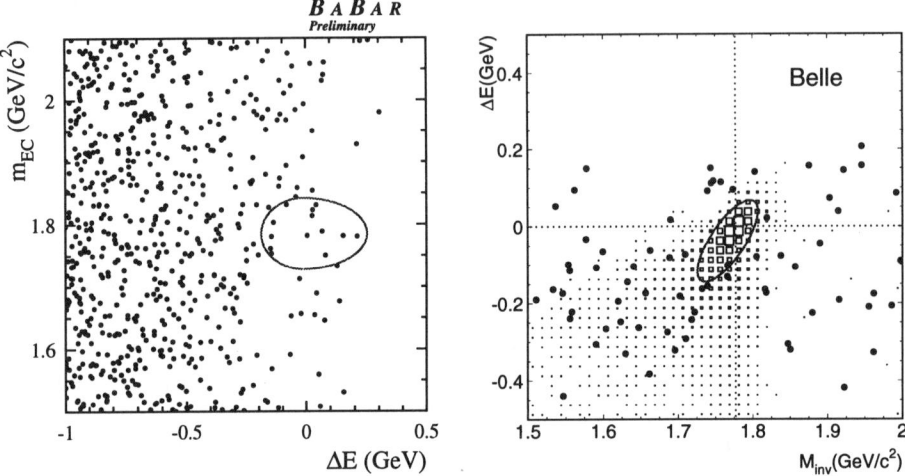

FIGURE 2. Left plot: Final sample of selected events in the BaBar $\tau \to \mu\gamma$ analysis, plotted as $M_{EC}(\mu\gamma)$ vs. ΔE, where $M_{EC}(\mu\gamma)$ is the beam energy constrained $\mu\gamma$ mass. The $\tau \to \mu\gamma$ signal region is indicated by the distorted ellipse. Right plot: The corresponding selected sample from Belle, plotted as ΔE vs. $M(\mu\gamma)$ (the 'raw' $\mu\gamma$ mass). The boxes represent the distribution from $\tau \to \mu\gamma$ Monte Carlo events, and the signal region is indicated by the ellipse. The different orientations of the ellipses in the two plots reflects the different correlations of the two variants of $\mu\gamma$ mass with ΔE for signal events.

A preliminary Belle analysis [8] of $\tau \to \mu\gamma$, based on $N_{\tau\tau} = 29.7$ million, was also presented at the Tau 2002 Workshop. The final event sample is shown as the right plot in Fig. 2. This analysis lacked some of the nice features of the BaBar analysis, namely the 'blind' approach and the attention to suppression of the $\mu\mu\gamma$ backgrounds. On the other hand, careful studies of the backgrounds, including the use of identified $\mu\mu\gamma$ events in the data to estimate this background, and a higher detection efficiency (9.0%) are strong elements of the Belle analysis.

Unlike BaBar, the Belle analysis was not 'unlucky' with regard to the observed yield in the signal region: one candidate event is observed while the background expectation was 2.5 ± 0.6 events. This leads to an upper limit on the $\tau \to \mu\gamma$ yield of 4.1 events and $\mathscr{B}(\tau \to \mu\gamma) < 6 \times 10^{-7}$, a considerable improvement over the CLEO limit.

Since this conference, Belle has presented an updated analysis [9], based on 86.3 fb^{-1} ($N_{\tau\tau} \sim 79$ million). In this update, Belle has adopted a blind analysis approach similar to that of BaBar, and also employs a likelihood analysis similar to that of CLEO. In an enlarged signal region (a box of dimension $\pm 3\sigma$ in both $M(\mu\gamma)$ and ΔE), 19 events are observed with an expected background of 20.2. The result is $\mathscr{B}(\tau \to \mu\gamma) < 3.2 \times 10^{-7}$.

TABLE 1. The 90% C.L. upper limits on branching fractions for τ lepton decays to the final states shown.

Decay Mode	CLEO limit [11, 12]	BELLE limit [13] (Preliminary)
$e^-e^+e^-$	2.9×10^{-6}	2.7×10^{-7}
$\mu^-\mu^+\mu^-$	1.9×10^{-6}	3.8×10^{-7}
$e^-\mu^+\mu^-$	1.8×10^{-6}	3.1×10^{-7}
$\mu^-e^+e^-$	1.7×10^{-6}	2.4×10^{-7}
$e^+\mu^-\mu^-$	1.5×10^{-6}	3.2×10^{-7}
$\mu^+e^-e^-$	1.5×10^{-6}	2.8×10^{-7}
$e^-K_s^0$	9.5×10^{-7}	2.9×10^{-7}
$\mu^-K_s^0$	9.5×10^{-7}	2.7×10^{-7}

3. SEARCHES FOR OTHER NEUTRINO-LESS DECAYS

Other neutrino-less channels of the type $\tau \to 3$ leptons (illustrated in Fig. 3) and $\tau \to e/\mu$+hadrons are also of great interest, potentially being sensitive to different interactions than $\tau \to \mu\gamma$. Like $\tau \to \mu\gamma$ these violate separate lepton number conservation, and in some cases (such as $\tau^- \to e^+\pi^-\pi^-$) fail to conserve total lepton number. The decays $\tau \to 3$ leptons are analogous to $\mu \to eee$, while the decays with hadrons are analogous to LFV decays of kaons as well as $\mu \to e$ conversion in a nuclear field, for all of which stringent limits exist [10].

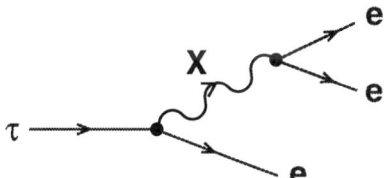

FIGURE 3. Possible diagram for $\tau \to eee$, mediated by a heavy neutral boson X.

Prior to the asymmetric B-factories, the most sensitive searches for many (~ 32) neutrino-less decay modes of the types described above had been carried out by CLEO [11, 12] yielding typical branching fraction limits in the range $1 - 8 \times 10^{-6}$. Belle has presented preliminary results [13] on eight channels based on 48.6 fb^{-1} of data, the results for which are shown in Table 1. Since this conference, Belle has also presented results [14] on the decay $\tau \to \mu\eta$, motivated by the suggestion [15] that the rate for this channel might be enhanced relative to other LFV τ decays. The 90% CL limit obtained is $\mathcal{B}(\tau \to \mu\eta) < 3.4 \times 10^{-7}$.

4. SUMMARY

The experimental search for τ lepton decays decays forbidden in the Standard Model continues to be an active area of research. The present scene is dominated by the B-factories, in particular by preliminary results from Belle where impressive sensitivities at the few $\times 10^{-7}$ level have already been attained. We can expect steady progress from both BaBar and Belle in the coming years. However, several points are worth noting:

- The $\tau \to \mu\gamma$ decay suffers serious background issues. With the hope that the two experiments will increase their data samples by at least a factor of five over the next several years, additional work will be needed if the sensitivity for this decay is to reach below the 10^{-7} level.
- The importance of a 'blind' analysis approach cannot be overstated, particularly in the situation where there are many decay modes being explored. I found it interesting to note that in the Belle analyses [8, 13] where such an approach was not used, the observed yields, summed over related channels, were consistently lower than the background expectations. For example, in the $\tau \to 3$ leptons samples the total number of events in each signal region was zero while the summed background expectation was 1.7 ± 0.6 events. In the $\tau \to K_S +$ lepton analysis, zero events were observed while 3.7 ± 0.7 were expected. Together, these results appear to be improbable, and so one's confidence in the integrity of the reported limits might have been buoyed had a blind approach been employed.

The question of the future beyond the current B-factory era is a more uncertain one. Substantial upgrades to the existing facilities to "super-B-factory" status are being considered, however the prospects for further progress in rare τ decays is not clear in light of the background issues. Detector and data analysis strategy issues are paramount in understanding these prospects.

Rare τ decay studies have so far come 'for free' as part of the larger program of heavy quark and lepton physics accessible with high-luminosity e^+e^- colliders. A dedicated rare τ decay experiment may have to be considered, as already has been done in the case of rare muon and kaon decay searches, if significant progress is to continue.

REFERENCES

1. D. Black, these proceedings.
2. MEGA Collaboration, M. L. Brooks et al., Phys. Rev. Lett. **83**, 1521 (1999).
3. See MEG Experiment web site: http://meg.web.psi.ch/.
4. R. Kitano, these proceedings.
5. CLEO Collaboration, A. Bean et al., Phys. Rev. Lett. **70**, 138 (1993).
6. CLEO Collaboration, S. Ahmed et al., Phys. Rev. D**61**, 071101 (2000).
7. C. Brown, presentation at the Seventh International Workshop on Tau Lepton Physics, Santa Cruz, Sept. 2002, arXiv:hep-ex/0212009.
8. K. Inami, T. Hokuue, and T. Ohshima, presentation at the Seventh International Workshop on Tau Lepton Physics, Santa Cruz, Sept. 2002, arXiv:hep-ex/0210036.
9. BELLE Collaboration, K. Abe et al., BELLE-CONF-0329.
10. For a summary and discussion of the planned MECO $\mu \to e$ conversion Experiment, see P. Yamin, these proceedings.
11. CLEO Collaboration, D. Bliss et al., Phys. Rev. D**57**, 5903 (1998).
12. CLEO Collaboration, S. Chen et al., Phys. Rev. D**66**, 071101 (2002).
13. Y. Yusa, T. Hayashii, T. Nagamine, and A. Yamaguchi, presentation at the Seventh International Workshop on Tau Lepton Physics, Santa Cruz, Sept. 2002, arXiv:hep-ex/0211017.
14. BELLE Collaboration, K. Abe et al., BELLE-CONF-0330.
15. M. Sher, Phys. Rev. D**66**, 057301 (2002).

QCD SPECTROSCOPY, STRUCTURE AND DYNAMICS

Observation of a Narrow Resonance in the $D_s^+ \pi^0$ System at 2.32 GeV/c^2 with BABAR

Ray F. Cowan

for the BABAR Collaboration

Massachusetts Institute of Technology, Cambridge, MA 02139, USA

Abstract. A new state with invariant mass near 2.32 GeV/c^2 has been observed by the BABAR Collaboration in the inclusive $D_s^+ \pi^0$ mass distribution using 91 fb^{-1} of e^+e^- annihilation data taken at the PEP-II asymmetric-energy storage ring at energies near 10.6 GeV. The new state is narrow, with an observed width that is consistent with the experimental resolution. This narrow width combined with the quantum numbers of the final state imply that the decay is isospin-violating. The state has natural spin-parity, and its low mass suggests a $J^P = 0^+$ assignment.

INTRODUCTION

Quark potential models have been successfully used to describe the properties of charm-strange mesons for some time. Two examples are the models of Godfrey, Isgur, & Kokoski [1, 2] and Di Pierro & Eichten [3]. Four of the states predicted by these models have been observed experimentally: the $D_s^{\pm 1}$, the $D_s^*(2112)^+$, the $D_{s1}(2536)^+$, and the $D_{s2}(2573)^+$ [4]. Two predicted but as yet unobserved states are expected to have masses in the 2.4–2.6 GeV/c^2 range. Expected to decay to $D^{(*)}K$, these states should be very broad, with widths on the order of a few hundred MeV, making them difficult to observe. However, if their masses should be below $D^{(*)}K$ threshold, they could be much narrower.

A new state, shown in Figure 1, decaying to $D_s^+ \pi^0$ with mass near 2317 MeV/c^2 and experimental width $\Gamma \approx 10$ MeV, was first noted by Antimo Palano, of the BABAR Collaboration, in early 2003 [5].

THE DATA SET, EVENT SELECTION, AND ANALYSIS

The BABAR detector is a general purpose, solenoidal, magnetic spectrometer, and is described in detail elsewhere [6]. A data set of 91.5 fb^{-1} of e^+e^- data taken near

---

[1] Charge-conjugate states are implied throughout.

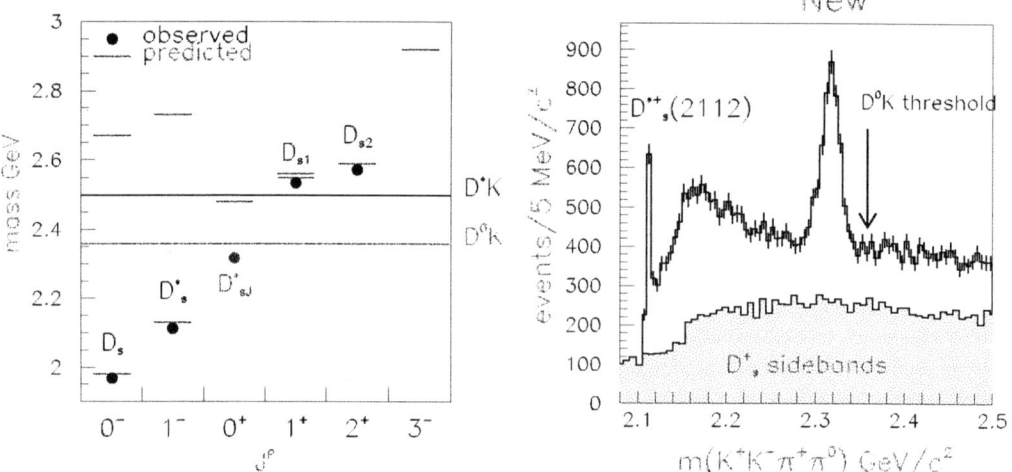

FIGURE 1. (Left) Spectroscopy of charm-strange mesons, showing known states (blue dots), potential model predictions (red lines), and the new $D_{sJ}(2317)^+$ state (red dot). (Right) $D_s^+\pi^0$ invariant mass distribution (white) for D_s^+ decay mode $D_s^+ \to K^+K^-\pi^+$, showing the known $D_s^*(2112)^+$ and the new state. The distribution for candidates in the D_s^+ sidebands is also shown (shaded).

10.58 GeV, on and off the $\Upsilon(4S)$ resonance, was used. Events from continuum $c\bar{c}$ production were selected and inclusive production of $D_s^+\pi^0$ studied, where $D_s^+ \to K^+K^-\pi^+(\pi^0)$. Combinations of three charged tracks were required to fit a common vertex consistent with production at the interaction region. Pairs of photons forming π^0 candidates were found and constrained to originate from the intersection of the $K^+K^-\pi^+$ candidate flight direction and the beam envelope.

Backgrounds were reduced by removing two-body D^0 decays, by selecting quasi two-body $\phi\pi^+$ and $\bar{K}^{*0}K^+$ decays of the D_s^+, and by helicity angle cuts on the ϕ and $\bar{K}^{*0}$ sub-decays, producing the distribution shown in Figure 1(right) with some 2200 events near 2.32 GeV. Additional checks showed that the signal was associated with both the D_s^+ and the π^0. The mass resolution was improved by use of a nominal mass [4] constraint

$$E_{D_s^+} = \sqrt{\mathbf{p}^2_{K^+K^-\pi^+} + m^2_{D_s^+}(PDG)}$$

The $D_s^+\pi^0$ mass distribution was studied in intervals of center-of-mass (CMS) momentum p^* from 2.5 GeV/c to 5 GeV/c. The signal was present in all intervals; events with $p^* > 3.5$ GeV/c were retained, further improving the signal-to-background and mass resolution. The $D_s^+\pi^0$ signal was also seen in another D_s^+ decay mode, $D_s^+ \to K^+K^-\pi^+\pi^0$, where the π^0 was required to originate from the $K^+K^-\pi^+$ vertex.

Fitting the mass spectrum yields a value for the mass of $m = 2316.8 \pm 0.4$ MeV/c^2 and a width of $\sigma = 8.6 \pm 0.4$ MeV/c^2 (errors are statistical). We call this state the $D_{sJ}(2317)^+$. Studies of detector response to a state at 2317 MeV/c^2 show that this width

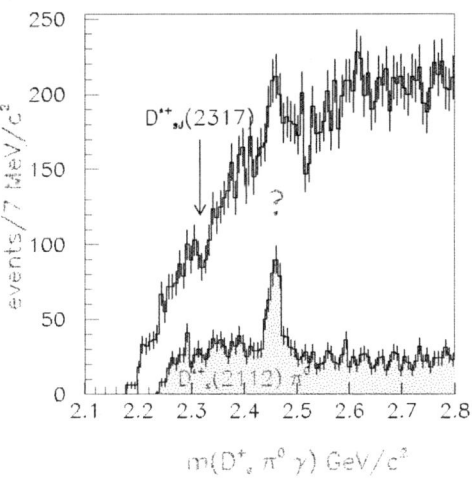

FIGURE 2. $D_s^+ \pi^0 \gamma$ invariant mass distribution (white) showing no obvious $D_{sJ}(2317)^+$ signal, but does show some structure near 2.46 GeV/c^2 is visible. The same distribution additionally requiring the sub-decay $D_s^*(2112)^+ \to D_s^+ \gamma$ shows an enhanced peak near 2.46 GeV/c^2 (shaded).

is consistent with the experimental resolution.

The decay angular distribution of the π^0 angle in the $D_s^+ \pi^0$ rest frame with respect to the $D_s^+ \pi^0$ flight direction in the CMS was consistent with being flat (43% probability). This would be expected for a spin-0 state or a higher-spin state produced unaligned.

Tests were performed to find reflections from known charm states and for particle identification problems by exchanging kaon and pion particle hypotheses. No significant signals near 2.32 GeV/c^2 were seen.

Searches for other decay modes of the $D_{sJ}(2317)^+$ were made: $D_s^+ \gamma$, $D_s^+ \gamma\gamma$, $D_s^*(2112)^+ \gamma$, $D_s^+ \pi^0 \pi^0$, and $D_s^+ \pi^0 \gamma$. No significant signals were seen for the first four modes. However, a small peak was observed in $m(D_s^+ \pi^0 \gamma)$, where the $D_s^*(2112)^+ \to D_s^+ \gamma$ and $D_{sJ}(2317)^+ \to D_s^+ \pi^0$ decays overlap. See Figure 2. A narrow state near 2.46 GeV/c^2 would produce a peak in $D_s^+ \pi^0$ near 2.32 GeV/c^2. However, simulations showed that the number of events in the small peak in the data could only explain $\approx 1/6$ of the observed $D_{sJ}(2317)^+$ signal, and therefore could not be its only source. The presence of the peak near 2.46 GeV/c^2 requires further study.

SUMMARY & CONCLUSIONS

BABAR has observed a narrow signal in the inclusive $D_s^+ \pi^0$ mass distribution with $m = 2316.8 \pm 0.4$ MeV/c^2 and has assigned a conservative systematic error on the mass of 3 MeV/c^2. The mass distribution has a gaussian width of $\sigma = 8.6 \pm 0.4$ MeV/c^2 which is consistent with the experimental resolution, implying a small intrinsic width $\Gamma <$ 10 MeV. Angular distribution studies show consistency with a spin-parity assignment $J^P = 0^+$, but other natural spin-parity possibilities are not excluded. A small peak in the $D_s^+ \pi^0 \gamma$ invariant mass distribution near 2.46 GeV/c^2 has also been observed and requires additional study[2].

If the $D_{sJ}(2317)^+$ is a $c\bar{s}$ state, it does not fit well into existing potential models, being too low in mass and, in particular, below DK threshold. The small intrinsic width is likely due to an isospin-violating decay (if $c\bar{s}$). Alternatively it could be a four-quark scalar state as proposed earlier by R.L. Jaffe [12, 13] and N. Isgur & H.J. Lipkin [14, 15, 16].

We expect interesting times ahead for experimentalists and theorists alike.

ACKNOWLEDGMENTS

We are grateful for the excellent luminosity and machine conditions provided by our PEP-II colleagues, and for the substantial dedicated effort from the computing organizations that support BABAR. The collaborating institutions wish to thank SLAC for its support and kind hospitality. This work is supported by DOE and NSF (USA), NSERC (Canada), IHEP (China), CEA and CNRS-IN2P3 (France), BMBF and DFG (Germany), INFN (Italy), FOM (The Netherlands), NFR (Norway), MIST (Russia), and PPARC (United Kingdom). Individuals have received support from the A. P. Sloan Foundation, Research Corporation, and Alexander von Humboldt Foundation.

REFERENCES

1. Godfrey, S., and Isgur, N., *Phys. Rev.*, **D32**, 189–231 (1985).
2. Godfrey, S., and Kokoski, R., *Phys. Rev.*, **D43**, 1679–1687 (1991).
3. Di Pierro, M., and Eichten, E., *Phys. Rev.*, **D64**, 114004 (2001).
4. Hagiwara, K., et al., *Phys. Rev.*, **D66**, 010001 (2002).
5. Palano, A., BABAR-COLL-0011, *Observation in the BaBar Experiment of a Narrow Meson in the $D_s^+ \pi^0$ System at 2.32 GeV/c^2*, SLAC Seminar presented on 28 April 2003.
6. Aubert, B., et al., *Nucl. Instrum. Meth.*, **A479**, 1–116 (2002).
7. Aubert, B., et al., *Phys. Rev. Lett.*, **90**, 242001 (2003).
8. Besson, D., et al., hep-ex/0305100 (2003).
9. Abe, K., hep-ex/0307041 (2003).
10. Abe, K., hep-ex/0307052 (2003).
11. Palano, A. (2003), XXIII International Conference on Physics in Collision, 26–28 June 2003.
12. Jaffe, R. L., *Phys. Rev.*, **D15**, 267 (1977).
13. Jaffe, R. L., *Phys. Rev.*, **D15**, 281 (1977).
14. Lipkin, H. J., *Phys. Lett.*, **B70**, 113 (1977).
15. Lipkin, H. J., *Phys. Lett.*, **B172**, 242 (1986).
16. Isgur, N., and Lipkin, H. J., *Phys. Lett.*, **B99**, 151 (1981).

---

[2] Much more has been learned since the time of this talk [7]. The observation of the $D_{sJ}(2317)^+$ has been confirmed by CLEO [8] and BELLE [9, 10], and the existence of a $D_{sJ}(2460)^+$ state established [8, 9, 11]. The latter has also been observed in B decays and via another decay mode $D_{sJ}(2460)^+ \to D_s^+ \gamma$.

A DK Molecule or other 4q model for the $D_s\pi$ resonance at 2.32 GeV

Harry J. Lipkin**

Department of Particle Physics Weizmann Institute of Science, Rehovot 76100, Israel
School of Physics and Astronomy, Raymond and Beverly Sackler Faculty of Exact Sciences, Tel Aviv University, Tel Aviv, Israel
High Energy Physics Division, Argonne National Laboratory, Argonne, IL 60439-4815, USA

Abstract. The recent reported charmed-strange resonance at 2.32 GeV/c suggests a possible multiquark state. Three types of multiquark bound states are reviewed. A previous model-independent variational approach considers a tetraquark with two heavy antiquarks and two light quarks as a heavy antidiquark with the color field of a quark bound to the two light quarks with a wave function like that of a heavy baryon. Results indicate that a charmed-strange tetraquark $\bar{c}\bar{s}ud$ or a bottom-strange tetraquark $\bar{b}\bar{s}ud$ with this "baryonium-type" wave function is not bound, in contrast to "molecular-type" $D-K$ and $B-K$ wave functions. However, a charmed-bottom tetraquark $\bar{c}\bar{b}ud$ might be bound with a very narrow weak decay mode. A "molecular-type" $D-B$ state can have an interesting $B_c\pi$ decay with a high energy pion.

The recent observation[1] of a charmed-strange state at 2.32 GeV that decays into $D_s\pi^o$ suggests a possible four-quark state (tetraquark). [2, 3, 4, 5, 6, 7]

Three different mass scales are relevant to the description of multiquark hadrons, the nuclear-molecular scale, the hyperfine or color-magnetic scale and the diquark scale.

The nuclear scale, charcterized by the deuteron, a binding energy of several MeV and a radius of $\approx M_\pi$ shows binding by interactions between color singlet states of two hadrons having a reduced mass of 500 MeV. The underlying quark structure of the hadrons plays no role. The kinetic energy of the state confined to this radius is

$$T_N = p^2/M_N \approx M_\pi^2/M_N \approx 20 \text{MeV} \tag{1}$$

The reduced mass of any two-meson state containing a pion is too small to be bound by such an interaction; its kinetic energy $T_\pi \approx M_\pi^2/M_\pi \approx 140\,\text{MeV}$ would be too high.

The two-kaon system with a reduced mass of 250 MeV seems to be on the borderline,

$$T_K = p^2/M_K \approx M_\pi^2/M_K \approx 40\,\text{MeV} \tag{2}$$

But $K\bar{K}$ couples to $\pi-\pi$ and $\eta-\pi$ and breaks up strongly. Suggestions that the f_o and a_o mesons are deuteron-like states or molecules are interesting, but controversial.

An attractive candidate for a molecular state[3, 4, 5, 6] suggested for the 2.32 GeV state[7] is the $D-K$ molecule with a kinetic energy

$$T_{DK} = p^2(M_D+M_K)/2M_DM_K \approx M_\pi^2(M_D+M_K)/2M_DM_K \approx 25\,\text{MeV} \tag{3}$$

The transition for the $I=0$ DK molecule to $D_s\pi$ is isospin forbidden thereby suggesting a narrow width. The isospin mixing is discussed in detail in ref.[7] .

The color-magnetic scale is characterized by a mass splitting of the order of 400 MeV; e.g. the $K^* - K$ splitting. Recoupling the colors and spins of a system of two color-singlet hadrons has been shown to produce a gain in color-magnetic energy[2, 3]. However, whether this gain in potential energy is sufficient to overcome the added kinetic energy required for a bound state is not clear without a specific model.

The diquark scale arises when two heavy quarks are bound in the well of the coulomb-like short-range potential required by QCD. A heavy $\bar{q}\bar{q}$ in a triplet of $SU(3)_c$ has the color field of a quark and can be bound to two light quarks with a wave function like that of a heavy baryon. The binding energy of two particles in a coulomb field is proportional to their reduced mass and all other interactions are mass independent. This diquark binding must become dominant at sufficiently high quark masses.

We apply the diquark-heavy-baryon model to new resonances containing heavy quarks. This model neglects the color-magnetic interactions of the heavy quarks, important for the charmed-strange four-quark system at the colormagnetic scale[2, 3] and is expected to overestimate the mass of a $\bar{c}\bar{s}ud$ state. The previous reults[3] at the color-magnetic or nuclear-molecular scale should be better. However, the bc system may already be sufficiently massive to lead to stable diquarks and the model predictions for the $\bar{c}\bar{b}ud$ state may suggest binding.

Our "model-independent" approach assumes that nature has already solved the problem of a heavy color triplet interacting with two light quarks and given us the answers; namely the experimental masses of the Λ, Λ_c and Λ_b. These answers provided by nature can now be used without understanding the details of the underlying theoretical QCD model. This approach was first used by Sakharov and Zeldovich[8] and has been successfully extended to heavy flavors[9] and applied to the $\Theta^+(1540)$.

The calculated mass can be interpreted as obtained from a variational principle with a particular form of trial wave function[4]. A mass value indicating an unbound state shows only that this type of diquark-heavy-baryon wave function does not produce a bound state; i.e that the heavy quark masses are not at the diquark scale.

We first consider the $\bar{c}\bar{s}ud$ state which may be relevant to the observed 2.32 GeV charmed-strange state. The wave function has a light ud pair seeing the color field of the $\bar{c}\bar{s}$ antidiquark like the field of a heavy quark in a heavy baryon, while the $\bar{c}\bar{s}$ antidiquark differs from the $c\bar{s}$ in the D_s by having a QQ potential which QCD color algebra requires[4] to have half the strength of the $Q\bar{Q}$ potential. in the D_s. The tetraquark mass is estimated by using the known experimental masses of the heavy baryons and heavy meson with the same flavors and introducing corrections for the difference between the heavy meson and the heavy diquark.

$$M(\bar{c}\bar{s}ud) = m_c + m_s + m_u + m_d + \langle H_{udQ}\rangle + \langle H_{ud}\rangle + \langle T_{cs}\rangle_{cs} + \langle V_{cs}\rangle_{cs} \tag{4}$$

$$M(cs) = m_c + m_s + \langle T_{cs}\rangle_{cs} + \langle V_{cs}\rangle_{cs} \tag{5}$$

$$M(D_s) = m_c + m_s + \langle T_{cs}\rangle_{c\bar{s}} + \langle V_{c\bar{s}}\rangle_{c\bar{s}} \tag{6}$$

$$M(\Lambda) = m_s + m_u + m_d + \langle H_{ud}\rangle + \langle H_{udQ}\rangle \tag{7}$$

$$M(\Lambda_c) = m_c + m_u + m_d + \langle H_{ud}\rangle + \langle H_{udQ}\rangle \tag{8}$$

where H_{ud} and H_{udQ} respectively denote the Hamiltonians describing the internal motions of the ud pair and of the three-body system of the ud pair and the antidiquark which behaves like a heavy quark, T_{cs} and V_{cs} denote the kinetic and potential energy operators for the internal motion of a cs diquark which is the same as that for a $\bar{c}\bar{s}$ antidiquark. The expectation values are taken with the "exact" wave function for the model, with the subscript cs indicating that it is taken with the wave function of a diquark and not of the D_s. The kinetic energy operator T_{cs} is the same for the cs diquark and the D_s but the potential energy operators V_{cs} and $V_{c\bar{s}} = 2V_{cs}$ differ by the QCD factor 2. This difference between cs diquark and D_s wave functions is crucial to our analysis. We can now write

$$M(\bar{c}\bar{s}ud) = (1/2) \cdot [M(D_s) + M(\Lambda) + M(\Lambda_c)] + \langle \delta H_{cs} \rangle \tag{9}$$

where $\langle \delta H_{cs} \rangle$ expresses the difference between the D_s and the $\bar{c}\bar{s}$ wave functions

$$\langle \delta H_{cs} \rangle = \langle T_{cs} \rangle_{cs} + \langle V_{cs} \rangle_{cs} - (1/2) \cdot [\langle T_{cs} \rangle_{c\bar{s}} + \langle V_{c\bar{s}} \rangle_{c\bar{s}}] \tag{10}$$

We improve on the treatment of ref[4] to calculate $\langle \delta H_{cs} \rangle$ by defining the Hamiltonian

$$H(\alpha) = \alpha T_{cs} + V_{cs} = \alpha T_{cs} + (1/2) \cdot V_{c\bar{s}} \tag{11}$$

This Hamiltonian $H(\alpha)$ is seen to describe both the cs diquark and the D_s

$$M(cs) = m_c + m_s + \langle H(\alpha) \rangle_{\alpha=1}; \quad M(D_s) = m_c + m_s + 2 \cdot \langle H(\alpha) \rangle_{\alpha=(1/2)} \tag{12}$$

$$\langle \delta H_{cs} \rangle = \langle H(\alpha) \rangle_{\alpha=1} - \langle H(\alpha) \rangle_{\alpha=(1/2)} \tag{13}$$

We use the Feynman-Hellmann theorem and the virial theorem to evaluate $\langle \delta H_{cs} \rangle$

$$\frac{d}{d\alpha} \cdot \langle H(\alpha) \rangle = \left\langle \frac{dH(\alpha)}{d\alpha} \right\rangle = \langle T_{cs} \rangle = \left\langle \frac{r}{2\alpha} \cdot \frac{dV_{cs}}{dr} \right\rangle_\alpha \tag{14}$$

$$\langle \delta H_{cs} \rangle = \int_{(1/2)}^{1} d\alpha \left\langle \frac{dH(\alpha)}{d\alpha} \right\rangle = \int_{(1/2)}^{1} d\alpha \left\langle \frac{r}{2\alpha} \cdot \frac{dV_{cs}}{dr} \right\rangle_\alpha \tag{15}$$

The Quigg-Rosner log potential[10] $V_{cs}^{QR} = (1/2) \cdot V_o \cdot log(r/r_o)$ with its parameter V_o determined by fitting the charmonium spectrum simplifies the expression (15) to

$$\langle \delta H_{cs} \rangle_{QR} = \frac{V_o}{4} \int_{(1/2)}^{1} \frac{d\alpha}{\alpha} = \frac{V_o}{4} log 2 = 126 \, MeV \tag{16}$$

This model must give a stable bound state in the high heavy quark mass limit. However, substituting experimental values[11] and eq.(16) into eq.(9) and a similar expression for $\bar{b}\bar{s}ud$ indicate that neither the cs diquark nor the bs diquark is heavy enough to produce a bound diquark-heavy-baryon state. The molecule[7] looks better.

$$M(\bar{c}\bar{s}ud) = 2685 + 126 = 2811 \, MeV; \quad M(D) + M(K) = 2361 \, MeV \ll 2811 \, MeV \tag{17}$$

$$M(\bar{b}\bar{s}ud) = 6180 \, MeV; \quad M(B) + M(K) = 5773 \, MeV \ll 6180 \, MeV \tag{18}$$

However, the bc diquark may be heavy enough to produce a bound four-quark state.

$$M(\bar{c}\bar{b}ud) = (1/2) \cdot [M(B_c) + M(\Lambda_b) + M(\Lambda_c)] + \langle \delta H_{cs} \rangle = 7280 \pm 200 \, \text{MeV} \quad (19)$$

$$M(D) + M(B) = 7146 \, \text{MeV} \quad (20)$$

Here the experimental error on the B_c mass is too large to enable any conclusions to be drawn. But if the bound state exists, it may produce striking experimental signatures.

A bound $\bar{c}\bar{b}ud$, $\bar{c}\bar{b}uu$ or $\bar{c}\bar{b}dd$ state would decay only weakly, either by b-quark decay into two charmed mesons (with the same sign of charm, so that there cannot be a J/psi decay mode), or a c-quark decay into a b meson and a strange meson. The signature with a vertex detector will see a secondary vertex with a multiparticle decay and one or two subsequent heavy quark decays and either one track or no track from the primary vertex to the secondary.

On the other hand, if the 2.32 GeV state seen by BaBar is really a DK $I = 0$ molecule[7] with an isospin violating $D_s - \pi$ decay, the analog for the bc system is a BD molecule with either $I = 1$ or $I = 0$ and a $B_c - \pi$ decay which is isospin conserving for $I = 1$ or isospin violating for $I = 0$.

Here the masses are very different and give a completely different signature with a high energy pion. $M(B) = 5279$ MeV, $M(D) = 1867$ MeV. This gives $M(B) + M(D) = 7146$ MeV, while $M(B_c) = 6400 \pm 400$ MeV. So a molecule just below BD threshold would just rearrange the four quarks into Bc-pi and fall apart, either with or without isospin violation, giving a neutral or charged pion having a well defined energy of 750 ± 400 MeV with the precision improved by better measurements[7].

In any case this is a striking signal which cannot be confused with a $q\bar{q}$ state. Experiments can look for a resonance with a pion accompanying any of B_c states.

ACKNOWLEDGMENTS

This research was supported in part by the U.S. Department of Energy, Division of High Energy Physics, Contract W-31-109-ENG-38.

REFERENCES

1. BABAR Collaboration: B. Aubert, et al, hep-ex/0304021
2. R.L.Jaffe, Phys. Rev. D15, 281 (1977).
3. H.J.Lipkin, Phys. Lett. 70B, 113 (1977);. N.Isgur and H.J.Lipkin, Phys. Lett. 99B, 151 (1981).
4. H. J. Lipkin, Phys. Lett. 172, 242 (1986).
5. T.Barnes, *The Status of Molecules*, Proc. of the XXIX Recontres de Moriond, (Meribel, France, 19-26 March 1994), hep-ph/9406215.
6. J.-M. Richard *Quasi-nuclear and quark model baryonium: historical survey*. Proc. of QCD99 (Montpellier, France, 7-13 July 1999), Nucl. Phys. Proc. Suppl. 86, 361 (2000), nucl-th/9909030.
7. T.Barnes, F. E. Close and H. J. Lipkin, hep-ph/0305025.
8. Ya. B. Zeldovich and A.D. Sakharov, Yad. Fiz 4, 395 (1966); Sov. J. Nucl. Phys. 4, 283 (1967)
9. M. Karliner and H.J. Lipkin, hep-ph/0307243
10. C. Quigg and J. L. Rosner, Phys.Rev. D17, 2364 (1978)
11. K.Hagiwara *et al.* (Particle Data Group), Phys. Rev. D66, 010001 (2002).

Observation of the $D_{sJ}(2463)$ and Confirmation of the $D^*_{sJ}(2317)$

Jon Urheim[a] and Sheldon Stone[b], for the CLEO Collaboration

[a] School of Physics and Astronomy, University of Minnesota, Minneapolis, MN 55455
[b] Physics Department, Syracuse, University, Syracuse, NY 13244-1130

Abstract. Using 13.5 fb^{-1} of e^+e^- annihilation data in the CLEO II detector at CESR, we have observed a new narrow state decaying to $D^{*+}_s\pi^0$, denoted the $D_{sJ}(2463)^+$. A possible interpretation holds that this is a $J^P = 1^+$ partner to the $D^*_{sJ}(2317)^+$ state recently discovered by the BaBar Collaboration which is consistent with $J^P = 0^+$. We have also confirmed the existence of the $D^*_{sJ}(2317)^+$ in its decay to $D^+_s\pi^0$. We have measured the masses of both states, accounting for the cross-feed background that the two states represent for each other, and have searched for other decay channels for both states. No narrow resonances are seen in $D^\pm_s\pi^\mp$ or $D^\pm_s\pi^\pm$ modes.

1. INTRODUCTION

Prior to this year, the spectrum of $c\bar{s}$ mesons was believed to be well-understood. The weakly-decaying ground state D^+_s meson was discovered by CLEO in 1983 with mass 1969 MeV and $J^P = 0^-$. The excited 1^- state at 2112.4 MeV, the D^{*+}_s meson, is also narrow, decaying to the D^+_s predominately via γ emission. It also has a 6% rate [1] for a strong transition via π^0 emission [2], which violates isospin symmetry since all $c\bar{s}$ mesons are isospin singlets while the pion is an isospin triplet. These states both have zero orbital angular momentum between the two quarks.

Four states with $L = 1$ are expected, corresponding to a spin singlet and triplet, giving one state with $J^P = 0^+$, two with 1^+, and one with 2^+. Considering the charm quark to be heavy, it is more natural to think of these as two doublets with $j = 1/2$ and $3/2$, where j is the angular momentum sum of L with the spin of the strange quark. The $j = 3/2$ states are expected to be narrow because their dominant (OZI- and isospin-favored) decays to $D^{(*)}K$ will proceed via D-wave. Indeed, the experimental observations of the $D_{sJ}(2573)^+$ (with J^P consistent with 2^+) and the $J^P = 1^+$ $D_{s1}(2536)^+$ were made feasible by the fact that these states are narrow. Most potential models expected the unobserved $j = 1/2$ states to have comparable masses, and to decay to same final states but with large widths, $\sim$200-300 MeV, since these decays would proceed via S-wave.

BaBar has recently reported the discovery of a new narrow state, the $D^*_{sJ}(2317)^+$, in its decay to $D^+_s\pi^0$ [3], its width consistent with experimental resolution. The low mass, below DK threshold, implies that despite its isospin violation, the observed channel is the most likely hadronic decay available, thus explaining the narrow width. The BaBar data are also consistent with a 0^+ spin/parity interpretation.

Various interpretations of this state have appeared in the literature. To give some examples: Barnes, Close and Lipkin speculate that this could be "baryonia" or a DK molecule [4]. Van Beveren and Rupp suggest a quasi bound scalar that arises due to coupling to the nearby DK threshold [5]. Cahn and Jackson formulate an acknowledgely poor explanation using non-relativistic vector and scalar exchange forces [6].

Bardeen, Eichten and Hill (BEH) [7] use HQET plus chiral symmetry to predict "parity doubling," where two orthogonal linear combinations of mesons transform as $SU(3)_L \times SU(3)_R$ and split into $(0^-, 1^-)$, $(0^+, 1^+)$ doublets. Assuming that the $D_{sJ}^*(2317)$ is the 0^+ state expected in the quark model, they predict that the mass splitting between the remaining 1^+ state and the 1^- should be the same as the $0^+ - 0^-$ splitting.

2. CONFIRMATION OF THE $D_{sJ}^*(2317)^+$

D_s^+ candidates are reconstructed in the $\phi\pi^+$ decay mode. The selection criteria are described in detail in Ref. [8]. The $D_s^+\pi^0$ mass distribution is shown in Fig. 1 for mass combinations with momenta above 3.5 GeV/c. Two peaks are evident: one with a mass difference (ΔM) near 0.1 GeV, due to the decay $D_s^{*+} \to D_s^+\pi^0$, and another, larger structure of 165±20 events near a ΔM of 0.35 GeV that confirms the existence of the $D_{sJ}^*(2317)^+$. The measured width of this peak is $8.0^{+1.3}_{-1.2}$ MeV, somewhat wider than the detector resolution of 6.0±0.3 MeV. The solid histogram shows our Monte Carlo simulation of the mass distribution, absolutely normalized, without the presence of any narrow states that decay into D_s^+ mesons. The CLEO Monte Carlo does an excellent job of reproducing the size and shape of our background π^0 candidates.

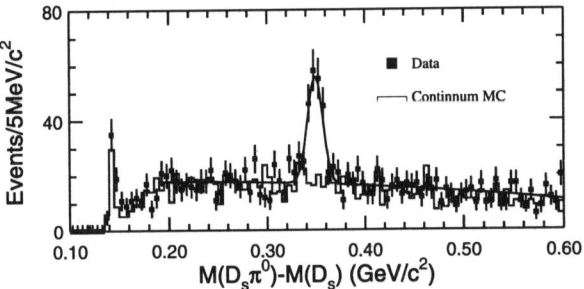

FIGURE 1. The $D_s^+\pi^0$ candidate mass distribution shown as the difference with respect to the D_s^+ mass. The solid histogram shows our Monte Carlo simulation of the spectrum, absolutely normalized, without narrow states decaying into D_s^+.

3. OBSERVATION OF THE $D_{SJ}(2463)^+$

We also looked for decays of the $D_{sJ}^*(2317)^+$ and possible additional narrow states in other channels, notably $D_s^{*+}\pi^0$. We use the $D^{*+} \to \gamma D_s^+$ decay mode. Photon candidates were selected from neutral energy clusters with lateral profiles consistent with electromagnetic showers and absolute energies above 50 MeV. Fig. 2 shows the mass difference distributions for both the peak and sideband regions of the D_s^{*+} signal.

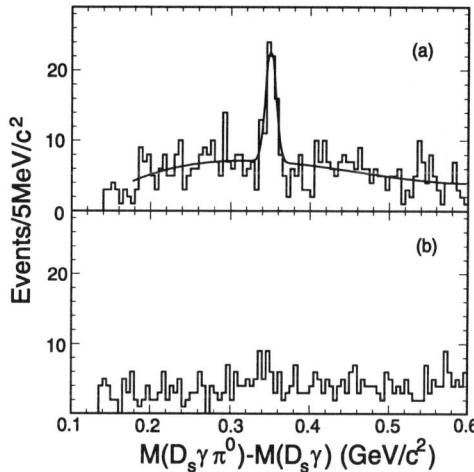

FIGURE 2. The $D_s^{*+}\pi^0$ candidate mass distribution shown as the difference with respect to the D_s^{*+} mass. (a) D_s^{*+} signal region; (b) D_s^{*+} sideband region.

We observe a peak consisting of 55±10 events, with a width of 6±1.0 MeV (r.m.s.) compared with the detector resolution of 6.6±0.5 MeV. The mass difference value is also about 0.35 GeV. The near equality of this ΔM with the previous one leads to the worry that there could be cross-contamination between the two final states.

4. ANALYSIS OF CROSS CONTAMINATION

Many studies were performed to see if these two states could arise from reflections of other known narrow states. These possibilities were excluded. It is possible, however, for a higher mass state decaying $D_s^{*+}\pi^0$ to be reconstructed as a lower mass state simply by ignoring the photon from the D_s^{*+} decay. In fact, taking the signal $D_s^{*+}\pi^0$ events and ignoring the photon from the D_s^{*+} decay causes a peak in the $D_s^+\pi^0$ spectrum at very nearly the same mass difference, but with a width of 14.9 MeV, considerably larger than our resolution. The efficiency of this process is rather high: (84±4±10)%.

It is also possible for the lower mass state to pick up a random photon, fake a D_s^{*+}, and thus be a candidate for the upper mass state. This is a much smaller probability, (9.0±0.7±1.5)% and can be estimated from the D_s^{*+} sidebands. The number of actual signal events can be estimated from these probabilities and the measured numbers of events in the peaks. Accounting for the background in this way, the peak in the $D_s^{*+}\pi^0$ sample corresponds to 41±12 signal events. The probability that this excess is due to a background fluctuation is in excess of 5σ. Thus CLEO has made the first observation of a new state near 2460 MeV. (Although the BaBar data also showed an excess of events in this mass region, the conclusion reached in Ref. [3] was that further study was needed to resolve whether the peak received contributions from a new state or was entirely due to a reflection of the $D_{sJ}^*(2317)$.)

5. MASS DETERMINATIONS

Because of the contamination of the lower mass state by the higher mass one, fitting the $D_s\pi^0 - D_s$ mass difference distribution to a single Gaussian could result in a biased mass determination. Making use of the excellent mass resolution of the CLEO CsI calorimeter we fit the $D_s^+\pi^0$ mass difference peak to two Gaussians whose means and widths are allowed to float. The fit determines one signal to be at a mean mass difference of 350.0±1.2 MeV with a width of 5.9±1.2 MeV and another wider Gaussian at 344.9±6.1 MeV with a width of 16.5±6.3 MeV, characteristic of the feed-down background. We assign a ±1.0 MeV systematic error to the fit result for the narrow Gaussian.

Since the feed-up from the first state to the second state is relatively small, ~20% of the signal of the higher mass state, we determine its mass by subtracting the D_s^* sidebands and performing a fit. The resulting mass difference is 351.2±1.7 MeV, to which we also assign a systematic error of ±1.0 MeV.

We note that a $D_s^+\pi^0$ system with $L = 0$ is a 0^+ state, and a $D_s^{*+}\pi^0$ system with $L = 0$ is a 1^+ state. If the $D_{sJ}(2463)$ were a 0^+ state, its dominant decay would be to DK as this is kinematically allowed. The narrowness of the $D_s^{*+}\pi^0$ peak excludes this possibility.

6. UPPER LIMITS ON OTHER DECAY MODES

6.1. Neutral and Doubly Charged Modes

In Fig. 3 we show the $D_s^\pm\pi^\mp$ and $D_s^\pm\pi^\pm$ mass difference distributions. No signals are visible and the production ratio times decay rate of any objects similar to the $D_{sJ}^*(2317)$ are lower by more than a factor of ten compared to the $D_s^{*+}\pi^0$ mode. This argues against a molecular interpretation.

6.2. Other Decay Modes of the $D_{sJ}^*(2317)^+$ and $D_{sJ}(2463)^+$

Upper limits on other decay modes of the $D_{sJ}^*(2317)^+$ relative to $D_s^+\pi^0$ are given in Table 1. Limits obtained on other decays of the $D_{sJ}(2463)^+$, relative to $D_s^{*+bb}\pi^0$, are summarized in Table 2.

The electromagnetic transition $D_{sJ}(2463)^+ \to D_{sJ}^*(2317)^+\gamma$ [9] presents a particularly difficult situation as the final state particles are again a D_s^+, a π^0, and a γ with momenta similar to that in the main $D_s^{*+}\pi^0$ mode. To reduce backgrounds from $D_{sJ}(2463)^+ \to D_s^{*+}\gamma$, we required that the $D_s\pi^0$ system be consistent with the decay of the $D_{sJ}^*(2317)$, namely that $|\Delta M(D_s\pi^0) - 350.0\,\text{MeV}/c^2| < 13.4\,\text{MeV}/c^2$ (~ 2σ based on Monte Carlo simulations). We also required that the $D_s\gamma$ system be inconsistent with D_s^* decay at the 1σ level (the corresponding $\Delta M(D_s\gamma)$ must deviate from the expected value for this decay by more than 4.4 MeV/c^2), and that the momentum of the π^0 be inconsistent with the $D_{sJ}(2463) \to D_s^*\pi^0$ transition, also at the 1σ level. Using these cuts, we see no evidence for a signal in this mode.

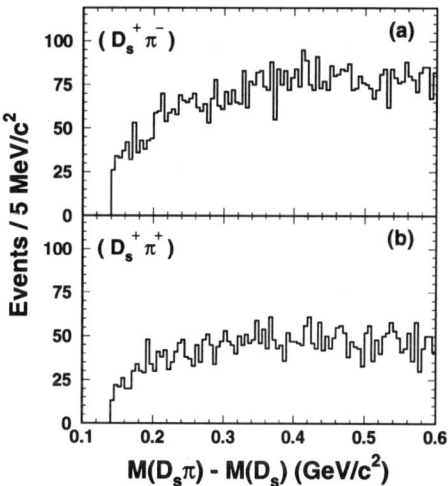

FIGURE 3. Mass difference distributions for $D_s^\pm \pi^\mp$ (top) and $D_s^\pm \pi^\pm$ (bottom) candidate samples.

TABLE 1. The 90% C.L. upper limits on the ratio of branching fractions for $D_{sJ}^*(2317)$ to the the channels shown relative to the $D_s^+ \pi^0$ state. Also shown are the theoretical expectations from Ref. [7], under the assumption that the $D_{sJ}^*(2317)$ is the lowest-lying 0^+ $c\bar{s}$ meson.

Final State	Yield	Efficiency	Ratio (90% C.L.)	Prediction
$D_s^+ \pi^0$	135 ± 23	$(9.7 \pm 0.6)\%$	—	
$D_s^+ \gamma$	-19 ± 13	$(18.1 \pm 0.1)\%$	< 0.052	0
$D_s^{*+} \gamma$	-6.5 ± 5.2	$(7.0 \pm 0.5)\%$	< 0.059	0.08
$D_s^+ \pi^+ \pi^-$	2.0 ± 2.3	$(19.8 \pm 0.8)\%$	< 0.019	0
$D_s^{*+} \pi^0$	-1.7 ± 3.9	$(3.6 \pm 0.3)\%$	< 0.11	0

We note that our upper limit for $D_{sJ}(2463)^+ \to D_s^+ \pi^+ \pi^-$ is considerably smaller than the BEH prediction. For this prediction they calculate both the isospin-violating $D_s^{*+} \pi^0$ rate and the decay into D_s^{*+} and a virtual σ meson that materializes as a $\pi^+ \pi^-$ pair. Although this is a difficult calculation, we should not be far from seeing this decay.

7. CONCLUSIONS

CLEO confirms the $c\bar{s}$ state near 2317 MeV discovered by BaBar, and measures a mass difference with respect to the D_s^+ of $350.0 \pm 1.2 \pm 1.0$ MeV. This state is likely to have $J^P = 0^+$.

CLEO has made the first observation of a new state near 2463 MeV and has measured $M(D_{sJ}(2463)^+) - M(D_s^+) = 351.2 \pm 1.7 \pm 1.0$ MeV. This is likely to be a 1^+ state. The mass splittings are consistent with being equal, as predicted by BEH; the difference

TABLE 2. The 90% C.L. upper limits on the ratio of branching fractions for $D_{sJ}(2463)$ to the the channels shown relative to the $D_s^{*+}\pi^0$ state. Also shown are the theoretical expectations from Ref. [7], under the assumption that the $D_{sJ}(2463)$ is the lowest-lying 1^+ $c\bar{s}$ meson.

Final State	Yield	Efficiency	Ratio (90% C.L.)	Prediction
$D_s^{*+}\pi^0$	41 ± 12	$(6.0 \pm 0.2)\%$	—	
$D_s^+\gamma$	40 ± 17	$(19.8 \pm 0.4)\%$	< 0.49	0.24
$D_s^{*+}\gamma$	-5.1 ± 7.7	$(9.1 \pm 0.3)\%$	< 0.16	0.22
$D_s^+\pi^+\pi^-$	2.5 ± 5.4	$(19.5 \pm 1.5)\%$	< 0.08	0.20
$D_{sJ}^*(2317)^+\gamma$	3.6 ± 3.0	$(2.0 \pm 0.1)\%$	< 0.58	0.13

$[(1^+ - 1^-) - (0^+ - 0^-)]$ being 1.2 ± 2.1 MeV. The two states are narrow and we limit the total decay widths of both of them to be $\Gamma < 7$ MeV.

We also do not see evidence for any narrow states in $D_s^\pm \pi^\mp$ or $D_s^\pm \pi^\pm$, which argues against a molecular interpretation.

Theoretical applications of QCD, including exploitation of lattice QCD, sum rules, and heavy quark and chiral symmetries, are necessary to extract information on fundamental parameters in the quark sector. By coupling HQET with chiral symmetry, the BEH model yielded predictions about masses, widths and decay modes that were in conflict with conventional thinking based on potential models. The experimental results reported here provide powerful support for the BEH approach.

ACKNOWLEDGMENTS

Support for this effort was provided by the U. S. National Science Foundation and the Dept. of Energy. D. Cinabro, S. Li, and J. C. Wang contributed greatly to this analysis. Useful conversations were held with W. Bardeen, T. Barnes, E. Eichten, C. Hill and J. Rosner.

REFERENCES

1. CLEO Collaboration, J. Gronberg et al., Phys. Rev. Lett. **75**, 3232 (1995).
2. P. Cho and M. B. Wise, Phys. Rev. D **49**, 6228 (1994).
3. BaBar Collaboration, B. Aubert et al., Phys. Rev. Lett. **90**, 242001 (2003).
4. T. Barnes, F. E. Close and H. J. Lipkin, hep-ph/0305025.
5. E. van Beveren and G. Rupp, hep-ph/0305035.
6. R. N. Cahn and J. D. Jackson, hep-ph/0305012.
7. W. A. Bardeen, E. J. Eichten, and C. T. Hill, hep-ph/0305049.
8. CLEO Collaboration, D. Besson et al., hep-ex/0305100, CLNS 03/1826 CLEO 03-09, to appear in Phys. Rev. D.
9. The calculation of cross feed backgrounds did not account for the possibility of $D_{sJ}(2463) \to D_{sJ}^*(2317)\gamma$ decays. However, based on our direct search for this decay, the impact of ignoring this channel is insignificant.

Photoproduction of Charm Pairs

Erik E. Gottschalk[1]

Fermilab, P.O. Box 500, Batavia, IL 60510, USA

Abstract.
A large sample of events containing fully and partially reconstructed pairs of charmed D mesons has been studied by the Fermilab photoproduction experiment FOCUS (FNAL-E831). Correlations between photoproduced D and $\overline{D}$ mesons are used to study heavy quark production dynamics. Correlation results are presented for fully and partially reconstructed pairs of charmed D mesons. The results are compared to Monte Carlo predictions based on a recent version of PYTHIA with default settings.

INTRODUCTION

A fundamental understanding of heavy-quark production has not yet been achieved. Quantum Chromodynamics (QCD) provides a theoretical framework, but perturbative QCD calculations can only be applied to some aspects of heavy-quark production while other aspects remain elusive. This is especially true for charm production, where perturbative QCD calculations have large uncertainties and non-perturbative effects must be included to adequately model physical observables. In the absence of a fundamental understanding of heavy-quark production, the best tools we have are models that are able to reproduce existing experimental data and make predictions about untested aspects of QCD.

One subject of considerable theoretical interest and ongoing experimental research is the study of correlations between heavy-quark pairs [1], in particular the study of correlations between D and $\overline{D}$ mesons. In this paper, highlights from a recent publication [2] on $D\overline{D}$ correlations in photoproduction are presented. The results are based on studies of fully and partially reconstructed charm-pair events, and are presented by comparing data distributions to predictions from a recent version of the Lund model [3] for photon-gluon fusion. The overall agreement between our data and the model is good, and significant improvements have been made compared to older versions of the model [4, 5]. In this paper we highlight noteworthy differences between our data and the model.

---

[1] On behalf of the FOCUS Collaboration: J. M. Link, P. M. Yager (**UC Davis**); J. C. Anjos, I. Bediaga, C. Göbel, J. Magnin, A. Massafferri, J. M. de Miranda, I. M. Pepe, A. C. Polycarpo, A. C. dos Reis (**CPBF, Rio de Janeiro**); S. Carrillo, E. Casimiro, E. Cuautle, A. Sánchez-Hernández, C. Uribe, F. Vázquez (**CINVESTAV, México City**); L. Agostino, L. Cinquini, J. P. Cumalat, B. O'Reilly, I. Segoni, K. Stenson, M. Wahl (**CU Boulder**); J. N. Butler, H. W. K. Cheung, G. Chiodini, I. Gaines, P. H. Garbincius, L. A. Garren, E. E. Gottschalk, P. H. Kasper, A. E. Kreymer, R. Kutschke, R. M. Wang (**Fermilab**); L. Benussi, M. Bertani, S. Bianco, F. L. Fabbri, A. Zallo (**INFN Frascati**); M. Reyes (**University of Guanajuato, Leon**); C. Cawlfield, D. Y. Kim, A. Rahimi, J. Wiss (**UI Champaign**); R. Gardner, A. Kryemadhi (**IU Bloomington**); Y. S. Chung, J. S. Kang, B. R. Ko, J. W. Kwak, K. B. Lee (**Korea University, Seoul**); K. Cho, H. Park (**Kyungpook National University, Taegu**); G. Alimonti, S. Barberis, M. Boschini, A. Cerutti, P. D'Angelo, M. DiCorato, P. Dini, L. Edera, S. Erba, M. Giammarchi, P. Inzani, F. Leveraro, S. Malvezzi, D. Menasce, M. Mezzadri, L. Moroni, D. Pedrini, C. Pontoglio, F. Prelz, M. Rovere, S. Sala (**INFN and Milano**); T. F. Davenport III (**UNC Asheville**); V. Arena, G. Boca, G. Bonomi, G. Gianini, G. Liguori, D. Lopes Pegna, M. M. Merlo, D. Pantea, S. P. Ratti, C. Riccardi, P. Vitulo (**INFN and Pavia**); H. Hernandez, A. M. Lopez, E. Luiggi, H. Mendez, A. Paris, J. Quinones, J. E. Ramirez, Y. Zhang (**Mayaguez, Puerto Rico**); J. R. Wilson (**USC Columbia**); T. Handler, R. Mitchell (**UT Knoxville**); D. Engh, M. Hosack, W. E. Johns, M. Nehring, P. D. Sheldon, E. W. Vaandering, M. Webster (**Vanderbilt**); M. Sheaff (**Wisconsin, Madison**)

CHARM-PAIR ANALYSIS

The data for our studies of photoproduced charm pairs were recorded by the FOCUS (FNAL-E831) experiment during the 1996–1997 fixed-target run at the Fermi National Accelerator Laboratory. The experiment ran with a photon beam and a BeO target. The average photon energy for the recorded data was ≈ 180 GeV with a width of ≈ 50 GeV.

A candidate-driven algorithm [2] was used to collect a sample of ≈ 7000 pairs of fully reconstructed D mesons: D^+D^-, $D^+\overline{D}^0$, D^0D^-, and $D^0\overline{D}^0$. For this sample we considered the decay modes $D^0 \to K^-\pi^+$, $D^+ \to K^-\pi^+\pi^+$, $D^0 \to K^-\pi^+\pi^+\pi^-$, and charged-conjugate modes. The algorithm found charm-pair events by reconstructing two D-meson candidates, and by using the two D's to reconstruct a primary vertex. The algorithm included other tracks in the primary vertex as long as the tracks satisfied confidence level cuts associated with the primary vertex. The number of tracks found in this manner together with the two D's define $N_{primary}$, the primary vertex multiplicity.

Figure 1 shows the $D\overline{D}$ signal that we obtain. The figure shows the normalized D invariant mass[2] $M_n(D)$ opposite the normalized $\overline{D}$ invariant mass $M_n(\overline{D})$. The $D\overline{D}$ yield that we obtain from a fit to the data is 7064 ± 119 (statistical error).

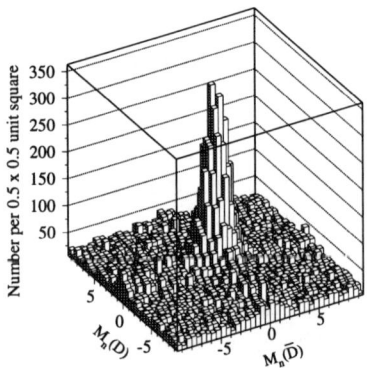

FIGURE 1. Normalized D mass vs. normalized $\overline{D}$ mass distribution for fully reconstructed charm pairs.

In addition to our sample of pairs of fully reconstructed D mesons, we have also obtained a sample of events where one D is fully reconstructed (referred to as the *recoil D*) and the other is kinematically tagged by a slow pion coming from the decay $D^{*+} \to \pi^+ D^0$. In these decays, the D^0 need not be reconstructed, and therefore we refer to this sample of events as *partially reconstructed charm pairs*[3]. For this sample we consider the same decay modes that were used for the fully reconstructed charm pairs. Figure 2a shows the invariant mass distribution that we obtain, with a total of $782\,630 \pm 1600$ candidates satisfying our selection criteria.

---

[2] The normalized mass, $M_n(D) = \Delta M/\sigma_M$, is defined as the difference between the reconstructed mass and the central value of the D^+ or D^0 mass distribution divided by the reconstructed-mass error σ_M.

[3] The partially reconstructed sample consists of $D^{*+}D^-$, $D^{*+}\overline{D}^0$, D^0D^{*-}, and D^+D^{*-} pairs.

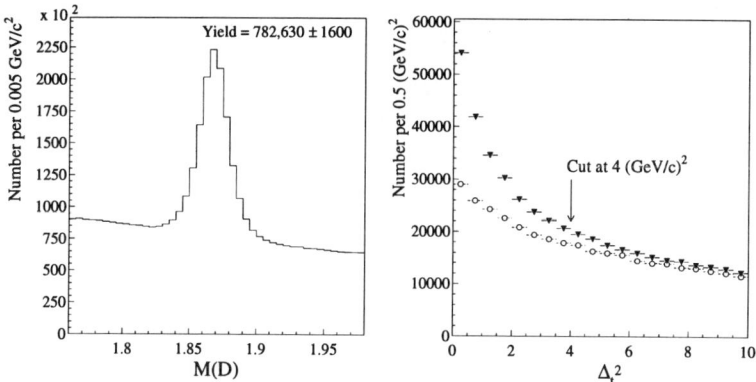

FIGURE 2. Invariant mass of the recoil D in the partially reconstructed charm-pair sample (left). The mass of charged D candidates is lowered by 3.74 MeV/c^2 to match the D^0 mass. The yield is a sum of individual yields for the three decay modes described in the text. Δ_t^2 distributions (right) for right-sign (filled triangles) and wrong-sign (open circles) combinations of partially reconstructed charm pairs.

For partially reconstructed charm pairs we perform a background subtraction that is based on the charge of the slow pion. This is done by treating each track that is assigned to the primary vertex (excluding the recoil D) as a slow-pion candidate from the decay $D^{*+} \to \pi^+ D^0$. If the charge of the slow pion is the same as the charge of the kaon from the recoil D, then the combination of the slow pion and recoil D is designated as a *right-sign* combination. Otherwise, it is a *wrong-sign* combination. This is used to subtract wrong-sign background from right-sign combinations.

To enhance the event selection procedure, a maximum cut of 4 (GeV/c)^2 is applied to $\Delta_t^2 = (p_x^{(r)} + 13.8 * p_x^{(\pi)})^2 + (p_y^{(r)} + 13.8 * p_y^{(\pi)})^2$, where $p_x^{(r)}, p_y^{(r)}$ and $p_x^{(\pi)}, p_y^{(\pi)}$ are transverse momentum components of the recoil D and slow pion [4], respectively. This cut enhances the selection of signal since genuine events balance Δ_t^2 (see reference [5] for more details). This is shown in Figure 2b, which shows a prominent excess of right-sign combinations close to $\Delta_t^2 = 0$ compared to the wrong-sign background. After applying the Δ_t^2 cut, we obtain a sample of $75\,160 \pm 1040$ partially-reconstructed charm pairs.

CHARM-PAIR PRODUCTION

Our study of charm-pair production compares FOCUS data to predictions from a Monte Carlo based on PYTHIA 6.203 [3] with default settings. We use PYTHIA with default settings (instead of using a Monte Carlo tuned to match our data) to facilitate comparisons with theoretical predictions and results from other experiments. To improve comparisons between data and model predictions we eliminate our lowest multiplicity

---

[4] The momentum of the soft pion approximates the momentum of the D^* (due to the low Q value of the D^* decay) when multiplied by the inverse of its energy fraction, which is ≈ 13.8.

charm-pair events by requiring $N_{primary} > 2$. This eliminates events that have only the two D mesons assigned to the primary and no additional tracks. The reason for imposing this cut when comparing data to PYTHIA is that the cut removes diffractively produced $\psi(3770)$ events, which are not present in PYTHIA. $\psi(3770)$ events are observed in FOCUS data (see Figure 3) as a threshold enhancement in the invariant $D\overline{D}$ mass for mass combinations with a net charge of zero (D^+D^- and $D^0\overline{D}^0$). This enhancement is especially evident for events with $N_{primary} = 2$ when we apply cuts that remove events with energy deposited in our electromagnetic calorimeters (see Figure 3b).

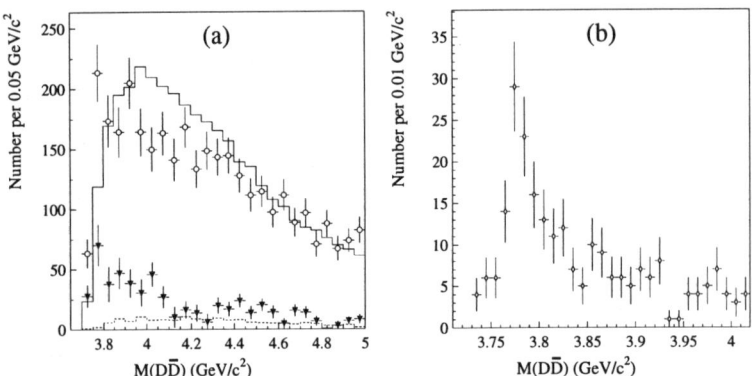

FIGURE 3. (a) Invariant $D\overline{D}$ mass for D^+D^- and $D^0\overline{D}^0$ mass combinations for background-subtracted FOCUS data (open circles), PYTHIA 6.203 (solid line), FOCUS data with an $N_{primary} = 2$ cut applied to the data (filled triangles), and PYTHIA 6.203 with $N_{primary} = 2$ cut (dashed line). (b) The $D\overline{D}$ mass for $N_{primary} = 2$ after removing events with energy deposited in electromagnetic calorimeters.

After requiring $N_{primary} > 2$ for fully reconstructed charm pairs, we compare FOCUS data to PYTHIA 6.203 for both the fully and partially reconstructed[5] charm pairs (see Figures 4 and 5). Although we observe some discrepancies, the PYTHIA model for charm photoproduction shows good agreement with the data. The distribution for the azimuthal angle, $\Delta\phi$, between the D and $\overline{D}$ momentum vectors in the plane transverse to the beam direction is reproduced by PYTHIA; however, we observe an enhancement in the first $\Delta\phi$ bin that may suggest the presence of a production mechanism that is not included in PYTHIA. The enhancement disappears for partially reconstructed charm pairs (see Figure 5a) due to resolution broadening and selection cuts used in the analysis. There is good agreement for the transverse momentum squared of the $D\overline{D}$ pair in Figures 4b and 5b, except that the data tend to have slightly larger values of $p_t^2(D\overline{D})$. Other comparisons between FOCUS data and PYTHIA are presented in reference [2].

In summary, FOCUS has extracted two large samples of charm pairs for studies of charm photoproduction. The agreement between data and a recent version of PYTHIA with default settings is good. Noteworthy differences are observed in the azimuthal angle between the D and $\overline{D}$ mesons, and in the production of $\psi(3770)$ events.

---

[5] Partially reconstructed charm-pair events satisfy an implicit $N_{primary} > 2$ cut, since these events have two D mesons and a slow pion that account for a minimum primary vertex multiplicity of three.

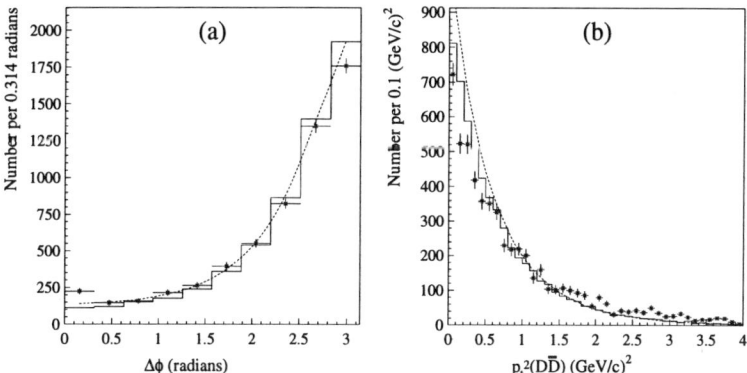

FIGURE 4. Charm-pair correlations for fully reconstructed charm pairs with $N_{primary} > 2$: (a) $\Delta\phi$ and (b) p_t^2 of the $D\overline{D}$ pair for background-subtracted FOCUS data (asterisks with error bars), PYTHIA 6.203 after detector simulation and data analysis cuts (solid line), and PYTHIA 6.203 parent distributions without acceptance or resolution effects (dashed line with arbitrary normalization).

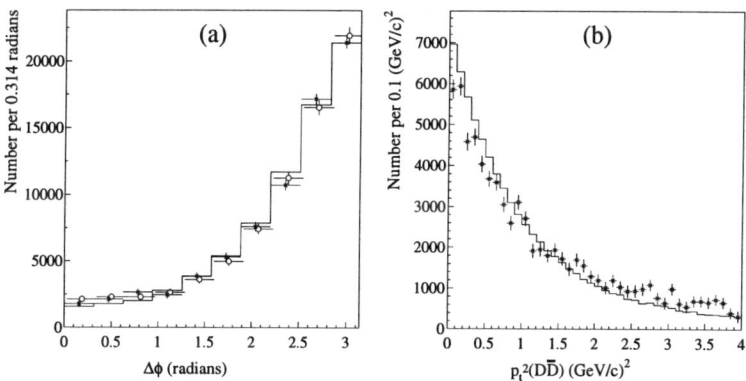

FIGURE 5. Charm-pair correlations for partially reconstructed charm pairs: (a) $\Delta\phi$ and (b) p_t^2 of the $D\overline{D}$ pair for background-subtracted FOCUS data (asterisks with error bars) and PYTHIA 6.203 after detector simulation and data analysis cuts (solid line). The $\Delta\phi$ distribution for fully-reconstructed charm pairs (open circles with error bars) is included for comparison in (a) after accounting for resolution broadening.

REFERENCES

1. S. Frixione, M.L. Mangano, P. Nason, and G. Ridolfi, *Adv. Ser. Direct. High Energy Phys.* **15** (1998) 609 [arXiv:hep-ph/9702287].
2. FOCUS Collaboration, J.M. Link *et al.*, *Phys. Lett.* **B 566** (2003) 51.
3. PYTHIA 6.203, T. Sjöstrand *et al.*, *Comput. Phys. Commun.* **135** (2001) 238.
4. PYTHIA 5.6 and JETSET 7.3, T. Sjöstrand *et al.*, *Comput. Phys. Commun.* **82** (1994) 74.
5. E687 Collaboration, P.L. Frabetti *et al.*, *Phys. Lett.* **B 308** (1993) 193.

Charm production asymmetries from heavy-quark recombination

Thomas Mehen

Department of Physics, Duke University, Durham NC 27708

Abstract. The large charm asymmetries in fixed-target hadroproduction experiments can be explained by an $O(\Lambda_{QCD}/m_c)$ power correction to the QCD factorization theorem called heavy-quark recombination. This mechanism explains D meson and Λ_c asymmetries with a minimal set of universal nonperturbative parameters.

Asymmetries in the production of charm particles in fixed-target hadroproduction experiments [1, 2, 3, 4, 5] are much larger than predicted by perturbative QCD, so they are a sensitive probe of nonperturbative aspects of heavy particle production. The cross-section for a charm particle, H, in the collision of two hadrons, A and B, is [6]

$$d\sigma[A+B \to H+X] = \sum_{i,j} f_{i/A} \otimes f_{j/B} \otimes d\hat{\sigma}[ij \to c\bar{c}+X] \otimes D_{c \to H} + \dots \quad (1)$$

Here $f_{i/A}$ is a parton distribution function, $D_{c \to H}$ is a fragmentation function and $d\hat{\sigma}[ij \to c\bar{c}+X]$ is a short-distance cross section. Corrections to the factorized form of the cross section are suppressed by Λ_{QCD}/m_c or $\Lambda_{QCD}/p_\perp$. At leading order in perturbation theory, charm particles and antiparticles are produced symmetrically because the partonic processes $gg \to c\bar{c}$ and $q\bar{q} \to c\bar{c}$ produce charm and anticharm symmetrically and $D_{c \to H} = D_{\bar{c} \to \bar{H}}$ due to charge conjugation invariance. At next-to-leading order, the asymmetry, $\alpha[H] = (\sigma[H] - \sigma[\bar{H}])/(\sigma[H] + \sigma[\bar{H}])$, is only a few percent [7].

In fixed-target hadroproduction charm hadrons that share a valence parton with the beam hadron are produced in much greater numbers in the forward direction of the beam than charm particles that do not share a valence quark with the beam [1, 2, 3, 4, 5]. This is known as the "leading particle effect". For example, experiments with a π^- beam incident on a nuclear target observe $\alpha[D^-] \approx 0.7$ (corresponding to $\approx 6 \, D^-$ for every D^+) at the highest x_F measured. In pN collisions $\alpha[\Lambda_c^+] \approx 1$ for all $x_F > 0.2$.

These asymmetries are usually explained by nonperturbative models of hadronization. A commonly used model is the Lund string fragmentation model [8] which can be implemented using PYTHIA [9]. (Another model of string fragmentation can be found in Ref. [10]). The asymmetry is generated by the "beam drag effect" [11] in which the charm quark binds to the remnants of the incident hadron via the formation of a color string. The PYTHIA Monte Carlo with default parameters rarely predicts the asymmetries correctly [1] and in the case of Λ_c asymmetries in πN collisions [4] gets the sign of the asymmetry wrong. Another approach is the recombination model first introduced in Ref. [12] (for recent analyses, see Refs. [13, 14, 15]). In this model the charm quarks coalesce with spectator partons in the beam hadrons whose momentum distribution is determined by double parton distributions. Finally there are models that generate the asymmetry from intrinsic charm in the incident hadron [16, 17, 18]. All models are sensitive to a number of poorly determined nonperturbative functions such as remnant parton distributions or recombination probabilities.

Since the observed asymmetries are much larger than predicted by perturbative QCD, they directly probe power corrections to the factorization theorem. Recently, Refs.[19, 20, 21, 22] have shown that an $O(\Lambda_{QCD}/m_c)$ power correction called heavy-quark recombination can explain charm hadron asymmetries in photo- and hadroproduction experiments. In this approach the asymmetry is generated in the short-distance process so cross sections are calculable up to an overall normalization set by a few universal nonperturbative parameters. For D mesons, the dominant contribution comes from a process in which a light antiquark participates in a hard scattering process that produces a charm-anticharm quark pair. The light antiquark emerges from the hard scattering with momentum of $O(\Lambda_{QCD})$ in the rest frame of the charm quark, then they hadronize into a final state that includes a D meson, giving the following contribution to the cross section:

$$d\hat{\sigma}[\bar{q}g \to D + X] = \sum_n d\hat{\sigma}[\bar{q}g \to c\bar{q}(n) + \bar{c}]\rho[c\bar{q}(n) \to D]. \qquad (2)$$

In this formula, $d\hat{\sigma}[\bar{q}g \to c\bar{q}(n) + \bar{c}]$ is a short-distance cross section for producing a $c\bar{q}$ with quantum numbers denoted by n and $\rho[c\bar{q}(n) \to D]$ parametrizes the hadronization of the $c\bar{q}$ into a state that includes a D meson.

Because the light antiquark is massless it is natural to expect the heavy-quark recombination contribution to be a convolution of a short-distance cross section with a distribution function that depends on the fraction of the light-cone momentum carried by the light quark. However, to lowest order in Λ_{QCD}/m_c only the leading moment of such a distribution contributes. Because the cross section is inclusive, the final state may include other soft quanta besides the D meson. Therefore the color and angular momentum quantum numbers of the $c\bar{q}$ can be different from the D meson. Amplitudes for production of $c\bar{q}$ in $L > 0$ partial waves are suppressed by additional powers of Λ_{QCD}/m_c relative to S-waves and can be neglected. Neglecting light quark-antiquark pair production, which is suppressed in the large-N_c limit of QCD, one finds that the D^+ cross section depends on four parameters which scale as Λ_{QCD}/m_c [19, 21]:

$$\rho_1 = \rho[c\bar{d}(^1S_0^{(1)}) \to D^+], \quad \tilde{\rho}_1 = \rho[c\bar{d}(^3S_1^{(1)}) \to D^+], \qquad (3)$$
$$\rho_8 = \rho[c\bar{d}(^1S_0^{(8)}) \to D^+], \quad \tilde{\rho}_8 = \rho[c\bar{d}(^3S_1^{(8)}) \to D^+].$$

Explicit expressions in terms of nonperturbative QCD matrix elements can be found in Ref. [23]. Analogous parameters for D^0, D^- and D^{*+} mesons are related to those in Eq. (3) by isospin, charge conjugation and heavy-quark spin symmetry, respectively.

The process which gives the leading contribution to charm baryon production is cq recombination [22]. The heavy-quark recombination contribution to Λ_c^+ production is

$$d\hat{\sigma}[qg \to \Lambda_c^+ + X] = \sum_n d\hat{\sigma}[qg \to cq(n) + \bar{c}]\eta[cq(n) \to \Lambda_c^+]. \qquad (4)$$

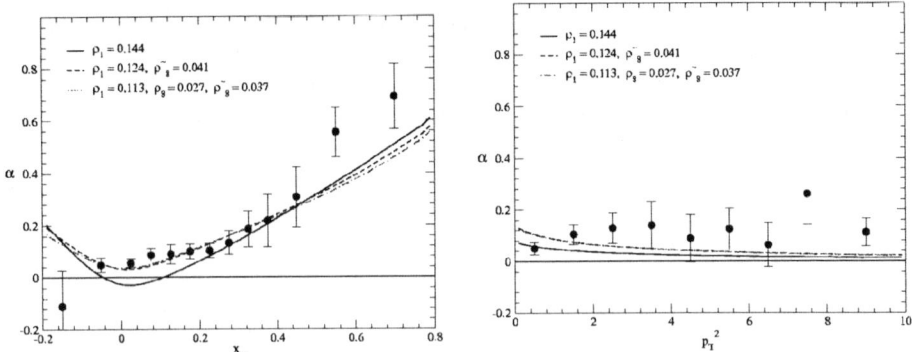

FIGURE 1. $\alpha[D^-]$ vs. x_F and $p_\perp^2$ for a 500 GeV π^- beam on a nuclear target [2]. One-, two- and three-parameter fits are solid, dashed and dot-dashed curves, respectively.

There are two possible color states, $\bar{3}$ and 6, and two possible spin states contributing at this order, for a total of four $O(\Lambda_{\text{QCD}}/m_c)$ parameters for Λ_c^+ production:

$$\eta_3 = \eta[cu(^1S_0^{(\bar{3})}) \to \Lambda_c^+], \quad \tilde{\eta}_3 = \eta[cu(^3S_1^{(\bar{3})}) \to \Lambda_c^+], \qquad (5)$$
$$\eta_6 = \eta[cu(^1S_0^{(6)}) \to \Lambda_c^+], \quad \tilde{\eta}_6 = \eta[cu(^3S_1^{(6)}) \to \Lambda_c^+].$$

Isospin symmetry requires $\eta[cu(n) \to \Lambda_c^+] = \eta[cd(n) \to \Lambda_c^+]$.

Techniques for calculating the short-distance cross sections $d\hat{\sigma}[\bar{q}g \to c\bar{q}(n)+\bar{c}]$ and $d\hat{\sigma}[qg \to cq(n)+\bar{c}]$ as well as explicit expressions can be found in Refs. [19, 22]. All these cross sections are strongly peaked in the forward direction of the initial light quark or antiquark. The cross section is larger for charm particles that share a valence quark with one of the colliding hadrons because the structure functions of the valence quarks are largest. Thus, heavy-quark recombination provides a natural explanation of the leading particle effect.

Charm hadrons can also be produced by ordinary fragmentation of charm quarks produced in $\bar{c}q$ or $\bar{c}\bar{q}$ recombination. This process, called "opposite side recombination", gives the following contributions to charm hadron production:

$$d\hat{\sigma}[qg \to H+X] = \sum_{n,\overline{D}} d\hat{\sigma}[qg \to \bar{c}q(n)+c]\,\rho[\bar{c}q(n) \to \overline{D}] \otimes D_{c \to H}, \qquad (6)$$

$$d\hat{\sigma}[\bar{q}g \to H+X] = \sum_{n,\overline{B}} d\hat{\sigma}[\bar{q}g \to \bar{c}\bar{q}(n)+c]\,\eta[\bar{c}\bar{q}(n) \to \overline{B}] \otimes D_{c \to H}. \qquad (7)$$

Here $\overline{B}$ is a charm antibaryon. The opposite side recombination mechanism can generate asymmetries even when there is no leading particle effect.

Fig. 1 compares calculations of $\alpha[D^-]$ using the heavy-quark recombination mechanism with data from the E791 experiment, which has a 500 GeV π^- beam [2]. The fragmentation functions are δ-functions times fragmentation probabilities taken from Ref. [25]. These fragmentation functions reproduce the single particle inclusive distributions at fixed-target energies better than Petersen fragmentation functions [24]. Other

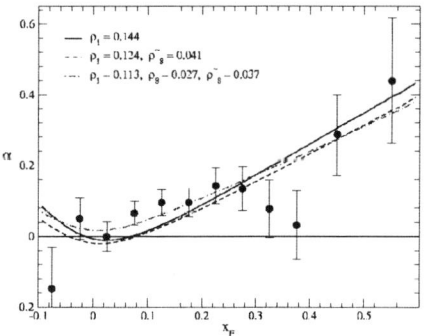 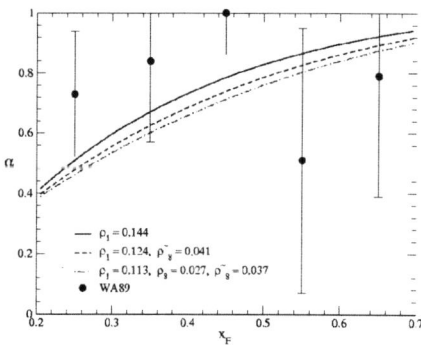

FIGURE 2. $\alpha[D^{*-}]$ vs. x_F for a 500 GeV π^- beam [2] (left) and 340 GeV Σ^- beam [3] (right). Solid, dashed and dot-dashed curves are the same as Fig. 1.

details can be found in Ref. [21]. The ρ parameters were determined by a global analysis of all measurements of D meson asymmetries in fixed-target hadroproduction experiments [26]. The result fits with one, two and three parameters are shown in Fig. 1. (A four-parameter fit did not yield significantly better results than the three-parameter fit.) Note that a one-parameter fit describes the data well in the forward region where heavy-quark recombination is most important. Inclusion of other parameters is necessary to obtain agreement in the central region $x_F \approx 0$. The universality of the ρ parameters can be tested with data on asymmetries of other D mesons. Fig. 2 compares predictions of the heavy-quark recombination mechanism with $\alpha[D^{*-}]$ from the E791 experiment [2] and $\alpha[D_s^-]$ from the WA89 experiment [3], which has a 340 GeV Σ^- beam, using the same parameter sets as in Fig. 1. The agreement with experiment is excellent.

Finally, in Fig. 3 we test the cq recombination mechanism for charmed baryons by comparing to measurements of $\alpha[\Lambda_c^+]$ in the E791 experiment [4] and experiments with 540 GeV p beams from SELEX [5]. The single nonvanishing ρ parameter is chosen to be consistent with the one-parameter fit to D meson asymmetries. Setting all the η parameters to zero gives the asymmetry shown by the dotted lines in Fig. 3, which is generated entirely by opposite side recombination. Though this is adequate for the π^- beam data, the p beam data clearly requires an additional mechanism. A one parameter fit with $\eta_3 = 0.22$ is shown by the solid lines in Fig. 3. The results of the calculation are in good agreement with both experiments.

ACKNOWLEDGMENTS

I thank Eric Braaten, Yu Jia and Masaoki Kusunoki for collaboration on the work presented in this talk. This research is supported by DOE Grants DE-FG02-96ER40945 and DE-AC05-84ER40150

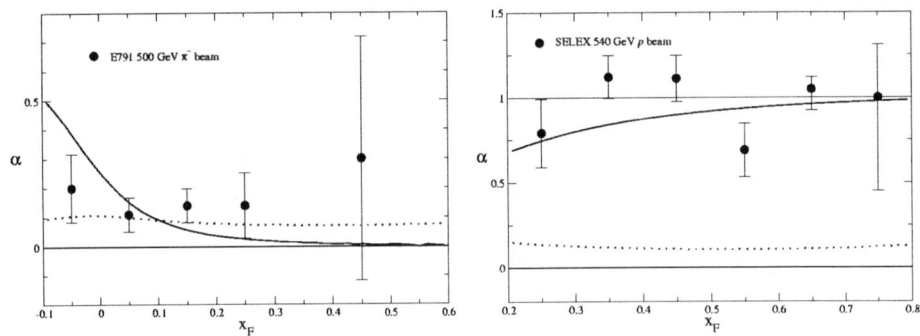

FIGURE 3. $\alpha[\Lambda_c^+]$ vs. x_F for a 500 GeV π^- beam [4] and 540 GeV p beam [5]. The solid curve is the best single-parameter fit with $\eta_3 = 0.22$, while the dotted curve is in the absence of cq recombination. The horizontal line at $\alpha = 1$ is the physical upper bound.

REFERENCES

1. E. M. Aitala et al. [E791 Collaboration], Phys. Lett. B **411**, 230 (1997);
   G. A. Alves et al. [E769 Collaboration], Phys. Rev. Lett. **72**, 812 (1994); ibid. **77**, 2392 (1996);
   M. Adamovich et al. [BEATRICE Collaboration], Nucl. Phys. B **495**, 3 (1997);
   M. Adamovich et al. [WA82 Collaboration], Phys. Lett. B **305**, 402 (1993).
2. E. M. Aitala et al. [E791 Collaboration], Phys. Lett. B **539**, 218 (2002); ibid. **371**, 157 (1996).
3. M. I. Adamovich et al. [WA89 Collaboration], Eur. Phys. J. C **8**, 593 (1999).
4. E. M. Aitala et al. [E791 Collaboration], Phys. Lett. B **495**, 42 (2000).
5. F. G. Garcia et al. [SELEX Collaboration], Phys. Lett. B **528**, 49 (2002).
6. J. C. Collins, D. E. Soper and G. Sterman, Nucl. Phys. B **263**, 37 (1986).
7. P. Nason, S. Dawson and R. K. Ellis, Nucl. Phys. B **327**, 49 (1989) [Erratum-ibid. B **335**, 260 (1989)].
   W. Beenakker, H. Kuijf, W. L. van Neerven and J. Smith, Phys. Rev. D **40**, 54 (1989).
   W. Beenakker, W. L. van Neerven, R. Meng, G. A. Schuler, J. Smith, Nucl. Phys. B **351**, 507 (1991).
8. H.-U. Bengtsson and T. Sjöstrand, Comput. Phys. Commun. **46**, 43 (1987).
9. T. Sjostrand, L. Lonnblad and S. Mrenna, hep-ph/0108264.
10. O. I. Piskounova, hep-ph/0202005.
11. E. Norrbin and T. Sjöstrand, Phys. Lett. B **442**, 407 (1998).
12. R. C. Hwa, Phys. Rev. D **51**, 85 (1995).
13. R. Rapp and E. V. Shuryak, Phys. Rev. D **67**, 074036 (2003).
14. A. K. Likhoded and S. R. Slabospitsky, Phys. Atom. Nucl. **65**, 127 (2002) [Yad. Fiz. **65**, 132 (2002)].
15. T. Tashiro, S. Nakariki, H. Noda and K. Kinoshita, Eur. Phys. J. C **24**, 573 (2002).
16. R. Vogt and S. J. Brodsky, Nucl. Phys. B **478**, 311 (1996).
17. E. Cuautle, G. Herrera and J. Magnin, Eur. Phys. J. C **2**, 473 (1998).
18. G. Herrera and J. Magnin, Eur. Phys. J. C **2**, 477 (1998).
19. E. Braaten, Y. Jia and T. Mehen, Phys. Rev. D **66**, 034003 (2002)
20. E. Braaten, Y. Jia and T. Mehen, Phys. Rev. D **66**, 014003 (2002)
21. E. Braaten, Y. Jia and T. Mehen, Phys. Rev. Lett. **89**, 122002 (2002)
22. E. Braaten, M. Kusunoki, Y. Jia and T. Mehen, hep-ph/0304280.
23. C. H. Chang, J. P. Ma and Z. G. Si, hep-ph/0301253.
24. S. Frixione, M. L. Mangano, P. Nason and G. Ridolfi, Nucl. Phys. B **431**, 453 (1994).
25. L. Gladilin, hep-ex/9912064.
26. E. Braaten, M. Kusunoki, Y. Jia and T. Mehen, in preparation.

Large-N_c selection rules for decay of J^{PC} exotic hybrid mesons

Philip R. Page

Theoretical Division, MS B283, Los Alamos National Laboratory, Los Alamos, NM 87545, USA

Abstract. The coupling of a neutral hybrid $\{1,3,5\ldots\}^{-+}$ exotic particle (or current) to two neutral (hybrid) meson particles with the same J^{PC} and $J = 0$ is proved to be sub-leading to the usual large-N_c QCD counting. The coupling of the same exotic particle to certain two - (hybrid) meson currents with the same J^{PC} and $J = 0$ is also sub-leading. The decay of a $\{1,3,5\ldots\}^{-+}$ hybrid to $\eta\pi^0$, $\eta'\pi^0$, $\eta'\eta$, $\eta(1295)\pi^0$, $\pi(1300)^0\pi^0$, $\eta(1440)\pi^0$, $a_0(980)^0\sigma$ or $f_0(980)\sigma$ is sub-leading, assuming that these final state particles are (hybrid) mesons in the limit of large N_c.

INTRODUCTION

States of Quantum Chromodynamics (QCD) can definitively be said not to be conventional mesons and baryons when these states have exotic J^{PC}, which cannot be constructed for conventional mesons in the quark model, or equivalently, cannot be built from local currents with only a quark and an antiquark field.

With the experimental discovery of isovector J^{PC} exotics, the question of their interpretation has come into focus. QCD with a large number of colours N_c offers a systematic expansion in $1/N_c$ with considerable phenomenological success [1, 2], which can address this question. This is because a glueball (built from only gluons) and a (hybrid) meson (quark-antiquark with additional gluons) do not mix in large-N_c [1]. Furthermore, four-quark states (two quark-antiquark pairs) are absent [2]. In large-N_c the isovector J^{PC} exotics *must* therefore be hybrid mesons, as glueballs are isoscalar. Here it is proved for the first time that certain decays of hybrid mesons that are allowed by the conserved quantum numbers of QCD are sub-leading to their usual large-N_c counting, providing a consistency check for the hybrid nature of the state.

Selection rules for J^{PC} exotic hybrid decays were noticed in non-field theoretic analyses [3]. A simple intuitive argument due to Lipkin of why a selection rule arises can be found in ref. [4]. In QCD it was found that these selection rules are really properties of certain three-point Green's functions [4, 5]. The first attempt to obtain hadronic properties from the Green's functions [5] contained some errors [4]. These properties were subsequently extracted in finite-N_c QCD, e.g. the physical $N_c = 3$ [4]. However, for technical reasons, the scope of the deductions was limited. These reasons disappear in large-N_c. The large-N_c treatment of the results of Ref. [4] is presented here.

COUPLING OF CURRENTS AND PARTICLES TO PARTICLES

In Eqs. 2, 3 and 14 of Ref. [4] it was proved (with no approximations) that

$$\int_{-\infty}^{\infty} dt\, e^{iEt}\, \hat{O}_\mathbf{p} \int d^3x\, d^3y\, e^{i(\mathbf{p}\cdot\mathbf{x}-\mathbf{p}\cdot\mathbf{y})} \langle 0| B(\mathbf{x},t)\, C(\mathbf{y},t)\, A_\mu(0) |0\rangle$$
$$= \sum_n (2\pi)^4\, \delta^3(\mathbf{p}_n)\, \delta(E_n-E)\, \hat{O}_\mathbf{p} \langle 0|\left(\int d^3x\, e^{i\mathbf{p}\cdot\mathbf{x}} B(\mathbf{x},0)\right) C(0) |n\rangle \langle n|A_\mu(0)|0\rangle. \quad (1)$$

The left-hand side (L.H.S.) of the equation contains the time integral and spatial Fourier transform of a three-point Green's function which describes the "decay" of A into B and C. The expression is in Minkowski (physical) space with E and $\mathbf{p}$ real numbers. The L.H.S. is expanded on the right-hand side (R.H.S.) by inserting an infinite set of asymptotic stable states n with energy E_n and momentum $\mathbf{p}_n$ in order to extract physical predictions. The delta functions indicate that the asymptotic states are at rest and have energy E. The gauge-invariant local currents B, C and A_μ have the flavour structure of a neutral (hybrid) meson (linear combinations of $\bar{u}u, \bar{d}d, \ldots$ quark fields), and can contain gluon fields [6]. The currents B and C both have the same colour-Dirac-derivative-gluon structure for a given flavour, a finite number of derivatives (when expanded as a power series) and $J=0$. Also, the currents $B(0)$ and $C(0)$ have equal P and C. The current $A_\mu(0)$ is assumed to have $P=-$ and odd J (with Lorentz indices denoted by μ). Conservation of charge conjugation then implies that this current is $J^{PC} = \{1,3,5\ldots\}^{-+}$ exotic, so that it should contain at least one gluon field: a hybrid meson current.

Eq. 1 would be of limited interest were it not for the fact that the action of the operator $\hat{O}_\mathbf{p}$ (containing a finite number of derivatives in powers of $\mathbf{p}$) allowed the demonstration that the L.H.S. contains only OZI rule forbidden contributions, and hence is $\mathscr{O}(1)$ to leading order in the large-N_c power counting, as opposed to the usual $\mathscr{O}(N_c)$. This behaviour of the L.H.S. is exploited to deduce the consequences for the R.H.S. The strategy is to keep only the leading contributions to the R.H.S. in large-N_c, and then to equate to the L.H.S.

The *first result*, derived in ref. [7], is that the coupling of currents to a particle

$$\langle 0| B(\mathbf{x},t)\, C(\mathbf{y},t) |\sigma \mathbf{0}\rangle = \mathscr{O}\left(\frac{1}{\sqrt{N_c}}\right), \quad (2)$$

where its usual counting is $\mathscr{O}(\sqrt{N_c})$. This holds for a neutral on-shell hybrid meson particle σ at rest with $J^{PC} = \{1,3,5\ldots\}^{-+}$. Also, B and C are neutral gauge-invariant local (hybrid) meson currents at space-time positions x and y at equal time with flavour structure a linear combination of $\bar{u}u, \bar{d}d,\ldots$, the same colour-Dirac-derivative-gluon structure for a given flavour, a finite number of derivatives and $J=0$. The currents $B(0)$ and $C(0)$ should have equal P and C.

The *second result*, derived in ref. [7], is that the coupling of particles to a current

$$\langle \sigma_1 k_1 \sigma_2 k_2 | A_\mu(z) |0\rangle = \mathscr{O}\left(\frac{1}{N_c}\right), \quad (3)$$

where its usual counting is $\mathscr{O}(1)$. This holds for neutral on-shell (hybrid) meson particles σ_1 and σ_2 with identical J^{PC} and $J=0$, and with arbitrary four-momenta k_1 and k_2. Also, $A_\mu(z)$ is a neutral gauge-invariant local hybrid meson $J^{PC} = \{1,3,5\ldots\}^{-+}$ current at space-time position z with flavour structure a linear combination of $\bar{u}u, \bar{d}d, \ldots$.

The *third result*, derived in ref. [7], is that the coupling of a particle to particles (the T-matrix expectation value)

$$\langle \sigma_1 k_1 \sigma_2 k_2 | T | \sigma \mathbf{0} \rangle = \mathscr{O}(\frac{1}{N_c^{\frac{3}{2}}}), \qquad (4)$$

where its usual counting is $\mathscr{O}(1/\sqrt{N_c})$. This holds for neutral on-shell (hybrid) meson particles σ_1 and σ_2 with identical J^{PC} and $J=0$; and for a neutral on-shell hybrid meson particle σ with $J^{PC} = \{1,3,5\ldots\}^{-+}$.

LARGE-N_C DECAY PHENOMENOLOGY

The three results are Eqs. 2, 3 and 4, including the discussion under each equation. These results are theorems of large-N_c QCD field theory with no approximations, and are valid within the generic large-N_c framework [1, 2, 8].

The first result (Eq. 2) implies that certain four-quark currents are not good interpolators for hybrid meson particles. This may have implications for Euclidean space lattice QCD, even though the result was derived only in Minkowski space. A special case of the second result (Eq. 3) was previously derived [4] for an $\eta\pi^0$ asymptotic state for certain quark masses within a certain kinematical range.

The third result (Eq. 4) is of more direct experimental relevance. For *example*, the decay amplitudes (couplings of a particle to particles) of a $\{1,3,5\ldots\}^{-+}$ hybrid to $\eta\pi^0$, $\eta'\pi^0$, $\eta'\eta$, $\eta(1295)\pi^0$, $\pi(1300)^0\pi^0$, $\eta(1440)\pi^0$, $a_0(980)^0\sigma$ or $f_0(980)\sigma$ are $\mathscr{O}(1/N_c^{\frac{3}{2}})$, while the usual counting is $\mathscr{O}(1/\sqrt{N_c})$, assuming that these final state particles are (hybrid) mesons in the limit of large N_c. Hence the widths of these decays are $1/N_c^2$ suppressed with respect to their usual counting. This is the same suppression that large-N_c predicts for decays forbidden by the OZI rule [1], implying that the suppressions predicted here should phenomenologically be similar. The selection rule is most useful when OZI allowed decay is expected to be important in the absence of the selection rule. In the example above, this is true for a hybrid composed dominantly of $u\bar{u}$ and $d\bar{d}$. An the other hand, it can be deduced from Eq. 4 that the coupling of a 1^{-+} hybrid to $\eta_c\eta$ is $\mathscr{O}(1/N_c^{\frac{3}{2}})$, but this is less useful as the OZI allowed coupling (of the $c\bar{c}$ component of the hybrid to η_c ($c\bar{c}$) and the $c\bar{c}$ component of the η) is not expected to be important. Interestingly, even in the unlikely case where the η' or σ is a pure glueball in the limit of large N_c, the decay amplitude of a $\{1,3,5\ldots\}^{-+}$ hybrids to $\eta'\pi^0, \eta'\eta$, $a_0(980)^0\sigma$ or $f_0(980)\sigma$ would be $\mathscr{O}(1/N_c)$ [1], which is still subdominant to the usual counting. Examples can also be given of 0^{+-} and 0^{--} exotic particles in the final state. Assuming isospin symmetry the results can also be extended to charged states by use of the Wigner-Eckart theorem.

Consider the decay of a 1^{-+} isovector hybrid with isospin symmetry. Decay to $\eta\pi$, $\eta'\pi$, $\eta(1295)\pi$, $\eta(1440)\pi$ and $a_0(980)\sigma$, which is ordinarily important, is suppressed. The experimental $\pi_1(1600)$ [9] is a candidate 1^{-+} exotic isovector resonance. It has not been seen in $\eta\pi$ and has prominently been seen in $\eta'\pi$ [9]. It has possibly been seen in $\eta(1295)\pi$: A 1.7 GeV π_1 was seen in $f_1\pi$ and $\eta(1295)\pi$ at a similar magnitude in $K^+\bar{K}^0\pi^-\pi^-$ [10]. However, more recent investigation indicates that only $f_1\pi$ is important [11]. If $\pi_1(1600)$ was found to have a large branching ratio to $\eta\pi'$ that would be inconsistent with large-N_c expectations which are otherwise consistent with its being a hybrid meson [12]. Recently a claim was made that a subset of the experimental data can be explained be generating the $\eta'\pi$ peak mostly via non-resonant mechanisms, with the possibility of an admixture of the resonance $\pi_1(1600)$ [13]. This mechanism would claim that the dominant part of the experimental π_1 wave is due to non-resonant $\eta'\pi$ production, which would tend to increase the consistency of experiment with large-N_c predictions.

This research is supported by the Department of Energy under contract W-7405-ENG-36.

REFERENCES

1. R.F. Lebed, *Czech. J. Phys.* **49**, 1273 (1999).
2. S. Coleman, *Aspects of Symmetry*, Cambridge University Press (Cambridge, UK, 1985), p. 351, and specifically pp. 377-378.
3. P.R. Page, *Phys. Lett.* **B401**, 313 (1997); H.J. Lipkin, *Phys. Lett.* **B219**, 99 (1989); F.E. Close and H.J. Lipkin, *Phys. Lett.* **B196**, 325 (1987); C.A. Levinson, H.J. Lipkin and S. Meshkov, *N. Cim.* **32**, 1376 (1964).
4. P.R. Page, *Phys. Rev.* **D64**, 056009 (2001).
5. F. Iddir et al., *Phys. Lett.* **B207**, 325 (1988).
6. The currents were previously taken to have the same colour-Dirac-derivative-gluon structure for all flavours (Eq. 1 of Ref. [4]). However, all derivations go through without this restriction (see Eq. 10 of Ref. [4] and discussion below it).
7. P.R. Page, hep-ph/0303170.
8. T.D. Cohen, *Phys. Lett.* **B427**, 348 (1998).
9. K. Hagiwara et al. (Particle Data Group), *Phys. Rev.* **D66**, 1 (2002).
10. J.H. Lee et al., *Phys. Lett.* **B323**, 227 (1994).
11. D. Weygand, *private communication*.
12. The partial widths of π_1 to $\eta\pi$ and $\eta'\pi$ have been addressed by A. Zhang and T.G. Steele, *Phys. Rev.* **D67**, 074020 (2003); *Phys. Rev.* **D65**, 114013 (2002); K.G. Chetyrkin and S. Narison, *Phys. Lett.* **B485**, 145 (2000); S. Narison, "QCD spectral sum rules", Lecture Notes in Phys. Vol. 26 (1989), p. 374; J.I. Latorre et al., *Z. Phys.* **C34**, 347 (1987).
13. A.P. Szczepaniak, M. Swat, A.R. Dzierba and S. Teige, hep-ph/030409.

Meson Spectroscopy in Photo-production at CLAS

M. Nozar – CLAS Collaboration

TJNAF, 12000 Jefferson Ave, MS 12H, Newport News, VA 23606

Abstract. Photo-production of excited mesons in the $1-2$ GeV mass range decaying via multi-pion or multi-kaon emission has been investigated at the TJNAF[1] experiment E01-017 (g6c) in the $4.8-5.4$ GeV photon beam energy range. The main objective of the experiment is to extract resonance parameters of the produced states by way of a Partial Wave Analysis (PWA) technique. In this paper, we will focus on the general characteristics of the data distributions in both the neutral and charged 3-pion final states, i.e. $\pi^+\pi^-\pi^0$ and $\pi^+\pi^+\pi^-$.

INTRODUCTION

The observation of an isovector exotic state with $J^{pc} = 1^{-+}$ quantum numbers at 1.6 GeV by the Brookhaven E852 experiment [1] provided the main motivation to search for such a state using a photon beam at JLab. This state is one of the signature hadronic states outside the constituent quark model, where either the excitation of the gluons within the hadron directly contributes to its degrees of freedom, giving a state configuration of ($q\bar{q}g$), or that the state is composed of more than one $q\bar{q}$ pair, i.e. $q\bar{q}q\bar{q}$. Such a state, if produced in a pion beam via ρ exchange, may also be produced in a photon beam, with the role of the beam and the exchange particle reversed and assuming Vector Meson Dominance (VMD) [2, 3, 4]. Photon beams, as probes for exotic meson production, have not been fully explored so far; the existing data on multi-particle final states are very sparse. The only relevant data comes from a SLAC bubble chamber experiment which employed a backscattered laser photon beam of 19.5 GeV average energy [5]. In the low mass region, the data is dominated by $a_2(1320)$ production with no clear evidence for $a_1(1260)$. In the high mass region, the group claimed evidence for a narrow state at 1.775 GeV, with possible $J^{pc} = 1^-, 2^-$, or 3^+ quantum numbers. Because of insufficient statistics, no PWA was performed on the data.

EXPERIMENTAL SETUP AND EVENT SELECTION.

The data for this analysis was collected during Aug.–Sep. of 2001. The 40 nA primary beam of 5.7 GeV electrons at 100% duty factor and was provided by CEBAF (Continuous Electron Beam Accelerator Facility). The secondary beam of photons was generated

---

[1] This work was supported by the U.S. Department of Energy and The U.S. National Science Foundation.

in Hall B via bremsstrahlung radiation, using a $2*10^{-4}$ r.l. radiator and a tagging system identifying photons in the [20% − 95%] range of the incident electron beam energy [6]. The high flux of $5x10^6$/sec photons (in the top 15% of the photon beam energy) was incident on an 18 cm long LH_2 target. The Hall B experimental hall houses the CLAS (CEBAF Large Acceptance Spectrometer) detector. CLAS covers a large solid angle with polar angle coverge in the range $8° \leq \theta \leq 145°$, and azimuthal angle coverage of 80%. The detector, composed of six independent sectors, provides a toroidal magnetic field, where positively charged particles bend outward and negatively charged tracks bend inward. Three sets of drift chambers in the radial direction are embedded in the space between the coils, and provide charged particle detection and track reconstruction. A set of time of flight scintillators are used for charged particle identification, and a set of electromagnetic calorimeters are used for neutral particle detection. Details of the CLAS detector design and performance are described elsewhere [7].

The lack of acceptance in the forward region ($\theta_{lab} < 8°$) presents a drawback to any meson spectroscopy program at CLAS in its current configuration. In addition, with the maximum available photon beam energy of 5.4 GeV, there is a significant contribution from t-channel baryon resonance production.

Both neutral and charged 3π final states were subjected to similar topological and geometrical cuts. The photon beam energy was selected to be higher than 4.8 GeV; vertex position cuts were applied to ensure the events originated within the target volume, and vertex timing constraints were imposed to reduce the accidentals between the CLAS and the tagging system.

DATA DISTRIBUTIONS: $\gamma p \to \pi^+ \pi^+ \pi^- (n)$

The π^+, π^+, and π^- were detected in CLAS, and the neutron was reconstructed by way of missing mass. In order to suppress the t-channel baryon background, we have employed two kinematical cuts, a low four-momentum transfer to the $\pi^+\pi^+\pi^-$ cut defined as $-t' \leq 0.4$ GeV^2, and the other, a small lab polar angle cut imposed on the two positively charged pions defined as $\theta_{lab} \leq 30°$. These two cuts, referred to as "baryon rejection cuts", discard mostly events associated with background processes. In all the plots shown for this final state, the shaded histograms represent the events that passed these two cuts collectively.

The left plot in Fig. 1 shows the missing mass off the $\pi^+\pi^+\pi^-$ for low $-t'$ events. The neutron peak sits on top of a negligible linearly increasing background, with a signal to background ratio of approximately 9/1. A Gaussian plus a 1st order polynomial fit to the peak gives a σ of 25 MeV. The area between the lines around the neutron peak represents the neutron selection cut. The $-t'$ distribution shown in the middle plot of Fig. 1 is fit to an exponential function of the form $f(t') = ae^{-b|t'|}$. The shape of the distribution is consistent with the characteristics of peripheral production, and the exponential constant of 4.4 GeV^{-2} is consistent with pion exchange [8]. In the 3π invariant mass spectrum shown in the right plot of Fig. 1 two enhancements are evident, one in the 1300 MeV region, and another in the 1600 − 1700 MeV range.

Figure 2 shows all three possible combinations of the $n\pi$ and $\pi\pi$ invariant mass distributions. In this analysis, the two positively charged pions were sorted based on mo-

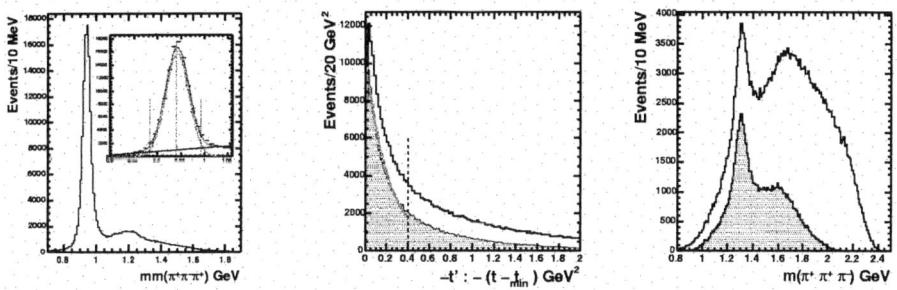

FIGURE 1. Left: Missing mass off of the $\pi^+\pi^+\pi^-$ for low $-t'$ events. Middle: four-momentum transfer to 3π for $\pi^+\pi^+\pi^-(n)$ events. Right: $\pi^+\pi^+\pi^-$ invariant mass distribution.

mentum, with the π_1^+ being the one with higher momentum. The two $n\pi^+$ combinations show peaks around the known baryon resonances, $\Delta(1232)$, $N^\star(1520)$, and $N^\star(1680)$, while the $n\pi^-$ shows a peak around the $\Delta(1232)$ only as is expected due to isospin considerations. It is clear from the shaded distributions that the baryon resonance peaks are greatly suppressed after the "baryon rejection" cuts. The neutral 2π effective mass distributions show signals around the mass of the $\rho(770)$ and the $f_2(1270)$, as well as a shoulder at the mass of the $f_0(980)$. The doubly-charged 2π combination doesn't show any distinct peak, indicative of the lack of an isospin $I=2$ state.

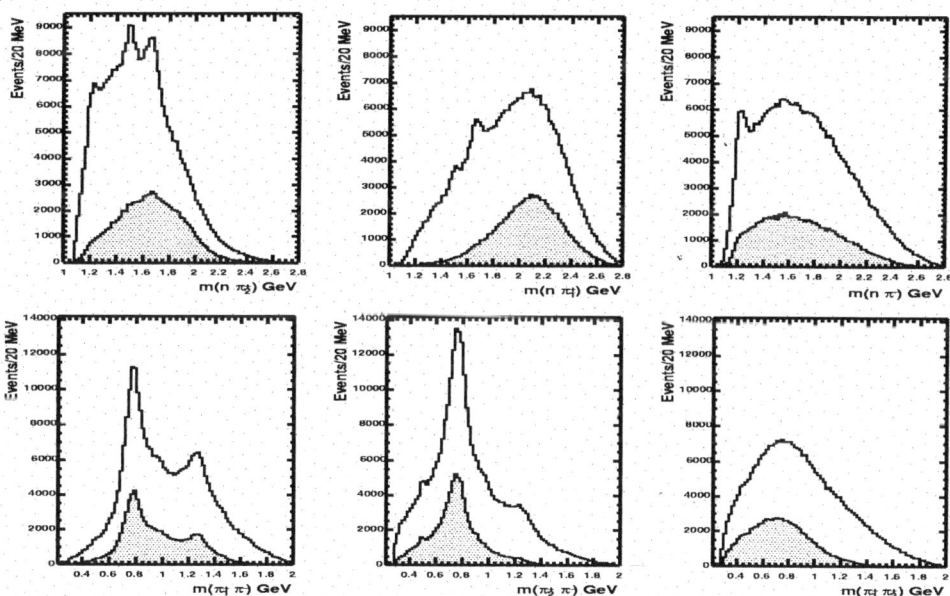

FIGURE 2. $n\pi$ (top row) and $\pi\pi$ (bottom row) invariant mass distributions for $\pi^+\pi^+\pi^-(n)$ events.

DATA DISTRIBUTIONS: $\gamma p \to p\pi^+\pi^-(\pi^0)$

The p, π^+, and π^- were detected in CLAS while the π^0 was reconstructed by way of missing mass. In order to enhance the 3π mesonic events, a combination of low $-t'$ and $p\pi$ mass cut was employed. The cut was defined as $0.1 \leq -t' \leq 1.0$ GeV^2, with the $p\pi$ effective masses for all three combination above 1.35 GeV. We'll refer to these two cuts, as "baryon rejection cuts". In all the plots shown for this final state, the shaded histograms represent the events that passed these two cuts.

The left plot in Fig. 3 shows the missing mass squared off the $p\pi^+\pi^-$. A Gaussian plus a 2nd order polynomial fit to the peak around the π^0 gives a σ of 41 MeV for the missing mass resolution. The region between the lines represents the π^0 selection cut; these events were subjected to kinematical fitting in a later stage. The $-t'$ distribution is shown in the middle plot of Fig. 3. The shaded distribution, after the $\Delta(1232)$ rejection cut is fit to an exponential function, $f(t') = ae^{-b|t'|}$, with the resulting $b = 1.3$ GeV^{-2}. The 3π invariant mass distribution is shown in the right plot of Fig. 3, with the shaded histogram showing the events which passed the "baryon rejection cuts". In the remaining events, one is able to see enhancement of events at 1300 MeV and in the 1600 − 1700 MeV region.

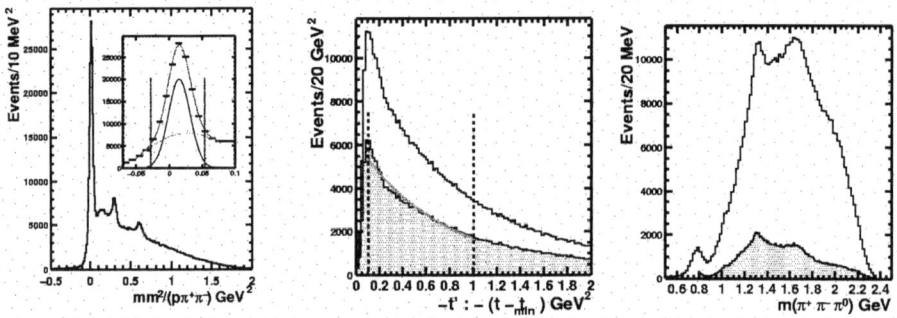

FIGURE 3. Left: Missing mass off of the $p\pi^+\pi^-$ for low $-t'$ events. Middle: four-momentum transfer to 3π for $p\pi^+\pi^-\pi^-(\pi^0)$ events. Right: $\pi^+\pi^-(\pi^0)$ invariant mass distribution.

Figure 4 shows all three possible combinations of the $p\pi$ and $\pi\pi$ invariant mass distributions. Before the "baryon rejection cuts", the most prominent feature observed is the peak around the $\Delta(1232)$ mass region. After the cuts, the peaks around the $N^*(1520)$ and $N^*(1680)$ still remain. The invariant mass distributions for the 2π charged combinations, $\pi^+\pi^0$ and $\pi^-\pi^0$, show strong peaks at the $\rho(770)$ mass, while the neutral 2π invariant mass distribution shows enhancements around the $\rho(770)$ and $f_2(1270)$ masses.

SUMMARY AND OUTLOOK

We have shown the data quality and the general characteristics of the distributions for both the neutral and the charged 3π final states from the data collected during the g6c

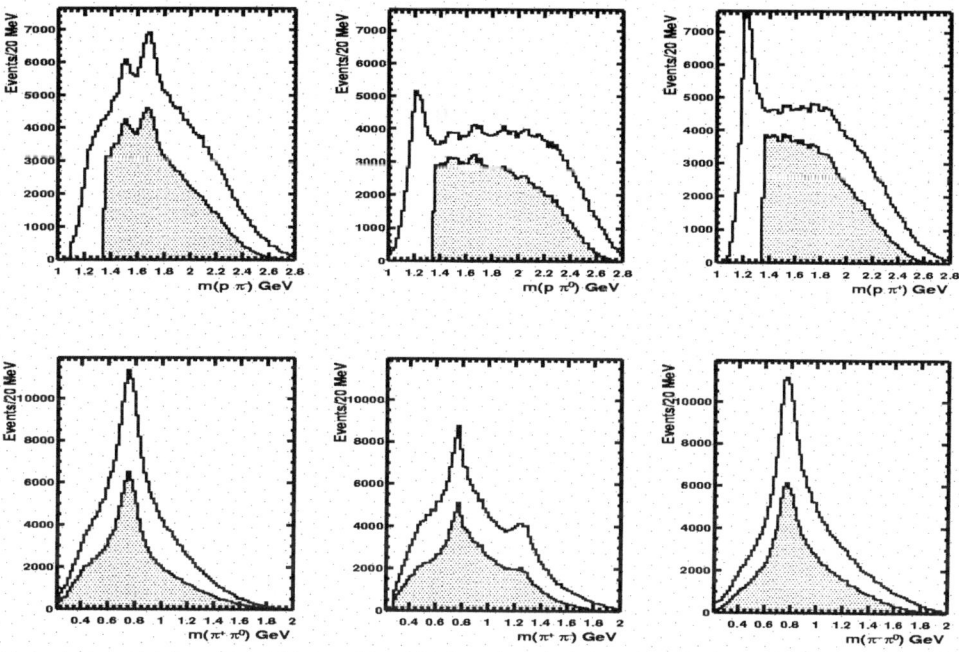

FIGURE 4. $p\pi$ (top row) and $\pi\pi$ (bottom row) invariant mass distributions for $p\pi^+\pi^-(\pi^0)$ events.

running of CLAS. The statistics for both sets of events is by a factor of few hundred higher than the existing data. Partial Wave Analysis on both systems are underway.

ACKNOWLEDGMENTS

The invaluable efforts of the Jlab Accelerator staff, the Physics Division, and the g6c group are greatly appreciated. I am especially grateful to Ji Li for his analysis of the neutral 3π final state and for providing the figures for this final state.

REFERENCES

1. S. U. Chung et al., *Physical Review D*, **65**, 1–16 (2002).
2. A. V. Afanasev and A. P. Szczepaniak, *Physical Review D*, **61** (2000).
3. A. V. Afanasev and P. R. Page, *Physical Review D*, **57** (1998).
4. F. E. Close and P. R. Page, *Physical Review D*, **52** (1995).
5. G. T. Condo et al., *Physical Review D*, **43** (1991).
6. D. I. Sober et al., *Nuclear Instruments and Methods in Physics Research A*, **440**, 263 (2000).
7. B. A. Mecking et al., *Nuclear Instruments and Methods in Physics Research A*, **503**, 513–553 (2003).
8. M. Guidal, J. M. Laget, M. Vanderhaeghen, *Nuclear Physics A*, **627**, 645–678 (1997).

Effect of light scalar mesons in $\eta \to 3\pi$ [1]

A. Abdel-Rehim*, Deirdre Black†, A.H. Fariborz** and J. Schechter*

Physics Department, Syracuse University, Syracuse, NY 13244
†*Jefferson Lab, 12000 Jefferson Ave, Newport News, VA 23606*
**Department of Mathematics/Science, State University of New York Institute of Technology, Utica, NY 13504*

Abstract. We discuss the decay $\eta \to 3\pi$, for which the observed branching ratio is larger than the simplest predictions from chiral symmetry. This process occurs primarily due to strong isospin violation and so is in principle a sensitive measure of the up-down quark mass difference. We study the role of a possible nonet of light scalar mesons in the $\eta \to 3\pi$ decay. Using a non-linear chiral Lagrangian approach which successfully describes several other strong processes, we find that the inclusion of the scalar mesons leads to a modest increase relative to the simplest prediction. The main effect of the scalar mesons is due to a light broad σ meson.

INTRODUCTION

The decay $\eta \to 3\pi$ has a rather interesting history. Current Algebra (CA) techniques, with the appropriate product of two weak currents, had proven very successful for describing the decay rate for $K \to 3\pi$, as well as the spectrum shape:

$$\sim 1 - \frac{2E_0}{m_K}, \qquad (1)$$

where E_0 is the energy of the neutral pion in the $\pi^+\pi^-\pi^0$ final state. The $\eta \to 3\pi$ decay violates isospin (an I=0=J three-pion state is ruled out by Bose statistics) and the current Particle Data Group average for the charged pion mode is [3]:

$$\Gamma(\eta \to \pi^0\pi^+\pi^-) = 267 \pm 25 \text{eV} \qquad (2)$$

In 1966 Sutherland [1] found that, using electromagnetic currents, the CA amplitude for this process vanished. Earlier that year, Bose and Zimmerman [2] found a non-zero result for the decay rate using current commutation relations based on the quark model (so a quark scalar density operator proportional to $u\bar{u} - d\bar{d}$). The origin of such an operator can be seen explicitly by writing the mass term for the up and down quarks as the sum of an isospin-conserving and a $\Delta I = 1$ piece:

$$m_u u\bar{u} + m_d d\bar{d} = \frac{m_u + m_d}{2}(u\bar{u} + d\bar{d}) + \frac{m_u - m_d}{2}(u\bar{u} - d\bar{d}). \qquad (3)$$

---

[1] Talk presented by D. Black.

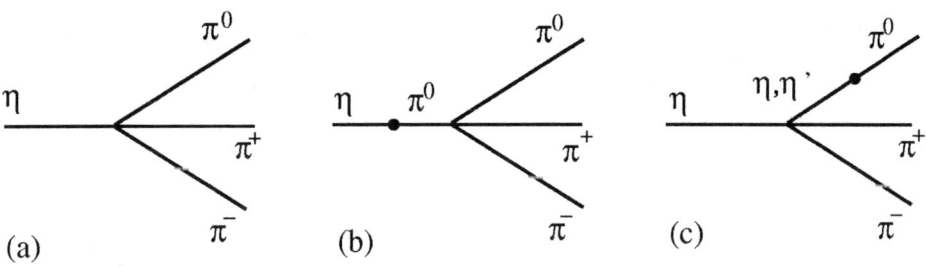

FIGURE 1. Diagrams contributing to the decay $\eta \to \pi^+\pi^-\pi^0$ at tree-level from the leading order U(3) chiral Lagragian given in Eq. (4). (a) is an intrinsic four-point isospin-violating vertex while (b) and (c) contain the two-point isospin-violating $\pi^0 - \eta$ transition, all constant and due to the second term in $\mathcal{L}_{LO}$.

Strong isospin-violating effects such as $\eta \to 3\pi$, which arise due to the $\Delta I = 1$ operator $u\bar{u} - d\bar{d}$, are small because the coefficient $(m_u - m_d)/2$ is small, but also very interesting as they are a sensitive probe of this up-down current quark mass difference.

CHIRAL LAGRANGIAN DESCRIPTION OF $\eta \to 3\pi$

The prediction using Current Algebra, or equivalently the leading order SU(2) chiral Lagrangian, is $\Gamma(\eta \to \pi^0\pi^+\pi^-) = 65\,\text{eV}$, much smaller than the experimental result in Eq. (2). This discrepancy was surprising since, given the low energy of the pions in this decay, one would expect the simplest predictions following from chiral symmetry to work reasonably well. The spectrum shape prediction is analogous to Eq. (1).

Next, we consider the effect of mixing of η with η' which is included in the leading order U(3) non-linear chiral Lagrangian:

$$\mathcal{L}_{LO} = \frac{F_\pi^2}{8}\text{Tr}(\partial_\mu U \partial^\mu U^\dagger) + \delta\text{Tr}[\mathcal{M}(U + U^\dagger)] + \frac{\kappa}{576}\ln^2\left(\frac{\det U}{\det U^\dagger}\right), \quad (4)$$

where $U = \exp(\frac{2i\phi}{F_\pi})$ and ϕ is the matrix of pseudoscalar mesons π, K, η and η'. The pseudoscalar masses are due to explicit symmetry breaking by the quark masses, contained in the spurion $\mathcal{M}$, and the last term in Eq. (4), which breaks $U(1)_A$. The resulting amplitude for the process $\eta \to 3\pi$ is given by the diagrams shown in Fig. 1.

$\mathcal{L}_{LO}$ in Eq. (4) has five free parameters, which we may choose to be F_π, δ, κ, $(m_u - m_d)/\hat{m}$ and $m_s/\hat{m}$, where $\hat{m}$ is the average of the up and down quark masses. The parameters in $\mathcal{L}_{LO}$ can be fixed by fitting to, for example, the measured values of F_π, m_π, m_K, $m_{\eta'}$ and $[\Delta m_K^2]_{\text{quark}}$, the piece of the $K^+ - K^0$ mass squared difference arising from the up-down quark mass difference. Once these parameters are fixed we obtain a prediction $\Gamma(\eta \to \pi^0\pi^+\pi^-) = 106\,\text{eV}$, which is larger than the SU(2) result, but still smaller than the experimental value.

A caveat in this prediction is that in order to extract $[\Delta m_K^2]_{\text{quark}}$ from the observed kaon mass difference, $[\Delta m_K^2]_{\text{phys}}$ we must subtract virtual photon effects:

$$[\Delta m_K^2]_{\text{quark}} = [\Delta m_K^2]_{\text{phys}} - [\Delta m_K^2]_\gamma = [\Delta m_K^2]_{\text{phys}} - [\Delta m_\pi^2]_{\text{phys}}.$$

To obtain the second equality we use Dashen's Theorem which states that $[\Delta m_K^2]_\gamma = [\Delta m_\pi^2]_\gamma$ and assume that the experimentally measured $\pi^+ - \pi^0$ mass difference arises solely from the photon diagrams. Thus the isospin-violation parameter $(m_u - m_d)/\hat{m}$, which is fixed from $[\Delta m_K^2]_{\text{quark}}$, is very sensitive to deviations from Dashen's Theorem.

Including scalar mesons

The light scalar mesons are the subject of considerable current discussion and our interest in [4] was to see what we could learn about them through their possible role in $\eta \to 3\pi$ decay. We add a nonet $[\sigma(550), \kappa(900), f_0(980), a_0(980)]$ of scalar mesons to $\mathcal{L}_{LO}$ in a chiral invariant framework [5] which is phenomenologically very successful in describing $\pi\pi$ and πK scattering and $\eta' \to \eta\pi\pi$. In this approach there are additional contributions, shown in Fig. 2, to the tree-level $\eta \to 3\pi$ amplitude. In Fig. 2, all of the trilinear scalar-pseudoscalar-pseudoscalar coupling constants and scalar meson masses are fixed from fitting to the strong scattering and decays just mentioned. Since σ is the lightest of the scalar mesons, one might expect diagrams (a) and (b) in Fig. 2 to give the largest scalar contribution to the amplitude. Individually this is indeed the case, but it turns out that these diagrams tend to cancel, which can be understood by noting that the π^0 and η, η' propagators in (a) and (b) have opposite signs. Nevertheless, the inclusion of the scalar mesons in this way increases the prediction to $\Gamma(\eta \to \pi^0\pi^+\pi^-) = 120\,\text{eV}$.

Next we considered the effect of adding higher order, i.e. $\mathcal{O}(m_q\partial^2, m_q^2)$, isospin-breaking terms to the lowest order Lagrangian of pseudoscalar and scalar mesons. These terms are given in [4]. It turns out that our prediction for the overall rate decreases to $\Gamma(\eta \to \pi^0\pi^+\pi^-) \approx 100\,\text{eV}$. However, interestingly, the prediction for the width due to the pseudoscalar mesons alone (diagrams in Fig. 1) decreases from 106 eV to 81 eV, meaning that the relative importance of the contribution of the scalar mesons increases. This is because there is an additional, momentum-dependent, contribution to the $\pi^0 - \eta$ mixing coefficient which means that the cancellation between diagrams Fig. 2 (a) and (b) is not so complete as at lowest order.

SUMMARY AND DISCUSSION

We discussed the decay $\eta \to 3\pi$ for which the simplest predictions based on chiral symmetry are too small. Experimentally, $\Gamma(\eta \to \pi^0\pi^+\pi^-)_{\text{expt}} = 267 \pm 25$ eV. We found in Ref. [4] that including the scalar mesons in a chiral invariant way boosts the initial theoretical prediction a little to $100 - 120$ eV (depending on the detailed treatment of the scalar widths and the vector mesons, which we do not discuss here). This may be compared with the one-loop Chiral Perturbation Theory result 160 ± 50 eV [6].

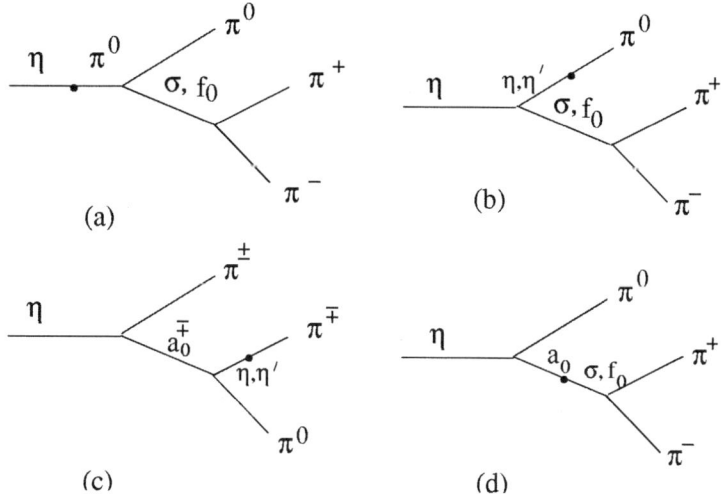

FIGURE 2. Diagrams involving a nonet of light scalar mesons, which contribute at tree-level to the $\eta \to 3\pi$ amplitude.

It will be interesting to include the effect of final state interactions in our model, since it has been found by other authors (see [4] for references) that these are significant. We also note that deviations from Dashen's Theorem increase the extracted value of $(m_u - m_d)/\hat{m}$ and so boost the $\eta \to 3\pi$ predictions by an overall factor. We also plan to study in detail the decay $\eta' \to 3\pi$, where the new scalar isospin violating $a_0 - f_0$ transition which occurs in Fig. 2 d) may be more important.

ACKNOWLEDGMENTS

The work of A. A-R. and J. S. has been supported in part by the US DOE under contract DE-FG-02-85ER 40231. D.B. wishes to acknowledge support from the Thomas Jefferson National Accelerator Facility operated by the Southeastern Universities Research Association (SURA) under DOE Contract No. DE-AC05-84ER40150. The work of A. H. F. has been supported by the 2003 grant from the State of New York/UUP Professional Development Committee, and the 2003 Faculty Grant from the School of Arts and Sciences, SUNY Institute of Technology.

REFERENCES

1. D.G. Sutherland, Phys. Lett **23**, 384 (1966).
2. S. Bose and A. Zimmerman, Nuovo Cimento A **43**, 1165 (1966).
3. Particle Data Group, K. Hagiwara *et al*, Phys. Rev. **D66**, 1-I 010001 (2002) and http://pdg.lbl.gov.
4. A. Abdel-Rehim, D. Black, A.H. Fariborz and J. Schechter, Phys. Rev. **D67**, 054001 (2003).
5. D. Black, A.H. Fariborz, F. Sannino and J. Schechter, Phys. Rev. **D59**, 074026 (1999).
6. J. Gasser and H. Leutwyler, Nucl. Phys. B250, 539 (1985).

Measurement of the angular distribution of $\psi'(3686) \to e^+e^-$ from $\bar{p}p$ annihilation

S. H. Seo

School of Physics and Astronomy, University of Minnesota
Minneapolis, MN 55455
(for E835 collaboration)

Abstract. E835 at Fermilab reports the angular distribution for the reaction $\bar{p}p \to e^+e^-$ at ψ' resonance leading to a measurement of the proton electromagnetic form factor ratio ($|\frac{G_E}{G_M}|$). Based on a clean sample of 6844 events for the exclusive reaction, we obtained a preliminary measurement of $\lambda = 0.67^{+0.15}_{-0.14} \pm 0.07$ and $|\frac{G_E}{G_M}| = 0.87^{+0.25}_{-0.24}$ at s = 13.59 GeV^2.

INTRODUCTION

One of the fundamental features of the proton is its structure which is described by the electromagnetic (Sachs) form factors, $G_E(Q^2)$ and $G_M(Q^2)$. In the space-like region ($Q^2 < 0$) the form factors are related to the charge and the magnetic moment distributions of the proton. The pioneering study of the Sachs form factors through elastic e^-p scattering, dates from the early 1950s by the Hofstadter group [1]. Later studies by many different groups led to an empirical relationship between the two form factors: $G_E = \frac{G_M}{\mu_p}$ where $G_M = (1 + \frac{|Q^2|}{0.71 GeV^2})^{-2}$. Traditionally the measurement of these two form factors are made using the Rosenbluth seperation technique [2]. Recently measurements based on the recoil polarization method [3, 4] have shown a breakdown of this empirical relationship [5]. Currently further measurements are underway at Jefferson Lab to understand this discrepancy [5].

The Sachs form factors have also been studied in the time-like region ($Q^2 > 0$). S-matrix unitarity requires the form factors to be complex functions in this region. Furthermore the measurements are more difficult due to the fall in the cross section at high Q^2 and the Rosenbluth seperation technique can not be used in this region. Therefore it is usually assumed that $|G_E|$ and $|G_M|$ are equal, which is only true by definition at the production threshold.

A measurement of $|G_M|$ at high time-like Q^2 [6] was performed for the first time by the Fermilab experiment E760. This was the predecessor of this experiment (E835) and subsequent measurements in the continuum by E835 have confirmed these results [7, 8]. Fig. 1(left) shows all the time-like measurements of $|G_M|$ made to date.

The ratio of $|G_E|$ to $|G_M|$ in the time-like region can be obtained by measuring the angular distribution of the process $\bar{p}p \to e^+e^-$, which is related via crossing symmetry to e^-p elastic scattering. However, practically this is feasable only at resonances like the

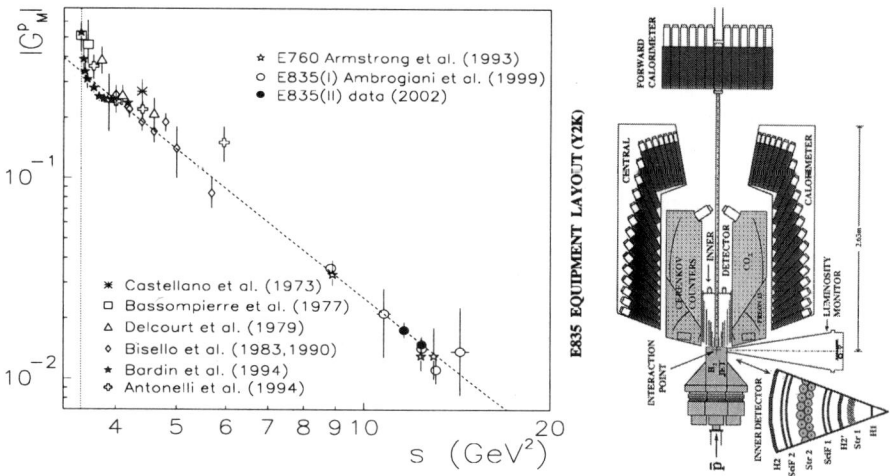

FIGURE 1. Left: Compilation of measurements of $|G_M|$ vs. Q^2 in the time-like region assuming $|G_E| = |G_M|$. Right: Cross sectional view of E835 detector for 2000 runs.

J/ψ and ψ', where there are enough statistics to measure the angular distribution. At the J/ψ or ψ' the form factors are dominated by the resonance. By making the assumption of one photon exchange, this angular distribution can be written as $\frac{dN}{d\cos\theta^*} \propto 1 + \lambda \cos^2\theta^*$ [9] with $\left|\frac{G_E}{G_M}\right| = \frac{\sqrt{s}}{2M_p}\sqrt{\frac{1-\lambda}{1+\lambda}}$, where θ^* is the angle between an electron and the $\overline{p}$ direction in the center-of-mass system. Equivalently, from this assumption the relative helicity amplitudes for $\overline{p}p$ annihilation can be obtained. The relation between the two possible relative helicities (C_0 and C_1) can be derived as $\left|\frac{C_0}{C_1}\right| = \sqrt{\frac{1-\lambda}{1+\lambda}}$ based on the helicity formalism [10] together with normalization condition ($2|C_1|^2 + |C_0|^2 = 1$). There have been several measurements of the angular distribution at the J/ψ but none at the ψ'.

EXPERIMENT

The experiment E835 was located in the Anti-proton Accumulator (AA) at Fermilab. A charmonium state is produced when the stochastically cooled $\overline{p}$ beam intersects a hydrogen gas-jet target. The small beam energy spread ($\frac{\Delta p}{p} \sim 10^{-4}$) is critical for the experiment especially for narrow resonances like the J/ψ and ψ'. The detector was a highly segmented non-magnetic spectrometer and is shown schematically in Fig. 1 (Right). It consisted of inner trackers, a Čerenkov detector, and electromagnetic calorimeters. The threshold Čerenkov counter was designed to distinguish electrons from pions. The central lead-glass calorimeter (CCAL) consisted of 1280 blocks and was used to meaure the energy and position of the electromagnetic final states (e^+, e^-, and γ). The resolutions of the CCAL are $\frac{\sigma(E)}{E} = \frac{6\%}{\sqrt{E(GeV)}} \pm 1.4\%$, 6 mrad in θ, and 11 mrad in ϕ. The total angular acceptance of the detector was $2^o < \theta_{lab} < 70^o$, with full coverage in the azimuth. Further details about the detector can be found in reference [11]. In the analysis of the data, collected energy deposits in 3x3 neighboring blocks were combined into a cluster. The timing of a cluster was determined by the timing of the highest or the second highest energy block in the cluster. The timing was classified as "in-time", "out-of-time", or "undetermined" according to whether the timing signal was "within", or "outside" a ± 10 ns window of the mean event time, or with "no information", respectively.

ANALYSIS AND RESULTS

Data sets from the two different data taking periods were used for this analysis. The 1996-1997 run collected 10.09 pb^{-1} at the ψ' resonance and the 2000 run collected 12.48 pb^{-1}.

The electron-positron candidate events were initially selected using the number of clusters, the timing of the events, the geometrical acceptance, and a four-constraint (4C) kinematic fit. For e^+e^- candidates, the two in-time clusters, that combined to form the highest invariant mass pair were considered. Two additional clusters were allowed to be present in the event to take into account either bremsstrahulung or secondary shower production in the CCAL. For the additional clusters, it was required that they were either in-time or undetermined with a minimum energy of 50 MeV and that the e^+e^- candidates fell within the geometrical acceptance of the Čerenkov detector. The kinematic fit probability required to be greater than 10^{-4}. Further cuts were made on the signal from the Čerenkov counter and on the structure of the clusters to ensure that these were not background events. The total number of events finally selected were 2391 from 1996-1997 data and 4453 from 2000 data in the range of $0 < \cos\theta^* < 0.58$. The background levels from mis-identified events were estimated using the GEANT MC. The channels which can potentially cause mis-identified events are $J/\psi\ \eta$, $J/\psi\ \pi^0\pi^0$, and $J/\psi\ \pi^+\pi^-$. The contamination levels from those channels were estimated as 0.05% and 0.14% from the 1996-1997 and 2000 MC, respectively. As these levels are negligible compared to the signal and its statistical significance, no background subtraction from the data was made.

Detection inefficiencies that can affect the angular distribution measurement are geometrical acceptance and cut efficiency. These efficiencies was studied using the GEANT MC and the overall efficiency was $\sim 90\%$. After the efficiency correction, a likelihood fit was applied to the angular distribution. The fit results were $\lambda = 0.59\ ^{+0.24}_{-0.23}$ and $\lambda = 0.71 \pm 0.18$ from 1996-1997 and 2000 data, respectively.

The systematic error was obtained by applying $\pm 10\%$ variations on the cuts made for the event selection. The values of the systematic error were 0.05 and 0.04 from 1996-1997 and 2000 data, respectively.

The Kolmogorov-Smirnov test between the two data sets resulted in 74.4% compatibility which allows the two to be combined. The likelihood fit values of the combined data were as $\lambda = 0.67\ ^{+0.15}_{-0.14}$ (stat.)± 0.04 (sys.) (see Fig. 2). Using the λ from the combined data, the Sachs form factor ratio of the proton and the relative helicity amplitude

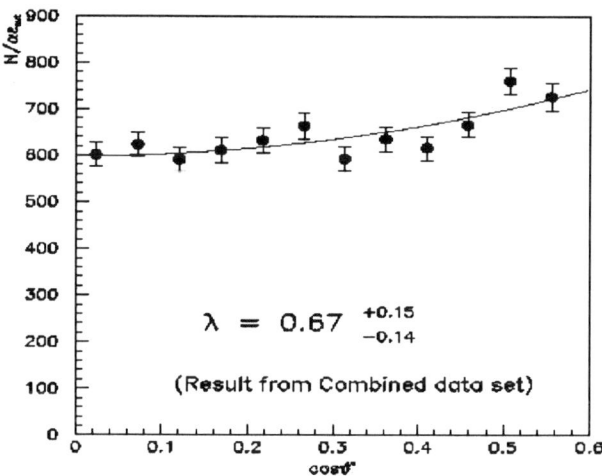

FIGURE 2. Angular distribution from the combined data set. (The line is the fit to the data.)

ratio were obtained as $\left|\frac{G_E}{G_M}\right| = 0.87^{+0.25}_{-0.24}$ and $\left|\frac{C_0}{C_1}\right| = 0.44^{+0.12}_{-0.11}$ at s = 13.59 GeV^2. The world average of λ at the J/ψ is 0.63 ± 0.08 [12], giving the ratios $\left|\frac{G_E}{G_M}\right| = 0.86 \pm 0.10$ and $\left|\frac{C_0}{C_1}\right| = 0.48 \pm 0.06$ at s = 9.61 GeV^2, which are consistent with those at the ψ'.

ACKNOWLEDGMENTS

I would like to thank E835 collaborators, scientists, engineers and technicians in the Fermilab beam's division and computing division. I also thank US Department of Energy and Italian Instito Nazionale di Fisica Nucleare for their grants. My personal thanks goes to Professor S. Rudaz for private communications regarding this research.

REFERENCES

1. Hofstadter, R., and McAllister, R.W., *Phys. Rev.*, **98**, pp. 217 - 218 (1955).
2. Rosenbluth, M.N., *Phys. Rev.*, **79**, pp. 615 - 619 (1950).
3. Arnold, R. *et al.*, *Phys. Rev.*, **23**, pp. 363 - 374 (1981).
4. Rekalo, M., and Tomasi-Gustafsson, E., nucl-th/0202025 (2002).
5. Arrington, J., nucl-ex/0305009 (2003).
6. Armstrong, T.A. *et.al.*, *Phys. Rev. Lett.*, **70**, pp. 1212 - 1215 (1993).
7. Ambrogiani, M. *et.al.*, *Phys. Rev.*, **D60**, pp. 032002-1 - 0302002-6 (1999).
8. Andreotti, M. *et.al.*, *Phys. lett.* ,**B559**, pp. 237 - 242(2003).
9. Zichichi, A. *et al.*, *Nuovo Cimento* **XXIV**, pp. 170 - 180 (1962).
10. Jacob, M., and Wick, G.C., *Ann. phys.*, **7**, pp. 404 - 428 (1959).
11. Andreotti, M. *et.al.*,*Nucl. Inst. Meth.*, **A**, to be published (2003).
12. Köpke, L., and Wermes, N., *Phys. Rep.*, **174**, pp. 67 - 227 (1989).

Baryon spectroscopy on the lattice: recent results

Colin Morningstar

Department of Physics, Carnegie Mellon University, Pittsburgh, PA, USA 15213-3890

Abstract. Progress in determining the baryon spectrum using computer simulations of quarks and gluons in lattice QCD are summarized and some future plans are outlined.

Baryon spectroscopy is plagued by numerous unresolved issues. The quark model predicts many more states[1, 2] than are currently known. Experiments in Hall B at Jefferson Laboratory are currently mapping out of the spectrum of N^* nucleon excitations, so the question of the so-called "missing resonances" should soon be resolved. A quark-diquark picture of baryons predicts a sparser spectrum[3]. Various bag and soliton models have also attempted to explain the baryon masses. The natures of the Roper resonance and the $\Lambda(1405)$ remain controversial. Experiment shows that the first excited positive-parity spin-1/2 baryon lies below the lowest-lying negative-parity spin-1/2 resonance, a fact which is difficult to reconcile in quark models. First principles studies of the baryon spectrum using lattice Monte Carlo methods are long overdue.

State of the art results for the low-lying hadron spectrum using the Iwasaki gauge action and a clover tadpole-improved fermion action are presented in Fig. 1. The quenched spectrum (upper left) deviates from experiment by under ten per cent. The inclusion of two flavors of light quark loops produces excellent agreement with experiment for the K and ϕ mesons, but the baryon masses show significant deviations from their experimental values. The authors suggest that finite volume errors will explain these discrepancies.

During the last few years, a handful of lattice studies have begun at last to focus attention on the excited baryon spectrum. Other than some preliminary unquenched results, all estimates to date have utilized the quenched approximation, and most have used unphysically heavy quarks. Systematic errors due to discretization and finite volume are not yet under control. Although the current status of such calculations is still embryonic, a renewed interest of the lattice community in such calculations promises substantial progress in the near future. For example, the Lattice Hadron Physics Collaboration (LHPC)[4] recently formed and one of its major objectives is the computation of the N^* spectrum. The formation of this collaboration was spearheaded by the late Nathan Isgur and is funded by the Department of Energy's Scientific Discovery through Advanced Computing (SciDAC) initiative. I shall outline the plans of this collaboration later in this talk. But first, results from four selected recent baryon studies are presented.

Latest results in the quenched approximation from the CSSM Lattice collaboration using an improved gauge field action and a fat-link irrelevant clover (FLIC) fermion action are shown in Fig. 2. Rather heavy quark masses were used, but the level orderings are in qualitative agreement with those observed in experiments.

The first-excited state in the positive-parity spin-1/2 sector is found to be significantly

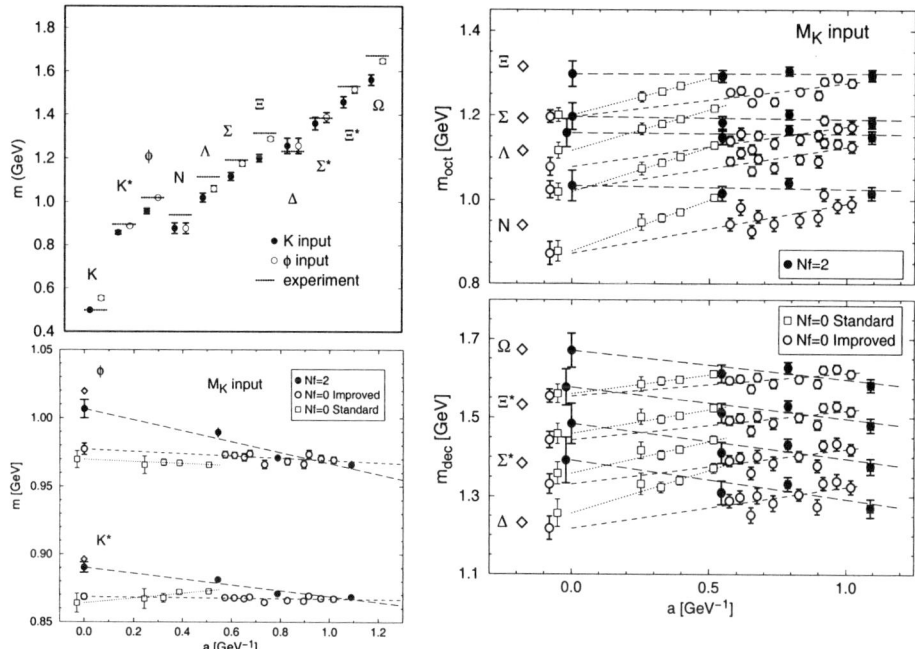

FIGURE 1. State of the art results from the CP-PACS collaboration for the low-lying hadron spectrum. Quenched results (upper left) in the continuum limit are from Ref. [5]. Solid symbols use the K to set the scale, and the hollow symbols use the ϕ to set the scale. The differences between the two sets measure the systematic errors from quenching. Agreement with experiment is remarkable, indicating that quenching errors in these observables are not large. In the lower right and the two leftmost plots, solid symbols indicate results from Ref. [6] including two flavors of light quark loops, whereas open symbols are quenched results shown against the lattice spacing a. For the K and ϕ mesons, one observes excellent agreement with experiment. However, the baryon results are problematic; the authors suggest that finite volume effects are to blame. Experimental results are indicated by the diamonds.

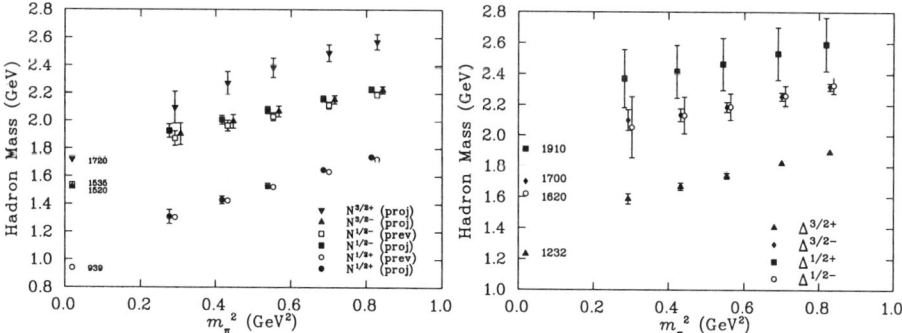

FIGURE 2. Mass estimates of the $J^P = \frac{1}{2}^\pm$ and $\frac{3}{2}^\pm$ N and Δ baryons from Ref. [7] against the square of the pion mass m_π in the quenched approximation. The results use an improved gauge field action and the fat-link irrelevant clover (FLIC) fermion action on a $16^3 \times 32$ lattice with spacing $a = 0.12$ fm, set by the string tension. Spin-projected results are compared with previous unprojected ones. Experimental values are shown near the vertical axis.

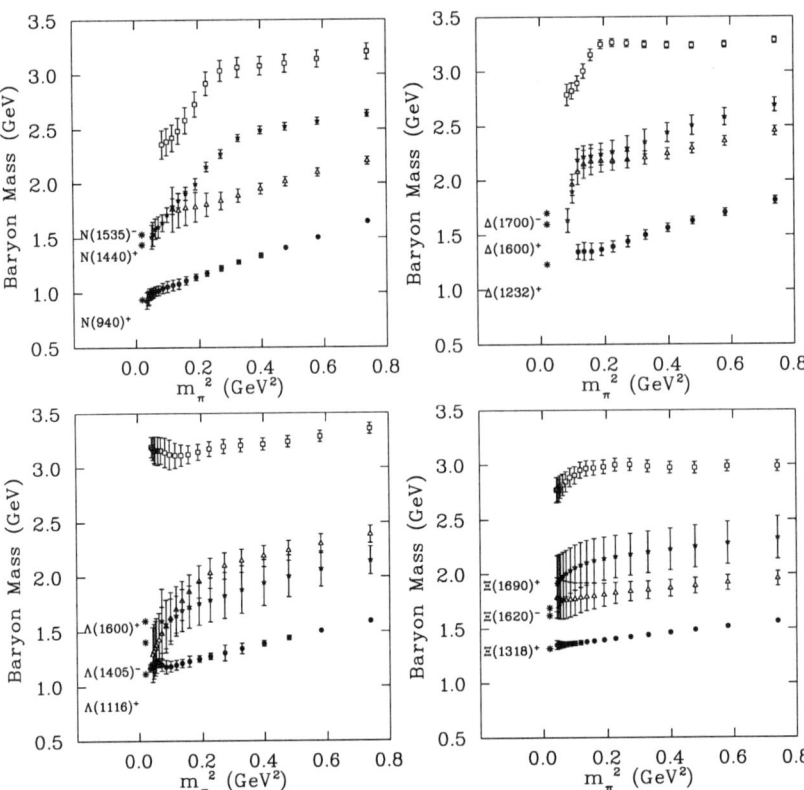

FIGURE 3. Two lowest-lying octet and decuplet baryon masses for both positive and negative parities in the quenched approximation from Ref. [12] at lattice spacing $a \approx 0.2$ fm against the square of the pion mass. Results were obtained on $16^3 \times 28$ lattices using an improved gauge action and overlap fermions for a large range of light quark masses. These figures emphasize the importance of simulating with sufficiently light quark masses. Experimental measurements are shown as bursts.

higher[8, 9, 10, 11] than the Roper mass when unphysically large quark masses are used. Recently, the use of overlap fermions has allowed quenched calculations[12] with realistically light quark masses, and this point appears to be crucial for identifying the Roper as a radial excitation of the nucleon. Fig. 3 shows dramatic changes in the quenched baryon masses as the pion mass drops below 300 MeV. An important note of caution concerning these findings is the use of empirical Bayesian constrained curve fitting to extract the excited state mass from a single correlation function. Although likely reliable, an analysis using several operators in a correlation matrix would be much preferred. Also, at such light quark masses, one must very carefully check finite volume errors (as evidenced in the next study described below). In a more recent paper[13], these authors have also addressed the issue of pollution of the first-excited state observed in Fig. 3 by unphysical $\eta'N$ quenched artifacts. Such ghost contributions were distinguished by obtaining results in two volumes $16^3 \times 28$ and $12^3 \times 28$, corresponding to lattice extents 3.2 and 2.4 fm, respectively. The conclusion, within the quenched approximation, that

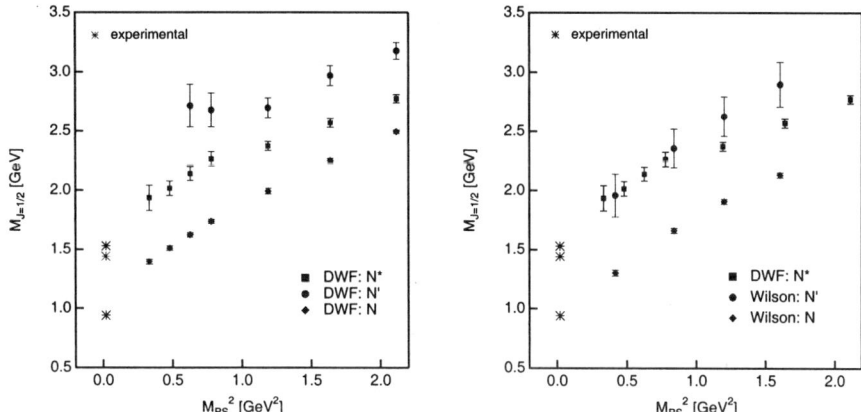

FIGURE 4. The low-lying nucleon masses in the quenched approximation from Ref. [8] (left) and Ref. [14] (right). Results on the left use domain wall fermions in a small volume with spatial extent $La \approx 1.5$ fm. On the right, results in three different volumes $La \approx 2.2 - 3.0$ fm were extrapolated using $1/L^3$ to the infinite volume limit. The Wilson gauge and Wilson fermion actions with $\beta = 6.0$ and spacing $a \sim 0.1$ fm were used with quark masses yielding pion masses in the range $m_\pi = 0.6 - 1.0$ GeV. Maximum entropy methods were employed. Experimental values are shown as bursts.

the Roper is a radial excitation of the nucleon with three valence quarks was confirmed.

A large sensitivity of the Roper resonance to finite volume errors in the quenched approximation has recently been reported in Ref. [14]. In Fig. 4, one sees that for $m_\pi^2 \approx 0.5$ GeV^2, the infinite volume results for the N' Roper are degenerate with the negative parity N^*, in disagreement with the results shown in Fig. 3. The lattice spacings and actions differ, so this discrepancy could simply be a discretization artifact. Also, maximum entropy methods are being employed, which further muddies the issue. Nevertheless, an important message seems clear: large volumes and small quark masses are especially important for reliable results in baryon spectroscopy.

Doubly charmed baryons, also of current experimental interest, have also been studied recently[15] in the quenched approximation using an improved gauge action on anisotropic lattices with the D234 action for the light quarks and a nonrelativistic (NRQCD) action for the heavy quarks. Two lattice spacings $a \sim 0.15, 0.22$ fm and four light quark masses were used. These authors found that mass splittings between $J = \frac{1}{2}$ and $\frac{3}{2}$ baryons from color hyperfine interactions were *not* suppressed, unlike the meson sector. Many charmed and bottom baryons were studied. No finite volume checks were done, and radiative corrections to the couplings in the NRQCD action were ignored. Results using a clover fermion action at $\beta = 6.2$ have also been presented[16] recently.

The Lattice Hadron Physics Collaboration is currently using large sets of extended operators with correlation matrix techniques to capture a significant portion of the baryon spectrum. Excited states will be extracted without resorting to maximum entropy methods. We believe that the construction of good operators is crucial: the operators have been designed with one eye towards maximizing overlaps with the low-lying states of interest, and the other eye towards minimizing the number of source needed in computing the required quark propagators. For example, the three-quark operators we plan to use are expressed in terms of smeared quark fields $\tilde{\psi}$, the covariant three-dimensional Laplacian $\tilde{\Delta}$, and the p-link covariant displacement $\tilde{D}_j^{(p)}$ by

$$(1): \phi_{ABC}^F \varepsilon_{abc} \Gamma_{\alpha\beta\gamma} (\tilde{\Delta}^{n_1}\tilde{\psi})_{Aa\alpha} (\tilde{\Delta}^{n_2}\tilde{\psi})_{Bb\beta} (\tilde{\Delta}^{n_3}\tilde{\psi})_{Cc\gamma},$$

$$(2): \phi_{ABC}^F \varepsilon_{abc} \Gamma_{\alpha\beta\gamma}^j (\tilde{\Delta}^{n_1}\tilde{\psi})_{Aa\alpha} (\tilde{\Delta}^{n_2}\tilde{\psi})_{Bb\beta} (\tilde{D}_j^{(p)}\tilde{\Delta}^{n_3}\tilde{\psi})_{Cc\gamma},$$

$$(3): \phi_{ABC}^F \varepsilon_{abc} \Gamma_{\alpha\beta\gamma}^{jk} (\tilde{\Delta}^{n_1}\tilde{\psi})_{Aa\alpha} (\tilde{D}_j^{(p_1)}\tilde{\Delta}^{n_2}\tilde{\psi})_{Bb\beta} (\tilde{D}_k^{(p_2)}\tilde{\Delta}^{n_3}\tilde{\psi})_{Cc\gamma},$$

where $n_1, n_2, n_3, p, p_1, p_2$ are positive integers, $j, k = \pm 1, \pm 2, \pm 3$ are spatial directions, α, β, γ are Dirac spin indices, A, B, C are quark flavors, and a, b, c indicate colors. Different powers of the spatial Laplacian $\tilde{\Delta}$ are utilized to build up radial structure, and the displacement operator $\tilde{D}_j$ is used to incorporate orbital structure. The group theoretical projections necessary to obtain operators transforming irreducibly under the symmetries of the lattice have been carried out using software written in Maple. Degeneracy patterns among the different irreducible representations of the cubic group must be exploited to identify angular momentum J eigenstates in the continuum limit. We hope to present our first results in the near future. Note that our operator construction approach can be easily adapted for mesons, pentaquark systems, and so on.

Regardless of how the operators are constructed, extracting the baryon spectrum in lattice simulations remains a challenge. It is especially important for baryons that the Monte Carlo calculations be done in large volumes with realistically light quark masses and without the quenched approximation. Furthermore, all baryon studies must ultimately confront the thorny issue of treating unstable resonances. The techniques for doing this are well known[17], but are untested in QCD. However, computing speeds continue to increase and large computer clusters dedicated to hadron physics are coming on-line. Given the renewed interest in baryon spectroscopy, substantial progress is inevitable. This work was supported by NSF award PHY-0099450.

REFERENCES

1. N. Isgur and G. Karl, Phys. Rev. D **18**, 4187 (1978); Phys. Rev. D **19**, 2653 (1979).
2. S. Capstick and N. Isgur, Phys. Rev. D **34**, 2809 (1986).
3. M. Oettel, G. Hellstern, R. Alkofer, and H. Reinhardt, Phys. Rev. C **58**, 2459 (1998).
4. The Lattice Hadron Physics Collaboration web site, http://www.jlab.org/~dgr/lhpc/.
5. S. Aoki et al. (CP-PACS collaboration), Phys. Rev. Lett. **84**, 238 (2000).
6. A. Ali Khan et al. (CP-PACS collaboration), Nucl. Phys. B (Proc. Suppl.) **94**, 229 (2001).
7. J.M. Zanotti et al., hep-lat/0304001.
8. S. Sasaki, T. Blum, and S. Ohta, Phys. Rev. D **65**, 074503 (2002).
9. C. Maynard and D. Richards, Nucl. Phys. B (Proc. Suppl.) **119**, 287 (2003).
10. D. Brömmel et al., hep-ph/0307073.
11. W. Melnitchouk et al., Nucl. Phys. B (Proc. Suppl.) **119**, 293 (2003).
12. F.X. Lee et al., Nucl. Phys. B (Proc. Suppl.) **119**, 296 (2003).
13. S.J. Dong et al., hep-ph/0306199.
14. S. Sasaki, Prog. Theor. Phys. Suppl. , to appear (nucl-th/0305014).
15. N. Mathur, R. Lewis, and R. Woloshyn, Phys. Rev. D **66**, 014502 (2002).
16. J Flynn, F. Mescia, and A. Tariq, JHEP **0307**, 066 (2003).
17. B. DeWitt, Phys. Rev. **103**, 1565 (1956).

An Analysis Of $\gamma p \to p\pi^+\pi^-$ Using The CLAS Detector

Matthew Bellis* and CLAS Collaboration[†]

Rensselaer Polytechnic Institute, Troy, NY 12180
[†]*Jefferson Lab, Newport News, VA*

Abstract. Two charged pion final states are studied for GeV photons incident on protons. The data come from Thomas Jefferson National Accelerator Facility using the CLAS detector. A tagged photon beam of 0.5-2.4 GeV/c was produced through bremsstrahlung radiation and was incident on a ℓH_2 target. This analysis looks at the reaction $\gamma p \to p\pi^+\pi^-$ using a partial wave analyis to identify intermediate baryon resonances. Total cross section is compared to previous experiments and preliminary differential cross sections for intermediate baryon resonance quantum numbers are shown.

INTRODUCTION

The consituent quark model does an excellent job of predicting the spectrum of the majority of baryons and mesons. Capstick, Isgur and others[1, 2, 3, 4] have augmented the quark model for baryons, including decays, with QCD-inspired corrections and get very good agreement with experimental data.

But it has been known since the 1960's that there are predicted baryon resonances which are not observed in the experimental data[5, 6]. Many of the models use a harmonic oscilliator basis, and it is found that these missing states all fall in the N=2 band. This prompted Lichtenberg[7] to propose the diquark model, where two of the three quarks become tightly bound, reducing the number of degrees of freedom. This constraint leads to a spectrum devoid of the missing resonances of the full model. There is nothing in QCD however, which would imply any sort of diquark coupling. Later calculations [1, 8, 2] suggest that these missing states may couple more strongly to $N\pi\pi$ final states than $N\pi$ final states. As the majority of relevant experiments involve $N\pi$ scattering, it may not be that surprising that we have not observed these missing resonances. JLab is in an excellent position to supplement the world's data with a large data set of $N\gamma$ scattering.

We perform a partial wave analysis on the reaction $\gamma p \to p\pi^+\pi^-$. By extracting the partial wave amplitudes it is hoped that any missing baryon states can be identified. Both intensity and relative phases should give us the handle needed to identify resonant states. In addition, this technique gives us the best description of the data and allows us to accurately calculate both the total and differential cross section. We show that this technique works and even in our preliminary analysis there appear to be some promising signals in the data.

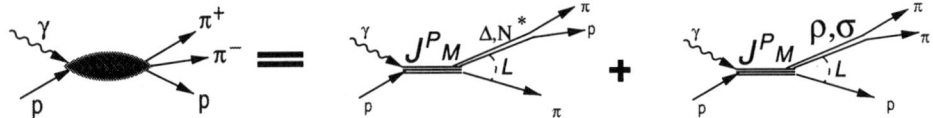

FIGURE 1. Representation of the decomposition of the scattering process into individual partial waves.

THE CLAS DETECTOR AND DATA SELECTION

The data was collected at the CLAS (CEBAF Large Acceptance Spectrometer) at Jefferson Lab in Newport News, VA. This sample of data is taken from the "g1c" running period which ran from Oct.-Nov. 1999. About 15% of the total run period is analyzed.

A 2.445 GeV electron beam was directed onto a thin foil radiator to produce a bremsstrahlung photon beam. CLAS is equipped with a hodoscope which allows tagging of photons with energies between 20% and 95% of the electron beam. This corresponds to a center-of-mass W from 1.3 to 2.3 GeV/c^2.

CLAS contains a large toroidal magnet to for determining the momentum of charged particles. A time-of-flight system is used for particle identification[9]. The particle identification is very clean for pions and protons. By making cuts on missing mass and missing z-component of momentum, we are able to identify exclusive events.

To simulate the detector a GEANT-based program was used. A detailed study of the acceptance was performed to check the agreement between the simulation and real-world data. This was used to identify our fiducial cuts. In the end, we have a very clean sample of 775,553 exclusive events.

PARTIAL WAVE ANALYSIS

The purpose of this analysis is to extract the partial waves amplitudes for this reaction. We want to expand the amplitude in some basis. For this study we use an s-channel decay basis, where we assume that all decays proceed through 2-body decays[10, 11]. Fig. 1 is a representation of how we label our basis states: J, P and M of the intermediate state, and the quantum numbers of the subsequent decays. We also sum over the initial and final state helicities and add them incoherently.

Slightly more formally we can represent the T-matrix in the following fashion.

$$\begin{aligned} T_{fi} &= \langle p\pi^+\pi^-; \tau_f | T | \gamma p; E \rangle \\ &= \sum_\alpha \langle p\pi^+\pi^-; \tau_f | \alpha \rangle \langle \alpha | T | \gamma p; E \rangle \\ &= \sum_\alpha \psi^\alpha(\tau_f) V^\alpha(E) \end{aligned}$$

We can calculate the decay amplitudes, $\psi^\alpha(\tau_f)$ and allow the fit to determine the production amplitudes, $V^\alpha(E)$. We use α to represent the quantum numbers of the intermediate "waves" and τ represents the kinematics of the reaction. This is an example of how our decay amplitudes are labeled.

TABLE 1. List of s-channel waves used in fit.

J^P	M	Isobars
$1/2^+$	$1/2$	$\Delta\pi$ ($\equiv \{\Delta^{++}\pi^-, \Delta^0\pi^+\}$)
$1/2^-$	$1/2$	$\Delta\pi, (p\rho)_{(s=1/2)}$
$3/2^+$	$1/2, 3/2$	$(\Delta\pi)_{(\ell=1)}, (p\rho)_{(s=1/2)}, (p\rho)_{(s=3/2;\ell=1,3)}, N^\star(1440)\pi$
$3/2^-$	$1/2, 3/2$	$(\Delta\pi)_{(\ell=0,2)}$
$5/2^+$	$1/2, 3/2$	$(\Delta\pi)_{(\ell=1)}, p\sigma$
$5/2^-$	$1/2, 3/2$	$(\Delta\pi)_{(\ell=2)}$

$$J^P, M = \frac{1}{2}^+, +\frac{1}{2} \rightarrow [\Delta^{++}\pi^-]_{\ell=1}, \lambda_{p_f} = +\frac{1}{2}$$

We remove the energy dependance of the production amplitudes by binning in W, the mass of the intermediate state. This method allows us to do an *energy independent* study of the scattering amplitudes.

As a starting point for fitting the production amplitudes we need some set of waves. The number of waves can quickly grow and become too cumbersome for the fitting routines. We started with waves that had been seen in a previous analysis of $\pi N \rightarrow N\pi\pi$[12]. These are shown in Table 1. This list represents 35 production amplitudes that we fit. We also include a non-interfering amplitude which represents t-channel ρ production.

This procedure allows us to perform a very good acceptance correction and so calculate a total cross section. Fig. 2 shows our calculated cross section plotted vs the results from the ABBHHM collaboration[13] as well as the CEA collaboration[14]. Their results are consistent with ours. In addition we plot the two strongest waves in the low and high mass region. In the low mass region, the reaction is dominated by $\frac{3}{2}^- \rightarrow \Delta^{++}\pi^- (\ell=0)$. This is to be expected, as some expect the "contact term" to show up in this wave[15]. The high mass region is dominated by our t-channel ρ wave.

We can look at individual wave intensities to see if there is evidence of resonance behavior. Fig. 3 shows the intensities of two waves and their relative phase. The $\frac{3}{2}^+$ wave shows an enhancement around 1600 MeV, while the $\frac{5}{2}^+$ wave shows an enhancement around 1650 MeV. The phase motion appears to qualitatively support the idea that these are resonant waves. One must remember that *each point is the result of an independant fit*. We are encouraged that the data seems to be consistent with some known states: $\Delta(1600)P_{33}$ and $N(1680)F_{15}$. In the future, we will explore different wave sets to find the best, stable description of the physics and perform a full mass dependant analysis.

The authors would like to thank the organizers of this conference for the opportunity to present our work.

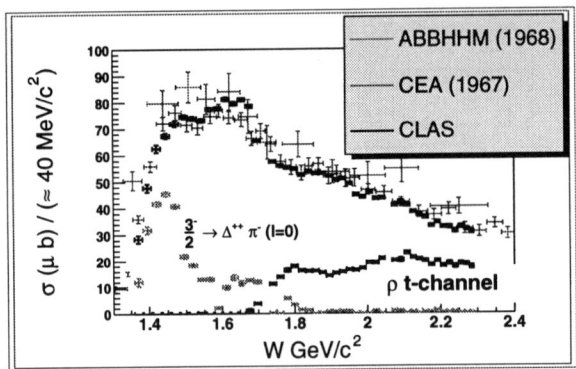

FIGURE 2. The total cross section for $\gamma p \to p\pi^+\pi^-$ is plotted for this analysis, the CEA[14] experiment, and the ABBHHM[13] experiment.

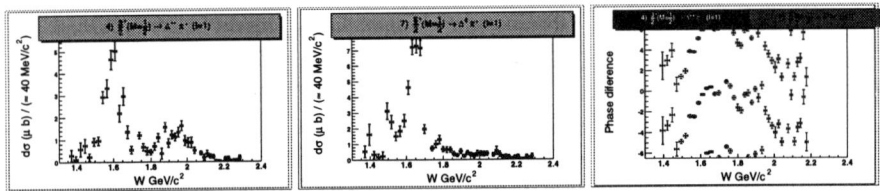

FIGURE 3. The intensities for two individual partial waves are plotted along with the relative phase for the same two partial waves. Beacause their phase difference is just an angle, it is plotted 3 times: ϕ, $\phi + 2\pi$ and $\pi - 2\pi$.

REFERENCES

1. Koniuk, R., and Isgur, N., *Phys. Rev.* **D21**, 1868 (1980).
2. Capstick, S., and Roberts, W., *Phys. Rev.*, **D47**, 1994–2010 (1993).
3. Capstick, S., and Roberts, W., *Phys. Rev.* **D49**, 4570–4586 (1994).
4. Capstick, S., and Isgur, N., *Phys. Rev.*, **D34**, 2809 (1986).
5. Faiman, D., and Hendry, A.W., *Phys. Rev.*, **173**, 1720–1729 (1968).
6. Faiman, D., and Hendry, A.W., *Phys. Rev.*, **180**, 1609–1610 (1969).
7. Lichtenberg, D.B., *Phys. Rev.*, **178**, 2197–2200 (1969).
8. Forsyth, C.P., and Cutkosky, R.E., *Z. Phys.*, **C18**, 219 (1983).
9. Mecking, B.A., et. al., *Nucl. Instrum. Meth.*, **A503**, 513–553 (2003).
10. Jacob, M. and Wick, G.C., *Ann. Phys.*, **7**, 404–428 (1959).
11. Chung, S.U., *Lectures given in Academic Training Program of CERN* (1969–1970).
12. Manley, D.M., and Saleski, E.M. *Phys. Rev.*, **D45**, 4002–4033 (1992).
13. *Phys. Rev.*, **175**, 1669–1696 (1968).
14. *Phys. Rev.*, **163**, 1510–1522 (1967).
15. Murphy, L.Y., and Laget, J.M., DAPNIA-SPHN-96-10 (1996).

Evidence for an exotic S=+1 baryon resonance at a mass of 1540 MeV

K.H. Hicks* and the LEPS Collaboration [†]

*Department of Physics, Ohio University, Athens, OH 45701
[†]email: hicks@ohio.edu

Abstract. A new baryon resonance with manifestly exotic strangeness quantum number $S = +1$ was first observed from the LEPS detector at the SPring-8 facility in Japan. The photoproduction reaction $\gamma n \to K^- \Theta^+$ was measured, where the neutron is bound in a ^{12}C nucleus and the Θ^+ is the new name of this resonance. The existence of the Θ^+ was inferred by observing a peak in the missing mass spectrum of the K^-, with the requirement of a coincident K^+ in the detector from the decay of the Θ^+. The mass of the resonance is 1.54 ± 0.01 GeV and a width of < 25 MeV.

INTRODUCTION

Our knowledge of bound quark configurations is determined largely by experiment because, until recently, non-perturbative calculations of quantum chromodynamics (QCD) were not possible [1]. Only baryons (qqq) and mesons ($q\bar{q}$) have been observed, even though QCD does not explicitly prohibit pentaquark $qqqq\bar{q}$ states. Extensive searches were carried out by experiments from over 30 years ago [2] for a baryon resonance with strangeness quantum number $S = +1$, but were fruitless. The lack of evidence for this exotic baryon led to a scientific bias against its existence that has deterred further experimental searches until recently.

The present results were motivated in part from a theoretical paper by Diakonov, Petrov and Polyakov [3], where a $S = +1$ baryon (originally called the Z^+) was predicted with a mass of 1.53 GeV and a narrow width of less than 15 MeV within the chiral soliton model. The prediction is based on group theory and symmetries within the chiral soliton model, and the tentative identification of the non-strange (N*) member of the $\overline{10}$ group as the known $P_{11}(1710)$ resonance. The new $S = +1$ resonance could be seen in reactions such as $\gamma n \to K^- Z^+$ where the mass of the hypothetical Z^+ is measured by the missing mass technique. Now, with experimental verification, the Z^+ has been renamed as the Θ^+, using a capital Greek letter like other baryon resonances.

---

[1] D.S. Ahn, J.K. Ahn, H. Akimune, Y. Asano, W.C. Chang, S. Daté, H. Ejiri, H. Fujimura, M. Fujiwara, K. Hicks, T. Hotta, K. Imai, T. Ishikawa, T. Iwata, H. Kawai, Z.Y. Kim, K. Kino, H. Kohri, N. Kumagai, S. Makino, T. Matsumura, N. Matsuoka, T. Mibe, K. Miwa, M. Miyabe, Y. Miyachi, M. Morita, N. Muramatsu, T. Nakano, M. Niiyama, M. Nomachi, Y. Ohashi, T. Ooba, H. Ohkuma, D.S. Oshuev, C. Rangacharyulu, A. Sakaguchi, T. Sasaki, P.M. Shagin, Y. Shiino, H. Shimizu, Y. Sugaya, M. Sumihama, H. Toyokawa, A. Wakai, C.W. Wang, S.C. Wang, K. Yonehara, T. Yorita, M. Yoshimura, M. Yosoi, and R.G.T. Zegers

At the time of this writing, four experiments have reported evidence for this pentaquark, all within the past year. The first evidence was presented by the LEPS collaboration at the 2002 PANIC conference [4]. A more complete description of this measurement is given in Ref. [5]. By early 2003, evidence from ITEP was made public [6]. Evidence from the CLAS collaboration [7, 8] followed closely. More recently, the SAPHIR collaboration has reported evidence [9] for the pentaquark. All experimental result give approximately the same mass (1.54 GeV) and a small width (< 9 MeV in [6]) for the resonance. Assuming baryon number and strangeness conservation, the peaks in the data are interpreted as an exotic $S = +1$ baryon, a pentaquark with $uudd\bar{s}$ constituents.

With the advent of confirmation of the SPring-8 result by several experiments, vigorous theoretical activity has ensued. Most of these models incorporate quark dynamics explicitly, rather than implicitly as in the chiral soliton model [3]. While the latter model was key as a motivation for experiments, it is not the only explanation for the results. The different theoretical models predict differences in the properties of the Θ^+, such as its isospin or its parity, and more data is needed to measure these properties. The result presented here is only the first step in an exciting path to discovery.

EXPERIMENTAL DETAILS

The experiment was carried out at the 8 GeV electron storage ring, SPring-8, in Japan. The photon beam was produced by back-scattering of laser light, with a maximum energy of 2.4 GeV for laser light of wavelength 351 nm used here. The scattered photon is boosted forward in the lab, and is naturally focused into a narrow beam onto an experimental target 70 m away. The struck electron is bent in the storage ring magnets onto a tagging detector consisting of several layers of scintillator strips and silicon microstrips. The tagging rate was $\sim 10^6$ per second, well within the capability of the tagger. The energy resolution of the photons is limited by the energy spread of the electron beam, not the tagger, which was about 15 MeV.

The LEPS detector is shown in Ref. [4]. The target was liquid hydrogen (LH_2) followed by a plastic scintillator known as the start counter (SC). Next is a Cerenkov counter with a index of refraction of 1.03, which is used to veto e^+e^- pairs from atomic events. This counter also removes high-energy pions from the trigger, which can sometimes be mis-identified as kaons. The particle tracks are measured before the dipole magnet using silicon vertex detectors and an upstream drift chamber, followed by two drift chambers after the magnet, giving the particle momentum. The particle velocity is found from the path-length and the time-of-flight (TOF) between the start counter and the plastic scintillators making up the TOF wall.

Data Analysis

The particle vertex position is shown in Fig. 1a) where a broad peak for the LH_2 target and a sharp peak at the position of the SC is seen. Small peaks from the windows of the LH_2 target can also be seen. Neutrons are present in the carbon nuclei in the SC, but

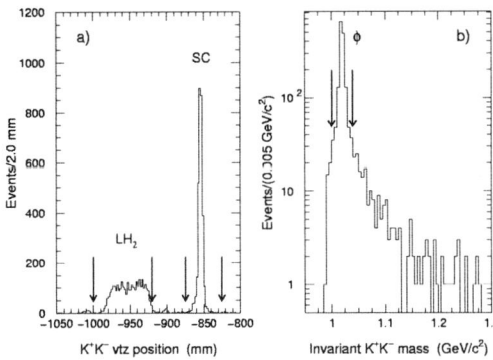

FIGURE 1. Event selection cuts for the data shown by a) the position along the beam direction of the K^+K^- vertex; b) the invariant mass of the K^+K^- to remove the ϕ-meson.

are absent in the LH$_2$ region. A clean test for whether events originate from a neutron can be done just by changing the software "cut" to select either the LH$_2$ or the SC, without changing any other characteristic of the beam or the detector. This provides a nice systematic check for the shape of background from events originating from protons. In particular, the reaction $\gamma n \to K^- \Theta^+$ cannot be produced in the LH$_2$.

The invariant mass of the detected K^+K^- pair is shown in Fig. 1b) where a strong peak is seen at the mass of the ϕ-meson. Production of the ϕ's is not expected to create a sharp peak in the missing mass spectrum of the K^- because the K^- comes from the decay in the ϕ center of mass frame. Cutting out the background from the ϕ is essential to increase the signal-to-background ratio in order to see the small Θ^+ signal.

In addition to event selection cuts at the vertex and removal of the ϕ, several other cuts are applied to the data. The missing mass of the K^+K^- pair forms a peak at the mass of the nucleon, which is required to be between 0.90 to 0.98 GeV. Events where a proton was found at the silicon vertex detector, within a given uncertainty of the calculated position for a proton struck by the photon, were removed since we desire events where a neutron was struck. Events outside the geometry of the vertex detector were removed, as were events with a momentum less that 0.35 GeV/c (due to the uncertainty in tracking such protons). The final sample had a total of 109 events from the SC vertex, with an integrated photon flux of 2.1×10^{12} over a six month period at SPring-8.

The final spectra of the K^+ and K^- missing mass are shown in Fig. 2. These spectra have been corrected for the Fermi momentum of the neutron as described in Ref. [5]. When no proton veto is applied, the K^+ missing mass shows a clear peak at the Λ^* mass of 1.52 GeV. Since the Λ^* is produced by $\gamma p \to K^+ \Lambda^*$, the peak is no longer seen when the proton veto is applied (solid histogram). In contrast, there is a prominent peak at a mass of 1.54 GeV in the K^- missing mass (with the proton veto) for a vertex cut on the SC (solid) but not for events from the LH$_2$ target (dotted) which has been normalized to the shape of the background above 1.6 GeV.

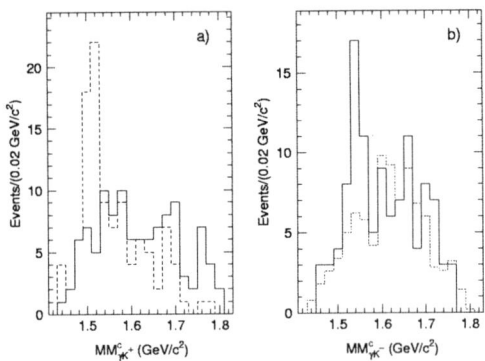

FIGURE 2. The missing mass of the K^+ and K^- events: a) with (solid) and without (dashed) the vertex proton veto; b) with all cuts for the SC (solid) and LH_2 (dotted) targets.

SUMMARY

An excess of 19 counts above an estimated background of 17 counts has been seen in the K^+ missing mass spectrum at a mass of 1.54 GeV. Assuming gaussian statistics, a 1-σ fluctuation in a background of 17 counts is about 4.1 counts. Hence, the peak at 1.54 GeV has a statistical significance of 4.6-σ. Extensive analysis and Monte Carlo [5] was done to show that this peak is not due to particle mis-identification or other instrumental effects. The SPring-8 result was the first of several experiments that have now confirmed the existence of the Θ^+ pentaquark at a mass of 1.54 GeV and a narrow width.

ACKNOWLEDGMENTS

We gratefully acknowledge the financial support of the Ministry of Education, Science, Sports and Culture of Japan, and also the support of the National Science Foundation.

REFERENCES

1. C. Morningstar, *Baryon Spectroscopy on the Lattice: Recent Results*, in these proceedings.
2. Particle Data Group, Phys. Lett. **B170**, 289 (1986).
3. D. Diakonov, V. Petrov and M. Polyakov, Z. Phys. A **359**, 305 (1997).
4. T. Nakano and the LEPS Collaboration, Nucl. Phys. **A721**, 112c (2003).
5. T. Nakano *et al.*, Phys. Rev. Lett. **91**, 012002, 2003.
6. V.V. Barmin *et al.*, e-print hep-ex/0304040.
7. V. Kubarovsky and S. Stepanyan, *Evidence for an Exotic Baryon State*, in these proceedings.
8. S. Stepanyan, K. Hicks *et al.*, *submitted to Phys. Rev. Lett.*, 2003; e-print hep-ex/0307018.
9. J. Barth *et al.*, e-print hep-ex/0307083.

Evidence for an Exotic Baryon State, $\Theta^+(1540)$, in Photoproduction Reactions from Protons and Deuterons with CLAS

Valery Kubarovsky*[†] and Stepan Stepanyan[†]

*Rensselaer Polytechnic Institute
[†]Jefferson Lab, 12000 Jefferson Ave., Newport News, VA 23606
The CLAS Collaboration

Abstract. CLAS photoproduction data on deuterium and hydrogen targets have been analyzed in a search for an exotic baryon state with strangeness $S = +1$, the Θ^+ (originally named the Z^+). This resonance was predicted recently in theoretical work based on the chiral soliton model as a lowest mass member of an anti-decuplet of 5-quark states. The reaction $\gamma d \to p K^- K^+ n$, which requires a final state interaction inside the deuteron, was used in the analysis of deuteron data. In the analysis of proton data, the reaction $\gamma p \to \pi^+ K^- K^+ n$ was studied. Evidence for the Θ^+ state is found in both analyzes in the invariant mass distribution of the nK^+. Our results are consistent with previously reported results by LEPS/Spring-8 collaboration (Japan), and by the ITEP (Moscow) group.

INTRODUCTION

Pentaquark resonances have been predicted decades ago and there have been experimental searches for many years. However, no significant signal was found in the early work. Recent theoretical work based on the chiral soliton model [1] made more quantitative predictions for the masses and widths of a spin $s = 1/2$ anti-decuplet of 5-quark states ($qqqq\bar{q}$). Using the $P_{11}(1710)$ resonance as the "anchor" for the masses of the anti-decuplet, the lowest lying member, Θ^+, is predicted to have a mass 1530 MeV/c^2 and a width of ~ 10 MeV/c^2. It is predicted to be an exotic baryon state with strangeness $S = +1$, and $I = 0$.

The LEPS collaboration at the SPring-8 facility in Japan recently reported [2] the observation of an $S = +1$ baryon at 1.54 GeV/c^2 with a width of < 25 MeV/c^2 from the reaction $\gamma n \to K^- K^+ n$ where the target neutron is bound in carbon, and the residual nucleus is assumed to be a spectator. This measurement reported a statistical significance of 4.6 ± 1.0 σ. Also, the DIANA collaboration at ITEP [3] recently announced results from an analysis of bubble-chamber data for the reaction $K^+ n \to K^0 p$, where the neutron is bound in a xenon nucleus, which shows a narrow peak at 1539 ± 2 MeV/c^2. The statistical significance of the ITEP result is 4.4 σ.

The data presented here were taken at the Thomas Jefferson National Accelerator Facility with the CLAS detector [5] and the photon tagging system [6] in Hall B. Data from two experiments have been used in these analyzes: i) photoproduction on deuterium using tagged photons produced by 2.478 and 3.115 GeV electrons; and ii) photoproduction on protons using tagged photons produced by 4.1 and 5.5 GeV electrons. The exclusive reaction $\gamma d \to pK^-K^+(n)$ was studied in the analysis of the deuteron data. A peak in the invariant mass distribution of nK^+ was found at 1.542 GeV/c^2 with a width of 21 MeV/c^2 and the statistical significance 5.3 ± 0.5 σ. The reaction $\gamma p \to \pi^+K^-K^+(n)$ was studied using the hydrogen data. A peak at 1.537 GeV/c^2 with a width of 31 MeV/c^2 in the invariant mass distribution of nK^+ was found in this reaction as well. The statistical significance of this peak is $4.8\pm0.4\sigma$.

PHOTOPRODUCTION ON DEUTERIUM

In the photoproduction on deuterium the Θ^+ can be produced directly on the neutron in the reaction $\gamma n \to \Theta^+ K^-$, similar to the reaction mechanism used by the LEPS collaboration. While the proton is a spectator in the direct production reaction and will not be detected in most cases, there are other ways to excite the Θ^+. Due to the final state interactions the proton can obtain high momentum and be detected. Fig.1 shows rescattering diagrams that may contribute to the production of the Θ^+ in the photoproduction on deuterium. For identification of such reactions the p, K^+, and K^- are detected, and the neutron is identified in the missing mass analysis. Although these reactions have a smaller cross section compared to the direct production because of an additional rescattering, they have the following advantages: i) the K^-s that are produced predominantly in the forward direction in the direct production mechanism will scatter at larger angles and will have higher probability of detection in CLAS; ii) the kinematics of such exclusive reactions puts additional constraints on the event selection that help to clean up the event sample significantly, and iii) due to the exclusive kinematics no Fermi momentum corrections are needed for the correct calculation of $M(nK^+)$.

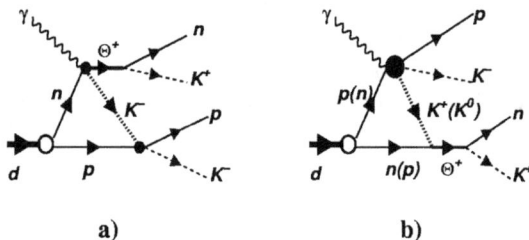

FIGURE 1. Feynman graphs for the Θ+ production in the exclusive reaction.

The analysis focused on events with detected proton, K^+ and K^- (and no other charged particles) in the final state. The missing mass (MM) distribution of selected events with the pK^+K^- final state shows a clear peak at the neutron mass. For further analysis events within $\pm 3\sigma$ of the neutron peak were kept. Estimated background due to the particle misidentification is $\sim 15\%$.

There are several known reactions, such as photoproduction of mesons (that decay into $K\bar{K}$) or excited hyperons (that decay into a pK^- or nK^-), that contribute to the same final state. The ϕ meson at $M(K^+K^-) = 1.02$ GeV/c^2, and the $\Lambda(1520)$ at $M(pK^-) = 1.518$ GeV/c^2 are cleanly seen in our event sample. Events from these resonances have been removed from the final sample.

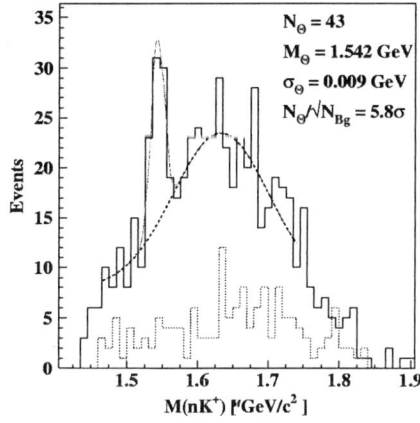 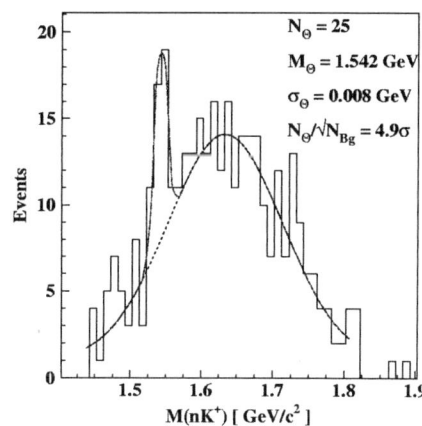

FIGURE 2. Invariant mass of the nK^+ system showing a sharp peak at the mass of 1.542 GeV/c². The panel on the left corresponds to all selected events, panel on the right corresponds to events with tight kaon timing cut. The dotted histogram shows the spectrum of events associated with $\Lambda(1520)$ production.

Two other event selection requirements are applied, based on kinematics. First, the missing momentum of the undetected neutron required to be greater than 80 MeV/c. For momenta below this value, the neutron is likely a spectator. Our studies show that increasing the value of this cutoff does not change the final results – in particular it does not eliminate the peak shown below – but does reduce the statistics in the $M(nK^+)$ spectrum. Second, events with K^+ momentum greater than 1.0 GeV/c were removed. This cutoff is based on Monte Carlo simulations of the 3-body kinematics, $\gamma d \rightarrow \Theta^+ p K^-$, uniform in phase space, which show that the K^+ momentum, from Θ^+ decay, rarely exceeds 1.0 GeV/c. This requirement reduces the background above 1.7 GeV/c².

The final $M(nK^+)$ spectrum is shown in Fig. 2, along with a fit to the peak at 1.542 GeV/c^2 and a Gaussian plus constant term fit to the background. The panel on the left shows the distribution for all selected events. The spectrum of events removed by the $\Lambda(1520)$ cut is shown in the dotted histogram, and does not appear to be associated with the peak at 1.542 GeV/c^2. The fit gives a Gaussian width for the peak consistent with the instrumental resolution of 21 MeV/c^2 (FWHM). The statistical significance of this peak is estimated to be 5.8σ, based on fluctuations of the background over a window of 36 MeV/c² centered on the peak. Different assumptions for the background shape lead to an additional uncertainty in the statistical significance, which is estimated at 5.3 ± 0.5 σ.

The panel on the right in Fig.2 corresponds to events with a tight timing cut on the kaon vertex time. The signal at 1.542 GeV/c^2 is clearly seen with a smaller number of events, and somewhat reduced significance. This result has since been submitted for publication in Phys. Rev. Lett. [4].

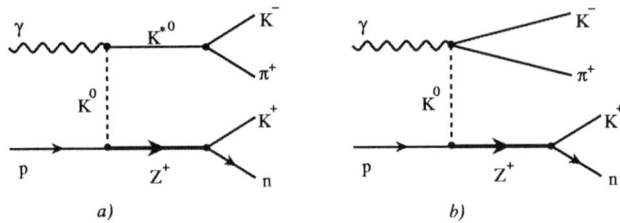

FIGURE 3. Feynman graphs for the Θ^+ photoproduction from a proton.

PHOTOPRODUCTION ON THE PROTON

In this analysis the reaction $\gamma p \to \pi^+ K^- K^+ n$ was studied. Possible diagrams contributing to the photoproduction of the Θ^+ from the proton are presented in Fig. 3. An estimate of the cross section for Θ^+ production in the reaction $\gamma p \to K^{*0} \Theta^+$ was made by M. Polyakov [7]. The $d\sigma/d\cos\theta_{cm}$ distribution (θ_{cm} is the angle between the $\pi^+ K^-$ momentum and photon beam in the center of mass system) peaks in the forward direction (small t region), as expected for the t-channel exchange mechanism (see Fig.3). About 80% of the events lie in the region with $\cos\theta_{cm} > 0.5$ (for a photon energy at 4 GeV), which appears to be a natural cut for the extraction of the Θ^+ signal from a proton target.

The reaction $\gamma p \to \pi^+ K^- K^+ n$ was studied at Jefferson Lab with photon energy from 3 to 5.25 GeV using an energy tagged photon beam. The final state particles, π^+, K^- and K^+, were detected in the CLAS detector[5], and the neutron was identified using the missing mass technique. There are 13.6K events, each having a positive pion and two kaons of opposite sign in the final state, which were selected for the analysis of the reaction $\gamma p \to \pi^+ K^- K^+ n$.

A cut on the $M_{K^+K^-}$ invariant mass with $M_{K^+K^-} > 1.040$ GeV/c^2 was applied to remove ϕ mesons. The M_{nK^+} invariant mass spectrum of the remaining 3699 events is presented in Fig.4.

To suppress background and extract the signal for the Θ^+ photoproduction, we used the general properties of the production mechanism of Θ^+ from a proton target (see Fig.3). In order to select the t-channel process illustrated in Fig.3, only events with $\cos\theta_{cm} > 0.5$ were taken for further analysis. Since we want to retain both the resonance K^{*0} production and non-resonance $K^- \pi^+$ continuum no cuts on $M_{K^-\pi^+}$ were applied.

Fig.4 (the right panel) presents the nK^+ invariant mass spectrum of the events with $\cos\theta_{cm} > 0.5$. This distribution was fitted by a Gaussian function and a smooth background. As suggested by data from this experiment the shape of the nK^+ invariant mass distribution in the reaction $\gamma p \to \pi^+ K^- K^+ n$ does not change significantly as a function of the θ_{cm} angle. For this reason the shape of the background was obtained from the full data set of the events (left panel in Fig.4). The resulting fit yields 27 counts in the peak with the mass $M = 1.54$ GeV/c^2 and width $FWHM = 32$ MeV/c^2. The mass scale uncertainty is estimated as ± 10 MeV/c^2. This uncertainty is mainly due to the energy

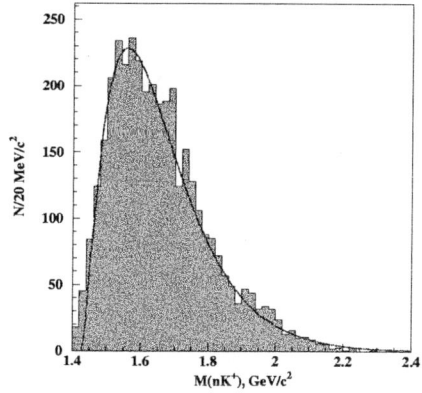
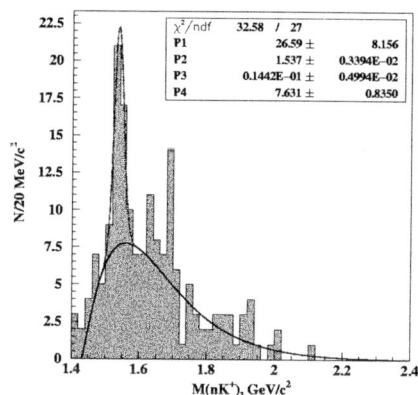

FIGURE 4. M_{nK^+} invariant mass spectrum in the reaction $\gamma p \to \pi^+ K^- K^+(n)$. Left panel is for all events. Right panel is for events with $cos\theta_{cm} > 0.5$. θ_{cm} is the angle between the $\pi^+ K^-$ system and the photon beam in the center of mass system.

calibration of the CLAS detector and the electron accelerator. The statistical significance of this peak is 4.8 ± 0.4 σ calculated over a window of 80 MeV/c^2. The mass resolution is close to the experimental resolution of CLAS.

SUMMARY

Analyzes of CLAS photoproduction data firmly establish the existence of a narrow $S = +1$ exotic baryon in the nK^+ system with a mass approximately at 1.54 GeV/c^2. The statistical significance of the peak in the invariant mass distribution of the nK^+ is 4.8σ for the analysis of the reaction $\gamma p \to \pi^+ K^- K^+ n$, and is 5.3σ for the analysis of the reaction $\gamma d \to p K^- K^+ n$. These results are consistent with the $S = +1$ state reported by LEPS and DIANA collaborations, and with the 5-quark ($uudd\bar{s}$) baryon predicted in the chiral soliton model.

REFERENCES

1. D. Diakonov et a., Z.Physics **A359** (1997), 305.
2. T. Nakano *et al.*, Phys. Rev. Lett. **91**, 012002 (2003).
3. V.V. Barmin *et al.*, e-print arXiv: hep-ex/0304040; V.V. Barmin *et al.* (to published in Yad. Fiz., vol. 66, issue 9 (2003)).
4. S. Stepanyan *et al.*, e-print arXiv: hep-ex/0307018, submitted to Phys. Rev. Lett.
5. B.A. Mecking *et al.*, Nucl. Instr. Meth. A **503**, 513 (2003).
6. D. Sober *et al.*, Nucl. Instr. Meth. A **440**, 263 (2000).
7. M.Polyakov. Photoproduction of Θ^1 on the proton target. Private communication (2003).

Excited $L=1$ baryons in large N_c QCD

Dan Pirjol* and Carlos Schat[†]

*Dept. of Physics and Astronomy, The Johns Hopkins University, 3400 N. Charles Street, Baltimore, MD 21218
[†]Department of Physics, Duke University, Durham, NC 27708

Abstract. The physics of the orbitally excited baryons simplifies drastically in the large N_c limit. The states are arranged into irreducible representations of the contracted $SU(4)_c$ symmetry, with mixing angles determined exactly. The ratios of the strong couplings $N^* \to [N\pi]_{S,D}$ are predicted in this limit, with results in agreement with those following from the quark model (with the large N_c mixing angles). We present a phenomenological analysis of the observed nonstrange baryons from the perspective of the $1/N_c$ expansion, including constraints from their masses and strong decays.

It has been known for some time that the large N_c limit of QCD [1, 2] can give a useful qualitative description of low energy hadronic physics. In the baryon sector this limit turns out to be considerably more predictive [3, 4, 5], and the $1/N_c$ expansion can be formulated in a systematic way allowing the treatment of power corrections and SU(3) breaking effects [6, 7, 8] (for a recent review see [9]).

The crucial point is the emergence of a new symmetry of QCD in the large N_c limit of the baryon sector - the contracted $SU(2n_f)_c$ symmetry (with n_f the number of light flavors) [6]. The physical states arrange themselves in irreducible representations of this symmetry group, which for $n_f = 2$ are labeled by $K = 0, \frac{1}{2}, 1, \ldots$ and contain all states satisfying $|I - J| \leq K$. The leading order predictions for masses and strong couplings recover the quark model with $SU(2n_f)$ spin-flavor symmetry. Using this approach, many applications have been discussed for the ground state baryons [9, 10, 11].

The large N_c expansion has been applied also to excited baryons, using different implementations of the idea [12, 13, 14, 15, 16, 17, 18, 19, 20, 21, 22]. In Refs. [14, 15] the contracted $SU(4)_c$ symmetry was found to extend also to these states, using consistency conditions for $N^{(*)}\pi \to N^{(*)}\pi$ scattering. However, because of the more complex mass spectrum of the excited states, the implications of this symmetry are more rich than in the ground state sector. In particular, the nonstrange negative parity $L = 1$ baryons fall into three irreducible representations of the $SU(4)_c$ symmetry

$$
\begin{aligned}
K = 0: \quad & N_{1/2}, \Delta_{3/2}, \ldots \\
K = 1: \quad & N_{1/2}, N_{3/2}, \Delta_{1/2}, \Delta_{3/2}, \Delta_{5/2}, \ldots \\
K = 2: \quad & N_{3/2}, N_{5/2}, \Delta_{1/2}, \Delta_{3/2}, \Delta_{5/2}, \Delta_{7/2}, \ldots
\end{aligned} \quad (1)
$$

This can be contrasted with the case of the corresponding ground state baryons, which include only one representation $K = 0 : N_{1/2}, \Delta_{3/2}, \ldots$. Thus, the large N_c limit implies a mass pattern for the excited baryons Eq. (1) which is very different from the quark model prediction of complete degeneracy into the **70** of $SU(6)$ [23, 24]. Still, the large N_c predictions for $N^* \to N\pi$ amplitude ratios are found to be again in agreement with those of the quark model with SU(4) spin-flavor symmetry [14]. [We note that very similar predictions are obtained for hybrid baryons in the large N_c limit [25].]

In a recent paper [26], the status of the $1/N_c$ expansion for the nonstrange $L = 1$ baryons was reexamined, working consistently at leading and subleading order in $1/N_c$. This was done using the operator approach proposed in [7, 6], and first applied to the excited states in [13, 16, 17]. The mass matrix of the $L = 1$ baryons can be written as a sum of operators acting on the quark basis as

$$\hat{M} = \sum_{k=0}^{N_c} \frac{1}{N_c^{k-1}} C_k \mathcal{O}_k \quad (2)$$

with $\mathcal{O}_k$ a k-body operator. Both the coefficients C_k and the matrix elements of the operators on baryon states $\langle \mathcal{O}_k \rangle$ have power expansions in $1/N_c$ with coefficients determined by nonperturbative dynamics

$$C_k = \sum_{n=0}^{\infty} \frac{1}{N_c^n} C_k^{(n)}, \qquad \langle \mathcal{O}_k \rangle = \sum_{n=0}^{\infty} \frac{1}{N_c^n} \langle \mathcal{O}_k \rangle^{(n)}. \quad (3)$$

The natural size for the coefficients $C_k^{(n)}$ is $\Lambda \sim 500$ MeV. A complete basis for the operators $\mathcal{O}_k^{(1,2)}$ has been constructed in [17], to which we refer for further details. At leading order in N_c only three operators contribute to the mass matrix, given by

$$O_1 = N_c \mathbf{1}, \qquad O_2 = l^i s^i, \qquad O_3 = \frac{3}{N_c} l^{(2)ij} g^{ia} G_c^{ja}. \quad (4)$$

At subleading order $O(N_c^{-1})$ five additional operators start contributing

$$O_4 = ls + \frac{4}{N_c+1} ltG_c, \quad O_5 = \frac{1}{N_c} lS_c, \quad O_6 = \frac{1}{N_c} S_c S_c, \quad O_7 = \frac{1}{N_c} sS_c, \quad O_8 = \frac{1}{N_c} l^{(2)} sS_c. \quad (5)$$

Their matrix elements on the excited baryon states can be found in the Appendix of [17]. These operators have a direct physical interpretation in the quark model in terms of one- and two-body quark-quark couplings.

Keeping only the operators $O_{1,2,3}$ contributing at $O(N_c^0)$, one finds by direct diagonalization of the mass matrix the mass eigenstates in the large N_c limit as linear combinations of the quark model $N_{1/2}, N'_{1/2}$ states

$$|K=0, J=\tfrac{1}{2}\rangle = \tfrac{1}{\sqrt{3}} N_{1/2} + \sqrt{\tfrac{2}{3}} N'_{1/2} \qquad \begin{cases} M_0^{(0)} = N_c C_1^{(0)} - C_2^{(0)} - \tfrac{5}{8} C_3^{(0)} \\ M_1^{(0)} = N_c C_1^{(0)} - \tfrac{1}{2} C_2^{(0)} + \tfrac{5}{16} C_3^{(0)} \end{cases} \quad (6)$$
$$|K=1, J=\tfrac{1}{2}\rangle = -\sqrt{\tfrac{2}{3}} N_{1/2} + \tfrac{1}{\sqrt{3}} N'_{1/2}$$

A similar diagonalization of the mass matrix for the $J = \tfrac{3}{2} N^*$ states gives the eigenstates

$$|K=1, J=\tfrac{3}{2}\rangle = \tfrac{1}{\sqrt{6}} N_{3/2} + \sqrt{\tfrac{5}{6}} N'_{3/2} \qquad \begin{cases} M_1^{(0)} = N_c C_1^{(0)} - \tfrac{1}{2} C_2^{(0)} + \tfrac{5}{16} C_3^{(0)} \\ M_2^{(0)} = N_c C_1^{(0)} + \tfrac{1}{2} C_2^{(0)} - \tfrac{1}{16} C_3^{(0)} \end{cases} \quad (7)$$
$$|K=2, J=\tfrac{3}{2}\rangle = -\sqrt{\tfrac{5}{6}} N_{3/2} + \tfrac{1}{\sqrt{6}} N'_{3/2}$$

TABLE 1. The four possible assignments of the observed nonstrange excited baryons into large N_c towers with $K = 0, 1, 2$.

	K = 0	K = 1	K = 2	ordering
#1	$N_{1/2}(1650)$	$\{N_{1/2}(1535), N_{3/2}(1520)\}$	$\{N_{3/2}(1700), N_{5/2}(1675)\}$	$\{M_0, M_2\} > M_1$
#2	$N_{1/2}(1535)$	$\{N_{1/2}(1650), N_{3/2}(1520)\}$	$\{N_{3/2}(1700), N_{5/2}(1675)\}$	$M_2 > M_1 > M_0$
#3	$N_{1/2}(1535)$	$\{N_{1/2}(1650), N_{3/2}(1700)\}$	$\{N_{3/2}(1520), N_{5/2}(1675)\}$	$M_1 > \{M_0, M_2\}$
#4	$N_{1/2}(1650)$	$\{N_{1/2}(1535), N_{3/2}(1700)\}$	$\{N_{3/2}(1520), N_{5/2}(1675)\}$	$M_0 > M_1 > M_2$

The $N_{5/2}$ state does not mix and has the mass $M_2^{(0)}$. These results make the tower structure in Eq. (1) explicit.

There is a discrete ambiguity in the assignment of the five observed N^* excited nucleons into the large N_c irreducible reps of $SU(4)_c$. The four possible ways of grouping them into multiplets are shown in Table 1. This implies a four-fold ambiguity in the coefficients of the mass operator $C_i^{(0)}$. In the following we extract these coefficients and attempt to resolve the discrete ambiguity by using experimental information on masses and strong decays of these states.

We start by determining the values of the coefficients $C_{1,2,3}^{(0)}$ in the large N_c limit, using the mass eigenvalues given in Eqs. (6) and (7). For each assignment, we fitted the coefficients $C_{1,2,3}^{(0)}$ to the observed N^* masses [26]. The results for $C_{2,3}^{(0)}$ are shown graphically in Figure 1 for each of the four possible assignments. The mixing angles θ_{N1}, θ_{N3} are fixed by Eqs. (6) and (7).

These results must satisfy an additional constraint, following from the no-crossing property of the eigenstates with the same quantum numbers. Consider the masses of the two $J = 1/2$ states as functions of $1/N_c$. They can not cross when N_c is taken from 3 to infinity. This means that the correspondence of the physical $N_{1/2}$ states with the large N_c towers is fixed by the relative ordering of the $K = 0, 1$ towers. This leads to a connection between the ordering of the tower masses and each of the 4 assignments, shown in the last column of Table 1. This constraint is also shown graphically in Fig. 1; it rules out the assignment No.4 and further restricts the solution for the assignment No.2.

We consider next also information from the strong decays $N^* \to [N\pi]_{S,D}$. The large N_c predictions for these decays were given in [14], where the consistency conditions for $N^{(8)}\pi \to N^{(*)}\pi$ scattering were solved exactly. Referring to Ref. [14, 15] for the full solution, we list in Eqs. (9) the results for the S- and D-wave reduced amplitudes A_{red} [defined up to spin and isospin CG coefficients as $A(N^* \to N\pi) = A_{\text{red}} \cdot CG_I \cdot CG_J$]. The 0's in these relations denote $1/N_c$ suppressed amplitudes. Including spin and isospin factors, these relations predict the large N_c partial width ratios shown in Eq. (10). In addition to constraining the masses of the tower states, the contracted $SU(4)_c$ symmetry relates also their strong decay widths which are predicted to be equal. This equality holds also for individual channels, which implies sum rules such as (for the $K = 2$ states)

$$\Gamma(N_{3/2} \to [N\pi]_D) + \Gamma(N_{3/2} \to [\Delta\pi]_D) = \Gamma(N_{5/2} \to [N\pi]_D) + \Gamma(N_{5/2} \to [\Delta\pi]_D). \quad (8)$$

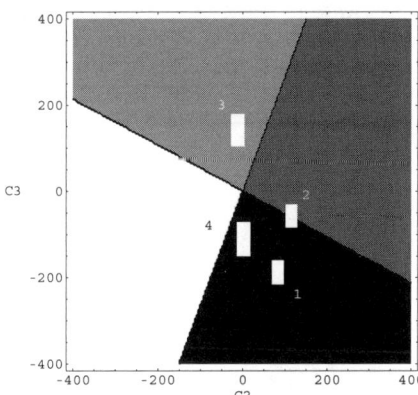

FIGURE 1. Fit results for the leading order coefficients $C_{2,3}^{(0)}$, corresponding to each assignment. The four wedgelike regions show the allowed values for each assignment following from the noncrossing argument.

$$
\begin{array}{lll}
K=0 & K=1 & K=2 \\
(N_{\frac{1}{2}} \to [N\pi]_S) = 0 & (N_{\frac{1}{2}} \to [N\pi]_S) = \sqrt{2}c_S & (N_{\frac{3}{2}} \to [\Delta\pi]_S) = 0 \\
 & (N_{\frac{3}{2}} \to [\Delta\pi]_S) = c_S & \\
\hline
(N_{\frac{1}{2}} \to [\Delta\pi]_D) = 0 & (N_{\frac{1}{2}} \to [\Delta\pi]_D) = c_{D1} & (N_{\frac{3}{2}} \to [N\pi]_D) = c_{D2} \\
 & (N_{\frac{3}{2}} \to [N\pi]_D) = -2c_{D1} & (N_{\frac{3}{2}} \to [\Delta\pi]_D) = -\frac{1}{2}c_{D2} \\
 & (N_{\frac{3}{2}} \to [\Delta\pi]_D) = -c_{D1} & (N_{\frac{5}{2}} \to [N\pi]_D) = \sqrt{\frac{2}{3}}c_{D2} \\
 & & (N_{\frac{5}{2}} \to [\Delta\pi]_D) = \frac{1}{2}\sqrt{\frac{7}{3}}c_{D2}
\end{array}
\quad (9)
$$

Note that these predictions depend crucially on the K assignment of the excited baryons. In particular, the strong couplings of the $K=0$ states are suppressed by $1/N_c$. Also, the $J=3/2$ $K=2$ state is predicted to decay in a pure D-wave. Therefore one expects these predictions to be useful for distinguishing among the possible assignments.

$$K=1: \Gamma(N_{\frac{1}{2}} \to [N\pi]_S) : \Gamma(N_{\frac{3}{2}} \to [\Delta\pi]_S) = 1:1 \tag{10}$$

$$K=1: \Gamma(N_{\frac{1}{2}} \to [\Delta\pi]_D) : \Gamma(N_{\frac{3}{2}} \to [N\pi]_D) : \Gamma(N_{\frac{3}{2}} \to [\Delta\pi]_D) = 2:1:1$$

$$K=2: \Gamma(N_{\frac{3}{2}} \to [N\pi]_D) : \Gamma(N_{\frac{3}{2}} \to [\Delta\pi]_D) : \Gamma(N_{\frac{5}{2}} \to [N\pi]_D) : \Gamma(N_{\frac{5}{2}} \to [\Delta\pi]_D)$$

$$= \frac{1}{2} : \frac{1}{2} : \frac{2}{9} : \frac{7}{9}.$$

TABLE 2. Large N_c predictions for the ratios of strong decay widths $R_{1,2}$, and their experimental values.

	R_1	R_2	$R_1 R_2$
#1	2.05	$O(1/N_c^2)$	$O(1/N_c^2)$
#2	$O(1/N_c^2)$	$O(N_c^2)$	2.4
#3	$O(1)$	$O(N_c^2)$	$O(N_c^2)$
#4	$O(N_c^2)$	$O(1/N_c^2)$	$O(1)$
exp	3.6-13.7	1.0-2.6	5.7-22.5

For this purpose, we consider the ratios of S−wave partial widths $R_1 = \frac{\Gamma(N_{1/2}(1535) \to N\pi)}{\Gamma(N_{3/2}(1520) \to [\Delta\pi]_S)}$ and $R_2 = \frac{\Gamma(N_{1/2}(1650) \to N\pi)}{\Gamma(N_{1/2}(1535) \to N\pi)}$. We present in Table 2 the large N_c predicted values for $R_{1,2}$ (including phase space factors), together with their experimental values.

Despite the large experimental errors, the combined constraints from masses and strong decays ($R_{1,2}$) appear to favor the assignment No.1 [14, 26]. In Ref. [26] the mass analysis presented here was extended to $O(1/N_c)$. The most important new point of this analysis is the appearance of a continuous set of solutions for the mass operator. Including also data from excited Δ states, it was found that the assignment No.3 is favored, in agreement with the analysis in [17], although No. 1 is still marginally allowed. More conclusive results will be possible once better data on the masses and decay widths of these states will become available.

Finally, we note that similar conclusions on the mass spectrum of these states (1) were also reached in [27, 28, 29] from a study of $N\pi$ scattering amplitudes in the Skyrme model.

ACKNOWLEDGMENTS

This work was supported by the DOE and NSF under Grants No. DOE-FG03-97ER40546 (D.P.), NSF PHY-9733343, DOE DE-AC05-84ER40150 and DOE-FG02-96ER40945 (C.S.).

REFERENCES

1. 't Hooft, G., *Nucl. Phys.*, **B 72**, 461 (1974).
2. Witten, E., *Nucl. Phys.*, **B 160**, 57 (1979).
3. Dashen, R. F., and Manohar, A. V., *Phys. Lett.*, **B 315**, 425 (1993).
4. Dashen, R. F., and Manohar, A. V., *Phys. Lett.*, **B 315**, 438 (1993).
5. Jenkins, E., *Phys. Lett.*, **B 315**, 441 (1993).
6. R. F. Dashen, E. J., and Manohar, A. V., *Phys. Rev.*, **D 49**, 4713 (1994).
7. Luty, M., and March-Russell, J., *Nucl. Phys.*, **B 426**, 71 (1994).
8. R. F. Dashen, E. J., and Manohar, A. V., *Phys. Rev.*, **D 51**, 3697 (1995).

9. Jenkins, E., "QCD Baryons in the $1/N_c$ Expansion," in *hep-ph/0111338*, 2001.
10. E. Jenkins, X. J., and Manohar, A., *Phys. Rev. Lett.*, **89**, 242001 (2002).
11. Jenkins, E., and Lebed, R., *Phys. Rev.*, **D 62**, 077901 (2000).
12. C. D. Carone, L. K., H. Georgi, and Morin, D., *Phys. Rev.*, **D 50**, 5793 (1994).
13. Goity, J. L., *Phys. Lett.*, **B 414**, 140 (1997).
14. Pirjol, D., and Yan, T. M., *Phys. Rev.*, **D 57**, 1449 (1998).
15. Pirjol, D., and Yan, T. M., *Phys. Rev.*, **D 57**, 5434 (1998).
16. C. E. Carlson, J. L. G., C. D. Carone, and Lebed, R. F., *Phys. Lett.*, **B 438**, 327 (1998).
17. C. E. Carlson, J. L. G., C. D. Carone, and Lebed, R. F., *Phys. Rev.*, **D 59**, 114008 (1999).
18. Carlson, C. E., and Carone, C. D., *Phys. Rev.*, **D 58**, 053005 (1998).
19. Carlson, C. E., and Carone, C. D., *Phys. Lett.*, **B 441**, 363 (1998).
20. C. Schat, J. L. G., and Scoccola, N. N., *Phys. Rev. Lett.*, **88**, 102002 (2002).
21. J. L. Goity, C. S., and Scoccola, N. N., *Phys. Rev.*, **D 66**, 114014 (2002).
22. J. L. Goity, C. S., and Scoccola, N. N., *Phys. Lett.*, **B 564**, 83 (2003).
23. Faiman, D., and Plane, D., *Nucl. Phys.*, **B 50**, 379 (1972).
24. A. J. Hey, P. J. L., and Cashmore, R. J., *Nucl. Phys.*, **B 95**, 516 (1975).
25. C. K. Chow, D. P., and Yan, T. M., *Phys. Rev.*, **D 59**, 056002 (1999).
26. Pirjol, D., and Schat, C., *Phys. Rev.*, **D 67**, 096009 (2003).
27. Cohen, T. D., and Lebed, R. F., *Phys. Rev. Lett.*, **91**, 012001 (2003).
28. Cohen, T. D., and Lebed, R. F., *Phys. Rev.*, **D 67**, 096009 (2003).
29. Cohen, T. D., and Lebed, R. F., *hep-ph/0306102* (2003).

An Experimental Overview of Gluonic Mesons

Curtis A. Meyer

Carnegie Mellon University, Pittsburgh, PA 15213

Abstract. In this paper, I review the experimental situation for hybrid mesons. Theoretical expectations are discussed, and a survey of what is known about hybrid mesons is undertaken. Good experimental evidence exists for states with exotic quantum numbers which is mixed with the nearby mesons, but a full understanding of these still requires additional information.

INTRODUCTION

Gluonic mesons are in the broadest sense a $q\bar{q}$ system in which the gluonic field contributes directly to the quantum numbers of the meson. In terms of the simple quark model, all quantum numbers of mesons are determined by the $q\bar{q}$ alone. However, Quantum Chromo Dynamics, (QCD) indicates that this picture is not complete. Lattice QCD calculations predict that both purely gluonic states, (glueballs), and states with the gluonic field carrying angular momentum, (hybrids) should exist. Beyond the lattice, most models which explain observed phenomena also predict such gluonic mesons to be present. A number of the hybrid states are predicted to have J^{PC} quantum numbers which are not accessible to simple $q\bar{q}$ systems, the so-called *exotics*.

This article will review the experimental situation for gluonic excitations. Of particular interest are states with exotic quantum numbers where two candidates exist. Hybrids with normal quantum numbers are more difficult to discern, as they are likely to mix with nearby normal mesons. It is only through detailed studies of decay and production that they can be identified. Finally, evidence exists for a $J^{PC} = 0^{++}$ glueball that is strongly mixed with the nearby scalar mesons.

Within the picture of the quark model, mesons are $q\bar{q}$ pairs which have been combined with spin, **S**, orbital angular momentum, **L**, and a possible radial excitation. S can be either 0 or 1, while **L** can be any non-negative integer. The quantum numbers of the allowed states which are conserved by the strong interaction can be built up from these as given as follows:

$$\text{Total Spin}: \mathbf{J} = \mathbf{L} \oplus \mathbf{S} = |\mathbf{L}-\mathbf{S}| \cdots |\mathbf{L}+\mathbf{S}| \quad \text{Parity}: \mathbf{P} = (-1)^{L+1}$$
$$\text{C}-\text{Parity}: \mathbf{C} = (-1)^{L+S} \quad \text{G}-\text{Parity}: \mathbf{G} = (-1)^{L+S+I}$$

The light-quark mesons are built up from u, d, s and their antiquarks. This yields nine possible $q\bar{q}$ combinations for each set of quantum numbers, (nonets). For the allowed values of **L** and **S**, If one looks at the J^{PC} quantum numbers, the following values are not allowed: $0^{--}, 0^{+-}, 1^{-+}, 2^{+-}, 3^{-+}, \ldots$. Anything identified with one of these quantum numbers falls outside of the normal $q\bar{q}$ picture of the quark model.

THE STATUS OF GLUONIC MESONS

Hybrid meson quantum numbers can be predicted within the *flux-tube model* [1]. In this picture, the gluonic field forms a flux-tube between the $q\bar{q}$ pair. In it ground state, the tube carries no angular momentum, but it can be excited. The lowest excitation is $L=1$ rotation which contains two degenerate states, (clock-wise and counter-clockwise rotations). Linear combinations of these can be taken such that the tube behaves as if it has $J^{PC}=1^{-+}$ or $J^{PC}=1^{-+}$. Adding these to the $L=0$ mesons, one obtains eight possible quantum numbers. What is of particular interest is that three of the J^{PC}s, 0^{+-}, 1^{-+} and 2^{+-} correspond to non $\bar{q}q$ combinations.

Within the flux-tube model, all eight hybrid nonets are degenerate. The Lattice also predicts the existence of a flux-tube forming between a heavy quark-antiquark pair. It is also possible to calculate the potentials for the ground state and excited states of the flux-tube [2]. Lattice predictions for hybrid mesons masses are shown in Table 1. The exotic 1^{-+} nonet is the lightest state with a mass in the range of 1.8 to $2 GeV/c^2$. The splitting between the the 1^{-+} and 0^{+-} nonets is predicted to be about $0.2 GeV/c^2$, (with large errors [3]).

TABLE 1. Recent results for 1^{-+} hybrid meson masses.

Light Quark 1^{-+}		Charmonium 1^{-+}	
Reference	Mass GeV/c^2	Reference	ΔM GeV/c^2
UKQCD [4]	1.87 ± 0.20	MILC [5]	$1.34 \pm 0.08 \pm 0.20$
MILC [5]	$1.97 \pm 0.09 \pm 30$	MILC [6]	1.22 ± 0.15
MILC [6]	2.11 ± 0.13	[7]	1.323 ± 0.130
LaSch [8]	1.9 ± 0.20	[9]	1.19
[10]	$2.013 \pm 0.026 \pm 0.071$		

Predictions for the widths and decays of hybrids are based on model calculations with the results of recent work [11]. The width predictions are fairly open. Most of the 0^{+-} exotic nonet are expected to be quite broad. However, both the 2^{+-} and the 1^{-+} nonets are expected to be much narrower. The non-exotic hybrids will be more difficult to disentangle as they are likely to mix with nearby normal $q\bar{q}$ States. The expected decay modes of hybrids involve daughters that in turn decay. This makes the overall reconstruction more complicated, with final states involving from four to seven pseudoscalar mesons.

However, these decays can be used as a guideline when looking for the states. Almost all models of hybrid mesons predict that the ground state ones will not decay to identical pairs of S-wave mesons, and that the decays to an $(L=0)(L=1)$ pair is favored. The one unit of angular momentum in the flux–tube remains in the internal orbital angular momentum of one of the daughter $q\bar{q}$ pairs.

The most striking experimental prediction for hybrid mesons is the fact that several of the nonets have non-$q\bar{q}$ quantum numbers, and the lightest of these will be 1^{-+}, or exotic. Over the last decade, several credible reports of such states have been published. An isospin 1 object, the $\pi_1(1400)$ was first reported in $\pi^- p \to \eta \pi^- p$ [14]. This state was quickly confirmed in antiproton-neutron annihilation [15]. Figure 1 shows the Dalitz plot from the latter analysis where the exotic signal is of the same strength as the $a_2(1320)$. The PDG [16] lists the mass as $m = 1.376 \pm 0.017$ GeV/c^2 and the width as $\Gamma = 0.300 \pm 0.040 \, GeV/c^2$ with observed decays to $\pi^-\eta$ and $\pi^0\eta$.

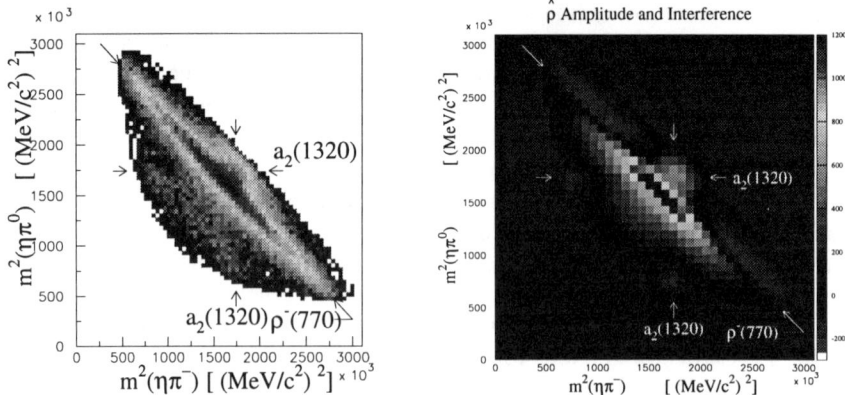

FIGURE 1. The left hand figure shows the Dalitz plot for $\bar{p}d \to \eta\pi^-\pi^0 p$ with both the $a_2(1320)$ and the $\rho(770)$ indicated. The right hand figure shows the contribution of the $\pi_1(1400)$ to the Dalitz plot.

A second such state, the $\pi_1(1600)$, was first observed in $\pi^- p \to \pi^+\pi^-\pi^- p$ [17]. The signal for the $\pi(1600)$ is shown in Figure 2. A latter observation reported the $\pi_1(1600) \to \eta'\pi$ [18] and various reports have been made at conferences about other observed decay modes. The VES experiment [19] reports the ratios of: $b_1\pi : \eta'\pi : \rho\pi = 1 : 1.0 \pm 0.3 : 1.6 \pm 0.4$. The PDG [16] lists the mass as $m = 1.596^{+0.025}_{-0.014} \, GeV/c^2$ and the width as $\Gamma = 0.312^{+0.064}_{-0.024} \, GeV/c^2$. Recently there has been a report of the $f_1(1285)\pi$ decay mode of the $\pi(1600)$ as well third π_1 state in the 1.9 GeV/c^2 region, also decaying to $f_1\pi$ [20].

The precise interpretation of these states is still open. The $\pi_1(1400)$ is significantly lighter than theoretical expectations, and its only observed decay mode, $\eta\pi$ is not expected for a hybrid. Recent work suggests that this state may actually be non-resonant scattering similar to the S-wave $\pi\pi$ scattering at low energy [21]. The same explanation in the $\pi\eta'$ system can also be invoked to explain a large part of the $\eta'\pi$ signal for the $\pi_1(1600)$ [22]. The $\pi_1(1600)$ as seen in $\rho\pi$ is still somewhat lower than theoretical expectations in mass, but could well be a hybrid meson. The open question now is what are its decay modes, and can we find any of its partner states, η_1 and η_1'?

There is also a more general issue of what is causing the over population of π_1 states? There is one 1^{-+} hybrid nonet, meaning that there should only be one π_1 state. While it is possible that the $\pi_1(1400)$ is just final state interactions, if there are really two states beyond this, ($\pi_1(1600)$ and $\pi_1(1900)$), it will be necessary to rethink what is happening.

Hybrids with non-exotic quantum numbers are more difficult to discern as they look like normal $q\bar{q}$ mesons. If one assumes that the $\pi_1(1600)$ sets the mass scale for the hybrids, then we are looking in the 1.6 to 2.2 GeV/c^2 mass range.

In the $J^{PC} = 0^{-+}$ system, we expect radial excitations of the pseudoscalar mesons as well as a glueball state. Three states of interest appear in this sector, the $\pi(1800)$, ($m = 1.8$, $\Gamma = 0.21$) has been observed with decays into $f_0(980)\pi$, $f_0(1370)\pi$, $\rho\pi$, $\eta\eta\pi$, $a_0(980)\pi$ and $f_0(1500)\pi$. and there has been speculation that due to its coupling

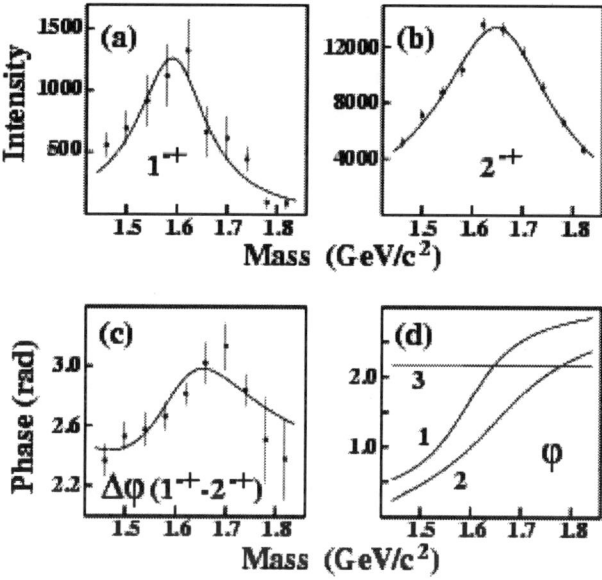

FIGURE 2. Data from the E852 experiment showing and exotic 1^{-+} signal in the $\rho\pi$ subsystem of $\pi^- p \to \pi^+\pi^-\pi^- p$. (a) shows the intensity of the 1^{-+} signal which interferes with (b) the 2^{-+} $\pi_2(1670)$. (c) shows the phase difference between the two waves, and (d) has the individual phases, with 1 corresponding to the 1^{-+}, 2 to the 2^{-+} and 3 to a background term.

to scalars, it may have a large hybrid component. The $\eta(1760)$, which decays into 4π, has only been observed in J/ψ decays. This is a likely partner for the $\pi(1800)$. Finally, the $\eta(2225)$, $(m = 2.2, \Gamma = 0.15)$ has been observed in J/ψ with decays into $\phi\phi$. This state is too high in mass to be the simple partner of the other two, but is consistent with what is expected of a glueball but needs confirmation.

In the $J^{PC} = 1^{--}$ system, we expect to see the radial excitations of the vector mesons as well as the $^3D_1$ nonet. The mass scale for the D-wave mesons are set by the $^3D_3$ nonet, $\rho_3(1690)$, $\omega_3(1670)$ and the $\phi_3(1850)$. There are a rather large number of known states in this region. The $\rho(1450)$, $\rho(1700)$, $\rho(1900)$, $\rho(2150)$, $\omega(1420)$, $\omega(1650)$, and $\phi(1680)$. This sector is probably completely mixed, so disentangling it is going to require a clear understanding of other sectors.

The $J^{PC} = 1^{+-}$ hybrid nonet is near the radial excitations of the b_1s nonet. One known state exists, the $h_1(1595)$, $(m = 1.6, \Gamma = 0.38)$, but little is known it. It is probably consistent with being a radial excitation of the $h_1(1170)$.

The $J^{PC} = 1^{++}$ nonet has the same quantum numbers as the radial excitations of the a_1s nonet. One known state exists, the $a_1(1640)$ which has been observed in a 3π decay. What little is known is consistent with this being a radial excitation of the $a_1(1260)$.

The $J^{PC} = 2^{-+}$ nonet can overlap with the D-wave nonet, $^1D_2$. There are a rather large number of candidates here. The $\pi_2(1670)$ and the $\eta_2(1645)$ are reasonably consistent with the D-wave mesons. There is a second η_2, the $\eta_2(1870)$ that mass-wise is consistent with being the η' of this nonet. However, its decay modes are consistent with it being composed of mostly non-strange light quarks. In fact, the $a_2\pi$ decay appears to be the largest mode for both the $\eta_2(1645)$ and the $\eta_2(1870)$. This sector has the strongest evidence for a hybrid state.

In any case, establishing the non-exotic hybrid nonets will almost certainly require more exotic states. These will allow us to both set the mass scale, and understand the actual decay patterns of the states.

SUMMARY

Clear experimental evidence exists for mesons with non-$q\bar{q}$ quantum numbers. Unfortunately, the exact nature of the exotic states remain unclear. Their observed mass and decay modes do not completely agree with theoretical expectations, which may well indicate that there are problems with the theory. In order to resolve these issues, it will be necessary to observe and measure both the partners of the existing states as well as states with other exotic quantum numbers. This will be studied in the light quark sector with the GlueX Experiment at Jefferson Lab [23].

REFERENCES

1. N. Isgur and J. Paton, Phys. Rev. D**31**, 2910, (1985).
2. C. J. Morningstar and M. Peardon, Phys. Rev. D**60**, 034509, (1999).
3. C. Morningstar, **The study of exotic hadrons in lattice QCD**.
4. P. Lacock *et al.*, Phys. Lett. B**401**, 309, (1997).
5. C. Bernard *et al.*, Phys. Rev. D**56**, 7039, (1997).
6. C. Bernard *et al.*, Nucl. Phys. B(Proc. Suppl.)**73**, 264, (1999).
7. T. Manke *et al.*, Phys. Rev. Lett.**82**, 4396, (1999).
8. P. Lacock and K. Schilling, Nucl. Phys. B (Proc. Suppl.)**73**, 261, (1999).
9. K. J. Juge, J. Kuti and C. Morningstar, Phys. Rev. Lett.**82**, 4400, (1999).
10. Zhong-Hao Mei and Xiang-Qian Luo, **Exotic mesons from quantum chromodynamics with improved gluon and quark actions on the anisotropic lattice**, (2002), (hep-lat/0206012).
11. P. R. Page, E. S. Swanson and A. P. Szczepaniak, Phys. Rev. D**59**, 034016, (1999).
12. N. Isgur, R. Kokoski and J. Paton, Phys Rev Lett.**54**, 869, (1985).
13. C. J. Morningstar and M. Peardon, Phys. Rev. D**56**, 4043, (1997).
14. D. R. Thompson *et al.* (E852 Collaboration) Phys. Rev. Lett. **79**, 1630, (1997) and S. U. Chung *et al.*, (E852 Collaboration) Phys. Rev. D**60**, 92001, (1999).
15. A. Abele *et al.* (Crystal Barrel Collaboration), Phys. Lett. B**423**, 175, (1998) and Phys. Lett. B**446**, 349, (1999).
16. K. Hagiwara *et al.*, **Review of Particle Physics**, Phys. Rev.D**66**, 010001, (2002).
17. G. S. Adams *et al.* (E852 Collaboration), Phys. Rev. Lett. **81**, 5760, (1998).
18. E. I. Ivanov *et al.* (E852 Collaboration), Phys. Rev. Lett. **86**, 3977, (2001).
19. Yu Khokhlov, *et al.* (VES Collaboration), Nucl. Phys. A**663**, 596 ,(2000).
20. Gary Adams, these proceeding, (2003).
21. A. R. Dzierba, Phys.Rev. D**67**, 094015, (2003).
22. Adam P. Szczepaniak, Maciej Swat, Alex R. Dzierba, Scott Teige, **Study of the $\eta\pi$ and $\eta'\pi$ spectra and interpetation of possible exotic $J^{PC} = 1^{-+}$ mesons**, hep-ph/0304095, (2003).
23. http://www.gluex.org.

Observation of a 1750 MeV/c^2 State in the Peripheral Photoproduction of K^+K^-

Ryan E. Mitchell[1]

University of Tennessee, 401 Nielsen Physics, Knoxville, TN 37996-1200

Abstract. The FOCUS collaboration at Fermilab has observed a clear enhancement at a mass of 1750 MeV/c^2 in the exclusive photoproduction of K^+K^-. The data were obtained using a photon beam with energies from 20 to 160 GeV incident on a BeO target. Mass and width measurements are presented, as well as a search in K^*K. This enhancement has not yet been correctly interpreted.

Peripheral photoproduction, where a photon beam softly glances off a nuclear target, is a unique and important way of studying the light meson spectrum [1]. Through the vector dominance model, photons in the beam can fluctuate into vector mesons such as the ρ, ω, and ϕ mesons, thus providing access to a range of final state mesons that differ from those accessible by the well-studied pion or kaon beams. Very few exclusively photoproduced mesons have been cleanly identified, however.

The $K^+K^-(1750)$ meson first appeared in photoprodution in the 1980's [2, 3, 4], and was immediately interpreted as the $\phi(1680)$, the radial excitation of the $\phi(1020)$ that has since become well established in e^+e^- annihilation [5, 6, 7]. Using data from the FOCUS experiment [10, 11, 12], we claim that the exclusively photoproduced $K^+K^-(1750)$ is not the $\phi(1680)$. This conclusion is based on two points: (1) FOCUS measures the mass of the $K^+K^-(1750)$ to be $1753.5 \pm 1.5 \pm 2.3$ MeV/c^2, clearly inconsistent with 1680 MeV/c^2; and (2) FOCUS finds the $K^+K^-(1750)$ to decay far more often to K^+K^- than K^*K, while the $\phi(1680)$ decays dominantly to K^*K [7, 8]. Details of this analysis have been published elsewhere [9].

Figure 1.b shows the K^+K^- mass spectrum in the region of 1750 MeV/c^2 for different requirements of p_T, the transverse momentum of the KK pair with respect to the beam direction. The $K^+K^-(1750)$ can clearly be seen in events with low p_T ($p_T < 0.15$ GeV/c). Fitting the 1750 MeV/c^2 mass region with a non-relativistic Breit-Wigner distribution and a quadratic background (Figure 1.a), we find: Yield = $11,700 \pm 480$ Events; M = $1753.5 \pm 1.5 \pm 2.3$ MeV/c^2; and $\Gamma = 122.2 \pm 6.2 \pm 8.0$ MeV/c^2. The systematic errors were determined by varying the p_T cut, the Cerenkov cuts, the form of the Breit-Wigner shape, and the form of the background shape, and include the sys-

---

[1] On behalf of the FOCUS collaboration: J. M. Link, P. M. Yager (**UC Davis**); J. C. Anjos, I. Bediaga, C. Göbel, J. Magnin, A. Massafferri, J. M. de Miranda, I. M. Pepe, E. Polycarpo, A. C. dos Reis (**CBPF, Rio de Janeiro**); S. Carrillo, E. Casimiro, E. Cuautle, A. Sánchez-Hernández, C. Uribe, F. Vázquez (**CINVESTAV, México City**); L. Agostino, L. Cinquini, J. P. Cumalat, B. O'Reilly, I. Segoni, M. Wahl (**CU Boulder**); J. N. Butler, H. W. K. Cheung, G. Chiodini, I. Gaines, P. H. Garbincius, L. A. Garren, E. Gottschalk, P. H. Kasper, A. E. Kreymer, R. Kutschke, M. Wang (**Fermilab**); L. Benussi, M. Bertani, S. Bianco, F. L. Fabbri, A. Zallo (**INFN Frascati**); M. Reyes (**Guanajuato, Leon**); C. Cawlfield, D. Y. Kim, A. Rahimi, J. Wiss (**UI Champaign**); R. Gardner, A. Kryemadhi (**IU Bloomington**); Y. S. Chung, J. S. Kang, B. R. Ko, J. W. Kwak, K. B. Lee (**Korea University, Seoul**); K. Cho, H. Park (**Kyungpook National University, Taegu**); S. Alimonti, S. Barberis, A. Cerutti, P. D'Angelo, M. DiCorato, P. Dini, L. Edera, S. Erba, M. Giammarchi, P. Inzani, F. Leveraro, S. Malvezzi, D. Menasce, M. Mezzadri, L. Moroni, D. Pedrini, C. Pontoglio, F. Prelz, M. Rovere, S. Sala (**INFN and Milano**); T. F. Davenport III, (**UNC Asheville**); V. Arena, G. Boca, G. Bonomi, G. Gianini, G. Liguori, M. M. Merlo, D. Pantea, S. P. Ratti, C. Riccardi, P. Vitulo (**INFN and Pavia**); H. Hernandez, A. M. Lopez, E. Luiggi, H. Mendez, A. Paris, J. Quinones, J. E. Ramirez, Y. Zhang (**Mayaguez, Puerto Rico**); J. R. Wilson (**USC Columbia**); T. Handler, R. Mitchell (**UT Knoxville**); D. Engh, M. Hosack, W. E. Johns, M. Nehring, P. D. Sheldon, K. Stenson, E. W. Vaandering, M. Webster (**Vanderbilt**); M. Sheaff (**Wisconsin, Madison**).

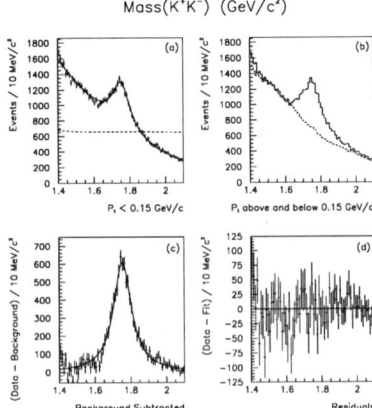

FIGURE 1. (a) The K^+K^- mass spectrum with the requirement that $p_T < 0.15$ GeV/c. The fit is described in the text. The dotted line is the Monte Carlo efficiency on a scale from 0 to 100%. (b) The solid line is the K^+K^- mass spectrum with the requirement that $p_T < 0.15$ GeV/c. The dotted line is the K^+K^- mass spectrum with $p_T > 0.15$ GeV/c scaled to the size of the low p_T spectrum for comparison. (c) The data and fit after subtracting the quadratic polynomial background shape. (d) The data minus the fit.

tematic uncertainty in our mass scale. The mass of $1753.5 \pm 1.5 \pm 2.3$ MeV/c^2 is in clear disagreement with 1680 MeV/c^2, but agrees well with earlier photoproduction measurements of the $K^+K^-(1750)$ mass [2, 3, 4]. Several interference senarios have also been studied, but in none of these does the mass of the $K^+K^-(1750)$ drop below 1747 MeV/c^2. Thus, the mass of the $K^+K^-(1750)$ substantially disagrees with the mass of the $\phi(1680)$.

Starting with a sample of $K_S K\pi$ events, two different K^*K combinations can be studied. After requiring $p_T < 0.15$ GeV/c (the same p_T cut imposed on the K^+K^- sample), and requiring a K^*, the two distinct K^*K spectra show no sign of a signal at 1750 MeV/c^2 (Figure 2). In order to place an upper limit on the number of $K^+K^-(1750)$ events in these K^*K modes, we have fit the mass spectrum with a parametrized background shape and a Breit-Wigner with mass and width fixed to the $K^+K^-(1750)$ values as determined in the fit to the K^+K^- mode. These fits provide an estimate of the number of events above background in the $K^+K^-(1750)$ region, -123 ± 120 events in the $K^{*0}K_S$ mode and 106 ± 117 in the $K^{*\pm}K^{\mp}$ mode. Using the relative detection effeciencies of the K^+K^- and K^*K final states, and correcting for the K_S unseen decay mode, we have found an upper limit on the following relative brancing ratios

$$\frac{\Gamma(K^+K^-(1750) \to \overline{K}^{*0}K^0 \to K^-\pi^+K_S + c.c.)}{\Gamma(K^+K^-(1750) \to K^+K^-)} < 0.065 \text{ at } 90\% \text{ C.L.}$$

$$\frac{\Gamma(K^+K^-(1750) \to K^{*+}K^- \to K_S\pi^+K^- + c.c.)}{\Gamma(K^+K^-(1750) \to K^+K^-)} < 0.183 \text{ at } 90\% \text{ C.L.}$$

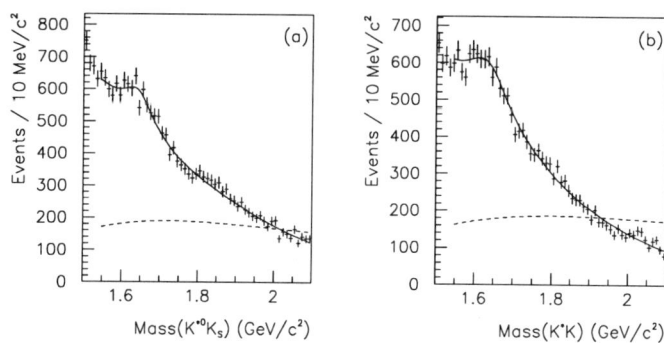

FIGURE 2. Fits to K^*K as described in the text. The dotted lines are the Monte Carlo efficiencies on a scale from 0 to 20%. (a) $K^{*0}K_S$ with K^{*0} to $K^{\pm}\pi^{\mp}$. (b) $K^{*\pm}K^{\mp}$ with $K^{*\pm}$ to $K_S\pi^{\pm}$.

The confidence limits were set using the Feldman-Cousins methodology [13]. The two relative branching ratios were measured to be -0.083 ± 0.081 and 0.065 ± 0.072, respectively. These limits are in stark contrast to the $\phi(1680)$ decay pattern, where the K^*K decay modes dominate.

The photoproduced $K^+K^-(1750)$ is not the $\phi(1680)$. The interpretation of the $K^+K^-(1750)$ remains unclear. An angular analysis must be conducted in order to determine the J^{PC} of the $K^+K^-(1750)$. In general, much is left to be understood in the peripheral photoproduction of mesons.

REFERENCES

1. S. Godfrey and J. Napolitano, Rev. Mod. Phys. 71 (1999) 1411.
2. D. Aston et al. Phys. Lett. B104 (1981) 231.
3. M. Atkinson et al., Z. Phys. C27 (1985) 233.
4. J. Busenitz et al., Phys. Rev. D40 (1989) 1.
5. A.B. Clegg, A. Donnachie, Z. Phys. C62 (1994) 455.
6. D. Bisello et al., Z. Phys. C52 (1991) 227.
7. J. Buon et al., Phys. Lett. B118 (1982) 221.
8. K. Hagiwara et al., Phys. Rev. D66, (2002) 1.
9. J. Link et al., Phys. Lett. B545 (2002) 50.
10. P.L. Fabretti et al., Nucl. Instrum. Methods A320 (1992) 519.
11. J. Link et al., Nucl. Instrum. Methods A484 (2002) 270.
12. J. Link et al., Nucl. Instrum. Methods A484 (2002) 174.
13. G.J. Feldman and R.D. Cousins, Phys. Rev. D57 (1998) 3873.

Confinement Dynamics

M. G. Olsson* and Theodore J. Allen[†]

*Department of Physics, University of Wisconsin,
1150 University Avenue, Madison, Wisconsin 53706 USA
[†]Physics Department, Hobart & William Smith Colleges
Geneva, New York 14456 USA

Abstract. We demonstrate the equivalence of the relativistic flux tube model of mesons to a simple potential model in the regime of large radial excitation. We make no restriction on the quark masses; either quark may have a zero or finite mass. Our primary result shows that for fixed angular momentum and large radial excitation, the flux tube/QCD string meson with a short-range Coulomb interaction is described by a spinless Salpeter equation with a time component vector potential $V(r) = ar - k/r$.

INTRODUCTION

Although Quantum Chromodynamics (QCD) is almost surely the correct theory of strong interactions, it remains difficult to explore its predictions in the non-perturbative regime. For hadron states the non-perturbative, or confinement, regime corresponds to large distances. For mesons, confinement dominates the dynamics of even the heaviest quark states. It has long been suspected that when the color sources are widely separated, the color electric fields collapse into relatively thin configurations known as flux tubes, or QCD strings. The evidence for such string-like configurations is primarily:

- universal linear Regge trajectories, reflecting a linear confining potential and relativistic kinematics [1];
- lattice simulation of the energy density [2];
- relativistic corrections of the flux tube model [3] agree with those of Wilson loop QCD [4];
- for heavy onia, the flux tube model reduces to the very successful linear confinement potential model;
- agreement of the vibrating string picture (at least at large source separations) [5] with lattice simulations of excited QCD states with fixed sources [6].

Recently, we have observed semi-classically and numerically that in the limit of large radial excitation [7] a simple time component vector potential (linear confinement) becomes identical with the string-flux tube model. This talk describes the complimentary observation that in the same high radial excitation regime the string equations, with arbitrary mass quarks, reduce to the simple time component vector (TCV) interaction [8].

THE QCD STRING WITH ARBITRARY QUARK MASSES

There are two equivalent methods of extracting the conserved quantities of the spinless quark-string system. The momentum-energy approach considers a straight color electric tube of energy a per unit length. From Lorentz boosting a string element perpendicular to its orientation, the momentum, angular momentum, and energy of the string are easily obtained [3, 9]. This intuitive construction is appealing for its simplicity. Using this approach the relativistic string/quark model for mesons can be exactly solved numerically for arbitrary quark masses [9]. One can also extract the conserved quantities with arbitrary quark masses more formally by using Noether's theorem [8].

By either method the conserved quantities are,

$$P^0 = W_{r1}\gamma_{\perp 1} + ar\left(\frac{r_1}{v_{\perp 1}}\arcsin(v_{\perp 1})\right) + (1 \leftrightarrow 2), \quad (1)$$

$$P_\perp = W_{r1}\gamma_{\perp 1}v_{\perp 1} - a\left(\frac{r_1}{v_{\perp 1}}\gamma_{\perp 1}^{-1}\right) - (1 \leftrightarrow 2), \quad (2)$$

$$L = W_{r1}\gamma_{\perp 1}v_{\perp 1}r_1 + \frac{a}{2}\frac{r_1^2}{v_{\perp 1}}\left(\frac{\arcsin(v_{\perp 1})}{v_{\perp 1}} - \gamma_{\perp 1}^{-1}\right) + (1 \leftrightarrow 2). \quad (3)$$

In the above we have used the identity,

$$m\gamma = \sqrt{p_r^2 + m^2}\,\gamma_\perp \equiv W_r\gamma_\perp. \quad (4)$$

THE HIGH RADIAL EXCITATION REGIME

We now make the crucial observation that, for fixed angular momentum and large radial excitation, we may assume the string velocity is small without changing the meson dynamics. This is because $v_\perp$ only becomes large near the inner turning point, where it reaches $v_\perp = 1$ in the case of a massless quark. However, near the inner turning point the string is short for very radial orbits so it carries little angular momentum or energy. Henceforth, for notational convenience, we generally suppress the $\perp$ subscript, and let v and γ denote $v_\perp$ and $\gamma_\perp$.

To demonstrate our approximation, we consider the equal quark mass case $m_1 = m_2$, from which follows $v_1 = v_2 \equiv v$, and $r_1 = r_2 = r/2$. We define

$$S(v) = \frac{\arcsin(v)}{v}, \quad (5)$$

$$f(v) = \frac{1}{2v}\left(S - \sqrt{1-v^2}\right). \quad (6)$$

We eliminate the radial energy term W_r between the energy and angular momentum equations to obtain a relation between the quark separation and transverse velocities

$$v\left(E - arS + \frac{k}{r}\right) + arf = \frac{2L}{r}. \quad (7)$$

To reach reasonably general numerical conclusions we define the dimensionless quantities

$$x = \frac{ar}{E}, \quad \beta = \frac{aL}{E^2}, \quad \kappa = \frac{ak}{E^2}. \tag{8}$$

In these dimensionless variables, Eq. (7) becomes

$$xv\left(1 - xS + \frac{\kappa}{x}\right) + x^2 f = 2\beta. \tag{9}$$

For fixed L and increasing E, we expect both β and κ to be small and $xv \approx 2\beta$. We thus expect v to be small unless x is small. In Fig. 1a we show the exact numerical solution of Eq. (9) for two values of β. For simplicity we consider no short range interaction ($\kappa = 0$).

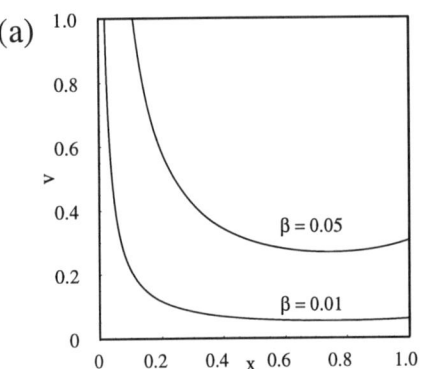
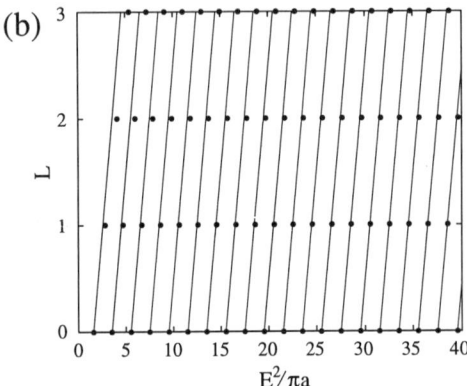

FIGURE 1. (a) Solution to Eq. (9) for the perpendicular velocity of the string end as a function of distance $x = ar/E$ for two values of $\beta = aL/E^2$ with no Coulomb potential; $\kappa = 0$. For any attractive Coulomb interaction ($\kappa > 0$) the velocities are lower, making the approximation $v \ll 1$ better. We observe that for radially excited states $\beta \ll 1$ and $v \to 0$ except at the inner turning point.
(b) The transition from a time component vector (TCV) confinement to string dynamics. The lines represent exact numerical solutions to the string equations in the case $m_1 = 0$, $m_2 = \infty$. The dots are exact numerical solutions to linear TCV confinement. The squared excitation energies $(E = M - m_2)$ of the TCV system converge to those of the string for large radial excitations with small angular momentum L.

Using the above relation between the transverse velocity and quark separation we can verify that the string plays a negligible role even near the inner turning point. Although this is expected due to the short string length it is numerically true [5] as shown in Figure 1b.

THE HIGH RADIAL EXCITATION REGIME AND REDUCTION TO A TCV POTENTIAL

We now make the crucial observation that, for fixed angular momentum and large radial excitation, we may assume the string velocity is small without changing the meson dynamics. This is because $v_\perp$ only becomes large near the inner turning point, where it reaches $v_\perp = 1$ in the case of a massless quark. However, near the inner turning point the string is short for very radial orbits so it carries little angular momentum or energy.

We now proceed to approximate the string velocity only to leading order while at the same time keeping full relativistic kinematics for the quarks. The details are straightforward but messy and are given in some detail in [8]. In the region of high radial excitation the string equations yield a spinless Salpeter wave equation

$$H\psi = M\psi, \qquad (10)$$

$$H = \sqrt{p^2 + m_1^2} + \sqrt{p^2 + m_2^2} + ar - \frac{k}{r}, \qquad (11)$$

which is the TCV Hamiltonian with a TCV potential

$$V(r) = ar - \frac{k}{r}. \qquad (12)$$

We note that at every step we have imposed the center of momentum rest condition and hence the Hamiltonian, Eq. (11), satisfies all conservation laws, even for arbitrary masses.

Finally, we verify numerically that for high radial excitations the string equation solutions become identical to the TCV spinless Salpeter solutions with linear confinement. The result is shown in Figure 1b.

ACKNOWLEDGMENT

This work was supported in part by the US Department of Energy under Contract No. DE-FG02-95ER40896.

REFERENCES

1. M. G. Olsson, Phys. Rev. D **55**, 5479 (1997).
2. G. Bali, Phys. Rept. **343**, 1 (2001).
3. C. Olson, M. G. Olsson, and Ken Williams, Phys. Rev. D **45**, 4307 (1992).
4. N. Brambilla and G. M. Prosperi, Phys. Lett. B **236**, 69 (1990); Phys. Rev. D **46**, 1096 (1992);
5. T. J. Allen, M. G. Olsson, and S. Veseli, Phys. Lett. B **434**, 110 (1998).
6. K. J. Juge, J. Kuti, and C. J. Morningstar, Proceedings of the XVth International Symposium on Lattice Field Theory, Nucl. Phys. B (Proc. Suppl.) **63** (1998) 326.
7. T. J. Allen, C. Goebel, M. G. Olsson, and S. Veseli, Phys. Rev. D **64**, 094011 (2001); T. J. Allen, Todd Coleman, M. G. Olsson, and S. Veseli, Phys. Rev. D **67**, 054016 (2003).
8. T. J. Allen, and M. G. Olsson, e-Print Archive: hep-ph/0306128 , Phys. Rev. D (in press).
9. D. La Course and M. G. Olsson, Phys. Rev. D **39**, 2751 (1989); M. G. Olsson and S. Veseli, Phys. Rev. D **51**, 3578 (1995).

Review of Two-Photon Interactions

David Urner

Cornell University

Abstract. Presented are recent results of two-photon interactions. Topics inlcude photon structure functions, inclusive hadron production, differential cross sections derived from tagged γγ fusion events and results in exclusive hadron production, particularly the observations of the η'_c.

Two-photon interactions provide a unique opportunity to study a large variety of physics topics. In electron positron machines two photons are emitted producing the photon-photon interaction. One can use this process to examine the structure and interaction of the photon. Two-photon data allow for a large number of tests of perturbative QCD and the exploration of non-perturbative phenomena in the light meson sector.

PHOTON STRUCTURE FUNCTION

Two photon interactions are instrumental to extract the hadronic structure function of the photon, which is related to the quark densities in the photon. It is measured in single tagged events, which means that one of the photons is real, and the other has a large momentum transfer much larger than Λ_{QCD}. This process can be viewed as deep inelastic electron e-γ scattering, where a quasi real photon is probed by a virtual photon with high Q^2. The differential cross section $F_2^\gamma(x, Q^2)$ is related to the hadronic structure function:

$\frac{d^2\sigma_{e\gamma \to eX}}{dxdQ^2} = [(1+1-y)^2)F_2^\gamma(x,Q^2) - y^2 F_L^\gamma(x,Q^2)]\frac{2\pi\alpha^2}{xQ^2}$ with $x \sim \frac{Q^2}{Q^2+W^2}$, $y = \frac{E_{tag}}{E_{beam}}cos^2\theta$.

The longitudinal term can usually be neglected since y^2 is small. The LEP2 data has considerably extended the reach towards small x. The present level of precision starts to challenge current structure function parameterizations. The newest result comes from ALEPH in the Q^2 range of 17.3 and 67.2 GeV^2 [1].

HEAVY QUARK PRODUCTION

Heavy quark production, provides important tests of perturbative QCD as well as probing the partonic densities in the photon. Recent results include a D* measurement of ALEPH [2] and a muon semileptonic measurement by DELPHI [3]. ALEPH shows that besides W, also the transverse momentum and the differential cross section in p_T and η are in good agreement with NLO QCD. DELPHI added a preliminary third measured point of the total bottom cross section, which agrees well with results from OPAL and L3. The large quark mass should allow good accuracy in perturbative calculations.

However the current status of NLO QCD underestimates the cross sections by about a factor of 3. This discrepancy cannot be tuned away by changing the b-quark mass.

INCLUSIVE MESON PRODUCTION

L3 has complemented their earlier result in inclusive π^0 production with an inclusive charged pion and kaon production measurement [4], using the LEP2 data in the region of $W_{\gamma\gamma}$ <5 GeV and Q^2 <8GeV^2, see Figure 1a). There is a good agreement between data and NLO QCD below a p_T of 3 GeV, but above about 5 GeV the data clearly surpass the expectation similar to the excess previously observed in the π^0 and K_s spectra. In the inclusive single jet production the data is compared to a NLO QCD calculation [5], which agrees well with many inclusive observables for the older OPAL data. However for the L3 data, see figure 1b), which go to a higher p_T, there is again a clear deviation observed [6].

EXCLUSIVE HADRON PAIR PRODUCTION

For small photon virtualities, large s and large momentum transfer from the photons to the hadrons, treating the $\gamma\gamma \to hh$ system in leading twist perturbation theory, the transition amplitude factorizes into a hard scattering amplitude $\gamma\gamma \to q\bar{q}q\bar{q}$ + a single hadron distribution for each hadron [7]. The hand bag model adds a soft 2-hadron distribution amplitude in what is basically a power correction.

Untagged $\gamma\gamma \to \pi\pi, KK$, Baryon-Antibaryon

The handbag model predicts the ratio $\frac{\gamma\gamma \to \pi\pi}{\gamma\gamma \to KK}$=1 [8], while the leading twist calculation would predict a ratio of 2. Aleph, Delphi [9], and with higher statistics, but still preliminary, BELLE [10] have measured these decays and find indeed that the ratio is around 1 over the full W range. The handbag model also predicts a $1/\sin^4(\theta)$ behavior for the differential cross section, which agrees well with the data.

Comparing predictions made in the framework of the hard scattering approach [7] with the $\gamma\gamma \to$ Baryon-Antibaryon cross section one is sensitive to the quark structure of the baryon. Reasonable agreement with the quark-diquark model [13] is found by the L3, OPAL, CLEO and BELLE [11] experiments in the $p\bar{p}$, $\Lambda\bar{\Lambda}$ and $\Sigma\bar{\Sigma}$ final states, see figures 1c)-1e), although the BELLE data starts to challenge the diquark model, while three-quark model [12] predictions are too low. The $\gamma\gamma \to p\bar{p}$ spectra can be used to fit the parameters of the handbag model, and hence make predictions for the cross sections of all other baryon octet members [14], using one additional parameter ρ - a ratio of form factors of the proton. They agree with the $\Lambda\bar{\Lambda}$ and $\Sigma\bar{\Sigma}$ measurement by CLEO and L3 [15], see figure 1c), 1d). One notes that for $W_{\gamma\gamma}$ below 2.6 GeV there are experimental discrepancies between the BELLE and CLEO experiments for the $p\bar{p}$ 1e) and between the L3 and CLEO experiments for the $\Lambda\bar{\Lambda}$ 1c) final states, which should be resolved.

$\gamma\gamma \to \rho\rho$

The L3 experiment measured $\gamma\gamma \to \rho^0\rho^0$ and $\gamma\gamma \to \rho^+\rho^-$ processes [17]. A simple partial wave analysis was performed on the 4-pion final states, including only $\rho\rho$ partial waves plus a 4 π isotropic background, fitting the data separately for each W bin. The only contributing waves have a J^P, J_Z of 0^+ and $2^+, 2$ as shown in figure 2. From this result one would conclude, that the isospin ratio of the $\rho\rho$ cross sections is incompatible with either I=0 or I=1.

The L3 experiment also analyzed the two-photon production of $\rho^0\rho^0$ in the single tagged mode at $\sqrt{s} = 89$-209 GeV [17], which allow a test of the qq, gg $\to$ meson-pair mechanism. The $\rho^0\rho^0$ signal was separated from the $\rho^0\pi^+\pi^-$ and 4π backgrounds with the box method. Figure 2c) shows the $\gamma^*\gamma$ differential cross section as a function of Q^2, which agrees well with the generalized vector dominance model.

Single Tagged $\gamma^*\gamma \to \pi^0\pi^0$

In tagged two-photon decays the process $\gamma^*\gamma \to \pi\pi$ in the region of large Q^2 but small W factorizes into a perturbatively calculable part dominated by short distance scattering [7]: $\gamma^*\gamma \to q\bar{q}$ or, $\gamma^*\gamma \to gg$ and non perturbative matrix elements measuring the transitions $q\bar{q} \to \pi\pi$ and $q\bar{q} \to gg$, called generalized distribution amplitudes [16]. CLEO has a preliminary measurement of the $\gamma^*\gamma \to \pi^0\pi^0$ cross sections for different Q^2 and W bins shown in figure 2d).

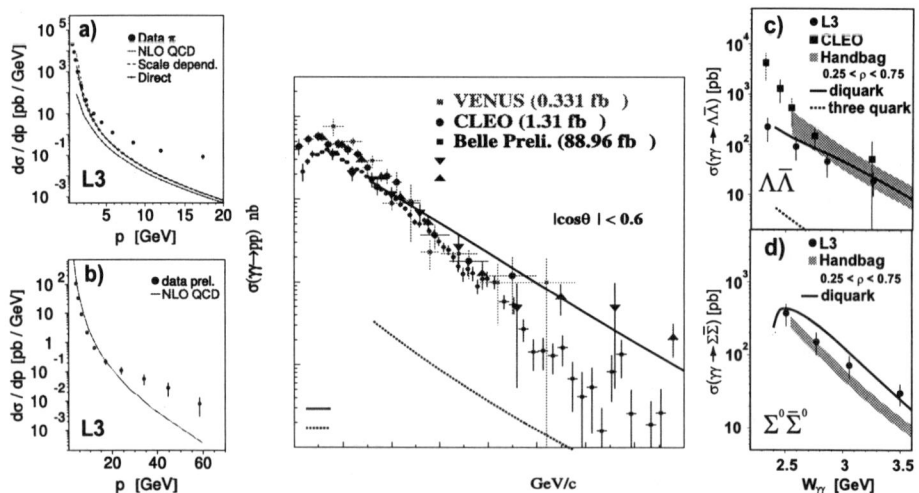

FIGURE 1. L3 experiment: Inclusive π^- (a) and single jet production (b) cross section. Exclusive $\gamma\gamma \to \Lambda\bar{\Lambda}$ (c) and $\gamma\gamma \to \Sigma\bar{\Sigma}$ (d) cross section and comparison with handbag and diquark models. BELLE experiment: Exclusive $\gamma\gamma \to p\bar{p}$ cross section (e) and comparison with diquark model.

$\gamma\gamma \to \eta(1440)$

Until recently $\eta(1440)$ has only been seen in gluon rich environments such as $\bar{p}p$ annihilation or J/ψ decay. Although quenched lattice calculations indicate a mass of the 0^{-+} glucball around 2GeV, some glueball content of the $\eta(1440)$ can presently not be excluded. The L3 experiment reported a first observation of the $\eta(1440)$ in two-photon collisions [18] with a $\Gamma_{\gamma\gamma}(\eta_{1440}) \cdot BR(\eta_{1440} \to K_s K^+ \pi^-) = 49 \pm 12$ eV. Since two-photon partial width of glueballs should be very small, this would indicate that the $\eta(1440)$ is mostly not a glueball.

CLEO has analyzed 13.8fb^{-1} of data collected around the $\Upsilon(4S)$ energies and searched for $\gamma\gamma \to \eta(1440) \to K_s K^\pm \pi^\mp$ decays [19]. There is no $\eta(1440)$ resonance observed and an upper limit for the two-photon partial width of 14.4 eV is obtained. This result, which includes all systematic errors, is 2.9 σ below the L3 result. Figure 3a) shows the CLEO data with the fit result and overlayed the signal (line) with errors (dashed), as expected from the L3 result.

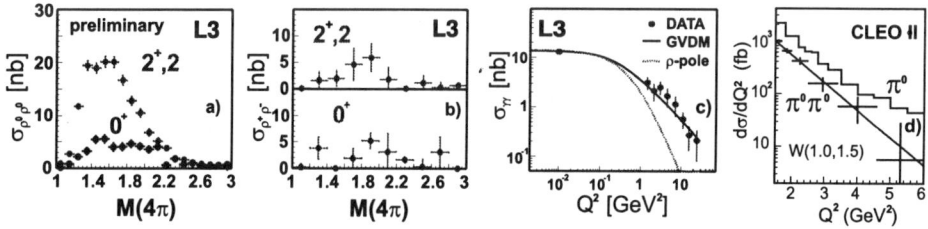

FIGURE 2. L3 experiment: cross section of contributing partial waves to untagged $\gamma\gamma \to \rho^0\rho^0$ (a) and $\gamma\gamma \to \rho^+\rho^-$ (b) decays. Single tagged $\gamma^*\gamma \to \rho^0\rho^0$ cross section (c). CLEO experiment: Single tagged $\gamma^*\gamma \to \pi^0\pi^0$ cross section compared to $\gamma^*\gamma \to \pi^0$ (d).

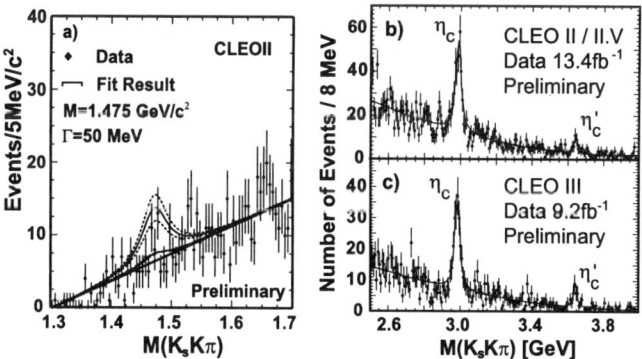

FIGURE 3. CLEO experiment: Search for $\gamma\gamma \to \eta(1440)(\eta(1440) \to K_s K \pi$ (a), the expected L3 signal (line) with errors (dashed) is superimposed. Observation of η'_c decaying to $K_s K \pi$ final state for CLEO II (b) and CLEO III (c) data sets.

η'_c OBSERVATION

The total existing experimental knowledge of the hyperfine splitting in any meson system is the $\Delta M = M(J/\psi)-M(\eta_c) = 117 \pm 2$ MeV. The measurement of $\Delta M = M(\psi')-M(\eta'_c)$ is important for the understanding of the spin-spin interaction in the confinement region. Theoretical predictions based on potential model calculations are $M(\eta'_c) \sim 3594$-3629 MeV, with a two-photon partial width ratio of $\Gamma_{\gamma\gamma}(\eta'_c)/\Gamma_{\gamma\gamma}(\eta_c) \sim 0.7$. A Crystal Ball measurement [20] at 3594 ± 5 MeV could not be confirmed by other searches [21]. Recently the BELLE experiment published an η'_c observation in $B \to K(K_s K^{\mp} \pi^{\pm})$ [22] with a mass of $3654 \pm 6 \pm 8$ MeV and a significance of more than 6σ. The BELLE experiment also observed a η'_c signal in the mode $e^+ e^- \to J/\psi(X)$ [23] with a mass of 3622 ± 12 MeV and a significance of 3.4σ. The BABAR experiment has presented a preliminary η'_c signal [24] in two-photon decays with a mass of $3633 \pm 5 \pm 1.8$ MeV.

The CLEO experiment has analyzed 13.9fb^{-1} taken with the CLEO II detector [25], see figure 3b) and finds a signal at a mass of 3642.7 ± 4.0 MeV with a significance of 4σ, a significance which assumes the mass or width of the resonance not to be known. CLEO confirmed this observation using 9.2fb^{-1} of data from the CLEOIII detector [25], see figure 3c), finding a mass of 3642.5 ± 3.4 MeV (no systematic error included) and a significance of 5.7σ. The ratio $\frac{\Gamma_{\gamma\gamma}(\eta'_c) x B(\eta'_c \to K_s K\pi)}{\Gamma_{\gamma\gamma}(\eta_c) x B(\eta_c \to K_s K\pi)} = 0.17 \pm 0.06$ (CLEO II) 0.29 ± 0.09 (CLEO III), with statistical errors only. All results are preliminary.

REFERENCES

1. ALEPH collaboration, paper in preparation.
2. ALEPH collaboration, *Eur. Phys. J.* **C28** 437-449 (2003).
3. DELPHI collaboration, presented at PHOTON 2003 in Frascati.
4. L3 collaboration *Phys. Letters.* **B554** 105-114 (2003).
5. L. Bertora *hep-ph/0306167* (2003).
6. L3 collaboration, presented at PHOTON 2003 in Frascati.
7. G. P. Lepage and S. J. Brodsky, *Phys. Rev. D* **22** 2157 (1980).
8. M. Diehl, et al. *Phys. Lett.* **B532** 99-110 (2002).
9. K. Grezlak Proceedings to PHOTON 2001 in Ascona *World Scientific*.
10. BELLE collabortion presented at PHOTON 2003 in Frascati.
11. L3 collaboration *hep-ex/0306017*, submitted to *Phys. Lett. B*, OPAL collaboration *hep-ex/0209052* (2002), CLEO collaboration *Phys. Rev. D* **50** 5484 (1994), VENUS *Phys. Lett B* **407** 185 (1997).
12. Farrar et al. *Nucl. Phys. B* **259** 702 (1985).
13. C. F. Berger and W. Schweiger, *hep-ph 0212066* (2002).
14. M. Diehl et al. hep-ph/0206288 (2002).
15. CLEO collaboration *Phys. Rev D* **56** 2485 (1997), L3 collaboration *Phys. Lett. B* **536** 24 (2002).
16. M. Diehl et al. *Phys Rev. D* **62** 073014 (2000).
17. L3 collaboration presented at PHOTON 2003 in Frascati.
18. L3 collaboration *Phys. Lett. B* **501** 1-11 (2001).
19. CLEO collaboration hep-ex/0212046 (2002).
20. Crystal Ball collaboration *Phys. Rev. Lett* **48** 70. (1982).
21. E760 Fermilab *Phys. Rev. D* **52** 4839 (1995), E835 Fermilab *Phys. Rev. D* **64** 052033 (2001).
22. BELLE collaboration *Phs. Rev. Lett.* **89** 102001 (2002).
23. BELLE collaboration *Phs. Rev. Lett.* **89** 142001 (2002).
24. BABAR collaboration G. Wagner *hep-ex/0305083* (2003).
25. CLEO collaboration *hep-ex/0306060* (2003).

An Interferometric Study of the $\chi_{c0}(1^3P_0)$ in the Reactions $\bar{p}p \to \pi^0\pi^0, \eta\eta, \pi^0\eta$

Jerome L. Rosen

(On behalf of the Fermilab E835 Collaboration.)

Northwestern University, Department of Physics and Astronomy

Abstract. Fermilab experiment E835 has observed $\bar{p}p$ annihilation production of the charmonium state χ_{c0} and its subsequent decay into $\pi^0\pi^0$ and $\eta\eta$. Although the resonant amplitude is an order of magnitude smaller than that of the non-resonant continuum production of $\pi^0\pi^0$, an enhanced interference signal is evident. A partial wave expansion is used to extract physics parameters. The amplitudes $J = 0$ and 2, of comparable strength, dominate the expansion.

The Fermilab E835 Collaboration has studied Charmonium production using a gas jet hydrogen target, the virtually monoenergetic stochastically cooled antiproton beam and a large acceptance shower spectrometer [1]. Data were recorded in year 2000, in particular 33 pb^{-1} of luminosity at the χ_{c0} energy in 17 energy points.

The angular distribution for $\bar{p}p \to \pi^0\pi^0, \eta\eta$, and $\pi^0\eta$ is:

$$\frac{d\sigma}{dz} = \underbrace{\left| \frac{-A_R}{x+i} + \sum_{J=0,2,\ldots}^{J_{max}} (2J+1)\, C_J(x)\, e^{i\delta_J(x)}\, P_J(z) \right|^2}_{\equiv A\, e^{i\delta_A}} + \underbrace{\left| \sum_{J=2,4,\ldots}^{J_{max}} \frac{(2J+1)}{\sqrt{J(J+1)}}\, C_J^1(x)\, e^{i\delta_J^1(x)}\, P_J^1(z) \right|^2}_{\equiv B\, e^{i\delta_B}} \quad (1)$$

where $x \equiv \frac{E_{CM}-M_{\chi_{c0}}}{\Gamma_{\chi_{c0}}/2}$ and $z \equiv |\cos\theta^*|$, with θ^* defined as the angle between the beam and the π^0 (or η) axes in the center of mass frame. The term $-A_R/(x+i)$ is the parameterization of an S-wave Breit-Wigner resonant amplitude. We can distinguish two contributions to the *non-resonant* cross section: $Ae^{i\delta_A}$, which *interferes* with the resonance, and $Be^{i\delta_B}$, which does not. No matter how many partial waves play a role, $Ae^{i\delta_A}$ and $Be^{i\delta_B}$ do not change markedly when the energy varies across the resonance.

Eq. 1 can be re-written as:

$$\frac{d\sigma}{dz} = \frac{A_R^2}{x^2+1} + A^2 + 2A_R A \frac{\sin\delta_A - x\cos\delta_A}{x^2+1} + B^2 \quad (2)$$

For fixed values of z, as the energy varies across the resonance (x passes through zero), even a very small A_R^2 (relatively to the non-resonant cross section) can give rise to a detectable signal thanks to the *interference-term* of Eq. 2.

In Fig. 1 the measured cross section is plotted versus z and versus the energy in the center of mass, E_{CM}. At off-resonance energy points, the cross section is *non-resonant*

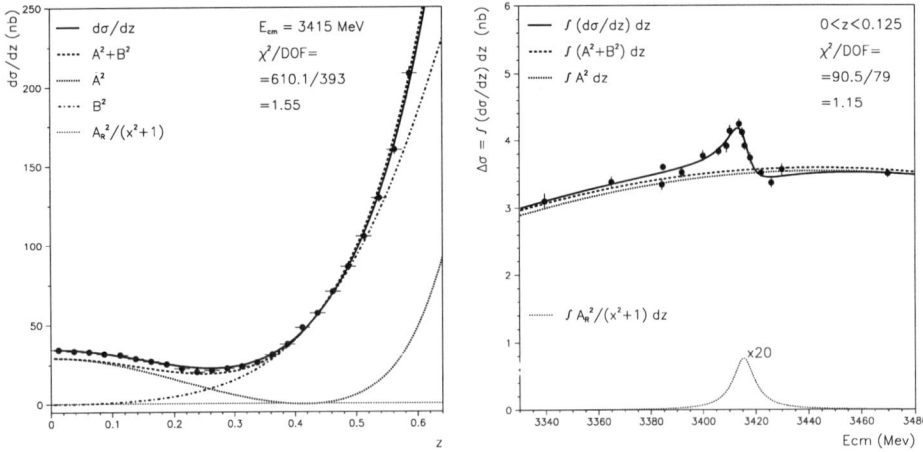

FIGURE 1. The $\bar{p}p \to \pi^0\pi^0$ cross section. Left: differential cross section versus z at $E_{CM} = 3415$ MeV. Right: integrated cross section (over $0 < z < 0.125$) versus E_{CM}.

production $\bar{p}p \to \pi^0\pi^0$ with a smooth dependence on the energy. The instrumental background (events from different channels, such as $\pi^0\pi^0\pi^0$ and $\pi^0\omega$, that are misidentified as $\pi^0\pi^0$) is $\sim 2\%$ at all energies and has been subtracted.

A maximum likelihood fit with the parameterization of Eq. 1, including partial waves up to $J_{max} = 4$, has been performed simultaneously on all energy points (Fig. 1-left). The number of free parameters is 15: the resonance amplitude A_R, the coefficients $C_{0,2,4}$ and $C^1_{2,4}$ (each of them is given a linear dependence on the energy), and the phases $\delta_{0,2,4}$ and $(\delta^1_4 - \delta^1_2)$; only the difference between δ^1_4 and δ^1_2 is measurable.

The line $A^2 + B^2$ shows the sum of the two contributions to the non-resonant cross section. The effect of the resonance, amplified by the interference, is seen in the gap (evident at small z) between $d\sigma/dz$ and $A^2 + B^2$ and is due almost entirely to the interference-term of Eq. 2. The gap decreases as z increases, following the trend of A. The term B^2 is small at small values of z, due to a factor z present in all the associate functions $P^1_J(z)$. The net suppression factor of B^2 with respect to A^2 is z^2 at small z-values. The contribution of the *pure* resonance $A_R^2/(x^2+1)$ is negligible.

We affirm that only $J = 0, 2, 4$ are significant [2]. Since the fit is dominated by the high statistics of the forward peak, where the resonant signal is not significant, we do not rely on the global fit to evaluate the magnitude of the resonance amplitude A_R.

A more reliable approach is to perform a different fit on a reduced range at small z (Fig. 1-right). In this range the resonance signal has a substantial size and, as observed above, the non-interfering part is very small (reducing therefore the uncertainty on the estimate of the ratio A^2/B^2, critical in determining the amplification effect of the interference). Notice that in this case we do not need to make any assumption on the number of partial waves involved in the reaction. We just perform a polynomial expansion on z about $z = 0$, exploiting the small extension of the range. A polynomial

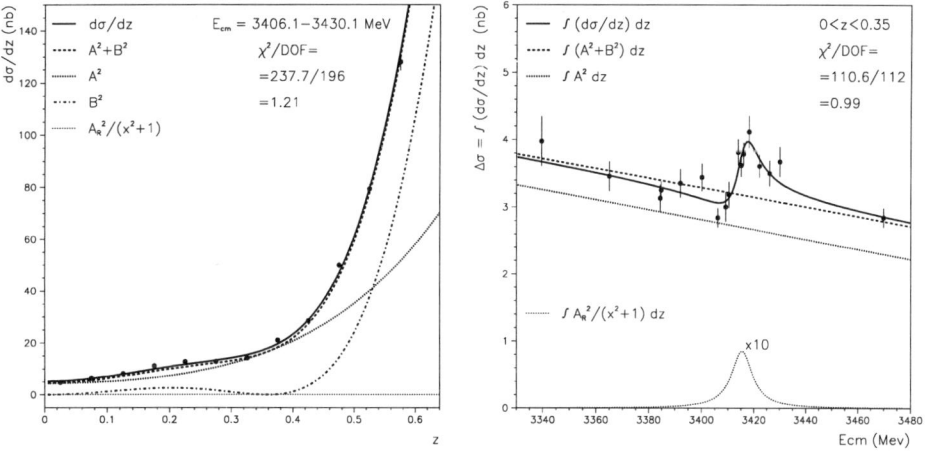

FIGURE 2. The $\bar{p}p \to \eta\eta$ cross section. Left: differential cross section versus z at $E_{CM} = 3406.1 - 3430.1$ MeV (the energies closer to the χ_{c0} resonance are merged in this plot to reduce fluctuations). Right: integrated cross section (over $0 < z < 0.35$) versus E_{CM}.

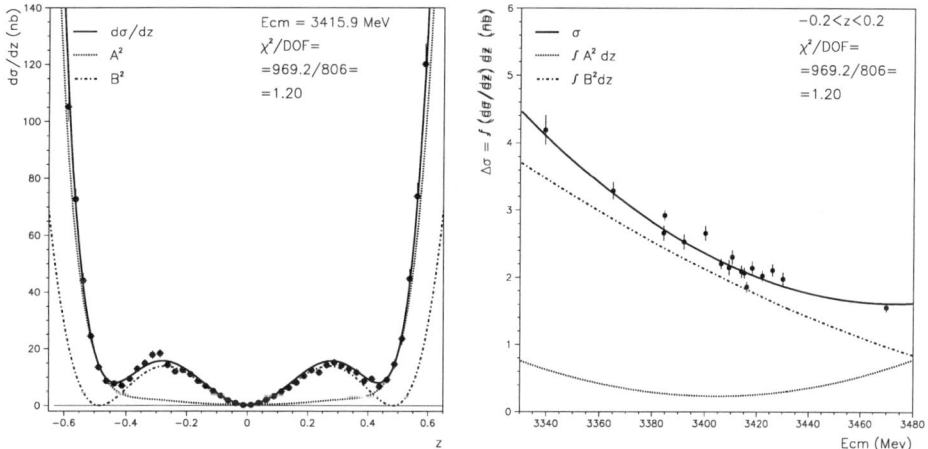

FIGURE 3. The $\bar{p}p \to \pi^0\eta$ cross section. Left: differential cross section versus z at $E_{CM} = 3415.9$ MeV. Right: integrated cross section (over $0 < z < 0.2$) versus E_{CM}.

expansion is used for the energy dependence as well.

It has to be stressed that the *pure* Breit-Wigner (the fictional cross section that would result if the non-resonant amplitudes could be turned off) is very small. In Fig. 1-right it is shown multiplied by 20 to make it comparable to the signal that we detect.

We observe a signal from the χ_{c0} in the $\eta\eta$ channel as well, as shown in Fig. 2. No signal from the χ_{c0} is observed in $\bar{p}p \to \pi^0\eta$ ($c\bar{c}$ is isospin-suppressed) and the fit shown in Fig. 3 is performed using Eq. 1 with A_R set to zero. Removing the constraint, the fit estimate of A_R is consistent with zero. The $\pi^0\eta$ channel provides a check on the systematics of the experiment.

E835 has significantly improved our knowledge of the χ_{c0} state, specifically $M_{\chi_{c0}}$, $\Gamma_{\chi_{c0}}$, $B(p\bar{p})$, $B(J/\psi\,\gamma)$, $B(\pi^0\pi^0)$, and $B(\eta\eta)$ (see Table 1). Accurate line widths for the 3 χ_{cJ} states are available in Table 2. The radiative transitions to the J/ψ are now in excellent agreement.

TABLE 1. E835 Results for the χ_{c0}.

$B_{in} \equiv$ $B_{out} \equiv$	$B(\chi_{c0} \to J/\psi\,\gamma)$ (^{a})	Common channel $B(\chi_{c0} \to \bar{p}p)$ $B(\chi_{c0} \to \pi^0\pi^0)$	$B(\chi_{c0} \to \eta\eta)$
$M_{\chi_{c0}}$ (MeV/c^2)	$3415.4 \pm 0.4 \pm 0.2$ (^{b})	$3414.7^{+0.7}_{-0.6} \pm 0.2$ (^{c})	$3412.2^{+2.1}_{-1.8} \pm 0.2$ (^{c})
$\Gamma_{\chi_{c0}}$ (MeV)	$9.8 \pm 1.0 \pm 0.1$ (^{b})	$8.6^{+1.7}_{-1.3} \pm 0.1$ (^{c})	$10.3^{+3.0}_{-3.1} \pm 0.1$ (^{c})
$B_{in} \times B_{out}$ (10^{-7})	$27.2 \pm 1.9 \pm 1.3$ (^{b})	$5.42^{+0.91}_{-0.96} \pm 0.22$ (^{c})	$4.1^{+1.2}_{-1.1}{}^{+0.5}_{-0.3}$ (^{c})
δ_A (degree)	–	$47 \pm 10 \pm 6$ (^{c})	$173^{+17}_{-19} \pm 6$ (^{c})
Final result for $B_{in} \times B_{out}$ (10^{-7}) and phase δ_A (degree)		$5.09 \pm 0.81 \pm 0.25$ (^{d}) $39 \pm 5 \pm 6$ (^{d})	$4.0 \pm 1.2^{+0.5}_{-0.3}$ (^{d}) $144 \pm 8 \pm 6$ (^{d})

(^{a}) The J/ψ was detected through its decay into e^+e^-.
(^{b}) From Ref. [3], where $B_{in} \times B_{out}$, $M_{\chi_{c0}}$ and $\Gamma_{\chi_{c0}}$ were free parameters.
(^{c}) This analysis with $B_{in} \times B_{out}$, $M_{\chi_{c0}}$ and $\Gamma_{\chi_{c0}}$ as free parameters.
(^{d}) This analysis with $M_{\chi_{c0}}$ and $\Gamma_{\chi_{c0}}$ fixed to values from Ref. [3].

TABLE 2. Comparison among the E1 radiative transitions of the χ_{cJ} states into J/ψ.

$c\bar{c}$	$B(\chi_{cJ} \to J/\psi\,\gamma)$ (%)	$\Gamma_{\chi_{cJ}}$ (MeV)	$\Gamma_{\chi_{cJ} \to J/\psi\gamma}$ (keV)	q_J (MeV)	$\left(\frac{q_J}{q_0}\right)^3$	$\frac{\Gamma_{\chi_{cJ} \to J/\psi\gamma}}{(q_J/q_0)^3}$ (keV)
χ_{c0}	1.33 ± 0.31	$9.8 \pm 1.0 \pm 0.1$	131 ± 33	303	1	131 ± 33
χ_{c1}	31.6 ± 3.2	0.92 ± 0.13	290 ± 50	389	2.10	138 ± 24
χ_{c2}	18.7 ± 2.0	2.08 ± 0.17	389 ± 52	430	2.86	136 ± 18

We have developed and proved the effectiveness of a technique for dealing with resonant/non-resonant interference and detecting a resonant signal in channels dominated by order-of-magnitude larger non-resonant cross section. Finally, we have gained insights into possible future strategies for attacking outstanding problems, namely the poor knowledge of the singlet $c\bar{c}$ states and the existence of hadromolecular $c\bar{c}q\bar{q}$ states.

This material is abstracted from the thesis of Dr. Paolo Rumerio. E835 thanks the staff of their respective institutions and the Antiproton Source Department of the Fermilab Beams Division. This research was supported by the US Department of Energy and the Italian Istituto Nazionale di Fisica Nucleare.

REFERENCES

1. M. Ambrogiani et al. [E835 Collab.], Phys. Rev. D **62**, 052002 (2000).
2. M. Andreotti et al. [Fermilab E835 Collaboration], "Interference Study of the $\chi_{c0}(1^3P_0)$ in the Reaction $\bar{p}p \to \pi^0\pi^0$," Accepted for publication on Phys. Rev. Lett. (July 2003).
3. S. Bagnasco et al. [E835 Collaboration], Phys. Lett. B **533**, 237 (2002).

CLEO Results in Upsilon Spectroscopy

Richard S. Galik

Newman Laboratory; Cornell University; Ithaca, NY 14853

Abstract. This presentation first updates several analyses of the CLEO Collaboration in the physics of the bound state Upsilon resonances: discovery and measurements of the 1^3D_J states, search for the singlets $1^0S_1(\eta_b)$, $2^0S_1(\eta_b')$, and $1^0P_1(h_b)$; and precision determination of Γ_{ee}. Then I present CLEO's preliminary results on di-pion transitions among the $^3S_1$ vector states and on the observation of the first hadronic transition in onia involving an ω, namely $\chi_b'(2^3P_J) \to \omega\Upsilon(1^3S_1)$.

INTRODUCTION AND DATA SAMPLES

As indicated in Fig. 1 the spectroscopy of the Υ system is potentially very rich. The depth of the QCD potential allows three n^3S_1 vector states below open-flavor threshold (compared to two in charmonium), includes a set "D" states (only stable L=2 $q\bar{q}$ states known), and has over 900 MeV of available energy for hadronic transitions. There are many lingering puzzles and undiscovered states. Further, analysis of the Υ system can provide tests of QCD predictions (especially lattice QCD), which could then enhance the usefulness of such theory inputs to the understanding of B physics.

The Cornell Electron Storage Ring (CESR) and the CLEO experimental apparatus have improved markedly since the last time data was taken by CLEO on the three narrow Υ vector resonances, which couple directly to the e^+e^- beams. The CLEO detector has evolved through CLEOII[1] to CLEOII.V[2] and on to CLEOIII[3], with which the data described here were collected. The total luminosity acquired was 1.2, 1.4 and 1.5 fb^{-1} at the $\Upsilon(1S)$, $\Upsilon(2S)$ and $\Upsilon(3S)$, respectively corresponding to roughly 20, 6, and 5 million produced events. At this point 100% of the $\Upsilon(1S)$ and $\Upsilon(3S)$ data have been processed as have 45% of the $\Upsilon(2S)$ data. Substantial data was also collected in the underlying four-quark continuum and in scans of the resonant line shapes.

WORK FIRST PRESENTED AT ICHEP'02

At the ICHEP meeting in Amsterdam, CLEO presented evidence for production of the 1^3D_J states via four photon cascade to the $\Upsilon(1S)$ (see Fig. 1) followed by its decay to lepton pairs.[4] The significance in the overall rate was over 9 standard deviations. The best fit was to a single state with mass of 10162.2 ± 1.6 MeV. The last 20% of the $\Upsilon(3S)$ data will be added and this analysis should be completed this summer.

Also at ICHEP'02 we showed[5] limits on the M1 radiative decays of the $\Upsilon(3S)$ to the singlets η_b and η_b'. These studies have been expanded to include searches (*i.e.*, only limits determined!) for the η_b in M1 decays of the $\Upsilon(2S)$ and transitions through the

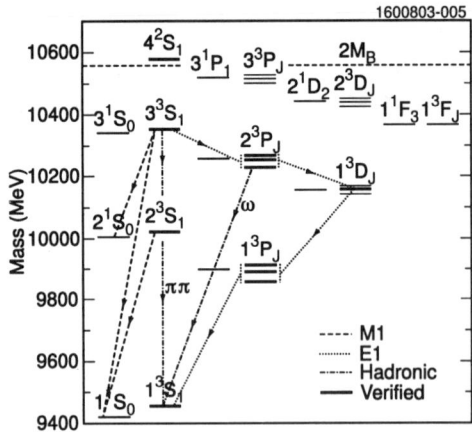

FIGURE 1. The $b\bar{b}$ spectrum, showing the transitions described in this report.

singlet "P" state, the h_b. These new results were reported at the 2003 APS meeting for a subset of the data; the full data set should be analyzed by Fall 2003.

MEASUREMENTS OF DI-LEPTON WIDTHS

The values of Γ_{ee} for each of the three vector bound-states will be a good test of the abilities of lattice QCD. Both theory and experiment should be able to push these widths, and even more so ratios among these widths, to the few percent level. Presently the PDG averages are known to 2%, 4%, and 9%, for the 1S, 2S and 3S, respectively. In collecting data at the three Υ energies, we carefully took multiple scans of the line shape, in that the *area* of the hadronic cross section in e^+e^- collisions is proportional to Γ_{ee}. These scans are shown in Fig. 2; statistical uncertainties are below 0.35% in each case.

The goal is to have overall systematic uncertainties below 2.5% for each resonance. Possible biases from backgrounds and energy calibration have been evaluated and found to contribute below 1% to the uncertainty. While not yet finished, it is believed that the acceptance can also be known to the 1% level. To achieve our goal the biggest effort will be in the systematic uncertainty on the luminosity; at present for CLEOIII this is at the 2.5% level, but was reduced in CLEOII to the 1% level, although with great effort.

DI-PION TRANSITIONS AMONG THE VECTOR STATES

There are four measured di-pion transitions among the states of quarkonia. When plotting the di-pion invariant mass, $m_{\pi\pi}$, the spectra for $\psi' \to \pi\pi(J/\psi)$, $\Upsilon(2S) \to \pi\pi\Upsilon(1S)$, and $\Upsilon(3S) \to \pi\pi\Upsilon(2S)$ all have a shape described nicely by multipole expansion of the gluon fields, peaking at high invariant mass. But the decay $\Upsilon(3S) \to \pi\pi\Upsilon(1S)$ is *different*, having two peaks in $m_{\pi\pi}$. Several explanations have been offered, but the experimental

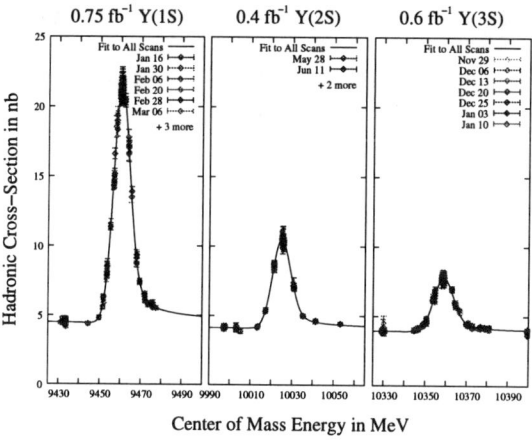

FIGURE 2. The hadronic cross section as a function of center-of-mass energy for each of the three narrow Υ resonances.

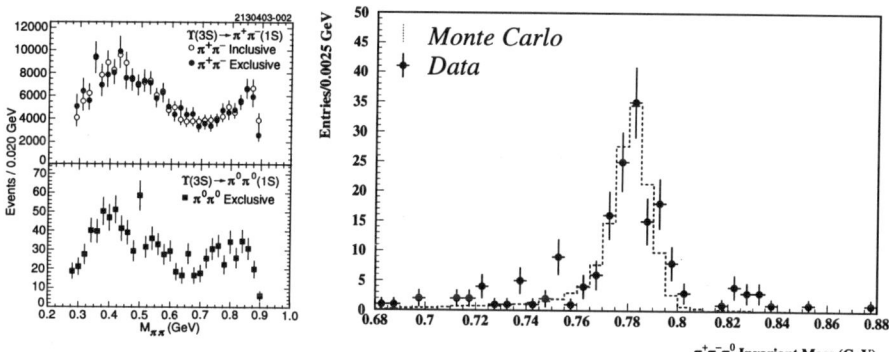

FIGURE 3. Left: The preliminary dipion invariant mass spectra for the decay $\Upsilon(3S) \to \pi\pi\Upsilon(1S)$. Right: The three-pion invariant mass for events consistent with $\Upsilon(3S) \to \gamma\pi\pi\pi\Upsilon(1S)$ showing evidence for the decay $\chi'_b \to \omega\Upsilon(1S)$.

data have been lacking the precision to distinguish among them.

Our present efforts on this $\Upsilon(3S)$ to $\Upsilon(1S)$ transition involve both *in*clusive and *ex*clusive (with $\Upsilon(1S) \to \ell^+\ell^-$) decays involving $\pi^+\pi^-$ and exclusive decays for the case of $\pi^0\pi^0$. Our preliminary spectra for $m_{\pi\pi}$ are shown in Fig. 3(a); these are corrected for detector acceptance. Work continues on the systematic uncertainties, which should considerably reduce the size of the error bars shown. We are also analyzing the angular distributions of the pions to determine the extent to which "D" waves contribute and whether the multipole expansion terms are sufficient to describe the decays.

FIRST OBSERVED TRANSITION INVOLVING AN ω MESON

CLEO has observed the first hadronic transition involving the $\chi_b(n^3P_J)$ states, the first hadronic transition in the $b\bar{b}$ system other than $\pi\pi$, and the first onium transition involving an ω meson. The mass splittings, as shown in Fig. 1, allow for the decay $\chi_{bJ} \to \Upsilon\omega$ for $J = 1, 2$ (but kinematically forbidden for $J = 0$).

The decay chain is $\Upsilon(3S) \to \gamma\chi'_{bJ}$, $\chi'_{bJ} \to \omega\Upsilon(1S)$, and $\Upsilon(1S) \to \ell^+\ell^-$. No particle identification is used. Events must have four or five tracks, with the two high momentum tracks assumed to be leptons and the others assumed to be pions (one spurious track is allowed to maximize efficiency). There must be at three or four showers, with one good π^0 found.

After demanding the di-lepton invariant mass be loosely consistent with that of the $\Upsilon(1S)$, we obtain the three pion invariant mass shown in Fig. 3(b). The data clearly show and ω with the proper mass and width as indicated from Monte Carlo simulations. Next, we form the mass recoiling against the three pions and the E1 photon and require this to again be consistent with the $\Upsilon(1S)$. Removing "cascade" events with requirements on the invariant mass of the two charged pions leaves us with 36 ± 6 signal events and backgrounds estimated at the level of one event.

The photon energy spectrum of these 36 events favors $J = 1$ over $J = 2$. Given that the efficiency in essentially independent of the parent J value leads us to a preliminary branching fraction of:

$$\mathcal{B}(\chi'_b \to \omega\Upsilon) = (2.30 \pm 0.40 \pm 0.16 \pm 0.14)\% \quad (1)$$

with the last uncertainty being from the other branching fractions in the decay chain. Publication is expected in late 2003.

ACKNOWLEDGMENTS

I wish to thank the many CLEO colleagues who helped me prepare this presentation, especially Tomasz Skwarnicki ("D" states), Hajime Muramatsu (M1 transitions), Jim Pivarski (Γ_{ee}), David Kreinick, Eric Engelson, Steve Pappas (all in the $\pi\pi$ transitions), and Todd Pedlar ($\pi\pi$ and ω transitions.)

REFERENCES

1. Y. Kubota et al., (CLEO Collaboration), *Nucl. Instrum. Meth.* A **320**, 66 (1992).
2. T. Hill, *Nucl. Instrum. Meth.* A **418**, 32 (1998).
3. G. Viehhauser, *Nucl. Instrum. Meth.* A **462**, 146 (2001); M. Artuso, et al., *Nucl. Instrum. Meth.* A **461**, 454 (2001); D. Peterson, et al., *Nucl. Instrum. Meth.* A **478**, 142 (2002).
4. S. E. Csorna et al., (CLEO Collaboration), ArXiv:hep-ex/0207060, submitted to ICHEP02.
5. A. H. Mahmood et al., (CLEO Collaboration), ArXiv:hep-ex/0207057, submitted to ICHEP02.

The Resummed Photon Spectrum in Radiative Upsilon Decays (And More)

Sean Fleming

Physics Department, Carnegie Mellon University, Pittsburgh PA 15213

Abstract. In this talk I present the results of two calculations that make use of Non-Relativistic QCD and the newly developed Soft-Collinear Effective Theory. The first process considered is inclusive radiative Υ decay. The second process considered is the leading color-octet contribution to $e^+e^- \to J/\psi + X$.

Bound states of heavy quarks and antiquarks have been of great interest since the discovery of the J/ψ [1, 2]. In particular the decay and production of quarkonium is an interesting probe of both perturbative and nonperturbative aspects of QCD dynamics. A systematic theoretical framework for handling the different scales characterizing both the decay and production of quarkonium is Non-Relativistic Quantum Chromodynamics (NRQCD) [3, 4]. NRQCD solves important conceptual as well as phenomenological problems in quarkonium theory. For instance, perturbative calculations of the inclusive decay rates for χ_c mesons in the color-singlet model suffer from nonfactorizable infrared divergences [5, 6, 7]. NRQCD provides a generalized factorization theorem so that infrared safe calculations of inclusive decay rates are possible [8]. In addition, color-octet production mechanisms are critical for understanding the production of J/ψ at large transverse momentum, $p_\perp$, at the Fermilab Tevatron [9, 10, 11]. There are still many challenging problems in quarkonium physics that remain to be solved [12]. One important problem is the polarization of J/ψ at the Tevatron. NRQCD predicts the J/ψ should become transversely polarized as the $p_\perp$ of the J/ψ becomes much larger than $2m_c$ [13, 14, 15, 16]. The theoretical prediction is consistent with the experimental data at intermediate $p_\perp$, but at the largest measured values of $p_\perp$ the discrepancies are at the 3σ level [17]. In this talk I present two additional puzzles where progress has been made lately: radiative Υ decay, and $e^+e^- \to J/\psi + X$.

Inclusive decays of quarkonium are understood in the framework of the operator product expansion (OPE), with power-counting rules given by NRQCD. The OPE for the direct photon spectrum of Υ decay is [3]

$$\frac{d\Gamma}{dz} = \sum_n C_n(M,z)\langle\Upsilon|\mathscr{O}_n|\Upsilon\rangle, \qquad (1)$$

where $z = 2E_\gamma/M$, with $M = 2m_b$. The C_i are short-distance coefficients, and the $\mathscr{O}$ are NRQCD operators. At leading order in v only one term in the sum must be kept, the so called color-singlet contribution.

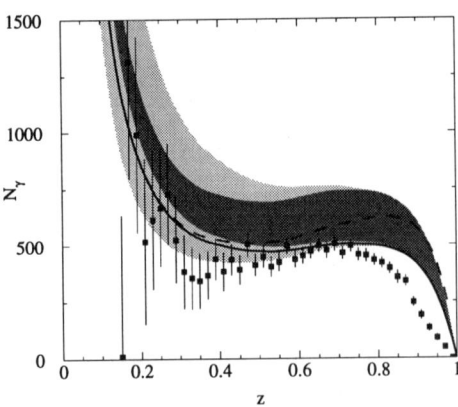

FIGURE 1. The inclusive radiative Υ photon spectrum, compared with data from CLEO [28].

This simple picture of the photon spectrum in inclusive Υ decays is only valid in the intermediate range of the photon energy spectrum ($0.3 \lesssim z \lesssim 0.7$). In the lower range, $z \lesssim 0.3$, photon-fragmentation contributions are important [18, 19]. At large values of the photon energy, $z \gtrsim 0.7$, both the perturbative expansion [19] and the OPE [20] break down.

The breakdown at the endpoint is a consequence of NRQCD not containing the correct low energy degrees of freedom. The effective theory which correctly describes this kinematic regime is a combination of NRQCD for the heavy degrees of freedom, and the soft-collinear effective theory (SCET) [21, 22, 23, 24] for the light degrees of freedom. In Refs. [25, 26, 27] SCET was applied to radiative Υ decay. A comparison of the calculation to CLEO [28] data is shown in Fig. 1.

The error bars on the data are statistical only. The dashed line is the direct tree-level and fragmentation result, and the solid curve is the sum of the interpolated resummed result and the fragmentation result. For these two curves we used the value of α_s extracted by CLEO from these data, $\alpha_s(M_\Upsilon) = 0.163$, which corresponds to $\alpha_s(M_Z) = 0.110$ [28]. We also show in this plot the interpolated resummed and fragmentation result, using the PDG value of $\alpha_s(M_Z)$, including theoretical uncertainties, denoted by the shaded region. The lighter band also includes the variation, within the errors, of the parameters for the quark to photon fragmentation function extracted by ALEPH [29].

New problems have arisen as a result of recent measurements of the spectra of J/ψ produced at the $\Upsilon(4S)$ resonance in e^+e^- collisions by the BaBar and Belle experiments [30, 31]. Leading order NRQCD calculations predict that for most of the range of allowed energies prompt J/ψ production should be dominated by color-singlet production mechanisms, while color-octet contributions dominate when the J/ψ energy is within a few hundred MeV of the maximum allowed. Furthermore, as pointed out in

Ref. [32], color-octet processes predict a dramatically different angular distribution for the J/ψ.

Experimental results do not agree with these expectations: the data does not exhibit any enhancement in the bins closest to the endpoint. However, the total cross section measured by the two experiments exceeds predictions based on the color-singlet model alone. The total prompt J/ψ cross section, which includes feeddown from ψ' and χ_c states but not from B decays, is measured to be $\sigma_{tot} = 2.52 \pm 0.21 \pm 0.21$ pb by BaBar, while Belle measures $\sigma_{tot} = 1.47 \pm 0.10 \pm 0.13$ pb. Estimates of the color-singlet contribution range from $0.4 - 0.9$ pb [33, 34, 35, 36]. Furthermore, the angular distribution disagrees with color-singlet result. These aspects of the data suggest that there is a substantial color-octet contribution which is not confined to the very endpoint.

In Ref. [37] the endpoint region is treated within the framework of NRQCD and SCET. The calculation depends on a nonperturbative function, and thus is not predictive. However, moments of the shape function are NRQCD operators whose size is constrained by the velocity scaling rules of NRQCD. Choosing a simple ansatz for the shape function whose moments are consistent with velocity scaling rules, one finds that the combined perturbative and nonperturbative effects lead to substantial broadening of the color-octet spectrum in a manner that is consistent with data.

In Fig. 2 I show the sum of the color-octet and color-singlet contributions as the upper line, and the color-singlet contribution only as the lower line. The color-octet matrix elements set the normalization. In the graph on the left they are chosen to be $\langle \mathscr{O}_8^{\psi}(^1S_0)\rangle = \langle \mathscr{O}_8^{\psi}(^3P_0)\rangle/m_c^2 = 1.3 \times 10^{-1}$ GeV^3. This is plotted against the BaBar data [31]. In the graph on the right they are chosen to be $\langle \mathscr{O}_8^{\psi}(^1S_0)\rangle = \langle \mathscr{O}_8^{\psi}(^3P_0)\rangle/m_c^2 = 6.6 \times 10^{-2}$ GeV^3, and is plotted against the Belle data [30].

While the calculations of Ref. [37] show that the leading color-octet contribution is broad enough to be compatible with the observed p_ψ distributions, other features of the e^+e^- data remain puzzling. In particular, Belle reports a large ratio of $J/\psi + c\bar{c}$ over inclusive J/ψ [38]. The predicted ratio from leading order color-singlet production mechanisms alone is at least a factor of three too small [33, 35] and a large color-octet contribution makes this ratio even smaller.

ACKNOWLEDGMENTS

I would like to thank my collaborators Adam Leibovich and Tom Mehen. This work is supported by Department of Energy grant number DOE-ER-40682-143.

REFERENCES

1. Aubert, J. J., et al., *Phys. Rev. Lett.*, **33**, 1404–1406 (1974).
2. Augustin, J. E., et al., *Phys. Rev. Lett.*, **33**, 1406–1408 (1974).
3. Bodwin, G. T., Braaten, E., and Lepage, G. P., *Phys. Rev. D*, **51**, 1125–1171 (1995).
4. Luke, M. E., Manohar, A. V., and Rothstein, I. Z., *Phys. Rev. D*, **61**, 074025 (2000).
5. Barbieri, R., Gatto, R., and Remiddi, E., *Phys. Lett. B*, **61**, 465 (1976).
6. Barbieri, R., Gatto, R., and Remiddi, E., *Phys. Lett.*, **B106**, 497 (1981).

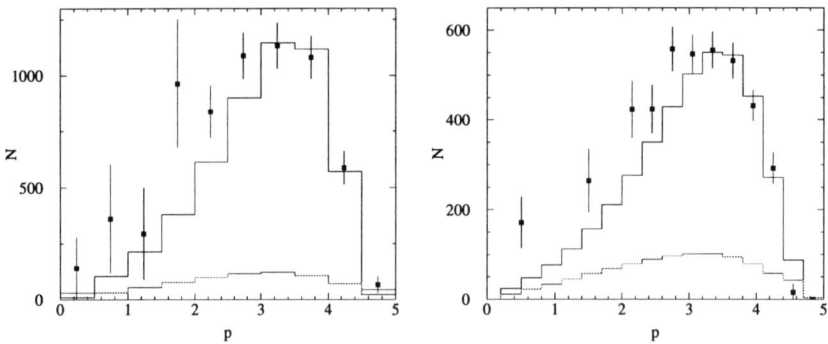

FIGURE 2. The sum of the color-octet and color-singlet contributions are plotted as the upper line. The lower line is the color-singlet contribution only. The graph on the left shows data from the BaBar collaboration [31]. The graph on the right shows data are from the Belle collaboration [30].

7. Barbieri, R., Caffo, M., Gatto, R., and Remiddi, E., *Nucl. Phys.*, **B192**, 61 (1981).
8. Bodwin, G. T., Braaten, E., and Lepage, G. P., *Phys. Rev.*, **D46**, 1914–1918 (1992).
9. Braaten, E., and Fleming, S., *Phys. Rev. Lett.*, **74**, 3327–3330 (1995).
10. Cho, P. L., and Leibovich, A. K., *Phys. Rev.*, **D53**, 150–162 (1996).
11. Cho, P. L., and Leibovich, A. K., *Phys. Rev.*, **D53**, 6203–6217 (1996).
12. Bodwin, G. T. (2002).
13. Cho, P. L., and Wise, M. B., *Phys. Lett.*, **B346**, 129–136 (1995).
14. Leibovich, A. K., *Phys. Rev.*, **D56**, 4412–4415 (1997).
15. Beneke, M., and Kramer, M., *Phys. Rev.*, **D55**, 5269–5272 (1997).
16. Braaten, E., Kniehl, B. A., and Lee, J., *Phys. Rev.*, **D62**, 094005 (2000).
17. Affolder, T., et al., *Phys. Rev. Lett.*, **85**, 2886–2891 (2000).
18. Catani, S., and Hautmann, F., *Nucl. Phys. Proc. Suppl.*, **39BC**, 359–363 (1995).
19. Maltoni, F., and Petrelli, A., *Phys. Rev.*, **D59**, 074006 (1999).
20. Rothstein, I. Z., and Wise, M. B., *Phys. Lett.*, **B402**, 346–350 (1997).
21. Bauer, C. W., Fleming, S., and Luke, M. E., *Phys. Rev.*, **D63**, 014006 (2001).
22. Bauer, C. W., Fleming, S., Pirjol, D., and Stewart, I. W., *Phys. Rev.*, **D63**, 114020 (2001).
23. Bauer, C. W., and Stewart, I. W., *Phys. Lett.*, **B516**, 134–142 (2001).
24. Bauer, C. W., Pirjol, D., and Stewart, I. W., *Phys. Rev.*, **D65**, 054022 (2002).
25. Bauer, C. W., Chiang, C.-W., Fleming, S., Leibovich, A. K., and Low, I., *Phys. Rev.*, **D64**, 114014 (2001).
26. Fleming, S., and Leibovich, A. K., *Phys. Rev. Lett.*, **90**, 032001 (2003).
27. Fleming, S., and Leibovich, A. K., *Phys. Rev.*, **D67**, 074035 (2003).
28. Nemati, B., et al., *Phys. Rev.*, **D55**, 5273–5281 (1997).
29. Buskulic, D., et al., *Z. Phys.*, **C69**, 365–378 (1996).
30. Abe, K., et al., *Phys. Rev. Lett.*, **88**, 052001 (2002).
31. Aubert, B., et al., *Phys. Rev. Lett.*, **87**, 162002 (2001).
32. Braaten, E., and Chen, Y.-Q., *Phys. Rev. Lett.*, **76**, 730–733 (1996).
33. Cho, P. L., and Leibovich, A. K., *Phys. Rev.*, **D54**, 6690–6695 (1996).
34. Yuan, F., Qiao, C.-F., and Chao, K.-T., *Phys. Rev.*, **D56**, 321–328 (1997).
35. Baek, S., Ko, P., Lee, J., and Song, H. S., *J. Korean Phys. Soc.*, **33**, 97–101 (1998).
36. Schuler, G. A., *Eur. Phys. J.*, **C8**, 273–281 (1999).
37. Fleming, S., Leibovich, A. K., and Mehen, T. (2003).
38. Abe, K., et al., *Phys. Rev. Lett.*, **89**, 142001 (2002).

The enhancement by the axial anomaly of the decay $\Upsilon(1D) \to \eta \, \Upsilon(1S)$

M.B. Voloshin

William I. Fine Theoretical Physics Institute, University of Minnesota, Minneapolis, MN 55455
and
Institute of Theoretical and Experimental Physics, Moscow, 117259

Abstract. It is shown that due to the enhancement by the anomaly in the flavor singlet axial current the rates of the decays $\Upsilon(1^3D_1) \to \eta \, \Upsilon(1S)$ and $\Upsilon(1^3D_2) \to \eta \, \Upsilon(1S)$ should be comparable to and likely exceed that of the recently discussed in the literature two-pion transition $\Upsilon(1D) \to \pi\pi \, \Upsilon(1S)$.

The D-wave states in the family of the $b\bar{b}$ resonances present a new interesting testing ground for the study of heavy quark dynamics. The ^3D_J states with $J = 1,2,3$ should form a closely spaced triplet of resonances, of which one (most likely the $^3D_2$) with the mass of about $10.16 \, GeV$ has been recently observed[1] in the CLEO experiment through the radiative transitions to and from the D wave state. It is clear however that similarly to other excited $b\bar{b}$ resonances there should also be strong-interaction transitions from the D states to lower resonances with emission of light mesons, i.e. of two pions, η, and also a weaker isospin violating transition with emission of π^0. In particular the transitions of the type $\Upsilon(1D) \to \pi\pi \, \Upsilon(1S)$ were discussed in the literature[2, 3, 4] in some detail. The amplitudes of such transitions between the $^3S_1$ states of heavy quarkonium and the pattern of the rates were understood long ago[5, 6, 7] within the general method of describing the hadronic transitions in heavy quarkonium using the multipole expansion[8, 9] in QCD for the interaction of the heavy quarkonium with soft gluonic field. In the transitions between the $^3S_1$ heavy resonances the relevant amplitudes for production of the light mesons are determined[5] by the low energy theorems arising from the quantum anomalies in QCD: the emission of an S wave pair of pions is dominated by the anomaly in the trace of the energy-momentum tensor[10], while the P wave emission of the η is regulated by the anomaly in the flavor singlet axial current[11]. The anomalous contribution greatly enhances both rates[5] in agreement with the available data.

The purpose of the present talk is to point out that the relation between the rates of the two-pion and η transitions from $\Upsilon(1^3D_{1,2})$ to $\Upsilon(1S)$ should be quite different from the pattern observed in the transitions between the $^3S_1$ states. Namely, the P wave emission of η in transitions to the $^3S_1$ state from the 1^3D_J resonances with $J = 1$ and $J = 2$, is enhanced by the axial anomaly in QCD, while the D wave emission of pion pairs decouples from the conformal anomaly. As a result the rate of the η transitions should be comparable to that of the two-pion ones, and in fact is quite likely to be the largest among the hadronic transitions from the ^3D_J states. A more definite quantitative estimate of the ratio of the rates, $\Gamma(1^3D_{1,2} \to \eta\, 1^3S_1)/\Gamma(1D \to \pi\pi\, 1S)$, is hindered by the present poor understanding of a parameter governing the non-anomalous amplitude of production of the pion pair by gluonic operators.

Within the QCD mltipole expansion the discussed two-pion transition arises in the second order in the $E1$ interaction with the chromoelectric gluon field $\vec{E}^a$ described by the Hamiltonian

$$H_{E1} = -\frac{1}{2}\xi^a \vec{r} \cdot \vec{E}^a(0)\,, \tag{1}$$

where $\xi^a = t_1^a - t_2^a$ is the difference of the color generators acting on the quark and antiquark (e.g. $t_1^a = \lambda^a/2$ with λ^a being the Gell-Mann matrices), and $\vec{r}$ is the vector for relative position of the quark and the antiquark.

The transitions of the type $^3D_J \to \eta\,^3S_1$ are induced by the interference of the $E1$ interaction in eq.(1) with the $M2$ term containing the chromomagnetic field $\vec{B}^a$ and described by the Hamiltonian

$$H_{M2} = -(4m_Q)^{-1}\xi^a S_j r_i \left(D_i B_j(0)\right)^a, \tag{2}$$

where $\vec{D}$ is the QCD covariant derivative, m_Q is the heavy quark mass, and $\vec{S} = (\vec{\sigma}_1 + \vec{\sigma}_2)/2$ is the operator of the total spin of the quark-antiquark pair. It should be noted that the $M1$ term, formally of a lower order in the multipole expansion, is proportional to the spin-flip operator $(\vec{\sigma}_1 - \vec{\sigma}_2)$ and thus does not contribute to transitions between states with the same total spin.

Using the expressions (1) and (2) the transition amplitudes are found in the standard way:

$$A_{\pi\pi} \equiv A(^3D_J \to \pi\pi\,^3S_1) = \langle \pi\pi | E_i^a E_j^a | 0 \rangle A_{ij}\,, \tag{3}$$

$$A_\eta^{(J)} \equiv A(^3D_J \to \eta\,^3S_1) = m_Q^{-1} \left\langle \eta \left| E_i^a (D_j B_k)^a + (D_j B_k)^a E_i^a \right| 0 \right\rangle A_{ijk}^{(J)}, \tag{4}$$

where A_{ij} and $A_{ijk}^{(J)}$ are the heavy quarkonium amplitudes, defined as

$$A_{ij} = \frac{1}{32}\langle 1S | \xi^a r_i \mathcal{G} r_j \xi^a | 1D \rangle \tag{5}$$

and

$$A_{ijk}^{(J)} = \frac{1}{64}\langle ^3S_1 | \xi^a r_i \mathcal{G} r_j \xi^a S_k |^3D_J \rangle \tag{6}$$

with $\mathcal{G}$ being the Green's function of the heavy quark pair in a color octet state.

Using the decoupling in a non-relativistic heavy quarkonium of the spin and orbital degrees of freedom, one can write the amplitudes A_{ij} and $A_{ijk}^{(J)}$ in a form with explicitly factorized spin and orbital components in the Cartesian coordinates. For this representation we denote as ζ_i and χ_i the spin polarization amplitude of respectively the initial ^3D state and the final $^3S_1$ state, and as ψ_{ij} the orbital polarization amplitude of the $L=2$ wave in the initial D states. The tensor ψ_{ij} is symmetric and traceless, as appropriate for an $L=2$ state. In this notation the amplitude A_{ij} is proportional to ψ_{ij} and can thus be written in terms of a scalar quantity A_2 as $A_{ij} = \psi_{ij} A_2$, while the amplitude $A^{(J)}_{ijk}$ is expressed in terms of the same A_2 as

$$A^{(J)}_{ijk} = \frac{i}{2} \varepsilon_{klm} \chi_l^* P^{(J)} \psi_{ij} \zeta_m A_2 , \qquad (7)$$

where $P^{(J)}$ is the projector on states with definite J, acting on the product of the spin and orbital polarization amplitudes $\psi_{ij} \zeta_m$.

The quantity A_2 depends on details of dynamics of heavy quarkonium, and at present is highly model-dependent. For this reason a prediction of the absolute rates of the discussed decays involves a considerable uncertainty. Clearly, however, A_2 cancels in the considered here ratio of the rates of the two-pion and η transitions, which is thus determined by the ratio of the matrix elements entering the equations (3) and (4), describing the production by the gluon operators of the corresponding light meson states.

The gluonic matrix element for the two-pion production in eq.(3) multiplies the traceless tensor ψ_{ij} and thus receives no contribution from the (enhanced) trace anomaly in QCD. Rather this matrix element is parameterized[6, 2] in terms of the QCD coupling α_s and the parameter ρ_G introduced in Ref.[6] as 'the fraction of the pion momentum carried by gluons'. Using this parameterization, one can write

$$A_{\pi^+\pi^-} = \langle \pi^+ \pi^- | E^a_i E^a_j | 0 \rangle \psi_{ij} (\chi_k^* \zeta_k) A_2 = 4\pi \alpha_s \rho_G p_i^+ p_j^- \psi_{ij} (\chi_k^* \zeta_k) A_2 , \qquad (8)$$

where the final state with charged pions is assumed for definiteness, and $p^\pm$ stand for the momenta of the pions in the heavy quarkonium rest frame.

The matrix element of the gluonic operators in eq.(4) can be found using the fact that the amplitude of general Lorentz structure $\langle \eta | G^a_{\mu\nu} (D_\rho G_{\lambda\sigma})^a | 0 \rangle$ is reduced[12] to the total derivative of the amplitude governed by the axial anomaly[11] $\langle \eta | G^a \tilde{G}^a | 0 \rangle = 8\pi^2 \sqrt{2/3} f_\eta m_\eta^2$:

$$i \langle \eta(p) | G^a_{\mu\nu} (D_\rho G_{\lambda\sigma})^a | 0 \rangle = -\frac{4\pi^2}{15} \sqrt{\frac{2}{3}} f_\eta m_\eta^2 \left[p_\rho \varepsilon_{\mu\nu\lambda\sigma} + \frac{1}{2} \left(p_\lambda \varepsilon_{\mu\nu\rho\sigma} - p_\sigma \varepsilon_{\mu\nu\rho\lambda} \right) \right] , \qquad (9)$$

thus determining the amplitude of the η transition as

$$A^{(J)}_\eta = \frac{8\pi^2}{15 m_Q} \sqrt{\frac{2}{3}} f_\eta m_\eta^2 p_i \varepsilon_{jlm} \chi_l^* P^{(J)} \psi_{ij} \zeta_m A_2 . \qquad (10)$$

The amplitude in eq.(10) for the P wave emission of the η naturally vanishes for the transitions from the $J = 3$ state, while after performing the projection on the states with $J = 1$ and $J = 2$ one finds the relation between the transition rates from these states:

$$\Gamma(^3D_1 \to \eta\, ^3S_1) = \frac{5}{9} \Gamma(^3D_2 \to \eta\, ^3S_1) . \qquad (11)$$

It should be noted that this relation is obtained in the limit where all effects of the spin-dependent interaction in the heavy quarkonium are neglected. The ignored effects include in particular the fine-structure splitting between the masses of the $^3D_2$ and $^3D_1$. However this splitting is expected to be quite small, not larger than about $10\,MeV$ (for a summary of potential model predictions see e.g. Ref.[13]). Thus if the difference in the kinematical factors p_η^3 in the decay rate is used as a representative measure of the contribution of the unaccounted corrections, one might expect that the accuracy of the relation (11) should be about 10%.

The relation between the rates of η and two-pion transitions is found from the equations (8) and (10), and after integration over the phase space one finds in particular

$$\frac{\Gamma(\Upsilon(1^3D_2) \to \eta\,\Upsilon)}{\Gamma(\Upsilon(1^3D_2) \to \pi^+\pi^-\Upsilon)} = \frac{448}{15} \frac{\pi^4}{0.44\,(\alpha_s\rho_G)^2} \frac{f_\eta^2 m_\eta^4 p_\eta^3}{m_b^2 \Delta^7} \approx \left(\frac{0.64}{\alpha_s\rho_G}\right)^2 \quad (12)$$

The numerical estimate corresponds to rather conservative values of f_η and m_b: $f_\eta \approx f_\pi \approx 130\,MeV$, $m_b \approx 5\,GeV$. Clearly, the main uncertainty in evaluation of the ratio of the decay rates comes from the poor knowledge of the dimensionless parameter $\alpha_s\rho_G$ with both factors normalized at a scale μ set by the characteristic size of the quarkonium. The estimates of the relevant value of this parameter range from $\alpha_s\rho_G \approx 0.2$[2] to $\alpha_s\rho_G \approx 0.59$[3], with a realistic value likely being close to 0.35. In either case, the numerical result in the equation (12) predicts that the η transition rate should be not smaller, but most plausibly larger, than the rate of the transition with the emission of two pions.

ACKNOWLEDGMENTS

A part of this text was prepared at the Aspen Center for Physics. This work is supported in part by the DOE grant DE-FG02-94ER40823.

REFERENCES

1. S.E. Csorna et. al. [CLEO Collaboration], CLEO Report No. CLEO-CONF-02-06, July 2002; [hep-ex/0207060].
2. P. Moxhay, Phys.Rev. **D37**, 2557 (1988).
3. P. Ko, Phys.Rev. **D47**, 208 (1993).
4. J.L. Rosner, U. Chicago report No. EFI 03-06, February 2003; [hep-ph/0302122].
5. M. Voloshin and V. Zakharov, Phys.Rev.Lett. **45**, 688 (1980).
6. V.A. Novikov and M.A. Shifman, Zeit.Phys. **C8**, 43 (1981).
7. B.L. Ioffe and M.A. Shifman, Phys.Lett. **95B**, 99 (1980).
8. K. Gottfried, Phys.Rev.Lett. **40**, 598 (1978).
9. M.B. Voloshin, Nucl.Phys. **B154**, 365 (1979).
10. R. Crewther, Phys.Rev.Lett. **28**, 1421 (1972);
    M. Chaniwitz and J. Ellis, Phys.Lett.**40B**, 397 (1972);
    J. Collins, L. Duncan, and S. Joglekar, Phys.Rev. **D16**, 438 (1977).
11. D.J. Gross, S.B. Treiman, and F. Wilczek, Phys.Rev. **D19**, 2188 (1979);
    V.A. Novikov et.al., Nucl.Phys. **B165**, 55 (1980).
12. M.B. Voloshin, Phys.Lett. **B562**, 68 (2003).
13. S. Godfrey and J.L. Rosner, Phys.Rev. **D64**, 097501b(2001); **D66** 059902 (2002) (E).

QCD Spectroscopy, Structure, and Dynamics

Parallel Session Summary for CIPANP 2003

Jim Napolitano* and James Russ[†]

*Department of Physics, Applied Physics, and Astronomy
Rensselaer Polytechnic Institute
Troy, NY 12180-3590
[†]Department of Physics
Carnegie Mellon University
Pittsburgh, PA 15213

Abstract. We summarize the parallel sessions on QCD Spectroscopy, Structure, and Dynamics, for the 2003 Conference on the Intersections of Particle and Nuclear Physics, held in New York City, 19-24 May 2003.

OVERVIEW

The parallel program on QCD Spectroscopy, Structure, and Dynamics for CIPANP03 covered six sessions, each about 100 minutes long. There were thirty presentations, most of which were contributed. Nineteen presentations were mostly concerned with experiment, while eleven were theoretical. We heard lots of exciting, new developments. It is unfortunate that we couldn't cover them all in the summary presentation or in this contribution to the proceedings, given time and page constraints.

This summary highlights only a few of the presentations. In all cases, refer to the other contributions to these proceedings for details, including references.

NEW NARROW STATES

Several new, narrow states were presented in these sessions. Just prior to the conference, BaBAR announced the discovery of a new meson decaying to $D_s \pi^0$. This discovery was quickly confirmed by CLEO, who went further and identified a second narrow state decaying to $D_s^* \pi^0$. Both results were confirmed by BELLE. These results were covered in talks by Ray Cowan, Jon Urheim, and Tom Browder, representing BaBAR, CLEO, and BELLE respectively. Possibly related to these new meson states are new doubly charmed baryons, discovered by SELEX and presented by Peter Cooper. Ken Hicks and Stepan Stepanyan showed tantalizing evidence for a new pentaquark baryon $\Theta^+(1540)$.

The signals for the new mesons are shown in Fig. 1. These mesons decay via isospin violation to $D_s \pi^0$ and $D_s^* \pi^0$ respectively, and this leads to their narrow widths. The $D_{sJ}^*(2317)$ lies just below the DK threshold. The $D_{sJ}^*(2463)$ lies just below D^*K threshold, and does not decay to DK presumably because it has unnatural parity. This leads one

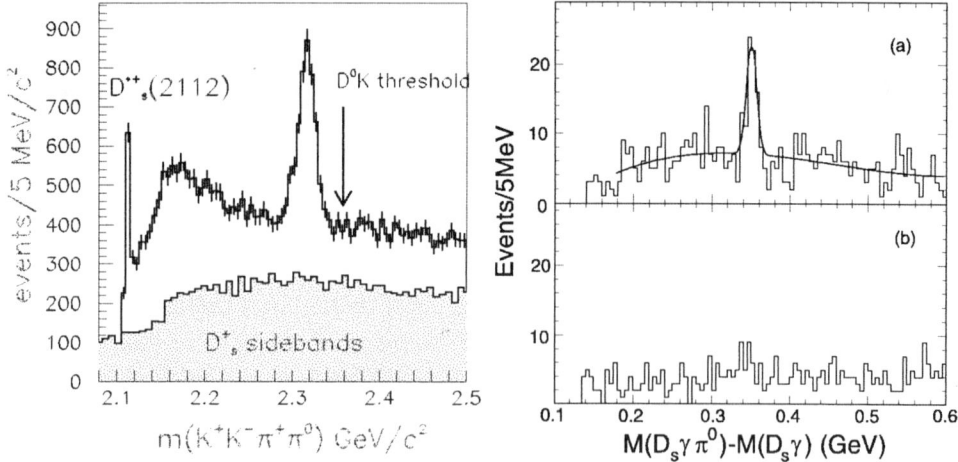

FIGURE 1. Observations of new narrow mesons with charm and strangeness, from BaBAR and CLEO. See talks by Ray Cowan and Jon Urheim respectively. On the left (from BaBAR) is the $K^+K^-\pi^+\pi^0$ mass for $D_s \to K^+K^-\pi^+$ and for the D_s sidebands. In addition to the well known $D_s^*(2112)$, a new state $D_{sJ}^*(2317)$ is observed. The right plot (from CLEO) shows the $D_s\gamma\pi^0$ mass for $D_s^*(2112) \to D_s\gamma$ (a) and in the $D_s^*(2112)$ sidebands (b). A second new state appears, the $D_{sJ}^*(2463)$.

to view the two new states as 0^+ and 1^+ "partners" of the D_s and D_s^*. Indeed, their mass difference (146 MeV) is very close to the D_s/D_s^* mass difference (142 MeV). (In fact, this leads to a potentially serious false $D_{sJ}^*(2463)$ signal, but CLEO investigated this in detail and concludes that such contamination cannot account for their observations.)

A variety of explanations for these states have been put forth. Harry Lipkin argued that they are DK and D^*K "molecular" states. A different explanation, from Bardeen, Eichten, and Hill (BEH), argues that the "partners" are required by an implementation of $SU(3)_L \times SU(3)_R$ chiral symmetry in heavy-light quark systems.

The BEH picture may also apply to new results from SELEX who observe "partners" of their previously reported doubly charmed baryons. (These states are naturally narrow since they cannot decay via the strong interaction.) Their results are summarized in Fig. 2. In Heavy Quark Effective Theory, the cc combination is equivalent to a heavy antiquark, so far as the symmetry is concerned, so these baryons behave like heavy/light $Q\bar{q}$ mesons, similar to the D_{sJ} system. The 78 MeV splitting in this case would not be the chromomagnetic fine structure splitting computed in potential models because the upper states also decay weakly, rather than by an $M1$ photon. The experimenters have some evidence that the splitting may be due to orbital angular momentum in the cc system.

A different, but probably unrelated, observation of a new narrow state $\Theta^+(1540)$ was reported by Ken Hicks for the SPRing-8 collaboration, with corroborating evidence given by Stepan Stepanyan for the CLAS collaboration, both using the reaction $\gamma n \to \Theta^+ K^-$. This state, which decays $\Theta^+ \to nK^+$, is noteworthy because it implies $uudd\bar{s}$, i.e. "pentaquark", degrees of freedom. The signals are shown in Fig. 3. Backgrounds are from competing reactions, in particular $\gamma p \to \Lambda(1520)K^+$ on protons in the nuclear target. Work is ongoing in the collaborations, and new results should be available shortly.

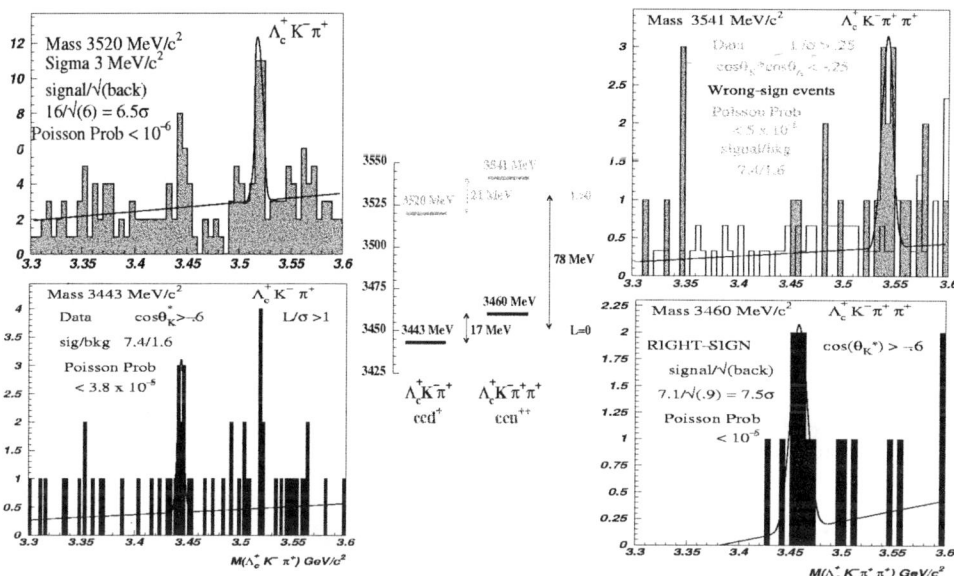

FIGURE 2. New results from SELEX on doubly charmed baryons, as reported by Peter Cooper. A pair of isodoublets is suggested, where in both the *ccd* and *ccu* cases, the first orbital excitation is at ~ 78 MeV. This may be consistent with the chiral symmetry mechanism put forth by Bardeen, Eichten, and Hill.

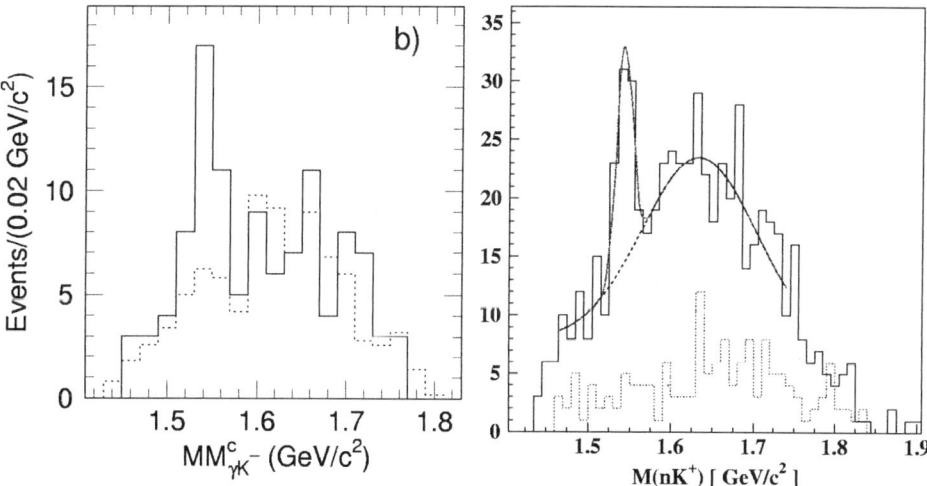

FIGURE 3. Evidence for the "pentaquark" $\Theta^+(1540)$ from the SPRing-8 collaboration (left), presented by Ken Hicks, and from the CLAS collaboration (right), presented by Stepan Stepanyan. In both cases, the signal for $\Theta^+ \to nK^+$ is derived from γn interactions. Backgrounds are determined by interactions with a hydrogen target (SPRing-8) and from events associated with $\Lambda(1520) \to pK^-$ production (CLAS).

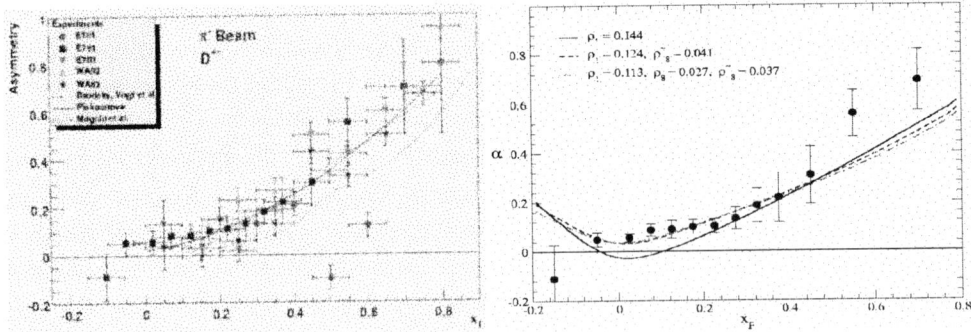

FIGURE 4. Data, presented by Maurizio Iori, and calculations by Tom Mehen on a Heavy Quark Recombination description of charm production dynamics. Plotted here is the $D^{\pm}$ asymmetry in the reaction $\pi^- p \to D^{\pm} X$ as a function of Feynman x from a variety of experiments, with a subset compared to the HQR prediction.

HEAVY QUARK DYNAMICS

Maurizio Iori and Erik Gottschalk gave us detailed summaries of heavy quark production data from hadro- and photo-production reactions, respectively. Figure 4 shows the inclusive $\pi^- p$ production asymmetry $\alpha \equiv [D^- - D^+]/[D^- + D^+]$ from a variety of experiments, along with a prediction by Tom Mehen, based on Heavy Quark Recombination (HQR) theory. Not only is the agreement quite good, but unlike Monte Carlo simulations such as PYTHIA, HQR attempts a more fundamental calculation from QCD. This calculation is based on an expansion in Λ_{QCD}/m_c, and leads to a more economical description of this phenomenon.

CHARMONIUM SPECTROSCOPY

New data on two $c\bar{c}$ states were also presented, summarized in Fig. 5. As part of a detailed review talk on two photon physics, David Urner showed results from CLEO on the long sought after η_c'. Produced through $\gamma\gamma \to \eta_c' \to K_S K^{\pm} \pi^{\mp}$, a signal appears above background with a (preliminary) mass of 3642.7 ± 4.0 MeV, in fair agreement with earlier measurements. This process is also observed by BaBAR.

Jerry Rosen, representing FNAL E835, showed data on the reaction $\bar{p}p \to \pi^0 \pi^0$ in the neighborhood of the $\chi_{c0}(3415)$. Although the resonant amplitude is much smaller than the nonresonant background, interference makes it possible to precisely extract parameters of the resonance using a partial wave expansion. (Also displayed in Fig. 5 is the fictional signal that would result if interference were not exploited, shown at 20 times its size.) E835 measurements significantly improve our knowledge of the basic χ_{c0} parameters M, Γ, $B(p\bar{p})$, $B(J/\psi\gamma)$, $B(\pi^0\pi^0)$, and $B(\eta\eta)$. Seon-Hee Seo also presented E835 results, on a measurement of the relative helicity amplitude ratio in the reaction $\bar{p}p \to \psi' \to e^+ e^-$.

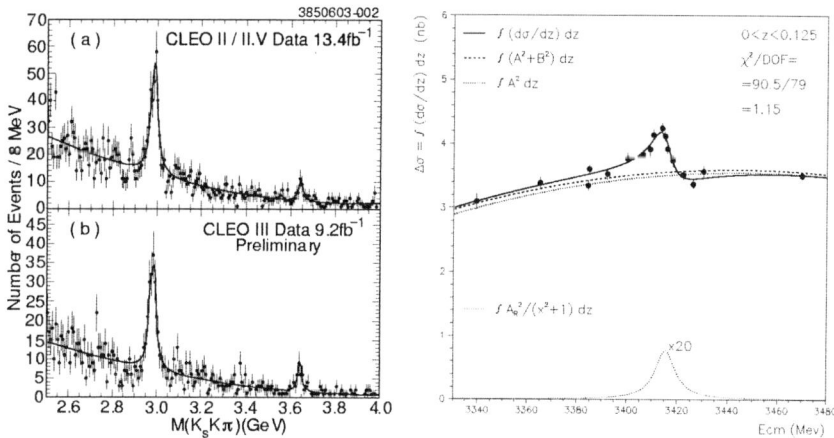

FIGURE 5. Results on charmonium spectroscopy. On the left, from David Urner's review of $\gamma\gamma$ processes, is the observation of the $\eta_c(2981)$ and the apparent discovery of the $\eta'_c(3643)$. On the right, data presented by Jerry Rosen, on an interferometric study of $\bar{p}p \to \pi^0\pi^0$ yielding data on the $\chi_{c0}(3415)$.

OTHER PRESENTATIONS

Curtis Meyer surveyed the status of mesons with exotic quantum numbers, including interpretations of $\eta\pi$ and $\eta'\pi$ signals. Gary Adams presented new results from BNL E852, and Mina Nozar showed work in progress on $\gamma p \to 3\pi N$ at Jefferson Lab. Ryan Mitchell discussed a new state from FOCUS decaying to K^+K^- with mass 1750 MeV. Philip Page gave new results on selection rules for QCD exotic mesons.

Mats Selen summarized hadronic D decays, including $D \to 3\pi$. FNAL E791 requires an isoscalar, scalar $\pi\pi$ component to fit this data, while CLEO sees no such term.

Rich Galik showed new results from CLEO on Υ spectroscopy. This was complemented by Mikhail Voloshin who discussed the role of the axial anomaly in the $\Upsilon(1D) \to \eta\Upsilon(1S)$ transition, and by Sean Fleming who presented calculations on Υ continuum radiative decay.

Colin Morningstar reviewed lattice QCD, with emphasis on the baryon spectrum and emerging unquenched calculations. Matthew Bellis told us of new measurements on the (light quark) baryon spectrum using the reaction $\gamma p \to p\pi^+\pi^-$ with CLAS.

Martin Olsson discussed reducing the QCD string to a time component vector potential. Dierdre Black gave us insight into the implementation of scalars in calculations of $\eta \to 3\pi$. Brett Van de Sande showed results on a transverse lattice treatment of QCD. Dan Pirjol and Boris Gelman discussed baryon spectroscopy calculated in large-N_c.

ACKNOWLEDGMENTS

Many thanks to all of our speakers and session convenors. We are also very grateful for all of the work of the conference organizers, especially Zohreh Parsa and Bill Marciano.

SPIN

Polarized Parton Distributions Measured at the HERMES Experiment

Jürgen Wendland (*for the HERMES collaboration*)

Department of Physics, Simon Fraser University, Burnaby BC V5A 1S6, Canada

Abstract. The HERMES collaboration has measured double spin asymmetries using polarized deep inelastic electron scattering. Inclusive and semi-inclusive asymmetries were measured from proton and deuteron targets with high precision. Asymmetries of pions from the proton and of pions and kaons from the deuteron were measured for the first time. Polarized parton densities of the u, $\bar{u}$, d, $\bar{d}$, $(s+\bar{s})$ flavours were extracted from the inclusive and semi-inclusive spin asymmetries in a LO analysis. The densities of the up (down) quarks were determined with good precision to be positive (negative). The polarized density of the sea quarks is consistent with zero.

Polarized deep inelastic scattering (DIS) is a powerful tool for the investigation of the nucleon spin structure. Measurements by the European Muon Collaboration (EMC) first indicated that only a small fraction of the nucleon spin is due to the spin of the quarks [1]. Experiments that followed [2, 3, 4, 5, 6] also found a significantly smaller value for the fraction of the nucleon spin carried by the quark spins than naively expected.

The HERMES experiment [7] at DESY was designed to perform precise measurements of the quark polarizations. The experiment uses the polarized 27.5 GeV electron or positron beam of the HERA accelerator in combination with an internal polarized gaseous target of 3-helium, hydrogen, or deuterium. With its large forward acceptance and good particle identification, the spectrometer is well suited to measure inclusive reactions where only the scattered positron is detected, and semi-inclusive processes where a hadron is detected in coincidence. A threshold Čerenkov detector provided identification of pions in the hydrogen data. A ring-imaging Čerenkov (RICH) detector installed in 1998 identified pions, kaons, and protons in the deuterium data.

In semi-inclusive DIS (SIDIS) the measurable kinematic variables are the momenta $k = (E, \vec{k})$ and $k' = (E', \vec{k}')$ of the incident and scattered lepton respectively, the momentum of the final state hadron $p_h = (E_h, \vec{p}_h)$. The initial momentum of the target nucleon is $P = (M, \vec{0})$ in the case of a fixed target. Common SIDIS variables are the negative squared invariant mass of the virtual photon $Q^2 = -(k-k')^2$, the Bjorken scaling variable $x = Q^2/(2M\nu)$, where $\nu = E - E'$, and the fractional energy of the hadron with respect to the virtual photon energy $z = (E_h/\nu)$.

The measured semi-inclusive lepton-nucleon asymmetry $A_{\parallel}^h$ of aligned and anti-aligned beam and target polarizations is related to the photon-nucleon asymmetry A_1^h,

$$\frac{A_{\parallel}^h(x,Q^2)}{D(1-\eta\gamma)} \simeq A_1^h(x,Q^2) \simeq \frac{1+R(x,Q^2)}{1+\gamma^2} \frac{\sum_q e_q^2 \Delta q(x,Q^2) \int_{z_{\min}}^{z_{\max}} dz\, D_q^h(z,Q^2)}{\sum_{q'} e_{q'}^2 q'(x,Q^2) \int_{z_{\min}}^{z_{\max}} dz\, D_{q'}^h(z,Q^2)}. \quad (1)$$

where D is the depolarization factor, and η and $\gamma = \sqrt{Q^2/v^2}$ can be calculated from the electron kinematics. The relation of the semi-inclusive asymmetries A_1^h to the parton densities in the quark parton model is given in the right hand side of Eq. (1). The unpolarized parton densities q and the polarized densities Δq are defined in terms of the parton densities q^+ and q^- for partons with spins aligned and anti-aligned with the nucleon spin respectively: $q \equiv (q^+ + q^-)$ and $\Delta q \equiv (q^+ - q^-)$. The factor $R = \sigma_L/\sigma_T$ is the ratio of the longitudinal to transverse photo-absorption cross sections. The fragmentation functions D_q^h give the probability that a parton of flavour q fragments into a final state hadron h.

The quark polarizations are isolated by rewriting Eq. (1) and introducing purities P_q^h,

$$A_1^h(x) = \frac{1+R(x)}{1+\gamma^2} \sum_q P_q^h(x) \frac{\Delta q}{q}(x), \quad P_q^h(x) \equiv \frac{e_q^2 \, q(x) \int_{z_{min}}^{z_{max}} dz \, D_q^h(z)}{\sum_{q'} e_{q'}^2 \, q'(x) \int_{z_{min}}^{z_{max}} dz \, D_{q'}^h(z)}, \quad (2)$$

where all quantities were integrated at each x over the corresponding range of Q^2. The purities P_q^h describe the probability that a hadron h originates from an event where a quark of flavour q was struck. They depend on the fragmentation functions, the unpolarized parton densities and in case of the deuterium target on the relative fluxes of hadrons originating from the two nucleons. To account for the limited acceptance of the spectrometer, the purities were extracted from Monte-Carlo simulations instead of using Eq. (2) directly. The fragmentation was modeled in the LUND string model implemented in the JETSET 7.4 package [8], which was tuned to fit the hadron multiplicities measured at HERMES. The CTEQ5L parton distributions [9] were used to model the unpolarized parton densities. Eq. (2) is generalized to include the inclusive asymmetries A_1 by defining inclusive purities $P_q \equiv e_q^2 \, q(x) / (\sum_{q'} e_{q'}^2 \, q'(x))$.

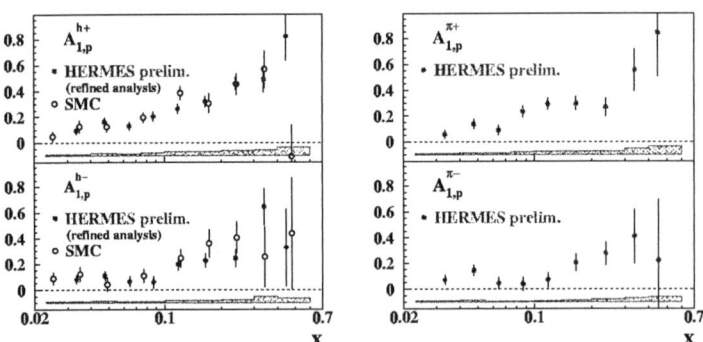

FIGURE 1. Preliminary HERMES results on the semi-inclusive hadron and pion asymmetries from the proton. The positive and negative hadron asymmetries measured by SMC [10] are also shown. The error bars represent the statistical uncertainties and the error band the systematic uncertainties.

The preliminary semi-inclusive asymmetries presented in Figs. 1 and 2 are based on 1.8 and 6.5 million DIS events respectively. The data were collected in the kinematic range $Q^2 > 1 \, \text{GeV}^2$ and $W^2 > 10 \, \text{GeV}^2$. Semi-inclusive hadrons were selected by requiring $0.2 < z < 0.8$ and $x_F \simeq 2p_L/W > 0.1$ where p_L is the longitudinal momentum of the hadron with respect to the virtual photon direction in the photon-nucleon center-of-

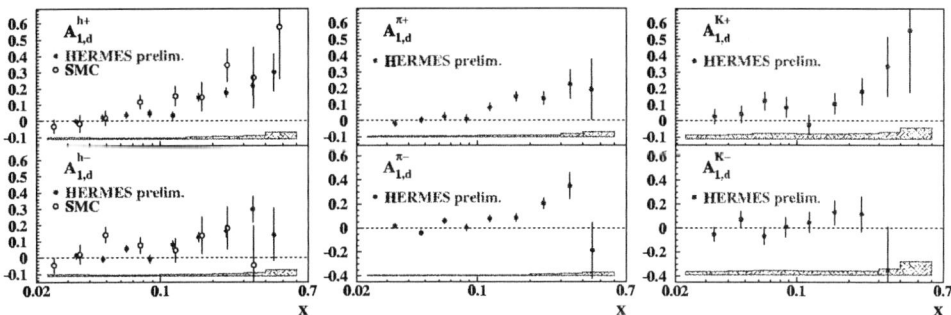

FIGURE 2. Preliminary HERMES results on the semi-inclusive hadron, pion, and kaon asymmetries on the deuteron. See Fig. 1 for more details.

mass frame. The lower limits suppress hadrons from the target fragmentation region. The upper cut on z rejects hadrons from exclusive events. In the hydrogen data the threshold Čerenkov detector identified pions with momenta above $\sim 4\,\mathrm{GeV}$. In the case of the deuterium data, the RICH detector provided kaon identification in addition. The asymmetries on the proton are positive over the entire range of x and largest at high x. The deuteron asymmetries are generally smaller. The pion and kaon asymmetries were measured for the first time by the HERMES collaboration.

The quark polarizations are obtained by combining the hydrogen and deuterium inclusive asymmetries (not shown) and the semi-inclusive asymmetries of positive and negative hadrons, pions, and kaons into an over-constrained system of linear equations and solving it for the quark polarizations $\Delta q/q$. This flavour decomposition was carried out with the parameters $\frac{\Delta u}{u}$, $\frac{\Delta \bar{u}}{\bar{u}}$, $\frac{\Delta d}{d}$, $\frac{\Delta \bar{d}}{\bar{d}}$, and $\frac{\Delta s}{s} \equiv \frac{\Delta \bar{s}}{\bar{s}}$. In contrast to earlier analyses [11, 10], the only symmetry assumption on the polarizations is that the strange sea be symmetric.

The polarizations and the polarized densities of the five quark flavours are shown in Fig. 3. The systematic uncertainties include those due to the asymmetry measurements, the unpolarized parton densities, and the fragmentation functions. At values larger than $x = 0.3$ the polarization of the sea flavours was fixed at zero. This results in small uncertainties for the non-sea flavours which were included in their systematic uncertainty. The polarized densities were obtained by multiplication of the polarizations with the CTEQ5L unpolarized PDFs at fixed $Q^2 = 2.5\,\mathrm{GeV}^2$. The polarizations are assmumed to be Q^2 independent. The density of the up quark is positive in the measured range of x and increases with x up to ~ 0.3 at $x = 0.47$. The down quark density is negative and shows only a small dependence on x. The densities of the light sea quarks are compatible with zero. The data favour a slightly positive strange quark density, in contrast to parameterizations of inclusive measurements [12, 13]. However, within the uncertainties Δs is also consistent with zero.

In conclusion the HERMES collaboration has collected a wealth of inclusive and semi-inclusive deep inelastic scattering data on polarized hydrogen and deuterium targets. Semi-inclusive hadron asymmetries on the proton and the deuteron were measured with good precision. Pion asymmetries on the proton and pion and kaon asymmetries on

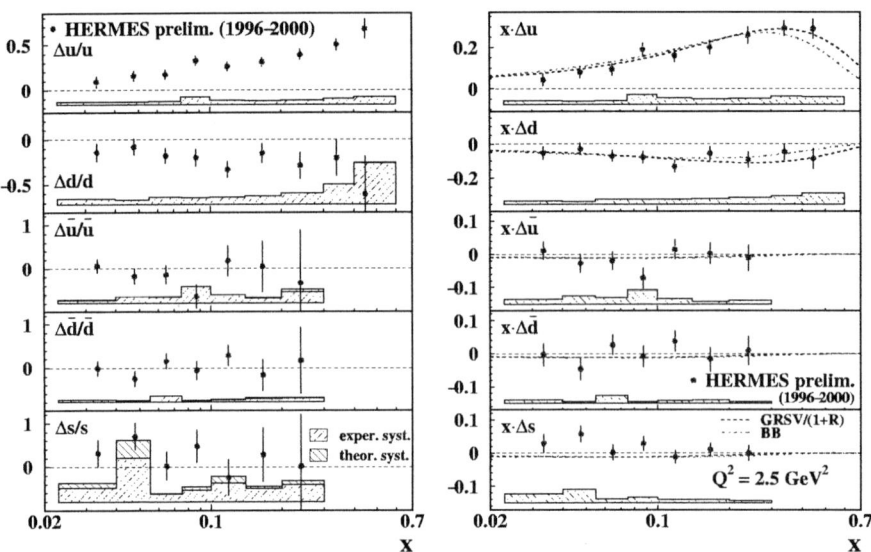

FIGURE 3. The left hand panel shows the measured quark polarizations as a function of x. The panel on the right shows the polarized parton densities at common $Q^2 = 2.5\,\text{GeV}^2$. The dashed lines and the dashed-dotted lines show parameterizations of Ref. [12, 13]. The uncertainty labelled "theor. syst." is due to the fragmentation tune and the unpolarized parton densities.

the deuteron were measured for the first time. Polarized parton densities were extracted from the semi-inclusive and inclusive asymmetries. The densities of the up and down quark were determined with good precision to be positive and negative respectively. The polarized quark sea was decomposed for the first time. Within the experimental uncertainty the densities of the sea quarks, $\Delta\bar{u}$, $\Delta\bar{d}$, and $\Delta s \equiv \Delta\bar{s}$, are compatible with zero.

REFERENCES

1. Ashman, J., et al., *Phys. Lett.*, **B 206**, 364 (1988).
2. Anthony, P. L., et al., *Phys. Rev.*, **D 54**, 6620 (1996).
3. Adeva, B., et al., *Phys. Lett.*, **B 412**, 414 (1997).
4. Abe, K., et al., *Phys. Rev. Lett.*, **79**, 26 (1997).
5. Abe, K., et al., *Phys. Rev.*, **D 58**, 112003 (1998).
6. Anthony, P. L., et al., *Phys. Lett.*, **B 463**, 339 (1999).
7. Ackerstaff, K., et al., *Nucl. Instrum. Meth.*, **A 417**, 230 (1998).
8. Sjostrand, T., *Comp. Phys. Comm.*, **82**, 74–90 (1994).
9. Lai, H. L., et al., *Eur. Phys. J.*, **C 12**, 375 (2000).
10. Adeva, B., et al., *Phys. Lett.*, **B 420**, 180 (1998).
11. Ackerstaff, K., et al., *Phys. Lett.*, **B 464**, 123 (1999).
12. Glück, M., Reya, E., Stratmann, M., and Vogelsang, W., *Phys. Rev.*, **D 63**, 094005 (2001).
13. Blümlein, J., and Böttcher, H., *Nucl. Phys.*, **B 636**, 225 (2002).

The status of the COMPASS experiment

E.M. Kabuß

Institut für Kernphysik, University of Mainz, Becherweg 45, 55099 Mainz, Germany
(supported by the BMBF)
on behalf of the COMPASS collaboration

Abstract. The COMPASS experiment at the SPS at CERN is investigating the spin structure of the nucleon via deep inelastic scattering of polarised muons on polarised nucleons. Currently, the main emphasis lies on the measurement of the gluon polarisation and a study of transverse quark distributions. For this purpose we use a newly built 2-stage large acceptance spectrometer supplemented by particle identification via RICH, calorimetry and muon identification. This setup allows to detect events in the quasi-real photoproduction regime which is required for the measurement of the gluon polarisation. First data were taken in 2002 with longitudinally and transversely polarised LiD targets. Currently the data are being analysed, the reconstruction is mostly done while improvements of the algorithms are still going on. The Analysis is concentrated on Λ polarisation, vector meson production, gluon polarisation, transverse asymmetries and semi-inclusive DIS.

INTRODUCTION

During the past two decades experiments at SLAC, CERN and DESY did a detailed investigation of inclusive deep inelastic scattering of polarized charged leptons off polarized protons and neutrons. The main aim of these investigations was the determination of the spin structure functions g_1 and g_2 using longitudinally and transversely polarized target nucleons. One important result is that the net contribution of the quarks to the nucleon spin is only about 25 – 30%. So the remaining part has to be carried by gluons and orbital angular momentum.

While it is very difficult to access the orbital angular momentum contribution e.g. via deeply virtual Compton scattering the gluon contribution to the nucleon helicity, ΔG, can be studied with the help of the photon gluon fusion process especially in the quasi-real photoproduction region ($Q^2 \approx 0$) where cross sections are reasonably high. To disentangle these processes from the normal DIS processes on quarks two strategies have been developed: The first uses the production of a heavy quark pair ($c\bar{c}$) which is signalled by open charm production, resulting in fragmentation mainly to D mesons. They can be reconstructed from their subsequent decay to e.g. a πK-pair [1]. The second method tries to use all produced quark flavours but enriches the photon gluon fusion process by selecting oppositely charged hadron pairs with high transverse momentum and opposite azimuth [2]. This selection yields much higher statistics than the open charm tagging but the signal is diluted by a considerable background mainly from the QCD Compton process and resolved photon processes at low Q^2.

THE COMPASS EXPERIMENT

The COMPASS experiment has been set up at the M2 muon beam line of the CERN SPS. Although it focuses on the measurement of ΔG in the initial phase it is a general purpose experiment to be used with hadron and muon beams to study a variety of physics items ranging from the Primakoff effect, glue balls, charmed baryons with hadron beams to the gluon polarisation, polarised quark distributions, transversity, lambda and vector meson production with the muon beam.

In 2002 the initial phase was completed and data taking for ΔG started. Fig. 1 shows the 2002 setup. In addition to the initial setup already some large-angle tracking (W4/5) is present to enlarge the kinematic range of the measurements.

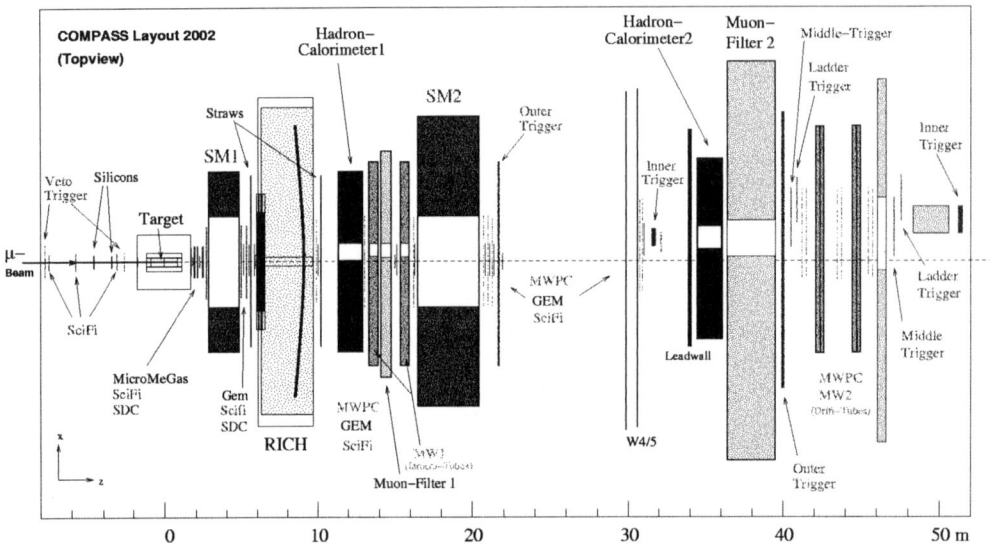

FIGURE 1. Schematic drawing of the COMPASS 2002 setup.

The COMPASS experiment consists of three main parts: the beam region, the target region and the spectrometer:

1) For the ΔG measurement a μ^+ beam of 160 GeV and about 80% polarisation is used. The muon beam has a momentum spread of about 5% and a large emittance of $\varepsilon > 6$ mm·mrad. Each incoming particle is measured in a series of scintillating fiber planes and silicon detectors interleaved with veto hodoscopes for trigger purposes.

2) The heart of the experiment is the polarised solid-state target which consists of two 60 cm long cells in a bath of ^{3}He/^{4}He mixture. The cells are filled with LID and are polarised oppositely. In 2002 target polarisations of up to 57% were obtained routinely.

3) The spectrometer is a large-acceptance two-stage spectrometer which covers a very large kinematic range from quasi-real photoproduction to the DIS region. A large-gap 1.5 Tm dipole magnet (SM1) in the first stage is used to measure low-momentum tracks, mainly hadrons produced in the target, whereas in the second stage a 5.2 Tm dipole magnet (SM2) is used for the scattered muons and high-momentum tracks. Both magnets are surrounded by a large number of tracking detectors to determine the trajectories

with high precision. Both stages use hadronic calorimetry for hadron and absorber walls (muon filter) for muon identification. In addition a RICH detector in stage 1 is used for π/K/p separation up to 50 GeV. At the end of the spectrometer most hodoscopes of the trigger system are located.

Depending on the distance to the muon beam different tracking chambers are used ranging from scintillating fibers in the beam region, GEM and MicroMega detectors close to the beam to MWPC, drift and straw chambers for large areas. Most of them performed very well in 2002 reaching the expected spatial and time resolution while maintaining high efficiencies above 95%.

Also first measurements with the RICH detector were obtained. The preliminary resolution of the measured Cherenkov angle distribution for ultrarelativistic particles is 0.4 mrad, corresponding to π/K separation at 3 sigma level up to 40 GeV/c.

THE 2002 DATA

In 2002 the measurements were split between longitudinally (57 days) and transversely (19 days) polarised targets yielding $3.8 \cdot 10^9$ and $1.2 \cdot 10^9$ events, respectively. The data analysis is in full swing, calibration and alignment are done and most data are reconstructed while improvement of the reconstruction algorithms is still going on.

For the beginning the analysis concentrated on lambda polarisation, vector meson production, ΔG from high p_T hadron pairs and transverse asymmetries. Results are shown from 16% of the longitudinal data. As an example fig. 2 (a) shows the invariant mass distribution for muon pairs in the J/ψ range from events with 3 tracks (among them 2 muons) depicting a clear peak for mainly elastic J/ψ. Fig. 2 (b) shows the invariant mass distribution for hadron pairs from events with 4 tracks (2 muons and 2 hadrons). Here a high statistics ($1.4 \cdot 10^6$) exclusive ρ^0 signal is seen. In this analysis already several kinematic cuts e.g. on the energy transfer and the missing mass were applied. With these ρ mesons analysis of the mass distribution and the angular distributions has been started [3].

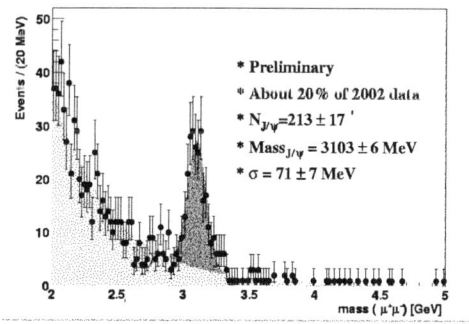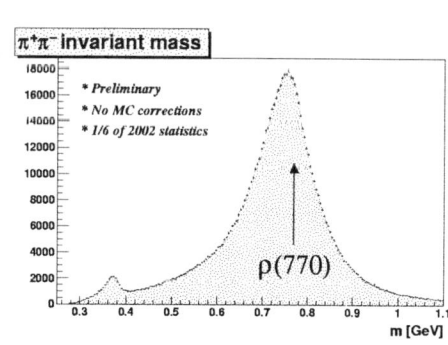

FIGURE 2. Invariant mass distribution for muon pairs (a) and hadron pairs (b)

Good prospects exist for the flavour separation of polarized quark distributions. Using the SMC proton data together with the COMPASS LiD data and making use of the K identification with the RICH a good handle will be obtained on the polarised strange quark distribution extending the HERMES measurement towards low values of x (see fig. 3 (a)).

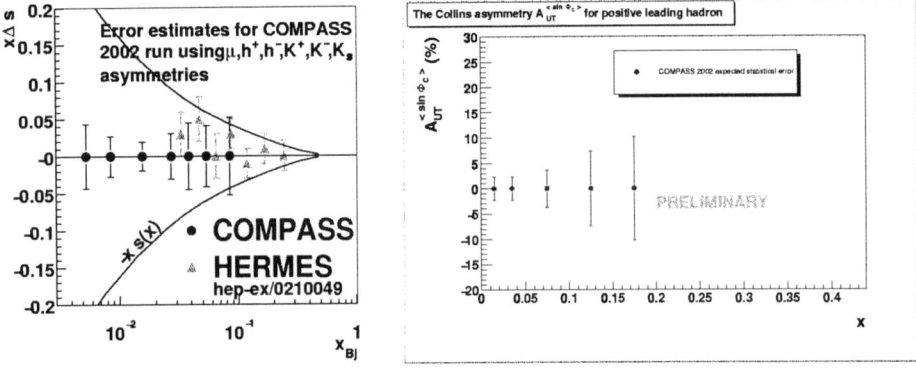

FIGURE 3. Estimated accuracy and kinematic range for the determination of the strange quark polarisation (a) and the transverse asymmetry (b).

Also work on high p_T hadron pairs, transverse asymmetries [4] (see fig. 3 (b)) and Λ polarisation [5] shows good perspectives for a first result with the 2002 data.

OUTLOOK

While the analysis of the 2002 data is still ongoing data taking has started in 2003 with the muon beam and the same setup as in 2002. During the shutdown a lot of work was done especially to improve the RICH radiator gas and the photon detectors. The planned hadron physics program using hadron beams will start later, its main topics being a search for glueballs, studies of charmed hadrons and measurements of the pion and kaon polarizabilities. This program will benefit from the planned completion of the spectrometer with a second RICH, full electromagnetic calorimetry and large-angle tracking. In parallel prospects for new topics like generalised parton distributions are being studied.

REFERENCES

1. COMPASS Collaboration, Compass proposal, *CERN/SPSLC 96-14*, *SPSC/P297*, *CERN/SPSLC 96-30* (1996).
2. A. Bravar, D. v. Harrach, and A. Kotzinian, *Phys. Lett. B*, **421**, 349 (1998).
3. A. Korzenev, Exclusive ρ and ϕ production from COMPASS, *Prodeedings of DIS03*, St. Petersburg, Russia (2003).
4. P. Pagano, Transversity at COMPASS, *Prodeedings of DIS03*, St. Petersburg, Russia (2003).
5. V. Alexakhin, Λ and $\bar{\Lambda}$ hyperon production by polarized muons at COMPASS, *Prodeedings of DIS03*, St. Petersburg, Russia (2003).

Fragmentation Functions and Implications for Spin Physics

Stefan Kretzer

Physics Department and RIKEN-BNL Research Center, Brookhaven National Laboratory, Upton, New York 11973, U.S.A.

Abstract. The present status of fragmentation function (FF) analysis is summarized and the role of FFs in QCD hard scattering phenomenology is outlined with emphasis on spin physics.

FF ↔ PDF

Parton distribution functions (PDFs) can be introduced through the local operator product expansion (OPE) in fully inclusive DIS – where PDFs emerge as the Mellin inverse of twist-2 operators. Fragmentation functions, on the other hand, relate to processes with one observed hadron in the final state (i.e. semi-inclusive) where the local OPE does not apply. FFs can, nevertheless, be defined in terms of the fields of the QCD Lagrangian within the generalized factorization theorems of QCD [1] or through cut vertices [2]. FFs are, therefore, part of a solid theory and no more model dependent than PDFs. Intuitively, the FF $D_f^h(z)$ represents a probability density that after a hard scattering event the parton f turns into hadron h with fractional momentum z of the parton [3].

OPERATIVE ROLE OF FFS FOR SPIN PHYSICS

The *fragmentation* process turns perturbatively produced partons into non-perturbative hadronic bound states. The structural wealth encoded in this process is very rich – in part because independent functions are generated through the relative spin directions of partons and hadrons and through the presence of transverse momentum when the hadron is not collinear to the fragmenting parton. We refer the interested reader to the database [4] as a point of entry into the corresponding literature. In this contribution to the proceedings, without any further specification we will use the term *fragmentation function* more traditionally – as describing the production of unpolarized (or spin-0) hadrons that are collinear with the fragmenting parton. These types of FFs are most important in their operative role of understanding hard scattering phenomena with identified hadrons in the final state at a quantitative level: Within factorized perturbative QCD we have that a hard scattering cross section σ with hadrons in the initial and final state can be written as

$$\sigma = \text{PDF} \otimes \hat{\sigma} \otimes \text{FF} \tag{1}$$

with the symbol ⊗ denoting a convolution integral and $\hat{\sigma}$ being a partonic hard scattering cross section. In spin physics, the strategy is to fix the FFs in unpolarized reactions where the unpolarized PDFs are well known by now - and then later to "divide out" the FFs in polarized measurements to asses the spin dependent PDFs.

RELEVANT PROCESSES: EXTRACTION AND APPLICATION

Below I will list some of the relevant process to analyze and apply FFs. The page limitation does not allow the inclusion of many figures so I will make some quantitative claims without proof, referring the reader to the literature for details.

e^+e^- Annihilations: $e^+e^- \to hX$

The extraction of FFs [5] comes dominantly from the QCD analysis of e^+e^- annihilations where the QCD of the final state (FF) is not intertwined with initial state (PDF) effects. A lot of information can be obtained from e.g. the high precision LEP data. Limitations of e^+e^- are, however:

- The leading order partonic process $e^+e^- \to q\bar{q}$ produces quarks only. Not much can be learnt about gluon fragmentation
- Statistics drops low toward the leading particle tail of the spectrum limiting information on large-z fragmentation.
- The internal light flavour structure cannot be disentangled. A systematic investigation [6] shows that mainly the flavour singlet (sum over flavours) FF is determined.

Because of these limitations FFs are not too well determined yet, even though high precision LEP e^+e^- data can be reproduced very successfully.

Gluon Jets in e^+e^-

Gluon jets can be identified by anti-tagging the "g" in three jet $e^+e^- \to b\bar{b}g$ configurations. Within the parton model, this process simply measures the gluon fragmentation function. A NLO QCD global analysis of anti-tagged gluon jets is a formidable task, though, which has not been achieved so far.

High $p_\perp$ Particle Production in Hadronic Collisions: $pp \to hX$

In figures 1, 2 we perform a decomposition of $pp \to h(p_\perp)X$ for (central rapidity) collisions at RHIC [9] into the contributing initial state PDFs (fig. 1) and final state FFs (fig. 2). In fig. 3 we plot the $p_\perp$ dependence of the averaged scaling variables: Initial state partons carry fractional momentum $x_{1,2}$ and final state partons transfer fractional momentum z in the hadronization process. Obviously, hadronic collisions

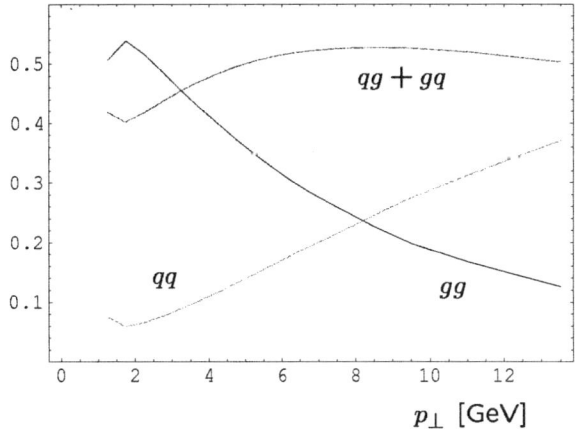

FIGURE 1. Initial state decomposition of (central rapidity) $pp \to \pi X$ at RHIC

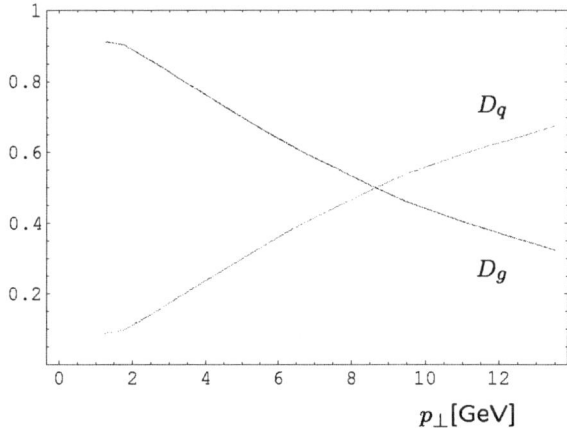

FIGURE 2. Final state decomposition of (central rapidity) $pp \to \pi X$ at RHIC

have the potential to provide information on large-z fragmentation as well as on gluon fragmentation, two shortcomings of the e^+e^- process.

Semi-Inclusive Deep Inelastic Scattering: $eN \to hX$

SIDIS has the potential of disentangling the light flavour sector of the fragmentation process [6]. Following their "operative role" outlined above, one can then apply the FFs to polarized SIDIS and disentangle the flavour structure of the longitudinal quark spin [7]. Uncertainties for the FFs (quantified in [6]) feed into uncertainties of the polarized PDFs. While SIDIS is a very promising process, a few obstacles of data at present

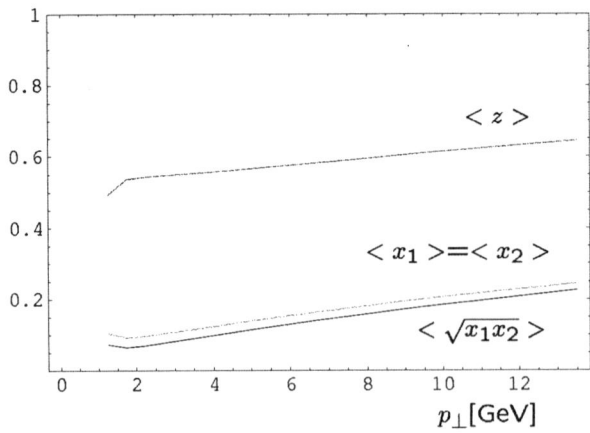

FIGURE 3. Average scaling variables

energies [8] will have to be better understood before it can be considered a bona fide lading twist perturbative reaction.

ACKNOWLEDGMENTS

It is a pleasure to thank E. Leader and E. Christova for collaboration and W. Vogelsang for discussions. I am grateful to RIKEN, Brookhaven National Laboratory and the U.S. Department of Energy (contract No. DE-AC02-98CH10886) for providing the facilities essential for the completion of this work.

REFERENCES

1. J. C. Collins, D. Soper and G. Sterman, in A. H. Mueller, ed., *Perturbative Quantum Chromodynamics* (World Scientific 1989); J.C. Collins and D.E. Soper, Nucl. Phys. **B193**, 381 (1981), **B213**, 545 (1983) (E); **B194** 445 (1982).
2. A.H. Mueller, Phys. Rev. **D18**, 3705 (1978).
3. R.D. Field and R.P. Feynman, Nucl. Phys. **B136**, 1 (1978).
4. http://www.pv.infn.it/radici/FFdatabase/
5. S. Kretzer, Phys. Rev. **D62** (2000) 054001; B. Kniehl, G. Kramer and B. Pötter, Nucl. Phys. **B582** (2000) 514; L. Bourhis, M. Fontannaz, J.P. Guillet, M. Werlen, Eur. Phys. J. **C19** (2001) 89; see these articles also for reference to the experimental data.
6. S. Kretzer, E. Leader, E. Christova, Eur. Phys. J. **C22**, 269 (2001).
7. See J. Wendtland's (for the HERMES collab.) contribution to these proceedings.
8. HERMES Collab. (A. Airapetian et al.), Eur. Phys. J. **C21** 599 (2001); M. Glück, E. Reya, hep-ph/0203063; G.A. Navarro, R. Sassot, Eur. Phys. J. **C28**, 321 (2003); A. Kotzinian, Phys. Lett. **B552**, 172 (2003).
9. PHENIX Collab. (S.S. Adler et al.), hep-ex/0304038.

Renormalons in exclusive meson electroproduction

A.V. Belitsky

*Department of Physics, University of Maryland
College Park, MD 20742-4111, USA*

Abstract.
We discuss the possibility of measuring generalized parton distributions in exclusive electroproduction of mesons off the nucleon and estimate the uncertainty from perturbatively induced higher-twist corrections. We find that, while the magnitude of the cross section changes significantly taking into account twist-four contributions modeled via renormalons, the transverse spin asymmetry is weakly sensitive to them and displays the precocious scaling.

Generalized parton distributions (GPDs) $F(x,\xi,\Delta^2)$ encode exhaustive information on one-particle correlations in the nucleon and thus carry the lore on its wave function and the phase structure of the latter. The quantum-mechanical wave function Ψ allows to predict expectation values of all observable for a given system. An identical description is achieved by means of the density matrix $\rho(x_1,x_2) = \Psi^*(x_1)\Psi(x_2)$. The latter can be used in turn to construct the quantum equivalent of the classical phase-space distribution, the primary example being known as the Wigner quasi-probability function $W(k,r) = \int \frac{dx}{2\pi\hbar} e^{-ikx/\hbar} \rho(r - \frac{1}{2}x, r + \frac{1}{2}x)$. Contrary to its classical counterpart it is not positive definite, — a hallmark of the interference. The marginals of $W(k,r)$ acquire however the probability interpretation as coordinate density $\rho(r) = \int dk W(k,r) = |\Psi(r)|^2$, or equivalently the Fourier transform of the atomic form factor, and momentum-space distribution $n(k) = \int \frac{dr}{2\pi\hbar} W(k,r) = |\widetilde{\Psi}(k)|^2$ with $\widetilde{\Psi}(k) = \int \frac{dx}{2\pi\hbar} e^{-ikx/\hbar} \Psi(x)$. The $W(k,r)$ is an analogue of a Fourier transformed one-dimensional GPD. The impact parameter-dependent parton distributions [1], related to GPDs again by a Fourier transform with respect to the momentum transfer $\Delta_\perp$, are transparently identified as relativistic nucleon Wigner distributions [2]. Thus, the studies of GPDs will shed the light on the phase-space distribution of quarks in the proton. GPDs are cleanly probed in deeply virtual Compton scattering involving only one hadron, the nucleon, whose structure is unraveled through electron scattering [3]. The same GPDs enter the amplitude of exclusive electroproduction of mesons in the asymptotic regime of large momentum transfer [4]. However due to the presence of an extra hadron in the final state and the specifics of perturbative QCD approach to such processes, one has to pose the question of applicability of hard gluon exchange mechanism at moderate photon virtualities. The cross section of meson photoproduction with longitudinally polarized γ^* is, see Fig. 1 (left),

$$\frac{d\sigma_L^M}{d|\Delta^2|d\varphi} = \frac{\alpha_{em}\pi}{\mathscr{Q}^6} \frac{f_M^2}{N_c^2} \frac{x_B^2}{(2-x_B)^2} \left\{ \sigma_M + \sigma_M^\perp \sin\Theta \sin(\Phi - \varphi) \right\}, \qquad (1)$$

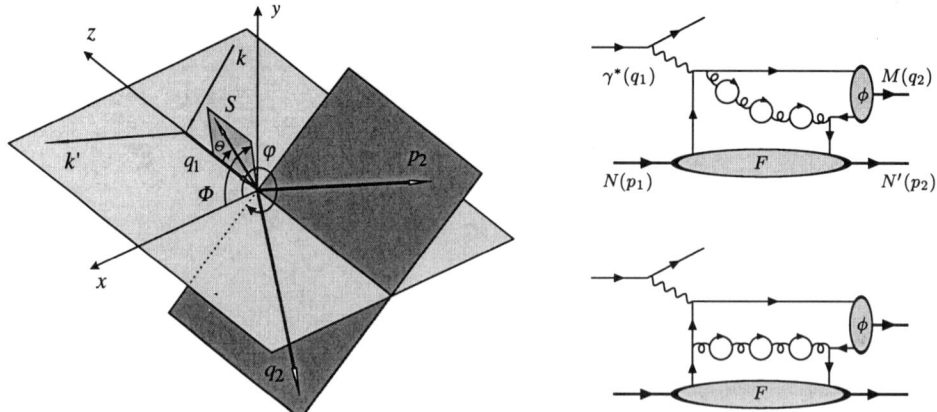

FIGURE 1. Kinematics of the exclusive meson electroproduction off the proton in its rest frame (left) and leading order perturbative diagrams in hard scattering approach (right) dressed by fermion bubble insertions which generate power corrections in the amplitude.

where $q_1^2 = -\mathcal{Q}^2$, $x_B = \mathcal{Q}^2/(2p_1 \cdot q_1)$, and M runs over mesons of different flavors. For the charged psedoscalar meson $P = \pi^+$ the decay constant is $f_\pi = 132$ MeV, while for the neutral vector mesons $V^0 = \rho^0, \omega$, they are $f_\rho = 153$ MeV and $f_\omega = 138$ MeV. The parts of the cross sections read for target polarization-independent [5, 6, 7]

$$\sigma_P = 8(1-x_B)|\widetilde{\mathcal{H}}_P|^2 - x_B^2 \frac{\Delta^2}{2M_N^2}|\widetilde{\mathcal{E}}_P|^2 - 4x_B^2 \, \Re\mathrm{e}\left(\widetilde{\mathcal{H}}_P^* \widetilde{\mathcal{E}}_P\right), \quad (2)$$

$$\sigma_V = 8(1-x_B)|\mathcal{H}_V|^2 - x_B^2 \left(2 + (2-x_B)^2 \frac{\Delta^2}{2M_N^2}\right)|\mathcal{E}_V|^2 - 4x_B^2 \, \Re\mathrm{e}\left(\mathcal{H}_V^* \mathcal{E}_V\right), \quad (3)$$

and target (transverse) polarization-dependent components

$$\sigma_P^\perp = -4x_B \sqrt{1-x_B} \sqrt{-\frac{\Delta^2}{M_N^2}} \sqrt{1 - \frac{\Delta_{\min}^2}{\Delta^2}} \, \Im\mathrm{m}\left(\widetilde{\mathcal{H}}_P^* \widetilde{\mathcal{E}}_P\right), \quad (4)$$

$$\sigma_V^\perp = 4(2-x_B)\sqrt{1-x_B} \sqrt{-\frac{\Delta^2}{M_N^2}} \sqrt{1 - \frac{\Delta_{\min}^2}{\Delta^2}} \, \Im\mathrm{m}\left(\mathcal{H}_V^* \mathcal{E}_V\right), \quad (5)$$

respectively. The generalized structure function $\mathcal{F} = \{\mathcal{H}, \mathcal{E}, \widetilde{\mathcal{H}}, \widetilde{\mathcal{E}}\}$ depends on the skewness $\xi = x_B/(2-x_B)$, the t-channel momentum transfer Δ^2 and resolution scale $\mathcal{Q}^2$. In leading-twist approximation, it is expressed as a convolution of the meson distribution amplitude $\phi(u)$, normalized to $\int_0^1 du \phi(u) = 1$, the quark or gluon GPD $F = \{H, E, \widetilde{H}, \widetilde{E}\}$ and, correspondingly, the quark or gluon coefficient function T via [4]

$$\mathcal{F}_M(\xi, \Delta^2; \mathcal{Q}^2) \quad (6)$$
$$\equiv \int_0^1 du \int_{-1}^1 dx \, \phi_M(u) \left\{ T_M(u, x, \xi; \mathcal{Q}^2) F_M(x, \xi, \Delta^2) + T_g(u, x, \xi) F_g(x, \xi, \Delta^2) \right\}.$$

The function T_M to lowest order approximation is given by the one-gluon exchange mechanism diplayed in Fig. 1 (right). The studies of higher-order perturbative corrections to the hard coefficient function in many physical observables have demonstrated that ambiguities generated by the perturbative resummation of fermion vacuum polarization insertions were of the same order of magnitude as available non-perturbative estimates of matrix elements of higher-twist operators. The development and sophistication of these ideas has led to some evidence that infrared renormalons might reflect the magnitude of higher-twist contributions and even their functional dependence on scaling variables and can thus be used as a rough estimate of power-suppressed effects. On the practical side to compute them in the present circumstances, one replaces the tree gluon propagator, in the single bubble-chain approximation, see Fig. 1 (right), by [in the Landau gauge]

$$\mathscr{D}_{\mu\nu}(k) = \frac{4\pi}{\alpha_s b} \int_0^\infty d\tau\, e^{-4\pi/(\alpha_s b)\tau} \left(\frac{\mu^2 e^C}{-k^2}\right)^\tau \frac{1}{k^2}\left(g_{\mu\nu} - \frac{k_\mu k_\nu}{k^2}\right),$$

where $C_{\overline{\mathrm{MS}}} = \frac{5}{3}$ in the $\overline{\mathrm{MS}}$ and $C_{\mathrm{MS}} = \frac{5}{3} - \gamma_E + \ln 4\pi$ in the MS scheme, and $b = \frac{11}{3}N_c - \frac{4}{3}T_F N_f$ is the first coefficient of the QCD beta-function and $\alpha_s = \alpha_s(\mu^2) = 4\pi/(b\ln\mu^2/\Lambda_{\overline{\mathrm{MS}}}^2)$, where the last equality hold to one-loop order. The functions F_M which enter the above structure functions are combinations of q-flavor quark GPDs

$$F_\pi = F_u - F_d, \qquad F_\rho = Q_u F_u - Q_d F_d, \qquad F_\omega = Q_u F_u + Q_d F_d, \qquad (7)$$

where the quark charges are $Q_u = \frac{2}{3}$ and $Q_d = -\frac{1}{3}$. For π^+ and V^0 only the polarized $F = \{\widetilde{H}, \widetilde{E}\}$ and, correspondingly, unpolarized GPDs $F = \{H, E\}$ enter the game. The quark coefficient function (with resummed renormalon chains) for the $M = \pi^+$ has the form

$$T_\pi(u,x,\xi;\mathscr{Q}^2) = \frac{4\pi C_F}{b}\int_0^\infty \frac{d\tau}{\xi} e^{-4\pi/(\alpha_s b)\tau}\left(\frac{2\mu^2 e^C}{\mathscr{Q}^2}\right)^\tau$$
$$\times \left\{\frac{Q_u}{[\bar{u}(1-\frac{x}{\xi}\ i0)]^{\tau+1}} - \frac{Q_d}{[u(1+\frac{x}{\xi}-i0)]^{\tau+1}}\right\}, \qquad (8)$$

with $C_F = (N_c^2 - 1)/2N_c$. The coefficient function for neutral vector mesons $M = V^0$, T_V, is obtained from this one by setting $Q_u, Q_d \to 1$ since the quark charges are included into flavor combinations of GPDs. Finally, for completeness we present the leading-order gluon coefficient function contributing to neutral vector meson production,

$$T_g(u,x,\xi) = \frac{\alpha_s}{\xi^2} \frac{4T_F \Sigma_q Q_q}{u\bar{u}(1-\frac{x}{\xi}-i0)(1+\frac{x}{\xi}-i0)}.$$

If one absorbs the dependence on the momentum fraction into the argument of the coupling, $\alpha_s(\frac{1}{2}u(1\pm\frac{x}{\xi})\mathscr{Q}^2 e^{-C})$, one explicitly sees that the end-point regions produce divergences. Infrared renormalons are caused by the end-point singularities [Feynman mechanism] in exclusive amplitudes [8], see also [9, 10]. This can be viewed as an estimate of the ambiguity in the resummation of higher-order perturbative corrections or, taken to the extreme, as a model of higher-twist contributions [11]. Convolution of the coefficient function with the distribution amplitude generates renormalon poles. For the asymptotic distribution amplitude $\phi_{asy}(u) = 6u\bar{u}$ one gets two poles $\tau = 1$ and $\tau = 2$, corresponding to ambiguities on the level of $\mathcal{Q}^{-2}$ and $\mathcal{Q}^{-4}$ power corrections. Since the latter receives extra contributions from higher order diagrams as well, we use only $\tau = 1$ pole for the estimates of the form of higher-twist corrections. Taking the imaginary part (divided by π) arising from the contour deformation around the renormalon poles as a measure of their magnitude, we get

$$\widetilde{\mathcal{H}}_\pi(\xi, \Delta^2; \mathcal{Q}^2) = \widetilde{\mathcal{H}}_\pi^{PV}(\xi, \Delta^2; \mathcal{Q}^2) + \theta \frac{\Lambda_{\overline{MS}}^2 e^{5/3}}{\mathcal{Q}^2} \int_{-1}^{1} dx \Delta_{\widetilde{H}}(x, \xi) \widetilde{H}_\pi(x, \xi, \Delta^2), \quad (9)$$

where $\theta = \pm 1$ comes from the ambiguity to go around the renormalon pole in the Borel plane. Here we have used the one-loop expression for the QCD coupling constant and

$$\Delta_{\widetilde{H}}(x, \xi) = 48 \frac{\pi C_F}{b\xi} \left\{ \frac{Q_u}{(1 - \frac{x}{\xi} - i0)^2} - \frac{Q_d}{(1 + \frac{x}{\xi} - i0)^2} \right\}, \quad (10)$$

Since the GPD and its first derivative are continuous functions at $x = \pm \xi$ [12] for models adopted below, the second integral is well-defined. In the first term of (9) one uses the principal value prescription to go around the poles in the Borel plane. For the $\widetilde{E}$ we use the pion-pole dominated form [7] and get

$$\widetilde{\mathcal{E}}_\pi(\xi, \Delta^2; \mathcal{Q}^2) = \widetilde{\mathcal{E}}_\pi^{PV}(\xi, \Delta^2; \mathcal{Q}^2) - \theta \frac{\Lambda_{\overline{MS}}^2 e^{5/3}}{\mathcal{Q}^2} \Delta_{\widetilde{E}}(\xi, \Delta^2; \mathcal{Q}^2), \quad (11)$$

where we have kept only the single and double poles at $\tau = 1$ in the second term, so that

$$\Delta_{\widetilde{E}}(\xi, \Delta^2; \mathcal{Q}^2) = 72 \frac{\pi C_F}{b\xi} F_\pi(\Delta^2) \left(2 + \ln \frac{\Lambda_{\overline{MS}}^2 e^{5/3}}{\mathcal{Q}^2} \right). \quad (12)$$

In the vicinity of the pion pole one can approximate $F_\pi(\Delta^2) = 4g_A M_N/(m_\pi^2 - \Delta^2)$.

In estimates, shown in Fig. 2, we relied on GPDs deduced from an ansatz based on modeling the double distribution as a product [12] $\Delta F(y, z, \Delta^2) = \pi(y, |z|) \Delta f(y, \Delta^2)$ of exclusive profile π and an inclusive parton distribution augmented to have an intrinsic momentum-transfer dependence $\Delta f_q(y, \Delta^2) = \eta_q A_q x^{a_q - \alpha'_q \Delta^2 (1-x)} (1-x)^{b_q} (1 + \gamma_q x + \rho_q \sqrt{x})$ with parameters fixed by the GSA forward densities [13] in $\Delta^2 = 0$ limit and slopes $\alpha'_u = 1.15 \, \text{GeV}^{-2}$, $\alpha'_d = 1.0 \, \text{GeV}^{-2}$ chosen to fit the dipole form of the axial form factor with the effective mass $m_A^2 = 0.9 \, \text{GeV}^2$. We give the cross section [at leading order compatible with earlier estimates [15, 6]] and transverse target-spin asymmetry $\mathcal{A}_P^\perp = (2\sigma_P)/(\pi\sigma_P^\perp)$. In our evaluations we set $\theta = 1$ and $\Lambda_{\overline{MS}} = 280 \, \text{MeV}$ for $N_f = 4$ and use the tree level result for $\mathcal{F}^{PV} \to \mathcal{F}^{LO}$. Note however that in calculations of higher-twist corrections via renormalons in deeply inelastic scattering in order to get the right magnitude of experimental data one has to take a larger value $|\theta| \approx 2 - 3$ [14]. The extremely large power corrections to the absolute cross section of pion leptoproduction are

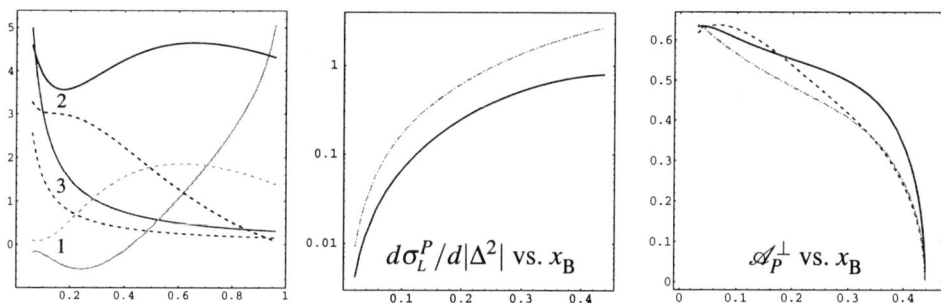

FIGURE 2. Generalized structure functions (left) in leading twist approximation (dashed) and including twist-four corrections (solid) as a function of x_B for $\Delta^2 = -0.3\,\text{GeV}^2$ and $\mathcal{Q}^2 = 10\,\text{GeV}^2$: (1) $\mathfrak{Re}\widetilde{\mathcal{H}}$, (2) $\mathfrak{Im}\widetilde{\mathcal{H}}$, and (3) $10^{-2}\cdot\widetilde{\mathcal{E}}$. The photoproduction cross section in units of nbarns (middle) without (solid) and with (dash-dotted) power suppressed contributions for the same values of the kinematical variables. The transverse spin asymmetry (right) at leading order (solid) and with twist-four power effects taken into account for $\Delta^2 = -0.3\,\text{GeV}^2$ and $\mathcal{Q}^2 = 4\,\text{GeV}^2$ (dashed) and $\mathcal{Q}^2 = 10\,\text{GeV}^2$ (dash-dotted). The maximal value of $x_{B,\max}$ is set by the kinematical constaint $|\Delta^2| > |\Delta^2_{\min}| = M_N^2 x_B^2/(1-x_B)$.

in qualitative agreement with the earlier consideration in Ref. [16]. As we observe, however, the renormalon model of higher-twist contributions affects in a marginal way the asymmetry and thus leads to the apparent conclusion of *the precocious scaling* in ratios of observables, — a fact pointed out previously in various circumstances [5, 6, 17].

We would like to thank D. Müller and A.V. Radyushkin for conversations and INT (Seattle) for its hospitality during the program "GPDs and hard exclusive processes". This work was supported by the U.S. DOE via grant DE-FG02-93ER-40762.

REFERENCES

1. D.E. Soper, Phys. Rev. D 15 (1977) 1141.
2. A.V. Belitsky, *New spin on the proton structure*, colloqium at College of William and Mary.
3. A.V. Belitsky, D. Müller, A. Kirchner, Nucl. Phys. B 629 (2002) 323.
4. J.C. Collins, L. Frankfurt, M. Strikman, Phys. Rev. D 56 (1997) 2982.
5. L. Frankfurt, P. Pobylitsa, M. Polyakov, M. Strikman, Phys. Rev. D 60 (1999) 014010.
6. A.V. Belitsky, D. Müller, Phys. Lett. B 513 (2001) 349.
7. K. Goeke, M. Polyakov, M. Vanderhaeghen, Prog. Part. Nucl. Phys. 47 (2001) 401.
8. S.S. Agaev, Phys. Lett. B 360 (1996) 117; Nucl. Phys. B Proc. Suppl. 74 (1999) 155.
9. P. Godzinski, N. Kivel, Nucl. Phys. B 521 (1998) 274.
10. A.I. Karanikas, N.G. Stefanis, Phys. Lett. B 504 (2001) 225.
11. M. Beneke, V.M. Braun, L. Magnea, Nucl. Phys. B 497 (1998) 297.
12. A.V. Radyushkin, Phys. Lett. B 449 (1999) 81.
13. T. Gehrmann, W.J. Stirling, Phys. Rev. D 53 (1996) 6100.
14. M. Maul, E. Stein, A. Schäfer, L. Mankiewicz, Phys. Lett. B 401 (1997) 100.
15. L. Mankiewicz, G. Piller, A.V. Radyushkin, Eur. Phys. J. C 10 (1999) 307.
16. M. Vänttinen, L. Mankiewicz, E. Stein, hep-ph/9810527.
17. A.V. Belitsky, X. Ji, F. Yuan, Phys. Rev. Lett. (2003) (in press), hep-ph/0212351.

Single Spin Asymmetries at CLAS

H.Avakian and L.Elouadrhiri for the CLAS Collaboration

Jefferson Lab, 12000 Jefferson Ave., Newport News, VA 23601.

Abstract. We present recent results from Jefferson Lab's CLAS detector on beam and target single-spin asymmetries in single pion electroproduction off unpolarized hydrogen and polarized NH_3 targets. Non-zero single-beam and single-target spin asymmetries are observed for the first time in semi-inclusive and exclusive pion production in hard-scattering kinematics.

Single-spin asymmetries (SSA) in hadronic reactions have been among the most difficult phenomena to understand from first principles in QCD. Large SSAs have been observed in hadronic reactions for decades [1, 2]. Recently, significant SSAs were reported in semi-inclusive DIS (SIDIS) by the HERMES collaboration at HERA [3, 4] for a longitudinally polarized target, by the SMC collaboration at CERN for a transversely polarized target [5], and by the CLAS collaboration at the Thomas Jefferson National Accelerator Facility (JLab) with a polarized beam [6]. In general, such single-spin asymmetries require a correlation of a particle spin direction and the orientation of the production or scattering plane. In hadronic processes, such correlations can provide a window into the physics of initial and final state interactions at the parton level.

As shown recently, partonic final state interactions give rise to interference effects in the DIS cross section[7, 8] and may be responsible for the observed SSAs. The non-trivial phase structure of QCD amplitudes, due to rescattering results in *time-reversal odd* (T-odd) effects and the appearance of single-spin asymmetries at leading twist[8, 9]. Furthermore, a nonzero orbital angular momentum of partons in the nucleon is crucial in forming the target SSA [12, 13, 14].

The orbital momentum of partons has been of central interest ever since the EMC [10] measurements implied that the constituent quarks account for only a fraction of the nucleon spin. Transverse momentum of quarks is a key to orbital angular momentum. In recent years parton distribution functions have been generalized to contain information not only on the longitudinal but also on the transverse distributions of partons in a fast moving hadron. With the transverse momentum k_T of partons included, the number of independent distribution functions at leading twist increases to eight [11, 15, 16]. This new degree of freedom makes possible studies of transitions of nucleons with one polarization state to a quark with another polarization state. In particular, the $f_{1T}^{\perp}$ known as the Sivers function [17, 18, 12, 9, 13, 14] describes unpolarized quarks in the transversely polarized nucleon. It is *time-reversal odd* (T-odd) and is only nonzero when final state interactions cause an interference between different helicity states. The counterpart of the Sivers function in the hadronization process is the Collins T-odd fragmentation function $H_1^{\perp}$ [7] describing fragmentation of transversely polarized quarks into unpolarized hadrons. As shown recently in Ref.[14], the interaction between the active parton in the hadron and the target spectators [8, 9, 13] leads to gauge-invariant transverse momentum dependent (TMD) parton distributions.

This paper presents measurements of single spin asymmetries in single pion inclusive leptoproduction using a 5.7 GeV electron beam and the CEBAF Large Acceptance Spectrometer (CLAS) [19] at JLab. Scattering of longitudinally polarized electrons off a liquid-hydrogen target and off a polarized NH_3 target was studied over a wide range of kinematics. The average beam polarization, frequently measured with a Møller polarimeter, was 0.73 ± 0.03 and the average target polarization was for NH_3 0.72 ± 0.05. The scattered electrons and pions were detected in CLAS [19]. The total number of electron-π^+ coincidences in the DIS range ($Q^2 > 1\,\text{GeV}^2$, $W^2 > 4\,\text{GeV}^2$) was $\approx 8 \times 10^6$.

For polarized targets, several azimuthal asymmetries already arise in leading order. The following contributions were investigated in Refs. [7, 15, 16, 8, 13, 24]:

$$\sigma_{UL}^{\sin 2\phi} \propto S_L 2(1-y) \sin 2\phi \sum_{q,\bar{q}} e_q^2 x h_{1L}^{\perp q}(x) H_1^{\perp q}(z), \tag{1}$$

$$\sigma_{UT}^{\sin \phi} \propto S_T(1-y) \sin(\phi+\phi_S) \sum_{q,\bar{q}} e_q^2 x h_1(x) H_1^{\perp q}(z), \tag{2}$$

$$+ S_T(1-y+y^2/2) \sin(\phi-\phi_S) \sum_{q,\bar{q}} e_q^2 x f_{1T}^{\perp q}(x) D_1^q(z), \tag{3}$$

where ϕ is the azimuthal angle between the scattering plane formed by the initial and final momenta of the electron and the production plane formed by the transverse momentum of the observed hadron and the virtual photon, ϕ_S is the azimuthal angle of the transverse spin in the scattering plane frame, the sum $\sum_{q,\bar{q}}$ is over quark flavors, y and z are fractions of electron energy carried by the virtual photon and the fraction of the virtual photon energy carried by the pion respectively. The $D_1^q(z)$ is the spin-independent fragmentation function. The subscripts in σ_{LU} specify the beam and target polarizations (L stands for longitudinally polarized, T for transversely polarized, and U for unpolarized), S_L and S_T are longitudinal and transverse components of the target polarization.

The SIDIS cross section with a longitudinally polarized target in sub-leading order contains an additional contribution to the $\sin\phi$ moment (σ_{UL}) [24, 25, 26]:

$$\sigma_{UL}^{\sin\phi} \propto S_L \sin\phi (2-y) \sqrt{1-y} \frac{M}{Q} \sum_{q,\bar{q}} e_q^2 x^2 h_L^q(x) H_1^{\perp q}(z). \tag{4}$$

Measurements of average moments $\langle W(\phi) \rangle_{UL} = \int \sigma_{UL}(\phi) W(\phi) d\phi / \int \sigma(\phi) d\phi$ ($W(\phi) = \sin\phi, \sin 2\phi$) of the cross section $\sigma_{UL}^{W(\phi)}$ will single out corresponding terms in the cross section. For spin-dependent moments this is equivalent to the corresponding spin asymmetries A_{UL}^W. Thus the $\sin\phi$ SSA of the cross section for longitudinally polarized beam and unpolarized target is defined as:

$$\frac{1}{2} A_{LU}^{\sin\phi} = \langle \sin\phi \rangle_{LU} = \frac{1}{P^{\perp} N^{\pm}} \sum_{i=1}^{N^{\pm}} \sin\phi_i, \tag{5}$$

where $P^{\pm}$ and $N^{\pm}$ are the polarization and number of events for $\pm$ helicity state, respectively. The final asymmetry is defined by the weighted average over two independent measurements for both helicity states.

Eq. 1 and Eq. 2 describe single-spin asymmetries involving the first moment of the Collins fragmentation function integrated over the transverse momentum of the final hadron $H_1^{\perp q}$. A unique feature of the Collins mechanism is the presence of a leading twist $\sin 2\phi$ SSA, for a longitudinally polarized target [24]. Measurements of the $\sin 2\phi$ SSA (see Fig. 1) thus allow a study of the Collins effect with no contamination from other mechanisms. Recent measurement of σ_{UL} by HERMES[3] is consistent with a $\sin 2\phi$ moment of zero. A sufficiently large effect has been predicted only at large x, a region well-covered by JLab [23]. The leading-twist distribution function $h_{1L}^{\perp}(x)$, accessible in this measurement, describes the transverse polarization of quarks in a longitudinally polarized proton.

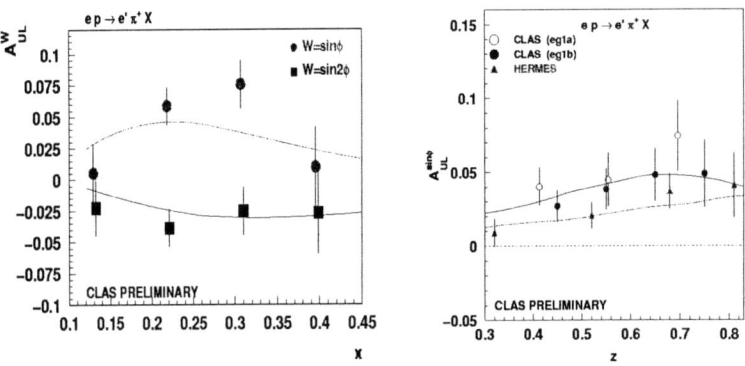

FIGURE 1. Azimuthal moments from CLAS 5.7 GeV polarized target data in a range $0.5 < z < 0.8$ (left panel). Curves on the left plot are the prediction from Ref. [23] for $\sin\phi$ and $\sin 2\phi$ SSA in the CLAS kinematics. The right panel shows the z dependence of A_{UL} compared with published HERMES data [3] and calculations from [8] for the CLAS (upper curve) and the HERMES (lower curve) kinematics.

The $\sin\phi$ moments of the cross section measured with CLAS at 5.7 GeV is in good agreement with the CLAS measurements at 4.3 GeV and the HERMES measurement at 27.5 GeV, which indicates that the asymmetry observables are not sensitive to the beam energy (see Fig.1). Curves for the $\sin\phi$ moment are calculations for HERMES and CLAS kinematics based on the BHS model [8] taking into account only the Sivers effect (right plot).

The $\sin\phi$ moments of the SIDIS cross section with a longitudinally polarized target contain contributions both from the Sivers effect (T-odd distribution) [17] and the Collins effect (T-odd fragmentation) [7] and additional measurements are required to separate them. They include SSA measurements with transversely polarized target and measurements with polarized beam and unpolarized target. The distribution and fragmentation functions responsible for a non-zero beam spin asymmetry in SIDIS were first identified by Levelt and Mulders [20]. They include the twist-3 unpolarized distribution function $e(x)$ introduced by Jaffe and Ji [21], and the polarized fragmentation function $H_1^{\perp}(z)$ first discussed by Collins [7]. The x dependence of beam SSA is defined by the ratio of the twist-3 unpolarized distribution function $e(x)$ and the leading twist distribution function $f_1(x)$ [20]. The expression for the beam SSA involves the convolution of the Collins function and the interaction dependent part $\tilde{e}(x)$ of the higher twist function $e(x)$:

$$\sigma_{LU}^{\sin\phi} \propto \lambda_e 2y\sqrt{1-y}\sin(\phi)\sum_{q,\bar{q}} e_q^2 x^2 \tilde{e}(x) H_1^{\perp q}(z). \qquad (6)$$

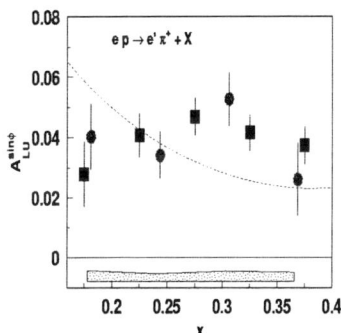 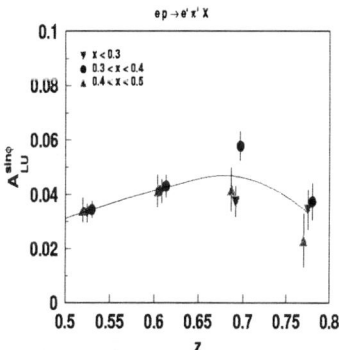

FIGURE 2. The beam-spin azimuthal asymmetry ($\sin\phi$ moment of the cross section) extracted from hydrogen data at 5.7GeV (red circles 4.3 GeV) as a function of x in a range $0.5 < z < 0.8$ (left panel) and z in a range $0.15 < x < 0.4$ (right panel).

The measured beam SSA $A_{LU}^{\sin\phi}$ is positive for a positive electron helicity in the range of $0.15 < x < 0.4$ (see Fig. 2). It is consistent with CLAS measurements at 4.3 GeV [6], increases with z, and shows no significant dependence on the x-range (Fig. 2, right panel). This behavior is consistent with factorization. The first extraction of the twist-3 distribution function from the CLAS beam SSA data has been reported recently by Efremov et al. [23] using a particular parametrization of the Collins fragmentation function. With a certain approximation for the twist-3 function $e(x)$, the beam SSA could become a major source of information on the T-odd polarized fragmentation function.

A beam spin asymmetry of this magnitude can be also obtained [22] using a mechanism similar to that proposed by Brodsky, Hwang and Schmidt for the target spin asymmetry case [8].

The beam $\sin\phi$ SSA for exclusive events in the hard scattering kinematics (see Fig. 3) unlike the case of the target $\sin\phi$ SSA [3, 27] is positive and compatible with the $A_{LU}^{\sin\phi}$ for the semi-inclusive sample at large z. Even though the power corrections for the absolute cross section of exclusive pion electroproduction analyzed in terms of generalized parton distributions are expected to be large, there are indications of a *precocious scaling* in ratios of observables[28].

In conclusion, significant SSAs have been observed in semi-inclusive and exclusive pion electroproduction in hard scattering kinematics. Current data are consistent with a partonic picture, and can be described by a variety of theoretical models. A global

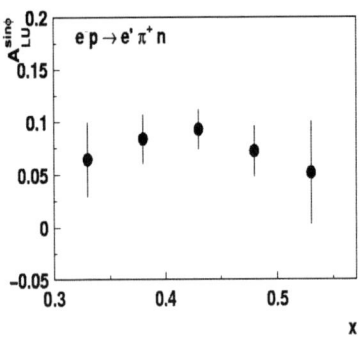
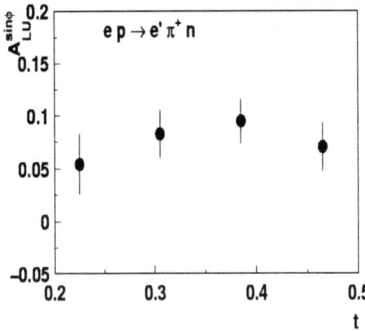

FIGURE 3. $A_{UL}^{\sin\phi}$ x and t dependences in hard scattering kinematics ($Q^2 > 2.5$ GeV^2, $W^2 > 5$ GeV^2).

analysis of available data on SSAs is needed to separate different contributions and to extract the underlying distribution and fragmentation functions.

REFERENCES

1. K. Heller et al. 'Proceedings of Spin 96',Amsterdam,Sep.1996,p23
2. Fermilab E704 collaboration (A. Bravar et al.), Phys.Rev.Lett. **77**, 2626 (1996).
3. HERMES collaboration (A. Airapetyan et al.), Phys.Rev.Lett. **84**, 4047 (2000).
4. HERMES collaboration (A. Airapetyan et al.), Phys.Rev. **D64**, 097101 (2001).
5. A. Bravar, Nucl. Phys. (Proc. Suppl.) **B79** 521 (1999).
6. CLAS Collaboration (H. Avakian et al.) hep-ex/0301005.
7. J. Collins, Nucl. Phys. **B396**, 161 (1993).
8. S. Brodsky et al. Phys. Lett. B **530**, 99 (2002).
9. J. Collins, Phys. Lett. B **536**, 43 (2002).
10. J. Ashman et al., Phys. Lett. B **206**, 364 (1988).
11. J. Ralston and D. Soper, Nucl. Phys. **B152**, 109 (1979).
12. S. Brodsky et al., Nucl. Phys. B **642**, 344 (2002).
13. X. Ji, F. Yuan, Phys. Lett. B **543**, 66 (2002).
14. A. Belitsky,X. Ji and F. Yuan, Nucl. Phys. **B 656**, 165 (2003).
15. P.J. Mulders and R.D. Tangerman, Nucl. Phys. **B461**, 197 (1996).
16. A. Kotzinian, Nucl. Phys. **B 441**, 234 (1995).
17. D. Sivers, Phys.Rev. **D43**, 261 (1991).
18. M. Anselmino and F. Murgia, Phys. Lett. B **442**, 470 (1998).
19. B. Mecking et al., Nucl. Instrum. Meth. A503, 513 (2003).
20. J. Levelt and P. J. Mulders, Phys. Lett. **B338**, 357 (1994).
21. R.L. Jaffe and X. Ji Nucl.Phys. **B375** (1992) 527.
22. A. Afanasev and C. Carlson, hep-ph/0308163.
23. A. Efremov et al. Phys. Rev. D67, 114014 (2003).
24. A.M. Kotzinian and P.J. Mulders, Phys. Rev. **D54** 1229 (1996).
25. D. Boer and P. Mulders, Phys.Rev. **D57**, 5780 (1998).
26. A. M. Kotzinian et al., Nucl.Phys. **A666**, 290-295 (2000).
27. H. Avakian, Proceedings of DIS-2000, Liverpool University 2000.
28. A. Belitsky hep-ph/0307256.

Novel Transversity Properties in SIDIS

Leonard Gamberg[*], Gary R. Goldstein[†] and Karo A. Oganessyan[**‡]

[*]*Division of Science, Penn State-Berks Lehigh Valley College, Reading, PA 19610, USA*
[†]*Department of Physics and Astronomy, Tufts University, Medford, MA 02155, USA*
[**]*INFN-Laboratori Nazionali di Frascati, Enrico Fermi 40, I-00044 Frascati, Italy*
[‡]*DESY, Notkestrasse 85, 22603 Hamburg, Germany*

Abstract. We consider a rescattering mechanism to calculate a leading twist T-odd pion fragmentation function, a candidate for filtering the transversity properties of the nucleon. We evaluate the single spin azimuthal asymmetry for a transversely polarized target in semi-inclusive deep inelastic scattering (for HERMES kinematics) and the double T-odd $\cos 2\phi$ asymmetry in this framework.

Introduction

The transversity distribution, h_1 which measures the probability to find a transversely polarized quark in the transversely polarized nucleon, is as important for the description of the internal nucleon spin structure as the more familiar helicity distribution function, g_1. However, it still remains unmeasured, unlike the spin-average and helicity distribution functions, which are known experimentally and extensively modeled theoretically. The difficulty is that h_1 is chiral odd, and consequently suppressed in inclusive deep inelastic scattering (DIS) processes [1]; it has to be accompanied by a second chiral-odd quantity. Semi-inclusive deep inelastic scattering (SIDIS) on polarized nucleons is one of several [2, 3, 4] promising methods proposed to access transversity. It relies on just such a quantity the so called Collins fragmentation function [5], which correlates the transverse spin of the fragmenting quark to the transverse momentum of the produced hadron. Beside being chiral-odd, this fragmentation function is also time-reversal odd (T-odd) [6, 7] which makes its calculation challenging. In this context, the non-zero single spin asymmetries in recent measurements [3] may signal the existence of a non-trivial T-odd effects which are intimately tied to our understanding of transversity. Here we explore [8] a one-gluon exchange mechanism, for the fragmentation of a transversely polarized quark into a spinless hadron similar to the approach we applied [9, 10] to the distribution of the transversely polarized quarks in the both unpolarized and transversely polarized nucleons. The non-perturbative information about the quark content of the target and the fragmentation of quarks into hadrons in SIDIS is encoded in the general form of the factorized cross sections in terms of the quark distributions $\Phi(p)$ and fragmentation functions $\Delta(k)$, entering the hadronic tensor. To leading order in $1/Q^2$ [11] the fragmentation functions are projected from

$$\Delta(k,P_h) = \sum_X \int \frac{d\xi^+ d^2\xi_\perp}{2z(2\pi)^3} e^{ik\cdot\xi} \langle 0|\mathscr{G}_{[\xi+,-\infty]}\psi(\xi)|X;P_h\rangle\langle X;P_h|\overline{\psi}(0)\mathscr{G}^\dagger_{[0,-\infty]}|0\rangle\Big|_{\xi^-=0}. \quad (1)$$

Here k quark fragmenting momenta and P_h is the fragmented hadron momenta. The path ordered exponential along the light like direction ξ^- is

$$\mathcal{G}_{[\xi^-,\infty]} = \mathcal{P}\exp\left(-ig\int_{\xi^-}^{\infty}d\xi^- A^+(\xi)\right).$$

In non-singular gauges [12, 13], the gauge link gives rise to initial and final state interactions which in turn provide a mechanism to generate leading twist T-odd contributions to both the distribution and *fragmentation* functions. The joint product of these functions enter novel azimuthal asymmetries and single spin asymmetries (SSAs) that have been reported in the literature [14, 15, 9, 10]. Such an analysis was recently applied to the T-odd $f_{1T}^\perp$ [14, 13, 16] and $h_1^\perp$ [15, 9, 10] distribution functions in addition to T-odd baryon fragmentation functions [17]. We apply [8] an analogous procedure to generate the T-odd pion fragmentation function, $H_1^\perp(z)$ (see also [18]).

Pion Fragmentation Function

The leading order contributions to the T-odd fragmentation functions come from the first non-trivial term in expanding the path ordered gauge link operator. The corresponding Feynman rules are those for interactions between an eikonalized struck quark and the remaining target [19] depicted in Fig. 1. In modeling the highly off-shell fragmenting quark we adopt a minimal spectator [20] approach. We couple the on-shell spectator, as a quark interacting with the produced pion through a Gaussian distribution in the transverse momentum dependence of the quark-spectator-pion vertex [10, 8] in order to address the log divergence arising in the moments of fragmentation functions. The leading order (in $1/Q$) one loop contribution which arises in the limit that the virtual photon's momentum becomes large corresponding to the rescattering of the initial state quark depicted in Fig. 1. The resulting twist 2, T-odd contribution the fragmentation function [8] projected from $\text{Tr}\left(\gamma^-\gamma^\perp\gamma_5\Delta\right)$ is

$$H_1^\perp(z,k_\perp) = \frac{\mathcal{N}'^2 f^2 g^2}{(2\pi)^4}\frac{1}{4z}\frac{(1-z)}{z}\frac{m}{\Lambda'(k_\perp^2)}\frac{M_\pi}{k_\perp^2}e^{-b'(k_\perp^2-\Lambda'(0))}\left[\Gamma(0,b\Lambda'(0))-\Gamma(0,b'\Lambda'(k_\perp^2))\right],$$

$\Gamma(0,z)$ is the incomplete gamma function and $\Lambda'(k_\perp^2) = k_\perp^2 + \frac{1-z}{z^2}M_\pi^2 + \frac{\mu^2}{z} - \frac{1-z}{z}m^2$. The average $<k_\perp^2> = 1/b'$ is a regulating scale which we fit to the expression for the integrated unpolarized fragmentation function

$$D_1(z) = \frac{\mathcal{N}'^2 f^2}{4(2\pi)^2}\frac{1}{z}\frac{(1-z)}{z}\left\{\frac{m^2 - \Lambda'(0)}{\Lambda'(0)} - \left[2b'\left(m^2 - \Lambda'(0)\right) - 1\right]e^{2b'\Lambda'(0)}\Gamma(0,2b'\Lambda'(0))\right\}$$

which is in good agreement with the distribution of Ref. [21]. In Fig. 1 the weighted the analyzing power, $H_1^{\perp(1)}(z)/D_1(z)$, is displayed. The resulting behavior is similar to a previous model ansatz proposed by Collins and calculated in Ref. [22]. The $\cos 2\phi$ asymmetry of SIDIS is projected out of the cross section and depends on a leading

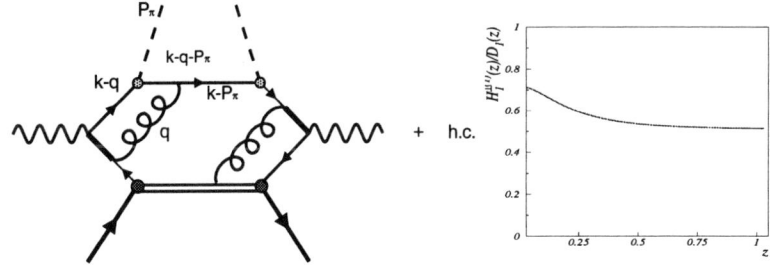

FIGURE 1. Left Panel: $h_1^\perp \star H_1^\perp \; \cos 2\phi$ asymmetry. Right Panel: The weighted analyzing power $H_1^{\perp(1)}(z)/D_1(z)$ as a function of z.

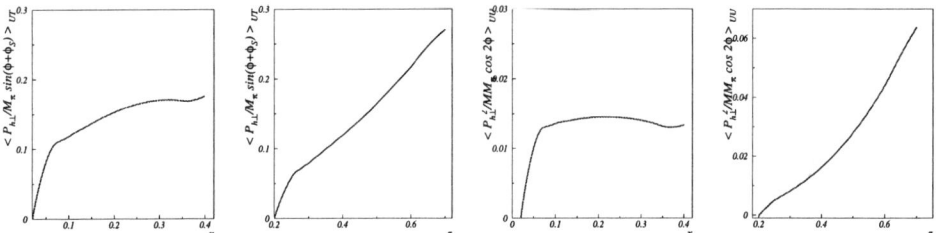

FIGURE 2. The $\langle \sin(\phi + \phi_s)\rangle_{UT}$ asymmetry for π^+ production as a function of x and z. The $\langle \cos 2\phi \rangle_{UU}$ asymmetry for π^+ production as a function of x and z.

double T-odd product,

$$\langle \frac{|P_{h\perp}^2|}{MM_\pi} \cos 2\phi \rangle_{UU} = \frac{8(1-y)\sum_q e_q^2 h_1^{\perp(1)}(x) z^2 H_1^{\perp(1)}(z)}{(1+(1-y)^2)\sum_q e_q^2 f_1(x) D_1(z)}. \quad (2)$$

UU indicates unpolarized beam and target and $h_1^{\perp(1)}(z)$ is the weighted moment of the distribution function [6, 10]. For a transversely polarized target nucleon, the $\sin(\phi + \phi_s)$ asymmetry[5, 11] can be similarly obtained yielding, the convolution of two chiral-odd structures,

$$\langle \frac{P_{h\perp}}{M_\pi} \sin(\phi + \phi_s) \rangle_{UT} = |S_T| \frac{2(1-y)\sum_q e_q^2 h_1(x) z H_1^{\perp(1)}(z)}{(1+(1-y)^2)\sum_q e_q^2 f_1(x) D_1(z)}. \quad (3)$$

The variable range to coincides with the HERMES kinematics [8]. In Fig. 2 the asymmetry of Eq. (3) for π^+ production on a proton target is presented as a function of x and z, respectively indicating approximately a $10-15\%$ $P_{h\perp}/M_\pi$ weighted $\sin(\phi + \phi_s)$ asymmetry. Also, in Fig. 3 the $P_{h\perp}^2/(MM_\pi)$ weighted $\cos 2\phi$ asymmetry of Eqs. (2) for π^+ production on an unpolarized proton target is presented as a function of x and z, respectively indicating a few percent effect.

Conclusion

A mechanism to generate the T-odd Collins fragmentation function that is derived from the gauge link has been considered. This approach complements the approach that was employed to generate the T-odd distribution functions, $f_{1T}^\perp$ and $h_1^\perp$ that fuel the Sivers and $\cos 2\phi$ asymmetries. The derivation of $H_1^\perp$ is consistent with the observation that intrinsic transverse quark momenta and angular momentum conservation are intimately tied with studies of transversity. Furthermore, this approach is interesting in that it does not suffer from the possible cancellation of the Collins effect cited in [23]. This effect is generated in the non-trivial phase associated with the gauge link operator [12, 13, 17, 9, 10, 24]. We have evaluated the analyzing power and predicted the $P_{h\perp}/M_\pi$ weighted $\sin(\phi + \phi_S)$ asymmetry at HERMES energies. Additionally, we predict that there is a non-trivial $\cos 2\phi$ asymmetry associated with the asymmetric distributions of transversely polarized quarks inside unpolarized hadrons. Generalizing from these model calculations, it is clear that initial and final state interactions can account for leading twist T-odd contributions to SSAs. Using rescattering as a mechanism to generate T-odd distribution and fragmentation functions opens a new window into the theory and phenomenology of transversity in hard processes.

Acknowledgments

L.G. thanks the organizers of CIPANP for the invitation to present this work.

REFERENCES

1. R. L. Jaffe and X. Ji, Phys. Rev. Lett. **67**, 552 (1991).
2. A. Bravar (Spin Muon Collaboration), Nucl. Phys. Proc. Suppl., **79** 520 (1999).
3. A. Airapetian *et al.*, Phys. Rev. Lett. **84**, 4047 (2000); Phys. Lett. B **562**, 182 (2003).
4. L. C. Bland, hep-ex/0212013; G. Rakness, hep-ex/0211068.
5. J.C. Collins, Nucl. Phys. **B396**,161 (1993).
6. D. Boer and P. J. Mulders, Phys. Rev. D **57**, 5780 (1998).
7. M. Anselmino and F. Murgia, Phys. Lett B **442**, 470 (1998).
8. L. P. Gamberg, G. R. Goldstein and K.A. Oganessyan, hep-ph/0307139, *To appear in Phys. Rev. D*.
9. L. P. Gamberg, G. R. Goldstein and K.A. Oganessyan, hep-ph/0211155, *Proceedings of the* 15th *International Spin Physics Symposium (SPIN 2002)*, Long Island, New York, September 2002.
10. L. P. Gamberg, G. R. Goldstein and K.A. Oganessyan, Phys. Rev. D **67**, 071504 (2003).
11. R. D. Tangerman and P. J. Mulders, Phys. Lett. B **352**, 129 (1995); Nucl. Phys. **B461**, 197 (1996).
12. J. C. Collins, Phys. Lett. B **536**, 43 (2002).
13. X. Ji and F. Yuan, Phys. Lett. B **543**, 66 (2002); A.V. Belitsky, X. Ji and F. Yuan, Nucl. Phys. B **656**, 156 (2003).
14. S. Brodsky, D.S. Hwang and I. Schmidt, Phys. Lett. B **530**, 99 (2002); D. Boer, S. Brodsky, D.S. Hwang, Phys. Rev D **67**, 054003 (2003).
15. G. R. Goldstein and L. P. Gamberg, hep-ph/0209085, *Proceedings of* 31st *International Conference on High Energy Physics (ICHEP 2002)*, Amsterdam, The Netherlands, Jul 2002.
16. D. Sivers, Phys. Rev D 41 (1990) 83 ; Phys. Rev. D **43**, 261 (1991).
17. A. Metz, Phys. Lett. B **549**, 139 (2002).
18. A. Bacchetta, A. Metz, and J. J. Yang, hep-ph/0307282.
19. J. C. Collins and D. E, Soper, Nucl. Phys. **B194**, 445 (1982).
20. P. Hoodbhoy, Phys.Rev. D **51**, 32 (1995).
21. S. Kretzer, Phys. Rev. D **62**, 054001 (2000).
22. A. M. Kotzinian and P. J. Mulders, Phys. Lett. B **406**, 373 (1997).
23. R. L. Jaffe, X. Jin, and J. Tang, Phys. Rev. Lett **80**, 1166 (1998).
24. D. Boer, P.J. Mulders, and F. Pijlman, hep-ph/0303034.

Single Spin Azimuthal Asymmetries and Transversity

Ralf Seidl
(on behalf of the HERMES Collaboration)

Physikalisches Institut II, University Erlangen-Nürnberg, Erwin-Rommel-Str.1, 91058 Erlangen

Abstract. Single Spin azimuthal Asymmetries (SSA) in semi-inclusive deep-inelastic lepton-nucleon scattering can be generally produced by time reversal odd (T-odd) distribution or fragmentation functions. In the case of an unpolarized positron beam and a longitudinally or transversely polarized target, SSAs can have two origins: the chiral odd transversity distribution h_1 in conjunction with the chiral odd Collins fragmentation function $H_1^\perp$ and the T-odd Sivers distribution function $f_{1T}^\perp$ in conjunction with a spin-independent fragmentation function. In the case of a longitudinally polarized target both mechanisms contribute to the same asymmetry, whereas in case of transverse target polarisation they give rise to different angular dependences and can therefore be measured separately. The present status of HERMES data taking with a transversely polarized proton target will be presented.

TRANSVERSE MOMENTUM DEPENDENT DISTRIBUTION AND FRAGMENTATION FUNCTIONS

In semi-inclusive deep-inelastic scattering (SIDIS) only 3 Parton-distribution functions (DF) remain after integration over the intrinsic transverse momentum of the quarks in the nucleon. Those are the well known momentum DF $f_1(x_B)$, the helicity DF $g_1(x_B)$ and the yet unmeasured transversity DF $h_1(x_B)$. The latter describes the probability to find a transversely polarized quark in a transversely polarized nucleon [1]. Here x_B is the Bjorken scaling variable, representing the fraction of the target momentum carried by the parton. In addition, also transverse momentum dependent DFs exist with its most prominent member beeing the Sivers Function $f_{1T}^\perp(x_B, p_T)$. This DF describes the probability to find an unpolarized quark in a transversely polarized nucleon [2]. Only recently it was realized that the transverse momentum dependent functions can be understood in terms of a final-state interaction. This creates an interference of amplitudes giving rise to the T-odd nature of these functions and resulting in observable Single Spin Asymmetries (SSA) [3, 4, 5, 6].

Since all hard interactions conserve chirality, the chiral odd transversity DF can only be probed in conjunction with another chiral odd object. In the Drell-Yan process this can be a second transversity distribution, whereas in SIDIS the 2nd chiral-odd structure has to be associated with the fragmentation process leading to the production of the detected hadron from the struck quark. The fragmentation function (FF) $H_1^\perp$, known as the Collins Function, is indeed chiral odd and it is also T-odd which results in a Single Spin Asymmetry.

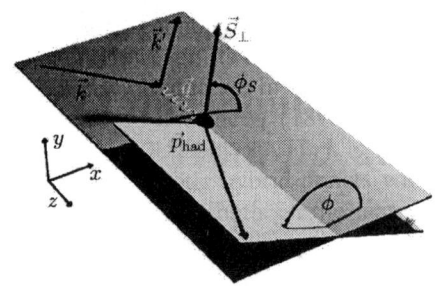

FIGURE 1. Definition of the azimuthal angles ϕ and ϕ_s.

LONGITUDINALLY POLARIZED VS. TRANSVERSELY POLARIZED TARGET

The aforementioned mechanisms both result in a SSA. These azimuthal asymmetries are characterized by the Fourier components of the involved azimuthal angles ϕ and ϕ_s, where ϕ describes the angle between the lepton scattering plane and the hadron production plane along the direction of the virtual photon and ϕ_s is the angle between the transverse component of the target spin, relative to the virtual photon, and the lepton scattering plane, as illustrated in Fig. 1. The $\sin(\phi + \phi_s)$ moment of the asymmetry A_{UT} of an unpolarized lepton beam and a transversely polarized target contains in leading twist $h_1(x_B)$ in conjunction with the Collins FF $H_1^\perp$, while its $\sin(\phi - \phi_s)$ moment contains the Sivers function $f_{1T}^\perp(x_B)$ in conjunction with the unpolarized FF D_1.
The previous measurements of HERMES made use of a longitudinally polarized target [7, 8, 9] and therefore the azimuthal target angle ϕ_s was always either 0 or π as the spin of the target was parallel to the incoming beam and thus its transverse component always stayed in the lepton scattering plane. For that reason both mechanism contributed to the same SSA. On the other hand, if a transversely polarized target is used, the two mechanisms can be distinguished as the $\sin(\phi - \phi_s)$ and $\sin(\phi + \phi_s)$ moments are different.

MOMENT RECONSTRUCTION, PRESENT STATUS OF THE HERMES RUN II

For the transversely polarized target, the cross section asymmetry containing two azimuthal angles is evaluated in 2 dimensions by

$$A_{UT}^h(\phi, \phi_s) = \frac{1}{|P_T|} \frac{N_h^\uparrow(\phi, \phi_s) - N_h^\downarrow(\phi, \phi_s)}{N_h^\uparrow(\phi, \phi_s) + N_h^\downarrow(\phi, \phi_s)} \quad (1)$$

where $|P_T|$ is the magnitude of the target polarization, and $N_h^{\uparrow(\downarrow)}(\phi, \phi_s)$ is the semi-inclusive hadron yield in that target spin state. The Collins and Sivers asymmetries,

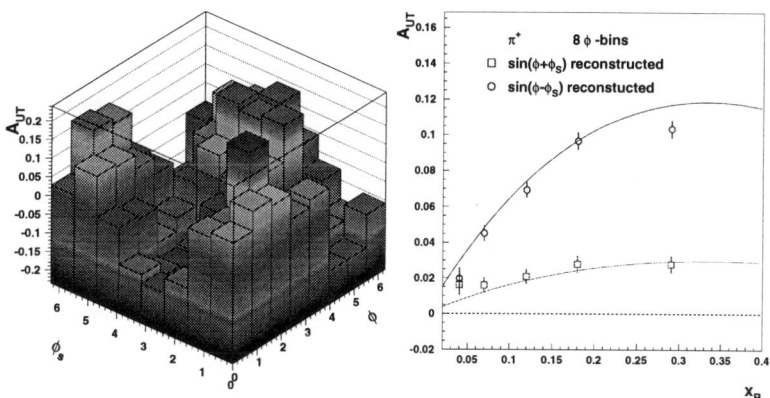

FIGURE 2. Asymmetries generated with the HERMES MC. Left:reconstructed 2-dimensional asymmetry, right: reconstructed (symbols) and generated (lines) asymmetries.

A_C^h and A_S^h are then extracted from a fit to the data using the following expression

$$A_{UT}^h(\phi,\phi_s) = A_C^h \sin(\phi + \phi_s) + A_S^h \sin(\phi - \phi_s). \tag{2}$$

The two linear combinations of the azimuthal angles are orthogonal such that they cannot interfere with each other but detector smearing and acceptance effects in the detector can cause such mixing effects. MC-studies have been performed, showing that the 2-dimensional fitting method reconstructs the generated asymmetries very well as illustrated in Fig.2.

In the first months of 2003 HERMES has collected about $9 \cdot 10^5$ DIS events from a transversely polarized hydrogen target which are under analysis. The results will be published soon.

ACKNOWLEDGMENTS

This work was supported by the German Bundesministerium für Bildung und Forschung, BMBF, contract nr. 06 ER 928I and nr. 06 ER 125I.

REFERENCES

1. Jaffe, R. L., and Ji, X.-D., *Nucl. Phys.*, **B375**, 527–560 (1992).
2. Sivers, D. W., *Phys. Rev.*, **D41**, 83 (1990).
3. Brodsky, S. J., Hwang, D. S., and Schmidt, I., *Phys. Lett.*, **B530**, 99–107 (2002).
4. Ji, X.-d., and Yuan, F., *Phys. Lett.*, **B543**, 66–72 (2002).
5. Collins, J. C., *Phys. Lett.*, **B536**, 43–48 (2002).
6. Belitsky, A. V., Ji, X., and Yuan, F., *Nucl. Phys.*, **B656**, 165–198 (2003).
7. Airapetian, A., et al., *Phys. Rev. Lett.*, **84**, 4047–4051 (2000).
8. Airapetian, A., et al., *Phys. Rev.*, **D64**, 097101 (2001).
9. Airapetian, A., et al., *Phys. Lett.*, **B562**, 182–192 (2003).

Single Transverse Spin Asymmetries

Dae Sung Hwang

Department of Physics, Sejong University, Seoul 143–747, Korea

Abstract. Recent measurements from the HERMES, SMC and CLAS collaborations show a remarkably large azimuthal single-spin asymmetries of the proton in semi-inclusive pion leptoproduction $\gamma^*(q)p \to \pi X$. The existence of such single-spin asymmetries requires a phase difference between two amplitudes coupling the proton target with $J_p^z = \pm\frac{1}{2}$ to the same final-state, the same amplitudes which are necessary to produce a nonzero proton anomalous magnetic moment. We show that the exchange of gauge particles between the outgoing quark and the proton spectators produces a Coulomb-like complex phase which depends on the angular momentum L^z of the proton's constituents and is thus distinct for different proton spin amplitudes. We then find that final-state interactions from gluon exchange between the outgoing quark and the target spectator system lead to single-spin asymmetries at leading twist in perturbative QCD; i.e., the rescattering corrections are not power-law suppressed at large photon virtuality Q^2 at fixed x_{bj}.

Single-spin asymmetries in hadronic reactions have been among the most difficult phenomena to understand from basic principles in QCD. The problem has become more acute because of the observations by the HERMES [1], SMC [2] and CLAS [3] collaborations of a strong correlation between the target proton spin $\vec{S}_p$ and the plane of the produced pion and virtual photon in semi-inclusive deep inelastic lepton scattering $\ell p^\uparrow \to \ell' \pi X$ at photon virtuality as large as $Q^2 = 6 \text{ GeV}^2$. Large azimuthal single-spin asymmetries have also been seen in hadronic reactions such as $pp^\uparrow \to \pi X$ [4, 5], where the target antiproton is polarized normal to the pion production plane, and in $pp \to \Lambda^\uparrow X$ [6], where the hyperon is polarized normal to the production plane.

In order to produce a correlation involving a transversely-polarized proton, there are two necessary conditions: (1) There must be two proton spin amplitudes $M[\gamma^* p(J_p^z) \to F]$ with $J_p^z = \pm\frac{1}{2}$ which couple to the same final-state $|F>$; and (2) The two amplitudes must have different, complex phases. The correlation is proportional to $\text{Im}(M[J_p^z = +\frac{1}{2}]^* M[J_p^z = -\frac{1}{2}])$. The analysis of single-spin asymmetries thus requires an understanding of QCD at the amplitude level. Since we need the interference of two amplitudes which have different proton spin $J_p^z = \pm\frac{1}{2}$ but couple to the same final-state, the orbital angular momentum of the two proton wavefunctions must differ by $\Delta L^z = 1$. The anomalous magnetic moment for the proton is also proportional to the interference of amplitudes $M[\gamma^* p(J_p^z) \to F]$ with $J_p^z = \pm\frac{1}{2}$ which couple to the same final-state $|F>$. Final-state interactions (FSI) in gauge theory can affect deep inelastic scattering reactions in a profound way, as has been demonstrated recently [7]. Here we shall show that FSI provide the required phase needed for single-spin asymmetries.

In Ref. [8] Brodsky, Schmidt and I calculated the single-spin asymmetry in semi-inclusive electroproduction $\gamma^* p \to HX$ induced by final-state interactions in a model of a spin-$\frac{1}{2}$ proton with mass M composed of charged spin-$\frac{1}{2}$ and spin-0 constituents

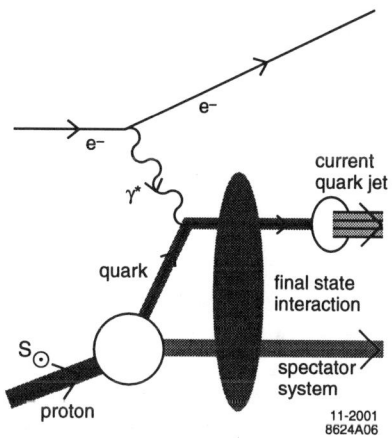

FIGURE 1. The final-state interaction in the semi-inclusive deep inelastic lepton scattering $\ell p^\uparrow \to \ell' \pi X$.

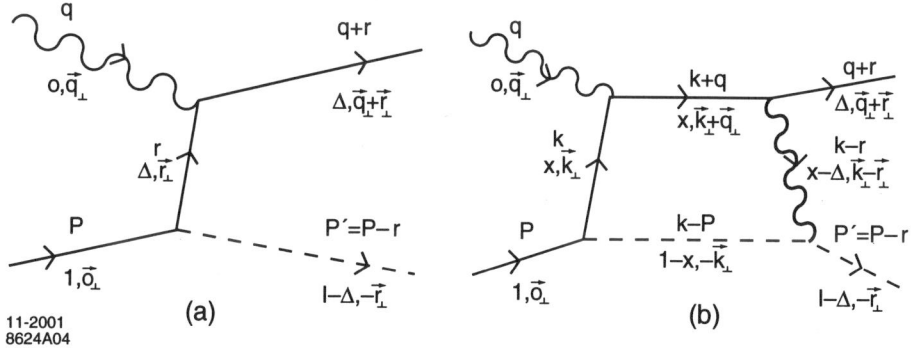

FIGURE 2. The tree (a) and one-loop (b) graphs for $\gamma^* p \to q(qq)_0$. The interference of the two amplitudes with $J_p^z = \pm 1/2$ provides the proton's single-spin asymmetry.

with respective mass m and λ, which is a QCD-motivated quark-scalar diquark model of a nucleon. The basic electroproduction reaction is then $\gamma^* p \to q(qq)_0$, as illustrated in Figs. 1 and 2. We shall take the case where the detected particle H is identical to the quark. One can take the asymmetry for a detected hadron by convoluting the jet asymmetry result with a realistic fragmentation function; e.g., $D_{q \to \pi X}(z, Q^2)$.

The amplitudes for $\gamma^* p \to q(qq)_0$ from Figs. 1 and 2 are given by

$$\mathscr{A}(\Uparrow \to \uparrow) = (M + \frac{m}{\Delta}) C \left(h + i\frac{e_1 e_2}{8\pi} g_1\right) \quad (1)$$

$$\mathscr{A}(\Downarrow \to \uparrow) = \left(\frac{+r^1 - ir^2}{\Delta}\right) C \left(h + i\frac{e_1 e_2}{8\pi} g_2\right) \quad (2)$$

$$\mathscr{A}(\Uparrow \to \downarrow) = \left(\frac{-r^1 - ir^2}{\Delta}\right) C \left(h + i\frac{e_1 e_2}{8\pi} g_2\right) \quad (3)$$

$$\mathcal{A}(\Downarrow \to \downarrow) = (M + \frac{m}{\Delta}) C (h + i\frac{e_1 e_2}{8\pi} g_1), \qquad (4)$$

where

$$C = -g e_1 P^+ \sqrt{\Delta} \, 2\Delta(1-\Delta), \quad h = \frac{1}{\vec{r}_\perp^2 + \Delta(1-\Delta)(-M^2 + \frac{m^2}{\Delta} + \frac{\lambda^2}{1-\Delta})}. \qquad (5)$$

The quark light-cone fraction $\Delta = \frac{k^+}{P^+}$ is equal to the Bjorken variable x_{bj} up to corrections of order $1/Q$. The label $\Uparrow / \Downarrow$ corresponds to $J_p^z = \pm\frac{1}{2}$. The second label $\uparrow / \downarrow$ gives the spin projection $J_q^z = \pm\frac{1}{2}$ of the spin-$\frac{1}{2}$ constituent. Here e_1 and e_2 are the electric charges of q and $(qq)_0$, respectively, and g is the coupling constant of the proton-q-$(qq)_0$ vertex. The first term in (1) to (4) is the Born contribution of the tree graph, and the second term is the imaginary part of the one-loop diagram 2b which are given by

$$g_1 = \int_0^1 d\alpha \frac{1}{\alpha(1-\alpha)\vec{r}_\perp^2 + \alpha\lambda_g^2 + (1-\alpha)\Delta(1-\Delta)(-M^2 + \frac{m^2}{\Delta} + \frac{\lambda^2}{1-\Delta})}, \qquad (6)$$

$$g_2 = \int_0^1 d\alpha \frac{\alpha}{\alpha(1-\alpha)\vec{r}_\perp^2 + \alpha\lambda_g^2 + (1-\alpha)\Delta(1-\Delta)(-M^2 + \frac{m^2}{\Delta} + \frac{\lambda^2}{1-\Delta})}. \qquad (7)$$

The crucial result is the fact that the contributions g_1 and g_2 from the one-loop diagram Fig. 2b are different, and that their difference is infrared finite. The virtual photon and produced hadron define the production plane which we will take as the $\hat{z} - \hat{x}$ plane. The azimuthal single-spin asymmetry transverse to the production plane is given by

$$\mathcal{P}_y = \frac{e_1 e_2}{8\pi} \frac{2(\Delta M + m) r^1}{\left[(\Delta M + m)^2 + \vec{r}_\perp^2\right]} \left[\vec{r}_\perp^2 + \Delta(1-\Delta)(-M^2 + \frac{m^2}{\Delta} + \frac{\lambda^2}{1-\Delta})\right]$$

$$\times \frac{1}{\vec{r}_\perp^2} \ln \frac{\vec{r}_\perp^2 + \Delta(1-\Delta)(-M^2 + \frac{m^2}{\Delta} + \frac{\lambda^2}{1-\Delta})}{\Delta(1-\Delta)(-M^2 + \frac{m^2}{\Delta} + \frac{\lambda^2}{1-\Delta})}. \qquad (8)$$

The linear factor of $r^1 = r^x$ reflects the fact that the single spin asymmetry is proportional to $\vec{S}_p \cdot \vec{q} \times \vec{r}$ where $\vec{q} \sim -v\hat{z}$ and $\vec{S}_p = \pm \hat{y}$. Here $\Delta = x_{bj}$.

Our analysis can be generalized to the corresponding calculation in QCD. The final-state interaction from gluon exchange has the strength $\frac{e_1 e_2}{4\pi} \to C_F \alpha_s(\mu^2)$. The scale of α_s in the $\overline{MS}$ scheme can be identified with the momentum transfer carried by the gluon $\mu^2 = e^{-5/3} (\vec{k}_\perp - \vec{r}_\perp)^2$ [9]. The matrix elements of the proton to its constituents will have the same numerator structure as the perturbative model since they are determined by orbital angular momentum constraints. The strengths of the proton matrix elements can be normalized by the anomalous magnetic moment and the total charge.

The azimuthal single-spin asymmetry $\mathcal{P}_y$ given in (8) transverse to the photon-to-pion production plane decreases as $\alpha_s(r_\perp^2) x_{bj} M r_\perp [\ln r_\perp^2]/r_\perp^2$ for large $r_\perp$, where $r_\perp$ is the magnitude of the momentum of the current quark jet relative to the virtual photon direction. The fall-off in $r_\perp^2$ instead of Q^2 compensates for the dimension of the $\bar{q}$-q - gluon correlation, i.e., the rescattering corrections are not power-law suppressed at large photon virtuality Q^2 at fixed x_{bj}. The mass M of the physical proton mass appears here since it determines the ratio of the $L_z = 1$ and $L_z = 0$ matrix elements. We estimate the scale of α_s as $\mathcal{O}(r_\perp^2)$. The nominal size of the spin asymmetry is thus $C_F \alpha_s(r_\perp^2) a_p$ where a_p is the proton anomalous magnetic moment.

There have been recently important developments in the study of the single-spin asymmetry. Collins demonstrated that the result (8) can be interpreted as the Sivers effect arising from a time-odd $k_\perp$-dependent distribution function [10, 11]. Belitsky, Ji and Yuan showed that the gauge invariance requires a transverse gauge link which is indispensible in the light-cone gauge [12, 13]. Then, Boer, Mulders and Pijlman extended the results to observables appearing in the first sub-leading order in $1/Q$ in the hadronic tensor [14]. By these works it has been understood that the time-odd distribution functions are allowed theoretically, and it became clear that there are two mechanisms for the target spin asymmetry, Collins and Sivers effects [15]. There are presently much efforts for distinguishing these two effects from the experimental results.

With the same mechanism presented in this proceeding, another time-odd distribution function $h_1^\perp$ (chiral-odd) was also calculated and it was applied to the $\cos 2\phi$ asymmetry in the Drell-Yan process [16, 17]. Futhermore, the time-odd fragmentation functions were calculated with a similar mechanism [18, 19, 20]. At the same time, the beam spin asymmetry was also measured [21] and it was calculated in the quark-scalar diquark model by Afanasev and Carlson [22].

The author wish to thank Stan Brodsky and Ivan Schmidt for the collaboration on the work presented here.

REFERENCES

1. HERMES Collaboration, A. Airapetian et al., Phys. Rev. Lett. **84**, 4047 (2000).
2. A. Bravar, for SMC Collaboration, Nucl. Phys. B (Proc. Suppl.) **79**, 520 (1999).
3. H. Avakian, for CLAS Collaboration, in these proceedings.
4. E704 Collaboration, A. Bravar et al., Phys. Rev. Lett. **77**, 2626 (1996).
5. L.C. Bland, for RHIC, hep-ex/0212013, in the proceedings of SPIN 2002.
6. K. Heller, in Proceedings of Spin 96, C. W. de Jager et al., Eds., World Scientific (1997).
7. S.J. Brodsky, P. Hoyer, N. Marchal, S. Peigne and F. Sannino, Phys. Rev. D **65**, 114025 (2002).
8. S.J. Brodsky, D.S. Hwang, and I. Schmidt, Phys. Lett. B **530**, 99 (2002).
9. S.J. Brodsky, A.H. Hoang, J.H. Kühn and T. Teubner, Phys. Lett. B **359**, 355 (1995).
10. J.C. Collins, Phys. Lett. B **536**, 43 (2002).
11. D.W. Sivers, Phys. Rev. D **43**, 261 (1991).
12. X. Ji and F. Yuan, Phys. Lett. B **543**, 66 (2002).
13. A. Belitsky, X. Ji, and F. Yuan, Nucl. Phys. B **656**, 165 (2003).
14. D. Boer and P.J. Mulders and F. Pijlman, hep-ph/0303034.
15. M. Anselmino, U. D'Alesio and F. Murgia, Phys. Rev. D **67**, 074010 (2003).
16. L. Gamberg, G.R. Goldstein and K.A. Oganessyan, Phys. Rev. D **67**, 071504 (2003).
17. D. Boer, S.J. Brodsky and D.S. Hwang, Phys. Rev. D **67**, 054003 (2003).
18. A. Metz, Phys. Lett. B **549**, 139 (2002).
19. L. Gamberg, G.R. Goldstein and K.A. Oganessyan, hep-ph/0307139.
20. A. Bacchetta, A. Metz and J.-J. Yang, hep-ph/0307282.
21. CLAS Collaboration, H. Avakian et al., hep-ex/0301005.
22. A. Afanasev and C.E. Carlson, hep-ph/0308163, in these proceedings.

Future Measurements of Spin Dependent Fragmentation Functions in e+e- Annihilation at Belle

Kazumi Hasuko[*†], Matthias Grosse Perdekamp[*], Akio Ogawa[***], Jens Söeren Lange[*‡] and Viktor Siegle[*]

[*]*RIKEN BNL Research Center, Upton, NY 11973-5000, USA*
[†]*RIKEN, Wako, Saitama 351-0198, Japan*
[**]*Brookhaven National Laboratory, Upton, NY 11973-5000, USA*
[‡]*University of Frankfurt, Frankfurt 60486, Germany*

Abstract. In unpolarized electron-positron annihilation, there may exist interesting and possibly non-zero azimuthal asymmetries, which measure novel chiral-odd fragmentation functions, such as the Collins-Heppelmann function, $H_1^\perp$, and the two-pion interference fragmentation function, $\delta \hat{q}^h$. We will present the experimental method to extract these functions using e^+e^- collision data from the Belle experiment at KEK B-factory (KEKB). In addition to the considerable interest in the properties of these new fragmentation functions, they are expected to be a powerful tool in accessing proton quark transversity distributions.

INTRODUCTION

In the study of transverse spin phenomena, much attention has been paid to understand the spin dependence in the fragmentation of a polarized quark. Particularly, nontrivial azimuthal angle asymmetries may arise from the non-perturbative dynamics in the fragmentation of a transversely polarized quark [1]. In this paper we discuss future measurements of such new fragmentation functions, so-called, spin-dependent fragmentation functions, from unpolarized e^+e^- annihilation at KEKB/Belle.

While there is substantial interest in the symmetry properties of these fragmentation functions, they are also useful tools to access proton quark transversity distributions, δq, in semi inclusive deep inelastic lepton scattering (SIDIS) experiments and polarized proton-proton scattering experiments. At leading twist transversity distributions remain the last unknown quark distribution functions and their knowledge is essential for a complete understanding of nucleon structure. Presently it is thought that transverse single spin asymmetries A_T in SIDIS and pp scattering offer the most practical way to measure transversity distributions. A_T is proportional to $\sum_q \delta q \times a_i^f \times FF$, where a_i^f is the transversity dependent partonic initial-final-state asymmetry which can be calculated from pQCD, and FF is the spin-dependent quark fragmentation function. The analyzing power in this process arises from the spin dependence of the partonic cross section as well as from the spin dependence of the fragmentation process. It is suggested that the quark spin direction might be reflected in the azimuthal distribution of a final state pion and the symmetry properties of the process do not require the proposed FF to be identical to zero [1]. Recent results in fact seem to suggest that these fragmentation functions and δq are different from zero [2, 3].

BELLE EXPERIMENT

KEKB is an asymmetric storage ring that collides 8 GeV electrons against 3.5 GeV positrons [4]. The experimental data are recorded at the (4S) resonance and in the continuum 60 MeV below the resonance, corresponding to integrated luminosities of more than 150 fb^{-1} on resonance and $\approx 10\%$ of this off-resonance.

The Belle detector is a general purpose, spectrometer based on a 1.5 T superconducting solenoid magnet. Charged particles are reconstructed with a three-layer double-sided silicon vertex detector (SVD) and a central drift chamber (CDC) that consists of 50-layer segmented into 6 axial and 5 stereo super-layers. The CDC covers the polar angle range between 17° and 150° in the laboratory frame, which corresponds to 92% of the full solid angle in the center of mass frame. Together with the SVD, a transverse momentum resolution of $(\sigma_{p_t}/p_t)^2 = (0.0019p_t)^2 + (0.0030)^2$ is achieved, where p_t is in GeV/c.

Particle identification (PID) for charged hadrons is provided by a combination of three sub-system devices: a sub-system of aerogel Čerenkov counters (ACC) covering the momentum range 1-3.5 GeV/c, a time-of-flight scintillation counter sub-system (TOF) for track momenta below 1.5 GeV/c, and dE/dx information from the CDC for particles with very low or high momenta. Information from these three devices is combined to give the likelihood of a particle, and hadron-type separation is then accomplished based on the likelihood ratio. The typical efficiency for identifying pions in the momentum region 0.5 GeV/c $< p <$ 4 GeV/c is (88.5±0.1)% and the systematic error in the PID is estimated to be 0.9% to 1.4% [5].

Surrounding the charged PID devices, a Cs(Tl) electromagnetic calorimeter and muon/K_L detector using iron plates interleaved with resistive plate counters are equipped. A detailed description of the Belle detector can be found elsewhere [6].

ANALYSIS METHOD AND SENSITIVITY

We intend to study two different spin-dependent fragmentation functions.

The one is called the Collins-Heppelmann function, $H_1^\perp$, which describes the fragmentation of a transversely polarized quark into a charged pion. It can be obtained using the recipe introduced in reference [7]: First identify a sample of two-jet events from light quark production. In each event, two unlike-sign pion tracks are selected, one in each event hemisphere. The fundamental observable is the angle between the two "pionplanes" formed by the pion momentum vectors and the jet axis. The pion planes include the angles ϕ_1 and ϕ_2 with the event plane defined by the beam axis and jet axis. The definition of the angles is shown in Figure 1-(a). The measured angular dependence in the angle $\phi_1 + \phi_2$ is then proportional to the product of the Collins functions on each side taken at their respective fractional pion energy z_1 and z_2: $A(\phi_1 + \phi_2) \propto H_1^\perp(z_1)H_1^\perp(z_2)$.

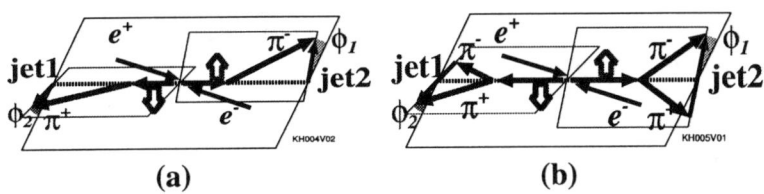

FIGURE 1. Kinematics of (a) $e^+e^- \longrightarrow \pi^+_{jet1}\pi^-_{jet2}X$ and (b) $e^+e^- \longrightarrow (\pi^+\pi^-)_{jet1}(\pi^+\pi^-)_{jet2}X$

The other is Interference fragmentation functions, $\delta\hat{q}^{h_1,h_2}$, which parameterizes the fragmentation of transversely polarized quarks into pairs of hadrons. In a model of $\delta\hat{q}$ [8], $\delta\hat{q}$ describes quark fragmentation into two pions in a state which is a linear superposition of s-wave and p-wave states. These two partial waves are active in the ρ-region. The effect is an s-p interference and the fragmentation function peaks just above and below the ρ resonance and is changing it's sign across the ρ. This sign change should help to identify the fragmentation function and discriminate against possible systematic effects. We follow the recipe provided by Atru and Collins [9] to extract $\delta\hat{q}$: In a sample of two-jet events, for each event $\pi^+\pi^-$ pairs are identified in each hemisphere of the event and the angle between the planes formed by the two pion pairs is measured. The distribution in this angle is the product of twice the interference fragmentation function evaluated at the invariant masses of the two pion pairs m_{12}, m_{34} and $z_{12,34}$, the longitudinal momentum fractions of the pairs. The kinematics is shown in Figure 1-(b). Schematically, the angular distribution is proportional to $f(z_{12}, m_{12}, Q^2) \times f(z_{34}, m_{34}, Q^2)$.

Figure 2-(a) shows the distribution of the angle between two unlike-sign hadron tracks for Belle continuum data and Monte Carlo generated data [10]. The distribution clearly displays jet-like correlations with peaks for the near side and away side jets. The jet axis can be measured by using the thrust axis and applying a thrust cut of $T > 0.85$ ensures the jetty events (two-jet events). The difference between the reconstructed thrust axis and the initial q ($\bar{q}$) direction in MC is shown in Figure 2-(b).

The essential experimental requirements for the fragmentation analysis are good particle identification in a wide momentum region and momentum resolution of about 100 MeV; these requirements are easily within the reach of the Belle detector [6, 11]. The fragmentation functions scale as $\log Q^2$. Subsequently, analyzing powers at Belle (at (4S), $Q^2 \approx 10^2$ GeV^2) are expected to be four times larger than ones at LEP (at Z^0, $Q^2 \approx 91^2$ GeV^2). By taking into account the difference in luminosity, the sensitivity $\Delta H_1^\perp/H_1^\perp$ will be higher by a factor of 20 at Belle than at LEP.

CONCLUSION

The existence of spin-dependence in fragmentation processes would be interesting and a powerful tool for the study of transverse nucleon quark spin structure. A recent

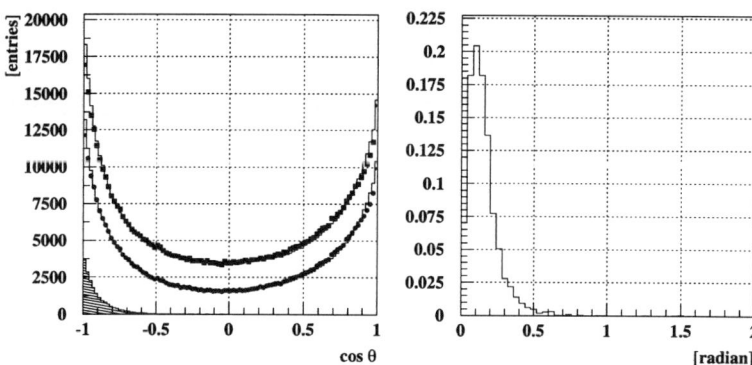

FIGURE 2. (a) The distribution of the angle between two unlike-sign charged hadron tracks. Upper distributions are all pairs for data (square) and MC (histogram); middle $T > 0.85$ for data (circle) and MC (histogram); bottom histogram shows data used for $H_1^\perp$ analysis ($T > 0.85, \cos\theta < 0, z_1 > 0.2$ and $z_2 > 0.2$). (b) The distribution of the angle between the thrust axis and the q ($\bar{q}$) direction with $T > 0.85$.

discussion of the prospects of future programs to access nucleon transversity using fragmentation function information from Belle can be found in reference [12].

The high luminosities at the KEK B-factory and the excellent momentum resolution and particle identification capabilities of the detector make Belle an ideal place for measurements of spin-dependent fragmentation functions.

Analysis has started using 32 million $q\bar{q}$ off-resonance events in the Belle experiment. Significance in the testing asymmetries will be at the level of a few percent. First results are expected within one year.

REFERENCES

1. Collins, J.C., *Nucl. Phys.* **B396**, 161 (1993); Collins, J.C., Heppelmann, S.F., and Ladinsky, G., *Nucl. Phys.* **B420**, 563 (1994).
2. Airapetian, A. *et al.*, *Phys. Rev. Lett.* **84**, 4047 (2000).
3. Bravar, A., *Nucl. Phys. (Proc. Suppl.)* **B79**, 520 (1999).
4. KEK B Factory Design Report No. 95-7, 1995 (unpublished).
5. Iijima, T. *et al.*, *Nucl. Instrum. Meth.* **A379**, 457-459 (1996).
6. Belle Collaboration, Mori, S. *et al.*, *Nucl. Instrum. Meth.* **A479**, 117-232 (2002).
7. Boer, D., Jakob, R., and Mulders, P.J., *Phys. Lett.* **B424**, 143-151 (1998).
8. Jaffe, R.L., Jin, X., Tang, J. *et al.*, *Phys. Rev. Lett.* **80**, 1166 (1998);
9. Artru, X., and Collins, J., *Z. Phys.* **C69**, 277(1996).
10. These MC events are generated with the CLEO group's QQ program;
    see http://www.lns.cornell.edu/public/CLEO/soft/QQ;
    The detector response is simulated using GEANT, R. Brun *et al.*, GEANT 3.21, CERN Report DD/EE/84-1 (1984).
11. Belle Collaboration, Gordon, A. *et al.*, *Phys. Lett.* **B542**, 183-192 (2002).
12. Boer, D., "Transversity Single Spin Asymmetries" in *9th International Workshop on Deep Inelastic Scattering (DIS 2001)*, edited by G. Bruni *et al.*, World Scientific, 2001.

Proton and Neutron Spin Structure Functions in and near the Resonance Region

R. C. Minehart

Physics Dept. U. of Virginia, Charlottesville, VA 22901

Abstract.
We present an overview of the measurements for polarized electron scattering from polarized NH_3, ND_3, and ^3He targets at JLab. All three experimental halls are involved in this program. The data span a range in Q^2 up to about 3.5 GeV^2 and a range in the γ^*N invariant mass W up to 3.5 GeV. The photon absorption function A_1 and the spin structure function g_1 for the proton and deuteron have been extracted from data obtained with the CLAS spectrometer. The photon absorption functions, A_1 and A_2, and the spin structure functions g_1 and g_2 for ^3He have been obtained from Hall A data.

An extensive program of double-polarization electron-nucleon scattering experiments is underway at Jefferson Lab(JLab). This report is limited to the inclusive scattering measurements. With a few exceptions, double polarization experiments have focused on deep inelastic scattering (DIS) using beam energies greater than 20 GeV. Lower energies are more suitable for probing QCD in the low Q^2 confinement region, where nucleon resonances play an important role.

Low Q^2 measurements can probe the transition region from confinement to DIS. At very low Q^2 the experiments can be used to test the GDH sum rule for real photons and chiral perturbation calculations of its generalization to non-zero Q^2. It is also possible that Bloom-Gilman duality may relate the average behavior of the spin structure functions measured at low Q^2 and high Bjorken x ($x = Q^2/(2M\nu)$) to their values in the DIS region. A complete isospin decomposition of the resonance transition form-factors requires scattering from both the proton and the neutron.

The spin dependence can be described in terms of two virtual photon absorption functions, A_1 and A_2:

$$A_1 \equiv \frac{\sigma_T^{1/2} - \sigma_T^{3/2}}{2\sigma_T} = \frac{g_1 - g_2/\tau}{F_1}, \quad A_2 \equiv \frac{\sigma_{LT}}{\sigma_T} = \frac{g_1 + g_2}{\sqrt{\tau}F_1} < \sqrt{\sigma_L/\sigma_T},$$

where $\tau = \nu^2/Q^2$ (ν is the electron energy loss), and $\sigma_T^{1/2}$ and $\sigma_T^{3/2}$ are transverse photon absorption cross sections for helicity 1/2 and 3/2 of the photon-nucleon system, The cross section $\sigma_T \equiv \frac{1}{2}(\sigma_T^{1/2} + \sigma_T^{3/2})$, and σ_{LT} arises from longitudinal-transverse interference. The nucleon structure function, F_1, is obtained from unpolarized scattering experiments. The functions $g_1(x,Q^2)$ and $g_2(x,Q^2)$ are the nucleon spin structure functions. In the scaling region, $g_1(x)$ is interpreted as a sum over polarizations of the various quark flavors in the nucleon, and $g_2 \to 0$.

The experiments are typically carried out with a longitudinally polarized electron beam and a target polarized either parallel or perpendicular to the beam, from which two asymmetries, $A_\|$ and $A_\perp$, linearly related to A_1 and A_2, can be formed:

$$A_\| = \frac{d\sigma^{\downarrow\Uparrow} - d\sigma^{\uparrow\Uparrow}}{d\sigma^{\downarrow\Uparrow} + d\sigma^{\uparrow\Uparrow}} = \frac{1}{P_e P_t} \frac{N^{\downarrow\Uparrow} - N^{\uparrow\Uparrow}}{N^{\downarrow\Uparrow} + N^{\uparrow\Uparrow}} = D(A_1 + \eta A_2),$$

$$A_\perp = \frac{d\sigma^{\downarrow\Rightarrow} - d\sigma^{\uparrow\Rightarrow}}{d\sigma^{\downarrow\Rightarrow} + d\sigma^{\uparrow\Rightarrow}} = \frac{1}{P_e P_t} \frac{N^{\downarrow\Rightarrow} - N^{\uparrow\Rightarrow}}{N^{\downarrow\Rightarrow} + N^{\uparrow\Rightarrow}} = d(A_2 - \zeta A_1).$$

The single arrow denotes the polarization of the electron, and the double arrow that of the nucleon. The horizontal arrow denotes polarization perpendicular to the beam direction. The terms N are the experimental counts normalized to the beam flux. The terms P_e and P_t specify the polarizations of the beam and target. The factors, D, d, η, and ζ depend on kinematical parameters and on the ratio, $R = \sigma_L/\sigma_T$.

The first moment, $\Gamma_1(Q^2) \equiv \int_0^1 g_1(x, Q^2) dx$ is constrained at high Q^2 by DIS. where it is positive for the proton. As $Q^2 \to 0$ the Gerasimov-Drell-Hearn [1, 2] sum rule for real photons requires that $\Gamma_1 \to 0$ with a negative slope. These two constraints imply that $\Gamma_1(Q^2)$ for the proton must change sign at some low Q^2 where it is expected to be dominated by the nucleon resonances [3, 4].

The electron beam at JLab has a polarization of 70-75%. In Halls B and C dynamically polarized solid NH_3 and ND_3 targets operating at T=1K in a homogeneous magnetic field of 5 T were used. A typical proton polarization of 70-90%, and a deuteron polarization up to 40% were obtained. Data were taken in the Resonance Spin Structure (RSS) experiment in Hall C [5] at one beam energy to measure both $A_\|$ and $A_\perp$ for $Q^2 \approx 1.3$ and $W < 2$ GeV. In the CEBAF Large Acceptance Spectrometer (CLAS) physical constraints have so far permitted measurements of only $A_\|$. However, with its large momentum and angle acceptance, the CLAS provides data over a continuous range in W and Q^2, obviating the necessity for interpolations between measured points. An extensive data set for several beam energies and Q^2 from 0.1 to 4. GeV^2 has been compiled with the CLAS.

Several experiments to measure both $A_\|$ and $A_\perp$ with beam energies up to 5.7 GeV have been carried out in Hall A using optically pumped gaseous ^{3}He targets with typical polarization of 40%. The scattered electrons are detected in two identical high resolution spectrometers, allowing simultaneous measurements at two electron angles. Recent results from the ^{3}He measurements were presented in the talk by Z. Meziani presented to the plenary session [6].

The experimental asymmetries, $A_\|$ and $A_\perp$, msut be corrected for the beam and target polarizations. Those measured with the ammonia targets must be converted to proton and deuteron asymmetries by correcting for the contribution of the unpolarized nitrogen, typically increasing the asymmetries by a factor of 7-8. The asymmetries must also be corrected for radiative effects. The photon asymmetries, A_1 and A_2, can be directly extracted from the two measured asymmetries. In the CLAS experiments only $A_\|$ is measured. When the complete data set is analyzed it may be possible to extract A_1 and A_2 from measurements of $A_\|$ at the same Q^2 and W but different beam energies (the "Rosenbluth technique"). In the meantime a model incorporating physical expectations

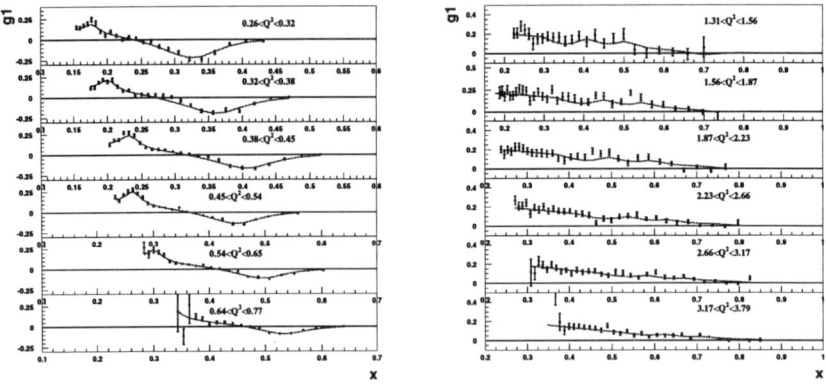

FIGURE 1. g_1 for the proton vs. Bjorken x at E=1.6 GeV(left) and 5.7 GeV(right). The solid lines are values obtained with the model used in the analysis.

and the world data was used to estimate F_2, ηA_2 and R, permitting A_1 and g_1 to be extracted from $A_{||}$. In general ηA_2 is small so that the extraction of A_1 is not very model dependent. The analysis of deuteron data from a first CLAS run has been published [7]. Results of the same run period for the proton have recently been submitted for publication [8]. Some preliminary results for g_1^p from a second CLAS run, using data taken with beam energies of 1.6 GeV and 5.6 GeV, are shown in Fig. 1. The solid line shows the predictions of the model used for extracting g_1. Systematic errors have not yet been determined. Results for g_1 for the deuteron are shown in Fig. 2. At the lower energy the prominent regions of negative values for both the proton and the deuteron are associated with excitation of the $\Delta(1232)$. At the higher energy the excitation of the $\Delta(1232)$ is much less prominent, and the results are close to those obtained with DIS. .

The contributions of the measured values of g_1^p to Γ_1^p are illustrated in Fig. 3. The points are the contributions of the measured values supplemented with the model calculations for low x values outside the measurement range. The published measurements are shown along with the new preliminary results. It can be seen that Γ_1^p crosses 0 around $Q^2 = 0.3$. The heavy dashed line near $Q^2 = 0$ shows the slope required by the GDH sum rule. The dashed line at $Q^2 > 1$ GeV is an evolution from DIS results. The upper solid line is a calculation of Soffer and Teryaev [9] and the lower solid line is a calculation of Burkert and Ioffe [10] using explicit resonance transition parameters.

We have presented a small sample of the high quality inclusive double-polarization data obtained at JLab by three large collaborations. Future activities include completion of the analysis of the existing CLAS data, the analysis of the RSS data, extraction of neutron structure functions using the combined data on the proton, deuteron and ^{3}He targets, and fits to moments of g_1 using the world data set. New measurements at very low Q^2 are planned for Halls A and B.

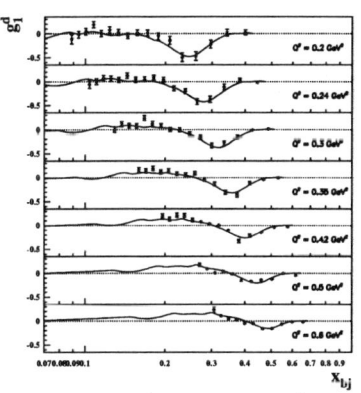
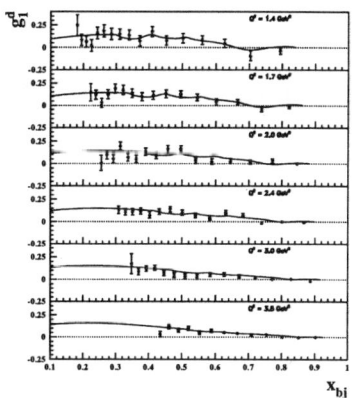

FIGURE 2. g_1 for the deuteron vs. Bjorken x at E=1.6 GeV (left) and E=5.6GeV (right). The solid lines are values obtained with the model used in the analysis.

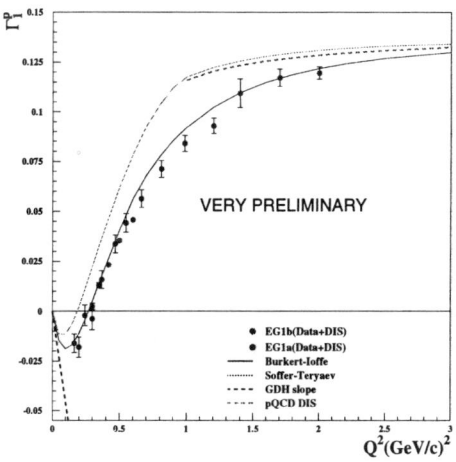

FIGURE 3. $\Gamma_1(Q^2)$ for the proton. See text for an explanation of the points and the curves.

REFERENCES

1. Gerasimov, S., *Yad. Fiz.*, **2**, 598 (1965).
2. Drell, S., and Hearn, A., *Phys. Rev. Lett.*, **16**, 908 (1966).
3. Burkert, V., and Li, Z., *Phys. Rev. D*, **47**, 46 (1993).
4. Drechsel, D., et al., *Phys. Rev.*, **D63**, 114010 (2001).
5. Rondon-Aramayo, O., *(Spokesperson),JLab Proposal E01-006* (2001).
6. Meziani, Z., *These proceedings* (2003).
7. Yun, J., et al., *Phys. Rev. C*, **67**, 055204 (2003).
8. Fatemi, R., et al., *arXiv: nucl-ex/0306091* (2003).
9. Soffer, J., and Teryaev, O. V., *Phys. Rev. D*, **51**, 25 (1993).
10. Burkert, V., and Ioffe, B., *J.E.T.P.*, **78**, 619 (1994).

Double-transverse spin asymmetries at NLO

A. Mukherjee*, M. Stratmann† and W. Vogelsang**

*Institut für Physik, Universität Dortmund, D-44221 Dortmund, Germany
†Institut für Theoretische Physik, Universität Regensburg, D-93040 Regensburg, Germany
**RIKEN-BNL Research Center and Physics Department, Brookhaven National Laboratory, Upton, New York 11973, U.S.A.

Abstract. We report on a next-to-leading order QCD calculation of the cross section and the spin asymmetry for isolated large-p_T prompt photon production in collisions of transversely polarized protons. Corresponding measurements may be used at RHIC to determine the transversity parton distributions of the proton.

The partonic structure of spin-1/2 targets at the leading-twist level is characterized by the unpolarized, longitudinally polarized, and transversely polarized distribution functions f, Δf, and δf, respectively [1, 2]. These non-perturbative parton densities can be probed universally in a multitude of inelastic scattering processes, for which it is possible to separate ("factorize") the long-distance physics relating to nucleon structure from a partonic short-distance scattering that is amenable to QCD perturbation theory.

In contrast to the f and Δf, the "transversity" distributions δf are unmeasured thus far. They are presently the focus of much experimental activity. For example, information should soon be gathered from transversely polarized proton-proton collisions at the BNL Relativistic Heavy Ion Collider (RHIC) [3]. The potential of RHIC in accessing transversity in measurements of transverse double-spin asymmetries A_{TT} was examined in [4] for high transverse momentum p_T prompt photon and jet production (for earlier studies, see [5, 6, 7]). All of these calculations were performed only at the lowest order (LO) approximation for the underlying partonic hard-scattering. As is well known, next-to-leading order (NLO) QCD corrections are generally indispensable in order to arrive at a firmer theoretical prediction for hadronic cross sections and spin asymmetries. The NLO calculation for A_{TT}^{γ} for isolated high-p_T prompt photon production, $pp \to \gamma X$, was recently completed [8]; here we give a brief report on those results.

Interesting new technical questions arise beyond the LO in case of transverse polarization. Unlike for longitudinally polarized cross sections where the spin vectors are aligned with momentum, transverse spin vectors specify extra spatial directions, giving rise to non-trivial dependence of the cross section on the azimuthal angle of the observed photon. As is well-known [2], for A_{TT} this dependence is always of the form $\cos(2\Phi)$, if the z axis is defined by the direction of the initial protons in their center-of-mass system (c.m.s.), and the spin vectors are taken to point in the $\pm x$ direction. Integration over the photon's azimuthal angle is therefore not appropriate. On the other hand, standard techniques developed in the literature for performing NLO phase-space integrations usually rely on integration over the full azimuthal phase space, and also on the choice of particular reference frames that are related in complicated ways to the one just specified. In [8], a new general technique was introduced which facilitates NLO calculations with transverse polarization by conveniently projecting on the azimuthal dependence of the matrix elements in a covariant way. The key point here is to recognize that the the factor $\cos(2\Phi)$ in the cross section actually results from the covariant expression

$$\mathscr{F}(p_\gamma, s_a, s_b) = \frac{s}{tu}\left[2(p_\gamma \cdot s_a)(p_\gamma \cdot s_b) + \frac{tu}{s}(s_a \cdot s_b)\right], \quad (1)$$

with s_a, s_b the initial spin vectors and p_γ the photon momentum. $\mathscr{F}$ reduces to $\cos(2\Phi)$ in the hadronic c.m.s. frame. One may thus integrate over all phase space without obtaining a vanishing result if one simply multiplies the squared matrix element by the factor $\mathscr{F}(p_\gamma, s_a, s_b)$. Integration over terms involving the s_a, s_b can be carried out in a covariant way by using standard tensor decompositions. After this step, there are no scalar products involving the s_i left in the squared matrix element. For the ensuing integration over all azimuthal phase space we can now employ techniques familiar from the corresponding calculations in the unpolarized and longitudinally polarized cases.

At NLO, there are two subprocesses that contribute for transverse polarization, $q\bar{q} \to \gamma X$ and $qq \to \gamma X$. The first one is already present at LO, where $X = g$. At NLO, one has virtual corrections to the Born cross section ($X = g$), but also $2 \to 3$ real emission diagrams, with $X = gg + q\bar{q} + q'\bar{q}'$. For the second subprocess, $X = qq$.

Owing to the presence of ultraviolet, infrared, and collinear singularities at intermediate stages of the calculation, it is necessary to introduce a regularization. Our choice is dimensional regularization, that is, the calculation is performed in $d = 4 - 2\varepsilon$ space-time dimensions. Ultraviolet poles in the virtual diagrams are removed by the renormalization of the strong coupling constant. Infrared singularities cancel in the sum between virtual and real-emission diagrams. After this cancelation, only collinear poles are left. These result for example from a parton in the initial state splitting collinearly into a pair of partons, corresponding to a long-distance contribution in the partonic cross section. From the factorization theorem it follows that such contributions need to be factored into the parton distribution functions. In our calculations [8], we have imposed on the photon the isolation cut proposed in [9]. All *final-state* collinear singularities then cancel. The isolation constraint was implemented analytically by assuming a narrow isolation cone.

For our numerical predictions we model the δf by assuming that the Soffer inequality [10] is saturated at some low input scale $\mu_0 \simeq 0.6\,\text{GeV}$. For $\mu > \mu_0$ the transversity densities $\delta f(x, \mu)$ are then obtained by solving the appropriate QCD evolution equations. Our numerical predictions apply for prompt photon measurements with the PHENIX detector at RHIC. Figure 1 shows our results for the transversely polarized prompt photon production cross sections at NLO and LO for two different c.m.s. energies. The lower part of the figure displays the so called "K-factor", $K = d\delta\sigma^{\text{NLO}}/d\delta\sigma^{\text{LO}}$. One can see that the NLO corrections are somewhat smaller for $\sqrt{S} = 500\,\text{GeV}$ and increase with p_T. The shaded bands in the upper panel of Fig. 1 indicate the uncertainties from varying the factorization and renormalization scales in the range $p_T/2 \leq \mu_R = \mu_F \leq 2p_T$. The solid and dashed lines are always for the choice where all scales are set to p_T, and so is the K factor underneath. One can see that the scale dependence becomes much weaker at NLO, as expected. The corresponding spin asymmetries $A_{TT}^\gamma = d\delta\sigma/d\sigma$ may be found in Fig. 2 of Ref. [8]; they are generally smaller at NLO than at LO.

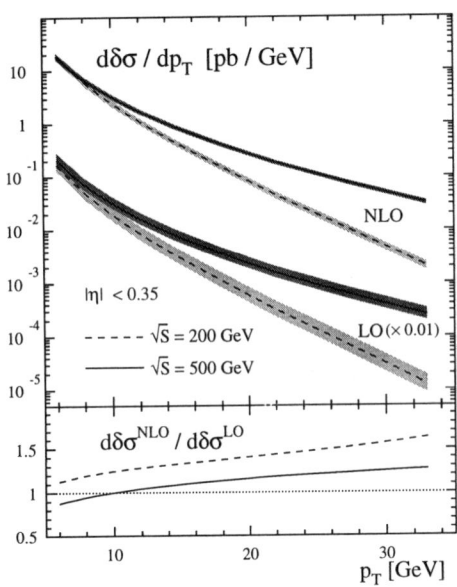

FIGURE 1. Predictions for the transversely polarized prompt photon production cross sections at LO and NLO, for $\sqrt{S} = 200$ and 500 GeV. The LO results have been scaled by a factor of 0.01. The shaded bands represent the theoretical uncertainty if μ_F $(= \mu_R)$ is varied in the range $p_T/2 \leq \mu_F \leq 2p_T$. The lower panel shows the ratios of the NLO and LO results for both c.m.s. energies.

ACKNOWLEDGMENTS

W.V. is grateful to RIKEN, Brookhaven National Laboratory and the U.S. Department of Energy (contract number DE-AC02-98CH10886) for providing the facilities essential for the completion of this work. This work is supported in part by the "Bundesministerium für Bildung und Forschung (BMBF)" and the "Deutsche Forschungsgemeinschaft (DFG)".

REFERENCES

1. R.L. Jaffe, X. Ji, Phys. Rev. Lett. **67**, 552 (1991); Nucl. Phys. **B375**, 527 (1992).
2. J.P. Ralston, D.E. Soper, Nucl. Phys. **B152**, 109 (1979).
3. G. Bunce, N. Saito, J. Soffer, W. Vogelsang, Annu. Rev. Nucl. Part. Sci. **50**, 525 (2000).
4. J. Soffer, M. Stratmann, W. Vogelsang, Phys. Rev. **D65**, 114024 (2002).
5. K. Hidaka, E. Monsay, D. Sivers, Phys. Rev. **D19**, 1503 (1979); X. Ji, Phys. Lett. **B284**, 137 (1992).
6. X. Artru, M. Mekhfi, Z. Phys. **C45**, 669 (1990).
7. R.L. Jaffe, N. Saito, Phys. Lett. **B382**, 165 (1996).
8. A. Mukherjee, M. Stratmann, W. Vogelsang, Phys. Rev. **D67**, 114006 (2003).
9. S. Frixione, Phys. Lett. **B429**, 369 (1998).
10. J. Soffer, Phys. Rev. Lett. **74**, 1292 (1995); D. Sivers, Phys. Rev. **D51**, 4880 (1995).

Status of the Michigan Ultra-Cold Spin-Polarized Hydrogen Jet [a]

K. Yonehara*, B.K. Harris*, M.C. Kandes*, B.H. Kienman*, A.D. Krisch*,
M.A. Leonova*, V.G. Luppov*, V.S. Morozov*, J.B. Olson*, C.C. Peters*,
R.S. Raymond*, D.L. Sisco*, N.S. Borisov[†], V.V. Fimushkin,[†] and
A.F. Prudkoglyad**

Spin Physics Center, University of Michigan, Ann Arbor, MI 48109-1120, USA
[†]*Joint Institute for Nuclear Research, RU-141980, Dubna, Russia*
**Institute for High Energy Physics, RU-142284, Protvino, Russia*

Abstract. Progress on the Michigan ultra-cold proton-spin-polarized atomic-hydrogen Jet target is presented. We describe the present status of the Jet and some beam test results.

We are developing an ultra-cold high-density Jet target of proton-spin-polarized hydrogen atoms (Michigan Jet) to study spin effects in high energy collisions. The Jet uses a very high magnetic field and an ultra-cold separation cell coated with a superfluid ^{4}He film to produce a slow monochromatic electron-spin-polarized atomic hydrogen beam. This beam is focused by a parabolic mirror coated with superfluid ^{4}He and a superconducting sextupole magnet. An rf transition unit will then convert this beam into a fully proton-spin-polarized beam [1].

A layout of the Michigan Jet is shown in Fig. 1. Atomic hydrogen is produced with a room-temperature rf dissociator and guided to the ultra-cold (170 mK) separation cell, which is coated with superfluid ^{4}He to suppress the surface recombination of hydrogen atoms, as shown in Fig. 2. The double walls of the cell form the mixing chamber of a dilution refrigerator. The cell's entrance and exit apertures are respectively located at about 95% and 50% of the superconducting solenoid's 12 Tesla magnetic field. After the hydrogen atoms are thermalized by collisions with the cell surface, the magnetic field gradient physically separates the atoms according to their electron-spin states. The "high-field-seeker" atoms in the two lowest hyperfine states ($|3\rangle$ and $|4\rangle$) are attracted up toward the high field region and most escape from the cell; they quickly recombine on bare surfaces and are cryopumped away. The "low-field-seeker" atoms in the two higher hyperfine states ($|1\rangle$ and $|2\rangle$) are accelerated toward the low field region and effuse from the exit aperture, forming a rather monochromatic electron-spin-polarized beam.

To increase the Jet density, we use a gold-plated copper focusing mirror with a polished surface covered with a ^{4}He superfluid film, which is similar to the prototype mirror [2]. After an rf transition unit [3], which will change state $|2\rangle$ atoms into state $|4\rangle$

---

[a] Supported by a research grant from the U.S. Department of Energy

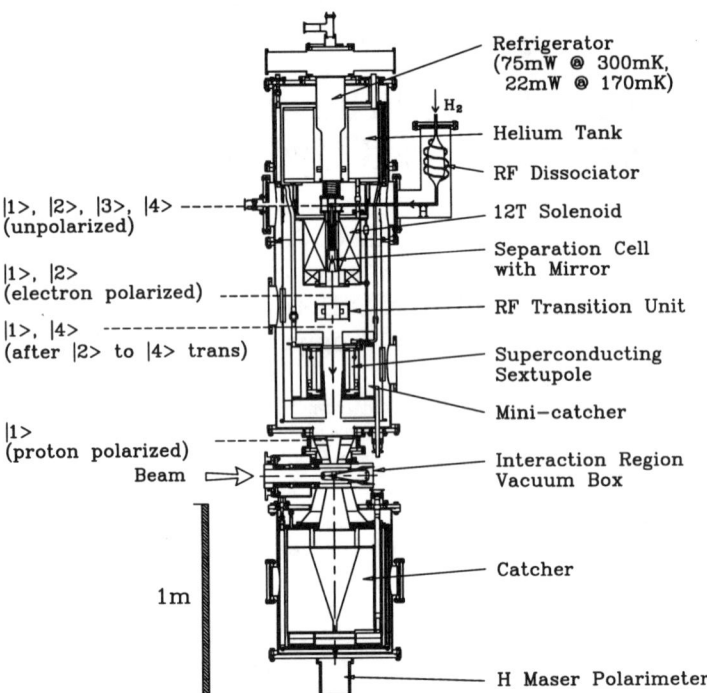

FIGURE 1. Layout of the Michigan ultra-cold Jet

atoms, the beam passes through the superconducting sextupole magnet. The sextupole focuses state $|1\rangle$ atoms into the interaction region and defocusses state $|4\rangle$ atoms, which are then cryopumped away. The proton-spin-polarized beam will then pass through the interaction region where it can collide with a proton beam in a high energy storage ring. The Jet beam is captured below by a huge cryopumping catcher [4] to keep the storage ring's vacuum uncontaminated. A maser polarimeter below the catcher monitors the Jet's proton polarization.

Most of the Michigan Jet parts have been fabricated, successfully tested, and used. This hardware includes a 12 Tesla superconducting solenoid with a very sharp gradient at its downstream end, a dilution refrigerator with a cooling power of about 22 mW at 170 mK, a 20 cm long superconducting sextupole magnet with a 3.1 kGauss field at its iron poles' 10.5 cm diameter bore, the catcher with a measured cryocondensation pumping speed of $1.2 \cdot 10^7$ liters s^{-1} [4], and a hydrogen maser polarimeter capable of monitoring the polarization with a precision of about $\pm$ 2% in a few minutes.

We studied a beam of polarized hydrogen atoms, which were focused by the superconducting sextupole into a compression tube detector that measured the polarized atoms' intensity. To avoid the superfluid ^{4}He film creeping out of the ultra-cold region, we installed three film burners above the mixing chamber, as shown in Fig. 2. Each film burner consists of a double-walled cell and a heater which is wound outside the cell. The superfluid ^{4}He film which creeps into the film burner cell is evaporated by the heater; the ^{4}He gas then recondences inside the the double-walled cell. By applying a higher heater voltage, we were able to build a thicker ^{4}He superfluid film, as shown in Fig. 3; this thicker film clearly increased the hydrogen beam intensity over some range, as shown in Fig. 3. The hydrogen intensity decreased when the film burner voltage was higher than 0.55 V since the heater power was then too high to recondense the ^{4}He gas. We concluded that the optimum film burner voltage was near 0.5 V.

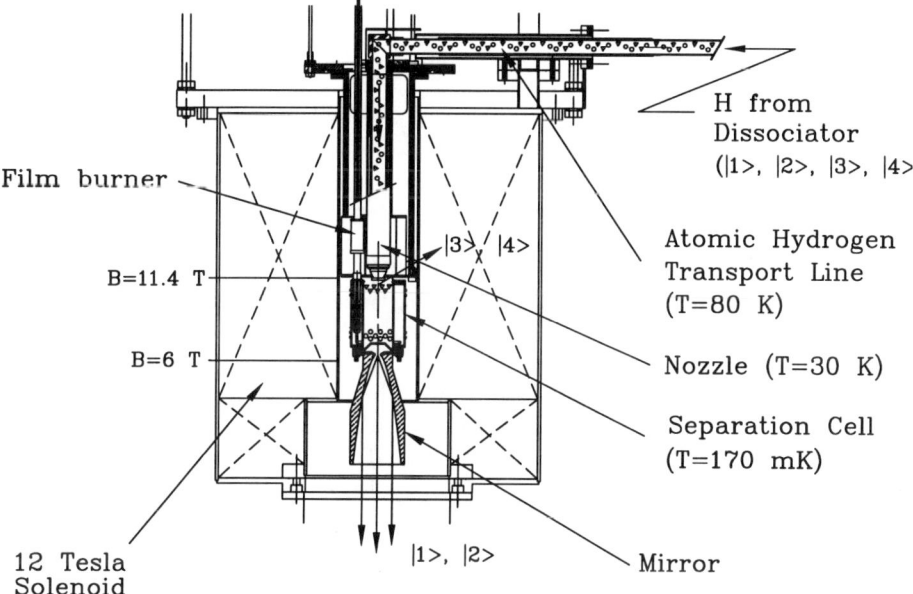

FIGURE 2. Details of the Michigan Jet's electron-spin separation region.

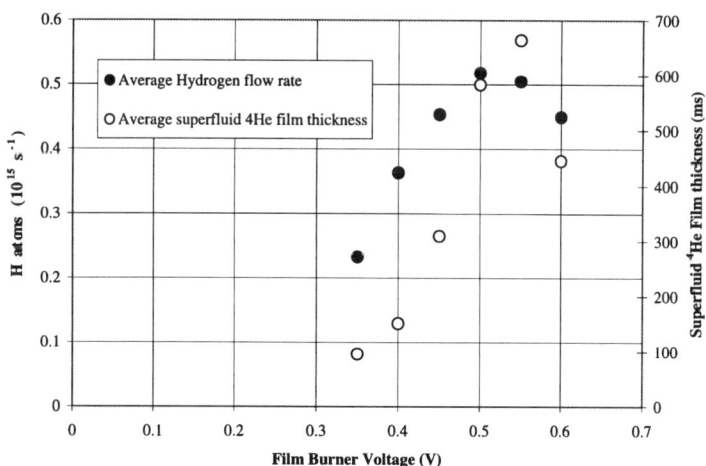

FIGURE 3. The observed hydrogen intensity entering the compression tube and superfluid ^{4}He film thickness as a function of the film burner voltage.

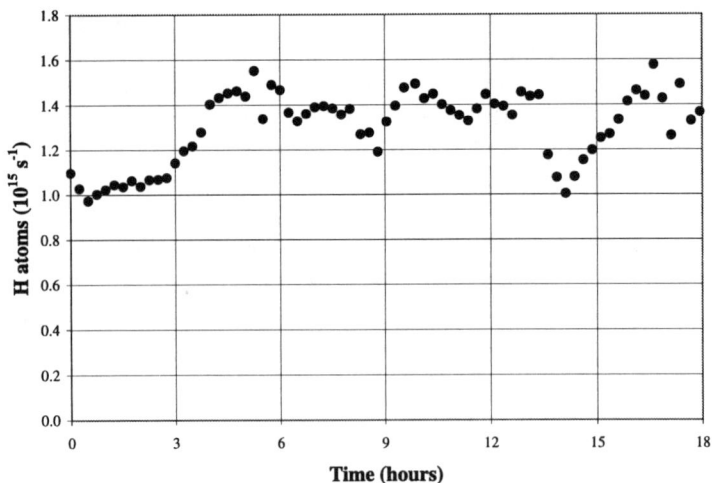

FIGURE 4. The observed hydrogen Jet intensity entering the compression tube during an 18-hour run.

By building a thick ^{4}He superfluid film, we were able to produce a high intensity spin-polarized hydrogen beam, which operated with good stability during an 18-hour run, until our liquid ^{4}He supply was depleted; the data is shown in Fig. 4. The average measured hydrogen intensity, into the 11 mm by 1.4 mm compression tube slot, was about $1.1 \; 10^{15}$ H s^{-1}. This intensity corresponds to a hydrogen Jet thickness of $0.8 \; 10^{12}$ H cm^{-2}; the maximum fluctuation was about ± 20 %. The Jet's highest measured intensity was $2.2 \; 10^{15}$ H s^{-1} [5]; this gave a Jet thickness of about $1.1 \; 10^{12}$ H cm^{-2}. We can heat the separation cell to about 40 K to remove the residual frozen H$_2$ molecules; this usually takes about 2 hours, but it was not needed during this 18-hour run.

We thank D. Kleppner and T. Roser for their earlier help.

REFERENCES

1. B.B. Blinov et al., "Michigan Ultra-Cold Polarized Atomic Hydrogen Jet", in Proc. of *14th International Spin Physics symposium, Osaka 2000*, edited by K. Hatanaka et al., American Institute of Physics, New York (2001), AIP Proceedings **570**, pp.856-860.
2. V.G. Luppov et al., *Phys. Rev. Lett.*, **71**, 2405 (1993).
3. R.S. Raymond, "Development of a Large-bore Cryogenic 2-4 Transition Unit", in Proc. of *International Workshop on Polarized Sources and Targets*, edited by A. Gute et al., Erlangen, Germany (1999).
4. V.G Luppov et al., "Status of the Mark-II Polarized Hydrogen Jet Target", in Proc. of *the 12th International Symposium on High-Energy Spin Physics*, edited by C. de Jager et al., New York (1997), World Scientific, pp. 434–437.
5. V.G. Luppov, "Status of the Michigan Ultra-Cold Polarized Hydrogen Target," in Proc. of *9th International Workshop on Polarized Sources and Targets*, edited by V.P. Derenchuk et al., Indiana (2001), World Scientific, pp. 32–36.

Spin Dependence in Polarized $pC \to pC$ Scattering at Low Momentum Transfer and Polarimetry at RHIC

A. Bravar[*], I. Alekseev[†], L. Ahrens[*], M. Bai[*], G. Bunce[*,‡],
S. Dhawan[§], H. Huang[*], V. Hughes[§], G. Igo[¶], O. Jinnouchi[‡],
K. Kurita[‖,‡], Z. Li[*], W.W. MacKay[*], S. Rescia[*], T. Roser[*],
N. Saito[**,‡], H. Spinka[††], D. Svirida[†], D. Underwood[††], C. Whitten[¶]
and J. Wood[¶]

[*]*Brookhaven National Laboratory, Upton, NY 11973, USA*
[†]*Institute for Theoretical and Experimental Physics, Moscow 117259, Russia*
[‡]*RIKEN BNL Research Center, Upton, NY 11973, USA*
[§]*Yale University, New Haven, CT 06511, USA*
[¶]*UCLA, Losa Angeles, CA 900095, USA*
[‖]*Rikkyo University, Toshima-ku, Tokyo 171-8501, Japan*
[**]*Kyoto University, Kyoto 606-8502, Japan*
[††]*Argonne National Laboratory, Argonne, IL 60439, USA*

Abstract. Preliminary results from the '03 RHIC Proton Spin Run on the spin dependence in the elastic scatterring of polarized protons off a carbon target at very low momentum transfer ($0.005 < |t| < 0.05$ GeV$^2/c^2$) are presented and discussed. Proton polarimeters based on this process are used in RHIC and AGS to measure reliably and in very short times the polarization of the proton beams. Polarimetry results from the just completed RHIC polarized proton run are also presented.

The analyzing power A_N is defined as the left-right asymmetry of the cross section in the scattering plane normal to the beam polarization. A_N arises from the interference between a spin-flip and spin-nonflip amplitude and thus provides important information on the spin dependence of the interaction. In high energy pp and pA elastic scattering at very low momentum transfer t, A_N originates from the interference between the imaginary electromagnetic (Coulomb) spin-flip amplitude, which is generated by the proton's anomalous magnetic moment, and the real hadronic (Nuclear) spin-nonflip amplitude (CNI - Coulomb Nuclear Interference). A_N has a maximum value of about 4% around a momentum transfer $-t$ of 3×10^{-3} (GeV/c)^2 and decreases with increasing $|t|$. The existence of a potential hadronic spin-flip amplitude interfering with a real electromagnetic spin-nonflip amplitude introduces a deviation from A_N calculated with no hadronic spin-flip amplitude. While the former contribution is fully calculable, the latter can be tackled only in Regge type phenomenological approaches and it is expected to diminish with increasing en-

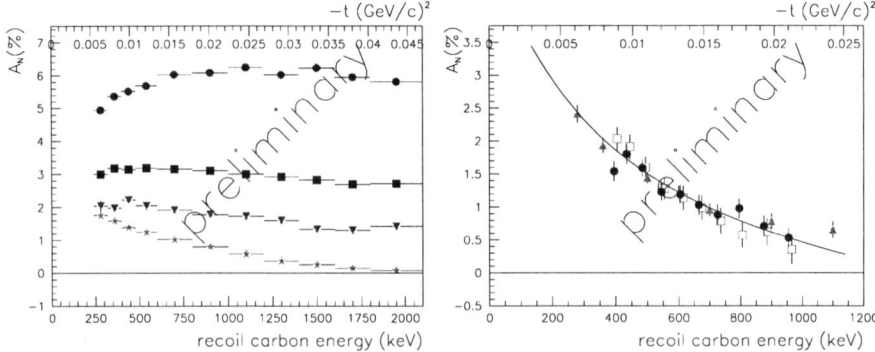

FIGURE 1. left: A_N^{pC} for $pC \to pC$ as function of the recoil carbon kinetic energy $T_{rec} = |t|/2M_C$ measured with the AGS polarized proton beam at 4 different energies; starting from top at 3.9 GeV, 6.5 GeV, 9.7 GeV, and 21.7 GeV. The displayed errors are statistical only. The estimated normalization error (i.e. ΔP_B) is $\sim 10\%$ for the lowest energy data points and increases to $\sim 20\%$ for the highest ones. P_B has been measured concurrently with an internal inelastic polarimeter ($pC \to p + p + X$). The systematic error, which comes mainly from backgrounds below the elastic pC peak, pileups and electronic noise, is estimated to be $< 20\%$.
right: A_N^{pC} as a function of T_{rec} at RHIC injection energy at 24.7 GeV measured with the AGS polarimeter (full triangles), and at the RHIC flattop at 100 GeV measured with the Yellow beam (open squares) and the Blue beam polarimeters (full circles). The line is a phenomenological fit [4] to the E950 A_N^{pC} data [2]. The measured asymmetries are normalized to this fit assuming the same A_N at 24.7 GeV and 100 GeV.

ergy [1]. The hadronic spin-flip amplitude carries important information on the static properties and on the constituent quark structure of the nucleon. Whitin Regge phenomenology, one can probe the long standing question of the magnitude of the Pomeron spin-flip amplitude through the study of A_N in the CNI region. A significant hadronic spin-flip amplitude, at a 15% level, was required to fit pC elastic scattering data at 22 GeV from experiment E950 at BNL [2].

pC elastic scattering events are identified by detecting the recoil carbon nuclei. Slow recoil carbon nuclei emerge at almost 90° w.r.t. the incident beam and are detected with silicon detectors, which provide energy and time of flight (t.o.f) information. On the basis of the energy – t.o.f correlation, carbon recoil events are identified and selected. An ultra thin carbon ribbon is used as target and is inserted into the beam during the measurement. A deadtimeless DAQ system, based on waveform digitizer modules with *on-board* event analysis, is used in order to handle the very high rates. The small value of A_N^{pC} makes it necessary to collect 2×10^7 events per measurement. A typical measurement in RHIC lasts only 30 sec. For more details on the RHIC pC CNI polarimeters see [3].

Figure 1 shows A_N^{pC} for elastic pC scattering as a function of the recoil carbon energy T_{rec}. At proton beam energies E_B below 10 GeV a very weak $|t|$ dependence is observed, which is quite different from the *typical* CNI-type behavior. In addition, A_N^{pC} decreases from $\sim 6\%$ to $\sim 2\%$ with E_B increasing from 4 GeV to 10 GeV.

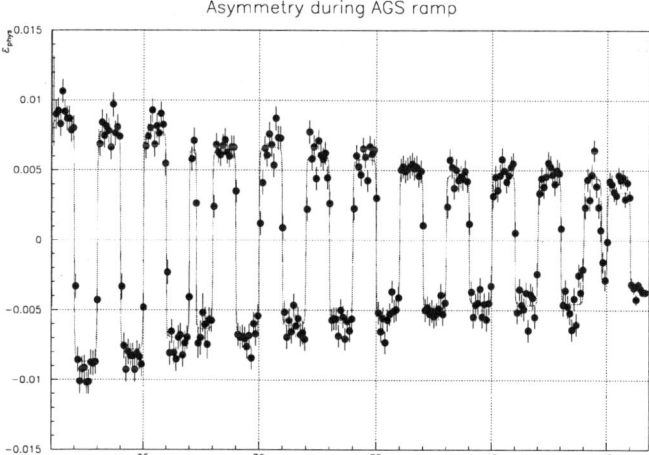

FIGURE 2. Measured asymmetry $\varepsilon_N = A_N \cdot P_B$ during the acceleration of the polarized proton beam in AGS as a function of $G\gamma = 1.91 \cdot E_B$ from $E_B \sim 11$ GeV to extraction at 24.7 GeV. Each point corresponds to an increase of E_B of ~ 50 MeV (1 msec measurement time). The continous line is an iterpolation to the measured points.

At higher E_B above 20 GeV the $|t|$ dependence resembles the CNI-type behavior consistent with a hadronic spin-flip amplitude at a 10–15% level. To be noted is the very similar t dependence of A_N^{pC} at E_B of 25 and 100 GeV.

The main motivation for studying A_N^{pC} in elastic pC scattering in the CNI region, however, comes from the RHIC polarized proton collider at BNL. This process is used for high energy proton polarimetry, since the acceleration of polarized proton beams and experiments with them require fast and reliable measurements of the beam polarization P_B. The RHIC polarimeters operate around $-t \sim 10^{-2}$ (GeV/c)^2, where $A_N^{pC} \sim 1$ %. The figure of merit, however, is very high, since the pC elastic cross section is very large in this kinematic region. The theoretical uncertainties from the unknown hadronic spin-flip amplitude limit the predictive power for this process, thus requiring that the pC CNI polarimeters be experimentally calibrated over the RHIC energy range before they can provide precise measurements of P_B. With the current knowledge on A_N^{pC} from the E950 experiment, P_B cannot be determined to better than ±30% (relative) at RHIC injection E_B of 25 GeV, while at the RHIC flattop of 100 GeV the unceratinty on P_B is even larger due to extrapolation uncertainties to the higher energy. For this purpose an absolute polarimeter using an internal polarized hydrogen gas jet target is being built [5]. The target polarization P_T will be measured to better than 3% using a Breit-Rabi polarimeter. P_T will be then transfered to P_B using pp elastic scattering in the CNI region. For pp elastic scattering A_N is identical, whether the beam or target is polarized. The jet target will be installed for the '04 run with the initial goal of determining P_B to about 10%. In 2005 a precision of 5% on P_B is expected. This setup will allow to study A_N^{pp} in polarized pp elastic scattering in

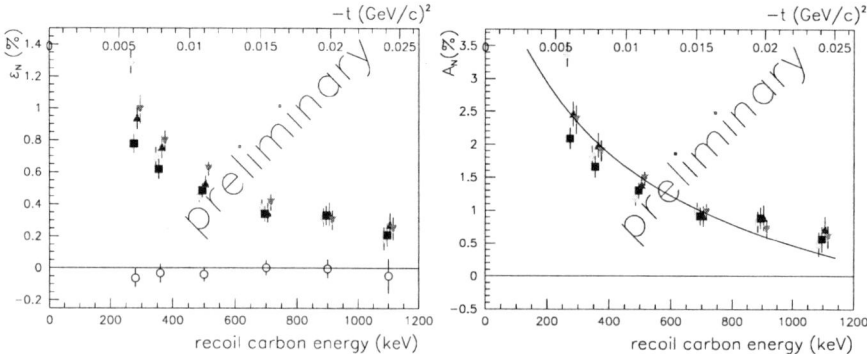

FIGURE 3. left: Several measurements of $\varepsilon_N = A_N \cdot P_B$ with the AGS CNI polarimeter as a function of T_{rec} at E_B of 24.7 GeV with similar P_B. Each measurement lasted about 5 minutes. The open circles are the asymmetries measured with an unpolarized beam.
right: A_N^{pC} for $pC \to pC$ at 24.7 GeV extracted from the measured ε_N. In order to determine A_N^{pC} from ε_N, P_B has been estimated by normalizing ε_N to a phenomenological fit to the E950 data (full line [4]). For these measurements $P_B \sim 0.4$. The overall error on P_B thus determined is around $\pm 30\%$ (relative) and comes mainly from the uncertainties in the E950 data.

the kinematical range of $0.001 < |t| < 0.02$ (GeV/c)^2 and $\sqrt{s}$ up to 22.4 GeV with a very high accuracy of $\Delta A_N^{pp} < 10^{-3}$. With both target and beam polarized also the double spin parameter A_{NN}^{pp} will be studied with similar precision.

Figure 2 shows the asymmetry $\varepsilon_N = A_N \cdot P_B$ measured during the acceleration of the polarized proton beam in the AGS as a function of $G\gamma$, where G is the proton anomalous magnetic moment and γ the Lorentz boost factor. $G\gamma$ is related to E_B by $G\gamma = 1.91 \cdot E_B$. $G\gamma$ is directly connected to the spin motion of particles in a storage ring and it gives the number of spin precessions during one orbit. During the acceleration, whenever the value of $G\gamma$ is an integer, an *imperfection* depolarizing resonance is crossed with consequent loss of beam polarization. To avoid such polarization losses, when crossing these resonances, the beam polariziation is reversed with the use of a partial solenoidal siberian snake. In addition to the *imperfection* resonances, two *intrinsic* resonances are visible in Figure 2 around $G\gamma$ of 27 and 45. When crossing these two resonances the beam polarization is reversed using an AC RF dipole (*spin flipper*). For more details on the spin motion see [6].

REFERENCES

1. B.Z. Kopeliovich and T.L. Trueman, Phys. Rev. D **64**, 034044 (2001).
2. J. Tojo et al., Phys. Rev. Lett. **89**, 052302 (2002).
3. O. Jinnouchi et al., Proc. 15th Int. Symp. on High-Energy Spin Physics SPIN2002.
4. T.L. Trueman, hep-ph/0305085.
5. A. Bravar et al., Proc. 15th Int. Symp. on High-Energy Spin Physics SPIN2002.
6. W.W MacKay, *these proceedings*.

A Study of Heavy Quark and Quarkonium Production in Polarized p-p collisions at RHIC

Ming Xiong Liu [1]

Los Alamos National Laboratory
Los Alamos, NM 87545

Abstract.
At RHIC, heavy quark production is sensitive to the gluon distribution inside the proton. The expected integrated luminosity of $3pb^{-1}$ from this year's first longitudinally polarized proton beams will provide an opportunity to study the helicity dependent heavy-quark and quarkonium production. This could lead to direct measurement of the gluon polarization in the polarized proton. I discuss the experimental sensitivity of such measurements with the PHENIX detector. New ideas to explore physics beyond the standard model with heavy quarks in the future high energy polarized p-p collisions are also discussed.

INTRODUCTION

The surprising result from the EMC experiment 15 years ago at CERN showed that the sum of the quark's spin only accounts for a small fraction of the proton's total spin[1]. Tremendous progress has been made in both experimental measurements and theoretical studies since then, however the origin of the proton's spin still remains an outstanding puzzle. One very likely candidate to solve this "missing" proton spin problem is the gluon polarization.

RHIC-SPIN is the first high-energy polarized proton-proton collider where hard gluon-gluon and gluon-quark scattering processes can be studied at leading order. It opens up a new frontier for the study of the gluon distribution inside the polarized proton.

PHENIX EXPERIMENT

The PHENIX experiment at RHIC is designed for a broad spectrum of physics involving both Heavy Ion collisions and Spin[2]. The excellent capability for lepton measurements allows the PHENIX experiment to have a rich program in heavy-quark physics, as has already been demonstrated in recent publications on open charm and J/ψ production through electron and muon measurements[3]. Here, I highlight selected topics from heavy-quark measurements with the PHENIX detector.

---

[1] for the PHENIX Collaboration

Heavy quark production and parton distribution

In the study of parton structure of the nucleon, perturbing QCD plays an critical role. The nucleon's structure functions measured in hard processes can be written as a sum of parton distributions $f_a(x,q^2)$ for each quark and gluon, where a is the index of parton flavor, x represents parton's momentum fraction in the nucleon, and q^2 is the energy scale at which it is measured. In the parton model, the heavy quark production process $a+b \to Q+\bar{Q}$ is given by,

$$E\frac{d^3\sigma}{d^3p} \propto f_a(x_a,q^2)f_b(x_b,q^2)\frac{d\hat{\sigma}(a+b \to Q\bar{Q})}{dt} \qquad (1)$$

where $\frac{d\hat{\sigma}(a+b \to Q+\bar{Q})}{dt}$, the parton scattering cross section, can be calculated with perturbative QCD. At RHIC energy, $\sqrt{s} = 200$ GeV, it is expected that charm and bottom quarks are predominately produced through the gluon fusion process. A simple simulation estimation with PYTHIA [4] shows that about 95% and 85% are from the gluon fusion process in open charm and open bottom production, respectively. So the the spin-dependent asymmetry A_{LL} from heavy-quark production in polarized p-p collisions, is given by,

$$A_{LL} = \frac{\sigma(++) - \sigma(+-)}{\sigma(++) + \sigma(+-)} \simeq \frac{\Delta G(x_1)}{G(x_1)}\frac{\Delta G(x_2)}{G(x_2)}\hat{a}_{LL}(g+g \to Q\bar{Q}) \qquad (2)$$

where $\Delta G(x) = G^+(x) - G^-(x)$ is the longitudinally polarized gluon distribution inside the proton, and $\hat{a}_{LL}$ is the asymmetry in the partonic process $g+g \to Q+\bar{Q}$. The latter has been computed at LO and NLO[5, 6].

Open heavy quark and quarkonium measurements in the PHENIX

In PHENIX, open heavy-quark production is measured with high P_T single leptons and dilepton pairs. For heavy quarkonium such as the J/ψ, it is fully reconstructed in its dilepton decay mode. PHENIX also has a plan to upgrade its inner tracking system to include a silicon vertex detector to explicitly detect heavy-quark decays.

By studying the spin-dependent asymmetry A_{LL} in heavy-quark production, the polarized gluon distribution $\frac{\Delta G(x)}{G(x)}$ can be extracted from equation (2). Fig. 1 show the expected asymmetry A_{LL} from single muon and J/ψ measurements in PHENIX for various projected luminosities.

Heavy quarks as a probe for new physics

Besides the importance in their own right in determining the spin structure of nucleon, study of the spin-dependent heavy-quark production will also provide a critical test of the QCD, and may eventually lead us to search for new physics beyond the standard

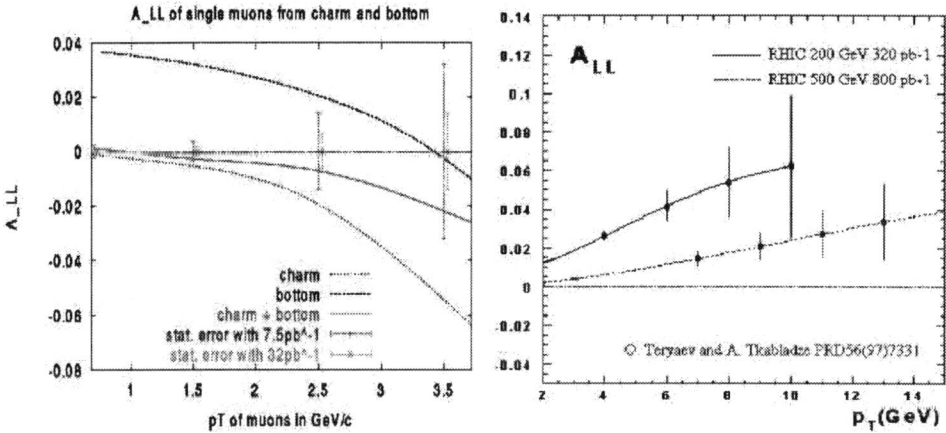

FIGURE 1. Left: Expected A_{LL} in open charm and open bottom production from single muon measurement. Right: Expected A_{LL} from J/ψ measurement.

model. The planned $\sqrt{s} = 500$ GeV polarized p-p collisions at RHIC will certainly open up a new window for such physics. Several ideas have already been proposed to probe new physics at RHIC-SPIN[7, 8].

Here we discuss one example where contributions from new heavier particles could be studied for bottom-quark production in polarized p-p collisions at RHIC. The observed large excess in bottom quark production compared to NLO pQCD prediction at Tevatron has triggered many theoretical discussions. Several models have been proposed to explain the long-standing discrepancy, including the use of improved B-hadron fragmentation functions and modified parton distribution functions[9]. There are also models based on physics beyond the standard model. In a recent paper[10], Berger et al. showed that the creation of light gluinos of mass ~ 15 GeV at the Tevatron would produce more bottom quarks and thus bring the theoretical expectation into good agreement with the experimental data. However, in general, it is hard to disentangle these different contributions in unpolarized collisions.

We argue that if indeed there exist such new particles of mass ~ 15 GeV that contribute significantly to the bottom quark production at high energy p-p collisions, it is possible to directly probe them at RHIC in the bottom-quark A_{LL} measurement if gluons have a sizable polarization inside the proton. The modified asymmetry A_{LL} is given by,

$$A_{LL} = \frac{\Delta\sigma_{SM} + \Delta\sigma_{NEW}}{\sigma_{SM} + \sigma_{NEW}} \quad (3)$$

where σ_{NEW} is the contribution from new processes such as light gluino production. Fig. 2 shows the LO calculation of A_{LL} as a function of $\sqrt{\hat{s}}/2M$. At high $P_T^* \geq 15$ GeV, the standard model process $g + g \to b + \bar{b}$ gives $A_{LL} \sim -1$ while the gluino processes $g + g \to \tilde{g} + \tilde{g}$, followed by $\tilde{g} \to b\tilde{b}$, will yield $A_{LL} \sim +1$ if the gluino has a large mass ~ 15 GeV. Detailed studies are underway to include NLO corrections.

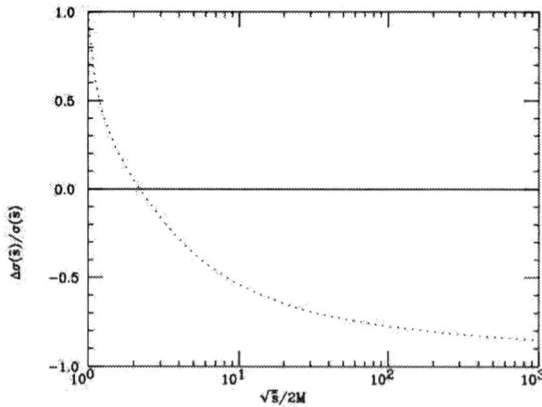

FIGURE 2. Parton level a_{LL} as a function of $\sqrt{\hat{s}}/2M$ in heavy-quark production.

SUMMARY

We discussed the possibility to use heavy-quark measurements to study the polarized gluon distribution inside the proton, and to probe new physics beyond the standard model. We expect with improved luminosity and beam polarization in the future, heavy-quark measurements will produce exciting results for the physics outlined above. The RHIC-SPIN program opens up a new frontier for nucleon structure physics and allows the search for new physics beyond the standard model.

ACKNOWLEDGMENTS

The author wishes to thank the members of PHENIX experiment for stimulating discussions.

REFERENCES

1. J.Ashman et al., Phys. Lett. B 206, 346 (1988), Nucl. Phys. B 328, 1(1989).
2. PHENIX Conceptual Design Report, BNL, 1993, unpublished.
3. D. Silyvermyr, see proceedings from this conference.
4. T. Sjostrand et al., Computer Physics Commun. 135 (2001) 238.
5. M. Karliner and R.W. Robinett, Phys. Lett. B 324 (1994) 209.
6. I. Bojak and M. Stratmann, hep-ph/0112276.
7. C. Kao et al, Phys. Lett. B 395 (1997) 327-333.
8. G. Bounce et al., Ann. Rev. Nucl. Part. Sci., 50 (2000) 525.
9. S.P. Baranov et al., hep-ph/0302171.
10. E.L. Berger et al., Phys. Rev. Lett. Vol. 86, 4231 (2001).

Single-spin asymmetries in exclusive electroproduction of pseudoscalar and vector mesons at HERMES

Delia Hasch
(On behalf of the HERMES Collaboration)

INFN - Laboratori Nazionali di Frascati, 00044 Frascati, Italy

Abstract. Results from the HERMES experiment are presented on target-related single-spin azimuthal asymmetries for exclusively produced π^+ and ρ^0 mesons in deep-inelastic scattering of 27.5 GeV positrons off a longitudinally polarized hydrogen and deuterium target, respectively. The azimuthal asymmetries are studied as a function of the relevant kinematic variables. The magnitude of the π^+ asymmetry is found to grow with decreasing Bjorken-x and increasing $-t$ (where t is the squared four-momentum transferred to the nucleon), while the ρ^0 asymmetry is found to be consistent with zero.

INTRODUCTION

Recent interest in exclusive processes has arisen with theoretical progress in the factorization for the non-forward regime for hard exclusive production of mesons by longitudinal virtual photons [1]. The Generalized Parton Distribution (GPD) functions appearing in the factorization scheme provide a unified description of exclusive and inclusive reactions [2]. GPDs provide access not only to longitudinal but also to transverse degrees of freedom. Moreover, it was shown [3] that the first moment of certain GPDs can be related to the total angular momentum of the quarks in the nucleon.

Since only a quadratic combination of GPDs appears in the unpolarized cross section for exclusive meson production, polarization is needed in order to disentangle the various distributions by accessing additional observables. It has been shown [4, 5] that for exclusive pseudoscalar production from a transversely polarized nucleon the interference between the two polarized GPDs leads to a large target-related single-spin asymmetry in the distribution of the angle ϕ. Here ϕ is the azimuthal angle of the meson around the virtual photon momentum relative to the lepton scattering plane. For exclusive vector meson production from a transversely polarized nucleon the interference between the two unpolarized GPDs leads to a single-spin asymmetry which depends *linearly* on the GPD E [2]. Therefore, the transverse spin asymmetry for vector mesons provides a unique observable to extract the GPD E and the quark total angular momentum J^q to the proton spin since J^q enters as a free parameter the parametrization of E.

As the asymmetries represent a ratio of cross sections the scaling region is reached at lower Q^2 of the virtual photon than for the absolute cross section. Moreover, corrections that are next-to-leading order in α_s cancel in the transverse asymmetry.

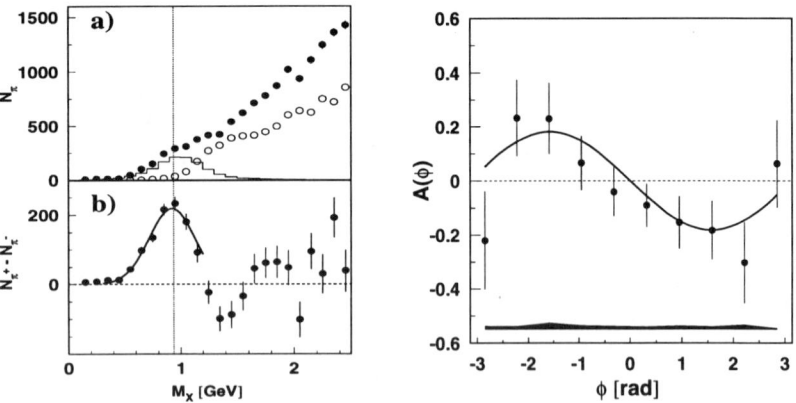

FIGURE 1. Left panel: a)Missing mass distribution for π^+ (filled circle) and π^- (empty circles) electroproduction on the proton. The histogram is a Monte Carlo prediction for exclusive π^+ production. b) Difference between the π^+ and normalized π^- distributions. The curve is a gaussian fit to the data, the dotted line indicates the nucleon mass. **Right panel:** Longitudinal target-spin asymmetry for exclusive π^+ production. The curve is a fit to the data with the function $A(\phi) = A_{UL}^{\sin\phi} \sin\phi$.

In this contribution, experimental data are presented for the measurement of target-related single-spin azimuthal asymmetries for exclusively produced π^+ and ρ^0 mesons in deep-inelastic scattering. The data has been collected by the HERMES experiment at DESY in Hamburg, Germany. A 27.5 GeV polarized electron or positron beam in the HERA storage ring is scattered off a longitudinally polarized hydrogen and deuterium gas target internal to the beam pipe. The HERMES forward spectrometer [6] features excellent particle identification capabilities which provide full separation between charged pions, kaons and protons over essentially the entire momentum range of the experiment.

PSEUDOSCALAR MESON PRODUCTION

HERMES has measured the single-spin asymmetry in the exclusive electroproduction of π^+ mesons using a longitudinally polarized hydrogen target [7]. With respect to the virtual photon, however, the target polarization vector has a certain transverse component as well.

Exclusive events were selected by requiring the missing mass (M_X) of the reaction $e^+ + \vec{p} \to e^+ + \pi^+ + X$ to correspond to the nucleon mass. With an experimental M_X resolution of about 230 MeV exclusive π^+ cannot be directly separated from non-exclusive channels. The non-exclusive channels are also present in the $e^+ + \vec{p} \to e^+ + \pi^- + X$ process while exclusive π^- production with a nucleon in the final state is forbidden by charge conservation. As for π^+, also non-exclusive π^- events are smeared into the exclusive region. Therefore, the non-exclusive π^+ background was estimated from the normalized number of π^- passing the same cuts as π^+. The left panel of Fig. 1 shows the difference between the π^+ and the normalized π^- missing mass distributions. A clear exclusive π^+-production peak is observed centered at the nucleon mass.

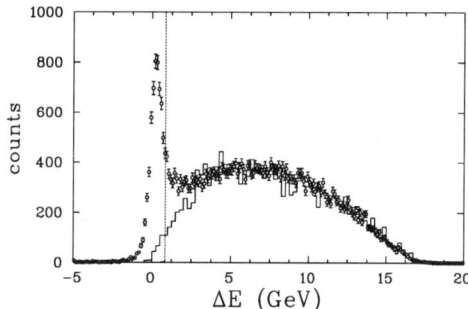

FIGURE 2. Missing energy distribution of the reaction $e^+ + \vec{d} \to e'^+ + \rho^0 + X \to e'^+ + \pi^+\pi^- + X$. The histogram shows a Monte Carlo simulation to describe the non-exclusive background. The dotted line indicates the cut used to select exclusive events.

The ϕ dependence of the polarized cross section appears in the cross section asymmetry defined by

$$A(\phi) = \frac{1}{|P_t|} \frac{N^+(\phi) - N^-(\phi) - |P_t| A_{bg}(\phi) N_{bg}(\phi)}{N^+(\phi) + N^-(\phi) - N_{bg}(\phi)} \quad (1)$$

Here, $N^+(-)$ is the number of events with the direction of the target spin antiparallel (parallel) to the positron beam momentum and P_t is the target polarization. N_{bg} is the number of events of the non-exclusive background estimated from π^- production and A_{bg} is the non-exclusive background asymmetry of π^+. The cross section asymmetry integrated over x, Q^2 and t is shown in the right panel of Fig. 1. The average values of the kinematic variables are $\langle x \rangle = 0.15$, $\langle Q^2 \rangle = 2.2$ GeV^2 and $\langle t \rangle = -0.46$ GeV^2. The data show a clear sinusoidal dependence of the polarized cross section with an amplitude of $A_{UL}^{\sin\phi} = -0.18 \pm 0.05(\text{stat.}) \pm 0.02(\text{syst.})$. The dependence of $A_{UL}^{\sin\phi}$ on the relevant kinematic variables has been investigated and is shown in Fig. 3. The absolute magnitude of the asymmetry is found to increase at low x and at large $-t$.

As the data are dominated by the contribution from the longitudinal target polarization component with respect to the virtual photon direction, a comparison to the predictions for a transversely polarized target would require a next-to-leading twist calculation [5]. Upcoming results from the measurement with a transversely polarized hydrogen target, which started in 2002, will allow a direct comparison with the predicted asymmetries.

VECTOR MESON PRODUCTION

HERMES has measured the single-spin asymmetry in the exclusive electroproduction of ρ^0 mesons using a longitudinally polarized deuteron target. The ρ^0 is identified by detecting its decay products, where the two opposite charged pions are required to have an invariant mass in the range $0.6 < M_{\pi\pi} < 1.0$ GeV. Exclusive events are selected by requiring the missing energy (ΔE) of the reaction $e^+ + \vec{d} \to e^+ + \rho^0 + X$ to be smaller

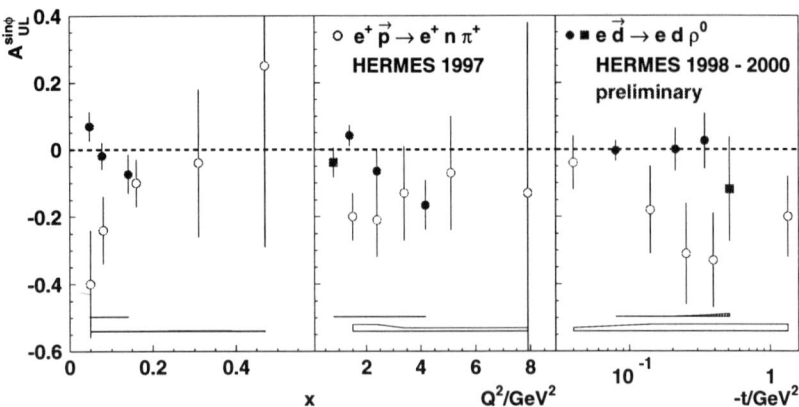

FIGURE 3. Kinematic dependence of $A_{UL}^{\sin\phi}$ on the relevant kinematic variables for exclusive π^+ production (open circles) and for exclusive ρ^0 production (filled circles).

than 0.6 GeV. As shown in Fig. 2 a clear peak of exclusive events appears in the missing energy distribution. Non-exclusive background can be well described by Monte Carlo and has also been estimated from a fit to the ΔE distribution.

The cross section asymmetry is calculated according to Equation 1. Integrated over x, Q^2 and t it amounts to $A_{UL}^{\sin\phi} = -0.003 \pm 0.027 \pm 0.001$; with average values of the kinematic variables $\langle x \rangle = 0.085$, $\langle Q^2 \rangle = 2.0$ GeV^2 and $\langle t' = t - t_{\min} \rangle = -0.12$ GeV^2. The dependence of the asymmetry on the relevant kinematic variables is shown together with the aforementioned results for exclusive π^+ production in Fig. 3.

While the separation of the longitudinal and transverse part of the cross section for exclusive π^+ production requires the measurement of the total cross section at different beam energies (Rosenbluth separation), the longitudinal part of the cross section for exclusive ρ^0 production can be obtained by extracting the spin density matrix elements from the decay angular distribution [8]. This analysis is under way and will provide first results on a single-spin asymmetry for exclusive production of longitudinal ρ_L^0.

REFERENCES

1. J. Collins, *Nucl. Phys.* B **396**, 161 (1993).
2. For a recent review see K. Goeke, M.V. Polyakov and M. Vanderhaegen, *Prog. Part. Nucl. Phys.* **47**, 401-515 (2001).
3. X. Ji, *Phys. Rev. Lett.* **78**, 610 (1997).
4. L.L. Frankfurt, P.V. Pobylitsa, M.V. Polyakov, and M. Strikman *Phys. Rev.* D **60**, 14010 (1999); L.L. Frankfurt, M.V. Polyakov, M. Strikman, and M. Vanderhaegen *Phys. Rev. Lett.* **84**, 2589 (2000).
5. A.V. Belitsky and D. Müller, *Phys. Lett.* B **513**, 349 (2001).
6. HERMES collaboration, *Nucl. Instrum. Methods* A **417**, 230 (1998).
7. HERMES collaboration, *Phys. Lett.* B **535**, 85 (2002).
8. HERMES collaboration, A. Airapetian et al., *Eur. Phys. J.* C **18**, 303 (2000).

Flavor Decomposition of Nucleon Spin Structure: A Proposed Experiment at Jefferson Lab Hall C

Xiaodong Jiang*, Donal B. Day[†] and Mark K. Jones**

*Department of Physics and Astronomy, Rutgers University, Piscataway, New Jersey.
[†]Department of Physics, University of Virginia, Charlottesville, Virginia.
**Thomas Jefferson National Accelerator Facility, Newport News, Virginia.

Abstract. We propose to measure the semi-inclusive double-spin asymmetries in deep-inelastic $\vec{p}(\vec{e},e'h)X$ and $\vec{d}(\vec{e},e'h)X$ reactions ($h = \pi^+, \pi^-, K^+$ and K^-) on longitudinally polarized NH_3 and LiD targets. The high statistic data will allow a leading order five-flavor decomposition (Δu, Δd, $\Delta \bar{u}$, $\Delta \bar{d}$, $\Delta s = \Delta \bar{s}$) of the nucleon spin structure in the region of $x = 0.12 \sim 0.43$ at a Q^2 of $1.21 \sim 3.22$ GeV^2. The combined spin dependent $\pi^+ + \pi^-$ yield ratios will be measured, since the acceptance and the hadron efficiencies of the detectors can be accurately determined. Independent of fragmentation functions and parton distributions, the z-dependence ($z = 0.38 \sim 0.71$) of the combined ratios provides a build-in test of factorization at the moderate Q^2. We estimate statistical uncertainties associated with such measurements.

INTRODUCTION

At the Thomas Jefferson National Accelerator Facility, the combination of a CW polarized electron beam and the use of polarized targets presents many new physics opportunities, especially in the measurements of spin observables of deep inelastic semi-inclusive scattering (SIDIS). If factorization between quark scattering and quark fragmentation can be clearly demonstrated, SIDIS can provide accesses to quark polarization distributions. In the current fragmentation regime, the leading hadron from the fragmentation is strongly correlated with the struck quark, and the virtual photon asymmetry A_1^h is related to quark polarization at the leading order as:

$$A_1^h(x, Q^2, z) = \frac{\sum_f e_f^2 \Delta q_f(x, Q^2) \cdot D_f^h(z, Q^2)}{\sum_f e_f^2 q_f(x, Q^2) \cdot D_f^h(z, Q^2)}. \tag{1}$$

Once factorization can be demonstrated at a reasonable level, flavor decomposition of nucleon spin structure can be achieved based on data sets of double-spin asymmetries [1]. In this proposed experiment, we will use asymmetries of $A_{1p}^{\pi^\pm}$, $A_{1p}^{K^\pm}$, $A_{1d}^{\pi^\pm}$, $A_{1d}^{K^\pm}$ and the better-known inclusive asymmetries A_{1p} and A_{1d}.

To quantitatively demonstrate factorization in SIDIS, it is crucial to perform precision measurements on observables that are sensitive to the violation of factorization. A schematic strategy of factorization tests was suggested [2] which requires no prior knowledge of fragmentation functions nor parton distributions. The major experimental challenge in this strategy is to observe the z-dependence of the combined $\pi^+ + \pi^-$ helicity asymmetries.

THE PROPOSED MEASUREMENTS

We propose to measure the $\vec{p}(\vec{e},e'h)X$ and $\vec{d}(\vec{e},e'h)X$ ($h=\pi^+,\pi^-,K^+$ and K^-) spin-dependent cross sections with longitudinally polarized NH_3 and LiD targets in Jefferson Lab Hall C with a 6 GeV polarized electron beam [3]. The existing HMS spectrometer will be located at 10.8° as the hadron arm detector. Three HMS momentum settings (2.00, 2.62 and 3.22 GeV/c) will be taken for both positive and negative polarities. A large calorimeter array, with a solid angle of 200 msr, will be centered at 30° as the electron arm in conjunction with a threshold gas Cherenkov detector. A polarized proton luminosity of 8.5×10^{34} cm^{-2}s^{-1} with 80% polarization can be achieved with the UVa polarized NH_3 target. For a total of 24 days of beam time, the expected statistical errors on A_{1p}^π and A_{1d}^π are shown in Fig. 1, and the expected errors on $A_{1p}^{\pi^++\pi^-}$ are shown in Fig. 2. The expected accuracies on the spin-flavor decomposition are shown in Fig 3.

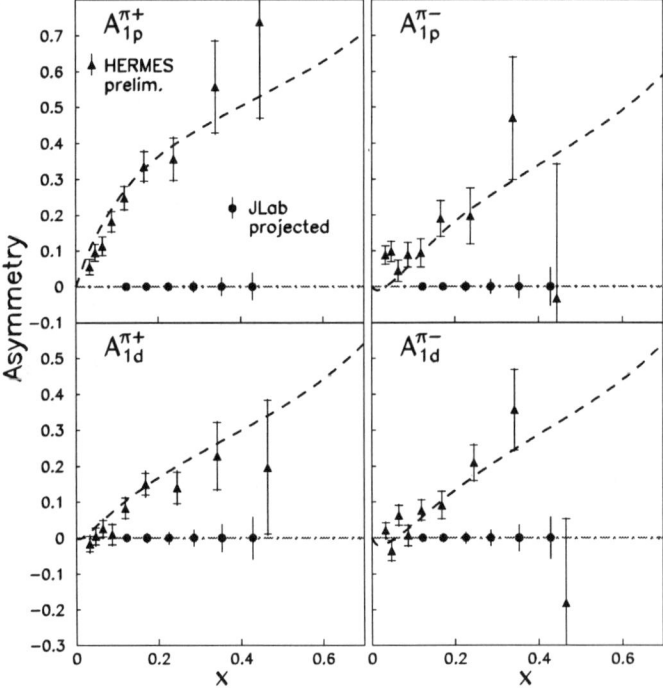

FIGURE 1. The expected statistical accuracy of $A_{1p}^{\pi^+}$, $A_{1p}^{\pi^-}$, $A_{1d}^{\pi^+}$ and $A_{1d}^{\pi^-}$ as functions of x. The HERMES preliminary results [1] are also shown. The dashed lines are the expected asymmetries for $\langle Q^2 \rangle = 2.5$ GeV^2 and $\langle z \rangle = 0.5$ using the CTEQ5 and the AAC quark distributions.

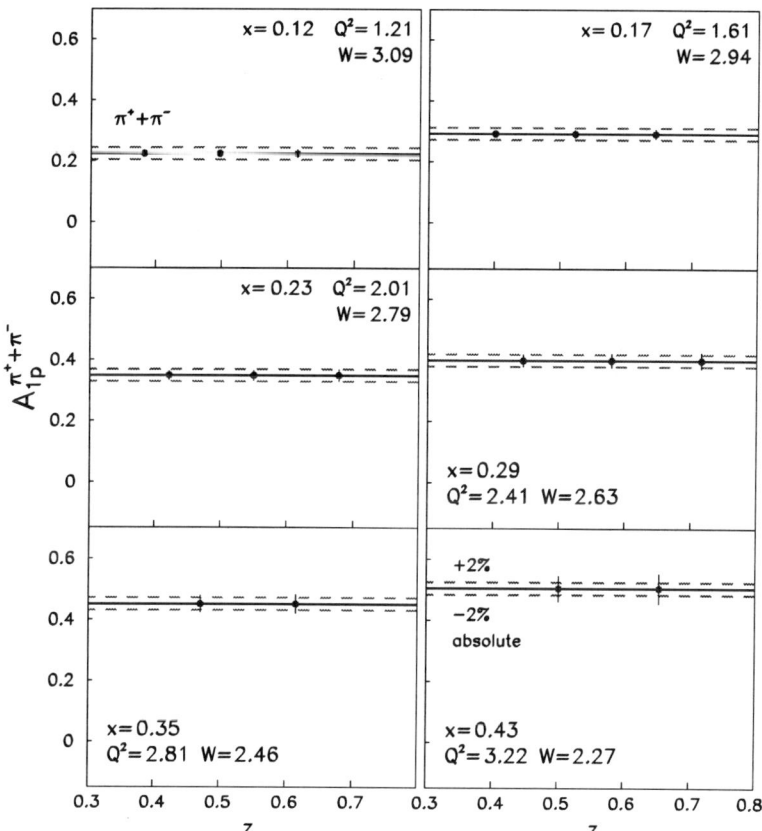

FIGURE 2. The expected statistical accuracy of spin-dependent yield ratio $A_{1p}^{\pi^++\pi^-}$ as functions of z are plotted for each x-bins. The solid line indicates the expected value from the factorization assumption, CETEQ5 and AAC parameterization. The dashed lines indicate an absolute variation of $\pm 2\%$.

ACKNOWLEDGMENTS

The authors thank R. Ent, J.-P. Chen, E.Christova, E. Leader for many discussions. This work is supported by the US Department of Energy and the National Science Foundation.

REFERENCES

1. HERMES collaboration, hep-ex/0307064; A. Miller, plenary talk at the SPIN2002 Conference, Sept 9-14, Long Island, New York, USA; B. Beckmann hep-ex/0210049.
2. E. Christova and E. Leader, Phys. Lett. **B468**, 299 (1999) and Nucl. Phys. **B607**, 369 (2001).
3. "Flavor Decomposition of Nucleon Spin Structure", Jefferson Lab proposal PR-03-114, D. B. Day, X. Jiang, M. K. Jones co-spokespersons.

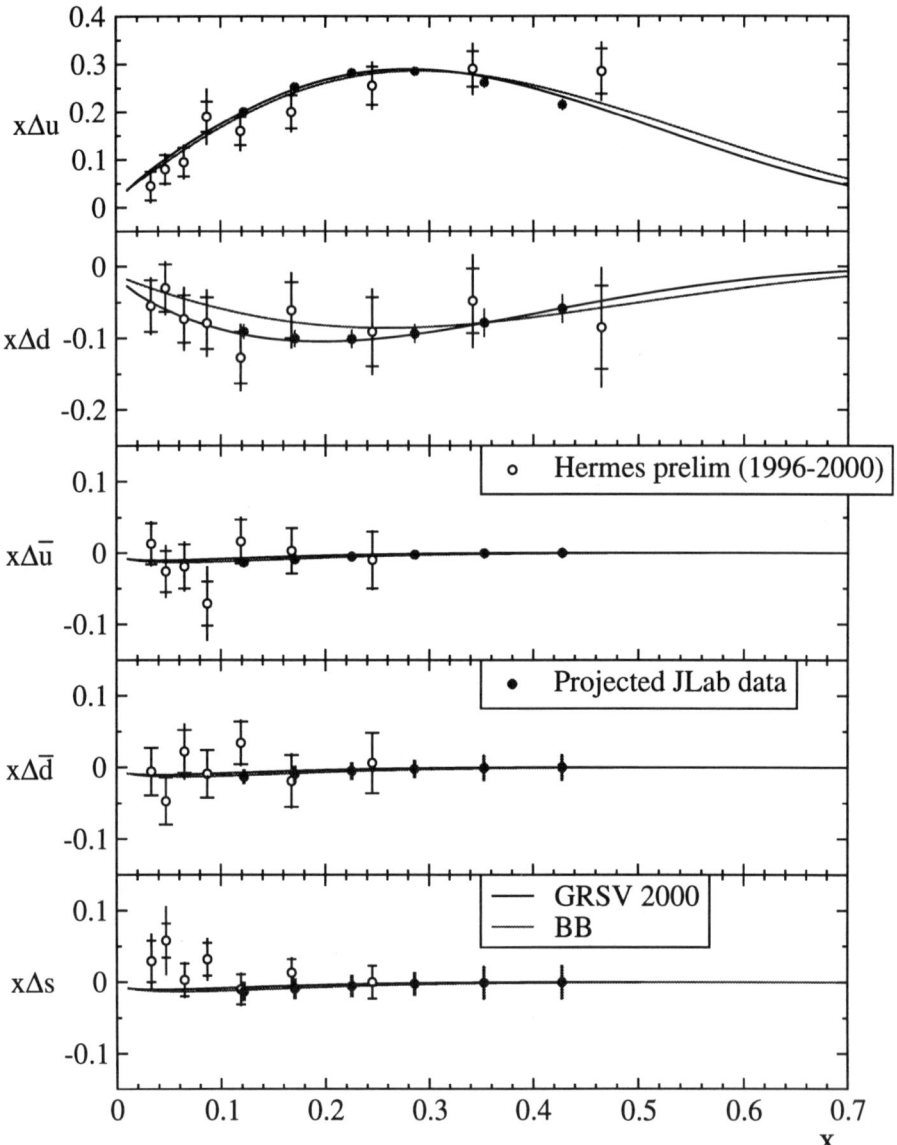

FIGURE 3. The expected statistical accuracy of 5-flavor decomposition compared with the preliminary HERMES data (1996-2000). The error bars on the projected JLab data are statistical only while the HERMES data includes both statistical (inner bars) and the systematic errors.

Summary of Spin Physics Parallel Sessions

Jianwei Qiu* and Matthias Grosse Perdekamp[†]

*Department of Physics and Astronomy, Iowa State University, Ames, IA 50011
[†]Department of Physics, University of Illinois Urbana Champaign, Urbana, IL 61801 and RIKEN BNL Research Center, Upton, NY 11973

Abstract. We summarize the activities in the spin physics parallel sessions of the 8th conference on intersections between particle and nuclear physics.

Introduction

For spin physics in this conference, we had five parallel sessions and over 500 minutes of presentations. We had twenty-six scheduled talks with twenty-four of them presented at the conference (fourteen in experiment, two in machine and instrumentation, and eight in theory). The talks reported the activities at almost all major high energy spin experiments around the world and covered a wild range of recent theoretical developments in this field. The spin physics parallel sessions had overwhelming participation and were full of exciting discussions.

Polarized parton distributions

The determination of polarized quark distributions, $\Delta q(x,\mu)$, and the gluon distribution, $\Delta g(x,\mu)$, is essential for testing QCD perturbation theory in its spin sector as well as for searching answers to the question on how the nucleon's spin is distributed among its constituents.

In the framework of QCD, the spin of a nucleon can be expressed as an expectation value of the QCD angular momentum operator in the nucleon state [1],

$$\frac{1}{2} = \frac{1}{2}\Delta\Sigma(\mu) + \Delta g(\mu) + L_q(\mu) + L_g(\mu), \tag{1}$$

where $\Delta\Sigma(\mu) \equiv \sum_q \int_0^1 dx[\Delta q(x,\mu) + \Delta\bar{q}(x,\mu)]$ and $\Delta g(\mu) = \int_0^1 dx \Delta g(x,\mu)$ are the total quark and gluon helicity, respectively; and $L_q(\mu)$ and $L_g(\mu)$ are quark and gluon orbital angular momentum (OAM), respectively. The scale μ indicates the momentum scale at which these quantities are measured. Combining all polarized deep inelastic scattering (DIS) measurements, the quark helicity contribution to the proton spin is found to be about $\Delta\Sigma \sim 0.2$ [1], which is much smaller than the unity expected from the naive quark model. In order to test the sum rule in Eq. (1), we need to measure the individual pieces on the right-hand-side.

The polarized parton distributions (pPDFs), $\Delta q(x,\mu)$ and $\Delta g(x,\mu)$ are nonperturbative but universal quantities, and are not direct physical observables. It is the QCD factorization theorem that connects them to polarized hadronic cross sections via perturbatively calculated partonic hard scattering processes. The pPDFs can be extracted from measurements of polarized cross sections with longitudinally polarized protons.

Until the advent of the RHIC spin program, high energy collisions with longitudinally polarized protons only took place in lepton-hadron DIS. By measuring the double longitudinal spin asymmetry of inclusive DIS cross sections, A_{LL}, one can extract the structure function, $g_1(x, Q^2)$, of the polarized proton,

$$g_1^p(x,Q^2) = \frac{1}{2}\left[\sum_{q,\bar{q}} \Delta C_q(x,\alpha_s) \otimes \Delta q(x,Q^2) + \Delta C_g(x,\alpha_s) \otimes \Delta g(x,Q^2)\right] \quad (2)$$

where the symbol $\otimes$ represents the convolution over the parton's momentum fraction and the coefficient functions, ΔC_i with $i = q, \bar{q}, g$, are perturbatively calculable. By fitting all available data on the g_1 structure function for proton and neutron at different values of Bjorken x and scale Q and using the DGLAP evolution equation to control the Q-dependence of pPDFs, good constraints on polarized valence quark distributions have been obtained [2, 3]. It was also found that the magnitude of polarized sea distributions are much smaller than that of polarized valence quark distributions and uncertainties on sea distributions are much larger.

The HERMES collaboration improved the separation of quark flavors and extracted the sea and strange quark distributions by the measurement of semi-inclusive DIS (SIDIS) production of $\pi^{\pm}$ and $K^{\pm}$. With the help of existing quark-to-hadron fragmentation functions, they reported the first determination of leading order (LO) $\Delta \bar{u}$, $\Delta \bar{d}$, and $\Delta \bar{s}$ with the assumption of $[\Delta s/s](x) = [\Delta \bar{s}/\bar{s}](x)$ [2]. The HERMES collaboration found no significant breaking of the flavor symmetry in the light sea and no indication of a negative strange sea contribution in fits to DIS data [2]. Kretzer in his talk emphasized the importance of good knowledge of the relevant fragmentation functions for the interpretation of these SIDIS measurements [4]. The HERMES collaboration also reports the effort to explore the use of SIDIS Λ production for improved sensitivity on Δs [5].

With the high luminosity at the Jefferson Laboratory (JLab), the double spin asymmetry A_{LL} in inclusive DIS was measured for the first time in and near the resonance region, from which spin structure functions were measured near $x = 1$ [6].

With a very successful 2002 run, the COMPASS collaboration recorded 1.2 fb^{-1} DIS data with a longitudinally polarized target and 0.3 fb^{-1} DIS data with a transversely polarized target. Reconstruction of events and data analysis are underway. The COMPASS collaboration expects to have first physics results soon [7].

The gluon helicity distribution, Δg, is a key quantity in our understanding of the proton spin. Unfortunately, because gluon contributions only enter the g_1 structure function at next-to-leading-order (NLO) order and indirectly via the DGLAP evolution, only limited information on Δg has been obtained from inclusive DIS measurements. The Asymmetry Analysis Collaboration (AAC) reports that global fits to all existing inclusive DIS data do not provide good constraints on $\Delta g(x, \mu)$ and do not even determine the sign [3]. With the fact that most hard processes at hadron colliders are dominated by gluon initiated subprocesses, the RHIC spin program will provide promising new information on Δg.

With the proton spin transversely polarized with respect to the collision axis, a novel helicity flip *chiral-odd* twist-2 quark distribution – known as transversity distribution,

$\delta q(x,\mu)$, is theoretically allowed [8]. However, there is no leading twist gluon transversity because it would require two units of helicity flip. Since perturbative hard processes conserve helicity, chiral-odd distributions must appear in pairs. That is, transversity can never be measured in polarized inclusive DIS cross sections.

Although transversity can be in principle extracted from the measurement of double transverse spin asymmetries, $A_{TT} \propto \delta q(x) \otimes \delta q(x')$, in jet- or inclusive hadron production in polarized hadronic collisions, the asymmetries are often too small to be extracted experimentally because of the dominance (or lack) of gluonic contribution to the unpolarized (or polarized) cross sections [9].

Therefore, $\delta q(x,\mu)$, is better determined from observables dominated by quark-initiated partonic subprocesses, like A_{TT} of Drell-Yan, which unfortunately suffers from the low rate at the luminosities presently available at RHIC. The single transverse spin asymmetry A_{UT} in SIDIS between a unpolarized lepton beam and a transversely polarized target is proportional to a combination of $\delta q(x) \otimes \Delta D(z,\mathbf{k}_T)$ with the Collins' $\mathbf{k}_T$-dependent fragmentation function $\Delta D(z,\mathbf{k}_T)$ [10]. Transversity distributions could also be accessed by measuring A_{UT} in SIDIS in combination with interference fragmentation functions [11].

Single spin asymmetries in SIDIS

Single spin asymmetries (SSA) in SIDIS between a unpolarized lepton beam and a polarized target can be achieved via the Collins [10] and Sivers [12] mechanisms at leading twist or the Qiu-Sterman mechanism [13] at twist-3. Because of Lorentz invariance of QCD, we need at least four vectors including the spin vector to construct a physically observed SSA. For example, SSAs in SIDIS should be proportional to $\varepsilon_{\mu\nu\alpha\beta} q^\mu P^\nu S^\alpha p^\beta$ where q is momentum of the virtual photon, P (S) are momentum (spin) of the polarized target, and p is the momentum of the observed final-state hadron. Consequently, SSAs have a unique $\sin(\phi)$ dependence with ϕ being the angle between the plane of the three four-vectors (q,P,S) and the plane of the four-vectors (q,P,p).

Let q and P define the collision z-axis and p_T be the transverse momentum of the observed hadron in this frame, then the SSA should be roughly proportional to the dimensionless coefficient: $p_T M/(p_T^2 + M^2)$ with the typical hadronic mass $M \ll Q$ [10]. Therefore, SSAs in general have two distinct regions of phase space: leading twist SSA when $p_T \ll M$ and twist-3 SSA when $p_T \gg M$. When p_T is small, while Q in SIDIS provides the hard scale required for the validity of the leading twist approximation, SSAs should be proportional to p_T/M; and the Collins and Sivers mechanisms lead to nonvanishing SSAs: $A_{UT} \propto \delta q(x) \otimes \Delta D(z,\mathbf{k}_T) + \Delta f(x,\mathbf{k}_T) \otimes D(z)$. The size and sign of the asymmetry are controlled by the nonperturbative $\mathbf{k}_T$-dependent Collins' fragmentation function $\Delta D(z,\mathbf{k}_T)$ and the $\mathbf{k}_T$-dependent Sivers' distribution function $\Delta f(x,\mathbf{k}_T)$. When p_T is of order of the hard scale Q, SSAs should be proportional to M/p_T, a typical twist-3 behavior; and Qiu and Sterman show that the SSA is proportional to new twist-3 tri-parton correlation functions and the corresponding partonic hard scattering pieces can be systematically calculated in pQCD [13]. In terms of final-state interaction between outgoing quark and target spectator, Brodsky et al. [14] explicitly demonstrated that the Sivers contribution to SSAs in SIDIS does not vanish.

Both the HERMES collaboration [15] and the CLAS collaboration [16] observed SSAs in $e+P(S) \to e' + \pi(p) + X$ when the target proton spins are oriented either along or perpendicular to the direction of incoming lepton, and verified that both A_{UL} and A_{UT} are proportional to $\sin(\phi)$. A_{UL} is mainly sensitive to the Collins effect while A_{UT} should include contributions from both the Collins and Sivers effects.

On the theory side, Gamberg [17] reported explicit model calculation of SSAs for both A_{UL} and A_{UT}. Afanasev [18] reported a calculation of a beam SSA. Instead of the target spin vector, the beam SSA is due to a polarized virtual photon on a unpolarized target in SIDIS. The beam SSA is suppressed by an extra power of $1/Q$ compared to the SSA from target spin.

Hasuko [19] reported the status of a program to measure spin dependent fragmentation functions at the Belle experiment in Japan. By measuring the final-state hadron azimuthal asymmetry and correlation, it will be possible to extract the Collins function and possibly other chiral-odd fragmentation functions.

Orbital angular momentum and generalized parton distributions

Asymptotically, the quarks, including both helicity and OAM, carry about 52% of total proton spin while the gluons carry the rest [20]. Since quark helicity contributes about 20% of the proton's spin, a significant quark contribution to the proton's spin must come from its OAM [1]. Therefore, knowing the OAM contribution of partons is crucial for understanding the decomposition of the nucleon's spin in Eq. (1).

Generalized parton distributions (GPDs) share the same operators as normal parton distributions, but are evaluated with a pair of nucleon states of different momenta. GPDs carry much more information than what parton distributions can provide. Belitsky [21] emphasized that parton orbital angular momentum is directly related to GPDs. He also pointed out that deep virtual Compton scattering (DVCS), recently measured at HERMES, H1 and CLAS, is the cleanest hadronic reaction that gives access to GPDs. The HERMES collaboration reported a measurement of GPDs from SIDIS meson production [22].

Ji et al. [23] introduced a phase-space distribution in terms of the GPDs to describe the probability to find a quark at a given momentum and position. Parton OAM plays an important role in determining the phase-space distribution because of its connection to GPDs. The phase-space distribution provides the possibility to study spatial distributions of partons inside a nucleon.

RHIC spin program

The RHIC spin program had its first polarized proton-proton collision two years ago. Two brief polarized proton runs in 2002 and 2003 have resulted in first physics results on transverse spin asymmetries and many important advances in accelerating high current polarized proton beams to high energies as well as proton beam polarimetry at high energies. The main goal of the program is to measure polarized parton distributions, in particular the polarized gluon distribution. However, the combination of available spin orientations and a broad range of experimental processes will also allow to probe the physics beyond the leading twist QCD dynamics as well as beyond the Standard Model.

The STAR Collaboration [24] has measured the single transverse spin asymmetry A_N for pion production both for positive and negative rapidity y. In the 2003 run, spin rotators were commissioned and for the first time, double spin asymmetries A_{LL} were measured in a hadron collider. The data analysis for A_{LL} in jet production is in progress and will provide first new information on the gluon helicity distribution [24]. With similar physics motivation, the PHENIX collaboration is carrying out analysis of double spin asymmetries in inclusive hadron production [25].

In order to improve the figure of merit for the asymmetry measurements, MacKay [26] argued that it will be possible for RHIC to raise the beam polarization from 35-40% to 70%. He discussed possible methods and plans to improve the luminosity of the polarized beam, which is now about a factor 30 below the design value. Bravar [27] showed that elastic $p^\uparrow C \to p^\uparrow C$ has been used successfully as a fast polarimeter at RHIC.

With three DIS machines (HERMES, Compass, JLab) running at different energies and RHIC spin program in a full swing, we will soon have more precise data to cover a wide spectrum of spin physics. A new experiment to test the factorization of SIDIS was just proposed to JLab [28]. Spin physics is getting more and more exciting now!

This work is supported by the U.S. Department of Energy under Grant No. DE-FG02-87ER40371.

REFERENCES

1. B. W. Filippone and X. D. Ji, Adv. Nucl. Phys. **26**, 1 (2001), and references therein.
2. J. Wendtland, HERMES Collaboration, in this proceedings.
3. M. Hirai *et al.*, AAC, Int. J. Mod. Phys. A **18**, 1203 (2003); and references therein.
4. S. Kretzer, in this proceedings.
5. N. Makins, HERMES Collaboration, in this proceedings.
6. R. Minehart, CLAS collaboration, in this proceedings.
7. E. Kabuss, Compass Collaboration, in this proceedings.
8. J. Ralston and D.E. Soper, Nucl. Phys. **B152**, 109 (1979).
9. W. Vogelsang, in this proceedings.
10. J. C. Collins, Nucl. Phys. B **396**, 161 (1993).
11. R. L. Jaffe, X. m. Jin and J. Tang, Phys. Rev. Lett. **80**, 1166 (1998).
12. D. W. Sivers, Phys. Rev. D **41**, 83 (1990); *ibid.* **43**, 261 (1991).
13. J. w. Qiu and G. Sterman, Phys. Rev. Lett. **67**, 2264 (1991); Nucl. Phys. B **378**, 52 (1992).
14. S. J. Brodsky, D. S. Hwang and I. Schmidt, Phys. Lett. B **530**, 99 (2002); D. S. Hwang, in this proceedings.
15. R. Seidel, HERMES Collaboration, in this proceedings.
16. H. Avakian, CLAS Collaboration, in this proceedings.
17. L. Gamberg, in this proceedings.
18. A. Afanasev, in this proceedings.
19. K. Hasuko, Belle collaboration, in this proceedings.
20. X. D. Ji, Phys. Rev. Lett. **78**, 610 (1997).
21. A.V. Belitsky, in this proceedings.
22. D. Hasch, HERMES Collaboration, in this proceedings.
23. X. Ji, in this proceedings; A. V. Belitsky, X. Ji and F. Yuan, arXiv:hep-ph/0307383.
24. G. Igo, STAR Collaboration, in this proceedings.
25. D. Field, PHENIX Collaboration, in this proceedings.
26. W. MacKay, in this proceedings.
27. A. Bravar, in this proceedings.
28. X. Jiang, in this proceedings.

RELATIVISTIC HEAVY IONS

Jets and high p_T hadrons in dense matter: recent results from STAR

Peter Jacobs*[†] and Jennifer Klay*[†]

*Lawrence Berkeley National Laboratory, 1 Cyclotron Road, Berkeley CA 94720
[†]for the STAR Collaboration

Abstract. We review recent measurements of high transverse momentum (high p_T) hadron production in nuclear collisions by the STAR Collaboration at RHIC. The previously observed suppression in central Au+Au collisions has been extended to much higher p_T. New measurements from d+Au collisions are presented which help disentangle the mechanisms responsible for the suppression. Inclusive single hadron spectra are enhanced in d+Au relative to p+p, while two-particle azimuthal distributions are observed to be similar in p+p, d+Au and peripheral Au+Au collisions. The large suppression of inclusive hadron production and absence of the away-side jet-like correlations in central Au+Au collisions are shown to be due to interactions of the jets with the very dense medium produced in these collisions.

1. INTRODUCTION

High energy partons propagating through matter are predicted to lose energy via induced gluon radiation, with the magnitude of the energy loss depending linearly on the color charge density of the matter[1]. This phenomenon may provide a sensitive probe of the medium generated in high energy heavy ion collisions, where a Quark-Gluon Plasma (QGP) is expected to form if sufficiently high energy density is achieved. The high energy partons in such collisions result from the hard scattering of quarks or gluons from the incoming nuclei and are observed experimentally as correlated "jets" of hadrons having large transverse momentum (p_T) with respect to the beam direction. The experimental challenge is to measure partonic energy loss ("jet quenching") in the extremely complex environment of high energy nuclear collisions.

In these talks we discuss recent progress towards the measurement of jet quenching in high energy nuclear collisions by the STAR experiment at the Relativistic Heavy Ion Collider (RHIC) at Brookhaven National Laboratory. At the values of jet E_T accessible at RHIC energies with the currently achieved integrated luminosities, full jet reconstruction with good energy resolution is difficult. We therefore utilize known features of jet fragmentation to study jet quenching, in particular the inclusive spectrum of high p_T ("leading") hadrons and the angular correlation of pairs of high p_T hadrons. We compare their distributions in centrality-selected Au+Au collisions to those in d+Au and non-singly diffractive (NSD) p+p collisions, all at $\sqrt{s_{NN}}$=200 GeV. Unless otherwise specified, the results reported here are for unidentified charged hadrons, measured in the large cylindrical STAR Time Projection Chamber with a 0.5T solenoidal magnetic field[2]. Related results from STAR are discussed in [3, 4].

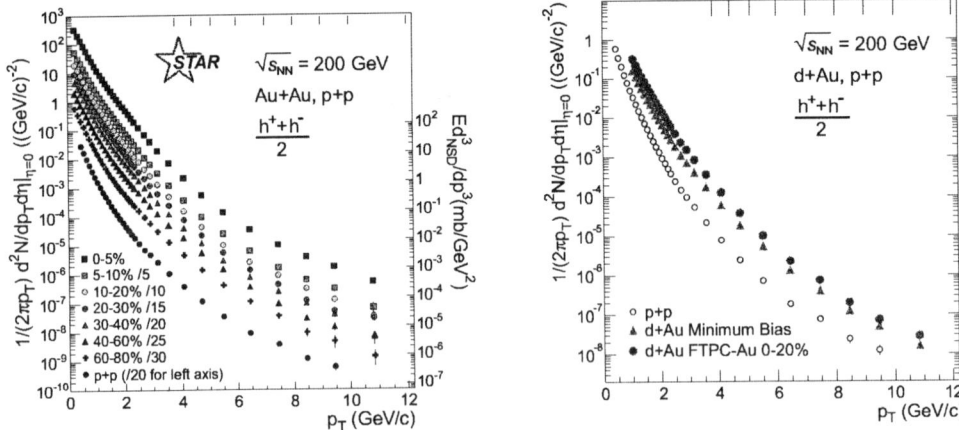

FIGURE 1. Invariant inclusive p_T distributions of charged hadrons at $\sqrt{s_{NN}}$=200 GeV. Left: NSD p+p and centrality-selected Au+Au collisions[5]. Right: central and minimum bias d+Au collisions, and same p+p spectrum as left panel[6].

2. SUPPRESSION OF INCLUSIVE CHARGED HADRONS

Jets occasionally fragment with a single hadron carrying a large fraction of the total jet energy. The resulting inclusive hadron distributions exhibit power-law $(1/p_T^n)$ shapes characteristic of the underlying perturbative QCD (pQCD) processes. Figure 1 shows the event-normalized invariant inclusive charged hadron distributions as a function of p_T for NSD p+p and centrality-selected Au+Au and d+Au collisions[5, 6]. The centrality bins correspond to the indicated percentiles of the total cross section: 0-5% for Au+Au indicates the most central and 60-80% the most peripheral collisions. Power-law shapes are clearly observed in all cases.

Nuclear effects on hadron production in d+Au and Au+Au collisions are measured through comparison to the p+p reference spectrum using the ratio

$$R_{AB}(p_T) = \frac{d^2N/dp_T d\eta}{T_{AB} d^2\sigma^{pp}/dp_T d\eta}, \qquad (1)$$

where $d^2N/dp_T d\eta$ is the differential yield per event in the nuclear collision $A+B$, $T_{AB}=\langle N_{bin}\rangle/\sigma^{pp}_{inel}$ describes the nuclear geometry, and $d^2\sigma^{pp}/dp_T d\eta$ for p+p inelastic collisions is determined from the p+p measurement. $\langle N_{bin}\rangle$ is the mean number of binary NN interactions for the given centrality class of $A+B$ collisions. In the absence of nuclear effects such as shadowing, the Cronin effect, or gluon saturation, hard process rates are expected to scale with $\langle N_{bin}\rangle$, and $R_{AB}(p_T)$=1.

Figure 2, left panel, shows $R_{AB}(p_T)$ for centrality-selected Au+Au relative to p+p collisions. The error bars indicate the statistical and systematic uncertainties of the spectra, while the bands indicate the uncertainty due to the geometrical scaling factor T_{AB}. At the highest p_T, hadron suppression of approximately a factor 5 is observed for the most central collisions: inclusive hadron production is strongly suppressed at high

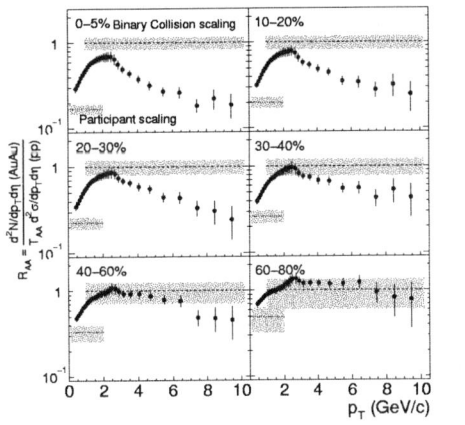 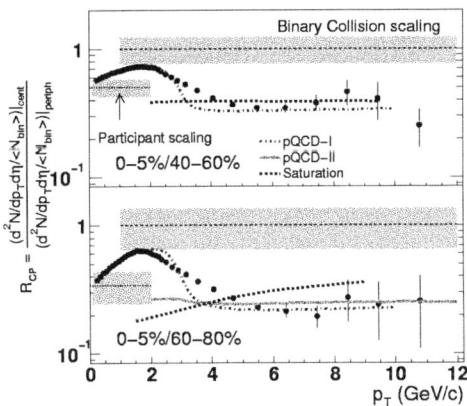

FIGURE 2. Left: $R_{AB}(p_T)$ (eq. 1) for Au+Au relative to p+p collisions[5]. Right: $R_{CP}(p_T)$ from Au+Au collisions[5].

p_T in central Au+Au collisions. For the most peripheral collisions $R_{AB}(p_T)$ is consistent with unity, while intermediate centralities interpolate smoothly between the extremes.

Figure 2, right panel, shows the related quantity $R_{CP}(p_T)$, the $\langle N_{bin}\rangle$-scaled ratio of particle yields in central relative to peripheral collisions, which also exhibits large suppression of inclusive hadron production in central collisions. Figure 2 also shows the results of three theoretical calculations: two models based on pQCD incorporating shadowing, the Cronin effect and jet quenching in dense matter (pQCD-I[7], pQCD-II[8]), and a model incorporating gluon saturation in the incoming Au nuclei[9]. The pQCD-based calculations contain one free parameter, the energy density for central collisions, which is fit to the data, yielding an initial density 30-50 times that of cold nuclear matter [7, 8]. These calculations then successfully describe the p_T and centrality dependence of the inclusive suppression for $p_T > 5$ GeV/c. The saturation model also describes the suppression and its p_T-dependence for the 0-5%/40-60% ratio for $p_T > 5$ GeV/c. These precision STAR data on inclusive charged hadron suppression, covering wide centrality and kinematic range, are well-described by widely differing models which attribute the suppression either to jet quenching in dense matter in the final state or to gluon saturation in the initial state.

In order to elucidate the particle-species dependence of the suppression, Fig. 3 shows $R_{CP}(p_T)$ for K_s^0 and $\Lambda + \bar{\Lambda}$ citeSTARStrangeRCP compared to the charged hadron $R_{CP}(p_T)$ already shown in Fig. 2. The mesons scale approximately as the charged hadrons throughout, while the baryons exhibit a pronounced enhancement in the region $2 < p_T < 4$ GeV/c. The origin of this behavior is at present not understood, though speculations include the Cronin effect and non-perturbative mechanisms such as baryon junctions. At $p_T = 5$ GeV/c, all particle species are strongly suppressed by the same amount in central collisions.

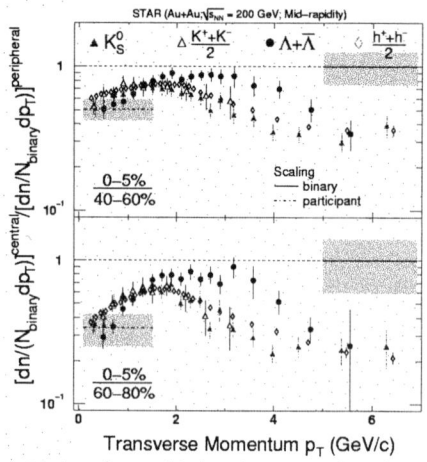

FIGURE 3. $R_{CP}(p_T)$ from Au+Au collisions for K_s^0 and $\Lambda + \bar{\Lambda}$[10], and for the charged hadron data also shown in Fig. 2, right panel.

3. CORRELATIONS

The angular correlations of pairs of high-p_T charged particles can be used to study jets in the complex environment of heavy ion collisions[3]. Figure 4 shows the two-particle azimuthal distribution $D(\Delta\phi)$, defined as

$$D(\Delta\phi) \equiv \frac{1}{N_{trigger}} \frac{1}{\varepsilon} \frac{dN}{d(\Delta\psi)}, \qquad (2)$$

for peripheral (left panel) and central (right panel) Au+Au collisions and for p+p collisions (both panels). Only particles within $|\eta|<0.7$ are included in the analysis. $N_{trigger}$ is the number of particles within $4<p_T(\text{trig})<6$ GeV/c, referred to as trigger particles. The distribution results from the correlation of each trigger particle with all associated particles in the same event having $2 < p_T < p_T(\text{trig})$, where ε is the tracking efficiency of the associated particles. The normalization uncertainties are less than 5%.

In order to compare correlations in Au+Au with p+p, the p+p correlations are scaled up to the same pedestal value at $|\Delta\phi| \sim \pi/2$ as Au+Au and superposed with a $\cos(2\Delta\phi)$ term that characterizes the azimuthal anisotropy in non-central Au+Au collisions (elliptic flow [4]). The magnitude of this additional correlation is measured independently and is given by the second coefficient v_2 of a Fourier expansion of the azimuthal distribution relative to the orientation of the reaction plane of the event, for particles having $p_T<2$ GeV/c [11]. Figure 4 shows that the correlation strength at small relative angle ($\Delta\phi \sim 0$) in peripheral and central Au+Au and at large relative angle ($\Delta\phi \sim \pi$) in peripheral Au+Au are very similar to the scaled correlations in p+p collisions.

The near-side peaks ($\Delta\phi \sim 0$) in all three collision systems are characteristic of jet fragmentation. The away-side peak ($|\Delta\phi| \sim \pi$) from the back-to-back partner jet, apparent in p+p and peripheral Au+Au collisions is strongly suppressed in central Au+Au

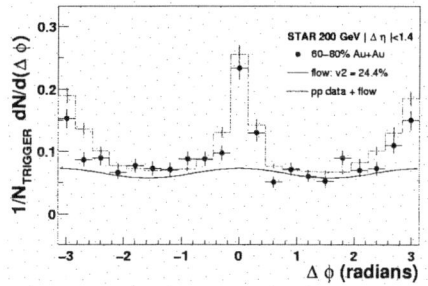 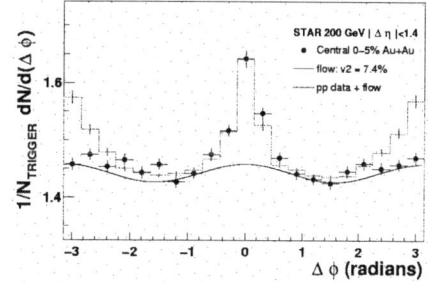

FIGURE 4. Azimuthal correlations for peripheral (left) and central (right) Au+Au collisions compared to the pedestal and flow-scaled correlations in p+p collisions[12].

collisions. These observations, together with the strong suppression of inclusive production and large elliptic flow at high p_T [4], suggest a picture in which jets traversing the bulk of the medium produced in Au+Au collisions are absorbed (strong jet quenching), and the observed jets are biased towards those generated on the surface and heading outwards. However, other explanations, in particular the gluon saturation picture, may also account for some or all of the observed phenomena. Experimentally, these very different scenarios can be discriminated through measurements of d+Au collisions, where initial state effects in the Au nucleus remain but no final state dense medium is generated.

4. D+AU COLLISIONS

pQCD-based models predict an enhancement in the production of high-p_T charged hadrons in d+Au collisions relative to binary-scaled p+p collisions and little change in the back-to-back correlation strength[13, 14], while one version of the saturation model[15] predicts an inclusive suppression in d+Au of about 30% and possibly also suppression of the back-to-back strength due to a mono-jet contribution [16].

Figure 5 shows $R_{AB}(p_T)$(left) and the pedestal subtracted two-particle azimuthal distributions (right) measured by STAR in minimum bias and central d+Au collisions[6]. $R_{AB}(p_T)$ exceeds unity for 2<p_T<7 GeV/c, consistent with expectations from the Cronin effect. However, no additional enhancement over p+p is observed for central relative to minimum bias d+Au collisions. In the top right panel, the azimuthal distributions are characterized by a fit to the sum of near-side and back-to-back Gaussian peaks plus a constant. The only significant difference between the p+p and d+Au correlations is the growth in the pedestal value [3]. The lower right panel shows the pedestal- and flow-subtracted azimuthal distribution from central Au+Au collisions along with the pedestal-subtracted distributions from central d+Au and p+p. The near-side peak is similar in all three collision systems, while the away-side peak is suppressed only in central Au+Au collisions.

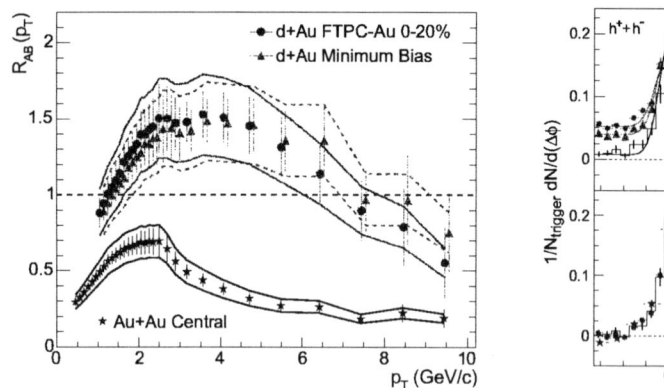 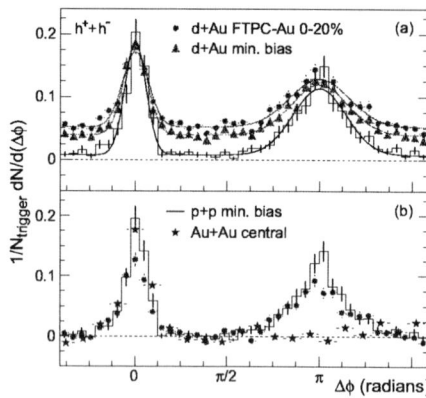

FIGURE 5. $R_{AB}(p_T)$ and two-particle azimuthal distributions in d+Au collisions[6]. The horizontal axes of the right panel are shifted by $\pi/2$ relative to Fig. 4.

5. SUMMARY

The strong suppression at high p_T of the inclusive hadron yield and back-to-back correlations in central Au+Au collisions at $\sqrt{s_{NN}}$=200 GeV are not observed in d+Au collisions. The inclusive yield in d+Au collisions is enhanced relative to binary-scaled p+p, consistent with expectations from the Cronin effect, and the back-to-back correlations show little variation relative to p+p. These results demonstrate conclusively that the striking suppression phenomena observed in central Au+Au collisions are due to the interaction of high energy partons or their fragmentation products in the dense medium created in such collisions.

REFERENCES

1. R. Baier, D. Schiff and B. G. Zakharov, Ann. Rev. Nucl. Part. Sci. **50**, 37 (2000); M. Gyulassy, I. Vitev, X.N. Wang, B. Zhang, nucl-th/0302077.
2. K.H. Ackermann *et al.*, Nucl. Instr. Meth. **A499**, 624 (2003).
3. M. Miller, these proceedings.
4. A. Tang, these proceedings.
5. J. Adams *et al.*(STAR Collaboration), nucl-ex/0305015.
6. J. Adams *et al.*(STAR Collaboration), Phys. Rev. Lett. **91**, 072304 (2003).
7. X.N. Wang, nucl-th/0305010; private communication. Calculations use model parameters $\mu_0 = 2.0$ GeV and $\varepsilon_0 = 2.04$ GeV/fm.
8. I. Vitev and M. Gyulassy, Phys. Rev. Lett. **89**, 252301 (2002).
9. D. Kharzeev, E. Levin, and L. McLerran, Phys. Lett. **561**, 93 (2003).
10. J. Adams *et al.*(STAR Collaboration), nucl-ex/0306007.
11. C. Adler *et al.*, Phys. Rev. Lett. **90**, 032301 (2003).
12. C. Adler *et al.*, Phys. Rev. Lett. **90**, 082302 (2003).
13. I. Vitev, Phys. Lett. **B562**, 36 (2003).
14. X.N. Wang, Phys.Rev. **C61**, 064910 (2000); nucl-th/0303004.
15. D. Kharzeev, E. Levin, and M. Nardi, hep-ph/0212316.
16. D. Kharzeev, Nucl. Phys. **A715**, 35c (2003).

PHENIX High Pt Results

L. Aphecetche for the PHENIX Collaboration

SUBATECH (EMN, CNRS-IN2P3, Université de Nantes) BP 20722 - 44307, Nantes, France

Abstract.
In heavy ion collisions, the high p_T observables convey information about the initial conditions of the system under study. The suppression of high p_T neutral pion and charged hadrons yields with respect to the nucleon-nucleon binary scaled p p collisions and the particle composition in Au+Au collisions (as measured by the PHENIX [1] experiment during the first two runs of the Relativistic Heavy Ion Collider (RHIC)) are briefly reviewed. Finally, the newest results from the RHIC Run 3 d+Au, which disentangle initial and final state effects, are presented.

Heavy-ion collisions at RHIC aim at the study of QCD matter at extreme energy densities where lattice [4] calculations predict a transition from hadronic matter to a deconfined plasma of quarks and gluons. High transverse momentum ($p_T > 2$ GeV/c) hadrons, arising from the fragmentation of partons (quarks and gluons) scattered with large momentum transfer in the initial parton-parton scattering[5], provide excellent probes of the matter thus created [2, 3].

A standard way to quantify nuclear medium effects on high p_T production is provided by the nuclear modification factor, which we define for A+B collisions as the ratio of invariant yield in A+B to that of p+p, scaled by the number of binary collisions.

$$R_{AB}(p_T) = \frac{(1/N_{AB}^{evt})\, d^2 N_{AB}/d\eta dp_T}{\langle N_{coll}\rangle / \sigma_{pp}^{inel}\, d^2 \sigma_{pp}/d\eta dp_T},$$

where $\langle N_{coll}\rangle$ is the average number of inelastic nucleon-nucleon (NN) collisions per event, and $\langle N_{coll}\rangle/\sigma_{pp}^{inel}$ is the average Glauber nuclear overlap function $\langle T_{AB}\rangle$.
For heavy ion collisions R_{AB} is expected to be below unity for $p_T < 2$ GeV/c, because the bulk of the particle production is due to soft processes which scale closer to the number of participant nucleons [6] than to $\langle N_{coll}\rangle$. On the contrary, for high p_T production, R_{AB} should be unity in absence of any nuclear medium effect. But it was in fact established[1] that R_{AuAu}, both for neutral pions and for charged hadrons, is significantly below unity for the most central collisions, which is now commonly referred to as "the high p_T suppression". The suppression is more or less independent of p_T above 4 GeV/c and increases as a function of the centrality of the collision, the most central collisions experiencing the most pronounced suppression (Fig. 1).

---

[1] first observed during RHIC Run I Au+Au at $\sqrt{s_{NN}} = 130$ GeV/c [7, 8, 9] and later on confirmed within an extended p_T range in Run II Au+Au at $\sqrt{s_{NN}} = 200$ GeV/c [10, 11]

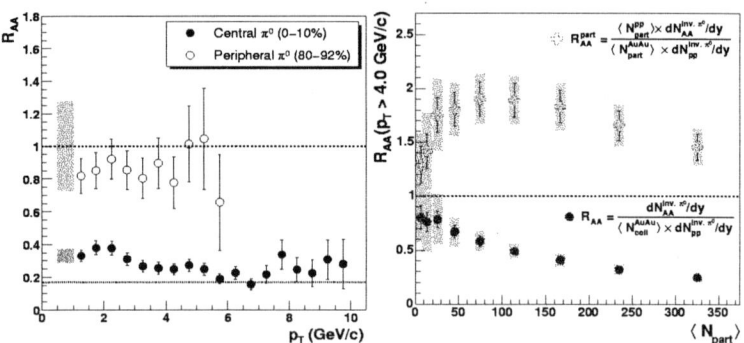

FIGURE 1. Left: nuclear modification factor $R_{AA}(p_T)$ for π^0 in central (closed circles) and peripheral (open circles) Au+Au collisions at $\sqrt{s_{NN}} = 200$ GeV as a function of p_T. Note how the π^0 production is suppressed by almost a factor 5 at high p_T in central collisions. Right: R_{AA} for $p_T > 4$ GeV/c as a function of centrality (both figures from [10]).

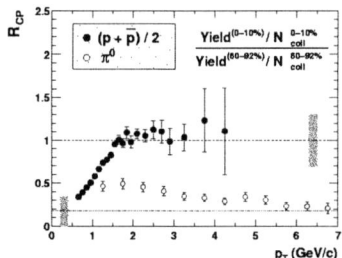

FIGURE 2. Nuclear modification factor R_{CP} for $(p+\bar{p})/2$ (filled circles) and π^0 (from [14]). Note how protons are not suppressed, in the accessible p_T range, in contrast to pions.

It should be noted that the p+p reference spectrum used to compute R_{AA} is also determined by the PHENIX experiment [12].

Another striking characteristics of the high p_T suppression is its dependence on particle type. It was observed that π^0 are more suppressed than charged hadrons [7] and that the yields of p and $\bar{p}$ near 2 GeV/c in central collisions are comparable to those of pions [13]. The nuclear modification factor for p and $\bar{p}$, expressed here in a slightly different way,

$$R_{CP} = \frac{yield^{0-10\%}/N_{coll}^{0-10\%}}{yield^{60-92\%}/N_{coll}^{60-92\%}},$$

i.e. as the N_{coll}-scaled ratio of the central yield over the peripheral yield, has been analyzed [14], and exhibit no suppression, in constrast to that of the pions (Fig. 2, see also [15]).

Shortly before this conference, the origin of the high p_T suppression could be roughly ascribed to two different kinds of processes. Theoretical studies of parton propagation

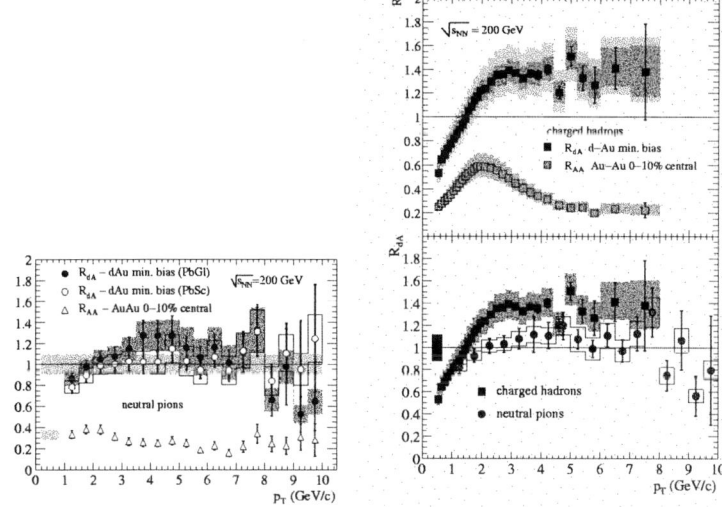

FIGURE 3. Nuclear modification factor for π^0 in d+Au (circles) and central Au+Au collisions (squares). Top right : nuclear modification factor for charged hadrons in d+Au and Au+Au, bottom right : comparison of R_{dA} for π^0 and $(h^+ + h^-/2)$ (from [24]). Note how R is different in d+Au and Au+Au collisions.

in high density matter suggest that partons lose a significant fraction of their energy through gluon bremsstrahlung [2, 3], reducing the parton momentum and depleting the yield of high p_T hadrons [16, 17, 18, 19, 20, 21]. This is a final-state effect in the spatially extended medium created in A+A collisions, and is the first possible explanation of the observed suppression. On the other hand, initial state effects (nuclear modifications to the parton momentum distributions -structure functions-, and soft scatterings experienced by incoming parton prior to its hard scattering) could also contribute. These initial effects should be both present in A+A and d+Au collisions. Interpretations of Au+Au collisions based on initial-state parton saturation effects [22] or final-state hadronic interactions [23] also predict a considerable suppression of the hadron production at high p_T. It was therefore of paramount interest to determine experimentally the modification, if any, of high p_T hadron yields due to initial state nuclear effects for a system in which a hot, dense medium is not produced in the final state. This was performed during the RHIC Run 3 (d+Au at $\sqrt{s_{NN}} = 200$ GeV/c) and the first results were presented at this conference, thus solving part of the puzzle.

The nuclear modification factor R_{dA} is plotted for three independent analyses in Fig. 3 : two neutral pion analyses making use of the two PHENIX electromagnetic calorimeters (PbSc and PbGl) and one charged hadrons analyses making use of the tracking capabilities of PHENIX [24]. The data clearly indicates that there is no suppression of high p_T particles in d+Au collisions. At variance, we do observe an enhancement for $p_T > 2$ GeV/c ("Cronin effect", already observed in p+A fixed target experiment [25]).

Those observations indicate that the suppression in central Au+Au collisions is not an initial state effect, but instead that it is a final state effect of the produced dense medium.

We thank the staff of the Collider-Accelerator and Physics Departments at BNL for their vital contributions. We acknowledge support from the Department of Energy and NSF (U.S.A.), MEXT and JSPS (Japan), CNPq and FAPESP (Brazil), NSFC (China), IN2P3/CNRS and CEA (France), BMBF, DAAD, and AvH (Germany), OTKA (Hungary), DAE and DST (India), ISF (Israel), KRF and CHEP (Korea), RMIST, RAS, and RMAE (Russia), VR and KAW (Sweden), U.S. CRDF for the FSU, US-Hungarian NSF-OTKA-MTA, and US-Israel BSF.

REFERENCES

1. K. Adcox et al., Nucl. Instrum. Methods **A499**, 469 (2003), see also E. O'brien, these proceedings.
2. M. Gyulassy and M. Plümer, Phys. Lett. **B243**, 432 (1990); X.N. Wang and M. Gyulassy, Phys. Rev. Lett. **68**, 1480 (1992).
3. R. Baier et al., Phys. Lett. **B345**, 277 (1995).
4. see e.g. E. Laermann and O. Philipsen, hep-ph/0303042 for a recent review.
5. J.F. Owens et al., Phys. Rev. **D18**, 1501 (1978).
6. K. Adcox et al., Phys. Rev. Lett. **86**, 3500 (2001).
7. K. Adcox et al., Phys. Rev. Lett. **88**, 022301 (2002).
8. K. Adcox et al., Phys. Lett. **B561**, 82 (2003).
9. C. Adler et al., Phys. Rev. Lett. **89**, 202301 (2002).
10. S.S. Adler et al., Phys. Rev. Lett. **91**, 072301 (2003)
11. J. Adams et al., nucl-ex/0305015
12. S.S. Adler et al., submitted to Phys. Rev. Lett. , hep-ex/0304038
13. K. Adcox et al., Phys. Rev. Lett. **88**, 242301 (2002)
14. S.S. Adler et al., submitted to Phys. Rev. Lett. , nucl-ex/0305036
15. T. Chujo, these proceedings.
16. M. Gyulassy and X.N. Wang, Nucl. Phys. **B420**, 583 (1994); X.N. Wang, Phys. Rev. **C58**, 2321 (1998).
17. I. Vitev and M. Gyulassy, Phys. Rev. Lett. **89**, 252301 (2002).
18. S. Jeon, J. Jalilian-Marian and I. Sarcevic, Phys. Lett. **B562**, 45 (2003).
19. G.G. Barnaföldi et al., nucl-th/0212111; V. Greco, C.M. Ko and P. Lévai, Phys. Rev. Lett. **90**, 202302 (2003).
20. X. N. Wang, hep-ph/0301196.
21. F. Arleo, J. High Energy Phys. **11**, 44 (2002).
22. D. Kharzeev, E. Levin and L. McLerran, Phys. Lett. **B561**, 93 (2003).
23. K. Gallmeister, C. Greiner and Z. Xu, Phys. Rev. **C67**, 044905 (2003).
24. S.S. Adler et al., nucl-ex/0306021, accepted for publication in PRL
25. D. Antreasyan et al., Phys. Rev. **D19**, 764 (1979).

First results on d+Au collisions from PHOBOS

B.B.Back*, M.D.Baker[†], M.Ballintijn**, D.S.Barton[†], B.Becker[†],
R.R.Betts[‡], A.A.Bickley[§], R.Bindel[§], A.Budzanowski[¶], W.Busza**,
A.Carroll[†], M.P.Decowski**, E.García[‡], T.Gburek[¶], N.George*[†],
K.Gulbrandsen**, S.Gushue[†], C.Halliwell[‡], J.Hamblen[∥], A.S.Harrington[∥],
C.Henderson**, D.J.Hofman[‡], R.S.Hollis[‡], R.Hołyński[¶], B.Holzman[†],
A.Iordanova[‡], E.Johnson[∥], J.L.Kane**, N.Khan[∥], P.Kulinich**, C.M.Kuo[††],
J.W.Lee**, W.T.Lin[††], S.Manly[∥], A.C.Mignerey[§], A.Noell[§], R.Nouicer[‡],
A.Olszewski[¶], R.Pak[†], I.C.Park[∥], H.Pernegger**, C.Reed**, L.P.Remsberg[†],
C.Roland**, G.Roland**, J.Sagerer[‡], P.Sarin**, P.Sawicki[¶], I.Sedykh[†],
W.Skulski[∥], C.E.Smith[‡], P.Steinberg[†], G.S.F.Stephans**, A.Sukhanov[†],
R.Teng[∥], M.B.Tonjes[§], A.Trzupek[¶], C.Vale**, G.J.van Nieuwenhuizen**,
R.Verdier**, G.I.Veres**, B.Wadsworth**, F.L.H.Wolfs[∥], B.Wosiek[¶],
K.Woźniak[¶], A.H.Wuosmaa*, B.Wysłouch** and J.Zhang**

Argonne National Laboratory, Argonne, IL 60439-4843, USA
[†]*Brookhaven National Laboratory, Upton, NY 11973-5000, USA*
**Massachusetts Institute of Technology, Cambridge, MA 02139-4307, USA*
[‡]*University of Illinois at Chicago, Chicago, IL 60607-7059, USA*
[§]*University of Maryland, College Park, MD 20742, USA*
[¶]*Institute of Nuclear Physics, Kraków, Poland*
[∥]*University of Rochester, Rochester, NY 14627, USA*
[††]*National Central University, Chung-Li, Taiwan*

Abstract.
We have measured transverse momentum distributions of charged hadrons produced in d+Au collisions at $\sqrt{s_{NN}} = 200$ GeV, in the range $0.25 < p_T < 6.0$ GeV/c. With increasing collision centrality, the yield at high transverse momenta increases more rapidly than the overall particle density, leading to a strong modification of the spectral shape. This change in spectral shape is qualitatively different from observations in Au+Au collisions at the same energy. The results provide important information for discriminating between different models for the suppression of high-p_T hadrons observed in Au+Au collisions.

Results from Au+Au collisions at $\sqrt{s_{NN}} = 130$ and 200 GeV have shown that the expected scaling of hadron production with the number of binary nucleon-nucleon collisions at p_T of 2–10 GeV/c is strongly violated [1, 2, 3, 4]. This effect had been predicted as a consequence of the energy loss of high-p_T partons in the hot and dense medium formed in Au+Au collisions [5]. However, the interpretation of the Au+Au data relies on the understanding of initial state effects, including gluon saturation [6], which can be investigated with the d+Au data presented here [7].

The data were collected using the PHOBOS silicon-based magnetic spectrometer [8, 9]. To characterize the collision centrality in d+Au collisions, we divided the total

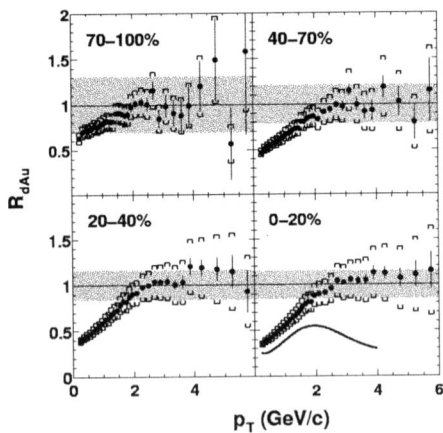

FIGURE 1. Nuclear modification factor R_{dAu} as a function of p_T for four bins of centrality. For the most central bin, the spectral shape for central Au+Au data relative to $p+\bar{p}$ is shown for comparison. The shaded area shows the uncertainty in R_{dAu} due to the systematic uncertainty in $\langle N_{coll} \rangle$ and the UA1 scale error (90% C.L.). The brackets show the systematic uncertainty of the d+Au spectra measurement (90% C.L.).

cross-section into four bins, and using a detector simulation and a Glauber model of the collision system, estimated the the average number of participating nucleons $\langle N_{part} \rangle$ and binary nucleon-nucleon collisions, $\langle N_{coll} \rangle$ in each bin [9].

We compare our d+Au data to results from UA1 for $p+\bar{p}$ collisions at the same energy [10] in terms of the nuclear modification factor defined as $R_{dAu} = \frac{\sigma_{p\bar{p}}^{incl}}{\langle N_{coll} \rangle} \frac{d^2 N_{dAu}/dp_T d\eta}{d^2\sigma(\text{UA1})_{p\bar{p}}/dp_T d\eta}$. The UA1 results have been modified to account for the difference in acceptance between UA1 and PHOBOS [9]. In Fig. 1 we present R_{dAu} as a function of p_T for each centrality bin. A value of $R_{dAu} = 1$ corresponds to scaling of the yield as an incoherent superposition of nucleon-nucleon collisions. For all centrality bins, we observe a rapid rise of R_{dAu} from low p_T, leveling off at $p_T \approx 2$ GeV/c. It is important to note when comparing to the results of the other RHIC experiments ([11, 12, 13]) that our acceptance is shifted by approximately one unit in pseudorapidity.

For comparison, we also plot the results from central Au+Au collisions at the same energy [4] in the lower right panel of Fig. 1. The average number of collisions undergone by each participating nucleon in the central Au+Au collision is close to 6, similar to that of each nucleon from the deuteron in a central d+Au collision. For central Au+Au collisions, the ratio of the spectra to $p+\bar{p}$ rises rapidly up to $p_T \approx 2$ GeV/c, but falls far short of collision scaling at larger p_T, in striking contrast to the behavior for central d+Au collisions.

Predictions for the evolution of R_{dAu} from semi-peripheral collisions with $\langle N_{coll} \rangle \approx 6$ to central collisions were made in two qualitatively different models. Perturbative QCD calculations [14] predict an increase in the maximum value of R_{dAu} at $p_T \approx 3.5$ GeV/c by 15%. In contrast, a decrease in R_{dAu} by 25–30% over the same centrality range is

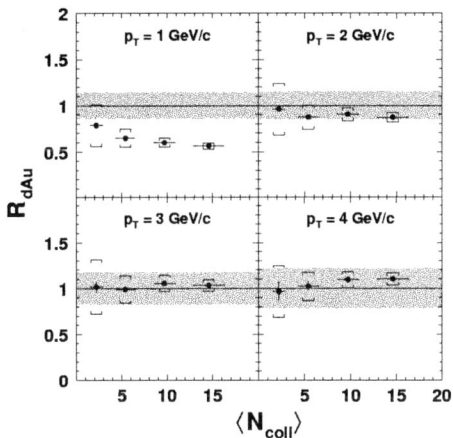

FIGURE 2. Nuclear modification factor R_{dAu} as a function of centrality in four bins of transverse momentum. The brackets indicate the point-to-point systematic error, dominated by the uncertainty in the number of collisions for each centrality bin. The grey band shows the overall scale uncertainty at each p_T. Systematic errors are at 90% C.L.

predicted in a parton saturation model [6]. The centrality evolution of R_{dAu} is shown in Fig. 2, where the points were obtained from a fit to the p_T dependence of R_{dAu} in each centrality bin.

Our data disfavor the prediction from the parton saturation model. This suggests that the observed suppression of high p_T hadrons in Au+Au collisions [1, 2, 3, 4] cannot be accounted for by initial state effects that should also be present in d+Au collisions.

This work was partially supported by US DoE grants DE-AC02-98CH10886, DE-FG02-93ER40802, DE-FC02-94ER40818, DE-FG02-94ER40865, DE-FG02-99ER41099, W-31-109-ENG-38, US NSF grants 9603486, 9722606, 0072204, Polish KBN grant 2-P03B-10323, and NSC of Taiwan contract NSC 89-2112-M-008-024.

REFERENCES

1. K. Adcox *et al.*, Phys. Rev. Lett. **88** (2002) 022301.
2. K. Adcox *et al.*, arXiv:nucl-ex/0207009.
3. C. Adler *et al.*, Phys. Rev. Lett. **89** (2002) 202301.
4. B. B. Back *et al.*, arXiv:nucl-ex/0302015, submitted to Phys. Lett. **B**.
5. M. Gyulassy and M. Plümer, Phys. Lett. **243** (1990) 432.
6. D. Kharzeev, E. Levin, L. McLerran, BNL preprint BNL-NT-02/22, arXiv:hep-ph/0210332.
7. X. N. Wang, Phys. Rev. **C61** (2000) 064910.
8. B. B. Back *et al.*, Nucl. Inst. Meth. **A499** (2003) 603.
9. B. B. Back *et al.*, Phys. Rev. Lett. **91** (2003) 072302.
10. C. Albajar *et al.*, Nucl. Phys. **B335**, (1990) 261.
11. S. S. Adler *et al.*, arXiv:nucl-ex/0306021.
12. J. Adams *et al.*, arXiv:nucl-ex/0306024.
13. I. Arsene *et al.*, arXiv:nucl-ex/0307003.
14. I. Vitev, arXiv:nucl-th/0302002.

Dynamics of Soft Particle Production in Heavy Ion Collisions

Peter Steinberg

Brookhaven National Laboratory
Upton, NY 11973

Abstract. The essential features of soft particle production data are discussed in the context of saturation models, hydrodynamical calculations, and empirical scaling rules.

INTRODUCTION

Heavy ion collisions are typically thought to proceed through several stages, each characterized by successively smaller energy scales. It is an open question how much we can understand about this complicated multi-stage dynamical scenario in detail. High-p_T hadrons may help probe the evolution using pQCD, but it is interesting to consider what we are learning by analyzing the bulk of produced particles with $p_T < 2$ GeV. In this work, three questions pertaining to soft particle production will be addressed:

- How do we understand the dynamics of the very early stage?
- Do we have evidence that reinteractions introduce additional dynamical considerations after the initial stage?
- What empirical constraints do we have on the validity of the theoretical models?

UNDERSTANDING THE EARLY STAGE

When a hadron is probed at high energy, the low-x part of the wave function dominates the interaction. This is seen most strikingly in the rapidly-rising gluon structure function extracted from HERA data. When considering a nucleus at high energy, one must consider the possibility that the low-x partons are coherent over distance scales $\Delta z \sim 1/x m_p$, where m_p is the proton mass [1]. In this case the gluons emitted from different nucleons can re-interact with each other, and even recombine. This naturally leads to the generation of a new energy scale $Q^2 \propto \nu \propto A^{1/3}$, where ν is the number of collisions per participant, and thus a measure of the nuclear thickness seen by a probe. Below this scale, the low-x gluons recombine and the cross section "saturates", taming the growth of the structure functions to enforce unitarity.

A large body of theoretical and phenomenological work has emerged describing the properties of this "Color Glass Condensate"(CGC), as discussed by Kovchegov in these proceedings [2]. It has had substantial success at describing the proton structure function at very low-x, as measured at HERA. Through this work, the important feature of "geometrical scaling"[3] has been explored and found to have possible consequences on heavy ion physics as well. Geometrical scaling is based on the observation that the cross section for a virtual photon to interact with a proton does not depend on Q^2 and x separately, but rather in the combination $x^\lambda/Q^2 = 1/\tau$, with λ the parameter that describes the rise of the structure function at very low-x. This leads to strong constraints on how Q_s^2 varies with energy and rapidity [4].

However, to apply saturation ideas to experimental data, several assumptions need to be made [4]: 1) Saturation models are formulated for small x. For larger values, one assumes the form found in pp collisions, $(1-x)^4$ [5], 2) It is assumed that the gluon distribution maps linearly onto the final state hadron distribution, a hypothesis found to be approximately true in e^+e^- collisions [6], and 3) A typical hadron mass must be chosen to convert the gluon rapidity to hadron pseudorapidity, which is a substantial effect, owing to the relation $dy = \beta d\eta$.

When this is done, impressive agreement with charged particle production over a large range in rapidity, energy, and centrality is achieved, as shown by Kharzeev [1]. The induced scale needed, $Q_s^2 \sim 2$ GeV^2, corresponds to a formation time for the bulk of the gluons of $\tau_0 \sim .2$ fm/c. This leads to an enormous energy density of almost 20 GeV/fm^3, or 40 times normal nuclear matter.

The CGC description of the initial state clearly captures the essential feature that at high energy, a nucleus is nearly "black" to oncoming nucleons, mainly due to the longitudinal coherence of the soft gluons. This leads to a dramatic reduction in the observed degrees of freedom, and thus in the expected multiplicities. The apparent success of parton-hadron duality naturally leads to the conclusion that we are observing a property of the initial state when we study the number of final state hadrons. However, since the assumed forward distributions are taken from pp physics, it is also not surprising that the pseudorapidity distributions in heavy ion collisions are quite similar to pp collisions at the same energy.

EVOLUTION FROM INITIAL TO FINAL STATE

If a QGP is formed in the early stage by sufficient reinteractions, thermodynamics and hydrodynamics should be applicable to the produced matter. Similar considerations emerged 50 years ago in the work of Landau, who solved 3+1D hydrodynamic equations with a massless equation of state (EOS) [7]. This was adapted by Bjorken [8] in the context of the emerging "boost invariant" plateau seen in ISR collisions and developed by many authors into a tool to explore the QGP and hadronic EOS by the study of soft particle production. The evidence for the applicability of statistical and hydrodynamic approaches in nuclear collisions has been mounting since the availability of the RHIC data, and is summarized by Kolb [9] in these proceedings.

While the successful phenomenological description of the data lends confidence to the validity of the approach, there are some outstanding issues. Even with the boost-invariant geometry, the hydrodynamic problem is under-constrained theoretically. Initial conditions are treated as input parameters for the calculations, the validity of which

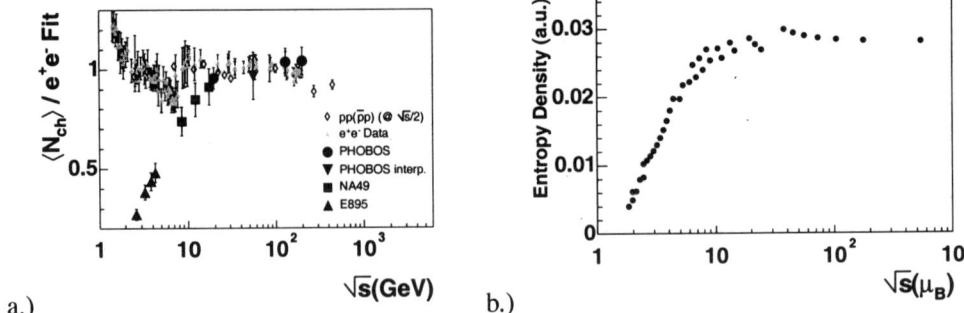

FIGURE 1. a.) Total charged particle multiplicities measured in e^+e^-, pp and central $A+A$ collisions vs. $\sqrt{s}$ divided by a fit to e^+e^-. b.) Entropy density (or entropy per participant) vs. $\sqrt{s}$ in the thermal model of Cleymans, et al.

are then evaluated on the basis of a successful phenomenological description. The conditions for the freezeout of the final state hadrons are also not understood from first principles. Typically, one identifies a space-time hypersurface on which the thermodynamic variables (e.g. temperature or energy density) reach a critical lower bound, beyond which hydrodynamics is not thought to apply. Finally, while the initial and final states are essentially boundary conditions applied at early and late times that set the size and energy scales for the evolution [10], the dynamical content is ultimately determined by the equation-of-state (EOS). Given the importance of the EOS, systematic studies are crucial to determine how uniquely it can be determined. In the meantime, it is growing clearer that only observables which are relatively immune to thermal fluctuations, like the flow of heavy particles, are the most sensitive to the details of the EOS.

However, there remain several important discrepancies between hydrodynamic calculations and experimental data. These are typically found in the forward directions, where longitudinal and transverse dynamics inevitably mix. It should be noted that the experimental data on rapidity distributions are manifestly in disagreement with the assumption of boost-invariance [11]. Boost-invariant calculations which assume no additional longitudinal expansion typically lead to a source which is over-extended in the R_{long} direction measured in HBT analyses. And yet, even calculations which do not even explicitly impose boost invariance are also unable to reproduce the rapid fall-off of the measured azimuthal asymmetry (v_2) at forward angles [12].

And yet, the hydrodynamical descriptions suggest that, in apparent contradistinction to the saturation assumptions, the final state particles are the result of strong reinteractions in the early phase. A natural resolution to this paradox is the fact that the evolution is isentropic, i.e. no new information (entropy) is introduced, but rather the new dynamics are encoded in the buildup of transverse expansion.

EMPIRICAL DESCRIPTION OF THE FINAL STATE

A typical example of a purely empirical approach to soft particle production is the study of the number of charged particles per participating nucleon pair $\langle N_{ch}\rangle/\langle N_{part}/2\rangle$. This tests the hypothesis that heavy ion collisions may be simply "constructed" out of a superposition of nucleon-nucleon collisions.

Interestingly, heavy ion data shows proportionality of N_{ch} with $N_{part}/2$ [13], as seen in pA collisions [14]. However, at high energies (above CERN SPS) the proportionality factor is not given by pp collisions, but rather the number of charged particles produced in e^+e^- annihilation to hadrons[13], as shown in Figure 1a. It also turns out that pp collisions follow the same systematic, once the energy of the leading particles is subtracted. This suggests that only the available energy controls the final-state multiplicity, and that AA collisions substantially reduce the leading particle effect. The agreement over an order of magnitude in $\sqrt{s}$ thus suggests a new type of observed "scaling". However, at lower energies this scaling is systematically violated, raising doubts about the presence of a truly simple behavior.

A resolution to this puzzle may be provided by insights gained from the success of statistical-thermal models in describing particle yields, described by Magestro [15] in these proceedings. While the analysis of elementary collisions has revealed that the freezeout temperature is approximately constant, heavy ion collisions show a tight coupling between the temperature (T) and the baryochemical potential (μ_B). This has been simply expressed by Cleymans and Redlich as the average energy per hadron $\langle E\rangle/\langle N\rangle \sim 1$ GeV, which appears to characterize all final states, from nuclei at rest to the recent RHIC data[16]. It also allows a simple parametrization of T and μ_B vs. $\sqrt{s}$, e.g. as in [17].

Landau and Belenkij noticed that an increasing net baryon number reduces the entropy density [7] as a consequence of energy conservation. This is illustrated using the model and parametrization of Cleymans, et al [17] in Figure 1b. It is observed that the observed deviation of the multiplicity per participant pair in A+A collisions with respect to e^+e^- is qualitatively similar to the suppression of the entropy density ($S/V \sim S/N_{part}$). This suggests that the similarities between A+A collisions and e^+e^- annihilations may be understandable over the whole energy range.

However, this also suggests that the dynamics which control how the initial baryons are stopped in the collision process play a more comprehensive role than might have been imagined in heavy ion collisions. The observed scaling of strangeness production in heavy ion collisions not with N_{part} or ν but rather with the fraction of multiply-struck participants [18, 19] also points to the importance of the initial excitation stage of the incoming baryons on the final state.

The key point gleaned from the multiplicity and strangeness data is that the distinguishing feature of AA collisions appears to be simply the multiple-excitations of the incoming baryons, which is dominantly a geometrical effect. This simultaneously leads to nearly-complete stopping and drives the system towards chemical equilibrium. It should be noted that neither of these observations would be made by simply assuming that simply the overall volume of the system (which scales as N_{part}) controls the relevant physics. Interestingly, recent work by Bass and Müller suggests that proton stopping may simply be the result of "destroying" the initial baryons entirely, the final rapidity distribution reflecting the x distribution of the quark PDFs. This itself is similar to Dokshitzer's observation that it only takes two collisions to fully dissociate a proton [20].

PUTTING IT ALL TOGETHER

While the variety of approaches and sheer amount of data may seem daunting, it is clear that the $A + A$ data is typically simpler than one would expect from the typical picture assuming simple pp collisions followed by several rescattering stages, none of which are constrained by data or theory. It may not be inconsistent to construct a physical scenario where saturation physics leads to N_{part} scaling and a large initial energy density from the rapid recombination process. After this, the energy density is propagated isentropically via hydrodynamic evolution. However, it is not obvious in such a model how to incorporate the apparent universality of particle multiplicities, which seems to depend more on the initially deposited energy rather than the induced momentum scale. Moreover, neither the saturation nor the hydrodynamic pictures to date lend substantial insight into the role of the initial baryon stopping (although progress is being made on this front [21]). These are clearly important issues illuminated by the phenomena at RHIC which will have to be addressed before a truly simple picture starts to emerge.

ACKNOWLEDGMENTS

I would like to thank the organizers of the CIPANP conference and especially the conveners of the heavy ion session, Jamie Nagle and Tim Hallman, for putting together a timely and well-organized set of talks.

REFERENCES

1. Kharzeev, D., These proceedings (2003).
2. Kovchegov, Y. V., These proceedings (2003).
3. Stasto, A. M., Golec-Biernat, K., and Kwiecinski, J., *Phys. Rev. Lett.*, **86**, 596–599 (2001).
4. Kharzeev, D., and Levin, E., *Phys. Lett.*, **B523**, 79–87 (2001).
5. Brodsky, S. J., and Farrar, G. R., *Phys. Rev. Lett.*, **31**, 1153–1156 (1973).
6. Azimov, Y. I., Dokshitzer, Y. L., Khoze, V. A., and Troian, S. I., *Z. Phys.*, **C27**, 65–72 (1985).
7. Belenkij, S. Z., and Landau, L. D., *Nuovo Cim. Suppl.*, **3S10**, 15 (1956).
8. Bjorken, J. D., *Phys. Rev.*, **D27**, 140–151 (1983).
9. Kolb, P. F., These proceedings (2003).
10. Carruthers, P, in *New York 1973, Annals Of The New York Academy Of Sciences, Vol.229 (1973),
11. Debbe, R., These proceedings (2003).
12. Back, B. B., et al., Phys. Rev. Lett., **89**, 222301 (2002).
13. Back, B. B., et al., nucl-ex/0301017 (2003).
14. Elias, J. E., et al., *Phys. Rev. Lett.*, **41**, 285 (1978).
15. Magestro, D., These proceedings (2003).
16. Cleymans, J., and Redlich, K., *Phys. Rev. Lett.*, **81**, 5284–5286 (1998).
17. Cleymans, J., *Pramana*, **60**, 787–794 (2003).
18. Hohne, C., et al., *Nucl. Phys.*, **A715**, 474–477 (2003).
19. Kampfer, B., Cleymans, J., Steinberg, P., and Wheaton, S., hep-ph/0304269 (2003).
20. Dokshitzer, Y. L., hep-ph/0306287 (2003).
21. Itakura, K., Kovchegov, Y. V., McLerran, L., and Teaney, D., hep-ph/0305332 (2003).

Saturation Physics in Heavy Ion Collisions

Yuri V. Kovchegov

Department of Physics, University of Washington, Box 351560, Seattle, WA 98195

Abstract. We discuss expectations of saturation physics for various observables in heavy ion collisions. We show how simple saturation-inspired assumptions about particle production in heavy ion collisions lead to Kharzeev-Levin-Nardi model. Comparing this model to RHIC data on particle multiplicities we conclude that saturation effects may play an important role in particle production and dynamics at the early stages of $Au - Au$ collisions already at RHIC energies. We then estimate the contribution of the initial state two-particle azimuthal correlations to elliptic flow observable v_2 in $Au - Au$ collisions by constructing a lower bound on these non-flow effects based on v_2 obtained from the analysis of proton-proton (pp) collisions.

INTRODUCTION

Saturation/Color Glass Condensate physics is based on the observation that the small-x wave functions of ultrarelativistic hadrons and nuclei are characterized by a hard scale Q_s, known as the saturation scale [1, 2, 3, 4]. The scale Q_s arises due to *saturation* of partonic densities at small-x and is an increasing function of energy and atomic number of the nucleus [5, 6, 4]. This large scale makes the strong coupling constant small $\alpha_s(Q_s) \ll 1$ leading to dominance of the classical gluonic fields in all high energy processes [4, 7]. Gluon production in high energy collisions is given by the classical field of the scattering color charges [8]. Corresponding gluon production cross section was found for pA collisions in [9] and the effects of quantum evolution [5] were included in it in [10]. The gluon production cross section for heavy ion collisions (AA) at the classical level has been studied both numerically [11] and analytically [12]. Since it is quite not clear at present how to include the effects of nonlinear quantum evolution [5] in the results of [11, 12], one has to construct models to describe the actual rapidity-dependent data produced in heavy ion collisions. Below we are going to show how some of these models, based on rather basic properties of saturation physics, provide a reasonably good description of RHIC data.

PARTICLE MULTIPLICITY FROM SATURATION MODELS

Multiplicity at Mid-Rapidity Versus Centrality

Classical field $A_\mu \sim 1/g$ leads to produced gluon multiplicity

$$\frac{dN}{d^2k\, d^2b\, dy} \sim \langle A_\mu A_\mu \rangle \sim \frac{1}{\alpha_s}. \tag{1}$$

Gluon transverse momentum spectrum described by a single scale Q_s can be written as

$$\frac{dN}{d^2k\,d^2b\,dy} = \frac{1}{\alpha_s} f(k_\perp/Q_s) \qquad (2)$$

with some unknown function $f(k_\perp/Q_s)$ to be determined by actual calculations. Integrating over k and b yields

$$\frac{dN}{dy} = \text{const}\,\frac{1}{\alpha_s}\pi R^2 Q_s^2. \qquad (3)$$

where the value of the constant is determined from $f(k_\perp/Q_s)$. Following [13, 14] we assume that the scale for the coupling constant in Eq. (3) is set by Q_s. (This step, of course, goes beyond the classical limit and assumes that Eq. (3) is valid even when running coupling corrections are included.) Then, as $R^2 \sim A^{2/3}$ and if $Q_s^2 \sim A^{1/3}$ [4, 7], together with running coupling

$$\alpha_s(Q_s) = \frac{1}{b \ln Q_s^2/\Lambda^2} \sim \frac{1}{\ln A} \qquad (4)$$

we conclude from Eq. (3) that

$$\frac{1}{A}\frac{dN}{d\eta} \sim \ln A. \qquad (5)$$

For heavy ion experiments at different collision centralities we substitute A by the number of participants N_{part} so that Eq. (5) becomes

$$\frac{1}{N_{part}}\frac{dN}{d\eta} \sim \ln N_{part}. \qquad (6)$$

Eq. (6) allowed the authors of [13] to correctly *predict* the particle multiplicity at mid-rapidity at RHIC as a function of centrality at $\sqrt{s} = 130$ GeV. A fit of particle multiplicity at other values of rapidity at $\sqrt{s} = 130$ GeV taken from [14] is shown in Fig. 1.

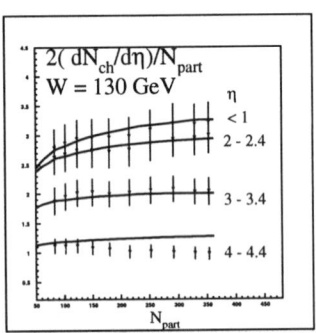

FIGURE 1. Saturation model fit of the PHOBOS data on total charged particle multiplicity at mid-rapidity as a function of centrality at $\sqrt{s} = 130$ GeV taken from [14].

Multiplicity as a Function of Energy

Eq. (3) allows one to test whether the scaling of total particle multiplicity with energy is consistent with saturation/Color Glass predictions. Using the fact that $Q_s^2 \sim 1/x_{Bj}^\lambda$ [5, 6] in Eq. (3) and, for the moment, dropping the slower Q_s-dependence in α_s leads to

$$\frac{dN/d\eta(\sqrt{s_1})}{dN/d\eta(\sqrt{s_2})} = \left(\frac{\sqrt{s_1}}{\sqrt{s_2}}\right)^\lambda. \tag{7}$$

Using PHOBOS data for total charge multiplicity at $\sqrt{s} = 130$ GeV for most central collisions

$$\frac{dN}{d\eta}(\sqrt{s} = 130\,\text{GeV}) = 555 \pm 12(stat) \pm 35(syst) \tag{8}$$

together with $\lambda = 0.25 \div 0.3$ obtained in [15] by analyzing HERA data, Kharzeev and Levin [14] *predicted* the total charge multiplicity at $\sqrt{s} = 200$ GeV to be

$$\frac{dN}{d\eta}(\sqrt{s} = 200\,\text{GeV}) = 616 \div 634 \tag{9}$$

in agreement with the later measured PHOBOS result

$$\frac{dN}{d\eta}(\sqrt{s} = 200\,\text{GeV}) = 650 \pm 35(stat). \tag{10}$$

$\frac{dN}{d\eta}$ Versus η and N_{part}

Describing rapidity distribution of the produced particles requires a little more modeling. Assuming k_T-factorization for particle production cross section

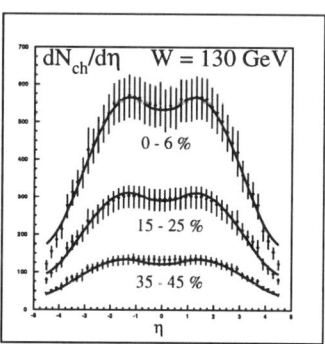

FIGURE 2. Saturation model fit of the PHOBOS data on total charged particle multiplicity as a function of rapidity and centrality at $\sqrt{s} = 130$ GeV taken from [14].

$$\frac{d\sigma^{AA}}{d^2k\,dy} = \frac{2\alpha_s}{C_F}\frac{1}{\underline{k}^2}\int d^2q\,\phi_A(\underline{q})\,\phi_A(\underline{k}-\underline{q}) \qquad (11)$$

along with saturation-inspired unintegrated gluon distribution functions ($\phi_A(\underline{k}) \sim \alpha_s/\underline{k}^2$ if $k_\perp > Q_s$ and $\phi_A(\underline{k}) \sim S_\perp/\alpha_s$ if $k_\perp < Q_s$) the authors of [14] produced an impressive fit of the PHOBOS data on charged particle multiplicities as functions of rapidity and centrality at $\sqrt{s} = 130$ GeV shown in Fig. 2. The *predictions* made in [14] for similar multiplicity data at $\sqrt{s} = 200$ GeV were also in good agreement with the later published BRAHMS data.

Phenomenological success of the saturation models presented above does not contradict the possibility of strong final state interactions leading to formation of quark-gluon plasma. As was argued in [16], thermalization in the saturation framework would not introduce any fundamentally new scales leaving Eqs. (3) and (6) practically unchanged. Late stage interactions are also not very likely to significantly modify the rapidity distribution of Fig. 2 due to causality constraints.

NON-FLOW CONTRIBUTION TO V_2

The contribution of non-flow two-particle azimuthal correlations from the early stages of heavy ion collisions to the elliptic flow observable v_2 has been estimated in [17] using a saturation-inspired model of particle correlations. Here we are going to construct a model-independent lower bound on these non-flow effects using v_2 extracted from the analysis of pp data. We start with the definition of $v_2(p_T)$ for pp corrected for uncertainty in the "reaction plane" definition (or, equivalently, defined through the two-particle correlation functions, such that $v_2(p_T) < v_2 > = < \cos 2(\phi_p - \phi_k) >_k$)

$$v_2^{pp}(p_T) = \frac{\int d^2k\,\frac{dN_{corr}^{pp}}{d^3p\,d^3k}\cos 2(\phi_p-\phi_k)}{\frac{dN^{pp}}{d^3p}\frac{dN^{pp}}{dy_k} + \int d^2k\,\frac{dN_{corr}^{pp}}{d^3p\,d^3k}} \sqrt{\frac{\frac{dN^{pp}}{dy_p}\frac{dN^{pp}}{dy_k} + \frac{dN_{corr}^{pp}}{dy_p\,dy_k}}{\int d^2p\,d^2k\,\frac{dN_{corr}^{pp}}{d^3p\,d^3k}\cos 2(\phi_p-\phi_k)}}. \qquad (12)$$

If we assume that the relative magnitude of the correlated terms in Eq. (12) compared to uncorrelated ones is roughly the same for all p_T, we can drop the former compared to the latter finding that Eq. (12) is approximately bounded from above by

$$v_2^{pp}(p_T) \leq \frac{\int d^2k\,\frac{dN_{corr}^{pp}}{d^3p\,d^3k}\cos 2(\phi_p-\phi_k)}{\frac{dN^{pp}}{d^3p}\frac{dN^{pp}}{dy_k}} \sqrt{\frac{\frac{dN^{pp}}{dy_p}\frac{dN^{pp}}{dy_k}}{\int d^2p\,d^2k\,\frac{dN_{corr}^{pp}}{d^3p\,d^3k}\cos 2(\phi_p-\phi_k)}}. \qquad (13)$$

We want to estimate the contribution to $v_2^{AA}(p_T)$ of the non-flow correlations of the same physical origin as the ones giving rise to $v_2^{pp}(p_T)$ in Eq. (12). The contribution is

$$v_2^{AA}(p_T)|_{non-flow} = \frac{\int d^2k\,\frac{dN_{corr}^{AA}}{d^3p\,d^3k}\cos 2(\phi_p-\phi_k)}{\frac{dN^{AA}}{d^3p}\frac{dN^{AA}}{dy_k}} \sqrt{\frac{\frac{dN^{AA}}{dy_p}\frac{dN^{AA}}{dy_k}}{\int d^2p\,d^2k\,\frac{dN_{corr}^{AA}}{d^3p\,d^3k}\cos 2(\phi_p-\phi_k)}}. \qquad (14)$$

Using the fact that, approximately, both in saturation models and in the data $dN/dy \sim N_{part}$ [13, 14], we rewrite the right hand side of Eq. (13) as

$$v_2^{pp}(p_T) \leq v_2^{AA}(p_T)|_{non-flow} \sqrt{\frac{N^{AA}}{N^{pp}}}, \qquad (15)$$

with N^{AA} and N^{pp} total particle multiplicities in AA and pp collisions, proportional to the average number of participants involved. Inverting Eq. (15) we obtain a lower bound

$$v_2^{AA}(p_T)|_{non-flow} \geq v_2^{pp}(p_T) \sqrt{\frac{N^{pp}}{N^{AA}}}. \qquad (16)$$

Preliminary analysis of pp data yields $v_2^{pp}(p_T) \approx 1$ at high-p_T [18]. To get a lower bound we use $N_{part}^{AA} = 394$ and $N_{part}^{pp} = 2$ in Eq. (16) obtaining $v_2^{AA}(p_T)|_{non-flow} \gtrsim 7\%$.

ACKNOWLEDGMENTS

This work was supported in part by the U.S. Department of Energy under Grant No. DE-FG03-97ER41014. The preprint number is NT@UW-03-020.

REFERENCES

1. L. V. Gribov, E. M. Levin and M. G. Ryskin, Phys. Rept. **100**, 1 (1983).
2. A. H. Mueller and J. w. Qiu, Nucl. Phys. B **268**, 427 (1986).
3. J. P. Blaizot and A. H. Mueller, Nucl. Phys. B **289**, 847 (1987).
4. L. D. McLerran and R. Venugopalan, Phys. Rev. D **49**, 2233 (1994) [arXiv:hep-ph/9309289]; Phys. Rev. D **49**, 3352 (1994) [arXiv:hep-ph/9311205]; Phys. Rev. D **50**, 2225 (1994) [arXiv:hep-ph/9402335].
5. I. Balitsky, Nucl. Phys. B **463**, 99 (1996) [arXiv:hep-ph/9509348]; Y. V. Kovchegov, Phys. Rev. D **60**, 034008 (1999) [arXiv:hep-ph/9901281].
6. E. Levin and K. Tuchin, Nucl. Phys. B **573**, 833 (2000) [arXiv:hep-ph/9908317]; E. Iancu, K. Itakura and L. McLerran, Nucl. Phys. A **708**, 327 (2002) [arXiv:hep-ph/0203137].
7. Y. V. Kovchegov, Phys. Rev. D **54**, 5463 (1996) [arXiv:hep-ph/9605446]; Phys. Rev. D **55**, 5445 (1997) [arXiv:hep ph/9701229]; J. Jalilian-Marian, A. Kovner, L. D. McLerran and H. Weigert, Phys. Rev. D **55**, 5414 (1997) [arXiv:hep-ph/9606337].
8. A. Kovner, L. D. McLerran and H. Weigert, Phys. Rev. D **52**, 6231 (1995) [arXiv:hep-ph/9502289].
9. Yu. V. Kovchegov and A. H. Mueller, Nucl. Phys. B **529**, 451 (1998) [arXiv:hep-ph/9802440].
10. Yu. V. Kovchegov and K. Tuchin, Phys. Rev. D **65**, 074026 (2002) [arXiv:hep-ph/0111362].
11. A. Krasnitz and R. Venugopalan, Phys. Rev. Lett. **84**, 4309 (2000) [arXiv:hep-ph/9909203].
12. Y. V. Kovchegov, Nucl. Phys. A **692**, 557 (2001) [arXiv:hep-ph/0011252].
13. D. Kharzeev and M. Nardi, Phys. Lett. B **507**, 121 (2001) [arXiv:nucl-th/0012025].
14. D. Kharzeev and E. Levin, Phys. Lett. B **523**, 79 (2001) [arXiv:nucl-th/0108006].
15. K. Golec-Biernat and M. Wusthoff, Phys. Rev. D **59**, 014017 (1999) [arXiv:hep-ph/9807513].
16. R. Baier, A. H. Mueller, D. Schiff and D. T. Son, Phys. Lett. B **539**, 46 (2002) [arXiv:hep-ph/0204211].
17. Y. V. Kovchegov and K. L. Tuchin, Nucl. Phys. A **708**, 413 (2002) [arXiv:hep-ph/0203213]; Nucl. Phys. A **717**, 249 (2003) [arXiv:nucl-th/0207037]; K. Tuchin, arXiv:hep-ph/0307097, these proceedings.
18. See talk by R. Snellings at the workshop on *Transverse Dynamics at RHIC*, BNL, March 6-8, 2003, URL: http://www.phenix.bnl.gov/phenix/WWW/publish/rak/workshop/int/program_TD.htm.

Particle production at RHIC energies

R. Debbe for the BRAHMS collaboration

Physics Dept. Brookhaven National Laboratory

Abstract.
This paper presents recent results from the BRAHMS experiment at RHIC; including results on particle production in rapidity space extending from y=0 to y ∼ 3 and on the transverse momentum distribution of fully identified charged particles. These results were obtained from the 5% most central Au-Au collisions recorded during RHIC Run-2 at $\sqrt{s_{NN}} = 200$ GeV.

INTRODUCTION

BRAHMS is the only RHIC experiment that is able to study fully identified particle production and energy flow over a wide range of rapidity (from y=0 to y=4 for pions). This coverage, which almost reaches the fragmentation regions, is ideal for studies of the bulk properties of the system formed in heavy ion collisions at RHIC energies. This work reports recent results obtained from the analysis of data collected with the BRAHMS spectrometers in Au-Au collisions at $\sqrt{s_{NN}} = 200$ GeV. A detailed description of the BRAHMS experimental setup can be found in [1]. All results shown here are preliminary and were obtained from a sample of the 5% most central events.

PARTICLE PRODUCTION

Momentum distributions of fully identified charged particles were obtained with conventional magnetic spectrometers instrumented with state-of-the-art time-of-flight and ring imaging and threshold Čerenkov detectors. For each particle type, the density in rapidity space is obtained by integration over the p_T dependence of these distributions. With the distributions dropping rapidly as p_T increases, a good fraction of the integral comes from unmeasured yields at low p_T. An extrapolation is thus necessary to cover the unmeasured regions. Empirically the pion yields were found to be best fitted by a power law, the kaon distributions by a single exponential in p_T, and the proton distributions with a single exponential in m_T.

The resulting rapidity density distributions are shown in panels a and b of figure 1. Panel a shows the densities for all charged particles. Because pions dominate this figure, panel b expands the view for the kaon, proton and antiproton distributions.

A remarkable feature of these distributions is their common bell shape character. (Pions, kaons and anti-proton distributions are fitted well with double Gaussians). This observation has a possible explanation based on the postulate that particle production in the momentum range measured in the present experiment is driven primarily by the distribution of partons in the colliding ions. As the energy of the collisions increases, the

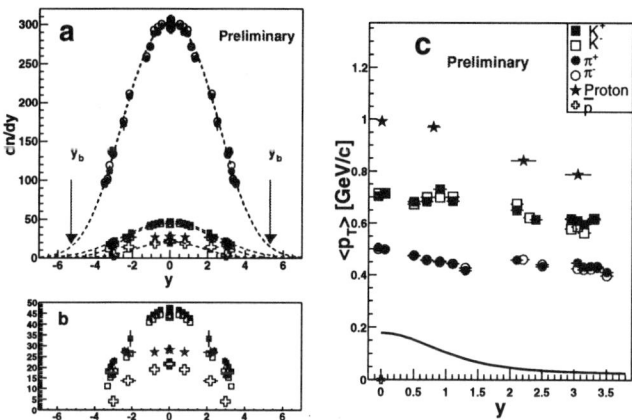

FIGURE 1. a) and b)Symmetrized rapidity density distributions for identified charged particles measured in the 5% most central events. The measurements were done for y>0. c) Mean ransverse momentum as function of rapidity

parton distributions can be resolved to smaller values of x (fraction of the total momentum of the hadron). The convolution of the left and right moving parton distributions leads to an initial bell-shaped dn/dy distribution for produced particles in symmetric system centered around y=0.

This distribution may evolve in later stages through secondary interactions, but it retains its bell shape. In this picture there is neither a wide plateau connecting the two fragmentation regions, as Feynman's intuition had it [2], nor an extended boost invariant longitudinal expansion, as proposed by Bjorken [3].

A good summary of all the p_T distributions extracted in this analysis is shown in panel c of figure 1; the average transverse momentum with which the detected particles are produced at different rapidities. Worth noting in this result is the small change of the pion and kaon average p_T as function of y. For comparison, a calculated average p_T for pions is also drawn in panel c. The curve was obtained with a single thermal source described by a Boltzmann distribution with a temperature of 200 MeV.

An inspection of the transverse momentum distributions shows that for the more massive particles there is a very clear curvature in the spectra as the value of p_T approaches zero (see Figure 2). This observation, together with the almost exponential shape of the distributions at higher p_T and for lower mass particles, is well reproduced by a functional form based on a thermalized system expanding radially [4].

After integration over y and azimuthal angle and assuming that a Boltzmann distribution describes the system, the following functional form is obtained, and is used to fit the spectra:

$$\frac{dn}{m_T dm_T} = Am_T \int_0^R rdr K_1\left(\frac{m_T \cosh\rho}{T}\right) I_0\left(\frac{p_T \sinh\rho}{T}\right)$$

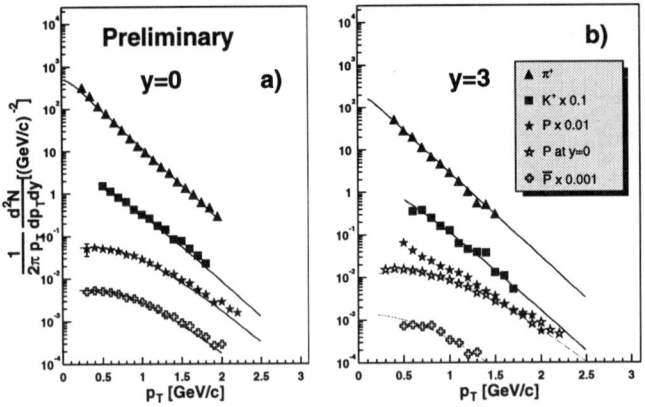

FIGURE 2. "Blast wave" fits to pions, kaons and protons at rapidity y=0 (panel a) and y=3 (panel b)

TABLE 1. Results of blast wave fits

Rapidity	Temperature [MeV]	Velocity
0	127 ± 2	0.57 ± 0.01
0.7	112 ± 1	0.60 ± 0.01
2.2	128 ± 3	0.50 ± 0.01
3	136 ± 4	0.44 ± 0.02

where $\beta_T = tanh\rho$ is the transverse velocity of the flow in units of c, and T is the decoupling temperature.

Several functions that describe the radial dependence of the flow velocity have been proposed. Here the simplest assumption of a transverse flow velocity that is constant at all radii is assumed. This choice may not be fully adequate, but does serve to highlight the rapidity dependence of the flow velocity. The table 1 summarizes the results.

The shapes of the distributions are well reproduced by the fits, with a 30 % reduction in transverse velocity found going from y-0 to y=3. This reduction is also suggested by the different shape of the mid-rapidity proton distributions as compared to that at the most forward rapidity, as shown in Fig. 2b where the y=0 points have been shifted down to facilitate the comparison.

The measured net proton at mid-rapidity [6],[7] was an early subject of much discussion in the community because it went against an expected baryon free region around $y \sim 0$; the energy of the colliding beams was so high that the initial baryon number should end up in the fragmentation regions if tied to the valence quarks . But the first results extracted at y=0 indicated otherwise, the net-proton number was not equal to zero. Some mechanism was transporting baryon number to mid-rapidity. D. Kharzeev [5] had predicted that the x distributions of gluons of the initial baryon extend to very small values of x. At the time of the collision these distributions would overlap and after "dressing" with quarks from the sea the low-x gluons of the initial baryons bring in

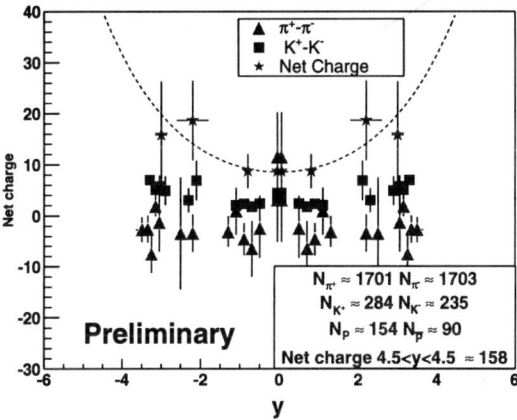

FIGURE 3. Net charge as function of rapidity (star symbols). The difference $K^+ - K^-$ is shown with squares, and the triangles show the difference $\pi^+ - \pi^-$. The dashed line is the fit to the function refered in the text.

effect net baryon number to mid-rapidity. This effect would have a rapidity dependence as $\frac{Zb}{sinh(y_{max}b)}cosh(yb)$ where Z is the charge of the ions, and b is related to nature of the system that brings baryon number to mid-rapidity and predicted to be close to 1/2. The parameter y_{max} is set to be equal to 4.5. Figure 3 shows the net charge measured up to rapidity 3 and a fit to the function mentioned above, The fit parameters are $b = 0.49 \pm 0.10$ with $\chi^2 = 1./7$.

SUMMARY

Analysis of the most central data sample collected during RHIC Run-2 (Au-Au at $\sqrt{s_{NN}} = 200$ GeV) is consistent with a thermalized system with a decoupling temperature around 120 MeV, and a strong radial flow $\beta_T \sim 0.6$ that diminishes by as much as 30% at the most forward rapidity measured.

Acknowledgments: This work was supported by the Division of Nuclear Physics of the Office of Science of the U.S. DOE, the Danish Natural Science Research Council, the Research Council of Norway, the Polish State Com. for Scientific Research, and the Romanian Ministry of Education and Research.

REFERENCES

1. M. Adamczyk et al, Nucl. Instr. and Meth. A499 437, (2003).
2. R. P. Feynman, Photon-Hadron Interactions, Addiso-Wesley, Massachusetts, 1998, p. 229.
3. J. D. Bjorken, Phys. Rev. D27 140, (1983).
4. E. Schnedermann et al Phys. Rev. C48 2462-2475, (1993).
5. D. Kharzeev Phys. Lett. B378 238-246, (1996).
6. C. Adler et al. Phys. Rev. Lett. 87 262302, (2001).
7. K. Adox et al. Phys. Rev. Lett. 88 242301 (2002).

What did we learn and what will we learn from hydrodynamics at RHIC?

Peter F. Kolb

Department of Physics and Astronomy, SUNY Stony Brook, Stony Brook, NY 11974, USA

Abstract.
A brief overview of the current status of hydrodynamic concepts applied to ultra relativistic heavy-ion collisions is presented. Special emphasis is placed on future prospects for extracting the thermodynamic properties and the bulk features of the created medium from experimental observations.

An extensive review of the application of hydrodynamics to describe the expansion stage of ultra relativistic heavy-ion collisions was recently given elsewhere [1]. Therefore I will be brief in reporting the highlights of past studies, but emphasize more what can be expected from the future application of hydrodynamics. For the basic concepts of hydrodynamics, its formalism and a complete list of references see also [1].

ANISOTROPIC FLOW AND HYDRODYNAMICS AT RHIC

One of the first and still most striking observations at RHIC is a strong event anisotropy in non-central collisions [2], which is generated through the elliptically deformed overlap region of the colliding nuclei, resulting in an eccentric distribution of matter and anisotropic pressure gradients in the early stages of the expansion [3]. From a microscopic point of view this strong collective anisotropy is best described under the assumption of extremely strong rescattering [4], strong enough in fact to reach the limit of continuum dynamics, whose predictions [5] were in quantitative agreement with first observations [2]. To achieve such a strong conversion of anisotropies from coordinate to momentum space, rescattering has to be strong at very early times and local thermalization has to occur while the geometric deformation of the source is still large [5]. More detailed subsequent hydrodynamic studies of non-central collisions gave additional predictions on characteristic features such as the mass dependence of elliptic flow [6] and investigated the influence of characteristics of the underlying equation of state [6, 7]. Results of such studies are given in Figure 1, where the left panel shows the momentum dependence of elliptic flow $v_2(p_T) = \langle \cos 2\varphi \rangle$, the second Fourier coefficient of the azimuthal dependence of the particle spectra $\frac{dN}{p_T dp_T dy d\varphi}$, as a function of transverse momentum for pions and protons. Experimental data from minimum bias collisions [8] are compared to hydrodynamic results [6, 9], once applying an equation of state including a phase transition to a plasma stage (solid lines), once using a soft resonance gas equation even at the earliest and hottest stages of the collision (dashed lines). The right panel of the figure shows the average elliptic flow of all charged particles as a function of centrality. Included are results of a hydrodynamic calculation assuming different latent heats Δe of the transition region (0.4, 0.8 and 1.6 GeV/fm^3) [7]. Clearly these results indicate that in order to describe the absolute magnitude, the mass splitting and the centrality dependence of elliptic flow, rapid thermalization with a strong push from a phase with a sufficiently hard equation of state (like the QCD plasma) and a fairly soft transition region (of width $\Delta e \sim 1$ GeV/fm^3) back to hadronic matter is required.

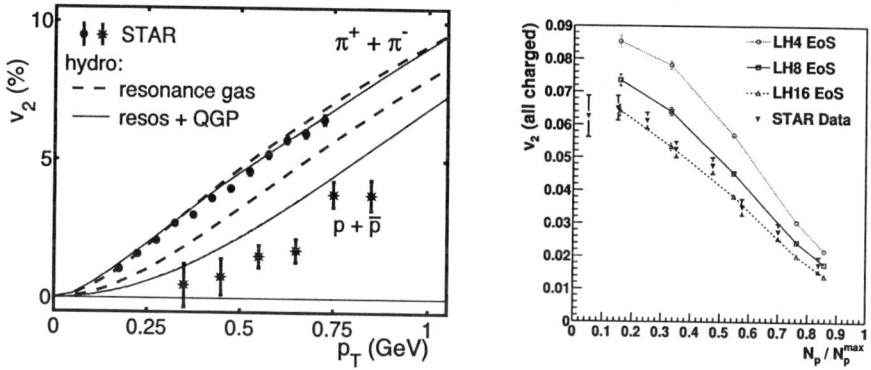

FIGURE 1. Experimental data of elliptic flow from Au+Au collisions at 130 AGeV [8, 2] in comparison to results from hydrodynamic calculations. Left: Elliptic flow as function of transverse momentum of pions and protons in minimum bias from a purely hydrodynamic calculation [6, 9], applying an equation of state with (solid) and without a phase transition (dashed). Right: centrality dependence of elliptic flow of a hydrodynamic calculation which treats the late hadronic stage within a microscopic picture. Results using different widths of the transition region (0.4, 0.8 and 1.6 GeV/fm^3) are shown [7].

THE ROAD AHEAD

With the prerequisites of a thermalized, adiabatically evolving source given, there is a wealth of topics that we can address within the hydrodynamic framework.

More quantitative extraction of the properties of the equation of state: Hydrodynamics is *the tool* to study how properties of the equation of state influence the dynamics of the system and final state observables, as the equation of state, which relates the local energy density to the pressure, explicitly enters the formalism in terms of the forces that drive the system apart. With the large variety of flow observables becoming currently available, we will be able to delimit parameters of the calculation, and particularly put stronger constraints on features of the nuclear equation of state by comparing experimental data with theoretical calculations. Helpful in this context are also finer details in the particle emission pattern, such as anisotropic components $v_n = \langle \cos n\varphi \rangle$ beyond the elliptic deformation, which may achieve significantly large values at intermediate to high transverse momenta [10]. The left panel of Fig. 2 shows the momentum dependence of the Fourier coefficients up to order 8 as expected from a hydrodynamic calculation to describe Au+Au collisions at $\sqrt{s_{NN}} = 200$ GeV.

Collectivity of which particles at what stages? The most abundant hadrons at RHIC – pions, kaons and protons – share signs of a strong common transverse expansion.

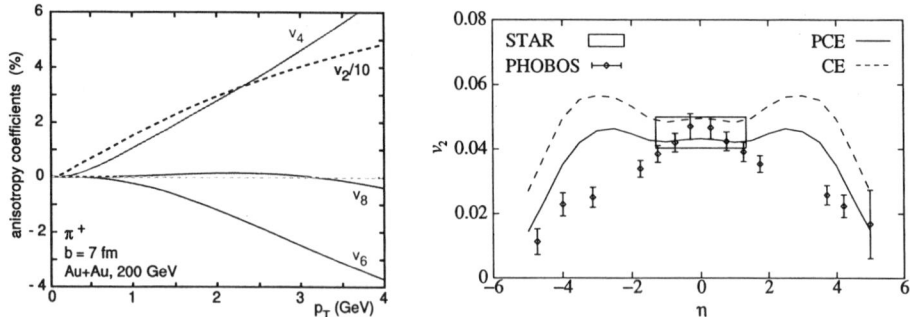

FIGURE 2. Left: Hydrodynamic expectations for higher flow anisotropies, calculated for pion spectra resulting from Au+Au collisions with $b = 7$ fm [10]. Right: rapidity dependence of elliptic flow from a hydrodynamic model [17] (with different degrees of chemical equilibration in the resonance stage), compared to experimental data from STAR and PHOBOS at 130 GeV [2, 18].

Newest results on particle spectra of strange and multi-strange baryons are yet inconclusive to which extent these species follow the collective dynamics. While the traditional 'blast wave fit' tends to assign them a higher temperature of origin [11], they appear to follow the common mass systematics of the full hydrodynamic calculation down to freeze-out temperatures of 100 MeV [1]. Signals attributed to decaying charmed mesons [12] do not yet prove or disprove whether even heavy flavors participate in the strong collectivity [13]. Whereas the question of flow of strangeness in the hadronic phase is still debatable, there is clear evidence from flow anisotropies of strange particles [14] that at hadronization all quark flavors share the same radial as well as anisotropic flow [6, 15], providing a clear sign of early thermalization and a common collective expansion of the fireball in the partonic stage.

Equation of state at larger baryon densities: Flow observables at forward/backward rapidities open the possibility to study the nuclear equation of state at larger baryon densities than found at central rapidity, and thus allow to scan a larger portion of the (T,μ)-plane of our nuclear equation of state, eventually bringing us into the realm of the expected tri-critical point [16]. This issue is however complicated through the breakdown of thermalization when moving toward the fragmentation region [17]. Before one can make quantitative statements about the equation of state in this realm, the exceedingly complicated interplay of initial conditions, viscous effects and the equation of state have to be well understood. Results from a hydrodynamic calculation that extends over a large rapidity window are shown in the right panel of Fig. 2 [17]. Deviations from ideal hydrodynamic behavior occur at rapidities $|\eta| > 1$ from where on the mentioned effects have to be taken into account.

Viscosity effects: Clearly the approach of ideal hydrodynamics works only under the stringent conditions of local thermalization followed by a 'gradual' adiabatic expansion. Deviations from this behavior are expected in the most peripheral collisions, when approaching the fragmentation region, in the late stages of the reaction and for the few particles emitted with large transverse momenta. Under these conditions, viscosity effects need to be considered [19]. Although the treatment of those effects within a full dynamical calculation is a very difficult task [20], it will eventually enrich our understanding of the transport properties of the quark gluon plasma and the hadron gas, and maybe resolve the persisting HBT-puzzle at RHIC [21].

Background medium for hard probes: Hydrodynamics has proved to be a great tool to study the properties of the bulk of the expanding matter. Even hard probes, although they do not follow the collective dynamics of the bulk, depend on the dynamical evolution of the fireball. The characteristics of energy-loss and jet-quenching should thus be folded into the hydrodynamic expansion, to get a reliable description of the net energy loss which the hard probes experience during the fireball evolution [22].

SUMMARY

To address the thermodynamic properties of the medium created at RHIC, it is essential that the system rapidly achieves local thermal equilibrium, which appears to happen within the first 1 fm/c after impact. The most natural language to study the nuclear equation of state, its influence on the dynamics of the system and the resulting observables is hydrodynamics. With the steady output of flow observables from RHIC we can constrain the parameters of the equation of state which is responsible for the observed strong collective expansion, but also address questions of viscosity and other non-ideal effects.

Acknowledgments: This work was supported in parts by the U.S. Department of Energy under Grant No. DE-FG02-88ER40388. Support from the Alexander von Humboldt Foundation in terms of a Feodor Lynen Fellowship is greatly appreciated.

REFERENCES

1. P.F. Kolb and U. Heinz, nucl-th/0305084.
2. K.H. Ackermann et al. (STAR Collaboration), Phys. Rev. Lett. 86 (2001) 402.
3. J.-Y. Ollitrault, Phys. Rev. D 46 (1992) 229.
4. D. Molnar and M. Gyulassy, Nucl. Phys. A 697 (2002) 495; Erratum-ibid. A 703 (2002) 893.
5. P.F. Kolb, J. Sollfrank and U. Heinz, Phys. Rev. C 62 (2000) 054909.
6. P.F. Kolb, P. Huovinen, U. Heinz, and H. Heiselberg, Phys. Lett. B 500 (2001) 232; P. Huovinen, P.F. Kolb, U. Heinz, P.V. Ruuskanen, and S. Voloshin, Phys. Lett. B 503 (2001) 58.
7. D. Teaney, J. Lauret, and E.V. Shuryak, Phys. Rev. Lett. 86 (2001) 3981, and nucl-th/0110037.
8. C. Adler et al. (STAR Collaboration), Phys. Rev. Lett. 87 (2001) 182301.
9. U. Heinz and P.F. Kolb, Nucl. Phys. A 702 (2002) 269, and hep-ph/0204061.
10. P.F. Kolb, nucl-th/0306081.
11. J. Adams et al. (STAR Collaboration), nucl-ex/0307024; J. Castillo et al. (STAR Collaboration), Nucl. Phys. A 715 (2003) 518; C. Suire et al. (STAR Collaboration), Nucl. Phys. A 715 (2003) 470.
12. K. Adcox et al. (PHENIX Collaboration), Phys. Rev. Lett. 88 (2002) 192303.
13. S. Batsouli, S. Kelly, M. Gyulassy, and J.L. Nagle, Phys. Lett. B 557 (2003) 26.
14. S.S. Adler et al. (PHENIX Collaboration) nucl-ex/0305013; J. Adams et al. (STAR Collaboration), nucl-ex/0306007.
15. R.J. Fries, B. Müller, C. Nonaka, and S.A. Bass, nucl-th/0301087, and nucl-th/0306027.
16. Z. Fodor and S.D. Katz, J. High Energy Phys. 03 (2002) 014.
17. T. Hirano, Phys. Rev. C 65 (2002) 011901; T. Hirano and K. Tsuda, Nucl. Phys. A 715 (2003) 821.
18. B. Back et al. (PHOBOS Collaboration), Phys. Rev. Lett. 89 (2002) 222301.
19. D. Teaney, nucl-th/0301099.
20. A. Muronga, Phys. Rev. Lett. 88 (2002) 062302; Erratum ibid, 89 (2002) 159901.
21. U. Heinz and P.F. Kolb, Nucl. Phys. A 717 (2003) 322; and nucl-th/0204061.
22. M. Gyulassy, I. Vitev, X.-N. Wang, and P. Huovinen, Phys. Lett. B 526 (2002) 301; T. Hirano and Y. Nara, nucl-th/0307015, and nucl-th/0307087.

Particle Correlations at High Partonic Density

Kirill Tuchin

*Institute for Nuclear Theory, University of Washington, Box 351550
Seattle, WA 98195, USA*

Abstract. We discuss manifestations of the particle correlations at high partonic density in the heavy-ion collisions at RHIC. In particular, we argue that the elliptic flow variable v_2 is dominated by particle correlations at high p_T.

Particle correlations at high partonic density (in Color Glass Condensate) are significantly different from those of the parton model. To illustrate this consider gluon production in Deep Inelastic Scattering on a heavy nucleus $A \sim 1/\alpha_s^6$ at high energies $x \sim e^{1/\alpha_s}$, Fig. 1. It was proved in [1] that the collinear factorization breaks down for this process. Instead $\gamma^* A$ cross section can be written in the k_T-factorized form [2]. This allows to introduce function $\phi(x, q_\perp)$ which encodes quantum evolution and multiple gluon rescattering in nucleus. $\phi(x, q_\perp)$ is simply related to the forward scattering amplitude which satisfies the QCD evolution equation for high partonic densities [3]. Solution to that equation implies that the scale inherent to function $\phi(x, q_\perp)$ is $q_\perp^2 \sim Q_s^2 = \Lambda_{\mathrm{QCD}}^2 A^{1/3} e^{4\alpha_s y}$. At high energies $Q_s^2 \gg \Lambda_{\mathrm{QCD}}^2$, therefore one cannot neglect the virtuality of the t-channel gluons compared to the momentum of the produced hard particle $k_T \gg 1$ GeV. As the result the transverse momentum conservation does not require anymore that the momentum of the hard jet be compensated by equally large momentum of another jet moving in the opposite direction in the transverse plane (in the center-of-mass frame).

The qualitative picture of the gluon production in heavy-ion collisions is pretty much the same as in DIS. Although the k_T factorization has not been proved in this case, there are reasons to believe that it is at least a fairly good approximation. Feynman diagram for the inclusive gluon production in AA collisions is shown in Fig. 2a. To study the gluon correlations in heavy-ion collisions we define the two-particle multiplicity distribution

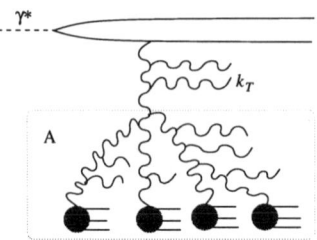

FIGURE 1. Gluon production in DIS on a heavy nucleus.

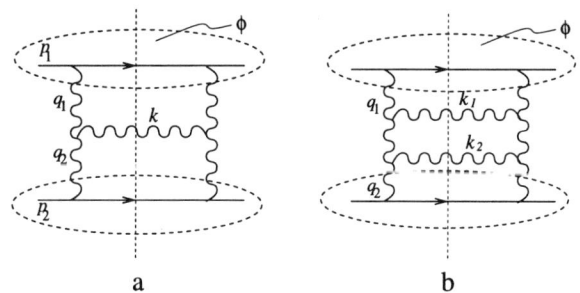

FIGURE 2. Inclusive (a) and double-inclusive (b) gluon production in AA

as
$$P(k_{1\perp},y_1;k_{2\perp},y_2) = \frac{dN}{d^2k_{1\perp}dy_1}\frac{dN}{d^2k_{2\perp}dy_2} + \frac{dN_{\text{corr}}}{d^2k_{1\perp}dy_1 d^2k_{2\perp}dy_2}, \qquad (1)$$

where the first term in the right-hand-side is just the square of the diagram Fig. 2a which gives the uncorrelated piece, and the second term is the diagram Fig. 2b which gives the correlated piece. The transverse momentum conservation applied to the later diagram gives $|k_{1\perp} + k_{2\perp}| \simeq NQ_s$, where N is the number of gluons in the nuclei wave functions, cf. Fig. 1. Since N can be large, the collinear factorization can be violated at $k_\perp^2 > Q_s^2$. It was argued in [5] that the collinear factorization is recovered when $k_\perp^2 \gtrsim \tilde{Q}^2 = Q_s^4/\Lambda_{\text{QCD}}^2 A^{1/3}$. The preliminary dA data at $\sqrt{s} = 200$ GeV suggest that $\tilde{Q}^2 \gtrsim Q_s^2$ at this energy. This means that at RHIC particles with momenta $k_\perp^2 \gg Q_s^2$ are correlated mostly back-to-back. It is important to emphasize, that bulk of particles is produced with $k_\perp < Q_s$ and therefore, saturation plays crucial role in understanding of the total multiplicity at RHIC.

Analysis of particle correlations with respect to the reaction plane azimuthal angle defined as
$$\tan 2\Psi_R = \frac{\sum_{i=1}^{'N}\sin 2\phi_i}{\sum_{j=1}^{'N}\cos 2\phi_i}, \qquad (2)$$

shows that the angular momentum distribution of the large multiplicity event $N \gg 1$ is given by [6]
$$\frac{dn}{d\phi_{p_T}d\Psi_R} = \frac{1}{(2\pi)^2}[1 + 2v_2(p_T,B)\Delta\cos 2(\phi_{p_T} - \Psi_R)], \qquad (3)$$

where Δ is the reaction plane resolution. The $\cos 2(\phi_{p_T} - \Psi_R)$ shape is the result of trivial autocorrelations of each particle with itself. However the coefficient v_2 carries information about the particle correlations. The elliptic flow variable v_2 is given by
$$v_2(p_T) = \frac{\langle\cos 2(\phi_1(p_T) - \phi_2)\rangle}{\sqrt{\langle\cos 2(\phi_1 - \phi_2)\rangle}}, \qquad (4)$$

where averaging over all events can be done using (1). The diagram Fig 2a entering the distribution function (1) can be calculated in a framework of the McLerran-Venugopalan

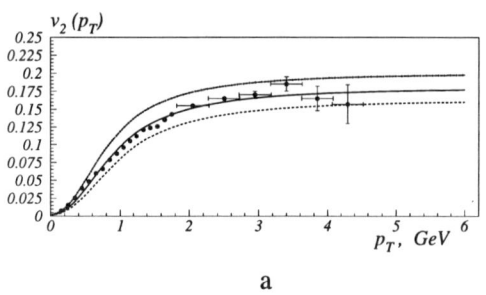
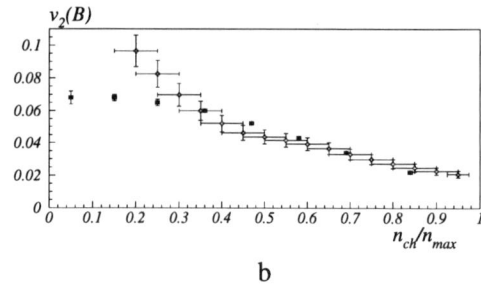

FIGURE 3. Elliptic flow variable as a measure of the particle correlations in gold-gold collisions at $\sqrt{s} = 130$ GeV vs STAR data.

model [7], i.e. treating the nuclear color field in the Weiszäcker-Williams approximation [8]. The diagram Fig. 2b requires the α_s quantum correction. For simplicity we assume that nuclei have cylindrical shape. We also neglect all finite state interactions. The result of calculation [8] is shown in Fig. 3. It is in a reasonable agreement with data.

We conclude that the particle correlations are essential to understand the behavior of the elliptic flow extracted from current flow analysis methods. It seems to account for most of the flow at large p_T. Elliptic flow appears to be sensitive to the saturation physics of the early stages of the collision.

ACKNOWLEDGMENTS

This work was sponsored in part by the U.S. Department of Energy under Grant No. DE-FG03-00ER41132.

REFERENCES

1. Y. V. Kovchegov and K. Tuchin, Phys. Rev. D **65**, 074026 (2002)
2. L. V. Gribov, E. M. Levin and M. G. Ryskin, Phys. Rept. **100**, 1 (1983).
3. I. Balitsky, Nucl. Phys. B **463**, 99 (1996); Y. V. Kovchegov, Phys. Rev. D **60**, 034008 (1999)
4. E. Levin and K. Tuchin, Nucl. Phys. B **573**, 833 (2000) Nucl. Phys. A **691**, 779 (2001) Nucl. Phys. A **693**, 787 (2001)
5. E. Iancu, K. Itakura and L. McLerran, Nucl. Phys. A **708**, 327 (2002); D. Kharzeev, E. Levin and L. McLerran, Phys. Lett. B **561**, 93 (2003)
6. Y. V. Kovchegov and K. L. Tuchin, Nucl. Phys. A **717**, 249 (2003)
7. L. D. McLerran and R. Venugopalan, Phys. Rev. D **50**, 2225 (1994) Phys. Rev. D **49**, 2233 (1994)
8. Y. V. Kovchegov and K. L. Tuchin, Nucl. Phys. A **708**, 413 (2002)

Elliptic Flow from Au+Au Collisions at $\sqrt{s_{NN}} = 200$ GeV

A. Tang for the STAR collaboration

Brookhaven National Laboratory, USA & The National Institute for Nuclear Physics and High Energy Physics, Netherlands

Abstract.
This paper presents results of elliptic flow measurements at moderate high transverse momentum in Au+Au collisions using the STAR detector at RHIC. Sizable v_2 is found up to 7 GeV/c in transverse momentum. Non-flow effects are discussed comparing correlations in p+p collisions and Au+Au collisions. v_2 from two-, four- and six-particle cumulant are shown and discussed.

INTRODUCTION

The azimuthal anisotropy of produced particles at large transverse momentum in non-central heavy ion collisions is one of several promising observables for probing the early partonic phase [1, 2], and in particular, this phenomenon can provide insights into partonic energy loss in the relevant medium [3]. Partonic energy loss increases systematically with increasing initial medium density and thus provides an important constraint on the initial parton densities. At moderate to large transverse momentum, the jet fragmentation process introduces a genuine correlation among fragmentation products, which is expected to be the dominant non-flow source [4] that can complicate the interpretation of flow analyzes that are based on two-particle correlations [5]. In order to investigate further the systematic uncertainties due to non-flow, comparison of azimuthal correlations is made between Au+Au and p+p collisions, in which only non-flow is expected to occur. Additionally, non-flow contributions to four-particle correlations, which are expected to be small [6], are investigated through comparison to six-particle correlations.

RESULTS

The left plot of figure 1 shows v_2 from the event plane (RP) and two- and four-particle cumulant ($v_2\{2\}$ and $v_2\{4\}$) methods. The difference between the RP and $v_2\{2\}$ methods may originate from the analytical extrapolation of event plane resolution from sub-events to full events. The difference between four-particle cumulant v_2 and the other two methods could be partially explained by non-flow effects and partially explained by the fluctuation of v_2 itself [7], but in either cases, 4-particle cumulant method gives a lower limit on v_2. From the plot we can see that in middle central events at $\sqrt{s_{NN}} = 200$ GeV, significant v_2 is found up to 7 GeV/c, which is the region where fragments of high p_t partons can be disentangled from the soft hydrodynamics component. The measurement will provide an important constraint on the initial parton densities in a "jet quenching" picture [3]. The finite $v_2\{4\}$ at high p_t seen in the figure demonstrates that v_2 is predominantly due to correlations relative to the the reaction plane rather than the intra-correlations of jet fragments [8]."

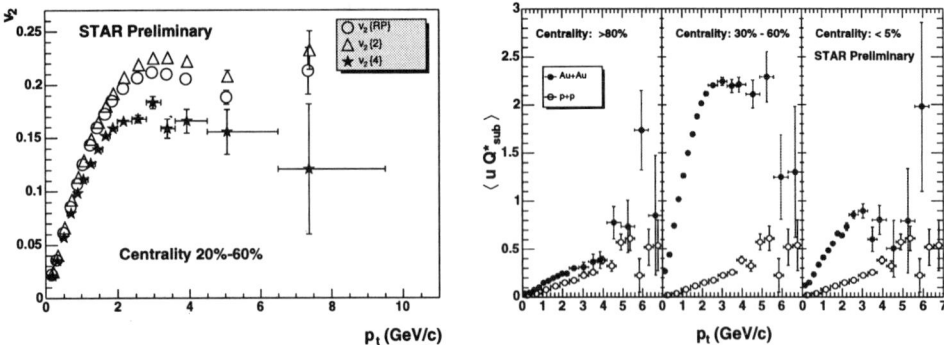

FIGURE 1. Left : v_2 as a function of transverse momentum from event plane method (circles), two-particle cumulant method (triangles) and four-particle cumulant method (stars). Right : Two-particle correlation in p+p (circles) and Au + Au (solid dots). Q_{sub}, which is the Q vector from randomly divided sub-events, is used in making this plot.

The four-particle cumulant v_2 analysis requires large statistics thus has limited power in terms of separation of non-flow effects at high p_t with currently available data. In order to get an insight to the problem, we can separate the two-particle correlation in Au+Au collisions as

$$\langle u_D Q^* \rangle^{AA} = M^{AA} v_D v_I + M^{AA} \delta_{DI}^{AA} , \qquad (1)$$

where $Q = \sum u_j$, $u_j = e^{2i\phi_j}$. v_D is differential flow and v_I is the integrated flow for particles used to define Q. δ_{DI}^{AA} is the two-particle non-flow correlation in Au+Au collisions and can be approximated to $\frac{\delta_{DI}^{AA}}{N_{collision}} \approx \frac{\delta_{DI}^{PP} M^{PP}}{M^{AA}} = \frac{\langle u_D Q^* \rangle^{PP}}{M^{AA}}$. M^{AA} and M^{PP} are multiplicities for Au+Au collisions and p+p collisions, respectively.

Here we have made assumptions, which are not necessarily true, about the similarity of non-flow for both p+p and Au+Au collisions, but this comparison is a very useful supplement to what is learned from four-particle cumulant studies. Rearranging terms, we have

$$\langle u_D Q^* \rangle^{AA} = M^{AA} v_D v_I + \langle u_D Q^* \rangle^{PP} . \qquad (2)$$

The right plot of figure 1 shows that the correlations have similar magnitude in p+p and the most central and peripheral Au+Au collisions, indicating that non-flow may dominate the correlations for these centrality classes. The correlations for mid-central Au+Au collisions are much stronger than those seen in p+p, so that true flow effects may dominate in this case.

In four-particle cumulant analysis, contributions of non-flow effect from four-particle correlations remains although non-flow from two-particle correlations are removed. To remove 4th order non-flow effect one needs to go to higher order cumulants. However this effect was expected to be small [6] and indeed confirmed by Figure 2. In the Figure

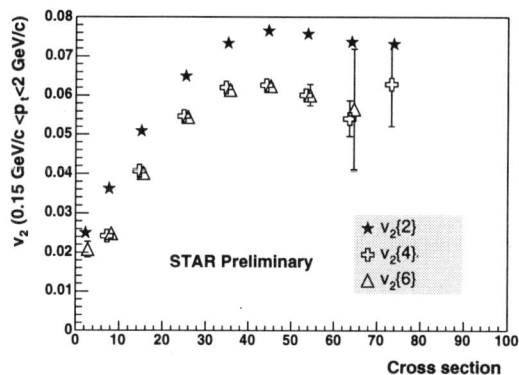

FIGURE 2. v_2 from two-(stars), four-(crosses) and six-(triangles) particle correlation, versus cross section at $\sqrt{s_{NN}} = 200$ GeV There is an overall 10% downward systematical uncertainty due to low p_t background contaminations in STAR detector.

$v_2\{4\}$ and $v_2\{6\}$ overlays on the top of each other very well, indicating that non-flow from four-particle correlation, which is shown by the difference between $v_2\{4\}$ and $v_2\{6\}$, is negligible.

SUMMARY

Sizable v_2 is found up to 7 GeV/c in p_t in Au + Au collisions at $\sqrt{s_{NN}} = 200$ GeV. Non-flow effect could be dominant at high p_t in peripheral and central events. Non-flow from pure four-particle correlation is negligible.

ACKNOWLEDGMENT

I thank S. Voloshin, K. Filimonov, A. Poskanzer and R. Snellings for their contributions.

REFERENCES

1. J.-Y. Ollitrault, Phys. Rev. **D46**, 229 (1992).
2. H. Sorge, Phys. Rev. Lett. **78**, 2309 (1997).
3. M. Gyulassy, I. Vitev and X.-N. Want, Phys. Rev. Lett. **86**, 2537 (2001).
   X.-N. Wang, Phys. Rev. **C63**, 054902 (2001).
4. N. Borghini, P. M. Dinh, and J.-Y. Ollitrault, Phys. Rev. **C63**, 054906 (2001).
5. A. M. Poskanzer and S. A. Voloshin, Phys. Rev. **C58**, 1671 (1998).
6. STAR collaboration, C. Adler *et al.*, Phys. Rev. **C66**, 034904 (2002).
7. M. Miller and R. Snellings, in preparation.
8. Y. V. Kovchegov and K. L. Tuchin, Nucl. Phys. **A708** 413-434 (2002); Y. V. Kovchegov and K. Tuchin, Nucl. Phys. **A717** 249-267 (2003).

Heavy Flavour Hadro-Production Cross-Sections

Hermine K. Wöhri and Carlos Lourenço

CERN/EP, Geneva, Switzerland

Abstract. Hadro-production data on charm and beauty absolute cross-sections, collected by experiments at CERN, DESY and Fermilab, are reviewed. The measurements, corrected for the 'time evolution' of the branching ratios, are compared to calculations done with Pythia, as a function of the collision energy, using the latest parametrizations of the parton densities. We then estimate some charm and beauty production cross-sections relevant for future measurements, including nuclear effectes in the PDFs. We finish by briefly addressing the relevance, in heavy-ion collisions, of beauty production as feed-down for J/ψ production.

In the near future, several heavy-ion experiments (at the SPS, RHIC and LHC) will be sensitive to the production of open charm and open beauty production. It is, therefore, timely to review what is known today about production yields and estimate what can be expected for these future measurements.

The data used in this study were obtained with proton and pion beams at energies from $E_{lab} = 200$ to 920 GeV and from $\sqrt{s} = 0.63$ to 1.96 TeV, being summarized in Table 1. The reported beauty production values refer to a global *mixture* of beauty hadrons, mostly measured by looking at high p_T single muons, dimuons or even trimuons, except for CDF which gives B^+ cross-sections. While most experiments had full coverage in the positive x_F region, the E789 experiment covered a rather small window. All experiments using nuclear targets assumed a *linear* dependence of heavy flavour production on the mass number of the target nucleus, to derive the cross-section per nucleon. The different measurements on charm production cross sections were normalized to common branching ratios, using the PDG 2002 tables [1], to remove the 'time evolution' of these values. We have also updated the systematic errors of the published values, to reflect the smaller uncertainties of the most recent branching ratios. If the D mesons were searched in more than one decay channel, the performed corrections were weighted according to the number of observed events in each of the channels. While the data on D meson production collected with pion beams were published for the positive x_F range, data obtained with proton beams were mostly published for the full x_F range. To ease their comparison, all measurements were normalized to the positive x_F range. The very recent CDF charm measurement was done for $p_T > 5.5$ GeV/c.

After having collected the available charm and beauty cross-sections we can compare their $\sqrt{s}$ dependence with calculations performed with Pythia [26], a Monte-Carlo event generator used by many experiments. The calculations used version 6.208, with its default settings. In particular, the c and b quark masses are 1.5 and 4.8 GeV/c^2, respectively. The neutral and charged D meson production cross-sections in π^-p and pp collisions are compared in Fig. 1 with the Pythia curves obtained with different PDFs [27]. After normalizing the curves to the data (not including the CDF point), fitting an empirical K-factor, none of the curves can be excluded.

While in pion induced collisions the D^+/D^- and $D^0/\overline{D^0}$ data require similar K-factors, of around 1.5, in p-A collisions the calculated charged D meson cross-sections are much lower than the measured values. Figure 2 (left) shows the charged to neutral D meson ratio measured in proton induced reactions. Their average is two times higher than the value given by Pythia, 0.32, and expected assuming equal production probabilities of the D and D* states and using known D* branching ratios. It is worth noting that the data collected with pion-induced collisions do not show such a strong discrepancy.

TABLE 1. Charm and beauty measurements used in this study.

	Exp. / Publication		Target(s)	$D^0 + \overline{D^0}$ events	$D^+ + D^-$ events	acc. window
p-A	NA16	[2] 1984	p	5	10	$x_F > -0.1$
	NA27	[3] 1988	p	98	119	$x_F > -0.1$
	E743	[4] 1988	p	10	46	$x_F > -0.1$
	E653	[5] 1991	emulsion	108	18	$x_F > -0.2$
	E789	[6] 1994	Be, Au	> 4000	—	$0.00 < x_F < 0.08$
	E769	[7] 1996	Be, Al, Cu, W	136	159	$x_F > -0.1$
p$\bar{\text{p}}$	CDF	[8] 2002	p$\bar{\text{p}}$	36804	28361	$p_T > 5.5$ GeV/c
π-A	NA16	[2] 1984	p	4	9	$x_F > -0.1$
	NA27	[9] 1986	p	49	14	$x_F > 0.0$
	NA32	[10] 1988	Si	75	39	$x_F > 0.0$
	NA32	[11] 1991	Cu	543	249	$x_F > 0.05$
	E653	[12] 1992	emulsion	328	351	$x_F > 0.0$
	E769	[7] 1996	Be, Al, Cu, W	62/353	73/414	$x_F > -0.1$
	WA92	[13] 1997	Cu, W	3873	3299	$x_F > 0.0$
	E791	[14] 1999	C, Pt	88990	—	$x_F > -0.1$
	E706	[15] 1997	Be, Cu	—	110	$x_F > -0.2$

	Exp. / Publication		Target(s)	$B\overline{B}$ events	acceptance window
π-A	NA10	[16] 1988	W	43	$x_F > 0$
	WA78	[17] 1989	U	12	$x_F > 0$
	E653	[18] 1993	emulsion	9 ± 3	$x_F > -0.3$
	E672/E706	[19] 1995	Be	8 ± 3.3	$x_F > 0$
	WA92	[20] 1998	Cu	26	$-0.5 < x_F < 0.6$
p-A	E789	[21] 1995	Au	19 ± 5	$0 < x_F^{J/\Psi} < 0.1$, $p_T^{J/\Psi} < 2$ GeV/c
	E771	[22] 1999	Si	15	$x_F > -0.25$
	HERA-B	[23] 2002	C, Ti	$10.5^{+13.1}_{-9.2}$	$-0.25 < x_F^{J/\Psi} < 0.15$
p$\bar{\text{p}}$	UA1	[24] 1991	p$\bar{\text{p}}$	2859	$|y| < 1.5$, $p_T^\mu > 6$ GeV/c
	CDF	[25] 2002	p$\bar{\text{p}}$	387 ± 32	$|y| < 1.0$, $p_T^\mu > 6$ GeV/c

Figure 3 shows the energy dependence of the total $c\bar{c}$ production cross-section, derived by adding the measured neutral and charged D meson values, and considering that the production of other charmed hadrons (Λ_c, D_s, etc) is 20 % of the total yield. Before being included in this figure, the CDF value was extrapolated down to zero p_T, using the distributions given by Pythia with the $\langle k_T^2 \rangle$ parameter varied between 1 and 2 (GeV/c)^2. An error of 40 % was added to account for the uncertainty in this extrapolation. Once Pythia is normalized to the available data we can estimate the charm production cross-sections

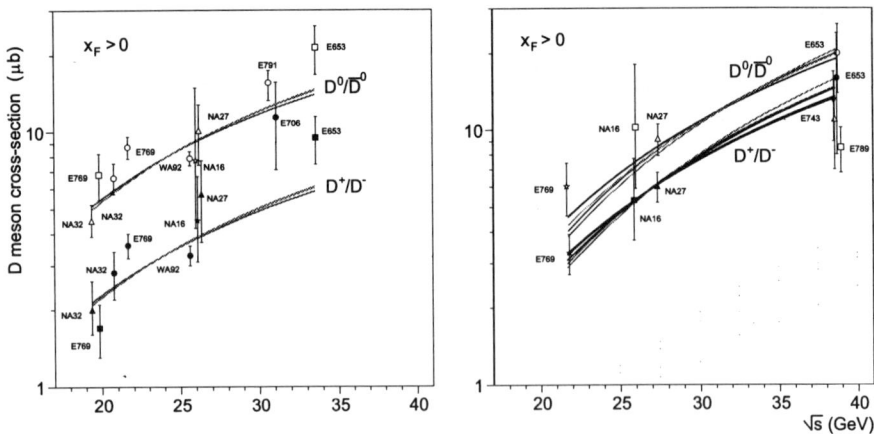

FIGURE 1. Energy dependence of the neutral (open symbols) and charged (closed symbols) D meson production cross-sections in $\pi^- p$ (left) and pp (right) collisions.

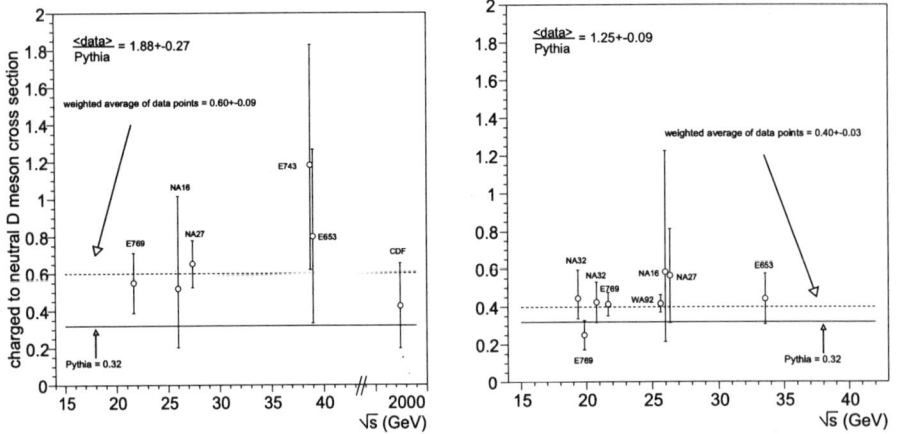

FIGURE 2. Ratio of charged to neutral D meson yields in proton (left) and pion (right) induced collisions.

for any other energies. The five curves that go below the CDF data point represent LO calculations (pure gluon fusion and $q\bar{q}$ annihilation) while the four higher curves include initial and final state parton showers, gluon splitting and flavour excitation, also simulated within Pythia.

For RHIC energies, $\sqrt{s} = 200$ and 500 GeV, the curves indicate $c\bar{c}$ cross-sections (per nucleon-nucleon collision) clustering around ~ 1 and ~ 3 mb, respectively, but with uncertainties around 100%. For the NA60 experiment [28], the spread in the curves is much smaller, giving ~ 5 μb and 20 μb, respectively for In-In collisions at $E_{lab} = 158$ GeV and p-A collisions at 400 GeV. Anti-shadowing of the PDFs was included using the EKS 98 model [29].

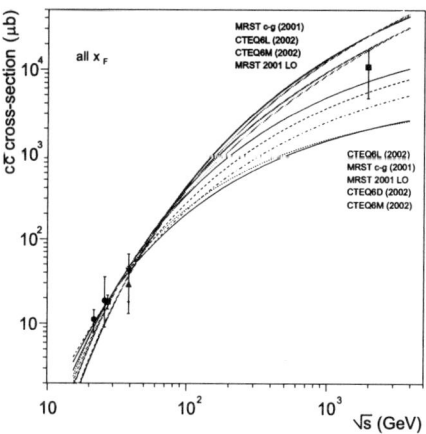

FIGURE 3. $c\bar{c}$ production cross-sections versus collision energy.

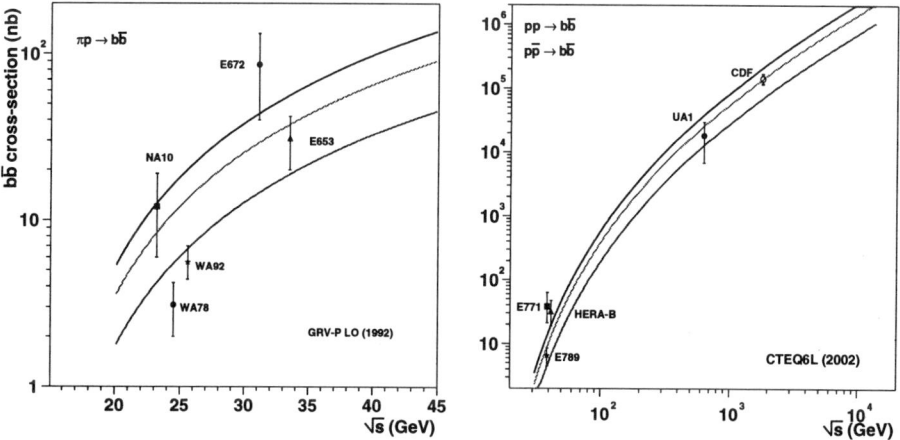

FIGURE 4. $b\bar{b}$ production cross-sections in pion (left) and proton (right) induced collisions.

Figure 4 shows the few available measurements of beauty production cross-sections in pion (left) and proton (right) induced collisions. Unfortunately, the data points are considerably spread around, with factors of 5 between measurements made at essentially the same energy (NA10 and WA78; E771 and E789). To minimize the number of standard deviations between the data points and the normalized curves, the pion induced collisions prefer K-factors below 1.4 while the proton data can accomodate values between 1 and 3. Therefore, rough estimates for the $b\bar{b}$ production cross-sections, per nucleon-nucleon collision, for the heavy-ion collider experiments are ~ 2 μb at RHIC and ~ 350 mb at the LHC. These values include nuclear effects in the PDFs, assuming Pb colliding nuclei. Clearly, better data is needed in the beauty sector, including a measurement of the nuclear effects in proton-nucleus collisions.

At collider energies, beauty production is a very important source of J/ψ mesons. CDF observed that beauty decays represent more than 15 % of the total J/ψ yield [30], at $\sqrt{s} = 1.8$ TeV, and for J/ψ mesons of p_T above 5 GeV/c. Since beauty production is expected to scale linearly with the mass number of the colliding nuclei, while J/ψ production scales as $A^{0.92}$, due to the normal nuclear absorption, in Au-Au collisions the relative fraction of J/ψ mesons resulting from beauty decays should be 2.3 times higher. If direct J/ψ production is further suppressed in heavy-ion collisions (NA50 measured a factor 2 in central Pb-Pb collisions at the SPS), beauty production might account for more than 50 % of the observed J/ψ yield, maybe already at RHIC energies. This observation underlines the importance of upgrading the RHIC experiments with vertexing detectors, in view of a proper interpretation of the measured J/ψ suppression pattern.

We thank A. David, A. Devismes and A. Morsch for considerable help in part of this work. We also thank M. Mangano, F. Mokler, L. Ramello, C. Salgado, J. Schukraft and P. Weilhammer for very useful discussions.

REFERENCES

1. K. Hagiwara et al. (PDG), Phys. Rev. **D66** (2002) 010001.
2. M. Aguilar-Benitez et al. (NA16 Coll.), Phys. Lett. **B135** (1984) 237.
3. M. Aguilar-Benitez et al. (NA27 Coll.), Z. Phys. **C40** (1988) 321.
4. R. Ammar et al. (E743 Coll.), Phys. Rev. Lett. **61** (1988) 2185.
5. K. Kodoma et al. (E653 Coll.), Phys. Lett. **B263** (1991) 573.
6. M.J. Leitch et al. (E789 Coll.), Phys. Rev. Lett. **72** (1994) 2542.
7. G.A. Alves et al. (E769 Coll.), Phys. Rev. Lett. **77** (1996) 2388.
8. D. Acosta et al. (CDF Coll.), hep-ex/0307080.
9. M. Aguilar-Benitez et al. (NA27 Coll.), Z. Phys. **C31** (1986) 491.
10. S. Barlag et al. (NA32 Coll.), Z. Phys. **C39** (1988) 451.
11. S. Barlag et al. (NA32 Coll.), Z. Phys. **C49** (1991) 555.
12. K. Kodoma et al. (E653 Coll.), Phys. Lett. **B284** (1992) 461.
13. M. Adamovich et al. (WA92 Coll.), Nucl. Phys. **B495** (1997) 3.
14. E.M. Aitala et al. (E791 Coll.), Phys. Lett. **B462** (1999) 225.
15. L. Apanasevich et al. (E706 Coll.), Phys. Rev. **D56** (1997) 1391.
16. P. Bordalo et al. (NA10 Coll.), Z. Phys. **C39** (1988) 7.
17. M.G. Catanesi et al. (WA78 Coll.), Phys. Lett. **B 231** (1989) 328.
18. K. Kodoma et al. (E653 Coll.), Phys. Lett. **B 303** (1993) 359.
19. R. Jesik et al. (E672 and E706 Coll.), Phys. Rev. Lett. **74** (1995) 495.
20. M. Adamovich et al. (WA92 Coll.) Nucl. Phys. **B519** (1998) 19.
21. D.M. Jansen et al. (E789 Coll.), Phys. Rev. Lett. **74** (1995) 3118.
22. T. Alexopoulos et al. (E771 Coll.), Phys. Rev. Lett. **82** (1999) 41.
23. I. Abt et al. (HERA-B Coll.), Eur. J. Phys. **C26** (2002) 345.
24. C. Albajar et al. (UA1 Coll.), Phys. Lett. **B256** (1991) 121.
25. D. Acosta et al. (CDF Coll.), Phys. Rev. **D65** (2002) 052005.
26. T. Sjöstrand et al., Comp. Phys. Comm. **135** (2001) 238.
27. H. Plothow-Besch, Comp. Phys. Comm. **75** (1993) 396; J. Pumplin et al., JHEP 0207 (2002) 012; A.D. Martin et al., Phys. Lett. **B531** (2002) 216; Eur. Phys. J. **C23** (2002) 73.
28. NA60 Proposal, CERN/SPSC 2000-010; http://cern.ch/na60
29. K.J. Eskola et al., Eur. Phys. J. **C9** (1999) 61.
30. F. Abe et al. (CDF Coll.), Phys. Rev. Lett. **79** (1997) 572.

J/Ψ and Open Charm Measurements at RHIC/PHENIX

D. Silvermyr for the PHENIX Collaboration

Los Alamos National Laboratory, P-25, MS H846, Los Alamos, NM 87545, USA

Abstract. PHENIX has measured inclusive single electron spectra at midrapidity in Au-Au collisions at $\sqrt{s_{NN}}$ = 130 and 200 GeV. PHENIX has also studied J/Ψ production at $x_F = 0$ via electrons and at forward and backward x_F via muons for p-p, d-Au and Au-Au collisions at $\sqrt{s_{NN}}$ = 200 GeV. An overview of some of the obtained results are presented here.

INTRODUCTION

Particles carrying charm or bottom quarks are sensitive probes to study the hot and dense matter created in the early stage of high-energy heavy-ion collisions. The PHENIX experiment at RHIC can study open charm production via single lepton spectra and charmonium (J/Ψ) via the dilepton channels. PHENIX is a high rate experiment, able to study rare signals via several different channels while RHIC is able to provide different colliding beam species at different energies. This combination results in a very powerful tool for understanding the relative importance of various phenomena relevant to high-energy heavy-ion collisions, such as effects of the initial gluon density and shadowing.

The PHENIX detector is composed of four spectrometer arms: two forward muon spectrometers covering the full azimuth and $1.1 < |\eta| < 2.4$ and two central spectrometer arms each covering $\pi/2$ in azimuth and $|\eta| < 0.35$. Electrons are identified in the central arms by matching charged particle tracks reconstructed with the Drift Chamber and Pad Chambers to deposited energy in the ElectroMagnetic Calorimeter and rings in the Ring Imaging Cherenkov detector. Muons are found in the forward arms by matching roads from the Muon Identifier planes with tracks from the Muon Tracker stations. The minimum momentum required for a muon to go deep enough to be reconstructed is about 2 GeV/c, while the required minimum transverse momentum allowing a track to be reconstructed in the central arms is about 0.2 GeV/c.

OPEN CHARM MEASUREMENTS IN AU-AU INTERACTIONS

The direct measurement of heavy-flavour is difficult in the high multiplicity heavy-ion collisions at RHIC (especially before the realization of the proposed silicon vertex tracker upgrade). So far, PHENIX has therefore used an alternative approach in which the contributions from heavy-flavor decays to the non-photonic part of the single electron spectra are determined. It is expected at RHIC that the charm decay contribution

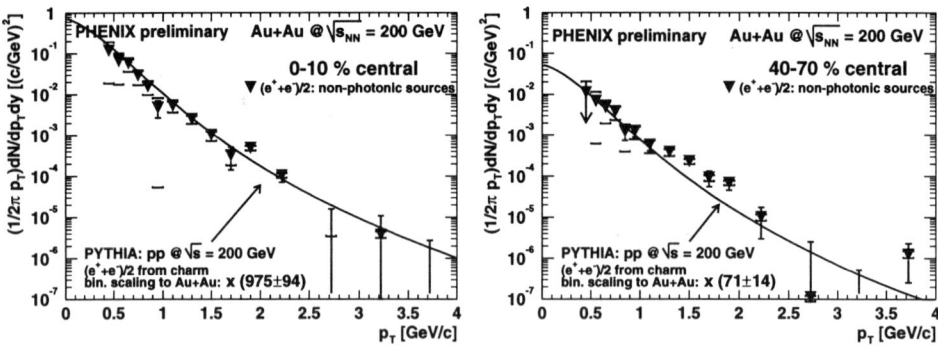

FIGURE 1. Invariant yield of electrons from non-photonic sources in Au-Au collisions, compared with the scaled PYTHIA yield, at $\sqrt{s_{NN}} = 200$ GeV for the 0-10% and the 40-70% central events, respectively.

would be the dominant component for single electrons with $p_T \sim > 1$ GeV/c, since the charm cross section is large and the production of high p_T pions has been found to be suppressed relative to binary collision scaling. In the analysis of the single electron spectrum from $\sqrt{s_{NN}} = 130$ GeV data the expected contributions from photonic conversions and light vector meson decays, were subtracted from the inclusive electron yield. An excess, increasing with p_T, was observed and interpreted as a charm signal [1]. The charm cross section at 130 GeV for the 10% most central collisions was found to be $380 \pm 60(stat) \pm 200(sys)\mu b$, and in reasonable agreement with PYTHIA [2], binary collision scaling, and results from experiments at lower energies.

For the 200 GeV run that followed, as an additional handle on the systematics, special runs with and without a 1% brass photon converter were studied. This approach allowed for a better determination of the conversion photon and Dalitz contributions to the spectra. The obtained non-photonic single electron p_T spectra for the most central events and for a semi-peripheral centrality bin compared with the PYTHIA yield, are shown in Fig. 1. The PYTHIA yield has been scaled with the average number of binary collisions for each centrality. There is agreement, within the relatively large uncertainties, between the result and the scaled PYTHIA prediction for all centralities, i.e. no obvious sign of a centrality dependent suppression, as has been observed for the high p_T pions. There are ongoing analyses to look at the single electron (and muon) spectra, also in p-p and d-Au interactions, to have a measured baseline for comparisons with the Au-Au results.

J/Ψ MEASUREMENTS

The original expectation for J/Ψ's in heavy-ion collisions was that due to color screening the production of heavy quarkonia should be suppressed in deconfined matter [3]. However, given the large charm cross section at RHIC, recombination may be important to the production of e.g. J/Ψ, which could instead be enhanced [4].

All the results that follow were obtained at $\sqrt{s_{NN}} = 200$ GeV.

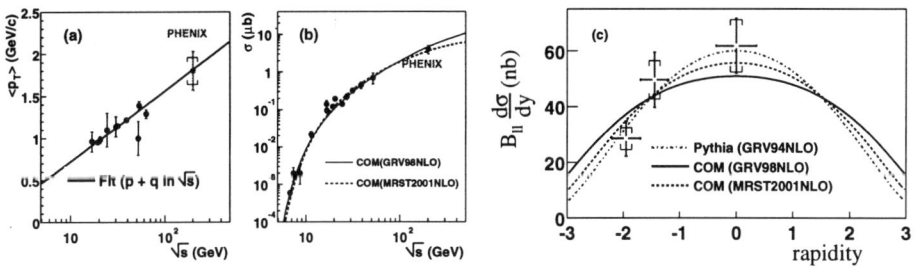

FIGURE 2. (a) J/Ψ mean p_T value and (b) J/Ψ total cross section ; both compared with measurements at lower energy. (c) J/Ψ rapidity distribution.

pp Results

In this analysis, both the dimuon and dielectron channels were studied. The measurements [5] are of intrinsic interest, but also establish a baseline for comparisons with p(d)-A and A-A collisions. The yields were obtained by comparing unlike-sign with like-sign (background) pairs. The muon arm results are divided into two forward rapidity bins, and the electron result gives a point at midrapidity. Together, these results provide information regarding the shape of the rapidity distribution: Fig. 2 (c). The values were fitted to several different shapes and the total J/Ψ cross section was estimated to be $3.99 \pm 0.61(stat) \pm 0.58(sys) \pm 0.40(abs) \mu b$. Also, by combining the dimuon and dielectron results and using several fits to the p_T spectra, a mean transverse momentum of $1.80 \pm 0.23(stat) \pm 0.16(sys)$ GeV/c was obtained. The obtained values are compared to results from experiments at lower energies in Fig. 2.

AuAu Results

In this measurement, which suffered from very limited statistics, the dielectron channel was studied. The results are reported in detail in [8]. The yields were obtained by comparing unlike-sign with like-sign (background) pairs. Three centrality bins were used: 0-20%, 20-40% and 40-90%. For each of these bins, the most probable signal as well as the 90% confidence limits were calculated. The results are summarized in Fig. 3 (a). The most probable values are indicated by square markers, while the arrows shows the 90% confidence limits, and the brackets include the estimated systematic errors. The grey band indicates the binary collision scaling of the p-p result.

Model comparisons are included in Fig. 3 (b). With the present statistical inaccuracy, models that predict enhancement are disfavored, but one can not distinguish between models that predict suppression relative to binary collision scaling.

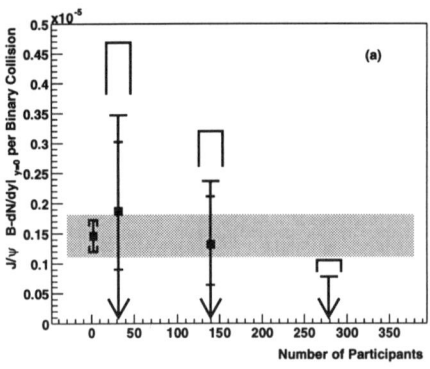 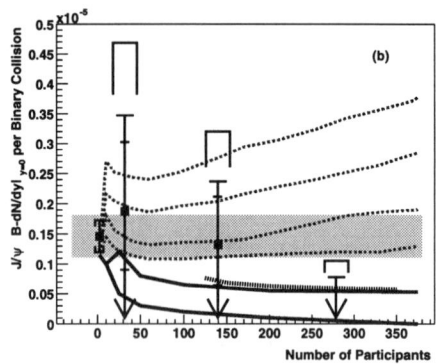

FIGURE 3. (a) The J/Ψ yields for p-p and for the three centrality bins. For additional information; see the text. (b) Model comparisons: The lowest two curves are from a calculation [6] that includes "normal" nuclear absorption, as well as absorption in a Quark-Gluon Plasma. The higher of the two includes recombination also. The next (dotted) curve is from a statistical model [7] and the top four curves is from a coalescence model [4], with different charm rapidity width values.

SUMMARY AND OUTLOOK

The early results from the recently completed (2003) d-Au and p-p runs, unfortunately had to be left out from these proceedings due to the length limits. Suffice it to say, that a larger J/Ψ sample was collected for the p-p part than in the previous p-p run, reported on above, and that the d-Au run resulted in the largest J/Ψ sample obtained so far at RHIC; of the order of 1000 J/Ψ's per muon arm. These data will help to determine the effects of shadowing in cold nuclear matter and will be essential in providing a baseline for the eagerly anticipated high statistics Au-Au run.

REFERENCES

1. K. Adcox et al. (PHENIX collaboration), Phys. Rev. Lett. 88 192303 (2002) and nucl-ex/0202002.
2. T. Sjöstrand et al., Comput. Phys. Commun. 135, 238 (2001).
3. T. Matsui and H. Satz, Phys. Lett B178, 416 (1986).
4. R. L. Thews et al., Phys. Rev. C63, 054905 (2001).
5. S. Adler et al. (PHENIX collaboration), submitted to Phys. Rev. Lett. and hep-ex/0307019.
6. L. Grandchamp and R. Rapp, Nucl. Phys. A 709, 415 (2002)
7. A. Andronic et al., nucl-th/0303036.
8. S. Adler et al. (PHENIX collaboration), submitted to Phys. Rev. Lett. and nucl-ex/0305030.

The NA60 Experiment: Results and Perspectives

J.M. Heuser[*†], R. Arnaldi[**], K. Banicz[‡], K. Borer[§], J. Buytaert[†], J. Castor[¶],
B. Chaurand[‖], W. Chen[††], B. Cheynis[‡‡], C. Cicalo[§§], A. Colla[**],
P. Cortese[**], A. David[¶¶], A. de Falco[§§], N. de Marco[**], A. Devaux[¶],
A. Drees[***], L. Ducroux[‡‡], H. En'yo[*], A. Ferretti[**], M. Floris[§§], P. Force[¶],
A. Grigorian[†††], J.-Y. Grossiord[‡‡], N. Guettet[¶], A. Guichard[‡‡],
H. Gulkanian[†††], M. Keil[‡], L. Kluberg[‖], Z. Li[††], C. Lourenço[‡], J. Lozano[¶¶],
F. Manso[¶], A. Masoni[§§], A. Neves[¶¶], H. Ohnishi[*], C. Oppedisano[**],
P. Parracho[¶¶], G. Puddu[§§], E. Radermacher[‡], P. Rosinsky[‡], E. Scomparin[**],
J. Seixas[¶¶], S. Serci[§§], R. Shahoyan[¶¶], P. Sonderegger[¶¶], R. Tieulent[‡‡],
G. Usai[§§], H. Vardanyan[†††], R. Veenhof[¶¶] and H. Wöhri[‡]

[*]*RIKEN - The Institute of Physical and Chemical Research, Wako, Saitama, Japan*
[†]*corresponding author for the NA60 collaboration, e-mail: Johann.Heuser@cern.ch*
[**]*University of Turin and INFN, Turin, Italy*
[‡]*CERN - European Organization for Nuclear Research, Geneva, Switzerland*
[§]*University of Bern, Bern, Switzerland*
[¶]*Université Blaise Pascal and CNRS-IN2P3, Clermont-Ferrand, France*
[‖]*LLR, Ecole Polytechnique and CNRS-IN2P3, Palaiseau, France*
[††]*Brookhaven National Laboratory, Upton, New York, USA*
[‡‡]*Université Claude Bernard Lyon-I and CNRS-IN2P2, Lyon, France*
[§§]*University of Cagliari and INFN, Cagliari, Italy*
[¶¶]*Instituto Superior Técnico, Lisbon, Portugal*
[***]*State University of New York at Stony Brook, Stony Brook, New York, USA*
[†††]*Yerevan Physics Institute, Yerevan, Armenia*

Abstract.
The NA60 experiment takes place at the SPS at CERN, to study the production of open charm and prompt dimuons in collisions induced by proton and heavy ion beams on nuclear targets. For this task, several novel detector systems were added to the dimuon spectrometer and zero-degree calorimeter, which were previously used in the NA50 experiment. The main upgrade is a new silicon pixel vertex spectrometer. It tracks the charged particles that are produced through the collisions in the target and allows us to match their trajectories and momenta with those of the muons that are measured behind the hadron absorber in the muon spectrometer. Besides improving considerably the dimuon mass resolution, the vertex telescope measures the offset of each muon track with respect to the interaction point. This allows us to select events where charm mesons were produced. We present in this article first results from dimuon data that were acquired in Summer 2002 with a 400 GeV proton beam. A silicon microstrip telescope was used at that time, since only a part of the pixel telescope was completed. The results include an improved dimuon mass resolution and an extended phase space coverage when compared with previous experiments. Data from Pb-Pb collisions at 20 and 30 GeV per nucleon were collected in October 2002 with three planes of pixel detectors. We show the very good accuracy with which we could reconstruct the interaction vertex. We conclude with a reminder of our physics programme and the ongoing completion of the pixel detector telescope for the Indium-Indium run at 158 GeV per nucleon, scheduled for September and October 2003.

1. PHYSICS MOTIVATION AND OVERVIEW OF THE NA60 EXPERIMENT

The NA60 experiment [1] at CERN's Super Proton Synchrotron addresses several questions that have been left open by previous experiments of the SPS heavy ion physics programme. NA60 aims at measuring prompt dimuon and open charm production in proton-nucleus and nucleus-nucleus collisions. Such measurements have not been performed to date because radiation hard detector technology capable of high-resolution tracking and vertex reconstruction in heavy ion collisions became available only recently. New measurements on open charm production are considered essential to the understanding of intriguing earlier observations. Is the origin of the intermediate mass dimuon excess, observed in the NA38 and NA50 experiments, due to the production of thermal dimuons from a plasma phase? Is the open charm yield enhanced in nucleus-nucleus collisions, and how does this compare with the observed anomalous suppression of charmonium production in central Pb-Pb collisions? Which physics variable drives the onset of ψ', χ_c and J/ψ suppression? What is the normal nuclear absorption pattern of the χ_c?

The concept of the NA60 experiment, taylored towards such measurements, is shown in Fig. 1. A newly instrumented target region complements the muon spectrometer and zero-degree calorimeter that were previously used in the NA50 experiment. A

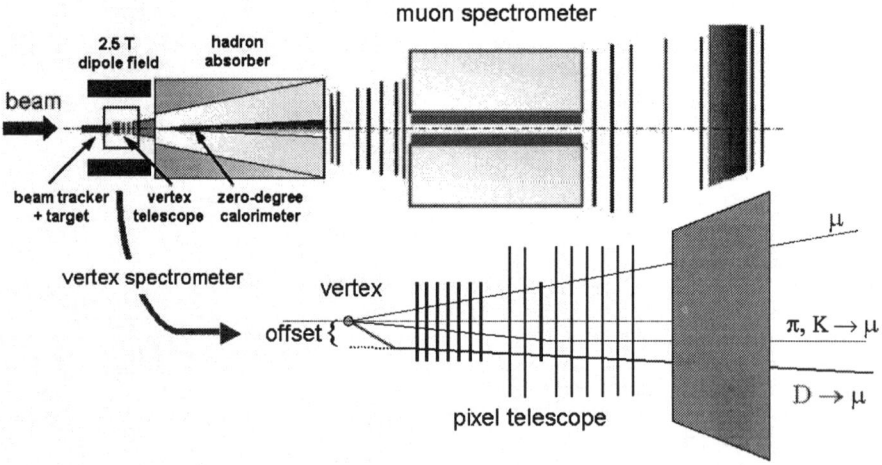

FIGURE 1. Concept of the NA60 experiment. The muon spectrometer and zero-degree calorimeter of NA50 are completed with vertex tracking capabilities for high track densities in heavy ion collisions.

beam tracker upstream and a vertex telescope downstream of the target, in front of the hadron absorber, allow us to determine the interaction point with high precision. The beam tracker [2] comprises two stations of silicon microstrip detectors that are operated at 130 K to increase its radiation tolerance. It tracks every beam ion and provides both the crossing time and the impact point on the target with good timing and spatial resolution. The vertex spectrometer consists of a multi-plane vertex tracking telescope, shown in Fig. 2, which is built from hybrid silicon pixel detectors that were developed at CERN [3]. The telescope is placed inside of a 2.5 T dipole magnetic field. Its modules are arranged to match the muon spectrometer's angular acceptance of $35 < \theta < 120$ mrad. The high granularity pixel detectors are essential to resolve the high track densities in heavy ion collisions. High-resolution track reconstruction and momentum measurement with the vertex spectrometer enable us to match tracks from the muon spectrometer through the hadron absorber into the vertex telescope. The dimuon mass resolution improves considerably, as well as the signal-to-background ratio through a rejection of pion and kaon decays into muons. Furthermore, a precise measurement of muon track offsets makes it possible to separate muons from charm decays or prompt sources.

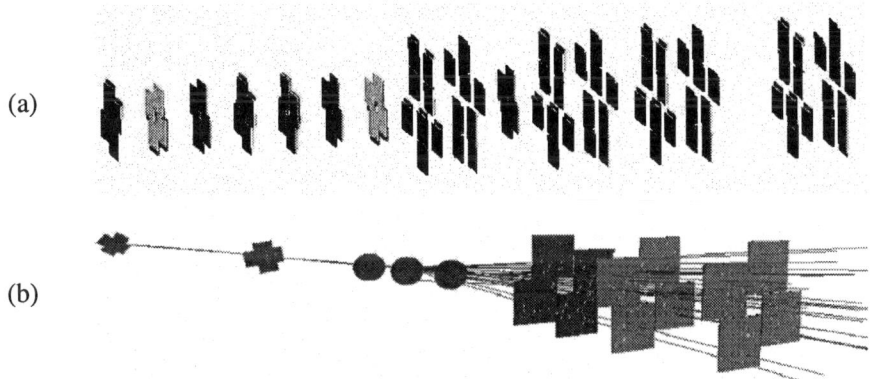

FIGURE 2. (a) Arrangement of pixel detector planes in the vertex spectrometer. Eight 4-chip and eight 8-chip modules are located between 7 cm and 32 cm distance from the target. Every hybrid pixel detector chip is segmented into 8192 pixels of 50 μm $\times$ 425 μm size.
(b) Operation of the vertex telescope: Average-multiplicity event from a Pb-Pb collision at 20 GeV per nucleon in the second target, reconstructed with the beam tracker (seen left from the targets) and three 4-chip pixel detector planes.

2. FIRST RESULTS FROM PROTON-NUCLEUS COLLISIONS

The first physics run of NA60 was performed in Summer 2002. Only one pixel detector module could be completed by then. A vertex telescope based on novel microstrip detectors was developed to perform track reconstruction in collisions of 400 GeV protons on six targets made from Be, In and Pb. The telescope comprised eight double-module tracking stations of sensors with variable strip pitch and length. First results from this run are shown in the three panels of Fig. 3. The track reconstruction and vertexing achieved a Z-vertex resolution of 900 μm, allowing to distinguish collisions in the different targets (Fig. 3a). The dimuon mass measurement was performed with a resolution better than 30 MeV at the ω and ϕ masses. Figure 3b shows the result for one of the Be targets. The good mass and vertex resolution will allow us to study with high precision the nuclear dependence of ω and ϕ production. In comparison with previous experiments, the NA60 experiment achieves a much more extended phase space coverage for those measurements, that includes very low transverse momenta (Fig. 3c).

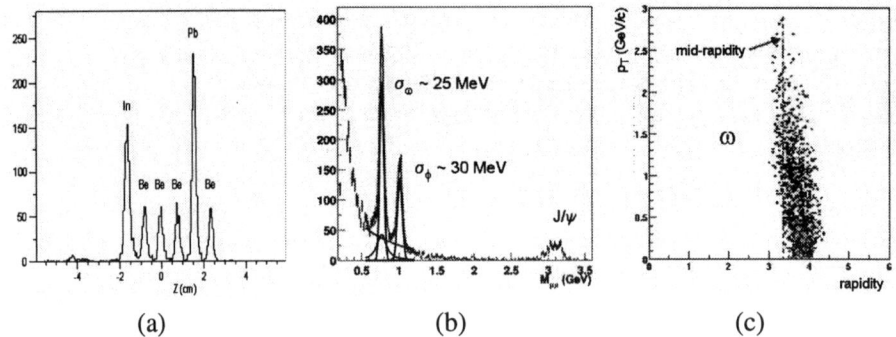

FIGURE 3. (a) Z-vertex reconstruction capabilities, (b) dimuon mass resolution, and (c) phase space coverage with the microstrip detector telescope in proton-nucleus collisions.

3. FIRST RESULTS FROM LEAD-LEAD COLLISIONS

A test run with three completed 4-chip pixel detector planes was performed in October 2002. Low-energy lead beams of 20 and 30 GeV per nucleon were available from the SPS at that time. The run was aimed at a thorough technical test of the detector components under real conditions in the experiment, including the tracking and vertexing algorithms that are used in the on-line and off-line data analysis. The target consisted of three Pb disks of 1.5, 1.0 and 0.5 mm thickness. During the run, the pixel detector modules performed very well and yielded efficient particle detection at low noise. Results from the track and vertex reconstruction are shown in Fig. 4. No dipole magnetic field was applied. Already with only three detector planes available, the positions of the vertices on the beam line, both in the targets and the steel windows of the target vacuum box, could be reconstructed with a Z-resolution better than 200 μm. The vertex resolution in the transverse direction reflects the combined track and vertex finding resolution of the pixel detectors, yielding approximately 20 μm, which is close to the value that is expected for the full vertex telescope. The correlation width of one of the transverse vertex coordinates from the pixel telescope with the measurement from the beam tracker is approximately 30 μm. These results confirm that we can reconstruct secondary vertices in heavy ion collisions.

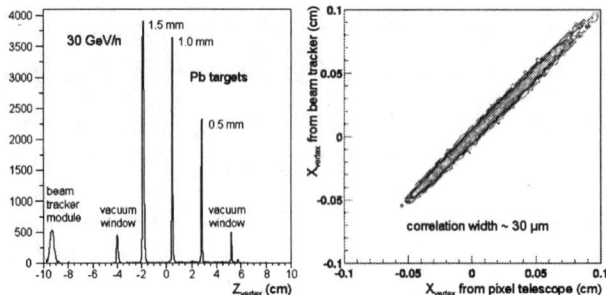

FIGURE 4. Vertex reconstruction performance of three pixel detector planes.

4. SUMMARY AND OUTLOOK

The NA60 collaboration has now fully developed their new detector systems. The completion of the pixel detector telescope for a physics run with heavy ion beams in Fall 2003 is ongoing. High-quality pixel detector components have been delivered from industry. About 75% of the pixel telescope's 16 tracking planes are already constructed and equipped with their final readout electronics. Further pixel detector planes are presently under construction.

Data collected in 2002 with the NA60 experiment, including a microstrip vertex telescope and a partially installed pixel detector telescope in the vertex spectrometer, confirm its feasibility:

- The dimuon mass resolution at the ω and ϕ mass is approximately 25-30 MeV.
- The phase space coverage extends down to low transverse momenta and masses.
- The resolution of the vertex measurement is about 20 μm in the transverse coordinates.

In Fall 2003, NA60 will collect data from In-In collisions at 158 GeV per nucleon, to study:

- The patterns of J/ψ and ψ' production to identify the physics mechanism driving charmonium suppression.
- ρ, ω and ϕ production.
- Open charm and thermal dimuon production.

REFERENCES

1. NA60 Collaboration, *Study of prompt dimuon and charm production with proton and heavy ion beams at the SPS*, CERN/SPSC 2000-010, March 2000.
2. V. Granata et al., *Cryogenic technology for tracking detectors*, Nucl. Instr. Meth. Phys. Res. **A461** (2001) 197; K. Borer et al., *Charge collection efficiency of irradiated silicon detectors operated at cryogenic temperatures*, Nucl. Instr. Meth. Phys. Res. **A440** (2000) 5; P. Rosinský et al., *The Cryogenic Silicon Beam Tracker of NA60 for Heavy Ion and Proton Beams*, Proc. 11th Int. Workshop on Vertex Detectors, Hawaii, Nov. 3-8, 2002, submitted to Nucl. Instrum. Methods Phys. Res. **A**.
3. W. Snoeys et al., Nucl. Instr. Meth. Phys. Res. **A465** (2001), 176; K. Wyllie et al., *A pixel readout chip for tracking at ALICE and particle identification at LHCb*, Proc. Fifth Workshop on Electronics for LHC Experiments, Snowmass, Colorado, 1999.

Charm Flow versus Fragmentation in RHIC

Sotiria Batsouli

Columbia University

Abstract. PHENIX data on single electron production from open charm decay in central and minimum bias $Au+Au$ collisions at $\sqrt{130}$ AGeV, yield important information on the kinematics of charm production [1]. The data appears to be consistent with perturbative QCD without final state interactions. However, the data are also consistent with thermal hydrodynamic models. The coincidence of D and B meson pQCD spectra with hydrodynamic flow spectra up to $p_T \approx 3$ and 5 GeV respectively, could imply that the initial charm and bottom quarks from pQCD may be produced following the collective flow of light quarks and gluons.

CHARM FLOW VERSUS FRAGMENTATION

Heavy quarks in nuclear collisions at RHIC energies are produced mainly via gluon fusion and they subsequently fragment into D and B mesons. It is thus possible that charm and bottom quark production can be calculated reasonably well by factorized perturbative QCD even in central $Au+Au$ collisions. In this case their yields should scale with the number of binary nucleon-nucleon collisions in the absence of medium effects. This assumption can be tested by comparing the PHENIX data [1] on "prompt" single electrons with the results of a pQCD based monte carlo, namely PYTHIA [2]. The PHENIX data in both central and "minimum bias" $Au+Au$ collisions at $\sqrt{130}$ AGeV were found to be in good agreement with PYTHIA [1].

However there are predictions that heavy ion collisions at RHIC energies create energy densities large enough for a transition from hadronic matter to a deconfined plasma of quarks and gluons (QGP) to be possible. If this is the case, then the fast partons going through the hot and dense matter could lose energy loss via induced gluon radiation [3, 4, 5] which would lead to a significant suppression of high p_T distributions relative to pQCD predictions. That is expected to occur mainly at central collisions. In PHENIX such suppression of the high p_T distribution of various hadrons has been observed [6].

So the obvious question is why there appears to be no indication of such a suppression in the case of the charm quarks. One possible reason is the so-called "dead cone" effect [7]. Gluon radiation from massive partons is suppressed at small angles, which is a manifestation of causality. Also the slower moving heavy quark samples a more dilute density profile as the medium expands. Recent calculations [8] however seem to indicate that the energy loss is not inhibited enough by the "dead cone" effect, for the produced medium to be transparent to heavy quarks. However the polarization of the QCD medium could lead to a further suppression of the radiation of soft gluons [8] in the medium as compared to the vacuum. The net result from the two effects would be equivalent to vacuum pQCD predictions, but only by coincidence. In other words the produced medium would appear essentially transparent to heavy quarks.

It would be informative however to consider a very different scenario for the interpretation of the charm results, namely thermal, hydrodynamic models. Hydrodynamic models assume that, at least for the light quark and gluons, the opacity of the produced QGP is high enough that local equilibrium is achieved early in the collision and maintained through hadronization. Such models have successfully reproduced a number of collective flow signatures observed at RHIC for light hadrons. Could it be possible then that heavy quarks exhibit such a behavior as well? It is possible that the heavy quarks undergo significant scattering in the partonic medium and thus do not reflect the initial hard scattering process in their momentum distribution. In this picture, we may want to consider that the charm quarks participate in a hydrodynamic type expansion. Thus the D mesons may exhibit a hydrodynamic momentum distribution consistent with the lighter hadrons. Furthermore if the heavy quarks fragment into D or B mesons inside the expanding fireball then the rescattering of these mesons of with surrounding particles may lead to thermalization and hydrodynamic flow in the hadronic phase. We can test this picture by computing the p_T spectra of D and B mesons using the same transverse boosted Bjorken model that fits the light hadron spectra up to 2 GeV [10]. The transverse velocity profile is assumed to be linear with radial position, $\beta_r(r) = \beta_s \times (\frac{r}{R})$, where β_s is the maximum surface velocity and R the maximum radius of the expanding source. Each fluid element is locally thermalized and boosted with a transverse angle $\rho = tanh^{-1}(\beta_r)$. The resulting spectrum is a superposition of individual thermal components and has the form:

$$\frac{dN}{m_t dm_t} = A \times \int_0^R rm_t I_0(\frac{p_t sinh\rho}{T_f}) K_1(\frac{m_t sinh\rho}{T_f}) dr \quad (1)$$

where T_f is the freeze out temperature.

In Figure ??, the hydrodynamic p_T distributions for π, D and B mesons are compared to the PYTHIA results for 10% central $Au+Au$ collisions at $\sqrt{130}$ GeV. The PYTHIA parameters for charm and bottom are taken from [1]. The PYTHIA π results use a K factor of 3.5, which agrees with the UA1 parametrization scaled to $\sqrt{130}$ GeV. For the hydrodynamic model, we use a fixed temperature T_f = 128 MeV and fit to the PHENIX π, K, p distributions to determine β_s = 0.65 which agrees with values previously derived [9]. The π hydrodynamic calculation is normalized to the PHENIX measured pion yield, and the D and B results are normalized to the PYTHIA pQCD dN/dy values. In the case of the pion the difference between pQCD and hydrodynamics is substantial. In the same figure the π^+, π^0 yields are also shown and it appears that the hydrodynamic model describes the data well up to 2 GeV but then the data falls below PYTHIA and UA1 parametrization at higher p_T, an indication of energy loss in the medium. However for the D and B mesons the two extreme dynamics are not substantially different well up to $p_T \approx 3$ and 5 GeV respectively.

In Figure ?? we show the hydrodynamic model and PYTHIA calculations for D mesons, B mesons and their resulting decay electron distribution for the 10% central $Au+Au$ collisions. Also shown is the PHENIX measured "prompt" single electron distribution. Both the zero mean free path hydrodynamic flow prediction and the infinite mean free path pQCD PYTHIA prediction reproduce the data equally well. Higher

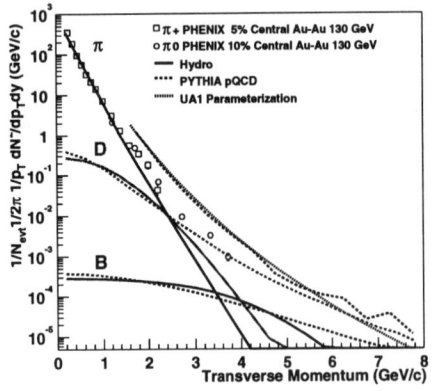
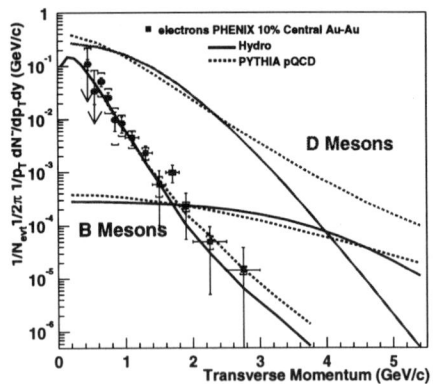

statistics single electron data are expected from PHENIX $Au+Au$ collisions at $\sqrt{200}$ GeV which will help distinguish between the two models at high p_T, but this still cannot rule out a role for hydrodynamics at lower p_T. The decisive test would be the observation or not of elliptic flow for D or B mesons, since in the hydrodynamic picture elliptic flow is expected unlike in the pQCD picture with a transparent medium. Predictions for the heavy quark v_2 according to a coalescence model is worth examining also [11].

CONCLUSIONS

The single "prompt" electron data in PHENIX can be described both by hydrodynamics that require the formation of a highly opaque medium and by pQCD in a completely transparent medium. Higher statistics and elliptic flow studies may help determine which of the two extreme scenarios is the one that applies to heavy quarks.

REFERENCES

1. K. Adcox *etal.*, Phys. Rev. Lett. 88, 192303 (2002)
2. T. Sjostrand *etal.*, Comput. Phys. Commun. 135, 238 (2001)
3. E. V. Shuryak *etal.*, Phys. Rev. C. 55, 961 (1997)
4. M. G. Mustafa *etal.*, Phys. Lett. B. 428, 234 (1998)
5. R. Vogt and X. N. Wang, Phys. Rev. C. 57, 899 (1998)
6. K. Adcox *etal.*, Phys. Rev. Lett. 88, 022301 (2002)
7. Y.L. Dokshitzer and D. Kharzeev, Phys. Lett. B 519, 1999 (2001). [hep-ph/0106202]
8. M. Djordjevic and M. Gyulassy, [nucl-th/0302069]
9. PHENIX Collaboration, Single Inclusive Hadron Spectra from $\sqrt{s_{NN}}=130$ GeV Au+Au Collisions at RHIC, in preparation
10. E. Schnedermann, J. Sollfrank and U. Heinz, Phys. Rev. C 48, 2462 (1993)
11. D. Molnar and Z. Lin, [nucl-th/0304045]

Multiparton Tomography of Hot and Cold Nuclear Matter

Ivan Vitev

Department of Physics and Astronomy, Iowa State University, Ames, IA 50010

Abstract. Multiple parton interactions in relativistic heavy ion reactions result in transverse momentum diffusion and medium induced non-Abelian energy loss of the hard probes traversing cold and hot nuclear matter. A systematic study of the interplay of nuclear effects on the $p_T \geq 2$ GeV inclusive hadron spectra demonstrates that the competition between nuclear shadowing, multiple scattering and jet quenching leads to distinctly different enhancement/suppression of moderate and high-p_T hadron production in $d+Au$ and $Au+Au$ collisions at RHIC. The associated increase of di-jet acoplanarity, measured via the broadening of the back-to-back di-hadron correlation function, provides an additional experimental tool to test the difference in the dynamical properties of the media created in such reactions.

Nuclear modification of hadron production

Particle production from a single hard scattering with momentum exchange much larger than 1/fm is localized in space-time. It is multiple parton scattering before or after the hard collision that is sensitive to the properties of the nuclear matter [1, 2]. By comparing the high-p_T observables in $p+p$, $p+A$ and $A+A$ reactions, we are able to study the strong interaction dynamics of QCD in the vacuum, cold nuclear matter and hot dense medium of quarks and gluons, respectively. So far the first two integral moments $\int dz\, z^n \rho(z)$ of the matter density in the interaction region can be deduced from experimental measurements since they are related to the broadening [3, 4] and energy loss [5, 6] of a fast parton traversing nuclear matter. From [3, 5] around midrapidity:

$$\langle \Delta \mathbf{k}_T^2 \rangle \approx 2\xi \int dz \frac{\mu^2}{\lambda_{q,g}} = 2\xi \int dz \frac{3C_R \pi \alpha_s^2}{2} \rho^g(z) = \begin{cases} 2\xi \frac{3C_R \pi \alpha_s^2}{2} \rho^g \langle L \rangle, & \text{static} \\ 2\xi \frac{3C_R \pi \alpha_s^2}{2} \frac{1}{A_\perp} \frac{dN^g}{dy} \ln \frac{\langle L \rangle}{\tau_0}, & 1+1D \end{cases} \quad (1)$$

$$\langle \Delta E \rangle \approx \int dz \frac{C_R \alpha_s}{2} \frac{\mu^2}{\lambda_g} z \ln \frac{2E}{\mu^2 \langle L \rangle} = \int dz \frac{9 C_R \pi \alpha_s^3}{4} \rho^g(z) \ln \frac{2E}{\mu^2 \langle L \rangle}$$

$$= \begin{cases} \frac{9 C_R \pi \alpha_s^3}{8} \rho^g \langle L \rangle^2 \ln \frac{2E}{\mu^2 \langle L \rangle}, & \text{static} \\ \frac{9 C_R \pi \alpha_s^3}{4} \frac{1}{A_\perp} \frac{dN^g}{dy} \langle L \rangle \ln \frac{2E}{\mu^2 \langle L \rangle}, & 1+1D \end{cases} \quad (2)$$

In Eq.(1) the factor 2 comes from 2D diffusion, $\xi \simeq \mathcal{O}(1)$ and ρ^g is the effective gluon density. For the 1+1D Bjorken expansion scenario $A_\perp$ is the transverse area of

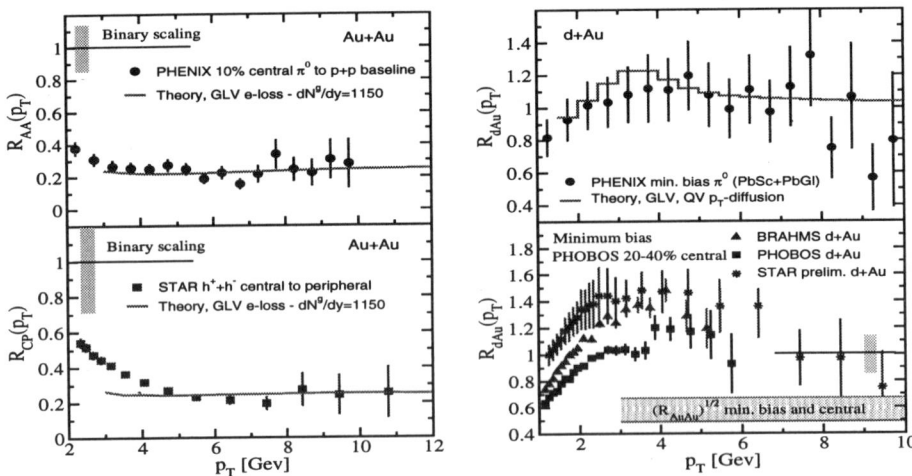

FIGURE 1. Left panel: predicted suppression for π^0 and $h^+ + h^-$ [1] in $Au + Au$ compared to PHENIX and STAR data [7]. Similar quenching is found by BRAHMS and PHOBOS [7]. Right panel: predicted small enhancement of neutral pions in $d + Au$ [8], data is from PHENIX. Left bottom panel: a test of a suggested interpretation of high p_T-hadron suppression as result of initial state wavefunction modification ($R_{dAu} \approx \sqrt{R_{AuAu}}$). Data is from BRAHMS, PHOBOS and STAR [8].

the interaction region, τ_0 is the initial equilibration time and dN^g/dy is the effective gluon rapidity density. In Eq.(2) the dominant logarithmically enhanced contribution to mean energy loss computed in the GLV approach [5] is shown. For further discussion on medium induced radiative energy loss in QCD see [5, 6].

Initial state parton broadening, nuclear shadowing and jet energy loss are incorporated in the lowest order pQCD hadron production formalism as in [1]. The interplay of dynamical nuclear effects can be studied through the nuclear modification ratio

$$R_{AB}(p_T) = \frac{dN^{AB}}{dyd^2\mathbf{p}_T} / \frac{T_{AB}(b)\, d\sigma^{pp}}{dyd^2\mathbf{p}_T}, \qquad \text{about impact parameter } b \text{ in } A+B. \quad (3)$$

Figure 1 compares the predicted [1] approximately constant suppression of π^0 and $h^+ + h^-$ in $\sqrt{s} = 200$ AGeV $Au + Au$ collisions at RHIC to PHENIX and STAR data [7]. The overall quenching magnitude and its centrality dependence are set by $(\langle L \rangle / A_\perp) dN^g/dy \propto N_{part}^{2/3}$, $dN^g/dy = 1150$. The shape of R_{AuAu} is a result of the interplay of all three nuclear effects. The full numerical calculation takes into account the dynamical Bjorken expansion of the medium, finite kinematic bounds, higher order opacity corrections and approximates multiple gluon emission by a Poisson distribution [1, 5]. In the right panel the Cronin enhancement, resulting from initial state parton broadening, $\langle \Delta k_T^2 \rangle - \langle \Delta k_T^2 \rangle_{vac} \sim (\mu^2/\lambda_g)_{eff} \langle L \rangle$ [1, 3] is seen to compare qualitatively to the shape of the PHENIX π^0 measurement [8] in $d + Au$. The calculations in Fig. 1 use the value $(\mu^2/\lambda_g)_{eff} = 2 \times 0.14$ GeV^2/fm for the cold nuclear matter transport coefficient constrained from existing Cronin data [1], although somewhat smaller scattering strength may be favored by PHENIX data. Larger enhancement of $h^+ + h^-$ production, consistent with results form low energy $p+A$ measurements, is also shown [8]. The lower right panel rules out the scenario for the initial wavefunction origin of moderate and high-p_T hadron suppression (see left panel of Fig. 1) since in this case $R_{dAu} \approx \sqrt{R_{AuAu}}$. For further discussion on the Cronin effect see [1, 9].

Broadening of the away-side di-hadron correlation function

The total vacuum+nuclear induced broadening for the two partons in a plane perpendicular to the collision axis in $p+A$ ($A+A$) reads [3]:

$$\langle \mathbf{k}_T^2 \rangle = \langle \mathbf{k}_T^2 \rangle_{vac} + \binom{1_{jet}}{2_{jets}} \left(\frac{\mu^2}{\lambda}\right)_{eff} \langle L \rangle_{IS} + 2_{jets} \left(\frac{1}{2}\right)_{projection} \left(\frac{\mu^2}{\lambda}\right)_{eff} \langle L \rangle_{FS} . \quad (4)$$

A typical range for the cold nuclear matter transport coefficient for gluons is given by $(\mu^2/\lambda_g)_{eff, IS \approx FS} = 2 \times 0.1$ GeV^2/fm - 2×0.15 GeV^2/fm. For final state scattering in a 1+1D Bjorken expanding quark-gluon plasma the broadening $((\mu^2/\lambda)_{eff}\langle L \rangle_{FS})$ can be evaluated from Eq.(1) with $dN^g/dy = 1150$, consistent with the inclusive hadron suppression pattern in Fig 1.

Figure 2 shows two measures of the predicted increase in di-jet acoplanarity for minimum bias $d + Au$ and central $Au + Au$ reactions [3]: $\langle |\mathbf{k}_{Ty}| \rangle = \sqrt{\langle \mathbf{k}_T^2 \rangle_{1\,parton}/\pi}$, $(\langle \mathbf{k}_T^2 \rangle_{1\,parton} = \langle \mathbf{k}_T^2 \rangle/2)$ and the away-side width σ_{Far} of the di-hadron correlation function $C(\Delta\phi) = N^{h_1,h_2}(\Delta\phi)/N_{tot}^{h_1,h_2}$. $C(\Delta\phi)$ is approximated by near-side and far-side Gaussians for a symmetric $p_T^{h_1} \approx p_T^{h_2}$ case and the vacuum widths are taken from PHENIX [10]. In the right panel of Fig. 2 di-hadron correlations in $d + Au$ are shown to be qualitatively similar to the $p + p$ case and in agreement with STAR measurements [8]. In $Au + Au$ reactions at RHIC di-jet acoplanarity is noticeably larger, but this effect alone does not lead to the reported disappearance of the back-to-back correlations [10]. To first approximation the coefficient of the away-side Gaussian (the area under $C(\Delta\phi)$, $\Delta\phi > \pi/2$), is determined by jet energy loss and given by $R_{AA} \propto N_{part}^{2/3}$. Broadening with and without away-side quenching is shown the bottom right panel of Fig. 2. Combined $d + Au$ and $Au + Au$ experimental data in Fig. 2 also rule out the existence of monojets at RHIC. For further discussion on di-hadron correlations see [3, 11].

In summary, evidence from jet tomography [1, 2], relativistic hydrodynamics [12] and parton cascade models [13] is in strong support of the creation of a deconfined phase of QCD at RHIC with initial energy density ~ 20 GeV/fm^3, more than 100 times the $1/7$ GeV/fm^3 density of cold nuclear matter.

ACKNOWLEDGMENTS

Useful discussion with J. Qiu is acknowledged. Thanks to I. Bearden, J. Dunlop, B. Jacak, J. Klay, M. Miller and G. Roland for help with experimental data. This work is supported by the U.S. Department of Energy under Contract No. DE-FG02-87ER40371.

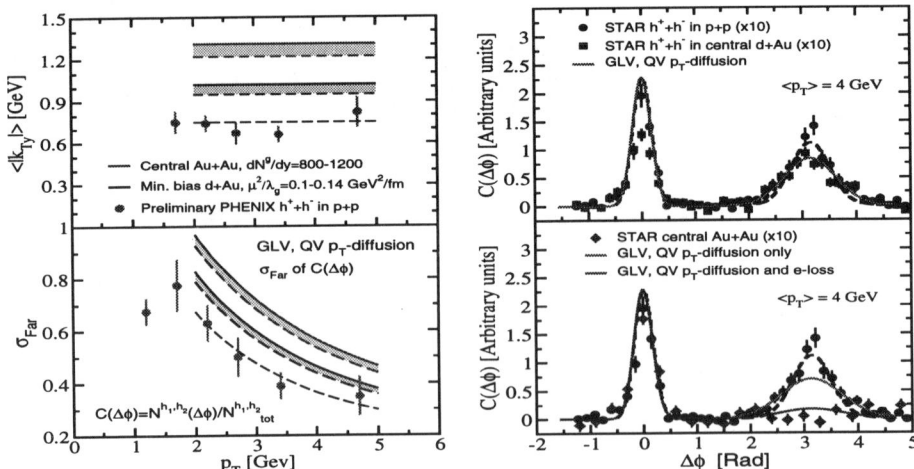

FIGURE 2. Left panel: predicted enhancement of $\langle |\mathbf{k}_{Ty}| \rangle$ and σ_{Far} in minimum bias $d+Au$ and central $Au+Au$ reactions at RHIC from p_T-diffusion [3]. Preliminary $p+p$ data is from PHENIX [10]. Right panel: the broadening of the far-side di-hadron correlation function in central $d+Au$ and $Au+Au$ compared to scaled (x10) STAR data [8]. In the bottom right panel the broadening with and without suppression, approximately given by R_{AA} from Fig. 1, are shown.

REFERENCES

1. I. Vitev and M. Gyulassy, Phys. Rev. Lett. **89**, 252301 (2002); M. Gyulassy, P. Levai and I. Vitev, Phys. Lett. B **538**, 282 (2002); I. Vitev, Phys. Lett. B **562**, 36 (2003).
2. F. Arleo, JHEP **0211**, 044 (2002); E. Wang and X. N. Wang, Phys. Rev. Lett. **89**, 162301 (2002); C. A. Salgado and U. A. Wiedemann, Phys. Rev. Lett. **89**, 092303 (2002); P. Levai et al., Nucl. Phys. A **698**, 631 (2002).
3. J. W. Qiu and I. Vitev, arXiv:nucl-th/0306039; M. Gyulassy, P. Levai and I. Vitev, Phys. Rev. D **66**, 014005 (2002).
4. M. Luo, J. W. Qiu and G. Sterman, Phys. Rev. D **49**, 4493 (1994); Phys. Rev. D **50**, 1951 (1994); J. W. Qiu and G. Sterman, Int. J. Mod. Phys. E **12**, 149 (2003).
5. M. Gyulassy, P. Levai and I. Vitev, Nucl. Phys. B **594**, 371 (2001); Phys. Rev. Lett. **85**, 5535 (2000); Nucl. Phys. B **571**, 197 (2000).
6. M. Gyulassy et al., arXiv:nucl-th/0302077; R. Baier, D. Schiff and B. G. Zakharov, Ann. Rev. Nucl. Part. Sci. **50**, 37 (2000); A. Kovner and U. A. Wiedemann, arXiv:hep-ph/0304151.
7. S. S. Adler et al., arXiv:nucl-ex/0304022; B. B. Back et al., arXiv:nucl-ex/0302015; J. Adams et al., arXiv:nucl-ex/0305015;
8. I. Arsene et al., arXiv:nucl-ex/0307003; S. S. Adler et al., arXiv:nucl-ex/0306021; B. B. Back, arXiv:nucl-ex/0306025; J. Adams et al., arXiv:nucl-ex/0306024.
9. X. f. Zhang and G. Fai, arXiv:hep-ph/0306227; J. Jalilian-Marian, Y. Nara and R. Venugopalan, arXiv:nucl-th/0307022; A. Accardi, arXiv:hep-ph/0212148; Y. Zhang, et al., Phys. Rev. C **65**, 034903 (2002).
10. C. Adler et al., Phys. Rev. Lett. **90**, 082302 (2003); J. Rak, arXiv:nucl-ex/0306031.
11. T. Hirano and Y. Nara, arXiv:nucl-th/0301042; X. N. Wang, arXiv:nucl-th/0305010.
12. P. Huovinen, arXiv:nucl-th/0305064; P. F. Kolb and U. Heinz, arXiv:nucl-th/0305084.
13. B. Zhang et al., Phys. Rev. C **61**, 067901 (2000); D. Molnar and M. Gyulassy, Nucl. Phys. A **697**, 495 (2002) [Erratum-ibid. A **703**, 893 (2002)].

The dA Collisions at Forward Rapidities at RHIC

Jamal Jalilian-Marian

Physics Department, Brookhaven National Laboratory, Upton, NY 11973-5000

Abstract.
We discuss the Color Glass Condensate model of high energy heavy ion collisions and argue that measuring particles at very forward rapidities in proton (deuteron)-nucleus collisions provides the best chance at RHIC to probe the Color Glass Condensate.

INTRODUCTION

The recent discovery of the high p_t suppression of hadron spectra in heavy ion collisions at the Relativistic Heavy Ion Collider (RHIC) at Brookhaven National Laboratory (BNL) has caused much excitement among the heavy ion physicists who may have realized their long awaited goals of creating the so called Quark Gluon Plasma (QGP), a state of deconfined quark and gluon matter which lives for times much longer than a Fermi. This is the state which must have existed in the very early universe, a few micro seconds after the big bang. The possibility to recreate the conditions of the early universe in a laboratory environment is truly an amazing accomplishment.

The colored quarks and gluons traversing this medium of quarks and gluons interact strongly with it and lose their energy through radiation. This was thought to cause the suppression of high p_t hadrons in heavy ion collisions at RHIC. However, an alternative idea was put forward which claimed that the suppression of the observed hadron spectra in heavy ion collisions at RHIC was due to the depletion of the partons in the wave function of the incoming nuclei prior to the collision (the Color Glass Condensate model). This would be an initial state effect as opposed to the energy loss paradigm which is a final state (medium) effect. The deuteron-nucleus experiment at RHIC was to decide whether the initial state or final state effects were the likely reason for the observed suppression of the hadronic spectra in heavy ion collisions at RHIC. While the preliminary data from the deuteron-gold collisions at **high p_t and mid-rapidity** seem to favor the energy loss scenario over the Color Glass Condensate model, one needs a more detailed analysis of deuteron-gold data at RHIC in order to ascertain the QGP paradigm.

It should be emphasized that at the mid-rapidity and high p_t region of the RHIC experiment ($\sqrt{s} = 200 GeV$), one probes values of the incoming parton momentum ratios (x's) which are quite large ($O \sim 0.1$). This is, almost certainly, too large of a x for the Color Glass Condensate physics to be important, let alone be dominant. Below, we argue that the best place to look for the manifestation of the Color Glass Condensate physics is in the very forward rapidity region of proton(deuteron)-nucleus collisions at high energy.

PROTON-NUCLEUS COLLISIONS AND THE COLOR GLASS CONDENSATE

In order to make the physics of Color Glass Condensate (the physics of the initial state wave function) the dominant physics in a given physical process, one needs to eliminate or at least minimize the final state (medium) effects as well as the contribution from the large x region. For sake of comparison, we loosely define large x as being larger than $x_0 = 0.01$. This value of x_0 is motivated by the success of the Color Glass Condensate model in description of the DIS ep data at HERA. For nuclear targets, the value of x_0 may be a little larger but we will not worry about that here.

To get a feeling for the kinematics of forward rapidities at RHIC, we recall that the beam rapidity at $\sqrt{s} = 200 GeV$ is about 5.3 and that the STAR and BRAHMS collaborations at RHIC measure particles at rapidities up to $y \sim 4$. For a produced particle with transverse momentum of $p_t = 1 GeV$, this roughly probes the target partons with momentum fraction $x_{target} = \frac{2p_t}{\sqrt{s}} e^{-y} \sim 10^{-4}$ at $y = 4$ while at mid rapidity, $x_{target} \sim 10^{-2}$ for $p_t = 1 GeV$. Furthermore, if a QGP medium is formed in a heavy ion collision, it will be in the mid rapidity region and not in the very forward region where produced multiplicities are very low (similar to a proton-proton collision) so that in the very forward rapidity region (in AA as well pA collisions) one does not have to worry about medium effects. An additional advantage of exploring the very forward rapidity region (proton fragmentation region) of proton-nucleus collisions over that of the very forward rapidity region of heavy ion collisions is that one does not have to worry about the possible nuclear modification of the projectile nucleus.

We now consider proton-nucleus collisions in the Color Glass Condensate model. In this formalism, the nucleus is treated as a Color Glass Condensate and described by a strong classical field while the proton is described as a collection of quarks and gluons with their distributions given by the standard parametrizations, as in the collinear factorization theory. One then considers scattering of quarks and gluons from the strong classical field of the nucleus (see [1] and the references therein). For simplicity, we consider the case of quark nucleus scattering as shown in Fig. (1). This leads to the following quark-nucleus scattering cross section

$$\frac{d\sigma^{qA \to qX}}{dq^- d^2q_t d^2b_t} \sim \tilde{\sigma}_{\text{dipole}}(x, q_t, b_t) \tag{1}$$

where the dipole cross section is defined as

$$\sigma_{\text{dipole}}(x, r_t, b_t) \equiv \frac{1}{N_c} Tr \left\langle 1 - V(b_t + \frac{r_t}{2}) V^\dagger(b_t - \frac{r_t}{2}) \right\rangle_\rho \tag{2}$$

and the matrix $V(z_t)$ defined as

$$V(z_t) \equiv \hat{P} \exp\left[-ig^2 \int_{-\infty}^{+\infty} dz^- \frac{1}{\partial_t^2} \rho_a(z^-, z_t) t_a\right] \tag{3}$$

is in the fundamental representation.

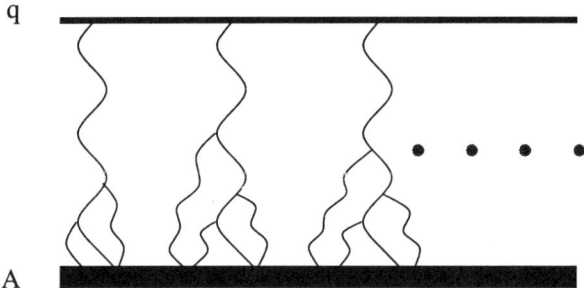

FIGURE 1. Quark-nucleus scattering at the classical level

This cross section has the following interesting properties; when $q_t^2 \gg Q_s^2$ where multiple scatterings can be ignored, the cross section is reduced to the standard leading twist pQCD cross section (with the nuclear A dependence given inside the parenthesis)

$$\frac{d\sigma}{d^2 q_t} \sim \frac{1}{q_t^4} \quad (\sim A) \qquad (4)$$

while in the kinematic region $q_t^2 > Q_s^2$, it is enhanced compared to the leading twist pQCD cross section,

$$\frac{d\sigma}{d^2 q_t} \sim \frac{\pi R^2 Q_s^2}{q_t^4}\left[1+\frac{Q_s^2}{q_t^2}\right] \quad (\sim A^{1+\alpha} \text{ with } \alpha > 0) \qquad (5)$$

This enhancement is known as the Cronin effect, first observed at low energy proton-nucleus collisions. In the transition region $q_t^2 \sim Q_s^2$, the cross section roughly goes like

$$\frac{d\sigma}{d^2 q_t} \sim \frac{\pi R^2}{q_t^2} \qquad (6)$$

while in the saturation region ($q_t^2 < Q_s^2$) where multiple scatterings are dominant, the cross section is almost independent of transverse momentum (up to possible logarithms)

$$\frac{d\sigma}{d^2 q_t} \sim \frac{\pi R^2}{Q_s^2} \quad (\sim A^{1-\alpha} \text{ with } \alpha > 0) \qquad (7)$$

One can also calculate the photon and dilepton production cross section in pA collisions in the Color Glass Condensate formalism. The universal component of these cross sections is the dipole cross section $\sigma_{dipole}(x, r_t, b_t)$ as defined in (2). It should be emphasized that the above is a classical calculation. Nevertheless, quantum corrections which are of the form $\alpha_s \log 1/x$ can be included through the use of renormalization group equations of small x QCD. In the large N_c limit, and for the dipole cross section this is the BK equation.

Inclusion of quantum effects in the quark-nucleus scattering cross section is straightforward. One needs to solve the non-linear RG equations (or the BK equation if the

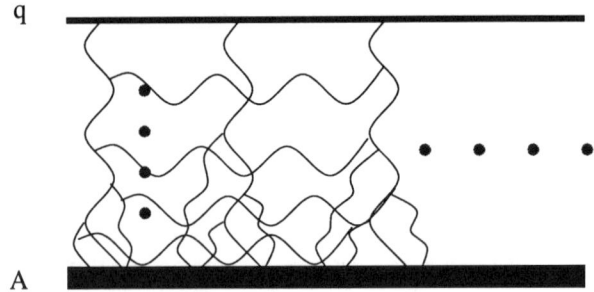

FIGURE 2. Quark-nucleus scattering including quantum evolution

large N_c approximation is used) for the dipole cross section and plug in the result into the quark-nucleus scattering cross section in (1). Diagrammatically this can be shown as in Fig. (2). It should be noted that inclusion of quantum effects ($\alpha_s \log 1/x$ terms) *preserves* the Cronin effect as can be seen by integrating the quark-nucleus cross section over q_t which gives a constant. This means that if the cross section is suppressed in some momentum range, it must be enhanced in another momentum range to keep the total cross section the same constant. This can be written formally as a sum rule which the quark-nucleus cross section satisfies. It can be shown that quantum evolution preserves this sum rule so that the Cronin effect survives the quantum evolution.

In order to relate the quark-nucleus scattering to the physically observed hadron spectra, needs to convolute the cross section in (1) with the quark distribution and fragmentation functions. This will nullify the formal derivation of the quark-nucleus scattering sum rule and may lead to disappearance of the Cronin effect at the hadronic level. This becomes more apparent when one considers gluon production in proton-nucleus collisions. Treating the proton as a weak classical field or as three valence quarks, one calculates the spectrum of produced gluons classically which again shown the Cronin enhancement at intermediate transverse momentum. However, explicit calculations show that when quantum evolution (or at least some of the diagrams) is included in the gluon production cross section, the Cronin effect, even if present in the classical calculation, is wiped out by quantum evolution. A detailed numerical study of both quark and gluon production including the non-linear quantum evolution in different kinematical regions is necessary to establish the prevalence of the Color Glass Condensate physics in proton-nucleus collisions.

ACKNOWLEDGMENTS

This presentation is based on work done in collaboration with A. Dumitru and F. Gelis.

REFERENCES

1. J. Jalilian-Marian, Y. Nara and R. Venugopalan, nucl-th/0307022.

Charged Particle Jet Studies in $\sqrt{s_{NN}}$=200 GeV p+p, d+Au, and Au+Au Collisions at RHIC

M.L. Miller*[†] and D.H. Hardtke**

Yale University, New Haven, Connecticut 06520
[†]*For the STAR Collaboration*
**Lawrence Berkeley National Laboratory, Berkeley, California 94720*

Abstract. Angular correlations for high transverse momentum ($p_\perp$) hadrons have been measured to study the fate of back-to-back jets in p+p, d+Au, and Au+Au collisions. We find no suppression of back-to-back jets in central d+Au collisions, thus implying that the large suppression of back-to-back jets in central Au+Au collisions is due to interaction of the hard scattered partons with the medium.

Quantum Chromo-Dynamics (QCD) makes two fundamental predictions that are currently being tested in ultra relativistic heavy ion collisions. First, at sufficiently high energy density, QCD predicts the existence of a color deconfined medium where the relevant degrees of freedom are quarks and gluons [1]. Second, QCD predicts that a high transverse momentum ($p_\perp$) parton traversing such a medium will lose energy via induced gluon radiation. The amount of energy loss per distance traversed (dE/dx) depends on the color charge and the gluon density of the medium [2] [3]. Partonic energy loss is predicted to soften the fragmentation of jets, thereby suppressing the yield of back-to-back high $p_\perp$ hadron pairs and single particle high $p_\perp$ spectra. Such suppression in the single particle high $p_\perp$ yields and back-to-back high $p_\perp$ hadron pairs have been recently reported in Au+Au collisions at $\sqrt{s_{NN}}$= 200 GeV [4] [5]. Before these results can be unambiguously interpreted as stemming from "final state" (post hard scattering) interactions such as partonic energy loss, alternative hypotheses must be considered. A primary but not extensive list of alternatives includes: initial state gluon radiation, initial state multiple scattering [6], and mono-jet production due to gluon saturation (three gluon coupling) effects [7]. If the observed high $p_\perp$ suppression can be attributed to such "initial state" (pre-hard scattering) effects, then one would predict similar suppression in d+Au collisions.

We present new results from d+Au collisions at $\sqrt{s_{NN}}$= 200 GeV. The yield of back-to-back high $p_\perp$ hadron pairs was studied and no significant suppression was measured. These data are compared to similar measurements in p+p and central Au+Au collisions at $\sqrt{s_{NN}}$= 200 GeV. From the d+Au results, we conclude that the high $p_\perp$ suppression in central Au+Au collisions is due to final state interactions of either hard scattered partons or their fragmentation products with the medium created in such collisions.

The data were collected using the Solenoidal Tracker at RHIC (STAR) detector, which is described in detail elsewhere [8]. STAR is a multipurpose system composed of trigger, tracking, and calorimeter detectors. These data depend primarily on charged particle momenta which were reconstructed in the main time projection chamber (TPC) which is located in a .5 Tesla magnetic field. The TPC has full azimuthal (ϕ) coverage and a large pseudo-rapidity range ($|\eta|<1.5$), with excellent momentum resolution. The specifics of the event trigger criteria can be found in Ref. [5] [9]. The p+p, d+Au, and Au+Au analyses consist of approximately 10^7, 10^7, and $3.2 \cdot 10^6$ events, respectively. Centrality (impact parameter) selection in the d+Au and Au+Au data is described in [9] and [5].

In central Au+Au events, full jet reconstruction has not yet been performed. It has been previously established in elementary collisions that the leading hadron in a jet is highly correlated with the parent parton. Using p+Nucleus collisions at $\sqrt{s_{NN}}$=38.8 GeV, the Fermilab E711 experiment found that leading hadrons of $3<p_\perp<8$ GeV/c carry $\simeq$ 75-85% of the jet energy and are highly collinear (within 5°) with the jet axis [10]. Using the p+p data presented here, jet reconstruction was performed using charged particles measured in the STAR TPC. A cone algorithm [11] was used with a cone radius of R=0.7 radians, as well as a k_T clustering algorithm [11] with an effective cone size of R=0.7 radians. Both methods showed results consistent with the E711 study. We therefore use $4 \leq p_\perp \leq 6$ GeV/c charged hadrons to tag the jet direction.

Events with at least one large transverse momentum hadron ($4<p_\perp^{trig}<6$ GeV/c), defined to be a *trigger* particle, are used in this analysis. There can be more than one trigger particle per event. For each of the trigger particles in the event, we increment the number $N(\Delta\phi, \Delta\eta)$ of *associated* tracks with 2 GeV/c$<p_\perp<p_\perp^{trig}$ as a function of their azimuthal ($\Delta\phi$) and pseudo-rapidity ($\Delta\eta$) relative to the trigger particle. We then construct an overall azimuthal pair distribution per trigger particle,

$$D(\Delta\phi) \equiv \frac{1}{N_{trigger}} \frac{1}{\varepsilon} \frac{dN}{d(\Delta\phi)} \qquad (1)$$

where $N_{trigger}$ is the observed number of tracks satisfying the trigger requirement. The efficiency ε for finding the associated particle is evaluated by embedding simulated tracks in real data. In order to have a high and constant tracking efficiency, the tracks are required to have $|\eta|<0.7$, which translates to a relative pseudo-rapidity acceptance of $|\Delta\eta|<1.4$. The single track reconstruction efficiency varies from 77% for the most central Au+Au collisions to 90% for d+Au and p+p collisions.

Fig. 1(a) shows the azimuthal distributions from p+p and d+Au collisions. The azimuthal distributions in d+Au collisions include a near-side($\Delta\phi \sim 0$) peak similar to that seen in p+p and Au+Au collisions [5] that is typical of jet production, and a back-to-back ($\Delta\phi \sim \pi$) peak similar to that seen in p+p and peripheral Au+Au collisions [9]. The azimuthal distributions are characterized by a fit to the sum of near-side (first term) and back-to-back (second term) Gaussian peaks and a constant:

$$D(\Delta\phi) = A_N \frac{e^{-(\Delta\phi)^2/2\sigma_N^2}}{\sqrt{2\pi}\sigma_N} + A_B \frac{e^{(-|\Delta\phi|-\pi)^2/2\sigma_B^2}}{\sqrt{2\pi}\sigma_B} + P \qquad (2)$$

Fit parameters are given in Table 1. The only significant difference between the p+p and d+Au fit parameters is the growth of the pedestal P. Both the near and back side jet widths σ_N and σ_B exhibit a small growth from p+p to central d+Au. A small growth in σ_B is expected from initial state multiple scattering [6]. A modest reduction in the

TABLE 1. Fit parameters from Eq. 2. Statistical error bars are shown. [9]

	p+p m.b.	d+Au m.b.	d+Au central
A_N	0.081±0.005	0.073±0.003	0.067±0.004
σ_N	0.18±0.01	0.20±0.01	0.22±0.02
A_B	0.119±0.007	0.097±0.004	0.098±0.007
σ_B	0.45±0.03	0.48±0.02	0.51±0.03
P	0.008±0.001	0.039±0.001	0.052±0.002

correlation strengths A_N and A_B from p+p to central d+Au is similar to that reported from peripheral Au+Au collisions [5].

Figure 1(b) shows the pedestal-subtracted azimuthal distributions for p+p and central d+Au collisions, as well as those from central Au+Au collisions after subtraction of the elliptic flow and pedestal contributions [5]. The near side peak is similar in all three collisions systems, while the back-to-back peak shows a significant suppression in central Au+Au compared to p+p and d+Au.

In conclusion, leading hadrons have been used to study the fate of back-to-back jets in p+p, d+Au, and Au+Au collisions. The suppression of back-to-back pairs found in central Au+Au collisions is not found in central d+Au collisions. We thus attribute the central Au+Au suppression to interactions of the hard scattered parton (or its fragmentation products) with the medium. Further studies are necessary to determine whether this suppression is due to parton-parton, parton-hadron, or hadron-hadron interactions.

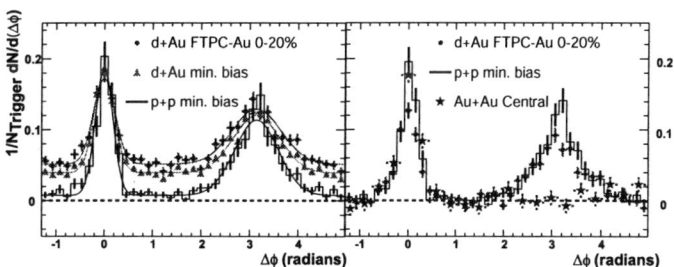

FIGURE 1. (a) Efficiency corrected two-particle azimuthal distributions for central and minimum bias d+Au collisions and minimum bias p+p collisions [5]. (b) Comparison of two-particle azimuthal distributions for central d+Au collisions to those seen in p+p and central Au+Au collisions [5]. The respective pedestals have been subtracted [9]. See text for details.

REFERENCES

1. Harris, J., and Muller, B., *Annu. Rev. Nucl. Part. Sci.* **46**, 71 (1996).
2. Baier, R., Schiff, D., and Zakharov, B. G., *Ann. Rev. Nucl. Part. Sci.*, **50**, 37–69 (2000).
3. Gyulassy, M., Vitev, I., Wang, X.-N., and Zhang, B.-W. (2003).
4. Adams, J., et al., *nucl-ex/0305015* (2003).
5. Adler, C., et al., *Phys. Rev. Lett.*, **90**, 082302 (2003).
6. Fields, T., and Corcoran, M., *Phys. Rev. Lett.*, **70**, 143 (1993).
7. Kharzeev, D., et al., *Phys. Lett.*, **561**, 93 (2003).
8. Ackermann, K., et al., *Nucl. Insr. Meth.*, **A499**, 624 (2003).
9. Adams, J., et al., *nucl-ex/0306024* (2003).
10. Boca, G., et al., *Z. Phys.*, **C49**, 543–554 (1991).
11. Blazey, G. C., et al., *hep-ex/0005012* (2000).

Particle Composition at High p_T in Au+Au Collisions at $\sqrt{s_{NN}} = 200$ GeV

Tatsuya Chujo

Brookhaven National Laboratory, Upton, NY 11973-5000, USA

Abstract. We report the recent results of proton and anti-proton yields as a function of centrality and p_T in Au+Au collisions at $\sqrt{s_{NN}} = 200$ GeV, measured by the PHENIX experiment at RHIC. In central collisions at intermediate transverse momenta ($1.5 < p_T < 4.5$ GeV/c) a significant fraction of all produced particles is protons and anti-protons. They show a different scaling behavior from that of pions. The $\bar{p}/\pi$ and p/π ratios are enhanced compared to peripheral Au+Au, p+p and e^+e^- collisions. This enhancement is limited to $p_T < 5$ GeV/c as deduced from the ratio of charged hadrons to π^0 measured in the range $1.5 < p_T < 9$ GeV/c.

Heavy-ion collisions at RHIC energies allow us to study the properties of nuclear matter at extreme energy densities. High p_T hadrons production originating in the fragmentation of partons with a large momentum transfer (hard processes) are sensitive probes of the hottest and densest stage of the collision. One of the most significant results from the first year of RHIC run was the suppression of yields both for charged and π^0 at high p_T in central Au+Au with respect to the number of nucleon-nucleon collisions (N_{coll}) [1, 2]. Moreover, it was found that π^0 yields are more strongly suppressed than for charged hadrons [1], and the yields of p and $\bar{p}$ near 2 GeV/c in central collisions are comparable to the yield of pions [3]. These observations suggest that a significant fraction of all particle yields is p and $\bar{p}$ at the intermediate p_T in central Au+Au collisions. We present here the results of p and $\bar{p}$ yields including their scaling properties and ratios of p/π, $\bar{p}/\pi$ as a function of centrality in Au+Au collisions at $\sqrt{s_{NN}} = 200$ GeV measured by the PHENIX experiment [4]. The detailed analysis methods and results are found in references [5, 6] for identified charged hadrons, in reference [7] for π^0, and in reference [8] for inclusive charged hadrons.

Figure 1 shows the p/π and $\bar{p}/\pi$ ratios as a function of p_T measured at mid-rapidity in central (0–10%), mid-central (20–30%), and peripheral (60–92%) Au+Au collisions at $\sqrt{s_{NN}} = 200$ GeV. For all centralities the ratios rise steeply at low p_T and then, at a value of p_T which increases from peripheral to central collisions, level off. In central collisions the ratios are a factor of ~ 3 larger than in peripheral events. At $p_T > 2$ GeV/c the peripheral Au+Au data agree well with the ratios observed in $p+p$ collisions at lower energies [9]. Above 3 GeV/c the p/π, $\bar{p}/\pi$ ratios in peripheral collisions are also consistent with gluon and quark jet fragmentation [10]. Deviations from jet fragmentation below 3 GeV/c indicate the absence of soft hadron production in the e^+e^- data. In Figure 2, we compare the N_{coll} scaled central to peripheral yield ratios, R_{CP}, for $(p+\bar{p})/2$ and π^0. In the p_T range from 1.5 to 4.5 GeV/c, p and $\bar{p}$ are not suppressed in

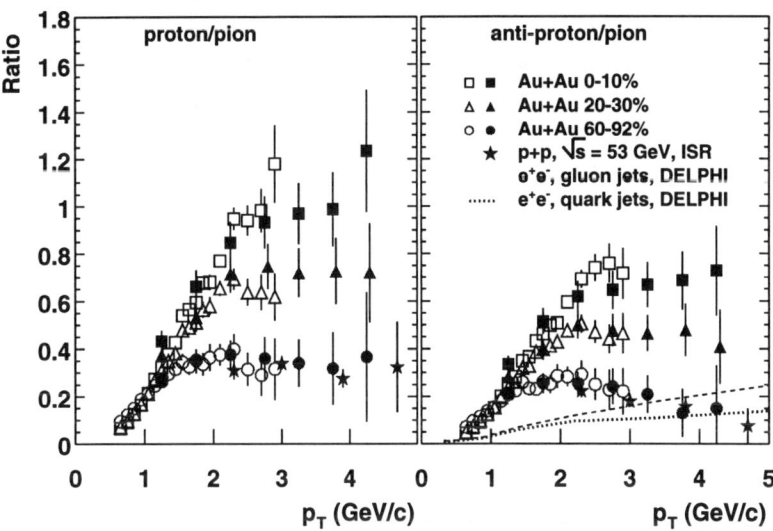

FIGURE 1. p/π (left) and $\bar{p}/\pi$ ratios for central(0-10%), mid-central(20-30%) and peripheral (60-92%) Au+Au collisions at $\sqrt{s_{NN}} = 200$ GeV [5, 6]. Open (filled) points are for charged (neutral) pions, respectively. Data from $\sqrt{s} = 53$ GeV $p+p$ collisions [9] are shown with stars. The dashed and dotted lines are $(\bar{p}+p)/(\pi^+ + \pi^-)$ ratio in gluon and in quark jets [10].

contrast to π^0 which are largely suppressed by a factor of 2-3. Moreover, this behavior holds for all centralities (see references [5, 6]), while the suppression in the π^0 yields increases from peripheral to central collisions [7].

It is interesting why the suppression for p and $\bar{p}$ is absent in central Au+Au collisions. Recently the observed abundance of protons yields relative to pions in central collisions has been attributed to the recombination of quarks, rather than fragmentation [11]. In this model, recombination for p and $\bar{p}$ is effective up to $p_T \simeq 5$ GeV above which fragmentation dominates for all particle species. Another explanation of the observed large baryon content invokes a topological gluon configuration: the baryon junction [12]. A centrality dependence, which is in qualitative agreement with the results presented here, has been predicted [13]. In both theoretical models, the baryon/meson enhancement is limited to $p_T < 5$–6 GeV/c. In order to test these theoretical predictions, we measure charged hadrons to π^0 measured in $1.5 < p_T < 9$ GeV/c (see references [5, 8]). It is found that in central collisions for $1 < p_T < 4.5$ GeV/c, h/π^0 ratio is enhanced by as much as 50% above the $p+p$ value. Above $p_T \simeq 5$ GeV/c, the particle composition is consistent with that measured in $p+p$ collisions. This indicates that the scaling of the proton yields should become consistent with that of pions at $p_T > 5$ GeV/c. Similar limiting behavior of baryon/meson enhancement is observed in Λ and K_S^0 by the STAR collaboration [14]. It is possible that nuclear effects such as the "Cronin effect" [15] contribute to the observed large (anti)proton/pion ratios. The recent results of inclusive charged hadrons and π^0 in d+Au at $\sqrt{s_{NN}} = 200$ GeV suggest that the Cronin effect in baryons is different from that in mesons [16]. Detailed studies of particle composition in d+Au collisions will help our understanding of the baryon production at the intermediate p_T region at

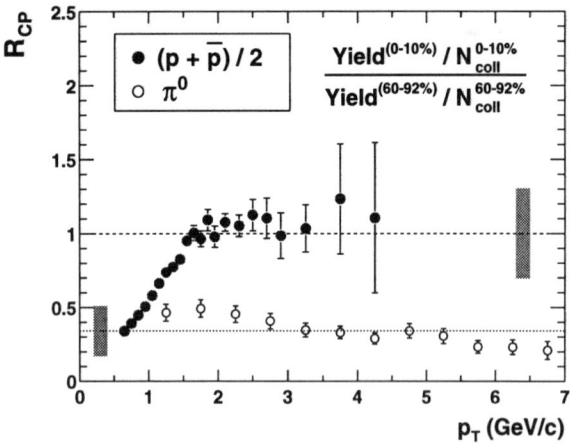

FIGURE 2. Nuclear modification factor R_{CP} for $(p+\bar{p})/2$ (filled circles) and π^0 [5, 6]. Dashed and dotted lines indicate N_{coll} and N_{part} (number of participant nucleons) scaling; the shaded bars show the systematic errors on these quantities.

RHIC.

We thank the staff of the Collider-Accelerator and Physics Departments at BNL for their vital contributions. We acknowledge support from the Department of Energy and NSF (U.S.A.), MEXT and JSPS (Japan), CNPq and FAPESP (Brazil), NSFC (China), CNRS-IN2P3 and CEA (France), BMBF, DAAD, and AvH (Germany), OTKA (Hungary), DAE and DST (India), ISF (Israel), KRF and CHEP (Korea), RMIST, RAS, and RMAE, (Russia), VR and KAW (Sweden), U.S. CRDF for the FSU, US-Hungarian NSF-OTKA-MTA, and US-Israel BSF.

REFERENCES

1. PHENIX Collaboration, K. Adcox *et al.*, Phys. Rev. Lett. **88**, 022301 (2002).
2. PHENIX Collaboration, K. Adcox *et al.*, Phys. Lett. **B561**, 82 (2003).
3. PHENIX Collaboration, K. Adcox *et al.*, Phys. Rev. Lett. **88**, 242301 (2002).
4. PHENIX Collaboration, K. Adcox *et al.*, Nucl. Instrum. Methods **A499**, 469-479 (2003).
5. PHENIX Collaboration, S. S. Adler *et al.*, submitted to Phys. Rev. Lett. , nucl-ex/0305036.
6. PHENIX Collaboration, S. S. Adler *et al.*, submitted to Phys. Rev. C, nucl-ex/0307022.
7. PHENIX Collaboration, S. S. Adler *et al.*, Phys. Rev. Lett. **91**, 072301 (2003), [nucl-ex/0304022].
8. PHENIX Collaboration, S. S. Adler *et al.*, submitted to Phys. Rev. C, nucl-ex/0308006.
9. B. Alper *et al.*, Nucl. Phys. **B100**, 237-290 (1975).
10. DELPHI Collaboration, P. Abreu *et al.*, EPJC**17**, 207 (2000).
11. R. C. Hwa *et al.*, Phys. Rev. C **67**, 034902 (2003); R. J. Fries *et al.*, nucl-th/0301087; V. Greco *et al.*, nucl-th/0301093.
12. G.C. Rossi *et al.*, Nucl. Phys. **B123**, 507 (1977); D. Kharzeev, Phys. Lett. **B378**, 238 (1996); S.E. Vance *et al.*, Phys. Lett. **B443**, 45 (1998).
13. I. Vitev *et al.*, Phys. Rev. **C65**, 041902 (2002); I. Vitev *et al.*, hep-ph/0109198.
14. STAR Collaboration, J. Adams *et al.*, submitted to Phys. Rev. Lett. , nucl-ex/0306007.
15. J. Cronin *et al.*, Phys. Rev. **D11**, 3105 (1975). D. Antreasyan *et al.*, Phys. Rev. **D19**, 764 (1979).
16. PHENIX Collaboration, S. S. Adler *et al.*, submitted to Phys. Rev. Lett. , nucl-ex/0306021.

Conference on the Intersections of Particle and Nuclear Physics 2003: Relativistic Heavy Ion Parallel Session Summary

J.L. Nagle* and T. Hallman[†]

*University of Colorado at Boulder
[†]Brookhaven National Laboratory

Abstract. The Relativistic Heavy Ion Collider (RHIC) came online in 2000, and the last three years have provided a wealth of new experimental data and theoretical work in this new energy frontier for nuclear physics. The transition from quarks and gluons bound into hadrons to a deconfined quark-gluon plasma is expected to occur at these energies, and the effort to understand the time evolution of these complex systems has been significantly advanced. The heavy ion parallel session talks from the Conference on the Intersections of Particle and Nuclear Physics (CIPANP) 2003 are posted at: "http://www.phenix.bnl.gov/WWW/publish/nagle/CIPANP/". We provide a brief summary of these sessions here.

INTRODUCTION

Many speakers separated the topics of heavy ion reactions into two broad categories. (1) The bulk system created which includes of order 5000 hadrons - with transverse momentum below 1.0 GeV/c. These hadrons result from the interaction of approximately 10,000 virtual gluons liberated from the nuclear wavefunction upon impact. These particles contain information about equilibration and thus the equation of state. (2) The probes of the bulk system, that are defined as calibrated colored systems such as hard scattered quarks and gluons that traverse the bulk media and tell us about its properties. Although these high p_T partons or heavy quark states are relatively rare, and thus do not impact the equation of state (they are very non-equilibrium), because they are accessible via pQCD or QCD phenomenology, they represent a key calibration of the system properties. M. Gyulassy referred to the importance of these pQCD probes as analogous to the "tail that wags the QGP dog."

BULK EFFECTS

P. Steinberg reviewed the contrasting pictures of bulk thermalization and initial state scaling behavior. The former implies a collective system whose properties are dictated by the dynamics after the first collision impact, while the latter are remnants of the initial parton distributions or first reaction mechanisms. P. Kolb pointed out that it appears for the first time in nuclear reactions, that hydrodynamics with no additional viscosity or multi-fluid expressions appears to describe the bulk momentum distribution of hadrons at low p_T, including the azimuthal anisotropy. However, the calculations appear to yield too long a lifetime for the system relative to two particle correlation measurements (HBT) or R_{out}. The calculations indicate enormous initial pressure, with a system decoupling rather quickly, of the time scale of 10 fm/c. P. Kolb gave a status report on the success of these calculations, and that some require a phase transition to extend the lifetime thus giving additional boost to the heavier particles.

M. Lisa gave a detailed presentation of the nearly complete experimental information on the phase space distribution - both momentum and position - for hadrons at the freeze-out (point of final interaction). Good progress has been made toward tracing back in time to the initial conditions and the equation of state. D. Magestro showed the outstanding agreement between statistical models and hadron ratios, including strange and multi-strange baryons. However, the suggestion that the fast time scale for thermalization implies that the system is "born into" into equilibrium was not a clear physical picture. Z. Xu and A. Tang presented data on flow and more exotic hadron production that help to complete the hadronic freeze-out picture. Results from STAR indicate modifications to the low mass vector mesons as observed via hadronic decay channels. Comparing these with reconstruction in the leptonic channels will be an interesting future measurement from PHENIX. R. Debbe showed that the BRAHMS hadron yields at forward rapidity appear as part of a "different source" than particles in the mid-rapidity region with a different net baryon density.

Y. Kovchegov and K. Tuchin presented global and azimuthal anisotropy observables that were generated with only initial state effects included - using a parton saturation picture as the calculation tool. If correct, this implies that the measured v_2 has no relation to a physical reaction plane. M. Lisa pointed out that the HBT results into and out of the reaction plane seem to disprove this assertion.

PROBES

An excellent probe of the medium is a hard scattered parton (quark or gluon) that transverses the medium. One can use QCD factorization to calculate the expected rate of these partons and their resulting fragmentation hadron products. It has already been observed by all four RHIC experiments in Au-Au reactions, that high p_T hadrons are suppressed relative to this particular expectation. Since a parton scattering through a dense gluonic medium loses additional energy via gluon bremsstrahlung, it can be used as a "gluonometer" - a calibratable probe of the color charge density of the medium. Another explanation for the lack of high p_T hadrons is that the high virtual parton density in the incoming nuclei saturates, and thus distinct partons are replaced by a gluon wavefunction that inherently break factorization. This saturation effect that might be probed in protons at $x \approx 10^{-4}$ (for example in HERA DIS), moves to $x \approx 10^{-2}$ in heavy nuclei due to the additional thickness. Many attempts to use this framework - sometimes referred to as the Color Glass Condensate (CGC) - to describe the bulk global observables have shown some success. More recently, D. Kharzeev and collaborators tried to extend the saturation regime via DGLAP evolution to higher Q^2, referred to as the Color Quantum Liquid (CQL) regime. This CQL region suppressed hard scattering for hadrons even up to $p_T \approx 8$ GeV/c - thus also providing a possible explanation of suppressed high p_T hadrons observed in Au-Au reactions.

This ambiguity led the RHIC experiments to propose a simple plan to resolve the question of whether the suppression is an initial state or final state effect. RHIC ran deuteron-Au reactions at the start of Run-3, since for the CQL picture, the suppression should still be present in the Au nucleus. In contrast, as pointed out by I. Vitev and others, the final state energy loss explanation would not be applicable with no dense medium created in deuteron-Au collisions.

Data from PHENIX, PHOBOS, and STAR as presented by L. Aphecetche, G. Roland, and P. Jacbos show an enhancement of high p_T hadrons (both unidentified and π^0) relative to factorization and nuclear thickness scaling in deuteron-Au reactions - as sometimes referred to as the Cronin effect. There is no observed suppression as predicted in the saturation model extension. Angular correlations also revealed very similar near-side and away-side jet structure as observed in proton-proton reactions. M. Miller and J. Rak emphasized that the disappearance of away side correlations in Au-Au reactions, is not seen in deuteron-Au reactions. Though there was much discussion - including talks by J. Jalilian Marion and L. McLerran - there was general agreement that the data rule out the saturation model extension - referred to as the Color Quantum Liquid regime, for $x \approx 10^{-1}$. This of course does not rule out saturation at lower x having an effect on the bulk dynamics as pointed out by L. McLerran and others.

Many speakers stated that with the extension to the saturation picture ruled our, parton energy loss in the dense gluonic medium is the only answer. L. McLerran stated that "if a 10 GeV quark loses 2-3 GeV of energy in medium, it seems clear that the bulk partonic matter is thermalized." However, challenging questions still remain that must be fully explored. T. Chujo and J. Klay pointed out that although light hadrons are suppressed, at least at intermediate p_T, the baryons - both protons and Λ are not. S. Bass and B. Muller both argued that while very high p_T hadrons may be dominated by partons that lose energy in medium and that fragment into hadrons in vacuum, intermediate p_T hadrons may come from parton recombination. This picture shows great promise in particular in the explanation of the v_2 scaling of different hadrons. However, the model is simple - not including gluon contributions - and further tests including ϕ mesons - large mass but $q\bar{q}$ state - are needed.

HERMES electron-nucleus deep inelastic scattering (DIS) data indicate modified fragmentation functions as shown by E. Kinney. These have been interpreted in the framework of fast hadronization or color dipole formation followed by re-interaction in the nucleas, and also in terms of multiple scattering from color charges in the nucleons of the nucleus inducing gluon bremsstrahlung. In Au-Au reactions, modification of QCD vacuum may also effect zeroth order gluon radiation, as described by M. Djordvic with regards to heavy quarks, and thus may also play a role for light partons. X.N. Wang argued that only the partonic energy loss explanation survives, but there was much debate and a well developed model including fast color dipole formation or hadronization is still lacking - that then needs to be confronted with all data. There is a wealth of new data that needs to be looked at in detail, and new ideas considered for a full description.

HEAVY FLAVOR PROBES

H. Woehre gave an overview of charm measurements at lower energies and detailed the constraints those place on heavy flavor production at higher energies. J. Heuser presented the detector performance of NA60, and it is clear that they have the ability to resolve the low mass vector meson states - down to low p_T - and the χ_c contribution to the J/ψ. Many expressed that it should be a high priority for the field that they get adequate running time. First charm result via "prompt" single electrons from PHENIX were shown by S. Batsouli and indicate a surprising feature. The spectra agree both with a unmodified charm quark fragmentation in vacuum and with a completely thermalized charm plus a hydrodynamic boost.

D. Silvermyr showed the first J/ψ results from at RHIC from PHENIX in proton-proton and Au-Au reactions. Despite low statistics, the Au-Au results appear to disfavor models of large J/ψ enhancement relative to binary scaling. However, future high statistics measurements are critical to utilitize these quarkonia probes. The J/ψ measurement in proton-proton collisions of the total cross section and mean p_T are the first at collider energies and should help address color-singlet versus color-octet production mechanisms.

SUMMARY OF THE SUMMARY

The heavy ion sessions included the presentation of a wealth of new data on high p_T probes, including the first deuteron-Au results. It is now clear that the suppression of high p_T hadrons is the result of a final state effect or some complete break-down of factorization - but not in the initial wavefunction. There was lively discussion of whether this is definitely partonic multiple scattering in the colored medium, or if contributions from fast hadronization, recombination, or zeroth order radiation modification may still play a role. New results on intermediate p_T baryons, low mass vector mesons, charm mesons, and quarkonia were shown and more data from these states will help to fill in the picture. For the bulk medium an overwhelming amount of data already exists and must be reconciled in terms of hydrodynamic pictures and observations of simple scalings.

We would like to thank all the speakers for their excellent presentations and all participants for spirited discussions.

ACKNOWLEDGMENTS

We would like to thank the organizers of the conference - Dr. W.J. Marciano, Dr. Z. Parsa and the organizing committee - for all their hard work in coordinating this event. JLN acknowledges support by the U.S. Department of Energy under Grant No. DE-FG02-00ER41152 and the Alfred P. Sloan Foundation.

ACCELERATORS, FACILITIES AND DETECTORS

The CMS Experiment - Status and Physics

Norbert Neumeister

Institute for High Energy Physics Vienna, Austria and CERN, Switzerland

Abstract. The Compact Muon Solenoid experiment will operate at the LHC, a high-energy, high-luminosity proton-proton collider that is under construction at CERN. The detector design and the status of the construction and installation of CMS is reviewed. The principal physics goals and the physics reach of CMS are discussed.

INTRODUCTION

The CMS (Compact Muon Solenoid) experiment [1] is one of the two general purpose experiments that will study proton-proton and heavy-ion collisions at the Large Hadron Collider (LHC). The accelerator is currently under construction at CERN, and is scheduled to start delivering collisions from mid 2007. With a center-of-mass energy for proton collisions of 14 TeV and a design luminosity of 10^{34} cm^{-2}s^{-1}, the LHC is a machine of unprecedented complexity and potential. The LHC bunch crossing frequency for proton-proton interactions is 40 MHz. At the design luminosity, about 20 interactions occur at each bunch crossing, corresponding to a total event rate of the order of 10^9 Hz. The rate of interesting physics like the production of the Higgs boson, however, is more than ten orders of magnitudes smaller.

DETECTOR

The design of the CMS detector is based on the choice of a large superconducting solenoid, which provides a strong 4 T magnetic field in the central part, ensuring high momentum resolution for charged particles. The magnetic flux is returned through a 1.5 m saturated iron yoke (1.8 T) instrumented with muon chambers.

A tracking system and both electromagnetic and hadron calorimeters are placed inside the magnet (bore radius $\sim$ 3 m), which leads to a very compact design. The tracking system [2] is based on silicon strip and silicon pixel detectors. The volume available for tracking is a cylinder of 6 m long and 1.2 m in radius. The whole volume is filled with 210 m^2 of silicon strip sensors, comprising 10M channels, and 40M silicon pixels close to the interaction region. The whole tracker volume will be maintained at $-10°$ C in dry atmosphere to minimize the effects of radiation damage of the silicon detectors.

The high precision electromagnetic calorimeter (ECAL) [3] is made of about 76,000 dense lead tungstate (PbWO$_4$) crystals. The scintillation light is detected by silicon avalanche photodiodes (APDs) in the barrel region ($|\eta| < 1.48$) and vacuum phototriodes (VPTs) in the endcap region ($1.48 < |\eta| < 3.0$). The ECAL surrounds the tracker

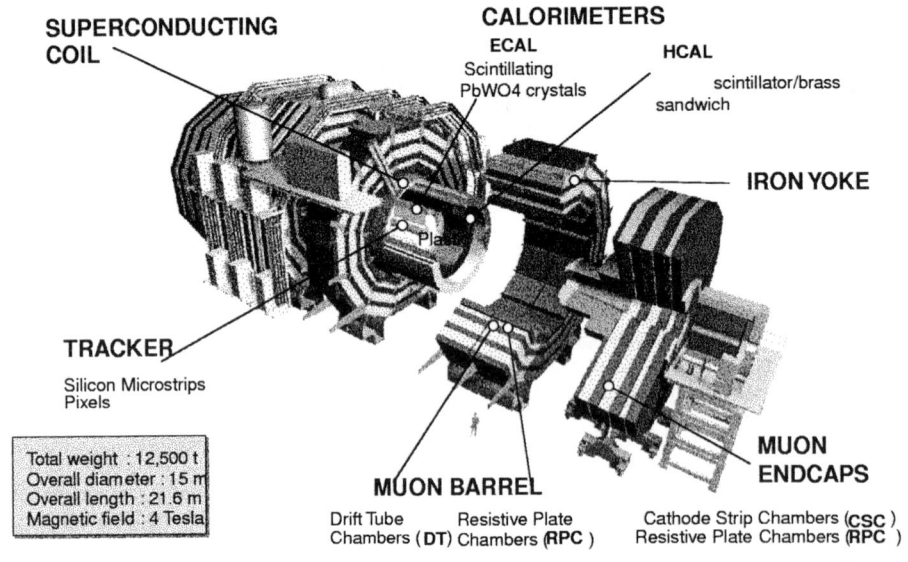

FIGURE 1. Overview of the CMS detector

volume and the photodetectors have to work in the 4 T magnetic field. A preshower system is installed in front of the endcap calorimeter ($1.6 < |\eta| < 2.6$).

The Hadron Calorimeter (HCAL) [4] in the central region ($|\eta| < 3$) is a copper (brass)/scintillator sampling calorimeter. It consists of a barrel part supported on rails by the vacuum tank and of two endcaps supported by the endcap magnet yoke. The copper absorber plates are interleaved with 4 mm thick plastic scintillator tiles. Coverage up to rapidities of 5.0 is provided by a steel/quartz fibre calorimeter. The Cerenkov light emitted in the quartz fibres is detected by photomultipliers.

The muon system [5] consists of four stations which are interleaved with the iron return yoke plates and are made up of tracking and triggering chambers. Three types of muon detectors are used: Drift Tube (DT) chambers in the central region (barrel), Cathode Strip Chambers (CSC) in the forward regions (endcaps), and Resistive Plate Chambers (RPC) in both barrel and endcaps. In total there are 250 drift tube chambers and 540 cathode strip chambers to be built.

The CMS detector has a length of 21.6 m, a diameter 14.6 m and a total weight of 12500 tons. Figure 1 shows the layout of the CMS detector.

The CMS trigger system consists of only two physical levels. The Level-1 trigger [6] is implemented on dedicated hardware and has access to data from the calorimeters and the muon detectors with coarse granularity. It has to reduce the input rate of 40 MHz up to a level acceptable for the Data Acquisition system (DAQ). At LHC startup, the DAQ system will be able to handle an event rate of up to 50 kHz, which will be increased to 100 kHz when the full LHC design luminosity is reached. The Level-1 selected events are passed to the High-Level Trigger (HLT) system [7], which will be implemented in software running on a farm of commercial processors. The HLT selection is carried out in successive steps and reduces the event rate to ~ 100 Hz at which the selected events (average event size ~ 1 MByte) are written to storage.

STATUS

The CMS collaboration consists currently of 1890 physicists and engineers from 152 institutes in 36 countries.

Excavation of the two underground caverns (for the experiment and the the counting rooms) was completed in September 2002. The caverns will be delivered to CMS in summer 2004, when the installation of infrastructure will start. The CMS experimental cavern will thus be ready to receive detector elements around mid 2005.

All big mechanical pieces have been assembled in the surface hall which is big enough (120 m long, 20 m high) to allow assembly and commissioning of big pieces of the CMS detector on the surface. The magnet yoke (12000 tons) is complete since mid 2002. Metallic structures for racks on the wheels and disks of the magnet yoke are ready since spring 2003. Gas and cooling pipes for the muon system are being installed. Installation and commissioning of the muon chambers on the magnet yoke will start in June 2003. Full test of the magnet on the surface will start in March 2005. Waiting for the underground caverns to be ready, installation and commissioning of the hadron calorimeter and of the muon system has started.

The assembly sequence followed by CMS is based on the completion of the full CMS detector, minus the staged items (outermost forward muon station, some RPC chambers at low angles, 50% DAQ online farm, and 3rd forward pixel disks), in time for physics in mid 2007.

PHYSICS

During the first physics run a data set corresponding to 5 to 10 fb^{-1} may be recorded. The most burning question surely revolves around the search for the Higgs boson. Between now and the LHC start-up the Higgs boson will be searched for by the experiments at the Tevatron, which are expected to increase the current Higgs mass limit of 114 GeV/c^2 from LEP. CMS has been designed to cover the entire Higgs mass range between 80 GeV/c^2 and the theoretical upper limit of ~ 1 TeV/c^2. The first physics run will provide statistics for discovery in channels with large branching ratios and/or a good signal-to-noise ratio. Taking into account the production cross section, the decay branching ratios and the expected signal-to-noise ratio for various selection criteria, the Higgs decay modes relevant to the discovery depend on the Higgs mass. Fig. 2 shows the expected discovery luminosity for various benchmark channels.

The search for supersymmetric (SUSY) particles is another high priority physics goal of CMS. If squarks and gluinos are light enough (below 1 to 2 TeV/c^2), their production dominates the total cross section for the direct production of supersymmetric particles at hadron colliders. This large cross section should allow the search for SUSY to lead to a discovery with the first few fb of data. Because squarks and gluinos decay in turn

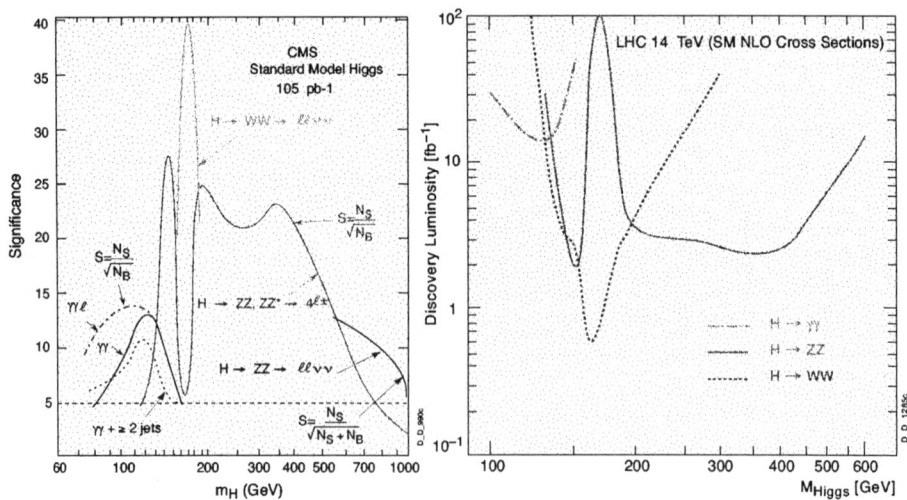

FIGURE 2. Signal significance and discovery luminosity for the Standard Model Higgs.

to lighter SUSY particles, the final states are in general characterized by many jets, leptons and large missing transverse energy. Gluino and squark masses up to 1.5 TeV/c^2 are accessible with about 1 fb. This requires efficient lepton identification and complete calorimeter coverage.

CONCLUSIONS

CMS is following an assembly sequence, that allows an initial detector to be ready for the first physics run starting in 2007. The proposed initial detector will be able to exploit the physics potential of the first physics run at medium luminosity (2×10^{33} cm^{-1}s^{-1}) for which an integrated luminosity of order 10 fb^{-1} could be accumulated in about 7 months (30% efficiency). With 10 fb^{-1} most of the Standard Model Higgs can be probed and SUSY squarks and gluinos with masses up to about 2 TeV can be discovered.

REFERENCES

1. CMS Collaboration, *CMS Technical Proposal*, CERN/LHCC 94–38 (1994).
2. CMS Collaboration, *The Tracker Project, Technical Design Report*, CERN/LHCC 98-06 (1998).
3. CMS Collaboration, *The Electromagnetic Calorimeter, Technical Design*, CERN/LHCC 97–33 (1997).
4. CMS Collaboration, *The Hadron Calorimeter Project, Technical Design Report*, CERN/LHCC 97–31 (1997).
5. CMS Collaboration, *The Muon Project, Technical Design Report*, CERN/LHCC 97–32 (1997).
6. CMS Collaboration, *CMS Level-1 Trigger Technical Design Report*, CERN/LHCC 2000–38 (2000).
7. The CMS Collaboration, *Data Acquisition and High-Level Trigger Technical Design Report*, CERN/LHCC 2002-26 (2002).

Overview of the PHENIX Experiment

Edward J. O'Brien
for the PHENIX Collaboration

Brookhaven National Lab, Upton, New York, 11973

Abstract. PHENIX is an large nuclear physics experiment at the Relativistic Heavy Ion Collider located at Brookhaven National Laboratory. The multi-kiloton detector is designed to address a spectrum of physics topics that range from quark-gluon plasma formation to spin structure of the nucleon and structure function physics. PHENIX is composed of a set of subdetectors that allow for a wide variety of spectrometry, calorimetry and event characterization measurements. The experiment has just completed its third year of data taking.

INTRODUCTION

PHENIX is a large, multi-purpose nuclear physics experiment located at one of the intersection regions of Brookhaven National Lab's Relativistic Heavy Ion Collider (RHIC). The 3.5 kiloton detector is composed of four spectrometer arms: two forward muon spectrometers covering the full azimuth and $1.1 \leq |\eta| \leq 2.4$ North (2.2 South) and two central spectrometer arms each covering $\pi/2$ in azimuth and $\eta \leq 0.35$ in pseudorapidity. Additional detectors at small radii cover a large rapidity range for both vertex determination and event characterization [1].

The experiment has been designed to work in conjunction with the RHIC accelerator to investigate a broad range of nuclear physics topics including the observation and characterization of the Quark-Gluon Plasma, spin structure of the nucleon, structure function physics, initial and final state cold nuclear effects. PHENIX has been designed, built and is operated by over 450 scientists and engineers from 56 institutions in 13 countries.

PHENIX resides in the 8 o'clock hall of the RHIC accelerator complex. The Relativistic Heavy Ion Collider has two counter-circulating rings 3.83 km in circumference. Each ring can be filled with a maximum of 120 equally spaced bunches of nuclei that cross the six machine intersection regions every 106 nanoseconds. The machine has the potential to collide any nuclear species with any other ranging from protons to fully stripped Au ions. The maximum machine energy is $\sqrt{s_{NN}}$ = 200 GeV for AuAu collisions and $\sqrt{s}$ = 500 GeV for p-p collisions. At the end of May 2003, RHIC had completed its third year of operation. During the machine's three physics runs it has delivered long data runs with beams of AuAu $\sqrt{s_{NN}}$ = 130 GeV, AuAu $\sqrt{s_{NN}}$ = 200 GeV, pp $\sqrt{s}$ = 200 GeV and dAu $\sqrt{s_{NN}}$ = 200 GeV.

PHYSICS PROGRAM

The PHENIX experiment has been designed with the ability to measure a large variety of physics signals over a broad momentum, energy and eta range [2]. The versatility of the RHIC accelerator to deliver collisions for a large variety of beam species challenges the experiment to handle luminosities and particle multiplicities ranging from AuAu ($\mathscr{L} = 2 \times 10^{26} cm^{-2} sec^{-1}$ with $dN_{charged}/dy \cong 700$) [3] to pp ($\mathscr{L} = 2 \times 10^{32} cm^{-2} sec^{-1}$ with $dN_{charged}/dy \cong 3$). In addition, the physics program of the experiment requires that p and p_T resolutions are maintained at the few percent level or less for identified hadrons (0.3 GeV/c $\leq p_T \leq 5.0$ GeV/c), inclusive charged hadrons (0.3 GeV/c $\leq p_T \leq 20$ GeV/c), and charged leptons ($p_T \leq 50$ GeV/c). Similarly the detector must maintain good M_{inv} resolution for pair reconstruction of particles with masses ranging from the π^o to the Υ.

The PHENIX physics program includes at least three major components.

1. The observation and characterization of the quark-gluon plasma (QGP) [4]: This is a deconfined state of quarks and gluons that is potentially produced when the high temperatures and matter densities created in a HI collision cause hadronic matter to go through a phase transition to a QGP. Initially the heaviest and most energetic beams at RHIC (AuAu at $\sqrt{s_{NN}} = 200\ GeV$) will be used in an attempt to observe this state of matter. Later systematic scans of collision energies and species will be used to detail physical characteristics of the phase transition.
2. Measurements of the spin structure of nucleons [5]: The spin of the proton has significant contributions from both the gluon and sea quarks. Colliding beams polarized in both the transverse and longitudinal direction will be used to measure the contributions of gluons and anti-quarks to nucleon spin.
3. Systematic studies of structure function physics and cold nuclear physics: These effects provide a nuclear physics baseline needed to disentangle QGP effects from more conventional nuclear physics. Nuclear physics topics including structure function saturation, gluon and quark shadowing, anti-shadowing, and Cronin effect. This physics is studied through pAu and dAu collisions, light ion collisions and energy scans of heavy ions.

PHENIX accesses these physics topics through a combination of spectrometer and calorimeter measurements. Physics quantities are derived from measurements using signals from photon, electron, muon, and hadron channels. A subset of the physics variables measured by the experiment is shown in table 1.

DESCRIPTION OF THE EXPERIMENT

The PHENIX detector is composed of two central spectrometer arms, two forward muon spectrometer arms and a set of subdetectors responsible for event characterization, triggering and global physics measurements .The central arms are optimized to measure photons,electrons and identified hadrons. They also have a good capacity for inclusive

TABLE 1. Physics Variables Measured by PHENIX

Signal	Physics Objective	Category
High pT π^o, hadrons from jets	Jet Quenching, parton dE/dx	QGP
$dN/dy, E_T$, inclusive hadron spectra	Temperature and Energy Density	QGP
Ident. hadron spectra, $\pi^{\pm}$,K,p,π^o,η..	Temperature and Energy Density	QGP
HBT($\pi\pi$,KK,pp), v2=Elliptic flow	Space-Time Evolution	QGP
Evt by evt fluctuations(p_T, E_T, net q)	Space-Time Evolution	QGP
$J/\psi,\psi' \to ee,\mu\mu$, $\Upsilon \to \mu\mu$	Deconfinement	QGP
$\phi \to ee$,KK, ϕ,ω,ρ width/shift	Chiral Symmetry Restoration	QGP
Dis. Chiral Condensate $(\pi^+\pi^-)/\pi^o$	Chiral Symmetry Restoration	QGP
anti-nucleon production	Chiral Symmetry Restoration	QGP
K/π, ϕ, $J/\psi,\psi'$,Υ, D, B mesons	Heavy Quark Production	QGP
Direct γ, $\gamma^* \to ee$, $\mu\mu$	Thermal Radiation	QGP
Direct γ, high pT π's	Gluon spin ΔG	Nucleon spin
$W^+/W^- \to e\nu, \mu\nu$ DY	Sea quark spin,$\Delta\bar{u}, \Delta\bar{d}$	Nucleon spin
π,K,p, charged hadron spectra	Cronin effect	Struct. function phys
$J/\psi,\psi'$, DY $\to ee,\mu\mu$, vs x, A	Shadowing, anti-shadowing, absorp.	Struct. function phys

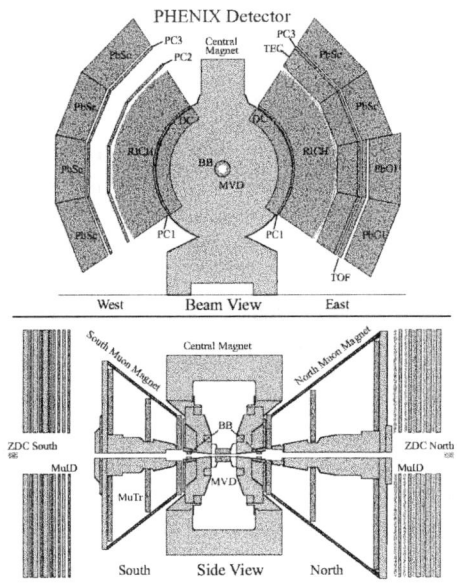

FIGURE 1. Top and side view of PHENIX detector.

hadron measurements. The forward muon arms are design to measure physics signals that produce either single muons or muon pairs such as open charm, ϕ, J/ψ, ψ' and Υ.

The two central arms are similar but not identical (Figure 1). The central arms contain subdetectors for charged particle tracking, particle ID devices and finely segmented calorimetry. The two central spectrometer arms are centered at y=0 and are built around an axial-field spectrometer magnet. The central magnet has a primary coil that can deliver a field integral of 0.78 T-m. A secondary inner coil also exists that allows the central magnet to be operated with an enhanced field integral 1.15 T-m or with the field bucked to provide essentially zero field integral within a radius of 70 cm from the first interaction point.

The charged particle tracking in the central arms is handled by three subsystems. The Drift Chamber is a multi-wire proportional chamber with its anode -cathode wires configured as a focusing jet-chamber. It provides a single-wire track resolution of better than 150 μm and a two-track separation of 1.5 mm. The Pad Chambers are cathode-readout wire chamber. Their fine-granularity pixel pad readout provides non-projective track information with a few mm precision. The Time Expansion Chamber is a radial drift wire chamber that measures charged track positions to 270 μm and dE/dx information in the chamber gas for particle ID.

Four different subsystems contribute to particle identification in the central arms. The PHENIX Time-of-Flight wall combines with a start counter, the Beam Beam, to obtain $\sigma_t < 96$ ps. The central arm ToF can separate $K/\pi \leq 2.5$ GeV/c, $p/K \leq 4.2$ GeV/c. The Ring Imaging Cerenkov Counter is optimized for e/π separation. The RICH uses CO_2 as a radiator gas and is readout with 5100 phototubes. The Time Expansion Chamber has been equipped with transition radiation radiator packs. When it is filled with a xenon gas mixture it will operate as a tracking Transition Radiation Detector.

The central spectrometer arms are instrumented with over 24k modules of electromagnetic calorimetry segmented 0.01 $\Delta\Phi$ x 0.01 $\Delta\eta$. There are two calorimeter technologies in the central arms. Pb-scintillator modules are arranged in a shishkabob-geometry with wave-length shifting fiber running longitudinally through the alternating Pb-scintillator layers and into the phototubes. The Pb-scintillator modules have a timing resolution $\sigma_t = 340$ ps and an energy resolution of $\sigma_E = 10\%\sqrt{E} + 6.5\%$. The Pb-glass modules make up 25% of the central arm calorimetry. The Pb-glass has an energy resolution of $\sigma_E = 8.5\%\sqrt{E} + 9.0\%$.

The forward muon spectrometer arms each have a piston-lampshade magnet that produces a radial magnet field and a 0.72 T-m field integral (at $\eta = 2.0$). The muon magnets are each instrumented with three stations of cathode-strip tracking chambers (CSCs). A CSC station has a charged track position resolution of 100 μm. Following the forward spectrometer magnet is the muon identifier. The MuID is 5 layers of steel absorber plate interleaved with 5 layers of Iarocci tubes (2X,2Y 4 planes/layer), 10 m × 10 m in active cross section. The MuID has a low energy cutoff of 1.9 GeV/c.

The PHENIX event characterization and global detectors consist of the beam-beam counters, zero-degree calorimeter, normalization trigger counter, multiplicity vertex detector and forward calorimeters. The beam-beam counters are two sets of quartz radiator-phototubes arranged symmetrically around the beam pipe and located 144 cm forward from the IR collision point. They provide input to the Level1 trigger and a start time for the PHENIX high-precision timing subsystems. The zero-degree calorimeters are tungsten calorimeters with fiber readout. They are an important Level1 trigger device and contribute to the absolute luminosity measurement. The normalization trigger counters are scintillator arrays which also contribute to the Level1 trigger. The MVD is a combination Si-strip/Si-pad detector which provides vertex, charged-track multiplicity and reaction plane information. The forward calorimeter is two arrays of Pb-scintillating fiber hadronic calorimeters located in the forward direction, 18m from the interaction region vertex point. They are used to characterize collision centrality in dAu and pAu running at RHIC by measuring forward protons near beam rapidity.

TABLE 2. Summary of the PHENIX Detector

Subsystem	$\Delta\eta$	$\Delta\phi$	#Chan.	Electronics	Characteristics
Central magnet	±0.35	2π			$Bdl \leq 1.15$ T-m
Muon magnet S	$-2.25 \leq \eta \leq -1.15$	2π			0.72 T-m $\eta=-2$
Muon magnet N	$1.15 \leq \eta \leq 2.44$	2π			0.72 T-m $\eta=2$
Silicon(MVD)	±2.5	2π	34,720	ADC	Si strip + pad
Beam Beam Counter	±(3.1-3.9)	2π	128	ADC/TDC	quartz + PMT
Normal. Trigger Countr	±(1.2-2.4)	2π	16	ADC/TDC	scin slat + PMT
Zero Degree Calorimtr	±2 mrad, $|\eta|>6$	2π	6	ADC/TDC	W+scin fiber
Forward Calorimeter	$|\eta| \geq 5.3$		180	ADC	Pb+scin fiber
Drift Chamber	±0.35	$2\times\pi/2$	12,800	TDC	MWPC Ar+C_2H_6
Pad Chamber	±0.35	$2\times\pi/2$	172,800	Discrim.	pixel cath readout
Time Expansion Chmbr	±0.35	$\pi/2$	20,480	FADC	MWPC P-10 or Xe
Ring Imaging Cerenkov	±0.35	$2\times\pi/2$	5,120	ADC/TDC	CO_2 + PMTs
Time of Flight	±0.35	$\pi/4$	1,920	ADC/TDC	scin slat+ PMT
EM Calorimeter	±0.35	$2\times\pi/2$	24,768	ADC/TDC	Pb Scin/Pb Glass
Muon Tracking S	$-2.25 \leq \eta \leq -1.15$	2π	21,984	ADC	Cath strip chambr
Muon Tracking N	$1.15 \leq \eta \leq 2.44$	2π	21,984	ADC	Cath strip chambr
Muon ID S	$-2.44 \leq \eta \leq -1.15$	2π	3,170	TDC	Iarocci tube+steel
Muon ID N	$1.15 \leq \eta \leq 2.44$	2π	3,170	TDC	Iarocci tube+steel

PHENIX OPERATIONS

When the experiment is operational, the detector timing, control, synchronization, monitoring, event selection, data collection, event building and archiving is all managed by the PHENIX Data Acquisition and Trigger systems [6]. RHIC design luminosities result in an interaction rate that varies from a few kHz in central AuAu to over 1MHz in pp minimum bias. The wide variety of data rates and event sizes is accommodated by a design that features a significant amount of parallelism and data buffering at every stage of the DAQ system. All data is digitized on the detector by front-end electronics that is pipelined with multi-event buffering. Zero-suppression is accomplished in the counting house at the data collection stage just prior to event building. The DAQ is fully partitioned with 32 partitions available.

PHENIX has two levels of triggering. A hardware LVL1 trigger operates as a synchronous pipeline with a latency of $\sim 40\mu s$. A software LVL2 trigger operates in a few dozen parallel cpu's on fully assembled events immediately after the event building stage. The prescales are adjusted for each type of LVL1 and LVL2 trigger to create a archived mix of min bias, large cross-section and rare triggered events. Events can be built by the PHENIX DAQ at a rate of 1.5 kHz ($\sim$100 kB/evt) and archived at a rate of 100 MB/s. Near term upgrades are expected to increase event building and archiving rates to 8 kHz and 250 MB/s. Summary of PHENIX runs to date is shown in table 3.

TABLE 3. PHENIX Data in the first 3 RHIC Runs

Run #	Year	Species	$\sqrt{s}$(GeV)	$\int \mathcal{L} dt$	N_{TOT}
Run1	2000	Au-Au	130	$1\,\mu b^{-1}$	10M
Run2	2001/02	Au-Au	200	$24\,\mu b^{-1}$	170M
		p-p	200	$0.15\,pb^{-1}$	3.7G
Run3	2002/03	d-Au	200	$2.74\,nb^{-1}$	5.5G
		p-p	200	$0.35\,pb^{-1}$	6.6G

SUMMARY

PHENIX has just completed its third year of operations at the RHIC accelerator. The experiment has a broad physics program encompassing the physics of the quark-gluon plasma, nucleon spin structure, structure functions, and cold nuclear physics topics. The program is carried out with a complex, multi-kiloton detector that combines excellent charged particle tracking, calorimetry, particle ID and event characterization subdetectors with a wide-bandwidth DAQ and multi-level triggering system. Even though RHIC has only operated for a few years, the experiment and accelerator have combined to produce a number of interesting results for a variety of running configurations in AuAu, pp and dAu. PHENIX has had significant physics production to date and intriguing physics effects are already being observed [7].

ACKNOWLEDGMENTS

We acknowledge support from the US Department of Energy and NSF, MEXT and JSPS(Japan), RAS, RMAE and RMIST(Russia), BMBF, DAAD and AvH(Germany), NSFC(China), ISF(Israel), VR and KAW(Sweden), CNPq and FAPESP(Brazil), IN2P3/CNRS and CEA(France), DAE and DST(India), OTKA(Hungary), KRF and CHEP(Korea) and the US-Israel BSF.

REFERENCES

1. K. Adcox et al., NIMA 499(2003)469.
2. PHENIX Conceptual Design Report, BNL, 1993, unpublished
3. K. Adcox et al., Phys. Rev. Lett. 86(2001)3500.
4. D.P. Morrison et al., Nucl. Phys. A 638(1998)565c.
5. N. Saito et al., Nucl. Phys. A 638(1998)575c.
6. S.S. Adler et al., NIMA 499(2003)560.
7. S.S. Adler et al., Phys. Rev. Lett. 91(2003)072303.

Recent Developments for Experiments in the MIT-Bates South Hall Ring

Wilbur A. Franklin

MIT-Bates Linear Accelerator Center, P.O. Box 846, Middleton, MA 01949

Abstract.
A program is underway at MIT-Bates to study nucleons and light nuclei using the recently installed Bates Large Acceptance Spectrometer Toroid (BLAST). These experiments are being carried out with polarized electron beams stored in the South Hall Ring (SHR) and polarized internal targets. Recent developments at the facility, including an upgrade of the polarized injector and the commissioning of a laser back-scattering polarimeter, permitted intense polarized electron beams to be reliably delivered at 850 MeV for initial tests of the BLAST Spectrometer and its detectors. In this paper, overviews of the SHR beam and BLAST are given, including selected results from the commissioning run.

Experiments with the BLAST Spectrometer address fundamental questions in fewbody physics using polarized beams and polarized targets. Electron scattering from light nuclei, such as ^{1}H, ^{2}H, and ^{3}He, can be simultaneously studied in multiple reaction channels over a large range of momentum transfer. Experiments with BLAST offer the promise of unprecedented statistical precision for spin observables in these systems. The measurements proposed are expected to improve substantially constraints on form factors of the nucleon (including G_E^n), providing important tests of models of nucleon structure [1, 2].

The statistical precision attainable in BLAST experiments results from a combination of the spectrometer's large acceptance and the high average currents achievable with stored electron beams in the South Hall Ring. The spectrometer subtends scattering angles between 20° and 80° in the lab frame. The BLAST detectors have a symmetric configuration with respect to the beam, making it an appropriate tool for semi-inclusive and exclusive measurements. In the SHR, stored electron beams now consistently meet the benchmarks of at least 80 mA average current with longitudinal polarization in excess of 0.6. With operation of an atomic beam source having been demonstrated in the BLAST magnetic field, luminosities in excess of 10^{31} cm^{-2}s^{-1} are available for experiments with hydrogen and deuterium.

POLARIZED ELECTRON BEAMS

Polarized electrons at Bates are produced by a laser-driven GaAs source [3]. Initially designed to deliver beam to parity-violating experiments which required high repetition rates and the removal of helicity-correlated position differences in the beam, the source has been reconfigured for storage mode operation to emphasize maximization of the beam polarization. A strained GaAsP photocathode[1] was installed allowing polarization to exceed the limit of 0.4 present in bulk GaAs sources [4]. Measurements with low energy Mott and transmission polarimeters indicate that beam polarization in excess of 0.7 can be reliably extracted from the polarized source.

To compensate for the reduction of quantum efficiency inherent in the use of strained photocathodes, a high-powered fiber-coupled diode array laser[2] system was implemented. The diode laser produces up to 200 W of power at 808 nm with an adjustable duty cycle allowing the peak current requirements for filling the South Hall Ring ($I_{peak} >$ 5 mA) to be consistently met. Requirements on the injector have been further eased by the installation of an additional bunching cavity which raises the fraction of electrons captured by the accelerator from 0.3 to 0.5. The polarized source has been able to reliably deliver beams for over one year without time-consuming replacement of the photocathode.

In the SHR, high average current is obtained through a combination of high injection current, long beam lifetime, and high duty cycle. Stored currents in excess of 300 mA have been measured following recent improvements in the SHR vacuum system. While such beams are short-lived and unsuitable for use during BLAST experiments, average currents of 120 mA can be injected reproducibly and tuned to reduce beam halo to low levels. Beam lifetimes in excess of 50 minutes have been observed without a target, although the actual lifetime is generally governed by the target thickness. An automated filling system has been created for the Ring, which includes ramping of detector voltages and setting of beam halo slits, to optimize the duty cycle for BLAST experiments. Typical filling time is less than one minute, an important consideration for a ring which is typically filled approximately every fifteen minutes.

BLAST experiments require longitudinal polarization at the internal target location. Electrons are injected into SHR polarized in the plane of circulation. The beam polarization is maintained by a superconducting solenoidal full Siberian Snake. Measurements of the stored beam polarization are made on a continuous basis by a laser backscattering Compton polarimeter [5] based on a similar device built for the AmPS Ring at NIKHEF [6]. The polarimeter detects gamma rays produced by interactions between circularly polarized 532-nm laser light and the stored beam. It is positioned upstream of the BLAST target to reduce sensitivity of the measurement to bremsstrahlung background. It is possible to operate the polarimeter continuously with currents in excess of 100 mA stored in the Ring. Results for the measured asymmetry from a typical run are shown in Fig. 1. The polarimeter has provided an important tool for avoiding depolarizing spin-orbit resonances in tuning the SHR [7]. Such considerations are particularly important for high current operation where betatron tune spreading can result in partial loss of the beam polarization. Although systematic uncertainties in the extraction of the absolute beam polarization are estimated to be approximately 5%, with a proper tune of the SHR, data taken over periods of weeks have yielded values for the beam polarization which consistently hover between 0.65 and 0.70. These data included a short run

---

[1] Bandwidth Semiconductor Inc., Bedford, NH
[2] Spectra Physics Laser, Mountain View, CA 94039

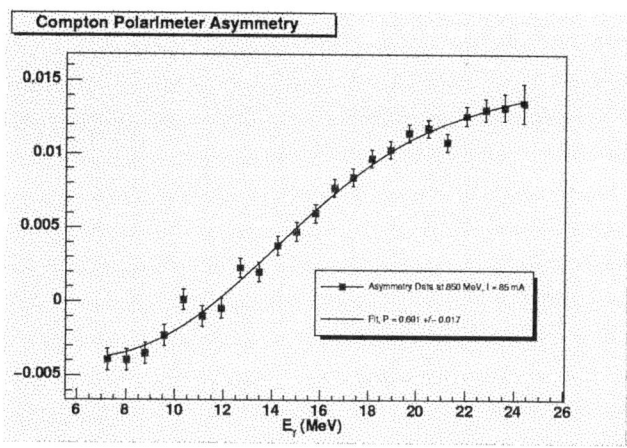

FIGURE 1. Asymmetry spectrum measured as a function of scattered gamma ray energy with the MIT-Bates Compton Polarimeter. The plot shows approximately 60 minutes of data with average current of 60 mA with beam polarization being measured to be $P = 0.691 \pm 0.017$.

in which dynamic reversal of the polarization of a beam in the South Hall Ring [8] was demonstrated using an rf dipole with a spin-flip efficiency of 0.98 ± 0.01.

BATES LARGE ACCEPTANCE SPECTROMETER TOROID

The BLAST Spectrometer consists of a set of eight copper coils carrying currents of nearly 7000 Amps. The resulting toroidal magnetic field has been mapped in detail and has strength of up to 0.4 T. Due to the symmetric design, the BLAST field exerts no force on the electron beam or on nuclei in the internal target region, although elements of the target system must reside in regions of strong magnetic field. The first internal target for BLAST is an atomic beam source [9] for experiments with polarized hydrogen and deuterium. RF transition units in the atomic beam line allow the polarization to be selected and reversed on a rapid time scale. Polarized gas is injected from above into a 400-mm long storage cell coated with drifilm to maintain polarization. A local magnetic holding field allows selection of the spin quantization axis, allowing great flexibility in the measurement of spin observables. Because the target is embedded in the BLAST magnetic field, it requires substantial magnetic shielding for vacuum equipment and to maintain proper focusing of the sextupole system.

The BLAST coils divide the spectrometer into eight sectors available for instrumentation. The two sectors in the horizontal plane have been outfitted with detector packages, each with an azimuthal acceptance of $35°$. Each detector pack presently consists of drift chambers for tracking, Cerenkov detectors for electron-pion identification, segmented timing scintillators for timing, and a wall of thick plastic scintillators for neutron detection. Silicon recoil detectors for the detection of low energy charged particles and forward angle lead glass calorimeters are also planned.

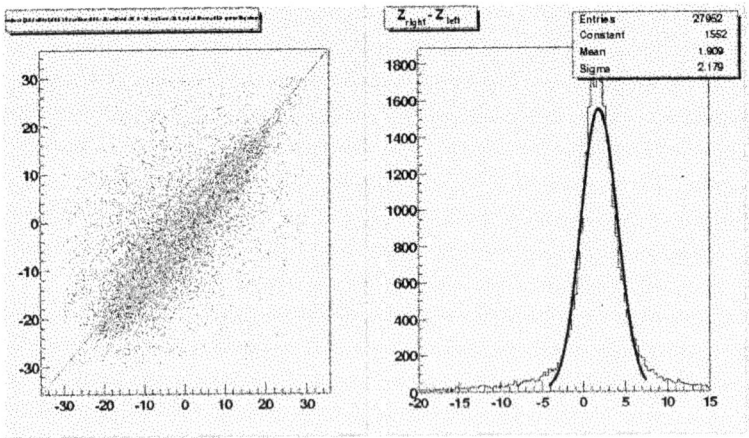

FIGURE 2. Vertex reconstruction in cm from target cell center.

The spectrometer is equipped with a versatile programmable trigger. During the initial run, the trigger was typically defined by particular combinations of two timing scintillators. For typical beam conditions, rates of 100 Hz were typically observed, which the CODA-based data acquisition system can handle comfortably. Under these conditions, the BLAST wire chambers are able to track of charged particles cleanly. Fig. 3 shows reconstruction of the scattering vertex with a profile which accurately reflects the dimensions of the target cell. Relative momentum resolution of about one percent appears to be a reachable goal based on analysis of $e - p$ elastic scattering data.

ACKNOWLEDGMENTS

This research has been supported in part by the U.S. Department of Energy under Cooperative Agreement #BEFC294ER40818. This article summarizes the contributions of many members of the MIT-Bates staff and user community, but I reserve particular thanks for H. Kolster, M. Farkhondeh, E. Tsentalovich, D. Hasell, and T. Lee who contributed material used in the presentation.

REFERENCES

1. Lomon, E., *Phys. Rev. C*, **64**, 035204 (2001).
2. Golak, J., et al., *Phys. Rev. C*, **65**, 044002 (2002).
3. Farkhondeh, M., et al., *AIP Conf. Proc.*, **570**, 955 (2001).
4. Maruyama, T., et al., *Phys. Rev. Lett.*, **66**, 2376 (1991).
5. Franklin, W., et al., *Prog. Part. Nucl. Phys.*, **44**, 61 (2000).
6. Passchier, I., et al., *Nucl. Inst. and Meth. A*, **414**, 446 (1998).
7. Montague, B., *Phys. Rep.*, **113**, 35 (1984).
8. Morozov, V., et al., *Phys. Rev. STAB*, **4**, 104002 (2001).
9. Szczerba, D., et al., *Nucl. Inst. and Meth. A*, **455**, 769 (2000).

Detection of High-Z Objects using Multiple Scattering of Cosmic Ray Muons

Gary E. Hogan, Konstantin N. Borozdin, John Gomez, Christopher Morris, William C. Priedhorsky, Alexander Saunders, Larry J. Schultz, Margaret E. Teasdale[1]

Los Alamos National Laboratory, Los Alamos NM 87545 USA

[1]*Present address: Department of Marine Science, University of Hawaii, 200 W. Kawili St., Hilo, HI 96720*

Abstract. Detection of high-Z material hidden inside a large volume of ordinary cargo is an important and timely task given the danger associated with illegal transport of uranium and heavier elements. Existing radiography techniques are inefficient for shielded material, often expensive and involve radiation hazards, real and perceived. We recently demonstrated that radiographs can be formed using cosmic-ray muons[1]. Here, we show that compact, high-Z objects can be detected and located in 3 dimensions with muon radiography. The natural flux of cosmic-ray muons[2], approximately 10,000 $m^{-2}min^{-1}$, can generate a reliable detection signal in a fraction of a minute, using large-area muon detectors as used in particle and nuclear physics.

Conventional radiography uses the absorption of x-ray or gamma radiation to provide image contrast. The flux in an image pixel is determined by Beer's Law, $N = N_0 \exp(-L/L_0)$, where L is the path length through the object and L_0 is the mean free path for scattering and/or absorption[3]. Radiographic measurements are limited by the Poisson counting statistics of the transmitted flux, $\frac{\Delta L}{L_0} = \frac{1}{\sqrt{N}}$, and the inability to penetrate dense materials. Even the most penetrating gamma rays (few MeV) are attenuated by an e-folding in 2 cm of lead. Objects much thicker than this can be penetrated only by a very large incident photon dose, which is harmful for living organisms or radiation sensitive cargo such as photographic film.

By contrast, our images are formed by the analysis of muon angular deflections caused by multiple Coulomb scattering. Muons undergo a random walk in direction, with a Gaussian width $\theta_0 = \frac{13.6 MeV}{\beta c p}\sqrt{\frac{L}{L_0}}[1 + 0.038 \ln(L/L_0)]$, where L_0 is the radiation length, p is the particle momentum in MeV/c and βc is its velocity[4]. Since radiation length drops rapidly with atomic number, in 10 cm a 3 GeV muon will scatter, on the average, 2.3 milliradians in water (L_0 = 36 cm), 11 milliradians in iron (L_0 = 1.76 cm), and 20 milliradians in lead. If the scattering angle in an object is measured, and the particle momentum is known, then the fractional path length,

$R = L/L_0$, can be determined to a precision of $\frac{\Delta R}{R} \cong \sqrt{\frac{2}{N}}$, where N is the number of transmitted particles. Thus each particle delivers, via its deflection, information on the density along the traversed path. In comparison, absorption radiography only contributes a binary count to a pixel.

The natural background of muons is free and available everywhere on the Earth. The muons are produced via interactions of primary cosmic rays in air. They are highly penetrating, having already traversed the equivalent of several meters of water by the time they reach sea level. Additionally, cosmic ray muons illuminate an object from a wide range of angles. This is helpful for 3D tomographic reconstruction.

Our newly developed technique allows the study of small and medium-sized objects such as passenger cars, sea containers, and commercial trucks. To obtain a reasonable cargo screening capability, we must produce a clean detection decision in a reasonable time. Complicating factors include the clustering of muons in showers and their distribution in momentum. The natural flux includes low momentum muons that scatter easily and high momentum muons that scatter less.

Consider, for example, a truck containing normal cargo, such as a flock of sheep (40-cm thick bags of water), in which is hidden a small (10-cm thick) volume of high-Z material such as lead. A muon that passes through a sheep will scatter with a mean angle, θ_0, of about 5 milliradians, but a muon that passes through the lead will scatter with a mean angle of about 20 milliradians. By measuring the actual scattering angles of all the muons that pass through the truck, and watching for an excess of muons that scatter through large angles, we can find the high-Z object.

To prove that these complications could be overcome, we developed an experimental detector system. The detector is a stack of delay line drift chambers that measure a total of eight X and eight Y locations for each muon. The active area of each chamber[5] is 60 x 60 cm^2. The top half of the stack measures the incident muon track, the bottom half measures the track after scattering. There is a 27 cm high volume in the middle for test objects. Calibrating the instrument with no scattering material in the object area, we determined that our position precision is about 400 μm (FWHM). Data is taken using a Windows® based data acquisition program developed at Los Alamos[6]. A pair of 30 cm square plastic scintillators placed below the lowest detector is used for triggering. Given the limited acceptance angle of this configuration, the expected trigger rate is about 850 counts•min^{-1}, consistent with the observed rate of 750 counts•min^{-1}. A tungsten cylinder of 5.5 cm radius and 5.7 cm height was used as a test object, supported by a Lexan plate and steel support beams.

The path of a charged particle through the test material is stochastic and can only be approximately reconstructed. We approximate multiply-scattered tracks as having only a single scattering event at their point of closest approach. The scattering signal from each muon is assigned to voxels (3-D pixels) along its track using a maximum likelihood technique that distributed the signal along the track according to the uncertainty in determining the point of scattering. Each scattering point is weighted by its scattering angle θ. We find that a $\theta^{1.5}$ weighting provides the highest contrast between high and low-Z objects in our images.

Figure 1 shows slices through the reconstructed, 3-dimensional image of the tungsten cylinder from a 2½ hour run. The tungsten is seen with sharp edges, and the steel support beams are visible as well, clearly distinguishable from the tungsten object. For a simple yes/no detection, considerably shorter runs will clearly do the job.

To compare these results with theory, and to establish a platform for investigating larger, more complex radiographic scenarios, we developed a Monte Carlo simulation code that i) generated cosmic ray muons with the appropriate distribution of energies and incident angles; ii) propagated them through a test volume, calculating their scattering in each 0.2 cm cubic resolution element; and iii) generated the positions at which they would be detected in four detector planes. The muon spectrum, angular distribution, and rate were appropriate for sea level[2]. Their scattering and propagation within the targets were calculated according to the multiple scattering law[4] described above. The geometry and densities of experimental test objects were incorporated in the simulation, and the same reconstruction technique used for the experimental data was applied to the simulated data. The simulation accurately matched the tungsten experiment, both in the reconstructed image, and the distribution of scatter angles.

We then simulated a hypothetical shipping container monitor. The overall geometry is similar to our experimental setup. The detector pairs, assumed to have our experimental resolution of 400 μm (FWHM), are spaced 1 meter apart (h) and span a 4.5 meter high (l) test volume. The resolution of the image would be improved if the pairs were more widely spaced (larger h); however, this might demand too large an instrument in a practical application. The simulation modeled a steel container (3 mm walls) that contained high-Z objects ("pigs", the informal name for shielded casks), surrounded by dozens of low-Z objects ("sheep"). The "pigs" (9 x 9 x 12 cm^3 bricks of uranium) are detected at high confidence in a 1-minute simulated exposure. The signal in the 3 x 3 x 3 cm voxel (3-D pixel) at the location of a brick is, in arbitrary units, 54 ± 24, while the background in an adjacent volume was 1.9 ± 1.1. Thus a threshold chosen to detect the object with 90% confidence will reject false positives from the background with 20σ confidence. Clearly, false positives are unlikely.

We improve the reconstruction algorithm by weighting the scattering signal by the approximate muon momentum. The signal-to-noise ratio is a function of our knowledge of the muon momentum. We have assumed a rough knowledge of the momentum (Δp/p of 50% in a log-normal distribution). This knowledge significantly improves the reconstruction. If the momentum is unknown, our 90% threshold will reject false positives from background with 3.5σ confidence. Momentum information can be obtained by measuring the multiple scattering in several layers of known material. This could be done by a scatter-detector sandwich below the lowest detector plane. We have determined via simulation that using two planes of scattering material in this sandwich provides the needed Δp/p of 50%. We reject muons with momentum greater than 20 GeV/c or with a scatter angle less than 5 milliradians to achieve higher contrast between high-scattering and low-scattering material in the image.

The results so far are encouraging, and there is room for further improvement in both detection and reconstruction techniques. For instance, the displacement of the muon track (i.e. the minimum distance between the incident and scattered path) also provides a measure of multiple scattering, because multiply-scattered muons are more likely to undergo a significant displacement, while a single-point scattering event necessarily has zero displacement. To achieve maximal contrast in images of different objects, one may vary weighting algorithms and apply different angular cuts.

We have confirmed that muon radiography of relatively large objects, on the order of the size of a commercial truck, can be performed in a reasonably short time (~1 minute). The method is particularly useful in detecting a high–Z target against a low-Z background. More sensitive reconstruction techniques are being developed.

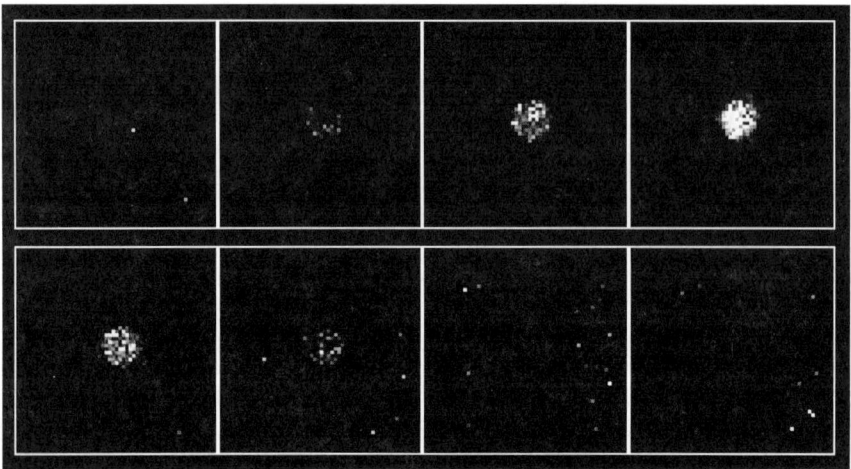

FIGURE 1. Reconstruction of test object based on an experimental run of 100,000 muons. The eight planes shown are horizontal slices through the test volume, moving top to bottom.

ACKNOWLEDGMENTS

This work was supported by the LDRD program at Los Alamos National Laboratory operated by the University of California for the U.S. Department of Energy. We would like to thank Val Armijo and Gary Blanpied (U. of South Carolina) for their help.

REFERENCES

1. Konstantin N. Borozdin, Gary E. Hogan, Christopher Morris, William C. Priedhorsky, Alexander Saunders, Larry J. Schultz, Margaret E. Teasdale, "Radiographic Imaging with Cosmic Ray Muons", Nature, 422, 277 (2003).
2. Grieder, P.K.F. Cosmic Rays at Earth, Elsevier Science (2001).
3. Beer, A. Bestimmung der absorption des rothen lichts in farbigen flussigkeiten, Ann. Physik. Chem. 86, 2 (1852).
4. Hagiwara, K., et al., Particle Data Group, Review of Particle Physics, Phys Rev D 66, 1 (2002).
5. Atencio, L. G., Amann, J. F., Boudrie, R. L., and Morris, C. L. Delay-Line Readout Drift Chambers. Nuclear Instruments & Methods In Physics Research 187, 381 (1981).
6. Hogan, G. E., "PC DAQ : a personal computer based data acquisition system," Los Alamos National Laboratory LA-UR-98-4531 (1998), http://lib-www.lanl.gov/cgi-bin/getfile?00418755.pdf.

Development of Photon Collider and Solid-State Modulator Technology for the NLC

Jeffrey Gronberg

LLNL,[1] 7000 East Ave., Livermore CA, 94550

Abstract. LLNL is part of the Next Linear Collider (NLC) R&D consortium. Among other projects, LLNL has been applying its core competencies in laser and induction modulator technology to the design of the NLC. Recent progress in the development of high average power, short-pulse lasers have made feasible the idea of a TeV-scale photon collider. A pre-conceptual design to integrate the laser and optics with the accelerator and detector components has been produced and test facilities based on the mothballed SLC are being examined. A prototype solid-state induction modulator has been created and is currently driving four XL4 klystrons at SLAC.

INTRODUCTION

Many NLC [1] parameters will extend previous experience with linear colliders by several orders of magnitude. LLNL has been applying it particular competencies to solving many of the issues facing the NLC. From precision machining of accelerator RF cavities to modeling of radiation damage in positron targets to the creation of ultra-stable mechanical supports for nanometer resolution beam position monitors, LLNL has contributed to a variety of efforts. This note will focus on the development of an induction modulator based on solid-state switches, which should be more efficient and robust than previous hydrogen thyratron based modulators. Also, the development of a new capability to produce photon-photon collisions will be detailed.

PHOTON COLLIDER

The basic idea for producing a photon collider by Compton scattering of a laser from an electron beam was first proposed by Ginzburg *et al.* in 1981 [2]. At a TeV scale linear collider photon-photon collisions with a significant fraction of the e^+e^- energy and luminosity are possible [3]. Such collisions are of great interest to the study of the Higgs boson [4] for several reasons. The CP nature of the Higgs boson can be uniquely studied since control of the laser polarization allows the CP of the initial state to be controlled. The production of Higgs through a loop diagram allows the existence of new

---

[1] This work was performed under the auspices of the U.S. Depatment of Energy by the University of California, Lawrence Livermore National Laboratory under Contract No. W-7405-Eng-48

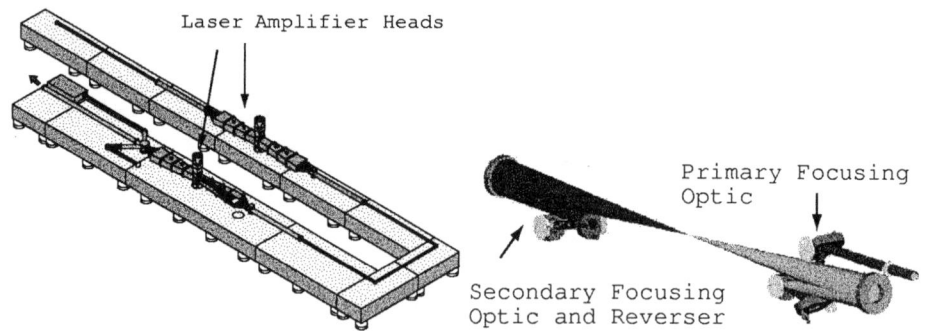

FIGURE 1. a) A diagram of the MERCURY laser layout is on the left. An initial laser pulse makes multiple passes through the laser gain medium before the full power of 100 Joules is extracted. Helium gas flow transverse to the crystal provides face cooling to minimize the impact of deposited power. b) On the right is the arrangement of the focusing mirrors within the detector beam pipe. The final focusing mirror has a hole to allow the electron beams to pass and all other mirrors are outside the path of the background particles.

particles to be probed and for SUSY Higgs the production mechanism allows the LHC blind spot in SUSY parameter space to be filled in.

While the basic idea is simple, the realization in practice requires designs for lasers and optics which push the state of the art in high average power, short-pulse lasers. Additionally, the laser light must be brought into focus with the confined space of the detector without degrading the performance of the detector or accelerator.

Over the past several years LLNL has been working to create a baseline design for a photon collider interaction region suitable for the NLC. The basis of the design is the MERCURY laser [5], shown in Figure 1a, which is designed to produce laser pulses of 100 Joules at 10Hz. The 100 Joule pulse is enough energy for a train of 95 NLC electron bunches. Since NLC trains are produced at 120 Hz, a battery of 12 lasers, firing sequentially, would be required to power a photon collider interaction region at the NLC. The MERCURY laser has been commissioning with a single amplifier head in place. It has reached steady-state operation of 20 Joule pulses at 10Hz. Currently, the experimental setup is being modified to include the second amplifier head and we expect operation at the design power of 100 Joules in the near future.

The task of bringing the laser light into the interaction point beam pipe and focusing it onto the electron beam a few millimeters from the interaction point is accomplished by a pair of four mirror telelscopes, as shown in Figure 1b. All mirrors are positioned so that they avoid the electron beams and the charged particle backgrounds produced in the electron beam collisions. This requires the final focusing mirror to have a large central hole. Since most of the laser light goes unused in the laser - electron beam interaction, the far set of mirrors catches the pulse, reflects and refocuses it onto the opposite electron beam. This allows each laser pulse to do double duty and cuts the required laser power in half. A set of half-size optics is currently assembled on a test bench along with an interferometric system, which is used for in situ alignment of the mirrors.

The design for the photon collider interaction region is sufficiently mature that test facilities can be considered. The Stanford Linear Collider (SLC) is currently the world's only e^+e^- linear collider. The SLC provides the opportunity to create the first photon collider and to gain experience with its operation before proposing an experiment at a future TeV-scale linear collider. An upgrade of the beam pipe to include focusing optics combined with a small laser would allow photon-photon luminosity to be generated. This idea has been studied and there are no technical show-stoppers to creating such a facility[2].

SOLID-STATE MODULATOR

A collaboration of SLAC, LLNL and Bechtel/Nevada have been working to design an inductive solid-state modulator, capable of driving eight klystrons, based on Insulated Gate Bipolar Transistors (IGBT). A simplified diagram is shown in Figure 2a. Capacitors are slowly charged and then discharged into the secondary of a transformer through an IGBT switch. The fast turn-on and turn-off times of the IGBT allow higher power conversion efficiency than in traditional pulse forming networks. Individual control of each IGBT provides increased robustness since spare circuits can be switched in when IGBTs fail. Additionally, the timing of an individual IGBT can be changed to improve the flat-top of the output pulse.

The prototype modulator has been installed in the SLAC endstation and is currently running a set of four XL4 klystrons. Figure 2b shows the ouput waveform. The modulator has a 1.6 microsecond flat-top voltage of 400kV. Several of the IGBTs have been fired late in order to eliminate ringing in the voltage waveform. The peak output current of the modulator is 1200A leading to a peak power of 0.5 GW. Currently, the modulator has been operated at 30Hz instead of the design of 120Hz due to heating limitations in the IGBTs. More robust heat dissipation in the next generation modulator should resolve that issue.

Initial operation of the modulator has yielded many lessons which are being incorporated into the next generation prototype. The next generation modulator will be designed to drive two klystrons. This reduction in the complexity of the circuit should make it more robust against damage to the IGBTs during transient faults. This is important since it allows the IGBTs to run faster leading to less power dissipation during switching and greater efficiency of the power conversion. The design of the new prototype is currently under way.

ACKNOWLEDGMENTS

The work reported in this paper is the result of the efforts of a broad collaboration of people. The modulator team is: R. Cassel, J. de Lamare, G. Leyh, M. Nguyen and C. Pappas from SLAC, E. Cook, J. Sullivan and S. Hawkins from LLNL and

---

[2] http://www-conf.slac.stanford.edu/lepcf

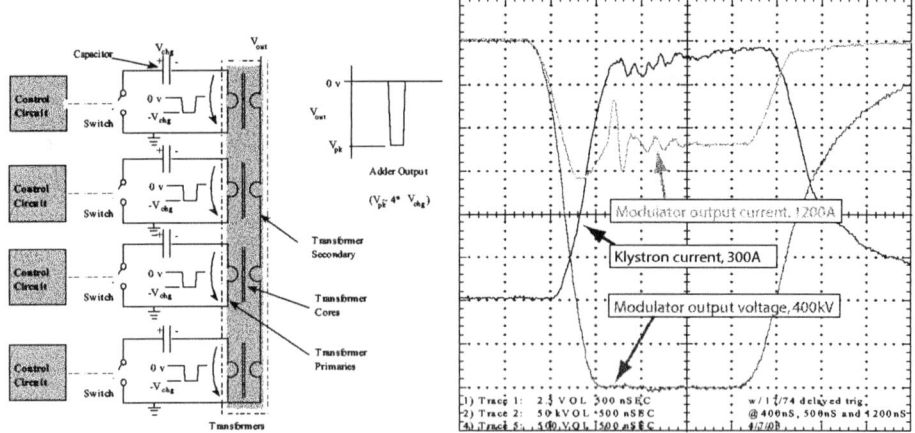

FIGURE 2. a) On the left is a block diagram of the transformer circuit with a few individual capacitor / IGBT cells shown. b) An output waveform from the modulator prototype is shown on the right. The modulator is running at 30Hz with a flat-top of 1.6 microseconds and driving 4 XL4 klystrons. The red curve is the output voltage. Modulator current is the green curve and the blue curve is an individual klystron current.

C. Brooksby and J. Yuhas from Bechtel / Nevada.

REFERENCES

1. The NLC Design Group, Zeroth Order Design Report for the Next Linear Collider, Tech. Rep. LBNL-5424, SLAC-474, UCRL-ID-124161, UC-414 (1996).
2. Ginzburg, I., Kotkin, G., Serbo, V., and Telnov, V., *Pizma ZhTEF*, **34**, 514 (1981).
3. Telnov, V., *Nucl. Instr. and Meth.*, **A355**, 3 (1995).
4. Asner, D., Gronberg, J., and Gunion, J., *Phys. Rev.*, **D67**, 035009 (2003).
5. Hackel, L., Tech. Rep. UCRL-TB-136126-01-02 (2001).

Spin-Flipping Polarized Deuterons At COSY [a]

K. Yonehara*, A.D. Krisch*, V.S. Morozov*, R.S. Raymond*, V.K. Wong*,
U. Bechstedt[†], R. Gebel[†], A. Lehrach[†], B. Lorenz[†], R. Maier[†], D. Prasuhn[†],
A. Schnase[†], H. Stockhorst[†], D. Eversheim**, F. Hinterberger**,
H. Rohdjess**, K. Ulbrich** and W. Scobel[‡]

*Spin Physics Center, University of Michigan, Ann Arbor, MI 48109-1120, USA
[†]Forschungszentrum Jülich, Inst. für Kernphysik, Postfach 1913, D-52425 Jülich, Germany
**Helmholtz-Inst. für Strahlen- und Kernphysik, Univ. Bonn, D-53115 Bonn, Germany
[‡]Inst. für Exper-physik, Univ. Hamburg, Luruper Chaussee 149, D-22761 Hamburg, Germany

Abstract. We recently stored a 1.85 GeV/c vertically polarized deuteron beam in the COSY Ring in Jülich; we then spin-flipped it by ramping a new air-core rf dipole's frequency through an rf-induced spin resonance to manipulate the polarization direction of the deuteron beam. We first experimentally determined the resonance's frequency and set the dipole's rf voltage to its maximum; then we varied its frequency ramp time and frequency range. We used the EDDA detector to measure the vector and tensor polarization asymmetries. We have not yet extracted the deuteron's tensor polarization spin-flip parameters from the measured data, since our short run did not provide adequate tensor analyzing-power data at 1.85 GeV/c. However, with a 100 Hz frequency ramp and our longest ramp time of 400 s, the deuterons' vector polarization spin-flip efficiency was 48±1%.

Recently polarized deuteron experiments have become an accessible and interesting area of subatomic physics. Studying the spin-flipping and polarization dynamics of polarized deuteron beams is an important step towards polarized deuteron, and thus polarized neutron, scattering experiments. Many polarized scattering experiments require frequent spin-direction reversals (spin-flips), while the polarized beam is stored, to reduce their systematic errors. Earlier we used an rf solenoid to study, for first time, spin flipping and spin manipulation of a simultaneously vector and tensor polarized 270 MeV deuteron beam stored in the IUCF Cooler Ring and obtained rather interesting results [1]. We recently studied spin flipping with an rf dipole of a simultaneously vector and tensor polarized 1.85 GeV/c deuteron beam at COSY's higher energy.

In any flat circular accelerator or storage ring, each deuteron's spin precesses around the Stable Spin Direction (SSD), which is defined by the ring's magnetic structure. With no horizontal magnetic fields in the ring, the SSD points along the vertical fields of the ring's bending magnets. The spin tune v_s, which is the number of spin precessions during one turn around the ring, is proportional to the deuteron's energy

$$v_s = G\gamma, \quad (1)$$

---

[a] Supported by research grants from the U.S. Department of Energy and the German BMBF Ministry.

where $G = (g-2)/2 = -0.1426$ is the deuteron's gyromagnetic anomaly and γ is its Lorentz energy factor.

The polarization can be perturbed by the horizontal rf magnetic field from either an rf-solenoid or an rf-dipole. This perturbation can induce an rf spin resonance, which can flip the spin of the stored polarized deuterons; the resonance's frequency is

$$f_r = f_c(k \pm v_s), \qquad (2)$$

where f_c is the deuteron's circulation frequency and k is an integer. Sweeping the rf magnet's frequency through f_r can flip the Ring's SSD [2]. For an adiabatic frequency sweep, each deuteron's vector polarization follows the SSD as it rotates by 180°; this causes a vector spin flip. The modified [3] Froissart-Stora formula [4] relates the beam's initial vector polarization P_i to its final polarization after crossing the resonance P_f

$$P_f = P_i \left\{ (1+\eta) \exp\left[-\frac{(\pi \varepsilon f_c)^2}{\Delta f/\Delta t}\right] - \eta \right\}, \qquad (3)$$

where η is the spin-flip efficiency, ε is the resonance strength, and $\Delta f/\Delta t$ is the resonance crossing rate, while Δf is the ramp's frequency range during the ramp time Δt.

In addition to the three ordinary vector polarization components, the polarization of a beam of spin-1 particles is also usually described by its tensor polarization. Unfortunately, there was no analyzing power data available at 1.85 GeV/c; thus, we were unable to extract the deuteron's tensor polarization from the measured asymmetries. Thus, we will only present results on spin-flipping the deuteron vector polarization. In future runs we plan to measure the analyzing powers and thus obtain the tensor polarization.

The apparatus used for this experiment is shown in Fig. 1; it includes the COSY storage ring [5], the EDDA detector, the Low Energy Polarimeter, the injector Cyclotron, the polarized ion source, and the rf dipole, which was two 6-turn air-core copper coils, installed around a fast quadrupole's ceramic chamber. As part of an LC resonant circuit, the rf dipole normally ran at about 4.3 kV rms producing an $\int B dl$

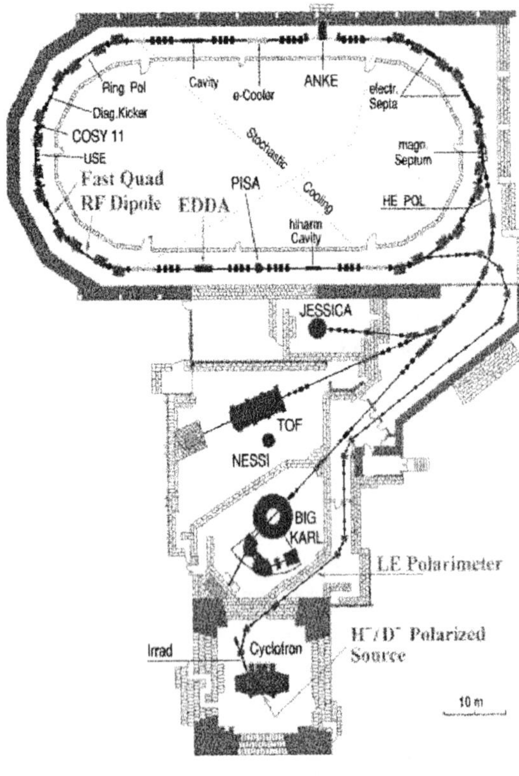

FIGURE 1. Layout of the COSY Storage Ring, with its injector Cyclotron, polarized ion source, rf dipole, EDDA detector, and Low Energy Polarimeter.

of about 0.15 T·mm rms. The beam emerging from the polarized D^- ion source was accelerated by the Cyclotron to COSY's injection energy. The Low Energy Polarimeter (LEP), which was between the Cyclotron and COSY, monitored the beam's polarization before injection into COSY. The LEP data was mostly used to check the stable operation of the ion source and Cyclotron.

To reduce our systematic errors, we cycled the polarized D^- source through five polarization states with nominal vector polarization values of $0, -2/3, -1/3, -1, +1$. Although no analyzing power data were available at 1.85 GeV/c, we measured the deuteron scattering asymmetries in the EDDA detector, which are related to the vector and tensor polarizations.

We first determined the spin resonance's approximate location $f_r = f_c(1 - |v_s|)$ by ramping the rf dipole's frequency through $\Delta f = 100$ Hz around the calculated frequency of $f_r = 917.4$ kHz; we then continued with 100 Hz ramps next to each previous frequency range until the beam was partly depolarized, as shown in Fig. 2. The source was cycled through only the $+1$ and -1 vector polarized states; after each 100 Hz ramp of duration $\Delta t = 19$ s, we measured the deuteron's left-right scattering asymmetry, which was proportional to its vector polarization. Fig. 2 shows the ratio P_f/P_i of the asymmetries measured before (P_i) and after (P_f) ramping the rf dipole. These data show that the resonance was at about 916.85 kHz, which is near the calculated f_r.

We then mapped the spin resonance by measuring the asymmetry ratio, with the rf dipole at different fixed frequencies near 916.85 kHz. These data, plotted in Fig. 3, have a wide and shallow dip centered near 916.85 kHz; we fit this data to a second-order Lorentzian. This wide and shallow shape could be due to a combination of a somewhat weak dipole and a finite spin tune spread.

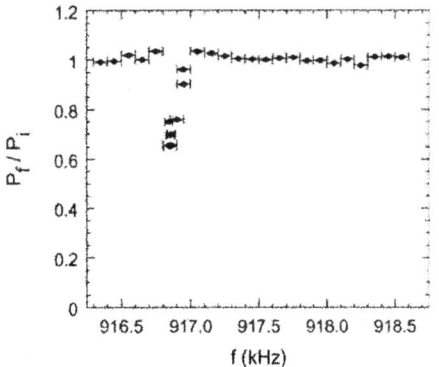

FIGURE 2. The measured deuteron's final-to-initial vector polarization ratio at 1.85 GeV/c is plotted against the range of each frequency ramp; a horizontal bar shows each ramp's Δf range.

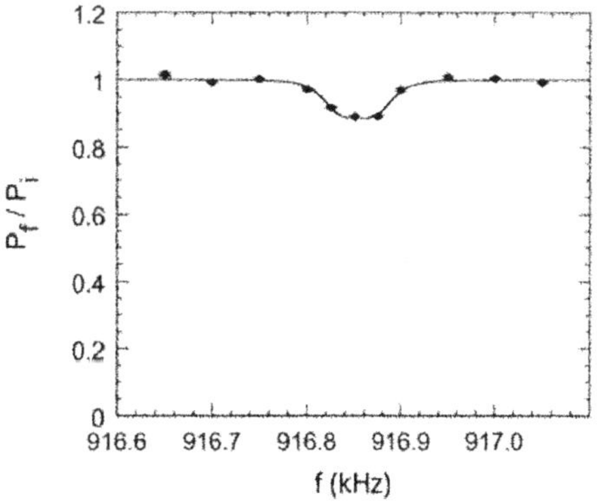

FIGURE 3. The measured deuteron's final-to-initial vector polarization ratio at 1.85 GeV/c is plotted against the rf-dipole's fixed frequency. The curve is a second-order Lorentzian fit.

After setting the rf dipole's voltage to its maximum and keeping the frequency range fixed at $\Delta f = 100$ Hz, we spin-flipped the beam by varying the frequency ramp time Δt. The measured left-right scattering asymmetries for the five polarization states are plotted in Fig. 4; the curve for each polarization state is a fit to Eq. (3). Notice that all five curves cross at the same non-zero point near 100 s. This indicates that, after subtracting the systematic offset of about 1.5%, each asymmetry was indeed linearly proportional to the state's vector polarization. Fig. 4 shows that the vector polarization was partially spin-flipped for the two longest ramp times of 200 and 400 s.

To further analyze the data, we first subtracted the unpolarized offset, shown in Fig. 4 by a solid line, from each other plotted asymmetry; we next divided the result by each corresponding initial asymmetry. Then, for each Δt, we averaged the four ratios and plotted the average P_f/P_i ratio in Fig. 5 vs. Δt; these data were fit to Eq. (3), which gave a spin-flip efficiency $\eta = 48 \pm 1\%$.

Fig. 5 suggests that increasing the ramp time even further would probably not significantly increase the spin-flip efficiency. Probably the $\Delta f = 100$ Hz frequency range did not cover the full resonance width, which limited the maximum spin-flip efficiency. We plan to increase the deuteron spin-flip efficiency by increasing the rf-dipole's strength, perhaps by increasing its input power and building a ferrite box around it.

We would like to thank the entire COSY staff for the successful operation of the Cooler Synchrotron, the injector Cyclotron, and the polarized ion source.

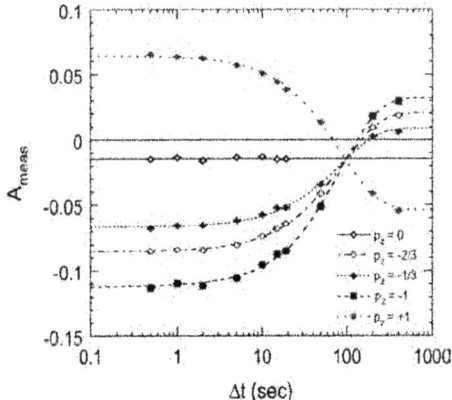
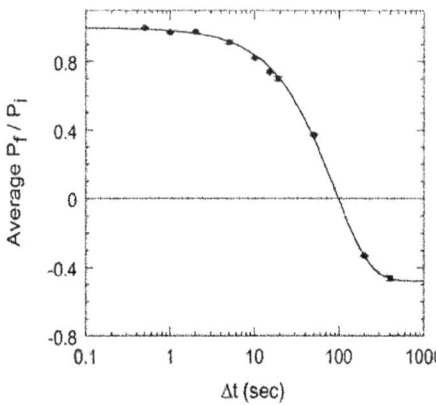

FIGURE 4. The measured deuterons' left-right scattering asymmetry for five polarization states at 1.85 GeV/c is plotted against the rf-dipole's ramp time Δt. The curves are fits using Eq. (3).

FIGURE 5. The 1.85 GeV/c deuterons' final-to-initial vector polarization ratio, averaged for the 4 polarized states, is plotted vs. the rf-dipole's ramp time Δt. The curve is a fit to Eq. (3).

REFERENCES

1. V.S. Morozov et al., "Spin flipping of a 270 MeV deuteron beam", submitted to Phys.Rev. Lett.
2. B.B. Blinov et al., *Phys. Rev. Lett.*, **88**, 014801 (2002).
3. V.S. Morozov et al., *Phys. Rev. ST-AB* **4**, 104002 (2001).
4. M. Froissart and R. Stora, *Nucl. Instrum. Meth.* **7**, 297 (1960).
5. R. Maier, *Nucl. Instrum. Meth.* **A390**, 1-8 (1997).

LENS – the Low-Energy Neutrino Spectrometer

Richard L. Hahn [+]

Chemistry Department, Brookhaven National Laboratory[*]
Upton, NY 11973-5000

Abstract. Despite the recent successes of SNO and KamLAND, the only detectors sensitive to <1 MeV-neutrinos have been the radiochemical chlorine and gallium solar neutrino experiments. LENS is being developed to measure in real time the charge-current fluxes and energy spectra of the lowest energy solar neutrinos, from the pp and ^{7}Be branches (91% and 7% of solar neutrinos). The detector medium will be a metal-loaded liquid scintillator (M-LS), with indium as the metal, which serves as the target for neutrino capture in the LS. ^{115}In has 95.71% natural abundance and a 114.2-keV neutrino-capture Q-value. Neutrino capture in ^{115}In produces a β particle and preferentially feeds an excited state of ^{115}Sn that de-excites by delayed emission of two γ rays. This triple-coincidence "tag" in time and space in principle is a very powerful tool to discriminate neutrino-capture events from backgrounds. The International LENS R&D Collaboration is working to develop an In-LS neutrino detector, with special properties, such as high indium content, high light yield, high optical transparency, long-term chemical stability, and good signal-to-background ratio. Success in preparing the In-LS has been achieved via the synthesis of In-carboxylates that are soluble in pseudocumene LS (note that these methods may also be applicable to the preparation of Gd-LS for reactor antineutrino experiments). Testing of some prototype LENS In-LS detector modules has begun recently at the Gran Sasso Laboratory.

SOLAR NEUTRINO DETECTION

To date, only the radiochemical neutrino detectors based on neutrino capture (via the Charge Current (CC) interaction) have had the capability of detecting (solar) neutrinos with energies < 1 MeV. The power of these radiochemical detectors comes from their great selectivity; <10 atoms of the radioactive products of neutrino capture are chemically separated from the ~10^{30} atoms in the neutrino detector and then purified, prior to being counted off-line in ultra-low background radiation detectors.

In the Homestake Chlorine radiochemical experiment, the CC reaction transmutes ^{37}Ar into radioactive ^{37}Cl, with a Q-value or threshold of 0.814 MeV. The detector is sensitive to solar neutrinos mainly from the ^{7}Be and ^{8}B solar branches. In the GALLEX and SAGE experiments, the CC reaction transmutes ^{71}Ga into radioactive ^{71}Ge, with a Q-value of 0.233 MeV. They are thus sensitive to a large portion of the pp solar spectrum plus the ^{7}Be, ^{8}B, and other solar branches. The main shortcoming of the radiochemical neutrino detectors is that their output is just one number, the Production Rate, which is the integral over energy (and over the exposure time) of the product of the energy-dependent neutrino flux and the energy-dependent CC cross section.

[+] For the International LENS R&D Collaboration
[*] Research sponsored by the Office of Science, Division of Nuclear Physics, of the U.S. Department Of Energy, under contract with Brookhaven Science Associates

In contrast to the radiochemical detectors, real-time neutrino detectors record each neutrino interaction as it occurs in the detector medium, so that information is obtained on the energy, direction, and time of occurrence of each event, as well as on the neutrino flux. To date, real-time neutrino detection has been successful in a few media: in H_2O in Kamiokande and Super-Kamiokande and in D_2O in SNO (these detectors have observed ^8B solar neutrinos and atmospheric neutrinos); and in liquid scintillators, e.g. in CHOOZ and KamLAND (which have observed antineutrinos from nuclear reactors). However, the operating thresholds of these detectors, at which the neutrino signal can be separated from the various backgrounds, are at best a few MeV.

Real-Time Neutrino Detectors at <1 MeV

Many creative ideas are being discussed for new real-time neutrino detectors of low-energy solar neutrinos, in particular those from the ^7Be and pp branches. These ideas have been discussed in detail in a recent series of "Low Nu" workshops [1].

One full-scale neutrino detector, Borexino [2], has actually been under construction at the Gran Sasso Laboratory for some time, with the goal of detecting ^7Be solar neutrinos in real time. The detection medium is a pure organic liquid scintillator, pseudocumene (trimethylbenzene). The neutrino interacts in the LS via Elastic Scattering (ES), which has some sensitivity to all three active neutrino flavors.

LENS, THE LOW-ENERGY NEUTRINO SPECTROMETER

Scientists from the USA, France, Germany, Italy, Russia, and Japan formed the LENS R&D Collaboration in 2000, to develop a new type of real-time low-energy neutrino detector [3]. Many of them had experience in solar neutrino experiments, such as SAGE, GALLEX, SNO, and Borexino. LENS has several important goals:

(a) To develop a new detector to measure in real time the Charge-Current (CC) fluxes and energy spectra of the lowest-energy, most abundant neutrinos (ν_e) from the pp and ^7Be solar branches.

(b) To use the pp solar flux as a "Standard Candle", for a model independent comparison with the measured LENS CC flux of pp neutrinos.

(c) To compare the results of the LENS CC pp flux measurement with the radiochemical gallium CC results from GALLEX and SAGE.

(d) To extract the ^7Be Neutral Current (NC) flux by comparing measurements of the LENS ^7Be CC flux with the Borexino ^7Be ES flux.

The LENS Concept

Liquid Scintillation (LS) counting is a well-established counting technique in nuclear physics that forms the basis of the LENS detector. Several important features of the LENS concept follow:

(i) Compared to previous LS detectors, LENS will contain a relatively high concentration of Metal (M-LS), ~8% (w/w) loading, to serve as the target for ν_e capture (the CC interaction).

(ii) Indium is the Metal favored for the M-LS (isotope ^{115}In, with 95.71% natural abundance and a half-life for β decay >10^{14} years). ^{115}In has a Q-value = 114.2 keV for ν_e-capture, way below the *pp* cutoff energy, 420 keV.

(iii) As shown in Figure 1, this ν_e capture produces a β particle and leads to an excited state in ^{115}Sn (lifetime τ = 4.7 μs) that decays by a cascade of two γ rays. This triple-coincidence "tag" of the β- and two γ-rays in time and in detector space will significantly reduce the backgrounds in the detector (even from the radioactive decay of the ^{115}In target).

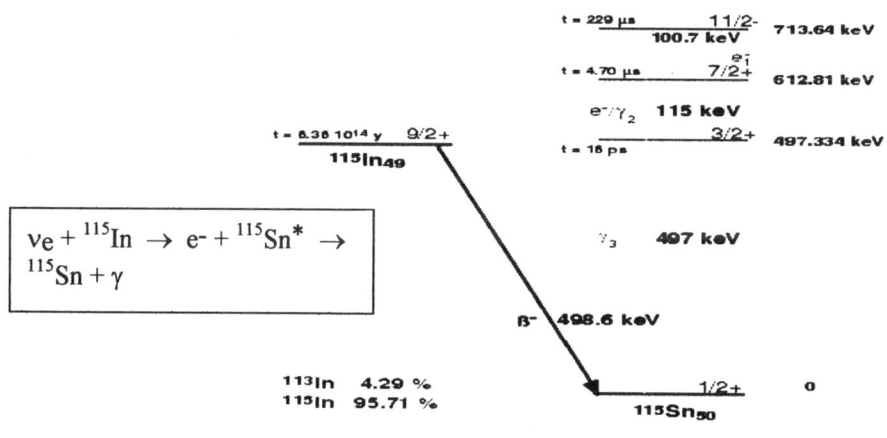

FIGURE 1. Nuclear level scheme of ^{115}In and ^{115}Sn, showing the 115-497 keV γ-ray cascade in delayed coincidence with the β particle from ν_e-capture in ^{115}In.

Monte Carlo simulations for a LENS In-LS containing 10 tons of indium in pseudocumene (PC) indicate that the energy spectrum will exhibit clear peaks for the *pp* and ^{7}Be CC interactions. With an assumed In-LS response of 300 photoelectrons per MeV, the MC predicts that one year of operation will yield 989 events detected for *pp* and 260 for ^{7}Be, with minor contributions from the ^{8}B, *pep*, and CNO, neutrinos.

Recent Progress in Preparing In-LS

The requirements for a satisfactory In-LS are straightforward in principle but not so easy to achieve in practice: (a) high indium content, on the order of several percent; (b) high light output from the In-LS, ~50% of that of the pure LS; (c) high optical transparency, i.e., an attenuation length >2 meters); (d) long-term chemical stability without formation of precipitates or gels; (e) stringent purity standards for chemical and for radioactive components; (f) reliable and reproducible preparation of tons of high-quality In-LS.

A key idea in meeting the above requirements is that certain organic complexes of indium, such as the carboxylates (and some organic phosphorus-oxygen compounds), are soluble in LS, such as in pseudocumene. Studies of a series of carboxylate complexes have led to the promising indium-methylvalerate complex, which is currently being studied with a large arsenal of chemical and physical methods at several LENS institutions, including the Brookhaven National Laboratory Chemistry Department [4]. The aim of this work is to understand the chemical processes and to optimize the In-LS product.

Recently, an extensive program of tests has begun at the Gran Sasso Laboratory of prototype LENS neutrino detector modules. The initial studies are being done with pure PC, to check electronics, data acquisition, and background levels. Then some modules will be filled with batches of In-carboxylates in PC to determine characteristics such as light yields and attenuation lengths, timing characteristics, and backgrounds.

Future Prospects

If all goes well with these prototype tests, the LENS Collaboration will consider the optimal design and associated costs of a full-scale solar neutrino LENS experiment. Such a detector will certainly contain >10 tons of indium, in >125 tons of In-LS.

It is worth noting that in addition to LENS, there are other neutrino experiments that are considering using a metal incorporated into LS. For example, some contemplated future reactor antineutrino detectors would involve loading gadolinium into the LS to enhance the detection efficiency for the product neutrons, via the $Gd(n,\gamma)$ reaction. However, the solubility of inorganic salts of Gd, such as chlorides or nitrates, in the LS is very small, and the reaction of the salts, e.g., of nitrates, with the organic liquid may lead to yellowing and decomposition of the LS. In principle, as was demonstrated in previous LENS R&D with the rare earth, Yb, the methods developed for LENS can be readily applied to Gd-LS at loading concentrations of several percent.

REFERENCES

1. See URL – http://cdfinfo.in2p3.fr/LowNu2003
2. See URL – http://borex.lngs.infn.it
3. See URL – http://lens.in2p3.fr
4. See URL – http://www.chemistry.bnl.gov/SciandTech/SN/default.htm

Future Neutrino Experiments at Nuclear Reactors

Jonathan M. Link

Columbia University, Dept. of Physics, New York, NY 10027, USA

Abstract. A next-generation neutrino oscillation experiment using reactor neutrinos could give important information on the size of the neutrino mixing angle θ_{13}. Considerations for designing a facility that optimizes experiment sensitivity are discussed, and potential experiment locations in the U.S. are summarized.

INTRODUCTION

The parameter θ_{13} is the last unmeasured neutrino mixing angle. Lately there has been a lot of interest in mounting experiments to measure θ_{13}. In particular the idea of performing the measurement using neutrinos from nuclear reactors shows great promise [1, 2, 3].

Previous reactor experiments have set a limit of $\sin^2 2\theta_{13} < 0.12$ at 90% CL at $\Delta m^2 = 2.5 \times 10^{-3}$ [4, 5]. Future experiments should attempt to improve on this sensitivity by at least an order of magnitude. Currently there are formal proposals for experiment in Russia [6] and Japan [7], and comprehensive searches for suitable reactor sites are ongoing in Europe and the U.S. (see Table 1).

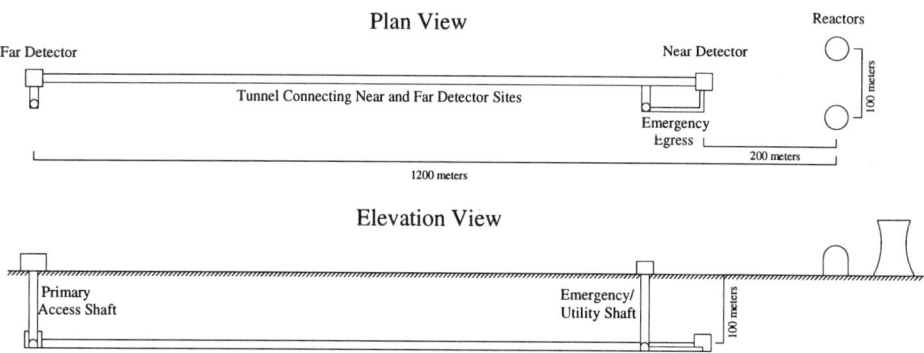

FIGURE 1. A conceptual design of the tunnel facility for a movable detector experiment at a two reactor site.

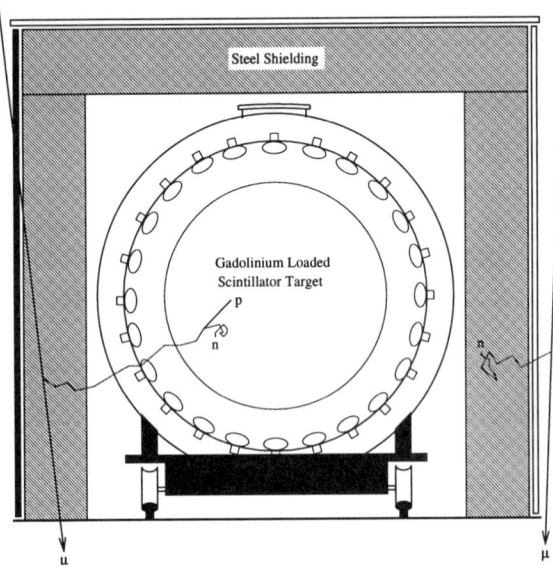

FIGURE 2. The Muon Veto Neutron Shield (MVNS) system will significantly reduce the correlated background event rate from fast neutrons by vetoing neutron producing cosmic rays in the shielding steel and ranging out fast neutrons produced outside the steel.

CONSIDERATION OF ERRORS

There is a general consensus that any future reactor experiment should use identical near and far detectors. The near detector measures neutrino flux before oscillation. Then neutrino oscillations are detected when, with sufficient confidence,

$$\frac{N_{far}L_{near}^2}{L_{far}^2 N_{near}} < 1 \qquad (1)$$

where N_{det} is the number of observed events in detector '*det*' and L_{det} is the distance from reactor core to detector. Uncertainties such as the neutrino cross section and reactor flux — limiting sources of systematic errors in the earlier single detector experiments — cancel in the far-near ratio of Eq. 1. The largest remaining systematic errors come from the uncertainty in the relative efficiency of the near and far detectors, and the uncertainty in the background rate.

To precisely determine the relative efficiency one needs a consistent source that spans the reactor neutrino energy range, and can be moved around inside the tank to simulate a sufficient set of event vertex locations. Alternatively the reactor neutrino flux itself can be used to determine the relative efficiency. By running about 10% of the time with both detectors in the large, unoscillated flux at the near detector location, a relative efficiency error that is small compared to the statistical error can be maintained. To facilitate this near calibration running the detectors must be easily movable, and, since the detectors

TABLE 1. List of the top 30 reactor sites in the U.S. by thermal power performance. The neutrino intensity is proportional to the reactor core thermal power.

Reactor Site	State	Cores	Avg MW_{th}*	Max MW_{th}
Palo Verde	AZ	3	10570	11552
South Texas Project	TX	2	6864	7600
Braidwood	IL	2	6491	7172
Vogtle	GA	2	6456	7130
Byron	IL	2	6442	7172
Browns Ferry	AL	2	6377	6916
Limerick	PA	2	6365	6916
Peach Bottom	PA	2	6290	6916
Sequoyah	TN	2	6209	6822
Oconee	SC	3	6204	7704
Susquehanna	PA	2	6161	6978
Catawba	SC	2	6116	6822
San Onofre	CA	2	6061	6876
Diablo Canyon	CA	2	6043	6749
Comanche Peak	TX	2	5986	6916
McGuire	NC	2	5880	6822
North Anna	VA	2	5129	5786
St. Lucie	FL	2	4925	5400
Edwin Hatch	GA	2	4901	5526
Arkansas Nuclear	AR	2	4844	5383
Calvert Cliffs	MD	2	4813	5400
Joseph Farley	AL	2	4801	5550
Dresden	IL	2	4779	5914
Brunswick	NC	2	4701	5116
Surry	VA	2	4664	5092
Nine Mile Point	NY	2	4500	5317
Quad Cities	IL	2	4481	5914
Indian Point	NY	2	4467	6096
La Salle	IL	2	4323	6978
Salem	DE	2	4281	6918

* Average taken over the years 1996 to 2002 [9, 10].

need to be under significant earth shielding to reduce cosmic backgrounds, the near and far detector sites should be connected by a tunnel (see Fig. 1).

The most difficult background in reactor experiments comes from cosmic muon induced fast neutrons. Fast neutrons can elastically scatter with a proton in the scintillator causing it to recoil. Following this the neutron thermalizes and captures. This sequence can mimic the two part coincidence signature of the inverse β-decay processed used by reactor neutrino experiments [8]. To address this background a Muon Veto Neutron Shield (MVNS) system can be used. The MVNS consists of a thick steel bunker surrounding the detector. This steel bunker sits inside a scintillating muon veto. Any muon that enters the system and creates a fast neutron inside the steel, is vetoed and any neutron created outside the system ranges out before entering the detector. Also, any fast neutrons that evade the system due to inefficiency or punch-through can be eliminated using the vetoed events. The proton recoil energy distribution for vetoed events can be matched to the prompt energy distribution for unvetoed events outside the reactor neutrino energy range. The veto distribution is then used to extrapolate into the signal region and the background is subtracted off.

CONCLUSION

By designing a reactor experiment that targets the limiting source of error from previous experiments it is possible to obtain sensitivities that improve on the existing limits by at least an order of magnitude.

REFERENCES

1. Shaevitz, M. H., and Link, J. M., e-print arXiv: hep-ex/0306031 (2003).
2. Huber, P., Lindner, M., Schwetz, T., and Winter, W., e-print arXiv: hep-ph/0303232 (2003).
3. Heeger, K.M., *in these proceedings* (2003).
4. Apollonio, M., et al., *Phys. Lett.*, **B420**, 397–404 (1998).
5. Boehm, F., et al., *Phys. Rev.*, **D62**, 072002 (2000).
6. Martemyanov, V., Mikaelyan, L., Sinev, V., Kopeikin, V., and Kozlov, Y., e-print arXiv: hep-ex/0211070 (2002).
7. Minakata, H., Sugiyama, H., Yasuda, O., Inoue, K., and Suekane, F., e-print arXiv: hep-ph/0211111 (2002).
8. Wang, Y. F., Miller, L., and Gratta, G., *Phys. Rev.*, **D62**, 013012 (2000).
9. *U.S. Nuclear Regulatory Commission Information Digest*, Office of the Chief Financial Officer, U.S. Goverment Printing Office (2002), volume 14.
10. *Nucleonics Week* (1997-2003), published by McGraw-Hill Companies, Inc.

Particle Identification in the PHENIX Experiment at RHIC (present and future)

Edouard Kistenev (for the PHENIX Collaboration)

Brookhaven National Laboratory, Upton, NY 11973-5000, USA

Abstract. The PHENIX[1] detector at the Relativistic Heavy Ion Collider, BNL is designed to perform a broad study of A-A, p-A and p-p collisions to investigate nuclear matter under extreme conditions. The charged particle identification is currently provided for kaons and pions to a momentum of 2.4 GeV/c and protons to 5 GeV/c. These limits would be greatly extended by the now under construction Aerogel Threshold Cherenkov detector (refractive index n=1.012).

PHENIX DETECTOR

The PHENIX detector (Fig. 1) comprises two central arms centered at zero rapidity and instrumented to detect electrons, photons and charged hadrons and two forward arms which have full azimuthal coverage and instrumented to detect muons. A number of global detectors (Zero Degree Calorimeters, Beam-Beam Counters and Multiplicity-Vertex Detector) measure the start time, vertex and multiplicity of the interactions.

In PHENIX semi-inclusive charged hadron spectra are measured in the central region over the whole kinematic range using data from a drift chambers (DC) and three segmented cathode pad chambers (PC) located outside of an axial magnetic field at a radial distances in the range of 2 to 5 m from the beam axis. About 25% of the azimuthal acceptance is covered by the time of flight system which allows k/p separation to 2.4 GeV/c and proton identification to $\sim 5 GeV/c$. When track momentum is known the electron identification is performed using Ring Imaging Cherenkov Counter (RICH) and the EMCal.

CHARGED PARTICLE IDENTIFICATION IN THE PHENIX CENTRAL ARMS

Time-of-Flight subsystem

The PHENIX ToF subsystem consists of 10 panels of ToF modules with dimensions of 38 cm x 200 cm, each panel houses 96 slats of 1.5 cm x 1.5 cm plastic scintillation counters (Bicron BC404) with 3/4" diameter PMT's (HAMAMATSU R3478S) attached at both ends[2, 3]. It is designed to have about 100 ps timing resolution in order to achieve p/K separation up to 2.4 GeV/c and K/p separation up to 5 GeV/c. Figure 2 shows a plot of time-of-flight as a function of the reciprocal momentum in minimum-bias

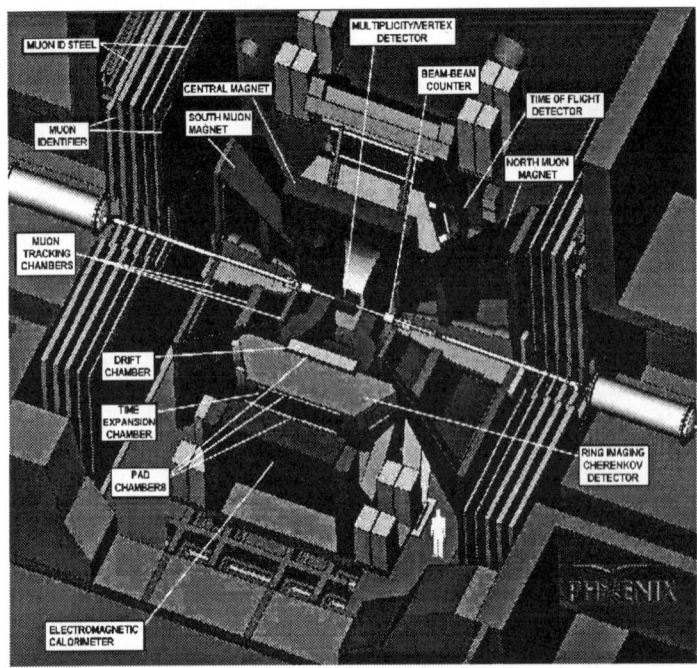

FIGURE 1. A cutaway drawing of the PHENIX detector. Labeled arrows point to the major detector subsystems.

Au-Au collisions. A resolution of $\sigma = 115 ps$ was obtained in the year-1 data analysis. Ultimately, 96 ps was achieved after further calibration, as reported in [1].

Ring Imaging Cherenkov Counter

The RICH (Fig. 3) is the primary device for electron identification[4, 5, 6] in PHENIX. It is a threshold gas Cherenkov detector with a high angular segmentation to cope with the high particle density. When an electron passes through the RICH gas volume, Cherenkov photons emitted from the electron are reflected by thin spherical mirrors and are focused on arrays of PMT's, forming a ring-shaped pattern. The direction of the electron track is then measured from the position of the Cherenkov ring. A charged particle track, measured by the tracking systemn of the central arm, is identified as an electron if its direction matches the direction measured from the ring.

The e/p discrimination capabilities of the RICH are determined by the value of the pion Cherenkov threshold, the statistical fluctuations in the number of photoelectrons in the ring and the background counting rates. The best compromise between photoelectron statistics and pion threshold is reached with ethane radiator (pion threshold of 3.5 GeV/c) which has a drawback of being flammable. An alternative radiator gas which is currently used in PHENIX RICH is CO_2 (pion threshold of 4.9 GeV/c). Using CO_2 results in 30%

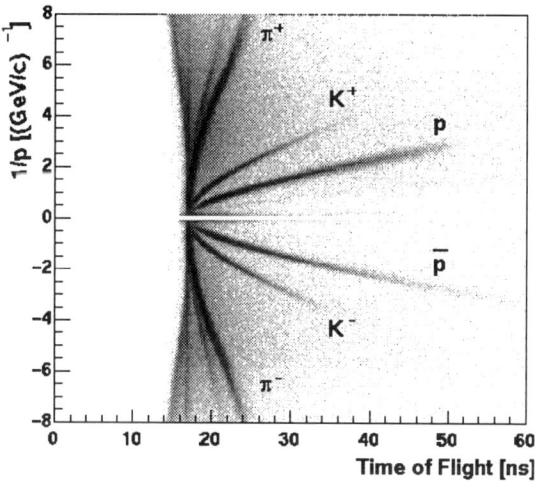

FIGURE 2. Scaled Time-of-Flight versus reciprocal momentum in minimum-bias Au+Au collisions at $\sqrt{s_{NN}} = 130$ GeV. The distribution demonstrates the particle identification capability using the TOF for the year-1 data taking period.

light loss and decreased diameter of Cherenkov ring but has only a minimal effect on the global PHENIX electron identification capabilities which rely heavily on using different detectors in conjunction.

Electromagnetic calorimeter subsystem

The EMCal system consists of a total of 24768 individual detector modules divided between the Pb-Scintillator calorimeter which provides 75% of azimuthal coverage and Pb-glass calorimeter in the rest of the acceptance. It measures energies of the impingent electrons and photons and plays a major role in particle identification and as an important part of the PHENIX trigger system.

The electromagnetic and hadronic particles produce quite different patterns of energy sharing between calorimeter towers what allows to use measurables dependent on the shower shape to differentiate between them[1]. PHENIX calorimeters also measure particle arrival times what provides an important extra tool for particle identification. In particular, timing is the only tool to reject neutral baryons and interacting antineutrons which are a major contributor to clusters with energy $\sim 2 GeV$. Slew corrected arrival times measured by exposing the calorimeter to electrons, pions and protons at 1 GeV/c momenta are shown in the top of Fig. 4. The distribution for electrons has a Gaussian shape with only a few events in the tails. The timing resolution curves (stochastic term) plotted in Fig. 4 combine the points measured by exposing the calorimeter to the particles in the 0.3-1.0 GeV/c momentum range.

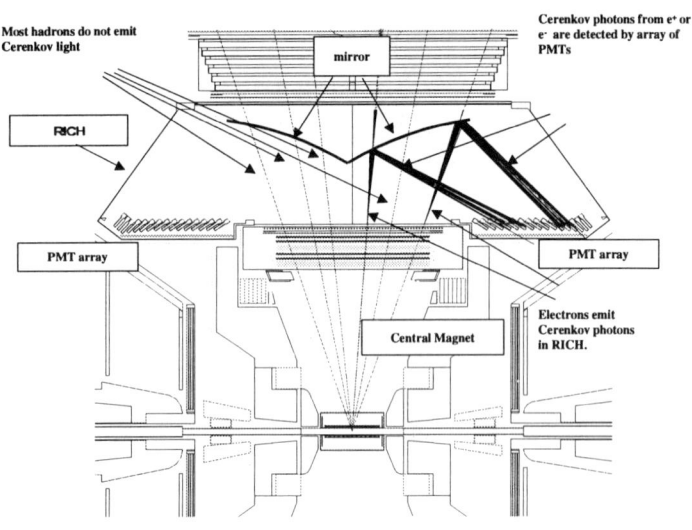

FIGURE 3. A cut-through view of PHENIX RICH detector

For energy deposits in the calorimeter $\geq 0.5 GeV$ the calorimeter timing resolution measured in the test beam is nearly constant at $\sim 120 ps$ for electrons and protons and $\sim 270 ps$ for pions where shower fluctuations are the major contributor to the measured resolution.

EXTENDED HIGH P_T HADRON IDENTIFICATION IN PHENIX

The discovery[7] of jet quenching[8] and the disappearance of back-to-back correlations[9] are the most direct evidence for the creation of dense QCD matter in nucleus-nucleus collisions. Further progress in this exiting new research program requires measuring transverse momentum spectra of identified hadron species in the transverse momentum range well above 2 GeV/c currently accessible to PHENIX. In 2001 the collaboration of PHENIX institutions began intense research and development program aimed to design, construct and install in the part of the PHENIX central Arm acceptance a new Čherenkov detector with highly segmented silica aerogel radiators. The low refractive index of the acrogel radiator makes it an ideal choice for the Cherenkov detector aimed at resolving remaining ambiguities in π/K identification

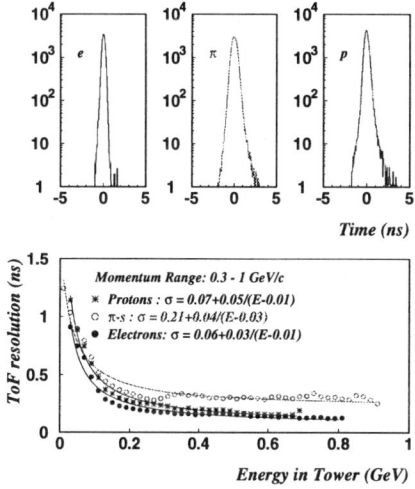

FIGURE 4. Pb-scintillator timing resolution for different particles. Top: lineshape for 1 GeV/c electrons, pions and protons. Bottom: resolution in the momentum range 0.3-1.0 GeV/c

in the few GeV/c momentum range, recent progress in manufacturing hydrophobic aerogels with improved optical quality[10] made it realistic to build and run the detector based on aerogel technology for many years without appreciable loss in the light yield.

Aerogel Detector Design

Aerogel Čerenkov detector covering the surface area of $\sim 4m^2$ will be installed into PHENIX in 2003. The detector consists of 160 identical cells of a $11 \times 22 \times 20 cm^3$ packaged with aerogel (refractive index of ~ 1.011) to the half depth. An empty space in the cell is designated an integrating box which works following the priciples similar to those of "integrating sphere" used in optical measurements. Cells are built of mylar-Goretex laminate, mylar provides for the stiffness while Goretex serves as a diffuse reflector. Aerogel cells are further loaded into 0.5 mm thick Al safety and light insulating boxes, the photomultipliers attached to the box lid (see Fig. 5).

To minimize loss of the space between boxes neighboring counters have different orientation with respect to direction towards collision point - with photomultipliers upstream and downstream of the aerogel layer. The detector topology is shown in Fig.6.

Performace of the Aerogel Detector (test beam results)

A number of prototype aerogel counters differed in the refractive indices of aerogel, configuration of aerogel cells, number and location of the photomultipliers and kind

FIGURE 5. Individual aerogel box. Lid with two photomultipliers, HV bases and LED is also shown.

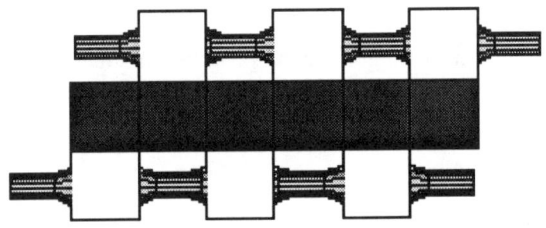

FIGURE 6. Topology of the PHENIX Aerogel Detector (view from above). Shaded - aerogel layer. PMT's viewing the neghbourring cells are upstream and downstream of the aerogel layer depending on the cell position in the layer.

of photomultipliers were tested in the test beam lines $\pi 2$ and T1 at the KEK proton synchrotron in Japan. The goal of the test beam work was to optimize the light yield and the uniformity of the light collection.

Light yield from aerogel counters

The light yield from aerogel counters was measured using cell configurations shown in the Fig.7. Block of aerogel tiles $12 \times 12 \times 12 cm^3$ in size was viewed by two 3" PMT's either in direct contact with aerogel or through the integrating box (in some cases equipped with spherical mirror). The total depth of the block (in the direction orthogonal to PMT axis) was adjustable. The refractive index of the aerogel used in this test was 1.017. The data presented below were recorded with R6233 3" HAMAMATSU

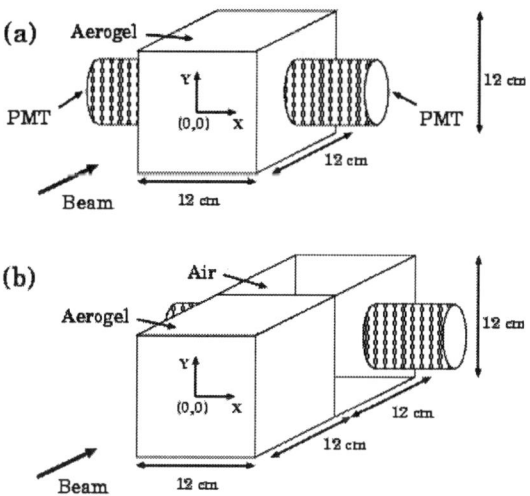

FIGURE 7. Schematic drawing of prototypes; (a)PMT's in direct contact with aerogel, (b)Light collection through integrating box.

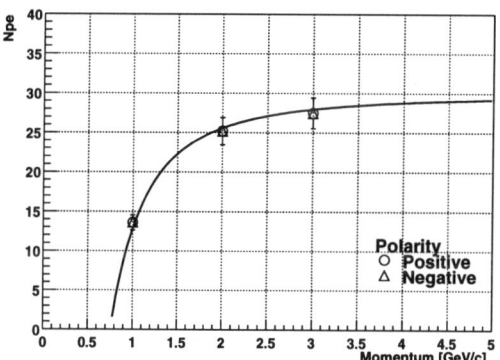

FIGURE 8. The threshold curve for pions in the n=1.017 aerogel counter. The fit as per Eq.1.

PMT's (S20 photocathode, 30% peak quantum efficiency at 420 nm). As expected(see Fig.8) the number of photoelectrons increases with the momentum, quantitatively the data agrees well with the function

$$N_{pe} \propto 1 - \frac{1}{n^2}\frac{m^2+p^2}{p^2}, \qquad (1)$$

which is characteristic of the cherenkov emission.

Light yield versus the aerogel block depth is shown in Fig.9. The PMT's were either in direct contact with aerogel at the center of aerogel block (depth direction) or the distance between PMT axis and the aerogel surface was maintained equal to halfsize of the aerogel tile (6 cm). The comparison is made for the light yield measured using single PMT (morror was installed into integrating box). When block is viewed at a center line the light yield was found proportional to the total depth of the block up to the maximum depth studied of 26 cm. When light is collected through the integrating box the saturation is reached at around 13 cm depth (light is absorbed due to Rayleich scattering). We measure only 20% difference in the maximum light yield between two configuration even if the total aerogel depth is different by factor two what is consistent with nearly 100% reflectivity of Goretex.

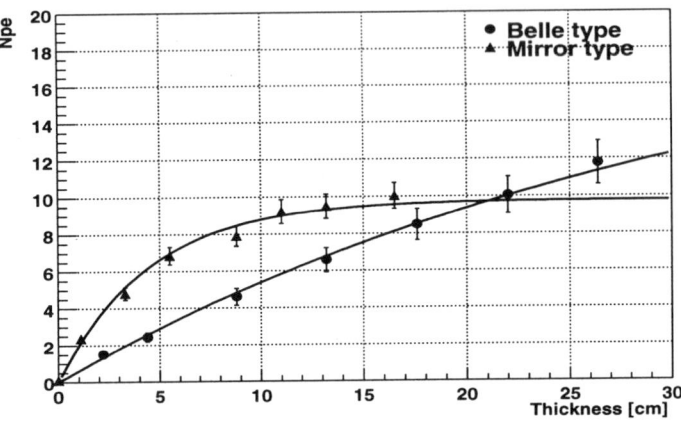

FIGURE 9. N_{pe} per PMT as a function of the aerogel block depth. Circles are for the PMT in direct contact with the block at half depth, triangles are for for mirrored integrating box. The solid curves are to guide the eyes.

The next two figures (Fig.10 and Fig.11) summarize uniformity data for the light yield in the geometry with two PMT's collecting light through the integrating box. Light yield per PMT varies for less then 30% over the whole cell surface while the total light yield stays constant to better then 15% (Fig.10). The comparison between two possible orintations (PMT's downstream or upstream) in Fig.11 shows that in the wave-length range covered by PMT's the effect of the orientation on the light yield is smaller then 30%.

The aerogel detector in PHENIX will be subjected to a very high background of secondary particles typical to heavy nuclei interractions. Only close to 5% of all charged hadrons will produce a Cherenkov light in aerogel, but a similar fraction of all charged hadrons will produce a scintillation light in the glass of the photomultiplier window. The cell design with two photomultipliers sharing in the light collection from the same cell helps to alleviate the background problem to large extent - requiring a coincidence between signals from both PMT's. An additional background rejection will be provided measuring time of the pulses from aerogel counters. Measurements made at KEK indicate that subnanosecond timing resolution can be reached triggering timing discriminator on the first photoelectron (no slewing correction). The measured dependence of the timing peak position in both PMT's on the impact point position is plotted in Fig.12.

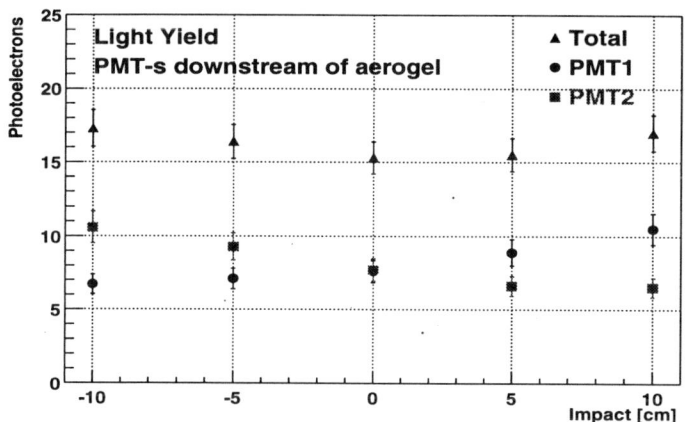

FIGURE 10. Light Yield from a single aerogel cell as a function of the impact position in the cell. Data for the cell with PMT's downstream of aerogel are shown.

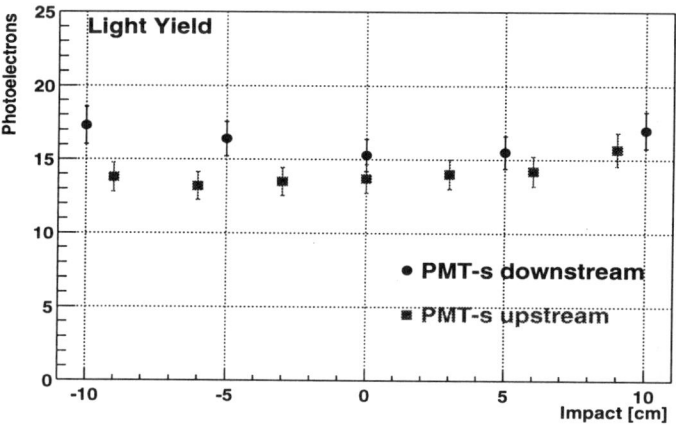

FIGURE 11. Total light yield from a single acrogcl ccll. Comparison is made between configurations with PMT's upstream and downstream with respect to particle direction.

A maximum timing difference 2 ns is observed between signals in two PMT's. If resolution can be improved - timing measurements in aerogel could be used for matching tracks in the spectrometer to hits in aerogel counters to better then counter size. On the other hand - if impact position is known, the appropriately corrected timing will be used to reject the soft electron background due to in flight decays.

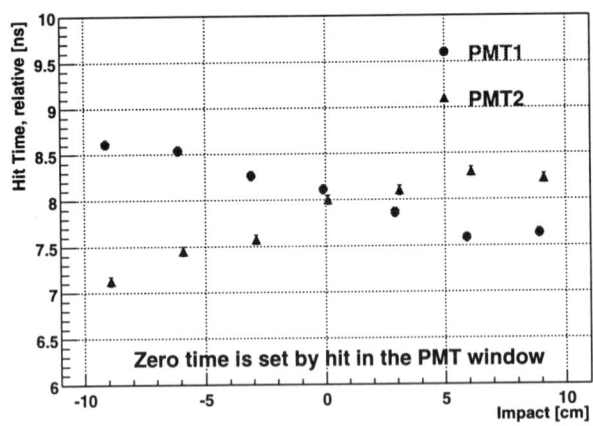

FIGURE 12. Timing difference between signals in two photomultipliers as function of the impact point position.

SUMMARY

Charged hadron identification in PHENIX is performed by combining the information from DC/PC, start counters and ToF. The time-of-flight resolution of about 100 ps allows 4σ π/K separation up to $2.4 GeV/c$. Upgrading the current configuration with high granularity Aerogel detector provides for the missing capability to separate π from kaons in the momentum range between 2 GeV/c and 4 GeV/c. To fully realize the potentials of this new detector PHENIX also plans to install in the same acceptance a Time-of-Flight detector with performance similar to that of described above. When project is completed the expected reach of PHENIX in identification of charged hadrons becomes complete for the whole range of kinematically accessible secondary momenta from magnetic field cutoff of $\sim 0.3 GeV/c$ to nearly 17GeV/c.

REFERENCES

1. K.Adcox et al., NIM A499 (2003) 469
2. Y. Miake, American Inst. Phys. 340 (1995) 78-89
3. L.Carlen et al., NIM A431 (1999) 123-133
4. Y.Akiba et al., NIM A433 (1999) 143-148
5. Y.Akiba et al., NIM A453 (2000) 279-283
6. T.Sakaguchi et al., NIM A453 (2000) 382-385
7. K.Adcox, et al., Phys.Rev.Lett. 88 (2002) 022301
   C.Adler, et al., Phys.Rev.Lett. 89 (2002) 202301
8. P.Levai, G.Papp, G.Fai, M.Gyulassy, G.G.Barnafoldi, I.Vitev, Y.Zhang, Nucl.Phys. A698 (2002) 631
9. C.Adler, et al., nucl-ex/0206006
10. Yokogawa, Yokogawa, J.Non-Cryst. Solids 186 (1995) 23

Silicon Vertex Detector Upgrade for the PHENIX Experiment at RHIC

Y. Akiba for the PHENIX Collaboration

RIKEN (Institute of Physical and Chemical Research), Wako, Saitama 351-0198, Japan

Abstract. We propose to build a Silicon Vertex Detector for PHENIX experiment at RHIC. The proposed detector covers $|\eta| < 2.7$ in pseudo-rapidity. With this upgrade, PHENIX will be able to improve its ability to measure heavy flavor production in $p+p$, $p+A$, and $A+A$ collisions, and will be able to measure gluon distribution $G(x)$ in much wider x range.

INTRODUCTION

We propose to construct a Silicon Vertex Detector (SVTX) for PHENIX experiment[1] at RHIC. The main physic motivation of the SVTX upgrade is measurement of heavy flavor production (charm and beauty) in $p+p$, $p+A$, and $A+A$ collisions. Since D and B mesons have finite decay length of order of 100 μm (for D) to 500 μm (for B), one can identify heavy flavor production by measuring displaced vertexes. With the present PHENIX detector, we have measured single electron from charm decay in Au+Au collision at RHIC[2]. The proposed detector extends our ability to measure heavy quark production in wider kinematic range. In addition, the large solid angle coverage of the proposed SVTX allows us to measure recoil jets in direct photon production in $p+p$ collisions. From the gamma-jets coincidence measurement, we can determine the gluon distribution in proton.

PROPOSED DETECTOR

The proposed detector is schematically shown in Figure 1. It is placed in the inner most part of the central magnet, and it is surrounded by the TPC/HBD detector[3], another proposed upgrade detector in PHENIX, as shown in the right part of the figure. The SVTX consists of the barrel part and the end-cap part. The barrel part covers $|\eta| < 1.2$ and almost 2 π in azimuth. Its acceptance extends much larger than that of the PHENIX central arms (two arms of $|\eta| < 0.35$ and $\Delta\phi = \pi/2$). It consists of one layer of Silicon Pixel detector at R = 2.5 cm from the beam line, and three concentric layers of Silicon Strip detector at R = 6, 8, and 10 cm. The pixel detector has a very small pixel size of 50 μm ×425 μm so that the occupancy is less than 1 % even in the central Au+Au collisions. The strip size is approximately 80 μm ×3 cm, and each layer has simultaneous x-u read-out. The thickness of each layer is kept to about 1 % raidation length to minimize the multiple scattering and photon conversion. The detector

Silicon Vertex Tracker in PHENIX

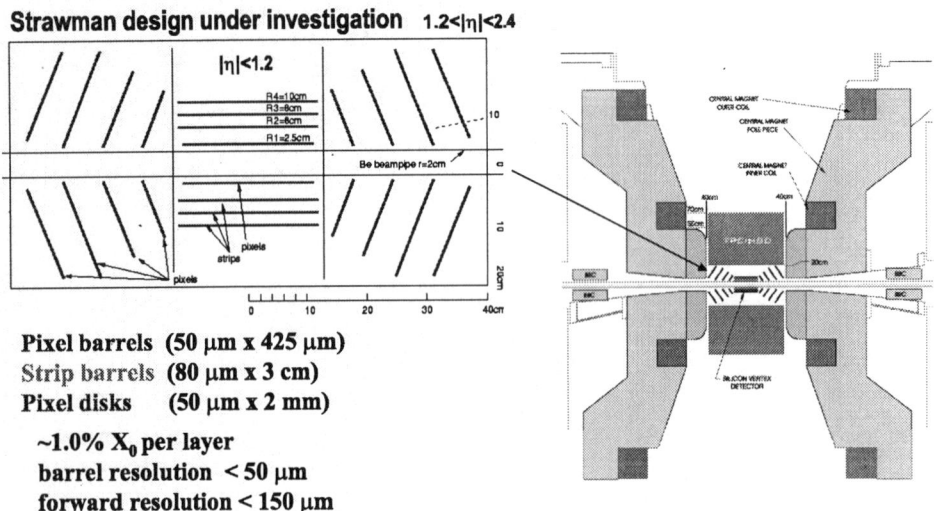

FIGURE 1. The straw-man model of the SVTX detector for PHENIX. The right panel shows its place in the PHENIX detector. The left part shows the schematic configuration of the SVTX.

has single track DCA (distance to the closest approach) resolution of about 50 μm. The end-cap part covers $1.2 < |\eta| < 2.7$, which matches the acceptance of the PHENIX muon arms ($1.2 < |\eta| < 2.4$). Each of the two end-cap detectors consists of four layers of mini strip detectors with pixel size of 50 μm $\times 2$ mm. The detector has single track DCA of about 150 μm

PHYSICS GOALS

The heavy flavor production will be studies in many channels by combining the measurement of the present PHENIX detector and the SVTX. The present PHENIX detector measure electrons in central rapidity and muons in forward rapidity. Combined with a high resolution DCA measurement of these leptonic tracks in the barrel and the end-cap SVTX, we can clearly identify single leptons from heavy flavor decays. This is illustrated by Figure 2. The figure shows that charm and beauty is the dominant source of electrons with the DCA larger than 200 μm in low p_T^e (left panel) and in the high p_T^e (right panel), respectively. In the present PHENIX, it is difficult to measure beauty production separately from charm. With the SVTX, B meson decay $B \to J/\psi + X$ is identified as $J/\psi \to \mu^+\mu^-$ and $J/\psi \to e^+e^-$ displaced from the collision vertex. At high p_T, charm meson decays $D \to K\pi, K\pi\pi$ from displaced vertexes can be directly reconstructed. The kinematic region that PHENIX can measure heavy flavor production

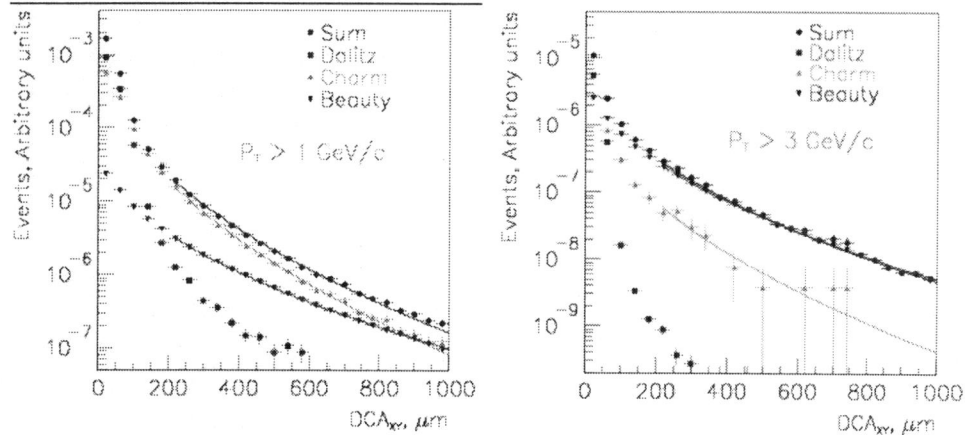

FIGURE 2. The DCA distribution of electrons from the various sources.

will be significantly extended with the SVTX.

There are many physics issues that can be addressed by measurements using the SVTX in $A+A$, polarized $p+p$, and $p+A$ collisions.

In heavy ion collisions, we can probe the early, highest energy density phase of the collision by measurement of the heavy flavor production. With the SVTX, PHENIX can measure charm production in heavy ion collision with improved accuracy, and PHENIX can measure beauty production separately from charm. One of the the important question that can be addressed is energy loss of heavy quark in the dense medium. Colored high momentum partons are predicted to lose energy as they propagate through the dense medium with medium induced gluon radiation[4]. The strong suppression of high p_T particle production observed in the central Au+Au collision at RHIC[5] is interpreted as due to this energy loss effect. The heavy quarks are predicted to lose less energy in the medium due to "dead cone effect"[6], and PHENIX measurement of single electron suggests that charm quark with moderate p_T indeed lose little or no energy[2]. Measurement of charm and beauty production in a very wide p_T range is needed to study energy loss of heavy quarks in the dense medium. Charm measurement with the present PHENIX is limited in $p_T^e < 4$ GeV/c since in the higher p_T^e the contribution from beauty decay becomes comparable or larger than that from charm and we can not distinguish electron from charm and beauty with the present PHENIX. With the SVTX, we can directly reconstruct D mesons, and therefore we can separately measure charm and beauty production in higher p_T region.

In polarized $p+p$ collisions, we can measure gluon polarization distribution $\Delta G(x)$ in a wide x range. The figure 3 shows the Bjorken x coverage of polarized gluon distribution that can be measured by the present PHENIX detector and with the SVTX upgrade. The x region extended by the SVTX is shown in the dashed line. The gluon distribution is determined by various physics processes including direct photon production and photon+jet measurement ($gq \to \gamma q$), charm production ($gg \to c\bar{c}$), beauty production ($gg \to b\bar{b}$). As shown in the figure, the x coverage of the measurement is extended much wider region with the SVTX.

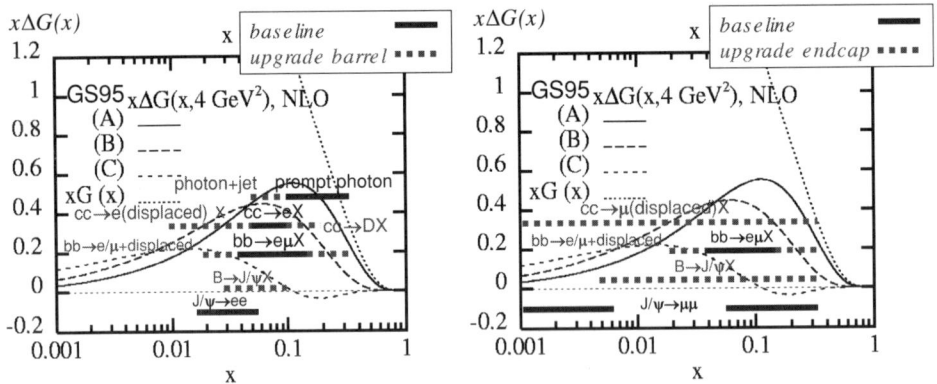

FIGURE 3. The Bjorken x coverage of gluon distribution with the present PHENIX experiment (solid line) and with the SVTX upgrade (dashed line) in various physics processes. The left plot shows the x coverage with central and barrel SVTX, and the right plot shows the x coverage with the muon arm and the end-cap.

In $p+A$ and $d+A$ collisions, nuclear shadowing of the gluon distribution $G(x)$ can be extensively studied. With the SVTX detector, the x range that we can measure G(x) through heavy flavor production is extended into very small x region ($x < 10^{-2}$), as shown in the right panel of Figure 3. This is the region where a strong nuclear shadowing is expected.

REFERENCES

1. E. O'Brien for the PHENIX Collaboration, these proceedings.
2. K. Adcox et al., Phys. Rev. Lett. 88, 19203 (2002); R. Averbeck for the PHENIX Collaboration, Nucl. Phys. A715, 695c (2003).
3. C. Aidala for PHENIX collaboration, these proceedings.
4. R. Baier, D. Schiff, B. Zakharov, Ann. Rev. Nucl. Sci 50, 37 (2000).
5. K. Adcox et al., Phys. Rev. Lett. 88, 022301 (2002); S. Adler et al., Phys. Rev. Lett. 91, 072301 (2003).
6. Y. L. Dokshitzer and D. E. Kharzeev, Phys. Lett. B519, 199 (2001).

A Large Tracking Detector In Vacuum Consisting Of Self-Supporting Straw Tubes

P. WINTZ for the COSY-TOF Collaboration

Institute for Nuclear Physics, Research Center Juelich, 52425 Juelich, Germany
p.wintz@fz-juelich.de

Abstract. A novel technique to stretch the anode wire simply by the gas over-pressure inside straw drift tubes reduces the necessary straw weight to an absolute minimum. Our detector will consist of more than 3000 straws filling up a cylindrical tracking volume of $1m$ diameter and $30cm$ length. The projected spatial resolution is $200 \mu m$. The detector with a total mass of less than $15kg$ will be operated in vacuum, but will have an added wall thickness of 3mm mylar, only. The detector design, production experience and first results will be discussed.

INTRODUCTION

The TOF-spectrometer at the Cooler Synchrotron accelerator COSY[1] [1] is a large 3m diameter, up to 8m long modular[2] barrel to measure time-of-flight and momentum vector of charged particles produced in nucleon-nucleon interactions. To avoid multiple scattering and secondary background production the barrel is evacuated to a vacuum pressure of a few $10^{-3} mbar$. Using the polarised proton or deuteron beam with momenta up to $3.5 GeV/c$ impinging on a tiny (few mm^3) target cell filled with liquid hydrogen or deuterium, fundamental nucleon-nucleon, nucleon-meson, and meson-meson interactions are studied [2]-[4]. In the center-of-mass system an almost complete phase space coverage of the detector allows to analyse dalitz plots of the reactions, e.g. to see a presence of N^* resonances. Total and differential cross-sections and various spin observables are measured. In particular, hyperon production, e.g. $pp \rightarrow pK^+\Lambda$ and the delayed decay $\Lambda \rightarrow p\pi^-$ can be studied exclusively at COSY-TOF.

STRAW TUBE DETECTOR

The detector will consist of more than 3000 straw tubes close-packed in 15 planar, vertical double-layers at three different φ-angles ($0°$, $\pm 120°$) for a 3-dimensional track reconstruction (fig. 1) close after the TOF-target. The overlapping detector volume has a cylindrical shape with $1m$ outer diameter and $30cm$ length. An inner diameter of $15mm$ avoids direct interactions of the beam halo. The tracking depth of at least $30cm$

---

[1] Address: D-52425 Juelich, Germany
[2] Depending on the reaction under study a barrel length of $1m$ up to $8m$ can be chosen.

FIGURE 1. Two ($1x1m^2$) straw double-layers each held by a lightweight rohacell carbon-fiber compound frame. The front frame is rotated by 120° relative to the back frame. The final detector will consist of a stack of 15 such frames alternating at the three φ-angles 0° and $\pm 120°$.

in all directions and the projected spatial resolution of $\sigma \simeq 200\mu m$ allows a precise reconstruction of the Λ decay ($c\tau \simeq 8cm$) and polarisation by its p and π^- decay tracks. The target interaction point will be resolved with sub-mm precision and the tracking resolution will be generally improved, in particular, if the shortest (1m) version of the TOF-spectrometer is used for studying multi-pion production in pp-scattering.

The straws have a length of 1050mm, 10mm outer diameter and a wall thickness of 30μm. The tubes are made of two spirally wound mylar strip films glued together and the inner film layer's inside being aluminised ($\simeq 0.2\mu m$) to be used as cathode. A gold plated tungsten-rhenium wire with 20μm diameter is used as anode. Cylindrical precision end plugs made from ABS[3] close the tube at both ends (fig. 2). They are glued to the mylar film leaving a small 1.5 mm film overlap on both ends. There, a gold-plated copper-beryllium spring wire is inserted to provide the electric cathode contacting (fig. 2). The springs allow 2mm elongation with a typical spring force equivalent to 10$gram$. The end plugs have a central hole with a 3mm thick cylindrical nose to insert and glue a crimp pin for the wire. A micro PVC tube is fed through another hole and glued in the end plugs to provide a gas flow through the tube.

The anode wire is stretched by a weight of 40 gram and crimped in the copper pins (0.1mm hole) at a gas over-pressure in the straw tube of 1.25$mbar$. At this wire tension the calculated gravitational sag is below 20μm, well below the projected spatial resolution of 200μm.

A double-layer contains 2×104 straws glued together and the second layer being staggered by one straw radius. Then, for an orthogonal track, lower detection efficiency close to the tube wall is always combined with high efficiency in the straw center in the next, staggered layer. Also, the tracks left/right ambiguity from the wire can be disentangled in the next layer.

The granularity (straw diameter) of 10mm allows a continuous tracking with up to 30

---

[3] Acrylonitrile-Butadiene-Styrene, density $\rho = 1.05 g/cm^3$, end plugs with 0.5mm wall thickness.

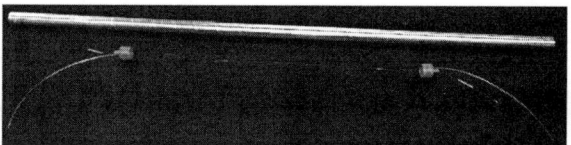

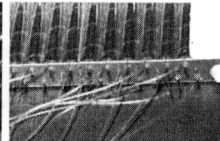

FIGURE 2. Left: A straw tube consisting of mylar film tube, end plugs, crimp pins to fix the anode wire and gas tubes to provide a gas flow through the tube. Center: Springs inserted in the small film overlap at the tube ends provide the electric cathode contacting. Right: The springs contact the copper coated side of a thin belt which also aligns the straw 2-layer.

hits per track, important to resolve complex track patterns. In the active detector part the added straw wall thickness of $3mm$ (mean) mylar for orthogonal tracks corresponds to 1% radiation lengths, only. This high transparency, i.e. less shadowing on the following scintillator hodoscopes allows a clean, background-free tracking close to the target.

The novel method to provide the wire tension by the gas over-pressure inside the straw tubes and a dedicated gluing method allows to build self-supporting double-layer packages. Therefore, a light-weight rectangular frame is sufficient to hold the straw layers. No heavy end or support structures are needed to maintain the added wire tension equivalent[4] to more than $120kg$. The frame bars are made of rohacell-foam ($\rho = 0.05g/cm^3$) reinforced by $0.3mm$ thick carbon fiber tape glued onto both sides. One double-layer including frame, front-end preamplifiers, gas and high voltage supply has a weight slightly below 1 kg. Thus, the final detector will have a weight of 15 kg.

Performance Tests

The measured tension for the first produced 1000 straws showed a very low spread (RMS) of $0.6g$ around the used stretching weight of $40gram$, mainly caused by temperature changes during production and tension measurement. The mylar film temperature coefficient ($\alpha \simeq 5 \times 10^{-5} K^{-1}$) amounts to a tube elongation of $\Delta L = 50\mu m$ per $\Delta T = 1K$ which corresponds to a change in the stretching weight of $0.5gram$.

The dependence of the gas over-pressure on the tension was also checked. A constant rise of $\simeq 10gram/1bar$ was measured for several straws up to $1.3bar$. At ZERO over-pressure a well-defined wire tension of $30gram$ was observed, maintained by the stiffness of the mylar tube alone. This allows to operate the straws even at a very low gas over-pressure, if needed.

With raising pressure a constant elongation of $\simeq 1.0mm/1bar$ was measured up to $4bar$ over-pressure. No damage of the tubes happened and no hysteresis was observed. But, during elongation a characteristic twisting angle of the mylar tubes of $\simeq 9°/1bar$ was seen. This up to now unknown phenomenon[5] of the mylar film tubes required a

---

[4] Needed weight to stretch all wires ($3000 \times 40gram$).

[5] We assume a difference in tension between the two mylar strips during winding of the tube showing up during elongation. No explanation from the manufacturing company about this effect.

modified gluing method of the double-layer packages.

A couple of gas-filled straws was put in a vacuum volume, pumps switched off and the vacuum pressure rise due to the straw leakage was recorded. Measurements with different gases (argon and CO_2), different gas over-pressure and different number of straws showed the characteristic dependence of gas leakage caused by permeation through the straws mylar film. The measured permeabilities for argon and CO_2 agreed with the numbers given by the manufacturer of the mylar film (DuPont Corporation). No real pore leaks or bad gluing were observed. Taking the existing pumping power the calculated vacuum end pressure worsen by the straws leakage in the TOF barrel will be still about a few $10^{-3} mbar$, low enough for the measurements.

The straws were operated with a premixed $Ar/CO_2 (82/18\%)$ gas mixture. To replace the gas loss of 3000 straws in the TOF vacuum an exchange of the gas bottle ($200 bar \times 20l$) every 200 days will be sufficient. But, the characteristic permeability difference of a factor 10 between argon and CO_2 will cause a de-mixing of the used gas, about -2% CO_2 per day. Therefore, the gas flow must be high enough to compensate this loss.

A vertical straw layer was illuminated by vertical cosmic tracks selected by an appropriate trigger scintillator setup. Relative and absolute efficiency checks were done by comparing the entry numbers in the straws recorded TDC spectra (variation within $\pm 2\%$, only) and by comparing them with the scintillator triggers, normalised to the active straw area (agreement better than 5%). The radial distance-drift time relation $R(t)$ was derived from the entry numbers in the recorded TDC-spectra bins. The obtained dependence was in accordance with simulation (GARFIELD) and the $R(t)$ spread of the individual straws was below $100 \mu m$.

During five days 50 straws were illuminated with a deuteron beam of intensity $\simeq 8 \times 10^6 s^{-1}$. No high voltage breakdowns or dark currents occurred. A clean ionisation beam current of $\simeq 80 \mu A$ was measured, corresponding to a gas gain of $5\text{-}10 \times 10^4$ and a total collected charge per straw of $1C$ per $2cm$ wire. The measured (discriminated) particle flux up to $6 \times 10^6 s^{-1}$ seen by each straw is more than factor of 100 higher than the expected particle rate during the TOF measurements, later.

CONCLUSIONS

Self-supporting straw double-layers reduce the overall detector weight to an absolute minimum, important for a clean and background-free tracking. The mechanical stability of this large, but thin-wall detector in vacuum is provided by the gas over-pressure inside the straw tubes. First measurements of efficiency and spatial resolution and with high particle rates were successful.

REFERENCES

1. R. Maier, Nucl. Instrum. Methods A **390**, 1997 (1).
2. S. Abd El-Samad et al., Phys. Lett. B **522**, 2001 (16).
3. M. Abdel-Bary et al., Eur. Phys. J. A **16**, 2003 (127).
4. W.K. Eyrich for the COSY-TOF Collaboration, Prog. Part. Nucl. Phys. **50**, 2003 (547).

Proposed Detector Upgrade for Measuring Low-Mass Lepton Pairs in PHENIX

Christine Aidala for the PHENIX Collaboration

Columbia University, New York, NY 10027

Abstract. The measurement of low-mass lepton pairs provides a unique opportunity to study in-medium effects of vector meson production and the possibility of chiral symmetry restoration in heavy ion collisions.However, this measurement is extremely difficult experimentally due to the large combinatorial background produced by Dalitz decays and conversions.The PHENIX collaboration has proposed constructing a new detector as part of its future upgrade program that would provide a high level of rejection against the Dalitz decay and conversion backgrounds, while at the same time preserving good efficiency for measuring vector mesons and low-mass electron pairs.This device would consist of a hadron-blind detector that would identify electrons by detecting Cherenkov light produced in a gas radiator, and could also incorporate the features of a Time Projection Chamber that would serve as a tracking detector to track low-momentum particles in a low-magnetic-field region produced in the central PHENIX spectrometer.The strategy for identifying and rejecting Dalitz pairs and conversions using this technique is discussed, along with some of the ongoing R&D efforts.

INTRODUCTION

In high energy heavy ion collisions, low-mass lepton pairs provide a clean signal for studying chiral symmetry restoration and in-medium effects on low-mass vector mesons, thermal radiation from the hadron gas, and strangeness production via the leptonic decay of the ϕ. We believe that the physics program at the Relativistic Heavy Ion Collider (RHIC) at Brookhaven National Laboratory (BNL) would be incomplete without a comprehensive measurement of low-mass electron pairs. At RHIC, the PHENIX experiment is unique in that it is the only one which has the capability of measuring the entire dilepton spectrum from the π^0 Dalitz decays up to invariant masses beyond that of the J/Ψ.

In the high-multiplicity environment of heavy ion collisions, electron pairs from Dalitz decays and photon conversions contribute to a formidable combinatorial background in the electron pair spectrum. The ability to identify and reject these pairs is essential to carrying out a measurement of this spectrum. An improvement in the signal-to-background ratio of at least two orders of magnitude is required, along with an electron detection efficiency of greater than 90%. A simulated combinatorial dilepton spectrum for central gold-gold collisions at $\sqrt{s_{NN}} = 200 GeV$ including electrons from conversions in the inner PHENIX detectors, charm decays, and meson decays is shown in Figure 1.

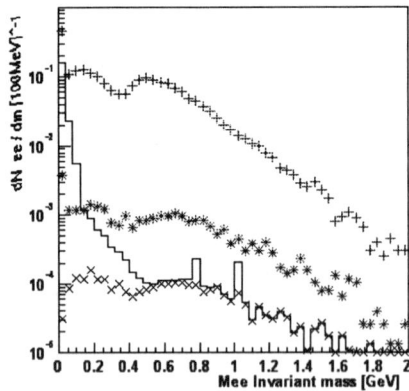

FIGURE 1. Simulated dilepton spectrum in PHENIX for 200 GeV central Au-Au collisions. The solid line represents the total of all correlated pairs; plus signs represent the total combinatorial background, including conversions, charm decays, and meson decays; asterisks represent the combinatorial background from charm decays alone; crosses represent the correlated charm signal.

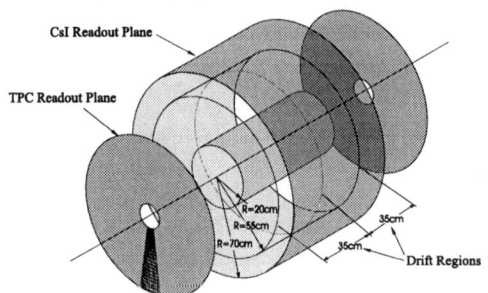

FIGURE 2. A schematic diagram of the proposed TPC/HBD.

PROPOSAL

To identify and reject Dalitz and conversion pairs, a hadron-blind detector (HBD), potentially combined with a fast, compact time projection chamber (TPC), has been proposed as an upgrade to the current PHENIX detector. A schematic diagram of the proposed TPC/HBD is shown in Figure 2. Information on the current PHENIX detector can be found in [1].

An inner magnet coil has been installed to create a low-field region in the central spectrometer, which will allow the measurement of low momentum electrons ($p < 200$ MeV/c) that would otherwise curl up and be lost in the normal magnetic field configuration. This lower magnetic field would also serve to better preserve pair opening angle, which would be important in the case where HBD measurements were available without any tracking information. Higher momentum electrons which pass through the stronger, outer magnetic field and reach the central spectrometer arms, such as those

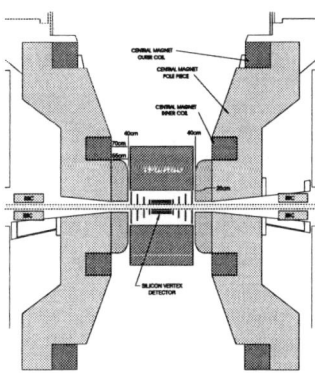

FIGURE 3. Inner region of the central spectrometer.

from the decay of low-mass vector mesons, can be identified and measured quite well via the full complement of PHENIX detector subsystems.

In Figure 3, the inner region of the central spectrometer is shown, with the inner and outer magnet coils, the TPC/HBD, and the proposed silicon tracker and vertex detector. Further information on the PHENIX silicon upgrade can be found in [2]. In this inner region, electrons would be tracked and identified by the TPC/HBD. The TPC would also provide full tracking coverage over 2π in azimuth and $|\eta| < 1.0$, while the current spectrometer arms cover π in azimuth and $|\eta| < 0.35$. Electron identification would be accomplished via detection of Cherenkov radiation by the HBD and dE/dx in the TPC. The TPC drift gas volume would double as a radiator for the HBD, and the HBD would detect the Cherenkov light from electrons using a large area photocathode such as CsI on its outer surface. The HBD would respond minimally to other particles, thus the term "hadron blind." The TPC would have a central, high-voltage plane and would drift charge axially toward readout planes on each end of the detector, with a drift distance of approximately 35 cm. As a readout device, both the TPC and HBD would use a micropattern detector such as a gas electron multiplier (GEM). A single fast, UV-transparent gas such as CF_4 could potentially be used as the TPC drift gas, the HBD radiator, and for operation of the micropattern readout detector.

With 2π azimuthal coverage, the HBD or HBD/TPC would see both partners for a significant fraction of the electron pairs produced. A cut on pair opening angle and/or invariant mass would be used to select out and reject pairs from Dalitz decays and conversions while preserving the majority of signal electrons. In this way the combinatorial background for the dilepton spectrum would be greatly reduced.

Aside from providing rejection against the Dalitz and conversion backgrounds, the TPC would enhance the capabilities of the PHENIX detector in other ways. Currently, PHENIX has no tracking within the magnetic field, and as such, decay and conversion backgrounds limit the high-p_T charged particle measurements. 2π tracking within the magnetic field by the TPC would eliminate much of this background. In addition, the TPC alone can provide a good momentum measurement over a large solid angle, thus permitting jet measurements in both heavy ion and proton-proton collisions. The TPC would also help to identify displaced vertices from charm production in conjunction with the silicon tracker and vertex detector upgrade as well as provide additional particle identification from dE/dx.

RESEARCH AND DEVELOPMENT

Research and development is ongoing, primarily at Brookhaven National Laboratory and the Weizmann Institute of Science. Some of this work is being performed jointly with the STAR Collaboration at RHIC, and related R&D is being performed by the LEGS group at BNL and for the NLC/TESLA.

Current R&D includes prototype hardware development, gas property studies, investigaton of GEMs as readout detectors, electronics development, and simulation. More detailed information on the HBD proposal and recent R&D can be found in [3] and [4].

SUMMARY

A hadron-blind detector combined with a time projection chamber has been proposed as an upgrade to the current PHENIX detector. Research and development is currently underway in a variety of areas. The proposed detector upgrade would allow PHENIX to identify and reject Dalitz and conversion pairs, thus reducing the combinatorial background in the dilepton spectrum and greatly improving the measurement of low-mass vector mesons as well as the low-mass dilepton continuum. In addition, in both heavy ion and proton-proton collisions, the 2π tracking within the magnetic field that would be provided by the TPC would improve charged particle measurements, allow the study of jets, and in conjunction with the proposed silicon upgrade, help to identify displaced vertices from charm production.

ACKNOWLEDGMENTS

The PHENIX Collaboration acknowledges support from the DOE and NSF (U.S.), MEXT and JSPS (Japan), CNPq and FAPESP (Brazil), NSFC (China), IN2P3/CNRS and CEA (France), BMBF, DAAD, and AvH (Germany), OTKA (Hungary), DAE and DST (India), ISF (Israel), KRF and CHEP (Korea), RMIST, RAS, and RMAE (Russia), VR and KAW (Sweden), U.S. CRDF for the FSU, US-Hungarian NSF-OTKA-MTA, and US-Israel BSF.

REFERENCES

1. E. J. O'Brien, these Proceedings.
2. Y. Akiba, these Proceedings.
3. C. Aidala et al., Nucl. Inst. Meth. A502, (2003) 200.
4. A. Kozlov et al., physics/0307101.

Electron Beam Polarimetry for EIC/eRHIC

W. Lorenzon

Randall Laboratory of Physics, University of Michigan, Ann Arbor, Michigan 48109-1120, USA

Abstract. Ideas are being presented about how to measure the electron beam polarization at the Electron Ion Collider (EIC) at RHIC with high precision. Based on lessons learned at the HERA storage ring it is discussed which design and physics processes might be most appropriate for EIC. This is a summary of the ideas and concepts discussed at the first workshop on EIC electron beam polarimetry at BNL in November 2002.

INTRODUCTION

The current design of the EIC project at RHIC foresees collisions of 5-10 GeV longitudinally polarized electrons on 25-250 GeV protons or 100 GeV/u heavy ions (such as gold). It is anticipated that the electron beam polarization is about 70% and that it needs to be measured with high precision ($\approx 1\%$ syst.). Unfortunately, a polarized electron bunch has no macroscopic properties that could be useful for measuring its polarization, as was pointed out by Ref. [1]. It is argued that a polarized electron bunch represent a very weak magnetic dipole which has a strength that is roughly seven orders of magnitude less than a piece of magnetized iron of comparable size. Therefore, one is inevitably lead to consider microscopic processes, i.e. spin-dependent scattering processes. The simplest such processes are the elastic processes which have three very useful properties: a) the cross sections for elastic scattering are usually large, b) elastic scattering processes have simple kinematical properties, and c) the physics or elastic electron (positron) scattering are quite well understood.

There are currently three different targets used to measure the polarization of electron (positron) beams: nuclei, electrons, and photons. Mott scattering, or e^- − nucleus scattering is mainly used at low energies (100 − 300 keV) to measure the polarization of electrons from polarized sources. Møller (Bhabha) scattering, or $e^-(e^+)$− electron scattering is widely used for polarized beams in the 100 MeV to many GeV energy range. Unfortunately, it is destructive to the beam and therefore not suitable for storage rings. In contrast, Compton scattering, or $e^\pm$− photon scattering, which is suitable for energies above 1 GeV, and ideal for energies above 10 GeV, is not destructive to beams in storage rings and is therefore the only choice to date for high energy storage rings, with the exception of a new idea discussed in Ref. [2].

There are many polarimeters that are, or have been, in use at various laboratories. Here, we describe a polarimeter that has taught as several lessons for how to design a polarimeter for EIC. It is the Longitudinal Polarimeter at the HERA accelerator at DESY [3].

THE LONGITUDINAL POLARIMETER AT HERA

At the HERA storage ring, the electron (positron) beam can become transversely polarized through the emission of synchrotron radiation (Sokolov-Ternov effect) with average asymptotic polarization values of 55% and typical rise times of 22 minutes. The Longitudinal Polarimeter is the second polarization monitor at HERA that is based on Compton back-scattering of circularly polarized photons from an laser beam [3]. In contrast to the Transverse Polarimeter, which measures the transverse polarization of the electron beam in the HERA West section [4], it provides a measurement of the longitudinal beam polarization in the East section of HERA between the spin rotators at the HERMES experiment, as shown in Fig. 1.

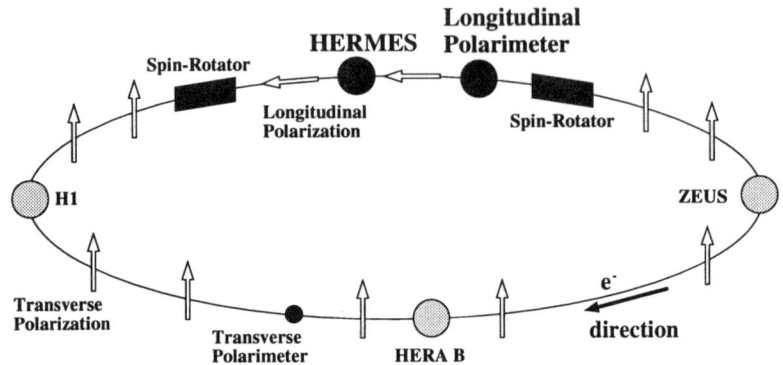

FIGURE 1. The electron ring layout at HERA.

The Longitudinal Polarimeter consists of two main components, a laser beam and a calorimeter that measures the back-scattered Compton photons. The longitudinal polarization can be obtained from the measurement of the energy-dependent asymmetry under reversal of the laser photon helicity using either the single-photon or the multi-photon mode. However, due to intense bremsstrahlung background from the HERMES target and the long straight section, the multi-photon mode is used [3]. In contrast to the single-photon mode, in which the energy of each individual Compton photon is analyzed, in the multi-photon mode one measures the total energy deposited in the detector by many Compton photons per laser pulse interaction with an electron bunch.

The advantages of running in single-photon mode would be twofold. The asymmetries are large, up to 60% for the conditions at HERA (532 nm laser light scattered off 27.5 GeV electrons), and the energy spectra can be compared to the Compton cross sections. Operating the Longitudinal Polarimeter in multi-photon mode has the advantage of being effectively independent of bremsstrahlung background in the HERA storage ring.

The longitudinal polarization of each electron bunch is determined by switching between the two light helicity states and analyzing the two resulting energy distributions. This calculation is provided every minute. The average polarization of the electron beam is finally computed as the mean of the individual bunch polarizations weighted by the corresponding time-averaged bunch currents. The Longitudinal Polarimeter can also measure the polarization of individual bunches with an absolute statistical accuracy of 3% in twenty minute long measurements.

POLARIMETRY AT EIC

Experience at the HERA storage ring has demonstrated that it is imperative to include polarization diagnostics and monitoring capabilities in the design of the electron beam lattice. The specifics depend on the design of the electron machine, and are much more crucial for a ring option than for a linac option. In either case, one has to ensure that the beam polarization can be measured continuously during data taking to minimize systematic uncertainties associated with the beam polarization. For a linac, destructive (Møller scattering with an iron foil) or non-destructive methods (Compton scattering or Møller scattering with an atomic beam source) can be employed.

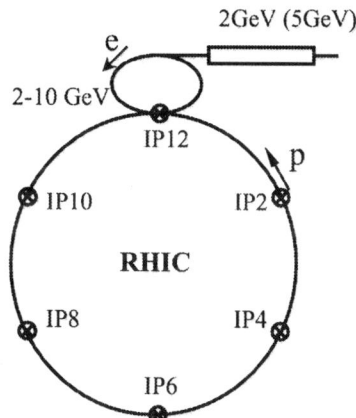

FIGURE 2. A possible layout for the EIC at Brookhaven National Laboratory.

The most recent design favors the ring option, as shown in Fig. 2. For a 5-10 GeV electron beam, there are essentially two options: either to measure the transverse polarization of the beam away from the target (as it is done at HERA with the Transverse Polarimeter), or to measure the longitudinal polarization close to the target (as it is done at HERA with the Longitudinal Polarimeter). The longitudinal polarization is measured via rate or energy asymmetries, which are generally much easier to measure than spatial asymmetries as in the case of transverse polarization. Nevertheless, having two independent measurements with vastly different systematic uncertainties should be strongly considered, if a precision of 1% (or better) needs to be achieved. If only one polarimeter is considered, it should be a longitudinal type polarimeter.

There are three components to be considered. The laser or laser transport system, the laser-electron interaction region, and the Compton detector.

Maybe the most important component is the location of the laser-electron interaction region. Its location has to be chosen to minimize bremsstrahlung and synchrotron radiation, and to optimize the rate of the back-scattered Compton photons versus the background rate. Minimizing bremsstrahlung background requires to have a short section of beam line, like introducing a chicane with soft bends to also minimize synchrotron background. This was done for the Transverse Polarimeter at HERA, and resulted in very small background rates, while for the Longitudinal Polarimeter, this was not possible (changing the lattice after it was built was too expensive) and resulted in much larger background rates. Maximizing the Compton rate means that the crossing angle between the laser beam and the electron beam should be as small as possible (of order 3-10 mrad), and the horizontal widths of the electron and laser beams should both be small (0.5 mm or smaller). In addition, the transverse spatial distribution of the back-scattered Compton photons due to the size and divergence of the electron beam has to be kept small if a) the Compton detector is far away (10s of meters), or b) if the transverse polarization needs to be measured, otherwise it will lead to a dilution of the spatial asymmetry.

One has to consider whether to build an optical cavity (which is expensive and delicate, and cannot be accessed during beam operation) or whether to build a more conventional laser (which can be accessed at all times) plus laser transport system. If rates, and therefore statistical precision allow it, a conventional laser transport system is more conservative. At low energies (<10 GeV) it is advantageous to use optical or maybe even UV laser light (as compared to IR light) because the analyzing power is proportional to the product of the electron and the photon energy. In addition, the energy of the Compton edge (which corresponds to the maximum energy of the back-scattered Compton photons) is proportional to the photon energy. The higher the Compton edge energy, the easier it is to detect it and to distinguish the Compton photons from background.

The Compton detector should be radiation hard and fast (< 35 ns) to be able to record signals for each individual bunch crossing. Calorimeters based on detecting Čerenkov light might be better suited than those based on detecting scintillation light, because they are inherently faster. The Compton detector should also be able to determine the impact position of the Compton photons. That could be achieved with the segmentation of the the detector or with an external position sensitive device.

Every effort should be made to keep the background conditions low. This allows to operate the Compton polarimeter in single-photon mode and to monitor the linearity of the detector, because it provides three energy calibration points, one at the beam energy (bremsstrahlung edge), one the Compton edge, and one at the zero crossing of the asymmetry (corresponding to 90° scattering in the electron rest frame).

ACKNOWLEDGMENTS

I wish to thank Abhay Deshpande, Bill Franklin, Eugene Chudakov and Peter Schüler for many fruitful discussions. The author's research is supported in part by the U.S. National Science Foundation, Intermediate Energy Nuclear Science Division under grant No. PHY-0072297 and PHY-0244842.

REFERENCES

1. Schwartz, M. L., Physics with polarized electron beams, Tech. Rep. SLAC-PUB-4656, Stanford Linear Accelerator Center, Stanford University, Stanford, CA 94309 (1988).
2. Chdukov, E. A., and Luppov, V. G., Møller polarimetry with atomic hydrogen tagets, Tech. rep., Ann Arbor, MI 48109 (2002).
3. Beckmann, M. et al., *Nucl. Instr. Meth. A*, **479**, 334 (2002).
4. Barber, D.P. et al., *Nucl. Instr. Meth. A*, **329**, 79 (1993).

Progress on the Concept and Design of the Rare Isotope Accelerator

Donald F. Geesaman

Argonne National Laboratory, Argonne, IL 60439

Abstract. The Rare Isotope Accelerator is a bold initiative for the U.S. nuclear science community that promises to revolutionize studies of nuclei far from stability. In this talk, a perspective on the role of RIA in enabling major progress in nuclear structure and nuclear reaction research and some important recent steps in technical progress for RIA are reviewed.

INTRODUCTION

Understanding atomic nuclei, their properties, reactions and how they are formed in the cosmic cauldrons of the universe, is a cornerstone of our knowledge of the natural world. The Rare Isotope Accelerator, RIA, is proposed as the next major step required to advance this understanding by producing beams of unstable nuclei in unprecendented variety and intensity. Today we have been able to study less than half of the nuclear isotopes that we believe can be bound by the strong interaction. Yet, these unknown isotopes hold the keys to a comprehenive picture of the structure of nuclei and the origins of the matter of which we are made.

The time is right for RIA due to a confluence of events. First, experimental progress in understanding the QCD substructure of hadrons and nuclei has validated hadron-based models of the nucleus to quite short distance scales, confirming the core of our picture of nuclear structure. At the same time advances in nuclear theory have made it clear that the solution to long-standing questions in nuclear structure lies in the many-body physics and has focused the physics discussions on issues such as: what is the isospin dependence of the nuclear three-body force? What are the dependences on proton-neutron asymmetry of the mean-field spin-orbit force and of the nuclear equation of state? RIA is designed to answer these questions. The astronomy and astrophysics communities are investing heavily in new generations of observatories. Interpreting their exciting results requires new insights from nuclear physics. What lets us be confident we can answer these questions is the rapid progress in: 1) accelerator and target technology and 2) experimental technique that makes a bold leap forward possible at this time. RIA will enable us to develop an overarching picture of the nature of nucleonic matter, understand the origin of the elements and test fundamental symmetries and search for physics beyond the standard model. This is why RIA received the highest priority for major new construction in the NSAC 2002 Long Range Plan.

In addition to vital advances in nuclear science, RIA will provide a copious bounty of isotopes for other basic research and societal applications. These major avenues include radioactive ion implantation for material science, isotope production for R&D by the biological and medical communities and stockpile stewardship applications.

THE RIA CONCEPT

The design of the Rare Isotope Accelerator evolved significantly in the late 1990's. Previously, e.g. in the 1996 NSAC Long Range Plan, it was widely accepted that there were two complementary techniques to produce rare isotope beams: 1) break up a fast heavy beam on a light target and perform experiments with the fast fragments or 2) use a high power light ion beam to break up a heavy target, collect and ionize the isotopes desired and accelerate them for experiments, the "isotope separation on-line" (ISOL) technique. In 1998 experiments carried out at Argonne to inject radioactive ions into the Canadian Penning trap demonstrated that one could stop fast ions, independent of the specific ion chemistry, in helium gas and the resulting 1^+ ions could be extracted rapidly (< 10 ms) and efficiently (up to 40%) by a combination of gas flow, DC and AC electric fields. This created a new paradigm for the production of rare isotopes and made it clear that an accelerator was needed that could accelerate all stable species from protons to uranium with high (>100 kW) beam powers. A superconducting linac was proposed to provide the optimum solution to accelerate these beams [1]. A key advantage is that the large acceptance of such a linac allows multiple charge states of ions, which emerge following passage through a stripper foil, to be simultaneously accelerated to the same energy [2], resulting in an order of magnitude increase in beam power available for the heaviest beams even with current ion source technology. New concepts had to be developed for high power targets to deal with these high (100 kW) beam powers. Here several ideas proved useful, including liquid lithium technology adapted from the fusion reactor community and two-step targets that separate the energy deposited by the beam from the energy released in fission. Finally, having worked so hard to produce radioactive beams, it is critical to develop techniques to reaccelerate them with optimum efficiency. This required the development of a new generation of accelerating structures to efficiently accelerate very low energy, high mass-to-charge-ratio 1^+ ions.

In 1999, the NSAC ISOL Task Force adopted each of these ideas as the basis for the Rare Isotope Accelerator. It concluded that the optimum performance for this facility required beams of all isotopes with energies of at least 400 MeV/u for uranium. Thus, the RIA concept was defined as illustrated in the schematic layout of Figure 1. Each of the four proposed production mechanisms, fast-fragmentation, fast-in-flight-fission of uranium, two-step neutron-induced fission and standard ISOL production, is needed to obtain the optimum yields [3]. For example, for high quality stopped and reaccelerated beams, ISOL and two-step fission provide orders of magnitude more beam intensity than fast beam techniques for the elements which diffuse rapidly through hot materials. For Sn isotopes, for example, these light ion induced techniques provide the optimum reaccelerated beams from ^{107}Sn to ^{139}Sn. When available, such high yields are very important not only for nuclear physics but for many of the applications that require harvesting or carefully controlled implantation of isotopes. Typically the fast, heavy-beam techniques are more efficient at the limit of proton and neutron rich isotopes, or

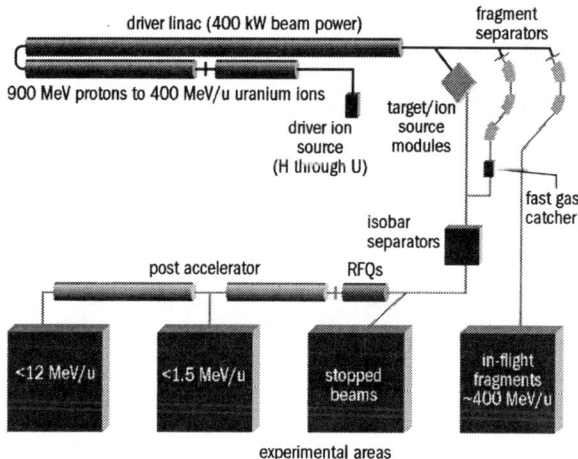

FIGURE 1. Schematic layout of the RIA facility.

for refractory elements that do not diffuse rapidly through materials.

Just as each of the production techniques is required to optimize the production of a specific rare isotope, each of the four experimental areas shown in Figure 1 has a special role to play in addressing the science case. Fast rare isotope beams provide the greatest reach at the limits of stability and in studies of the nuclear equation of state. The higher resolution of reaccelerated beams offers major advantages in heavy nuclei with high densities of states. The lower velocities make them ideal for studying the interplay between collective and single particle degrees of freedom. One important example in this regard is reactions to produce the shell-stabilized superheavy elements. The range of predictions for the location of the center of the island (or peninsula) of superheavy elements is precisely a result of our lack of quantitative understanding of this interplay of collective and single particle degrees of freedom that RIA will resolve. While there are many indirect techniques for determining the reaction rates for nucleosynthesis, as the current situation for the ^{7}Be(p,γ) reaction makes clear, presently there is still no substitute, when feasible, for the direct measurement of low energy reaction rates. Finally, the continued development of ion and atom traps offers great potential in the study of fundamental interactions and the high-precision determination of masses required for both astrophysics and fundamental interactions studies. Recent measurements using a Cf source, a gas-stopping cell and the Canadian Penning trap at ANL have determined the masses of 20 neutron-rich nuclei in the Ba-Pr region to typically better than 100 keV. With RIA the number of isotopes which could be addressed with this technique is in the thousands. New experiments, for example a measurement of the electric dipole moment of ^{225}Ra with trapped atoms, are pioneering the technology for precision searches at RIA for physics beyond the standard model with unstable isotopes.

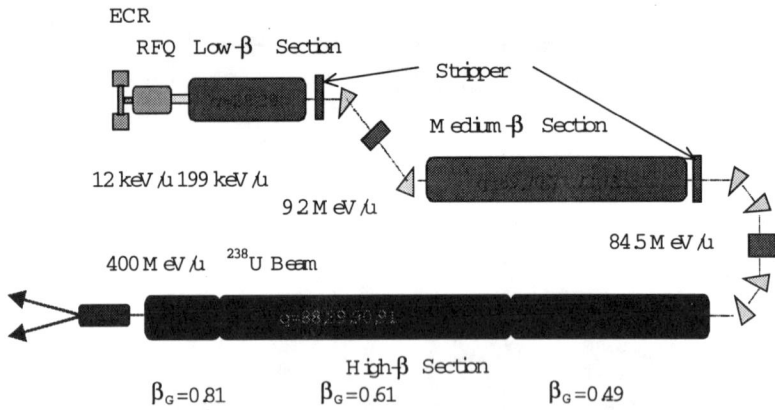

FIGURE 2. A typical configuration of the RIA Driver superconducting linac illustrating the two stripper locations and the range of charge states of uranium ions simultaneously accelerated through each section.

TECHNICAL PROGRESS

Currently ten U.S. laboratories: ANL, BNL, CSM, JLab, LANL, LBNL, LLNL, NSCL/MSU, ORNL and Texas A&M are participating in the DOE-funded RIA R&D that received over $3M in FY2003. Major progress is also being made internationally as new facilities are planned and constructed. Reference [4] provides an extensive list of references to this work. To make sure that RIA is choosing the optimum technology, there is close cooperation on an international scale. In May 2002, the 2nd RIA Driver workshop brought experts together to review the concepts for the superconducting driver linac (Figure 2). The conclusion of this workshop was that the RIA technology is sound; there are no hard failure modes. A number of areas were identified for optimization and cost minimization as the project planning progresses. I would like to review here some of the important technical developments of the past year.

The work on the accelerating cavities and the accelerator beam optics must proceed hand in hand to achieve a robust and reliable design. New classes of superconducting cavities have been developed to cover the entire velocity range from $\beta = 0.02$ to $\beta = 0.8$. Capitalizing on the ongoing development work, the superconducting cavities designed for the Oak Ridge Spallation Neutron Source have been adopted as the baseline choice for the high velocity particles. Considerable progress has also been made in achieving higher accelerating field gradients in superconducting RF cavities at a number of laboratories over the past few years using electro-polishing and high-pressure rinse techniques. For example, performance measurements of cavities at ANL, JLab, Legnaro and MSU surpass the RIA design specifications. Potentially, this enhanced performance could result in significant cost savings for the project.

Decisions about how to implement this technology rest heavily on detailed calculations of the accelerator beam optics. It is especially important in such a high power machine to minimize beam losses to ensure that one can properly perform accelerator maintenance. Extensive calculations show that one can maintain the high standards required by careful construction of the lattice and optimizing the phase space acceptance of the accelerator [5].

Targets that can handle the high beam power are a major challenge for RIA, both for the ISOL and fragmentation production mechanisms. Liquid lithium has the thermal properties and low Z to be ideal for this latter application. A prototype of a RIA windowless flowing liquid lithium is operating at ANL and will be tested under high power beam conditions later this year. Full RIA-energy tests of the helium gas stopper cell are being performed by an international collaboration at GSI. In the first measurements, it was demonstrated that the energy spread of the slowed beam could be controlled to the level required through the use of energy dispersion and precision graded absorbers. A full RIA scale, 0.5 Atm-m gas cell has been constructed at Argonne and undergone detailed performance tests at ATLAS. It will be transported to GSI this year for the full RIA-energy measurements. This sophisticated device contains over 7400 parts, over 4000 of which must be prepared to ultra-high vacuum specifications to maintain the clean environment required for the gas cell to operate properly.

SUMMARY

In summary, the Rare Isotope Accelerator will open exciting new opportunities in nuclear science and a broad range of applications. Significant progress has occurred on all the technical aspects of the machine concept which have validated the design and costing assumptions. Many of these developments have important consequences for other future accelerator projects. RIA will produce rare isotope beams of unprecented variety and intensity. With these capabilities, RIA will transform our understanding of nuclei, the cosmos and the world in which we live.

ACKNOWLEDGMENTS

This work is supported in part by the U. S. Department of Energy under contract No. W-31-109-ENG-38.

REFERENCES

1. K. W. Shepard et al., "SC Driver Linac for a Rare Isotope Accelerator Facility" in the Proc. of the 9th Int'l Workshop of RF Superconductivity, Santa Fe, New Mexico, 1999 ed. B. Rusnak p. 345.
2. P. Ostroumov et al., Phys. Rev. Lett. **86**, 2798 (2000).
3. C.-L. Jiang et al., Nucl. Intr. Meth. A **492**, 57 (2002).
4. J. A. Nolen, "The U. S. Rare Isotope Acclerator Project" in the Proc. of the XXI Int'l Linac Conf., Gyeongju, Korea, 2002.
5. P. Ostroumov, "Design Features of High-Intensity Medium-Energy Superconducting Heavy-Ion LINAC" in the Proc. of the XXI Int'l Linac Conf., Gyeongju, Korea, 2002 .
6. M. P. Kelly et al., Proc. of the 2001 Particle Accelerator Conf., June 18-22, Chicago, IL, 2001, p. 506.

The Physics Of eRHIC

Richard G. Milner

MIT-Bates Linear Accelerator Center
Massachusetts Institute of Technology
Cambridge, MA 02139

Abstract. Over the last five years, there has been substantial international interest in a high luminosity (approx. $10^{33} cm^{-2} s^{-1}$) polarized electron-ion collider covering a CM energy range from about 30 to 100 GeV. Workshops have taken place in Europe in 1997 [1] and in the United States at IUCF (1999) [2], BNL (1999) [3], Yale (2000) [4] and MIT (2000) [5]. An Electron-Ion Collider (EIC) collaboration was formed in the fall of 2000, and in the United States, the 2001 Long Range Plan favorably endorsed the science of EIC and urged R&D support as a high priority. In March 2002, a series of workshops at BNL involving over 150 physicists from around the world agreed to produce a conceptual design of an electron-ion collider within three years. Further, it was agreed to develop this design around the existing Relativistic Heavy Ion Collider (RHIC) at BNL. This design is known as eRHIC. In this paper the scientific highlights of eRHIC are described.

SCIENTIFIC HIGHLIGHTS

The primary scientific motivation for eRHIC is to probe the fundamental quark and gluon structure of strongly interacting matter. The great advantage of the collider is the ability to probe large momentum transfer (Q^2) and small parton momentum fraction (x). In addition, the collider enables complete detection of the hadronic final-state in hard scattering processes. Experiments on fixed targets at higher energies suffer from the fact that many of the reaction products end up at ever decreasing forward angles. Further, an electron-nucleon collider will allow high polarization of both electron and nucleon beams and so the ability to probe completely uncharted aspects of nucleon structure using spin-dependent scattering. Finally, eRHIC with its nuclear beams can probe the partonic structure of nuclei with unprecedented scope and precision. Partons in nuclei are essential for understanding relativistic heavy ion collisions. In addition, there are predictions of new phenomena in nuclei at low parton momentum. The scientific opportunities provided by eRHIC are outlined in the February 2002 white paper[6].

Quark And Gluon Distributions In The Nucleon

eRHIC offers a unique capability for measuring 'flavor tagged' structure functions by providing access to a wide range of final states. The collider geometry makes measurement of semi-inclusive reactions very efficient so that quark and gluon distributions in nucleons, nuclei, and possibly even mesons can be mapped in a flavor tagged mode.

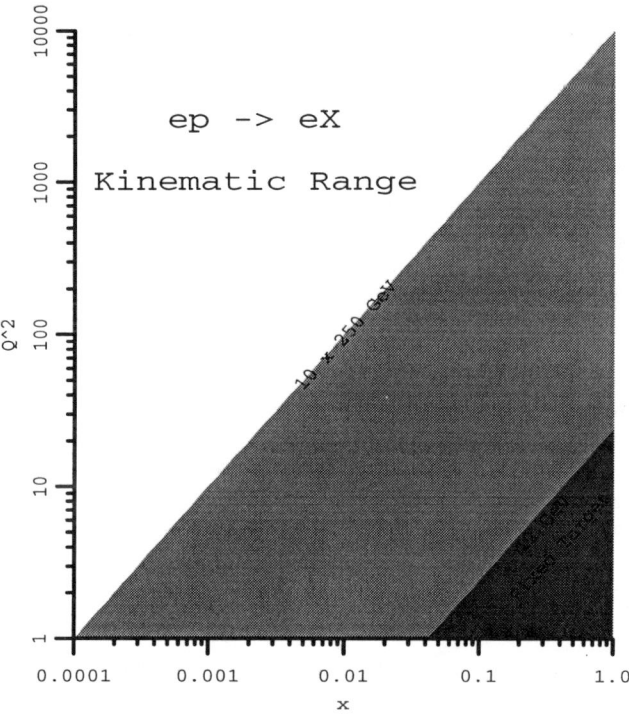

FIGURE 1. The kinematic range in parton momentum fraction x and 4-momentum transfer Q^2 made possible by a 10 GeV electron beam on a 250 GeV proton. For comparison, the kinematic range accessible with a 12 GeV electron beam on a fixed target is shown.

Quark And Gluon Distributions In The Nucleon

eRHIC offers a unique capability for measuring 'flavor tagged' structure functions by providing access to a wide range of final states. The collider geometry makes measurement of semi-inclusive reactions very efficient so that quark and gluon distributions in nucleons, nuclei, and possibly even mesons can be mapped in a flavor tagged mode.

Spin Structure Of The Nucleon

eRHIC, operating at the highest center-of-mass energy, will provide crucial data on the proton's spin-dependent structure function at lower x than possible in any previous experiment. This is urgently required to verify the QCD prediction of a dramatic decrease in the proton's spin-structure function, based on an evolution analysis using existing data. In addition, eRHIC will provide precision measurement of the polarization of the sea quarks, currently a matter of controversy within sophisticated and successful models of the nucleon. Further, eRHIC will provide the possibility with transversely polarized beams to explore the leading-twist transversity distribution in the nucleon.

Correlations Between Partons

A complete characterization of the partonic substructure of the nucleon requires a description of the correlations between the parton densities. Progress in this area can be realized by measuring hard exclusive processes expressed, as a result of the new QCD factorization theorems, through a new class of parton distributions termed Generalized Parton Distributions. eRHIC kinematics are optimal for measuring these processes.

The Role Of Quarks And Gluons In The Nuclei

eRHIC opens new horizons fundamental to nuclear physics by determining the quark and gluon momentum distributions in the nuclei. In particular, the gluon distribution in nuclei is at present almost undetermined. Further, its determination is vital to understanding the complicated processes underway in relativistic heavy ion collisions.

Hadronization In Nucleons And Nuclei

How do the colored quarks and gluons struck by the virtual photon in deep inelastic scattering evolve into the colorless hadrons that must eventually appear? This process is one of the clearest manifestations of confinement: the asymptotic physical states must be color neutral. Hadronization is a complex process that involves both the structure of hadronic matter and the long range nonperturbative dynamics of confinement. eRHIC will make it possible to observe the complete array of decay products in hard scattering from the nucleon and nuclei. Recent observation of jet quenching in heavy ion collisions underscores the necessity to understand how fast partons traverse nuclei.

Partonic Matter Under Extreme Conditions

Very high energy deep inelastic scattering on nuclear targets offers opportunities to study partonic matter under extreme conditions. Measurements of the proton structure function at low x show that the gluon distribution grows rapidly with decreasing x. When the gluon density becomes large, it may saturate and give rise to a new form of matter known as a *color glass condensate*. eRHIC will enable the search for this exotic component of the proton structure by measurement of gluon observables in nuclei.

DETECTORS

The design of an optimized suite of detectors for eRHIC is underway. This involves iteration between the machine design and the detailed simulation of the physics measurements. In particular, the detailed design of the electron-ion interaction region is closely coupled to the design of the eRHIC detectors. However, it is clear that at least one central detector will be required. This will require tracking of the scattered electron and hadronic final state, calorimetery to determine the final state energies, particle identification for both electron and hadrons and jet reconstruction. For example, the ZEUS detector at the electron-proton collider HERA contains many of these characteristics for high center-of-mass energies.

In addition to a central detector, many proposed measurements will require additional instrumentation at small angles in the forward or rear directions. For example, the detection of the complete final state in electron-nucleus collisions (as proposed by M. W. Krasny in [6]) will require additional magnetic elements and detectors at small angles. The measurement of hard exclusive processes, e.g. deeply virtual Compton scattering, requires detection of forward hadrons with high momentum.

SUMMARY

eRHIC is essential for the study of the fundamental structure of matter. It will provide a new vista into the role and behavior of quarks and gluons in atomic nuclei. All future planning for the field of hadron structure worldwide involves serious consideration of an electron-ion collider. Further, eRHIC has sparked considerable synergy among the different sub-fields of nuclear physics – electromagnetic, hadronic and heavy ion communities. In 2003, eRHIC has been judged to be absolutely essential to the future of nuclear physics in the United States. It is anticipated that a conceptual design for eRHIC will be realized in 2004.

ACKNOWLEDGMENTS

The work reported here is the fruit of a collaboration involving many people. In particular, I would like to acknowledge discussions with A. Bruell, J. Cameron, A. Deshpande, G. Garvey, R. Holt, V. Hughes, R. Jaffe, T. Londergan, L. McLerran, P. Paul, J.-C. Peng, D. van Harrach, and R. Venugopalan on the scientific case for eRHIC. I have learned much from discussions with D. Barber, I. Ben-Zvi, I. Koop, S. Peggs, T. Roser, Y. Shatunov and C. Tschalaer on the machine design. The author's work is supported by the United States Department of Energy under Cooperative Agreement #DE-FC02-94ER40818.

REFERENCES

1. Slide report of the Joint DESY/GSI/NuPecc *Workshop on Electron-Nucleon/Nucleus Collisions*, March 3-4 1997, Lufthansa-Zentrum Seeheim, Germany, GSI REPORT 97-04.

2. Proceedings of the *Workshop on Physics with a High Luminosity Polarized Electron Ion Collider* (**EPIC99**), .April 8-11 1999, Bloomington, Indiana, USA, Editors L.C. Bland, J.T. Londergan, and A.P. Szczepaniak, World Scientific.
3. Proceedings of the *eRHIC Workshop,* December 3-4 1999, Brookhaven National Laboratory.
4. Proceedings of the *Second eRHIC Workshop,* April 6-8 2000, Yale University, New Haven, Connecticut, USA, BNL Report 52592.
5. Proceedings of the *Second Workshop on Physics with an Electron Polarized Light Ion Collider-EPIC 2000*, September 14-16 2000, MIT, Cambridge, Massachusetts, USA, Editor R.G. Milner, AIP Conference Proceedings No. 588.
6. The Electron Ion Collider: A high luminosity probe of the partonic substructure of nucleons and nuclei , BNL Report 68933, February 2002.

ELIC: A High Luminosity And Efficient Spin Manipulation Electron-Light Ion Collider Based At CEBAF

Lia Merminga and Yaroslav Derbenev

Center for Advanced Studies of Accelerators,
Jefferson Laboratory, Newport News, VA 23606

Abstract. Electron-light ion colliders with center of mass energy between 20 and 100 GeV, luminosity between 10^{33} and 10^{35} cm^{-2} sec^{-1}, and polarization of both beams at or above 80% have been proposed for the study of hadronic structure. The Electron-Light Ion Collider (ELIC) facility would require the upgrade of CEBAF to 5-7 GeV energy recovering linac and the realization of an ion storage ring complex, accelerating and storing light ions of up to 150 GeV. In this report several innovative features of electron and ion beam designs and their advantages in delivering the luminosity and spin are described. These features include: electron circulator ring to reduce electron polarized source and energy recovering linac requirements, twisted spin booster and collider ring; interaction points with low beta-star and crab-crossing using the short, cooled ion bunches. Accelerator physics and technology issues for both protons/ions and electrons are presented. The feasibility of an integrated fixed target program at 25 GeV and collider program with center of mass energy between 20 and 65 GeV is explored.

INTRODUCTION

Thirty years after the establishment of QCD as the theory of the strong nuclear interaction, and despite significant achievements in our knowledge of the structure of hadronic matter in the intervening decades, understanding how QCD works in practice remains one of the great puzzles in nuclear physics. Some crucial questions in the field remain open and involve the role and behavior of quarks and gluons in atomic nuclei. There are still gaps in our understanding of how quarks and gluons provide the binding and the spin of the nucleon, of how quarks and gluons evolve into hadrons via the dynamics of confinement, and of how the nuclear binding arises from QCD. The nuclear physics community worldwide has suggested that a high luminosity, polarized electron-ion collider would be a powerful new microscope to probe the hadronic structure of matter, and provide answers to these questions. The luminosity of this collider should be greater than 10^{33} cm^{-2} sec^{-1} with both beams longitudinally polarized at 80% in the interaction region. The center-of-mass energy should be variable between 20 and 100 GeV. Spin-flip of both beams is extremely desirable for exclusive measurements.

Two accelerator design scenarios are being explored for a high luminosity, high polarization collider: the scheme of colliding storage rings and the linac-ring scheme, with the electron beam generated from a polarized source and accelerated in a superconducting, energy recovering linac (ERL) [1]. A critical challenge of the storage ring scenario is related to the complexity of preserving and manipulating the electron spin, while the challenge of the linac-ring option stems from the necessity to generate, accelerate and energy recover a high bunch charge, high average current polarized electron beam. However, there are also significant advantages in the latter scheme, which make further exploratory studies compelling. In the linac-ring scenario, higher luminosities appear feasible and flexibility with respect to spin manipulations is greatly enhanced.

The Electron – Light Ion Collider (**ELIC**) facility [2] is a high luminosity, polarized electron – light ion collider in the center of mass energy range of 20-65 GeV. It would require the upgrade of CEBAF to 5-7 GeV energy recovering linac and the realization of a storage ring complex, accelerating and storing light ions of up to 150 GeV. Preliminary design studies indicate that the luminosity of the colliding electron and light ion beams can be as high as 10^{35} cm^{-2} sec^{-1} with an arbitrary polarization direction of either particle for up to four interaction regions. The upgrade of CEBAF could also result in a 25 GeV fixed-target facility.

In the following, we describe the proposed CEBAF-based electron-ion collider and discuss the integration of the collider with a 25 GeV fixed target program. We conclude with the accelerator physics and technology challenges that must be resolved in order for the feasibility of this proposal to be established.

THE ELIC PROPOSAL

A schematic layout of the ELIC collider is shown in Figure 1. Longitudinally polarized electrons generated from a high current polarized source are injected into the CEBAF accelerator, and accelerated to 5-7 GeV in a single pass through the accelerator. To ease the high current polarized photoinjector and ERL requirements, a circulator ring [3] may be used. In this scenario, the electrons are injected into the circulator ring, after acceleration in the ERL, and circulate for approximately 100 revolutions while they continuously collide with the ions. They are subsequently extracted, transported back to CEBAF for deceleration and energy recovery, and are dumped at approximately their injection energy. If the polarized source development should become so advanced that the circulator ring does not offer any advantages, the electrons can still follow the same transport, but now circulate only once before being reinjected to CEBAF for energy recovery.

All ion species are injected longitudinally polarized and accelerated in a conventional ion RF linac with maximum energy of 200 MeV. Currently, two options are being considered for the ion beam stacking and acceleration after the linac. The first option comprises the conventional scheme of a 2-3 GeV kinetic energy "Figure-8" [3] pre-booster ring (approximately 70 m in circumference, not shown in Fig. 1) followed by a large booster (coincident with the "Figure-8" electron circulator ring). Stripping injection of the polarized negative ions can be used to accumulate the polarized protons and deuterons in the pre-booster. However, this method is not

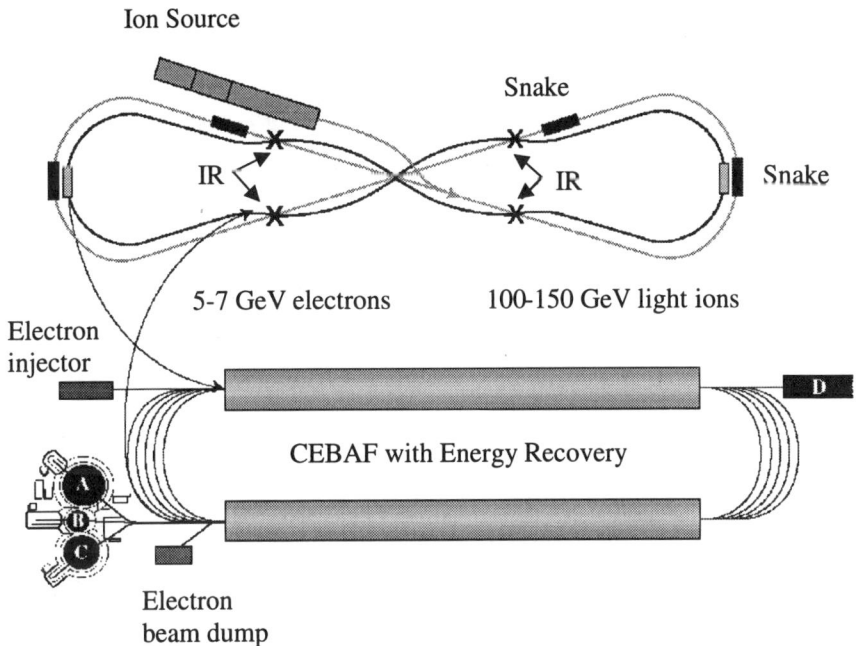

FIGURE 1. ELIC Layout.

applicable to the polarized He^3 (positive ion source). To make the accumulation of polarized He^3 possible, we rely on the experience of accumulation of polarized protons in the IUCF Cooler Ring with electron cooling. We consider an alternate scheme that includes a small, up to ~15 m in circumference, 200 MeV accumulator ring, with conventional electron cooling, following the linac and before the pre-booster. In this scheme, all ion species can be accumulated using the positive polarized sources. Another important advantage of this option is the possibility to cool the ion beam after acceleration in the linac in order to reduce the transverse emittance, if needed.

As mentioned earlier, the circulator ring can also be used as the booster ring bringing the ion energy up to ~20 GeV. The ions are then injected and stored in the "Figure 8" storage ring housed in the same tunnel with the CR. "Figure-8" rings, including pre-booster, booster and storage ring, are used for the ions for spin preservation and flexible manipulation of all species of interest. Specifically, "Figure-8" rings have zero spin tune, as a result of which intrinsic spin resonances and spin resonance-crossing are avoided. In the "Figure-8" storage ring, longitudinal polarization for all ion species at all energies is possible by introducing solenoids in the straight sections or horizontal dipoles in the arcs. Spin rotators around the interaction points would not be needed. For protons, up to 4 simultaneous interaction regions (IRs) can exist with longitudinal or transverse polarization. Compact, non-expensive full Siberian Snakes (shown in the Figure) can be used to provide longitudinal polarization for the proton beam in all 4 IRs simultaneously, while helping to stabilize the spin. For D and He^3 up to 2 simultaneous IRs can exist with longitudinal or transverse spin. With "Figure-8" storage ring, spin-flip will work reliably without loss of polarization. To ensure that the electron spin remains longitudinal at the IRs, a Wien filter in the injector or one Siberian snake (i.e. superconducting solenoid) is required for two IRs, and a Wien filter plus 2 Siberian snakes, or three Siberian snakes without the Wien filter for 4 IRs.

A series of evolutionary upgrades to the collider facility are envisioned [2]. A first milestone may be at a luminosity of 10^{33} cm^{-2} sec^{-1}, which could be reached if the polarized electron injector current is 2.4 mA, assuming that we use the circulator ring configuration. A next milestone, at a luminosity near 10^{34} cm^{-2} sec^{-1} may be reached by utilizing the extremely short, of order 1 cm, ion bunches, resulting from electron cooling [4], which allow the use of "crab crossing" [5] resulting in increase of the collision frequency and reduction of parasitic collisions. With the use of the circulator ring, the polarized electron source is required to provide about 16 mA. From there, the maximally attainable luminosity, is obtained at the beam-beam limit of the ion ring, together with maximum collision frequency, equal to the rf frequency. Luminosity of 10^{35} cm^{-2} sec^{-1} or greater is feasible when the ion beam energy is 100 GeV or above.

In every stage electron cooling is introduced to suppress ion beam blowup due to IBS and its impact on the luminosity. At energies above the transition values, energy exchange between two particles at intra-beam collisions leads to horizontal emittance increase due to the energy-orbit coupling and vertical emittance increase due to x-y coupling. In the limit of $\gamma \gg Q$, where Q is the betatron tune, and when the x-y coupling parameter κ is small, then the equilibrium criterion obtained by setting the IBS rate equal to the cooling rate, leads to flat beams and maximum cooling rate [4]. Thus by reducing the transverse coupling to a minimum, while conserving the beam area (which determines the luminosity), we are able to decrease the impart of IBS on luminosity. Electron cooling then leads to a flat equilibrium with large aspect ratio. Table 1 shows ELIC parameters with flat beams. Note that the beam-beam parameter has been set to the modest value of 0.01 per interaction point. The use of strong superconducting rf, which leads to a rather high value of synchrotron tune (ν_s=0.06) [4], helps stabilize the short ion bunches against microwave instabilities and beam-beam interaction.

When the equilibrium described above between cooling and IBS is established, the horizontal emittance will be determined by the multiple IBS while the vertical emittance will be determined by the beam-beam interaction. The luminosity lifetime is determined by the single or Touschek scattering, which is IBS at large momentum transfer driving particles out of the core, and thus limiting the luminosity lifetime. A phenomenological model, which includes single scattering and cooling time of the scattered particles has been used to estimate an optimum set of parameters for maximum luminosity, at a given luminosity lifetime [4]. Table 1 parameters incorporate this optimization, for a luminosity lifetime of 20 hours.

Another potential limitation on the luminosity lifetime is due to the interactions between colliding beams. If the electron scatters elastically from the oncoming proton at sufficiently large angles, both of these particles are lost from the acceleration cycle. We evaluated this effect assuming the value of critical angle to be 0.3 mrad (which is the design angular spread at the interaction point), and found the proton beam half-life to be 5 days for 5 GeV electron on 50 GeV proton beams, and 20 days for 100 GeV protons. Contributions from inelastic processes, including electron collisions with light ions, appear to have an effect of a factor of ~10 smaller.

The same electron accelerator that is used in the collider mode can also provide up to 25 GeV electrons for fixed target experiments for physics. This scheme would require the implementation of a 5-pass recirculator at 5 GeV per pass as in present CEBAF. It is a subject of further investigation whether the collider and fixed target modes could run simultaneously or in alternating modes. The emittance growth due to synchrotron radiation in the CEBAF arcs at the higher energies has been addressed. A novel optics design for the higher arcs results in reduction in emittance growth by a factor of 10 compared to the present optics. The beta functions in arc 9 are ~ 70 m and the spot sizes at the IP are between 0.3-0.5 mm at 25 GeV [6].

The ELIC proposal is at an early stage of development. A number of technical challenges must be resolved, and several R&D projects have been started. These include development of a high average current polarized electron source with a high bunch charge, electron cooling of protons/ions, energy recovery at high current and high energy, and the design of an interaction region and detector that support the combination of high luminosity and high detector acceptance and resolution, which are essential to carry out the proposed physics program.

This work was supported by the U.S. Department of Energy under Contract No DE-AC05-84ER40150.

TABLE 1. ELIC Parameters With Flat Beams.

Parameter	Unit	Value (Protons/Electrons)
Beam energy	GeV	150/7
Energy of cooling beam	MeV	75
Bunch rep rate	GHz	1.5
Particles/bunch	10^{10}	0.2/1
Beam current	A	0.5/2.5
Cooling current	A	2.5
Horizontal emittance (norm, rms)	μm	1/100
Vertical emittance (norm, rms)	μm	0.01/1
Number of interaction points		4
Total beam-beam tune shift		0.04/0.16
Laslett's tune shift in p-beam		0.02
Luminosity overall IP (10^{35})	$cm^{-2}s^{-1}$	2
Cooling/IBS time in p-beam core	min	5
Luminosity Touschek's lifetime	h	20

REFERENCES

1. Merminga, L., Douglas, D.R., and Krafft, G.A., *Annu. Rev. Nucl. Part. Sci.* 53:387-429 (2003).
2. Merminga, L., et al., Proc. European Particle Accelerator Conference (2002).
3. Derbenev, Ya., Proc. European Particle Accelerator Conference (2002)
4. Derbenev, Ya., Proc. COOL'03 Workshop (2003).
5. Derbenev, Ya., Proc. Particle Accelerator Conference (2003)
6. Chao, Y.-C. JLab Technical Note TN-99-037.

A New Detector for Physics at HERA

I. Abt for the HERA-III Detector Group

MPI für Physik, Föhringer Ring 6, 80805 München, Germany

Abstract.
A new detector[1] designed to enlarge the kinematic range to study the structure functions F_2 and F_l at the e[27.5GeV] p[920GeV] collider HERA at DESY is presented. The kinematic region of low x and low Q^2 is opened by extending the acceptance of the device to rapidities up to 8. The detector also allows studies of vector meson production and forward jet physics through a symmetric forward-backward spectrometer arrangement based on silicon tracking stations and silicon-tungsten calorimetry. GEANT studies detailing the projected capabilities are presented.

PHYSICS PROGRAM

The new detector will facilitate a huge variety of measurements in the field of e p[A] collisions. The highlights of the proposed program are:

- The measurement of F_2 from $Q^2 = 0.1$ GeV^2 to $Q^2 = 10$ GeV^2 with high precision to better understand the observed transition of the cross sections from partonic to hadronic behavior.
- The measurement of the longitudinal structure function, F_L, particularly at Q^2 values below 10 GeV^2, where present theoretical and experimental uncertainties are very large.
- The measurement of diffractive and exclusive reactions (VM production and DVCS) over the full W range, and to values of $|t| \leq 1.5$ GeV^2, with no proton dissociation background, to perform a three dimensional mapping of the proton and perform first extractions of generalized parton distributions.
- The measurement of forward jets and forward particle production up to pseudorapidities of at least $\eta = 6$ to test in a direct way our understanding of parton branching in strong interactions and to see the onset of collective phenomena. Acceptance for forward jets will also allow the measurement of F_2 to $x = 1$ at moderate Q^2.

All of the above measurements should be performed with protons and with at least two nuclear targets (e.g., deuterons and Hg nuclei) in order to search for the gluon condensate and understand nuclear effects in parton distributions.

EXPERIMENTAL REQUIREMENTS

The **luminosity requirements** for the e p program are set by exclusive cross-section measurements at high t, F_2 measurements near $x = 1$, and by the F_L measurement. The need for data sets at three different proton energy settings ($E_P = 460, 690, 920$ GeV) with integrated luminosities of about 50 pb^{-1} is anticipated. The luminosity required for the e A program has been estimated at 2 pb^{-1} per nucleon. We do not anticipate needing electron beams - positron beams will suffice - since we are concerned primarily with the Q^2 region where photon exchange dominates.

The **detector requirements** resulting from the goals of the physics program are

- dipole magnets to separate scattered positrons from the beam
- precision forward and backward tracking up to rapidities of 8
- good electromagnetic calorimetry, matching the tracking acceptance
- forward [proton direction] hadron calorimetry

The **dipole** field facilitating the extension in rapidity range is oriented vertically. In both the electron and proton hemispheres it extends from 0.3m to 4.8m from the interaction point and in the two hemispheres it has opposite polarity.

The **interaction region** can be designed using existing magnets. The first machine elements are positioned at ±5.8m.

DETECTOR CONCEPT

The main idea is to build a compact detector with tracking and central electromagnetic calorimetry inside a magnetic dipole field of 0.3T and calorimetric end-walls outside. The detector inside the magnet[s] should not exceed a radius of 80 cm. The coordinate system has the z-axis parallel to the proton beam, the x-axis horizontal and the y-axis vertical. The positrons thus point towards negative z.

The **tracking** focuses on the forward and backward tracks. The calorimetry is to show the best performance in the central region where momentum measurements are intrinsically less precise due to the field configuration and thus e-π separation more difficult. Tracking for $|\eta| < 0.5$ is currently not foreseen.

The precision tracking required translates into hit resolutions of less than approximately 50 μm. This makes silicon strip detectors an obvious choice of technology. As the material budget will be important, double-sided detectors are desirable. A read-out pitch between 50 μm and 100 μm will be adequate. Two double-sided detectors with appropriate stereo angle design can yield unambiguous space-points. For the baseline design we assume silicon tracking stations with two double-sided, 300 μ thick silicon strip detectors and support structures with material equivalent to 1.2 mm of carbon fiber. This results in stations with a material budget equivalent to ≈ 1 % of a radiation length X_0.

The complete tracker is composed of planes oriented perpendicular to the beam. The planes are centered around the proton beam line and measure approximately 40×40 cm^2. They have a central cut-out that follows the beam-pipe design. Each plane is composed

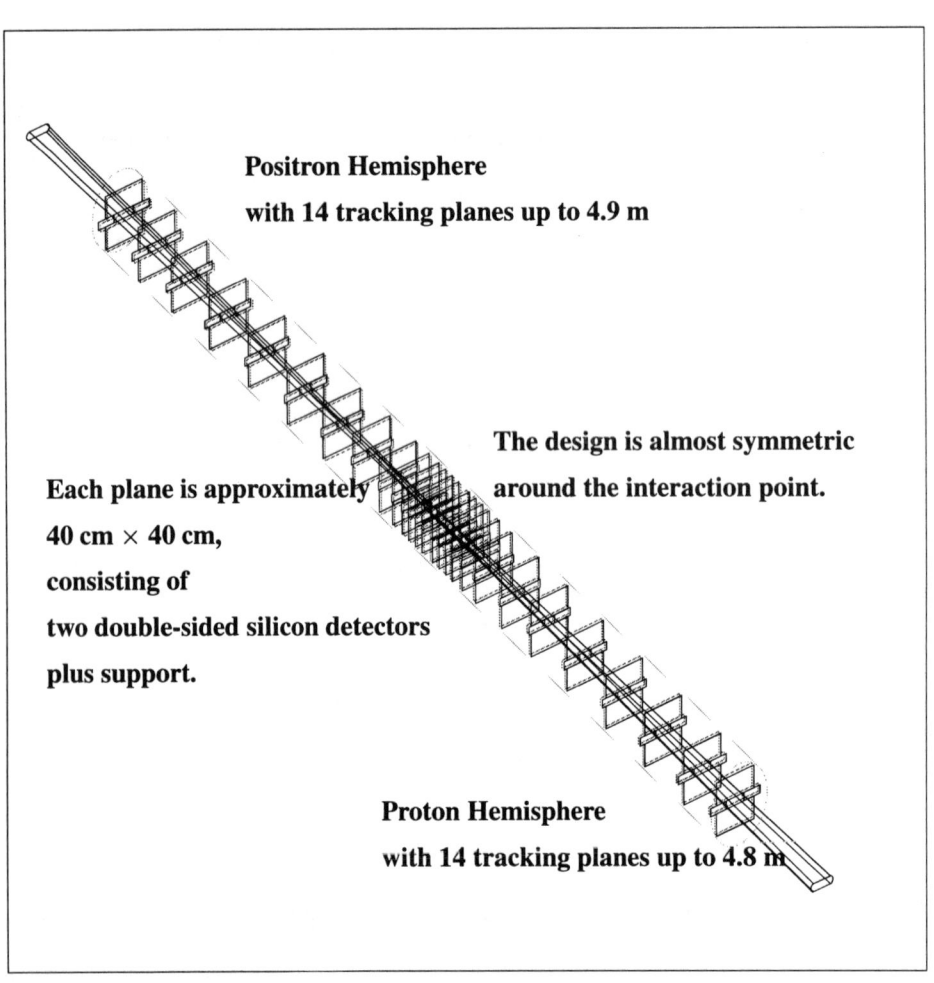

FIGURE 1. Overview over the silicon tracker. The system is confined to cylinder with a radius of 30 cm. Also visible is the widening beam-pipe.

of a top and a bottom half and two horizontal plugs that are adjusted to the required cut-out. In order to leave space for the end-wall calorimetry the silicon planes extend to 4.9 m in the electron and 4.8 m in the proton hemisphere. Figure 1 gives an overview of the design.

The **electromagnetic calorimetry** is required to match the forward tracking acceptance and to cover the central region. The goal of a compact detector leads to silicon-tungsten as the choice of technology. The active components are confined to a tube with a radius of 60 cm. In order to cover 4π in an elongated design as this, multiple structures have to be adapted. Figure 2 gives an overview over the full detector, Fig. 3 depicts the central region.

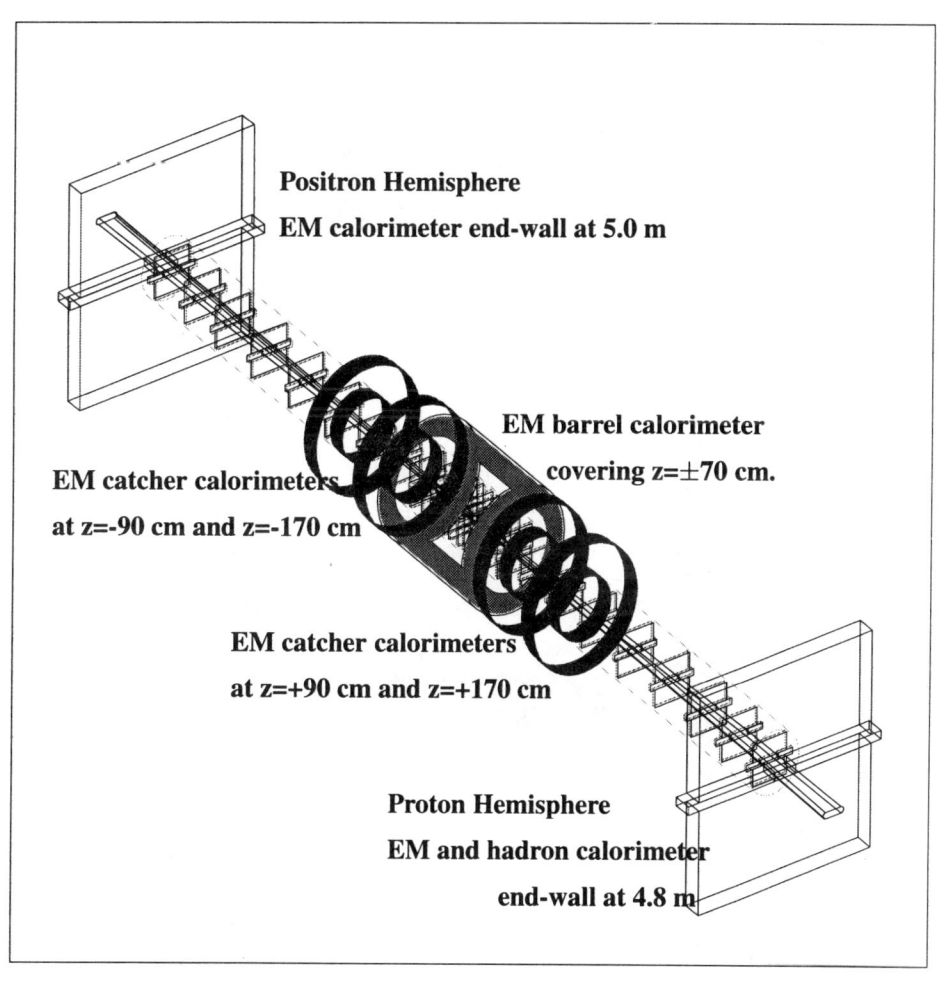

FIGURE 2. Schematic overview over the detector components within ≈ ±6 m of the interaction point. The silicon planes are visible in the center. The calorimeter system consists of a central barrel, 2 catcher rings on each side and 2 end-walls.

The central region in rapidity $[|\eta| < 1.3]$ is covered by a barrel structure with an inner radius of 40 cm. It extends to ±70 cm in z. For a simple, non pointing layer geometry this results in a doubling of the path-length through a layer for a particle coming from the interaction point. The tungsten layers have a thickness of 1.75 mm which is equivalent to half a radiation length $[X_0]$ per layer. For particles hitting the end of the barrel the effective layer thickness is 1 X_0. There are 50 layers resulting in an overall thickness of 25 X_0 at the center. The active medium are 500 μm thick silicon pads. The pad size will be around 1cm×1cm and there will be three longitudinal sections.

High rapidities, $|\eta| > 2.4$, are covered by end-walls. These end-walls cannot be

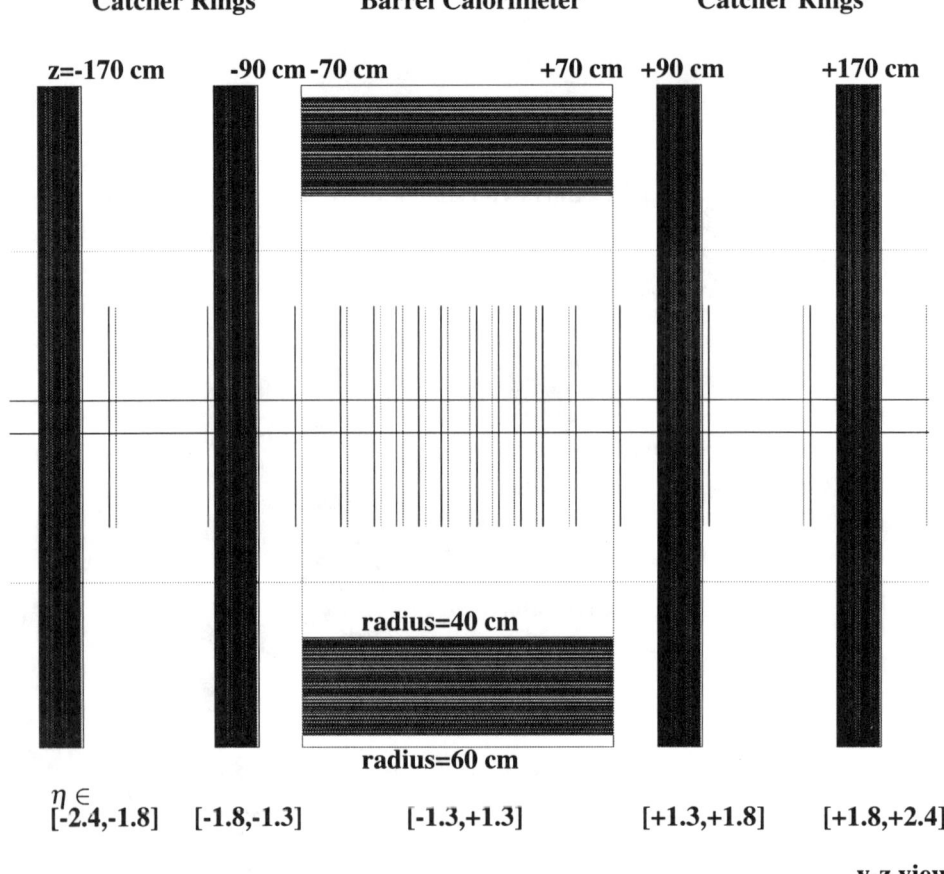

FIGURE 3. Overview over the central part of the em calorimeter system. The system is confined to a cylinder with a radius of 60 cm. Visible are the central barrel and the 4 catcher rings. The sandwich structures have active silicon planes and tungsten absorber plates.

large enough to cover all the non-barrel rapidity range, because of the certainly limited aperture of the magnets. Therefore the intermediate range in rapidity is covered by "catcher rings". Their absorber plates are perpendicular to the beam-line and cover radii from 30 cm to 60 cm. Two catchers are placed on each side of the interaction point. They start at $z=\pm 90$ cm and $z=\pm 170$ cm and cover the rapidity ranges from 1.3 to 1.8 and 1.8 to 2.4 with some reasonable additional overlap of about 10 % of a unit in rapidity.

Both catchers and the end-walls have tungsten absorber plates of 3.5 mm, i.e. $1X_0$, thickness. They have 25 layers and again silicon pads of 500 μm thickness as active elements. In addition there is a need for hadronic calorimetry in the proton hemisphere.

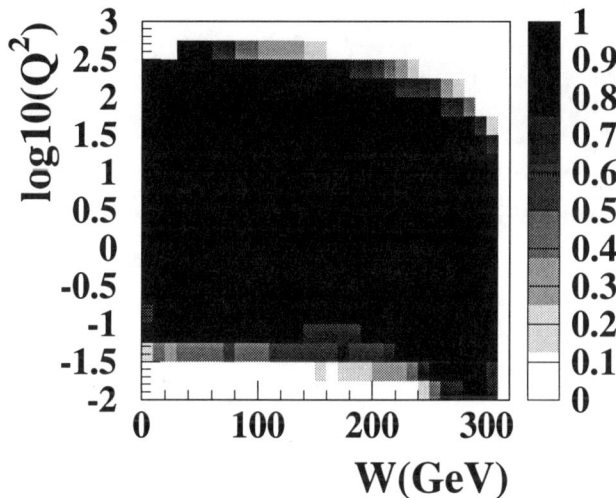

FIGURE 4. Acceptance of the tracking system vs. W and Q^2. A track has to pass 3 stations to count as accepted.

There are possibilities to use parts of existing calorimeters in this design. The requirements of the positron hemisphere end-wall could be met by the HERA-B tungsten shashlik calorimeter. The ZEUS uranium scintillator forward calorimeter with its electromagnetic and hadronic sections could be used as the proton hemisphere end-wall.

DETECTOR PERFORMANCE

The detector performance was studied using a full GEANT3 simulation. The **acceptance** of the tracking system is depicted in Fig. 4. The excellent coverage of low Q^2 events should be noted.

The **momentum** resolution was studied using different assumptions about the material in the detector, as shown in Table 1. These assumptions do not correspond to real technical designs; they are merely used to study the influence of material to guide later designs.

TABLE 1. Material budgets used to study the momentum resolution. 600 μm silicon represent 2 silicon wafers, 1.2 mm CF assume a uniform support structure of 1.2 mm carbon fiber and the beam-pipe is assume to have a thickness of 500μ m aluminium.

Scenario	silicon	support	beam-pipe
standard	600 μm	1.2 mm CF	500μ m Al
light	600 μm	1.2 mm CF	none
extra-light	600 μm	none	none

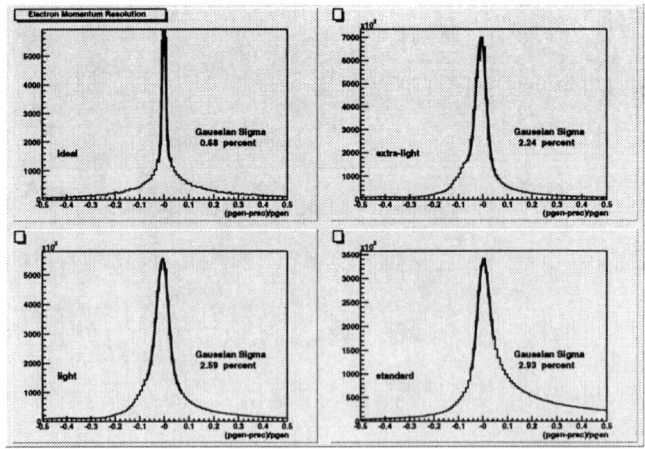

FIGURE 5. Distribution of the normalized difference between generated and reconstructed momentum for an ideal detector, top left, the extra-light, top right, the light, bottom left, and the standard, bottom right, scenario [see also Table 1].

Single track events are used to determine the resolution. All impact points on the planes created in GEANT by a test particle are handed to a fit program. This program treats the components perpendicular and parallel to the magnetic field separately and then recombines the projections. It also smears the projected coordinates, simulated hits, by an assumed hit resolution. Two physics processes directly influence the resolution. One is multiple scattering, the other one Bremsstrahlung. While multiple scattering changes the track in a random way, Bremsstrahlung always causes an energy loss and thus a bias. Some Bremsstrahlung photons convert and cause extra hits. In addition, some of the primary positrons interact and create a shower. All these effects make some pattern recognition, even for a preliminary study, indispensable. Normally a hit resolution σ_h of 20 μm is assumed. After a first fit using all simulated hits, the hit with the largest residual is taken out, if it has a residual larger than 3 σ_h. This procedure is continued until either all hits have a residual of less than 3 σ_h or there are only 3 hits left. A track is removed from the sample if the average residual is larger then 2 σ_h. This affects approximately 4 % of the tracks.

In Fig. 5 the overall distribution of the normalized difference between reconstructed and generated momentum is shown for the scenarios from Table 1 and for an ideal detector with no material. Obviously the material is important for the central part of the distribution which is dominated by multiple scattering. However, it is even more important for the tails caused by Bremsstrahlung. These tails contains up to 20 % of the tracks. The distributions are fit with a central Gaussian and an exponential tail. The central Gaussian part is taken as the resolution $\delta p/p$.

The hit resolution is not critical below at least 50 μm when reasonable amounts of material for a silicon tracker are assumed. This is understandable, as hits in tracking stations with 1% X_0 and a lever arm of about 50 cm, as is the situation for forward tracks in the current design, are subject to distortions of about 70 μm from multiple

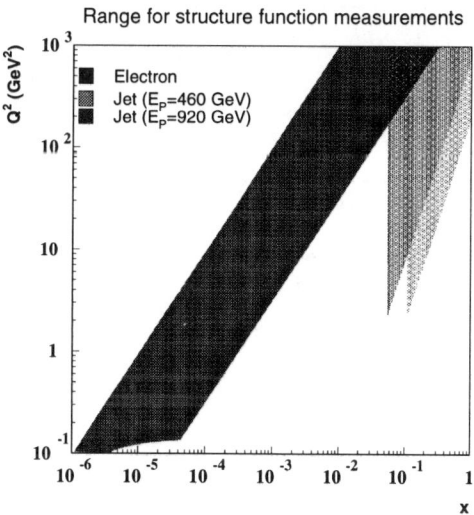

FIGURE 6. The kinematic coverage with the electron method assuming a 1 % electron momentum measurement at small-y (area shaded dark grey) and with a combination of the electron and jet information (area shaded medium grey: $E_P = 920$ GeV, area shaded light grey: $E_P = 460$ GeV). The intermediate region requires a measurement of y from the hadronic system.

scattering alone. However, hit resolutions become more important, if the the material can be reduced. A study of alternative silicon designs based on pixel detectors is underway.

The **energy resolution** for positrons hitting the barrel depends on the rapidity of the particle, i.e. on the impact location. Between the center and the end of the barrel the effective plate thickness doubles. The resolution is

$$14\%\sqrt{E} \oplus 2\% \quad \text{and} \quad 15\%\sqrt{E} \oplus 2\%$$

for the central and end part of the barrel respectively. The situation in the catcher rings and end-wall[s] is similar to that at the end of the barrel, as the plate-thickness was chosen to also correspond to $1X_0$ there.

e-π Separation is straight-forward for particle energies above 5 GeV. Therefore the current studies focus on the most critical case of the lowest energy accepted, i.e. 2 GeV. In areas where the momentum is measured it is possible to achieve rejection factors of larger than 50 with efficiencies above 95% by using the particle momentum and the longitudinal and transverse shower profile. In the central region where no momentum measurement is planned the rejection factor drops to around 25. This is still acceptable, as the total number of pion tracks in the positron hemisphere is generally less than 4.

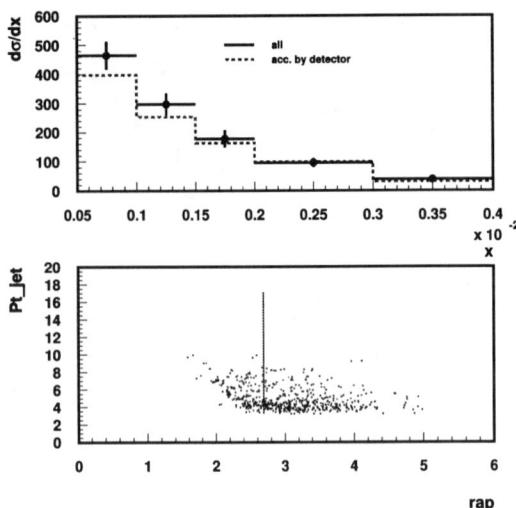

FIGURE 7. A comparison between generated and accepted cross-section for events with a forward jet is shown in the top figure. The bottom figure depicts the acceptance for forward jets in the rapidity η and transverse momentum p_T plane. The range covered by H1 and ZEUS is to the left of the vertical line.

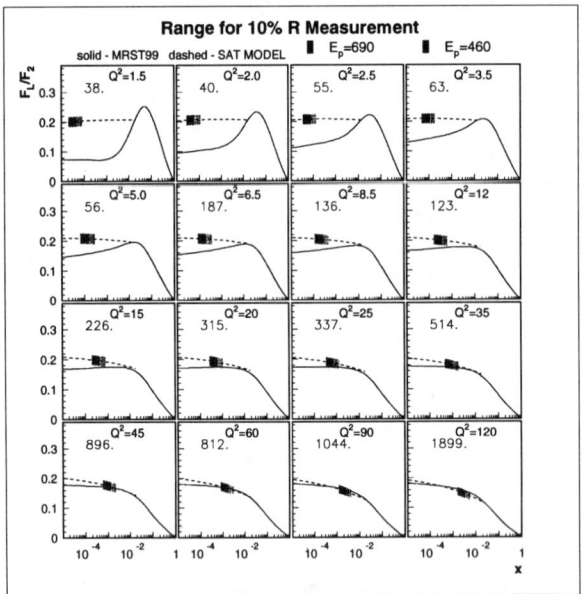

FIGURE 8. The x-range where a 10 % or better measurement of F_L is possible is shown in different Q^2 bins. The area shaded in light grey is deduced from data with $E_p = 920$ and $E_p = 460$ GeV, while the area shaded in dark grey is determined by the combination of $E_p = 920$ and $E_p = 690$ GeV. The required luminosity is given in each bin in the upper left corner (in pb^{-1}). Note that bins will be combined at high Q^2 to obtain statistically significant measurements. Plausible predictions are indicated by the solid and the dashed lines.

PHYSICS PERFORMANCE

The kinematic range targeted for F_2 is depicted in Fig. 6. It should be noted that though the main focus is to extend the accessible range to lower x, there are also opportunities at large x by using jet information. Figure 7 demonstrates the good acceptance for events with a forward jet. In addition Fig. 7 provides a comparison between the jet acceptance of the detector proposed here and the existing detectors ZEUS and H1.

Figure 8 demonstrates the power of this detector to distinguish between different predictions for F_L in the low x, low Q^2 regime where theoretical prediction have a large variation.

CONCLUSIONS

A new detector especially focused on the low x, low Q^2 kinematic regime of e p physics at HERA would be able to make a major contribution towards the better understanding of the proton structure. A detector based on silicon tracking and silicon-tungsten-calorimetry within a dipole magnetic field can provide the required performance.

REFERENCES

1. H. Abramowicz et al, A New Detector foe HERA Letter of Intent submitted to DESY PRC, May 2003, MPI-PhE/2003-06

CIPANP2003 PLENARY TALK PROGRAM

CIPANP2003 Conference on the Intersections of Particle and Nuclear Physics

Grand Hyatt Hotel – New York City

May 19–24, 2003

Tuesday May 20	*8:20–8:30 a.m.*	Welcome	
	8:30–9:10 a.m.	Electroweak Physics	P. Langacker
	9:15–9:55 a.m.	Dipole Moments	B. L. Roberts
	10:00–10:30 a.m.	Coffee Break	
	10:30–11:10 a.m.	Strangeness	R. Jaffe
	11:15–11:55 a.m.	Electron-Nucleus Scattering	E. Beise
	12:00–1:40 p.m.	Lunch	
	1:40–3:20 p.m.	Parallel Talks Session 1	
	3:20–3:50 p.m.	Coffee Break	
	3:50–5:30 p.m.	Parallel Talks Session 2	
	6:15–7:15 p.m.	Reception in Grand Hyatt	
Wednesday May 21	*8:30–9:10 a.m.*	Structure Functions	J. C. Peng
	9:15–9:55 a.m.	RHIC Results	B. Jacak
	10:00–10:30 a.m.	Coffee Break	
	10:30–11:10 a.m.	QCD	D. Kharzeev
	11:15–11:40 a.m.	Recent Neutron Spin Results	Z. Meziani
	11:45–12:10 p.m.	Recent E158 Results	P. Souder
	12:15–1:40 p.m.	Lunch	
	1:40–3:20 p.m.	Parallel Talks Session 1	
	3:20–3:50 p.m.	Coffee Break	
	3:50–5:30 p.m.	Parallel Talks Session 2	

Thursday May 22	8:30–9:10 a.m.	Neutrino Physics	S. Freedman
	9:15–9:55 a.m.	Particle and Nuclear Astrophysics	J. Beacom
	10:00–10:30 a.m.	Coffee Break	
	10:30–11:10 a.m.	Theoretical Perspective on WMAP	R. Easther
	11:15–11:40 a.m.	Recent Results from KLOE	M. Moulson
	11:45–12:10 p.m.	Recent Results NA48	P. Rubin
	12:15–1:40 p.m.	Lunch	
	1:40–3:20 p.m.	Parallel Talks Session 1	
	3:20–3:50 p.m.	Coffee Break	
	3:50–5:30 p.m.	Parallel Talks Session 2	
	6:30–7:30 p.m.	Reception	
	7:30–10:00 p.m.	Conference Dinner	
Friday May 23	8:30–9:10 a.m.	Lattice Gauge Theory	R. Mawhinney
	9:15–9:55 a.m.	Heavy Quark Physics	S. Stone
	10:00–10:30 a.m.	Coffee Break	
	10:30–11:10 a.m.	Hadron Collider Results	T. Liss
	11:15–11:40 a.m.	Time Varying Fundamental Constants	V. Flambaum
	11:45–12:25 p.m.	Extra Dimensions	J. Lykken
	12:30–1:50 p.m.	Lunch	
	1:50–3:30 p.m.	Parallel Talks Session 1	
	3:30–4:00 p.m.	Coffee Break	
	4:00–5:40 p.m.	Parallel Talks Session 2	

Saturday
May 22

8:20–10:00 a.m.	Parallel Summary Talks Sessions 1 to 5	
10:00–10:30 a.m.	Coffee Break	
10:30–12:10 p.m.	Parallel Summary Talks Sessions 6 to 10	
12:15–1:40 p.m.	Lunch	
1:40–2:20 p.m.	Future Accelerators	J. Womersley
2:25–3:05 p.m.	National Underground Lab	J. Wilkerson
3:10–4:00 p.m.	Outlook	F. Wilczek

CIPANP2003 List of Registered Participants*

Abt, Iris D.	MPI Fuer Physik
Adams, Gary S.	Rensselaer Polytechnic Institute
Afanasev, Andrei	Jefferson Lab.
Aidala, Christine A.	Columbia University
Akiba, Yasuyuki	RIKEN
Aleem, Fazal	Centre for High Energy Physics
Anatoly, Maltsev A.	JINR
Aoki, Yasumichi	Brookhaven National Laboratory
Aphecetche, Laurent B	Subatech
Arakelyan, Gevorg H.	Yerevan Physics Institute
Aronson, Samuel H.	Brookhaven National Laboratory
Arrington, John R.	Argonne National Laboratory
Artuso, Marina	Syracuse University
Aslanyan, Petros Z.	JINR/LHE
Avagyan, Harut	Jefferson Lab.
Baker, Mark D.	Brookhaven National Laboratory
Baltz, Anthony J.	Brookhaven National Laboratory
Baltz, Edward A.	Columbia Astrophysics Laboratory
Barannikova, Olga	Purdue University
Barenboim, Gabriela	Fermi National Accelerator Laboratory
Barish, Kenneth N.	University Of California-Riverside
Barnes, Peter D.	Los Alamos National Laboratory
Bass, Steffen A.	Duke University
Batsouli, Sotiria	Columbia University
Beacom, John F. .	Fermi National Accelerator Laboratory
Beck, Douglas H.	University Of Illinois-Urbana Champaign
Beck, Marcus	K. U. Leuven
Beise, Elizabeth J.	University Of Maryland
Belitsky, Andrei	University Of Maryland
Belli, Pierluigi	INFN-Sezione Roma 2
Bellis, Matthew	Rensselaer Polytechnic Institute
Berger, Bruce E.	Lawrence Berkeley National Laboratory
Berger, Edmond L.	Argonne National Laboratory
Bhuyan, Bipul	Brookhaven National Laboratory
Bijnens, Johan L.	Lund University
Black, Deirdre M.	Jefferson Lab.
Blanc, Frederic	University Of Colorado
Blum, Tom	Brookhaven National Laboratory
Bodek, Arie	University Of Rochester
Borean, Cristiano	University And INFN Of Trieste
Bowers, Jeffrey A.	Massachusetts Institute Of Technology
Bowman, David J.	Los Alamos National Laboratory
Boyd, Steven B.	University Of Pittsburgh
Bravar, Alessandro	Brookhaven National Laboratory

Brooker, Carl	Los Alamos National Laboratory
Browder, Thomas E.	University Of Hawaii
Bueltmann, Stephen	Brookhaven National Laboratory
Bugg, William M.	University Of Tennessee
Bunce, Gerry	Brookhaven National Laboratory
Burkert, Volker	Jefferson Lab.
Burles, Scott M.	Massachusetts Institute Of Technology
Butler, Malcolm N.	Saint Mary's University
Canelli, Florencia	University Of Rochester
Chatterjee, Lali	Institute Of Physics Publishing
Chattopadhyay, Swapan	Jefferson Lab.
Chen, Mu-Chun	Brookhaven National Laboratory
Chen, Zhang	Manhattanville College
Chitwood, Daniel B.	University Of Illinois-Urbana
Choudhury, Deo C.	Polytechnic University
Chujo, Tatsuya	Brookhaven National Laboratory
Cinabro, David	Wayne State University
Coan, Thomas E.	Southern Methodist University
Cooper, Martin D.	Los Alamos National Laboratory
Cooper, Peter S.	Fermi National Accelerator Laboratory
Cowan, Ray F.	Massachusetts Institute Of Technology
Cranshaw, Jack M.	Texas Tech University
Cummings, John P.	Rensselaer Polytechnic Institute
Czarnecki, Andrzej	University Of Alberta
D'Enterria, David G.	Columbia University
De Jager, Kees	Jefferson Lab.
Debbe, Ramiro R.	Brookhaven National Laboratory
Derevianko, Andrei	University Of Nevada-Reno
Deshpande, Abhay L.	Brookhaven National Laboratory
Diwan, Milind V.	Brookhaven National Laboratory
Djordjevic, Magdalena R.	Columbia University
Dokshitzer, Yuri L	Univ. Paris-6
Easther, Richard J.	Iscap, Columbia University
Echmalian, Melanie T.	Brookhaven National Laboratory
Ecklund, Karl M.	Cornell University
Egiyan, Kim S.	Yerevan Physics Institute
Eitel, Klaus G.	Forschungszentrum Karlsruhe
Empl, Anton	University Of Houston
Ent, Rolf	Jefferson Lab.
Ferroglia, Andrea	Phys. Inst. Albert-Ludwigs-Univ. Freiburg
Fields, Douglas E.	University Of New Mexico
Flambaum, Victor	University Of New South Wales
Fleming, Bonnie T.	Fermi National Accelerator Laboratory
Fleming, Sean	Carnegie Mellon University
Fox, Brendan D.	Brookhaven National Laboratory

Fraenkel, Zeev	Weizmann Institute Of Science
Franklin, Wilbur A.	Mit-Bates Linear Accelerator Center
Freedman, Stuart J.	University Of California At Berkeley
Gadiyak, Valeriya G.	Columbia Univesrity
Gagliardi, Carl A.	Texas A&M University
Gainutdinov, Renat	Kazan State University
Galik, Richard S.	Cornell University/Cleo
Gallagher, Hugh	Tufts University
Gamberg, Leonard P.	Penn State Berks
Geesaman, Donald F.	Argonne National Laboratory
Gainutdinov, Renat	Kazan State University
Galik, Richard S.	Cornell University/Cleo
Gallagher, Hugh	Tufts University
Gamberg, Leonard P.	Penn State Berks
Geesaman, Donald F.	Argonne National Laboratory
Gelman, Boris	University Of Arizona
Gialas, Ioannis	University Of Aegean
Goldhaber, Maurice	Brookhaven National Laboratory
Goldin, Daniel	University Of Basel
Goldman, Terrance J.	Los Alamos National Laboratory
Gondolo, Paolo	Case Western Reserve University
Gottschalk, Erik	Fermi National Accelerator Laboratory
Goulianos, Konstantin	Rockefeller University
Groer, Leslie S.	Columbia University/Dzero
Gronberg, Jeffrey B.	Lawrence Livermore National Laboratory
Grosse Perdekamp, Matthias	UIUC/RBRC
Hahn, Richard L.	Brookhaven National Laboratory
Hallman, Timothy J.	Brookhaven National Laboratory
Han, Tao	University Of Wisconsin-Madison
Hardy, John C.	Texas A&M University
Harrison, Michael	Brookhaven National Laboratory
Hart, Terrence L.	University Of Colorado
Hasch, Delia	LNF-INFN
Hasuko, Kazumi	RIKEN
Hawker, Eric A.	University Of Cincinnati
Heeger, Karsten M.	Lawrence Berkeley National Laboratory
Henderson, Stuart D.	Oak Ridge National Laboratory
Hertzog, David W.	University Of Illinois-Urbana Champaign
Heuser, Johann M.	RIKEN
Hicks, Kenneth H.	Ohio University
Hiejima, Hiroyoshi	University Of Illinois-Urbana Champaign
Hogan, Gary E.	Los Alamos National Laboratory
Holstein, Barry R.	University Of Massachusetts
Honscheid, Klaus	Ohio State University
Huang, Xiaobo	Boston University

Huentemeyer, Petra H.	University Of Utah
Huseinov, Vali A.	Nakhchivan State University
Hyde-Wright, Charles E.	Old Dominion University
Igo, George J.	University Of California-Los Angeles
Islam, Muhammad M.	University Of Connecticut
Ito, Takeyasu	California Institute Of Technology
Jacak, Barbara	SUNY Stony Brook
Jacobs, Peter M.	Lawrence Berkeley National Laboratory
Jaffe, David E.	Brookhaven National Laboratory
Jaffe, Robert L.	Massachusetts Institute Of Technology
Jalilian-Marian, Jamal	Brookhaven National Laboratory
Jeffrey, Ruth M.	Brookhaven National Laboratory
Jia, Jiangyong	SUNY Stony Brook
Jia, Yu	Michigan State University
Jiang, Xiaodong	Rutgers University
Johnson, Brant M.	Brookhaven National Laboratory
Joo, Kyungseon	University Of Connecticut
Kabuss, Eva-Maria	University Of Mainz
Kainnady, Jacob V.	Kingborough Community College
Kakuno, Hidekazu	Tokyo Institute Of Technology
Karle, Albrecht	University Of Wisconsin-Madison
Kawall, David M.	Yale University
Kelly, James J.	University Of Marylanad
Keppel, Cynthia E.	Hampton University/Jefferson Lab.
Kezerashvili, Roman Y.	New York City College Of Technology
Kharzeev, Dmitri E.	Brookhaven National Laboratory
Kilminster, Benjamin J.	University Of Rochester Cdf
Kinney, Edward R.	University Of Colorado
Kirk, Thomas B.	Brookhaven National Laboratory
Kistenev, Edouard P.	Brookhaven National Laboratory
Kitano, Ryuichiro	Institute For Advanced Study
Klay, Jennifer L.	Lawrence Berkeley National Laboratory
Kobayashi, Kazuyoshi	SUNY Stony Brook
Kolb, Peter F.	SUNY Stony Brook
Komatsu, Masahiro	Nagoya University
Kopytine, Mikhail L.	Kent State University
Koubarovski, Valeri	Rensselaer Polytechnic Institute
Kovchegov, Yuri V.	University Of Washington
Kowalski, Stanley	Massachusetts Institute Of Technology
Kraft, David W.	U.S. Military Academy
Kretzer, Stefan	Brookhaven National Laboratory
Krisch, Alan D.	University Of Michigan
Krutenkova, Anna	Inst. Of Theoretical And Experimental Physics
Kumar, Krishna S.	University Of Massachusetts-Amherst
Kumar, Sushil S.	Research Fellow

Kuno, Yoshitaka	Osaka University
Kurylov, Andriy	California Institute Of Technology
Laiho, John W.	Princeton University
Lamoreaux, Steven K.	Los Alamos National Laboratory
Langacker, Paul	University Of Pennsylvania
Lauss, Bernhard	University Of California-Berkeley
Lee, Christopher	California Institute Of Technology
Lin, Hueywen	Columbia University
Link, Jonathan M.	Columbia University
Lipkin, Harry J.	Weizmann Institute Of Science
Liss, Tony M.	University Of Illinois
Liu, Ming X.	Los Alamos National Laboratory
Lorenzon, Wolfgang B.	University Of Michigan
Lykken, Joseph D.	Fermi National Accelerator Laboratory
Magestro, Daniel	Ohio State University
Makins, Naomi C.	University Of Illinois-Urbana Champaign
Malek, Matthew S.	SUNY Stony Brook
Marciano, William J.	Brookhaven National Laboratory
Maurizio, Iori	University Of Rome
Mawhinney, Robert D.	Columbia University
May, Morgan	Brookhaven National Laboratory
Mccauley, Neil K.	University Of Pennsylvania
Mcdonald, Kirk T.	Princeton University
Mcfarland, Kevin	University Of Rochester
Meadows, Brian T.	University Of Cincinnati
Mehen, Thomas C.	Duke University
Merminga, Lia	Jefferson Lab.
Mesropian, Christina	The Rockefeller University
Meyer, Curtis A.	Carnegie Mellon University
Meziani, Zein-Eddine	Temple University
Miller, James P.	Boston University
Miller, Michael L.	Yale University
Milner, Richard G.	Massachusetts Institute Of Technology
Minehart, Ralph C.	University Of Virginia
Mintz, Stephan L.	Florida International University
Mitchell, Gregory S.	Los Alamos National Laboratory
Mitchell, Jeffery T.	Brookhaven National Laboratory
Mitchell, Ryan E.	University Of Tennessee
Mocioiu, Irina	University Of Arizona
Mohapatra, Rabindra N.	University Of Maryland
Morningstar, Colin J.	Carnegie Mellon University
Moulson, Matthew	Laboratori Nazionali Di Frascati/INFN
Mueller, Berndt	Duke University
Muto, Ryotaro	RIKEN
Nagle, James L.	University Of Colorado-Boulder

Napolitano, Jim	Rensselaer Polytechnic Institute
Navas, Sergio S.	University Of Granada
Neubert, Matthias	Cornell University
Neumeister, Norbert	CERN/HEPHY
Niebuhr, Carsten	DESY
Noaki, Jun-Ichi	Brookhaven National Laboratory
Norman, Eric B.	Lawrence Berkeley National Laboratory
Nozar, Mina	Jefferson Lab.
O'brien, Edward J.	Brookhaven National Laboratory
Oganessyan, Karo A.	INFN-LNF
Ohta, Shigemi	IPNS/KEK and RBRC/BNL
Ojha, Indrani D.	Vanderbilt University
Olsson, Martin G.	University Of Wisconsin
Orrell, John L.	Cen. For Exp. Nucl. Phys. And Astrophysics
Ossola, Giovanni	New York University
Oyunzendem, Handmaa	Natl. Univ. Of Mongolia Nuclear Res. Centre
Page, Philip R.	Los Alamos National Laboratory
Pak, Robert	Brookhaven National Laboratory
Parrish, Sabrina L.	Brookhaven National Laboratory
Parsa, Zohreh	Brookhaven National Laboratory
Patel, Yogeshbhai S.	Research In Applied Physics
Paul, Peter	Brookhaven National Laboratory
Peng, Jen-Chieh	University Of Illinois
Peters, Klaus J.	Ruhr-University Bochum
Petrov, Alexey A.	Wayne State University
Pile, Philip H.	Brookhaven National Laboratory
Pirjol, Dan	Johns Hopkins University
Pitzl, Daniel	DESY
Polly, Charles C.	University Of Illinois-Urbana Champaign
Ptitsyn, Vadim	Brookhaven National Laboratory
Qiu, Jianwei	Iowa State University
Raby, Stuart	The Ohio State University
Rak, Jan	Iowa State University
Ramsey-Musolf, Michael J,	California Institute Of Technology
Redlinger, George	Brookhaven National Laboratory
Reimer, Paul E.	Argonne National Laboratory
Relyea, David R.	Stanford Linear Accelerator Center
Roberts, B. Lee	Boston University
Roland, Gunther M.	Massachusetts Institute Of Technology
Rosen, Jerome L.	Northwestern University
Rosenberg, Leslie J.	Lawrence Livermore National Laboratory
Rowson, Peter	Stanford Linear Accelerator Center
Rubin, Philip D.	Univ. Of Richmond/Univ. Of Edinburgh/CERN
Rumerio, Paolo	Northwestern University
Russ, James S.	Carnegie Mellon University

Samios, Nicholas P.	Brookhaven National Laboratory
Saunders, Alexander	Los Alamos National Laboratory
Schubnell, Michael	University Of Michigan
Schwartz, Alan J.	University Of Cincinnati
Seidl, Ralf	University Of Erlangen
Selen, Mats A.	University Of Illinois-Urbana Champaign
Semertzidis, Yannis K.	Brookhaven National Laboratory
Seo, Seon-Hee	University Of Minnesota
Seth, Kamal K.	Northwestern University
Shebeko, Alexandr V.	NSC Kharkov Institute Of Physics & Technology
Sheldon, Paul	Vanderbilt University
Sher, Alexander	Santa Cruz Inst. For Particle Phys.-Uc Santa Cruz
Shul'ga, Nikolai F.	NSC Kharkov Institute Of Physics & Technology
Silvermyr, David O.	Los Alamos National Laboratory
Silvestrini, Luca	INFN-Rome
Simone, James N.	Fermi National Accelerator Laboratory
Singh, Jaivir	Panjab University-Chandigarh
Sirlin, Alberto	New York University
Smith, Elton	Jefferson Lab.
Sobel, Henry	University Of California-Irvine
Soni, Amarjit	Brookhaven National Laboratory
Sorel, Michel	Columbia University
Souder, Paul	Syracuse University
Stanev, Todor S.	Bartol Research Institute/Univ. Of Delaware
Stein, Jeffrey S.	American Physical Society Member
Steinberg, Peter A.	Brookhaven National Laboratory
Stepanyan, Stepan	Jefferson Lab.
Stoler, Paul	Physics Dept., Rensselaer Polytechnic Inst.
Stone, James	Boston University
Stone, Sheldon	Syracuse University
Su, Shufang	California Institute Of Technology
Sulak, Lawrence R.	Boston University
Swain, John D.	Northeastern University
Takahashi, Jun	University Of Sao Paulo
Tang, Aihong	Brookhaven National Laboratory
Tannenbaum, Michael J.	Brookhaven National Laboratory
Tojo, Junji	RIKEN
Trodden, Mark	Syracuse University
Trueman, Larry	Brookhaven National Laboratory
Tuchin, Kirill	University Of Washington
Tuts, Philip M.	Columbia University
Urheim, Jon	University Of Minnesota
Urner, David C.	Cornell University
Vainshtein, Arkady	William I. Fine Theoretical Physics Institute
Van De Sande, Brett D.	Geneva College

Van Kolck, Ubirajara	University Of Arizona
Variola, Alessandro	INFN-Genova
Vitev, Ivan M.	Iowa State University
Vogelsang, Werner	Brookhaven National Laboratory
Volmer, Jochen	DESY-Zeuthen
Voloshin, Mikhail	William I. Fine Theoretical Physics Institute
Vorov, Oleg K.	Grand Accelerateur National D'ions Lourds
Wascko, Morgan O.	Lousiana State University
Weinstock, Steffen M.	Brookhaven National Laboratory
Wendland, Juergen	Simon Fraser University
Westerhoff, Stefan	Columbia University
Wilczek, Frank A.	Massachusetts Institute Of Technology
Wilkerson, John F.	University Of Washington
Wintz, Peter	Research Centre Juelich
Woehri, Hermine K.	CERN/EP
Womersley, John	Fermi National Accelerator Laboratory
Wong, Tim K.	Researcher
Xu, Zhangbu.	Brookhaven National Laboratory
Yamada, Norikazu	Brookhaven National Laboratory
Yamin, Peter	Brookhaven National Laboratory
Yokoya, Hiroshi	Brookhaven National Lab./Hiroshima Univ.
Yonehara, Katsuya	University Of Michigan
Yoshida, Rik	Argonne National Laboratory
Young, Albert R.	North Carolina State University
Zadorozhny, Vladimir F.	Institute Of Cybernetics
Zmeskal, Johann	Institute For Medium Energy Physics

---

*The total attendance at this conference was 415. This list only includes those who have paid the registration fee in full.

Author Index

A

Abdel-Rehim, A., 522
Abt, I., 816
Adams, T., 95
Ahrens, L., 643
Aidala, C., 793
Akiba, Y., 785
Alekseev, I., 643
Allen, T. J., 562
Alton, A., 95
Ames, F., 180
Amoretti, M., 205
Andersen, V., 349
Aoki, A., 196
Aoki, M., 200
Aoki, Y., 243
Aphecetche, L., 673
Arnaldi, R., 713
Arrington, J., 148
Auzinsh, M., 196, 200
Avakian, H., 612
Avvakumov, S., 95
Awes, T. C., 100

B

Back, B. B., 677
Bai, M., 643
Baker, M. D., 677
Balakin, V., 196, 200
Ballarini, F., 349
Ballintijn, M., 677
Baltz, E.A., 374
Banicz, K., 713
Barton, D. S., 677
Batsouli, S., 718
Battistoni, G., 349
Bay, F., 192
Bazhan, A., 196, 200
Bechstedt, U., 763
Beck, D., 180
Beck, M., 180
Becker, B., 677
Beddo, M. E., 100
Beise, E. J., 23

Belitsky, A. V., 607
Belli, P., 328
Bellis, M., 535
Bennett, G. W., 196, 200
Berger, B. E., 258, 314
Bernabei, R., 328
Bernstein, R. H., 95
Betts, R. R., 677
Bickley, A. A., 677
Bickman, S., 192
Bijnens, J., 407
Bindel, R., 677
Black, D., 478, 522
Blanc, 417
Bodek, A., 95, 148
Bolton, T., 95
Bonomi, G., 205
Borer, K., 713
Borisov, N. S., 639
Borozdin, K. N., 755
Boutcha, A., 205
Bowe, P., 205
Boyd, S., 95
Brau, J., 95
Bravar, A., 643
Brooks, M. L., 100
Brown, C. N., 100
Buchholz, D., 95
Budd, H., 95, 148
Budzanowski, A., 677
Bugel, L., 95
Bunce, G., 643
Bush, J. D., 100
Busza, W., 677
Butler, M. N., 278
Buytaert, J., 713

C

Campanella, M., 349
Canelli, M. F., 469
Cappella, F., 328
Carboni, M., 349
Carey, R. M., 196, 200
Carey, T. A., 100
Carraro, C., 205

Carroll, A., 677
Castor, J., 713
Cerulli, R., 328
Cerutti, F., 349
Cesar, C. L., 205
Chang, T. H., 100
Charlton, M., 205
Chaurand, B., 713
Chen, M.-C., 222
Chen, W., 713
Cheynis, B., 713
Chiba, J., 138
Choudhury, D. C., 345
Chujo, T., 732
Cicalo, C., 713
Cinabro, D., 452
Ciuchini, M., 422
Coan, T. E., 461
Coeck, S., 180
Colla, A., 713
Conrad, J., 95
Cooper, W. E., 100
Cortese, P., 713
Cowan, R. F., 489
Cranshaw, J., 474
Cushman, P., 196, 200

D

Dai, C. J., 328
David, A., 713
Day, D. B., 655
de Barbaro, L., 95
de Barbaro, P., 95
Debbe, R., 690
Debevec, P. T., 196, 200
Decowski, M. P., 677
de Falco, A., 713
Delahaye, P., 180
Delauré, B., 180
de Marco, N., 713
DeMille, D., 192
Derbenev, Y., 811
Devaux, A., 713
Dhawan, S., 643
Doser, M., 205
Drees, A., 713
Drucker, R. B., 95
Ducroux, L., 713

Dudnikov, A., 196, 200

E

Easther, R., 64
Ecklund, K. M., 432
Eitel, K., 265
Elouadrhiri, L., 612
Empl, A., 349
En'yo, H., 138, 713
Eversheim, D., 763

F

Fariborz, A. H., 522
Farley, F. J. M., 196, 200
Fassò, A., 349
Ferrari, A., 349
Ferretti, A., 713
Filippini, V., 205
Fimushkin, V. V., 639
Fleming, B. T., 95, 314
Fleming, S., 579
Floris, M., 713
Fontana, A., 205
Force, P., 713
Formaggio, J., 95
Franco, E., 422
Franklin, W. A., 751
Freedman, S. J., 303
Frey, R., 95
Fujiwara, M., 205
Fukao, Y., 138
Funahashi, H., 138
Funakoshi, R., 205

G

Gadioli, E., 349
Gagliardi, C. A., 100, 234
Galik, R. S., 575
Gallagher, H., 153, 293
Gamberg, L., 617
García, E., 677
Garvey, G. T., 100
Garzelli, M. V., 349
Gburek, T., 677

Gebel, R., 763
Geesaman, D. F., 100, 801
Genova, P., 205
George, N., 677
Gialas, I., 115
Goldhaber, M., 390
Goldin, D., 134
Goldman, J., 95
Goldstein, G. R., 617
Golovko, V. V., 180
Gomez, J., 755
Goncharov, M., 95
Gottschalk, E. E., 503
Goulianos, K., 110
Grigorian, A., 713
Groer, L. S., 238
Gronberg, J., 759
Grosse Perdekamp, M., 628, 659
Grossiord, J.-Y., 713
Guenaut, C., 180
Guettet, N., 713
Guichard, A., 713
Gulbrandsen, K., 677
Gulkanian, H., 713
Gushue, S., 677

Hertzog, D. W., 196, 200
Heuser, J. M., 713
Hicks, K. H., 539
Hinterberger, F., 763
Hofman, D. J., 677
Hogan, G. E., 755
Hollis, R. S., 677
Holstein, B. R., 176
Hołyński, R., 677
Holzman, B., 677
Huang, H., 643
Hughes, V., 643
Hüntemeyer, P., 341
Hwang, D. S., 624

I

Ieiri, M., 138
Igo, G., 643
Incicchitti, A., 328
Iordanova, A., 677
Isenhower, L. D., 100
Ishino, M., 138
Islam, M. M., 142
Ito, T. M., 161
Iwasaki, M., 196, 200

H

Hahn, R. L., 767
Halliwell, C., 677
Hallman, T., 735
Hall Reno, M., 165
Hamagaki, H., 138
Hamblen, J., 677
Han, T., 209
Hangst, J. S., 205
Hardtke, D. H., 729
Hardy, J. C., 184
Harrington, A. S., 677
Harris, B. K., 639
Harris, D. A., 95
Hart, T., 270
Hasch, D., 651
Hasuko, K., 628
Hawker, E. A., 100, 283
Hayano, R. S., 205
He, X. C., 100
Heeger, K. M., 303
Henderson, C., 677

J

Jacobs, P., 667
Jalilian-Marian, J., 725
Jia, Y., 465
Jiang, X., 655
Jiang, Y., 192
Jinnouchi, O., 643
Johnson, E., 677
Johnson, R. A., 95
Jones, M. K., 655
Jorgensen, L. V., 205
Jungmann, K., 196, 200

K

Kabuß, E. M., 599
Kanda, H., 138
Kandes, M. C., 639
Kane, J. L., 677

851

Kaplan, D. M., 100
Kaufman, S. B., 100
Kawall, D., 192, 200
Keil, M., 713
Kelly, J. J., 393
Kezerashvili, R. Y., 353
Khan, N., 677
Kharzeev, D., 44
Khazin, B., 196, 200
Khriplovich, I. B., 196, 200
Kienman, B. H., 639
Kilminster, B., 218
Kim, J. H., 95
Kirk, B., 200
Kirk, P. N., 100
Kistenev, E., 775
Kitaguchi, M., 138
Klay, J., 667
Kluberg, L., 713
Koetke, D. D., 100
Kolb, P. F., 694
Koutsoliotas, S., 95
Kovchegov, Y. V., 685
Kozlov, V. Y., 180
Kraev, I. S., 180
Kraft, D. W., 345
Kretzer, S., 165, 603
Krisch, A. D., 639, 763
Kuang, H. H., 328
Kubarovsky, V., 543
Kulinich, P., 677
Kuno, Y., 196, 200
Kuo, C. M., 677
Kurita, K., 643
Kurylov, A., 168, 188
Kyle, G., 100

L

Lagomarsino, V., 205
Lamm, M. J., 95
Landua, R., 205
Langacker, P., 1
Lauss, B., 230
Lazarus, D. M., 196, 200
Lee, D. M., 100
Lee, J. W., 677
Lee, K., 349
Lee, W. M., 100

Lehrach, A., 763
Leipuner, L. B., 196, 200
Leitch, M. J., 100
Leonova, M. A., 639
Li, Z., 643, 713
Lin, W. T., 677
Lindelof, D., 205
Lindroth, A., 180
Link, J. M., 771
Lipkin, H. J., 493
Liu, M. X., 647
Lodi Rizzini, E., 205
Logashenko, V., 196, 200
Lorenz, B., 763
Lorenzon, W., 119, 797
Lourenço, C., 704, 713
Lozano, J., 713
Luddy, R. J., 142
Luk, K.-B., 303
Luppov, V. G., 639
Lynch, K. R., 196, 200

M

Ma, J. M., 328
MacKay, W. W., 643
Macrì, M., 205
Madsen, N., 205
Mahanthappa, K. T., 222
Maier, R., 763
Makins, N., 100
Manly, S., 677
Manso, F., 713
Manuzio, G., 205
Marciano, W. J., 196, 200
Marsh, W., 95
Masiero, A., 422
Mason, D., 95
Masoni, A., 713
McCauley, N., 253
McDonald, J., 95
McFarland, K. S., 95
McGaughey, P. L., 100
McNabb, R., 196, 200
McNulty, C., 95
Mehen, T., 508
Meng, W., 196, 200
Merminga, L., 811
Mesropian, C., 105

Meyer, C. A., 554
Meziani, Z.-E., 54
Mignerey, A. C., 677
Mihara, M., 138
Miller, J. P., 196, 200
Miller, M. L., 729
Milner, R. G., 806
Minehart, R. C., 632
Mintz, S. L., 411
Mitchell, G. S., 172
Mitchell, R. E., 559
Miwa, K., 138
Miyashita, T., 138
Montagna, P., 205
Montecchia, F., 328
Morningstar, C., 530
Morozov, V. S., 639, 763
Morris, C., 755
Morse, W. M., 196, 200
Moss, J. M., 100
Moulson, M., 74
Mueller, B. A., 100
Mukherjee, M., 636
Murakami, T., 138
Muto, R., 138

N

Nagle, J. L., 735
Nakura, T., 138
Naples, D., 95
Napolitano, J., 587
Naruki, M., 138
Navas-Concha, S., 298
Neubert, M., 427
Neumeister, N., 741
Neves, A., 713
Nienaber, P., 95
Noaki, J., 247
Noell, A., 677
Nomachi, M., 138
Nord, P. M., 100
Nouicer, R., 677
Nozar, M., 517
Nozzoli, F., 328

O

O'Brien, E. J., 745
Oganessyan, K. A., 617
Ogawa, A., 628
Ohnishi, H., 713
Ohta, S., 398
Olson, J. B., 639
Olsson, M. G., 562
Olszewski, A., 677
Onderwater, C. J. G., 196, 200
Oppedisano, C., 713
Orlov, Y. F., 196, 200
Orrell, J. L., 275
Ottolenghi, A., 349
Ozawa, K., 138
Ozben, C. S., 196, 200

P

Page, P. R., 513
Pak, R., 677
Papavassiliou, V., 100
Park, B. K., 100
Park, I. C., 677
Parracho, P., 713
Parsa, Z., 307
Pelliccioni, M., 349
Peng, J. C., 33, 100
Pernegger, H., 677
Peters, C. C., 639
Petitt, G., 100
Petrov, A. A., 456
Phalet, T., 180
Pinsky, L. S., 349
Pirjol, D., 548
Polly, C. C., 226
Prasuhn, D., 763
Priedhorsky, W. C., 755
Prigl, R., 196, 200
Prokudin, A. V., 142
Prosperi, D., 328
Prudkoglyad, A. F., 639
Pruys, H., 205
Puddu, G., 713

Q

Qiu, J., 659
Quint, W., 180

R

Radermacher, E., 713
Radescu, V., 95
Ramsey-Musolf, M. J., 168, 188
Ranft, J., 349
Raymond, R. S., 639, 763
Reed, C., 677
Regenfus, C., 205
Reimer, P. E., 100
Remsberg, L. P., 677
Rescia, S., 200, 643
Riedler, P., 205
Roberts, B. L., 13, 196, 200
Roesler, S., 349
Rohdjess, H., 763
Roland, C., 677
Roland, G., 677
Romosan, A., 95
Rosen, J. L., 571
Rosenberg, L. J., 332
Roser, T., 643
Rosinsky, P., 713
Rotondi, A., 205
Rubin, P. D., 80
Russ, J., 587

S

Sadler, M. E., 100
Sagerer, J., 677
Saito, N., 643
Sakuma, F., 138
Sakumoto, W. K., 95
Sala, P. R., 349
Sarin, P., 677
Sasaki, O., 138
Sato, A., 196
Saunders, A., 755
Sawicki, P., 677
Schat, C., 548
Schechter, J., 522
Schellman, H., 95
Schnase, A., 763
Scholberg, K., 362
Schubnell, M., 323
Schultz, L. J., 755
Schwartz, A. J., 440
Scobel, W., 763

Scomparin, E., 713
Sedykh, I., 677
Seidl, R., 621
Seixas, J., 713
Sekimoto, M., 138
Semertzidis, Y. K., 196, 200
Seo, S. H., 526
Serci, S., 713
Severijns, N., 180
Shaevitz, M. H., 95
Shafer-Ray, N., 196, 200
Shahoyan, R., 713
Sher, A., 381
Siegle, V., 628
Silenko, A., 196, 200
Silvermyr, D., 709
Silvestrini, L., 422
Sirlin, A., 385
Sisco, D. L., 639
Skulski, W., 677
Smith, E. S., 129
Smith, P., C. E., 677
Söeren Lange, J., 628
Sonderegger, P., 713
Sondheim, W. E., 100
Souder, P. A., 59
Spentzouris, P., 95
Spinka, H., 643
Stanev, T., 357
Stankus, P. W., 100
Steinberg, C. E., 677
Steinberg, P., 680
Stepanyan, S., 543
Stephans, G. S. F., 677
Stephenson, E. J., 196, 200
Stern, E. G., 95
Stockhorst, H., 763
Stone, J., 374
Stone, S., 497
Stratmann, M., 636
Su, S., 168, 188
Sukhanov, A., 677
Suwonjandee, N., 95
Svirida, D., 643
Swain, J., 366
Sylvestrov, G., 196

T

Tabaru, T., 138
Tanaka, K. H., 138
Tang, A., 701
Teasdale, M. E., 755
Teng, R., 677
Testera, G., 205
Thompson, T. N., 100
Tieulent, R., 713
Tobien, N., 95
Togawa, M., 138
Tonjes, M. B., 677
Towell, R. S., 100
Tribble, R. E., 100
Trzupek, A., 677
Tuchin, K., 698

U

Ulbrich, K., 763
Underwood, D., 643
Urheim, J., 482, 497
Urner, D., 566
Usai, G., 713

V

Vainshtein, A., 403
Vaitaitis, A., 95
Vakili, M., 95
Vale, C., 677
Van der Werf, D. P., 205
van Nieuwenhuizen, G. J., 677
Vardanyan, H., 713
Variola, A., 205
Vasiliev, M. A., 100
Veenhof, R., 713
Verdier, R., 677
Veres, G. I., 677
Versyck, S., 180
Vitev, I., 721
Vogelsang, W., 636
Volmer, J., 124

Voloshin, M. B., 436, 583

W

Wadsworth, B., 677
Wang, Y. C., 100
Wang, Z. F., 100
Webb, J. C., 100
Wendland, J., 595
Westerhoff, S., 370
Whitten, C., 643
Willis, J. L., 100
Wilson, T. L., 349
Wintz, P., 789
Wise, D. K., 100
Wöhri, H. K., 704, 713
Wolfs, F. L. H., 677
Womersley, J., 85
Wong, V. K., 763
Wood, J., 643
Wosiek, B., 677
Woźniak, K., 677
Wuosmaa, A. H., 677
Wysłouch, B., 677

Y

Yamada, S., 138
Yang, U. K., 95
Ye, Z. P., 328
Yokkaichi, S., 138
Yonehara, K., 639, 763
Yoshimura, K., 196, 200
Yoshimura, Y., 138
Young, A. R., 288
Young, G. R., 100
Yu, J., 95

Z

Zeller, G. P., 95
Zhang, J., 677
Zimmerman, E. D., 95

Previous Proceedings in the Series of Conferences on Intersections of Particle and Nuclear Physics

	Year	Held in	Publisher	ISBN
7th	2000	Quebec City, Canada	AIP Conf. Proceedings vol. 549	1-56396-978-5
6th	1997	Big Sky, Montana	AIP Conf. Proceedings vol. 412	1-56396-712-X
5th	1994	St. Petersburg, Florida	AIP Conf. Proceedings vol. 338	1-56396-335-3
4th	1991	Tucson, Arizona	AIP Conf. Proceedings vol. 243	0-88318-950-X
3rd	1989	Rockport, Maine	AIP Conf. Proceedings vol. 176	0-88318-376-5
2nd	1986	Lake Louise, Canada	AIP Conf. Proceedings vol. 150	0-88318-349-8
1st	1984	Steamboat Springs, CO	AIP Conf. Proceedings vol. 123	0-88318-322-6

Other Related Titles from AIP Conference Proceedings

688 Scalar Mesons: An Interesting Puzzle for QCD
Edited by Amir H. Fariborz, November 2003, 0-7354-0159-4

687 High Energy Physics: The 25th Annual Montreal-Rochester-Syracuse-Toronto Conference on High Energy Physics, MRST 2003: A Tribute to Joe Schechter
Edited by Amir H. Fariborz, November 2003, 0-7354-0161-6

660 Hadron Physics: Effective Theories of Low Energy QCD; Second International Workshop on Hadron Physics
Edited by A. H. Blin, B. Hiller, A. A. Osiplov, M. C. Ruivo, and E. van Beveren, March 2003, 0-7354-0120-9

656 Frontiers of Nuclear Structure
Edited by Paul Fallon and Rod Clark, March 2003, 0-7354-0116-0

642 High Intensity and High Brightness Hadron Beams: 20th ICFA Advanced Beam Dynamics Workshop on High Intensity and High Brightness Hadron Beams; ICFA-HB2002
Edited by Weiren Chou, Yoshiharu Mori, David Neuffer, and Jean-François Ostiguy, November 2002, CD-ROM included, 0-7354-0097-0

619 Hadron Spectroscopy: Ninth International Conference on Hadron Spectroscopy; HADRON2001
Edited by Dmitry Amelin and Alexander M. Zaitsev, June 2002, 0-7354-0067-9

610 Nuclear Physics in the 21st Century: International Nuclear Physics Conference, INPC 2001
Edited by Eric Norman, Lee Schroeder, and Gordon Wozninak, April 2002, 0-7354-0056-3

To learn more about these titles, or the AIP Conference Proceedings Series, please visit the webpage http://proceedings.aip.org/proceedings